Algorithms and Combinatorics 25

Springer-Verlag Berlin Heidelberg GmbH

Boris Aronov • Saugata Basu
János Pach • Micha Sharir

Editors

Discrete and Computational Geometry

The Goodman-Pollack Festschrift

With 287 Figures

Springer

Editors

Boris Aronov

Department of Computer
and Information Science
Polytechnic University
Six MetroTech Center
Brooklyn, NY 11201, USA
e-mail: aronov@poly.edu

Saugata Basu

School of Mathematics
Georgia Institute of Technology
Atlanta, GA 30332-0160, USA
e-mail: saugata@math.gatech.edu

János Pach

City College
and Courant Institute
251 Mercer St.
New York, NY 10012, USA
e-mail: pach@cims.nyu.edu

Micha Sharir

School of Computer Science
Tel Aviv University
Tel Aviv 69978, Israel
e-mail: michas@tau.ac.il

Cataloging-in-Publication Data applied for

Bibliographic information published by Die Deutsche Bibliothek

Die Deutsche Bibliothek lists this publication in the Deutsche Nationalbibliografie;
detailed bibliographic data is available in the Internet at <http://dnb.ddb.de>.

Mathematics Subject Classification (2000):
52-XX, 68-XX, 05-XX, 14-XX, 57-XX, 60-02

ISSN 0937-5511
ISBN 978-3-642-62442-1 ISBN 978-3-642-55566-4 (eBook)
DOI 10.1007/978-3-642-55566-4

http://www.springer.de

Originally published by Springer-Verlag Berlin Heidelberg New York in 2003
Softcover reprint of the hardcover 1st edition 2003

Typeset by the authors. Edited and reformatted by Kurt Mattes, Heidelberg, using a Springer LaTeX macro package
Cover design: *design & production* GmbH, Heidelberg

Printed on acid-free paper 46/3142/LK - 5 4 3 2 1 0

Preface

Eli Goodman and Ricky Pollack were born at about the same time and brought up in the same city, within the same social milieu. They had many musical, literary, and scientific interests in common that drew them to the same events and locations. Their orbits were bound to intersect.

Precisely where and when it happened, they disagree. Was it at a concert in the late fifties? In a group of bridge enthusiasts? Or perhaps a little later, at the Heights campus of New York University, where each of them taught? In fact, Eli and Ricky disagree about a great many little things and details: this seems to be their working method, the secret of their relationship – carefully testing commonly accepted views and concepts, no matter how self-evident they might appear, by exposing them to each other's critical eyes and words.

Both of them received a first-class mathematical education. Ricky's thesis was in number theory, under the supervision of Harold Shapiro at New York University, while Eli became an algebraic geometer under the direction of Heisuke Hironaka and Steven Kleiman at Columbia University. Both of them published remarkable results in their respective research areas and had a taste of scientific success. In the mid 1970s, apparently on divergent paths, both of them took sabbaticals, and soon after each decided to try something new.

Ricky spent a semester at McGill University, collaborating with a group of excellent combinatorists. In particular, he learned about and was fascinated by the Erdős–Szekeres conjecture, which states that in every set of more than 2^{n-2} points in general position in the plane one can find n points that form the vertex set of a convex n-gon. He spent two months trying to come up with a proof. Unbeknownst to him, Eli had been working in topological graph theory and had recently turned his attention to some problems in combinatorial geometry. Shortly after Ricky's return from Montreal, he bumped into Eli at Courant Institute, at a seminar talk on graph algorithms. "What are you doing here?" – they asked each other in surprise. A few weeks later Eli called Ricky: – "Have you ever heard of the Erdős–Szekeres problem? I have some ideas ..."

They started meeting once or twice a week and, based on Eli's preliminary ideas, they soon developed a natural combinatorial encoding of a planar point configuration. This later became known as an "allowable sequence," and turned out to be an important tool in combinatorial geometry that led to spectacular breakthroughs such as Ungar's "book" proof of Scott's conjecture, according to which every set of n noncollinear points in the plane determines at least $2\lfloor n/2 \rfloor$ distinct directions. Dozens of joint papers evolved from this first fruitful collaboration, some of them developing theoretical aspects of the material, others solving problems that had seemed insurmountable by other methods. To their surprise, Eli and Ricky found themselves drawn more to combinatorial geometry than to their original fields. They loved this newly-found playground, and made it their own. They made a permanent move and never looked back. Gradually their joint work spread out to encompass various related topics in discrete geometry, and later included additional authors as well. Of their many ground-breaking discoveries we mention two. (1) They revitalized geometric transversal theory by showing that, using the notion of "order types," one can establish many exciting Helly-type theorems in d-space for the existence of a k-dimensional flat intersecting all members of a family of convex bodies, for $0 < k < d$. (2) They were among the first mathematicians to use real algebraic geometry to enumerate order types, simplicial polytopes, and other geometric objects. Their surprising results found numerous applications in ray shooting, in computer graphics, and elsewhere.

In 1984, after it had become clear that their work had implications not only in the well-established discipline of discrete geometry but also in the then nascent field of computational geometry, the opportunity arose of founding a journal devoted to both fields. Three publishers were interested, but Eli and Ricky decided on Springer-Verlag (whose New York office was then under the leadership of Walter Kaufmann-Bühler) as the best of the choices, and the journal *Discrete & Computational Geometry* was born the following year. Its first issue appeared in 1986, and the journal has thrived since then, having grown from under 400 pages a year at the start to nearly 1300 pages currently. At the same time the field has grown enormously. It is hard to overestimate the role Eli, Ricky, and their journal have played in these developments. The marriage of the two fields proved to be fruitful and long-lasting. DCG is now the leading journal in a fast growing area along the borderline of pure mathematics and computer science. Eli and Ricky co-organized the first Special Year at DIMACS (1989–90) and several major conferences dedicated to the subject (Santa Cruz 1986, Mount Holyoke 1996, Ascona 1999, and Oberwolfach 2000), as well as a number of special sessions at meetings of the American Mathematical Society. The investment of seemingly infinite resources on their part has paid off: they have built bridges between remote groups of researchers working on similar problems as well as bridges between various subdisciplines. In this sense, they are perhaps the two most conspicuous "founding fathers" of the field. As such, their names have become inseparable.

The idea of editing a collection of papers dedicated to Eli and Ricky was proposed by N. Prabhu, a former student of Ricky's, to whom we are grateful. We were soon overwhelmed by an avalanche of excellent manuscripts submitted by leading researchers in discrete and computational geometry. They contributed state-of-the-art results covering virtually every aspect of the field. In the selection process, we followed the simple rule that each paper should be refereed according to the high standards adopted by Eli's and Ricky's journal. It is most appropriate that the publisher of this volume is Springer-Verlag, under whose auspices the journal is produced. We are very grateful for their enthusiastic support.

Each editor of this volume regards himself as a student, a close friend, a collaborator, and a co-author of Eli and Ricky – all in one. They have shaped our lives, our mathematical thinking, our views of the world. We wish that their golden age last forever to the benefit of our community.

Boris Aronov
Saugata Basu
János Pach
Micha Sharir

Contents

Are Your Polyhedra the Same as My Polyhedra?

Branko Grünbaum

1 Introduction

"Polyhedron" means different things to different people. There is very little in common between the meaning of the word in topology and in geometry. But even if we confine attention to geometry of the 3-dimensional Euclidean space – as we shall do from now on – "polyhedron" can mean either a solid (as in "Platonic solids", convex polyhedron, and other contexts), or a surface (such as the polyhedral models constructed from cardboard using "nets", which were introduced by Albrecht Dürer [17] in 1525, or, in a more modern version, by Aleksandrov [1]), or the 1-dimensional complex consisting of points ("vertices") and line-segments ("edges") organized in a suitable way into polygons ("faces") subject to certain restrictions ("skeletal polyhedra", diagrams of which have been presented first by Luca Pacioli [44] in 1498 and attributed to Leonardo da Vinci). The last alternative is the least usual one – but it is close to what seems to be the most useful approach to the theory of *general* polyhedra. Indeed, it does not restrict faces to be planar, and it makes possible to retrieve the other characterizations in circumstances in which they reasonably apply: If the faces of a "surface" polyhedron are simple polygons, in most cases the polyhedron is unambiguously determined by the boundary circuits of the faces. And if the polyhedron itself is without selfintersections, then the "solid" can be found from the faces. These reasons, as well as some others, seem to warrant the choice of our approach.

Before deciding on the particular choice of definition, the following facts – which I often mention at the start of courses or lectures on polyhedra – should be considered. The regular polyhedra were enumerated by the mathematicians of ancient Greece; an account of these five "Platonic solids" is the final topic of Euclid's "Elements" [18]. Although this list was considered to be complete, two millennia later Kepler [38] found two additional regular polyhedra, and in the early 1800's Poinsot [45] found these two as well as two more; Cauchy [7] soon proved that there are no others. But in the 1920's Petrie and Coxeter found (see [8]) three new regular polyhedra, and proved the completeness of that enumeration. However, in 1977 I found [21] a whole lot of new regular polyhedra, and soon thereafter Dress proved [15], [16] that one needs to add just one more polyhedron to make my list complete. Then, about ten years ago I found [22] a whole slew of new regular polyhedra, and so far nobody claimed to have found them all.

How come that results established by such accomplished mathematicians as Euclid, Cauchy, Coxeter, Dress were seemingly disproved after a while? The answer is simple – all the results mentioned are completely valid; what changed is the meaning in which the word "polyhedron" is used. As long as different people interpret the concept in different ways there is always the possibility that results true under one interpretation are false with other understandings. As a matter of fact, even slight variations in the definitions of concepts often entail significant changes in results.

In some ways the present situation concerning polyhedra is somewhat analogous to the one that developed in ancient Greece after the discovery of incommensurable quantities. Although many of the results in geometry were not affected by the existence of such quantities, it was philosophically and logically important to find a reasonable and effective approach for dealing with them. In recent years, several papers dealing with more or less general polyhedra appeared. However, the precise boundaries of the concept of polyhedra are mostly not explicitly stated, and even if explanations are given – they appear rather arbitrary and tailored to the needs of the moment [12] or else aimed at objects with great symmetry [40]. The main purpose of this paper is to present an internally consistent and quite general approach, and to illustrate its effectiveness by a number of examples.

In the detailed discussions presented in the following sections we shall introduce various restrictions as appropriate to the classes of polyhedra considered. However, I believe that in order to develop any general theory of polyhedra we should be looking for a definition that satisfies the following (admittedly somewhat fuzzy) conditions.

(i) *The generality should be restricted only for very good reasons, and not arbitrarily or because of tradition.* As an example, there is no justification for the claim that for a satisfactory theory one needs to exclude polyhedra that contain coplanar faces. (Thus, if we were to interpret the two regular star-polyhedra found by Kepler as solids – the way they are usually shown – each would be bounded by 60 congruent triangles. Since quintuplets of triangles are coplanar, these "polyhedra" would be inadmissible.) In particular, the definition should not be tailored to fit a special class of polyhedra (for example, the regular ones, or the uniform polyhedra), in such a way that it is more or less meaningless in less restricted situations (such as the absence of high symmetry).

(ii) *The combinatorial type should remain constant under continuous changes of the polyhedron.* This is in contrast to the situation concerning the usual approach to convex polyhedra, where the combinatorial type is easily seen to be discontinuous. The point is illustrated in Figure 1, where the first three diagrams show pentagonal dodecahedra that are becoming

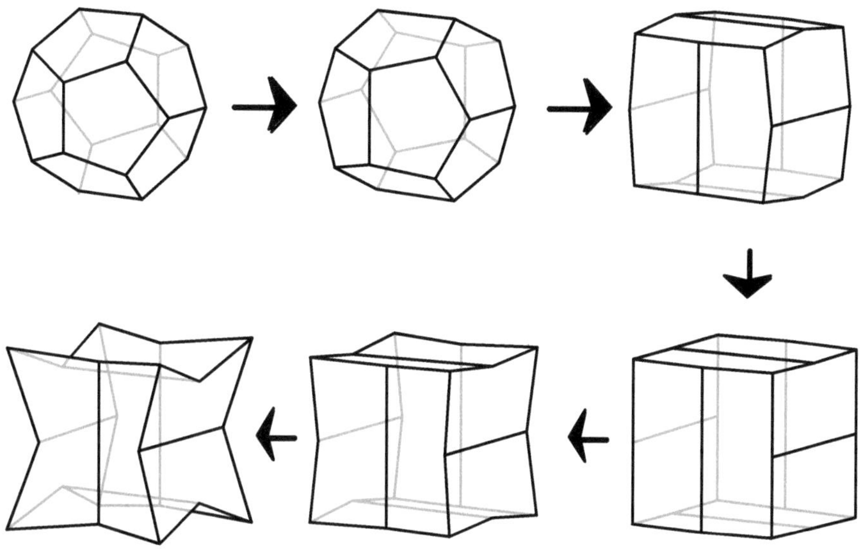

Fig. 1. The polyhedron with pairs of coplanar faces (at right in bottom row) is not a cube (even though the set of its points coincides with that of a cube) but is a pentagonal dodecahedron. It marks the transition between convex and nonconvex realizations of the same combinatorial type. Realizations of two polyhedra that have different combinatorial structure but coincide as sets of points (such as the cube and the above dodecahedron) are said to be *isomeghethic* (from Greek $\mu\varepsilon\gamma\varepsilon\theta o\sigma$ – extent, bulk).

more box-like, till pairs of faces turn coplanar for the fourth polyhedron. We wish to consider this 12-faced polyhedron as distinct from the cube, and as just a transitional step from convex to nonconvex polyhedra (as shown in the later parts of Figure 1), – all with the same combinatorial structure .

(iii) *To every combinatorial type of polyhedron there should correspond a dual type.* This is a familiar condition which is automatically satisfied by convex polyhedra, and is frequently stated as valid in all circumstances – although in fact it fails in some cases. We shall discuss this in Section 3.

Preparatory for the discussion of polyhedra, in Section 2 we consider polygons. Our working definition of "polyhedron" is presented and illustrated in Section 3. Sections 4 to 8 are devoted to the analysis of some specific classes of polyhedra that have been discussed in the literature and for which we believe the present approach provides a better and more consistent framework than previously available.

2 Polygons

Since we consider polyhedra as families of points, segments and polygons (subject to appropriate conditions), it is convenient to discuss polygons first.

Like "polyhedron", the word "polygon" has been (and still is) interpreted in various ways.

A *polygon* (specifically, an *n-gon* for some $n \geq 3$) is a cyclically ordered sequence of arbitrarily chosen points $V_1, V_2, \dots, V_n$, (the *vertices* of the polygon), together with the segments E_i determined by pairs of vertices V_i, V_{i+1} adjacent in the cyclic order (the *edges* of the polygon). Each vertex V_i is said to be *incident* with edges E_{i-1} and E_i, and these edges only. Here and in the sequel subscripts should be understood mod n. Polygons were first considered in close to this generality by Meister [41] nearly 250 years ago. (The assertion by Günther [30, p. 25] and Steinitz [52, p. 4] that already Girard [20] had this perception of "polygon" seems unjustified.) As explicitly stressed by Meister, this definition implies that distinct vertices of a polygon may be represented by the same point, without losing their individuality, and without becoming incident with additional edges even if the point representing a vertex is situated on another edge. Hence the definition admits various unexpected possibilities: edges of length 0; collinear edges – adjacent or not; edges overlapping or coinciding in pairs or larger sets; the concurrence of three or more edges. In order to simplify the language, the locution "vertices coincide" is to be interpreted as "the *points* representing the *distinct* vertices coincide"; similarly for edges.

Most of the important writings on polygons after Meister (such as Poinsot [45], Cauchy [7], Möbius [42], Wiener [55], Steinitz [52], Coxeter [9]) formulate the definition in the same way or equivalent ones, even though in some cases certain restrictions are added; for example, Möbius insists that the polygon not be contained in a line. However, all these writers *tacitly* assume that no two vertices fall on coinciding points. It is unfortunate that Günther [30, pp. 44ff] misunderstands Meister and imputes to him the same restriction. Other authors (for example, Brückner [3, p. 1], Hess [32, p. 611]) insist explicitly in their definitions that no two vertices of a polygon are at the same point; but later Brückner [3, p. 2] gives another definition, that coincides with ours, apparently written under the impression that it has the same meaning as his earlier one. However, disallowing representation of distinct vertices by one point is a crippling restriction which, I believe, is one of the causes for the absence of an internally consistent general theory of polyhedra. (The present definition coincides with what were called *unicursal polygons* in [23], where more general objects were admitted as "polygons".)

It should be mentioned that all these authors define a polygon as being a single circuit; Möbius explicitly states that it would be contrary to the customary meaning of the word if one were to call "hexagon" the figure formed by two triangles. This seems to have had little influence on later writers. For example, without any formalities or explanations, Brückner [3, p. 6] introduces such figures as "discontinuous polygons", in contradiction to his own earlier definition of "polygon". Hess [32] has a more vague definition of polygons, and explicitly allows "discontinuous" ones.

Since the number of "essentially different" shapes possible for n-gons increases very rapidly with increasing n, it is reasonable and useful to consider various special classes which can be surveyed more readily. The historically and practically most important classes are defined by symmetries, that is, by isometric transformations of the plane of the polygon that map the polygon onto itself. In case some of the vertices coincide, symmetries should be considered as consisting of an isometry paired with a permutation of the vertices. Thus, the *quadrangle* in Figure 2 does not admit a 120° rotational symmetry, but it admits a reflection in a vertical mirror paired with the permutation (12)(34). Clearly, all symmetries of any polygon form a group, its *symmetry group*.

A polygon is called *isogonal* [*isotoxal, regular*] provided its vertices [edges, flags] form a single orbit under its symmetry group. (A flag is a pair consisting of a vertex and one of the edges incident with it.) It is easily proved that a polygon is regular if and only if it is both isogonal and isotoxal. Moreover, if $n \geq 3$ is odd, every isogonal n-gon is regular, as is every isotoxal one. The more interesting situation of even n is illustrated for $n = 6$ in Figure 3. Similar illustrations of the possibilities for other values of n appear in [23], [24], [25].

Two consequences of the above definition of polygons deserve to be specifically mentioned; both are evident in Figure 3, and become even more pronounced for larger n. First, all isogonal n-gons fit into a small number of continua, and so do all isotoxal n-gons. If polygons having some coinciding vertices were excluded, the continua would be artificially split into several components, and the continuity would largely disappear. Second, the number of regular polygons would be considerably decreased. Under our definitions, for every pair of integers n and d, with $0 \leq d \leq n/2$, there exists a regular n-gon, denoted by its *Schläfli symbol* $\{n/d\}$. The construction of polygons $\{n/d\}$ inscribed in a unit circle can be described as follows (this was first for-

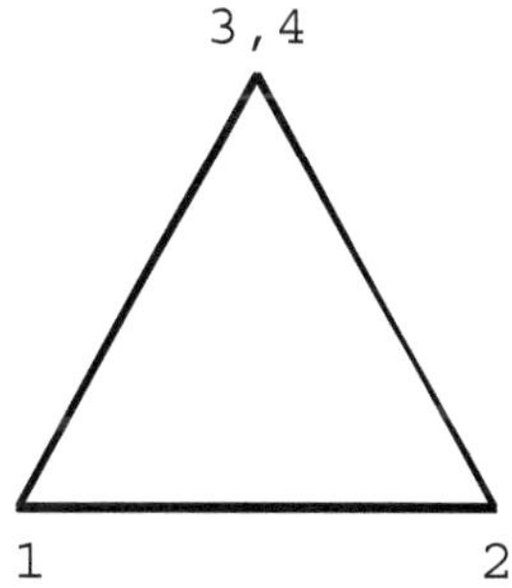

Fig. 2. This polygon looks like an equilateral triangle, but is in fact a quadrangle with two coinciding vertices. Besides the identity, the only symmetry it admits is a reflection (in a vertical mirror through the coinciding vertices 3 and 4) paired with the permutation (12)(34).

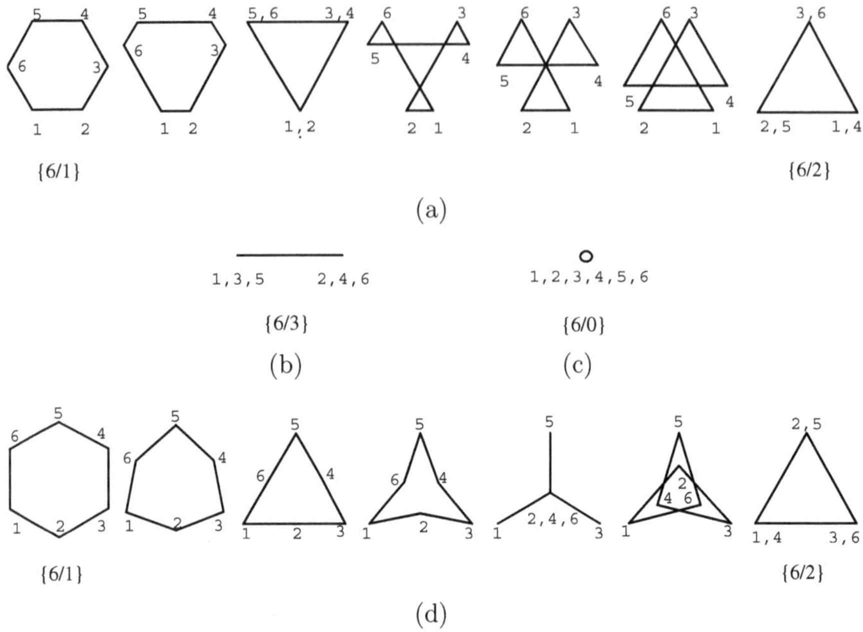

Fig. 3. Illustration of some of the polygons with high symmetry; shown is the case of hexagons. Isogonal polygons are shown in parts (a), (b), (c). Isotoxal polygons (in which all edges form one orbit under symmetries) are shown in parts (b), (c), (d). Representatives of the various shapes that isogonal or isotoxal hexagons can assume are illustrated. Regular polygons are indicated by their Schläfli symbol $\{n/d\}$.

mulated by Meister [41], and was also stated by Poinsot [45]): From a point on the circle, taken as the first vertex, advance to the next vertex by turning through an angle of $2\pi d/n$, and repeat this procedure from each resulting point till the starting point is reached at the nth step. Naturally, depending on the values of n and d, some of the intermediate vertices may coincide, but their identities are determined by the number of steps that led to them. Thus, for example, $\{6/0\}$ has six coinciding vertices, $\{6/2\}$ has three pairs of coinciding vertices, and $\{6/3\}$ has two triplets of coinciding vertices; in contrast, all vertices of $\{6/1\}$ are distinct. It takes no effort to realize that all vertices of $\{n/d\}$ will be distinct if and only if n and $d > 0$ are relatively prime. This connection between geometry and number theory was stressed by Poinsot and all writers following his example, and seems to have been one of the reasons why they banished from consideration all polygons with coinciding vertices. This was done despite the fact that the various results concerning angles, areas and other properties of regular polygons remain valid regardless of the relative primeness of n and d. Moreover, allowing polygons with coinciding vertices is essential if one wishes to have continuity in the combinatorial types of polyhedra.

One other consideration requires admitting polygons – and in particular, regular polygons – with coinciding vertices. In the present paper we are concerned with unoriented polygons; however, in some situations it is convenient or necessary to assign to each polygon an orientation. This yields two oriented polygons for each unoriented $\{n/d\}$ (except if $d = 0$ or $d = n/2$). Among regular polygons it is convenient to understand that the rotations through $2\pi d/n$ yielding $\{n/d\}$ are taken in the positive orientation; then the polygon oppositely oriented to $\{n/d\}$ is $\{n/e\}$, where $e = n - d$. Thus, oriented regular polygons $\{n/d\}$ exist for all $n > d \geq 0$, and these n polygons are all distinct. The appropriateness of such a convention is made evident by its applicability in many results concerning arbitrary polygons. It would lead us too far to describe these results, which can be interpreted as consequences of the possibility of expressing every n-gon as a weighted sum (in an appropriate sense) of regular polygons. The results range from Napoleon-type theorems to the elucidations of limits of iterations of various averaging operations on polygons. Detailed information about such applications, which would not be possible under the Poinsot restriction, may be found in [2], [13], [14], [19], [39], [43], [46], [47], [48], and in their references.

3 Definition of Polyhedron

In my opinion, the most satisfying way to approach the definition of polyhedra is to distinguish between the **combinatorial structure** of a polyhedron, and the **geometric realizations** of this combinatorial structure. We start by listing the conditions under which a collection of objects called *vertices*, *edges*, and *faces* will be called an *abstract polyhedron*. The conditions involve a (primitive) relation of *incidence*, and a (derived) relation of *adjacence*. In an abstract way of thinking, an edge is a pair of vertices, and a face is a circuit of edges. More specifically, in an abstract polyhedron we have to have:

(P1) Each edge is incident with precisely two distinct vertices and two distinct faces.

Each of the two vertices is said to be *incident* (via the edge in question) with each of the two faces. Two vertices incident with an edge are said to be *adjacent*; also, two faces incident with an edge are said to be *adjacent*.

(P2) For each edge, given a vertex and a face incident with it, there is precisely one other edge incident to the same vertex and face.

This edge is said to be *adjacent* to the starting edge.

(P3f) For each face there is an integer k, such that the edges incident with the face, and the vertices incident with it via the edges, form a circuit in the sense that they can be labeled as $V_1E_1V_2E_2V_3E_3 \ldots V_{k-1}E_{k-1}V_kE_kV_1$, where each edge E_i is incident with vertices V_i and V_{i+1}, and adjacent to edges E_{i-1} and E_{i+1}. All edges and all vertices of the circuit are distinct, all subscripts are taken mod k, and $k \geq 3$.

(P3v) For each vertex there is an integer j, such that the edges incident with the vertex, and the faces incident with it via the edges, form a circuit in the sense that they can be labeled as $F_1E_1F_2E_2F_3E_3 \ldots F_{j-1}E_{j-1}F_jE_jF_1$, where each edge E_i is incident with faces F_i and F_{i+1}, and adjacent to edges E_{i-1} and E_{i+1}. All edges and all faces of the circuit are distinct, all subscripts are taken mod j, and $j \geq 3$.

Thus, each face corresponds to a simple circuit of length at least 3, and similarly for the circuits that correspond to the vertices; the latter circuits are known as *vertex stars*.

(P4) If two edges are incident with the same two vertices [faces], then the four faces [vertices] incident with the two edges are all distinct.

(P5f) Each pair F, F* of faces is connected, for some j, through a finite chain $F_1E_1F_2E_2F_3E_3 \ldots F_{j-1}E_{j-1}F_j$ of incident edges and faces, with $F_1 = F$ and $F_j = F^*$.

(P5v) Each pair V, V* of vertices is connected, for some j, through a finite chain $V_1E_1V_2E_2V_3E_3 \ldots V_{j-1}E_{j-1}V_j$ of incident edges and vertices, with $V_1 = V$ and $V_j = V^*$.

It should be noted that with this definition, the duality requirement is satisfied in an essentially trivial way: Given an abstract polyhedron, a dual abstract polyhedron is obtained by interchanging "vertices" and "faces". The formulation of the conditions (P1) to (P5) shows that they will be satisfied after such an exchange.

A *symmetry* of an abstract polyhedron is an automorphism induced by incidence-preserving permutations of the vertices, the edges, and the faces. In most cases we shall encounter, such an automorphism is already determined by a permutation of the vertices.

It is clear that the above definition could have been formulated as pertaining to a special class of cell-complexes representing 2-dimensional closed manifolds. In fact, each face may be understood as the boundary of a 2-dimensional topological disk, and the identifications determined by incidences and adjacencies determine the cell-decomposition of a manifold, which we shall call the *associated manifold* of the polyhedron. For a given abstract polyhedron we shall often refer to its associated manifold and we shall assign to the abstract polyhedron as its *genus*, or its *Euler characteristic*, the values of these functions for the associated manifold. On the other hand, cell-decompositions in general admit features that cannot occur in polyhedra; for example, our definition does not admit monogons or digons.

Equally obvious is the fact that the conditions listed above (hence the definition of an abstract polyhedron) could have been formulated in terms of lattices. Such an approach is taken by McMullen and Schulte [40], to define not only objects more general than our polyhedra, but also the analogous higher-dimensional *abstract polytopes*.

A *geometric polyhedron* or *polyhedron* for short is an image of an abstract polyhedron under a mapping in which vertices go to points, edges to segments (possibly of length 0) and faces to polygons (which are understood as circuits of incident vertices and edges). *Incidence* means that the point representing a vertex is an endpoint of a segment representing an edge, and that a segment (which represents an edge) is a member of the cycle which defines a polygon (representing a face). We say that the polyhedron is a *realization* of the *underlying* abstract polyhedron.

If all faces of a geometric polyhedron are simple polygons, we may interpret each face as a topological disk. Their totality forms a surface which may have selfintersections or overlaps. Best known examples of this kind are the two regular polyhedra first discovered by Poinsot [45] – the great icosahedron $\{3, 5/2\}$ and the great dodecahedron $\{5, 5/2\}$.

Polyhedra with the same underlying abstract polyhedron are said to be *combinatorially equivalent*, or to have the same *combinatorial type*. Realizations of two polyhedra that have different combinatorial types but coincide as sets of points are said to be *isomeghethic* (from Greek $\mu\varepsilon\gamma\varepsilon\theta o\sigma$ – extent, bulk). This term may be used in cases where we interpret the polyhedra as surfaces (such as the cube and the fourth dodecahedron in Figure 1), as well as in cases in which selfintersecting polygons necessitate interpreting faces as circuits of vertices and edges. In this sense the regular dodecahedron is isomeghetic with the uniform polyhedron in Figure 15(c).

A *symmetry* of a (geometric) polyhedron is a pairing of an isometric mapping of the polyhedron onto itself with an automorphism of the underlying abstract polyhedron. The polyhedron is *isogonal* [*isohedral*, *regular*] if its vertices [faces, flags] form one orbit under its symmetries. (A *flag* of a polyhedron is a triplet consisting of a vertex, an edge, and a face, all mutually incident.) A polyhedron is *noble* if it is both isogonal and isohedral.

Every abstract polyhedron has realizations: Nothing in the definition prevents all vertices to be represented by the same point. Clearly, such *trivial* realizations are usually of little interest, but in some contexts they need to be considered. Also, abstract polyhedra may have other *subdimensional* realizations – that is, the affine hull of a realization may well be 1- or 2-dimensional. An example of a noble polyhedron that realizes the Klein bottle in the plane is shown in Figure 4. In the remaining part of the paper we shall concentrate on full-dimensional polyhedra, that is, polyhedra with 3-dimensional affine hull.

For a given geometric polyhedron the construction of a dual polyhedron is most often carried out by applying to its faces and vertices a *polarity* (that is, a reciprocation in a sphere). From properties of this operation it follows at once that the polar of a given polyhedron is a realization of the abstract polyhedron dual to the given one. However, the possibility of carrying out the polarity depends on choosing a sphere for the inversion in such a way that its center is not contained in the plane of any face. While this is easy to accomplish in any case, the resulting shape depends strongly on the position of that center. The main problem arises in connection with polyhedra with high symmetry (for example, isogonal or uniform polyhedra) if it is desired to find a dual with the same degree of symmetry: If the only position for the center is at the centroid of the polyhedron, and the polyhedron has some faces that contain the centroid – then it is not possible to

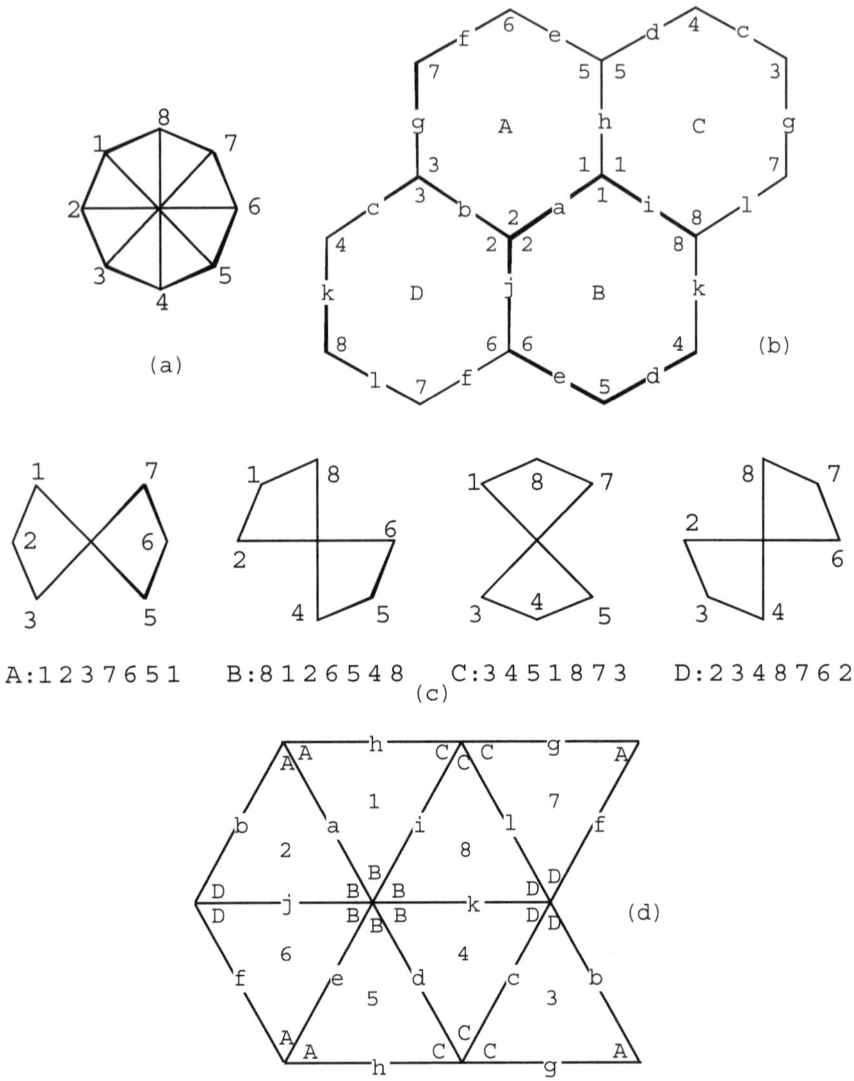

Fig. 4. A subdimensional noble polyhedron is shown in (a). The associated cell complex representing the underlying abstract polyhedron is shown in (b); the manifold is the Klein bottle. Each of the four faces of the geometric realization in (a) is shown separately in (c). Note that any two faces are both incident with two edges, but these have distinct vertices as required by condition (P4). The cell complex representing the abstract polyhedron dual to the one in (a) and (b) is shown in (d); as is easily verified, although the abstract polyhedron is noble, it admits no nontrivial realization as a noble geometric polyhedron.

find a polar polyhedron with the same symmetry. Moreover, if a polyhedron has coplanar faces [coinciding vertices] then any polar polyhedron will have coinciding vertices [coplanar faces]. All these possibilities actually occur for various interesting polyhedra. Clearly, duality-via-polarity is uninteresting for subdimensional polyhedra – it yields only trivial ones.

Our definitions are applicable to finite as well as infinite polyhedra; this enables one to include tilings and honeycombs among the objects studied. However, for the present discussion we shall restrict attention to finite polyhedra, that is, we shall assume the cardinalities of the sets of vertices, edges, and faces to be finite.

Despite the adaptability of the "skeletal" approach to such topics as polyhedra with skew polygons as faces, in the present work we shall consider only polyhedra with planar faces.

4 Regular Polyhedra

We shall now present constructions that lead to some "new" regular polyhedra. One construction which may be applied to polyhedra in general, is by the following *vertex-doubling*. Start with any abstract or geometric polyhedron. Replace each vertex by a pair of vertices, for example a green one and a red one. For each face, as you go around it, alternate between red and green vertices. If the face is an n-gon for some odd n, then there will now be a $(2n)$-gon in its place; if n is even, the vertices along the n-gon will have alternating colors – but there will be another n-gon, with vertices of the opposite colors. The collection of these new faces will be an (abstract or geometric) polyhedron if and only if there is at least one odd-sided face in the original polyhedron. If there is no such face, the resulting family of polygons does not satisfy condition (P.5f) of the definition is Section 3; instead of a polyhedron, a *compound* of two polyhedra is obtained. Dually, one can start with any polyhedron, replace each face by a pair of faces of different "colors", and take as adjacent those faces which arise from adjacent faces of the original and have different colors. This *face-doubling* gives rise to a new polyhedron if and only is there is at least one vertex of odd valence in the original polyhedron.

It should be stressed that the above comments *do not* mean that if all faces are even-sided, then there is no polyhedron in which the vertices are doubled up and all new faces have double the number of sides of the original ones. It means *only* that the above method of replacing one vertex by two vertices represented by the same point does not lead to such new polyhedra. At the end this section we shall encounter an example that illustrates this comment.

A general property of the vertex-doubling construction considered here is that it transforms regular polyhedra into regular ones, and isogonal or isohedral polyhedra into isogonal or isohedral ones, respectively. Analogously for the face-doubling construction. Probably the most interesting instances to which the vertex-doubling procedure can be applied are eight of the nine regular polyhedra (five convex and four Kepler-Poinsot) – all except the cube. The resulting "new" polyhedra are regular and can be denoted by their Schläfli symbols $\{6/2,3\}$, $\{6/2,4\}$, $\{6/2,5\}$, $\{10/2,3\}$, $\{6/2,5/2\}$, $\{10/2,5/2\}$, $\{10/4,3\}$, $\{10/4,5\}$. The face-doubling construction can be applied to all regular polyhedra except the octahedron, and yields "new" regular polyhedra $\{3,6/2\}$, $\{4,6/2\}$, $\{5,6/2\}$, $\{3,10/2\}$, $\{5/2,6/2\}$, $\{5/2,10/2\}$, $\{3,10/4\}$ and $\{5,10/4\}$. Clearly, these sixteen polyhedra form eight pairs of dual polyhedra. Moreover, the duality can be effected by a polarity (that is, by reciprocation in a suitable sphere). It should be noted that the number of combinatorial types of these regular polyhedra is smaller. Just as the combinatorial types of the icosahedron $\{3,5\}$ and the great icosahedron $\{3,5/2\}$ coincide, so do pairs of polyhedra $\{6/2,5\}$ and $\{6/2,5/2\}$, $\{10/2,3\}$ and $\{10/4,3\}$, $\{10/2,5/2\}$ and $\{10/4,5\}$, $\{5,6/2\}$ and $\{5/2,6/2\}$, $\{3,10/2\}$ and $\{3,10/4\}$, $\{5/2,10/2\}$ and $\{5,10/4\}$; each pair represents a single combinatorial type.

The polyhedra $\{3,6/2\}$ and $\{6/2,3\}$ are shown in Figure 5, where lower and upper case characters are used instead of different colors. All the other regular polyhedra listed above would appear, analogously, like their counterparts among the convex or Kepler-Poinsot polyhedra to which they are isomeghethic; however, their combinatorial structure – determined by the underlying abstract polyhedron – is different from that of the traditional ones.

A natural question that arises from these constructions is whether it is possible to perform **vertex k-tupling**, that is replace each face of a polyhedron by a polygon having k times as many sides or by a family of k polygons with the same number of sides, where $k \geq 3$. While we have seen that cases in which the doubling operation yields a polyhedron are rather easily characterized, no corresponding general result is known for ktupling. However, in case the operation is performed on a tetrahedron, there is an affirmative answer, as follows.

For a given $k \geq 2$ we may replace each vertex of the tetrahedron by k vertices; if these are denoted $\mathrm{a}_1, \ldots, \mathrm{a}_k, \mathrm{b}_1, \ldots, \mathrm{b}_k, \mathrm{c}_1, \ldots, \mathrm{c}_k, \mathrm{d}_1, \ldots, \mathrm{d}_k$, then the four faces are given by $[\mathrm{d}_1\mathrm{c}_1\mathrm{b}_1\ \mathrm{d}_2\mathrm{c}_2\mathrm{b}_2\ \mathrm{d}_3\mathrm{c}_3\mathrm{b}_3 \ldots \mathrm{d}_k\mathrm{c}_k\mathrm{b}_k\ \mathrm{d}_1]$, $[\mathrm{c}_1\mathrm{d}_1\mathrm{a}_1\ \mathrm{c}_k\mathrm{d}_k\mathrm{a}_k\ \mathrm{c}_{k-1}\mathrm{d}_{k-1}\mathrm{a}_{k-1} \ldots \mathrm{c}_2\mathrm{d}_2\mathrm{a}_2\ \mathrm{c}_1]$, $[\mathrm{b}_1\mathrm{a}_1\mathrm{d}_1\ \mathrm{b}_k\mathrm{a}_k\mathrm{d}_k\ \mathrm{b}_{k-1}\mathrm{a}_{k-1}\mathrm{d}_{k-1} \ldots \mathrm{b}_2\mathrm{a}_2\mathrm{d}_2\ \mathrm{b}_1]$, $[\mathrm{a}_1\mathrm{b}_1\mathrm{c}_1\ \mathrm{a}_2\mathrm{b}_2\mathrm{c}_2\ \mathrm{a}_3\mathrm{b}_3\mathrm{c}_3 \ldots \mathrm{a}_k\mathrm{b}_k\mathrm{c}_k\ \mathrm{a}_1]$; each face is of type $\{3k/k\}$. (Here, and throughout the paper, the first vertex of each face is repeated at the end to make the checking of incidences simpler; each face is described by listing its vertices in a cyclic order, and spaces are inserted to facilitate understanding the structure.) This determines an orientable polyhedron $\mathrm{P}(k)$ with $4k$ vertices, $6k$ edges, and 4 faces; hence the associated map has genus $g = k-1$. For $k = 2$ the polyhedron has as its map the regular one in Figure 5(b); no $\mathrm{P}(k)$ with $k \geq 3$ is regular, as can be checked easily. The map corresponding to $\mathrm{P}(3)$ is the only possible map of type $\{9/3,3\}$, hence these parameters do not admit any regular map or polyhedron. On the other hand, at least

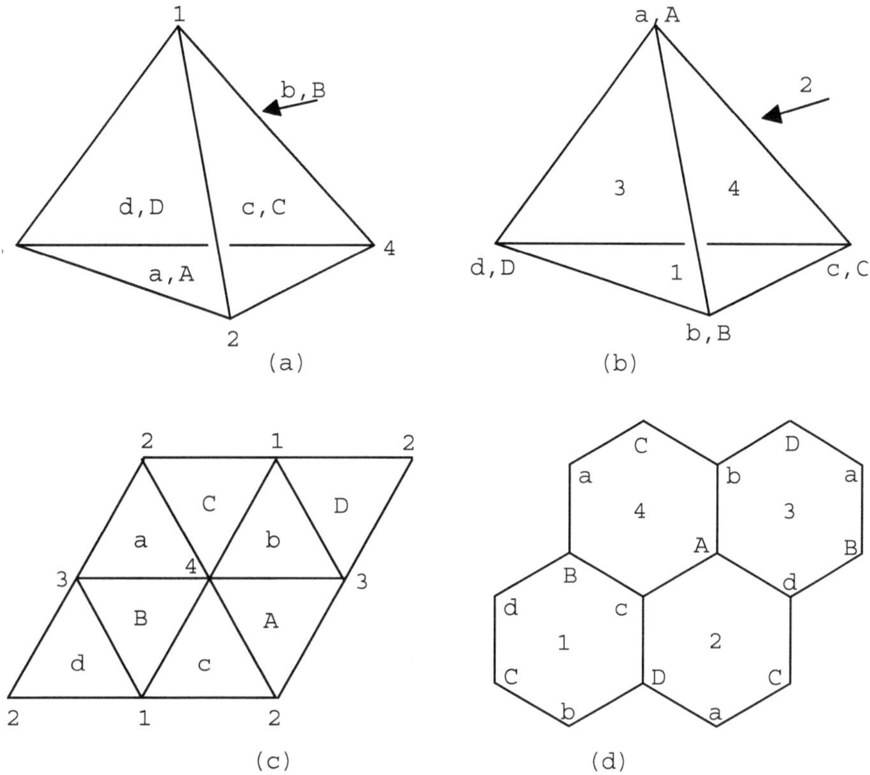

Fig. 5. The regular polyhedron in (a) was obtained by face-doubling from the regular tetrahedron, the regular polyhedron in (b), dual to it, resulted from vertex-doubling of the regular tetrahedron. The underlying abstract polyhedra are indicated by the cell complexes representing them in (c) and (d), respectively.

for $k = 4$ there is one other polyhedron $P^{\#}$ with four faces of type $\{3k/k\}$, and it is regular (map W#24.22 in Wilson's catalog [56]). To distinguish between the two polyhedra we note that the faces of P(4) are $[d_1c_1b_1\ d_2c_2b_2\ d_3c_3b_3\ d_4c_4b_4\ d_1]$, $[c_1d_1a_1\ c_4d_4a_4\ c_3d_3a_3\ c_2d_2a_2\ c_1]$, $[b_1a_1d_1\ b_4a_4d_4\ b_3a_3d_3\ b_2a_2d_2\ b_1]$ and $[a_1b_1c_1\ a_2b_2c_2\ a_3b_3c_3\ a_4b_4c_4\ a_1]$, while the faces of the $P^{\#}$ are $[d_1c_1b_1\ d_2c_2b_2\ d_3c_3b_3\ d_4c_4b_4\ d_1]$, $[c_1d_1a_3\ c_2d_2a_4\ c_3d_3a_1\ c_4d_4a_2\ c_1]$, $[b_1a_1d_3\ b_2a_2d_4\ b_3a_3d_1\ b_4a_4d_2\ b_1]$ and $[a_1b_1c_1\ a_2b_2c_2\ a_3b_3c_3\ a_4b_4c_4\ a_1]$. The fact that these polyhedra are not combinatorially equivalent can most easily be established from their maps, see Figure 6.

For $k \geq 5$ there probably exist polyhedra different from P(k), having faces of type $\{3k/k\}$ and $4k$ trivalent vertices. It may be conjectured that none of these polyhedra is regular. For $k \leq 16$ the validity of this can be inferred from the fact that there are no regular maps satisfying these conditions in Wilson's catalog [56], where all regular maps with at most 100 edges are listed.

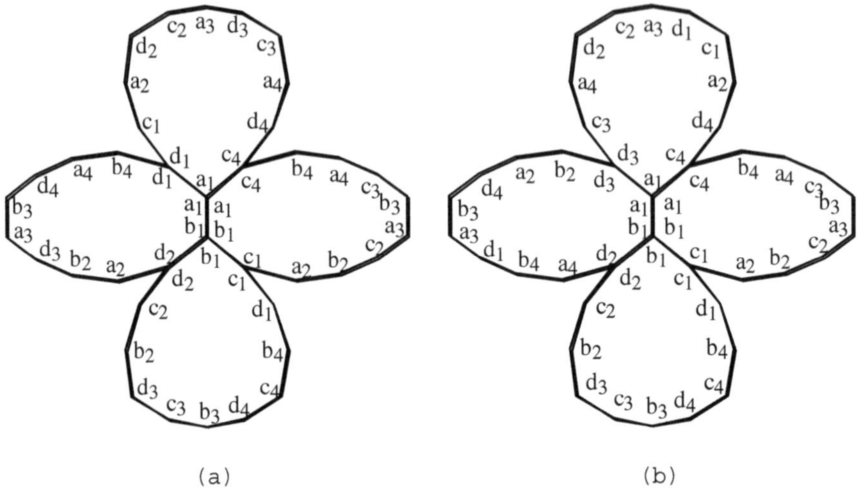

Fig. 6. The maps underlying two polyhedra, each with four faces of type $\{12/4\}$ and sixteen 3-valent vertices. The map in (b), and the corresponding polyhedron $P^{\#}$, are regular. The map of P(4) shown in (a) is not combinatorially equivalent to the map in (b). One way to see this is to observe that since (b) is regular, all its flags are equivalent. The labels of the two maps coincide for all vertices on the flags associated with vertex a_1, the edge $[a_1,b_1]$, and the face to the right of it. Hence, if the map in (a) were regular, the only possible isomorphism would preserve all labels – but this is clearly not the case with the faces on top or on the left.

Although the general vertex-doubling construction described above does not apply to the cube, there is a polyhedron obtained by doubling the vertices on the cube. This polyhedron Q and its map are shown in Figure 7. Both are regular; the polyhedron has the Schläfli symbol $\{8/2, 3\}$, and the map is W#24.21 in [56].

A construction of two infinite families of regular polyhedra should be mentioned here. The first arises from a certain vertex k-tupling of the octahedron. It is most simply described by saying that one of the triangular faces is replaced by a polygon $\{3k/k\}$, and the other triangular faces are replaced by suitable reflections of this face. The resulting polyhedron $\{3k/k, 4\}$ is easily seen to be regular; it is orientable, with 6k vertices, 12k edges and 8 faces, hence has genus $g = 3k-3$. The second family is polar to this; it arises by face k-tupling of the cube. Its Schläfli symbol is, accordingly, $\{4, 3k/k\}$. The case $k = 2$ of both families is illustrated in [27], where they are obtained by a different construction.

It may be conjectured that there is only a finite number of infinite families of full-dimensional regular polyhedra.

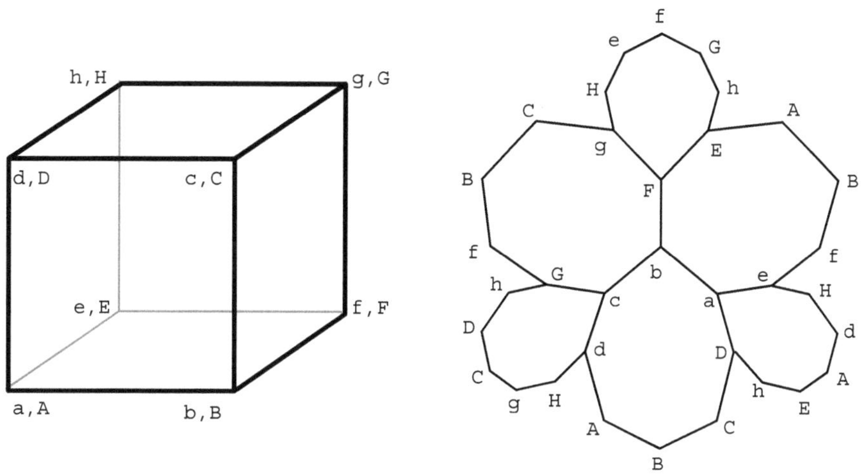

Fig. 7. Doubling-up vertices of the cube, in the way indicated in the map, yields a regular polyhedron $\{8/2, 3\}$ with 16 vertices, 24 edges and six faces. It is orientable, of genus 2. Its map is W#24.21 in the catalog [56].

5 Noble Polyhedra

Polyhedra that are noble (that is, isogonal as well as isohedral) have been studied considerably less than the slightly more symmetric regular ones. Nevertheless, they seem quite interesting. In particular, it seems that beyond finitely many infinite families of full-dimensional noble polyhedra, there exists only a finite number of individual polyhedra of this kind. This conjecture is one of the intriguing open questions concerning symmetric polyhedra.

The study of noble polyhedra was begun by Hess in the 1870's (see [33], [34], [35], [36]), and continued by Brückner [3], [4], [5], [6]. In the early papers (for example, in [33]), Hess considered noble polyhedra as generalization of regular ones. However, there seems to be no mention of these polyhedra in the literature after the works of Hess and Brückner, until [23], close to a century later. This may in part be due to a general neglect of nonconvex polyhedra during most of the 20th century, and in part to the inconsistent and clumsy exposition by Hess and Brückner of their own results.

It is obvious that all regular polyhedra are noble. It is well known that among convex polyhedra the only nonregular ones are *sphenoids*, that is, tetrahedra with congruent faces, different from equilateral triangles. From now on we shall discuss nonconvex noble polyhedra only.

We have encountered one subdimensional noble polyhedron in Figure 4, and many other examples of this kind are possible. However, the possibilities are much more restricted if we are interested in full-dimensional noble polyhedra.

Several infinite families of such polyhedra have symmetries of prisms or of anti-prisms. One family consists of the remarkable prismatic and antiprismatic *crown* polyhedra, discovered by Hess [35]; he called them *stephanoids* (from the Greek for "crown"). Their faces are selfintersecting quadrangles. Detailed descriptions and illustrations can be found in [23], where also the prismatic and antiprismatic *wreath* polyhedra (with triangular faces) and *V-faced* polyhedra are introduced and illustrated. The faces of V-faced polyhedra are full-dimensional quadrangles with vertices V_1,V_2,V_3,V_4, in which V_2 and V_4 are represented by the same point. (It should be noted that, contrary to the impression given by the illustrations in [23], the quadrangles in V-faced noble polyhedra need not be equilateral.) Hess and Brückner showed considerable ingenuity in discovering noble polyhedra, and one may wonder why they did not find the (rather simple) wreath polyhedra and V-faced polyhedra. One possible reason is that they were ignoring polyhedra with coplanar faces or coinciding vertices. However, rather inconsistently, in other instances (as mentioned below) they did allow such polyhedra, even in the same publications in which they state that vertices have to be distinct.

Probably more interesting are some of the noble polyhedra with octahedral or icosahedral symmetries. One polyhedron, a version of which was described by Hugel [37] but recognized as noble by Hess [35] and Brückner [3, p. 215], is shown in Figure 8. Its 20 vertices are the same as those of a regular dodecahedron, and its 20 faces are selfintersecting hexagons; one is indicated in Figure 8 by heavy lines. The polyhedron is *autopolar*, in the sense that its polar with respect to a suitable sphere coincides with the polyhedron itself. It is remarkable that the underlying abstract polyhedron is regular; its map is the one listed as W#60.57 in [56].

The situation is more complex concerning a noble polyhedron described by Hess [34] and Brückner [3, p. 215]. It is supposed to look like the object in Figure 9, that is, to be isomeghethic with the regular polyhedron $\{5, 5/2\}$. However, each of the twelve faces is formed by *all* segments determined by its five vertices, that is, each face is supposed to look like the union of a pentagon with a pentagram, shown by heavy lines in Figure 9. This is quite appropriate provided one allows a polygon to revisit vertices (as was the setup in [23]). In fact, in this case there are two distinct possibilities of turning the figure into an isogonal polygon (see [23]). Hess and Brückner seem to have noticed only one of these – but even this is contrary to their usual (and explicitly stated) requirement that a polygon cannot revisit any vertex. In any case, under the definitions we adopted in the present paper these are not polyhedra. However, doubling-up the vertices leads to polyhedra that are noble (as noted in [23]). One face of the first is the polygon [d F b D e B c E f C d] in the notation of Figure 9; the other 23 faces result by the application of symmetries of the icosahedron, and the interchange of lower and upper case vertices. In contrast, one face of the other polyhedron is [b C f B e F d E c D b], and the other faces are obtained analogously.

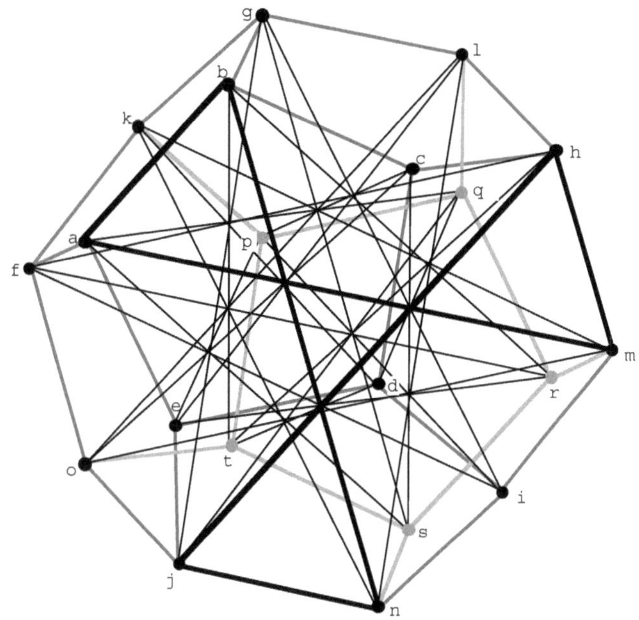

Fig. 8. An orientable selfpolar polyhedron recognized as noble by Brückner [2]. It has 20 vertices and 20 faces, one of which is emphasized; its map is not only noble, but regular, of genus 9, as observed by Prof. J. Wills. The dodecahedron serves only to guide the construction and recognition of the faces of this polyhedron.

Another remarkable invention of Brückner [3] is shown in Figure 10, which is meant to illustrate the construction of two noble polyhedra. The idea is to start with a uniform rhombicuboctahedron (shown in gray lines); the five points a, e, u, w, h are coplanar, and according to Brückner they determine two distinct polygons: [a e w a u h a] and [a e w a h u a]. The other faces are obtained by applying to each of these two the symmetries of the rhombicuboctahedron. However, each of these polygons revisits a vertex (as was allowed in [23] but not here); therefore these objects are not polyhedra in the present sense, or in the sense generally accepted by Brückner. On the other hand, as described in [23], vertex-doubling produces in each of the two cases an acceptable noble polyhedron with 48 vertices and 48 faces.

In Figure 11 are shown two additional noble polyhedra, which seem to have escaped Brückner's attention. They are generated in the same way from the uniform quasirhombi-cuboctahedron (−3.4.4.4) (see [W, p. 132]) as the ones described above from the rhombicuboctahedron. Although the noble polyhedra arising from Figure 11 are combinatorially equivalent to the ones in Figure 10, their metric difference can most simply be established by observing that the ratio of lengths of the longest diagonal of each face to the shortest is $2 + \sqrt{2} = 3.14142\ldots$ in Figure 11, but $1 + 1/\sqrt{2} = 1.7071\ldots$ in Figure 10.

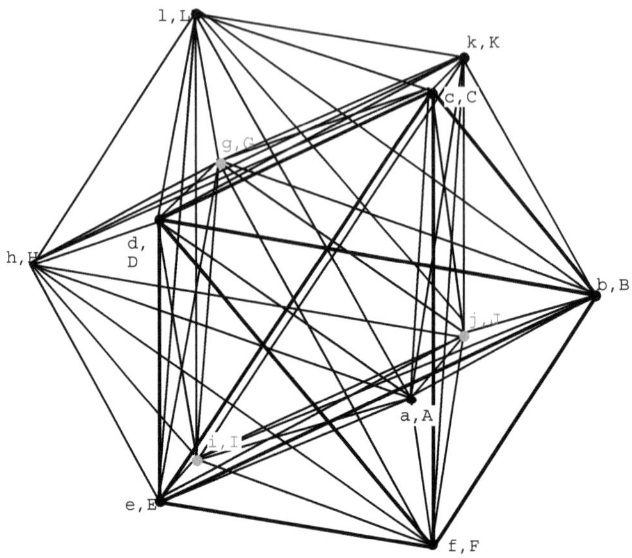

Fig. 9. The construction of noble polyhedra isomeghethic with the regular polyhedron $\{5, 5/2\}$. The emphasized decagon can be interpreted in two isogonal ways – either as [d F b D e B c E f C d] or as [b C f B e F d E c D b]; the other 23 faces result by the application of symmetries of the icosahedron, and the interchange of lower and upper case vertices. Two distinct noble polyhedra are obtained. However, if the pairs of coinciding vertices are simply identified (as done by Hess [34] and Brückner [2]), the resulting object is not a polyhedron in our sense, or in the sense ostensibly accepted by Hess and Brückner.

In view of the ingenuity with which Hess and Brückner pursued noble polyhedra, and their willingness to stretch their own rules in order to admit the ones they found, it is strange that they never mention two rather simple polyhedra (and their polars), which are shown and explained in Figure 12. The probable reason for this omission is, again, the shying away from coplanar faces or coinciding vertices.

Another way of constructing certain noble polyhedra will be mentioned in the next section.

6 Uniform Polyhedra

Uniform polyhedra are defined as isogonal polyhedra with all faces regular. They are closely related to the *Archimedean* polyhedra, studied since antiquity; the older concept requires congruence of the vertex stars instead of the more restrictive isogonality. The convex uniform polyhedra are all well known since the work of Kepler [38], but the determination of nonconvex ones was done piecemeal by many people, over close to a century, – and that only in the traditional understanding of what is a polyhedron. An illustrated list of

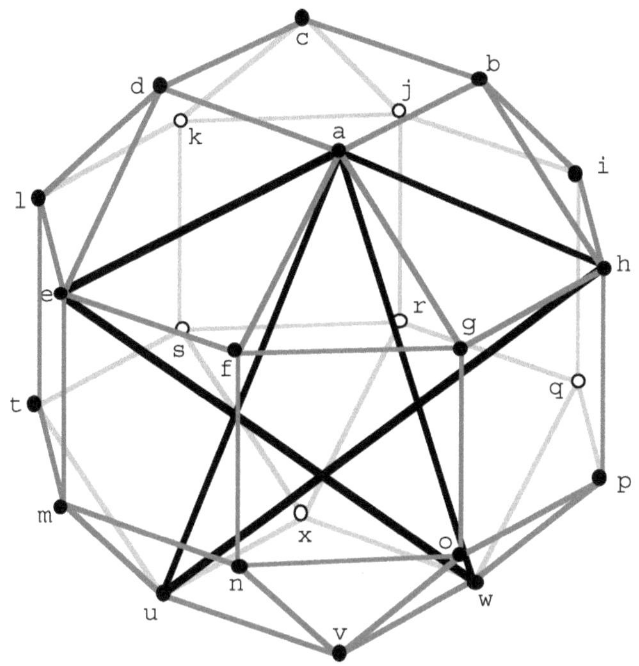

Fig. 10. The construction of two noble polyhedra, according to Brückner [2]: one of these has the face [a e w a u h a] and the other faces obtaineble by symmetries, while the other has the face [a e w a h u a] and those in its orbit. Since this involves revisiting a vertex, these are not polyhedra in our sense. However, by doubling-up vertices each leads to a noble polyhedron. The rhombicuboctahedron (3.4.4.4) serves only to guide the construction and recognition of the faces of this polyhedron.

such uniform polyhedra appears in Coxeter *et al.* [12], and the fact that this list is complete was established by Sopov [51] and Skilling [50]. Additional illustrations of all these polyhedra can be found in Wenninger [53] and Har'El [31].

It should not come as a surprise that with our definition of polyhedra there are many new possibilities for the formation of uniform ones. As with "new" regular polyhedra, the visual appearance of many "new" uniform polyhedra is somewhat disappointing – they look exactly like appropriate "old" uniform polyhedra since they are isomeghethic with them. However, their inner structure (the underlying abstract polyhedron) is different. In many cases, the abstract polyhedron admits a continuum of non-uniform isogonal realizations which do seem interesting, and in the limit become uniform; in some instances, this approach to visualization of the structure of polyhedra works for regular ones as well. Examples of both possibilities appear in [27]; one of the uniform ones is (3 . 6/2 . 3 . 6/2), another is (3 . 6/2 . 6/2), and a

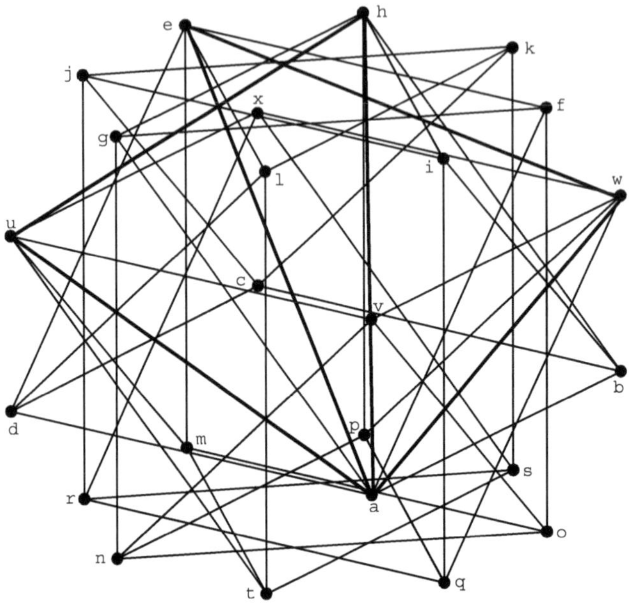

Fig. 11. Another pair of noble polyhedra, combinatorially equivalent to the ones in Figure 10, can be obtained using the vertices of the uniform quasirhombicuboctahedron (−3.4.4.4). This uniform polyhedron is combinatorially equivalent to the rhombicubocta-hedron, under the correspondence between their vertices indicated by the labels. Although the resulting noble polyhedra obtainable by doubling-up vertices are combinatorially equivalent to the the ones described in the caption of Figure 10, they are metrically distinct from them.

regular one is a 24-faced $\{4, 6/2\}$. More interesting are cases in which something genuinely new occurs. One example is (3 . 6 . 4 . 6/2 . 4 . 6), which is presented in figure 13 and its caption. Two other uniform polyhedra, with symbol (8 . 8 . 8 . 8), are shown in Figure 14. They are representative of several others that can be obtained analogously from uniform polyhedra by deleting one transitivity class of faces, and doubling-up the remaining faces. Some – though not all – of such polyhedra are noble.

Another example concerns two polyhedra the existence of which under the traditional concept of polyhedron was rejected in [12]. Discussing the possibility of existence of polyhedra with symbols $t\{5/2, 5\}$ and $t\{5/2, 3\}$ in the notation of [12], the authors say (on page 411) that "... $t\{5/2, 5\}$ consists of three coincident dodecahedra, while $t\{5/2, 3\}$ consists of two coincident great dodecahedra along with the icosahedron that has the same vertices and edges ...". (The construction in question consists of truncating the regular polyhedra to the extent of completely cutting off their "points".) While the non-acceptance of the resulting object among the uniform polyhedra in the traditional meaning is fully justified (even if not for the reason stated),

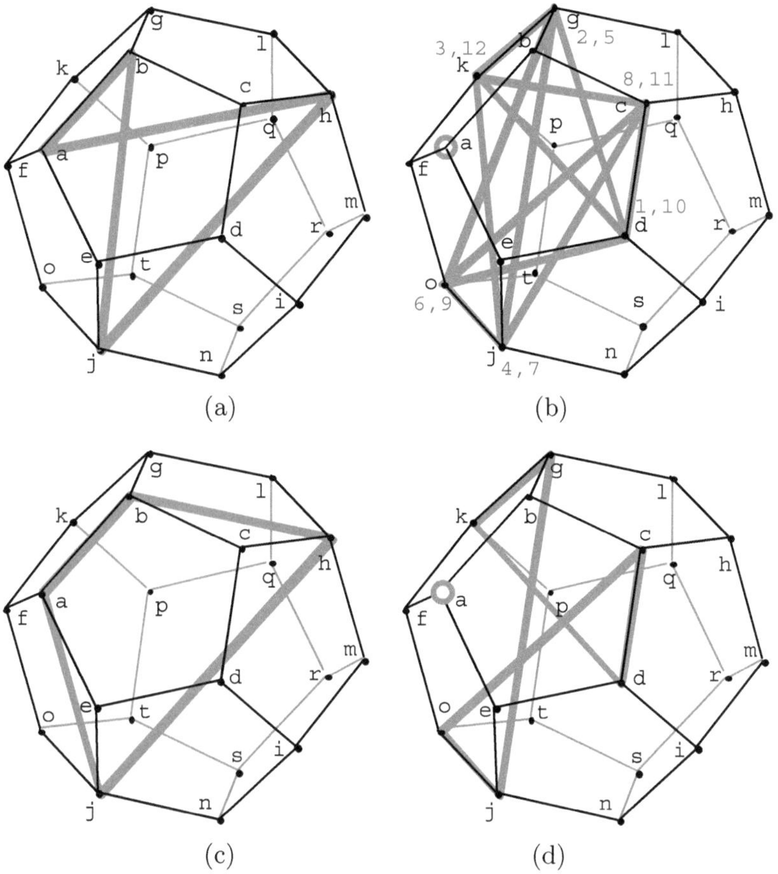

Fig. 12. Four noble polyhedra. The hexecontahedron in (a) consists of the sixty quadrangles congruent to the one emphasized, that can be inscribed in the regular dodecahedron. Its polar is an icosahedron, with twenty 12-gonal faces. The diagram in (b) shows the face which is the polar of the vertex a. Each of the 20 vertices of the icosahedron in (b) represents three coinciding vertices, while each face meets six pairs of coinciding vertices. Clearly, the coincidences here are no more against the traditional grain than the ones in the polyhedron in Figure 9. The other hexecontahedron is obtained similarly from the quadrangle in (c), while (d) shows the face of a noble icosahedron polar to the polyhedron in (c). The diagram in (d) shows the face which is the polar of the vertex a. Each of the 20 vertices of the icosahedron in (d) represents three coinciding vertices, while each face meets six pairs of coinciding vertices, each pair determining an edge of zero length. In all diagrams the dodecahedra serve only to guide the construction and recognition of the faces of the polyhedra described.

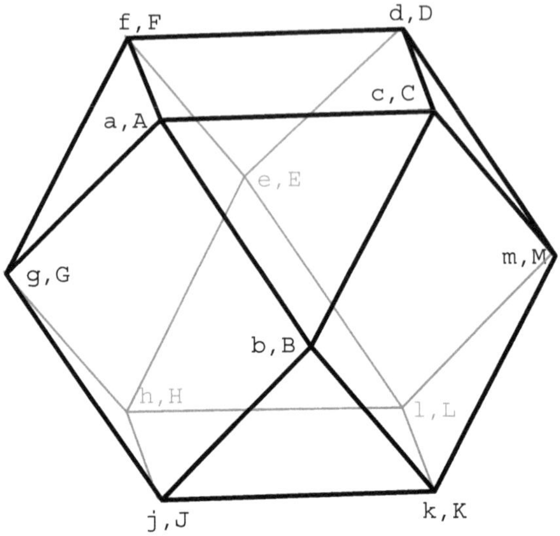

Fig. 13. A nontrivial uniform polyhedron (3 . 6 . 4 . 6/2 . 4 . 6) with 24 vertices, coinciding in pairs with the vertices of a cuboctahedron. Its faces are eight triangles: (a b c a), (A B C A), (d e f d), (D E F D), (g h j g), (G H J G), (k l m k), (K L M K); twelve squares: (a g J B a), (A G j b A), (b k M C b), (B K m c B), (c d F A c), (C D f a C), (D M l e D), (d m L E d), (E H g f E), (e h G F e), (H L k j H), (h l K J h); eight hexagons {6}: (a f e l k b a), (A F E L K B A), (b j h e d c b), (B J H E D C B), (a c m l h g a), (A C M L H G A), (d f g j k m d), (D F G J K M D); and four hexagons {6/2}: (a B c A b C a), (d E f D e F d), (g H j G h J g), (k L m K l M k). The polyhedron is orientable and of genus 9; since some of the faces pass through the center, no density at the center can be defined, and there is no polar polyhedron with the same degree of symmetry.

the uniform polyhedra $t\{5/2, 5\}$ and $t\{5/2, 3\}$ exist in our interpretation of "polyhedron". Indeed, as is best seen from the illustration in Figure 15, the truncation of $t\{5/2, 5\}$ leads to a uniform polyhedron (5 . 10/2 . 10/2) with sixty vertices. In a similar way, the truncation of $t\{5/2, 3\}$ yields a uniform polyhedron (3 . 10/2 . 10/2) with sixty vertices.

As a final example in this section, we recall that in the process of verification of the completeness of the enumeration of the uniform polyhedra in [12], Skilling [50] found one extraordinary object, the *great disnub dirhombidodecahedron*, which would have qualified as a uniform polyhedron in every respect except that it has four faces incident with some edges. However, as Skilling points out on p. 123, this object *is* a polyhedron if the exceptional edges are interpreted as two distinct edges which happen to be represented by the same segment although they are determined by different pairs of faces; in other words, it is a polyhedron in the sense adopted here. This and other "new" uniform polyhedra are discussed in greater detail in [29].

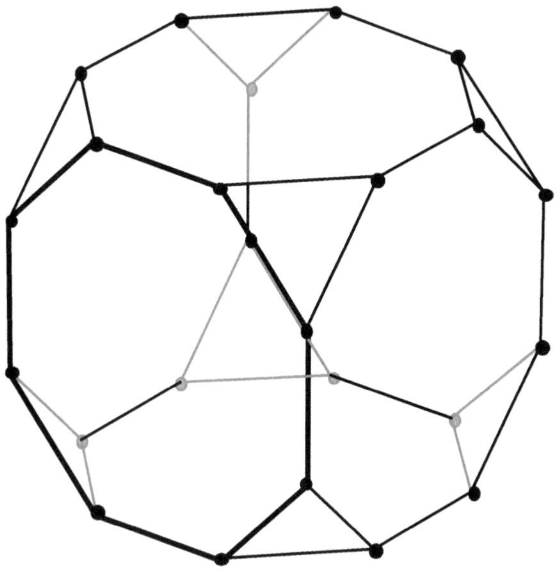

Fig. 14. Deleting all triangles from the uniform truncated cube (3 . 8 . 8), and then face-doubling the octagons, leads to two "new" polyhedra (8 . 8 . 8 . 8), which are not only uniform, but noble; moreover, they are isomeghetic. In both, each octagon has been replaced by one "red" and one "green" octagon. The two are adjacent along the four edges previously adjacent to triangles. In the first polyhedron the remaining edges are adjacent to octagons of the same color, in the second to differently colored ones. The difference between the two is that in the first, each triangular hole is surrounded by two circuits of three octagons each, while in the second one it is surrounded by one circuit of six octagons.

7 Isohedral Polyhedra with Regular Vertices

The polars of uniform polyhedra (with respect to a sphere whose center coincides with the centroid of the polyhedron) are isohedral polyhedra with regular vertex-stars. In the case of convex uniform polyhedra these isohedral polyhedra are often called *Catalan polyhedra*, although the historical justification for this seems to be ambiguous. There has been no name proposed for the general case, and, in fact, there appears to have been avoidance in considering such polyhedra. There are several reasons for this situation.

To begin with, in a number of uniform polyhedra some of the faces pass through the centroid of the polyhedron; therefore there is no polar polyhedron in either the traditional sense or in the meaning of "polyhedron" accepted here. Brückner [3, p. 191] ignores the question of polars of such polyhedra, although he claims to be systematically discussing the isogonal polyhedra *and* their polar isohedral ones. Wenninger [54] and Har'El [31] solve the problem of polars of some of the uniform polyhedra by admitting unbounded faces.

Fig. 15. The truncation of the regular polyhedron $\{5/2, 5\}$. (a) shows an early stage of the truncation; one of the pentagonal faces, and one of the decagonal faces are emphasized. (b) shows an almost complete truncation, illustrating the proximity of the emphasized pentagon and decagon. (c) is the complete truncation, in which each face of the "dodecahedron" represents one pentagon $\{5\}$ and one decagon $\{10/2\}$. Each dodecahedral vertex represents three vertices of the uniform polyhedron (5 . 10/2 . 10/2), the truncation of $\{5/2, 5\}$. Continuation of this sequence leads to several interesting polyhedra; they will be described in detail elsewhere.

While such an approach is interesting, it certainly does not fall within the usual scope of the meaning of "isohedral polyhedron".

Another difficulty for the traditional approach is that some of the uniform polyhedra have pairs of coplanar faces; hence the polar polyhedra must have pairs of coinciding vertices – which would make them unacceptable under the traditional definition of polyhedra. However, neither in [54] nor in [31] is any mention made of this fact. The vertices which are incident with two cycles of faces are neither noticed nor explained, nor is any mention made of the fact that, for example, the uniform polyhedron (3.3.3.3.3.5/2) has 112 faces, but the purported polar shown in [54] and [31] has only 92 vertices. On the other hand, in our interpretations of polyhedra there is no problem in such cases: the two vertices of each pair are distinct, and only in the realization they happen to be represented by a single point.

8 Other Polyhedra

There are several other classes of polyhedra for which the definition of polyhedra as presented here is useful – either in clarifying and eliminating what seemed to be unexpected exceptional cases, or in enabling a complete and unambiguous determination of all members of the class.

One example of the former kind concerns the recent study by Shephard [49] of isohedral *deltahedra* (polyhedra all faces of which are equilateral triangles). After explaining one of the constructions of such polyhedra – the replacement of each face of a regular polyhedron by the mantle of a pyramid (with equilateral triangles) erected over the face as basis – the claim is made that this construction works on eight of the regular polyhedra but not on $\{5, 5/2\}$. In fact, the construction works in this case as well, and results in an isohedral hexecontahedron of type [5 . 10/2 . 10/2] that looks like the regular icosahedron to which it is isomeghethic, but has three (combinatorially distinguishable) faces over each icosahedral face. A similar construction consisting of "excavating" the pyramids is said in [49] to fail when applied to the tetrahedron "... since the construction leads to a set of twelve equilateral triangles which coincide in four sets of three." In our interpretation, the resulting polyhedron is combinatorially equivalent to the one obtained by erecting the pyramids, except that in this realization each triplet of (distinguishable) faces is represented by one triangle.

Settling on a particular definition of polyhedra makes possible the completion of enumeration of several classes of polyhedra. The determination of all face-transitive polyhedra with rectangular faces, started in [10] and [11] is being carried out in work in preparation. Also in preparation are enumerations of rhombic or parallelogram-faced isohedra (extending work in [26] and [28]) and on simplicial isohedra.

While definitions of the polyhedral concept different from the one adopted here are certainly possible, and possibly useful, at the moment there seems to be no better alternative available that is both general and internally consistent, and also satisfies the criteria set out in Section 1.

References

[1] A. D. Aleksandrov, Convex Polyhedra. [In Russian]. Moscow 1950. German translation: A. D. Alexandrow, Konvexe Polyeder, Akademie-Verlag, Berlin 1958.

[2] E. R. Berlekamp, E. N. Gilbert and F. W. Sinden, A polygon problem. Amer. Math. Monthly 72(1965), pp. 233 – 241; reprinted in *Selected Papers on Algebra*, Mathematical Association of America, Washington, D.C., 1977.

[3] M. Brückner, *Vielecke und Vielflache. Theorie und Geschichte.* Teubner, Leipzig 1900.

[4] M. Brückner, Über die diskontinuierlichen and nicht-konvexen gleicheckig-gleichflächigen Polyeder. *Verh. des dritten Internat. Math.-Kongresses* Heidelberg 1904. Teubner, Leipzig 1905, pp. 707 – 713.

[5] M. Brückner, Über die gleicheckig-gleichflächigen, diskontinuirlichen und nichkonvexen Polyeder. Nova Acta Leop. 86(1906), No. 1, pp. 1 – 348 + 29 plates.

[6] M. Brückner, Zur Geschichte der Theorie der gleicheckig-gleichflächigen Polyeder. Unterrichtsblätter für Mathematik und Naturwissenschaften, 13(1907), 104 – 110, 121 – 127 + plate.

[7] A. L. Cauchy, Recherches sur les polyèdres; Premier mèmoire. J. École Polytech. 9(1813), 68 – 98. German translation by R. Haußner, with comments, as

"Abhandlung über die Vielecke und Vielflache". Pages 49 – 72 and 121 – 123 in *Abhandlungen über die regelmäßigen Sternkörper*, Ostwald's Klassiker der exakten Wissenschaften, Nr. 151. Engelmann, Leipzig 1906.

[8] H. S. M. Coxeter, Regular skew polyhedra in three and four dimensions. Proc. London Math. Soc. (2) 43(1937), 33 – 62.

[9] H. S. M. Coxeter, Regular Polytopes. 3rd ed. Dover, New York 1973.

[10] H. S. M. Coxeter and B. Grünbaum, Face-transitive polyhedra with rectangular faces. Math. Reports Acad. Sci. Canada 20(1998), 16 – 21.

[11] H. S. M. Coxeter and B. Grünbaum, Face-transitive polyhedra with rectangular faces and icosahedral symmetry. Discrete & Comput. Geometry 25(2001), 163 – 172.

[12] H. S. M. Coxeter, M. S. Longuet-Higgins and J. C. P. Miller, Uniform polyhedra. Philos. Trans. Roy. Soc. London (A) 246(1953/54), 401 – 450.

[13] P. J. Davis, *Circulant Matrices*. Wiley-Interscience, New York 1979.

[14] J. Douglas, Geometry of polygons in the complex plane. *J. Math. Phys.* 19(1940), 93 – 130.

[15] A. W. M. Dress, A combinatorial theory of Grünbaum's new regular polyhedra, Part I: Grünbaum's new regular polyhedra and their automorphism group. Aequationes Math. 23(1981), 252 – 265.

[16] A. W. M. Dress, A combinatorial theory of Grünbaum's new regular polyhedra, Part II: Complete enumeration. Aequationes Math. 29(1985), 222 – 243.

[17] A. Dürer, *Unterweisung der Messung mit dem Zirkel und Richtscheit*. 1525. English translation with commentary by W. L. Strauss: *The Painter's Manual*. Abaris, New York 1977.

[18] Euclid, *The Thirteen Books of Euclid's Elements*, translated and edited by T. L Heath. Vol. III, Books X – XIII, Cambridge Univ. Press 1926. There are many other editions and translations.

[19] J. C. Fisher, D. Ruoff and J. Shilleto, Polygons and polynomials. In: *The Geometric Vein: the Coxeter Festschrift*, C. Davis, B. Grünbaum, F. A. Scherk, eds., pp. 165 – 176. Springer, New York, 1981.

[20] A. Girard, *Table des sines, tangentes & secantes, selon le raid de 100000 parties. Avec un traicté succinct de la trigonometrie tant des triangles plans, que sphericques. Où son plusieurs operations nouvelles, non auparavant mises en lumiere, tres-utiles & necessaires, non seulment aux apprentifs; mais aussi aux plus doctes practiciens des mathematiques*. Elzevier, à la Haye 1626.

[21] B. Grünbaum, Regular polyhedra – old and new. Aequationes Math. 16(1977), 1 – 20.

[22] B. Grünbaum, Regular polyhedra. In *Companion Encyclopaedia of the History and Philosophy of the Mathematical Sciences,* I. Grattan-Guinness, ed. Routledge, London 1994. Vol. 2, pp. 866 – 876.

[23] B. Grünbaum, Polyhedra with hollow faces. In *POLYTOPES: Abstract, Convex and Computational*, Proc. NATO – ASI Conference, Toronto 1993. T.

Bisztriczky, P. McMullen, R. Schneider and A. Ivic' Weiss, eds. Kluwer Acad. Publ., Dordrecht 1994, pp. 43 – 70.

[24] B. Grünbaum, Metamorphoses of polygons. In *The Lighter Side of Mathematics, Proc. Eugène Strens Memorial Conference*, R. K. Guy and R. E, Woodrow, eds. Math. Assoc. of America, Washington, D.C. 1994, pp. 35 – 48.

[25] B. Grünbaum, Isogonal decagons. In: *The Pattern Book. Fractals, Art, and Nature*, C. A. Pickover, ed. World Scientific, Singapore 1995, pp. 251 – 253.

[26] B. Grünbaum, Still more rhombic hexecontahedra. Geombinatorics 6(1997), 140 – 142.

[27] B. Grünbaum, Realizations of symmetric maps by symmetric polyhedra. Discrete Comput. Geom. 20(1998), 19 – 33.

[28] B. Grünbaum, Parallelogram-faced isohedra with edges in mirror-planes. Discrete Math. 221(2000), 93–100.

[29] B. Grünbaum, "New" uniform polyhedra. In: *Discrete Geometry: In Honor of W. Kuperberg's 60th Birthday*, A. Bezdek, ed. Dekker, New York 2003, pp. 331–350.

[30] S. Günther, *Vermischte Untersuchungen zur Geschichte der mathematischen Wissenschaften*. Teubner, Leipzig 1876.

[31] Z. Har'El, Uniform solutions for uniform polyhedra. Geometriae Dedicata 47(1993), 57 – 110.

[32] E. Hess, Über gleicheckige und gleichkantige Polygone. Schriften der Gesellschaft zur Beförderung der gesammten Naturwissenschaften zu Marburg, Band 10, Abhandlung 12, pp. 611 – 743, 29 figures. Th. Kay, Cassel 1874.

[33] E. Hess, Ueber zwei Erweiterungen des Begriffs der regelmässigen Körper. Sitzungsberichte der Gesellschaft zur Beförderung der gesammten Naturwissenschaften zu Marburg 1875, pp. 1 – 20.

[34] E. Hess, Ueber die zugleich gleicheckigen und gleichflächigen Polyeder. Schriften der Gesellschaft zur Beförderung der gesammten Naturwissenschaften zu Marburg, Band 11, Abhandlung 1, pp. 1 – 97, 11 figures. Th. Kay, Cassel 1876.

[35] E. Hess, Ueber einige merkwürdige nichtkonvexe Polyeder. Sitzungsberichte der Gesellschaft zur Beförderung der gesammten Naturwissenschaften zu Marburg 1877, pp. 1 – 13.

[36] E. Hess, *Einleitung in die Lehre von der Kugelteilung*. Teubner, Leipzig 1883.

[37] T. Hugel, *Die regulären und halbregulären Polyeder.* Gottschick-Witter, Neustadt a. d. H., 1876.

[38] J. Kepler, *Harmonices mundi*. J. Planck, Linz 1619. Also in *Opera omnia*, Vol. V, Frankfurt 1864, pp. 75 – 334. German translation in *Gesammelte Werke*, Vol. 6, Beck, Munich 1940, pp. 3 – 337. There are many other editions and translations.

[39] H. Martini, On the theorem of Napoleon and related topics. Math. Semesterber. 43(1996), 47 – 64.

[40] P. McMullen and E. Schulte, *Abstract Regular Polytopes*. Cambridge Univ. Press 2002.

[41] A. L. F. Meister, Generalia de genesi figurarum planarum et inde pendentibus earum affectionibus. Novi Comm. Soc. Reg. Scient. Gotting. 1(1769/70), pp. 144 – 180 + plates.

[42] A. F. Möbius, Ueber die Bestimmung des Inhaltes eines Polyëders. Ber. Verh. Sächs. Ges. Wiss, math.-phys. Kl. 17(1865), 31 – 68. (= *Ges. Werke*, Vol.2 pp. 473 – 512. Hirzel, Leipzig 1886.)

[43] B. H. Neumann, Plane polygons revisited. In *Essays in Statistical Science: Papers in Honour of P. A. P. Moran*, J. Gani and E. J. Hannan, eds. Also in J. of Appl. Prob., Special volume 19A(1982), pp. 113 – 122.

[44] L. Pacioli, *De divina proportione*. 1498.

[45] L. Poinsot, Mémoire sur les polygones et les polyèdres. J. École Polytech. 10(1810), 16 – 48. German translation by R. Haußner, with comments, as "Abhandlung über die Vielecke und Vielflache". Pages 3 – 48 and 105 – 120 in *Abhandlungen über die regelmäßigen Sternkörper*, Ostwald's Klassiker der exakten Wissenschaften, Nr. 151. Engelmann, Leipzig 1906.

[46] I. J. Schoenberg, The finite Fourier series and elementary geometry. Amer. Math. Monthly 57(1950), pp. 390 – 404.

[47] W. Schuster, Polygonfolgen und Napoleonsätze. Math. Semesterberichte 41(1994), 23 – 42.

[48] W. Schuster, Regularisierung von Polygonen. Math. Semesterberichte 45(1998), 77 – 94.

[49] G. C. Shephard, Isohedral deltahedra. Periodica Math. Hungar. 39(1999), 83 – 106.

[50] J. Skilling, The complete set of uniform polyhedra. Philos. Trans. Roy. Soc. London (A) 278(1975), 111 – 135.

[51] S. P. Sopov, Proof of the completeness of the enumeration of uniform polyhedra. [In Russian] Ukrain. Geom. Sbornik 8(1970), 139 – 156.

[52] E. Steinitz, Polyeder und Raumteilungen. *Encykl. Math. Wissenschaften* 3(1922), Geometrie, Part 3AB12, pp. 1 – 139.

[53] M. J. Wenninger, *Polyhedron Models*. Cambridge University Press 1971.

[54] M. J. Wenninger, *Dual Models*. Cambridge University Press 1983.

[55] C. Wiener, *Über Vielecke und Vielflache*. Teubner, Leipzig 1864.

[56] S. E. Wilson, New techniques for the construction of regular maps. Ph.D. thesis, University of Washington, Seattle 1976.

Branko Grünbaum
University of Washington
Seattle, WA 98195-4350
grunbaum@math.washington.edu

Some Algorithms Arising in the Proof of the Kepler Conjecture

Thomas C. Hales

Abstract

By any account, the 1998 proof of the Kepler conjecture is complex. The thesis underlying this article is that the proof is complex because it is highly under-automated. Throughout that proof, manual procedures are used where automated ones would have been better suited. This paper gives a series of nonlinear optimization algorithms and shows how a systematic application of these algorithms would bring substantial simplifications to the original proof.

1 Introduction

In 1998 a proof of the Kepler conjecture was completed [8]. By any account, that solution is complex (300 pages of text, 3 GB stored data on the computer, computer calculations taking months, 40K lines of computer code, and so forth). The thesis underlying this article is that the 1998 proof is complex because it is highly under-automated. Throughout that proof, manual procedures are used where automated ones would have been better suited.

Ultimately, a properly automated proof of the Kepler conjecture might be short and elegant. The hope is that the Kepler conjecture might eventually become an instance of a general family of optimization problems for which general optimization techniques exist. Just as today linear programming problems of a moderate size can be solved without fanfare, we might hope that problems of a moderate size in this family might be routinely solved by general algorithms. The proof of the Kepler conjecture would then consist of demonstrating that the Kepler conjecture can be structured as a problem in this family, and then invoking the general algorithm to solve the problem.

As a step toward that objective, this article frames the primary algorithms of that proof in sufficient generality that they may be applied to much larger families of problems. The algorithms are arranged into four sections: *Quantifier Elimination, Linear Assembly Problems, Automated Inequality Proving, Plane Graph Generation.*

We do not claim any originality in the algorithms. In fact, the purpose is just the contrary: to exhibit the proof of the Kepler conjecture insofar as possible as an instance of standard optimization techniques. To keep things as general as possible, the algorithms we present here will make no mention of the particulars of the Kepler conjecture. A final section lists parts of the 1998 proof that can be structured according to these general algorithms.

2 Quantifier Elimination

We might try to structure the Kepler conjecture as a statement in the elementary theory of the real numbers. Tarski proved the decidability of this theory, through quantifier elimination. G. Collins and others have developed and implemented concrete algorithms to perform quantifier elimination [5]. The Kepler conjecture, as formulated in [6], is not a statement in this theory, because the transcendental arctangent function enters into the statement.

However, it seems that the arctangent is not essential to the formulation of the Kepler conjecture, and that it enters only because no attempt was made to do without it. For example, it is plausible that it can be replaced with a close rational approximation, without doing violence to the proof. In fact, the computer calculations in that proof are already based on rational approximations with explicit error bounds, and on its rational derivative $1/(1+x^2)$.

Assuming this can be done, quantifier elimination gives a procedure to solve the Kepler conjecture. Unfortunately, these algorithms are prohibitively slow (exponential, or doubly exponential in the number of variables).

Section 3 of this article proposes a different family of optimization problems for which algorithmic performance is satisfactory. These are called linear assembly problems.

Although quantifier elimination is too slow to be of practical value as a 1-step solution to the Kepler conjecture, it can be of great value in proving intermediate results. Recent algorithms are able to solve problems nearly at the level of difficulty of intermediate results in the Kepler conjecture [2], [13].

Instances of the following families of problems arise as intermediate steps in the Kepler conjecture. Each instance of the following families must provide an explicit set of parameters r_k (or **dmin**, **dmax**), and the problem becomes to show that configurations of points in $\mathbf{R}^3$ with the given parameters do not exist. In theory, the problems are all amenable to solution by quantifier elimination.

Problem 2.1. *Let S be a simplex whose edges all have length at most given parameter values r_i. Show that there is no point in the interior of the simplex with distance at least r from every vertex.*

Problem 2.2. *Show that there does not exist a triangle of circumradius at most r_1, and a segment of length at most r_2 such that the segment passes* *through the interior of the triangle and such that each endpoint of the segment has distance at least* r_3 *from each vertex of the triangle.*

Problem 2.3. *Show that there does not exist a configuration of* 5 *points* $0, p_1, p_2, p_3, q$ *with given minimum* $\mathbf{dmin}(p,q)$ *and maximum* $\mathbf{dmax}(p,q)$ *distances between each pair* (p,q) *of points. The line through* $(0,q)$ *must link the triangle with vertices* p_i.

Problem 2.4. *Show that there does not exist a configuration of* 6 *points* 0, $p_1, \ldots, p_4$, q, *with given minimum and maximum distances between each pair* (p,q) *of points. The line* $(0,q)$ *must link the skew quadrilateral with vertices* q_i *(ordered according to subscripts).*

Problem 2.5. *Show that there does not exist a configuration of* 7 *points* 0, $p_1, \ldots, p_4$, q_1, q_2 *with given minimum* $\mathbf{dmin}(p,q)$ *and maximum* $\mathbf{dmax}(p,q)$ *distances between each pair* (p,q) *of points. For* $j = 1, 2$, *the line* $(0, q_j)$ *must link the skew quadrilateral with vertices* q_i *(ordered according to subscripts).*

Although we hope that one day these problems will all be amenable to direct solution by quantifier elimination, in practice, we did not try to apply quantifier elimination directly without preprocessing them. The idea of preprocessing is that if a configuration exists, then the points can be moved in such a way to make various upper and lower bound constraints on distances bind. With a large number of binding constraints, the dimension of the configuration space becomes smaller and the problem easier to solve. In some cases, preprocessing reduces the configuration space to a single configuration, so that the existence of the configuration can be tested by choosing coordinates and calculating whether all the metric constraints are satisfied.

These five families of quantifier elimination problems have a similar feel to them. Let us give a preprocessing algorithm in general enough terms that it applies uniformly to all five problem families.

The primary preprocessing of the configurations is a deformation that we call *pivoting*. Fix any three of the points p_1, p_2, and q of the configuration. We call a *pivot through axis* (p_1, p_2) the continuous motion of q in the perpendicular bisecting plane B of (p_1, p_2) at constant distance from p_1 and p_2. Thus, the pivot moves q in a circular path in the plane B.

Usually, the plane $P = (p_1, p_2, q)$ through the three points is chosen to have the property that the entire configuration lies in a half-space through P. If q moves away from the half-space containing P, the distances from q to the other points of the configuration increase or remain the same. If q moves into the half-space, the distances decrease or remain the same.

More generally, we allow the plane P to separate the points of the configuration into two groups, such that the lower distance bounds from q to the first group do not bind, and such that the upper distance bounds from q to the second group do not bind.

To apply pivots, we must prescribe their directions, whether into the half-space or away from it. To do this, we give a *model*, which is a configuration

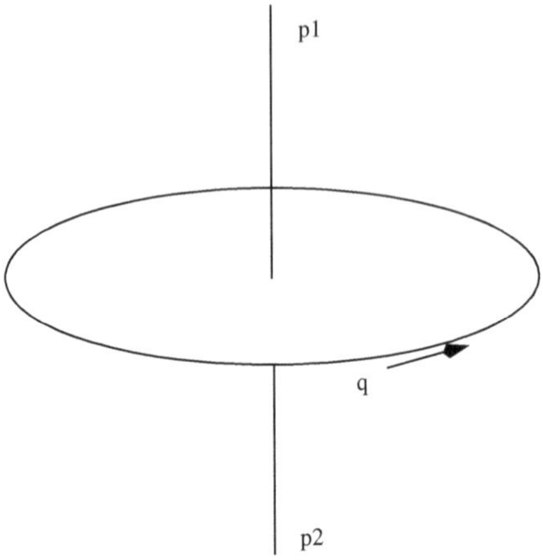

Fig. 1. A pivot is the circular motion of a point around a fixed axis.

that exists, of the form indicated in the problem family, but which is not required to satisfy the various constraints. Various edges in the model are marked with a strut (indicating a lower bound) or a cable (indicating an upper bound). If the model has a cable, then preprocessing pivots are applied to increase the corresponding distance in the configuration space, until the given upper bound is reached. Where the model has a strut, pivots are applied so as to decrease distances to the lower bound.

(Bob Connelly has pointed out that some of these problems can be viewed as tensegrity problems, but we do not see how to treat them all as tensegrities, so we do not pursue this point of view here. Globally rigid tensegrities are analogous to our models. However, our models are not claimed to be rigid.)

Example 2.6. *In Problem 2.1, let the upper bounds on the edges of the simplex be $\sqrt{8}$, and let $r = 2$. We take the model to be a regular tetrahedron with edges marked as cables. Mark the edges from the circumcenter to the vertices as struts. We apply pivots to the simplex to increase its edges to $\sqrt{8}$. Move the interior point by a sequence of pivots so that it has distance exactly 2 to three of the four vertices of the simplex.*

After these pivots are completed, the configuration is uniquely determined, and a calculation with explicit coordinates shows that the configuration does not exist, because the distance from the interior point to the fourth vertex of the simplex is too small.

In these problems, when we pivot in the correct direction, the distance constraints between points take care of themselves. However, some of the problems impose additional constraints. In Problem 2.1, the point is constrained to lie in the simplex. In the other problems, lines are required to be linked with various space polygons. A separate verification is required to see that pivots do not violate these additional constraints. *These separate ver-*

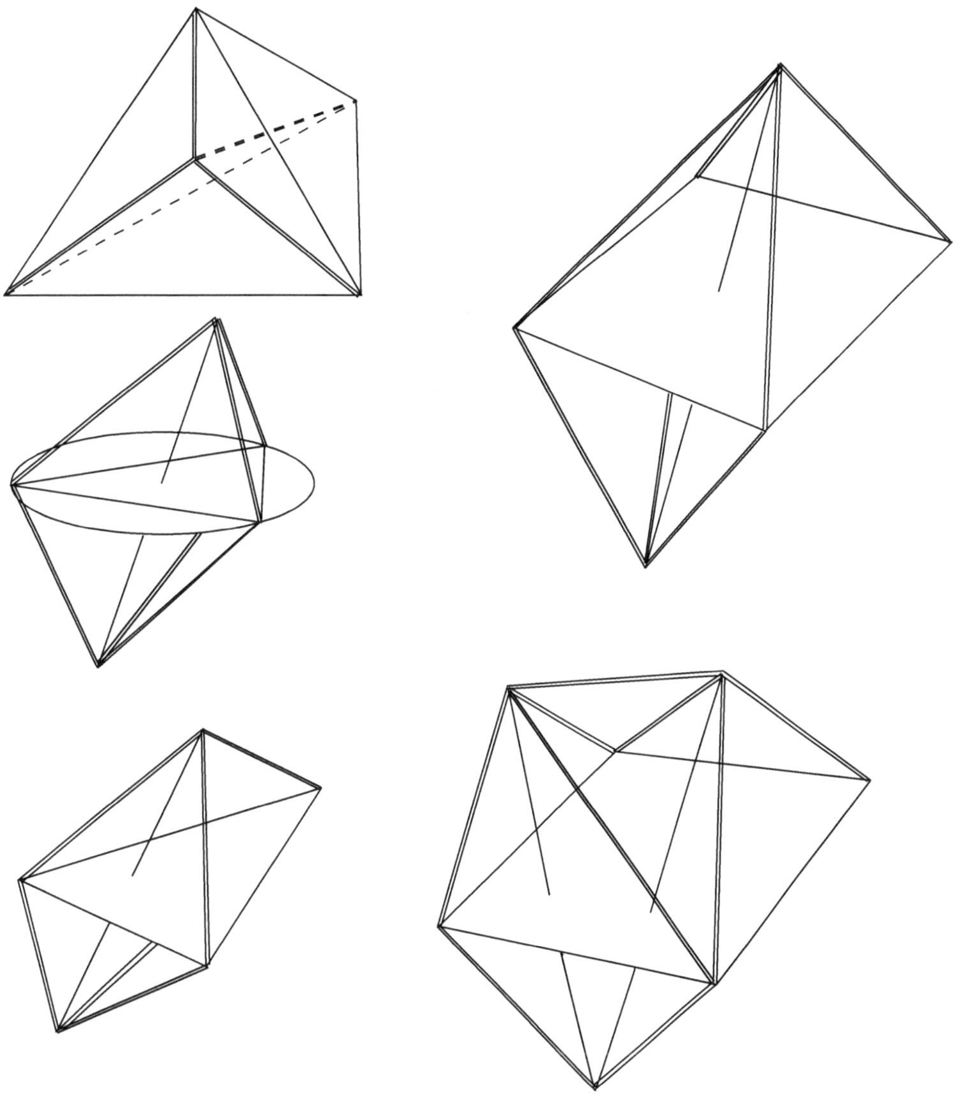

Fig. 2. Some models for the sample quantifier elimination problems. Struts are doubled lines.

ifications are again quantifier elimination problems, of a smaller magnitude than the original problem.

For example, in Problem 2.1, we verify that the point cannot be an interior point of a face of the simplex. This insures that the point does not escape from the interior of the simplex during the sequence of pivots. The argument that there does not exist a point in a face, under the given metric constraints, is similar to Example 2.6, but all the arguments are reduced to two dimensions, instead of three.

The preprocessing in most other cases is similar to Example 2.6, and can be reconstructed without difficulty from the models. The two exceptions are Problem 2.4 and Problem 2.5, which require substantial preprocessing and a lemma to insure that the pivots can be carried out. (I would much prefer to have arguments based on a pure quantifier elimination algorithm and bypass this lemma entirely, but the current quantifier elimination algorithms do not seem up to the task.)

Lemma 2.7. *Fix constants ℓ_i, k_i, ϵ, h_i, and ℓ subject to the constraints*

$$\ell_i < \sqrt{8}, \quad k_i \in [2, 2.61], \quad \epsilon \geq 2, \quad h_i \in [2, \sqrt{8}], \quad \ell \in [2, \sqrt{8}]. \tag{1}$$

Pick the following parameters in Example 2.5

$$\begin{array}{lll} \mathbf{dmin}(p, q_j) = h_i & \mathbf{dmin}(q_1, q_2) = \epsilon & \mathbf{dmin}(others) = 2 \\ \mathbf{dmax}(0, p_i) = \ell_i & \mathbf{dmax}(0, q_j) = \ell & \mathbf{dmax}(p_i, p_{i+1}) = k_i \end{array}$$

and let the other values of **dmax** *be $+\infty$. If a configuration of 7 points exists with these parameters, then a configuration also exists with these parameters and the additional constraints that*

$$|p_i| = 2, \quad |p_i - p_{i+1}| = k_i, \quad |q_j| = \ell.$$

Furthermore, the same lemma and conclusion holds in the context of the 6-point configuration of Example 2.4, if we take $q_1 = q_2 = q$ and **dmin**$(q_1, q_2) =$ 0.

Proof. This is Lemma 4.3 of [7]. Some of the constants have been relaxed in a way that affects the proof in a very minor way. (Two modifications must be made to the proof. The assertion that the circumradius of a triangle of sides $2.1, 2.51, 2.51$ is less than $\sqrt{2}$ of the original must be replaced with the assertion that there exists an $x > 2$ such that the circumradius of the triangle of sides $x, 2.61, 2.61$ is less than $\sqrt{2}$. Also, an instance of Problem 2.1 is needed, with $r = 2$ and the length-bounds of the sides of the simplex 2.61, 2.61, $\sqrt{8}$ at one vertex, and lengths at most $\sqrt{8}$ at the edges opposite the edges of length at most 2.61.) □

3 Linear Assembly Problems

In this section we define a class of nonlinear optimization problems that we call *linear assembly problems.*

Assume given a topological space X, and a finite collection of topological spaces, called *local domains.* For each local domain D there is a map $\pi_D : X \to D$. There are functions u_i, $i = 1, \ldots, N$, each defined on some local domain $D_i = \mathbf{dom}(u_i)$, and we let x_i denote the composite $x_i = \pi_{D_i} \circ u_i$.

On each local domain D, the functions u_i are related by a finite set of nonlinear relations

$$\phi(u_i : \mathbf{dom}(u_i) = D) \geq 0, \quad \phi \in \Phi_D. \tag{2}$$

We use vector notation $x = (x_1, \ldots, x_N)$, with constant vectors c, b, and matrix A given.

The problem is to maximize $c \cdot x$ subject to the constraints

$$A\,x \leq b, \tag{3}$$

and to the nonlinear relations 2. A problem of this form is called a linear assembly problem. (Intuitively, there are a number of nonlinear objects D, that form the pieces of a jigsaw puzzle that fit together according to the linear conditions 3.)

Example 3.1. *Assume a single local domain D, and let $\pi_D : X = D$ be the identity map. The function $f = c \cdot x$ is nonlinear. The problem is to maximize f over D subject to the nonlinear relations Φ_D. This is a general constrained nonlinear optimization problem.*

Example 3.2. *Assume that each u_i has a distinct local domain $D_i = \mathbf{R}$. Let $X = \mathbf{R}^N$, let π_D be the projection onto the ith coordinate, and let x_i be the ith coordinate function on $\mathbf{R}^n$. Assume that Φ_D is empty for each D. The problem becomes the general linear programming problem*

$$\max c \cdot x$$

such that $Ax \leq b$.

These two examples give the nonlinear and linear extremes in linear assembly problems. The more interesting cases are the mixed cases which combine nonlinear and linear programming. Example 2.3 gives one such case.

Example 3.3. *(2D Voronoi cell minimization). Take a packing of disks of radius 1 in the plane. Let Λ be the set of centers of the disks. Assume that the origin $0 \in \Lambda$ is one of the centers. The truncated Voronoi cell at 0 is the set of all $x \in \mathbf{R}^2$ such that $|x| \leq t$, and x is closer to the origin than to any other center in Λ. We assume $t \in (1, \sqrt{2})$.*

Only the centers of distance at most $2t$ affect the shape and area of the truncated Voronoi cell. For each $n = 0, 1, 2, \ldots$, we have a topological space of all truncated Voronoi cells with n nonzero disk centers v_i at distance at most $2t$. Fix n, and let X be the topological space.

Let $D = D_i$, $i = 1, \ldots, n$, be the sectors lying between consecutive segments $(0, v_i)$. Each sector is characterized by its angle α and the lengths y_a and y_b of the two segments $(0, v_i)$, $(0, v_j)$ between which the sector lies. The part A in D of the area of the truncated Voronoi cell is a function of the

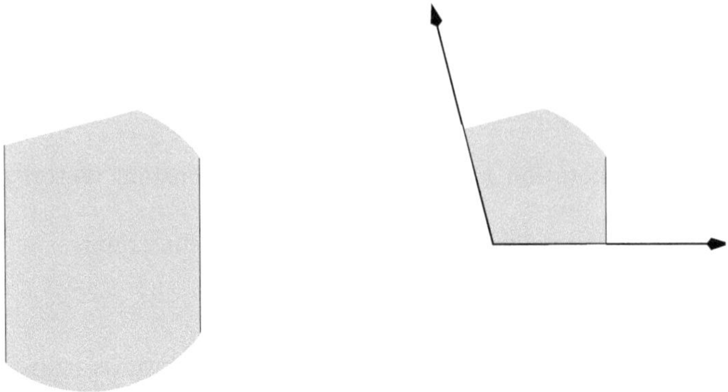

Fig. 3. A truncated Voronoi cell and a subset of the cell lying in a sector

variables α, y_a, y_b. *A nonlinear implicit equation* $\phi = 0$ *relates* A, α, y_a, *and* y_b *on* D. *The variables* u_i *of the linear assembly problem for the local domain* D *are* A, y_a, y_b, α.

We have a linear assembly problem. The function $c \cdot x$ *is the area of the truncated Voronoi cell, viewed as a sum of variables* A, *for each sector* D *(or rather, their pullbacks to* X *under the natural projections* $X \to D$*).*

The assembly constraints are all linear. One linear relation imposes that the angles of the n *different sectors must sum to* 2π. *Other linear relations impose that the variable* y_a *on* D *equals the variable* y_b *on* D' *if the two variables represent the length of the same segment* $(0, v_i)$ *in* X.

3.1 Solving linear assembly problems

In this section we describe how various linear assembly problems are solved in the proof of the Kepler conjecture in terms sufficiently general to apply to other linear assembly problems as well.

Let us introduce some general notation. Let $x_D = (x_i : \mathbf{dom}(u_i) = D)$ be the vector of variables with local domain D. Write $c{\cdot}x$ in the form $\sum_D c_D{\cdot}x_D$ and the assembly conditions as

$$Ax = \sum_D A_D x_D,$$

according to the local domain of the variable.

3.1.1 Linear relaxation

The first general technique is *linear relaxation*. We replace the nonlinear relations $\phi(x_D) \geq 0, \phi \in \Phi_D$ with a collection of linear inequalities that are true whenever the constraints Φ_D are satisfied: $A'_D x_D \leq b_D$. A linear program is obtained by replacing the nonlinear constraints Φ_D with the linear constraints. Its solution dominates the nonlinear optimization problem. In this way, the nonlinear maximization problem can be bounded from above.

Let us review some constructions that insure rigor in linear programming solutions. We assume general familiarity with the basic theory and terminology of linear programming. It is well-known that the primal has a feasible solution iff the dual is bounded. We will formulate our linear programs in such a way that both the primal and the dual problems are feasible and bounded.

We use vector notation to formulate a primal problem as

$$\max c \cdot x \tag{4}$$

such that $Ax \leq b$, where x is a column vector of free variables (no positivity constraints), A is a matrix, c is a row vector, and b is a column vector.

We can insure that this primal problem is bounded by bounding each of the variables x_i. (This is easily achieved considering the geometric origins of our problem, which provides interpretations of variables as particular dihedral angles, edge lengths, and volumes.) We assume that these bounds form part of the constraints $Ax \leq b$.

The linear programs we consider have the property that if the maximum is less than a constant K, the solution does not interest us. (For instance, in the dodecahedral conjecture, Voronoi cell volumes are of interest only if the volume is less than the volume of the regular dodecahedron.) This observation allows us to replace the primal problem with one having an additional variable t:

$$\max c \cdot x + K t \tag{5}$$

such that $Ax + bt \leq b$, and $0 \leq t \leq 1$. This modified primal is bounded for the same reasons that the original primal is. It has the feasible solution $x = 0$ and $t = 1$.

Lemma 3.4. *If the maximum M of the original primal is greater than K, then the optimal solution of the modified primal has $t = 0$, and hence its maximum is also M*

Proof. Assume that (x_0, t_0) gives an optimal solution to the modified problem for some $1 > t_0 > 0$, with $c \cdot x_0 + Kt_0 > M$. Then $(x_1, t_1) = (x_0/(1 - t_0), 0)$ is also a feasible solution and it beats the optimal solution

$$c \cdot x_1 + Kt_1 > c \cdot x_0 + Kt_0. \tag{6}$$

This contradiction proves $t_0 = 0$. □

The output from linear program that is solved by numerical methods can be transformed into a rigorous bound as follows. Based on the preceding remarks, we assume that these linear programs are feasible and bounded. The dual is then also feasible and bounded. We assume that the numerical solutions are carried out with sufficient accuracy to insure bounded feasible approximations to the true optima.

To explain the rigorous verification, we separate the equality constraints from the inequality constraints, and rewrite the problem as

$$\max c \cdot x \tag{7}$$

such that $A'x = b'$, $Ax \leq b$, with x free. The dual problem yields a solution to

$$\min yb' + zb, \tag{8}$$

such that $yA' + zA = c$, with $z \geq 0$ and y free. Let (y_0, z_0) be a numerically obtained approximation to the dual solution. The vector z_0 will be approximately positive, and by replacing negative coefficients by 0, we may assume $z_0 \geq 0$. Let $\delta = c - yA' - zA$ be the error row vector resulting from numerical approximations. Then for any feasible solution x of the primal, we have

$$c \cdot x = (\delta + y_0 A' + z_0 A)x \leq \delta \cdot x + y_0 \cdot b' + z_0 \cdot b. \tag{9}$$

Using the bounds of the variables x_i, we bound $\delta \cdot x \leq D$, and thus obtain the rigorous upper bound $c \cdot x \leq D + y_0 \cdot b' + z_0 \cdot b$ on the primal.

3.1.2 Implementation details

The linear programs are solved numerically using a commercial package (CPLEX). The input and output to these numerical programs are processed by a custom java program, which is linked to CPLEX with a java API provided by the software manufacturer. Each bound is calculated with interval arithmetic to insure that it is reliable. (We use a simple implementation of interval arithmetic in java based on the math.BigDecimal implementation of arbitrary precision arithmetic.)

3.2 Nonlinear duality

The second general technique is nonlinear duality. Suppose that we wish to show that the maximum of the primal problem 4 is at most M.

Let $x^* = (x^*_D)$ be a guess of the solution to the problem, obtained for example, by numerical nonlinear optimization. We relax the nonlinear optimization by dropping from the matrix A and the vector b those inequalities that are not binding at x^*. With this modification, we may assume that $Ax^* = b$. Let m be the size of the vector b, that is, the number of binding linear conditions. Let d be the number of local domains D.

We introduce a linear dual problem with real variables t, $r_\phi : \phi \in \Phi_D$, and $w \in \mathbf{R}^m$. The variables r_ϕ and w are constrained to be non-negative.

We consider the linear problem of maximizing t such that

$$M + dt - c \cdot x^* \geq 0 \tag{10}$$

and such that for each x_D in each D the linear inequality

$$c_D \cdot (x_D - x_D^*) + \sum_{\Phi_D} r_\phi \phi(x) + w A_D(x_D^* - x_D) + t < 0 \tag{11}$$

is satisfied.

There is no guarantee that a feasible solution exists to this system of inequalities. However, any feasible solution gives an upper bound M. Indeed, let $x = (x_D)$ be any feasible argument to the primal, and let t, r_ϕ, w be a feasible solution to the dual. Taking the sum of the linear inequalities 11, over D at x, we have (recall $\phi \geq 0$ and $Ax \leq b$):

$$\begin{aligned} M &\geq M + c \cdot (x - x^*) + \textstyle\sum_D \sum_{\Phi_D} r_\phi \phi(x) + wA(x^* - x) + d\,t, \\ &\geq c \cdot x + (M + d\,t - c \cdot x^*) + w(b - Ax), \\ &\geq c \cdot x. \end{aligned}$$

Since the dual problem has infinitely many constraints (because of constraints for each $x \in D$), we solve the dual problem in two stages. First, we approximate each D by a finite set of test points, and solve the finitely constrained linear programming problem for t, r_ϕ, and w.

We replace t with $t_0 = (-M + c \cdot x^*)/d$ (to make the constraint 10 bind). It follows from the feasibility of t that $t \geq t_0$, and that t_0, r_ϕ, w is also feasible on the finitely constrained problem. To show that t_0, r_ϕ, w satisfies all the inequalities 11 (under the substitution $t \mapsto t_0$), we use interval arithmetic to show that each of these inequalities hold. (To make these interval arithmetic verifications as easy as possible, we have chosen the solution t_0, r, w to make the closest inequality hold by as large a margin $t - t_0$ as possible. This is the meaning of the maximization over t in the dual problem.) The next section will give further details about interval arithmetic verifications.

3.3 Branch and bound

The third technique is branch and bound. When no feasible solution is found in step (2), it may still be possible to partition X into finitely many sets $X = \coprod X_i$, on which feasible solutions to the dual may be found. Although this is an essential part of the solution, the rules for branching in the Kepler conjecture follow the structure of that problem, and we do not give a general branching algorithm.

4 Automated Inequality Proving

What we would like is a general, efficient algorithm for proving inequalities of several real variables. Each inequality $f < 0$ of a continuous function on a compact domain can be expressed as a maximization problem:

$$\max f < 0. \tag{12}$$

Generally efficient algorithms are not possible because NP hard problems can be encoded as optimization problems of quadratic functions [10].

This section describes an inequality proving procedure that has worked well in practice, and which could be automated to provide a method of general interest. This section assumes some general familiarity with issues of floating-point and interval arithmetic, such as can be found in [1], [3]. Our methods are similar to those in [12].

To prove $f < 0$, it is enough to show that the maximum of f is less than 0. For this reason, we use interval arithmetic to bound the maximum of functions. Through interval arithmetic, an interval $[a, b]$ containing the range of f can be obtained. By verifying that $b < 0$, it follows that the range of f is negative, and hence that $f < 0$.

All our functions can be built from arithmetic operations. (Transcendental functions are replaced with explicit rational approximations with known error bounds.)

Often, the functions f are twice continuously differentiable. To obtain additional speed and accuracy, we use interval arithmetic to obtain rigorous bounds on the second partial derivatives of f. (We obtain formulas for the second partials through symbolic and automatic differentiation of the function f). With bounds on the second partials, we obtain rigorous bounds on f through its Taylor approximation.

The accuracy of the Taylor approximation improves as the domain shrinks in size. We chop the domain into a collection of small rectangles and check on each rectangle whether the Taylor bound implies $f < 0$. If Taylor bound is too crude to give $f < 0$, we divide it into smaller rectangles and recompute the Taylor bounds. By a process of adaptive subdivision of rectangles, the inequality $f < 0$ is eventually established.

Derivative information can be used to speed up the algorithm. Taylor bounds can also be applied to the first partial derivatives of f. If a partial derivative of a variable x is of fixed sign on a rectangle, then the function is maximized along an edge $x = a$ of the rectangle. If this edge is shared with an adjacent rectangle, the maximization of f is pushed to an adjacent rectangle. If this edge lies on the boundary of the domain, the dimension of the optimization problem is reduced by one.

The method outline above works extremely well for simple functions in a small number of variables. The complexity grows rapidly with the number of variables. We are able to obtain satisfactory results for many inequalities that depend on a single simplex S, that is, functions of six variables parameterized by the edge lengths of a simplex.

4.1 Generative Programming

Most of the computer code for the proof of the Kepler conjecture implements the Taylor approximations of the nonlinear functions. The computer code for proving $f < 0$ is obtained as follows.

First, an expression for f is derived. The formulas for the first and second partial derivatives of the function are obtained (say by a symbolic algebra system) from the expressions for f.

These symbolic expressions are then converted to an interval arithmetic format. In a language such as C++ with operator overloading, this can be achieved by defining a class for intervals and overloading arithmetic operations so that they may be applied to instances of the class. In languages without operator overloading, the conversion from the symbolic expression to computer source code is more involved.

There are other considerations to bear in mind in producing the interval code. In practice, there is a substantial degradation of performance when the rounding mode on the computer is frequently switched, and often it is necessary to rearrange the code substantially to reduce the number of changes in rounding mode. Also, floating point arithmetic is not associative, so that in order to obtain rigorous results based on interval arithmetic, great care must be paid to the placement of parentheses. Another issue is the input of floating point constants. In C++, the line of code in 13 sets $x = 1.0$, no matter the rounding flags. (The constant is parsed at compile time and truncated to 16 digits, and there is no control over rounding modes until later, when the program executes.) The code must insure that no errors are introduced through compiler constant truncation.

```
x = 1.000000000000000000001; // set x=1.0, regardless of rounding flags
```
(13)

There are many such perils in the production of reliable interval arithmetic code. Overall, a great deal of effort must be expended to produce the computer code for rather simple inequalities. This effort must be expended every time a new function is introduced into an inequality. This simple fact has kept the inequality-proving software developed for the proof of the Kepler conjecture from having more widespread applicability to more general inequality proving.

Figure 4 shows a snippet of C++ code that computes the arctangent of a linear germ of a function.

If the Kepler conjecture is eventually to be proved by generic tools, we must find a less cumbersome way to produce the computer code. Indeed, a fundamental principle of software design is that there should be no manual procedures (Pragmatic Programmer, Tip 61) [11]. Generative programming gives methods to automate the production of computer code [4]. There is nothing about the interval arithmetic computer code for a new function that requires human thought or effort in an essential way. For example, an examination of the code for the arctangent in Figure 4 reveals that it is a shallow reformatting of the formula for the derivative of the arctangent, combined with the quotient rule in calculus. Why should the code be produced by hand, if it the process is entirely mechanical?

```
/**
 * A lineInterval is an interval version of a linear approximation to a
 * function in 6 variables. The linear
 * approximation is +f + Df[0] x0 + Df[1] x1 + Df[2] x2 +...+ Df[5] x5.
 */
class lineInterval {
   public:
      interval f,Df[6];
      // rest of class omitted
};

/**
 * Sample implementation of the arctangent function.
 * This computes the linear approximation only.  The second derivatives
 * are much more involved.
 */
static lineInterval atan(lineInterval a,lineInterval b) // atan(a/b);
{
   static const interval one("1");
   lineInterval temp;
   temp.f = interMath::atan(a.f/b.f); // computes interval-valued
   arctangent interval rden = one/(a.f*a.f+b.f*b.f);
   for (int i=0;i<6;i++) temp.Df[i]= rden*(a.Df[i]*b.f-b.Df[i]*a.f);
   return temp;
}
```

Fig. 4. Code to calculate an interval version of the arctangent function

A generative program could be written that takes as its input a function and produces as output the interval arithmetic computer code for the Taylor series bounds of that function. The program would parse the definition of the function, generate symbolic derivatives of the function, convert the derivative information to computer code for calculating the derivatives, and so forth.

What advantages would this bring? First of all, it would no longer be necessary to read 40K lines of check to check the correctness of the proof of the Kepler conjecture. It would be enough to check the code on the much smaller generator. Also, the same generator could be used to prove many other inequalities.

4.1.1 Implementation details

The generative program has not been written. Some feasibility tests have been made with javaCC for parsing and XSLT for abstract syntax tree transformations.

5 Plane Graph Generation

A sphere graph is a graph together with an embedding of it into the unit sphere. We discuss a simple sphere graph generating algorithm in this section.

Figure 5 shows a sequence of sphere graphs, giving a sequence of faces that are added to a square to produce the graph dual to the edge graph of the rhombic dodecahedron. We can represent the sequence abstractly as a directed graph with vertices $v_1, \ldots, v_{11}$ with edges from v_i to v_{i+1}. Figure 6 shows that the sequence of faces can be generated in different orders, and that all different sequences can be represented as a directed graph whose root is the square v_1. Call this the derivation graph. The terminal vertices of the directed graph represent sphere graphs isomorphic to v_{11}.

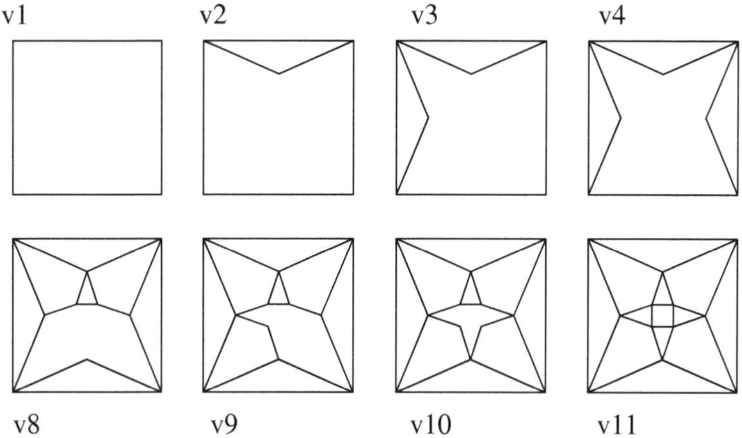

Fig. 5. The first few stages and the final few stages of drawing the graph dual to the rhombic dodecahedron

In going from a parent to a child, we always add one face. Each sphere graph in the derivation graph has two types of faces – those such as the pentagon in v_2 that does not survive unmodified in the terminal nodes, and those such as the triangle in v_2 that do. Call these two types of faces *modifiable* and *unmodifiable* respectively.

Let us generalize this construction to generate all the sphere graphs that are needed for the proof of the Kepler conjecture. Fix a natural number N.

Consider the set V_0 of nonempty sphere graphs with at most N vertices, and no loops or multiple joins. All faces of the sphere graph are assumed to be polygons, and all polygons are assumed to be simple. We give each face one of the two attributes *modifiable* or *unmodifiable* and call a graph with these attributes a *decorated* graph. Let V be the set of all decorated graphs of V_0.

Let P be a polygon. We say that a simple polygon Q is an *admissible refinement* of P if every vertex of Q is either a vertex of P or an interior point

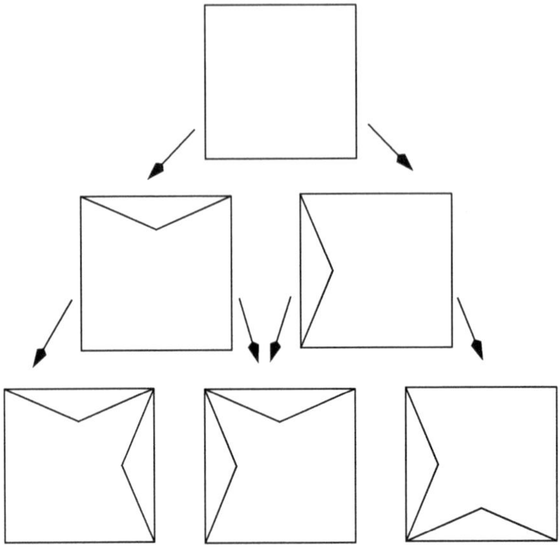

Fig. 6. The same graphs can be generated by different sequences of adding faces

of P and if Q shares at least one edge with P. An admissible refinement of P partitions the region P into the region Q and finitely many other polygons.

Let v be a decorated sphere graph in V. We say that v' is an *admissible refinement* of v if there is a modifiable face P of v' and an admissible refinement Q of P such that the graph obtained by adding Q is v' that is compatibly decorated. We say that v' is compatibly decorated if the unmodifiable faces of v' are Q together with the unmodifiable faces of v. A special case occurs, when $P = Q$, and in this case, we simply change the attribute of P to unmodifiable.

The set V becomes a directed graph Γ with an edge from each v to all of the admissible refinements of v. The root of the directed graph is an empty node. The children of the root are graphs consisting of a single polygon dividing the sphere into an interior and exterior, one side modifiable and the other not. The terminal vertices in this directed graph are decorated graphs, with no modifiable faces. Thus, terminal vertices are in natural bijection with V_0.

If we take any graph in V_0, it can be reached from the root as follows. Pick a face of V_0 and draw its edges as the initial polygon. Make the interior of the polygon modifiable. Then continue to pick faces that have at least one edge already drawn, and draw all remaining edges of that face, marking the completed face as unmodifiable. This corresponds to following a edge in the directed graph Γ.

This gives us an algorithm to generate all sphere graphs in V_0: begin with the children of the root vertex (polygons with at most N vertices) and generate all admissible refinements (that is, follow all possible directed edges) until terminal vertices are reached.

We can improve on this algorithm by fixing for each $v \in \Gamma$ a modifiable face P and an edge on that face, and then taking only admissible refinements Q of P that share the given edge. Each sphere graph in V_0 is still generated under this restriction. We can also assume without loss of generality that the initial polygon is chosen to be one with the most edges.

There is an enormous combinatorial explosion as all admissible refinements are generated. As fortune has it, we are not interested in all sphere graphs V_0, but rather only those that arise as a potential counterexample to the Kepler conjecture. Let $V_1 \subset V_0$ denote this smaller set of relevant graphs. This allows us to combine the general graph-generating algorithm with pruning operations that keep the combinatorics from getting out of hand.

What is needed are criteria on $v \in \Gamma$ that allows us to conclude that v has no no admissibly refined descendents in V_1, that is, to conclude there is no directed path from v to $v_1 \in V_1$. The proof of the Kepler conjecture gives a long list of properties of graphs in V_1 and this avoids the combinatorial explosion. The implementation of the graph generator includes many other minor tricks to keep the execution time manageable. Pruning and the other tricks are rather mundane, and we refer the reader to the source code for details.

6 Conclusion

We will not try to list all of the places where the algorithms of this article would bring a simplification of the 1998 proof of the Kepler conjecture. The list would be extensive. To give a rough indication, we list a few places these algorithms are relevant to the two shortest articles of the proof [6] and [9].

In the article [6] alone, low-dimensional quantifier elimination problems are the subject of Lemma 1.2, Lemma 1.3, Lemma 1.4, Lemma 1.5, Lemma 1.6, Lemma 1.7, Lemma 1.8, Lemma 1.9, Lemma 1.11, Lemma 2.1, and Lemma 2.2. In [9], an additional low-dimensional quantifier elimination problem appears in Lemma 2.2. In general, the parts of the 1998 proof that rely on what that proof calls *geometric considerations* are amenable to preprocessed quantifier elimination.

Linear assembly algorithms generalize the algorithm presented in [9], Appendix 2. Some examples where linear assembly would simplify the 1998 proof are [6] Section 4, and [9] Proposition 4.1, Proposition 4.2, Proposition 5.2, Proposition 5.3, Appendix 1 (A.5 and A.7).

Interval arithmetic inequalities are used throughout the 1998 proof, in sections such as [6] Appendix 3.13.1–4.7.5 and [9] Section 10, Appendix 1. In the early articles in the series, interval arithmetic Taylor approximations are not used [7]. As a result, these early papers only prove very limited types of inequalities. The entire strategy of the proof of the Kepler conjecture in [7] is shaped by these algorithmic limitations. This profoundly affects the structure of the optimization problem in [6], because a scoring function within the reach of the early primitive algorithms was chosen, although such a function was highly suboptimal. With improved automated inequality proving algorithms, it should be possible to make a fresh start and devise a much more efficient scoring function.

Graph generation is carried out in [9] Section 8.

We have not yet reached the fundamental objective of avoiding all manual procedures; some parts of the proof remain hand-made (even after taking account of the algorithms of this article). A fully automated proof would have to develop additional algorithms to prove these estimates. Nevertheless, this article brings us one step closer to that objective.

References

[1] Anonymous, What every computer scientist should know about floating-point arithmetic, http://docs.sun.com/htmlcoll/coll.648.2/iso-8859-1/NUMCOMPGD/ncg_goldberg.html

[2] S. Basu, R. Pollack, and M.-F. Roy, On the combinatorial and algebraic complexity of Quantifier Elimination, In Procedings of the Foundations of Computer Science, pp. 632–641, 1994.

[3] Goetz Alefeld and Juergen Herzberger, Introduction to Interval Computations, Academic Press, N.Y., 1983.

[4] Krzysztof Czarnecki and Ulrich Eisenecker, Generative Programming: Methods, Tools, and Applications, Addison-Wesley, 2000.

[5] Bob F. Caviness (Editor), J. R. Johnson (Editor), Quantifier Elimination and Cylindrical Algebraic Decomposition (Texts and Monographs in Symbolic Computation), Springer-Verlag, 1998.

[6] S. Ferguson and T. Hales, A formulation of the Kepler conjecture, preprint, 1998.

[7] T. Hales, Sphere Packings I, Disc. Comp. Geom, 1997, 17:1-51.

[8] T. Hales, An overview of the Kepler conjecture, preprint, 1998.

[9] T. Hales, Sphere Packings III, preprint, 1998.

[10] Reiner Horst, P. M. Pardalos, and Nguyen V. Thoai, Introduction to Global Optimization - Second Edition (Nonconvex Optimization and its Applications, Volume 48), Kluwer, 2000.

[11] Andrew Hunt, David Thomas, and Ward Cunningham Pragmatic Programmer: From Journeyman to Master, Addison-Wesley, 1999.

[12] R. Baker Kearfott : Rigorous Global Search: Continuous Problems Kluwer Academic Publishers, Dordrecht, Netherlands, 1996.

[13] Bhubaneswar Mishra, Computational Real Algebraic Geometry, Handbook of Discrete and Computational Geometry, CRC Press, 1997.

About Author

Thomas C. Hales is at the Department of Mathematics, University of Pittsburgh.

Acknowledgments

The interval-arithmetic algorithms were developed in collaboration with S. Ferguson.

The Minimal Number of Triangles Needed to Span a Polygon Embedded in $\mathbb{R}^d$

Joel Hass
Jeffrey C. Lagarias

Abstract

Given a closed polygon P having n edges, embedded in $\mathbb{R}^d$, we give upper and lower bounds for the minimal number of triangles t needed to form a triangulated PL surface embedded in $\mathbb{R}^d$ having P as its geometric boundary. More generally we obtain such bounds for a triangulated (locally flat) PL surface having P as its boundary which is immersed in $\mathbb{R}^d$ and whose interior is disjoint from P. The most interesting case is dimension 3, where the polygon may be knotted. We use the Seifert surface construction to show that for any polygon embedded in $\mathbb{R}^3$ there exists an embedded orientable triangulated PL surface having at most $7n^2$ triangles, whose boundary is a subdivision of P. We complement this with a construction of families of polygons with n vertices for which any such embedded surface requires at least $\frac{1}{2}n^2 - O(n)$ triangles. We also exhibit families of polygons in $\mathbb{R}^3$ for which $\Omega(n^2)$ triangles are required in any immersed PL surface of the above kind. In contrast, in dimension 2 and in dimensions $d \geq 5$ there always exists an embedded locally flat PL disk having P as boundary that contains at most n triangles. In dimension 4 there always exists an immersed locally flat PL disk of the above kind that contains at most $3n$ triangles. An unresolved case is that of embedded PL surfaces in dimension 4, where we establish only an $O(n^2)$ upper bound. These results can be viewed as providing qualitative discrete analogues of the isoperimetric inequality for piecewise linear (PL) manifolds. In dimension 3 they imply that the (asymptotic) discrete isoperimetric constant lies between 1/2 and 7.

1 Introduction

Given a closed polygon P in $\mathbb{R}^d$ having n edges, we consider the problem of giving upper and lower bounds for the minimal number of triangles t needed to form a triangulated PL surface in $\mathbb{R}^d$ having P as its geometric boundary. This has a well known answer in $\mathbb{R}^2$, which is

$$t = n - 2,$$

in which t is the number of triangles in any triangulation of the (convex or nonconvex) polygon that adds no extra vertices, see [18, Theorem 23.2.1]. Such triangulations minimize the number of triangles over all possible triangulations, in which extra vertices might be added.

What happens in higher dimensions? We consider two versions of the problem, in both of which extra vertices are permitted, in the surface and added to the polygonal boundary.

(1) The surface is an embedded oriented PL surface with no restriction on its genus.

(2) The surface is an immersed oriented PL disk, with the extra restriction that the interior of the surface cannot cut through its boundary.

Recall that a surface is *embedded* if it does not intersect itself, and is *immersed* if it is locally embedded, i.e. if each point on the surface has an embedded neighborhood. More precisely, in the second case we allow only a *complementary immersed surface*, by which we mean an immersed surface (of any genus) whose interior does not intersect its boundary, and whose boundary is embedded [1]. Complementary immersed surfaces allow extra freedom over embedded surfaces, but in dimension $d = 3$ they still detect knottedness; a polygon P in $\mathbb{R}^3$ has a complementary immersed surface that is a topological disk if and only if P is unknotted. If we were to allow general immersed surfaces rather than restricting to complementary immersed surfaces, then the two-dimensional construction above works in all higher dimensions, to produce an immersed disk having $n - 2$ triangles (whose interior in general will intersect its boundary.) With the extra restrictions given above on the surfaces the answer becomes non-trivial. We show that the minimal number of triangles grows like n^2 in dimension 3, is $O(n)$ in dimensions 5 and above, and also is $O(n)$ in dimension 4 for complementary immersed surfaces.

We first consider dimension 3, where the polygon may be knotted. It is known that the Seifert surface construction leads to an $O(n^2)$ algorithm for constructing oriented embedded surfaces, see Vegter [19, p. 532]. Using such a construction we obtain the following explicit upper bound for a triangulated oriented embedded surface having the polygon as boundary,

Theorem 1.1. *For each closed polygonal curve P with n line segments embedded in $\mathbb{R}^3$, there exists an embedded oriented triangulated PL surface, having a PL subdivision of P as boundary, that has at most $7n^2$ triangles.*

In this result, the polygonal curve P is the geometric boundary of the surface, but extra vertices may have to be added to P to get the boundary of the surface as a PL manifold.

Theorem 1.1 shows that for embedded surfaces of unspecified genus the upper bound is polynomial in n. This result contrasts with bounds for the size of an embedded oriented triangulated PL surface of minimal genus spanning the curve P. Hass, Snoeyink and Thurston [11] show in the unknotted case

[1] This condition rules out a surface whose boundary is a multiple covering of the given boundary curve.

(minimal genus = 0) that $(C_1)^n$ triangles are sometimes required as $n \to \infty$, where $C_1 > 1$ is a fixed constant. Complementing this result, Hass, Lagarias and Thurston [10] show that in the unknotted case there always exists an embedded minimal genus PL surface (an embedded disk) spanning the curve P, that contains at most $(C_2)^{n^2}$ triangles, for a fixed constant $C_2 > 1$.

In §3 we show that the upper bound of $O(n^2)$ in Theorem 1.1 is the correct order of magnitude. We present two constructions based on different principles, giving $\Omega(n^2)$ lower bounds.

The first method involves the genus $g(K)$ of the knot K. The *genus* of a knot is the minimal genus of any orientable embedded surface that has the knot as its boundary. The lower bound is

$$t \geq 4g(K) + 1,$$

and it applies to embedded orientable PL surfaces having K as boundary. This lower bound depends only on the (ambient isotopy) type of the knot K, so one gets the best result by minimizing the number of edges in a polygon representing the knot, which is called the *stick number* of the knot. Using the $(n, n-1)$ torus knot, we obtain the following result.

Theorem 1.2. *There exists an infinite sequence of values of $n \to \infty$ with closed polygonal curves P_n having n line segments embedded in $\mathbb{R}^3$, for which any embedded triangulated PL surface, that is oriented and that has a PL subdivision of P_n as boundary, requires at least $\frac{n^2}{2} - 3n + 5$ triangles.*

The second method uses an invariant of a knot diagram K, the *writhe* $w(K)$. The writhe of a knot diagram K is obtained by assigning an orientation (direction) to the knot diagram, and then assigning a sign of ± 1 to each crossing, with $+1$ assigned if the two directed paths of the knot diagram at the crossing have the undercrossing oriented by the right hand rule relative to the overcrossing, and -1 if not. The writhe $w(K)$ of the oriented diagram is the sum of these signs over all crossings; it is independent of the orientation chosen.

The lower bound is

$$t \geq |w(K)|, \tag{1}$$

and it applies to complementary immersed surfaces. The writhe is not an ambient isotopy invariant, but is an invariant of a of knot diagram under Reidemeister moves of types II and III only, with type I moves forbidden.

We apply this bound to show that there is an infinite family of polygonal curves P_n in $\mathbb{R}^3$ having a quadratic lower bound for the number of triangles in a complementary immersed surface, of unrestricted genus.

Theorem 1.3. *There exists an infinite sequence of closed polygonal curves P_n in $\mathbb{R}^3$ having n line segments, with values of $n \to \infty$, for which any complementary immersed triangulated PL surface that has a PL subdivision of P_n as boundary, requires at least $\frac{n^2}{36}$ triangles.*

The writhe bound (1) implies that a polygonal knot that has a large writhe in one direction must have a large number of crossings in any projection direction (Theorem 3.3).

In §4 we consider the combinatorial isoperimetric problem for embeddings of a curve in dimensions $d \geq 4$. We obtain two $O(n)$ upper bounds. In these dimensions we construct PL surfaces spanning the polygon which are locally flat, as defined at the beginning of §4. The local flatness condition is a restriction on how the surface is situated in $\mathbb{R}^d$. It is known that local flatness always holds for an embedded PL surface in codimension 3 or more (see [15, Corollary 7.2]), hence requiring local flatness puts a constraint only in dimension $d = 4$.

We first treat dimension $d = 4$, and construct a complementary immersed surface that is locally flat and has $O(n)$ triangles.

Theorem 1.4. *Let P be a closed polygonal curve embedded in $\mathbb{R}^4$ consisting of n line segments. Then there exists a complementary immersed triangulated PL disk, which is locally flat, has P as its PL boundary, and contains $3n$ triangles.*

Second, in dimensions $d \geq 5$, by coning the polygon P to a suitable point we obtain an embedded PL disk with n triangles.

Theorem 1.5. *Let P be a closed polygonal curve embedded in $\mathbb{R}^d$, with $d \geq 5$, consisting of n line segments. Then there exists an embedded triangulated PL disk which is locally flat, has P as its PL boundary, and contains n triangles.*

To summarize, these results establish that the complexity of the spanning surface is $O(n^2)$ in dimension 3, and is $O(n)$ in all other dimensions, except possibly in dimension 4 for embedded surfaces. The increased complexity in dimension 3 might be expected, since dimension 3 is the only dimension in which knotting is possible for curves. As far as we know, the remaining unresolved case of embedded surfaces in $\mathbb{R}^4$ might conceivably have super-linear complexity; if so, this would represent a new phenomenon peculiar to the discrete case. For this case we establish only an $O(n^2)$ upper bound, as explained at the end of §5, while an $\Omega(n)$ lower bound is immediate.

Our motivation for study of these questions comes from an analogy with isoperimetric inequalities, which we considered in [11]. The classical isoperimetric inequality asserts that for a simple closed curve γ of length L in $\mathbb{R}^2$, the area A that it encloses satisfies

$$A \leq \frac{1}{4\pi} L^2,$$

with equality only in the case of a circle. This inequality generalizes to all higher dimensions, where we allow either immersed surfaces, which can be restricted to be disks, or embedded surfaces of arbitrary genus, as follows. For a closed C^2-curve γ of length L embedded in $\mathbb{R}^d$ there exists an immersed disk of area A having γ as boundary, as well as an embedded orientable surface of area A having γ as boundary, such that in either case

$$A \leq \frac{1}{4\pi} L^2.$$

The first of these d-dimensional results traces back to Beckenbach and Rado [3], while the second traces back to Blaschke [4], see Osserman [13, p. 1202]. The problems we consider here are discrete analogues of these two variants of the isoperimetric inequality. The discrete measure of "length" of the polygon is the number of line segments n in its boundary, and the discrete measure of the "area" of a triangulated surface is the number of triangles t that it contains. This type of combinatorial minimal area problem is associated to affine geometry because these measures of "length" and "area" are both affine invariant. It follows that our results are most appropriately viewed as results concerning d-dimensional affine space $\mathbb{A}^d$ without a metric structure, rather than $\mathbb{R}^d$ with its Euclidean structure. However for convenience we formulate all results in $\mathbb{R}^d$.

Our results determine the order of growth of the discrete isoperimetric bounds as a function of n. The discrete problem has some differences from the classical problem, in that its bounds grow linearly in n rather than quadratically as in the classical isoperimetric inequality, except in dimension 3, and possibly dimension 4 for embedded surfaces. Our bounds are qualitative, so are not a perfect analogue of the classical isoperimetric inequality which gives an exact constant. For exact answers in the discrete case there are an infinite number of cases, one for each value of the number of edges n in the polygon. It therefore seems more natural to consider a notion of asymptotic isoperimetric constant as $n \to \infty$, We formulate this in the most interesting case of dimension 3. For each $n \geq 3$ we define the *discrete isoperimetric constant* $\gamma(n)$ by

$$\gamma(n) = \max_{P_n} \left(\min_{\Sigma \text{ spans } P_n} \frac{1}{n^2} t(\Sigma) \right),$$

in which P_n runs over all polygons with n edges embedded in $\mathbb{R}^3$, and all surfaces Σ are embedded surfaces. We define the *asymptotic discrete isoperimetric constant* Γ in $\mathbb{R}^3$ to be

$$\gamma := \limsup_{n \to \infty} \gamma(n).$$

Combining Theorems 1.1 and 1.2 implies that the asymptotic isoperimetric constant τ must lie between 1/2 and 7. It would be interesting to determine whether the constant γ is a limiting value rather than a lim sup, and to determine its exact value.

A further direction for such PL isoperimetric problems would be to establish isoperimetric bounds for higher-dimensional submanifolds. Consider a k-dimensional triangulated closed PL-manifold M embedded in $\mathbb{R}^d$, where $k \geq 2$, and ask: what is the minimal number of $(k+1)$-simplices in an embedded triangulated PL $(k+1)$-dimensional manifold having a PL-subdivision of M as its boundary?

Earlier work on the complexity of embedded surfaces bounding unknotted curves in $\mathbb{R}^3$ under various restrictions includes Almgren and Thurston [2]. Connections between combinatorial complexity of such surfaces and the computational complexity of problems in knot theory appear in [8], [9].

2 Upper Bound

We establish Theorem 1.1 by a straightforward analysis of the construction due to Seifert [16] of an orientable surface having a given knot as boundary. A general description of Seifert surfaces and their construction appears in Rolfsen [14, Chapter 5].

Proof of Theorem 1.1: Given a closed polygon P in $\mathbb{R}^3$ having n line segments, we first choose an orientation for it. We obtain a knot diagram by orthogonally projecting it onto a plane. Fix once and for all a projection direction in "general position", so that the projections of any two line segments in P intersect in at most one point, and if the two segments in P are disjoint then this point must correspond to interior points of the two segments. Without loss of generality we may rotate the polygon so that the projection direction is in the z-direction and the projected plane is $z = 0$, and we may translate it in the z-direction so that it lies in the half-space $z \geq 1$. The projected image of the polygon in the plane has n vertices and c crossing points, where

$$c \leq n(n-3)/2,$$

since an edge cannot intersect its two adjacent edges or itself. We make the projection into a planar graph by marking vertices at each crossing point, which we call *crossing vertices.* This graph is a directed graph, with directed edges obtained by projection of the orientation assigned to the polygonal knot P, and is regarded as sitting in the plane $z = 0$. Each vertex of this planar graph has either two or four edges incident on it, so the faces of the graph can be two-colored; call the colors white and black. The graph has a single unbounded face, which we consider colored white; it also has at least one bounded region colored black. We denote this directed colored graph $\mathcal{G}$; it has $n + c$ vertices, which we call *initial vertices* in what follows.

We now add new vertices to this graph as follows: At each edge containing a crossing vertex we insert new vertices very close to each of its crossing vertex endpoints; call these *interior vertices.* The resulting graph has $n + c$ *initial vertices* and $4c$ interior vertices. Each crossing vertex has four edges incident to it.

Near each crossing vertex we now add edges connecting pairs of interior vertices on adjacent edges in cyclic order around the crossing point. There are four such edges which form a small quadrilateral enclosing the crossing vertex. We choose the interior points close enough to the crossing vertex so that the interior of each edge in the boundary of this quadrilateral does not intersect any other edge of the graph, and so lies entirely inside one of the colored polygons of the original graph. We assign that color to the edge. These four edges form two white edges and two black edges; we discard the black edges and add the two white edges only, to form an augmented planar graph.

The example of the trefoil knot is pictured in Figure 1; part (b) shows the added vertices of the augmented planar graph, and the white edges are indicated by dotted lines in (b).

The added white-colored edges create two white-colored triangular regions adjacent to each crossing vertex, which together form a "bow-tie" shaped region. If one now deletes all crossing vertices and the four edges incident to them (which have as other endpoint an interior vertex), and adds in the white-colored edges only, then one obtains a new planar graph $\mathcal{G}'$, that has $n+4c$ vertices. This new graph may be disconnected, and consists of a union of simple closed polygons. Its regions are two-colorable, with the coloring obtained from that of $\mathcal{G}$ by changing the color of the bow-tie shaped regions from white to black. For the trefoil knot this is pictured in Figure 1(c).

The graph $\mathcal{G}'$ is a union of simple closed curves C which we call *circuits*, some of which may be nested inside others. To each circuit we assign an integer level that measures its nesting. An *innermost circuit* is one that contains no other circuit: we assign these innermost circuits level 1. We now inductively define a level for each other circuit, to be one more than the maximal level of any circuit they contain; the maximal level is at most n.

We now construct a triangulated embedded spanning surface Σ for γ as follows.

(1) For each circuit C of level k in $\mathcal{G}'$ we make a copy $\tilde{C}$ of this circuit in the plane $z=-k$, i.e. we translate it from the plane $z=0$ by the vector $(0,0,-k)$. It forms a simple closed polygon whose interior in this plane will form part of the surface Σ. If it has m sides, then we may triangulate it using $m-2$ triangles lying in the plane $z=-k$.

(2) We next add vertical faces connecting the circuit $\tilde{C}$ to the the polygon P lying above it. More precisely, we must enlarge P to a one-dimensional simplicial complex P' which includes preimages of the white edges. We first add new vertices on P which lie vertically above the endpoints of the white edges; these points are unique because the projection is one-to-one off crossing points; this gives a subdivision of P. We next add new edges (not on P) connecting these points, which project to the white edges; P' is the one-dimensional simplicial complex in $\mathbb{R}^3$ resulting from adding these points and edges to P. To each edge of the circuit $\tilde{C}$ there is a unique edge of P' that projects vertically onto it. We take the convex hull of these two edges, that

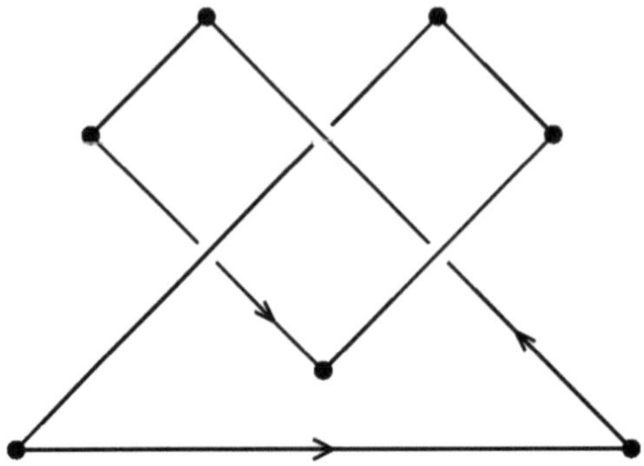

(a) Trefoil knot diagram ($n = 7$, $c = 3$)

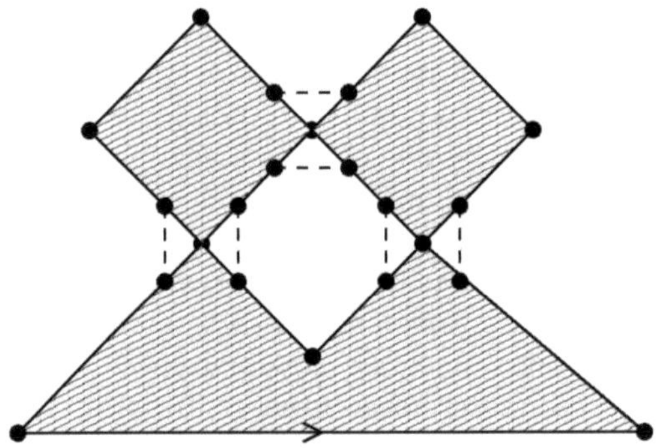

(b) Augmented graph

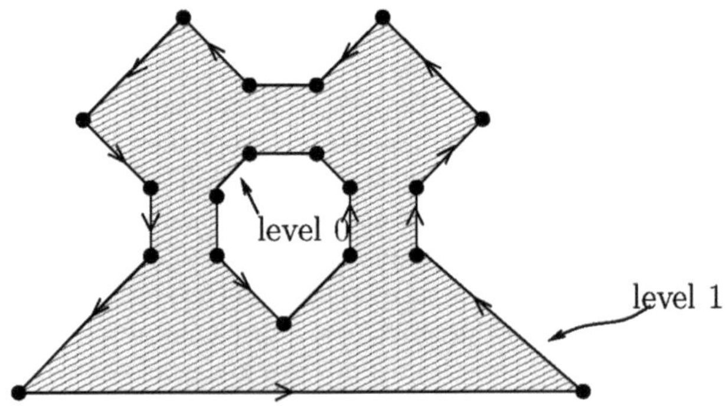

(c) Graph $\mathcal{G}'$

Fig. 1. Trefoil knot

forms a trapezoid with two vertical edges, plus the two edges we started with. These trapezoids will form part of the surface Σ. Each of them may be triangulated by adding a diagonal to the trapezoid. See Figure 2.

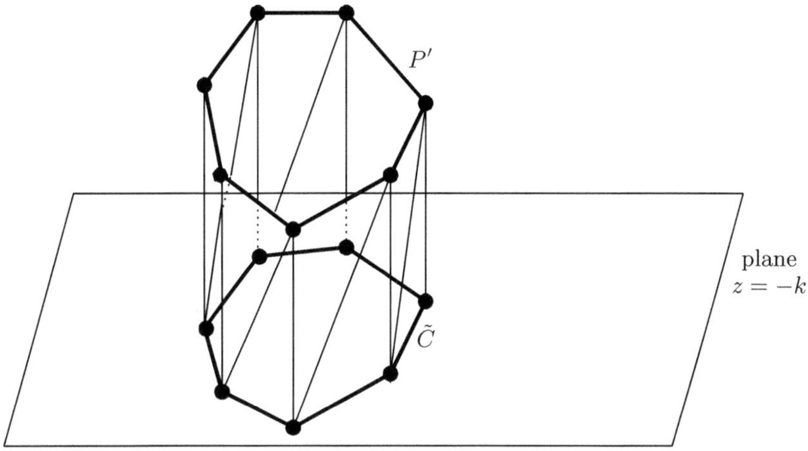

Fig. 2. Triangulated bowl-shaped region for cycle C

The part of the surface Σ associated to the circuit $\tilde{C}$ in steps (1) and (2) forms a bowl-shaped region whose base is a polygon in the plane $z = -k$ and part of P' above $\tilde{C}$ as its lip.

(3) Above each bow-tie shaped region of $\mathcal{G}'$ containing a crossing vertex and two white edges, there lie four edges of P', two edges of which project to the white edges on the plane $z = 0$, and the other two of which are part of edges of P whose projections on the plane $z = 0$ are disjoint except at the crossing vertex. Let the vertices of the two edges of P' lying above the white edges be labeled $[\mathbf{x}_1, \mathbf{x}_2]$ and $[\mathbf{y}_1, \mathbf{y}_2]$ respectively, with the black edges (not part of P') being $[\mathbf{x}_1, \mathbf{y}_2]$ and $[\mathbf{x}_2, \mathbf{y}_1]$, and with the line segments $[\mathbf{x}_1, \mathbf{y}_1]$ and $(\mathbf{x}_2, \mathbf{y}_2)$ being subsets of the original polygon P. We then form the two triangles $[\mathbf{x}_1, \mathbf{x}_2, \mathbf{y}_1]$ and $[\mathbf{x}_2, \mathbf{y}_1, \mathbf{y}_2]$, that share a common black edge $[\mathbf{x}_2, \mathbf{y}_1]$, and add them to Σ. See Figure 3.

We claim that Σ forms a triangulated surface embedded in $\mathbb{R}^3$, which is orientable and has a subdivision of P as its boundary.

To see that Σ is embedded in $\mathbb{R}^3$, note that the triangulated pieces (1)-(3) of Σ are embedded, and when projected to the z-axis have disjoint interiors. Thus these pieces can only overlap along their boundary edges.

We use Seifert's argument to show Σ is orientable. The contribution of (1) and (2) corresponding to each circuit is a bowl-shaped surface that is topologically a disk, with the crossing points located near the lip of the bowl. At each crossing point the cup is attached to another bowl by a rectangular strip with a half-twist in it, twisting through an angle of π. Since it is constructed from disks attached along boundary intervals, Σ is topologically a 2-manifold with boundary. In addition, the construction connects a bowl

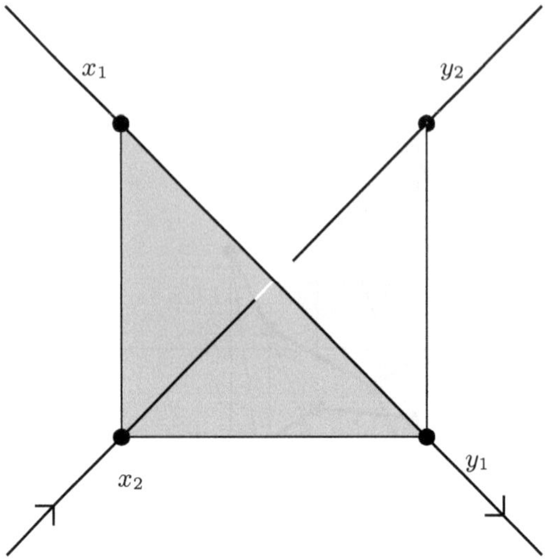

Fig. 3. Triangulated "bow-tie" region

at level j only to bowls at level $j \pm 1$. A compatible orientation then takes as one side the upper (inside) surface of bowls at level $2j$ and the lower (outside) surface of bowls at level $2j+1$, plus corresponding sides of the strips connecting them. Thus Σ is orientable.

We bound the number of triangles in Σ. The totality of triangles produced in step (1) above is at most the number of edges in all the circuits; this is at most the number of edges in P' after adding internal vertices, and is at most $n+4c$. In step (2) each trapezoid is associated to one of the at most $n+4c$ edges of the circuits, and has two triangles; thus these contribute at most $2n+8c$ triangles. In step (3) there are two triangles added for each crossing vertex, which totals $2c$. Thus the total is at most $3n+14c$, which is at most $7n^2-18n$ triangles. This gives the desired upper bound $7n^2$. □

Remark. It is possible to modify the construction in Theorem 1.1 to further improve the asymptotic upper bound to cn^2 with a constant c smaller than 7. We leave the problem of obtaining the optimal constant unanswered.

3 Lower Bound Constructions

We present two different constructions giving quadratic lower bounds for triangulated surfaces.

Lemma 3.1. *Let P be a closed polygonal curve embedded in $\mathbb{R}^3$ whose associated knot type K has genus $g(K)$. If t is the number of triangles in a triangulated oriented PL surface Σ which is embedded in $\mathbb{R}^3$ and has a subdivision of P as boundary, then*

$$t \geq 4g(K)+1. \tag{1}$$

Proof: Without loss of generality we may assume that the surface Σ is connected, by discarding any components having empty boundary; this only decreases t. If V, E and F denote the number of vertices, edge and faces in the triangulated orientable surface Σ , then its Euler characteristic is

$$\chi(\Sigma) = V - E + F.$$

By definition of knot genus this surface is of genus $g \geq g(K)$. Recall that the genus of a surface with boundary Σ is the smallest genus of a connected surface Σ' without boundary in which it can be embedded; In this case Σ' is obtained by gluing in a disk attached to the (topological) boundary P. This adds one face, and no new edges or vertices, hence we obtain

$$\chi(\Sigma) = -1 + \chi(\Sigma') = 1 - 2g \geq 1 - 2g(K).$$

We have $F \geq t$, and since all faces in the surface are triangles, we obtain

$$3t = 2E - m,$$

where m is the number of edges on the boundary of the surface. Counting the number of edges on the boundary gives

$$V \geq m.$$

From these bounds follows

$$\chi(\Sigma) \geq 1 - 2g(K) = V - E + F \geq m - (\frac{3}{2}t + \frac{m}{2}) + t,$$

which simplifies to

$$\frac{t}{2} \geq 2g(K) - 1 + \frac{m}{2} \geq 2g(K) + \frac{1}{2},$$

since $m \geq 3$. $\square$

Remark. Define the *unoriented genus* $g^*(K)$ of a knot K to be the minimal value possible of $1 - \frac{1}{2}\chi(\Sigma)$ taken over all embedded connected surfaces Σ, orientable or not, having K as boundary. Then $g^*(K)$ is an integer or half-integer, and the same reasoning as above shows that

$$t \geq 4g^*(K) + 1$$

for the number of triangles in any triangulated PL surface bounding a polygon P of knot type K.

Proof of Theorem 1.2:
We consider the $(m, m-1)$ torus knot $K_{m,m-1}$. This has a polygonal representation using $n = 2m$ line segments, given in Adams et al [1, Lemma 8.1].

They also show that a polygonal realization of this knot requires at least $2m$ segments [1, Theorem 8.2].

In 1934 Seifert [16, Satz 4] showed that the (p, q)-torus knot $K_{p,q}$ has genus

$$g(K_{p,q}) = \frac{(p-1)(q-1)}{2}.$$

Thus we have $g(K_{m,m-1}) = \dfrac{m^2 - 3m + 2}{2}$.

We apply Lemma 3.1 to $K_{m,m-1}$ and obtain $t \geq 2m^2 - 6m + 5 \geq \frac{1}{2}n^2 - 3n + 5$, as asserted. □

We next obtain a lower bound in terms of the *writhe* (or *Tait number*) of a knot diagram K associated to P by planar projection. The writhe of a knot diagram K is calculated by assigning an orientation (direction) to the knot diagram, then associating a sign of ± 1 to each crossing, using $+1$ if the two directed paths of the knot diagram at the crossing have the undercrossing oriented by the right hand rule relative to the overcrossing, and -1 if not. The writhe $w(K)$ of the oriented diagram is the sum of these signs over all crossings. The quantity $w(K)$ is independent of the orientation, but depends on the direction of projection.

Lemma 3.2. *Let P be a closed polygonal curve embedded in $\mathbb{R}^3$ that has an orthogonal planar projection K that has writhe $w(K)$. If t is the number of triangles in a complementary immersed PL surface in $\mathbb{R}^3$ which is triangulated and has a subdivision of P as boundary, then*

$$t \geq |w(K)| + 1. \tag{2}$$

Proof: The quantity $|w(K)|$ reflects the amount of twisting between two different longitudes of the knot, the z-pushoff and the preferred longitude. The *preferred longitude* is a longitude on the knot defined by an embedded two-sided surface bounding the knot P. The preferred longitude is defined intrinsically as the two primitive homology classes $\pm[\tau]$ of a peripheral torus of the knot that are annihilated on injecting into $H^1(\mathbb{R}^3 - P, \mathbb{Z})$ (a peripheral torus is the boundary of an embedded regular neighborhood of the knot). The z-pushoff is the curve on the peripheral torus directly above K in the z direction.

Any triangulated complementary immersed surface Σ having a subdivision of P as boundary necessarily defines a curve on the peripheral torus in the class $\pm[\tau]$. The total twisting about K of the boundary of a smooth surface Σ having P as boundary, relative to the z-direction, is given by $\pi w(K)$. In the case of a PL surface this total twisting can be computed by adding successive jumps in a normal vector to P pointing along the surface as one travels along the (subdivided) curve P. Since triangles are flat, twisting occurs only at the boundaries of two adjacent triangles, and can be no more than π at such a point. While it is possible that a triangle meets P at as many as three points, the total twisting contributed by any triangle at all points where it meets P is at most π, the sum of its interior angles. It follows that there must be more than $|w(K)|$ triangles in the surface that meet the polygon P. □

Proof of Theorem 1.3:
There exists a family of polygonal curves P_m for $m \geq 1$ having $n = 6m + 3$ segments and with writhe $w(P_m) = m(m+1)$. The knot diagram for the polygon P_3 is pictured in Figure 4 below. The construction for general m consists of adding more parallel strands to the pattern.

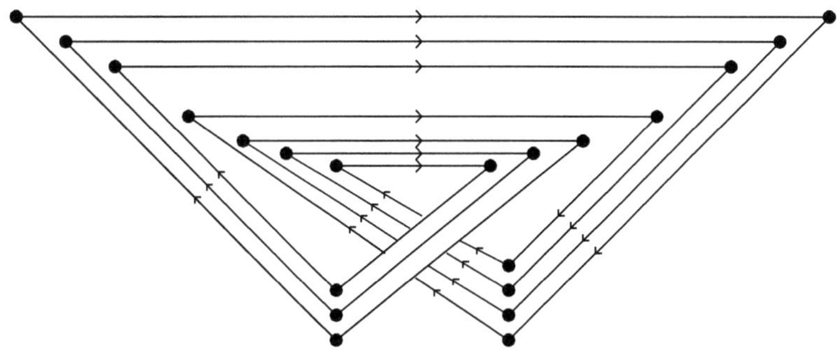

Fig. 4. Knot diagram of polygon P_3.

The theorem now follows by applying the bound of Lemma 3.2, namely $t \geq \dfrac{n^2}{36} + \dfrac{3}{4}$. □

Combining Lemma 3.2 with the construction of a surface in Theorem 1.1 yields a new result in knot theory. It says that if a polygon P embedded in $\mathbb{R}^3$ has a large writhe in some projection direction, relative to its number of edges n, then in all projection directions it has a large number of crossings.

Theorem 3.3. *Let P be a polygonal knot embedded in $\mathbb{R}^3$. If $w(K)$ is the writhe of one projection of P, then the number of crossings c of any projection K' of P satisfies*

$$c \geq \frac{1}{16}(|w(K)| - 3n).$$

Proof: By Lemma 3.2 one has $t \geq |w(K)| + 1$, However if a knot projection K has c crossings, then by the proof of Theorem 1.1 one can construct an oriented, embedded, PL triangulated surface having P as boundary with $t \leq 3n + 14c$ triangles. Combining these estimates gives the lemma. □

As an example, the polygons P_m in $\mathbb{R}^3$ given in the proof of Theorem 1.3 (see Figure 4) must have crossing number $c \geq (m^2 - 17m - 9)/16$ in any projection.

Remarks. (1) The polygonal curves P_m used in the proof of Theorem 1.3 are knotted. We do not know how large $|w(P)|$ can be for an unknotted polygon with n edges. One can easily construct representatives of the unknot having writhe $|w(P)| > cn$, for a positive constant c and $n \to \infty$, but we do not know whether it is possible to get representations P of the unknot having n crossings and writhe $|w(P)| > cn^2$ with $n \to \infty$.

(2) The proofs of Theorem 1.3 use topological invariants associated to knottedness. It may be that a quadratic lower bound can hold for strictly geometric reasons. A relevant geometric construction was given by Chazelle [6], who used it to give examples of polyhedra with n faces which require $\Omega(n^2)$ tetrahedra in any triangulation. A reviewer has suggested that this approach might conceivably produce a family of unknotted polygons with a $\Omega(n^2)$ lower bound. Chazelle's construction takes a hyperbolic paraboloid, H and makes two parallel translates H^+ and H^- just above it and just below it. Now H is a ruled surface, and one takes n line segments connecting close points in a ruling of H, and n segments each in the conjugate ruling of H^+ and H^-, so that each pair of segments from opposite rulings cross in vertical projection, so there are $\Omega(n^2)$ crossings under vertical projections. Then one connects the $3n$ segments in a zigzag manner to produce a polygon with at most $9n$ segments. (There is some freedom of choice in how to make the connections, allowing the construction of polygons of various knot types.) The geometric principle to exploit is that the projection of any triangle with one edge on a segment with endpoints in H cannot cross more than a small number of projections of segments in H^+ and H^-, unless the triangle is very narrow. It seems plausible that an $\Omega(n^2)$ lower bound can be proved for such a construction, but we leave this as an open problem. This approach, if successfully carried out, would give lower bounds that apply to the complexity of unoriented spanning surfaces.

4 Higher Dimensions

In this section we consider polygons P embedded in $\mathbb{R}^d$, for dimensions $d \geq 4$, and construct locally flat PL surfaces having P as boundary. Recall that a surface Σ is *locally flat* if at each point $\mathbf{x}$ of the surface Σ there is a neighborhood in $\mathbb{R}^d$ homeomorphic to $D \times I^{d-2}$ in $\mathbb{R}^d$, where D is a topological 2-disk in the surface (or is a half 2-disk with boundary for a boundary point $\mathbf{x}$ of Σ) and $I = [-1, 1]$ and $D \times 0^{d-2}$ is part of Σ, see [14, p. 36], [15, p. 50]. For immersed surfaces we interpret local flatness to apply to each sheet of the surface separately.

Proof of Theorem 1.4: We are given a closed polygon P with n edges embedded in $\mathbb{R}^4$, and vertices $\mathbf{v}_1, ..., \mathbf{v}_n$. We can always pick a point $z \in \mathbb{R}^4$ such that coning the polygon P to the point z will produce a complementary immersed surface. However this surface need not be locally flat at the cone point. To circumvent this problem, we replace the cone point with a convex planar polygon Q having n vertices and produce a triangulated immersed (polygonal) annulus connecting P to Q. Combining this with a triangulation of Q to its centroid will yield the desired locally flat immersed surface.

Given a convex planar polygon Q with vertices $\mathbf{w}_1, ..., \mathbf{w}_n$ we form the (immersed) triangulated annulus Σ_1 between P and Q with triangles $[\mathbf{v}_j, \mathbf{v}_{j+1}, \mathbf{w}_{j+1}]$ and $[\mathbf{v}_j, \mathbf{w}_j, \mathbf{w}_{j+1}]$, for $1 \leq j \leq n$, using the convention that $\mathbf{v}_{n+1} := \mathbf{v}_1$ and $\mathbf{w}_{n+1} := \mathbf{w}_1$. Then we triangulate Q to its centroid vertex $\mathbf{v}_0$, obtaining a triangulated disk Σ_2. For "general position" Q (described below) this construction produces an immersed surface $\Sigma = \Sigma_1 \cup \Sigma_2$ consisting of n triangles from triangulating Q and $2n$ triangles in the annular part, for $3n$ triangles in all. In general the surface Σ_1 is immersed rather than embedded, because the triangles in it may intersect each other.

We show that Q can be chosen so that Σ is a complementary immersed surface. The main problem is to ensure that Σ_1 intersects P only in its boundary $\partial\Sigma_1$. We wish Σ_1 to contain no line segment in any of its triangles which intersects P in two or more points. We define a "bad set" to avoid. Take the polygon P, extend its line segments to straight lines l_j in $\mathbb{R}^4$, and call a point "bad" if it is on a line connecting any two points on the extended polygon. The "bad" set B consists of a union of $n(n+1)/2$ sets B_{ij}, in which B_{ij} is the union of all lines connecting a point of line l_i to a point of line l_j, with $1 \leq i < j \leq n$. Each B_{ij} is either a hyperplane (codimension 1) in $\mathbb{R}^4$ or is a plane (codimension 2 flat) in $\mathbb{R}^4$. If a plane F (codimension 2 flat) is picked in "general position" in $\mathbb{R}^4$ it will intersect B in at most $n(n-1)/2$ lines and points. In particular, such an F contains a (two-dimensional) open set U not intersecting B and disjoint from the convex hull of P. We choose Q to lie in this open set. Now Σ_2 is the convex hull of Q, which lies in U, so does not intersect P. We claim that Σ_1 intersects P in $\partial\Sigma_1$. Indeed any point $\mathbf{x}$ in Σ_1 not on P lies on a line connecting a point of P to a point of Q, and this line contains at most one point of P because the point in Q is not in the "bad set" B. Since we already know of one point on P on this line, which is not $\mathbf{x}$, the claim follows. We conclude that Σ is a complementary immersed surface.

We next show that Σ is a locally flat (immersed) surface. We need only verify this at the vertices of Σ. At the vertices $\mathbf{v}$ of P three triangles meet, so locally the configuration is three-dimensional, and local flatness holds in the three-dimensional subspace around $\mathbf{w}$ determined by the edges, (as it does for any embedded polyhedral surface in $\mathbb{R}^3$) and this extends to local flatness in $\mathbb{R}^4$ by taking a product in the remaining direction. At a vertex $\mathbf{w}$ of Q five triangles meet. However two of these triangles lie in the plane F of the polygon Q, hence for determining local flatness we may disregard the edge into the interior of the polygon, and treat the vertex as having four incident triangles. Suppose the remaining 4 edge directions leaving $\mathbf{w}$ span a four dimensional space. Take an invertible linear transformation L that maps these vectors to $\mathbf{x}_1 = (1, 1, 0, 0), \mathbf{x}_2 = (1, -1, 0, 0), \mathbf{x}_3 = (-1, -1, \epsilon, 0)$, and $\mathbf{x}_4 = (-1, 1, 0, \epsilon)$, in cyclic order. Because each angle in the convex polygon is less than π, we conclude that the interiors of all four triangles project onto the positive linear combinations of consecutive vectors, e.g. the first triangle maps into the region $\lambda_1 \mathbf{x}_1 + \lambda_2 \mathbf{x}_2$ with $\lambda_1, \lambda_2 \geq 0$. Now projection on the first two coordinates in this new coordinate system extends to a local homeomorphism $U_1 \times I^2$ in a neighborhood of the vertex, and pulling back by L^{-1} gives the required local flat structure in a neighborhood of the vertex. If instead the four edge directions span a three-dimensional space, then the argument used for a vertex of P applies. Thus Σ is locally flat at each vertex of Q. Finally the centroid vertex added to the polygon Q is obviously locally flat, and we conclude that Σ is locally flat.

Finally we note that Σ is a topological disk, since it is two-sided and topologically is an annulus glued onto a disk. □

Proof of Theorem 1.5: Given the polygon P in $\mathbb{R}^d$, for $d \geq 5$ we cone it to a suitably chosen point $z \in \mathbb{R}^d$, chosen so that the coning is an embedding. It suffices to choose a "general position" point, because two planes (codimension $d-2$ flats) in $\mathbb{R}^d$ generically have empty intersections. The resulting surface Σ has n triangles, and is a topological disk.

By a standard result, see Rourke and Sanderson [15, Corollary 5.7 and Corollary 7.2], this embedded surface is locally flat. □

We conclude with some remarks concerning the unresolved case of embedded surfaces in $\mathbb{R}^4$ having a given polygon P with n edges as boundary. First, one can always find such a triangulated surface using at most $21n^2$ triangles, as follows. Take a projection of the polygon P into a hyperplane H, resulting in a polygon P^* in H, picking a projection direction such that the vertical surface Σ_1 connecting P to P^* is embedded. Theorem 1.1 gives a surface Σ_2 that uses at most $7n^2$ triangles which lies entirely in H, and has a subdivision of P^{**} of P^* as boundary and with P^{**} having at most $7n^2$ vertices. We obtain a triangulated vertical surface Σ_1 connecting P to P^{**} using at most $14n^2$ triangles, and $\Sigma = \Sigma_1 \cup \Sigma_2$ is the required surface. It can be checked that this surface is locally flat.

Second, the immersed surface constructed in Theorem 1.4 can be converted to an embedded surface of higher genus by cut-and paste, but we show that for some P such a surface must contain $\Omega(n^2)$ triangles. Recall that the *4-ball genus* of a knot embedded in a hyperplane in $\mathbb{R}^4$ is the smallest genus of any spanning surface of it that lies strictly in a half-space of $\mathbb{R}^4$ on one side of this hyperplane. If we start with a polygon P in $\mathbb{R}^4$ that lies in a hyperplane, the construction of Theorem 1.4 will (in general) produce an immersed surface lying in a half-space on one side of the hyperplane, and a cut-and-paste construction will preserve this property. (Note that cut-and-paste in 4-dimensions to replace two triangles intersecting in an interior point with a non-intersecting set may result in eight triangles.) As noted earlier, the $(2n, 2n-1)$ torus knot has a polygonal representation P_n in a hyperplane using $4n$ line segments (Adams et al [1, Lemma 8.1]) while a result of Shibuya [17] implies that its 4-ball genus is at least $2n(2n-1)/8$. Applying

Lemma 3.1, we conclude that the number of triangles needed in an embedded orientable PL surface of this type spanning P_n must grow quadratically in n. This can be taken as (weak) evidence in support of the possibility that for embedded surfaces in $\mathbb{R}^4$ the best combinatorial isoperimetric bound may be $O(n^2)$.

References

[1] C. Adams, B. M. Brennan, D. L. Greilsheimer and A. K. Woo, Stick numbers and composition of knots and links, J. Knot Theory Ramifications **6** (1997), 149–161.

[2] F. J. Almgren and W. P. Thurston, Examples of unknotted curves that only bound surfaces of high genus within their convex hulls, *Ann. Math.* **105** (1977), 527–538.

[3] E. F. Beckenbach and T. Rado, Subharmonic functions and surfaces of negative curvature, *Trans. Amer. Math. Soc.* **35** (1933), 662–682.

[4] W. Blaschke, *Vorlesungen über Differentialgeometrie I*, Third Ed., Springer-Verlag: Berlin 1930.

[5] Yu. D. Burago and V. A. Zalgaller, *Geometric Inequalities*, Springer-Verlag: Berlin 1988.

[6] B. Chazelle, Convex partitions of polyhedra: a lower bound and worst case optimal algorithm, SIAM J. Comput. **13** (1984), 488–507.

[7] E. Furstenber, Jie Li and J. Schneider, Stick knots, Chaos, Solitons and Fractals **9** (1998), 561–568.

[8] J. Hass and J. C. Lagarias, The number of Reidemeister moves needed for unknotting, *J. Amer. Math. Soc.* 14 (2001), 399-428.

[9] J. Hass, J. C. Lagarias and N. Pippenger, The computational complexity of knot and link problems, *J. ACM* 46 (1999), no. 2, 185–211.

[10] J. Hass, J. C. Lagarias and W. P. Thurston, Area inequalities for embedded disks bounding unknotted curves, in preparation.

[11] J. Hass, J. S. Snoeyink and W. P. Thurston, The size of spanning disks for polygonal knots, Discrete & Computational Geometry **29** (2003) no. 1, 1–17. eprint: `arXiv math.GT/9906197`.

[12] J. H. Hubbard, On the convex hull genus of space curves, Topology **19** (1980), 203–208.

[13] R. Osserman, The isoperimetric inequality, Bull. Amer. Math. Soc. **84** (1978), 1182–1238.

[14] D. Rolfsen, *Knots and Links,* Math. Lecture Series No. 7, Publish or Perish: Berkeley, Cal. 1976.

[15] C. P. Rourke and B. J. Sanderson, *Introduction to Piecewise-Linear Topology,* Springer-Verlag: New York-Heidelberg 1982.

[16] H. Seifert, Über das Geschlecht von Knoten, Math. Ann. **110** (1934), 571–592.

[17] T. Shibuya, A lower bound of 4-genus for torus knots, Mem. Oaska Inst. Tech. Ser. A **31** (1986), 11–17.

[18] S. Suri, Polygons, in: *Handbook of Discrete and Computational Geometry*, (J. E. Goodman and J. O'Rourke, Eds.), Chapter 23, CRC Press, Boca Raton, FL 1997, pp. 429–444.

[19] G. Vegter, Computational Topology, in: *Handbook of Discrete and Computational Geometry*, (J. E. Goodman and J. O'Rourke, Eds.), Chapter 28, CRC Press, Boca Raton, FL 1997, pp. 517–536.

About Authors

Joel Hass is at the Department of Mathematics, University of California, Davis, CA 95616; *hass@math.ucdavis.edu.*

Jeffrey C. Lagarias is at AT&T Labs-Research, 180 Park Avenue, Florham Park, NJ 07932-0971; *jcl@research.att.com.*

Acknowledgments

The authors are indebted to the reviewers for several useful remarks and suggestions, and for bringing some references to our attention.

Part of the work on this paper was done by the authors during a visit to the Institute for Advanced Study. The first author was partially suppored by NSF grant DMS-0072348, and by a grant to the Institute of Advanced Study by AMIAS.

Jacobi Decomposition and Eigenvalues of Symmetric Matrices

W. He
N. Prabhu

Abstract

We show that every n-dimensional orthogonal matrix can be factored into $O(n^2)$ Jacobi rotations (also called Givens rotations in the literature). It is well known that the Jacobi method, which constructs the eigen-decomposition of a symmetric matrix through a sequence of Jacobi rotations, is slower than the eigenvalue algorithms currently used in practice, but is capable of computing eigenvalues, particularly tiny ones, to a high relative accuracy. The above decomposition, which to the best of our knowledge is new, shows that the infinite-precision nondeterministic Jacobi method (in which an oracle is invoked to obtain the correct rotation angle in each Jacobi rotation) can construct the eigen-decomposition with $O(n^2)$ Jacobi rotations. The complexity of the nondeterministic Jacobi algorithm motivates the efforts to narrow the gap between the complexity of the known Jacobi algorithms and that of the nondeterministic version. Speeding up the Jacobi algorithm while retaining its excellent numerical properties would be of considerable interest. In that direction, we describe, as an example, a variant of the Jacobi method in which the rotation angle for each Jacobi rotation is computed (in closed-form) through 1-dimensional optimization. We also show that the computation of the closed-form optimal solution of the 1-dimensional problems guarantees convergence of the new method.

1 Introduction

The symmetric eigenvalue problem has been investigated extensively over the years and the research has yielded several efficient algorithms [4–6, 8, 9, 12, 13, 18, 20, 28, 31, 34, 36, 42, 44] . At present, for $n \times n$ symmetric matrices with $n \leq 25$, most commercial software packages usually use the QR method [13, 16, 27, 47, 48]; for n between 25 and a few thousands they switch to divide-and-conquer methods [8, 10, 12, 20, 42] and for $n \gg 10^3$ the packages usually resort to one of the iterative methods based on Krylov subspaces, such as the Lanczos algorithm [15, 28, 34, 37, 44]. However, for certain matrices whose eigenvalues are very tiny, or when the eigenvalues need to be computed to high relative accuracy, Jacobi's method – the oldest method for computing eigenvalues – devised in 1846 by Jacobi [25] continues to be the algorithm of choice. Given a symmetric matrix A_0, Jacobi's method constructs a sequence of orthogonally similar matrices $A_0, A_1, \ldots$ which converge to a diagonal matrix Λ. A_{i+1} is constructed from A_i using a Jacobi rotation

$$J_i = I + (\cos(\theta) - 1)(e_r e_r^T + e_s e_s^T) + \sin(\theta)(e_r e_s^T - e_s e_r^T)$$

as $A_{i+1} = J_i^T A_i J_i$, where e_r is the rth column of an $n \times n$ identity matrix. No upper bound is known on the number of Jacobi rotations needed to diagonalize a given symmetric matrix to a pre-specified accuracy. However, it is known that the Jacobi method is asymptotically quadratically convergent [2,21,36,39,45,49]. The numerical stability of the Jacobi method comes from the usual floating point error analysis of matrix multiplication which shows that if Q is an orthogonal matrix and $\tilde{Q}$ its floating point representation, then for any matrix A, $\tilde{Q}A = Q(A+E)$ where $\|E\|_2 \leq O(\epsilon)\cdot\kappa_2(Q)\cdot\|A\|_2$; ϵ represents the machine precision, $\kappa_2(Q)$, the condition number of Q (in the 2-norm) and $\|\cdot\|_2$, the 2-norm. Given A and ϵ, $\|E\|_2$ is minimized when Q is orthogonal matrix in which case $\kappa_2(Q) = 1$. In fact, it is known that for certain symmetric matrices [11] every method that involves reducing the matrix to a bidiagonal form loses all significant digits in all but the largest singular value while Jacobi's method calculates all the singular values to full machine precision. For further discussion of Jacobi's method and its numerical properties see [16,35,49,50].

As we remarked above, Jacobi's method constructs the eigen-decomposition of a symmetric matrix one Jacobi rotation at a time. We call the decomposition of an orthogonal matrix into a product of Jacobi rotations, the *Jacobi decomposition of the orthogonal matrix.* To understand the complexity of the best Jacobi-like algorithm for eigen-decomposition we then need to determine an upper bound on the number of Jacobi rotations needed to construct a Jacobi decomposition of an arbitrary orthogonal matrix.

Observe that a Jacobi rotation can be written as

$$J = \exp(\alpha E_{ij}); \quad \text{where } E_{ij} = e_i e_j^T - e_j e_i^T.$$

Further, the space of $n \times n$ antisymmetric matrices form the Lie algebra of the Lie group SO(n). Hence any orthogonal matrix Q can be written as

$$Q = \exp\left(\sum_{i<j} \alpha_{ij} E_{ij}\right).$$

Writing Q as a product of factors of the form $\exp(\beta_{ij} E_{ij})$ is complicated by the fact that

$$[E_{ij}, E_{kl}] = f_{ij,kl}^{rs} E_{rs}, \text{ where } f_{ij,kl}^{rs} \triangleq \delta_{il}\delta_j^r\delta_k^s + \delta_{jk}\delta_i^r\delta_l^s - \delta_{jl}\delta_i^r\delta_k^s - \delta_{ik}\delta_j^r\delta_l^s,$$

that is, the structure constants $f_{ij,kl}^{rs}$ of the Lie algebra are not zero. It is not straightforward to try to factor the matrix Q using the Baker-Campbell-

Hausdorff formula

$$\exp(A)\exp(B) = \exp\left(A + B + \frac{1}{2}[A,B] + \frac{1}{12}([A,[A,B]] - [B,[B,A]]) + \dots\right)$$

either since one has an infinite series of commutators. With the hindsight of the Decomposition Theorem, which we prove in the next section, it appears unlikely that Q has a decomposition

$$Q = \Pi_{i<j} \exp(\beta_{ij} E_{ij}),$$

where each pair (i, j) occurs exactly once, although not necessarily in any definite order; the preceding remark further suggests that factoring Q into Jacobi rotations using Baker-Campbell-Hausdorff formula is likely infeasible. We prove the sought Decomposition Theorem for orthogonal matrices using a different parametrization of the orthogonal group.

We show that any $n \times n$ orthogonal matrix can be written as a product of at most $\frac{n(n-1)}{2}$ Jacobi rotations. Therefore, a nondeterministic Jacobi algorithm, which makes the right choice of rotation angle for each Jacobi rotation, can diagonalize an n-dimensional orthogonal matrix with $O(n^2)$ Jacobi rotations. While one may not expect to find a deterministic Jacobi algorithm that converges in $O(n^2)$ steps, the upper bound on the time complexity of the nondeterministic algorithm certainly motivates the search for faster variants of the Jacobi algorithm. In that direction, we outline a new method for computing an eigenvalue and prove the convergence of the new algorithm. The algorithm we outline does not generate the optimal sequence of Jacobi rotations. However it does generate a sequence of locally optimal Jacobi rotations, exploiting the closed form solutions of cubic and quartic equations [3].

The paper is organized as follows. In Section 2 we prove the Decomposition Theorem. In Section 3 we present an algorithm based on the Decomposition Theorem and prove that it converges. The standard methods for solving cubic and quartic equations are summarized in Appendices A and B.

2 Decomposition Theorem

In this section we establish a decomposition theorem for orthogonal matrices. The theorem could be of independent interest in applications involving the *Special Orthogonal Group* $SO(d)$. The d-dimensional representation of $SO(d)$ is

$$SO(d) = \{M \in \mathcal{M}_{d\times d} \mid M^T M = I;\ \det(M) = 1\},$$

where $\mathcal{M}_{d\times d}$ is the group of all $d \times d$ real matrices and $\det(M)$ denotes the determinant of M. The group $SO(d)$ represents the set of all rotations in R^d. The Lie algebra of $SO(d)$ is the set of $d \times d$ antisymmetric matrices as the following calculation shows. If A is an antisymmetric $d \times d$ matrix

$(A^T = -A)$, then

$$\begin{aligned} e^A &\equiv \sum_{n=0}^{\infty} \frac{A^n}{n!}, \\ \left(e^A\right)^T &= \left(\sum_{n=0}^{\infty} \frac{A^n}{n!}\right)^T = \sum_{n=0}^{\infty} \frac{\left(A^T\right)^n}{n!} = \sum_{n=0}^{\infty} \frac{(-A)^n}{n!} = e^{-A}. \end{aligned}$$

Therefore

$$\left(e^A\right)^T e^A = e^{-A} e^A = I.$$

Further, since for any $M \in \mathcal{M}_{d\times d}$,

$$\begin{aligned} \det(M) &= e^{\text{trace}(\log(M))}, \\ \det\left(e^A\right) &= e^{\text{trace}(A)} = 1. \end{aligned}$$

Besides the exponentiation, which is a computationally expensive operation, there is yet another remarkable representation of SO(d) matrices in terms of antisymmetric matrices. Again if A is any antisymmetric matrix then set

$$B \equiv (I + A)(I - A)^{-1}. \tag{1}$$

The definition is valid since all the eigenvalues of an antisymmetric matrix are zeros and pure imaginary; hence $(I - A)$ is invertible. One can easily verify that $B^T B = I$. Further, since for any matrix Q, $\det(Q) = \det\left(Q^T\right)$, we have

$$\det(I + A) = \det\left((I + A)^T\right) = \det(I - A).$$

Therefore,

$$\begin{aligned} \det(B) &= \det\left((I + A)(I - A)^{-1}\right) = \det(I + A)\det\left((I - A)^{-1}\right) \\ &= \det(I + A) \times \frac{1}{\det(I + A)} \\ &= 1, \end{aligned} \tag{2}$$

which shows that B is an SO(d) matrix. However, the converse is not true. That is, given an SO(d) matrix B, it is not always possible to find an antisymmetric matrix A to satisfy equation (1). This fact is significant and we elaborate on it further. If for a given B, there exists an antisymmetric A such that equation (1) holds, then

$$B = (I + A)(I - A)^{-1} \Rightarrow B(I - A) = (I + A) \Rightarrow (B - I) = (B + I)A \tag{3}$$

The system (3) is not solvable for every SO(d) matrix B. The reason, of course, is that since both $+1$ and -1 are possible eigenvalues of B, both

$B+I$ and $B-I$ could be rank-defective matrices, but with different rank-defects. For example, consider,

$$B = \begin{pmatrix} 1 & & \\ & -1 & \\ & & -1 \end{pmatrix}.$$

Then the system (3) becomes

$$\begin{pmatrix} 0 & & \\ & -2 & \\ & & -2 \end{pmatrix} = \begin{pmatrix} 2 & & \\ & 0 & \\ & & 0 \end{pmatrix} A$$

which, clearly, has no solution for A. However the system

$$(B-I) = (B+I)A \tag{4}$$

would have a solution for A, if $B+I$ is invertible, or in other words, if -1 is not an eigenvalue of B. Thus, if -1 is not an eigenvalue of B,

$$A = (B+I)^{-1}(B-I) \tag{5}$$

We'd still need to verify if the A computed in (5) is antisymmetric. Let

$$\begin{aligned} M = A + A^T &= (B+I)^{-1}(B-I) + \left[(B+I)^{-1}(B-I)\right]^T \\ &= (B+I)^{-1}(B-I) + (B^T - I)(B^T+I)^{-1} \end{aligned}$$

or

$$\begin{aligned} (B+I)M(B^T+I) &= (B-I)(B^T+I) + (B+I)(B^T-I) \\ &= 0. \end{aligned}$$

Since $\det(B+I) \neq 0$, by assumption, both $B+I$ and (B^T+I) are invertible, which implies that $M=0$, or that A is antisymmetric.

To summarize, given any antisymmetric A equation (1) enables us to compute the corresponding $\mathrm{SO}(d)$ matrix. However, given an $\mathrm{SO}(d)$ matrix B, we cannot in general find an antisymmetric matrix to satisfy (1). If -1 is not one of the eigenvalues of B however one can find an antisymmetric matrix corresponding to B. Our objective is to decompose a general $\mathrm{SO}(d)$ matrix into factors of the form

$$(I + \alpha E_{ij})(I - \alpha E_{ij})^{-1}$$

where $\alpha \in R$, and

$$E_{ij} \equiv e_i e_j^T - e_j e_i^T;$$

e_i is the $n \times 1$ column vector whose j^{th} component is

$$(e_i)_j = \begin{cases} 1 & \text{if } j = i, \\ 0 & \text{otherwise.} \end{cases}$$

$\{E_{ij}\}_{i<j}$ form a basis for the vector space of antisymmetric matrices. The following Lemma is needed for the proof of the Decomposition Theorem that follows.

Lemma 2.1. *Let S be an anti-symmetric matrix, then there exists an orthogonal matrix P such that*

$$P^T SP = \begin{pmatrix} 0 & & & & & \\ & \ddots & & & & \\ & & 0 & & & \\ & & & A_1 & & \\ & & & & \ddots & \\ & & & & & A_s \end{pmatrix},$$

where $A_i = \begin{pmatrix} 0 & a_i \\ -a_i & 0 \end{pmatrix}$, *and a_i is a nonzero real number, $i = 1, 2, \cdots, s$.*

Proof: Since the eigenvalues of an antisymmetric matrix are either zero or pure imaginary numbers, from the real Schur decomposition [10, Theorem 4.3]) we know that there exists an orthogonal matrix P such that

$$U = P^T SP = \begin{pmatrix} 0 & & & & & \\ & \ddots & & & & \\ & & 0 & & * & \\ & & & A_1 & & \\ & & & & \ddots & \\ & & & & & A_s \end{pmatrix}.$$

where U is block upper triangular – *i.e.*, $A_1, \ldots, A_s$ are 2×2 matrices.

Since S is antisymmetric, it commutes with its transpose and $[S, S^T]$=0 and therefore $[U, U^T] = 0$ as well. Writing

$$U = \begin{pmatrix} B & C \\ 0 & D \end{pmatrix}$$

we see that $[U, U^T] = 0$ implies $BB^T + CC^T = B^T B$ and $DD^T = C^T C + D^T D$. Hence, trace$(CC^T) =$ trace$(B^T B - BB^T) = 0$ which implies that $C = 0$. Therefore B and D must also commute with their respective transposes. Applying the preceding argument recursively, we conclude that

$$P^T SP = \begin{pmatrix} 0 & & & & & \\ & \ddots & & & & \\ & & 0 & & & \\ & & & A_1 & & \\ & & & & \ddots & \\ & & & & & A_s \end{pmatrix}.$$

where $A_1, \ldots, A_s$ are 2×2 matrices. Since each A_j, $1 \leq j \leq s$ corresponds to a pair of conjugate pure imaginary eigenvalues of S, each A_j must be of the form $A_j = \begin{pmatrix} 0 & a_j \\ -a_j & 0 \end{pmatrix}$. □

Lemma 1 is used in the proof of the following decomposition theorem.

Theorem 2.2 (Decomposition Theorem). *Let* $E_{ij} \equiv e_i e_j^T - e_j e_i^T$ *and*

$$\begin{aligned}
Z_1(n) &\equiv \{A \in \mathcal{M}_{n\times n} \mid A = (I+S)(I-S)^{-1};\ S^T = -S\}, \\
Z_2(n) &\equiv \{B \in \mathcal{M}_{n\times n} \mid B = \prod_{1\leq i<j\leq n} (I+\theta_{ij}E_{ij})(I-\theta_{ij}E_{ij})^{-1};\ \theta_{ij} \in R\}, \\
Z_3(n) &\equiv \{C \in \mathcal{M}_{n\times n} \mid \\
&\qquad C = \prod_{\substack{1\leq i,j\leq n \\ 1\leq k\leq \frac{n(n-1)}{2}}} (I+\theta_{ij}^k E_{ij})(I-\theta_{ij}^k E_{ij})^{-1}\ ;\ \theta_{ij}^k \in R\}, \\
\bar{Z}_3(n) &\equiv \{C \in \mathcal{M}_{n\times n} \mid \\
&\qquad C = MD;\ M \in Z_3(n);\ D \in \mathcal{M}_{n\times n};\ D = Diag(1,\ldots,1,-1)\}, \\
Z_4(n) &\equiv \{Q \in \mathcal{M}_{n\times n} \mid Q^T Q = I\}.
\end{aligned}$$

Then

1. $Z_1(n), Z_2(n) \subset Z_3(n) \subset Z_4(n)$ *for* $n \geq 2$.

2. $Z_4(n) = Z_3(n) \cup \bar{Z}_3(n)$ *for* $n \geq 2$.

Proof: First we show that

$$Z_1(n) = \{M \in \mathrm{SO}(n) | -1 \notin Spectrum(M)\} \tag{6}$$

where *Spectrum*(M) is the set of eigenvalues of M. To show (6) observe that if S be an anti-symmetric matrix, then there exists an orthogonal matrix P such that

$$P^T SP = \begin{pmatrix} 0 & & & \\ & B_1 & & \\ & & \ddots & \\ & & & B_l \end{pmatrix} \equiv H,$$

where $B_i = \begin{pmatrix} 0 & b_i \\ -b_i & 0 \end{pmatrix}$, and b_i is a nonzero real number, $i = 1, 2, \cdots, l$. Hence,

$$\begin{aligned}(I-S)^{-1} &= (P(I-H)P^T)^{-1} \\ &= P\begin{pmatrix} I_s & & & \\ & \begin{pmatrix} 1 & -b_1 \\ b_1 & 1 \end{pmatrix}^{-1} & & \\ & & \ddots & \\ & & & \begin{pmatrix} 1 & -b_l \\ b_l & 1 \end{pmatrix}^{-1} \end{pmatrix} P^T.\end{aligned}$$

Therefore,

$$\begin{aligned}A &= (I+S)(I-S)^{-1} \\ &= P\begin{pmatrix} I_s & & & \\ & \begin{pmatrix} \frac{1-b_1^2}{1+b_1^2} & \frac{2b_1}{1+b_1^2} \\ -\frac{2b_1}{1+b_1^2} & \frac{1-b_1^2}{1+b_1^2} \end{pmatrix} & & \\ & & \ddots & \\ & & & \begin{pmatrix} \frac{1-b_l^2}{1+b_l^2} & \frac{2b_l}{1+b_l^2} \\ -\frac{2b_l}{1+b_l^2} & \frac{1-b_l^2}{1+b_l^2} \end{pmatrix} \end{pmatrix} P^T.\end{aligned}$$

Since $\frac{1-b_i^2}{1+b_i^2} \neq -1$, $1 \leq i \leq l$, we conclude that $-1 \notin \mathit{Spectrum}(A)$.

Next we show that

$$Z_3(n) = \{A \in \mathcal{M}_{n\times n} | A^T A = I; \mathit{Spectrum}(A) \text{ has an even number of -1's}\} \quad (7)$$

The proof of (7) is by induction on n. Consider $n = 2$. First we show that the set on the right is a subset of $Z_3(2)$. Any 2×2 matrix A satisfying $A^T A = I$ and having an even number of -1's in its spectrum can be written as

$$A = \begin{pmatrix} \cos(\alpha) & \sin(\alpha) \\ -\sin(\alpha) & \cos(\alpha) \end{pmatrix}$$

for some α. Further observe that

$$\begin{aligned}(I+\theta E_{12})(I-\theta E_{12})^{-1} &= \begin{pmatrix} 1 & \theta \\ -\theta & 1 \end{pmatrix}\begin{pmatrix} 1 & -\theta \\ \theta & 1 \end{pmatrix}^{-1} \\ &= \frac{1}{1+\theta^2}\begin{pmatrix} 1-\theta^2 & 2\theta \\ -2\theta & 1-\theta^2 \end{pmatrix}\end{aligned}$$

We set $\theta^2 = \frac{1-\cos(\alpha)}{1+\cos(\alpha)}$, if $\alpha \neq \pi$. If $\alpha = \pi$, then $A = -I$ and we can write $A = B^2$ where

$$B = \begin{pmatrix} 0 & -1 \\ 1 & 0 \end{pmatrix} = \begin{pmatrix} 1 & -1 \\ 1 & 1 \end{pmatrix}\begin{pmatrix} 1 & 1 \\ -1 & 1 \end{pmatrix}^{-1} = (I-E_{12})(I+E_{12})^{-1}.$$

Therefore, the set on the right hand side of (7) is a subset of $Z_3(2)$. Conversely, consider a 2×2 matrix $A \in Z_3(2)$. By definition,

$$A = (I + \theta_{12}E_{12})(I - \theta_{12}E_{12})^{-1}$$

which implies that

$$\det(A) = \det((I + \theta_{12}E_{12})(I - \theta_{12}E_{12})^{-1}) = 1.$$

Hence A cannot have an odd number of -1's in its spectrum and (7) holds for $n = 2$.

Now assume that (7) holds for $n = m - 1 \geq 2$. We'll prove it for $n = m$. Consider an $m \times m$ matrix A, such that $A^T A = I$ and *Spectrum(A)* has an even number of -1's. We'll show that there exists a matrix M such that $M^T \in Z_3(m)$ and AM is of the form

$$AM = \begin{pmatrix} A_1 & \\ & 1 \end{pmatrix}$$

We can then use induction on A_1. We present the argument through two cases.

Case 1: $a_{mm} = -1$

If $a_{mm} = -1$ and $A^T A = I$ implies that A must be of the form

$$A = \begin{pmatrix} \tilde{A} & \\ & -1 \end{pmatrix}$$

Let $M \equiv Diag_m(-1, 1, 1, \ldots, 1, -1)$. Then M can be written as

$$M = M^T = \left[(I + E_{1m})(I - E_{1m})^{-1}\right]^2. \tag{8}$$

Clearly $M^T \in Z_3(m)$ and AM is of the form

$$AM = A = \begin{pmatrix} A' & \\ & 1 \end{pmatrix} \text{ where } A' = \tilde{A} \times Diag_{m-1}(-1, 1, \ldots, 1).$$

Case 2: $a_{mm} \neq -1$ In this case there must exist $\alpha_{1m}, \alpha_{2m}, \cdots, \alpha_{m-1,m}$ satisfying $\cos\alpha_{im} \neq -1$ for $i = 1, 2, \cdots, m-1$ and

$$\begin{aligned}
a_{m1} &= -\sin\alpha_{1m}, \\
a_{m2} &= -\cos\alpha_{1m}\sin\alpha_{2m}, \\
&\vdots \\
a_{m,m-1} &= -\cos\alpha_{1m}\cdots\cos\alpha_{m-2,m}\sin\alpha_{m-1,m}, \\
a_{mm} &= \cos\alpha_{1m}\cdots\cos\alpha_{m-2,m}\cos\alpha_{m-1,m}.
\end{aligned}$$

The α_{im} that solve the above system are given by

$$\sin\alpha_{im} = -\frac{a_{mi}}{\sqrt{1 - a_{m1}^2 - \ldots - a_{m,i-1}^2}}, \quad 1 \leq i \leq m-1$$

If $a_{mi} = 0$, set $\sin\alpha_{im} = 0$ and $\cos\alpha_{im} = 1$. So we can write A as

$$A = \begin{bmatrix} a_{11} & a_{12} & \dots & a_{1m} \\ a_{21} & a_{2m-1} & \dots & a_{2m} \\ \vdots & \vdots & & \vdots \\ a_{m-1,1} & a_{m-1,2} & \dots & a_{m-1,m} \\ -\sin\alpha_{1m} & -\cos\alpha_{1m}\sin\alpha_{2m} & \dots & \cos\alpha_{1m}\cdots\cos\alpha_{m-2,m}\cos\alpha_{m-1,m} \end{bmatrix}$$

Next we show that if

$$R_m = \prod_{i=1}^{m-1} (I + \theta_{im}E_{im}) \cdot (I - \theta_{im}E_{im})^{-1} \tag{9}$$

then the last row of R_m must be of the form

$$(-\sin\beta_{1m}, -\cos\beta_{1m}\sin\beta_{2m}, \cdots,$$
$$\cdots, -\cos\beta_{1m}\cdots\cos\beta_{m-2,m}\sin\beta_{m-1,m}, \cos\beta_{1m}\cdots\cos\beta_{m-2,m}\cos\beta_{m-1,m})$$

where $\cos\beta_{im} = \frac{1-\theta_{im}^2}{1+\theta_{im}^2} \neq -1$ for $1 \leq i \leq m$.

Again the proof is by induction on m. From definition of R_n it follows that

$$R_2 = (I + \theta_{12}E_{12})(I - \theta_{12}E_{12})^{-1} = \begin{pmatrix} \cos\beta_{12} & \sin\beta_{12} \\ -\sin\beta_{12} & \cos\beta_{12} \end{pmatrix}$$

where $\cos\beta_{12} = \frac{1-\theta_{12}^2}{1+\theta_{12}^2} \neq -1$. Hence, the claim is correct for $m = 2$. Now suppose that it is correct for $m = r - 1$. Then for $m = r$, we have, from definition,

$$\begin{aligned} R_r &= \prod_{i=1}^{r-1} (I + \theta_{ir}E_{ir}) \cdot (I - \theta_{ir}E_{ir})^{-1} \\ &= \begin{pmatrix} \cos\beta_{1r} & 0 & \cdots & 0 & \sin\beta_{1r} \\ 0 & 1 & \cdots & 0 & 0 \\ \vdots & \vdots & \ddots & \vdots & \vdots \\ 0 & 0 & \cdots & 1 & 0 \\ -\sin\beta_{1r} & 0 & \cdots & 0 & \cos\beta_{1r} \end{pmatrix} \cdot \prod_{i=2}^{r-1} (I + \theta_{ir}E_{ir}) \cdot (I - \theta_{ir}E_{ir})^{-1} \end{aligned}$$

where $\cos\beta_{1r} = \frac{1-\theta_{1r}^2}{1+\theta_{1r}^2} \neq -1$. Note that $\prod_{i=2}^{r-1}(I+\theta_{ir}E_{ir})\cdot(I-\theta_{ir}E_{ir})^{-1}$ has the form $\begin{pmatrix} 1 & 0 \\ 0 & \tilde{R}_{r-1} \end{pmatrix}$. By induction, the last row of $\tilde{R}_{r-1}$ is

$$(-\sin\beta_{2r}, -\cos\beta_{2r}\sin\beta_{3r}, \cdots,$$
$$\cdots, -\cos\beta_{2r}\cdots\cos\beta_{r-2,r}\sin\beta_{r-1,r}, \cos\beta_{2r}\cdots\cos\beta_{r-2,r}\cos\beta_{r-1,r}).$$

where $\cos\beta_{ir} = \frac{1-\theta_{ir}^2}{1+\theta_{ir}^2} \neq -1$, $2 \leq i \leq r$. Hence the last row of R_r is

$$(-\sin\beta_{1r}, -\cos\beta_{1r}\sin\beta_{2r}, \cdots,$$
$$\cdots, -\cos\beta_{1r}\cdots\cos\beta_{r-2,r}\sin\beta_{r-1,r}, \cos\beta_{1r}\cdots\cos\beta_{r-2,r}\cos\beta_{r-1,r})$$

where $\cos\beta_{ir} = \frac{1-\theta_{ir}^2}{1+\theta_{ir}^2} \neq -1$, for $1 \leq i \leq r$.

From the preceding argument we see that we can choose θ_{ij} in the design of R_m such that β_{im} in R_m coincide with α_{im} of A for $1 \leq i \leq m$. For such an R_m, from the orthogonality of R_m and A it follows that AR_m^T must be of the form

$$AR_m^T = \begin{pmatrix} A_1 & \\ & 1 \end{pmatrix} \tag{10}$$

Therefore in this case ($a_{mm} \neq -1$) we set $M = R_m^T$. We observe that $M^T = R_m \in Z_3(m)$.

In summary, if *Spectrum(A)* has an even number of -1's and $A^TA = I$, then we can find a matrix M such that $M^T \in Z_3(m)$ and

$$AM = \begin{pmatrix} A_1 & \\ & 1 \end{pmatrix}. \tag{11}$$

Since *Spectrum(A)* has an even number of -1's, *Spectrum(A_1)* must have an even number of -1's as well. By induction $A_1 \in Z_3(m-1)$ which implies that $AM \in Z_3(m)$. Since $AM, M^T \in Z_3(m)$, so does $AMM^T = A$ and we have shown that the set on the right hand side of (7) is contained in $Z_3(m)$.

To show the converse that $Z_3(m)$ is contained in the set on the right hand side of (7), consider an $m \times m$ matrix $A \in Z_3(m)$. One can easily verify that $A^TA = I$. Further, from the arguments presented in equations (1) and (2) it follows that $\det(A) = 1$. If λ is an eigenvalue of A and V_λ the corresponding eigenvector then

$$(\lambda^* V_\lambda^H)(\lambda V_\lambda) = (AV_\lambda)^H(AV_\lambda) = V_\lambda^H A^H A V_\lambda = V_\lambda^H A^T A V_\lambda = V_\lambda^H V_\lambda \tag{12}$$

which implies that $|\lambda|^2 = \lambda^*\lambda = 1$ where $*, H$ denote complex and Hermitian conjugation respectively. Therefore, $\det(A) = -1$ if and only if *Spectrum(A)* has an odd number of -1's. Since $\det(A) = 1$, we conclude that *Spectrum(A)* must have an even number of -1's.

Finally, we show that $1 \leq k \leq \frac{m(m-1)}{2}$ for $m \geq 3$. Consider equation (11) and the statement preceding it. If $a_{mm} = -1$, then we showed in (8) that the M in (11) can be written as a product of two factors of the form $(I+\theta E_{ij})(I-\theta E_{ij})^{-1}$. On the other hand if $a_{mm} \neq -1$, then the M in (11) can be written as a product on $m-1$ factors of the form $(I+\theta E_{ij})(I-\theta E_{ij})^{-1}$ as shown in equations (9), (10) and the discussion between them. Thus in either case, $a_{mm} = -1$ or $a_{mm} \neq -1$, the M in equation (11) can be written as a product of at most $m-1$ elementary rotations ($(I+\theta E_{ij})(I-\theta E_{ij})^{-1}$). Now observe that A_1 in equation (11) belongs to $Z_3(m-1)$ and hence we can find an

$$M_1 = \begin{pmatrix} \tilde{M}_1 & \\ & 1 \end{pmatrix}$$

such that

$$\begin{pmatrix} A_1 & \\ & 1 \end{pmatrix} \begin{pmatrix} \tilde{M}_1 & \\ & 1 \end{pmatrix} = \begin{pmatrix} A_2 & & \\ & 1 & \\ & & 1 \end{pmatrix}$$

where M_1 is a product of at most $m-2$ elementary rotations. Thus we have

$$A \times M \times M_1 \times \ldots \times M_{m-2} = I$$

which implies that

$$A = [M \times M_1 \times \ldots \times M_{m-2}]^{-1} \tag{13}$$

If we define $U \equiv (I - \theta E_{ij})$, then $U^T U^{-1} = U^{-1} U^T$ as the following calculation shows.

$$\begin{aligned} UU^T U^{-1} &= (I - \theta E_{ij})(I + \theta E_{ij})(I - \theta E_{ij})^{-1} \\ &= (I + \theta E_{ij})(I - \theta E_{ij})(I - \theta E_{ij})^{-1} \\ &= (I + \theta E_{ij}) \\ &= U^T \end{aligned}$$

Therefore we have, for all $-\infty < \theta < \infty$ and all $1 \leq i < j \leq m$,

$$(I - \theta E_{ij})(I + \theta E_{ij})^{-1} = (I + \theta E_{ij})^{-1}(I - \theta E_{ij}) \tag{14}$$

Therefore if we define $R(\theta) \equiv (I - \theta E_{ij})(I + \theta E_{ij})^{-1}$ then from (14) it follows that

$$\begin{aligned} [R(\theta)]^T = \left[(I - \theta E_{ij})(I + \theta E_{ij})^{-1}\right]^T &= (I - \theta E_{ij})^{-1}(I + \theta E_{ij}) \\ &= (I + \theta E_{ij})(I - \theta E_{ij})^{-1} \\ &= R(-\theta). \end{aligned}$$

Further

$$[R(\theta)]^T R(\theta) = (I + \theta E_{ij})(I - \theta E_{ij})^{-1}(I - \theta E_{ij})(I + \theta E_{ij})^{-1} = I$$

That is, the transpose of an elementary rotation $R(\theta)$, namely $[R(\theta)]^T = R(-\theta) = [R(\theta)]^{-1}$ is also an elementary rotation. Since each of the M_i, $1 \leq i \leq m-2$ and M in (13) is a product of elementary rotations M_i^{-1} is also a product of the same number of elementary rotations as M_i (and M^{-1}

is a product of the same number of elementary rotations as M). Therefore, from (13) we have

$$A = M_{m-2}^{-1} \times \ldots, M_1^{-1} \times M^{-1}$$

is a product of at most $\frac{m(m-1)}{2}$ elementary rotations.

Equations (6) and (7) imply that $Z_1(m) \subset Z_3(m)$. Next we show that $Z_2(m) \subset Z_3(m)$. Clearly $Z_2(m) \subseteq Z_3(m)$. One can verify that $-I_2 = \begin{pmatrix} -1 & \\ & -1 \end{pmatrix} \notin Z_2(2)$ but as we showed earlier $-I_2 \in Z_3(2)$.

Finally, to show that $Z_4(m) = Z_3(m) \cup \bar{Z}_3(m)$, we note that if $A \in Z_4(m)$, then $\det(A) = \pm 1$. From equation (12), the discussion following it, equation (7) and the definitions of $Z_3(m)$, $\bar{Z}_3(m)$ and $Z_4(m)$ it follows that we can rewrite $Z_3(m)$ and $\bar{Z}_3(m)$ as

$$\begin{aligned} Z_3(m) &= \{C \in Z_4(m) | \det(C) = +1\} \\ \bar{Z}_3(m) &= \{C \in Z_4(m) | \det(C) = -1\} \end{aligned}$$

which shows that $Z_4(m) = Z_3(m) \cup \bar{Z}_3(m)$. □

As a corollary of the Theorem we have

Corollary 2.3. *Given an oracle for specifying the correct rotation angle for each Jacobi rotation, the (nondeterministic) Jacobi algorithm can diagonalize any $n \times n$ real symmetric matrix with at most $\frac{n(n-1)}{2}$ Jacobi rotations.*

Proof: Let S be an $n \times n$ real symmetric and Q the orthogonal matrix that diagonalizes S; $S = Q^T \Lambda Q$, $\Lambda = diag(\lambda_1, \ldots, \lambda_n)$. Q has an odd number of -1's in its spectrum if and only if $\det(Q) = -1$. Let $D = diag(-1, 1, \ldots, 1)$. Then $Q^T D^T \Lambda D Q = Q^T \Lambda Q = S$, which shows that if Q diagonalizes S, then so does DQ. Clearly DQ is orthogonal and $\det(DQ) = -\det(Q)$. In other words, there exists a matrix $\bar{Q} \in Z_3$ ($\bar{Q} = Q$ or DQ) such that $\bar{Q} S \bar{Q}^T = \Lambda$. From Theorem 1 the claim follows. □

Theorem 1 (and Corollary 1) reduce the problem of diagonalizing a symmetric matrix to that of computing the optimal rotation angles of $O(n^2)$ Jacobi rotations. One may not expect to devise an algorithm that computes the optimal sequence of Jacobi rotations (as stated in Corollary 1) since that would yield a finitely-convergent algorithm for computing roots of the characteristic polynomials of symmetric matrices. However, Corollary 1 certainly motivates one to attempt speeding up the Jacobi method through a more judicious choice of rotation angles. In the next section we outline a variant of the Jacobi method in which the rotation angles are computed using closed form solutions of certain 1-dimensional optimization problems. The construction of global optima of the 1-dimensional optimization problems is also shown to yield a proof of convergence of the algorithm.

3 Rotation Method

In this section we outline a Jacobi-like algorithm for computing an eigenvalue of a given symmetric matrix A. The sequence of Jacobi rotations is constructed through 1-dimensional optimization guided by the objective to drive the commutator $[A, xx^T]$ to zero, where x is a column vector. We also show that the closed-form optimization implies convergence of the algorithm. We start with an outline of the algorithm. We'll use the following notation.

$$\begin{aligned} Q_{ij}(\theta) &\equiv (I + \theta E_{ij})(I - \theta E_{ij})^{-1} \\ f_{ij}^k(\theta) &\equiv \left[Q_{ij}(\theta)x^k\right]^T A \left[Q_{ij}(\theta)x^k\right] \end{aligned}$$

Also, the jth row of A will be denoted A_j, $1 \leq j \leq n$.

Rotation Algorithm

Step 1: Select a random initial unit vector x^0. Set $k \leftarrow 0$.

Step 2: Compute

$$v_k \equiv \max_{1 \leq i,j \leq n} |(A_j x^k)x_i^k - (A_i x^k)x_j^k|.$$

Let (i_k, j_k) be the subscript pair at which the maximum is attained. That is,

$$v_k = |(A_{j_k} x^k)x_{i_k}^k - (A_{i_k} x^k)x_{j_k}^k|.$$

Step 3: If $v_k = 0$,

then $\left[x^k\right]^T Ax^k$ is an eigenvalue of A. Stop.

else
- solve

$$\max_\theta f_{i_k j_k}^k(\theta) \tag{15}$$

  to obtain a *global* optimal solution θ_k. The optimization problem (15) has closed form solutions, making it possible for us to compute the global maximum as described below.
- set

$$\begin{aligned} x^{k+1} &\leftarrow Q_{i_k j_k}(\theta_k)x^k \\ k &\leftarrow k+1 \end{aligned}$$

- Go to Step 2.

End of algorithm

The convergence of the algorithm is established in the theorem below.

Theorem 3.1. *The sequence of vectors $x^1, x^2, \ldots$ generated by the Rotation Algorithm converges to an eigenvector of the symmetric matrix A.*

Proof: If a nonzero x is an eigenvector of A with eigenvalue λ, then,

$$Ax = \lambda x \Rightarrow A_i x = \lambda x_i \Rightarrow x_j A_i x = x_i A_j x$$

where A_i denotes the ith row of A and x_j the jth component of x, $1 \leq i, j \leq n$. Conversely, if $x_i A_j x = x_j A_i x$ for all $1 \leq i, j \leq n$, then we have for some $x_k \neq 0$,

$$x_k \begin{pmatrix} A_1 \\ A_2 \\ \vdots \\ A_n \end{pmatrix} x = (A_k x) \begin{pmatrix} x_1 \\ x_2 \\ \vdots \\ x_n \end{pmatrix}.$$

That is, $x_k Ax = (A_k x)x$ or $Ax = \frac{A_k x}{x_k} x$ which shows that x is an eigenvector of A. In short, x is an eigenvector of a symmetric matrix A, if and only if

$$Axx^T = xx^T A \tag{16}$$

From (16) it follows that if x is a unit vector satisfying (16), then not only is x an eigenvector of A but $x^T Ax$ is the eigenvalue of A corresponding to x. (Multiply both sides of equation (16) with x^T and take transposes.) Therefore, if

$$V^k \equiv Ax^k [x^k]^T - x^k [x^k]^T A \tag{17}$$

and

$$v_k \equiv \max_{1 \leq i,j \leq n} |(V^k)_{ij}|$$

then to show that the sequence $x^1, \ldots, x^k, \ldots$ converges to an eigenvector it suffices to show that either

$$\begin{aligned} v_k &= 0 \text{ for some finite } k \text{ or} \\ \lim_{k \to \infty} v_k &= 0. \end{aligned} \tag{18}$$

To establish (18) we assume, on the contrary, that there exists a $v^* > 0$, such that $v_k \geq v^*$ for all $k = 1, 2, \ldots$. To obtain a contradiction we need to derive an expression for $f'_{ij}(\theta)$ where $f_{ij}(\theta) \equiv [Q_{ij}(\theta)x]^T A [Q_{ij}(\theta)x]$ and the f'_{ij} denotes the derivative with respect to θ. Let $y \equiv Q_{ij}(\theta)x$. Then

$$f_{ij}'(\theta) = \frac{4}{1+\theta^2}[(A_i y)y_j - (A_j y)y_i] \tag{19}$$

From the definition of $Q_{ij}(\theta)$, we have

$$\begin{aligned} y_k &= x_k \text{ for } k \neq i, j, \\ y_i &= x_i \cos\alpha + x_j \sin\alpha, \\ y_j &= -x_i \sin\alpha + x_j \cos\alpha \end{aligned} \tag{20}$$

where $\cos\alpha = \frac{1-\theta^2}{1+\theta^2}$ and $\sin\alpha = \frac{2\theta}{1+\theta^2}$. Hence

$$\begin{aligned} y'_k &= 0 \text{ for } k \neq i,j, \\ y'_i &= \frac{2}{1+\theta^2}(-x_i \sin\alpha + x_j \cos\alpha) = \frac{2}{1+\theta^2} y_j, \\ y'_j &= \frac{2}{1+\theta^2}(-x_i \cos\alpha - x_j \sin\alpha) = -\frac{2}{1+\theta^2} y_i; \end{aligned} \tag{21}$$

Note that $a_{ij} = a_{ji}$ and

$$f_{ij}(\theta) = \sum_{k=1}^{n}\sum_{l=1}^{n} a_{kl} y_k y_l \tag{22}$$

A straightforward calculation using (20), (21) and (22) shows that

$$f_{ij}{}'(\theta) = \frac{4}{1+\theta^2}[(A_i y) y_j - (A_j y) y_i].$$

Since $\|y\| = 1$ and A is a given constant matrix, (19) implies that for all θ, and all $1 \leq i,j \leq n$, we have

$$f'_{ij}(\theta) \leq M, \tag{23}$$

where M is a constant that depends on the matrix A. Also,

$$\begin{aligned} f_{ij}{}''(\theta) &= -\frac{2\theta}{1+\theta^2} f_{ij}{}'(\theta) \\ &\quad + \frac{8}{(1+\theta^2)^2}[(a_{ii}y_j - a_{ij}y_i)y_j - (A_i y)y_i - (a_{ij}y_j - a_{jj}y_i)y_i - (A_j y)y_j] \\ &= -\frac{2\theta}{1+\theta^2} f_{ij}{}'(\theta) \\ &\quad + \frac{8}{(1+\theta^2)^2}[a_{ii}y_j^2 - 2a_{ij}y_i y_j + a_{jj}y_i^2 - (A_i y)y_i - (A_j y)y_j]. \end{aligned}$$

Again, since $\|y\| = 1$, there exists a constant ξ such that for all θ and for all $1 \leq i,j \leq n$

$$f''_{ij}(\theta) \leq \xi. \tag{24}$$

We also note that if λ_1 and λ_n are the smallest and the largest eigenvalues of A respectively then $\lambda_1 \leq f_{ij}(\theta) \leq \lambda_n$ for all θ and all $1 \leq i,j \leq n$. Since, $f_{ij}(\theta)$ is not a constant function, it must have at least one maximum and one minimum on the real line. Further, $f_{ij}(+\infty) = f_{ij}(-\infty)$ and f'_{ij} vanishes at $\pm\infty$. Hence we conclude that $f_{ij}(\theta)$ has at least one critical point at a finite θ. Let θ_k be a critical point of $f^k_{i_k j_k}$ computed in Step 3 of the algorithm; we don't assume that θ_k is a finite number. Then since $f^k_{i_k j_k}{}'(\theta_k) = 0$, from (24) it follows that

$$4v^* \leq 4v_k = |f^k_{i_k j_k}{}'(\theta_k) - f^k_{i_k j_k}{}'(0)| = |f^k_{i_k j_k}{}''(\eta)\theta_k| \leq \xi|\theta_k| \tag{25}$$

where $\eta \in [0, \theta_k]$. Since the constant ξ in (25) does not depend on either θ_k or i_k, j_k, (25) also shows that at any critical point $\bar{\theta}$ of f_{ij}, for any $1 \le i, j \le n$, $|\bar{\theta}| \ge \frac{4v^*}{\xi}$.

Observe that $\theta = 0$ is not a critical point of $f^k_{i_k j_k}$ in Step 3 of the algorithm, unless x^k is an eigenvector. Let $\tilde{\theta_k}$ be a critical point of $f^k_{i_k j_k}$ which is closest to the point 0 and $f^k_{i_k j_k}(\tilde{\theta_k}) \ge f^k_{i_k j_k}(0)$; once again we do not assume that $\tilde{\theta_k}$ is a finite number. Set $\theta_0 = \frac{4v^*}{\xi}$. Then since θ_k computed in Step 3 of the algorithm is a global maximum, we have

$$
\begin{aligned}
|f^k_{i_k j_k}(\theta_k) - f^k_{i_k j_k}(0)| &\ge |f^k_{i_k j_k}(\tilde{\theta}_k) - f^k_{i_k j_k}(0)| \\
&\ge |f^k_{i_k j_k}(\theta_0) - f^k_{i_k j_k}(0)| \\
&= |f^k_{i_k j_k}{}'(0)\theta_0 + \frac{1}{2} f^k_{i_k j_k}{}''(\eta)\theta_0^2| \\
&\ge 4v_k|\theta_0| - \frac{1}{2}\xi\theta_0^2 \\
&\ge 4v^*|\theta_0| - \frac{1}{2}\xi\theta_0^2 \\
&= \frac{8v^*}{\xi}.
\end{aligned}
\tag{26}
$$

Recall that $f^k_{i_k j_k}(\theta_k) = [x^{k+1}]^T A x^{k+1}$ and $f^k_{i_k j_k}(0) = [x^k]^T A x^k$. Also, the sequence $F_k \equiv [x^k]^T A x^k$ increases monotonically and for all $k \ge 1$, $\lambda_1 \le F_k \le \lambda_n$, where λ_1 and λ_n are the smallest and largest eigenvalues of A. Hence, v^* in (26) cannot be nonzero as we had assumed earlier. □

The proof of convergence relies on the algorithm's ability to compute the global maximum in Step 3. We show below that there are closed form expressions for the critical points of $f^k_{i_k j_k}$ which enable us to compute the global maximum.

Recall from equation (19) that in order to compute the finite critical points of $f^k_{i_k j_k}$, it suffices to solve the equation

$$(A_{i_k} y) y_{j_k} - (A_{j_k} y) y_{i_k} = 0 \tag{27}$$

where $y = Q_{i_k j_k}(\theta) x^k$. In order to simplify the notations, we use the expression for Q_{ij} instead of $Q_{i_k j_k}$. A lengthy but straightforward calculation shows that equation (27) can be written as

$$(S - U)\theta^4 + 2(V - T)\theta^3 - 6S\theta^2 + 2(T + V)\theta + (S + U) = 0 \tag{28}$$

where

$$\begin{aligned} S &= a_{ij}x_i^2 + (a_{jj} - a_{ii})x_ix_j - a_{ij}x_j^2, \\ T &= (a_{ii} - a_{jj})x_i^2 + 4a_{ij}x_ix_j + (a_{jj} - a_{ii})x_j^2, \\ U &= (\sum_{l \neq i,j} a_{jl}x_l)x_i - (\sum_{k \neq i,j} a_{ik}x_k)x_j, \\ V &= (\sum_{l \neq i,j} a_{jl}x_l)x_j + (\sum_{k \neq i,j} a_{ik}x_k)x_i. \end{aligned}$$

That equation (27) reduces to a quartic equation is to be expected given the form of $Q_{ij}(\theta)$. Appendices A and B outline the standard methods for obtaining the closed form solutions of cubic and quartic equations in terms of the coefficients of the equations. Thus we can compute all the solutions of (27) and hence all the critical points of $f^k_{i_k j_k}$, evaluate the function at each of the critical points and pick the global maximum.

At each step the Rotation Method performs $n^2 + 4n + 64$ multiplications, $\frac{n^2+9n}{2} + 18$ additions and computes two square roots, one cube root. In contrast, the classical Jacobi Method performs $8n + 4$ multiplications, $4n + 3$ additions and computes two square roots per iteration. The Jacobi Method would hence be inferior to the Rotation Method only if it performs at least $O(n)$ times as many iterations as the Rotation Method. We are currently engaged in extensive numerical experiments to compare the efficiency of Jacobi Method with a large family of methods based on Theorem 1 (including the Rotation Method described above). Some of the methods we are studying numerically, require significantly less computation per iteration than the Rotation Method. The relative performance of the family of algorithms vis-a-vis the classical Jacobi Method and QR Method will be reported elsewhere.

4 Concluding Remarks

The Decomposition Theorem presented in Section 2 shows that the nondeterministic complexity of eigen-decomposition of a general symmetric $n \times n$ real matrix is $O(n^2)$. An interesting direction for further work is to investigate if one can derive a better nondeterministic complexity for special factorizations such as QR factorization or for the diagonalization of special forms of the symmetric matrices such as the symmetric tridiagonal matrices. Computational studies of a family of eigenvalue algorithms based on the Decomposition Theorem as well as non-deterministic complexity of eigen-decomposition of special forms of symmetric matrices and of QR decomposition will be presented elsewhere.

5 Appendix A: Solving Cubic Equations – Cardano's Method

We recall the closed form solutions of cubic and quartic equations in this and the next Appendix. The discussion is taken from [3] A general cubic equation can be written in the form

$$ax^3 + 3bx^2 + 3cx + d = 0 \tag{29}$$

where a, b, c and d are real numbers. Setting

$$z = ax + b \tag{30}$$

we get, using (29),

$$\begin{aligned} z^3 &= a^3x^3 + 3a^2x^2b + 3ab^2x + b^3 \\ &= a^2[ax^3 + 3bx^2] + 3ab^2x + b^3 \\ &= a^2[-3cx - d] + 3ab^2x + b^3 \\ &= ax[3b^2 - 3ac] + b[3b^2 - 3ac] - b[3b^2 - 3ac] - a^2d + b^3 \\ &= 3(b^2 - ac)z + [3abc - a^2d - 2b^3] \\ &= -3Gz - H \end{aligned}$$

where $G \equiv ac - 3b^2$ and $H \equiv 2b^3 + a^2d - 3abc$. Thus equation (29) can be written as

$$z^3 + 3Gz + H = 0 \tag{31}$$

Now set

$$z = p^{\frac{1}{3}} + q^{\frac{1}{3}} \tag{32}$$

Cubing (32) we get

$$\begin{aligned} z^3 &= p + q + 3p^{\frac{1}{3}}q^{\frac{1}{3}}(p^{\frac{1}{3}} + q^{\frac{1}{3}}) \\ &= 3p^{\frac{1}{3}}q^{\frac{1}{3}}z + (p + q) \end{aligned} \tag{33}$$

Comparison of equations (31) and (33) suggests the following identifications

$$G \equiv -p^{\frac{1}{3}}q^{\frac{1}{3}}; \quad H \equiv -(p + q)$$

or

$$pq = -G^3; \;\; p + q = -H \tag{34}$$

Solving (34) we get

$$p = \frac{1}{2}[-H \pm \sqrt{H^2 - 4G^3}]; \; q = \frac{1}{2}[-H \mp \sqrt{H^2 - 4G^3}]$$

The two different sets of solutions of (34)

$$(p,q) = (\tfrac{1}{2}[-H+\sqrt{H^2-4G^3}], \tfrac{1}{2}[-H-\sqrt{H^2-4G^3}])$$

and

$$(p,q) = (\tfrac{1}{2}[-H+\sqrt{H^2-4G^3}], \tfrac{1}{2}[-H-\sqrt{H^2-4G^3}])$$

yield the same value of $z = p^{\frac{1}{3}} + q^{\frac{1}{3}}$. Thus, if the cube roots of unity are 1, ω and ω^2, then the three roots of equation (31) are

$$\begin{aligned} z_1 &= \left[\frac{-H+\sqrt{H^2-4G^3}}{2}\right]^{\frac{1}{3}} - G^3\left[\frac{-H+\sqrt{H^2-4G^3}}{2}\right]^{-\frac{1}{3}} \\ z_2 &= \omega\left[\frac{-H+\sqrt{H^2-4G^3}}{2}\right]^{\frac{1}{3}} - \frac{G^3}{\omega}\left[\frac{-H+\sqrt{H^2-4G^3}}{2}\right]^{-\frac{1}{3}} \\ z_3 &= \omega^2\left[\frac{-H+\sqrt{H^2-4G^3}}{2}\right]^{\frac{1}{3}} - \frac{G^3}{\omega^2}\left[\frac{-H+\sqrt{H^2-4G^3}}{2}\right]^{-\frac{1}{3}} \end{aligned} \tag{35}$$

Using (30) and (35) we get the three roots of (29) to be

$$x = \frac{z_1 - b}{a}, \frac{z_2 - b}{a}, \frac{z_3 - b}{a} \tag{36}$$

where z_1, z_2, z_3 are given by (35) and $G = ac - 3b^2$ and $H = 2b^3 + a^2d - 3abc$.

6 Appendix B: Solving Quartic Equations

Consider a quartic equation of the form

$$ax^4 + 4bx^3 + 6cx^2 + 4dx + e = 0 \tag{37}$$

Making the substitution $z = ax + b$ equation (37) can be written as

$$z^4 + 6Hz^2 + 4Gz + a^2I - 3H^2 = 0 \tag{38}$$

where $H \equiv ac - b^2$, $G \equiv 2b^3 + a^2d - 3abc$ and $I \equiv ae - 4bd + 3c^2$.

Assume that a root of (38) is of the form

$$z = \sqrt{p}\sqrt{q} + \sqrt{p}\sqrt{r} + \sqrt{q}\sqrt{r} \tag{39}$$

Then squaring (39) twice and simplifying we get

$$z^4 - 2(qr + pr + pq)z^2 - 8pqrz + (qr + rp + pq)^2 - 4(p + q + r)pqr = 0 \tag{40}$$

Comparing (39) and (38) we get

$$pq + qr + pr = -3H; \; pqr = -\frac{G}{2}; \; p + q + r = \frac{a^2I - 3H^2}{2G}$$

which suggests that p, q and r are the roots of the cubic equation

$$2Gt^3 + (3H^2 - a^2I)t^2 - 6HGt + G^2 = 0 \tag{41}$$

Using Cardano's method outlined in Appendix A, we can solve (41) to obtain p, q and r and hence the roots of the equation (37). The identity

$$2(\sqrt{p}\sqrt{q} + \sqrt{p}\sqrt{r} + \sqrt{q}\sqrt{r}) = (\sqrt{p} + \sqrt{q} + \sqrt{r})^2 - p - q - r$$

shows that the different sign choices for the square roots of p, q and r generate only four different values of z as required.

References

[1] Arnoldi W.E. *The principle of minimized iteration in the solution of the matrix eigenproblem.* Quart. Appl. Math. 9 (1951) 17-29.

[2] Brent R.P. and Luk F.T. *The solution of singular value and symmetric eigenvalue problems on multiprocessor arrays.* SIAM J. Sci. Statist. Comput. 6 (1985) 69-84.

[3] Burnside W.S. and Panton A.W. *The Theory of Equations*, Dublin University Press Series, Dublin, Ireland, 1892.

[4] Crouzeix M., Philippe B. and Sadkane M. *The Davidson method.* SIAM J. Sci. Comput. 15 (1994) 62-76.

[5] Cullum J. and Willoughby R.A. *Lanczos Algorithms for Large Symmetric Eigenvalue Computations.* Vol. I, Birkhäuser, Boston, MA, 1985.

[6] Cullum J. and Willoughby R.A. *Lanczos Algorithms for Large Symmetric Eigenvalue Computations.* Vol. II, Birkhäuser, Boston, MA, 1985.

[7] Cullum J. and Willoughby R.A. *A practical procedure for computing eigenvalues of large sparse nonsymmetric matrices.* In: Cullum J., Willoughby R.A. (Eds.), Large Scale Eigenvalue Problems, Amsterdam, North-Holland, 1986, pp. 193-240.

[8] Cuppen J.J.M. *A divide and conquer method for the symmetric eigenproblem.* Numer. Math. 36 (1981) 177-195.

[9] Davidson E.R. *The iterative calculation of a few of the lowest eigenvalues and corresponding eigenvectors of large real symmetric matrices.* J. Comput. Phys. 17 (1975) 87-94.

[10] Demmel J.W. *Applied Numerical Linear Algebra.* SIAM, Philadelphia, PA, 1997.

[11] Demmel J.W. and Veselić K. *Jacobi's method is more accurate than QR.* SIAM J. Matrix Anal. Appl. 13 (1992) 1204-1246.

[12] Dongarra J.J. and Sorensen D.C. *A fully parallel algorithm for the symmetric eigenvalue problem.* SIAM J. Sci. Statist. Comput. 8 (1987) 139-154.

[13] Francis J.G.F. *The QR transformation: a unitary analogue to the LR transformation.* Parts I and II Comput. J. 4 1961 265-272, 332-345.

[14] Givens W. *Numerical computation of the characteristic values of a real symmetric matrix.* Oak Ridge National Lab., ORNL-1574, 1954.

[15] Golub G.H. *Some uses of the Lanczos algorithm in numerical linear algebra.* In: Miller J.J.H. (Ed.), Topics in Numerical Analysis, Academic Press, New York, 1973, pp. 173-184.

[16] Golub G.H. and Van Loan C.F. *Matrix Computations.* The Johns Hopkins University Press, Baltimore, 1996.

[17] Golub G.H. and van der Vorst H.A. *Eigenvalue computation in the 20th century.* J. Comput. Appl. Math. 123 (2000) 35-65.

[18] Greenstadt J. *A method for finding roots of arbitrary matrices.* Math. Tables Aids Comput. 9 (1955) 47-52.

[19] Gregory R.T. *Computing eigenvalues and eigenvectors of a symmetric matrix on the ILLIAC.* Math. Tables Aids Comput. 7 (1953) 215-220.

[20] Gu M. and Eisenstat S.C. *A divide-and-conquer algorithm for the symmetric tridiagonal eigenproblem.* SIAM J. Matrix Anal. Appl. 16 (1995) 172-191.

[21] Henrici P. *On the speed of convergence of cyclic and quasicyclic Jacobi methods for computing the eigenvalues of Hermitian matrices.* SIAM J. Appl. Math. 6 (1958) 144-162.

[22] Hessenberg K. *Auflösung linearer Eigenwertaufgaben met Hilfe der Hamilton-Cayleyschen Gleichung.* Ph.D. Thesis, Technische Hochschule, Darmstadt, 1940.

[23] Hoffmann W. and Parlett B.N. *A new proof of global convergence for the tridiagonal QL algorithm.* SIAM J. Numer. Anal. 15 (1978) 929-937.

[24] Householder A.S. *The Theory of Matrices in Numerical Analysis.* Dover, New York, 1964.

[25] Jacobi C.G.J. *Über ein leichtes Verfahren, die in der Theorie der Säcularstörungen vorkommenden Gleichungen numerisch aufzulösen.* J. Reine Angew. Math. 30 (1846) 51-94.

[26] Krylov A.N. *O cislennom resenii uravnenija, kotorym v techniceskih voprasah opredeljajutsja castoty malyh kolebanii material'nyh sistem.* Izv. Akad. Nauk SSSR. Ser. Fiz.-Mat. 4 (1931) 491-539.

[27] Kublanovskaja V.N. *On some algorithms for the solution of the complete eigenvalue problem.* Zh. Vych. Mat. 1 (1961) 555-570.

[28] Lanczos C. *An iteration method for the solution of the eigenvalue problem of linear differential and integral operators.* J. Res. Natl. Bur. Stand 45 (1950) 225-280.

[29] Li T.-Y. and Zeng Z. *Homotopy-determinant algorithm for solving nonsymmetric eigenvalue problems.* Math. Comput. 59 (1992) 483-502.

[30] Liu S.H. and Golub G.H. *Homotopy method for the numerical solution of the eigenvalue problem of self-adjoint partial differential operators.* Numer. Algorithms 10 (1995) 363-378.

[31] Morgan R.B. and Scott D.S. *Generalizations of Davidson's method for computing eigenvalues of sparse symmetric matrices.* SIAM J. Sci. Statist. Comput. 7 (3) (1986) 817-825.

[32] Olsen J., Jorgensen P. and Simons J. *Passing the one-billion limit in full configuration-interaction (FCI) calculations.* Chem. Phys. Lett. 169 (1990) 463-472.

[33] Ostrowski A.M. *On the convergence of the Rayleigh quotient iteration for the computation of characteristic roots and vectors I-VI.* Arch. Rational. Mesh. Anal. I 1:233-241, II 2:423-428, III 3:325-340, IV 3:341-347, V 3:472-481, VI 4:153-165, 1958-1959.

[34] Paige C.C. *The computation of eigenvalues and eigenvectors of very large matrices.* Ph.D. Thesis, University of London, 1971.

[35] Parlett B.N. *The Symmetric Eigenvalue Problem.* Prentice-Hall, Englewood Cliffs, NJ, 1980.

[36] Pope D.A. and Tompkins C. *Maximizing functions of rotations: experiments concerning speed of diagonalization of symmetric matrices using Jacobi's method.* J. ACM 4 (1957) 459-466.

[37] Ruhe A. *Implementation aspects of band Lanczos algorithms for computation of eigenvalues of large sparse symmetric matrices.* Math. Comput. 33 (1979) 680-687.

[38] Saad Y. *Variations on Arnoldi's method for computing eigenelements of large unsymmetric matrices.* Linear Algebra Appl. 34 (1980) 269-295.

[39] Schönhage A. *Zur Konvergenz des Jacobi-Verfahrens.* Numer. Math. 3 (1961) 374–380.

[40] Sleijpen G.L.G. and van der Vorst H.A. *A Jacobi-Davidson iteration method for linear eigenvalue problems.* SIAM J. Matrix Anal. Appl. 17 (1996) 401-425.

[41] Stewart G.W. and Sun Ji-Guang *Matrix Perturbation Theory.* Academic Press, San Diego, CA, 1990.

[42] Tisseur F. and Dongarra J. *A parallel divide and conquer algorithm for the symmetric eigenvalue problem on distributed memory architectures.* SIAM J. Sci. Comput. 20 (1999) 2223-2236.

[43] Trefethen L.N. and Bau D. III, *Numerical Linear Algebra.* SIAM, Philadelphia, PA, 1997.

[44] Underwood R.R. *An iterative block Lanczos method for the solution of large sparse symmetric eigenproblems.* Technical Report STAN-CS-75-495, Department of Computer Science, Stanford University, Stanford, CA, 1975.

[45] van Kempen H.P.M. *On quadratic convergence of the special cyclic Jacobi method.* Numer. Math. 9 (1966) 19-22.

[46] von Mises R. and Pollaczek-Geiringer H. *Praktische Verfahren der Gleichungsauflösung.* Z. Angew. Math. Mech. 9 (1929) 58-77, 152-164

[47] Watkins D.S. *Some perspectives on the eigenvalue problem.* SIAM Rev. 35 (1993) 430-471.

[48] Watkins D.S. and Elsner L. *Chasing algorithms for the eigenvalue problem.* SIAM J. Matrix Anal. Appl. 12 (1991) 374-384.

[49] Wilkinson J.H. *The Algebraic Eigenvalue Problem.* Clarendon Press, Oxford, 1965.

[50] Wilkinson J.H. and Reinsch C. (Eds.) *Handbook for Automatic Computation.* Vol. 2, Linear Algebra, Springer, Heidelberg, 1971.

About Authors

W. He is at the School of Industrial Engineering, Purdue University, West Lafayette, IN 47907.

N. Prabhu is at the School of Industrial Engineering, Purdue University, West Lafayette, IN 47907.

Acknowledgments

W. He was s supported in part by ONR grant N00014-96-1-0281. N. Prabhu was supported in part by ONR grant N00014-96-1-0281 and NSF grant 9800053CCR.

Discrete Geometry on Red and Blue Points in the Plane – A Survey –

Atsushi Kaneko
M. Kano

Abstract

In this paper, we give a short survey on discrete geometry on red and blue points in the plane, most of whose results were obtained in the past decade. We consider balanced subdivision problems, geometric graph problems, graph embedding problems, Gallai-type problems and others.

1 Notation and Definitions

In this paper, we give a short survey on discrete geometry on red and blue points in the plane, most of whose results were obtained in the past decade. We consider two disjoint sets R and B of red points and of blue points in the plane, respectively, such that no three points of $R \cup B$ lie on the same line. Throughout this paper, R and B always denote the sets mentioned above unless noted otherwise.

We begin with some notation and definitions, which will be used throughout this paper. A *(directed) line* l trisects the plane into three pieces: l, $right(l)$ and $left(l)$, where $right(l)$ and $left(l)$ denote the *open half-planes* which are on the right side and on the left side of l, respectively (Figure 1). Let r_1 and r_2 be two rays emanating from the same point p. Then we denote by $right(r_1) \cap left(r_2)$ the open region that is swept by the ray being rotated clockwise around p from r_1 to r_2 (Figure 1). The open region $left(r_1) \cap right(r_2)$ is similarly defined. Then $r_1 \cup r_2$ trisects the plane into three pieces: $r_1 \cup r_2$ and two open regions $right(r_1) \cap left(r_2)$ and $left(r_1) \cap right(r_2)$. If the internal angle $\angle r_1pr_2 = \angle r_1r_2$ of $right(r_1) \cap left(r_2)$ is less than π, then we call $right(r_1) \cap left(r_2)$ the *wedge* defined by r_1 and r_2, and denote it by $wedge(r_1r_2)$, $wedge(r_2r_1), wedge(r_1pr_2)$ or $wedge(r_2pr_1)$.

Given a point p and a line l passing through p, let l^* denote the line lying on l and having the opposite direction of l. We define the rays r and r^* to be the two rays emanating from p and lying on the line l, while having the direction of l and l^* respectively (Figure 1). Conversely, given a ray r, we can define r^*, l and l^*.

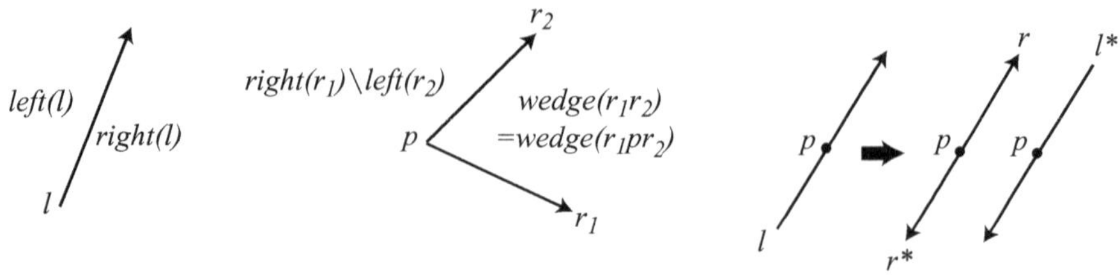

Fig. 1. Open regions $right(l)$, $left(l)$, $left(r_1) \cap right(r_2)$ and $wedge(r_1r_2) = wedge(r_1pr_2)$, the rays r and r^*, and the line l^*.

For a set X of points in the plane, let $\mathrm{conv}(X)$ denote the *convex hull* of X, which is the smallest convex set containing X. For convenience, a region in the plane whose boundary consists of straight-line segments is called a *polygon* even if it is an infinite region. For example, Figure 2 (b) illustrates a subdivision of the plane into seven convex polygons.

2 Balanced Subdivisions

In this section we consider problems of subdivisions of the plane into convex polygons such that (i) each polygon contains exactly a red points and b blue points, (ii) each polygon contains an_i red points and bn_i blue points, or (iii) polygons satisfy other prescribed conditions, where a and b are constant positive integers independent of polygons and n_i is a positive integer depending on each polygon. A famous classical theorem on this problem is the following Ham-sandwich Theorem.

Theorem 2.1 (Ham-sandwich Theorem [17]). *For given R and B, there exists a line l such that $|left(l) \cap R| = |right(l) \cap R|$, $|l \cap R| \leq 1$, $|left(l) \cap B| = |right(l) \cap B|$ and $|l \cap B| \leq 1$ (Figure 2 (a)). If both $|R|$ and $|B|$ are even, then l passes through no red points and no blue points, and thus l bisects R and B. Moreover, there exists a linear time $O(|R|+|B|)$ algorithm for finding line l [31].*

The line l given in the above Ham-sandwich Theorem is called a *bisector line*, and the subdivision of the plane by a bisector line is called a *Ham-sandwich cutting.*

Before considering balanced subdivisions, let us give a result on balanced undirected lines. When $|R| = |B|$, an undirected line l is said to be *balanced* if each open half-plane bounded by l contains precisely the same number of red points as blue points. Then if $|R| = |B|$, an undirected bisector line is balanced. The following theorem gives an lower bound of the number of balanced undirected lines, and thus settles a conjecture of G. Baloglou.

Theorem 2.2 (Pach and Pinchasi [36]). *If $|R| = |B| = n$, then there exist at least n balanced undirected lines each of which passes through one red point and one blue point.*

Note that if $|R| = |B| = n$ and R and B are separated by a line, then there exists exactly n balanced undirected lines determined by them, and so the above lower bound n is best possible.

The following theorem, which was conjectured in [21] and proved for $a = 1, 2$ in [21] and [26], was completely proved by Bespamyatnikh, Kirkpatrick and Snoeyink [7], Sakai [39] and by Ito, Uehara and Yokoyama [19] independently. Note that for $g = 2$, it is equivalent to the Ham-Sandwich Theorem above.

Theorem 2.3 (The Equitable Subdivision Theorem [7], [39], [19]). *Let $a \geq 1$, $b \geq 1$ and $g \geq 2$ be integers. If R contains ag red points and B contains bg blue points, then there exists a subdivision $X_1 \cup X_2 \cup \cdots \cup X_g$ of the plane into g disjoint convex polygons such that every X_i contains exactly a red points and b blue points (Figure 2 (b)).*

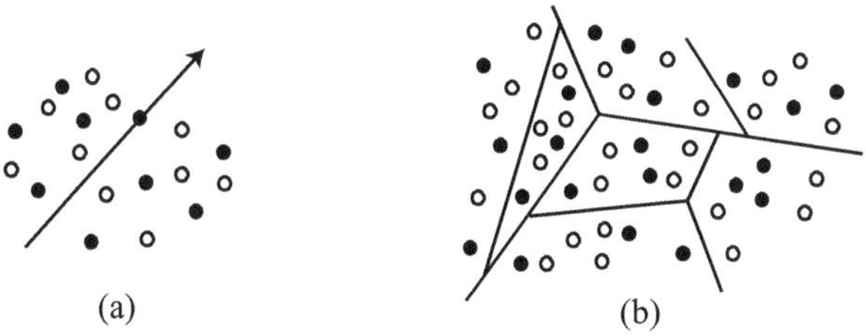

Fig. 2. (a) A Ham-sandwich cutting; (b) An equitable subdivision.

Such a subdivision of the plane is called an *equitable subdivision* of the plane. An algorithm for finding an equitable subdivision of the plane with time complexity $O(n^{4/3} \log^3 n \log g)$, where $n = (a + b)g$, was obtained by Bespamyatnikh, Kirkpatrick and Snoeyink [7].

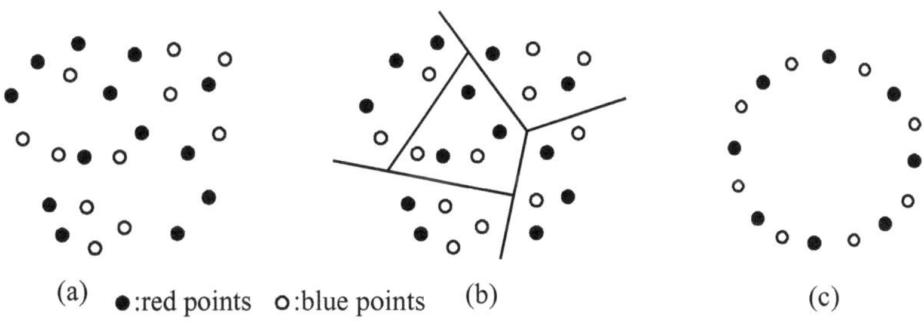

Fig. 3. (a), (b) $R \cup B$ and a subdivision from Theorem 2.4; (c) A configuration having no subdivision with $a = 1$, $b = 3$ and $g = h = 2$..

Theorem 2.4 (Kaneko, Kano and Suzuki [27]). *Let $a \geq 1$, $g \geq 0$ and $h \geq 0$ be integers such that $g+h \geq 1$. If R contains $ag+(a+1)h$ red points and B contains $(a+1)g+ah$ blue points, then there exists a subdivision $X_1 \cup \cdots \cup X_g \cup Y_1 \cup \cdots \cup Y_h$ of the plane into $g+h$ disjoint convex polygons such that every X_i $(1 \leq i \leq g)$ contains exactly a red points and $a+1$ blue points and every Y_j $(1 \leq j \leq h)$ contains exactly $a+1$ red points and a blue points (Figure 3 (b)).*

The above theorem naturally suggests the following problem. Let $a \geq 1$ and $b \geq a+2$ be integers. If R contains $ag+bh$ red points and B contains $bg+ah$ blue points, then does there exist a subdivision $X_1 \cup \cdots \cup X_g \cup Y_1 \cup \cdots \cup Y_h$ of the plane into $g+h$ disjoint convex polygons such that every X_i contains exactly a red points and b blue points and every Y_j contains exactly b red points and a blue points? However this statement does not hold since for a configuration in which $ag+bg$ red points and $bg+ag$ blue points alternately lie on a circle, there exists no subdivision mentioned above (Figure 3 (c)).

The Ham-sandwich Theorem, the Equitable Subdivision Theorem and Theorem 2.4 may be applied to some problems on geometry on red and blue points in the plane. We give such an example.

It is well-known ([30]) that given $R \cup B$ with $|R| = |B|$, there exists a non-crossing geometric perfect matching on $R \cup B$, that is, there exist $|R|$ disjoint straight-line segments which connect red points and blue points and have no crossings. This result can be easily proved by using the Ham-sandwich Theorem and induction on $|R|$. That is, if $|R|$ is odd, then a straight-line segment lying on the bisector line l whose endpoints are the red point and the blue point on l is chosen to be an element of a perfect matching, and it is obvious how to apply the induction hypotheses in the case where $|R|$ is even.

We can generalize this problem to the following problem, which is called a *non-crossing geometric alternating path-covering problem* (Figure 4 (a)). We denote by P_k the path with k vertices and length $k-1$ that passes through red points and blue points alternately. For a constant integer n and for given R and B, do there exist disjoint non-crossing geometric alternating paths P_n which cover $R \cup B$? If $P_n = P_{2m}$ and $|R| = |B| = mg$, then we first apply the Equitable Subdivision Theorem to $R \cup B$, and obtain the equitable subdivision of the plane in which each convex polygon contains m red points and m blue points. Thus if we can show that for every arrangement of m red points and m blue points in the plane in general position, there exists a non-crossing geometric alternating path P_{2m} that passes through all these red and blue points, the above problem is affirmatively solved.

When we consider the above problem with path P_{2m+1}, R must contain $mg+(m+1)h$ red points and B must contain $(m+1)g+mh$ blue points, where integers g and h correspond to the numbers of paths P_{2m+1} whose both endpoints are blue points or red points, respectively. In this case, we first obtain a subdivision of the plane by Theorem 2.4, and if we can show that for every arrangement of m red points and $m+1$ blue points in the

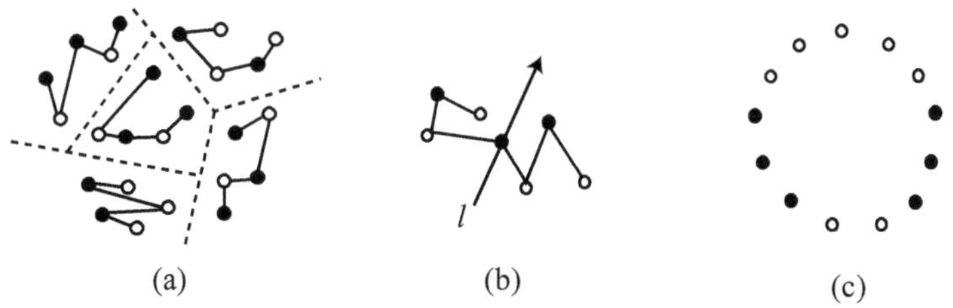

Fig. 4. (a) A non-crossing geometric alternating P_5-covering (b) A non-crossing geometric alternating P_7 and a bisector line l; (c) A configuration having no non-crossing geometric alternating path.

plane in general position, there exists a non-crossing geometric alternating path P_{2m+1} that passes through all these red and blue points, then the above problem is affirmatively solved.

It is shown that if R and B are separated by a line and $|R| = |B|$, then there exists a non-crossing geometric alternating path that passes through all the points in $R \cup B$ [1]. On the other hand, as we show in Figure 4 (c), there exists a configuration consisting of six red points and seven blue points in the plane for which there exists no non-crossing geometric alternating path P_{13} that passes through all these 13 points. For $n \geq 15$, there exists such a configuration of $\lfloor n/2 \rfloor$ red points and $\lceil n/2 \rceil$ blue points. Considering this observation, we obtain the following theorem, which is easily proved for $2 \leq n \leq 7$ by the Ham-sandwich Theorem (Figure 5 (b)).

Theorem 2.5 ([27]). *Suppose that $2 \leq n \leq 12$ or $n = 14$. Then for any given $\lfloor n/2 \rfloor$ red points and $\lceil n/2 \rceil$ blue points in the plane in general position, there exists a non-crossing geometric alternating path P_n that passes through all these red and blue points.*

We now consider balanced subdivisions of the plane, in which the numbers of red points and blue points are different depending on convex polygons.

Theorem 2.6 ([28]). *Let $m \geq 1, n \geq 1$ and $g \geq 2$ be integers such that $m + 2n = 2g + 1$. If R contains $2g + 1$ red points and B contains $(2g + 1)b$ blue points, then there exists a subdivision $X_1 \cup X_2 \cup \cdots \cup X_{m+n}$ of the plane into $m + n$ disjoint convex polygons such that every X_i $(1 \leq i \leq m)$ contains one red point and b blue points and every X_j $(m < j \leq m + n)$ contains two red points and $2b$ blue points (Figure 5 (a)).*

It is easy to see that if the above theorem holds for $m = 1$, then the theorem holds for every m, since each convex polygon containing two red points and $2b$ blue points can be bisected by the Ham-sandwich Theorem.

Theorem 2.7 ([24]). *Let $2 \leq n \leq 8$, $1 \leq b$ and $2 \leq g$ be integers. If R contains n red points and B contains bn blue points, then for every integer-partition $n = n_1 + n_2 + \ldots + n_g$ such that $1 \leq n_i \leq n/2$ for every $1 \leq i \leq g$,*

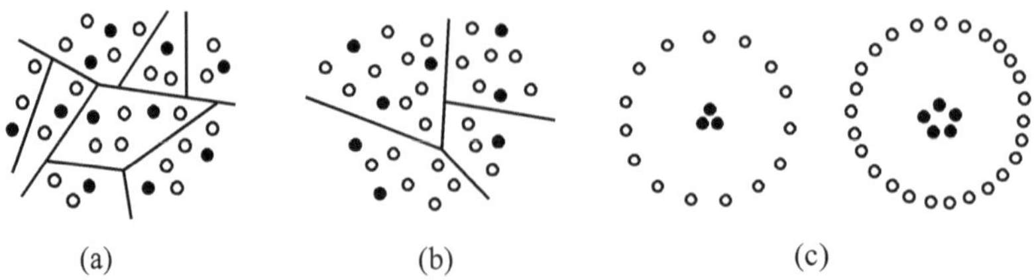

Fig. 5. (a) A balanced subdivision from Theorem 2.6; (b) A balanced subdivision from Theorem 2.7; (c) Configurations having no balanced subdivisions with $(n_1, n_2) = (2, 1)$ and $(n_1, n_2) = (3, 2)$, respectively.

there exists a subdivision $X_1 \cup X_2 \cup \cdots \cup X_g$ of the plane into g convex polygons such that each X_i $(1 \le i \le g)$ contains exactly n_i red points and bn_i blue points.

Figure 5 (b) shows a balanced subdivision with $(n_1, n_2, n_3, n_4) = (3, 2, 2, 1)$, Figure 5 (c) gives two configurations $R \cup B$ having no balanced subdivision with $(n_1, n_2) = (2, 1)$ and $(n_1, n_2) = (3, 2)$, respectively. Similarly, if $n/2 < n_1 < n$, then there exist configurations $R \cup B$ having no balanced subdivision with $n = n_1 + n_2$. Based on Theorems 2.6 and 2.7 and by some other configurations given in [24], we propose the following conjecture.

Conjecture 2.8 ([24]). *Let $n = n_1 + n_2 + \cdots + n_g$ be an integer partition such that $1 \le n_i \le n/3$ for every $1 \le i \le g$. If $|R| = an$ and $|B| = bn$, then there exists a subdivision $X_1 \cup X_2 \cup \cdots \cup X_g$ of the plane into g disjoint convex polygons such that each X_i $(1 \le i \le g)$ contains exactly an_i red points and bn_i blue points.*

The following theorem gives a subdivision, which might be called a semi-balanced subdivision.

Theorem 2.9 ([26]). *Let $m \ge 0$ be an integer, and let $R = R_1 \cup R_2$ be a partition of R with $|R_1| = s \ge 0$ and $|R_2| = t \ge 0$. If $|B| = ms + (m+1)t$, then there exists a subdivision $X_1 \cup X_2 \cup \cdots \cup X_{s+t}$ of the plane into $s + t$ convex polygons such that each X_i $(1 \le i \le s)$ contains one red point of R_1 and m blue points and each $X_j (s < j \le s + t)$ contains one red point of R_2 and $m + 1$ blue points.*

We can consider a generalization of the above theorem by taking a partition $R = R_1 \cup R_2 \cup R_3$ and three integers $m, m+1, m+2$. But this generalization does not hold as shown in [26].

Bárány and Matoušek [5] consider balanced subdivisions of the plane by k-fan, and obtain similar results. Though they consider t Borel probability measure in the plane, here we consider t sets $X_1, X_2, \ldots, X_t$ of points in the plane in general position. A *k-fan* is defined to be a point x in the plane and k rays emanating from x, and by this k-fan, the plane is subdivided into k open regions $W_1 \cup W_2 \cup \cdots \cup W_k$, one of which may not be convex

(Figure 6 (a)). A k-tuple of parallel lines is also considered to be a k-fan (this is a limit case for x receding to infinity). For a vector $\alpha = (\alpha_1, \alpha_2, \ldots, \alpha_k)$ of non-negative real numbers whose components sum up to 1, we say that a k-fan *simultaneously α-partitions* $X_1, X_2, \ldots, X_t$ if

$$|X_i \cap W_j| = \alpha_j |X_i| \qquad \text{for all } \ 1 \le i \le t \ \text{ and } \ 1 \le j \le k.$$

Note that when we consider a simultaneous α-partition, we always assume that $\alpha_j |X_i|$ is an integer for all i and j. In particular, a k-fan simultaneously $(\frac{1}{k}, \frac{1}{k}, \ldots, \frac{1}{k})$-partitions $X_1, X_2, \ldots, X_t$ if

$$|X_i \cap W_j| = \frac{1}{k}|X_i| \qquad \text{for all } \ 1 \le i \le t \ \text{ and } \ 1 \le j \le k.$$

The following theorem was proved by using topological methods, and so it would be interesting to have an elementary proof for it.

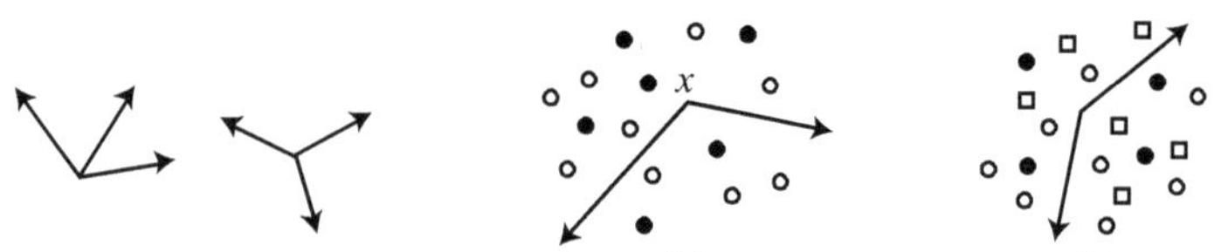

(a) (b) (c)

Fig. 6. (a) Two 3-fans; (b) R and B is simultaneously $(\frac{1}{3}, \frac{2}{3})$-partitioned by a 2-fan with center x. (c) R, B and Y is simultaneously $(\frac{1}{2}, \frac{1}{2})$-partitioned by a 2-fan.

Theorem 2.10 (Bárány and Matoušek [5]). *Let R, B, Y be three disjoint sets of red points, blue points and yellow points in the plane, respectively, such that no three points of $R \cup B \cup Y$ lie on the same line. Then the following statements hold.*

(i) For any $\alpha = (\alpha_1, \alpha_2)$ and any point x in the plane, R and B can be simultaneously α-partitioned by a 2-fan with center x.

(ii) For $\alpha = (\frac{1}{2}, \frac{1}{2})$ and for $\alpha = (\frac{2}{3}, \frac{1}{3})$, any R, B and Y can be simultaneously α-partitioned by a 2-fan.

(iii) Every R and B can be simultaneously $(\frac{1}{3}, \frac{1}{3}, \frac{1}{3})$-partitioned by a 3-fan, or $(\frac{1}{2}, \frac{1}{4}, \frac{1}{4})$-partitioned by a 3-fan. They can also be $(\frac{2}{5}, \frac{1}{5}, \frac{1}{5}, \frac{1}{5})$-partitioned by a 4-fan.

(iv) For any $k \ge 3$ and any α, there exist R, B and Y that can not be simultaneously α-partitioned by a k-fan.

Theorem 2.11 (Bárány and Matoušek [6]). *Every R and B can be simultaneously $(\frac{1}{4}, \frac{1}{4}, \frac{1}{4}, \frac{1}{4})$- partitioned by a 4-fan.*

3 Proof Techniques

In this section we give some theorems and lemmas which are used in the proofs of some theorems on balanced subdivisions from Section 2 and other geometric problems. We begin with the following lemma, whose distinct proofs can be found in [7] and [24].

Lemma 3.1 (The Intermediate Line Lemma). *Let $k \geq 0$ be an integer. If there exist two lines l_1 and l_2 such that $|left(l_1) \cap R| = |left(l_2) \cap R|$ and $|left(l_1) \cap B| \leq k \leq |left(l_2) \cap B|$, then there exists a line l_3 such that $|left(l_3) \cap R| = |left(l_1) \cap R|$, $|left(l_3) \cap B| = k$ and l_3 passes through no point in $R \cup B$.*

We now prove the Ham-sandwich Theorem by using the above Intermediate Line Lemma. For simplicity, we assume that $|R|$ and $|B|$ are even. It is easy to find a line l with $|left(l) \cap R| = |right(l) \cap R|$, which implies $|left(l) \cap R| = |left(l^*) \cap R| = |R|/2$. Without of loss generality, we may assume that $|left(l) \cap B| < |B|/2$. Then

$$|left(l) \cap B| < |B|/2 < |left(l^*) \cap B|.$$

Hence by the Intermediate Line Lemma, there exists a line l_3 satisfying $|left(l_3) \cap R| = |R|/2$ and $|left(l_3) \cap B| = |B|/2$, and thus l_3 is the desired bisector line.

The following theorem, called the 3-cutting Theorem, is very powerful when we consider balanced subdivision problems and other related problems. This theorem was originally proved by Bespamyatnikh, Kirkpatrick and Snoeyink [7] under the assumption that

$$\frac{g_1}{h_1} = \frac{g_2}{h_2} = \frac{g_3}{h_3}.$$

But as mentioned in [24,27], this condition can be removed without changing the arguments in the proof given in [7]. Similar results, which seems to be essentially equivalent to the following 3-cutting Theorem, were obtained in [39] and [19], respectively.

Theorem 3.2 (The 3-cutting Theorem). *Let $g_1, g_2, g_3, h_1, h_2, h_3$ be positive integers such that $|R| = g_1 + g_2 + g_3$ and $|B| = h_1 + h_2 + h_3$. Suppose that one of the following statements (i) and (ii) holds:*

(i) For every integer $i \in \{1, 2, 3\}$ and for every line l with $|left(l) \cap R| = g_i$, $|left(l) \cap B| < h_i$.

(ii) For every integer $i \in \{1, 2, 3\}$ and for every line l with $|left(l) \cap R| = g_i$, $|left(l) \cap B| > h_i$.

Then there exists three rays emanating from a common point such that the three open regions W_i $(1 \leq i \leq 3)$ defined by these three rays are wedges, and each wedge W_i $(1 \leq i \leq 3)$ contains exactly g_i red points and h_i blue points. Moreover, one of the three rays can be chosen to be a vertically downward ray (Figure 7).

Now we prove the Equitable Subdivision Theorem using Lemma 3.1 and the 3-cutting Theorem.

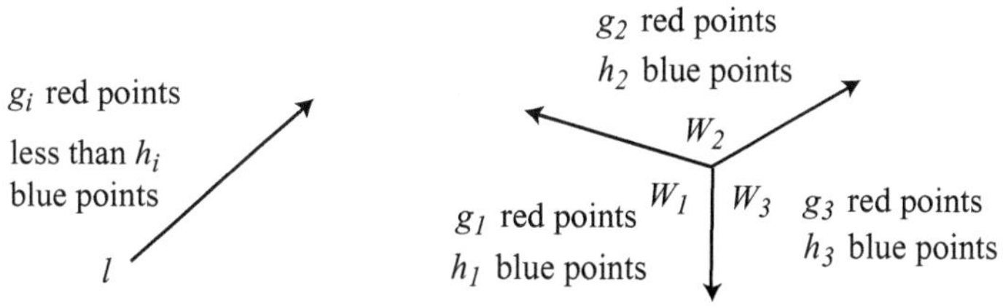

Fig. 7. A subdivision given in the 3-cutting theorem.

Proof of the Equitable Subdivision Theorem [7]. We prove the theorem by induction on g. Let m be an integer such that $1 \leq m \leq g-1$. If there exists a line l such that $left(l)$ contains exactly am red points and bm blue points, then by applying the inductive hypotheses to $left(l)$ and $right(l)$ respectively, we can obtain the desired subdivision of the plane. Hence we may assume that for every line l with $|left(l) \cap R| = am$, it follows that $|left(l) \cap B| \neq bm$. By Lemma 3.1, this fact implies that for every line l with $|left(l) \cap R| = am$, either always $|left(l) \cap B| < bm$ holds or always $|left(l) \cap B| > bm$ holds. Hence we can define $sign(m)$ as

$sign(m) = +$ if $|left(l) \cap B| > bm$ for every line l with $|left(l) \cap R| = am$;
$sign(m) = -$ if $|left(l) \cap B| < bm$ for every line l with $|left(l) \cap R| = am$.

Without loss of generality, we may assume $sign(1) = -$ since otherwise by exchanging the colors red and blue, we have $sign(1) = -$. It is clear that $sign(1) = -$ implies $sign(g-1) = +$ since $|left(l) \cap R| = a$ and $|left(l) \cap B| < b$ implies that $|left(l^*) \cap R| = (g-1)a$ and $|left(l^*) \cap B| > (g-1)b$. Hence there exists an integer $2 \leq k \leq g-1$ such that

$$sign(1) = \cdots = sign(k-1) = - \qquad \text{and} \qquad sign(k) = +.$$

Since $sign(k) = +$, we have $sign(g-k) = -$, and hence

$$sign(1) = sign(k-1) = sign(g-k) = - \qquad \text{and} \qquad 1 + (k-1) + (g-k) = g.$$

Therefore by the 3-cutting Theorem, the plane can be divided into three wedges $W_1 \cup W_2 \cup W_3$ such that W_1, W_2, W_3 contain a red points and b blue points, $a(k-1)$ red points and $b(k-1)$ blue points, and $a(g-k)$ red points and $b(g-k)$ blue points, respectively. By applying the inductive hypothesis to each W_i, we can obtain the desired equitable subdivision of the plane.

The next lemma is used in the proofs of Theorems 4.4 and 4.5, and can be proved by induction on $|R|$.

Lemma 3.3. *Let $|R| \geq |B| \geq 2$ and let p be a point in the plane not contained in* $\mathrm{conv}(R \cup B)$ *such that no three points of $R \cup B \cup \{p\}$ lie on the same line. Suppose that two vertices of* $\mathrm{conv}(R \cup B \cup \{p\})$ *adjacent to p are blue points. Then there exist two rays r_1 and r_2 emanating from p such that (i) each r_i passes through exactly one red point, (ii) $right(r_1) \cap left(r_2)$ contains no points in $R \cup B$, and (iii) $|(left(l_1) \cup l_1) \cap R| = |left(l_1) \cap B|$. (Figure 8).*

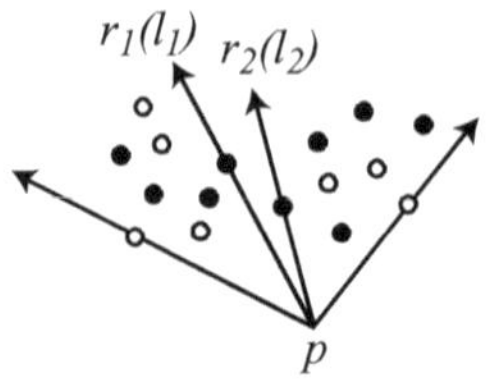

Fig. 8. Two rays r_1 and r_2 from Lemma 3.3.

4 Geometric Graphs

In this section we consider some problems on geometric graphs on red and blue points in the plane. Readers who are interested in geometric graphs in the plane should refer to the book [34] by Pach and Agarwal and the survey [33] by Pach.

A *geometric graph* is a graph drawn in the plane whose edges are straight-line segments. Let X be a set of points in the plane in general position. Then a *geometric spanning tree* on X, denoted by $tree(X)$, is defined to be a spanning tree on X whose edges are straight-line segments connecting two points in X. A geometric Hamiltonian path, $path(X)$, on X and a geometric Hamiltonian cycle, $cycle(X)$, on X are defined analogously (Figure 9 (a),(b)). The *geometric complete graph* $K(X)$ on X is defined to be a graph whose vertex set is X and whose edge set consists of all the straight-line segments connecting any two points of X. Then $tree(X)$, $path(X)$ and $cycle(X)$ are a spanning tree, a Hamiltonian path and a Hamiltonian cycle of $K(X)$, respectively. For two disjoint sets X and Y of points in the plane, we define the *geometric complete bipartite graph* $K(X, Y)$ as a graph whose vertex set is $X \cup Y$ and whose edge set consists of all the straight-line segments connecting any point in X and any point in Y. Thus it follows that the disjoint union $K(X) \cup K(Y) \cup K(X, Y)$ is the geometric complete graph $K(X \cup Y)$.

We first consider the following problem. When we wish to draw red and blue geometric spanning trees $tree(R)$ and $tree(B)$ on given $R \cup B$ so that the number of crossings in $tree(R) \cup tree(B)$ is as small as possible, how do we draw such spanning trees and what is the minimum number of crossings? This problem is solved in the following theorem.

Theorem 4.1 (Tokunaga [42]). *Let $\tau(R, B)$ denote the number of unordered pairs $\{x, y\}$ of vertices of* conv$(R \cup B)$ *such that one of $\{x, y\}$ is red and the other is blue, and xy is an edge of* conv$(R \cup B)$*. Then $\tau(B, R)$ is an even number, and the minimum number of crossings in $tree(R) \cup tree(B)$ among all pairs $(tree(R), tree(B))$ is equal to*

$$\max\{\frac{\tau(R, B) - 2}{2},\ 0\}.$$

In particular, we can draw red and blue geometric spanning trees without crossings if and only if $\tau(B, R) \leq 2$ (Figure 9 (a)). Furthermore, the proof *gives a polynomial time algorithm for drawing* $tree(R) \cup tree(B)$ *with minimum number of crossings.*

Theorem 4.2 (Tokunaga [42]). *For given* R *and* B*, there exists a pair* $(path(R), path(B))$ *of red and blue geometric Hamiltonian paths such that each edge of* $path(R)$ *intersects at most one edge of* $path(B)$ *and vice versa. In particular, the number of crossings in* $path(R) \cup path(B)$ *is less than or equal to* $\min\{|B|, |R|\} - 1$ *(Figure 9 (b)).*

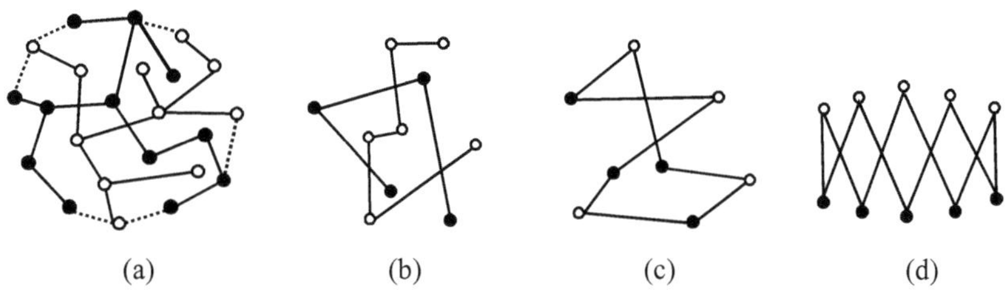

Fig. 9. (a) $tree(R)$, $tree(B)$ and $\tau(R, B) = 6$; (b) $path(R)$ and $path(B)$; (c) A geometric alternating Hamiltonian cycle with two crossings; (d) A geometric alternating Hamiltonian cycle with $|R| - 1$ crossings.

Theorem 4.3 (Dumitrescu and Kaye [11]). *Let* $n = |R| + |B|$*. Then the following two statements hold.*

(i) For given R *and* B*, there exists a non-crossing matching in* $K(R) \cup K(B)$ *which covers at least* $0.8571n$ *points of* $R \cup B$*. There exists an algorithm for finding such a matching in* $O(n^2)$ *time.*

(ii) There exists a configuration $R \cup B$ *for which every non-crossing matching in* $K(R) \cup K(B)$ *covers at most* $0.9871n$ *points of* $R \cup B$*.*

We next consider geometric alternating spanning graphs on $R \cup B$, whose edges connect red points and blue points, that is, hereafter we deal with spanning subgraphs of $K(R, B)$. For example, a geometric alternating Hamiltonian cycle passes through all the points of $R \cup B$ and through red points and blue points alternately.

Akiyama and Urrutia [4] considered a geometric alternating Hamiltonian cycle on a set of n red points and n blue points lying on the same circle, and gave a $O(n^2)$ time algorithm for finding such a geometric alternating Hamiltonian cycle (if it exists). An upper bound of the number of crossings of geometric alternating Hamiltonian cycle is given in the following theorem.

Theorem 4.4 (Kaneko, Kano, Yoshimoto [25]). *For* R *and* B *with* $|R| = |B|$*, there exists a geometric alternating Hamiltonian cycle on* $R \cup B$ *that has at most* $|R| - 1$ *crossings (Figure 9 (c)). Moreover there exist configurations* $R \cup B$ *for which this upper bound* $|R| - 1$ *is best possible (Figure 9 (d)).*

As shown in Theorem 4.1, for sets R and B with $\tau(R, B) \geq 4$, every $tree(R) \cup tree(B)$ contains at least one crossing. On the other hand, we can easily show by induction that for every sets R and B, there exists a geometric alternating spanning tree on $R \cup B$ that has no crossings. Namely, we can show that $K(R, B)$ contains a spanning tree without crossings. In 1996, Albellanas et al. [1] proved that if $|R| = |B|$, then there exists a non-crossing geometric alternating spanning tree on $R \cup B$ whose maximum degree is at most $O(\log |R|)$. Recently, Kaneko obtained the best upper bound for the maximum degree of the tree mentioned above. In his proof, Lemma 3.3 plays an important role.

Theorem 4.5 (Kaneko [22]). *For R and B with $|R| = |B|$, there exists a non-crossing geometric alternating spanning tree on $R \cup B$ whose maximum degree is at most three (Figure 10 (a)). This upper bound is sharp.*

The following theorem, which was conjectured in [22], has been recently proved by Kaneko.

Theorem 4.6. *Let $|R| \leq |B|$ and $k = \lceil |B|/|R| \rceil$. Then there exists a non-crossing geometric alternating spanning tree on $R \cup B$ whose maximum degree is at most $k + 2$.*

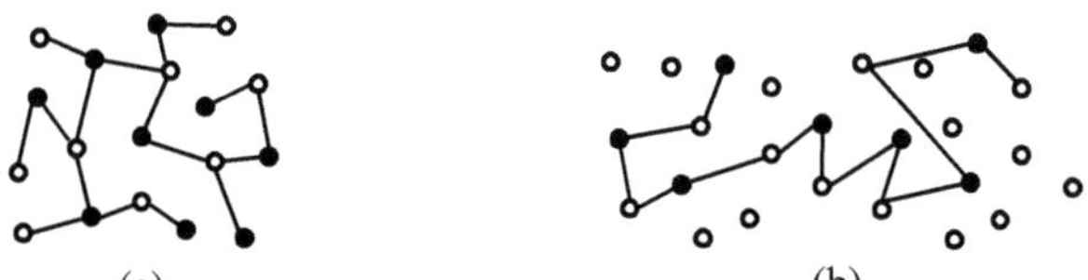

Fig. 10. (a) A non-crossing spanning tree of $K(R, B)$ with maximum degree at most three; (b) A non-crossing Hamiltonian path of $K(R, B')$ for some $B' \subset B$.

As stated in Theorem 4.4, geometric alternating Hamiltonian cycle and path may have many crossings for some $R \cup B$. However, if the number of blue points is much bigger than that of red points, can we then choose a subset B' of B such that there exists a non-crossing geometric alternating Hamiltonian path (or cycle) on $R \cup B'$? The following theorem gives a partial answer to this question on a Hamiltonian path.

Theorem 4.7 (Kaneko and Kano [22]). *Let $|R| = n \geq 3$. If $|B| \geq (n+1)(2n-4)+1$, then we can find a subset B' of B with n points such that there exists a non-crossing geometric alternating Hamiltonian path on $R \cup B'$ (Figure 10 (b)). Moreover, there exists a configuration of $R \cup B$ with $|B| = \frac{n^2}{16} + \frac{n}{2} - 1$ for which no such subset B' of B exists.*

5 Graph embeddings

For a graph G, we denote by $|G|$ the number of vertices of G. Given a planar graph G, let X be a set of $|G|$ points in the plane in general position. Then

G is said to be *line embeddable onto* X if G can be embedded in the plane so that every vertex of G corresponds to a point of X, every edge corresponds to a straight-line segment, and no two straight-line segments intersect except possibly having a common endpoint. A graph is called an *outerplanar graph* if it can be embedded in the plane so that every vertex of G lies on the exterior region.

The following theorem is a basic result on line embedding problems. Note that its proof is not found in [14], but Lemma 1 in Section 3 implies this theorem ([42]).

Theorem 5.1 (de Fraysseix, Pach and Pollack [14]). *A planar graph G can be line embedded onto an arbitrarily given set of $|G|$ points in the plane in general position if and only if G is an outerplanar graph.*

We now consider a line embedding on red and blue points. Let G be a planar graph with n specified vertices $v_1, v_2, \ldots, v_n$, and X a set of $|G|$ points in the plane in general position which contains n red points $p_1, p_2, \ldots, p_n$ and $|G| - n$ blue points. Then we say that G is *strongly line embeddable onto* X or G has a *strong line embedding onto* X if G can be line embedded onto X so that for every $1 \leq i \leq n$, v_i corresponds to the red point p_i (Figure 11). A tree with one specified vertex v is called a *rooted tree* with root v. Given n disjoint rooted trees T_i with root v_i, $1 \leq i \leq n$, the union $T_1 \cup T_2 \cup \cdots \cup T_n$, whose vertex set is $V(T_1) \cup V(T_2) \cup \cdots \cup V(T_n)$ and whose edge set is $E(T_1) \cup E(T_2) \cup \cdots \cup E(T_n)$, is called a *rooted forest* with roots $v_1, v_2, \ldots, v_n$, which are specified vertices of it.

Before giving our results, we remark that for a cycle C having two specified adjacent vertices and for a set X of $|C| - 2$ blue points and two red points in the plane such that all the points of X lie on a circle, and the two red points are not adjacent in $conv(X)$, C cannot be strongly line embedded onto X (Figure 12 (a)). Another such example is given in Figure 12 (b), where a graph H consisting of a cycle C_5 and a path P_2 cannot be strongly line embedded onto Y. Hence, when we consider a strong line embedding problem, we may restrict ourselves to rooted forests.

The following theorem, conjectured by Perles [38], was partially solved by Pach and Törőcsik [37], and completely proved by Ikebe, Perles, Tamura and Tokunaga [20]. A simpler proof of it can be found in Tokunaga [41]. Another related result can be found in [14].

Theorem 5.2. *A rooted tree T can be strongly line embedded onto every set of $|T|$ points in the plane in general position containing one red point.*

Theorem 5.3 ([23]). *A rooted forest F consisting of two rooted trees can be strongly line embedded onto every set of $|F|$ points in the plane in general position containing two red points (see Figure 11 (a), (b)). There exists an $O(|F|^2 \log |F|)$ time algorithm for finding a strong line embedding.*

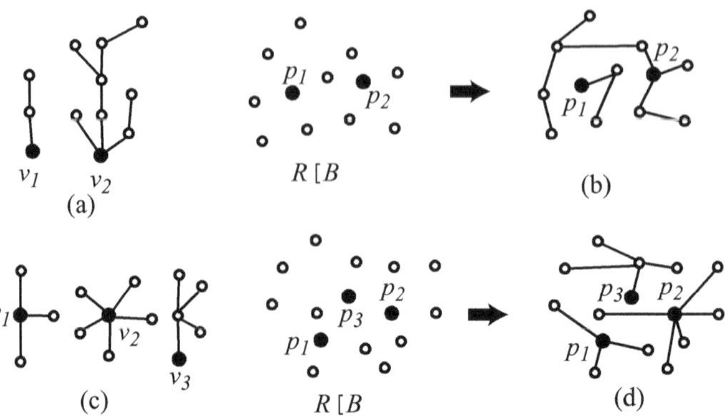

Fig. 11. (a) A rooted forest F with two components; (b) A strong line embedding of F; (c) A rooted star forest G; (d) A strong line embedding of G.

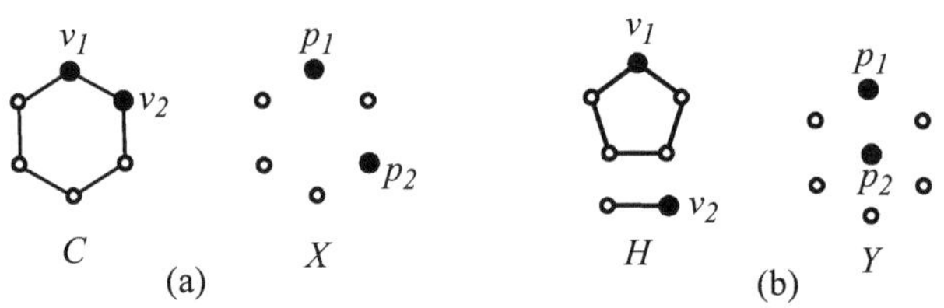

Fig. 12. (a) A cycle C cannot be strongly line embedded onto X; (b) A graph H cannot be strongly line embedded onto Y.

It was conjectured that a rooted forest F consisting of three rooted trees can be strongly line embedded onto every set of $|F|$ points in the plane in general position containing three red points ([22]). But a counter-example to this conjecture was found in [29].

A *star* $K(1,k)$ $(k \geq 1)$ consists of the vertex set $\{x, y_1, y_2, \ldots, y_k\}$ and the edge set $\{xy_1, xy_2, \ldots, xy_k\}$, where x is called its center and y_i is called its end-vertex. The union of stars is called a *star forest*, and the union of rooted stars, some of whose roots may be end-vertices, is called a *rooted star forest* (Figure 11 (c)).

Theorem 5.4. *A rooted star forest F consisting of n rooted stars can be strongly line embedded onto every set of $|F|$ points in the plane in general position containing n red points (see Figure 11 (c), (d)).*

By combining Theorem 2.9 and Theorem 5.2, we can obtain the next theorem.

Theorem 5.5. *Let m be a positive integer, and let $T_1, T_2, \cdots, T_n$ be n disjoint rooted trees such that $|T_i| \in \{m, m+1\}$ for all $1 \leq i \leq n$. Then the rooted forest $F = T_1 \cup T_2 \cup \cdots \cup T_n$ can be strongly line embedded onto every set of $|F|$ points in the plane in general position containing n red points (Figure 13).*

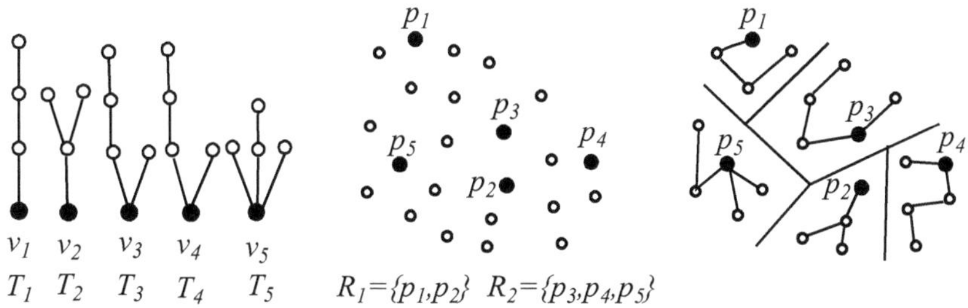

Fig. 13. A rooted forest F, and its strong line embedding onto $R \cup B$.

6 Gallai-type and Other Problems

In this section we consider $R \cup B$ not necessarily in general position.

A well-known theorem of Gallai ([16], [40]) states that for any set S of points in the plane all of which do not lie on the same line, there exists a line that passes through exactly two points of S. Such a line is called an *ordinary line*. Csima and Sawyer [9] improved this result by showing that there are at least $\lceil 6|S|/13 \rceil$ ordinary lines.

We now consider a similar problem on $R \cup B$. A line that passes through at least two red points and no blue points or through at least two blue points and no red points is called a *monochromatic line*. On the other hand, a line that passes through at least one red point and at least one blue point is called a *bichromatic line*. A *bichromatic ordinary line* is a line that passes through exactly one red point and one blue. It is easy to show that there exist configurations $R \cup B$ for which there exists no bichromatic ordinary lines (Figure 14 (a)). We beign with a result on monochromatic lines.

Theorem 6.1 (Chakerian [8], [2]). *If $R \cup B$ does not lie on a line, then there exists a monochromatic line.*

The next theorem deals with the number of bichromatic lines.

Theorem 6.2 (Pach and Pinchasi [35]). *Suppose $|R| = |B| = n$. Then there exists more than $n/2$ bichromatic lines that pass through at most two red points and at most two blue points. Furthermore, the number of bichromatic lines passing through at most six points is at least $m/10$, where m is the total number of connecting lines.*

Conjecture 6.3 (Fukuda [15], [10]). *If (i) R and B are separated by a line, and (ii) $|R|$ and $|B|$ differ by at most one, then there exists a bichromatic ordinary line.*

Note that the two conditions in the above conjecture are necessary. Namely, Figure 14 (a) and (b) show that condition (i) is necessary, and (c) shows that the condition (ii) is necessary ([35]). Recently, Finschi and Fukuda [13]

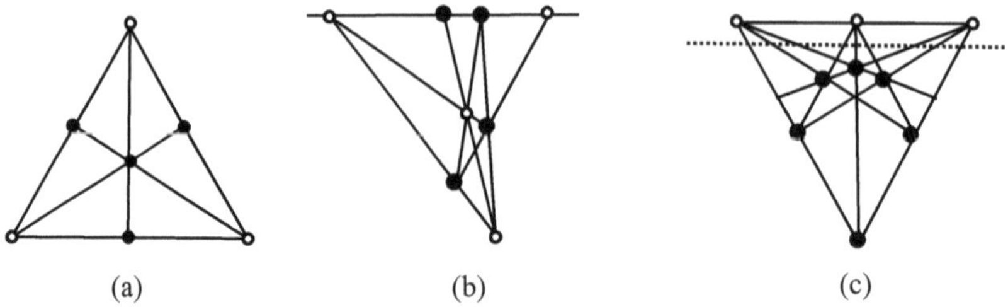

Fig. 14. Configurations having no bichromatic ordinary lines.

showed that the above Conjecture 6.3 is true for $|R \cup B| \le 8$, and found a counter-example to the conjecture, which consists of five red points and four blue points.

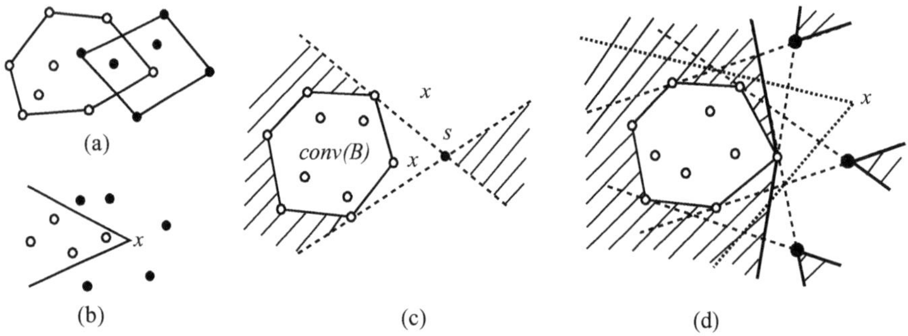

Fig. 15. (A) $R \cup B$ that cannot be separated by a wedge; (b) A point x for which there exists a wedge separating B and R; (c) The shaded region $Shade(s)$ with respect s; (d) The region of points x for which there exists a wedge separating B and R.

We next consider a separation problem. It is obvious that R and B can be separated by a line if and only if $conv(R) \cap conv(B) = \emptyset$. It is also easy to see that if $conv(B)$ contains a red point and if $conv(R)$ contains a blue point, then R and B cannot be separated by a wedge (Figure 15 (a)). Thus when we consider a wedge-separation, we may assume that $conv(B)$ contains no red points. Then for a red point s, if a point x is not contained in the shaded region in Figure 15 (c), then there exists a wedge with top x that separates $conv(B)$ and s. By this observation, we can obtain the following theorem.

Theorem 6.4 (Hurtado, Noy, Ramos and Seara [18]). *Suppose that $conv(B) \cap R = \emptyset$. For any red point $s \in R$, we denote by $Shade(s)$ the shaded region with respect s. Then there exists a wedge with top x that separate R and B if and only if*

$$x \notin \bigcup_{s \in R} Shade(s) \cup conv(B)$$

(Figure 15 (d)).

We now turn our attention to a monochromatic partitioning problem. Given a $S = R \cup B$ in general position, let $p(S)$ be the minimum number k of monochromatic subsets $S = X_1 \cup X_2 \cup \cdots \cup X_k$ such that $conv(X_i) \cap conv(X_j) = \emptyset$ for all $i \neq j$. Let

$$p(n) = \max\{p(S) \mid |S| = |R \cup B| = n\}.$$

Then $p(n)$ was recently determined as below.

Theorem 6.5 (Dumitrescu and Pach [12]).

$$p(n) = \left\lceil \frac{n+1}{2} \right\rceil.$$

The following proposition was posed as a problem in the 27th International Mathematics Olympiad, 1986.

Proposition 6.6. *Let P be n points with integer coordinates. Then each point of P can be colored red or blue so that for every row and column, the number of red points on it differs from the number of blue points on it by at most one. This coloring is called a balanced coloring.*

This proposition can be proved by using graph theory as follows. Let $X = \{x_1, \ldots, x_n\}$ and $Y = \{y_1, \ldots, y_m\}$ be the sets of integers such that for every $1 \leq i \leq n$ and $1 \leq j \leq m$, P contains a point with coordinate (x_i, y_t) and a point with coordinate (x_s, y_j) for some y_t and x_s. Then we construct the bipartite graph $B = B(X, Y)$ with bipartition $X \cup Y$ in which $x_i \in X$ and $y_j \in Y$ are joined by an edge if and only the point with coordinate (x_i, y_j) is contained in P. In particular, the exists a one to one correspondence between the edges of B and the points of P. Then by a well-known theorem [32], every edge of B can be colored red and blue so that for every vertex v of B, the number of red edges incident with v differs from the number of blue edges incident with v by at most one. This implies the above Proposition 6.6.

Akiyama and Urrutia [3] generalized the above proposition by considering an m-coloring, where $m \geq 3$. This generalization also can be obtained by the same arguments as above since a similar result hold for any m-edge-coloring of a bipartite graph.

References

[1] M. Abellanas, J. García, G. Hernández, M. Noy and P. Ramos, Bipartite embeddings of trees in the plane, *Discrete Appl. Math.* **93** (1999) 141–148.

[2] M. Aigner and G.M. Ziegler, *Proofs from the book*, 2nd edition, Springer, Berlin, (2001) 63.

[3] J. Akiyama and J. Urrutia, A note on balanced colorings for lattice points, *Discrete Math.* **83** (1990) 123–126.

[4] J. Akiyama and J. Urrutia, Simple alternating path problem, *Discrete Math.* **84** (1990) 101–103.

[5] I. Bárány and J. Matoušek, Simultaneous partitions of measures by k-fans, *Discrete Comput. Geom.* **25** (2001)317–334.

[6] I. Bárány and J. Matoušek, Equipartition of two measures by a 4-fan, preprint.

[7] S. Bespamyatnikh, D. Kirkpatrick and J. Snoeyink, Generalizing ham sandwich cuts to equitable subdivisions, *Discrete Comput. Geom.* **24** (2000) 605–622.

[8] G.D. Chakerian, Sylvester's problem on collinear points and a relative, *Amer. Math. Monthly* **77** (1970) 164–167.

[9] J. Csima and E.T. Sawyer, There exist $6n/13$ ordinary points, *Discrete Comput. Geom.* **9** (1993) 87–202.

[10] P.F. Da Silva and K. Fukuda, Isolating points by line in the plane, *J. Geom.* **62** (1998) 48–65.

[11] A. Dumitrescu and R. Kaye, Matching colored points in the plane: Some new results, *Comput. Geom.* **19** (2001) 69–85.

[12] A. Dumitrescu and J. Pach, Partitioning colored point sets into monochormatic parts, LNCS **2125** (2001), 264-275.

[13] L. Finschi and K. Fukuda, Complete combinatorial generation of small point configurations and hyperplane arrangements, *Discrete and Computational Geometry: The Goodman-Pollack Festschrift* (in the Series Algorithms and Combinatorics), Springer Verlag, preprint.

[14] H. de Fraysseix, J. Pach and R. Pollack, How to draw a planar graph on a grid, *Combinatorica* **10** (1990) 41–51.

[15] K. Fukuda, Question raised at the problem session, in "AMS-IMS-SIAM Joint Summer Research Conference on Discrete and Computational Geometry: Ten Years Later", Mount Holyoke College, South Hadley, Massachusetts, (1996).

[16] T. Gallai, Solution of problem 4065, *Amer. Math. Monthly* **51** (1944) 179–171.

[17] *Handbook of Discrete and Computational Geometry*, edited by J. Goodman and J. O'Rourke, CRC Press, (1997) 211

[18] F. Hurtado, M. Noy, P.A. Ramos and C. Seara, Separating objects in the plane by wedges and strips, *Discrete Math.* **109** (2001) 109-138.

[19] H. Ito, H. Uehara, and M. Yokoyama, 2-dimensional ham-sandwich theorem for partitioning into three convex pieces, *Discrete Comput. Geom.* LNCS **1763** (2000) 129-157.

[20] Y. Ikeba, M. Perles, A. Tamura and S. Tokunaga, The rooted tree embedding problem into points on the plane, *Discrete Comput. Geom.* **11** (1994) 51–63.

[21] A. Kaneko and M. Kano, Balanced partitions of two sets of points in the plane, *Comput. Geom.* **13** (1999) 253–261.

[22] A.Kaneko and M. Kano, Straight-line embeddings of two rooted trees in the plane, *Discrete Comput. Geom.* **21** (1999) 603–613.

[23] A.Kaneko and M. Kano, Straight line embeddings of rooted star forests in the plane, *Discrete Appl. Math.* **101** (2000)167–175.

[24] A. Kaneko and M. Kano, Generalized balanced partitions of two sets of points in the plane, *Discrete Comput. Geom.* LNCS **2098** (2001) 176-186.

[25] A. Kaneko, M. Kano and K. Yoshimoto, Alternating Hamiltonian cycles with minimum number of crossings in the plane, *Internat. J. Comput. Geom. Appl.*, **10** (2000) 73-78.

[26] A. Kaneko and M. Kano, A balanced partition of points in the plane and tree embedding problems, preprint.

[27] A. Kaneko, M. Kano and K. Suzuki, Path Coverings of Two Sets of Points in the Plane, preprint.

[28] A. Kaneko and M. Kano Certain Balanced Partitions of Two Sets of Points in the Plane, preprint.

[29] A. Kaneko, M. Kano and S. Tokunaga, Straight-line embeddings of three rooted trees in the plane, in Tenth Canadian Conference on Computational Geometry, Extended abstract (1998).

[30] L.C. Larson, *Problem-Solving Through Problems*, (Springer, New York), (1983) 200–201.

[31] C.-Y. Lo, J. Matoušek, and W. Steiger, Algorithms for ham-sandwich cuts, *Discrete Comput. Geom.*, **11** (1994) 433–452.

[32] L. Lovász, *Combinatorial Problems and Exercises*, North-Holland, Amsterdam (1979) p.50, Problem 11.

[33] J. Pach, Geometric Graph Theory, *Surveys in Combinatorics, 1999*, London Math. Soc. Lecture Note Series **267**, (1999) 167–200.

[34] J. Pach and P.K. Agarwal, *Combinatorial Geometry*, Wiley, (1995) 223–242.

[35] J. Pach and R. Pinchasi, Bichromatic lines with few points, *J. Combinatorial Theory Ser. A*, **90** (2000) 326–335.

[36] J. Pach and R. Pinchasi, On the number of balanced lines, *Discrete Comput. Geom.*, **25** (2001) 611–628.

[37] J. Pach and J. Törőcsik, Layout of rooted tree, *Planar Graphs (DIMACS Series in Discrete Math. and Theoritical Comput. Sci.)* **9** (1993) 131–137.

[38] M. Perles, Open problem proposed at the DIMACS Workshop on Arrangements, Rutgers University, 1990.

[39] T. Sakai, Balanced Convex Partitions of Measures in R^2, to appear in *Graphs and Combinatorics.*

[40] J. Sylvester, Mathematical question 11851, *Educational Times* **59** (1893) 98–99.

[41] S. Tokunaga, On a straight-line embedding problem of graphs, *Discrete Math.* **150** (1996) 371–378.

[42] S. Tokunaga, Intersection number of two connected geometric graphs *Information Proc. Discrete Math.* **150** (1996) 371–378.

About Authors

Atsushi Kaneko is at the Department of Computer Science and Communication Engineering, Kogakuin University, Nishi-Shinjuku, Shinjuku-ku, Tokyo 163–8677 Japan; *kaneko@ee.kogakuin.ac.jp*.

M. Kano is at Department of Computer and Information Sciences, Ibaraki University, Hitachi 316–8511 Japan; *kano@cis.ibaraki.ac.jp*.

Acknowledgments

Work on this paper has been supported by the Ministry of Education of Japan (the grant-in-aid to Kaneko and Kano). The authors would like to thank a referee for his very careful and helpful suggestions, and the editors for their helpful suggestions and nice contribution.

Configurations with Rational Angles and Trigonometric Diophantine Equations

M. Laczkovich

Abstract

A subset E of the plane is said to be a *configuration with rational angles* (CRA) if the angle determined by any three points of E is rational when measured in degrees. We prove that there is a constant C such that whenever a CRA has more than C points, then it can be covered either by a circle and its center or by a pair of points and their bisecting line. The proof is based on the description of all rational solutions of the equation

$$\sin \pi p_1 \cdot \sin \pi p_2 \cdot \sin \pi p_3 = \sin \pi q_1 \cdot \sin \pi q_2 \cdot \sin \pi q_3 .$$

1 Introduction and main results

We shall say that a subset E of the plane is a *configuration with rational angles* (CRA) if the angle determined by any three points of E is rational when measured in degrees; that is, a rational multiple of π when measured in radians. Obviously, every collinear set is a CRA. To present a less trivial example, let

$$E_1 = \{U,\ V\} \cup \{(\tan \pi r, 0) : r \in \mathbf{Q},\ |r| < \pi/2\},$$

where $U = (0,1)$ and $V = (0,-1)$. It is easy to see that for every $|r| < 1/2$ the angles of the triangle with vertices U, V and $(\tan \pi r, 0)$ are $\pi|r|$, $\pi|r|$ and $\pi - 2\pi|r|$, and for every $0 \le r < s < 1/2$ the angles of the triangle with vertices U, $(\tan \pi r, 0)$ and $(\tan \pi s, 0)$ are $\pi(s-r)$, $(\pi/2) + \pi r$ and $(\pi/2) - \pi s$ (see Figure 1). Therefore E_1 is a CRA. We shall say that a set E is a *configuration of type I*, if E is similar to a subset of E_1.

Our next example is

$$E_2 = \{O\} \cup \{(\cos \pi r, \sin \pi r) : r \in \mathbf{Q}\},$$

where $O = (0,0)$. If $0 < s - r < 1$ then the angles of the triangle with vertices O, $(\cos \pi r, \sin \pi r)$ and $(\cos \pi s, \sin \pi s)$ are $\pi(s-r)$, $\pi(1-(s-r))/2$ and $\pi(1-(s-r))/2$, and for $r < s < t$, $t - r < 1$, the angles of the

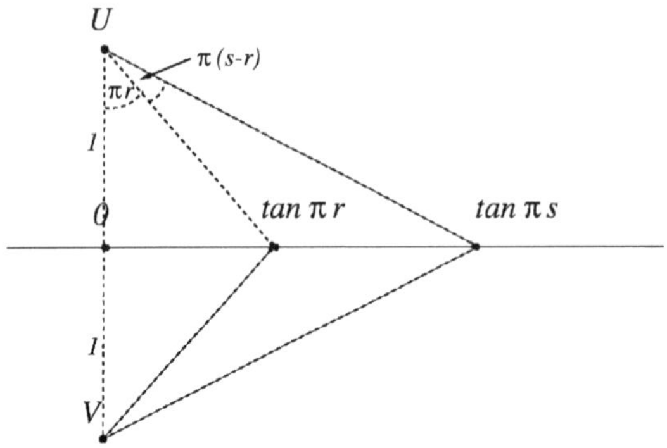

Fig. 1. E_1 is a CRA

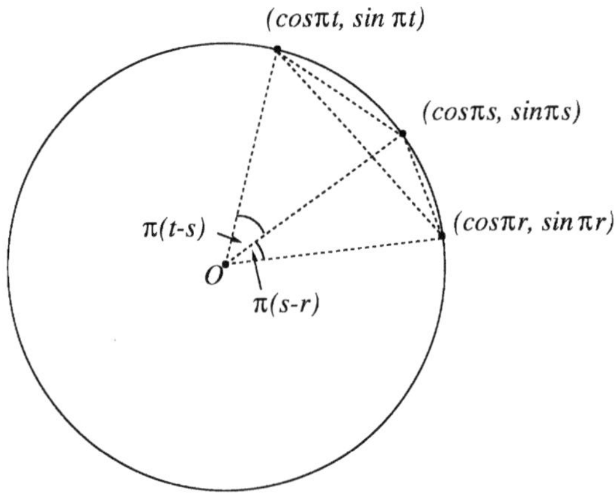

Fig. 2. E_2 is a CRA

triangle with vertices $(\cos \pi r, \sin \pi r)$, $(\cos \pi s, \sin \pi s)$ and $(\cos \pi t, \sin \pi t)$ are $\pi(s-r)/2$, $\pi(t-s)/2$ and $\pi(1-(t-r)/2)$. (See Figure 2.) Therefore E_2 is also a CRA. We shall say that a set E is a *configuration of type II,* if E is similar to a subset of E_2.

Our aim is to show that, apart from finite sets, every nontrivial CRA is of type I or II. More precisely, we shall prove the following.

Theorem 1. *Let E be a configuration with rational angles. Then one of the following statements is true.*

(i) *E is collinear.*

(ii) *E is a configuration of type I or II.*

(iii) *E is finite and has at most C elements, where C is an absolute constant.*

The following statement is an immediate corollary.

Corollary 2. *There is an absolute constant C such that whenever a configuration with rational angles has more than C elements then it can be covered either by a circle and its center or by a pair of points and their bisecting line.*

Let $ABC\triangle$ be a triangle. We shall say that P is a *regular point of* $ABC\triangle$ if the set $\{A, B, C, P\}$ is a configuration of type I or II.

The regular points of a triangle $ABC\triangle$ can be located as follows. First, we may assume that $ABC\triangle$ has rational angles, since otherwise there are no regular points. Then each of the lines AB, AC, BC contains infinitely many regular points. For example, if $A = U = (0, 1)$ and BC is the real axis then each point $\tan \pi r$ ($r \in \mathbf{Q}$) is a regular point of $ABC\triangle$. If $ABC\triangle$ is isosceles then we also find infinitely many regular points on its axis of symmetry. It is also easy to check that there are infinitely many regular points on the circle circumscribed to $ABC\triangle$. If $ABC\triangle$ is isosceles and, say, $\overline{AB} = \overline{AC}$ then there are infinitely many regular points on the circle with radius $\overline{AB}$ and center A. One can easily prove that the only regular points not covered by these lines and circles are the center of the circumscribed circle and the image of each vertex under the reflection about the opposite side. In other words, the following is true.

Lemma 3. *The set of regular points of a non-isosceles triangle $ABC\triangle$ are covered, with the exception of four points, by the lines AB, AC, BC, and the circumscribed circle. If $ABC\triangle$ is isosceles then we have to add the symmetry axes of $ABC\triangle$ and the circles having one of the vertices A, B, C as centers and containing the other two vertices.* □

The point P is said to be a *singular point of* $ABC\triangle$ if the set $\{A, B, C, P\}$ is a CRA, but P is not a regular point. If $ABC\triangle$ has rational angles and is not a right triangle then $ABC\triangle$ has at least five singular points: the four intersections of the inner and outer bisectors of the angles, and the orthocenter. One can easily construct triangles having other singular points; in the example below we shall construct one with (at least) 10 singular points. Our main result is the following.

Theorem 4. *Every triangle has at most K singular points, where K is an absolute constant.*

Now we prove Theorem 1, supposing the truth of Theorem 4. Let K be as in Theorem 4. We claim that if E is a CRA, E is not collinear and has more than $11K + 84$ elements, then E is of type I or II. Let A, B, C be fixed non-collinear elements of E. Then, for every $P \in E$, the set $\{A, B, C, P\}$ is a CRA, and thus every element of E is either a regular or a singular point of $ABC\triangle$. Therefore, by Theorem 4, more than $10K + 84$ elements of E are regular points of $ABC\triangle$. Since, by Lemma 3, the regular points of $ABC\triangle$ can be covered by six lines, four circles and four points, we obtain that more than $K + 8$ elements of E are covered by a line or by a circle.

First suppose that the line ℓ contains more than $K + 8$ elements of E. We claim that $E \setminus \ell$ is either a singleton or contains two points, and in the latter case these points are symmetric about ℓ. It is clear that in this case E must be a configuration of type I. Suppose the statement is not true. Then there are elements P, Q of $E \setminus \ell$ such that ℓ is not the perpendicular bisector of the segment PQ. Since more than five points of E lie on ℓ, we can choose a point $R \in E \cap \ell$ such that the triangle $PQR\triangle$ is not isosceles. Now every point of $E \cap \ell$ is either a regular or a singular point of $PQR\triangle$ and thus, by Theorem 4, there are more than 8 regular points of $PQR\triangle$ lying on ℓ. By Lemma 3, the set of regular points of $PQR\triangle$ can be covered by the lines PQ, PR, QR, the circle circumscribed to $PQR\triangle$, and at most four points. Therefore ℓ contains at most 7 regular points of $PQR\triangle$, which is impossible.

Next suppose that the circle c contains more than $K + 8$ elements of E. We claim that $E \setminus c$ is either empty or equals $\{O\}$, the center of c. It is clear that in this case E must be a configuration of type II. If the statement is not true, then the set $E \setminus (c \cup \{O\})$ is nonempty. Let $P \in E \setminus (c \cup \{O\})$, and let Q be an arbitrary point of $E \cap c$. Since more than five points of E lies on c, we can choose a point $R \in E \cap c$ such that $PQR\triangle$ is not isosceles. Every point of $E \cap c$ is either a regular or a singular point of $PQR\triangle$ and thus, by Theorem 4, there are more than 8 regular points of $PQR\triangle$ lying on c. By Lemma 3, the set of regular points of $PQR\triangle$ can be covered by the lines PQ, PR, QR, the circle circumscribed to $PQR\triangle$, and at most four points. Therefore c contains at most 8 regular points of $PQR\triangle$, which is a contradiction. □

Sections 2–5 will be devoted to the proof of Theorem 4, based on the solution of a certain trigonometric Diophantine equation. Although the proof gives an explicit upper bound for K, it is much bigger than the expected value. As for the constant C, the argument above gives $C \leq 11K + 84$. However, it is very likely that $C < K < 100$. In order to find lower bounds, one has to construct large nontrivial CRA's and triangles with many singular points. The best we can offer is a nontrivial CRA having 10 points and a triangle having 10 singular points.

Example. Let r be a rational number satisfying $\frac{5}{12} < r < \frac{1}{2}$ and put $\alpha = \pi r$. Let $ABC\triangle$ be a regular triangle with center O. Let $AEB\triangle$, $BFC\triangle$, $CDA\triangle$ be nonoverlapping isosceles triangles such that

$$BAE\measuredangle = ABE\measuredangle = FBC\measuredangle = BCF\measuredangle = DCA\measuredangle = DAC\measuredangle = \alpha$$

(see Figure 3). The points O, C, and the middle point of the segment DF (denoted by M) are collinear. Since

$$OCD\measuredangle = \alpha + \frac{\pi}{6} > \pi - \alpha > \frac{\pi}{2} = OMD\measuredangle,$$

there exists a point G in the segment CM such that $OGD\measuredangle = \pi - \alpha$. Rotating the segment OG about the endpoint O with angles $2\pi/3$ and $4\pi/3$ we obtain the points H and I.

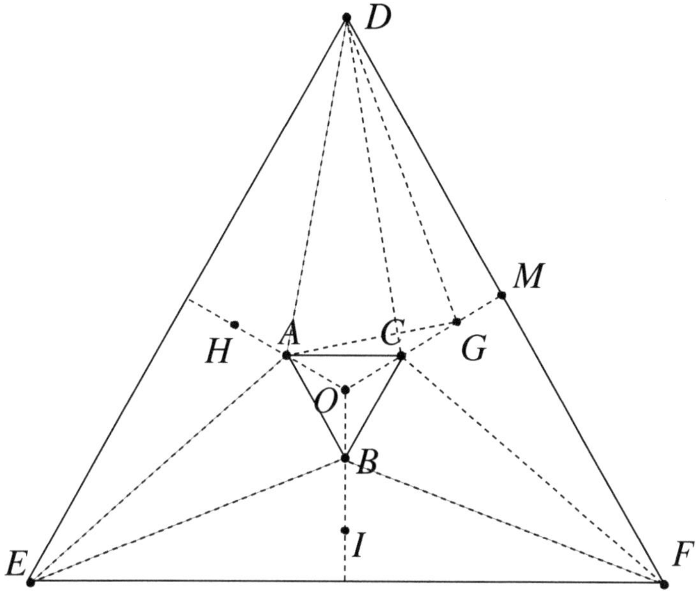

Fig. 3. a nontrivial CRA having 10 points

We claim that the set $\Gamma = \{A, B, C, G, H, I, D, E, F, O\}$ is a CRA. First we show that the quadrilateral $ACGD$ is a CRA. Since $CGD\measuredangle = OGD\measuredangle = \pi - \alpha$, it follows that $ACGD$ can be inscribed in a circle, and thus $CGA\measuredangle = CDA\measuredangle = \pi - 2\alpha$ is a rational multiple of π. Therefore $CAG\measuredangle = \pi - (5\pi/6) - CGA\measuredangle$ is also a rational multiple of π. The same is true for $GDA\measuredangle = \pi - ACG\measuredangle = \pi/6$ and $CDG\measuredangle = CAG\measuredangle$, which proves that $ACGD$ is a CRA.

By the *slope* of a line ℓ we shall mean the angle between the lines ℓ and EF. Since AC is parallel to EF and $ACGD$ is a CRA, it follows that the slope of every line going through two points of $ACGD$ is a rational multiple of π. Based on this observation it is easy to check that for arbitrary points $P, Q \in \Gamma$ the slope of the line PQ is a rational multiple of π. If P, Q, $R \in \Gamma$ then $PQR\measuredangle$ is the difference between the slopes of PQ and QR, which proves that Γ is a CRA.

Now we claim that the triangle $AGD\triangle$ has at least 10 singular points. If $P \in \Gamma$ then $\{A, G, D, P\}$ is a CRA. It is easy to see that no point of $\{O, B, E, I, F\}$ lies on any of the lines AG, GD, DA or on the circumscribed circle of $AGD\triangle$, and does not coincide with the reflection of any of the vertices of $AGD\triangle$ about the opposite side. That is, each point of $\{O, B, E, I, F\}$ is a singular point of $AGD\triangle$. One can also check that the four intersections of the inner and outer bisectors of $AGD\triangle$ and the orthocenter of $AGD\triangle$ are distinct from the points of Γ. Therefore $AGD\triangle$ has at least $5 + 5 = 10$ singular points.

2 Reduction to a Diophantine sine equation

Let A, B, C, P be distinct points of the plane, and put

$$\begin{aligned} CAB\measuredangle &= \alpha_1, \ CBA\measuredangle = \alpha_2, \\ PAC\measuredangle &= \beta_1, \ PBC\measuredangle = \beta_2, \\ APC\measuredangle &= \gamma_1, \ BPC\measuredangle = \gamma_2. \end{aligned} \tag{1}$$

These angles satisfy two relations. First, they satisfy one of the relations

$$\alpha_1 + \alpha_2 \pm \beta_1 \pm \beta_2 \pm \gamma_1 \pm \gamma_2 = \pm\pi, \tag{2}$$

where the signs depend on the location of the point P. In order to find the second relation, observe that

$$\frac{\sin\alpha_2}{\sin\alpha_1} = \frac{\overline{AC}}{\overline{BC}} = \frac{\overline{AC}/\overline{PC}}{\overline{BC}/\overline{PC}} = \frac{\sin\gamma_1/\sin\beta_1}{\sin\gamma_2/\sin\beta_2}.$$

Therefore we obtain

$$\sin\alpha_2 \sin\beta_1 \sin\gamma_2 = \sin\alpha_1 \sin\beta_2 \sin\gamma_1. \tag{3}$$

We shall say that a solution of (3) is *Diophantine* if all angles involved are rational multiples of π. Clearly, if $\{A, B, C, P\}$ is a CRA then the angles defined above constitute a Diophantine solution of (3). In the next section we shall prove that, apart from a bounded number of exceptions, every Diophantine solution of (2) and (3) determines a regular point of $ABC\triangle$. This will prove Theorem 4 subject to Theorem 6 below which describes all Diophantine solutions of (3). We shall prove Theorem 6 in Sections 4 and 5.

Trigonometric Diophantine equations have been investigated by many authors. For a short survey of the topic up to 1976 see Section 7 of the paper [2] by J. H. Conway and A. J. Jones. The Diophantine solutions of the equation $\sin\alpha \sin\beta = \sin\gamma \sin\delta$ were found by G. Myerson in [5]. As Myerson remarks in [5, p. 81], it would be very difficult, if not impossible, to give an exact list of all rational solutions of similar equations containing more factors. However, for our purposes it is sufficient to describe the structure of the solutions, and such a description is possible. We start with the following general result saying that the rational solutions of every trigonometric Diophantine equation can be given by a finite system of identities.

Theorem 5. *Let $P \in \mathbf{Q}[x_1, \ldots, x_n]$ be a nonzero polynomial with rational coefficients. Then there is a finite set S of maps $\Lambda = (\Lambda_1, \ldots, \Lambda_n) : \mathbf{Q}^n \to \mathbf{Q}^n$ with the following properties.*

(i) *Each component $\Lambda_i(t)$ of every $\Lambda \in S$ is a linear function of the form $m_1 t_1 + \ldots + m_n t_n + c$, where the coefficients m_i are integers and the constant c is rational;*

(ii)
$$P(\sin \pi\Lambda_1(t), \dots, \sin \pi\Lambda_n(t)) = 0 \tag{4}$$
for every $\Lambda \in S$ *and for every* $t = (t_1, \dots, t_n)$; *and*

(iii) *all rational solutions of the equation*
$$P(\sin \pi p_1, \dots, \sin \pi p_n) = 0 \tag{5}$$
are given by (4) *in the following sense: if* $(p_1, \dots, p_n)$ *is a rational solution of* (5) *then there is a* $\Lambda \in S$ *and there are rational numbers* $t_1, \dots, t_n$ *such that* $p_i \equiv \Lambda_i(t) \pmod 2$ *for every* $i = 1, \dots, n$.

This theorem is essentially contained in the paper of Conway and Jones (cf. [2, Theorem 3]). Considering, however, that the result is not stated explicitly in [2] and that the proof of [2, Theorem 3] can be simplified, we shall prove Theorem 5 in Section 4.

M. Newman proved in [7] that the equation $\sin \pi x_1 \cdots \sin \pi x_n = r$ only has a finite number of rational solutions for every n and for every nonzero rational r. (For a simple proof see [5, Theorem 1].) We remark that this is an immediate consequence of Theorem 5. Moreover, the following is true.

Let $Q \in \mathbf{Q}[x_1, \dots, x_n]$, *and let* r *be a nonzero rational number. Then in the rational solutions of*
$$\sin \pi x_1 \cdot Q(\sin \pi x_1, \dots, \sin \pi x_n) = r$$
the factor $\sin \pi x_1$ *only can take a finite number of values.* Proof: Applying Theorem 5 with $P = x_1 \cdot Q - r$ we find that for every $\Lambda \in S$ the component Λ_1 must be constant. Indeed, since
$$\sin \pi\Lambda_1(t) \cdot Q(\sin \pi\Lambda_1(t), \dots, \sin \pi\Lambda_n(t)) = r$$
is an identity and $r \neq 0$, it follows that $\sin \pi\Lambda_1(t) \neq 0$ for every t, which is possible only if Λ_1 is constant.

In Section 4 we shall also prove that in the special case of the equations
$$\sin \pi p_1 \cdots \sin \pi p_s = r \cdot \sin \pi q_1 \cdots \sin \pi q_t \tag{6}$$
the maps Λ can be chosen in such a way that each component of every Λ only depends on one variable (see Theorem 12). In other words, all rational solutions of (6) can be obtained by multiplying a finite system of one-variable identities.

Making use of these results, we can describe the Diophantine solutions of (3). In order to simplify the description we shall use the following terminology. We say that the equations $x_1 \dots x_n = y_1 \dots y_n$ and $u_1 \dots u_n = v_1 \dots v_n$ are *equivalent* if, with a suitable permutation of the factors and changing the sides of one of the equations if necessary, we obtain $x_i = \pm u_i$ and $y_i = \pm v_i$ for every $i = 1, \dots, n$.

Theorem 6. *There are finite sets $F_3 \subset \mathbf{Q}^3$, $F_4 \subset \mathbf{Q}^4$ and $F_6 \subset \mathbf{Q}^6$ such that whenever the rational numbers p_i, q_i $(i = 1, 2, 3)$ satisfy*

$$\sin \pi p_1 \cdot \sin \pi p_2 \cdot \sin \pi p_3 = \sin \pi q_1 \cdot \sin \pi q_2 \cdot \sin \pi q_3 \tag{7}$$

then one of the following statements is true.

(i) *Both sides of* (7) *are zero.*

(ii) (7) *is equivalent to an equation of the form $uvw = uvw$.*

(iii) (7) *is equivalent to an equation of the form*

$$\sin 2\pi x \cdot \sin \pi y \cdot \cos \pi y = \sin \pi x \cdot \cos \pi x \cdot \sin 2\pi y. \tag{8}$$

(iv) (7) *is equivalent to an equation of the form*

$$u \cdot (1/2) \cdot \sin 2\pi x = u \cdot \sin \pi x \cdot \cos \pi x.$$

(v) *There are nonnegative integers n_i $(i = 1, \dots, 6)$ and rational numbers a_i $(i = 1, \dots, 6)$ such that*

$$\begin{aligned} &\sin \pi(n_1 x + a_1) \sin \pi(n_2 x + a_2) \sin \pi(n_3 x + a_3) = \\ &\sin \pi(n_4 x + a_4) \sin \pi(n_5 x + a_5) \sin \pi(n_6 x + a_6) \end{aligned} \tag{9}$$

is an identity, $n_i \le 24$ $(i = 1, \dots, 6)$, n_4, n_5, n_6 are positive, the numbers $120 \cdot a_i$ are integers for every $i = 1, \dots, 6$ and (7) *is equivalent to* (9) *for a suitable rational value of x.*

(vi) (7) *is equivalent to an equation of the form*

$$\sin \pi a_1 \cdot \sin \pi a_2 \cdot \sin 2\pi x = \sin \pi b_1 \cdot \sin \pi x \cdot \cos \pi x, \tag{10}$$

where $(a_1, a_2, b_1) \in F_3$.

(vii) (7) *is equivalent to an equation of the form*

$$u \cdot \sin \pi a_1 \cdot \sin \pi a_2 = u \cdot \sin \pi b_1 \cdot \sin \pi b_2, \tag{11}$$

where $(a_1, a_2, b_1, b_2) \in F_4$.

(viii) $$(p_1, p_2, p_3, q_1, q_2, q_3) \in F_6.$$

Before turning to the proof of these theorems we make some comments on the cases (v)–(viii).

In [5, Theorem 4, p. 80] Myerson lists all solutions of $\sin \pi a_1 \cdot \sin \pi a_2 = \sin \pi b_1 \cdot \sin \pi b_2$. As it turns out, there are 15 nontrivial solutions. Thus F_4 contains 15 quadruples. On the other hand, F_3 only contains 4 triples.

Indeed, (10) holds if and only if $2\sin\pi a_1 \cdot \sin\pi a_2 = \sin\pi b_1$. It follows from Myerson's theorem that the only nontrivial solutions of this equation are

$$\begin{aligned} a_1 &= 1/10,\ a_2 = 3/10,\ b_1 = 1/6; \\ a_1 &= 1/12,\ a_2 = 5/12,\ b_1 = 1/6; \\ a_1 &= 1/15,\ a_2 = 4/15,\ b_1 = 1/10; \\ a_1 &= 2/15,\ a_2 = 7/15,\ b_1 = 3/10. \end{aligned}$$

We note that this fact also follows from a theorem by Mann [4, Theorem 6, p. 114] and from a result by W. J. R. Crosby [3].

We do not have any reasonable estimate of the size of the set F_6. Although the proof gives an explicit upper bound for the denominators, it is enormous, and a computer search for listing F_6 does not seem to be feasible.

The identities that occur in (v) are special cases of the more general identities

$$\begin{aligned} &\sin\pi(n_1x + a_1)\cdots\sin\pi(n_sx + a_s) \\ &\qquad = \lambda\sin\pi(m_1x + b_1)\cdots\sin\pi(m_tx + b_t). \end{aligned} \tag{12}$$

In the next theorem (which is a generalization of [1, Theorem 1]) we shall describe these identities. Let $\mathcal{A} = (A_1, \dots, A_s)$ and $\mathcal{B} = (B_1, \dots, B_t)$ be finite sequences of sets. We shall say that the pair $(\mathcal{A};\mathcal{B})$ is *exact* if $\sum_{i=1}^{s} \chi_{A_i} = \sum_{j=1}^{t} \chi_{B_j}$, where χ_H denotes the characteristic function of H.

Theorem 7. *Let n_i, m_j be positive integers, a_i, b_j be real numbers, and put $\mathcal{A} = (A_1, \dots, A_s)$ and $\mathcal{B} = (B_1, \dots, B_t)$, where*

$$A_i = \left\{\frac{k - a_i}{n_i} : k \in \mathbf{Z}\right\} \quad (i = 1, \dots, s),$$

and

$$B_j = \left\{\frac{k - b_j}{m_j} :\ k \in \mathbf{Z}\right\} \quad (j = 1, \dots, t).$$

(i) *Suppose that (12) is an identity, where λ is a real number. Then $\lambda = \pm 2^{t-s}$, and the pair $(\mathcal{A};\mathcal{B})$ is exact.*

(ii) *If the pair $(\mathcal{A};\mathcal{B})$ is exact then, with a suitable choice of the sign in the right hand side,*

$$\begin{aligned} &\sin\pi(n_1x + a_1)\cdots\sin\pi(n_sx + a_s) \\ &\qquad = \pm 2^{t-s}\sin\pi(m_1x + b_1)\cdots\sin\pi(m_tx + b_t) \end{aligned} \tag{13}$$

is an identity.

Proof. First we prove (ii). Suppose that $(\mathcal{A};\mathcal{B})$ is exact, and let $N = n_1 + \ldots + n_s$, $M = m_1 + \ldots + m_t$. Since $\operatorname{card}(A_i \cap [0,1)) = n_i$, $\operatorname{card}(B_j \cap [0,1)) = m_j$

and the pair $(\mathcal{A};\mathcal{B})$ is exact, it follows that

$$N=\sum_{i=1}^{s}\operatorname{card}\left(A_i\cap[0,1)\right)=\sum_{j=1}^{t}\operatorname{card}\left(B_j\cap[0,1)\right)=M.$$

It is not difficult to prove, using the identities $\sin\pi x=\left(e^{i\pi x}-e^{-i\pi x}\right)/(2i)$ and $x^n-1=\prod_{\nu=0}^{n-1}\left(x-e^{2\pi i\nu/n}\right)$ that

$$\sin\pi nx=2^{n-1}\sin\pi x\cdot\sin\pi\left(x+\frac{1}{n}\right)\cdots\sin\pi\left(x+\frac{n-1}{n}\right)\tag{14}$$

for every $n=1,2,\ldots$ and for every x. Applying (14) s times we obtain

$$\sin\pi(n_1x+a_1)\cdots\sin\pi(n_sx+a_s)=\pm2^{N-s}\prod_{i=1}^{s}\prod_{a\in A_i\cap[0,1)}\sin\pi(x-a).$$

We have a similar representation of $\sin\pi(m_1x+b_1)\cdots\sin\pi(m_tx+b_t)$. Since $(\mathcal{A};\mathcal{B})$ is exact and $N=M$, this proves (13).

Next suppose that (12) is an identity. The sets $\bigcup_{i=1}^{s}A_i$ and $\bigcup_{j=1}^{t}B_j$ are the sets of the roots of the left hand side and of the right hand side of (12), respectively. Taking the multiplicities of the roots into account, we find that the pair $(\mathcal{A}:\mathcal{B})$ is exact. As we proved above, this implies (13), and thus λ must be equal to $\pm2^{t-s}$. □

Using Theorem 7 we may list all identities that occur in (v) of Theorem 6. Some of them are as follows:

$\sin(\pi/6)\cdot\sin(\pi/6)\cdot\sin3\pi x=\sin\pi x\cdot\sin\pi\left(x+\frac{1}{3}\right)\cdot\sin\pi\left(x+\frac{2}{3}\right);$

$\sin(\pi/6)\cdot\sin3\pi x\cdot\sin\pi\left(x+\frac{1}{2}\right)=\sin2\pi x\cdot\sin\pi\left(x+\frac{1}{3}\right)\cdot\sin\pi\left(x+\frac{2}{3}\right);$

$\sin(\pi/6)\cdot\sin(\pi/6)\cdot\sin4\pi x=\sin2\pi x\cdot\sin\pi\left(x+\frac{1}{4}\right)\cdot\sin\pi\left(x+\frac{3}{4}\right);$

$\sin(\pi/6)\cdot\sin\pi\left(3x+\frac{1}{4}\right)\cdot\sin\pi\left(3x+\frac{3}{4}\right)=$
$\sin\pi\left(2x+\frac{1}{2}\right)\cdot\sin\pi\left(2x+\frac{1}{6}\right)\cdot\sin\pi\left(2x+\frac{5}{6}\right);$

$\sin3\pi x\cdot\sin\pi\left(x+\frac{1}{2}\right)\cdot\sin\pi\left(x+\frac{1}{6}\right)=\sin2\pi x\cdot\sin\pi\left(2x+\frac{1}{3}\right)\cdot\sin\pi\left(x+\frac{1}{3}\right);$

$\sin4\pi x\cdot\sin\pi\left(x+\frac{1}{3}\right)\cdot\sin\pi\left(x+\frac{2}{3}\right)=\sin3\pi x\cdot\sin\pi\left(2x+\frac{1}{2}\right)\cdot\sin\pi\left(x+\frac{1}{2}\right).$

3 Proof of Theorem 4

Let $ABC\triangle$ be a fixed triangle, and let P be a point distinct from A, B, C. Recall that if P is a singular point of $ABC\triangle$ then P does not lie on any of the lines AB, AC, BC or on the circle circumscribed to $ABC\triangle$. We shall denote the interior of $ABC\triangle$ by T. The open angular domain bounded by the rays $\overrightarrow{CA}$ and $\overrightarrow{CB}$ will be denoted by U_C. The reflection of U_C about the vertex C is denoted by V_C. The line AB partitions U_C into two parts; we denote the (open) unbounded component by W_C. (See Figure 4.) By changing the labelling of the vertices if necessary, we may assume (and shall assume throughout the proof) that P belongs to one of the domains T, $V=V_C$

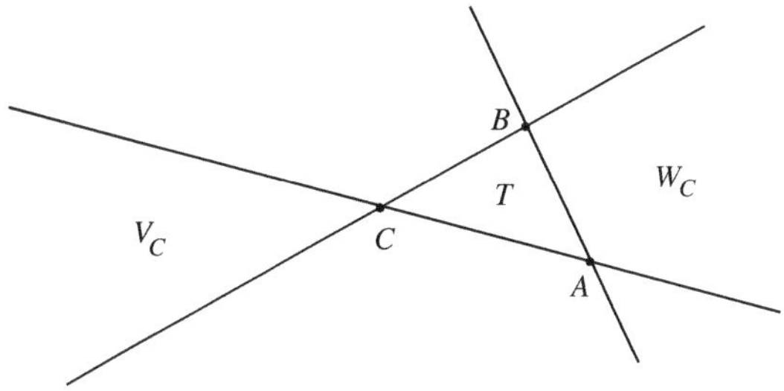

Fig. 4. the domains T, W_C and V_C

and $W = W_C$. We shall apply the notation introduced in (1). Note that the angles α_1, α_2 are fixed, while $\beta_1, \beta_2, \gamma_1, \gamma_2$ are functions of P. It is easy to check that the angles $\alpha_i, \beta_i, \gamma_i$ satisfy the following inequalities and equations (see Figures 5-7):

If P $\in$ T then
$$\beta_1 < \alpha_1 < \gamma_2, \quad \beta_2 < \alpha_2 < \gamma_1,$$
$$\beta_1 + \gamma_1 < \pi, \quad \beta_2 + \gamma_2 < \pi,$$
$$\alpha_1 + \alpha_2 = \beta_1 + \beta_2 + \gamma_1 + \gamma_2 - \pi;$$

If P $\in$ W then
$$\alpha_1 < \beta_1, \; \alpha_2 < \beta_2, \quad \gamma_1 + \gamma_2 < \pi,$$
$$\beta_1 + \gamma_1 < \pi, \quad \beta_2 + \gamma_2 < \pi,$$
$$\alpha_1 + \alpha_2 = \beta_1 + \beta_2 + \gamma_1 + \gamma_2 - \pi;$$

If P $\in$ V then
$$\alpha_1 + \alpha_2 + \beta_1 + \beta_2 + \gamma_1 + \gamma_2 = \pi.$$

Lemma 8. *For an arbitrary pair of angles δ_1, δ_2 there are at most 64 singular points P such that two different terms of the quadruple $(\beta_1, \beta_2, \gamma_1, \gamma_2)$ equal δ_1 and δ_2 respectively.*

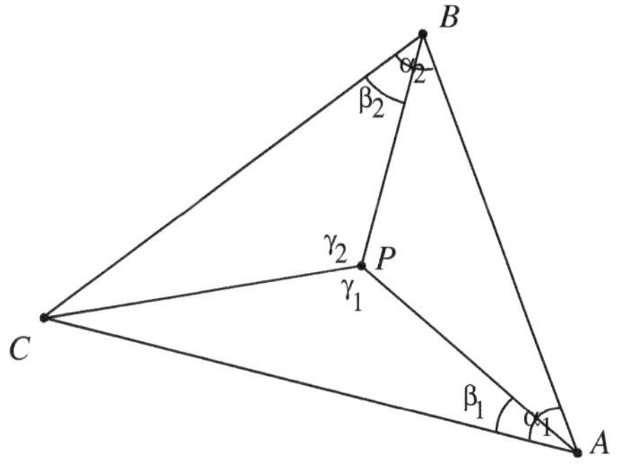

Fig. 5. the case $P \in T$

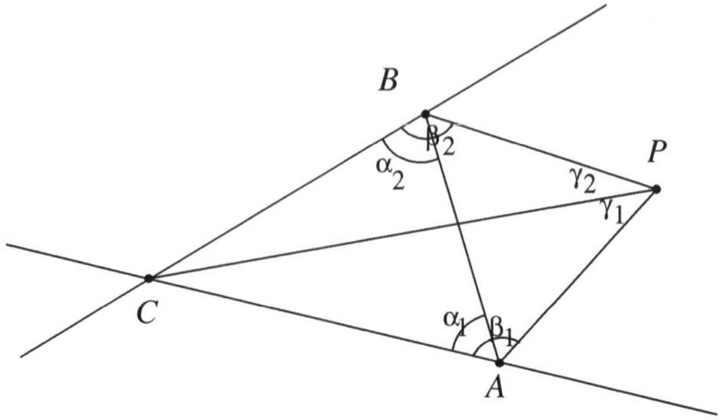

Fig. 6. the case $P \in W$

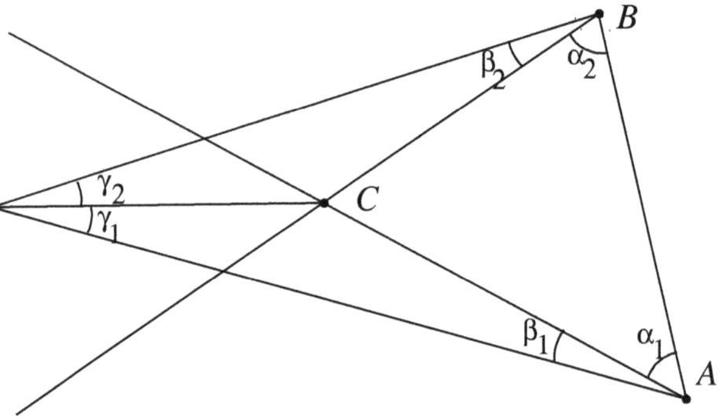

Fig. 7. the case $P \in V$

Proof. If $\beta_1 = \delta$ (δ is fixed) then P belongs to the union of the two lines going through the vertex A and making the angle δ with the line AB. Similarly, the condition $\beta_2 = \delta$ implies that P belongs to the union of two fixed lines going through B. Consequently, if $\beta_1 = \delta_1$ and $\beta_2 = \delta_2$ then P is one of the intersections of the corresponding lines. Since P is not on the line AB, we find that there are at most four such points. That is, the number of singular points P satisfying $\beta_1 = \delta_1$ and $\beta_2 = \delta_2$ is at most four.

If $\gamma_1 = \delta$ then P must belong to the union of two fixed circles going through C and A. Therefore, the condition $\beta_1 = \delta_1$, $\gamma_1 = \delta_2$ implies that P equals one of the intersections of two given lines going through A with two given circles containing A and C, allowing 4 possible points. We have 8 possible points if $\beta_1 = \delta_1$ and $\gamma_2 = \delta_2$. Similarly, the condition $\beta_2 = \delta_1, \gamma_1 = \delta_2$ gives 8 points, while $\beta_2 = \delta_1, \gamma_2 = \delta_2$ gives 4. If $\gamma_1 = \delta_1$ and $\gamma_2 = \delta_2$ then P is one of the intersections of two given circles going through A and C with two other given circles going through B and C. Being a singular point, P does not lie on the circle circumscribed to $ABC\triangle$, and thus the circles in question do not coincide, and give at most 4 possible points. Altogether we find 32 possible points; but we also have to take into consideration the cases when δ_1 and δ_2 are interchanged, amounting 64 possibilities. □

Since $\{A, B, C, P\}$ is a CRA, the angles $\alpha_i, \beta_i, \gamma_i$ $(i = 1, 2)$ are rational multiples of π, and satisfy (3). Therefore, by Theorem 6, one of (i)–(viii) holds (with (3) instead of (7)). We shall consider the cases in the order (i), (viii), (vii), (v), (ii), (vi), (iv), (iii). In the sequel, when we say that a certain set is finite, we shall always mean that the number of elements of the set in question is, in fact, bounded by an explicit bound.

First note that each of the angles $\alpha_i, \beta_i, \gamma_i$ $(i = 1, 2)$ is strictly between 0 and π, and thus each factor in (3) is positive. In particular, (i) of Theorem 6 cannot occur. If (viii) of Theorem 6 holds then each of $\alpha_i, \beta_i, \gamma_i$ $(i = 1, 2)$ belongs to a given finite set of angles and thus, by Lemma 8, the number of corresponding singular points is finite. A similar argument works in case of (vii). Indeed, in this case the numbers $\sin \pi a_i$, $\sin \pi b_i$ $(i = 1, 2)$ belong to a given finite set S, and thus at least two of the numbers $\sin \beta_i$, $\sin \gamma_i$ $(i = 1, 2)$ also belong to S. Since $\beta_i, \gamma_i \in (0, \pi)$ we conclude that two of β_i, γ_i $(i = 1, 2)$ belong to a given finite set and thus, by Lemma 8, the number of corresponding singular points is finite.

Next suppose (v). The number of possible identities (9) is finite. Fix any of them, and suppose that (3) is equivalent to it. Then $\sin \alpha_1$ or $\sin \alpha_2$ must be equal to one of the numbers $\pm \sin \pi(n_i x + a_i)$ $(i = 4, 5, 6)$, and thus the number of possible values of $x \pmod 1$ is finite. Therefore β_i, γ_i $(i = 1, 2)$ belong to a given finite set, and we may conclude, as before, that the set of corresponding singular points is finite.

Now suppose that (ii) of Theorem 6 holds. Then we have

$$\{\sin \alpha_2, \sin \beta_1, \sin \gamma_2\} = \{\sin \alpha_1, \sin \beta_2, \sin \gamma_1\}.$$

If $\sin \alpha_1 \neq \sin \alpha_2$ then two of $\beta_1, \beta_2, \gamma_1, \gamma_2$ equal some of $\alpha_1, \alpha_2, \pi - \alpha_2, \pi - \alpha_2$, and then, by Lemma 8, the number of corresponding singular points P is finite. Therefore we may assume $\sin \alpha_1 = \sin \alpha_2$. Since $\alpha_1 + \alpha_2 < \pi$, this implies $\alpha_1 = \alpha_2$. Thus $ABC\triangle$ is isosceles, and $\overline{AC} = \overline{BC}$. We claim that in this case P must be a regular point. If $\beta_1 = \beta_2$ then P lies on the symmetry axis of $ABC\triangle$, hence P is regular. (At this point we used the assumption that $P \in T \cup V \cup W$.) Therefore we may assume that $\beta_1 \neq \beta_2$.

First suppose $\sin \beta_1 = \sin \beta_2$ and $\sin \gamma_1 = \sin \gamma_2$. Then we have $\beta_1 + \beta_2 = \pi$. If $P \in T$ then $\beta_1 < \alpha_1$ and $\beta_2 < \alpha_2$ imply $\beta_1 + \beta_2 < \pi$, which is impossible. If $P \in W$ then $\gamma_1 + \gamma_2 < \pi$ gives $\gamma_1 = \gamma_2$. Since

$$\alpha_1 + \alpha_2 = \beta_1 + \beta_2 + \gamma_1 + \gamma_2 - \pi = \gamma_1 + \gamma_2$$

and $\alpha_1 = \alpha_2$, $\gamma_1 = \gamma_2$, we obtain $\alpha_1 = \alpha_2 = \gamma_1 = \gamma_2$. In particular, we have $\gamma_2 = \alpha_1$. Then P lies on the circle circumscribed to $ABC\triangle$, and thus P must be regular. In the case $P \in V$ we have $\beta_1 + \beta_2 < \pi$, which is impossible.

Next suppose $\sin\beta_1 = \sin\gamma_1$ and $\sin\beta_2 = \sin\gamma_2$. If $P \in T$ then $\beta_1 + \gamma_1 < \pi$, $\beta_2 + \gamma_2 < \pi$ give $\beta_1 = \gamma_1$, $\beta_2 = \gamma_2$. Then we have $\beta_1 < \gamma_2 = \beta_2 < \gamma_1 = \beta_1$, which is impossible. If $P \in W$ then $\beta_1 + \gamma_1 < \pi$, $\beta_2 + \gamma_2 < \pi$ give $\beta_1 = \gamma_1$, $\beta_2 = \gamma_2$. Then the triangle ACP is isosceles and $\overline{PC} = \overline{AC} = \overline{BC}$. Thus P lies on the circle with center C and radius $\overline{AC} = \overline{BC}$, hence P is regular. Finally, if $P \in V$ then $\beta_1 + \beta_2 + \gamma_1 + \gamma_2 < \pi$ implies $\beta_1 = \gamma_1$, $\beta_2 = \gamma_2$, and we have the same conclusion. This completes the investigation of case (ii).

In order to deal with the remaining cases (vi), (iv) and (iii) we shall need three lemmas.

Lemma 9. (i) *If* $\sin\alpha_1 = 2\sin\alpha_2 \cdot \sin\beta_1$ *and* $\beta_2 + \gamma_1 + \gamma_2 = \frac{3}{2}\pi$ *then the point* P *is regular.*

(ii) *If* $\sin\alpha_1 = 2\sin\alpha_2 \cdot \sin\gamma_2$ *and* $\beta_1 + \beta_2 + \gamma_1 = \frac{3}{2}\pi$ *then the point* P *is regular.*

Proof. Let $\eta = \beta_1$ if (i) holds and $\eta = \gamma_2$ in case (ii); then $\sin\alpha_1 = 2\sin\alpha_2 \cdot \sin\eta$. Since, in both cases, $\beta_1 + \beta_2 + \gamma_1 + \gamma_2 > \pi$, $P \in V$ is impossible. Therefore we have

$$\alpha_1 + \alpha_2 = \beta_1 + \beta_2 + \gamma_1 + \gamma_2 - \pi = \eta + \frac{\pi}{2}, \tag{15}$$

and thus $\sin\eta = -\cos(\alpha_1 + \alpha_2)$. Then

$$\sin\alpha_1 = -2\cos(\alpha_1 + \alpha_2) \cdot \sin\alpha_2 = -2\cos\alpha_1 \cos\alpha_2 \sin\alpha_2 + 2\sin\alpha_1 \sin^2\alpha_2,$$

$$\sin\alpha_1 \cdot (1 - 2\sin^2\alpha_2) = -\cos\alpha_1 \cdot \sin 2\alpha_2,$$

and hence

$$\sin\alpha_1 \cdot \cos 2\alpha_2 = -\cos\alpha_1 \cdot \sin 2\alpha_2. \tag{16}$$

If $\cos 2\alpha_2 = 0$ then $\sin 2\alpha_2 = 0$ by (16) which is impossible. Thus $\cos 2\alpha_2 \neq 0$, and then (16) gives $\tan\alpha_1 = -\tan 2\alpha_2$. Since $\alpha_1 + 2\alpha_2 < 2(\alpha_1 + \alpha_2) < 2\pi$, we obtain $2\alpha_2 = \pi - \alpha_1$, and thus

$$BCA\measuredangle = \pi - \alpha_1 - \alpha_2 = \alpha_2.$$

Therefore $ABC\triangle$ is isosceles and $\overline{AB} = \overline{AC}$. We also have, by (15),

$$\eta = \alpha_1 + \alpha_2 - \frac{\pi}{2} = \alpha_1 + \left(\frac{\pi}{2} - \frac{\alpha_1}{2}\right) - \frac{\pi}{2} = \frac{\alpha_1}{2}.$$

Since $\eta \in \{\beta_1, \gamma_2\}$ by assumption, we find that either $\beta_1 = \alpha_1/2$ or $\gamma_2 = \alpha_1/2$.

Suppose (i). Then $\beta_1 = \alpha_1/2 < \alpha_1$ and thus $P \in W$ is impossible. Since, as we noted already, $P \in V$ is also excluded, we have $P \in T$. Then $\beta_1 = \alpha_1/2$ implies that P lies on the symmetry axis of $ABC\triangle$, and thus P is regular.

Now suppose (ii). Then $\gamma_2 = \alpha_1/2 < \alpha_1$, and hence $P \in T$ is impossible. Thus $P \in W$. Since

$$CPB\measuredangle = \gamma_2 = \alpha_1/2 = \frac{1}{2}CAB\measuredangle,$$

we conclude that P lies on the circle with center A and radius $\overline{AB} = \overline{AC}$. Hence P is regular again. □

Lemma 10. *Let u, v, w, x be real numbers such that*

$$u \in \{2x,\ \pi - 2x\},\quad v \in \{x,\ \pi - x\},\quad w \in \left\{\frac{\pi}{2} - x, \frac{\pi}{2} + x\right\}.$$

Then either $u+v+w = \frac{3}{2}\pi$ or $u+v+w = a\cdot\frac{\pi}{2}+b\cdot x$, where a, b are integers, $|a|$, $|b| \le 5$ and $b \neq 0$.

Proof: by inspection of all cases. □

Lemma 11. *To every finite set S of real numbers there corresponds a finite subset S' of the plane with the following property: whenever P is a singular point of $ABC\triangle$ such that (3) is equivalent to an equation*

$$u \cdot v \cdot \sin 2\pi x = w \cdot \sin \pi x \cdot \cos \pi x$$

such that u, v, $w \in S$, then $P \in S'$.

Proof. Replacing x by an $x' \in (0, 1/2)$ such that $x \equiv \pm x' \pmod 1$, we may assume that $x \in (0, 1/2)$, and thus the numbers $\sin 2\pi x$, $\sin \pi x$, $\cos \pi x$ are positive. If any of $\sin 2\pi x$, $\sin \pi x$, $\cos \pi x$ equals any of the numbers $\sin \alpha_1$, $\sin \alpha_2$, then the number of possible values of x is finite. Then the number of possible equations is finite, and we may apply Lemma 8. Therefore we may assume that each of the numbers $\sin 2\pi x$, $\sin \pi x$, $\cos \pi x$ equals one of $\sin \beta_i$ or $\sin \gamma_i$ $(i = 1, 2)$. By symmetry we may assume that

$$\{\sin \pi x,\ \cos \pi x\} = \{\sin \beta_2,\ \sin \gamma_1\}.$$

Then we have either $\sin \beta_1 = \sin 2x$ and $\sin \alpha_1 = 2 \sin \gamma_2 \cdot \sin \alpha_2$, or $\sin \gamma_2 = \sin 2x$ and $\sin \alpha_1 = 2 \sin \beta_1 \cdot \sin \alpha_2$. We only consider the first case, the other being similar.

If $\sin \beta_1 = \sin 2x$ then we have $\beta_1 \in \{2x,\ \pi - 2x\}$, one of β_2 and γ_1 belongs to $\{x,\ \pi - x\}$, and the other belongs to $\left\{\frac{\pi}{2} - x, \frac{\pi}{2} + x\right\}$. Then, by Lemma 10, $\beta_1 + \beta_2 + \gamma_1$ equals either $\frac{3}{2}\pi$ or $a \cdot \frac{\pi}{2} + b \cdot x$, where a, b are integers, $|a|$, $|b| \le 5$ and $b \neq 0$. In the first case we may apply (ii) of Lemma 9, and obtain that P is regular. In the second case we find that

$$\begin{aligned} a \cdot \frac{\pi}{2} + b \cdot x &= (\beta_1 + \beta_2 + \gamma_1 + \gamma_2) - \gamma_2 \\ &= [\pm(\alpha_1 + \alpha_2) + \pi] - \gamma_2. \end{aligned} \tag{17}$$

Now $\sin \alpha_1 = 2 \sin \gamma_2 \cdot \sin \alpha_2$ and $\pm \sin \alpha_1, \pm \sin \alpha_2 \in S$ imply that the set of possible values of γ_2 is finite. Therefore, by (17), the same is true for x, and then an application of Lemma 8 completes the proof. □

Now we consider the cases (vi), (iv) and (iii). First suppose (vi). Since the set of numbers $\sin a_i$, $\sin b_i$ $(i = 1, 2)$ is finite, Lemma 11 takes care of this case.

Next suppose that (iv) holds; that is, (3) is equivalent to the equation $u \cdot (1/2) \cdot \sin 2\pi x = u \cdot \sin \pi x \cdot \cos \pi x$. If any of the numbers $\sin 2\pi x$, $\sin \pi x$, $\cos \pi x$ equals any of the numbers $\sin \alpha_1$, $\sin \alpha_2$, then the number of possible values of x is finite. Since at least two of the numbers $\sin \beta_i$, $\sin \gamma_i$ $(i = 1, 2)$ equals some of the numbers $\frac{1}{2}$, $\sin 2\pi x$, $\sin \pi x$, $\cos \pi x$, we find that in this case two of the numbers β_i, γ_i $(i = 1, 2)$ belong to a given finite set, and we may apply Lemma 8. Otherwise we have $u = \pm \sin \alpha_1$ or $u = \pm \sin \alpha_2$ and then we may apply Lemma 11.

Finally, suppose that (iii) holds. In this case $\sin \alpha_1$ equals one of the factors in (8). Therefore either x or y must belong to a given finite set (mod 1). Then another application of Lemma 11 completes the proof. □

4 Trigonometric Diophantine equations

As we mentioned in Section 2, Theorem 5 is essentially contained in [2] and, in fact, it is an easy consequence of [2, Theorem 3]. The proof given in [2] depends on [2, Lemma 1] and [2, Theorem 2]. However, the proof of [2, Lemma 1] is not correct (see the remark following Lemma 3 in [5, p. 73]), and we have to replace this lemma by Mann's theorem [4, Theorem 1]. On the other hand, as we shall see, [2, Theorem 2] is not needed for the proof. We shall use the notation $e(x) = e^{2\pi i x}$. The product of primes up to k will be denoted by P_k. We shall say that $\sum_{i=1}^k c_i$ is a *minimal vanishing sum* if, for every $\varepsilon_i \in \{0, 1\}$ we have $\sum_{i=1}^n \varepsilon_i c_i = 0 \iff \varepsilon_1 = \varepsilon_2 = \ldots = \varepsilon_k$.

Suppose that $S = \sum_{\nu=1}^k a_\nu \zeta_\nu$ is a minimal vanishing sum, where the a_ν are rational numbers and the ζ_ν are roots of unity. Then Mann's theorem [4, Theorem 1] states that there are roots of unity ω and η_ν $(\nu = 1, \ldots, k)$ such that each η_ν is a P_kth root of unity, and $\zeta_\nu = \omega \eta_\nu$ $(\nu = 1, \ldots, k)$. Let $\eta_\nu = e(c_\nu)$, where $P_k c_\nu \in \mathbf{Z}$ $(\nu = 1, \ldots, k)$. Since $\sum_{\nu=1}^k a_\nu e(c_\nu) = \sum_{\nu=1}^k a_\nu \zeta_\nu / \omega = S/\omega = 0$, we obtain the following corollary. *If $\sum_{\nu=1}^k a_\nu \zeta_\nu$ is a minimal vanishing sum, where the a_ν are rational numbers and the ζ_ν are roots of unity, then there are rational numbers c_ν such that $P_k c_\nu \in \mathbf{Z}$ for every ν, $\sum_{\nu=1}^k a_\nu e(z + c_\nu) = 0$ for every z, and $\zeta_\nu = e(z + c_\nu)$ for every $\nu = 1, \ldots, k$ for a suitable rational value of z.*

Proof of Theorem 5. Let $P \in \mathbf{Q}[x_1, \ldots, x_n]$ be a nonzero polynomial with rational coefficients. We shall consider the equation

$$P(\cos 2\pi x_1, \ldots, \cos 2\pi x_n) = 0 \tag{18}$$

instead of (5). Of course, this change is irrelevant, as the substitution $x \mapsto \frac{1}{2} - 2x$ turns $\sin \pi x$ into $\cos 2\pi x$. Applying the identities $\cos 2\pi x = (e(x) + e(-x))/2$ and $e(x) \cdot e(y) = e(x+y)$, we obtain that

$$P(\cos 2\pi x_1, \ldots, \cos 2\pi x_n) \equiv \sum_{\nu=1}^{N} a_\nu e(A_\nu(x)), \tag{19}$$

where $a_1, \ldots, a_N$ are nonzero rational numbers and $A_\nu(x) = A_\nu(x_1, \ldots, x_n)$ are linear forms with integer coefficients. Suppose that $p = (p_1, \ldots, p_n)$ is a rational solution of (18). Then, by (19) we obtain $\sum_{\nu=1}^{N} a_\nu \zeta_\nu = 0$, where $\zeta_\nu = e(A_\nu(p))$ $(\nu = 1, \ldots, N)$. Fix a partition of the sum $\sum_{\nu=1}^{N} a_\nu \zeta_\nu$ into minimal vanishing sums, and let $S_1, \ldots, S_K$ be an enumeration of these sums. Let I_k denote the set of indices ν for which $a_\nu \zeta_\nu$ belongs to S_k $(k = 1, \ldots, K)$. As we saw above, there are rational numbers $c_1, \ldots, c_N$ and $z_1, \ldots, z_K \in [0,1)$ with the following properties.

$$P_N c_\nu \in \mathbf{Z} \quad (\nu = 1, \ldots, N); \tag{20}$$

for every $k = 1, \ldots, K$ we have $\sum_{\nu \in I_k} a_\nu e(z + c_\nu) = 0$ for every z; and $\zeta_\nu = e(z_k + c_\nu)$ for every $\nu \in I_k$. Since $\zeta_\nu = e(A_\nu(p))$, we obtain $A_\nu(p) \equiv z_k + c_\nu \pmod 1$ for every $\nu \in I_k$. By adding suitable integers to c_ν, we may assume that

$$A_\nu(p) = z_k + c_\nu \quad (\nu \in I_k,\ k = 1, \ldots, K). \tag{21}$$

Let $y = (y_1, \ldots, y_K)$, and put $B_\nu(y) = y_k$ for every $k = 1, \ldots, K$ and $\nu \in I_k$. Then we have

$$\sum_{\nu=1}^{N} a_\nu e(B_\nu(y) + c_\nu) = 0 \tag{22}$$

for every $y \in \mathbf{Q}^K$.

Let $A = (A_1, \ldots, A_N)$ and $B = (B_1, \ldots, B_N)$, then $A : \mathbf{Q}^n \to \mathbf{Q}^N$ and $B : \mathbf{Q}^K \to \mathbf{Q}^N$ are linear maps. Put $X = \{x \in \mathbf{Q}^n : \exists\, y \in \mathbf{Q}^K,\ A(x) = B(y)\}$; then X is a linear subspace of $\mathbf{Q}^n$ as a vector space over $\mathbf{Q}$. Then there are linear maps $L : \mathbf{Q}^n \to \mathbf{Q}^n$ and $M : \mathbf{Q}^n \to \mathbf{Q}^K$ such that $A \circ L = B \circ M$, and $X = \operatorname{Im} L$.

Indeed, let $u^1, \ldots, u^n$ be a generating system of X, and choose vectors $v^1, \ldots, v^n$ such that $A(u^j) = B(v^j)$ $(j = 1, \ldots, n)$. It is easy to check that $L(t) = u^1 t_1 + \ldots + u^n t_n$ and $M = v^1 t_1 + \ldots + v^n t_n$ satisfy the requirements. Multiplying L and M with the common denominator of the coordinates of u^j and v^j we may assume that the coefficients of each component of L and M are integers.

Put $c = (c_1, \ldots, c_N)$, and fix a pair of elements $x^0 \in \mathbf{Q}^n$ and $y^0 \in \mathbf{Q}^K$ such that $A(x^0) = B(y^0) + c$. (It follows from (21) that there are elements with this property.) Then we have

$$A(L(t) + x^0) = B(M(t) + y^0) + c \tag{23}$$

for every $t \in \mathbf{Q}^n$. Let $L = (L_1, \dots, L_n)$. Then it follows from (19), (23), and (22) that

$$P(\cos 2\pi(L_1(t) + x_1^0), \dots, \cos 2\pi(L_n(t) + x_n^0)) = 0$$

for every t. Since $p - x_0 \in X$ by (21) and by $A(x^0) = B(y^0) + c$, we find that $p = L(t) + x^0$ for a suitable t. That is, the map $\Lambda = L + x^0$ satisfies the conditions.

In this way we constructed a map Λ starting from an arbitrary rational solution p of (18). In order to complete the proof, we have to check that the number of maps constructed is finite, at least if we restrict ourselves to those solutions $p = (p_1, \dots, p_n)$ for which $p_i \in [0, 1)$ for every $i = 1, \dots, n$.

Note that the linear forms A_ν are fixed (they only depend on the polynomial P). Therefore, supposing $p_i \in [0, 1)$ for $i = 1, \dots, n$, the numbers $A_\nu(p)$ are bounded. Then (21) implies that the numbers c_ν are bounded as well, since $z_k \in [0, 1)$. Then it follows from (20) that the set of possible N-tuples $c = (c_1, \dots, c_N)$ is finite. The map B only depends on the partition $S_1, \dots, S_K$, and thus the set of possible maps B is also finite. Since x^0, y^0 and L only depend on c and B, we conclude that the number of maps $\Lambda = L + x^0$ is finite. □

Now we turn to the Diophantine solutions of the equation

$$\sin \pi x_1 \cdots \sin \pi x_s = r \cdot \sin \pi x_{s+1} \cdots \sin \pi x_{s+t}, \tag{24}$$

where s, t are nonnegative integers and r is a nonzero rational number. We put $s + t = n$ and $u = (u_1, \dots, u_n)$.

Theorem 12. *There exists a finite set T of maps $\Lambda = (\Lambda_1, \dots, \Lambda_n) : \mathbf{Q}^n \to \mathbf{Q}^n$ with the following properties.*

(i) *Each component Λ_i of every $\Lambda \in T$ is a linear function of the form $mu_\nu + c$, where m is a nonnegative integer, c is rational and $1 \le \nu \le n$;*

(ii) *$p_1 = \Lambda_1(u), \dots, p_n = \Lambda_n(u)$ is a solution of (24) for every $\Lambda \in T$ and $u \in \mathbf{Q}^n$; and*

(iii) *whenever $(p_1, \dots, p_n)$ is a rational solution of (24) then there is a $\Lambda \in T$ and there are rational numbers $u_1, \dots, u_n$ such that $p_i \equiv \Lambda_i(u) \pmod 2$ for every $i = 1, \dots, n$.*

Proof. We put

$$P(x_1, \dots, x_n) = x_1 \cdots x_s - r \cdot x_{s+1} \cdots x_n \tag{25}$$

and apply Theorem 5. Then we obtain a finite set S of maps Λ satisfying (i)–(iii) of Theorem 5. We shall prove that for every $\Lambda \in S$ there is a map

Λ' such that $\operatorname{Im}\Lambda' \supset \operatorname{Im}\Lambda$ and Λ' satisfies (i) and (ii) of Theorem 12. Then the set T of these maps Λ' will satisfy the requirements.

Let $\Lambda \in S$ be fixed. Let $\Lambda = (L_1 + c_1, \ldots, L_n + c_n)$, where $L_1, \ldots, L_n$ are linear forms with integer coefficients involving the variables $u_1, \ldots, u_n$. We put $A_i = \{u \in \mathbf{Q}^n : L_i(u) + c_i \in \mathbf{Z}\}$; then A_i is the set of rational roots of the map $\sin \pi(L_i(u) + c_i)$ for every $i = 1, \ldots, n$. Since

$$\prod_{i=1}^{s} \sin \pi(L_i(u) + c_i) = r \cdot \prod_{i=s+1}^{n} \sin \pi(L_i(u) + c_i) \tag{26}$$

for every u, we have

$$\bigcup_{i=1}^{s} A_i = \bigcup_{i=s+1}^{n} A_i. \tag{27}$$

It is clear that $A_i = \emptyset$ if $L_i \equiv 0$ and $c_i \notin \mathbf{Z}$; $A_i = \mathbf{Q}^n$ if $L_i \equiv 0$ and $c_i \in \mathbf{Z}$; and A_i is a countable union of hyperplanes if $L_i \not\equiv 0$. Note that in the latter case these hyperplanes form a discrete system; that is, every point of $\mathbf{Q}^n$ has a neighbourhood that intersects at most one of these hyperplanes. Since $\mathbf{Q}^n$ cannot be covered by a discrete system of hyperplanes, it follows from (27) that if $L_i \equiv 0$ and $c_i \in \mathbf{Z}$ for some $i \le s$ then there is a $j > s$ such that $L_j \equiv 0$ and $c_j \in \mathbf{Z}$. Suppose, for example, that $L_1 \equiv 0$, $c_1 \in \mathbf{Z}$ and $L_{s+1} \equiv 0$, $c_{s+1} \in \mathbf{Z}$. Then we define $\Lambda'_1 \equiv \Lambda'_{s+1} \equiv 0$ and $\Lambda'_i = u_i$ for every $i \ne 1, s+1$. It is obvious that $\Lambda' = (\Lambda'_1, \ldots, \Lambda'_n)$ satisfies the requirements.

Therefore we may assume that, for every i, either $L_i \not\equiv 0$ or $c_i \notin \mathbf{Z}$. Let $\mathcal{A} = (A_1, \ldots, A_s)$ and $\mathcal{B} = (A_{s+1}, \ldots, A_n)$. We prove that the pair $(\mathcal{A}; \mathcal{B})$ is exact. (See the definition preceding Theorem 7.) Let $u^0 \in \mathbf{Q}^n$ be arbitrary, and choose a vector $v^0 \in \mathbf{Q}^n$ such that $L_i(v^0) \notin \mathbf{Z}$ for every i for which $L_i \not\equiv 0$. Then the function

$$x \mapsto f_i(x) = \sin \pi(L_i(x \cdot v^0 + u^0) + c_i) \quad (x \in \mathbf{R})$$

has a root at $x = 0$ if and only if $u^0 \in A_i$. Also, if $f_i(0) = 0$ then 0 is a simple root. Since $f_1 \cdots f_s = r \cdot f_{s+1} \cdots f_n$ by (26), we conclude that the number of indices $i \le s$ satisfying $u^0 \in A_i$ equals the number of indices $i > s$ with $u^0 \in A_i$. In other words, the pair $(\mathcal{A}; \mathcal{B})$ is exact.

Let $L(u)$ be a nonzero linear form, and suppose that at least one of the forms L_i is a nonzero constant multiple of L. Using a suitable permutation of the indices we may assume that there are indices $s_1 \le s$ and $t_1 \le t$ such that L_i is a nonzero constant multiple of L if and only if i belongs to the set $I_1 = \{1, \ldots, s_1, s+1, \ldots, s+t_1\}$. We claim that the sequences $\mathcal{C} = (A_1, \ldots, A_{s_1})$ and $\mathcal{D} = (A_{s+1}, \ldots, A_{s+t_1})$ form an exact pair $(\mathcal{C}; \mathcal{D})$. Indeed, let $u^0 \in \mathbf{Q}^n$ be arbitrary. Choose a vector $w^0 \in \mathbf{Q}^n$ such that $L(w^0) = 0$, but $L_i(w^0) \ne 0$ whenever $1 \le i \le n$, $i \notin I_1$, and $L_i \not\equiv 0$. If $u^0 \in A_i$ where $i \notin I_1$ then $u^0 + y \cdot w^0 \notin A_i$ if $y \in \mathbf{Q}$ is positive and small enough. (Note that if $L_i \equiv 0$ then $A_i = \emptyset$.) On the other hand, if $u^0 \in A_i$ where $i \in I_1$, then $u^0 + y \cdot w^0 \in A_i$ for every y. Since $(\mathcal{A}; \mathcal{B})$ is exact, the sets of indices $\{i \le s : u^0 + y \cdot w^0 \in A_i\}$ and $\{s < i \le n : u^0 + y \cdot w^0 \in A_i\}$ have the same number of elements for every $y \in \mathbf{Q}$. Choosing a small positive y we obtain that the same is true for the sets $\{i \le s_1 : u^0 \in A_i\}$ and $\{s < i \le s + t_1 : u^0 \in A_i\}$. That is, the pair $(\mathcal{C}; \mathcal{D})$ is exact.

Let $L_i = m_i \cdot L$ for every $i \in I_1$. Dividing L by a suitable integer we may assume that the numbers m_i are nonzero integers for every $i \in I_1$. We prove that

$$\begin{aligned} \sin \pi(m_1 x + c_1) \cdots \sin \pi(m_{s_1} x + c_{s_1}) = \\ \pm 2^{t_1 - s_1} \sin \pi(m_{s+1} x + c_{s+1}) \cdots \sin \pi(m_{s+t_1} x + c_{s+t_1}) \end{aligned} \tag{28}$$

is an identity (with a suitable choice of the sign). By Theorem 7 it is enough to show that the pair

$$((E_1, \ldots, E_{s_1}); (E_{s+1}, \ldots, E_{s+t_1})) \tag{29}$$

is exact, where $E_i = \{(k - c_i)/m_i : k \in \mathbf{Z}\}$ $(i \in I_1)$. Now for every $u \in \mathbf{Q}^n$ and $i \in I_1$ we have $L(u) \in E_i \iff u \in A_i$. Since $(\mathcal{C}; \mathcal{D})$ is exact and L is nonzero, we conclude that the pair in (29) is exact. Therefore, (28) is an identity.

Suppose that $L_i \not\equiv 0$ for some $i \notin I_1$. We may assume that there are indices $s_1 < s_2 \le s$, $t_1 < t_2 \le t$ and there is a nonzero linear form M such that L_i is a nonzero constant multiple of M if and only if $i \in I_2 = \{s_1 + 1, \ldots, s_2, s + t_1 + 1, \ldots, s + t_2\}$, and that $L_i = m_i \cdot M$ for every $i \in I_2$, where the m_i are integer. Repeating the argument above we obtain that

$$\begin{aligned} \sin \pi(m_{s_1+1} x + c_{s_1+1}) \cdots \sin \pi(m_{s_2} x + c_{s_2}) = \\ \pm 2^{t_2 - s_2} \sin \pi(m_{s+t_1+1} x + c_{s+t_1+1}) \cdots \sin \pi(m_{s+t_2} x + c_{s+t_2}) \end{aligned} \tag{30}$$

is an identity.

We continue this way until we exhaust all indices i with $L_i \not\equiv 0$. Suppose we reach this stage at the kth step. Then we have indices $0 = s_0 < s_1 < \ldots < s_k \le s$, $0 = t_0 < t_1 < \ldots < t_k \le t$ and nonzero linear forms M_j such that

(i) $L_i \equiv 0$ for every $s_k < i \le s$ and $s + t_k < i \le s + t$;

(ii) $L_i = m_i \cdot M_j$ for every $j = 1, \ldots, k$, $s_{j-1} < i \le s_j$ and $s + t_{j-1} < i \le s + t_j$; and

(iii) an identity analogous to (28) and (30) holds for every $j = 1, \ldots, k$.

Now we define $\Lambda'_i(u) = m_i \cdot u_j + c_i$ for every $j = 1, \ldots, k$, $s_{j-1} < i \le s_j$ and $s + t_{j-1} < i \le s + t_j$; and $\Lambda'_i(u) = c_i$ for every $s_k < i \le s$ and $s + t_k < i \le s + t$. We claim that $p_1 = \Lambda'_1(u), \ldots, p_n = \Lambda'_n(u)$ is a solution of (24) for every $u \in \mathbf{Q}^n$. Indeed, it follows from the identities (28), (30), etc. that

$$\frac{\sin \pi \Lambda'_1(x) \cdots \sin \pi \Lambda'_n(x)}{\sin \pi \Lambda'_{s+1}(x) \cdots \sin \pi \Lambda'_n(x)}$$

is constant; that is, has the same value for every $x = (x_1, \ldots, x_n)$. Substituting $x_j = M_j(u)$ we find that this constant must be r. Therefore the map $\Lambda' = (\Lambda'_1, \ldots, \Lambda'_n)$ satisfies the requirements, apart from the fact that the coefficients m_i are not necessarily nonnegative. But this is irrelevant, since replacing the pair (m_i, c_i) by $(-m_i, c_i + 1)$, the value of the function $\sin \pi(m_i x + c_i)$ does not change. □

5 Proof of Theorem 6

First we need some simple facts about arithmetical progressions. The set $\{dk + b : k \in \mathbf{Z}\}$ will be denoted by $(d : b)$. A subset A of $\mathbf{R}$ is called an *arithmetical progression* (AP) if $A = (d : b)$ for some $d > 0$ and $b \in \mathbf{R}$. The smallest positive d such that $A = (d : b)$ for some b is the *difference* of A, denoted by $d(A)$. We say that the nonzero real number a *divides* the number b, and write $a \mid b$, if $b/a \in \mathbf{Z}$.

Lemma 13. *Let A, B, C be AP's such that $A \subset B \cup C$, but A is not covered by any of B, C. Then $d(B) \mid 2d(A)$ and $d(C) \mid 2d(A)$.*

Proof. We may assume that $A = \mathbf{Z}$ and that $0 \in B$. Since $1 \in B$ would imply $A \subset B$, we have $1 \in C$. By the same argument, $2 \in B$ and $3 \in C$. Then $0, 2 \in B$ gives $d(B) \mid 2$, and $1, 3 \in C$ implies $d(C) \mid 2$. □

Suppose that the pair $(\mathcal{A}; \mathcal{B})$ is exact, where $\mathcal{A}$ and $\mathcal{B}$ are finite sequences of sets. We shall say that the pair $(\mathcal{A}; \mathcal{B})$ is *indecomposable* if no subsequences $\mathcal{A}_1 \subset \mathcal{A}$, $\mathcal{B}_1 \subset \mathcal{B}$ form an exact pair $(\mathcal{A}_1; \mathcal{B}_1)$ except for $\mathcal{A}_1 = \mathcal{A}$, $\mathcal{B}_1 = \mathcal{B}$ and $\mathcal{A}_1 = \mathcal{B}_1 = \emptyset$.

Lemma 14. *Let $(\mathcal{A}, \mathcal{B})$ be an indecomposable exact pair, where*

$$\mathcal{A} = (A_1, \ldots, A_s), \; \mathcal{B} = (B_1, \ldots, B_t),$$

$s, t \le 3$, and the sets A_i, B_j are AP's. Then one of the following statements is true.

(i) *$s = t = 1$ and $A_1 = B_1$.*

(ii) *$s = 1$, $t = 2$, and there are $d \neq 0$ and b such that $A_1 = (d : b)$, $\mathcal{B} = \{(2d : b), (2d : d + b)\}$.*

(iii) *The same as (ii) with the roles of $\mathcal{A}$ and $\mathcal{B}$ interchanged.*

(iv) *$\max(s, t) = 3$, and there is an AP $(d : b)$ such that*

$$\bigcup_{i=1}^{s} A_i = \bigcup_{j=1}^{t} B_j \subset (d : b),$$

and each of the differences $d(A_i)$ and $d(B_j)$ divides $24d$ for every $i = 1, \ldots, s$ and $j = 1, \ldots, t$.

Proof. We may assume that $d(A_1)$ is (one of) the smallest among the numbers $d(A_i)$ and $d(B_j)$. We may also suppose that $A_1 = \mathbf{Z}$; therefore $d(A_i) \geq 1$ and $d(B_j) \geq 1$ for every i and j. Since $(\mathcal{A}, \mathcal{B})$ is exact, we have $A_1 \subset B_1 \cup \ldots \cup B_t$.

If $A_1 \subset B_j$ then, by $d(B_j) \geq 1$, we have $A_1 = B_j$. Since $(\mathcal{A}, \mathcal{B})$ is indecomposable, this implies (i). Therefore we may assume that A_1 is not covered by any of the AP's $B_1, \ldots, B_t$. Then $d(B_j) > 1$ whenever $B_j \cap A_1 \neq \emptyset$.

Suppose that $A_1 = \mathbf{Z}$ is covered by two of the B_j's, say, $A_1 \subset B_1 \cup B_2$. Then, by Lemma 13, $d(B_j) \,|\, 2$ $(j = 1, 2)$. However, $B_j \cap A_1 \neq \emptyset$ implies $d(B_j) > 1$, and thus $d(B_j) = 2$ $(j = 1, 2)$. Then A_1 is the disjoint union of B_1 and B_2. Since $(\mathcal{A}, \mathcal{B})$ is indecomposable, we find that $s = 1,\ t = 2$, and (ii) holds.

Therefore we may assume that $t = 3$ and that $A_1 = \mathbf{Z}$ is not covered by any of the sets $B_j \cup B_k$ $(j \neq k)$. Let $|H|$ denote the cardinality of the set H. We claim that $|\mathbf{Z} \cap B_j| \geq 2$ for every $j = 1, 2, 3$. It is clear that $\max_{1 \leq j \leq 3} |\mathbf{Z} \cap B_j| \geq 2$; we may assume that $|\mathbf{Z} \cap B_3| \geq 2$. Then $\mathbf{Z} \cap B_3$ is an AP. Since $\mathbf{Z} \cap B_3 \neq \mathbf{Z}$, it follows that $\mathbf{Z} \setminus B_3$ is infinite. As $\mathbf{Z} \setminus B_3 \subset B_1 \cup B_2$, we have $\max_{1 \leq j \leq 2} |\mathbf{Z} \cap B_j| \geq 2$. We may assume that $|\mathbf{Z} \cap B_2| \geq 2$, and thus $\mathbf{Z} \cap B_2$ is an AP. Then the set $D = \mathbf{Z} \cap (B_2 \cup B_3)$ is periodic. Since $D \neq \mathbf{Z}$, the set $\mathbf{Z} \setminus D$ is infinite and then so is $\mathbf{Z} \cap B_1 \supset \mathbf{Z} \setminus D$.

Then each of the sets $\mathbf{Z} \cap B_j$ $(j = 1, 2, 3)$ is an AP. Let $d_j = d(A_1 \cap B_j)$ $(j = 1, 2, 3)$. Then d_1, d_2, d_3 are integers, they are greater than 1, and satisfy $(1/d_1) + (1/d_2) + (1/d_3) \geq 1$. We may assume $d_1 \leq d_2 \leq d_3$. Then either $d_1 = d_2 = d_3 = 3$ or $d_1 = 2$. In the latter case $A_1 \setminus B_1$ is an AP covered by $B_2 \cup B_3$ but not covered by any of B_2, B_3. By Lemma 13 we find that in this case $d(B_2)$ and $d(B_3)$ both divide $2 \cdot d(A_1 \setminus B_1) = 4$. Since $d(B_j) \,|\, d(A_1 \cap B_j)$, we obtain that either $d(B_j) \,|\, 3$ $(j = 1, 2, 3)$ or $d(B_j) \,|\, 4$ $(j = 1, 2, 3)$. Now, $B_j \cap A_1 \neq \emptyset$ implies $d(B_j) > 1$, and then $d(B_j) \,|\, 3$ gives $d(B_j) \in \{3/2,\ 3\}$, while $d(B_j) \,|\, 4$ yields $d(B_j) \in \{4/3,\ 2,\ 4\}$. We conclude that either $d(B_j) \in \{3/2,\ 3\}$ for every $j = 1, 2, 3$, or $d(B_j) \in \{4/3,\ 2,\ 4\}$ $(j = 1, 2, 3)$.

We shall only consider the case $d(B_j) \in \{3/2,\ 3\}$ $(j = 1, 2, 3)$, the other case being analogous. Then we have $\bigcup_{i=1}^{s} A_i = \bigcup_{j=1}^{3} B_j \subset ((1/2) : 0)$. We shall prove that each of the numbers $d(A_i)$ divides 6, and thus (iv) holds with $d = 1/2$ and $b = 0$.

If $s = 1$ then we are done, since $d(A_1) = 1$. Suppose $s \geq 2$. Since

$$\left(\sum_{j=1}^{3} \chi_{B_j} \right) - \chi_{A_1} = \sum_{i=2}^{s} \chi_{A_i} \tag{31}$$

and each of the functions χ_{A_1} and χ_{B_j} $(j = 1, 2, 3)$ is periodic mod 3, it follows that $\sum_{i=2}^{s} \chi_{A_i}$ is also periodic mod 3. Let $x_0 \in A_2$. Then $\sum_{i=2}^{s} \chi_{A_i}(x_0) > 0$, and thus $\sum_{i=2}^{s} \chi_{A_i}(x) > 0$ for every $x \in (3 : x_0)$. Hence $(3 : x_0) \subset \bigcup_{i=2}^{s} A_i$. Therefore, by Lemma 13, one of the following statements is true.

- $s = 3$ and $d(A_i) \,|\, 6$ $(i = 2, 3)$. In this case the proof is finished.

• $(3 : x_0) \subset A_i$ for one of $2 \le i \le s$; we may assume that $(3 : x_0) \subset A_2$. Then $d(A_2) \mid 3$. If $s = 2$ then we are done. Suppose $s = 3$. Since

$$\chi_{A_3} = \sum_{j=1}^{3} \chi_{B_j} - (\chi_{A_1} + \chi_{A_2})$$

and each of the sets A_1, A_2, B_j $(j = 1, 2, 3)$ is periodic mod 3, the same is true for A_3. Therefore $d(A_3) \mid 3$, and the proof is complete. □

Now we turn to the proof of Theorem 6. Applying Theorem 12 with $s = t = 3$ and $r = 1$ we obtain a finite set T of maps Λ satisfying (i)–(iii) of Theorem 12. Let F denote the set of the constant terms of the components of all $\Lambda \in T$. Then F is a finite set of rationals. We put $F_3 = F^3$, $F_4 = F^4$ and $F_6 = F^6$. It is enough to show that for every $\Lambda \in T$ and $u \in \mathbf{Q}^6$, the equation

$$\begin{aligned}\sin \pi\Lambda_1(u) \cdot \sin \pi\Lambda_2(u) \cdot \sin \pi\Lambda_3(u) &= \\ &\sin \pi\Lambda_4(u) \cdot \sin \pi\Lambda_5(u) \cdot \sin \pi\Lambda_6(u)\end{aligned} \tag{32}$$

satisfies one of (i) - (viii) of Theorem 6 with (32) instead of (7).

Let $\Lambda \in T$ be fixed. If one of the factors $\sin \pi\Lambda_i(u)$ is identically zero then (i) holds. Therefore we may assume that this is not the case. Let $\Lambda_i(u) = m_i u_{\nu_i} + c_i$ $(1 \le i \le 6)$, and put $A_i = \{(k - c_i)/m_i : k \in \mathbf{Z}\}$ for every i with $m_i > 0$. Then, for every such index, A_i is an AP with $d(A_i) = 1/m_i$.

Let I_ν denote the set of indices i such that Λ_i depends on the variable u_ν. We may fix the other variables in such a way that each factor corresponding to the indices $i \notin I_\nu$ becomes a nonzero constant, and thus (32) becomes an identity in u_ν. Then, by Theorem 7, the sequences $\mathcal{A}_\nu = (A_i : i \le 3,\ i \in I_\nu)$ and $\mathcal{B}_\nu = (A_i : 4 \le i \le 6,\ i \in I_\nu)$ form an exact pair $(\mathcal{A}_\nu; \mathcal{B}_\nu)$. This is true for every ν, and thus the nonconstant factors of (32) can be partitioned into disjoint groups such that the factors of each group depend on the same variable, and the corresponding pair $(\mathcal{A}_\nu; \mathcal{B}_\nu)$ is exact. Partitioning each group further, if necessary, we may assume that each pair $(\mathcal{A}_\nu; \mathcal{B}_\nu)$ is exact and indecomposable. Now the cases (ii)–(viii) of Theorem 6 correspond to the possible partitions as follows.

Each pair $(\mathcal{A}_\nu; \mathcal{B}_\nu)$ satisfies one of (i)–(iv) of Lemma 14. We shall say that $(\mathcal{A}_\nu; \mathcal{B}_\nu)$ is of type (x), if it satisfies (x) of Lemma 14.

If $(\mathcal{A}_\nu; \mathcal{B}_\nu)$ is of type (i) and $\mathcal{A}_\nu = (A_i)$, $\mathcal{B}_\nu = (B_j)$, then $A_i = B_j$. It is easy to see that $A_i = B_j$ implies $\sin \pi\Lambda_i(u) = \pm \sin \pi\Lambda_j(u)$. If there are three groups then necessarily each pair is of type (i) and then (ii) of Theorem 6 holds. The same is true if there are two groups and both of the corresponding pairs is of type (i).

Suppose that $(\mathcal{A}_\nu; \mathcal{B}_\nu)$ is of type (ii), and let $\mathcal{A}_\nu = (A_i)$, $\mathcal{B}_\nu = (B_j, B_k)$. Then $A_i = B_j \cup B_k$ and $B_j \cap B_k = \emptyset$. In addition, $m_j = m_k$ and, denoting the common value by m we have $m_i = 2m$. Also, we have either $c_j \equiv c_i/2$ and

$c_k \equiv (c_i/2) + (1/2) \pmod 1$ or the other way around. Putting $x = mu_\nu + (c_i/2)$ we find that $\sin \pi\Lambda_i(u) = \sin 2\pi x$, one of $\sin\pi\Lambda_j(u)$ and $\sin\pi\Lambda_k(u)$ equals $\pm\sin\pi x$, and the other is $\pm\sin\pi(x+\frac{1}{2}) = \pm\cos\pi x$.

Therefore, if there are two groups and each pair is of type (ii) or (iii) then we have (iii) of Theorem 6. If one of the pairs is of type (i) and the other is of type (ii) or (iii) then we obtain (iv) of Theorem 6.

Next suppose that there is only one group. If the pair $(\mathcal{A}_\nu; \mathcal{B}_\nu)$ is of type (i) then we have (vii) of Theorem 6. If it is of type (ii) or (iii) then we get (vi) of Theorem 6. Suppose that $(\mathcal{A}_\nu; \mathcal{B}_\nu)$ is of type (iv). We may assume that $t = 3$; that is, each factor of the right hand side of (32) belongs to the group. Thus $m_i > 0$ for $i = 4, 5, 6$. We may also assume $m_1 > 0$. By (iv) of Lemma 14, there is an AP $(d : b)$ such that $A_i \subset (d : b)$ and $1/m_i = d(A_i) \mid 24d$ for every i for which A_i belongs to the group; that is, whenever $m_i > 0$. Putting $n_i = 24dm_i$, we find that $n_i \in \mathbf{Z}$ for every $i = 1, \ldots, 6$. If $m_i > 0$ then $A_i \subset (d : b)$ gives $d \mid d(A_i) = 1/m_i$, and thus $n_i \mid 24$. We put

$$x = \left(u_\nu + \frac{c_1}{m_1}\right) / (24d).$$

If $m_i > 0$ then, by $A_i \subset (d : b)$ and $A_1 \subset (d : b)$ we have

$$\frac{c_i}{m_i} - \frac{c_1}{m_1} = k_i d,$$

where $k_i \in \mathbf{Z}$. Therefore

$$m_i u_\nu + c_i = m_i\left(24dx - \frac{c_1}{m_1}\right) + c_i = n_i x + \frac{n_i k_i}{24}.$$

Thus we have (v) of Theorem 6 if $m_i > 0$ for every i. If $m_3 = 0$ but all other m_i are positive then it follows from Theorem 7 that $\sin\pi c_3 = \pm 1/2$. Thus $c_3 \equiv \pm 1/6 \pmod 1$, and (v) of Theorem 6 holds in this case as well. Next suppose $m_2 = m_3 = 0$. Then, by Theorem 7, $\sin\pi c_2 \cdot \sin\pi c_3 = \pm 1/4$. Therefore, by a theorem of M. Newman [6], c_2 and c_3 satisfy one of the following congruences (mod 1):

$$\begin{aligned} &c_2 \equiv \pm 1/6, \quad c_3 \equiv \pm 1/6; \\ &c_2 \equiv \pm 1/12, \ c_3 \equiv \pm 5/12; \\ &c_2 \equiv \pm 1/10, \ c_3 \equiv \pm 3/10; \end{aligned}$$

and the same relations with c_2 and c_3 interchanged. Then (v) of Theorem 6 holds again.

Finally, when each factor of (32) is constant, then we obtain (viii) of Theorem 6. □

References

[1] J. Beebee, Some trigonometric identities related to exact covers, *Proc. Amer. Math. Soc.* 112 (1991), 329-338.

[2] J. H. Conway and A. J. Jones, Trigonometric diophantine equations, *Acta Arithmetica* 30 (1976), 229-240.

[3] W. J. R. Crosby, Solution to problem 4136, *Amer. Math. Monthly* 53 (1946), 103-107.

[4] H. B. Mann, On linear relations between roots of unity, *Mathematika* 12 (1965), 107-117.

[5] G. Myerson, Rational products of sines of rational angles, *Aequationes Math.* 45 (1993), 70-82.

[6] M. Newman, A diophantine equation, *J. London Math. Soc.* 43 (1968), 105-107.

[7] M. Newman, Some results on roots of unity, with an application to a diophantine problem, *Aequationes Math.* 2 (1969), 163-166.

About the Author

M. Laczkovich is at the Department of Analysis of the Eötvös Loránd University, Budapest, Pázmány Péter sétány 1/C, Hungary 1117, and at the Department of Mathematics of the University College London, Gower Street, London WC1E 6BT, England; *laczko@renyi.hu*.

Acknowledgments

This work was completed when the author had a visiting position at the Mathematical Institute of the Hungarian Academy of Sciences. Also supported by the Hungarian National Foundation for Scientific Research Grant No. T032042.

Reconstructing Sets From Interpoint Distances

Paul Lemke
Steven S. Skiena
Warren D. Smith

Abstract

Which point sets realize a given distance multiset? Interesting cases include the "turnpike problem" where the points lie on a line, the "beltway problem" where the points lie on a loop, and multidimensional versions. We are interested both in the algorithmic problem of determining such point sets for a given collection of distances and the combinatorial problem of finding bounds on the maximum number of different solutions. These problems have applications in genetics and crystallography.

We give an extensive survey and bibliography in an effort to connect the independent efforts of previous researchers, and present many new results. In particular, we give improved combinatorial bounds for the turnpike and beltway problems. We present a pseudo-polynomial time algorithm as well as a practical $\mathbf{O}(2^n n \log n)$-time algorithm that find all solutions to the turnpike problem, and show that certain other variants of the problem are NP-hard. We conclude with a list of open problems.

1 Introduction

A set of n points in some space defines a set of distances between all pairs of points. In this paper we consider the inverse problem of constructing all point sets which realize a given distance multiset. The complexity of an algorithm to generate all such point sets depends upon the number of solutions, and so we are also interested in bounds on the maximum number of distinct solutions in a given space, as a function of n.

The problem dates back to the origins of X-ray crystallography in the 1930's [patt35] [picc39] [patt44]. More recently it has arisen in restriction site mapping of DNA, and was independently posed by M.I. Shamos [sham77] as a computational geometry problem. We encourage the reader to consult the recent thesis by Dakic [daki00] for the most recent results, including efforts to apply semi-definite programming to the problem. Pandurangan and Ramesh [pand01] have recent work on a variant of our problem which assumes additional information.

Spaces of particular interest include restricting the points to a line or a circular loop. The analogy of these points as exits on a road lead us to call these cases the "turnpike" and "beltway" problems, respectively. A turnpike problem instance consists of a multiset of $\binom{n}{2}$ distances; a beltway instance consists of a list of $(n-1)n$ distances.

It should be made clear that the correspondences between the distances and point pairs are *not* known and the entire difficulty of the reconstruction problem is to deduce such labeling information. If the labels are known, then given the $\binom{n}{2}$ labeled distances among n points in d-space, a suitable set of coordinates may be determined in $\mathbf{O}(n^2 d)$ time. Let the nth point lie at $\vec{0}$ and let the coordinates of the $n-1$ nonzero points be given by the columns of a $d \times (n-1)$ upper triangular matrix A. Define B by $B = A^T A$. Then $B_{ij} = \frac{1}{2}(q_{0i} + q_{0j} - q_{ij})$ for $1 \leq i, jn$, and $B_{ii} = q_{0i}$ where q_{ij} is the squared distance between points i and j. We may solve for A in terms of B consecutively column by column in time $\mathbf{O}(dn)$ per column, for a total runtime $\mathbf{O}(dn^2)$. If $d = n$, this algorithm is called the "Cholesky factorization" [golu83]. Alternately, we may find the "eigendecomposition" of $B = Q^T \Lambda Q$ where Λ is the diagonal matrix of $n-1$ real eigenvalues of B (in decreasing order; only the first d can be nonzero) and the columns of Q are the eigenvectors of B [golu83]. Q has orthonormal rows and columns. Then $\Lambda^{1/2} Q$ has d nonzero rows. Its $n-1$ columns give coordinates for our $n-1$ points. This approach has numerical advantages in situations in which our distances are contaminated by noise or roundoff error, because, e.g. the best approximation, in the Frobenius norm, of a symmetric matrix by a rank-d positive definite symmetric matrix is precisely its eigendecomposition with all eigenvalues besides the d largest ones, artificially zeroed ([golu83]; this is the symmetric case of the SVD approximation theorem).

1.1 Notation

$\mathbf{Z}$ is the set of integers and $\mathbf{R}$ the set of real numbers, while $\mathbf{S}^d$ denotes the unit d-sphere $\{\vec{x} \in \mathbf{R}^{d+1}\colon\ |\vec{x}| = 1\}$. We will use $(\mathbf{R}/\mathbf{Z})^d$ to denote a *flat d-torus,* i.e. the d-cube $[0,1]^d$ with opposite faces equivalenced. In X-ray crystallography, one must reconstruct a point set from its vector-differences modulo some d-parallelipiped unit cell; by taking an affine transformation this parallelipiped may be transformed to $(\mathbf{R}/\mathbf{Z})^d$. Our asymptotic notation $\mathbf{O}()$, $o()$, $\theta()$, $\sim$ follows [knut76].

An algorithm is said to run in *pseudo-polynomial time* if it runs in time polynomial in the size of its input, when this input consists of integers written as unary numbers [gare78]. Similarly, a problem is *strongly NP-complete* if it is still NP-complete even if the input is required to consist of unary integers. For convenience, we adopt the "real RAM" [prep85] as our model of computation in this paper, although we have taken care not to abuse its excessive power. All of our lower bounds, NP-completeness proofs, and undecidability proofs have been obtained by the explicit use of weaker models of computation polynomially equivalent to a Turing machine.

Two noncongruent n-point sets are *homometric* if the multisets of $\binom{n}{2}$ distances they determine are the same. A set of points in $\mathbf{R}^d$ are in *general position* if their coordinates do not satisfy any nontrivial algebraic equation with integer coefficients.

The function $H_d(n)$ denotes the maximum possible number of mutually noncongruent and homometric n-point sets which can exist in $\mathbf{R}^d$. $S_d(n)$ denotes the maximum possible number of such sets for which all points must lie on the sphere $\mathbf{S}^d$. $H_d^*(n)$ and $S_d^*(n)$ are defined in the same way, except that we allow 0-distances or, equivalently, overlapping points. Thus $H_d(n) \leq H_d^*(n)$.

1.2 Summary of results

Our results on the worst-case computational complexity of n-point reconstruction problems may be summarized as follows. All d-dimensional vector difference turnpike reconstructions can be found in $\mathbf{O}(2^n n \log n)$ time and all d-torus vector difference beltway reconstructions in $\mathbf{O}(dn^n \log n)$ time. These include the one-dimensional turnpike and beltway problems respectively as special cases. The turnpike problem is also soluble in pseudo-polynomial time by a different algorithm. The *decision* problem of whether a multiset of $\binom{n}{2}$ distances is realized by n points in $\mathbf{R}^d$ is NP-complete. Also shown NP-complete is the question of whether there exist n points in $\mathbf{R}$ whose distances are contained in $\binom{n}{2}$ given distance intervals.

Our bounds on $H_d(n)$ are summarized below. The lower bounds hold for an infinite number of values of n, while the upper bounds are valid for all n.

$$\frac{1}{2}n^{0.8107144} \leq H_1(n) \leq \frac{1}{2}n^{1.2324827}, \qquad H_1^*(n) \leq \frac{1}{2}n^{2.4649654}$$

$$n^{(d/2-1-\epsilon)n} \leq H_d(n) \leq H_d^*(n) \leq n^{(2d-1)n}, \quad d \geq 2,\ \epsilon > 0$$

Further, $H_1(n)$ is always a power of 2. The maximal order of $\log H_d(n)$ is known to within a constant factor as $n \to \infty$ except when $d = 2$. We also have similar results on the number of solutions for the case when the points lie on a d-dimensional sphere or torus. If the requirement that the points be distinct is relaxed, then we can show

$$\exp\left\{\theta(n \exp[-0.7044985\sqrt{\ln n}])\right\} \leq H_2^*(n)$$

Concerning the maximum possible number $S_1(n)$ of n-point beltway reconstructions, the upper bound $S_1(n) \leq H_2(n)$ can be shown by embedding a circle in the plane. We also have

$$\exp(2^{\frac{\ln n}{\ln \ln n}+o(1)}) \leq S_1(n) \leq S_1^*(n) \leq \frac{1}{2}n^{n-2};$$

this lower bound is far stronger asymptotically. But it is valid only for a certain infinite set of n, while the upper bound is valid for all n.

A preliminary version of this paper appears in [skie90]. This paper is organized as follows. Section 1.3 discusses applications of reconstruction problems to DNA sequencing and x-ray crystallography. Section 2 discusses bounds on $H_d(n)$ and related functions. Section 3 presents reconstruction algorithms and their analysis. Section 4 4 presents computational complexity results including NP-completeness proofs. We conclude by presenting a list of open problems in Section 5.

1.3 Application in DNA sequencing

The genetic code of an organism can be thought of as a string on an alphabet $\{A, C, G, T\}$ (adenine, cytosine, guanine, or thymine). *Sequencing* is the process of determining this pattern for a given strand of DNA. *Restriction enzymes* are chemicals which recognize and cut DNA molecules at particular patterns. For example, the enzyme *Eco RI* cuts at $GAATTC$. Over 100 restriction enzymes are known and each may cut a given strand of DNA many times. The lengths of DNA fragments may be measured by means of differing forced diffusion speeds in electrophoresis; length measurement accuracies approaching $\approx 0.1\%$ are feasible.

Restriction site mapping is the process of determining where all the restriction sites lie by measuring the lengths of DNA fragments. By partially digesting DNA with some restriction enzymes, fragments of all possible lengths are produced. Additional information may be obtained by using different combinations of enzymes, for intervals containing sites for more than one enzyme, although we not consider this possibility. Then determining where the restriction sites for those enzymes lie is a turnpike or beltway problem, depending on whether the original DNA was linear or circular. In this paper, we give a backtracking algorithm for the turnpike problem that in practice is capable of handling almost all distance sets of up to several hundred points.

Additional work extending the backtracking algorithm discussed in this paper to biological data is reported in [skie94] [zhan94]. Related combinatorial bounds on the probed partial digestion problem are reported in [newb93]. For additional information on restriction site mapping, see [ally88] [bell88] [dixk88] [rhee88] [stef78] [tuff88].

1.4 Applications in X-ray crystallography

In Fraunhofer monochromatic X-ray diffraction [hose62] [patt44] [patt35] there is a simple correspondence, up to certain angle-dependent factors, between the far-field amplitude pattern of X-rays scattered from the sample, and the modulus

$$\left|\int\int\int_{\text{sample}} e^{i\vec{k}\cdot\vec{x}} \rho(\vec{x})\, \mathrm{d}^3\vec{x}\right|$$

of the Fourier transform of the X-ray scattering density $\rho(\vec{x})$ within the sample. Our object is to determine the function $\rho(\vec{x})$ from this scattering data. This problem would be solvable directly by an inverse Fourier transform, except that only the *modulus* and not the (complex) phase of the Fourier transform is known. By the Fourier convolution identity, inverse Fourier transforming the square of this modulus gives us the autocorrelation function or *Patterson function*

$$Q(\vec{y}) = |\int\int_{\text{sample}} \rho(\vec{x})\rho(\vec{y}-\vec{x})\, \mathrm{d}^3\vec{x}|$$

which constitutes exactly the information that is recoverable from the scattering data.

If the atoms are modeled as identical Dirac-delta masses, then Q gives us exactly the vector-difference multiset for the n atoms in the sample. Reconstructing the coordinates of the atoms from this multiset is a three-dimensional vector-difference problem, while considering each coordinate separately gives a turnpike problem. Generally one is dealing with a crystal, in which the atoms lie in an essentially infinite periodic pattern. X-ray scattering then gives the vector differences among the atoms in a single (parallelepiped) unit cell, *modulo* fundamental translations. To reconstruct each cell, a beltway problem must be separately solved in each coordinate, while the entire problem is a vector difference problem in a 3-torus.

Other instances of distance reconstruction problems occur in astronomy [fink83], pattern recognition [mcla61], and the psychology of vision [gilb74].

2 Bounds on the Number of Homometric Sets

In this section we analyze the behavior of the functions $H_d(n)$, $S_d(n)$, $H_d^*(n)$, $S_d^*(n)$. We remark that the maximum number of n-point sets in $\mathbf{R}^d$ (respectively a d-torus) having the same *vector-difference* multiset, is exactly $H_1(n)$ (resp. $S_1(n)$) by projection onto a general rational line, so that we need not consider this problem separately.

2.1 Homometric Sets for the Turnpike Problem

We will prove power law upper and lower bounds on $H_1(n)$.

Lemma 2.1 *If* $n \leq 5$, *then* $H_1(n) = 1$, *but if* $n \geq 6$, *then* $H_1(n) \geq 2$.

Proof: For $n \geq 6$, the following construction by Lemke and Werman [lemk88], based on a 6-point pair found by Hosemann and Bagchi [hose62] gives a homometric pair. Observe that for all $n \geq 6$, the two n-point sets X and X' given by $X = n+1, n+3 \cup S \cup T$ and $X' = 2, n+2 \cup S \cup T$ where S is the set of integers i with $5 \leq i \leq n-2$, and $T = \{0, 1, n, n+5\}$, are homometric and noncongruent.

Table 1. Some examples of homometric sets.

n	the sets	source
6	{0,1,2,6,8,11}, {0,1,6,7,9,11}	[hose54], Lemma 2.1
6	{0,1,4,10,12,17}, {0,1,8,11,13,17}	[bloo77]
7	{0,1,5,7,8,10,12}, {0,1,2,5,7,9,12}	[ross82]
8	{0,1,5,6,8,9,11,13}, {0,1,2,5,6,8,10,13}	Lemma 2.1
9	{0,1,2,3,4,6,7,8,11}, {0,1,4,5,6,7,8,9,11}	Lemma 2.2 and {0,1,4}, {0,2,7}
13	{0,2,3,5,7,9,10,13,16,17,18,22,28}, {0,2,7,9,13,14,15,17,18,19,22,25,28}, {0,2,6,11,12,13,15,17,18,20,21,25,28}, {0,2,3,5,6,9,11,13,15,16,20,21,28}	computer search
14	{0,1,5,10,11,12,13,15,17,19,20,22,23,26}, {0,1,3,4,7,9,10,11,12,14,16,21,22,26} {0,1,3,4,5,8,10,11,13,14,15,20,22,26}, {0,1,7,9,11,12,13,16,18,19,21,22,23,26}	computer search
15	{0,1,4,5,8,9,10,11,12,13,19,23,25,26,28}, {0,1,2,3,10,13,15,16,17,19,20,21,24,25,28} {0,1,2,3,4,5,6,10,13,14,18,19,21,25,28}, {0,1,7,8,10,11,15,19,20,22,23,24,25,26,28}	computer search

We now show that $H_1(n) = 1$ if n =2, 3, 4, or 5. The cases $n = 2$ and $n = 3$ are trivial. For the case $n = 4$, place the two furthest-apart points $x_1 = 0$, x_4; then the second-largest distance determines the third point (up to a mirror reflection). Finally the fourth point is then completely determined by the three remaining distances.

For the case $n = 5$, place the two furthest apart points $x_1 = 0$ and $x_5 = 1$ (say). Of the remaining 9 distances, there are the three pairs of distances each of which sum to 1; call this set of 6 distances S and the remaining 3 distances T. The set T ={a, b, c} must have the property that $a+b = c$. This partitioning into sets S and T is necessarily unique because since $a + b = c$, if $a + c = 1$ then b uniquely determines T, while if $a + c \neq 1$ (and $b + c \neq 1$), then S is determined uniquely. By the case $n = 3$, T determines the three remaining points x_2, x_3, x_4 uniquely up to a translation and a reflection. The set S then determines the entire 5-point set up to a reflection about the midpoint $\frac{1}{2}$. Note that this idea will not suffice to prove uniqueness for $n \geq 6$, because there are then enough degrees of freedom to make the selection of the $(n - 2)$-pair set S ambiguous. ∎

Some examples of homometric sets appear in Table 1. Homometric sets of larger multiplicity can be constructed with the following observation:

Lemma 2.2 *If* $b \geq a \geq 3$, *then* $H_1(ab) \geq 2\, H_1(a)\, H_1(b)$.

Proof: To each n-point set a_i-$1 \leq i \leq n$, we associate the generating function

$$P(x) = \sum_i x^{a_i} \tag{1}$$

There is a correspondence between the distance multiset $|a_i - a_j|$, $1 \leq ij \leq n$ and the *distance generating function* $P(x)P(1/x)$. Any two sets with the same distance generating function must be homometric. Given a set of $H_1(a)$ homometric a-point sets whose generator polynomials are $P_i(x)$, $1 \leq i \leq H_1(a)$, and a set of $H_1(b)$ homometric b-point sets whose generator polynomials are $Q_j(x)$, $1 \leq j \leq H_1(b)$, we can construct $2H_1(a)H_1(b)$ mutually homometric ab-point sets as follows. Construct the sets whose generating functions are $P_i(x)Q_j(x)$ and $P_i(x)Q_j(1/x)$. If the a-point sets have been appropriately pre-scaled, e.g. by dilation with some constant incommensurable with the b-point sets, then no point overlaps can occur. ∎

In the proof above, we have implicitly assumed that the original sets from $P_i(x)$ and $Q_j(x)$ were not reflection symmetric; this assumption is justified because:

Lemma 2.3 *Lemma 1 Any point set in* $\mathbf{R}$ *that is invariant under a reflection, cannot be homometric with any incongruent set.*

Proof: Follows immediately from either of the algorithms of Section 3 for finding all point sets with a given distance set (see second paragraph of Section 3.3). ∎

Theorem 2.4 *For an infinite number of values of* n, $\frac{1}{2}n^{\alpha} \leq H_1(n)$, *where*

$$\alpha = \ln(8)/\ln(13) \approx 0.8107144 .$$

Proof: Table 1 proves that $H_1(13) \geq 4$. By iterative application of Lemma 2.2, $H_1(13^k) \geq 2^{3k-1}$ for all $k \geq 1$. If $n = 13^k$, this may be rewritten $2H_1(n) \geq n^{\ln 8/\ln 13}\ n^{0.8107144}$, giving the result. More generally, if $H_1(a) \geq r$, then whenever $n = a^k$,

$$H_1(n) \geq \frac{1}{2} n^{\log_a(2r)} .$$

∎

This result provides incentive to determine the least n such that $H_1(n) \geq 2^r$, which is open for $r \geq 2$. In particular, demonstrating that $H_1(n) = 8$ for some $n \leq 30$ would improve the result of Theorem 2.4.

Before presenting our asymptotic upper bound on $H_1(n)$, let us first present a few results concerning the structure of homometric sets in $\mathbf{R}^1$.

Consider the set of all equations of the form $d_1 + d_2 = d_3$ satisfied by the distances d_i. Two distance sets are called *equivalent* if they satisfy exactly same system of equations of this form (up to some relabeling), because any solution of the turnpike problem for one distance set, is immediately converted into a solution for an equivalent set by substituting corresponding distances. In theory, it is possible to characterize all possible inequivalent types of distance sets for n-point sets, and hence determine any desired value of $H_1(n)$, in $\mathbf{O}(n^{6n})$ time. This is by investigating every possible set of $\leq n$ linearly independent equations (in addition to the usual set of triangle equalities) and solving the resulting linear systems for the point sets. Another result along these lines is

Lemma 2.5 *Given any n-point turnpike problem, there is an equivalent turnpike problem whose distances are all integers bounded by $6^{(n-2)/2}$. This reduction may be carried out in polynomial time.*

Proof: Consider the set of all equations of the form $d_1 + d_2 = d_3$ satisfied by the distances d_i, where d_i is the ith largest distance. Since its coefficients are integers, this system must have a rational solution – indeed an integer solution by scaling – satisfying no other linearly independent equations of this form. Simply find such a solution in polynomial(n) operations.

This linear system has rank $\leq n-2$ because all the distances are determined as differences among the coordinates of the n points and we may fix the two furthest-apart points at 0 and 1 without loss of generality. We may use the coordinates of the middle $n-2$ points as (at least) a basis. When all equations are rewritten in these variables, every equation has norm (sum of the squares of its coefficients) ≤ 6. Hence Hadamard's inequality and Cramer's rule, applied to a spanning set of $\leq n-2$ equations, shows that the numerator and (common) denominator of each (rational) coordinate is $\leq 6^{(n-2)/2}$. The bound follows upon removing the denominators. ▮

A theorem given by Rosenblatt and Seymour [ross82] nicely characterizes homometric pairs using generating functions.

Theorem 2.6 *Two point sets with generator polynomials $P(x)$ and $Q(x)$ are homometric if and only if there exist generating functions $A(x)$ and $B(x)$ and numbers μ and ν such that $P(x) = x^{\mu}A(x)B(x)$ and $Q(x) = x^{\nu}A(x)B(1/x)$.*

Theorem 2.7 *The number of finite subsets of $\mathbf{R}^1$ that have a given distance multiset, is always a power of 2. Consequently, $H_1(n)$ is always a power of 2.*

The proof will require a few intermediate lemmas.

Lemma 2.8 *If $F(x)$ is a polynomial with integer coefficients which divides a polynomial whose coefficients are 0's and 1's only, then $F(x)$ has first and last coefficients that are either both $+1$ or both -1.*

Proof: Clearly the first and last coefficients are ± 1. If their signs differed, then $F(x)$ would have a positive real root, and 0-1 polynomials cannot. ▮

Lemma 2.9 *If $F(x)$ and $G(x)$ are monic polynomials with integer coefficients, and if $F(x)F(\frac{1}{x}) = G(x)G(\frac{1}{x})$, then if either $F(x)$ or $G(x)$ is 0-1, then the other is also.*

Proof: Let $F(x) = \sum_k f_k x^k$, $G(x) = \sum_k g_k x^k$. We have $F(1) = \pm G(1)$; assume the $+$ sign for the moment. It follows that $\sum_k g_k = \sum_k f_k$. Equating the constant terms in $F(x)F(\frac{1}{x}) = G(x)G(\frac{1}{x})$ yields $\sum_k (g_k)^2 = \sum_k (f_k)^2$. Hence

$$\sum_k g_k - (g_k)^2 = \sum_k f_k - (f_k)^2 . \qquad (2)$$

If all f_k are 0 or 1, then both sides of this equation are zero; but if any g_k were not 0 or 1, then the left hand side of (2) would be negative – a contradiction. If, on the other hand, $F(1) = -G(1)$, then by the same argument $-G(x)$ would be 0-1, contradicting the assumption that it is monic. ∎

A corollary is that if $P(x) = F(x)G(x)$ is 0-1, then so is $Q(x) = F(x)G(\frac{1}{x})$, because $P(x)P(\frac{1}{x}) = Q(x)Q(\frac{1}{x})$. An alternate proof follows from Filastta's discussion of the factorization of 0-1 polynomials into reciprocal and nonreciprocal parts [fila99].

A polynomial $P(x)$ of degree k is said to be *reciprocal* if $P(x) = x^k P(\frac{1}{x})$, meaning it corresponds to a palindromic point set.

Lemma 2.10 *If $F(x)$ is a non-reciprocal polynomial with integer coefficients, then $F(x)^2$ does not divide any 0-1 polynomial.*

Proof: Let $F(x) = \sum_{k=0}^{D} a_k x^k$. Then let j be the smallest index such that $a_j \neq a_{D-j}$. By Lemma 2.8, we may assume (by negating F if necessary) that $a_0 = a_D = 1$, hence $j \geq 1$. Let $F(x)^2 = \sum_{k=0}^{D} A_k x^k$, where $A_0 = a_0^2$, $A_1 = 2a_0 a_1$, $A_2 = a_1^2 + 2a_0 a_2$, and so on. It is also the case that A_k and A_{2D-k} first differ when $k = j$; specifically $A_j - A_{2D-j} = 2(a_j - a_{D-j})$. Now assume that a polynomial $G(x)$ with integer coefficients exists, such that $F(x)^2 G(x)$ is 0-1. Let $G(x) = \sum_{k=0}^{E} g_k x^k$, and without loss of generality $g_0 = 1$. Then the coefficients C_k of $F(x)^2 G(x)$ again first differ from C_{E+2D-k} when $k = j$, specifically

$$C_j - C_{E+2D-j} \quad = \quad (A_j - A_{2D-j})g_0 \quad = \quad 2(a_j - a_{D-j})g_0 .$$

But the fact that two coefficients differ by ≥ 2 contradicts the assumption that this polynomial is 0-1. ∎

Theorem 2.7 now follows from applying Theorem 2.6 in every possible way to the factorization of the generating function of a point set. Lemma 2.10 shows that a power of two generating functions must be obtained in this way; Lemma 2.9 and its corollary show that all generating functions obtained are 0-1, i.e. actually correspond to legitimate point sets.

We will now prove an asymptotic upper bound on $H_1(n)$ [lemk88]. This proof will involve several kinds of polynomial norms, which we will now define. Given a polynomial

$$P(x) = a_0 + a_1 x^1 + \ldots + a_k x^k ,$$

with integer coefficients, define the L_2-*norm*

$$L_2(P) = \left(\sum_{0 \le i \le k} a_i^2 \right)^{1/2}$$

and the *Mahler measure* $M(P)$ [mahl76 p. 5]

$$M(P) = |a_k| \prod_{1 \le i \le k} \max(|\alpha_i|, 1)$$

where α_i are the roots of P. Two properties of the Mahler measure are multiplicativity: $M(AB) = M(A)M(B)$, and "Specht's inequality" [spec49] $M(P) \le L_2(P)$.

An important property of the Mahler measure, due to Smyth [smyt71], is that $M(P) \ge M(x^3 - x - 1) \approx 1.32472$ if P is a non-reciprocal polynomial.

Theorem 2.11 *For all n, $H_1(n) \le \frac{1}{2} n^{\beta}$ and $H_1^*(n) \le \frac{1}{2} n^{2\beta}$, where $\beta = \frac{\ln(2)}{2\ln(\phi)} \approx 1.2324827$, and $\phi \approx 1.32472$ satisfies $\phi^3 - \phi = 1$.*

Proof: Since noncongruent homometric sets are determined by different permutations of the same set of distances, it is clear that for any given n, $H_1(n)$ is finite. By Lemma 2.5 we may assume all the points have integer coordinates, so the generating functions may be assumed to be polynomials with integer coefficients. By Theorem 2.6, the number of point sets homometric to a set with generator polynomial $P(x)$ cannot exceed 2^{F-1}, where $P(x)$ has F irreducible nonreciprocal factors, since we do not count mirror images twice. From Specht's inequality, Smyth's lower bound on $M(P)$, and the fact that Mahler measures are multiplicative, $F \le \ln(L_2(P))/\ln(\phi)$, and we obtain the result since $L_2(P) = \sqrt{n}$ if P is the generator function of a set of n distinct points. Even if the points are not required to be distinct, $L_2(P) \le n$ so $H_1^*(n) \le \frac{1}{2} n^{2.4649657}$. ∎

It has been conjectured that almost all polynomials with 0-1 coefficients, or *Newman polynomials,* are irreducible. This conjecture is supported by statistical evidence, and is also known to be true for trinomials, since there is an exact formula [fila95,ljun60] for the number of divisors of a Newman trinomial $1 + x^A + x^B$.

If this conjecture is true, then by Theorem 2.6, almost all integer point sets are uniquely determined by their distance multisets, and may be determined by factoring the distance generating function. We conducted an exhaustive search among the n-element subsets of $\{1,2,..,M\}$ for homometric sets. The number $f(M,n)$ of homometric pairs appears in Table 2. Only subsets including 1 and M were considered, and of these, only those that were lexicographically larger than their reflections. Thus if M is even, exactly $\frac{1}{2}\binom{M-2}{n-2} - \frac{1}{2}\binom{(M-2)/2}{(n-2)/2}$ candidate subsets were considered, and among these, $f(M,n)$ homometric pairs were found.

Table 2. A partial census of homometric pairs among n-element subsets of $\{1, 2, \ldots, n\}$. The starred entries do not count any pairs arising inside the unique homometric quadruplet with these parameters. For these quadruplets, see Table 1.

M, n	6	7	8	9	10	11	12	13	14	15	16	17
12	1	0	0	1								
13	0	2	0	4	0							
14	1	2	2	7	0	0						
15	1	1	2	11	1	0	0					
16	0	4	6	14	8	4	1	0				
17	0	2	5	25	10	6	7	0	0			
18	1	3	6	40	16	11	27	2	0	0		
19	1	2	3	44	33	16	45	9	2	5	0	
20	1	3	9	63	38	32	99	15	12	16	0	0
21	1	2	4	78	39	43	148	36	21	50	2	0
22	1	2	11	95	68	78	227	69	65	106	14	2
23	1	2	9	104	62	70	316	88	107	186	27	11
24	2	3	9	144	89	99	541	164	169	405	84	15
25	0	4	5	142	109	123	618	268	313	648	189	61
26	1	4	8	186	109	161	909	364	498	1144	369	154
27	2	4	11	196	112	164	1100	381	681*	1639	601	248
28	1	3	10	232	153	194	1368	529	987	2539	1082	512
29	2	3	10	270	143	220	1720	635*	1070	3702*	1443	886
30	2	4	2	4	13	319	167	247	2246	812	1484	5461

2.2 Homometric Sets for the Beltway Problem

Let $S_d(n)$ be the maximum number of mutually homometric sets on the d-sphere $\mathbf{S}^d$. Since three points on a circle are uniquely determined up to a dihedral symmetry by their distances, $S_1(n) = 1$ for $n \leq 3$. For $n \geq 4$, the homometric pair given by Patterson [patt44]

$$\{0, t, 1{+}t, 2, 4, \ldots, 2(n{-}3)\} \text{ and } \{0, t, 2, 4, \ldots, 2(n{-}3), 2n{-}5{+}t\} \bmod 2(n{-}2)$$

(where t is a free parameter) proves that $S_1(n) \geq 2$.

The examples in Table 3 prove that $S_1(7) \geq 6$ and $S_1(13) \geq 19$, refuting a claim by Bullough [bull61, pp. 265] that $S_1(n) \leq n - 2$.

We have conducted an exhaustive search for all homometric sets that are n-element subsets of a regular M-gon, $M \leq 31$. The results of our census are in Table 4. By the use of distance generating functions modulo $x^M - 1$, it may be readily shown (see also [buer77] [chie79]) that two subsets of a regular n-gon are homometric if and only if their complements are. For this reason, we have only tabulated n for $4 \leq n \leq M/2$.

As may also be shown using distance generating functions, every n-point subset of the regular $2n$-gon is homometric to its complement. Of course, it is usually also incongruent to its complement. Hence with probability $\Omega(n^{-1/2})$, a random subset of a regular $2n$-gon is homometric to at least one other set. For a deeper investigation along these lines, see [rau80]. Thus while homometric turnpike pairs appear exponentially rare, beltway pairs are common.

2.2.1 Constructing Homometric Beltway Sets with Singer Difference Sets

Using Singer difference sets, we can construct beltway instances with a quadratic number of non-isomorphic reconstructions. *Singer difference sets* [sing38]

Table 3. Some n-point h-way-homometric beltway examples on a circle of perimeter M. Examples are included for all record-breaking parameter sets (n, M), $M \leq 30$, i.e. such that q is larger than for any smaller M with that n.

n	h	M	h sets as binary (hexadecimal) M-bit numbers, each having n ones corresponding to points
4	2	8	E4 D8 (= 11100100 and 11011000 in binary = {0,1,2,5} and {0,1,3,4} mod 8 explicitly)
5	2	10	3B0,3C8
5	3	18	34880,34084,32500
6	2	12	F90,F60
6	3	16	F40C,F908,F620
6	4	24	D11800,CC1400,E40220,E22400
6	5	31	68441000,604A2000,68110010,64142000,65020200
7	2	14	3EC0,3F20
7	3	18	3D820,3E410,3CD00
7	6	24	D0C810,D08184,CA4060,D04818,E24840,D18480
8	4	16	F42C,F948,F4C2,ED60
8	6	24	D4C180,D8C140,E46820,E26840,E4C280,E8C240
9	4	18	3EB04,3DD20,3EC14,3ED10
9	8	24	EA48C0,D4C580,E86242,EA4260,E842C8,D584C0,E26A40,E8CA40
10	6	20	F530C,EC31A,F6518,F350C,F3944,EB21C
10	8	24	F51A40,F205A8,EB1680,F50B04,ED02B0,F40B14,F60950,F50348
11	6	22	3F9460,3EE340,3DD380,3F8D10,3F4E20,3E7580
11	8	24	F91298,F584C8,E846D8,EC7242,F33520,ED068C,EE5260,EB4CC0
11	12	30	35982920,3A086494,36904C28,39205348,3A4A6410,36944C20 39A45240,3A486414,35902930,35024B0C,39A05248,36296410
12	12	24	F68722,F90AD8,F60E4A,F684E8,FB12B0,F2B720 FC8D28,F95B10,EC0EB4,F485D8,F72560,F62E12
13	8	26	3DB8740,3F47890,3F8D1A0,3E9D380,3E74B80,3F8B460,3F488F0,3C3BB40
13	19	28	ED60CC8,EB4CCC0,EC58C86,F448D8C,F662650,F9894C8,E606CD4,F46684C,E8D909C F94CC48,ED891C8,ECC5A0C,F623464,EC6CD08,F6468C4,F333520,E656CC0,EC5C90C, ECD1C84
14	16	28	ED05C74,F9CA9A0,FA6AC60,F125770,F4CF054,F6862E8,F2CEAC0,F90ACB8 F115B70,FA468E2,FAB0B30,F60E2CA,F48745C,F44D43C,F537430,FCA8CB0
15	20	30	3F929328,3B2DAE40,3DB51B04,3BB60B48,3D9215E4,3EB213A4,3E8266B4, 3D941B46,3D9ACB02,3ACDAD80,3B02B6D8,3EC24AE4,3F532584,3DB21B0A, 3F426534,3E96E224,3D8159B4,3F5249C4,3F6292C8,3ECCA582

[hall67] are n-element subsets of $\mathbf{Z}_M$, where $M = (q^3-1)/(q-1)$ and $q = n-1$ is a prime power, with the property that the $n(n-1) = M-1$ differences they determine are all of the nonzero elements of $\mathbf{Z}_M$, each one occurring exactly once. Singer showed that at least one example always exists for each prime power q.

To construct other point sets homometric to Singer sets, we observe that multiplying a Singer set by any element r of $\mathbf{Z}_M$ (r relatively prime to M) preserves the distance set. Unfortunately, as a consequence of Hall's multiplier theorem [hall56] [hall67], if r divides q, then multiplying by r will merely translate, reflect, and/or permute the elements of the difference set.

But it seems likely that upon multiplying a Singer set by any element r of $\mathbf{Z}_M$ such that r is relatively prime to M and $\pm r$ does not divide q, a noncongruent set will be obtained. Let us specifically examine the case when M and q are prime. In this case, $q^3 \equiv 1 \ mod\ M$, so that q and -1 multiplicatively generate the order-6 dihedral group D_3 (inside $\mathbf{Z}_{M-1}$) of trivial multiplicative symmetries. Therefore if a Singer set with these parameters exists with *no* further multiplicative symmetries, then it would have $(M-1)/6$ equivalence classes of noncongruent homometric multiples, each equivalence class of size 6.

By the method outlined in [Hall67 end of Section 11.3], it is easy to generate Singer sets by computer. One may then find all their multiples. After translating and reflecting each such set to reduce it to a canonical (least-lexicographic) form, throwing out degenerate multiples, and sorting to remove redundant sets, one has a large collection of noncongruent homometric

Table 4. A partial census of homometric n-point subsets of the regular M-gon. To explain the entries by example: The $M = 18$, $n = 9$ entry indicates that there are 512 homometric pairs composed of 9-subsets of the regular 18-gon, 6 such homometric triplets, and 54 such homometric quadruplets.

M, n	4	5	6	7	8	9	10	11
8	1							
9	0							
10	0	3						
11	0	0						
12	1	3	15					
13	1	0	2					
14	0	6	6	48				
15	0	5	25	10				
16	2	10	28,3	40,4	177,0,3			
17	0	0	16	24	52			
18	0	13,1	56,6	118,16	139,11	512,6,54		
19	0	0	21	57	90	156		
20	2	22	96,2	180,11	491,12,32	535,14,16	1973,1,130,0,2	
21	0	0	96	220	276,6	1032,23,7	568,45	
22	0	20	55,5	310,25	540,35	1300,125	1430, 120	6985,5,390,0,10
23	0	0	33	110	429	803,11	1144	1342,33

Table 5. Constructions of n-point h-way-homometric beltway examples on a circle of perimeter $M = \frac{q^3-1}{q-1}$.

n	q	h	M	n	q	h	M
4	3	2	13	8	7	6	57
6	5	5	31	12	11	18	133
18	17	51	307	14	13	20	183
42	41	287	1723	20	19	42	381
60	59	590	3541	24	23	78	553
72	71	852	5113	30	29	132	871
90	89	1335	8011	32	31	110	993

sets which may be used to verify this construction. Explicitly, the first three of these examples are:

$\{0,1,3,9\} \times$ (1 or 2) mod 13 [2 homometric 4-point sets are given]

$\{0,1,3,8,12,18\} \times$ (1,2,3,4, or 8) mod 31 [5 homometric 6-point sets]

$\{0,1,3,30,37,50,55,76,98,117,129,133,157,189,199,222,293,299\} \times$ (1,13,14,15,16,20,21,103,104,106,108,109,112,113,116,122,125,126,128, 129,131,135,136,137,138,140,141,142, 143,150,154,156,158,159,160,168,175,183,184,186,193,200,202,218,219,220 ,222,239,256,269,273) mod 307 [51 homometric 18-point sets]

with each incarnation of each set having the distance (mod M) set $\{1,2,..M\}$.

Table 5 shows this construction yields a $(q+1)q/6$ homometric sets for every prime q with $q < 100$ and q^2+q+1 prime. The table gives n,q,h, and M, meaning there are h homometric noncongruent n-point sets on a circle of perimeter M. The right hand column of Table 5 contains some examples of noncongruent homometric multiples of Singer sets for prime q such that q^2+q+1 is not prime. In these cases, less than $q(q+1)/6$ sets are obtained, but still, the number is often quite large.

We conjecture that the Singer set construction generates $\geq (q+1)q/6$ homometric $(q+1)$-point beltway sets for every prime q such that q^2+q+1 is prime. This would follow from showing that for each such q at least one Singer set with no nontrivial symmetries exists. This seems likely, considering that no Singer set with *any* nontrivial symmetries is known. That there are an infinite number of such q is a standard Hardy-Littlewood conjecture.

2.2.2 A Lower bound on $S_1(n)$

We now present a construction yielding an asymptotically better lower bound on $S_1(n)$. This construction uses the monic cyclotomic polynomial [bate49]

$$\Phi_k(z) \equiv \prod_{\substack{1 \leq m \leq k \\ \gcd(m,k)=1}} (z - \exp(\frac{2\pi i m}{k})) = \prod_{d|k} (1-z^d)^{\mu(k/d)} \tag{3}$$

Here $\mu(x)$ is the Moebius function, which is zero if x has a square factor, is $(-1)^c$ if x is the product of c different primes, and $\mu(1) = 1$. $\Phi_k(z)$ is the unique monic irreducible polynomial with integer coefficients whose roots include the *kth* root of unity. The fact which we will need about the cyclotomic polynomials is that there exist distinct irreducible polynomials $\Phi_1(z)$, $\Phi_2(z)$,... with integer coefficients such that

$$z^n - 1 = \prod_{d|n} \Phi_d(z).$$

Theorem 2.12 $S_1(n) \geq \exp(2^{\frac{\ln n}{\ln \ln n}+o(1)})$ *for infinitely many* n.

Proof: Let p_i denote the ith odd prime, $K \geq 5$, and $Q \equiv 2^K$. Let $R \equiv p_1 p_2 \ldots p_K Q$ and $n \equiv (p_1+2)(p_2+2)..(p_K+2)$. We will construct $\geq 2^Q/4R$ mutually homometric and incongruent n-element subsets of the regular $2R$-gon.

Throughout the proof, $[K]$ will denote the set $\{1, 2, \ldots, K\}$ and U and T will denote subsets of $[K]$. ($\bar{T}$ means $[K] - T$.) The quantities

$$P_T \equiv \prod_{i \in T} p_i \,, \qquad z(T) \equiv (\sum_{j \in T} 2^{j-1})$$

each uniquely specify such a set T. (In place of $z(T)$, we could have used any 1-1 mapping between subsets of $[K]$ and the integers m with $0 \leq m2^K$.) We also define the following polynomials.

$$F_T(x) = \frac{x^{2R}-1}{x^{2P_T Q}-1} = 1 + x^{2P_T Q} + x^{4P_T Q} + \ldots + x^{2R-2P_T Q}$$

$$G_T(x) = (x^{P_T Q}+1) \prod_{i \in T} (x^{2P_T Q/p_i}+1)$$

$$H_T(x) = (x^{P_T Q} - 1) \prod_{i \in T} (x^{2P_T Q/p_i} - 1)$$

The point sets all correspond to generating polynomials $S(x)$ where

$$S(x) = \sum_T x^{z(T)} F_T(x) \frac{G_T(x) \pm H_T(x)}{2} \tag{4}$$

where all Q of the $\pm$ signs in the sum are independent, hence we have defined 2^Q possible point sets $S(x)$. We will now make a succession of claims, each of which are readily verified by use of the definitions above and the preceding claims, and from which the proof will follow.

1. $F_T(x)H_T(x)$ is divisible by $\Phi_d(x)$ for any d dividing $2R$, except that it is not divisible by $\Phi_d(x)$ with $d = 2P_T Q$.

2. $F_T(x)G_T(x)$ is divisible by $\Phi_d(x)$ for all d which divide $2R$ but do not divide R. In particular, it is divisible by $\Phi_d(x)$ for any d of form $d = 2P_U Q$, $U \subseteq [K]$.

3. Hence any two $S(x)$ are different modulo $\Phi_d(x)$ if $d = 2P_T Q$ and the two $S(x)$ differ in term T. Hence all the point sets $S(x)$ differ modulo $x^{2R} - 1$.

4. The difference generating function $S(x)S(\frac{1}{x})x^{2R}$ is invariant (with respect to the choice of signs in the sum (4) modulo $\Phi_d(x)$ for all d dividing $2R$, hence is invariant modulo $x^{2R} - 1$, hence all these sets $S(x)$ are homometric. This is because all of the "cross terms" of form $H_T(x)G_U(x)F_T(x)F_U(x)$ and $H_U(x)H_T(x)F_T(x)F_U(x)$, the latter with $U \neq T$, are zero modulo $\Phi_d(x)$ for every d dividing $2R$.

5. $G_T(x)$ is a polynomial with 0-1 coefficients, in fact it has exactly $2^{|T|+1}$ 1's. $H_T(x)$ is a polynomial with 0 and ± 1 coefficients, in fact it has exactly $2^{|T|+1}$ nonzero terms all of which coincide with nonzero terms of $G_T(x)$; exactly half of the nonzero coefficients on $H_T(x)$ are -1's and half are $+1$'s. Hence $[G_T(x) \pm H_T(x)]/2$ has only 0-1 coefficients, and has exactly $2^{|T|}$ 1's. $F_T(x)$ times this has only 0-1 coefficients (since a sum of reciprocals of distinct p_i cannot be an integer, no overlap leading to non 0-1 coefficients is possible) and has exactly $2^{|T|} P_{\bar{T}}$ 1's.

6. If a is such that the coefficient of x^a in term T of the sum is non-zero, then $a = z(T)$ *mod* Q. Since all of the $z(T)$'s are distinct and obey $0 \leq z(T)Q$, none of the 1-coefficients in different terms of the sum coincide, hence $S(x)$ has only 0-1 coefficients. Therefore every $S(x)$ really does represent a valid set of distinct points.

7. The total number of points in each set $S(x)$ is, as was claimed earlier

$$n = \sum_T 2^{|T|} P_{\bar{T}} = \prod_{i=1}^{K} (2 + p_i) .$$

8. The total number of distinct, incongruent homometric n-point sets generated (after removal of a possible $4R$ dihedral symmetries) is at least $2^Q/4R$.

Upon applying the Prime Number Theorem $p_i \sim i \ln i$ and approximating various products using Riemann sums of logarithms, we see that

$$n = \left(\frac{(1+o(1))K\ln K}{e}\right)^K = o(R)$$

$$R = \left(\frac{2(1+o(1))K\ln K}{e}\right)^K = n^{1+o(1)}$$

$$K = \frac{\ln n}{\ln\ln n} + o(1) = \frac{\ln R}{\ln\ln R} + o(1)$$

from which the theorem follows. ∎

As a direct consequence of the fact that $R = n^{1+o(1)}$, there cannot be a polynomial time algorithm to find all beltway reconstructions of a given distance multiset, even if all the input distances are required to be unary integers.

The first time the construction of Theorem 2.12 beats the quadratic construction with Singer sets is with $n = 1,167,075$, $R = 16,336,320$, $K = 6$, when there are at least 282,296,503,645 homometric incongruent n point subsets of the regular $2R$-gon. The next instance, $K = 7$, involves $n = 24,508,575$, $R = 620,780,160$, and 10^{29} sets. The Singer set construction gives more spectacular and frequent homometric examples for $n < 1,167,075$. It is still unclear what the asymptotic behavior of $S_1(n)$ is, although a $\mathbf{O}(n^{n-2})$ upper bound follows from the algorithm of Theorem 3.5.

2.3 Homometric sets in higher dimensions

In $\mathbf{R}^d$, $d \geq 2$, at least three noncongruent n point sets exist whenever $n \geq 4$. Gilbert [gilb74] gives the following 2-parameter family of three homometric noncongruent 4-point planar sets. The three sets are $\{A,B,C,D\}$, $\{A,B,C,D'\}$, and $\{A,B,C,D''\}$. Given a general triangle ΔABC, let a be the midpoint of BC and b be the midpoint of AC. Let K be the line through a perpendicular to the line Aa. Let L be the line through b perpendicular to the line Bb. Then: D is the intersection of L and K. D' is the point of L such that $\text{dist}(D',b) = \text{dist}(D,b)$, and D'' is the point of K such that $\text{dist}(D'',a) = \text{dist}(D,a)$.

Since any sufficiently small perturbation of a regular d-simplex may be reconstructed with all possible assignments of edge lengths,

$$H_d(d+1) \geq \left(\frac{(d+1)d}{2}\right)! \,/\, (d+1)! \tag{5}$$

mutually noncongruent homometric $(d+1)$-point sets exist in $\mathbf{R}^d$, $d \geq 2$. Thus $3 \leq H_2(4) \leq H_3(4) = 30$.

T. Caelli [cael80] claimed to have a complete characterization of all 4-point homometric pairs in $\mathbf{R}^2$, but unfortunately his result is incorrect. Hence the value of $H_2(4)$ is still open within the bounds [3,30].

We will now briefly survey some known results on *few-distance sets,* in order to explain the connections with homometric sets.

Erdös [erd46] defined the function $F_d(n)$ to be the minimum possible number of distinct distances determined by n points in $\mathbf{R}^d$. One may define the similar function $G_d(n)$ for the case when the n points must lie on the sphere $\mathbf{S}^d$. It is easy to see that $F_1(n) = n - 1$ and $G_1(n) = \lfloor n/2 \rfloor$.

Any set of n points, all lying on the unit sphere $\mathbf{S}^d$ in $(d+1)$-dimensional space and determining s distances, must obey [dels77] (see also [koor76])

$$n \le \binom{d+s}{d} + \binom{d+s-1}{d}, \tag{6}$$

and any set of n points in $\mathbf{R}^d$ determining s distances must obey [bann83]

$$n \le \binom{d+s}{d}. \tag{7}$$

The d-dimensional integer grid $\{0, 1, 2, ..n^{1/d}\}^d$ proves the upper bound $F_d(n) \le dn^{2/d}$. Thus

$$(\frac{d}{e}n)^{1/d} - d < F_d(n) < dn^{2/d}, \quad d \ge 3. \tag{8}$$

In two dimensions [soli01]

$$\Omega(n^{6/7}) \le F_2(n) \le 0.7044984310\frac{n}{\sqrt{\ln n}}; \tag{9}$$

the upper bound arises from the points of the equilateral triangle lattice lying inside a circle. P.Erdös offered \$500 for bounds on $F_2(n)$ tight to within a factor of $o(n^\epsilon)$.

In three dimensions [clar90]

$$\Omega(\sqrt{n}(\frac{n}{\lambda_6(n)})^{1/4}) \le F_3(n) \lesssim \frac{5}{12}(\frac{3}{\pi})^{2/3}n^{2/3}$$

where $\lambda_6(n)/n$ is an extremely slowly growing function related to Davenport-Schinzel sequences [agar89].

Theorem 2.13 *For any fixed d with $d \ge 2$, and for any desired function $g(n)$,*

$$H_d(n) \ge \frac{\binom{g(n)}{n}}{\binom{\binom{n}{2} + F_d(g(n)) - 1}{F_d(g(n)) - 1}} \frac{1}{nQ_d(n)}$$

$$S_d(n) \geq \frac{\binom{g(n)}{n}}{\binom{\binom{n}{2}+G_d(g(n))-1}{G_d(g(n))-1}} \frac{1}{Q_{d+1}(n)}$$

Here $Q_d(n)$ represents the maximum size of any finite group of origin-preserving reflections or rotations preserving the set achieving $H_d(n)$ or $S_{d-1}(n)$; specifically $Q_d(n) \leq 600\, d!\, 2^d n$.

Proof: From the $g(n)$-point set in $\mathbf{R}^d$ defining the minimum possible number of distances, choose a random n-element subset. The numerator of the bound gives the number of such subsets, while the denominator dominates the number of possible distance multisets. The theorem now follows by the pigeonhole principle; the $nQ_d(n)$ and $Q_{d+1}(n)$ terms remove all possible overcounting due to reflection, rotation, and translation symmetries. ∎

Theorem 2.14 *For $d \geq 3$ and for any $\epsilon > 0$, there are an infinite number of values of n such that $H_d(n) \geq n^{(d/2-1-\epsilon)n}$.*

Proof: Using the grid-set bound $F_d(n) \leq dn^{2/d}$ and choosing $g(n) = n^{d/2-\epsilon}$ in Theorem 2.13 yields

$$H_d(n) \geq \frac{\binom{n^{d/2-\epsilon}}{n}}{\binom{\binom{n}{2}+d\,o(n)}{d\,o(n)}} \frac{1}{Q_d n} = e^n n^{(d/2-1-\epsilon)n} n^{d\,o(n)} .$$

∎

Theorem 2.15 *For $n > d \geq 2$, $H_d(n) < n^{2dn}$.*

Proof: Without loss of generality, we may assume that the point set achieving $H_d(n)$ spans $\mathbf{R}^d$. Hence, $d+1$ of the point sites must define some nondegenerate d-simplex. We will call some particular $(d-1)$-simplex defined by d of the point sites the *central simplex.* Consider the following rigid structure defined by $(n-d)d + \binom{d}{2}$ line segments: Every pair of the vertices in the central simplex is joined by a line segment, and each of the remaining $n-d$ sites is joined to each of the d vertices of the central simplex by a line segment. If distances chosen from the $\binom{n}{2}$ available are assigned to each line segment in this structure, and also the $n-d$ points not on the central simplex are assigned $\pm$ signs depending on which side of the hyperplane (defined by the central simplex) they lie on, then clearly the point set is completely defined. The number of possible ways to accomplish these sign and distance assignments, is

$$< \binom{n}{2}^{(n-d)d+\binom{d}{2}} 2^{n-d}.$$

We may remove a factor $d!\,(n-d)!$ due to automorphism symmetries of the graph structure. $\lfloor d/2 \rfloor$ of the distances in the central simplex (plus one external edge, if d is odd) may be assumed without loss of generality to be the $\lceil d/2 \rceil$ largest distances available, so that we may remove an additional factor of $\binom{n}{2}^{\lfloor d/2 \rfloor}$ and replace it with a factor

$$(1 + (d-1)d/2)^{\lfloor d/2 \rfloor} \; .$$

The three reasons that we obtain an upper bound are: firstly, we are counting impossible distance assignments, secondly we are overestimating falling factorials by using powers, and thirdly, if any distances in our set of $\binom{n}{2}$ were equal, then we could have improved our bound even further.

The result follows by combining these terms:

$$\begin{aligned} H_d(n) & \quad \frac{1}{d!\,(n-d)!} \binom{n}{2}^{(n-d)d + \binom{d}{2} - \lfloor d/2 \rfloor} 2^{n-d} \left(1 + (d-1)d/2\right)^{\lfloor d/2 \rfloor} \\ & < \; n^{(2d-1)n - 3d^2/2} e^n 2^{d^2/2 + (1-d)n} \, . \end{aligned}$$

Some simpler but weaker bounds are $n^{(2d-1)n} e^n$ and n^{2dn}. ∎

In three or higher dimensions, the *logarithm* of the upper and lower bounds of Theorems 2.14 and 2.15 are tight to within a multiplicative factor of 4. Unfortunately in *two* dimensions, the best lower bound we have on $H_2(n)$ is subexponential (from Theorem 2.12), while the best upper bound we have is superexponential. It appears difficult to tighten these bounds, since a significant improvement would resolve Erdös's \$500 problem. Specifically: If there is some positive δ such that n-point planar sets can determine $\mathbf{O}(n^{1-\delta})$ distances, then there are

$$\geq \; n^{\frac{\delta}{1-\delta} n - o(n)}$$

(a superexponential number) of mutually noncongruent and homometric n-point planar sets. Conversely, if $H_2(n)$ grows more slowly than this, then $H_2(n) = \Omega(n^{1-\delta})$ for all $\delta > 0$. A new attack on the Erdös problem may be possible by using this fact.

A slightly better lower bound may be obtained if we allow overlapping points.

Theorem 2.16 *For all sufficiently large n, there are at least*

$$H_2^*(n) \; \geq \; \exp\left\{\theta(n \exp[-0.7044985\sqrt{\ln n}])\right\}$$

incongruent homometric sets of n (not necessarily distinct) points in the plane.

Proof: Assume that there is a sequence of n-point planar sets $S[n]$ with $\lesssim \kappa n/\sqrt{\ln n}$ distances. To prove Theorem 2.16, we choose n points at random (only with replacement this time) $S[g(n)]$ where

$$g(n) \; = \; \theta(n \exp[-\kappa\sqrt{\ln n}])\, .$$

Note $g(n) = o(n)$. The number of point sets that may be obtained in this way (after removal of at most $8n$ rotation/reflection equivalences) is at least

$$P = \frac{1}{8n}\binom{n+g(n)-1}{g(n)}.$$

The number of possible distance multisets that can arise from the random n-point sets is at most

$$D = \binom{\binom{n}{2} + \frac{g(n)}{\sqrt{\ln g(n)}} - 1}{\frac{g(n)}{\sqrt{\ln g(n)}}}.$$

Using Stirling's formula

$$\ln\binom{A}{B} \sim B\ln\frac{A}{B} + B - \frac{1}{2}\ln B + \mathbf{O}(1)$$

to asymptotically analyze $\ln\frac{P}{D}$ gives

$$\ln\frac{P}{D} = \theta(g(n)).$$

Since $H_2^*(n) \geq \frac{P}{D}$ by the pigeonhole principle, using $\kappa \leq c_\Delta$ gives the result. ∎

Examples of incongruent sets of points in higher dimensions with stronger kinds of homometricity properties may be obtained from [lind74] [doyv71].

3 Algorithms for Reconstructing Point Sets from their Distance Multiset

In this section we present a pseudo-polynomial time algorithm [gare78] for finding all n-point turnpike reconstructions, as well as an $O(2^n n\log n)$ algorithm which in practice shows excellent behavior, despite having exponential worst-case runtime.

3.1 Turnpike reconstruction by polynomial factoring

The algorithm for turnpike reconstruction which we discuss is due to Rosenblatt and Seymour [ross88], although the analysis showing that it runs in pseudo-polynomial time is due to Lemke and Werman [lemk88]. Given a set of distances $d_1, d_2, \ldots, d_N$, where $N = \binom{n}{2}$, we form the distance generating function

$$Q(x) = n + \sum_i (x^{d_i} + x^{-d_i}).$$

We factor this polynomial into irreducibles over the ring of polynomials with integer coefficients, using a factoring algorithm with runtime polynomial in $\max_i d_i$ [land87] [lens82]. It is convenient to write this factorization as

$$Q(x) = \prod_{i=1}^{F} P_i(x)P_i(\frac{1}{x})\, R(x)R(\frac{1}{x})$$

where the $P_i(x)$'s are monic, irreducible, and non-reciprocal, while all the reciprocal factors are collected in the polynomial $R(x)$. If Q does not have a factorization of this form then no turnpike reconstruction is possible. The point set $a_1, a_2, \ldots, a_n$ must have a generating function (1) obeying

$$Q(x) = P(x)P(\frac{1}{x}).$$

Therefore, we simply try all 2^F possible subsets S of $\{1, \ldots, F\}$ as putative factors of $P(x)$, i.e.

$$P_S(x) = \prod_{i \in S} P_i(x) \prod_{i \in \overline{S}} P_i(\frac{1}{x})\, R(x),$$

finding a superset of all the possible point sets $P(x)$ in time polynomial in N and 2^F. To avoid finding reflected solutions we may assume $1 \in S$ and translate the point sets to assure the leftmost point lies at the origin. After eliminating the point sets from generating functions with negative coefficients and sorting to remove possible redundant copies of point sets, we are done.

The runtime claim follows from Theorems 2.6 and 2.11 that if a turnpike reconstruction is possible, then $2^F \leq n^{2.4649657}$. If 2^F does not obey this bound, then no reconstruction is possible, and so we may terminate the algorithm early.

As an example, consider Bloom's distance set {1,2,3,4,5,6,7,8,9,10,11,12, 13,16,17}, whence

$$\begin{aligned} Q(x) &= x^{17} + x^{16} + x^{13} + .. + x + 6 + x^{-1} + .. + x^{-13} + x^{-16} + x^{-17} \\ &= P_1(x)P_1(\frac{1}{x})P_2(x)P_2(\frac{1}{x}) \end{aligned}$$

where $P_1(x) = x^6 + x + 1$, $P_2(x) = x^{11} - x^5 + x^4 + 1$ and the point sets {0,1,4,10,12,17} and {0,1,8,11,13,17} arise from $P_1(x)P_2(x)$ and $P_1(x)P_2(\frac{1}{x})$ respectively.

The factoring algorithm can be made to work even if it is given non-integer real distances by the use of Lemma 2.5, although then "pseudo-polynomial time" will no longer be an applicable notion.

3.2 A Backtracking Algorithm for Turnpike Reconstruction

The turnpike reconstruction problem has a combinatorial flavor which is not reflected by the polynomial factorization algorithm. In this section, we take

```
        17                      17
      12  16                  13  16
    10  11  13              11  12   9
   4   9   8   7          8   10   5   6
 1   3   6   2   5      1   7   3   2   4
```

Fig. 1. Distance Pyramids for Bloom's 6-point Homometric Example

a different look at the problem which leads to a practical algorithm for reconstruction.

Let d_{ij} represent the distance between the ith and jth points on a line, ordered from the left. We shall assume each distance to be a real number, $d_{ij} \geq 0$. For convenience, we shall represent these distances by a vector v sorted in increasing order, so that $v[i] \leq v[j]$, $1 \leq ij \leq \binom{n}{2}$, The complete set of distances and their initially unknown assignments can be represented as a triangular matrix or *pyramid* as in Figure 1. The set of distances d_{ij} such that $j - i = l$ defines row l of the pyramid.

As with Pascal's triangle, several identities can be observed on this structure. For example, there is a simple relationship between any four elements forming a "parallelegram" in the pyramid.

Lemma 3.1 $d_{ij} + d_{kl} = d_{il} + d_{kj}$, *where* $i \leq k \leq l \leq j$.

The relationship between the sums of rows is particularly pretty.

Theorem 3.2 *For any pyramid, the sum of distances on the kth row equals the sum of the distances on the $(n-k)$th row. Formally:*

$$\sum_{i=1}^{n-k} d_{i(i+k)} = \sum_{i=1}^{k} d_{i(i+n-k)} .$$

Proof: We will use the triangle equality and count how many times each base distance $d_{i,i+1}$ is added to the row.

$$\begin{aligned}\sum_{i=1}^{n-k} d_{i(i+k)} &= d_{1(k+1)} + d_{2(k+2)} + \ldots d_{(n-k)k} \\ &= \sum_{i=1}^{k-1} i \cdot d_{i,i+1} + k \sum_{i=k}^{n-k} d_{i,i+1} + \sum_{i=1}^{k-1} i \cdot d_{(n-i)(n-i+1)} \\ &= d_{1(n-k+1)} + d_{2(n-k+1)} + \ldots + d_{kn} = \sum_{i=1}^{k} d_{i(i+n-k)}\end{aligned}$$

∎

One implication of Theorem 3.2 is that reconstructing a pyramid with an even number of rows (ie. $n - 1 = 2k$) requires solving a partition problem, since the sum of the $(n^2 - 2n)/8$ distances in the top half of the pyramid equals those of the $(3n^2 - 2n)/8$ in the bottom half.

Based on the triangle equality, we can make certain assignments between distances and pairs of cities. Clearly, the largest distance $v[\binom{n}{2}]$ represents d_{1n}. The reflection of any pyramid is also a pyramid. To eliminate double counting of reflections, we make the convention that a pyramid is in canonical order if it is lexicographically less than its reflection. Thus $d_{2n} = v[\binom{n}{2} - 1]$, which implies $d_{12} = v[\binom{n}{2}] - v[\binom{n}{2} - 1]$. Unfortunately, this represents the extent of our ability to assign distances absolutely.

Theorem 3.3 *There is a* $\mathbf{O}(2^n n \log n)$*-time algorithm for finding all possible reconstructions of an* n*-point set from its* $\binom{n}{2}$*-element distance multiset.*

Proof: We shall use a backtracking procedure, and repeatedly position the *largest remaining* unpositioned distance. These elements will be filled in from the top of the pyramid, using backtracking to select either the left or right side. Because we are placing the largest available distance in the pyramid, there will be only two possible locations to choose from. The key to making this procedure efficient is to immediately fill in the pyramid any values which are determined by our previous (non-deterministic) selections.

Assume that we have thus far placed l elements along the left-hand side of the pyramid and r on the right, as shown in Figure 2.

All the positions in the shaded regions are determined by the $l + r - 1$ distances d_{1j} and d_{ir}, $(n - l + 1) \leq j \leq n$, $1 \leq i \leq r$. The largest remaining distance must be associated with either $d_{1(n-l)}$ or $d_{(r+1)n}$.

Suppose that the backtrack procedure selects the left side to receive the next element. Thus $d_{1(n-l)}$ is assigned the largest remaining distance. This determines $d_{i(n-l)}$, $2 \leq i \leq r$, since $d_{i(n-l)} = d_{1(n-l)} - d_{1i}$, which are already in the pyramid, as well as $d_{(n-l+1)(n-j)}$, $0 \leq j \leq l-2$, since $d_{(n-l+1)(n-j)} =$

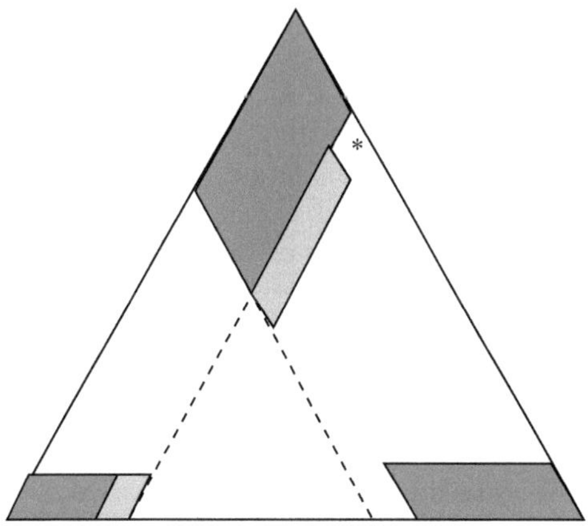

Fig. 2. The Effect of Choosing $d_{1(n-l+1)}$.

$d_{1n} - d_{1(n-l+1)} - d_{(n-j)n}$. If any of these $l + r - 1$ determined values is not a remaining distance, the partial reconstruction cannot be extended into a pyramid and we backup. If they are all remaining distances, we mark them as used and advance to the next level. Since the ith choice determines i values, $n - 1$ choices are sufficient to determine any pyramid. On the ith choice, we will perform i binary searches in a sorted list of the distances to test whether the ones we need are available.

The space complexity of this procedure is optimal - $\mathbf{O}(n^2)$ - since we need maintain only one pyramid if we remove the i elements we positioned when we backtrack from the ith level. (The state information needed at each level is $\mathbf{O}(1)$.) By assigning $d_{1n} = v[\binom{n}{2}]$, $d_{1(n-1)} = v[\binom{n}{2} - 1]$, and $d_{(n-1)n} = d_{1n} - d_{1(n-1)}$, the search is initialized with $l = 2$ and $r = 1$. ∎

While this algorithm takes exponential time in the worst case, notice that the exponent is n instead of $\binom{n}{2}$, the total number of distances. Further, it is independent of the magnitude of the distances, unlike the factoring algorithm.

3.3 Performance in Practice

Since the search is pruned whenever one of the i derived values determined on the ith level does not appear in the set of remaining distances, we usually get much more efficient behavior in practice than suggested by the worst-case analysis. For example, if the distance set arises from n real points in general position, then one of the two choices will be pruned immediately with probability 1, so that the procedure will use $\mathbf{O}(n^2 \log n)$ operations.

Another interesting aspect of our algorithm is that if the point set being reconstructed is reflection invariant, then the (unique) solution will be found in $\mathbf{O}(n^2 \log n)$ time, since the entire pyramid is reflection invariant and thus it doesn't matter which side the largest remaining distance is positioned.

The backtrack algorithm was implemented by Chi-long Lin and Yaw-ling Lin to help assess its performance in practice. This implementation includes a simple ordering heuristic which significantly improves performance on randomly generated problems: Since the span of a position in the pyramid overlaps that of its immediate ancestor in all but one base segment, it stands to reason that of the two possible positions for the largest remaining unpositioned distance, the position with the larger parent value is more likely to be correct. Therefore we investigate this choice first.

The algorithm was run on a series of randomly generated examples with n points, such that the distances between neighboring points on the line were independent uniform integer deviates in $[1, m]$. The backtracker usually takes longer when n increases or m decreases, although the run-time for large n and small m tends to fluctuate drastically. Table 6 presents the average size of the backtrack trees arising from 100 random examples for each (n, m) entry - with the exception of starred entries which summarize 25 random examples.

Table 6 shows that so long as the multiplicity of distances is not too excessive, the ordering heuristic almost always makes the correct decision at

Table 6. Mean Sizes of Backtrack Trees Produced by Reconstruction Algorithm.

number of	Maximum Neighbor Distance m						
points n	100	50	25	15	10	5	2
50	49.04	49.10	49.59	50.86	57.26	134.13	3064.21
100	99.18	99.21	99.65	102.43	113.35	269.05	10012.31
150	149.01	149.37	149.95	152.84	171.65	411.89	16730.36
200	199.08	199.46	199.73	206.12	233.74	559.25	7930.00
250	249.02	249.35	249.47	255.15	281.23	1452.03	31272.60*
300	299.10	299.34	299.95	307.65	348.84	826.69	76930.00*
350	349.01	349.12	349.98	358.80	421.56	1563.34	14822.76*
400	399.21	399.23	399.81	408.25	457.21	1172.73	83016.56*

each step, and false steps are quickly detected. These results show that almost all instances with up to several hundred points can be solved in essentially real-time.

We note that the backtracking algorithm can be modified to reconstruct data with experimental errors, by using interval arithmetic [moor66] instead of testing directly for equality. Skiena and Sundaram [skiena94] demonstrated both analytically and experimentally that the algorithm runs in polynomial-time with high probability if the relative errors of the fragment size are bounded by $O(1/n^2)$. Zhang [zhang94] constructed pathological input that causes the exact-data backtrack algorithm to take exponential time.

3.4 Vector differences in higher dimensions

Crystallographic applications require reconstructing sets from the set of pairwise vector differences. As with the beltway problem, this generates $n(n-1)$ distinct differences.

A simple general method for solving the vector-difference problem uses a subroutine to solve the one-dimensional turnpike problem. With probability 1, projecting the vector differences onto a random direction yields a 1-1 correspondence between the vector differences and the projected distances, i.e. no repeated distances will occur unless the corresponding vector differences are also repeated. (We remove vectors with negative projections.) Therefore, finding all solutions of the projected turnpike problem, and then replacing each distance by its vector equivalent, gives a superset of the possible vector difference reconstructions. Similar ideas also work for the d-vector difference beltway problems modulo a torus.

This leads to an algorithm for vector difference reconstruction running in

$$\mathbf{O}(2^n n \log n + dn^2 A)$$

time where A is the number of solutions found for the projected problem, $An^{1.23}$. It is also possible to generalize the backtrack approaches directly to vector-differences in d-space, using the additional restrictions that the parallelegram identities of Lemma 3.1 must hold as vector identities. The backtrack trees thus generated will never be larger, and indeed usually will be smaller, than for each coordinatewise 1D projected problem.

The Rosenblatt-Seymour polynomial-factoring algorithm may be generalized for the vector-difference problem by using multivariate polynomials. Instead of polynomials $P(\frac{1}{x})$, $P(x)$ and $Q(x)$, we use a d-variate argument

$$\vec{x} = (x_1, x_2, \ldots, x_d)$$

and

$$\frac{1''}{\vec{x}} = (\frac{1}{x_1}, \frac{1}{x_2}, .., \frac{1}{x_d})$$

The exponents are also vector valued – we use the notation

$$\vec{x}^{\vec{e}} = x_1^{e_1} x_2^{e_2} \ldots x_d^{e_d}.$$

With these multidimensional conventions, the entire algorithm now works exactly as in one-dimension, using a routine which factors d-variate polynomials with integer coefficients symbolically into irreducibles. Since this is possible in time polynomial in the number of bits in the input coefficient array if the degree d of the polynomial is fixed [lens82], we conclude that there is a pseudo-polynomial algorithm for solving the vector-difference turnpike problem in fixed dimensions.

3.5 A backtracking algorithm for distance reconstruction in d-dimensional space

The "central simplex" method used in the upper bound proof for Theorem 2.15 may be used to construct a backtracking algorithm for finding all n-point sets in d-space with a given distance multiset.

Theorem 3.4 *If $n > d \geq 2$, then all noncongruent n-point sets in $\mathbf{R}^d$ having a given distance multiset may be found in time*

$$\mathbf{O}(n^{(2d-1)n} e^n)$$

Proof: First, construct the "central simplex" in all possible non-isomorphic ways, using at least $\lfloor d/2 \rfloor$ of the largest distances. Let this $(d-1)$-simplex lie in the hyperplane $x_1 = 0$ without loss of generality. Then for each central simplex, by backtracking consider the addition of one point at a time. Each time a point is added – by selection of the d distances to the central simplex from among the unused distances and the selection of the sign of its x_1-coordinate – one may prune the backtrack search if the selected distances and the central simplex do not form a valid d-simplex or if the point thus embedded does not assume valid distances to every other currently embedded point. Pruning also occurs if the d-tuple of distances from the new point to the central simplex, is lexicographically larger than some previous d-tuple.

This eliminates $(n-d)!$ automorphisms and is easily implemented by considering candidate d-tuples in lexical order. Each embedding of a point may be accomplished in $\mathbf{O}(d^3)$ time and then all the needed distances may be found in $\mathbf{O}(n \log n)$ time by n binary searches. Every time all points have been successfully embedded, an output occurs. The bound on the run-time of this algorithm now follows from the proof of Theorem 2.15. ∎

In fact, if the points lie in general position, then with probability 1 there will be a unique reconstruction and only one branch of the backtracking will not immediately be pruned once the correct central simplex is found. In this case, the algorithm will run in time

$$\mathbf{O}(n^2 \log n \,+\, n^{d^2-d-2\lfloor d/2 \rfloor} \,+\, n^{2d}).$$

3.6 Backtrack solution of the Beltway Problem

Let d_{ij} be the distance going clockwise from point i to point j. The points are numbered clockwise and the numbers are taken mod n. These satisfy the triangle equality

$$d_{ij} + d_{jk} = d_{ik}$$

and the complement distance identity

$$d_{ij} + d_{ji} = L$$

where L is the perimeter of the beltway loop. The identities

$$k\,L \;=\; \sum_{0 \le in} d_{i(i+k)} \quad \text{and} \quad \binom{n}{2} L \;=\; \sum_{0 \le i,jn} d_{ij}$$

enable one to determine L.

Theorem 3.5 *There is a $\mathbf{O}(n^n \log n)$-time algorithm for finding all possible reconstructions of an n-point circular set from its $(n-1)n$-element difference multiset. This algorithm runs in optimal $\mathbf{O}(n^2)$ space.*

Proof: The $n(n-1)$ distances must be assigned places in a $(n-1) \times n$ cylindrical "tableaux" to solve the problem. The the *kth* row of the "tableaux" consists of the distances $d_{i(i+n-k)}$ for $i = 0, \ldots, n-1$. We will use the partial order inequalities

$$d_{(i-1)\,(j+1)} \;\ge\; d_{ij} \;\ge\; d_{(i-1)\,j}$$

which state that a tableaux element is at least as large as the one lying directly below it and also the one below and to the right of it.

Now we describe a $\mathbf{O}(n^n \log n)$-time reconstruction algorithm. As usual we fill in the distances largest first so that at any time we need only choose which of the n columns to place the next distance in. After each such choice is made, the triangle equalities allow one to fill in various other distances.

In fact, all distances that follow from repeated application of the triangle equality to a new distance may be determined in $\mathbf{O}(\log n)$ time per distance filled in, in the case of Theorem 3.3. We will see that the total number of automatic fill-ins is always $(n-1)^2$, so that only $n-1$ decisions need be made by the backtrack algorithm. Of these, the first choice may always be forced to avoid constructing each of n cyclic shifts of a point set, so we really only have to make $n-2$ decisions. Since each decision involves at most n choices, and each choice may be made (along with all the fill-ins it causes) in $\mathbf{O}(n^2 \log n)$ time, we may conclude that the total run time is $\mathbf{O}(n^n \log n)$. ∎

We may also conclude, upon removing mirror symmetries, that $S_1^*(n) \leq \frac{1}{2}n^{n-2}$, as was claimed in the introduction.

The complexity of the beltway problem is at least as great as that of the turnpike problem, as may be shown by a simple reduction; in view of Theorems 2.11 and 2.12, it is probably greater.

4 The complexity of distance embedding problems

The algorithms discussed in Section 3 all had worst-case exponential time behavior, but we have been unable to prove that the problems are intractable. In fact, there is evidence that the turnpike problem is not NP-complete, as a consequence of the fact that the maximum number of solutions is polynomially bounded.

In this section, we prove some related distance embedding problems are intractable.

The *turnpike problem with experimental error bounds,* which one might also call the *distance-interval reconstruction problem,* is as follows.

Instance: A multiset of $\binom{n}{2}$ distance-intervals and a positive number d. These closed intervals are specified by their integer endpoints.
Question: Is there an n-point set in $\mathbf{R}^d$ which satisfies the distance intervals. In other words, is there a 1-1 correspondence between the distances determined by this putative set and the distance-intervals, with each distance lying inside its corresponding interval.

See [papa76] [saxe79] for some similar NP-complete problems.

To prove its NP-completeness, we will use the following "modified 3-Partition problem" (M3PP): Given a set S of $3K$ positive integers $A_1, A_2, \ldots, A_{3K}$, is there a way to partition S into K disjoint triples such that two of the elements in each triple add up to the third. This problem is strongly NP-complete, as is readily shown by a reduction from "numerical matching with target sums," which is problem SP17 in [gare88].

Theorem 4.1 *The distance-interval reconstruction problem is strongly NP-complete, even if the distance intervals all have arbitrarily small width compared to the value at their midpoints and $d = 1$.*

Proof: Clearly the problem is in NP. Our reduction from M3PP uses the additive constraints to specify K clusters of three points each, with the clusters widely but regularly spaced apart.

Specifically, we claim that if $K \geq 30$, the following set of $(n-1)n/2$ distances arise from a collinear set of $n = 3K$ points if and only if the modified 3-partition problem is solvable

set number	distance interval midpoint	distance interval width	distance multiplicity	total number of distance-intervals
1	$\lvert p-q\rvert B\,K,\ 1 \leq pq \leq K$	$5B$	9	$9(K-1)K/2$
2	$A_m,\ 1 \leq m \leq 3K$	0	1	$3K$
total				$(3K-1)3K/2$

where $B = \sum A_m$. ∎

The *arbitrary-dimensional distance reconstruction problem* is as follows.

Instance: a multiset of $(n-1)n/2$ distances and a number $d > 0$.
Question: Is there an n-point set in d-space which realizes the distances.

For the purposes of the proof, we will assume that the *squares* of the distances are to be specified rationals.

Our reduction uses the original version of 3-partition [gare88, problem SP15] where we are given a set S of $3K$ integers A_m, $1 \leq m \leq 3K$, summing to KB and with $B/4 < A_m < B/2$, and we are to determine whether S may be partitioned into K disjoint triples such that each triple sums to B. We will actually merely require that each triple sums to $\leq B$, which of course implies equalities.

Theorem 4.2 *The arbitrary-dimensional distance reconstruction problem is strongly NP-complete, even if the input distances determine a simplex.*

Proof: The fact that the problem is in NP follows from the fact that we may guess the correspondences between distances and point-pairs, and then deducing the embedding may be accomplished via a matrix (Cholesky) factorization, by using a sequence of $\mathbf{O}(n^3)$ rational operations and n extractions of square roots. Each of the coordinates of the points will be expressible in the form $a + \sqrt{b}$ where a and b are rationals, since the Cholesky factorization does not introduce any numbers involving more than one square root. By the usual subdeterminant arguments [papa82] the total number of binary bits in these numbers will be a polynomial, as will also be the case for all the intermediate quantities arising in the computation.

Define $b_g \equiv K^6 + gK^4 + g^2K$, $1 \leq g \leq K$. We claim if $K \geq 2$, then the following set of $(n-1)n/2$ distances, $n = 3K+2$, is embeddable as a simplex in $\mathbf{R}^d$, $d = 3K+1$, if and only if the 3-partition problem below is solvable.

set number	squares of distances	distance multiplicity	total number of distances
1	$(2K^8)^2$	1	1
2a	$1+(K^8+b_g)^2$, $1 \le g \le K$	3	$3K$
2b	$1+(K^8-b_g)^2$, $1 \le g \le K$	3	$3K$
3	$(b_p-b_q)^2+2$, $1 \le p < q < K$	9	$9(K-1)K/2$
4	$2-2\underline{\cos}(2\pi A_m/B)$, $1 \le m \le 3K$	1	$3K$
total			$(3K+2)(3K+1)/2$

Here $\underline{\cos}(x)$ denotes the best rational approximation p/q to $\cos(x)$ subject to the restrictions that $\underline{\cos}(x) = \cos(y)$ for some y with $0 \le y < x$, and $|p| \le q \le 10B$. (This function may be computed quickly using Taylor series and regular continued fractions [hard79] and has the property that $x - y < 1/(10B)$ if $0 < x \le \pi$.) Note that if the inputs A_m and B to the 3-partition problem are expressed as unary integers, we may still generate this squared-distance set, and output it in the form of unary rationals, in polynomial time. If both the input and the output are expressed using binary integers, the reduction is even easier.

To prove the claim, regard $\mathbf{R}^d$ as $\mathbf{R} \times \mathbf{R}^{3K}$ for convenience. The largest distance is in set 1, so without loss of generality we may take the two furthest-apart points (the "endpoints") to lie at $(\pm K^8; \vec{0})$. Now no distance from sets numbered ≥ 3 can be added to any distance from sets numbered ≥ 2 to obtain a distance $\ge 2K^8$, so by the triangle inequality the distances from sets 2a, 2b are precisely the distances involving one endpoint and one other point.
In fact these distances must be paired – each distance from set 2a represents the distance from a point to one endpoint and the corresponding (same m) distance from 2b is the distance to the other endpoint – since any other pairing would entail some distance-pair sum $2K^8$. Therefore we see that the $3K$ non-endpoints have coordinates $(b_m; \vec{x}_{m,i})$, $1 \le m \le K$, $1 \le i \le 3$, and each $\vec{x}_{m,i}$ is a $3K$-vector with unit norm. (The fact that we may use b_m instead of $\pm b_m$ is because none of the distances in sets numbered ≥ 3 have magnitude as large as $\min_{p,q} b_p + b_q$. Now we see the distances in set 3 must be precisely the distances between points with coordinates $(b_p, \vec{x}_{p,i})$ and $(b_q, \vec{x}_{q,j})$, $p \ne q$, which forces the orthogonality of $\vec{x}_{p,i}$ and $\vec{x}_{q,j}$. We may now assume without loss of generality that $\vec{x}_{p,j}$ is of the form $(0, 0, \ldots, 0, e_{p,j}, f_{p,j}, g_{p,j}, 0, 0, .., 0)$ where the e, f,and g are in the $3p^{\text{th}}$, $3p+1^{\text{th}}$, and $3p+2^{\text{th}}$ positions respectively and $e_{p,j}^2 + f_{p,j} + g_{p,j}^2 = 1$. This is since the three points $\vec{x}_{1,i}$ must lie in some 3-space, without loss of generality [by a rotation of the coordinate system, if necessary] the one spanned by

$$(1,0,0,0,..,0), (0,1,0,0,..,0), \text{ and } (0,0,1,0,..,0),$$

and then the three points $\vec{x}_{2,j}$ must lie in an orthogonal 3-space, without loss of generality the one spanned by

$$(0,0,0,1,0,0,0,..,0), (0,0,0,0,1,0,0,..,0), \text{ and } (0,0,0,0,0,1,0,..,0),$$

and so on. Now, all this is possible if and only if the 3 angles determined by the 3 points $(b_m; \vec{x}_{m,i})$, $1 \le i \le 3$, in each orthogonal sphere sum to $\le 2\pi$.

Such an assignment of angles is possible (considering set 4 and the law of cosines) if and only if the 3-partition problem is solvable. ∎

5 Conclusions

We have presented several new results concerning algorithms for turnpike reconstruction and the number of homometric sets. In particular, our backtracking algorithm is sufficiently simple and fast in practice that it should be suitable for almost all turnpike problems arising in applications. Several outstanding open problems remain:

1. Improve our bounds on the constant $0.810 < C < 1.233$ defined by

$$C = \limsup_{n\to\infty} \frac{\ln H_1(n)}{\ln n}$$

2. What is the least value of n such that there are at least 2^r non-congruent homometric n-point sets? Improved bounds for small r can tighten lower bound of problem 1 via Lemma 2.2.

3. Find n-point homometric pairs, $n \geq 7$, with no repeated distances. Piccard [picc39] believed she had shown that no such pair could exist, but Bloom [bloo77] found a counterexample for $n = 6$.

4. What is the asymptotic behavior of $S_1(n)$? It could range from subexponential to superexponential.

5. Find a strongly-polynomial algorithm for turnpike reconstruction. No NP-complete problem is known with the property that the number of solutions is $2^{o(n^{o(1)})}$. Therefore, in light of Theorem 2.4 it seems doubtful that the turnpike problem is NP-hard. Further, find a reasonable algorithm for beltway reconstruction.

References

[agar89] P. Agarwal, M. Sharir, P. Shor: *Sharp upper and lower bounds on the length of general Davenport-Schinzel sequences,* J. Comb. Theory A, 52 (1989) 228-274.

[ally88] L. Allison and C. N. Yee: *Restriction Site Mapping is in Separation Theory,* Comput. Appl. Biol. Sci. 4,1 (1988) 97-101

[bann83] E.Bannai, E.Bannai, D.Stanton: *An upper bound for the cardinality of s-distance subset of Euclidean space II,* Combinatorica 3,2 (1983) 147-152.

[bate49] P.T. Bateman: *Note on the coefficients of the cyclotomic polynomial,* Bull. AMS 55 (1949) 1180-1181

[bell88] Bernard Bellon: *Construction of Restriction Maps,* Comput. Appl. Biol. Sci. 4,1 (1988) 111-115

[bloo77] G.S. Bloom: *A counterexample to a theorem of Piccard,* J. Comb. Theory A22 (1977) 378-379

[bull61] R.K. Bullough: *On homometric sets,* I. Acta Cryst. 14 (1961) 257-268 II. Acta Cryst. 17 (1964) 295-308

[buer77] M.J. Buerger: *Exploration of cyclotomic point sets in tautoeikonic complementary pairs,* Z. Kristallogr. 145 (1977) 371-411

[cael80] T. Caelli: *On generating spatial configurations with identical interpoint distance distributions,* Proc. 7th Australian Conf. Combinatorial Math. Newcastle Autralia August 1979 = Springer Lecture Notes in Math. 829 (1980) 69-75

[chie79] C. Chieh: *Analysis of cyclotomic sets,* Z. Kristallogr. 150 (1979) 261-277

[clar90] K.L.Clarkson, H. Edelsbrunner, L.J. Guibas, M.Sharir, E.Welzl: *Combinatorial complexity bounds for arrangements of curves and spheres,* Discrete & Comput. Geom. 5,2 (1990) 99-160.

[daki00] T. Dakic: *On the Turnpike Problem* PhD Thesis, Simon Fraser University, 2000.

[dels77] P.Delsarte, J.Goethals, J.J.Seidel: *Spherical codes and designs,* Geometraie Dedicata 6,3 (1977) 363-388

[dixk88] T.I. Dix and D.H. Kieronska: *Errors between sites in restriction site mapping,* Comput. Appl. Biol. Sci. 4,1 (1988) 117-123

[doyv71] J. Doyen & M. Vandensavel: *Non-isomorphic Steiner quadruple systems,* Bull. Soc. Math. Belgium 23 (1971) 393-410

[erdo46] P. Erdös: *On sets of distances of* n *points,* AMM 53 (1946) 248-250

[fila95] M. Filaseta, The irreducibility of all but finitely many Bessel polynomials. Acta Math. 174 (1995), no. 2, 383–397.

[fila99] M. Filaseta: *On the factorization of polynomials with small Euclidean norm,* pp. 143-163 in *Number theory in progress* 1 (volume 2 of 2, e.d K.Györy et al.) W. de Gruyter 1999

[fink83] A.M. Finkelstein, O. M. Kosheleva, and V. JA. Kreinovic: *On the uniqueness of image reconstruction from the amplitude of radiointerferometric response.* Astro. Sp. Sci. 92 (1983) 31-36

[gare78] M.R. Garey & D.S. Johnson: *Computers and Intractibility: a guide to the theory of NP-completeness,* Freeman 1978

[gilb74] E.N. Gilbert & L.A. Shepp: *Textures for discrimination experiments,* Bell Laboratories Murray Hill NJ, TM-74-1218-6, TM-74-1215-15 April 15,1974. Filing case 20878.

[golu83] G.H. Golub & C. Van Loan: *Matrix Computations,* Johns Hopkins University Press 1983

[hall56] M. Hall Jr.: *A survey of difference sets,* Proc. AMS 7 (1956) 975-986

[hall67] M. Hall Jr.: *Combinatorial Theory,* Wiley 1967

[hard79] G.H. Hardy & E.M. Wright: *Introduction to the theory of numbers,* Oxford University Press 1979

[hose54] R. Hosemann & S.N. Bagchi: *On homometric structures,* Acta. Cryst. 7 (1954) 237-241

[hose62] R. Hosemann & S.N. Bagchi: *Direct analysis of diffraction by matter,* North-Holland 1962

[humn01] Special issues on human genome: *Nature* 409 (15 Feb 2001); *Science* 291, 5507 (16 Feb 2001).

[knut76] D.E. Knuth: *Big omicron and big omega and big theta,* SIGACT News 8,2 (April-June 1976) 18-24

[koor76] T.M.Koornwinder: *A note on the absolute bound for systems of lines,* Indag. Math. 38,2 (1976) 152-153.

[land87] Susan Landau: *Factoring polynomials quickly,* Notices Amer. Math. Soc. 34,1 (1987) 3-8

[lemk88] P. Lemke and M. Werman, *On the complexity of inverting the autocorrelation function of a finite integer sequence, and the problem of locating n points on a line, given the unlabeled distances between them,* manuscript, 1988.

[lens82] Arjen K. Lenstra, H.W. Lenstra Jr., Lazslo Lovasz: *Factoring polynomials with rational coefficients,* Math.Ann. 261 (1982) 515-534

[lind74] C.C. Lindner: *On the construction of non-isomorphic Steiner quadruple systems,* Colloq. Math. 29 (1974) 303-306

[ljun60] W. Ljungren: *On the irreducibility of certain trinomials and quadrinomials* Math. Scandinav. 8 (1960) 65-70; also see H. Tverberg 121-126 same issue.

[mahl76] K. Mahler: *Lectures on transcendental numbers,* Lecture Notes in Math. 546 Springer-Verlag 1976

[mcla61] Dan McLachlan, Jr.: *Similarity function for pattern recognition,* J. Appl. Phys. 32 (1961) 1795-1796

[moor66] R.E. Moore: *Interval Analysis,* Prentice-Hall 1966

[newb93] L. Newberg and D. Naor, *A Lower Bound on the Number of Solutions to the Probed Partial Digest Problem* Advances in Applied Mathematics 14 (1993).

[pand01] G. Pandurangan and H. Ramesh, *The Restriction Mapping Problem Revisited,* Journal of Computer and System Sciences (JCSS), to appear

[papa76] C.H. Papadimitriou: *The NP-completeness of the bandwidth minimization problem,* Computing 16 (1976) 263-270

[patt35] A. L. Patterson: *A direct method for the determination of the components of interatomic distances in crystals,* Zeitschr. Krist. 90 (1935) 517-542

[patt44] A.L. Patterson: *Ambiguities in the X-ray analysis of crystal structures,* Phys. Review 65 (1944) 195-201.

[picc39] Sophie Piccard: *Sur les ensembles de Distances des Ensembles de points d'un Espace Euclidean,* Mem. Univ. Neuchatel 13 (Neuchatel Switzerland 1939)

[prep85] F. Preparata and M. Shamos, *Computational Geometry,* Springer-Verlag, New York, 1985.

[rau80] V.G. Rau, L.G. Parkhomov, V.V. Ilyukhin, N.V.Belov: *On the calculation of possible Patterson cyclotomic sets,* I: Doklady Akademii Nauk SSSR 255,4 (1980) 859-861; II: ibid. 255,5 (1980) 1110-1113. (Both in Russian.)

[rhee88] Gwangsoo Rhee: *DNA restriction mapping from random-clone data,* Technical Report WUCS-88-18, Department of Computer Science, Washington University St. Louis MO 1988

[ross82] Joseph Rosenblatt & Paul Seymour: *The structure of homometric sets,* SIAM J. Alg. Disc. Methods 3,3 (1982) 343-350

[saxe79] James B. Saxe: *Embeddibility of weighted graphs in k-space is strongly NP-hard,* Proc. 19th Allerton Conference on Computers, Controls, and Communications 19, Urbana, IL (1979) 480-489

[sham77] M.I. Shamos: *Problems in computational geometry,* Unpublished manuscript, Carnegie Mellon University, Pittsburgh, PA 1977.

[sing38] James A. Singer: *A theorem in finite projective geometry and some applications to number theory,* Trans. Amer. Math. Soc. 43 (1938) 377-385

[skie90] S. S. Skiena, W. D. Smith, and P. Lemke: *Reconstructing sets from interpoint distances (extended abstract)* Proc. Sixth ACM Symposium on Computational Geometry (1990) 332-339.

[skie94] S. S. Skiena and G. Sundaram: *A Partial Digest Approach to Restriction Site Mapping*
Bulletin of Mathematical Biology, 56 (1994) 275-294.

[smyt71] C. J. Smyth: *On the product of the conjugates outside the unit circle of an algebraic integer,* Bull. London Math. Soc. 3 (1971) 169-175

[soly01] J.Solymosi & Cs. D. Toth: *Distinct distances in the Euclidean plane,* Discr. & Comput. Geom. 25,4 (2001) 629-634.

[spec49] W. Specht: *Abschätzungen der Wurzeln albebraischer Gleichungen,* Math. Zeit. 52 (1949) 310-321

[stef78] Mark Stefik: *Inferring DNA structures from segmentation data,* Artificial Intelligence 11 (1978) 85-114

[tuff88] P. Tuffery, P. Dessen, C. Mugnier, S. Hazout: *Restriction Map Construction Using a 'Complete Sentences Compatibility' Algorithm,* Comput. Appl. Biol. Sci. 4,1 (1988) 103-110

[zhan94] Z. Zhang: *An exponential example for partial digest mapping algorithm,* J. Computational Biology 1,3 (1994) 235-239.

About Authors

Paul Lemke did this work as a member of the Department of Mathematical Sciences, Troy NY 12180-3590.

Steven Skiena (to whom correspondence should be addressed) is at the Department of Computer Science, SUNY Stony Brook, Stony Brook, NY 11794-4400; *skiena@cs.sunysb.edu*.

Warren D. Smith did most of this work as a member of AT&T Bell Laboratories and the NEC Research Laboratory. His current affiliation is DIMACS, Rutgers University, 96 Frelinghuysen Road, Piscataway, NJ 08854-8018.

Acknowledgments

We thank Bernard Chazelle for first bringing this problem to our attention. A. Blokhuis, Dan Gusfield, A.M. Odlyzko, J.C. Lagarias, N.J.A. Sloane, Pong-Chi Chu, Arch Robison, and Carla Savage provided insights and helpful comments. Discussions with Gene Myers, Webb Miller, and John Turner helped clarify the connection to restriction site mapping.

Dense Packings of Congruent Circles in Rectangles with a Variable Aspect Ratio

Boris D. Lubachevsky
Ronald Graham

Abstract

We use computational experiments to find the rectangles of minimum area into which a given number n of non-overlapping congruent circles can be packed. No assumption is made on the shape of the rectangles. Most of the packings found have the usual regular square or hexagonal pattern. However, for 1495 values of n in the tested range $n \leq 5000$, specifically, for $n = 49, 61, 79, 97, 107, ...4999$, we prove that the optimum cannot possibly be achieved by such regular arrangements. The evidence suggests that the limiting height-to-width ratio of rectangles containing an optimal hexagonal packing of circles tends to $2 - \sqrt{3}$ as $n \to \infty$, if the limit exists.

1 Introduction

Consider the task of finding the smallest area rectangular region that encloses a given number n of circular disks of equal diameter. The circles must not overlap with each other or extend outside the rectangle. The aspect ratio of the rectangle, i.e., the ratio of its height to width, is variable and subject to the area-minimizing choice as well as the positions of the circles inside the rectangle.

Packing circles in a square has been the subject of many investigations [GL], [NO1], [NO2], [NO3]. Because the aspect ratio is not fixed in our present problem, the solutions are typically different from the dense packings in a square. For example, the density $\pi/4 = 0.785...$ of the proved optimum (see [NO3]) for packing 25 congruent circles in a square (see Figure 1a) can be increased to $25\pi/(26(2+\sqrt{3})) = 0.809...$, if we let the rectangle assume its best aspect ratio (see Figure 1b). Our experiments offer only three values of n for which the dense packing in a square is also a solution to our present problem: $n = 4, 9$, and of course, $n = 1$.

Similarly, dense packings in long rectangles with a fixed aspect ratio usually do not yield solutions for our problem. According to a long-standing conjecture attributed to Molnar (see [Füredi]), dense packings in long fixed rectangles tend to form periodic up and down alternating triangular "teeth,"

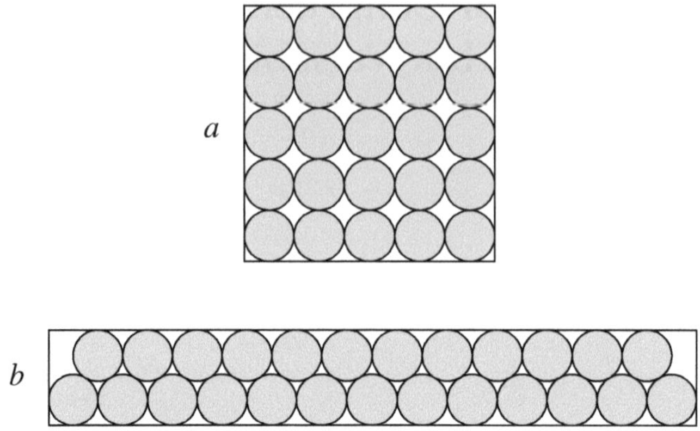

Fig. 1. The best packings found for 25 circles *a*) in a square, and, *b*) in a rectangle with variable aspect ratio.

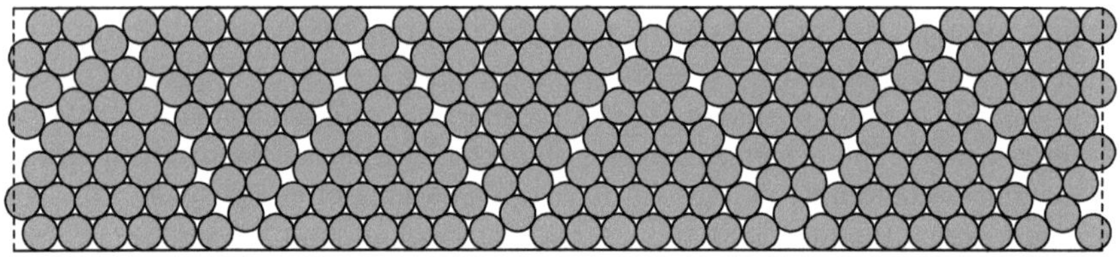

Fig. 2. A 224-circle fragment of the packing with the highest density found in a long rectangle with a fixed aspect ratio.

as in the example shown in Figure 2. We can usually increase the density of such packings by slightly changing the aspect ratio of the rectangle.

The present problem has been occasionally mentioned among packing problems ; for example, Web enthusiasts have recently began discussing it. However, the problem was considered as early as in 1970 in [Ruda], where the optimum packings were determined for all $n \leq 8$ and conjectured for $9 \leq n \leq 12$. We learned about Ruda's results after we performed our computations and made our conjectures. Fortunately, our conjectures restricted to the values $n \leq 12$ agree with the results and conjectures in [Ruda]!

It appears that the complexity of the proofs of optimality in the packing-circles-in-a-square problem increases exponentially with n; computer generated/assisted proofs of optimality of such packings do not reach $n = 30$ (see [NO3] ; proofs that do not utilize computers have not been found except for quite small values of n), while conjectures extend to $n = 50$ and even much larger values (see [NOR]).

Similarly, it seems to be difficult for the present problem to prove optimality for the configurations we find. This paper describes an experimental approach, where good packings are obtained with the help of a computer.

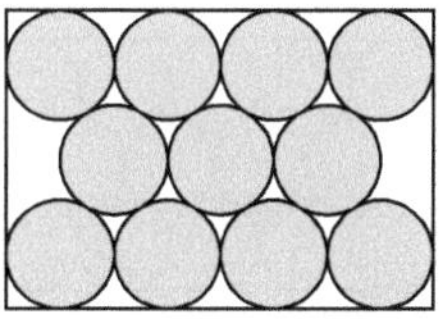

Fig. 3. The best packing found for 11 circles in a rectangle.

Some of these packings hopefully will be proved optimal in the future. At present, most of the statements about the packings made in this paper are only conjectures, except for a few which we explicitly claim as proven. For instance, we prove that the best packings found are better than any other in their class, e.g., that the packing of 79 circles with a monovacancy (see Figure 5a), while non-optimal, is better than any hexagonal or square grid packing of 79 circles without a monovacancy. Also, we prove that $n = 11$ is the smallest n for which a hexagonal packing as in Figure 3 is better than any square grid packing of n circles.

2 A priori expectations and questions about best packings

Since it is well known that the hexagonal arrangement of congruent circles in the infinite plane has the highest possible density, (see [Th], [Ft], [FG],[Oler]), before plunging into our experiments, we expected to obtain good finite packings by "carving" rectangular subsets out of this infinite packing. It will be useful to classify here such finite arrangements. For that we will discuss packings in Figures 1b, 3, 4a, 4b, and 5a, irrespective of their optimality (which will be discussed in the following sections). For the purpose of exposition, we will pretend in this section that there are no holes in the arrangements in Figures 4b and 5a, that is, the question mark in the figures is covered with an additional circle of the common radius.

Under such an assumption, the arrangement in Figure 5a consists of $h = 5$ alternating rows with $w = 16$ circles in each, a total of $n = w \times h = 80$ circles. Arrangements as in Figures 1b, 3, 4a, and 4b, are obtained by depleting such $w \times h$ arrays of some circles along one side of the array. The ways of performing the depletion depend on the parity of h.

If h is even, then the depletion can be done in only one way by removing $h/2$ circles. Figure 1b presents an example with $h = 2$.

The case of an odd h presents two possibilities: we can remove $\lfloor h/2 \rfloor$ circles on a side as in the example of Figure 4b, or $\lfloor h/2 \rfloor + 1$ circles on a side as in the example of Figure 4a.

Alternatively, one can consider rectangular square grid arrangements as candidates for the best packings, e.g., see those in Figure 6. The density of a square grid arrangement is fixed at $\pi/4$ independent of n. When both

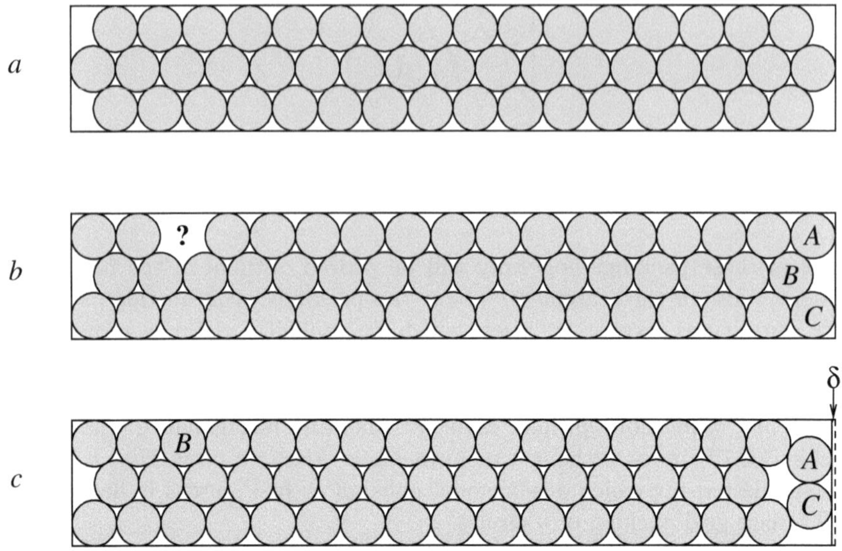

Fig. 4. Packings of 49 circles in a rectangle: *a*) the best in the class of hexagonal packings, *b*) a best in the class of hexagonal packings with monovacancies (one of 17 equally dense packings with the hole), *c*) the best we found.

sides h and w of the carved rectangle tend to infinity, the hexagonal packing density tends to $\pi/(2\sqrt{3})$ which is the density of the hexagonal packing in the infinite plane. Since $\pi/(2\sqrt{3}) > \pi/4$, one naturally wonders for which n the best hexagonal grid arrangement becomes better than the square grid arrangement.

A natural question is whether or not both arrangements exhaust all possibilities for the best packing. In other words, does there exist an optimal packing of n disks in a rectangle such that it *cannot* be represented as being carved out by a rectangle from either square grid or hexagonal grid packing of the infinite plane?

3 Results of compactor simulations

To tackle the problem by computer, we developed a "compactor" simulation algorithm. The simulation begins by starting with a random initial configuration with n circles lying inside a (large) rectangle without circle-circle overlaps. The starting configuration is feasible but is usually rather sparse. Then the computer imitates a "compactor" with each side of the rectangle pressing against the circles, so that the circles are being forced towards each other until they "jam." Possible circle-circle or circle-boundary conflicts are resolved using a simulation of a hard collision so that no overlaps occur during the process.

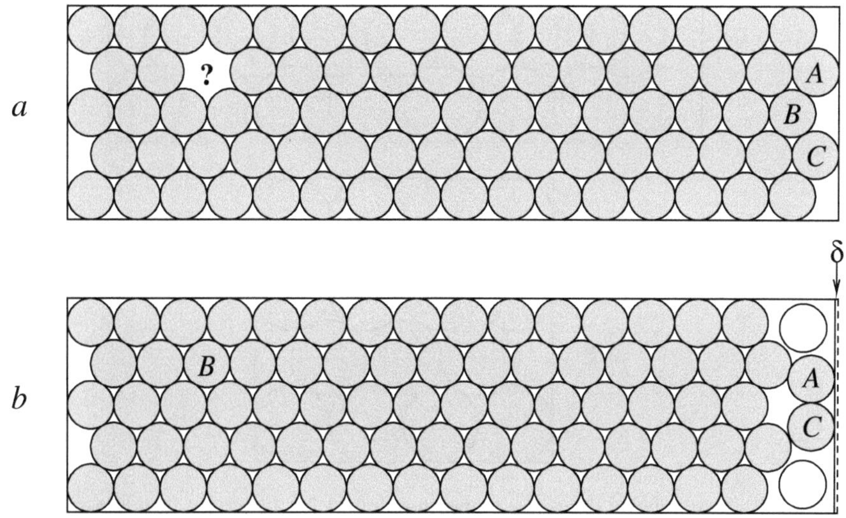

Fig. 5. Packings of 79 circles in a rectangle: *a*) a best in the class of hexagonal packings with possible monovacancies, and, *b*) the best we found.

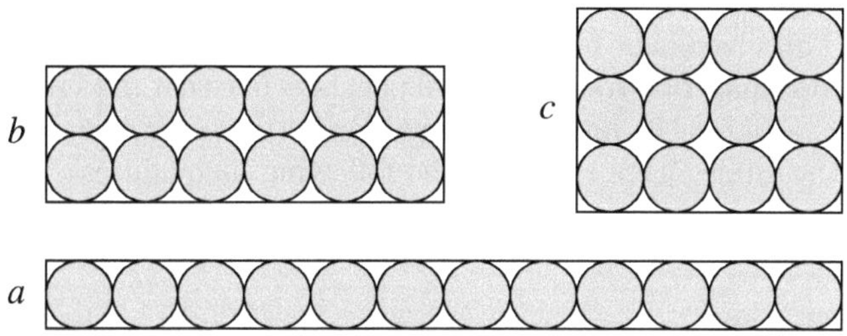

Fig. 6. Best packings found for 12 circles in a rectangle.

The simulation for a particular n is repeated many times, with different starting circle configurations. If the final density in a run is larger than the record achieved thus far, it replaces this record. Eventually in this process, the record stops improving up to the level of accuracy induced by the double precision accuracy of the computer. The resulting packing now becomes a candidate for the optimal packing for this value of n.

In this way we found that the square grid pattern with density $\pi/4$ supplies the optimum for $n = 1, ..., 10$, and that $n = 11$ is the smallest number of circles for which a density better than $\pi/4$ can be reached; the density of this packing is $11\pi/(16(1+\sqrt{3}) = 0.790558..$ The corresponding conjectured optimum pattern for $n = 11$ is shown in Figure 3. The pattern is hexagonal. For $n = 12$ and 13 the square grid pattern briefly takes over as the experimental optimum again (see Figure 6 for the case of $n = 12$). Remarkably,

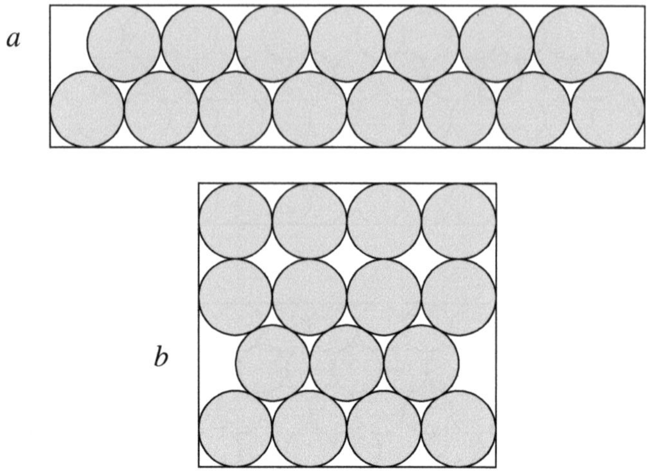

Fig. 7. Two equally dense best packings found for 15 circles in a rectangle.

Ruda [Ruda] *proves* that for $n \leq 8$, the square-grid packings are optimal. His conjectures for $9 \leq n \leq 12$ also agree with our findings.

Our simulation results show that for $n \geq 14$, the best densities are larger than $\pi/4$. This statement is also easy to prove, considering examples of (not necessarily optimal) two-row hexagonal packings like that in Figure 1b for odd $n = 2m + 1 > 14$, or the ones with equal length rows for even $n = 2m \geq 14$. Assuming the circles have radius 1, the following inequalities

$$2(m+1)(2+\sqrt{3}) < 4(2m+1) \tag{1}$$

for $n = 2m + 1$ and

$$(2m+1)(2+\sqrt{3}) < 8m \tag{2}$$

for $n = 2m$ have to be satisfied. The left-hand sides in (1) and (2) are the areas of the enclosing rectangles for the two-row hexagonal packings, and the right-hand sides are the areas for the corresponding square grid packings. It is easily seen that all integers $m \geq 7$ satisfy either inequality. These correspond to all integer values of $n \geq 14$.

A surprise awaited us at the value $n = 15$. We found two different equally dense packings. One, shown in Figure 7a, is of the expected hexagonal type. However, the other, shown in Figure 7b, is not, nor can it be carved out of the square grid. It is easy to verify that two equally dense packings a and b of these types also exist for any n of the form $n = 15 + 4k$, $k = 1, 2, 3, \ldots$. Packing a, such as that in Figure 7a, has two alternating rows, with one row one circle shorter than the other, and with the longer row consisting of $w = 8 + 2k$ circles. Packing b, such as that in Figure 7b, has 4 rows, the longest row having $w = 4 + k$ circles, three bottom rows alternate, the middle one having $w - 1$ circles, and the 4th row of w circles stacked straight on the top (or on the bottom; these would produce the same configuration). Our experiments suggest that for $k = 1$ and 4, i.e., for $n = 19$ and 31, these pairs of configurations might also be optimal; however, they are provably not optimal for other values of $k > 0$. The packings of pattern b for $n = 15$, 19, and 31, if they are proved to be optimal, answer positively the second question posed in Section 2.

Thus, the optimal container for 15 bottles apparently can have two different shapes, as can the optimal containers for 19 and 31 bottles! In an obvious way, more than one optimum container shape also exists for square grid packings of n circles if n is not a prime. See Figure 6 for the example of $n = 12$; there are three different rectangles which are equally good and probably optimal for $n = 12$. Apparently, three is the maximum number of rectangular shapes, any number n of circles can optimally fit in. It seems that for $n = 4, 6, 8, 9, 10, 12, 15, 19$ and 31, there are exactly two shapes. For all other n tested, only one best aspect ratio was found experimentally.

Another packing surprise awaited us at $n = 49$. Figures 4a and 4b show two among several equivalent packings achieved in our simulation experiments. The configuration a is hexagonal, while configuration b contains a monovacancy, that is, a hole than can accommodate exactly one circle. Monovacancies in hexagonal arrays of congruent circles appear often in the simulation experiments for $14 \leq n < 49$ but only in packings of inferior quality, so that a better quality hexagonal packing without monovacancies could always be found. No higher density hexagonal packing without monovacancies was found for $n = 49$.

Such large n already present substantial difficulties for our simulation procedure. The procedure failed to produce a packing which is better than those in Figures 4a and 4b. However, it is easy to prove that the density of a hexagonal packing in a rectangle with a monovacancy can be increased. For example, to improve the packing in Figure 4b we relocate circle B into the vacancy and then rearrange circles A and C along the side. This reduces the width of the rectangle by a small but positive δ as shown in Figure 4c, where $\delta = 2 - \sqrt{2\sqrt{3}} = 0.13879..$ of the circle radius. The density of c is $49\pi/(2(1 + \sqrt{3})(34 - \delta)) = 0.83200266...$

It is not known whether or not the resulting configuration c can be further improved. However, using the exhaustive search we show (see next section) that the optimum packing for 49 circles in a rectangle, whatever it might be, cannot be purely hexagonal. The value $n = 49$ is apparently the smallest one for which neither square-grid nor perfect hexagonal pattern delivers the optimum.

4 Exhaustive search

The simulation method becomes progressively slower for increasing n. However, by exercising the simulation for smaller n we are able to refine the idea

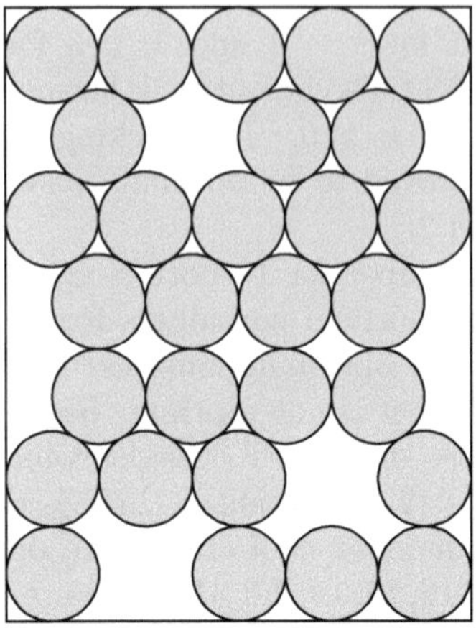

Fig. 8. A "general case" member of the class of packings among which we search for the optimum; here $w = 5$, $h = 5$, $h_- = 2$, $s = 2$, $s_- = 1$, and $d = 3$.

of the class of packings which might deliver the optimum. We have chosen the class that consists of square grid and hexagonal packings and their hybrids; we also allow for monovacancies in configurations of the class, although these configurations are never optimal. We have extended our computing experiments to larger values of n using exhaustive search. For a given n, the method simply evaluates each candidate in the class by computing the area of the rectangle and selects the one with the smallest area. Note that for each n in this class, there are only a finite number of candidates.

A general packing in the class (see Figure 8), consists of $h+s$ rows and has d monovacancies. The h rows are hexagonally alternating, and the s rows are stacked directly on top of the previous row as in square grid packings. The longest row consists of w circles. Assuming that all monovacancies are filled with circles, among the h hexagonal rows, h_- rows consist of $w-1$ circles each, the remaining $h - h_-$ rows consist of w circles each, and among the s square grid rows, s_- rows consist of $w-1$ circles each, and the remaining $s - s_-$ rows consist of w circles each. Then the number of circles with the configuration is

$$n = w(h+s) - h_- - s_- - d \tag{3}$$

Table 1 lists the best packings found of n circles in a rectangle with variable aspect ratio for n in the range $1 \leq n \leq 53$, and Table 2 continues this list for $54 \leq n \leq 213$. The packings in Table 1 are obtained by the simulation procedure described in Section 3 and verified by the exhaustive search. Most packings in Table 2 were generated only by the exhaustive search. An entry in either table consists of the n, followed by the set of integers which represent the parameters of the packing structure as explained above:

w, the number of circles in the longest row,

Table 1. Packings of n circles in a rectangle for $1 \leq n \leq 53$ with w circles in a row, h hexagonal rows, h_- of which consist of $w-1$ circles each, and s square grid rows of w circles each. Except for the case $n = 49$ marked with a star, all packings are the best we could find.

n	w	h	h_-	s	n	w	h	h_-	s	n	w	h	h_-	s
1	1	0	0	1	15	8	2	1	0	33	7	5	2	0
2	2	0	0	1		4	3	1	1	34	9	4	2	0
3	3	0	0	1	16	8	2	0	0	35	12	3	1	0
4	4	0	0	1	17	6	3	1	0	36	12	3	0	0
	2	0	0	2	18	9	2	0	0	37	19	2	1	0
5	5	0	0	1	19	10	2	1	0	38	13	3	1	0
6	6	0	0	1		5	3	1	1	39	13	3	0	0
	3	0	0	2	20	7	3	1	0	40	10	4	0	0
7	7	0	0	1	21	7	3	0	0	41	14	3	1	0
8	8	0	0	1	22	11	2	0	0	42	11	4	2	0
	4	0	0	2	23	8	3	1	0	43	9	5	2	0
9	9	0	0	1	24	8	3	0	0	44	15	3	1	0
	3	0	0	3	25	13	2	1	0	45	15	3	0	0
10	10	0	0	1	26	9	3	1	0	46	12	4	2	0
	5	0	0	2	27	9	3	0	0	47	16	3	1	0
11	6	2	1	0	28	6	5	2	0	48	10	5	2	0
12	12	0	0	12	29	10	3	1	0	*49	17	3	2	0
	6	0	0	2	30	10	3	0	0	50	17	3	1	0
	4	0	0	3	31	16	2	1	0	51	17	3	0	0
13	13	0	0	1		8	3	1	1	52	13	4	0	0
14	5	3	1	0	32	11	3	1	0	53	11	5	2	0

h, the number of rows arranged in a hexagonal alternating pattern,
h_-, the number of rows that consist of $w-1$ circles each,
s, the number of rows, in addition to h rows, that are stacked in the square grid pattern.

Column s is absent in Table 2, because, as explained in Section 3, no optimum packing found for $n > 31$ has s positive. Also, for any n, no optimum packing was found with $s_- > 0$; thus, the s_- column is omitted.

In most packings presented in these two tables, the number of monovacancies $d = 0$, and so the d column is omitted. A few entries where monovacancies are possible are marked with stars. The non-starred entries describe the best packings found. The entries marked with stars cannot be optimal, as explained in Section 3.

An improved packing for the marked case of 49 circles can be obtained as described in Section 3 (see Figure 4). The next marked case $n = 61$ is similar; Table 1 lists the variant a for it with $h = 3$ and $h_- = \lfloor h/2 \rfloor + 1 = 2$. The following marked case is $n = 79$. Here too we improve the packing with the hole by relocating side circles. A, B, C, D, and E as shown in Figure 5, where the value of improvement δ is the same as in the case $n = 49$.

This method of relocation applies to all marked cases in the tables when h is odd. For example, for the case of $h = 5$, when $h_- = 3$ the relocation can be done as illustrated in Figure 10. This applies to the cases $n = 97$, 107, and 142 in Table 2. In all these cases, before using this method, we should replace listed in the tables variant that has $h_- = \lfloor h/2 \rfloor = 2$ with the variant that has $h_- = \lfloor h/2 \rfloor + 1 = 3$. This would produce a configuration with the pattern of the side as in Figure 10a. We would then improve this configuration by relocating circles A, B, C, D, and E to yield a configuration as in Figure 10b. The value of the improvement here is $\delta = 2 - 0.5\sqrt{3} - 3^{1/4}(2\sqrt{3}-1)/(2\sqrt{4-\sqrt{3}}) = 0.05728...$ of the circle radius. A similar method works for even h.

Sometimes during these improvements, some circles become the so-called *rattlers*, i.e., they become free to move and hit their neighbors, like the two unshaded circles in Figure 5. For glass bottles tightly packed in an empty box, packings with rattlers should definitely be avoided even though the box area was minimal!

The case $n = 79$ and several others are marked in Table 2 with two stars to signify that the packing not only *may* contain a monovacancy, but in fact, it *must*: any hexagonal packing of n circles without a monovacancy is provably worse than the one represented in the table when n is marked with two stars.

All such cases in the table also happen to have $h_- = 0$. Also, a double-star marked entry in Table 2 always has a discrepancy between the n computed from the parameters of the packing structure using (3),(here it would be $n = wh$), and actual n. In all the other cases, the n computed from the packing structure parameters using (3) will always match the correct n because there are no monovacancies. However, no entry without a monovacancy is possible in the double-star marked cases. Another observation: an entry $n+1$ that follows an entry n marked by one or two stars, always has the same w and h, and hence its packing fits in the same rectangle as that of the entry n.

5 Best packings found for larger n

The exhaustive search procedure of Section 4 produces the best packings in the class defined in Section 4 for values of n on the order of several thousands. The features we observed for $n \leq 213$ hold until $n = 317$. Namely, only one best size rectangle exists for each n. Most of the best packings in the search class are perfectly hexagonal and unique for their n. Each such packing can be described by the set w, h, and h_-, with the latter parameter taking on up to three possible values:

$$\begin{array}{lllllll} \text{if } h \text{ is even} & \text{then} & h_- = 0 & \text{or} & h_- = h/2 & & \\ \text{if } h \text{ is odd} & \text{then} & h_- = 0 & \text{or} & h_- = \lfloor h/2 \rfloor & \text{or} & h_- = \lfloor h/2 \rfloor + 1 \end{array} \quad (4)$$

Table 2. Packings of n circles in a rectangle for $54 \leq n \leq 213$ with w circles in a row, h hexagonal rows, h_- of which consist of $w - 1$ circles each. Except for the cases marked with stars, all packings are the best we could find.

n	w	h	h_-	n	w	h	h_-	n	w	h	h_-	n	w	h	h_-
54	14	4	2	94	24	4	2	134	34	4	2	174	29	6	0
55	11	5	0	95	14	7	3	135	23	6	3	175	25	7	0
56	19	3	1	96	16	6	0	136	17	8	0	176	20	9	4
57	19	3	0	*97	20	5	3	137	20	7	3	177	30	6	3
58	12	5	2	98	20	5	2	138	28	5	2	178	36	5	2
59	20	3	1	99	17	6	3	*139	16	9	5	179	26	7	3
60	9	7	3	100	20	5	0	140	16	9	4	180	23	8	4
*61	21	3	2	101	34	3	1	141	24	6	3	**181	26	7	0
62	21	3	1	102	15	7	3	*142	29	5	3	182	26	7	0
63	13	5	2	103	21	5	2	143	29	5	2	183	37	5	2
64	16	4	0	104	12	9	4	144	21	7	3	184	23	8	0
65	22	3	1	105	18	6	3	145	29	5	0	185	21	9	4
66	17	4	2	106	27	4	2	146	37	4	2	186	27	7	3
67	10	7	3	*107	22	5	3	147	21	7	0	187	17	11	0
68	14	5	2	108	22	5	2	148	30	5	2	188	24	8	4
69	12	6	3	109	16	7	3	149	17	9	4	189	27	7	0
70	14	5	0	110	22	5	3	150	25	6	0	190	19	10	0
71	24	3	1	111	19	6	3	151	22	7	3	**191	24	8	0
72	18	4	0	112	16	7	0	152	19	8	0	192	24	8	0
73	15	5	2	113	23	5	2	153	31	5	2	193	28	7	3
74	11	7	3	114	19	6	0	154	22	7	0	194	22	9	3
75	15	5	0	115	23	5	0	155	31	5	0	195	20	10	5
76	19	4	0	116	17	7	3	156	20	8	4	196	25	8	4
77	26	3	1	117	20	6	3	*157	23	7	4	**197	22	9	0
78	16	5	2	118	24	5	2	158	23	7	3	198	22	9	0
**79	16	5	0	119	17	7	0	159	27	6	3	*199	29	7	4
80	16	5	0	120	20	6	0	160	20	8	0	200	29	7	3
81	12	7	3	*121	14	9	5	161	23	7	0	201	34	6	3
82	21	4	2	122	14	9	4	162	27	6	0	202	16	13	6
83	17	5	2	123	18	7	3	163	33	5	2	203	23	9	4
84	14	6	0	124	16	8	4	164	21	8	4	204	19	11	5
85	17	5	0	125	25	5	0	165	24	7	3	205	21	10	5
86	22	4	2	126	21	6	0	*166	19	9	5	*206	30	7	4
87	15	6	3	127	12	11	5	167	19	9	4	207	30	7	3
88	18	5	2	128	26	5	2	168	34	5	2	208	26	8	0
89	30	3	1	129	22	6	3	169	13	13	0	209	19	11	0
90	18	5	0	130	19	7	3	170	17	10	0	210	30	7	0
91	13	7	0	131	15	9	4	171	16	11	5	*211	24	9	5
92	23	4	0	132	22	6	0	172	25	7	3	212	24	9	4
93	19	5	2	133	27	5	2	173	35	5	2	213	36	6	3

A few exceptional cases have a single monovacancy. These are similar to the one-star marked cases with odd h and with two options for h_- (the choice of the option defines the presence or absence of the vacancy), or they are similar to the two-star marked cases with $h_- = 0$.

The case $n = 317$ deviates from this pattern. Here, the best packing in the class has parameters $w = 27$, $h = 12$, $h_- = 6$, and $d = 1$. Unlike the one-star marked cases, h is even, and unlike the two-star marked cases, h_- is non-zero. Such a new type of packing recurs with one hole for $n = 334$ ($w = 34$, $h = 10$, and $h_- = 5$) and then for $n = 393$ ($w = 40$, $h = 10$, and $h_- = 5$), but in the latter case with *two* monovacancies. In other words, the best possible packing for 393 circles in a rectangle in the class of hexagonal packings with possible monovacancies must have two of them.

With two holes, we have more freedom to improve the quality of the packing than having just one hole. Several ways are possible, including combining the ones described before for cases of smaller n. Same as for the other packings with monovacancies, the optimal packing for $n = 393$ is not known, but the exhaustive search in our experiments proves that it cannot be a hexagonal packing.

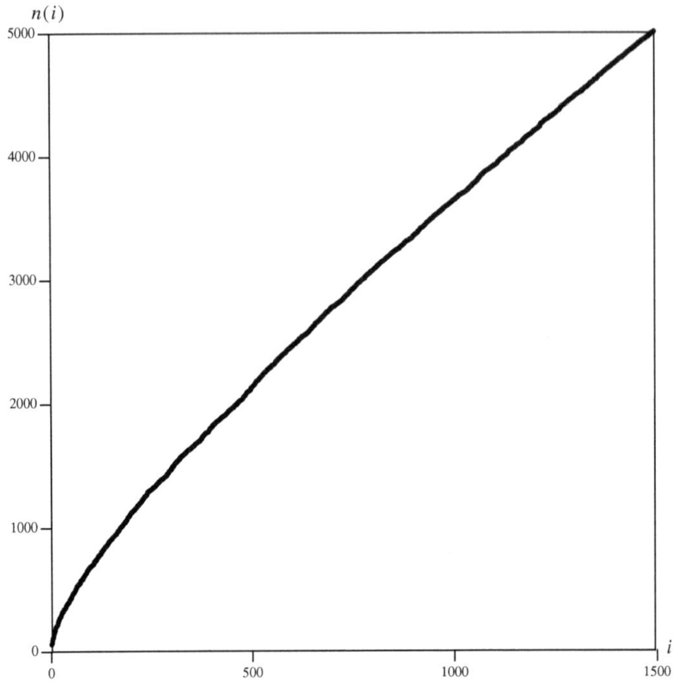

Fig. 9. Numbers of circles $n(i)$ for which the optimum packing in a rectangle is not perfectly hexagonal for $49 \leq n(i) \leq 5000$ plotted versus the rank i.

Naturally, the best packing (in the search class) of $n = 394$ circles also has parameters $w = 40$, $h = 10$, $h_- = 5$ and $d = 1$, i.e., a single monovacancy. The same rectangle also accommodates the best packing in the class of $n = 395$ circles, with the same $w = 40$, $h = 10$, $h_- = 5$. Since the latter packing is without holes, it is conceivable that it is optimal.

The next outstanding case is $n = 411$ where $w = 38$, $h = 11$, $h_- = 6$, and $d = 1$. This is the smallest n where the h is odd and $h_- = \lfloor h/2 \rfloor + 1$, that is, the packing looks like the one in Figure 4a, but unlike the latter it has a hole. An equivalent packing that looks like the one in Figure 4b, where $h_- = \lfloor h/2 \rfloor = 5$ also exists, but unlike the latter it has not one but two holes. For next value $n = 412$, we have a standard one-star marked situation with the same sizes of the rectangle, i.e., $w = 38$, $h = 11$ and $h_- = \lfloor h/2 \rfloor = 5$ for the variant with one hole and $h_- = \lfloor h/2 \rfloor + 1 = 6$ for the variant without holes.

The smallest n for which as many as three monovacancies exist in the best packing in the class is $n = 717$. The corresponding packing parameters are $w = 48$, $h = 15$, $h_- = 0$ and no hexagonal packing with a smaller number of holes can be better or even as good as this packing. Similarly, with four monovacancies, the smallest n is $n = 2732$ ($w = 86$, $h = 32$, and $h_- = 16$), and for five monovacancies, $n = 2776$ is the smallest ($w = 103$, $h = 27$, and

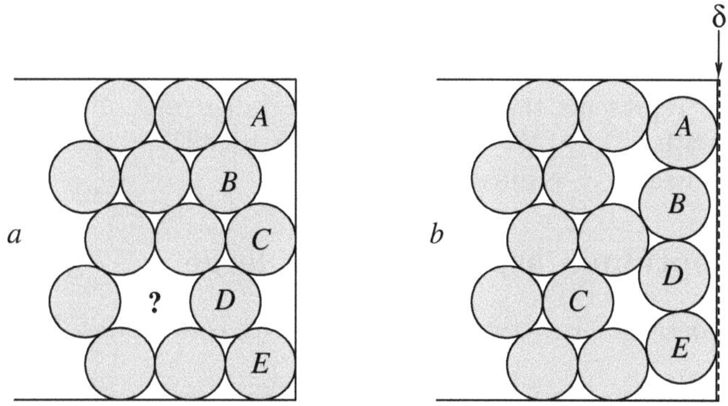

Fig. 10. Improving a packing with a monovacancy when $h = 5$, $h_- = 2$: *a*) an original hexagonal packing with a monovacancy, *b*) the transformed packing.

$h_- = 0$). No optimum packing in the class for $n \leq 5000$ has six or more monovacancies.

In all, we found 1495 values of n on the interval $1 \leq n \leq 5000$ for which the best packing in the class must or can have monovacancies. As explained above, for each such n the best packing provably cannot be of a pure square-grid or hexagonal pattern. It is not proven though, but we believe it to be also true, that those 1495 values of n are *all* such irregular values among the considered 5000 values.

The chance to encounter such an irregular n seems to increase with n. For example, here are all the experimentally found irregular values of n among 100 consecutive values on the intervals $401 + 1000k \leq n \leq 500 + 1000k$ for $k = 0, 1, 2, 3, 4$:

17 values for $401 \leq n \leq 500$:
409, 411, 412, 421, 422, 433, 439, 453, 454, 461, 463, 467, 471, 478, 487, 489, 499.

24 values for $1401 \leq n \leq 1500$:
1401, 1402, 1405, 1409, 1412, 1414, 1423, 1427, 1429, 1434, 1446, 1447, 1451, 1453, 1457, 1459, 1466, 1468, 1477, 1483, 1486, 1487, 1489, 1497.

33 values for $2401 \leq n \leq 2500$:
2401, 2402, 2406, 2411, 2419, 2421, 2423, 2428, 2429, 2435, 2437, 2439, 2441, 2443, 2446, 2452, 2454, 2455, 2456, 2458, 2462, 2467, 2469, 2474, 2476, 2477, 2479, 2481, 2487, 2491, 2493, 2495, 2497.

33 values for $3401 \leq n \leq 3500$:
3407, 3409, 3411, 3412, 3414, 3415, 3418, 3421, 3425, 3428, 3431, 3433, 3436, 3442, 3446 3447, 3453, 3455, 3459, 3461, 3464, 3467, 3469, 3473, 3476, 3479, 3481, 3487, 3489, 3490, 3493, 3494, 3499.

38 values for $4401 \leq n \leq 4500$:

4401, 4404, 4405, 4409, 4411, 4414, 4417, 4419, 4421, 4426, 4430, 4434, 4436, 4438, 4441, 4443, 4447, 4450, 4453, 4456, 4457, 4458, 4461, 4462, 4467, 4468, 4474, 4476, 4479, 4483, 4486, 4487, 4491, 4492, 4493, 4495, 4497, 4499.

Figure 9 represents these 1495 irregular values $n(i)$, $1 \leq n(i) \leq 5000$, beginning with $n(1) = 49$ and ending with $n(1495) = 4999$, by plotting points {abscissa $= i$, ordinate $= n(i)$}.

6 The optimum aspect ratio for large n

Figure 11 contains for each n, a data point with coordinates

$$(\text{abscissa} = n, \text{ ordinate} = \text{best aspect ratio found for this } n).$$

All $n \leq 5000$ are represented with the exception of a few n where the best packings found are of the square grid type. Also, the hybrid packings like those in Figure 7b are excluded. In this way we assure the uniqueness of the best aspect ratio for each n represented. The aspect ratios for the best packings in the search class are described in Section 4. The changes to the aspect ratios are due to the δ shrinkages of the rectangles in those cases with monovacancies, such as those in Figures 4, 5, and 10. Presumably these are small in number and should not be noticeable in Figure 11.

The data points tend to form patterns of descending and ascending "threads." To examine the threads in detail, a rectangular box which is close to the y-axis in Figure 11 is magnified and represented in Figure 12. All data points present in the box can be found in Tables 1 and 2.

The steep downward threads correspond to configurations with fixed height, and with widths increasing by 1 from one point to the next as we move down the thread. Each such thread eventually terminates in either direction which means that the optimum rectangle cannot be too flat or too tall. The less steep upward and downward threads (they are dotted) correspond to sequences (w, h), where from one point to the next w increases by 3 or 6 and h increases by 1 or 2, respectively.

To compute the optimum aspect ratio of the rectangle that encloses the optimum packing for large n, we analyze the area wasted along the rectangle sides. Note that in the infinite hexagonal packing, the uncovered area is $s = \sqrt{3} - \pi/2$ per each $\pi/2$ of the covered area. This is obvious from examining a single triangle XYZ in Figure 13 and observing that the entire infinite packing is composed of such triangles. The waste s here is the area of the central triangle formed by three circular arcs and it is equal to the full area $\sqrt{3}$ of triangle XYZ (assuming circle radius is 1) minus $\pi/2$, which is the total area of the three $\pi/6$ sectors covered. (By the way, we remind the reader that the density of the infinite hexagonal packing is $(\sqrt{3} - s)/(\pi/2) = \pi/(2\sqrt{3}) = 0.90689968....$)

The structure of the uncovered area in the infinite hexagonal packing can be also understood if we think of each covered circle bringing with it two

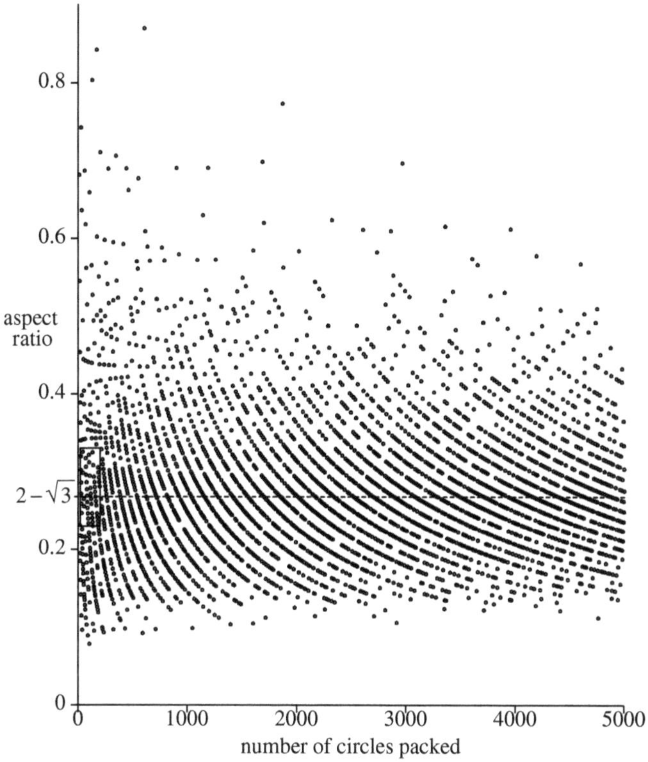

Fig. 11. Values of the aspect ratio for optimal hexagonal packings.

curved uncovered triangles s, those adjacent to this circle on its right-hand side. With such bookkeeping, each triangle s will be counted exactly once. We conjecture that these two triangles s per each circle is the unavoidable, i.e., fixed, waste in any finite hexagonal packing carved out by a rectangle. There will be additional, i.e., variable, waste along the rectangle sides and we are trying to minimize this variable waste.

Along the bottom and top sides, for each 2 units of length, like the side AD of rectangle $ABCD$ in Figure 13, the additional waste is the area of rectangle $ABCD$ minus two covered quarter-circle areas and minus s. This s represents one of the two curved triangles attached to the right-hand side of the circle with the center at B and as an unavoidable waste should not be included. Hence the waste per unit length of the top and bottom sides is $a = (2 - \pi/2 - s)/2 = (2 - \sqrt{3})/2$.

The waste along the left- or right-hand sides per two alternating rows, i.e., per $2\sqrt{3}$ units of length, is the area of the semicircle that is cut in half by the side plus or minus additional area depending on the side. Along the left-hand side, all additional uncovered area is additional waste, since we

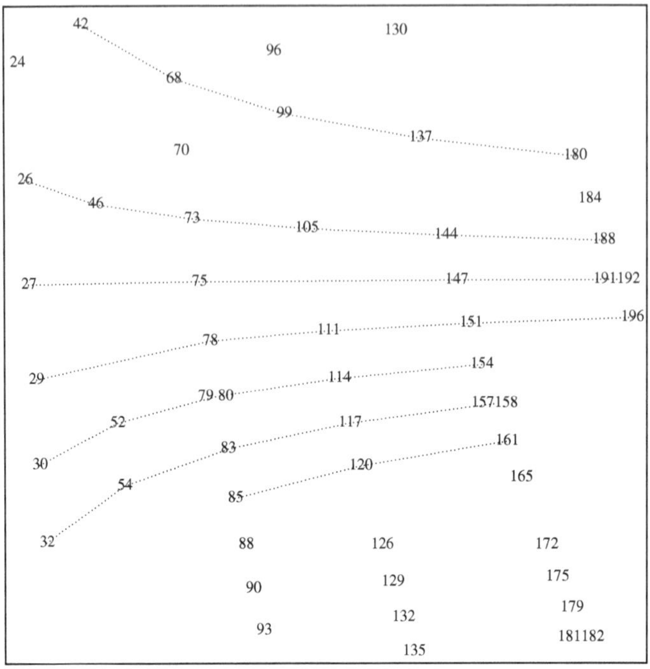

Fig. 12. A box selected in Figure 11 magnified. Each data point in Figure 11 is replaced here with the corresponding number of circles.

assume that the curved triangles s are attached to the right-hand side of the covered circles. The addition consists of two triangles s and two halves of triangles $s/2$ as shown in Figure 13. The left-hand side waste per $2\sqrt{3}$ of length is thus $\pi/2 + 3s$. Along the right-hand side we subtract from the area of the semicircle the outstanding two half triangles $s/2$ because these are necessary in the infinite packing. The right-hand side waste per $2\sqrt{3}$ units of length is thus $\pi/2 - s$. Averaging this for both sides and dividing by $2\sqrt{3}$, we have $b = (\pi/2 + s)/(2\sqrt{3}) = 1/2$ as the additional waste per unit length of left-hand or right-hand sides.

Now we should take $a/b = 2 - \sqrt{3}$ as the ratio of height over width that yields the optimal balance between the waste along the sides of the enclosing rectangle and hence the minimum area. Figure 11 includes a horizontal line at the level of the aspect ratio $2 - \sqrt{3}$.

7 Concluding remarks

As pointed out in the beginning of the paper, we hope that these computational results will lead to proofs of optimality for larger (or even infinite classes of) n.

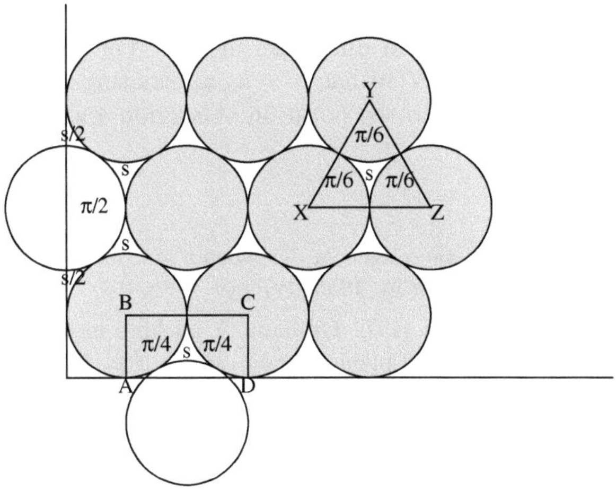

Fig. 13. Calculation of the waste area adjacent to the sides of the rectangle.

Our experiments suggest that the frequency of occurrence of non-hexagonal best packings increases with n. It is not clear whether or not the frequency has a limit as $n \to \infty$ and if it has, whether of not this limit is smaller than 1. While it is not easy to find small $n \geq 14$ for which the best packing is not perfectly hexagonal, it might be more difficult to find large n for which the best packing *is* perfectly hexagonal. Do there exist infinitely many n, for which the densest packing of n circles in a rectangle is hexagonal?

A related phenomenon is conjectured to hold for $n = N(k)$ of the form $N(k) = \frac{1}{2}(a_k + 1)(b_k + 1)$ where a_k and b_k are given by:

$$a_1 = 1, a_2 = 3, a_{k+2} = 4a_{k+1} - a_k$$

$$b_1 = 1, b_2 = 5, b_{k+2} = 4b_{k+1} - b_k$$

so that, $N(2) = 12, N(3) = 120, N(4) = 1512$, etc. The fractions $\frac{a_k}{b_k}$ are actually (alternate) convergents to $\frac{1}{\sqrt{3}}$, and it has been conjectured by Nurmela et al. [NOR] that for these n, a "nearly" hexagonal packing of n circles in a square they describe is in fact optimal.

In our case, one should seek alternate convergents to $\sqrt{3} + 3/2$ which yields sequences a_k and b_k given by:

$$a_k = 2v_{k+1} - v_k, \quad b_k = 2v_k,$$

where

$$v_0 = 0, \quad v_1 = 1, \quad v_{k+2} = 4v_{k+1} - v_k, \quad k = 2, 3, ...$$

Thus,

$$(a_1, b_1) = (7, 2), \quad (a_2, b_2) = (26, 8), \quad (a_3, b_3) = (97, 30)....$$

so that $N(2) = 208$, $N(3) = 2910$, etc. No such $N(k)$ for $k = 2, 3, ...$ is indeed found to be irregular in our experiments. The best packing found experimentally for such an $N(k)$ has $h = b_k$ alternating rows of full length $w = a_k$ with $h_- = 0$, $s = 0$, in the notation of Section 4 and Tables 1 and 2.

References

[Ft] L. Fejes-Toth, Lagerungen in der Ebene, auf der Kugel und im Raum, *Springer-Verlag*, Berlin, 1953, 197 pp.

[FG] J. H. Folkman and R. L. Graham, A packing inequality for compact convex subsets of the plane, *Canad. Math. Bull.* **12** (1969), 745–752.

[Füredi] Z. Füredi, The densest packing of equal circles into a parallel strip. *Discr. Comp. Geom.* **6** (1991), no. 2, 95–106.

[GL] R. .L. Graham and B. D. Lubachevsky, Repeated Patterns of Dense Packings of Equal Disks in a Square, *Electr. J. Combin.* **3(1)** (1996), #R16.

[L] B. D. Lubachevsky, How to simulate billiards and similar systems, *J. Comp. Phys.* **94** (1991), 255–283.

[NO1] K. J. Nurmela and P. R. J. Östergård, Packing up to 50 equal circles in a square, *Discr. Comp. Geom.* **18** (1997), 111-120.

[NO2] K. J. Nurmela and P. R. J. Östergård, Optimal Packings of Equal Circles in na Square, in: *Combinatorics, Graph Theory, and Algorithms*, Vol. II, Y. Alavi, D.R. Lick, and Schwenk (eds.), New Issues Press, Kalamazoo 1999

[NO3] K. J. Nurmela and P. R. J. Östergård, More optimal packings of equal circles in a square, *Discr. Comp. Geom.* **22** (1999), 439-457.

[NOR] K. J. Nurmela, P. R. J. Östergård and R. aus dem Spring, Asymptotic behavior of optimal circle packings in a square, *Canad. Math. Bull.* **42** (1999), 380-385.

[Oler] N. Oler, A finite packing problem, *Canad. Math. Bull.* **4** (1961), 153-155.

[Ruda] M. Ruda, The packing of circles in rectangles. (Hungarian. English summary) *Magyar Tud. Akad. Mat. Fiz. Tud. Oszt., Közl.*, **19** (1970), 73–87.

[Th] A. Thue, Über die dichteste Zussamenstellung von kongruenten Kreisen in einer Ebene, *Christiania Vid. Selsk. Skr.* no. 1 (1910), 3-9.

Colorings and Homomorphisms of Minor Closed Classes

Jaroslav Nešetřil
Patrice Ossona de Mendez

Abstract

We relate acyclic (and star) chromatic number of a graph to the chromatic number of its minors and as a consequence we show that the set of all triangle free planar graphs is homomorphism bounded by a triangle free graph. This solves a problem posed in [15]. It also improves the best known bound for the star chromatic number of planar graphs from 80 to 30. Our method generalizes to all minor closed classes and puts Hadwiger conjecture in yet another context.

1 Introduction

Denote by $\chi_a(G)$ the acyclic chromatic number of a graph G, i.e. the minimum number of colors which are sufficient for a (proper) coloring of the vertices of G so that every cycle in G gets at least 3 colors. It is known that χ_a is bounded for graphs of bounded genus and also for bounded degree graphs, see [2,3] for the best known bounds. Similarly, let $\chi_{\mathrm{st}}(G)$ denote the *star chromatic* number of a graph G, i.e. the minimal number of colors which are sufficient for a (proper) coloring of the vertices of G so that 4 vertices of every path of length 3 in G get at least 3 colors. Clearly $\chi_{\mathrm{st}}(G) \geq \chi_a(G)$. It is known that χ_{st} is bounded whenever χ_a is bounded (folklore, see e.g. [6]).

It is also well known that χ_a differs arbitrarily from χ as it is unbounded for bipartite graphs and even for bipartite 2-degenerated graphs: It suffices to consider the graph $K_n^\star$ which we get from K_n by subdividing every edge by a single vertex. We complement these results by the following result:

Theorem 1.1. *There exists a function $f : \mathbf{N} \to \mathbf{N}$ such that for any graph G holds*

$$\chi_a(G) \leq \chi_{\mathrm{st}}(G) \leq f(\max \chi(H))$$

where the maximum is taken over all minors H of G. In fact $f(n) \leq cn^2$ where c is a constant.

(The proof is in Section 3.)

Recall, that G is a *minor* of H if G can be obtained from a subgraph of H by a sequence of edge-contractions. We consider only loopless simple graphs.

We denote by $\preceq$ the minor relation for graphs. A class $\mathcal{K}$ is said to be *minor closed* if $H \in \mathcal{K}$ and $G \preceq H$ implies $G \in \mathcal{K}$. The class $\mathcal{K}$ is said to be *proper* if it does not contain all graphs. It follows that for any minor closed class $\mathcal{K}$, $\{\chi_a(G), G \in \mathcal{K}\}$ is bounded iff $\{\chi(G), G \in \mathcal{K}\}$ is bounded. It follows that this is equivalent to bounded *oriented chromatic number* $\vec{\chi}$ ([12, 18]) and to bounded colorings of mixed graphs with colored edges ([1, 19]), see Theorem 6.1.

We shall make use of the following two (we believe) well known results. We sketch a proof of Lemma 2 for completeness.

Lemma 1. *For each k, there exists some natural number $h(k) < \gamma k\sqrt{\log k}$ (for some constant γ) such that every graph of minimum degree at least $h(k)$ contains K_k as a minor.*

This Lemma 1 has been proved by Kostochka [11] and Thomason [21] (extending earlier work of Mader [13]). The constant γ is well understood, [22].

Lemma 2. *For any minor closed class $\mathcal{K}$, $\{\chi(G), G \in \mathcal{K}\}$ is bounded if and only if $\mathcal{K}$ is proper (i.e. different from the class of all graphs).*

Proof. If $\{\chi(G), G \in \mathcal{C}\}$ is unbounded then the vertex critical graphs in $\mathcal{C}$ have unbounded minimal degree and Lemma 1) applies. □

We discovered Theorem 1.1 in the context of graphs and their homomorphisms. Recall: A *homomorphism* $f : G \to H$ is a mapping $f : V(G) \to V(H)$ satisfying $\{f(x), f(y)\} \in E(H)$ whenever $\{x, y\} \in E(G)$. We also write $G \leq H$ if there exists a homomorphism $G \to H$. This quasiorder is denoted by $\mathcal{C}$. It is called *homomorphism* or *coloring order*. By the well known *Grötsch's Theorem* (see e.g. [10] and [23] for a simple proof), every triangle free planar graph is 3-colorable. Using homomorphism order $\mathcal{C}$ this means that $G \leq K_3$ for any triangle free planar graph G. In the other words K_3 is an upper bound (in the coloring order $\mathcal{C}$) of the class $\mathcal{P}_3$ of all triangle free planar graphs. We can also say that the class $\mathcal{P}_3$ of all K_3-free graphs is bounded by K_3 (in $\mathcal{C}$). The following problem has been formulated in [15]:

Problem 1. *Does there exist a triangle free graph H such that $G \leq H$ for any triangle free planar graph G? In other words, is the class $\mathcal{P}_3$ (of all planar triangle free graphs) bounded by a triangle free graph H?*

Here we give an affirmative answer to this problem. In fact one can prove that the class $\mathcal{P}_3$ is bounded by a 3-colorable triangle free graph and we also prove an analogous result for any minor closed class $\mathcal{K}$ with bounded chromatic number:

Theorem 1.2. *Let $\mathcal{C}$ be a minor closed class of graphs all of which are k-colorable. Then the class $\mathcal{C}_3$ of all triangle free graphs in $\mathcal{C}$ is bounded by a k-colorable triangle free graph.*

Related results were obtained recently in [14] and we use one of the constructions of [14] here (Section 4). However the case of planar triangle-free graphs was left open. Here we treat the problem in a more general context. All results may be seen as an evidence for the following general conjecture.

Let $\mathcal{A}, \mathcal{B}$ be classes of graphs, $\mathcal{A} \subseteq \mathcal{B}$. We say that the class $\mathcal{A}$ is *bounded in* $\mathcal{B}$ if there exists a graph $H \in \mathcal{B}$ such that $G \leq H$ for any $G \in \mathcal{A}$. (Thus $\mathcal{A}$ is bounded in $\mathcal{B}$ if the class $\mathcal{A}$ is bounded by a graph in $\mathcal{B}$.) Note that $\mathcal{A}$ is bounded in $\mathcal{A}$ iff $\mathcal{A}$ has the greatest element (with respect to $\leq$). The study of boundedness phenomena is one of the basic problems and we are pleased that in our setting it relates questions like Hadwiger conjecture to the mainstream mathematics (see Remarks).

Given a finite set $\mathcal{F}$ of graphs we denote by $\mathrm{Forb}_h(\mathcal{F})$ the class of all graphs G for which there is no homomorphism $F \to G$ for every $F \in \mathcal{F}$. Equivalently and more formally, $\mathrm{Forb}_h(\mathcal{F}) = \{G; F \in \mathcal{F} \Rightarrow F \not\leq G\}$. As an example, note that $\mathrm{Forb}_h\{K_3\}$ is the class of all triangle free graphs.

Conjecture 1. *Let $\mathcal{F}$ be any finite set of graphs. Let $\mathcal{K}$ be a proper minor closed class of graphs. Then the class $\mathcal{K} \cap \mathrm{Forb}_h(\mathcal{F})$ is bounded in* $\mathrm{Forb}_h(\mathcal{F})$.

The results of this paper verify the conjecture for $\mathcal{F} = \{K_3\}$. It has been proved in [4, 9] (see also [14] for a different proof) that for the class $\mathcal{C}_d$ of all graphs with all their vertices having a degree bounded by d and for any finite set of graphs $\mathcal{F}$ the analogous conjecture holds. Note that for graphs in general and even classes of degenerated graphs the analogous statement fails to be true: For $\mathcal{F} = \{K_3\}$ consider graphs $K_n^{\star\star}$ formed from K_n by subdividing each edge by two new vertices. All the graphs $K_n^{\star\star}$ are 2-degenerated yet they are not bounded by a finite triangle-free graph. Note also that this cannot be saved by (large) girth: Let G be a graph of girth ℓ with chromatic number k. Then the graph $G^{\star\star}$ has girth 3ℓ and there is no homomorphism of G into a triangle free graph with at most k vertices.

This paper is organized as follows: In Section 1 we prove results on star chromatic number of graphs. In Section 3 we prove Theorem 1.1. In Section 4 we prove the following result which is perhaps of an independent interest:

Theorem 1.3. *For any minor closed class $\mathcal{C}$ with bounded chromatic number there exists an integer $k = k(\mathcal{C})$ such that any graph $G \in \mathcal{C}$ has a proper k-coloring with the property that any odd cycle of length ≥ 5 get at least 4 different colors.*

Note that, again, an analogous statement fails to be true in general: In any k-coloring of the graph $K_n^{\star\star}$, n sufficiently large, there exist cycles of length ≥ 6 which are colored by at most 3 colors. Also the graphs $G^{\star\star}$ (see above) have girth 3ℓ and in any k-coloring contain cycles colored by at most 3 colors.(More complicated examples are provided by Ramsey theory.)

The key notion for the proof of Theorem 1.3 is the notion of *folding* and using that we prove in Section 5 Theorem 1.2 by an universal construction similar to those given in [1, 14, 19]. Section 5 contains concluding remarks and open problems.

2 Star chromatic number bounded by density

Recall that the (hereditary) density $m(G)$ of a graph G is $\max_{H \subseteq G} \frac{|E(H)|}{|V(H)|}$.

Theorem 2.1. *Let G be a simple graph with density α, and let $\beta \geq \alpha$ be the maximum density over the minors of G which are simple (including G itself). Then, the star chromatic number $\chi_{\mathrm{st}}(G)$ is bounded by:*

$$\chi_{\mathrm{st}}(G) \leq \lceil \alpha \rceil (2\lceil \beta \rceil + \lceil \alpha \rceil - 1) + \lfloor 2\alpha \rfloor + 1.$$

Proof. First, we shall find an acyclic orientation $\vec{G}$ such that each vertex has indegree at most $\lceil \alpha \rceil$. (It is well known that a graph G with density $t = \max_{H \subseteq G} \frac{|E(H)|}{|V(H)|}$ may be oriented in such a way that any vertex v of G has indegree $d^-(v)$ at most $\lceil t \rceil$ (see for instance, Hakimi [8].) The acyclic orientation $\vec{G}$ leads to a linear ordering $x_1, \dots, x_n$ of vertices of $\vec{G}$ with the property that $(x_i, x_j) \in E(\vec{G})$ implies $i < j$. Proceeding for vertices $x_1, x_2, \dots$ we can obviously color each arc e of $\vec{G}$ with a color $c(e) \in \{1, \dots, \lceil \alpha \rceil\}$, in such a way that all the arcs incoming a same vertex have different colors.

Then, let $\widetilde{G}$ be the graph $(V(G), E(G) \cup E_1 \cup E_2)$, where (denoting (x, y) an arc and $\{x, y\}$ an edge):

- $(x, y) \in E_1$ if $\{x, y\} \notin E(G)$ and $\exists z \in V(G), \{(x, z), (z, y)\} \subseteq E(\vec{G})$,
- $(x, y) \in E_2$ if $\{x, y\} \notin E(G)$ and $\exists z \in V(G), \{(x, z), (y, z)\} \subseteq E(\vec{G})$ with $c((x, z)) < c((y, z))$.

Thus the graph $\widetilde{G}$ is the graph G together with potentially conflicting pairs of vertices for a star coloring. We now estimate the density of $\widetilde{G}$. Let A be any subset of $V(G)$ and let B be the set of neighbors of A. Formally, B is the subset of $V(G) \setminus A$ defined by:

$$B = \{x \in V(G) \setminus A; \quad \exists v \in A, \{v, x\} \in E(G)\}$$

The edges of the subgraph $\widetilde{G}_A$ of $\widetilde{G}$ induced by A may be partitioned as follows: $E(\widetilde{G}_A) = E(G_A) \cup E_1^{\mathrm{i}} \cup E_1^{\mathrm{e}} \cup E_2^{\mathrm{i}} \cup E_2^{\mathrm{e}}$, where:

- $(x, y) \in E_1^{\mathrm{i}}$ if $\{x, y\} \notin E(G_A)$ and $\exists z \in A, \{(x, z), (z, y)\} \subseteq E(\vec{G})$,
- $(x, y) \in E_1^{\mathrm{e}}$ if $\{x, y\} \notin E(G_A)$ and $\exists z \in B, \{(x, z), (z, y)\} \subseteq E(\vec{G})$,
- $(x, y) \in E_2^{\mathrm{i}}$ if $\{x, y\} \notin E(G_A)$ and $\exists z \in A, \{(x, z), (y, z)\} \subseteq E(\vec{G})$ with $c((x, z)) < c((y, z))$.
- $(x, y) \in E_2^{\mathrm{e}}$ if $\{x, y\} \notin E(G_A)$ and $\exists z \in B, \{(x, z), (y, z)\} \subseteq E(\vec{G})$ with $c((x, z)) < c((y, z))$.

Let H be the graph obtained from $G_{A\cup B}$ by deleting any edges having both incidences in B, and let H_i be the graph obtained from H by deleting all the vertices of B having no incoming edge of color i and contracting all the arcs $(x,y), y \in B$ of color i. Moreover, $V(H_i)$ may be identified with A and the edge set of H_i is given by:

$$\begin{aligned} E(H_i) =& E(G_A) \\ & \cup \{(x,y) \in E_1^{\mathrm{e}}, \exists z \in B, \{(x,z),(z,y)\} \subseteq E(H) \text{ and } c((x,z)) = i\} \\ & \cup \{(x,y) \in E_2^{\mathrm{e}}, \exists z \in B, \{(x,z),(y,z)\} \subseteq E(H) \text{ and } c((x,z)) = i\} \end{aligned}$$

Hence, we have, as H_i being a minor of H an thus a minor of G, has density at most β (and order $|A|$):

$$\lceil\alpha\rceil\lceil\beta\rceil|A| \geq \sum_{i=1}^{\lceil\alpha\rceil} |E(H_i)| \geq \lceil\alpha\rceil|E(G_A)| + |E_1^{\mathrm{e}}| + |E_2^{\mathrm{e}}|$$

It follows that

$$\lceil\alpha\rceil\lceil\beta\rceil|A| - \lceil\alpha\rceil|E(G_A)| \geq |E_1^{\mathrm{e}}| + |E_2^{\mathrm{e}}|$$

Moreover, we obviously have $|E_1^{\mathrm{i}}| \leq \lceil\alpha\rceil|E(G_A)|$ and $|E_2^{\mathrm{i}}| \leq \binom{\lceil\alpha\rceil}{2}|A|$. Summarizing, we get:

$$|E(\widetilde{G}_A)| \leq |E(G_A)| + \left(\lceil\alpha\rceil\lceil\beta\rceil + \binom{\lceil\alpha\rceil}{2}\right)|A|$$

and thus, the density of $\widetilde{G}$ is at most:

$$\max_{A\subseteq V(G)} \frac{|E(\widetilde{G}_A)|}{|A|} \leq \lceil\alpha\rceil\lceil\beta\rceil + \binom{\lceil\alpha\rceil}{2} + \alpha$$

It follows that $\chi(\widetilde{G}) \leq 2\lceil\alpha\rceil\lceil\beta\rceil + \lceil\alpha\rceil(\lceil\alpha\rceil - 1) + \lfloor 2\alpha \rfloor + 1$.

Now, consider a coloring of $\widetilde{G}$ with $\chi(\widetilde{G})$ colors and think of this as a coloring of G. Then, for any path P of length 3 in $\vec{G}$, P either contains a subpath of length 2 whose internal vertex is a sink or it contains a directed subpath of length 2. In both cases, according to the definition of $\widetilde{G}$, this path induces a triangle in $\widetilde{G}$ and thus is 3-colored. As a consequence, no bi-colored subgraph of G may include a path of length 3 and, as the coloring is obviously a proper coloring of G (G is a partial graph of $\widetilde{G}$), we get $\chi_{\mathrm{st}}(G) \leq \chi(\widetilde{G})$, which concludes the proof. □

Corollary 1. *The star chromatic number of every planar graph is at most 30.*

This improves bounds given in [6].

Proof. For any planar graph G of order $n \geq 3$, we get $\alpha \leq \beta \leq 3 - \frac{6}{n} < 3$. □

Here is an example where we use $\alpha \neq \beta$:

Corollary 2. *The star chromatic number of a bipartite planar graph is at most 18, The star chromatic number of a planar graph with the degree of all its vertices ≤ 4 is at most 19.*

Proof. For any bipartite planar graph G of order $n \geq 3$, we get $\alpha \leq 2 - \frac{4}{n} < 2$ and $\beta < 3$ (as any minor of a bipartite planar graph is planar). For any planar graph with the degree of all its vertices ≤ 4 holds $\alpha \leq 2$ and $\beta < 3$ (as previously). □

Let us remark that the above proof of Theorem 2.1 gives a structural characterization of star chromatic number in terms of chromatic number. Given an orientation $\vec{G}$ of G we define an undirected graph $\breve{G}$ (similarly as the graph $\widetilde{G}$ in the above proof): $V(\breve{G}) = V(G)$ with the edges of $E(\breve{G})$ being all edges of G together with all pairs $\{x, y\}$ for which there exists a vertex $z \in V(G)$ such that either $(x, z), (z, y) \in E(\vec{G})$ or $(x, z), (y, z) \in E(\vec{G})$.

Corollary 3. $\chi_{st}(G) = \min \chi(\breve{G})$ *where minimum is taken over all graphs $\breve{G}$ which correspond to an orientation $\vec{G}$ of G.*

Proof. Clearly any orientation $\vec{G}$ of G and any proper coloring of $\breve{G}$ gives a star coloring of G. Conversely, given any star coloring c of G define orientation $\vec{G}$ as follows: for an edge $e = \{x, y\}$ of G with colors $c(x) = i, c(y) = j$ the subgraph induced by all the vertices with colors i and j is a star forest and we orient e from the center of the star (in the case that the star has 2 vertices we take any orientation). One can check that c is a proper coloring of the graph $\breve{G}$ corresponding to this orientation $\vec{G}$. □

3 Acyclic chromatic number bounded by chromatic number

In this section we prove Theorem 1.1. First, we shall relate the density of a graph with the maximum value of the chromatic numbers of its minors.

Lemma 3. *There exists a function $g : \mathbb{N} \to \mathbb{N}$, such that any graph G has density bounded as follows:*

$$\max_{H \subseteq G} \frac{|E(H)|}{|V(H)|} \leq g(\max_{H \preceq G} \chi(H)) \tag{1}$$

Proof. According to Lemma 1, there exists a natural number $h(k)$, such that every graph of minimum degree at least $h(k)$ contains K_k as a minor.

Thus, assume G has density $\alpha = \max_{H \subseteq G} \frac{|E(H)|}{|V(H)|}$. Then, it includes as a subgraph a graph H with minimum degree at least α and thus G has K_p as a minor with $h(p) \geq \alpha$. Thus,

$$\max_{H \subseteq G} \frac{|E(H)|}{|V(H)|} \leq h^{-1}(\max_{H \preceq G} \chi(H)) \tag{2}$$

□

Now we can apply

$$\chi_a \leq \chi_{st} \leq \lfloor 2\alpha \rfloor + 3\lceil \alpha \rceil^2 - \lceil \alpha \rceil + 1$$

and we get Theorem 1.1.

4 Foldings

Definition 1. Let G and H be graphs and let $f : V(G) \to V(H)$ be a homomorphism from G to H. Then f is a *folding* of G in H if

$$\forall x, y, z \in V(G), \quad \big((x,z) \in E(G) \wedge (y,z) \in E(G)\big) \implies f(x) \neq f(y) \tag{3}$$

The following result generalizes [18] where a similar result was obtained for (no partially colored) degree 3 planar graphs.

Proposition 1. *Let G be an undirected graph with acyclic chromatic number $\chi_a(G)$. Assume G is partially oriented and let Δ^- be the maximum indegree of the partial orientation. Then, there exists an extension of the partial orientation of G into a full orientation and a folding, with respect to this orientation, from G to K_p (with some orientation), where*

$$p \leq \chi_a(G)(\Delta^- + 2)^{\chi_a(G)-1}.$$

Proof. Consider an acyclic coloring of G with $\chi_a(G)$ colors $\{1, \ldots, \chi_a(G)\}$. Let $1 \leq i < j \leq \chi_a(G)$. According to the definition of an acyclic coloring, the subgraph $G_{i,j}$ induced by the vertices colored i or j is a forest. Consider an extension of the orientation of the edges of $G_{i,j}$ such that the originally non-oriented edges of any tree not reduced to an isolated vertex is oriented from a root. Then, each vertex has indegree at most $\Delta^- + 1$. Then, compute a function $\mu_{i,j} : V(G_{i,j}) \to \{0, 1, \ldots, \Delta^- + 1\}$ as follows: For any isolated vertex v, $\mu_{i,j}(v) = 0$. For any non trivial tree Y, let r be a vertex of the tree of color i and let $X = \{r\}$. While $X \neq Y$, we consider a vertex v in $Y \setminus X$ having a neighbor u in X. Then, we compute $\mu_{i,j}(v)$ as follows:

- if $\{u, v\}$ is oriented from u to v and u has color i, then $\mu_{i,j}(v) = \mu_{i,j}(u)$;
- if $\{u, v\}$ is oriented from u to v and u has color j, then $\mu_{i,j}(v) \equiv \mu_{i,j}(u) + 1 \pmod{\Delta^- + 2}$;
- if $\{u, v\}$ is oriented from v to u and u has color i, then choose $\mu_{i,j}(v) \not\equiv \mu_{i,j}(u) \pmod{\Delta^- + 2}$ while avoiding the values given to the predecessors of u (that is: the neighbors z of u, such that $\{z, u\}$ is oriented from z to u) already in X (at most Δ^- such vertices exist);

- if $\{u, v\}$ is oriented from v to u and u has color j, then choose $\mu_{i,j}(v) \not\equiv \mu_{i,j}(u)+1 \pmod{\Delta^- + 2}$ while avoiding the values given to the predecessors of u already in X (at most Δ^- such vertices exist).

If we consider the union of the preceding partial orientations for all the possibles values of i and j $(1 \le i < j \le \chi_a(G))$ and if we recolor any vertex v of color c with the t-tuple $(c, \mu_{1,c}(v), \dots \mu_{c-1,c}(v), \mu_{c,c+1}(v), \dots, \mu_{c,\chi_a(G)}(v))$, we obtain a natural folding of G in $K_{\chi_a(G)(\Delta^-+2)^{\chi_a(G)-1}}$. □

We prove Theorem 1.3 in the following more technical form. Together with the previous Proposition and Theorem 1.1 this implies Theorem 1.3.

Theorem 4.1. *Let G be a graph and f a folding of G in a complete graph K_p. Then, there exists a coloring of G with at most q colors, such that any odd cycle of G of length at least 5 has vertices of at least 4 colors, where $q \le \left(\max_{H \preceq G} \chi(H)\right)^{\binom{p}{3}}$.*

Proof. Consider any numbering of the vertices of K_p with integers $1, \dots, p$ and let $i, j, k \in \{1, \dots, p\}$ be three distinct integers. Then, f induces a folding of the subgraph $G_{i,j,k}$ of G induced by the vertices mapped by f into one of i, j, k in a directed triangle, whose vertices are i, j, k. Two cases may then occur:

- The triangle (i, j, k) is directed as a circuit.

  In such a case, any vertex in $G_{i,j,k}$ has at most one incoming edge and any cycle of $G_{i,j,k}$ is thus oriented as a circuit. Moreover, no vertex may belong to more than one circuit and every circuit has a length which is a multiple of 3, with vertices successively colored i, j, k. Then, we assign a mark from $\{0, 1\}$ to the vertices: 0 for all the vertices, but one in each circuit. This way, in each cycle of length bigger than 3 there exist at least 3 vertices of different colors with mark 0 and one vertex with mark 1.

- The triangle (i, j, k) is acyclically oriented.

  Then, we may also assume it is directed as $i \to j \to k$. In such a case, no cycle of $G_{i,j,k}$ may be directed as a circuit and hence every cycle γ of $G_{i,j,k}$ includes at least one sink v. As the sink has to have at least two incoming edges, it is colored k and its neighbors have respectively color i and j. Let H be the minor of G obtained by contracting all the edges incident to a vertex colored i and call $g : V(G) \to V(H)$ the identification mapping of the contraction. The loops in H then correspond to initial cycles of length 3. Let $\hat{H}$ be the graph obtained from G by removing these loops (we identify $V(\hat{H})$ with $V(H)$) and let $\chi(\hat{H})$ be its chromatic number. Color $\hat{H}$ with $\chi(\hat{H})$ colors and call $\mu : V(H) \to \{1, \dots, \chi(\hat{H})\}$ the coloring. Then, consider an odd cycle γ of $G_{i,j,k}$ having length at least 5. We will prove that this cycle contains four vertices having distinct $(f(v), \mu \circ g(v))$ pairs. This cycle corresponds, in $\hat{H}$ to an union of cycle (as some paths joining two vertices of γ may be contracted to a single vertex of $\hat{H}$). As, in any cycle, the number of edges contracted is even (2 per vertex colored i in the cycle), one of the elementary cycle γ' resulting from γ will have an odd length. Three cases may occur:

- γ' is a loop.
  Then, three vertices respectively colored i, j, k have been contracted into a single vertex. These three vertex and the neighbor of the vertex colored j in γ which is not colored i will get four distinct $(f(v), \mu \circ g(v))$ pairs.
- γ' has length at least 3 and γ includes no vertex colored i mapped to a vertex of γ'.
  Then, γ' will get at least 3 colors in the $\hat{H}$. Moreover, as notice before, γ includes at least one vertex v_0 colored i. The vertex v_0 and three vertices of γ mapped to vertices of γ' having 3 different colors in the coloration of $\hat{H}$ will have four distinct $(f(v), \mu \circ g(v))$ pairs.
- γ' has length at least 3 and γ includes a vertex v_0 colored i mapped to a vertex of γ'. Let v_1 be a neighbor of v_0 ($g(v_1) = g(v_0)$) and let v_2, v_3 be two vertices of γ, such that $g(v_0), g(v_2)$ and $g(v_3)$ get 3 distinct colors in $\hat{H}$. Then, v_0, v_1, v_2, v_3 will have four distinct $(f(v), \mu \circ g(v))$ pairs.

□

5 Triangle free bounds

In this section we prove Theorem 1.2. Let $\mathcal{K}$ be a minor closed class of graphs. We assume that $\chi(G), G \in \mathcal{K}$ is a bounded set. By Theorems 1.1 and 1.3 we know that there exists a positive q such that any graph $G \in \mathcal{K}$ may be proper colored by q colors in such a way that any odd cycle of G of length ≥ 5 gets at least 4 colors. We construct graph $H = (V, E)$ as follows: The vertices V are all pairs of the form (i, ϕ) where $1 \leq i \leq q$ and ϕ is a function which assigns to every triple $T = \{i, j, k\}; 1 \leq j < k < l \leq q$ value $\phi(T) \in \{0, 1\}$ with $\phi(T) = 1$ whenever $i \notin T$. The edges E are all pairs of the vertices of the form $\{(i, \phi), (i', \phi')\}$ where $i \neq i'$ and $\phi(T) \neq \phi'(T)$ whenever triple T contains both i and i'. We shall prove that the graph H has all the properties claimed by Theorem 1.2.

The graph H has clearly no triangle as if $(i, \phi), (i', \phi'), (i'', \phi'')$ are 3 vertices of H then considering the triple $T = \{i, i', i''\}$ we see that at least two of the values $\phi(T), \phi'(T), \phi''(T)$ coincide and thus the corresponding vertices do not form an edge of H. Next we prove that H is a bound for the class $\mathcal{K}_3$ of all triangle free graphs in $\mathcal{K}$. Towards this end let $G \in \mathcal{K}_3$ and let

$c : V(G) \to \{1, \dots q\}$ be a proper coloring of G guaranteed by Theorem 1.3. Given a triple $T \subset \{1, \dots q\}$ the subgraph G_T of G induced by the set $c^{-1}(T)$ is bipartite and thus there exists a homomorphism (coloring) $\phi_T : G_T \to K_2$. Define the mapping $f : V(G) \to V(H)$ as follows: $f(v) = (c(v), \phi)$ where $\phi(T)(v) = \phi_T(v)$ providing $v \in T$ and $\phi(T)(v) = 1$ otherwise. It is easy to check that this is a homomorphisms $G \to H$. Finally suppose that all graphs from the class $\mathcal{K}$ are k-colorable. Then the graph $H \times K_k$ is k-colorable triangle free bound for $\mathcal{K}_3$. This completes the proof of Theorem 1.2. □

From Theorem 1.2, one can deduce the following:

Corollary 4. *There exists a function $f : \mathbb{N} \to \mathbb{N}$ such that, for any minor closed class of graph $\mathcal{C}$ with maximum chromatic number k, any triangle free graph $G \in \mathcal{C}$ may be properly colored in $f(k)$ colors, in such a way that any subgraph H of G gets a number of colors at least equal to the minimum number of vertices of a triangle free graph with chromatic number $\chi(H)$.*

Proof. According to Theorem 1.2, there exists a triangle free graph U with order $f(k)$, such that $G < U$, for any $G \in \mathcal{C}$. Color the vertices of G according to their image by a homomorphism from G to U. Then, any subgraph H of G is mapped into a subgraph of U_H of U. The graph H has $|V(U_H)|$ colors and chromatic number $\chi(H) \le \chi(U_H)$. □

Corollary 5. *There exists a constant c, such that, for any triangle free graph G, there exists a coloring of G with $f(\max_{H \preceq G} \chi(H))$ colors, such that any subgraph H of G with chromatic number k gets at least $ck^2 \log k$ colors.*

Proof. According to Erdős and Hajnal [5], a triangle free graph with chromatic number k has order at least $ck^2 \log k$. □

6 Remarks

1. The notion of folding is an interesting notion as it is sandwiched between locally injective homomorphisms (studied e.g. in [16] and [7] from complexity point of view) and homomorphisms of bounded degree graphs. Because of this we formulated Proposition 1 in a greater generality (than necessary for our proof of Theorem 1.2. More results are going to appear in our forthcoming paper [17].

2. Our Conjecture 1 may be seen as a finitary approximation to Hadwiger conjecture, see e.g. [10]. In our language Hadwiger conjecture may be expressed as follows:

Conjecture 2. *(Hadwiger)*

Any minor closed class $\mathcal{K}$ with bounded chromatic number has a greatest element which is a complete graph.

In other words, if a minor closed class $\mathcal{K}$ is proper then it is bounded (by a finite graph, for example by a large complete graph) and Hadwiger conjecture asserts that then it has the greatest element which is a complete graph. In this context one may see our Conjecture 1 as an approximation to Hadwiger conjecture: instead of asking for the greatest element of class $\mathcal{K}$ we ask for a bound with local properties similar to those in $\mathcal{K}$ (such as not containing a given complete graph). On the other hand the following naturally arises as a weaker form of Hadwiger conjecture:

Conjecture 3. *Any proper minor closed class $\mathcal{K}$ has a greatest element.*

3. The following is a consequence of Theorem 1.1

Theorem 6.1. *(Characterization of bounded minor closed classes)*
Let $\mathcal{K}$ be a a minor closed class of graphs. The following statements are equivalent

i. the acyclic chromatic number $\chi_a(G)$ is bounded for $G \in \mathcal{K}$;
ii. the oriented chromatic number $\vec{\chi}(G)$ is bounded for $G \in \mathcal{K}$;
iii. the star chromatic number $\chi_{st}(G)$ is bounded for $G \in \mathcal{K}$;
iv. colored mixed graphs in $\mathcal{K}$ may be properly colored by a fixed number of colors (in the sense of [1, 19]);
v. the chromatic number $\chi(G)$ is bounded for all $G \in \mathcal{K}$;
vi. the clique number $\omega(G)$ is bounded for all $G \in \mathcal{K}$;
vii. the edge density of all graphs G is bounded for all $G \in \mathcal{K}$;
viii. the average degree of all graphs G is bounded for all $G \in \mathcal{K}$;
ix. the minimal degree of a vertex is bounded for all $G \in \mathcal{K}$;
x. $\mathcal{K}$ is proper *minor closed class of graphs (i.e. $\mathcal{K}$ is not the class of all graphs).*

We shall comment on this more extensively in [17].

4. The following is an interesting consequence of Theorem 1.2 presented as a problem in [14]: For a graph $G = (V, E)$ and a positive integer t define the graph $G^{(t)} = (V, E^{(t)})$ by $\{x, y\} \in E^{(t)}$ iff the vertices x and y are joined in G by a path of length t. Note that for an even t the graph $G^{(t)}$ may contain an arbitrarily large complete graph even for a tree G (consider subdivision of a star). However for $t = 3$ we have the following, perhaps surprising, general result:

Corollary 6. *For every minor closed class $\mathcal{K}$ the following two statements are equivalent:*

i. The chromatic number of triangle free graphs from $\mathcal{K}$ is bounded;
ii. The chromatic number of graphs $\chi(G^{(3)}), G \in \mathcal{K}, G$ triangle free, is bounded.

Proof. Any homomorphism $f : G \to H$ for a triangle free graph H may be viewed as a coloring of $G^{(3)}$ by $|V(H)|$ colors. Thus *i.* implies *ii.* by Theorem 1.2. In the reverse direction suppose contrary: let $\chi(G^{(3)}) \leq k$ for any triangle free $G \in \mathcal{K}$ and assume that there exists a triangle free graph G in $\mathcal{K}$ with $\chi(G) \geq 3k$. Let $V(G) = V_1 \cup \ldots \cup V_k$ be a coloring of the graph $G^{(3)}$. It follows that there exists $i, 1 \leq i \leq k$ such that the subgraph G_i of G induced by the set V_i is not bipartite. As G_i is triangle free and as V_i is color class of a coloring of $G^{(3)}$ G_i does not contain an induced copy of the path of length 3. However it is well known (see [20, 24]) that then G_i is a perfect graph and thus a bipartite graph. This is a contradiction. □

(Note that the implication *ii.* ⇒ *i.* holds for any class of graphs. The reverse implication does not hold generally. This proposition may be formulated in terms of multiplication of incidence matrices.) Clearly many further questions may be asked. We decided to stop here (and continue in [17]).

5. It is a bit surprising that in order to prove the main Theorem 1.2 for planar graphs one can proceed in purely combinatorial way for all minor closed classes. This leads to the following:

Conjecture 4. *For any $k \geq 3$ there exists a function f_k such that every graph G has a proper coloring with at most $f_k(\max\{\chi(H); H \preceq G\})$ colors such that any k-color critical subgraph G' of G is either isomorphic to K_k or gets at least $k+1$ colors.*

This is true for $k = 3$ (by Theorem 1.2). It is proved in [14] that the the validity of Conjecture 4 implies the existence of a K_k-free bound. For the completeness we give here a simple proof of this particular case.

Lemma 4. *Let $\mathcal{K}$ be a minor closed class of graphs, such that $\chi(G), G \in \mathcal{K}$, is bounded by q. Let $3 \leq k \leq q$ be an integer. Assume there exists a function f such that any graph $G \in \mathcal{K}$ has a coloring with $f(q)$ colors such that any k color-critical subgraph of G is either isomorphic to K_k or gets at least $k+1$ colors.*

Then, the class $\mathcal{K}_k$ of all K_k free graphs in $\mathcal{K}$ is bounded by a q-colorable K_k free graph.

Proof. We construct graph $U = (V, E)$ as follows: The vertices V are all pairs of the form (i, ϕ) where $1 \leq i \leq f(q)$ and ϕ is a function which assigns to every k-uple $X = \{i_1, \ldots, i_k\}; 1 \leq i_1 < \cdots < i_k \leq f(q)$ value $\phi(X) \in \{1, \ldots, k-1\}$. The edges E are all pairs of the vertices of the form $\{(i, \phi), (i', \phi')\}$ where $i \neq i'$ and $\phi(X) \neq \phi'(X)$ whenever k-tuple X contains both i and i'. We shall prove that the graph U has all the properties claimed by Lemma 4.

The graph U has clearly no subgraph isomorphic to K_k as if $(i_1, \phi_1), \ldots, (i_k, \phi_k)$ are k vertices of U, then considering the k-uple $X = \{i_1, \ldots, i_k\}$ we see that at least two of the values $\phi_1(X), \ldots, \phi_k(X)$ coincide and thus the corresponding vertices do not form an edge of U. Next we prove that U is a bound for the class $\mathcal{C}$. Towards this end let $G \in \mathcal{K}_k$ and let $c : V(G) \to \{1, \ldots f(q)\}$ be a proper coloring of G guaranteed by the assumptions of the lemma. Given a k-tuple $X \subset \{1, \ldots f(q)\}$ the subgraph G_X of G induced by the set $c^{-1}(X)$ may not have chromatic number k. Otherwise, G_X would have a k color-critical subgraph isomorphic to K_k. Hence, G_X is $(k-1)$-colorable and thus there exists a homomorphism (coloring) $\phi_X : G_X \to K_{k-1}$. Define the mapping $g : V(G) \to V(U)$ as follows: $g(v) = (c(v), \phi)$ where $\phi(X)(v) = \phi_X(v)$ providing $v \in X$ and $\phi(X)(v) = 1$ otherwise. It is easy to check that this is a homomorphisms $G \to U$. Then the graph $U \times K_q$ is q-colorable K_k free bound for $\mathcal{K}_k$. This completes the proof of Lemma 4. □

References

[1] N. Alon, T. H. Marshall: Homomorphisms of edge-colored graphs and Coxeter groups, J. Algebraic Comb.8 (1998), 5-13.

[2] N. Alon, C. J. H. McDiarmid, B. A. Reed: Acyclic Coloring of Graphs, Random Structures and Algorithms 2 (1991), 343-365.

[3] N. Alon, B. Mohar, D. P. Sanders: On acyclic colorings of graphs on surfaces, Israel J. Math. 94 (1996), 273-283.

[4] P. Dreyer, Ch. Malon, J. Nešetřil: Universal H-colorable graphs without a given configuration, Discrete Math.(in press)

[5] P. Erdös, A. Hajnal: Chromatic number of finite and infinite graphs and hypergraphs, Discrete Math. 53 (1985), 281-285.

[6] G. Fertin, A. Raspaud, B. Reed: On Star Coloring of graphs. In: Proccedings of GW'01, LNCS, Springer Verlag

[7] J. Fiala, J. Kratochvíl, A. Proskurowski: Partial covers of graphs, to appear in Discussiones Mathematicae Graph Theory.

[8] S.L. Hakimi: On the degree of the vertices of a directed graph, J. Franklin Inst. 279 (1965), 4.

[9] R. Häggkvist, P. Hell: Universality of A-mote graphs, European J. Combinatorics 14 (1993), 23-27.

[10] T. Jensen, B. Toft: Graph coloring problems, Willey, 1995.

[11] A. Kostochka: On the minimum of the Hadwiger number for graphs with given average degree, *Metody Diskret. Analiz.*, 38(1982), Novosibirsk, 37-58, in Russian,
English translation: *AMS Translations (2)*, 132(1986), 15-32.

[12] A. Kostochka, E. Sopena, X. Zhu: Acyclic and oriented chromatic number of graphs, J. of Graph Th. 24 (1997), 331-340.

[13] W. Mader: Homomorphiesätze für Graphen, Math. Ann. 178 (1968), 154–168.

[14] T. H. Marshall, R. Nasraser, J. Nešetřil: Homomorphism Bounded Classes of Graphs (to appear in European J. Comb.)

[15] J. Nešetřil: Aspects of Structural Combinatorics, Taiwanese J. Math. 3, 4 (1999), 381 - 424.

[16] J. Nešetřil: Homomorphisms of derivative graphs, Discrete Math., 1, 3 (1971), 257-268.

[17] J. Nešetřil, P. Ossona de Mendez: Foldings (submitted)

[18] E. Sopena, A. Raspaud: Good and semi-strong colorings of oriented planar graphs, Inf. Processing Letters 51 (1994), 171-174.

[19] J. Nešetřil, A. Raspaud: Colored homomorphisms of colored mixed graphs, J. Comb. Th. B, 80 (2000), 147-155.

[20] D. Seinsche: On a property of the class of n-colorable graphs, J. Comb. Th. B, 16 (1974), 191–193.

[21] A. Thomason: An extremal function for contractions of graphs, Math. Proc. Cambridge Philos. Soc. 95 (1984), 261–265.

[22] A. Thomason: The extremal function for complete minors, J. Comb. Th. B 81 (2001), 318–338.

[23] C. Thomassen: Grötsch's 3-color theorem and its counterparts for torus and the projective plane, J. Comb. Th. B, 62 (1994), 268-279.

[24] E. S. Wolk: A note on the comparability graph of a tree, Proc. Amer. Math. Soc. 16 (1965), 17–20.

About Authors

Jaroslav Nešetřil is at the Department of Applied Mathematics and Institute of Theoretical Computer Science (ITI), Charles University, Malostranské nám. 25, 11800 Praha 1, Czech Republic; *nesetril@kam.ms.mff.cuni.cz*

Patrice Ossona de Mendez is at the Centre d'Analyse et de Mathématiques Sociales (UMR 8557), CNRS, 54 Bd Raspail, 75006 Paris, France; *pom@ehess.fr*

Acknowledgements

Work by Jaroslav Nešetřil has been partially supported by the ITI Grant LN00A56 of the Czech Ministry of Education and CRM, Barcelona, Spain.

We thank A. Raspaud who informed us about star colorings, Jakub Nešetřil for a help, and Tim Marshall for having pointed out mistakes in a first version of Theorem 4.1.

Conflict-free Colorings

János Pach
Géza Tóth

Abstract

A coloring of the elements of a planar point set P is said to be *conflict-free* if, for every closed disk D whose intersection with P is nonempty, there is a color that occurs in $D \cap P$ precisely once. We solve a problem of Even, Lotker, Ron, and Smorodinsky by showing that any conflict-free coloring of every set of n points in the plane uses at least $c \log n$ colors, for an absolute constant $c > 0$. Moreover, the same assertion is true for homothetic copies of any convex body D, in place of a disk.

1 Introduction

Motivated by a frequency assignment problem in cellular telephone networks, Even, Lotker, Ron, and Smorodinsky [ELRS02] studied the following question. Given a set P of n points in the plane, what is the smallest number of colors in a coloring of the elements of P with the property that any closed disk D with $D \cap P \neq \emptyset$ has an element whose color is not assigned to any other element of $D \cap P$. We refer to such a coloring as a *conflict-free* coloring of P with respect to disks.

In the specific application, the points correspond to *base stations* interconnected by a fixed backbone network. Each *client* continuously scans frequencies in search of a base station within its (circular) range with good reception. Once such a base station is found, the client establishes a radio link with it, using a frequency not shared by any other station within its range. Therefore, a conflict-free coloring of the points corresponds to an assignment of frequencies to the base stations, which enables every client to connect to a base station without interfering with the others.

Even et al. proved that any set of n points in the plane has a conflict-free coloring with $O(\log n)$ colors, and they exhibited an example showing that this bound cannot be improved. The aim of the present note is to show that in fact *any* set of n points requires at least constant times $\log n$ colors for a conflict-free coloring.

Theorem 1. *Every conflict-free coloring of every set of n points in the plane uses at least $\log_8 n$ colors.*

In Section 3, we show that a similar result holds for conflict-free colorings with respect to non-circular ranges (see Theorem 2), and we discuss some related questions.

2 Proof of Theorem 1

Throughout this section, we fix a set P of n points in the plane and a conflict-free coloring of P with k colors.

It is sufficient to establish the following

Lemma 2.1. *For any $0 \leq i < k$, there is a closed axis-parallel square $\boldsymbol{S}_i$ with circumscribing circle $\boldsymbol{C}_i$ such that*
(a) $|P \cap \boldsymbol{S}_i| \geq \frac{n}{8^i}$, and
(b) the elements of P belonging to the interior of $\boldsymbol{C}_i$ are colored with at most $k-i$ colors.

Applying the lemma with $i = k-1$, we obtain a circle $\boldsymbol{C}_{k-1}$ containing $m \geq \frac{n}{8^{k-1}} - 4$ points of P in its interior (not counting the corners of the inscribed square $\boldsymbol{S}_{k-1}$, which may belong to P). By (ii), all of these points must have the same color. Using the property that the coloring is conflict-free, we have that $m \leq 1$, so that $k-1 \geq \log_8(n/5)$ and Theorem 1 follows.

It remains to prove Lemma 2.1. We use induction on i. For $i = 0$, let $\boldsymbol{S}_0$ be any axis-parallel closed square containing P. Suppose that, for some $0 \leq i \leq k-2$, we have already found a square $\boldsymbol{S}_i$ with circumscribing circle $\boldsymbol{C}_i$, satisfying the requirements of the lemma. Denote the vertices of $\boldsymbol{S}_i$ by V_1 (upper left), V_2 (upper right), V_3 (lower right), and V_4 (lower left). The coloring of P is conflict-free, so among the elements of P in the interior of $\boldsymbol{C}_i$ there is one with a unique color. Pick such a point and denote it by O.

We distinguish two cases.

CASE A: $O \notin \boldsymbol{S}_i$.

For any $1 \leq j \leq 4$, define $\boldsymbol{S}^j = \boldsymbol{S}_i^j$ to be the *largest* axis-parallel closed square with circumscribing circle $\boldsymbol{C}^j = \boldsymbol{C}_i^j$ satisfying the following three conditions:

(i) $\boldsymbol{S}^j \subseteq \boldsymbol{S}_i$,
(ii) V_j is a vertex of $\boldsymbol{S}^j$,
(iii) O does not lie in the interior of $\boldsymbol{C}^j$.

Claim 2.2. *For any $1 \leq j \leq 4$,*
(a) circle $\boldsymbol{C}^j$ is inside $\boldsymbol{C}_i$,
(b) $\bigcup_{j=1}^4 \boldsymbol{S}^j = \boldsymbol{S}_i$.

Proof. Part (a) follows directly from the definition.

Suppose without loss of generality that O belongs to the part of the disk enclosed by $\boldsymbol{C}_i$ that is cut off by the segment V_1V_2 (see Fig. 1, left). For any square $\boldsymbol{S}$, let $l(\boldsymbol{S})$ denote its side length. Draw a vertical line through O, and let O' denote its intersection with V_1V_2.

The square $\overline{\boldsymbol{S}}^1$ with upper left corner V_1 and upper right corner O' and its circumscribing circle $\overline{\boldsymbol{C}}^1$ satisfy conditions (i)–(iii) with $j = 1$. Therefore, by the maximality of $\boldsymbol{S}^1$, we have $l(\boldsymbol{S}^1) \geq l(\overline{\boldsymbol{S}}^1) = V_1O'$. Similarly, we obtain $l(\boldsymbol{S}^2) \geq V_2O'$, whence

$$l(\boldsymbol{S}^1) + l(\boldsymbol{S}^2) \geq V_1V_2 = l(\boldsymbol{S}_i). \tag{1}$$

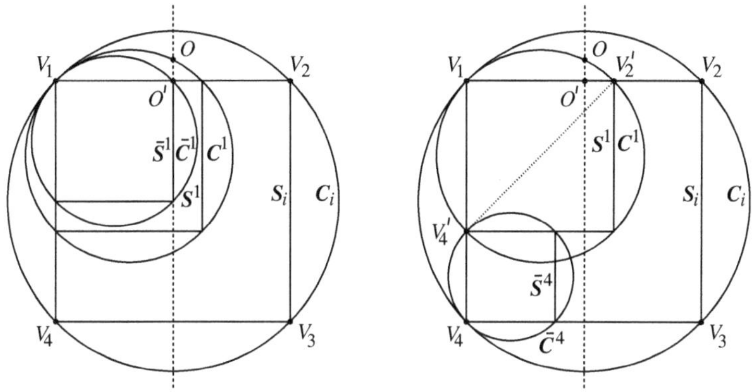

Fig. 1.

Let V_4' and V_2' denote the lower left and the upper right corners of $\boldsymbol{S}^1$, respectively. Consider the closed square $\overline{\boldsymbol{S}}^4$ with lower left corner V_4 and upper left corner V_4', and let $\overline{\boldsymbol{C}}^4$ denote its circumscribing circle (see Fig. 1, right). Notice that the line $V_4'V_2'$ is tangent to $\overline{\boldsymbol{C}}^4$ and it separates $\overline{\boldsymbol{C}}^4$ from O' and O. Therefore, $\overline{\boldsymbol{S}}^4$ and $\overline{\boldsymbol{C}}^4$ satisfy conditions (i)–(iii) with $j = 4$. Hence, by the maximality of $\boldsymbol{S}^4$, we have $l(\boldsymbol{S}^4) \geq l(\overline{\boldsymbol{S}}^4) = V_4V_4'$ so that

$$l(\boldsymbol{S}^1) + l(\boldsymbol{S}^4) \geq V_1V_4 = l(\boldsymbol{S}_i). \tag{2}$$

By symmetry, we also have

$$l(\boldsymbol{S}^2) + l(\boldsymbol{S}^3) \geq l(\boldsymbol{S}_i). \tag{3}$$

Now let $\widehat{\boldsymbol{S}}^4$ denote the closed square whose lower left corner is V_4 and whose size is the same as that of $\boldsymbol{S}^1$, and let $\widehat{\boldsymbol{C}}^4$ denote its circumscribing circle (see Fig. 2). Again, it is easy to check that $\widehat{\boldsymbol{S}}^4$ and $\widehat{\boldsymbol{C}}^4$ satisfy conditions (i)–(iii) with $j = 4$. Thus, we obtain $l(\boldsymbol{S}^4) \geq l(\widehat{\boldsymbol{S}}^4) = l(\boldsymbol{S}^1)$ and, analogously, $l(\boldsymbol{S}^3) \geq l(\boldsymbol{S}^2)$. Therefore, we have

$$l(\boldsymbol{S}^3) + l(\boldsymbol{S}^4) \geq l(\boldsymbol{S}^2) + l(\boldsymbol{S}^1) \geq l(\boldsymbol{S}_i). \tag{4}$$

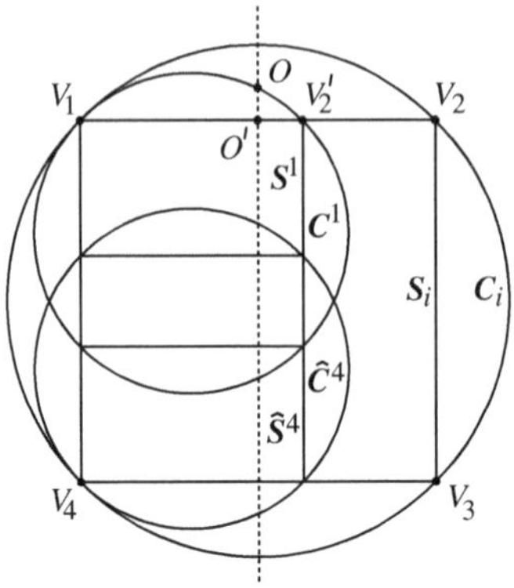

Fig. 2.

Equations (1)–(4) immediately imply part (b) of the claim. □

CASE B: $O \in \boldsymbol{S}_i$.

For any $1 \le j \le 4$, define the closed square $\boldsymbol{S}^j$ with circumscribing circle $\boldsymbol{C}^j$, as in Case A. Furthermore, for any $5 \le j \le 8$, let $\boldsymbol{S}^j$ denote the *largest* axis-parallel closed square with circumscribing circle $\boldsymbol{C}^j$ satisfying the following three conditions:

(i) $\boldsymbol{S}^j \subseteq \boldsymbol{S}_i$;

(ii) O is the lower right corner of $\boldsymbol{S}^5$, the lower left corner of $\boldsymbol{S}^6$, the upper left corner of $\boldsymbol{S}^7$, and the upper right corner of $\boldsymbol{S}^8$;

(iii) O does not lie in the interior of $\boldsymbol{C}^j$.

Claim 2.3. *For any* $1 \le j \le 8$,

(a) circle $\boldsymbol{C}^j$ *is inside* $\boldsymbol{C}_i$,

(b) $\bigcup_{j=1}^{8} \boldsymbol{S}^j = \boldsymbol{S}_i$.

Proof. Part (a) follows from the fact that each $\boldsymbol{S}^j$ is contained in and can be obtained from $\boldsymbol{S}_i$ by shrinking it from one of its points q. Therefore, q must lie on or inside of $\boldsymbol{C}_i$, and the same shrinking transformation will take $\boldsymbol{C}_i$ into $\boldsymbol{C}^j$, proving that $\boldsymbol{C}^j \subset \boldsymbol{C}_i$.

Draw a vertical and a horizontal line through O, and denote their intersections with the four sides of $\boldsymbol{S}_i$ by V^{up}, V^{down}, V^{left}, and V^{right}. These two lines divide $\boldsymbol{S}_i$ into four rectangles. To prove part (b) of the claim, it is enough to show, by symmetry, that the upper left rectangle, $\boldsymbol{R} = V_1 V^{\mathrm{up}} O V^{\mathrm{left}}$ is covered by $\boldsymbol{S}^1 \cup \boldsymbol{S}^5$ (see Fig. 3, left).

Suppose without loss of generality that $V_1 V^{\mathrm{up}} \ge V^{\mathrm{up}} O$, so that $l(\boldsymbol{S}^5) = V^{\mathrm{up}} O$. Let V_1' denote the upper left corner of $\boldsymbol{S}^5$.

Assume first that $V_1 V^{\mathrm{left}} \ge V_1 V_1'$. Let $\overline{\boldsymbol{S}}^1$ be the closed square whose upper left and lower left corners are V_1 and V^{left}, resp., and let $\boldsymbol{C}$ denote its circumscribing circle. By our assumption, we have $\overline{\boldsymbol{S}}^1 \cup \boldsymbol{S}^5 = \boldsymbol{R}$. On

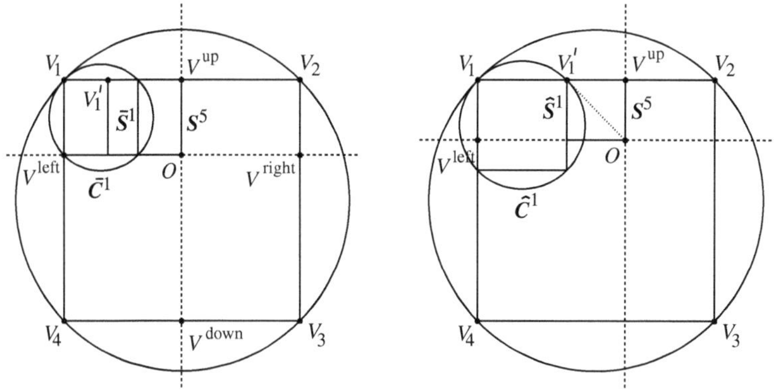

Fig. 3.

the other hand, $\overline{\boldsymbol{S}}^1$ and $\overline{\boldsymbol{C}}^1$ obviously satisfy conditions (i)–(iii) with $j = 1$. Thus, by the maximality of $\boldsymbol{S}^1$, we have $\boldsymbol{S}^1 \supseteq \overline{\boldsymbol{S}}^1$, yielding that $\boldsymbol{S}^1 \cup \boldsymbol{S}^5 \supseteq \boldsymbol{R}$.

Assume next that $V_1 V^{\text{left}} \leq V_1 V_1'$. Now let $\widehat{\boldsymbol{S}}^1$ denote the closed square with upper left corner V_1 and upper right corner is V_1', and let $\widehat{\boldsymbol{C}}^1$ be its circumscribing circle. Line $V_1'O$ is tangent to $\widehat{\boldsymbol{C}}^1$, so O cannot lie inside $\widehat{\boldsymbol{C}}^1$ (see Fig. 3, right). Therefore, by our assumption, $\widehat{\boldsymbol{S}}^1 \cup \boldsymbol{S}^5 \supseteq \boldsymbol{R}$. Since $\widehat{\boldsymbol{S}}^1$ and $\widehat{\boldsymbol{C}}^1$ satisfy conditions (i)–(iii) with $j = 1$, we obtain $\boldsymbol{S}^1 \supseteq \widehat{\boldsymbol{S}}^1$. Consequently, we also have $\boldsymbol{S}^1 \cup \boldsymbol{S}^5 \supseteq \boldsymbol{R}$. This completes the proof of part (b) of Claim 2.3. □

Now we are in a position to complete the induction step in the proof of Lemma 2.1. By the induction hypothesis, we have $|P \cap \boldsymbol{S}_i| \geq \frac{n}{8^i}$. In both cases (A and B), condition (iii) guarantees that, for every j, all points in the interior of $\boldsymbol{C}^j$ are colored with at most $k - i - 1$ colors, because the color of O cannot be used. On the other hand, by parts (b) of Claims 2.2 and 2.3, there *exists* a j ($1 \leq j \leq 8$) such that $|P \cap \boldsymbol{S}^j| \geq |P \cap \boldsymbol{S}_i|/8 \geq \frac{n}{8^{i+1}}$. Setting $\boldsymbol{S}_{i+1} := \boldsymbol{S}^j$ and $\boldsymbol{C}_{i+1} := \boldsymbol{C}^j$, the assertion of the lemma follows for $i + 1$.

This concludes the proof of Lemma 2.1 and hence the theorem.

3 A generalization and concluding remarks

For any compact convex body D in the plane, a coloring of the elements of a point set P is said to be *D–conflict-free* if, for any *homothetic* (i.e., translated and similar) copy of D, whose intersection with P is nonempty, there is a color that occurs in $D \cap P$ precisely once.

Even et al. [ELRS02] extended their result on disks by showing that, for any given D, any set of n points permits a D–conflict-free coloring with

$O(\log n)$ colors. They gave an example of n points requiring $\Omega(\log n)$ colors. The argument presented in the previous section easily generalizes to this case.

Theorem 2. *For any compact convex body D, every D–conflict-free coloring of every set of n points in the plane uses at least $\log_8 n$ colors.*

Proof. (Sketch) Let X and Y be two points on the boundary of D at maximum distance from each other, i.e., let XY be a *diameter* of D. Let ℓ_1 and ℓ_2 be two lines parallel to XY such that the lengths of the segments $X_1Y_1 = D \cap \ell_1$ and $X_2Y_2 = D \cap \ell_2$ are equal to half of the length of XY. Applying a proper affine transformation to the plane (including D and the point set), the parallelogram $X_1Y_1X_2Y_2$ becomes an axis-parallel square. So, without loss of generality, we can assume that D has an inscribed square. Now one can repeat the proof of Theorem 1 with the only difference that, instead of axis-parallel squares with circumscribing circles, we have to use axis-parallel squares with circumscribing homothetic copies of D. □

In the same spirit, for any family of sets, $\mathcal{F}$, a coloring of a point set P is said to be *$\mathcal{F}$–conflict-free* or *conflict-free with respect to $\mathcal{F}$* if, for every member $F \in \mathcal{F}$ whose intersection with P is nonempty, there is a color that appears in $F \cap \mathcal{F}$ precisely once.

It was pointed out in [HS02] that every set of n points in general position in the plane permits a conflict-free coloring using $O(\sqrt{n})$ colors, with respect to all axis-parallel closed rectangles. This can be slightly improved, as follows.

Proposition 3.1. *Every set of n points in general position in the plane permits a conflict-free coloring using $O(\sqrt{n \log\log n / \log n})$ colors, with respect to the family of all axis-parallel closed rectangles.*

Proof. Given a set P of n points in general position (i.e., no two of them have the same x-coordinate or y-coordinate), define a graph G on the vertex set P by connecting two points with an edge if and only if the smallest axis-parallel closed rectangle containing both of them has no element of P in its interior. It is easy to verify that G is "uncrowded:" it has no complete subgraph with 5 vertices.

We *claim* that G has an independent set of size $\Omega\left(\sqrt{n \log n / \log\log n}\right)$. For any $p \in P$, the vertical and horizontal lines passing through p divide the plane into four quadrants. Obviously, the neighbors of p lying in each of these quadrants form either a monotone increasing or a monotone decreasing subsequence, i.e., all slopes determined by their point pairs have the same sign. Taking every other element of each of these sequences, we obtain an independent set, so that the neighborhood of p can be decomposed into at most *eight* (in fact, at most *four*) independent sets. Therefore, if there exists a point $p \in P$ whose degree is at least $\Omega\left(\sqrt{n \log n / \log\log n}\right)$, we are done. Otherwise, the maximum degree D in G satisfies $D = O\left(\sqrt{n \log n}\right)$. Now we can apply an extension of a result of Ajtai, Komlós, and Szemerédi [AKS80] on uncrowded graphs, due to Shearer [S95]. This implies that G has an independent set of size $\Omega\left((n/D)(\log D/\log\log D)\right) = \Omega\left(\sqrt{n\log n/\log\log n}\right)$.

We follow the approach of [ELRS02] to argue that Proposition 3.1 is an easy consequence of the above *claim*. Pick an independent set $S_1 \subseteq P$ of size $\Omega\left(\sqrt{n\log n/\log\log n}\right)$ in G. Color all elements of S_1 with color 1, and use the claim to find a large independent set S_2 in the subgraph of G induced by $P - S_1$. Color all elements of S_2 with color 2. Continue like this until no points are left. The resulting coloring will meet the requirements of Proposition 3.2. □

References

[AKS80] M. Ajtai, J. Komlós, and E. Szemerédi, A note on Ramsey numbers, *J. Combin. Theory Ser. A* **29** (1980), 354–360.

[ELRS02] G. Even, Z. Lotker, D. Ron, and S. Smorodinsky, Conflict–free colorings of simple geometric regions with applications to frequency assignment in cellular networks, in: *Proceedings of 43rd Annual IEEE Symposium on the Foundations of Computer Science*, 691–700, 2002.

[HS02] S. Har-Peled and S. Smorodinsky, On conflict-free coloring of points and simple regions in the plane, manuscript.

[S95] J. B. Shearer, On the independence number of sparse graphs, *Random Structures and Algorithms* **7** (1995), 269–271.

About Authors

János Pach is at the Courant Institute of Mathematical Sciences, New York University, New York, NY 10012, USA and the A. Rényi Institute of Mathematics, Hungarian Academy of Sciences, Budapest, POB 127, 1364 Hungary; *pach@cims.nyu.edu.*

Géza Tóth is at the A. Rényi Institute of Mathematics, Hungarian Academy of Sciences, Budapest, POB 127, 1364 Hungary. *geza@renyi.hu.*

Acknowledgments

János Pach has been supported by NSF grant CR-00-98246, PSC-CUNY Research Award 63382-0032 and OTKA-T-032452. Géza Tóth has been supported by OTKA-T-038397, OTKA-T-032452 and the New York University Research Challenge Fund.

We are indebted to Shakhar Smorodinsky for bringing the above problems to our attention and for making many interesting remarks. He informed us that Proposition 3.1 was also proved by Noga Alon and Timothy Chan.

New Complexity Bounds for Cylindrical Decompositions of Sub-Pfaffian Sets

Savvas Pericleous
Nicolai Vorobjov

Abstract

Tarski-Seidenberg principle plays a key role in real algebraic geometry and its applications. It is also constructive and some efficient quantifier elimination algorithms appeared recently. However, the principle is wrong for first-order theories involving certain real analytic functions (e.g., an exponential function). In this case a weaker statement is sometimes true, a possibility to eliminate *one sort* of quantifiers (either $\forall$ or $\exists$). We construct an algorithm for a cylindrical cell decomposition of a closed cube $I^n \subset \mathbb{R}^n$ compatible with a semianalytic subset $S \subset I^n$, defined by analytic functions from a certain broad finitely defined class (Pfaffian functions), modulo an oracle for deciding emptiness of such sets. In particular the algorithm is able to eliminate one sort of quantifiers from a first-order formula. The complexity of the algorithm and the bounds on the output are doubly exponential in $O(n^2)$.

1 Introduction

Semianalytic sets are defined as subsets of points in $\mathbb{R}^n$ satisfying Boolean combinations of atomic formulae of the kind $f > 0$, where f's are real analytic functions defined in a common open domain $G \subset \mathbb{R}^n$. Subanalytic sets are defined as images of relatively proper real analytic maps of semianalytic sets. If functions f are polynomials, then these two classes of sets coincide (Tarski-Seidenberg principle). An equivalent formulation of this statement is that the first-order theory of reals admits quantifier elimination. It plays a key role in many aspects and applications of real algebraic geometry. However, Tarski-Seidenberg principle is wrong already if one of atomic f's is an exponential function, in which case a subanalytic set may not be semianalytic [16]. Thus, the quantifier elimination is not generally possible in a first-order theory with real analytic functions. A theorem due to Gabrielov [6, 7] shows however that at least one sort of quantifiers (either $\forall$ or $\exists$) *can* be eliminated. This is equivalent to saying that the complement to a subanalytic set is subanalytic.

It is well-known that Tarski-Seidenberg principle is constructive. Original Tarski's proof provided an algorithm for quantifier elimination (computing of a projection) with a non-elementary complexity. In mid-70-s, [3] and [20] proposed elementary algorithms (doubly-exponential in the number of variables n). In recent years the problem had received a significant attention, had attracted a number of powerful mathematical techniques, and as a result some very efficient quantifier elimination algorithms were designed (see [1, 11, 12, 17]).

Attempting to extend the complexity results from algebraic to real analytic case, we have firstly to restrict the class of real analytic functions to a finitely defined subclass which would include as many as possible important analytic functions (for example, all algebraic functions, exponentials, logarithms, etc.), and for whose members a natural concept of a *size* or *format* would be definable. A suitable class of such kind is formed by *Pfaffian* functions. Pfaffian functions are solutions of triangular systems of first order partial differential equations with polynomial coefficients. *Semi-Pfaffian* sets, defined by systems of equations and inequalities between these functions, are characterized by global finiteness properties [13, 14] (formal definitions are given in Section 2). This means that their basic geometric and topological characteristics can be explicitly estimated in terms of formats of their defining formulae. *Sub-Pfaffian* sets are relatively proper images of semi-Pfaffian sets, and their complements are also sub-Pfaffian.

A common technique for proving quantifier elimination results is constructing a *cylindrical cell decomposition* of the set defined by the quantifier-free part of a given formula (see definition in the next section), i.e. representing this set as a disjoint union of geometrically simple *cells*, homeomorphic to open balls of some dimensions, which induces (via projections) similar decompositions on a certain filtration of subspaces. This method was used in [3,20] to obtain doubly-exponential upper complexity bounds for algebraic case (more efficient modern algorithms [1, 11, 12, 17] don't use cylindrical decomposition).

The technique of cylindrical cell decomposition was applied to Pfaffian case in the context of model-theoretic study of *o-minimality* (see [5,19]). The complexity estimates which can be extracted from these works are apparently non-elementary.

Recently Gabrielov and Vorobjov in [9] suggested an algorithm which produces cylindrical cell decompositions of sub-Pfaffian sets in $\mathbb{R}^n$. In particular, this algorithm finds complements to sub-Pfaffian sets, in other words eliminates one sort of quantifiers from prenex first-order formulae involving Pfaffian functions. As a model of computation [9] uses a *real numbers machine* (Blum-Shub-Smale model) [2] equipped with an *oracle* for deciding the feasibility of any system of Pfaffian equations and inequalities. The complexity bound of this algorithm, the number and formats of cells are doubly exponential in $O(n^3)$ (assuming that each oracle call has a unit cost).

In the present paper we obtain a new upper complexity bound by using a very different and more elementary technique. As in [9], this bound is doubly exponential in the number of variables and, being formally incomparable with the one from [9], is better for a long Pfaffian chain for defining functions. We rely on the two known results: the Khovanskii's upper bound on the number of connected components of a semi-Pfaffian set [13,14] and Gabrielov's algorithm for computing the closure of a semi-Pfaffian set [8]. Unlike [9], we do not use a stratification algorithm.

2 Pfaffian functions and sub-Pfaffian sets

Definition 2.1. (See [13,14] , and [10].) A *Pfaffian chain* of the order $r \geq 0$ and degree $\alpha \geq 1$ in an open domain $G \subset \mathbb{R}^n$ is a sequence of real analytic functions $f_1, \ldots, f_r$ in G satisfying Pfaffian equations

$$df_j(X) = \sum_{1 \leq i \leq n} g_{ij}(X, f_1(X), \ldots, f_j(X))dX_i$$

for $1 \leq j \leq r$. Here $g_{ij}(X, Y)$ are polynomials in $X = (X_1, \ldots, X_n)$ and $Y = (Y_1, \ldots, Y_j)$ of degree not exceeding α. A function

$$f(X) = P(X, f_1(X), \ldots, f_r(X))$$

where $P(X, Y_1, \ldots, Y_r)$ is a polynomial of degree not exceeding $\beta \geq 1$ is a *Pfaffian function* of order r and degree (α, β).

Example 2.2.

1. *Pfaffian functions of order 0 and degree $(1, \beta)$ are polynomials of degree not exceeding β.*

2. *The exponential univariate function $f(X) = e^{aX}$ is a Pfaffian function of order 1 and degree $(1,1)$ in $\mathbb{R}$, due to the equation $df(X) = af(X)dX$.*

3. *The function $f(X) = 1/X$ is a Pfaffian function of order 1 and degree $(2,1)$ in the domain $X \neq 0$, due to the equation $df(X) = -f^2(X)dX$.*

4. *The logarithmic function $f(X) = \ln(|X|)$ is a Pfaffian function of order 2 and degree $(2,1)$ in the domain $X \neq 0$, due to the equations $df(X) = g(X)dX$ and $dg(X) = -g^2(X)dX$ with $g(X) = 1/X$.*

For more examples of Pfaffian functions see [10,14].

Lemma 2.3. *(See [10, 14])*

1. *The sum (resp. product) of two Pfaffian functions, f_1 and f_2, of orders r_1 and r_2 and degrees (α_1, β_1) and (α_2, β_2), is a Pfaffian function of the order $r_1 + r_2$ and degree $(max(\alpha_1, \alpha_2), max(\beta_1, \beta_2))$ (resp. $(max(\alpha_1, \alpha_2), \beta_1 + \beta_2)$). If the two Pfaffian functions are defined by the same Pfaffian chain of order r, then the order of the sum and product is also r.*

2. *A partial derivative of a Pfaffian function of order r and degree (α, β) is a Pfaffian function of order r and degree $(\alpha, \alpha + \beta - 1)$.*

In what follows we only consider the "restricted" case in which Pfaffian functions are defined also on the boundary of the domain.

Let $I^k = [0,1]^k$ denote the unit cube in $\mathbb{R}^k$.

Definition 2.4. (*Semi- and sub-Pfaffian set.*)

1. A set $S \subset \mathbb{R}^s$ is called *semi-Pfaffian* in an open domain $G \subset \mathbb{R}^s$ if it consists of points from G satisfying a Boolean combination of atomic equations and inequalities $f = 0, g > 0$, where f, g are Pfaffian functions having a common Pfaffian chain defined in the domain G.

2. Consider $I^{m+n} \subset G$, where $G \subset \mathbb{R}^{m+n}$ is an open domain, and the projection map $\pi : \mathbb{R}^{m+n} \to \mathbb{R}^n$. A subset $W \subset \mathbb{R}^n$ is called *(restricted) sub-Pfaffian* if $W = \pi(S)$ for semi-Pfaffian set $S \subset I^{m+n}$.

According to [6, 7], the complement $I^n \setminus W$ in $I^n = \pi(I^{n+m})$ of a sub-Pfaffian set W is also sub-Pfaffian.

Definition 2.5. (*Format.*) For a semi-Pfaffian set

$$S = \bigcup_{1 \le l \le M} \{f_l = 0, g_{l1} > 0, \dots, g_{lJ_l} > 0\} \subset G \subset \mathbb{R}^s, \tag{1}$$

where f_i, g_{ij} are Pfaffian functions with a common Pfaffian chain, of order r and degree (α, β), defined in an open domain G, its *format* is a quintuple (N, α, β, r, s), where $N = 1 + \sum_{1 \le l \le M} (J_l + 1)$. Let $D = \alpha + \beta$. For $s = m + n$ and a sub-Pfaffian set $W \subset \mathbb{R}^n$ such that $W = \pi(S)$, its *format* is the format of S. In the sequel we will use the notation $g_l > 0$ for the system of inequalities $g_{l1} > 0, \dots, g_{lJ_l} > 0$.

Proposition 2.6. *([13, 14]) The number of connected components of a semi-Pfaffian set S with the format (N, α, β, r, s), does not exceed*

$$2^{r^2} s^{O(r)} (ND)^{O(r+s)}. \tag{2}$$

Corollary 2.7. *The number of connected components of a sub-Pfaffian set $W = \pi(S)$, with format (N, α, β, r, s), defined by a formula having only existential quantifiers, does not exceed bound (2).*

As a model of computation we use a *real numbers machine* (Blum-Shub-Smale model) [2] equipped with an *oracle* for deciding the feasibility of any system of Pfaffian equations and inequalities. An oracle is a subroutine which can be used by the algorithm any time the latter needs to check feasibility. We assume that this procedure always gives the correct answer though we do not specify how it actually works. For some classes of Pfaffian functions the feasibility problem is decidable on real numbers machines or Turing machines with explicit (singly-exponential) complexity bounds. Apart from polynomials, such class form, for example, terms of the kind $P(e^h, X_1, \dots, X_n)$ where h is a fixed polynomial in $X_1, \dots, X_n$ and P is an arbitrary polynomial in $X_0, X_1, \dots, X_n$ (see [18]). For such classes the oracle can be replaced by a deciding procedure, and we get an algorithm in the usual sense. As far as the computational complexity is concerned, we assume that each oracle call has the unit cost.

Definition 2.8. The *closure* $cl(S)$ of a sub-Pfaffian set S in an open domain G is an intersection with G of the usual topological closure of S:

$$cl(S) = \{x \in G \,:\, \forall \varepsilon > 0\, \exists z \in S\, (|x - z| < \varepsilon)\}.$$

The *frontier* ∂S of S is $cl(S) \setminus S$.

Lemma 2.9. *Let S be a semi-Pfaffian set in an open domain $G \subset \mathbb{R}^s$, of format (N, α, β, r, s), defined by (1) where $s = n + m$ and the variables are $X = (X_1, \dots, X_n), Y = (Y_1, \dots, Y_m)$. There is an algorithm which produces a Boolean formula $F(X, Y)$ in a disjunctive normal form with atomic Pfaffian functions such that for any fixed $y \in \mathbb{R}^m$ the closure $cl(S \cap \{Y = y\}) \subset \mathbb{R}^n$ coincides with $\{F(X, y)\}$. The format of $\{F(X, Y)\}$ is*

$$((Nd)^{O((s+r)s)}, \alpha, d^{O(s)}, r, s),$$

where $d = 2^{r^2}(sD)^{s+r}$ and $D = \alpha + \beta$. The complexity of the algorithm does not exceed $(Nd)^{O((s+r)s)}$.

Proof. The proof of this lemma is a straightforward parameterization of the proof of Theorem 1.1 from [8]. □

If a set S is defined by a formula Ψ, then $F(X, Y)$ from the proof of Lemma 2.9 will be sometimes denoted by $cl(\Psi)$.

Definition 2.10. ([5, 19]) Cylindrical cell is defined as follows.

1. Cylindrical 0-cell in $\mathbb{R}^n$ is an isolated point.
2. Cylindrical 1-cell in $\mathbb{R}$ is an open interval $(a, b) \subset \mathbb{R}$.
3. For $n \geq 2$ and $0 \leq k < n$, a cylindrical $(k+1)$-cell in $\mathbb{R}^n$ is either a *section over* C, i.e., a graph of a continuous bounded function $f : C \to \mathbb{R}$ where C is a cylindrical $(k+1)$-cell in $\mathbb{R}^{n-1}$ equipped with coordinates $X_2, \dots, X_n$, or else a *sector over* C, i.e., a set of the form

$$(f, g) \equiv \{(x_1, \dots, x_n) \in \mathbb{R}^n : \ (x_2, \dots, x_n) \in C \ \text{ and }$$
$$f(x_2, \dots, x_n) < x_1 < g(x_2, \dots, x_n)\}$$

where C is a cylindrical k-cell in $\mathbb{R}^{n-1}$ and $f, g : C \longrightarrow \mathbb{R}$ are continuous bounded functions satisfying $f(x_2, \dots, x_n) < g(x_2, \dots, x_n)$ for all points $(x_2, \dots x_n) \in C$.

Clearly, a cylindrical k-cell is homeomorphic to an open k-dimensional ball, and its closure is homeomorphic to a closed k-dimensional ball.

Definition 2.11. Cylindrical cell decomposition, say $\mathcal{D}$, of a subset $A \subset \mathbb{R}^n$ is defined as follows.

1. If $n = 1$, then $\mathcal{D}$ is a finite family of pair-wise disjoint cylindrical cells (i.e., isolated points and intervals) whose union is A.
2. If $n \geq 2$, then $\mathcal{D}$ is a finite family of pair-wise disjoint cylindrical cells in $\mathbb{R}^n$ whose union is A and there is a cell decomposition $\mathcal{D}'$ of $\pi(A)$ such that for each cell C of $\mathcal{D}$, the set $\pi(C)$ is a cell of $\mathcal{D}'$, where $\pi : \mathbb{R}^n \longrightarrow \mathbb{R}^{n-1}$ is the projection map onto the coordinate subspace of $X_2, \dots, X_n$. We say that $\mathcal{D}'$ is induced by $\mathcal{D}$.

Definition 2.12. If $A \subset \mathbb{R}^n$, $B \subset \mathbb{R}^n$ and $\mathcal{D}$ is a cylindrical cell decomposition of A, then $\mathcal{D}$ is *compatible* with B if for all $C \in \mathcal{D}$ either $C \subset B$ or $C \cap B = \emptyset$ (i.e. some $\mathcal{D}' \subset \mathcal{D}$ is a cylindrical cell decomposition of $B \cap A$).

3 The main result

We describe an algorithm for producing a cylindrical cell decomposition of a semi-Pfaffian set S in the closed unit cube $I^n = [0,1]^n \subset \mathbb{R}^n$. By the definition, this decomposition induces a cylindrical decomposition of the projection W of S onto a subspace $\mathbb{R}^m$, $m \leq n$.

More precisely, an input of the algorithm is a semi-Pfaffian set S defined by (1) with $s = n$, and we assume that S is contained in $I^n \subset G$.

Let

$$\pi : \mathbb{R}^n \to \mathbb{R}^m,$$

$$\pi : (X_1, \dots, X_n) \mapsto (X_{n-m+1}, \dots, X_n)$$

be the projection map with $\pi(S) = W$.

The output of the algorithm is a cylindrical cell decomposition $\mathcal{D}_n$ of I^n compatible with S. Each cell is described by a formula of the type

$$\pi'\Big(\bigcup_{1 \leq i \leq M'} \ \bigcap_{1 \leq j \leq M''} \{h_{ij} *_{ij} 0\} \Big),$$

where h_{ij} are Pfaffian functions in $n' \geq n$ variables, π' is the projection map $\pi' : \ \mathbb{R}^{n'} \to \mathbb{R}^n$, $*_{ij} \in \{= \ , >\}$, and M', M'' are certain integers. By the definition of a cylindrical cell decomposition, $\mathcal{D}_n$ induces a cylindrical decomposition $\mathcal{D}_m$ of the cube $I^m = \pi(I^n) \subset \mathbb{R}^m$ compatible with W. Using an oracle the algorithm can then decide which cells from $\mathcal{D}_m$ belong to W and which to its complement $I^m \setminus W$.

We prove that the number of cells in $\mathcal{D}_n$, the components of the format and the complexity of the algorithm are less than

$$(\alpha + \beta N)^{r^{O(n)} 2^{O(n^2)}}.$$

Note that in [9] the bound for these parameters is

$$N^{(r+m+n)^{O(d)}}(\alpha+\beta)^{r^{O(d(m+dn))}},$$

where $d = \dim(W) \leq n$. Note that if S is *semialgebraic*, then our bound is essentially the same as the best known upper bound in a cylindrical cell decomposition for the polynomial case [3,20]. Recall also result of Davenport and Heintz [4], that real quantifier elimination is doubly exponential (and hence any cylindrical algebraic decomposition algorithm should have the same complexity).

4 Description of a cell decomposition

Let $S \subset I^n \subset G \subset \mathbb{R}^n$ be a semi-Pfaffian set, defined by (1) with $s = n$ and having format (N, α, β, r, n).

Firstly, we reduce the formula defining set S to a simple special form which is essentially a single Pfaffian equation. Introduce a new variable X_0 and the function

$$f \equiv \prod_{1 \leq i \leq M} \Big((f_i^2 + (X_0 - iN)^2) \cdot \prod_{1 \leq j \leq J_i} (g_{ij}^2 + (X_0 - iN - j)^2) \Big).$$

Notice that f is a Pfaffian function of order r and degree $(\alpha, 2\beta N)$.

Let $\mathcal{D}$ be a cylindrical decomposition of $I^n \times [0, N^2] \subset \mathbb{R}^{n+1}$ compatible with $\{f = 0\} \cap (I^n \times [0, N^2])$, and $\mathcal{D}'$ be the cylindrical decomposition of I^n induced by $\mathcal{D}$.

By the definition of the cylindrical decomposition, $\mathcal{D}'$ is compatible with $\pi(\{f = 0\})$ where $\pi : \mathbb{R}^{n+1} \to \mathbb{R}^n$ is the projection map onto the subspace $\{X_0 = 0\}$. Generally, $\pi(\{f = 0\}) \neq S$.

Lemma 4.1. *The cylindrical decomposition $\mathcal{D}'$ is compatible with S.*

Proof. We need to prove that for any cell C' of $\mathcal{D}'$ either $C' \subset S$ or $C' \cap S = \emptyset$. Suppose that contrary to this, for a cell C' of $\mathcal{D}'$, there are points $x, y \in C'$ such that $x \in \{f_i = 0, g_{i1} > 0, \dots, g_{iJ_i} > 0\}$ for a certain i, and $y \notin S$. In particular, $y \notin \{f_i = 0, g_{i1} > 0, \dots, g_{iJ_i} > 0\}$, i.e., either $g_{ij}(y) \leq 0$ for some $1 \leq j \leq J_i$ or $f_i(y) \neq 0$. In the case $g_{ij}(y) \leq 0$, since C' is connected and $g_{ij}(x) > 0$, there is a point $z \in C' \cap \{g_{ij} = 0\}$ and therefore a point $(z, iN + j) \in \{f = 0\}$. The point $(z, iN + j)$ belongs to a cell, say C, of $\mathcal{D}$. Note that $\pi(C) = C'$. Clearly, $C \subset \{g_{ij} = 0\} \cap \{X_0 = iN + j\}$. It follows that $C' \subset \{g_{ij} = 0\}$ which contradicts to $x \in C'$ and $g_{ij}(x) > 0$. In the case of $f_i(y) \neq 0$, the point $(y, iN) \notin \{f = 0\}$. The point (y, iN) belongs to a cell, say C'', of $\mathcal{D}$, and $\pi(C'') = C'$. But $C'' \cap \{f = 0\} = \emptyset$, since $\mathcal{D}$ is compatible with $\{f = 0\}$, which is a contradiction, since $x \in C'$ and $f_i(x) = 0$. □

We proved that it is sufficient to construct a cylindrical decomposition of the intersection $\{f = 0\} \cap (I^n \times [0, N^2])$. For simplicity of notations, in what follows we assume that the function f has just n variables $X_1, \dots, X_n$ and $S = \{f = 0\} \cap I^n$. In the remaining part of this section we give a non-constructive description of a certain cylindrical cell decomposition of I^n compatible with $\{f = 0\}$.

Define:

- $I^k_m(a) \equiv I^k \cap \{X_m = a\}$, $1 \le m \le k$, $a \in [0, 1]$.
- $I^k_m \equiv I^k_m(0) \cup I^k_m(1)$, $1 \le m \le k$.
- $L^k_0(a) \equiv I^k$ and $L^k_m(a) \equiv \bigcap_{1 \le i \le m} I^k_i(a)$, $1 \le m \le k$, $a \in [0, 1]$.
- $L^k_m \equiv \bigcap_{1 \le i \le m} I^k_i$, $1 \le m \le k$.

Without loss of generality assume that $\{f = 0\} \cap I^n_1 = \emptyset$. Set $V \equiv (\{f = 0\} \cap I^n) \cup I^n_1$. Note that the format of V is $(O(n2^n), \alpha, 2\beta N, r, n)$.

The description of a cell decomposition proceeds by induction on n.

Definition 4.2. For a subanalytic curve Γ (a subanalytic set of dimension at most 1) in $\mathbb{R}^n$, define:

1. $E_k(\Gamma)$ to be the set of all points of local extrema of X_k-coordinate on $cl(\Gamma)$.
2. $R(\Gamma)$ to be the set of all ramification points of $cl(\Gamma)$. (A point x is called a ramification point of $cl(\Gamma)$, if for all sufficiently small $0 < t \in \mathbb{R}$ the intersection of $cl(\Gamma)$ with a sphere of the radius t centered at x, contains at least three different points.)
3. $B(\Gamma) \equiv \partial\Gamma = cl(\Gamma) \setminus \Gamma$.
4. The set of all *special* points relative to X_k-coordinate to be $S_k(\Gamma) \equiv E_k(\Gamma) \cup R(\Gamma) \cup B(\Gamma)$.

Observe that an isolated point in Γ is a special point.

We make several initial steps of the induction.

Let $n = 1$. Then let $\Omega^{(0)}_s = S_1(V)$ and define $\Omega^{(0)}_0 = \Omega^{(0)}_s$. For all pairs of points $x, y \in \Omega^{(0)}_s$ consider the set $V \cap \{1/2(x + y)\}$ and denote by $\Omega^{(0)}_m$ the union of all these sets. Notice that if $\{f = 0\}$ is finite, then $\Omega^{(0)}_m = \Omega^{(0)}_0$. Each member of $\Omega^{(0)}_0$ is a zero-dimensional cylindrical cell. A cylindrical cell decomposition $\mathcal{D}$ of I^1 compatible with V and therefore with $\{f = 0\} \cap I^1$ consists of these points and open intervals on the line between them. One can enumerate alternatively these points and intervals by successive non-negative integers j_1 in the ascending along X_1 order by assigning index $j_1 = 0$ to 0, index $j_1 = 1$ to its neighbouring interval, and so on. Notice that $|\mathcal{D}| < 2|\Omega^{(0)}_0|$.

Let $n = 2$. Then for every fixed value ω of X_2-coordinate define finite sets $\Omega_0^{(0)}(\omega)$ and $\Omega_m^{(0)}(\omega)$ as in case $n = 1$ by restricting V to $\{X_2 = \omega\}$. Let

$$\Gamma_0^{(1)} \equiv \bigcup_{\omega \in L_1^2(0)} \Omega_0^{(0)}(\omega), \quad \Gamma_m^{(1)} \equiv \bigcup_{\omega \in L_1^2(0)} \Omega_m^{(0)}(\omega).$$

Clearly, $\Gamma_0^{(1)}$, $\Gamma_m^{(1)}$ are 1-dimensional (not necessarily closed) subsets of I^2. Observe that $L_1^2 \subset \Gamma_0^{(1)} \subset \Gamma_m^{(1)} \subset V$.

Let

$$\Omega_s^{(1)} = S_2(\Gamma_0^{(1)}) \cup S_2(\Gamma_m^{(1)}).$$

For all $x = (x_1, x_2) \in \Omega_s^{(1)}$ denote by $\Omega_0^{(1)}$ the union of finite sets of the kind $\Gamma_0^{(1)} \cap \{X_2 = x_2\}$. For all pairs of points $x = (x_1, x_2), y = (y_1, y_2) \in \Omega_s^{(1)}$ denote by $\Omega_m^{(1)}$ the union of finite sets of the kind $\Gamma_m^{(1)} \cap \{X_2 = 1/2(x_2 + y_2)\}$.

Let $\omega_1 < \omega_2$ be two neighbouring X_2-coordinates of points from $\Omega_0^{(1)}$ (that is, there are no X_2-coordinates ω of points from $\Omega_0^{(1)}$ such that $\omega_1 < \omega < \omega_2$). Then for each $\omega \in (\omega_1, \omega_2)$, the set $\Omega_0^{(0)}(\omega) \subset \{X_2 = \omega\}$ consists of the same finite number of points. Let us enumerate these points and intervals between them, as we did in case $n = 1$, by successive non-negative integers in the ascending along X_1 order. It is clear that the set of all points having the same index for all $\omega \in (\omega_1, \omega_2)$ is an open interval of the curve $\Gamma_0^{(1)}$, which is a one-dimensional cylindrical cell being a graph of a continuous function defined on an interval in the 1-dimensional set $L_1^2(0)$. The set of all intervals having the same index for all $\omega \in (\omega_1, \omega_2)$ is an open 2-dimensional cylindrical cell being the set of points strictly between the non-intersecting graphs of two continuous functions defined on an interval in $L_1^2(0)$.

Now we can describe all zero-, one-, and two-dimensional cells of the cylindrical decomposition of I^2 that is compatible with V. Enumerate each cell by a 2-multi-index (j_1, j_2) in a following way. Index j_2 enumerates (by successive non-negative integers starting from zero) alternatively points in $\Omega_0^{(1)} \cap L_1^2(0)$ and intervals between these points in $L_1^2(0)$ in the ascending along X_2 order. For a fixed value of j_2, index j_1 enumerates points in $\Omega_0^{(0)}(\omega) \subset \{X_2 = \omega\}$ and intervals between them (as in case $n = 1$), where ω is either the X_2-coordinate of the point in $\Omega_0^{(1)} \cap L_1^2(0)$ having index j_2, or the X_2-coordinate of a point in the interval between two neighbouring points of $\Omega_0^{(1)} \cap L_1^2(0)$ having index j_2.

It is easy to see that the defined family of the cylindrical cells is a cylindrical cell decomposition of I^2 compatible with V and therefore with $\{f = 0\} \cap I^2$. A cell having index (i, j) is cylindrical over the cell with index $(0, j)$ that belongs in the decomposition of $L_1^2(0)$. Observe that the number of cells in this decomposition is $O(|\Omega_0^{(1)}|)$.

We proceed to the description of a general induction step.

For every fixed value ω of X_n-coordinate finite sets of points of the kind $\Omega_0^{(n-2)}(\omega)$ and $\Omega_m^{(n-2)}(\omega)$ can be defined by applying the inductive hypothesis to $V \cap \{X_n = \omega\}$. An important property of these sets, is that there are formulae (with quantifiers) $\Phi_0^{(n-2)}(X_1, \dots, X_{n-1}, X_n)$ and $\Phi_m^{(n-2)}(X_1, \dots, X_{n-1}, X_n)$ having free variables $X_1, \dots, X_n$ and not depending on ω, such that the replacement of the variable X_n by ω gives formulae $\Phi_0^{(n-2)}(X_1, \dots, X_{n-1}, \omega)$ and $\Phi_m^{(n-2)}(X_1, \dots, X_{n-1}, \omega)$ in free variables $X_1, \dots, X_{n-1}$ defining the sets $\Omega_0^{(n-2)}(\omega)$ and $\Omega_m^{(n-2)}(\omega)$ respectively for the section $\{X_n = \omega\}$. Let

$$\Gamma_0^{(n-1)} \equiv \{\Phi_0^{(n-2)}(X_1, \dots, X_{n-1}, X_n)\}, \quad \Gamma_m^{(n-1)} \equiv \{\Phi_m^{(n-2)}(X_1, \dots, X_{n-1}, X_n)\}.$$

Clearly, $\Gamma_0^{(n-1)}$, $\Gamma_m^{(n-1)}$ are 1-dimensional (not necessarily closed) subsets of I^n.

Observe that $L_{n-1}^n \subset \Gamma_0^{(n-1)} \subset \Gamma_m^{(n-1)} \subset V$. Moreover, for any $k = 2, \dots, n-1$ by the definitions of $\Gamma_*^{(n-1)}$, where $* \in \{0, m\}$, we have the inclusions $\pi_k(\Gamma_*^{(n-1)}) \subset \Gamma_*^{(n-1)}$, where π_k denotes the projection on the subspace of coordinates $X_k, X_{k+1} \dots, X_n$.

Let

$$\Omega_s^{(n-1)} = S_n(\Gamma_0^{(n-1)}) \cup S_n(\Gamma_m^{(n-1)}).$$

For all points $x = (x_1, \dots, x_n) \in \Omega_s^{(n-1)}$ denote by $\Omega_0^{(n-1)}$ the union of finite sets of the kind $\Gamma_0^{(n-1)} \cap \{X_n = x_n\}$. For all pairs of points $x = (x_1, \dots, x_n), y = (y_1, \dots, y_n) \in \Omega_s^{(n-1)}$ denote by $\Omega_m^{(n-1)}$ the union of finite sets of the kind $\Gamma_m^{(n-1)} \cap \{X_n = 1/2(x_n + y_n)\}$.

Let the index j_n enumerate in the ascending along X_n order alternatively points in $\Omega_0^{(n-1)} \cap L_{n-1}^n(0)$ and intervals between these points on $L_{n-1}^n(0)$. Let $\omega_1 < \omega_2$ be two neighbouring X_n-coordinates of points from $\Omega_0^{(n-1)}$.

Assume that the interval (ω_1, ω_2) is indexed by j_n and a point $\omega \in (\omega_1, \omega_2)$. It follows from the inductive hypothesis that there is a certain cylindrical cell decomposition of the intersection $I_n^n(\omega) = I^n \cap \{X_n = \omega\}$ compatible with $V \cap \{X_n = \omega\}$ and all cells are enumerated by $(n-1)$-multi-indices. In the next section we will prove that for all $\omega \in (\omega_1, \omega_2)$ the sets of multi-indices coincide. Moreover, we will prove that any fixed multi-index corresponds to cells of the same dimension and finally that the union of all p-cells for $p = 0, 1, \dots, n-1$ having the same multi-index $(j_1, \dots, j_{n-1})$ for all $\omega \in (\omega_1, \omega_2)$ is a cylindrical $(p+1)$-cell to which we will assign multi-index $(j_1, \dots, j_{n-1}, j_n)$.

Let ω be the X_n-coordinate of the point in $\Omega_0^{(n-1)} \cap L_{n-1}^n(0)$ having index j_n. By the inductive hypothesis there is a cylindrical cell decomposition of $I_n^n(\omega)$. All cells of this decomposition are also the elements of a cell decomposition of I^n. If a cell in $I_n^n(\omega)$ has a multi-index $(j_1, \dots, j_{n-1})$, then considering it as a cell in I^n we assign to it the multi-index $(j_1, \dots, j_{n-1}, j_n)$.

We prove in the next section that the described decomposition $\mathcal{D}$ is compatible with V and therefore with $\{f = 0\} \cap I^n$. Observe that its total number of cells is $O(|\Omega_0^{(n-1)}|)$.

5 Cell decomposition is well defined

Let $\omega_1 < \omega_2$ be two neighbouring X_n-coordinates of points from $\Omega_s^{(n-1)}$ and $\alpha_1,\ \alpha_2$ be any two numbers such that $\omega_1 < \alpha_1 < \alpha_2 < \omega_2$. According to the inductive hypothesis (of the induction described in previous section), on both $I_n^n(\alpha_1)$ and $I_n^n(\alpha_2)$ certain cylindrical cell decompositions compatible with V are defined. Let $(i_1, \dots, i_{n-1})$ be the multi-index of a cylindrical p-cell C_1 in $I_n^n(\alpha_1)$.

Lemma 5.1. *There exists a cylindrical p-cell C_2 in $I_n^n(\alpha_2)$ having the same multi-index $(i_1, \dots, i_{n-1})$.*

Proof. According to the definition of ω_1, ω_2, the set $\Omega_s^{(n-1)} \cap \{\omega_1 < X_n < \omega_2\} = \emptyset$. It follows that

$$\Gamma_m^{(n-1)} \cap \{\alpha_1 \le X_n \le \alpha_2\}$$

consists of a finite number of disjoint curve segments each of which is homeomorphic to $[0,1]$, and on each of which X_n monotonously increases. Therefore, the sets of points of the kind $\Omega_*^{(n-2)}(\alpha_1)$ for $I_n^n(\alpha_1)$ and $\Omega_*^{(n-2)}(\alpha_2)$ for $I_n^n(\alpha_2)$, where $* \in \{0, m\}$, are in a natural bijective correspondence: $x \in \Omega_*^{(n-2)}(\alpha_1)$ corresponds to $y \in \Omega_*^{(n-2)}(\alpha_2)$ if and only if x and y belong to the same connected component of the intersection $\Gamma_*^{(n-1)} \cap \{\alpha_1 \le X_n \le \alpha_2\}$.

Let $x^{(1)} = (x_1^{(1)}, \dots, x_n^{(1)}),\ x^{(2)} = (x_1^{(2)}, \dots, x_n^{(2)}) \in \Omega_*^{(n-2)}(\alpha_1)$, and j, $1 \le j \le n-1$, be such that

(1) $x_n^{(1)} = x_n^{(2)} = \alpha_1, x_{n-1}^{(1)} = x_{n-1}^{(2)}, \dots, x_{j+1}^{(1)} = x_{j+1}^{(2)}$ and $x_j^{(1)} < x_j^{(2)}$,

(2) in the interval $(x_j^{(1)}, x_j^{(2)})$ there is no j-coordinate of any point from $\Omega_*^{(j-1)}$ with respect to $\{X_n = x_n^{(1)}, \dots, X_{j+1} = x_{j+1}^{(1)}\}$.

There exist certain $y^{(1)} = (y_1^{(1)}, \dots, y_n^{(1)}),\ y^{(2)} = (y_1^{(2)}, \dots, y_n^{(2)}) \in \Omega_*^{(n-2)}(\alpha_2)$ which are the images of $x^{(1)}, x^{(2)}$ respectively under the bijective correspondence described above.

To prove the lemma, it is sufficient to show that $y^{(1)}, y^{(2)}$ satisfy the conditions similar to (1), (2) for $x^{(1)}, x^{(2)}$. Namely:

(i) $y_n^{(1)} = y_n^{(2)} = \alpha_2,\ y_{n-1}^{(1)} = y_{n-1}^{(2)}, \dots, y_{j+1}^{(1)} = y_{j+1}^{(2)}$ and $y_j^{(1)} < y_j^{(2)}$;

(ii) in the interval $(y_j^{(1)}, y_j^{(2)})$ there is no j-coordinate of any point from $\Omega_*^{(j-1)}$ with respect to $\{X_n = y_n^{(1)}, \dots, X_{j+1} = y_{j+1}^{(1)}\}$.

Indeed, suppose that (i), (ii) are established. Index i_{n-1}, by definition, enumerates either the X_{n-1}-projection of a point $x \in \Omega_0^{(n-2)}(\alpha_1)$ or an interval between two neighbouring X_{n-1}-projections of some points $x^{(1)}, x^{(2)} \in$

$\Omega_0^{(n-2)}(\alpha_1)$. In the first case, according to the properties (i), (ii), the image point $y \in I_n^n(\alpha_2)$ of x has the same last index i_{n-1}. In the second case, by the same argument, the interval between points $y^{(1)}, y^{(2)}$ has number i_{n-1}. Repeating this argument by induction for indices $i_{n-2}, \dots, i_1$ we prove that there exists a p-cell on $I_n^n(\alpha_2)$ with multi-index $(i_1, \dots, i_{n-1})$.

Now we proceed to the proof of conditions (i), (ii). Let $\Gamma_{x^{(1)},y^{(1)}}$ (respectively, $\Gamma_{x^{(2)},y^{(2)}}$) be the connected component of the intersection $\Gamma_*^{(n-1)} \cap \{\alpha_1 \le X_n \le \alpha_2\}$ realizing the bijection between $x^{(1)}, y^{(1)}$ (respectively, between $x^{(2)}, y^{(2)}$). Suppose that (i) is false because there exists $k \in \{j+1, \dots, n-1\}$ such that $y_k^{(1)} \ne y_k^{(2)}$. Let $s \in \{j+1, \dots, n-1\}$ be the maximum among such numbers k and $y_s^{(1)} < y_s^{(2)}$. According to the construction of the set $\Gamma_*^{(n-1)}$, we have that $\pi_s(\Gamma_*^{(n-1)}) \subset \Gamma_*^{(n-1)}$, where π_s denotes the projection onto the subspace $\{X_1 = \cdots = X_{s-1} = 0\}$ equipped with coordinates $X_s, X_{s+1} \dots, X_n$. In particular, $\pi_s(\Omega_*^{(n-2)}(\alpha)) \subset \Omega_*^{(n-2)}(\alpha) \subset I_n^n(\alpha)$, where $\alpha = \alpha_1$ or α_2.

Define the points $y^{(3)} = \pi_s(y^{(1)}), y^{(4)} = \pi_s(y^{(2)})$ which belong to

$$\Omega_*^{(n-2)}(\alpha_2) \cap \{X_{s+1} = y_{s+1}^{(1)}, \dots, X_n = y_n^{(1)}\}$$

and the point $x^{(3)} = \pi_s(x^{(1)})$, which belongs to

$$\Omega_*^{(n-2)}(\alpha_1) \cap \{X_{s+1} = x_{s+1}^{(1)}, \dots, X_n = x_n^{(1)}\}.$$

Thus, $y_i^{(3)} = y_i^{(4)} = x_i^{(3)} = 0$ for every $1 \le i < s$, $y_s^{(3)} = y_s^{(1)}$, $y_s^{(4)} = y_s^{(2)}$ (so, $y_s^{(3)} < y_s^{(4)}$) and $x_s^{(3)} = x_s^{(1)} = x_s^{(2)}$, since by assumption $\pi_s(x^{(1)}) = \pi_s(x^{(2)})$.

Define curves $\Gamma' = \pi_s(\Gamma_{x^{(1)},y^{(1)}})$ and $\Gamma'' = \pi_s(\Gamma_{x^{(2)},y^{(2)}})$ which both are connected and contained in $\Gamma_*^{(n-1)}$. We have: $y^{(3)} \ne y^{(4)}$, $y^{(3)} \in \Gamma'$, $y^{(4)} \in \Gamma''$, and $x^{(3)} \in \Gamma' \cap \Gamma''$. It follows that $x^{(3)} \in S_n(\Gamma_*^{(n-1)})$, which contradicts to the choice of α_1, α_2.

We have proved that for all $k \in \{j+1, \dots, n\}$ the equality $y_k^{(1)} = y_k^{(2)}$ is true. The inequality $y_j^{(1)} < y_j^{(2)}$ (see (i)) can be proved by a symmetric argument. The property (ii) can be proved similarly. □

Let C_1, C_2 be p-cells in $I_n^n(\alpha_1)$ and $I_n^n(\alpha_2)$ respectively, having the same multi-index $(i_1, \dots, i_{n-1})$. By inductive hypothesis for any $j = 1, 2$, either $C_j \subset V$ or $C_j \cap V = \emptyset$.

Lemma 5.2. *$C_1 \subset V$ if and only if $C_2 \subset V$.*

Proof. One can show by induction that there exists a point $x \in C_1 \cap \Gamma_m^{(n-1)}$. Since $\Omega_s^{(n-1)} \cap \{\omega_1 < X_n < \omega_2\} = \emptyset$, there exists $y \in C_2 \cap V$ such that $y \in \Gamma_m^{(n-1)}$. Thus, $C_2 \subset V$. □

For any α such that $\omega_1 < \alpha < \omega_2$ denote by C_α the cylindrical p-dimensional cell in $I_n^n(\alpha)$ having a multi-index $(i_1, \dots, i_{n-1})$.

Lemma 5.3. *The union $C = \bigcup_{\alpha \in (\omega_1, \omega_2)} C_\alpha$ is a cylindrical $(p+1)$-dimensional cell in I^n.*

Proof. We prove by induction on n and p that C satisfies Definition 2.10. The base case of $n = 1$ is trivial. According to the construction and the inductive hypothesis, for each α there exists a cell C'_α in $I^n_n(\alpha) \cap \{X_1 = 0\}$ having the multi-index $(0, i_2, \dots, i_{n-1})$. In particular, $\pi_2(C_\alpha) = C'_\alpha$. By the inductive hypothesis, the union $C' = \bigcup_{\alpha \in (\omega_1, \omega_2)} C'_\alpha$ is a cylindrical cell in $I^n_1(0)$. Also, by the inductive hypothesis, C_α is either a graph of a continuous bounded function $h_\alpha : C'_\alpha \to I^1$, or a sector between two such functions (see Definition 2.10). Let, for definiteness, C_α be the graph of some h_α. Then C is the graph of the continuous bounded function $h : C' \to I^1$ such that $h(x_2, \dots, x_{n-1}, x_n) = h_{x_n}(x_2, \dots, x_{n-1})$. □

Lemma 5.3 implies that either $C \subset V$ or $C \cap V = \emptyset$, i.e., $\mathcal{D}$ is compatible with V.

Lemma 5.4. *The decomposition $\mathcal{D}$ is a cylindrical cell decomposition.*

Proof. Straightforward. □

Remark 5.5. Let in the induction step n of the description of $\mathcal{D}$, the definition of the set $\Omega_s^{(n-1)}$ be modified by adding an arbitrary point $x \in cl(\Gamma_m^{(n-1)})$. Thus,

$$\Omega_s^{(n-1)} \equiv S_n(\Gamma_0^{(n-1)}) \cup S_n(\Gamma_m^{(n-1)}) \cup \{x\}.$$

Clearly, this leads to another cylindrical cell decomposition of I^n compatible with V which is a refinement of $\mathcal{D}$.

6 Constructing a cell decomposition: stage I

The algorithm recursively constructs the decomposition $\mathcal{D}$ described in previous sections. The idea is to find the finite set $\Omega_0^{(n-1)}$ and thereby the cell decomposition $\mathcal{D}^{(n-1)}$ induced by $\mathcal{D}$ on $I^1 \subset \{X_1 = \dots = X_{n-1} = 0\}$. For each cell C of $\mathcal{D}^{(n-1)}$ the algorithm finds the finite set $\Omega_0^{(n-2)}$ parameterized by the points of C, and thereby the cell decomposition $\mathcal{D}^{(n-2)}$ induced on $I^2 \subset \{X_1 = \dots = X_{n-2} = 0\}$. On the last step of this recursion the parameterized set $\Omega_0^{(0)}$ and the cell decomposition $\mathcal{D} = \mathcal{D}^{(0)}$ of I^n will be found, that is compatible with V.

A straightforward representation of a finite (generally parametric) set $\Omega_0^{(i)}$, by means of formula $\Phi_0^{(i)}$ (see Section 4) would require quantifier alternation which we clearly want to avoid. In the following algorithm, at step i of the induction, for fixed values of coordinates $X_{i+1}, \dots, X_n$, we represent $\Omega_0^{(i-1)}$ in the following three stages. Firstly, we approximate $\Omega_s^{(i-1)}$ by a finite (parametric) set of points defined by an *existential* formula. Then we define $\Omega_s^{(i-1)}$ by passing to limit with a help of Lemma 2.9. Finally, we define the set $\Omega_0^{(i-1)}$ itself by an existential formula which involves the formula for the set $\Omega_s^{(i-1)}$.

Define recursively the sequence of integers $s_1, \ldots, s_n$ by setting $s_1 = 2$ and $s_{i+1} = 5s_i + 2i + 2$ for $1 \leq i \leq n-1$. Introduce new variables $Y_1, \ldots, Y_{s_n}, Z$.

Let $D_i \equiv (X_i \geq 0) \wedge (X_i \leq 1)$, so that $\{\bigwedge_{1 \leq i \leq k} D_i\} = I^k$. Denote $T^{(m)} = (T_1, \ldots, T_m)$, the m-tuple of variables T_i, $i \leq m$.

Let $X = X^{(n)}$. We now introduce formulae $G_0^{(i)}, G_m^{(i)}, G^{(i)}$ by induction on i. In cases $i = 1, 2$ we include comments explaining some non-trivial parts of the construction. Note that at a step i we treat $X_{i+1}, \ldots, X_n$ as parameters of formulae.

Step i = 1.

$f^{(0)}(X) \equiv (f(X))^2$

In case when $f \not\equiv 0$ the equation $f^{(0)}(X) = 0$ defines the set $\Omega_0^{(0)} \setminus \{0, 1\}$ and possibly some points outside $[0, 1]$.

$h_Z^{(0)}(X, Z) \equiv (f^{(0)}(X) - Z)^2$

The points satisfying $f^{(0)}(X) = 0$ are perturbed by Z.

$H_Z^{(0)} \equiv \left(h_Z^{(0)} = 0\right)$

$\Theta_0^{(0)} \equiv \left(cl\left(H_Z^{(0)} \wedge (Z > 0)\right) \wedge (Z = 0)\right) \vee (X_1 \cdot (X_1 - 1) = 0)$

This formula defines the limits of perturbed points as $Z \to +0$ with added $\{0, 1\}$, i.e, the set $\Omega_0^{(0)}$ and possibly some points outside $[0, 1]$. The only purpose of perturbation is to start the pattern which is meaningful on further induction steps.

$\Theta_m^{(0)} \equiv \big(X_1 = 1/2(Y_1 + Y_2) \wedge \Theta_0^{(0)}(Y_1, X_2, \ldots, X_n) \wedge \Theta_0^{(0)}(Y_2, X_2, \ldots, X_n) \wedge (f^{(0)}(X) = 0)\big)$

Defining the set $\Omega_m^{(0)}$ and possibly some points outside $[0, 1]$.

$G_0^{(1)}(Y^{(2)}, X) \equiv \Theta_0^{(0)} \wedge (Y_1 = Y_2 = 0) \wedge D_1$

Defining the set $\Omega_0^{(0)}$ and the (parametric) curve $\Gamma_0^{(1)}$ as projections along variables Y_1, Y_2.

$G_m^{(1)}(Y^{(2)}, X) \equiv \Theta_m^{(0)} \wedge D_1$

Defining the set $\Omega_m^{(0)}$ and the (parametric) curve $\Gamma_m^{(1)}$ as projections along variables Y_1, Y_2.

$G^{(1)}(Y^{(2)}, X) \equiv G_0^{(1)}(Y^{(2)}, X) \vee G_m^{(1)}(Y^{(2)}, X)$

Defining $\Gamma_0^{(1)} \cup \Gamma_m^{(1)}$.

Step i = 2.

$G_*^{(1)}(Y^{(2)}, X) \equiv \bigvee_{1 \leq l \leq M_1} \left((f_{l*}^{(1)}(Y^{(2)}, X) = 0) \wedge (g_{l*}^{(1)}(Y^{(2)}, X) > 0)\right)$ for $* \in \{0, m\}$.

Representing each $G_*^{(1)}(Y^{(2)}, X)$ as a Boolean combination of atomic equations and inequalities.

For each l, $1 \le l \le M_0$, define :
$h_{l,Z,*}^{(1)}(Y^{(4)}, X, Z) \equiv (f_{l*}^{(1)}(Y^{(2)}, X) - Z)^2 + \left(\frac{\partial f_{l*}^{(1)}}{\partial X_1}\right)^2 + \left(\frac{\partial f_{l*}^{(1)}}{\partial Y_1}\right)^2 + \left(\frac{\partial f_{l*}^{(1)}}{\partial Y_2}\right)^2 + Y_3^2 + Y_4^2$
For small values of Z the equation $f_{l*}^{(1)}(X) = Z$ defines a smooth hypersurface. Then $h_{l,Z,*}^{(1)}(Y^{(4)}, X, Z) = 0$ defines the set of all critical points of the coordinate function X_2 on this hypersurface. The purpose of introducing variables Y_3, Y_4 will be explained below.

$H_{Z*}^{(1)} \equiv \bigvee_{1 \le l \le M_1} \left((h_{l,Z,*}^{(1)} = 0) \wedge (g_{l*}^{(1)} > 0)\right)$
Collecting together the critical points on $f_{l*}^{(1)}(X) = Z$ for all l, $1 \le l \le M_1$ and selecting the ones which are relevant. Note that for small values of $Z > 0$ all points of local extrema of the coordinate function X_2 on $\{G_*^{(1)}(Y^{(2)}, X)\}$, except possibly the ones with $X_2(X_2 - 1) = 0$, are close to corresponding critical points.

$\Theta_{e*}^{(1)}(Y^{(4)}, X) \equiv (cl\big(H_{Z*}^{(1)} \wedge (Z > 0)\big) \wedge (Z = 0)) \vee (G_*^{(1)} \wedge (X_2(X_2 - 1) = 0))$
Passing to limit as $Z \to +0$ and adding $G_*^{(1)} \wedge (X_2(X_2 - 1) = 0)$ produces a finite (parameterized) set of points on $\{G_*^{(1)}\}$ which includes all points of local extrema of X_2 on $\{G_*^{(1)}\}$. The projection of $\{\Theta_{e*}^{(1)}(Y^{(4)}, X)\}$ along variables Y_1, Y_2, Y_3, Y_4 contains all points of local extrema of X_2 on $\Gamma_*^{(1)}$.

$\Theta_{\partial*}^{(1)}(Y^{(4)}, X) \equiv \partial(G_*^{(1)}(Y^{(2)}, X)) \wedge (Y_3 = Y_4 = 0)$
Defining a finite set of frontier points of $\{G_*^{(1)}(Y^{(2)}, X)\}$. The projection of $\{\Theta_{\partial*}^{(1)}(Y^{(4)}, X)\}$ along variables Y_1, Y_2, Y_3, Y_4 contains $B(\{\Gamma_*^{(1)}\})$.

$G_{1*}^{(1)} \equiv G_*^{(1)}(Y^{(2)}, X_1 - Z, X_2, \ldots, X_n)$
This defines a curve obtained from $\{G_*^{(1)}\}$ by shifting it along the coordinate axis X_1 by Z.

$Q_{Z*}^{(1)}(Y^{(4)}, X_1 - Z, X) \equiv G_{1*}^{(1)}(Y^{(2)}, X_1 - Z, X_2, \ldots, X_n) \wedge G_*^{(1)}(Y_3, Y_4, X)$
Intersecting the projection of $\{G_*^{(1)}\}$ with the projection of its shift produces a finite (parameterized) subset of $\Gamma_*^{(1)}$. Note that we need two additional variables Y_3, Y_4. Observe that for a small value $|Z|$ each ramification point of $\{\Gamma_*^{(1)}\}$ is close to the projection along Y_1, Y_2, Y_3, Y_4 of some of the points from $\{Q_{Z*}^{(1)}\}$.

$\Theta_{r*}^{(1)}(Y^{(4)}, X_1 - Z, X) \equiv cl\big(Q_{Z*}^{(1)} \wedge (Z > 0)\big) \wedge (Z = 0)$
Passing to limit as $Z \to +0$ produces a finite (parameterized) set of points on $\{G_*^{(1)}\}$ such that its projection along variables Y_1, Y_2, Y_3, Y_4, X_1 contains all X_2-coordinates of ramification points of $\Gamma_*^{(1)}$.

$\Theta_s^{(1)}(Y^{(4)}, X) \equiv \Theta_{e0}^{(1)} \vee \Theta_{\partial 0}^{(1)} \vee \Theta_{r0}^{(1)} \vee \Theta_{em}^{(1)} \vee \Theta_{\partial m}^{(1)} \vee \Theta_{rm}^{(1)}$

Defining a set whose projection along variables $Y_1, \dots, Y_4, X_1$ is a finite set containing all X_2-coordinates of points from $S_2(\Gamma_0^{(1)}) \cup S_2(\Gamma_m^{(1)})$.

$\Theta_0^{(1)}(Y^{(7)}, X) \equiv G_0^{(1)}(Y^{(2)}, X) \wedge \Theta_s^{(1)}(Y_3, Y_4, \dots, Y_7, X_2, \dots, X_n)$
Defining a set whose projection along variables $Y_1, \dots, Y_7$ contains $\Omega_0^{(1)}$. Note that in the expression $\Theta_s^{(1)}(Y_3, Y_4, \dots, Y_7, X_2, \dots, X_n)$ variables $Y_3, \dots, Y_6$ stand for $Y^{(4)}$ in the definition of $\Theta_s^{(1)}$ while Y_7 stands for X_1. For any fixed values of parameters $X_3, \dots, X_n$ the set $\Theta_s^{(1)}$ is finite and therefore the set $\{G_0^{(1)} \wedge \Theta_s^{(1)}\}$ reduces to an intersection of two finite unions of affine subspaces of complementary dimensions in 8-dimensional space. It follows that $\Theta_0^{(1)}(Y^{(7)}, X)$ is finite.

$$\Theta_m^{(1)}(Y^{(14)}, X) \equiv \big(X_2 = 1/2(Y_8 + Y_{14}) \wedge$$
$$G^{(1)}(Y^{(2)}, X) \wedge \Theta_s^{(1)}(Y_3, Y_4, \dots, Y_8, X_3, \dots, X_n) \wedge$$
$$\wedge \Theta_s^{(1)}(Y_9, \dots, Y_{14}, X_3, \dots, X_n)\big)$$

Defining a finite set of points whose projection along variables $Y_1, \dots, Y_{14}$ contains $\Omega_m^{(1)}$.

$G_0^{(2)}(Y^{(14)}, X) \equiv \Theta_0^{(1)}(Y^{(7)}, X) \wedge (Y_8 = \dots = Y_{14} = 0) \wedge D_2$
$G_m^{(2)}(Y^{(14)}, X) \equiv \Theta_m^{(1)}(Y^{(14)}, X) \wedge D_2$
$G^{(2)}(Y^{(14)}, X) \equiv G_0^{(2)}(Y^{(14)}, X) \vee G_m^{(2)}(Y^{(14)}, X)$

General step. Assume that on step i, $i \leq n-1$, the expression
$G^{(i)}(Y^{(s_i)}, X) \equiv G_0^{(i)} \vee G_m^{(i)}$
was defined. The interpretations of the following formulae are analogous to the ones provided in **step 2**.

Step (i + 1).

$G_*^{(i)}(Y^{(s_i)}, X) \equiv \bigvee_{1 \leq l \leq M_i} \big((f_{l*}^{(i)}(Y^{(s_i)}, X) = 0) \wedge (g_{l*}^{(i)}(Y^{(s_i)}, X) > 0)\big)$, where $* \in \{0, m\}$.

For each l, $1 \leq l \leq M_i$, define :
$h_{l,Z,*}^{(i)}(Y^{(2s_i)}, X, Z) \equiv \big(f_{l*}^{(i)} - Z\big)^2 + \sum_{1 \leq j \leq i+1} \Big(\frac{\partial f_{l*}^{(i)}}{\partial X_j}\Big)^2 + \sum_{1 \leq j \leq s_i} \Big(\frac{\partial f_{l*}^{(i)}}{\partial Y_j}\Big)^2 + \sum_{s_i+1 \leq j \leq 2s_i} \Big(Y_j\Big)^2$

$H_{Z*}^{(i)} \equiv \bigvee_{1 \leq l \leq M_i} \big((h_{l,Z,*}^{(i)} = 0) \wedge (g_{l*}^{(i)} > 0)\big)$

$\Theta_{e*}^{(i)} \equiv cl\big(H_{Z*}^{(i)} \wedge (Z > 0)\big) \wedge (Z = 0) \vee (G_*^{(i)} \wedge (X_{i+1}(X_{i+1} - 1)) = 0)$

$\Theta_{\partial *}^{(i)} \equiv \partial(G_*^{(i)}(Y^{(s_i)}, X)) \wedge (Y_{s_i+1} = \dots = Y_{2s_i} = 0)$
$G_{1*}^{(i)} \equiv G_*^{(i)}(Y_{1+s_i}, \dots, Y_{2s_i}, X_1, \dots, X_{i-1}, X_i - Z, X_{i+1}, \dots, X_n)$

$G_{2*}^{(i)} \equiv G_*^{(i)}(Y_{1+s_i}, \dots, Y_{2s_i}, X_1, \dots, X_{i-2}, X_{i-1} - Z, X_i, \dots, X_n)$

......

$G^{(i)}_{j*} \equiv G^{(i)}_{*}(Y_{1+s_i}, \dots, Y_{2s_i}, X_1, \dots, X_{i-j}, X_{i+1-j} - Z, X_{i+2-j}, \dots, X_n)$

......

$G^{(i)}_{i*} \equiv G^{(i)}_{*}(Y_{1+s_i}, \dots, Y_{2s_i}, X_1 - Z, X_2, \dots, X_n)$

$Q^{(i)}_{1,Z,*}(Y^{(2s_i)}, X_i - Z, X) \equiv G^{(i)}_{1*} \wedge G^{(i)}_{*}$

$Q^{(i)}_{2,Z,*}(Y^{(2s_i)}, X_{i-1} - Z, X) \equiv G^{(i)}_{2*} \wedge G^{(i)}_{*}$

......

$Q^{(i)}_{j,Z,*}(Y^{(2s_i)}, X_{i+1-j} - Z, X) \equiv G^{(i)}_{j*} \wedge G^{(i)}_{*}$

......

$Q^{(i)}_{i,Z,*}(Y^{(2s_i)}, X_1 - Z, X) \equiv G^{(i)}_{i*} \wedge G^{(i)}_{*}$

$Q^{(i)}_{Z*}(Y^{(2s_i)}, X_1 - Z, \dots, X_i - Z, X) \equiv \bigvee_{1 \le j \le i} (Q^{(i)}_{j,Z,*})$

$\Theta^{(i)}_{r*} \equiv cl\left(Q^{(i)}_{Z*} \wedge (Z > 0)\right) \wedge (Z = 0)$

$\Theta^{(i)}_{s}(Y^{(2s_i)}, X) \equiv \Theta^{(i)}_{e0} \vee \Theta^{(i)}_{\partial 0} \vee \Theta^{(i)}_{r0} \vee \Theta^{(i)}_{em} \vee \Theta^{(i)}_{\partial m} \vee \Theta^{(i)}_{rm}$

$\Theta^{(i)}_{0}(Y^{(3s_i+i)}, X) \equiv G^{(i)}_{0}(Y^{(s_i)}, X) \wedge$
$\quad \wedge \Theta^{(i)}_{s}(Y_{s_i+1}, \dots, Y_{3s_i}, Y_{3s_i+1}, \dots, Y_{3s_i+i}, X_{i+1}, \dots, X_n)$

$\Theta^{(i)}_{m}(Y^{(s_{i+1})}, X) \equiv (X_{i+1} = 1/2(Y_{3s_i+i+1} + Y_{5s_i+2i+2}) \wedge G^{(i)}(Y^{(s_i)}, X) \wedge$
$\quad \wedge \Theta^{(i)}_{s}(Y_{s_i+1}, \dots, Y_{3s_i+i+1}, X_{i+2}, \dots, X_n) \wedge$
$\quad \wedge \Theta^{(i)}_{s}(Y_{3s_i+i+2}, \dots, Y_{5s_i+2i+2}, X_{i+2}, \dots, X_n)$

$G^{(i+1)}_{0}(Y^{(s_{i+1})}, X) \equiv \Theta^{(i)}_{0}(Y^{(3s_i+i)}, X) \wedge (Y_{3s_i+i+1} = \dots = Y_{s_{i+1}} = 0) \wedge D_{i+1}$

$G^{(i+1)}_{m}(Y^{(s_{i+1})}, X) \equiv \Theta^{(i)}_{m}(Y^{(s_{i+1})}, X) \wedge D_{i+1}$

$G^{(i+1)}\left(Y^{(s_{i+1})}, X\right) \equiv G^{(i+1)}_{0} \vee G^{(i+1)}_{m}$

End of the general step.

For each i, $1 \le i \le n$ let $\rho_i : \mathbb{R}^{s_i+i} \to \mathbb{R}^i$ be the projection map along $Y^{(s_i)}$ onto the subspace with coordinates $X_1, \dots, X_i$. Consider a vector $(\omega_{i+2}, \dots, \omega_n)$ such that $0 \le \omega_j \le 1$ for all $j = i+2, \dots, n$. For any $* \in \{0, m\}$ let $\Omega^{(i)}_{*}(\omega_{i+2}, \dots, \omega_n)$ denote the set $\Omega^{(i)}_{*}$ for $V \cap \{X_{i+2} = \omega_{i+2}, \dots, X_n = \omega_n\}$ in the cube I^{n-i+1} identified with $I^n \cap \{X_{i+2} = \omega_{i+2}, \dots, X_n = \omega_n\}$.

Lemma 6.1. *For any* $* \in \{0, m\}$ *the projection* $\rho_{i+1}(\{\Theta^{(i)}_{*}(Y^{(3s_i+i)}, X)\} \cap \{X_{i+2} = \omega_{i+2}, \dots, X_n = \omega_n\})$ *is a finite set of points containing* $\Omega^{(i)}_{*}(\omega_{i+2}, \dots, \omega_n)$.

Proof. The proof is a straightforward routine using induction on i. For $i = 2$ it is actually contained in the comments to formulae defining $\Theta^{(1)}_{0}$, $\Theta^{(1)}_{m}$ (see **step i = 2**). □

Lemma 6.2. *For each $k \in \{1, \ldots, n\}$ the format of the sets defined by $G_*^{(k)}$ (for $* \in \{0, m\}$) is $(N_k, \alpha, \beta_k, r_k, m_k)$, where*

$$N_k = (\alpha + \beta N)^{(r+n)^{O(k)} 2^{O(k^2)}}, \beta_k = (\alpha + \beta N)^{(r+n)^{O(k)} 2^{O(k^2)}},$$

$$r_k = r 5^{k-1}, m_k = O(n 5^k).$$

Proof. Recall that the semi-Pfaffian set $V = (\{f = 0\} \cap I^n) \cup I_1^n$ has format $(O(n2^n), \alpha, 2\beta N, r, n)$. Introduce notations $N_V = O(n2^n)$ and $D_V = O(\alpha + \beta N)$.

At step k, $1 \leq k \leq n$, let $m_k = s_k + n$ be the total number of variables and r_k the size of the Pfaffian chain for the functions in $G_*^{(k)}$. Recall that $s_1 = 2$ and $s_k = 5s_{k-1} + 2k$. Then $s_k = O(k5^k)$ and $m_k = O(n5^k)$.

For $k = 1$ the order $r_1 = r$ since the operation of taking closure (see Lemma 2.9) leaves the size of a Pfaffian chain unchanged. Suppose that $F(Y^{(s_{k-1})}, X) = \left(f_1(Y^{(s_{k-1})}, X), \ldots, f_{r_{k-1}}(Y^{(s_{k-1})}, X)\right)$ is the Pfaffian chain of the set defined by $G_*^{(k-1)}$. Notice that the order of the Pfaffian chain of the set defined by $Q_{j,Z,*}^{(k-1)}$, $1 \leq j \leq k-1$ is $2r_{k-1}$ since we need to add in this chain the same functions as before but with variables $Y^{(s_{k-1})}, X_{k+1-j}$ replaced by $Y_{s_{k-1}+1}, \ldots, Y_{2s_{k-1}}, X_{k+1-j} - Z$ respectively. Thus, the size of the Pfaffian chain of $Q_{Z*}^{(k-1)}$ is equal to kr_{k-1}. According to Lemma 2.9, there is a Boolean formula, say $Q_*^{(k-1)}$, with atomic Pfaffian functions in variables $Y^{(2s_{k-1})}, X_1 - Z, \ldots, X_{k-1} - Z, X, Z$, having the same common Pfaffian chain as $Q_{Z*}^{(k-1)}$ such that $\{Q_*^{(k-1)}\} = cl\{Q_{Z*}^{(k-1)} \wedge (Z > 0)\}$. The formula $\Theta_{r*}^{(k-1)} \equiv Q_*^{(k-1)} \wedge (Z = 0)$ is equivalent to an expression involving Pfaffian functions only in variables $Y^{(2s_{k-1})}, X$. Substituting the value 0 for the variable Z in every function present in the Pfaffian chain of $Q_*^{(k-1)}$ we see that the Pfaffian chain of $\Theta_{r*}^{(k-1)}$ is $F(Y^{(s_{k-1})}, X), F(Y_{s_{k-1}+1}, \ldots, Y_{2s_{k-1}}, X)$. Similarly, formulae $\Theta_{e*}^{(k-1)}$ and $\Theta_{\partial*}^{(k-1)}$ are equivalent to expressions involving Pfaffian functions only in variables $Y^{(2s_{k-1})}, X$, having common Pfaffian chain $F(Y^{(s_{k-1})}, X)$. Thus, the Pfaffian chain of $\Theta_s^{(k-1)}$ is again $F(Y^{(s_{k-1})}, X), F(Y_{s_{k-1}+1}, \ldots, Y_{2s_{k-1}}, X)$. It follows that the Pfaffian chain of $\Theta_0^{(k-1)}$ is

$$F(Y^{(s_{k-1})}, X), F(Y_{s_{k-1}+1}, \ldots, Y_{2s_{k-1}}, Y_{3s_{k-1}+1}, \ldots, Y_{3s_{k-1}+k-1}, X_k, \ldots, X_n),$$

$$F(Y_{2s_{k-1}+1}, \ldots, Y_{3s_{k-1}}, Y_{3s_{k-1}+1}, \ldots, Y_{3s_{k-1}+k-1}, X_k, \ldots, X_n)$$

and the common Pfaffian chain of $\Theta_m^{(k-1)}$ and $G_*^{(k)}$ is

$$F(Y^{(s_{k-1})}, X), F(Y_{s_{k-1}+1}, \ldots, Y_{2s_{k-1}}, Y_{3s_{k-1}+1}, \ldots, Y_{3s_{k-1}+k}, X_{k+1}, \ldots, X_n),$$

$$F(Y_{2s_{k-1}+1}, \ldots, Y_{3s_{k-1}}, Y_{3s_{k-1}+1}, \ldots, Y_{3s_{k-1}+k}, X_{k+1}, \ldots, X_n),$$

$$F(Y_{3s_{k-1}+k+1}, \ldots, Y_{4s_{k-1}+k}, Y_{5s_{k-1}+k+1}, \ldots, Y_{5s_{k-1}+2k}, X_{k+1}, \ldots, X_n),$$

$$F(Y_{4s_{k-1}+k+1},\dots,Y_{5s_{k-1}+k},Y_{5s_{k-1}+k+1},\dots,Y_{5s_{k-1}+2k},X_{k+1},\dots,X_n).$$

We conclude that the order of the set defined by $G_*^{(k)}$ is $r_k = 5r_{k-1} = r5^{k-1}$.

Let

$$p_k = \prod_{1\le j\le k-1} m_j = n^{k-1}2^{O(k^2)}, \quad q_k = \prod_{1\le j\le k-1} (m_j + r_j) = (r+n)^{k-1}2^{O(k^2)}.$$

For $k=1$, applying the bounds from Lemma 2.9 we get

$$N_1 = (2^{r^2} N_V)^{O(m_1(m_1+r_1))}(m_1 D_V)^{O(m_1(m_1+r_1)^2)}$$

and

$$\beta_1 = 2^{m_1 r_1^2}(m_1 D_V)^{O(m_1(m_1+r_1))}$$

Note that at each other step we perform two iterations of the closure operation. It can be seen that

$$\beta_k = (2^r p_k D_V)^{O(p_k^2 q_k^2)} = D_V^{(r+n)^{O(k)}2^{O(k^2)}}$$

and

$$N_k = (2^{r^2} p_k N_V \beta_k^{2k})^{O(p_k^2 q_K^3)} = D_V^{(r+n)^{O(k)}2^{O(k^2)}}.$$

□

The algorithm writes out formulae $G_*^{(k)}$, $G^{(k)}$ for all $1\le k\le n$ using the described recursive formulae. The complexity of this stage of the algorithm does not exceed $(\alpha+\beta N)^{r^{O(n)}2^{O(n^2)}}$.

7 Constructing a cell decomposition: stage II

The second stage of the algorithm consists of the following recursive procedure. For $0\le i\le n-1$ and for fixed values of coordinates $X_{i+2},\dots,X_n$ let $\Omega_L^{(i)} = \Omega_0^{(i)} \cap L_i^n(0) \subset \Omega_0^{(i)}$. This is the projection of $\Omega_0^{(i)}$ on $\{X_1 = \cdots = X_i = 0\}$. According to Lemma 6.1, $\Omega_L^{(i)} \subset \rho_{i+1}(\{G_0^{(i+1)}(Y^{(s_i)},X) \wedge (X_1 = \cdots = X_i = 0)\})$. Introduce the notation

$$\Psi_L^{(i)}(X) \equiv (\exists Y^{(s_i)})(G_0^{(i)}(Y^{(s_i)},X) \wedge (X_1 = \cdots = X_i = 0)).$$

Thus, $\Omega_L^{(i)} \subset \{\Psi_L^{(i)}(X)\}$.

For each $i\in\{0,\dots,n\}$ we define as follows a cylindrical cell decomposition $\mathcal{D}^{(i)}$ of $L_i^n(0)$ which is compatible with the projection of V onto the subset $L_i^n(0)$ of $\mathbb{R}^{n-i}$ equipped with coordinates $X_{i+1},\dots,X_n$.

For each $k\in\{0,\dots,n\}$ let $i=n-k$. We proceed by induction on k.

For $k=0$ set $\alpha=(0,\dots,0)$ and $C_\alpha^{(n)} = \{(0,\dots,0)\}$ to be the only cell of the cylindrical cell decomposition $\mathcal{D}^{(n)}$ of $L_n^n(0)$.

Suppose that on the step k a cylindrical cell decomposition $\mathcal{D}^{(i)}$ of the cube $L_i^n(0)$ was defined. On the next step $k+1$ (with $i = n-k-1$) the input is the cylindrical cell decomposition $\mathcal{D}^{(i+1)}$ of the cube $L_{i+1}^n(0)$ obtained on the previous step. For each cell $C_\alpha^{(i+1)} \in \mathcal{D}^{(i+1)}$ denote by $Z(C_\alpha^{(i+1)})$ the set $C_\alpha^{(i+1)} \times [0,1]$, which is the bounded cylinder over $C_\alpha^{(i+1)}$ and along X_{i+1}, contained in the cube $L_i^n(0)$. The algorithm constructs a cell decomposition of $Z(C_\alpha^{(i+1)})$ in the following way. Observe that for any point $z = (z_{i+2}, \ldots, z_n) \in C_\alpha^{(i+1)}$, the cardinality of $\{\Psi_L^{(i)}(X_1, \ldots, X_{i+1}, z)\}$ is finite and constant over $C_\alpha^{(i+1)}$. Corollary 2.7 implies that this number does not exceed

$$M_i = (\alpha + \beta N)^{(r+n)^{O(i)} 2^{O(i^2)}}.$$

Introduce new variables $x_{j,i+1}^{(i)}$, $1 \le j \le M_i$, and denote $x_j^{(i)} = (0, \ldots, 0, x_{j,i+1}^{(i)}, z)$.

The algorithm finds the exact number K_α of distinct points in

$$\{\Psi_L^{(i)}(X_1, \ldots, X_{i+1}, z)\}$$

by testing with the oracle for each m, $1 \le m \le M_i$, whether the statement

$$(\exists z)(\exists x_{1,i+1}^{(i)}) \cdots (\exists x_{m,i+1}^{(i)}) \Bigg[\Big(\bigwedge_{1 \le j \le m} \Psi_L^{(i)}(x_j^{(i)}) \Big) \wedge$$

$$\Big(\bigwedge_{1 \le r \le m-1} \bigwedge_{1 \le j \le r} (x_{r+1,i+1}^{(i)} \ne x_{j,i+1}^{(i)}) \Big) \Bigg]$$

is true. The number K_α is the maximal m for which the statement holds.

We now describe all cells in $Z(C_\alpha^{(i+1)}) \subset L_i^n(0)$ by the following formulae.

- Sections over $C_\alpha^{(i+1)}$: for $1 \le j \le K_\alpha$

$$C_{\alpha'}^{(i)} = \Bigg\{ (0, \ldots, 0, z_{i+1}, \ldots, z_n) \in Z(C_\alpha^{(i+1)}) \ :$$

$$(\exists x_1^{(i)}) \cdots (\exists x_{K_\alpha}^{(i)}) \Big(\bigwedge_{1 \le j \le K_\alpha} \Psi_L^{(i)}(x_j^{(i)}) \wedge$$

$$\wedge x_{1,i+1}^{(i)} < \cdots < x_{j-1,i+1}^{(i)} < x_{j,i+1}^{(i)} = z_{i+1} < x_{j+1,i+1}^{(i)} < \cdots < x_{K_\alpha,i+1}^{(i)} \Big) \Bigg\},$$

where the index $\alpha' = (0, \ldots, 0, 2j-2, \alpha_{i+2}, \ldots, \alpha_n)$.

- Sectors over $C_\alpha^{(i+1)}$: for $1 \le j \le K_\alpha - 1$

$$C_{\alpha'}^{(i)} = \Bigg\{ (0, \ldots, 0, z_{i+1}, \ldots, z_n) \in Z(C_\alpha^{(i+1)}) \ :$$

$$(\exists x_1^{(i)}) \cdots (\exists x_{K_\alpha}^{(i)}) \Big(\bigwedge_{1 \le j \le K_\alpha} \Psi_L^{(i)}(x_j^{(i)}) \wedge$$

$$\wedge x_{1,i+1}^{(i)} < \cdots < x_{j-1,i+1}^{(i)} < x_{j,i+1}^{(i)} < z_{i+1} < x_{j+1,i+1}^{(i)} < \cdots < x_{K_\alpha,i+1}^{(i)}) \Big\},$$

where the index $\alpha' = (0, \ldots, 0, 2j-1, \alpha_{i+2}, \ldots, \alpha_n)$.

Combining the cell decompositions of $Z(C_\alpha^{(i+1)})$ for all cells $C_\alpha^{(i+1)}$ in $\mathcal{D}^{(i+1)}$, we obtain a cylindrical cell decomposition $\mathcal{D}^{(i)}$ of the cube $L_i^n(0)$. This concludes the description of the algorithm.

We conclude by estimating the complexity of the algorithm. On each recursion step $k = 0, \ldots, n$ of the second stage of the algorithm, the degrees of Pfaffian functions remain the same as in $G_0^{(n-k)}$, while the numbers of variables, orders and atomic formulae are multiplied by at most $M_{n-k}+1$ each. The number of cells is increased on a recursion step according to the formula $|\mathcal{D}^{(n-k)}| \le O(M_{n-k})|\mathcal{D}^{(n-k+1)}|$ and $|\mathcal{D}^{(n)}| = 1$, so $|\mathcal{D}^{(0)}| \le O(M_n)^n$.

It follows that the described two-stage algorithm produces a cylindrical cell decomposition $\mathcal{D}^{(0)}$ of I^n, compatible with the semi-Pfaffian set $S \cap I^n$, consisting of $\Lambda = (\alpha + \beta N)^{r^{O(n)} 2^{O(n^2)}}$ cells. Each cell is a sub-Pfaffian set having a format $(\Lambda, \alpha, \Lambda, \Lambda, \Lambda)$. The complexity of the algorithm does not exceed Λ.

References

[1] Basu, S., Pollack, R., Roy, M.-F., On the combinatorial and algebraic complexity of quantifier elimination, *Journal of the ACM,* 43, 1996, 1002–1045.

[2] Blum, L., Cucker, F., Shub, M., Smale, S., *Complexity and Real Computation,* Springer-Verlag, 1997.

[3] Collins, G.E., Quantifier elimination for real closed fields by cylindrical algebraic decomposition, *Lecture Notes in Computer Science*, 33, 1975, 134–183.

[4] Davenport, J.H., Heintz, J., Real quantifier elimination is doubly exponential, *J. Symbolic Comput.,* 5, 1988, 29–35.

[5] Dries van den, L., *Tame Topology and O-minimal Structures, LMS Lecture Note Series*, 248, Cambridge University Press, Cambridge, 1998.

[6] Gabrielov, A., Projections of semi-analytic sets, *Functional Anal. Appl.,* 2, 1968, 282–291.

[7] Gabrielov, A., On complements of subanalytic sets and existential formulas for analytic functions, *Invent. Math.,* 125, 1996, 1–12.

[8] Gabrielov, A., Frontier and closure of a semi-Pfaffian set, *Discrete Comput. Geom.,* 19, 1998, 605–617.

[9] Gabrielov, A., Vorobjov, N., Complexity of cylindrical decompositions of sub-Pfaffian sets, *J. Pure and Appl. Algebra,* 164 (2001), 179–197.

[10] Gabrielov, A., Vorobjov, N., Complexity of stratifications of semi-Pfaffian sets, *Discrete Comput. Geom.,* 14, 1995, 71–91.

[11] Grigoriev, D., Complexity of deciding Tarski algebra, *J. Symbolic Comput.,* 5, 1988, 65–108.

[12] Grigoriev, D., Vorobjov, N., Solving systems of polynomial inequalities in subexponential time, *J. Symbolic Comput.*, 5, 1988, 37–64.

[13] Khovanskii, A., On a class of systems of transcendental equations, *Soviet Math. Dokl.*, 22, 1980, 762–765.

[14] Khovanskii, A., *Fewnomials,* AMS Transl. Math. Monographs, 88, AMS, 1991.

[15] Milnor, J. W. *Topology from the differentiable viewpoint*, Univ. Pres, Virginia, 1965

[16] Osgood, W. F., On functions of several complex variables, *Trans. Amer. Math. Soc.*, 17, 1916.

[17] Renegar, J., On the computational complexity and the first order theory of reals. Parts I–III, *J. Symbolic Comput.*, 13, 1992, 255–352.

[18] Vorobjov, N., The complexity of deciding consistency of systems of polynomial in exponent inequalities, *J. Symbolic Comput.*, 13, 1992, 139–173.

[19] Wilkie, A., A general theorem of the complement and some new o-minimal structures, *Selecta Mathematica, New Series,* 5, 1999, 397–421.

[20] Wüthrich, H.R., Ein Entscheidungsvefahren für die Theorie der reell-abgeschlossenen Körper, *Lecture Notes in Computer Science*, 43, 1976, 138–162.

About Authors

Savvas Pericleous (*mapsp@cs.bath.ac.uk*) and
Nicolai Vorobjov (*nnv@cs.bath.ac.uk*)
are at the Department of Computer Science, University of Bath, Bath, BA2 7AY, UK.

Acknowledgments

This work was supported in part by EPSRC grant GR/L77928. We are grateful to Andrei Gabrielov and Dan Richardson for useful discussions.

Note on the Chromatic Number of the Space

Radoš Radoičić
Géza Tóth

Abstract

The chromatic number of the space is the minimum number of colors needed to color the points of the space so that every two points unit distance apart have different colors. We show that this number is at most 15, improving the best known previous bound of 18.

1 Introduction

The *unit distance graph* G_d of the space $\Re^d$ is the graph whose vertices correspond to the points of $\Re^d$ and two vertices are connected if and only if the corresponding points have distance 1. The classical Hadwiger-Nelson problem asks for the chromatic number of the plane, or more precisely the chromatic number of the unit distance graph of the plane. The best known bounds are four and seven, due to Nelson and Isbell, respectively.

In three dimensions, Raiskii [R70] proved that $\chi(G_3) \geq 5$ which was recently improved by Nechushtan [N00] to 6. For the upper bound, Székely and Wormald [SW89] (see also [BT96]) proved $\chi(G_3) \leq 21$ which was improved to $\chi(G_3) \leq 18$ by Coulson [C97].

Theorem. *The chromatic number of the space $\Re^3$ is at most 15.*

In higher dimensions, the best known bounds are

$$(1 + o(1)) \cdot 1.2^d \leq \chi(G_n) \leq (3 + o(1))^d$$

due to Frankl and Wilson [FW81] and Larman and Rogers [LR72] respectively. For a survey on this problem see [S91] and [C93].

2 Proof of the Theorem

Let Π_3 be the convex hull of all vectors that are obtained by permuting the coordinates of the vector $(1, 2, 3, 4)$. This is a three dimensional polytope, called the permutahedron (see [Z98]). It has 24 vertices and 14 facets, 8

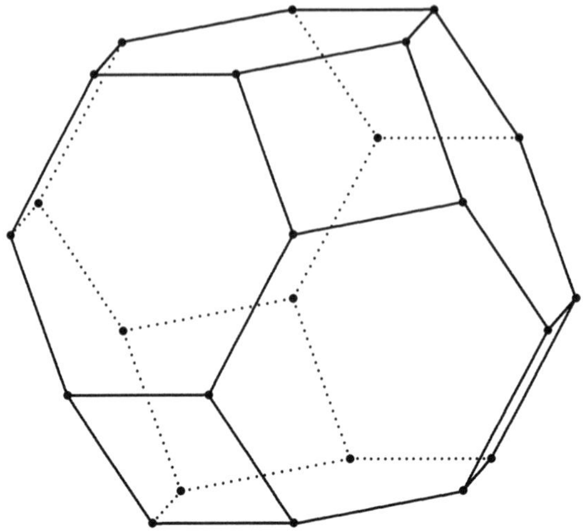

Fig. 1. The permutahedron Π_3

regular hexagons and 6 squares (see Figure). Our 15-coloring of the space is based on a tiling of the space with copies of Π_3.

Let $\mathcal{T}$ be isometric to Π_3, with center at the origin, and edge length $\frac{\sqrt{10}}{10}$. The vertices of $\mathcal{T}$ are all vectors that are obtained by permuting the coordinates of the vectors

$$\left(\pm\frac{\sqrt{20}}{20}, \pm\frac{\sqrt{20}}{10}, 0\right).$$

It is easy to check that $\mathcal{T}$ has diameter 1, the distance between any pair of opposite vertices. Remove those vertices of $\mathcal{T}$ which have $-\frac{\sqrt{20}}{20}$ as a coordinate. For simplicity, we still call the resulting body $\mathcal{T}$. Now the distance 1 is not realized within $\mathcal{T}$.

Let Λ be the lattice generated by

$$\vec{a} = \left(\frac{\sqrt{20}}{5}, 0, 0\right), \quad \vec{b} = \left(0, \frac{\sqrt{20}}{5}, 0\right), \quad \vec{c} = \left(\frac{\sqrt{20}}{10}, \frac{\sqrt{20}}{10}, \frac{\sqrt{20}}{10}\right).$$

Then $\{\mathcal{T} + \vec{u} \mid \vec{u} \in \Lambda\}$ is a tiling of $\Re^3$. Let $\mathcal{T}_{i,j,k} = \mathcal{T} + i\vec{a} + j\vec{b} + k\vec{c}$ for any i, j, k integers. Note that the boundary points of any $\mathcal{T}_{i,j,k}$ are covered twice.

Now we define a coloring of the space, using colors $0, 1, \ldots 14$. For any point $p \in \mathcal{T}_{i,j,k}$ let the color of p be $5i + 3j + k \pmod{15}$. For multiply covered points choose any of the resulting colors. We show that it is a proper coloring, that is, the unit distance is not realized between points of the same color.

A triple (i, j, k) is called *dangerous* if there are two points, $p \in \mathcal{T}_{0,0,0}$ and $q \in \mathcal{T}_{i,j,k}$ at unit distance.

Claim. If (i, j, k) is dangerous, then $5i + 3j + k \not\equiv 0 \pmod{15}$.

Proof of Claim. Clearly, (i, j, k) is dangerous if and only if $(-i, -j, -k)$ is. And $5i + 3j + k \equiv 0 \pmod{15}$ if and only if $5(-i) + 3(-j) - k \equiv 0 \pmod{15}$. Therefore, it is enough to check check those triples where $k \geq 0$.

Here is the list of all such dangerous triples:

$(1,0,0)$,	$(0,1,0)$,	$(-1,0,0)$,	$(0,-1,0)$,	$(1,1,0)$,	$(-1,1,0)$,
$(-1,-1,0)$,	$(1,-1,0)$,	$(2,0,0)$,	$(0,2,0)$,	$(-2,0,0)$,	$(0,-2,0)$,
$(0,0,1)$,	$(-1,0,1)$,	$(-1,-1,1)$,	$(0,-1,1)$,	$(1,0,1)$,	$(0,1,1)$,
$(1,-1,1)$,	$(0,-2,1)$,	$(-1,-2,1)$,	$(-2,-1,1)$,	$(-2,0,1)$,	$(-1,1,1)$,
$(0,0,2)$,	$(-1,0,2)$,	$(-2,0,2)$,	$(-2,-1,2)$,	$(-1,-1,2)$,	$(-2,-2,2)$,
$(-1,-2,2)$,	$(0,-2,2)$,	$(0,-1,2)$			
$(-1,-1,3)$,	$(-1,-2,3)$,	$(-2,-1,3)$,	$(-2,-2,3)$,	$(-2,-2,4)$	

Finally, it is easy to check that none of the triples satisfy $5i + 3j + k \not\equiv 0 \pmod{15}$. □

Return to the proof of the Theorem. Suppose that there are two points, p and p' of the same color and at unit distance, $p \in \mathcal{T}_{i,j,k}$ and $p' \in \mathcal{T}_{i',j',k'}$. Since the unit distance is not realized in $\mathcal{T}$, $(i, j, k) \neq (i', j', k')$. Let $I = i' - i$, $J = j' - j$, $K = k' - k$. Since p and p' have the same color, $5I + 3J + K \equiv 0 \pmod{15}$. Let $q = p - i\vec{a} - j\vec{b} - k\vec{c}$, $q' = p' - i\vec{a} - j\vec{b} - k\vec{c}$. Then $q \in \mathcal{T}_{0,0,0}$ and $q' \in \mathcal{T}_{I,J,K}$, q and q' have the same color they are unit distance apart. But then (I, J, K) is a dangerous triple so by the Claim q and q' have different colors, a contradiction. This concludes the proof of the Theorem. □

Remarks. 1. If we add the missing vertices to each $\mathcal{T}_{i,j,k}$, then we have two problems. The unit distance would be realized within each tile, and we would get an additional set of dangerous neighbors:

$(1,2,0)$,	$(2,1,0)$,	$(2,-1,0)$,	$(1,-2,0)$,	$(-1,-2,0)$,	$(-2,-1,0)$,
$(-2,1,0)$,	$(-1,2,0)$,	$(1,-1,2)$,	$(-1,-3,2)$,	$(-3,-1,2)$,	$(-1,1,2)$,
$(-1,1,-2)$,	$(1,3,-2)$,	$(3,1,-2)$,	$(1,-1,-2)$,	$(-1,-2,4)$,	$(-3,-2,4)$,
$(-2,-1,4)$,	$(-2,-3,4)$,	$(1,2,-4)$,	$(3,2,-4)$,	$(2,1,-4)$,	$(2,3,-4)$

If we want a modular coloring where these tiles also have different colors than $\mathcal{T}_{0,0,0}$, we have to use 24 colors. In other words, our construction is "rigid", the tiling can not be scaled even a little bit.

2. Since the permutahedron used in the construction has 14 facets, we can not have a similar proper coloring with 14 colors.

3. We conjecture that there is a proper modular coloring based on the lattice tiling of $\Re^d$ with d-dimensional permutahedra, that uses asymptotically fewer than $(3 + o(1))^d$ colors. With this method we found a proper 54-coloring of $\Re^4$.

Added in proof. Very recently, Coulson [C02] has independently found a very similar 15-coloring of 3-space.

Acknowledgement. We are grateful to János Pach for many helpful remarks, and to François Blanchette for technical help.

References

[BT96] M. Bóna, G. Tóth: A Ramsey-type problem on right-angled triangles in space, *Discrete Mathematics* Selected papers in honour of Paul Erdös on the occasion of his 80th birthday (Keszthely, 1993) **150** (1996), 61–67.

[C93] K. B. Chilakamarri: The unit-distance graph problem: a brief survey and some new results, *Bull. Inst. Combin. Appl.* **8** (1993), 39–60.

[C97] D. Coulson: An 18-colouring of 3-space omitting distance one, *Discrete Mathematics* **170** (1997), 241–247.

[C02] D. Coulson: An 15-colouring of 3-space omitting distance one, *Discrete Mathematics* **256** (2002), 83–90.

[FW81] P. Frankl, R. M. Wilson: Intersection theorems with geometric consequences, *Combinatorica* **1** (19981), 357–368.

[LR72] D. G. Larman, C. A. Rogers: The realization of distances within sets in Euclidean space, *Mathematika* **19** (1972), 1–24.

[M80] P. McMullen: Convex bodies which tile space by translation, *Mathematika* **27** (1980), 113–121.

[N00] O. Nechushtan: A note on the space chromatic number, *Discrete Mathematics* **256** (2002), 499–507.

[R70] D. E. Raiskii: The Realization of all distances in a decomposition of the space R^n into $n+1$ parts, *Math. Notes* **7** (1970), 194–196.

[SW89] L. A. Székely, N. C. Wormald: Bounds on the measurable chromatic number of R^n, *Discrete Mathematics* **75** (1989), 343-372.

[S91] A. Soifer: Chromatic number of the plane: A historical essay, *Geombinatorics* **1** (1991), 13–15.

[Z98] G. Ziegler: *Lectures on Polytopes,* Springer-Verlag, New York, 1998.

About Authors

Radoš Radoičić is at the Department of Mathematics, Massachusetts Institute of Technology, Cambridge, MA 02139, USA; *rados@math.mit.edu.*

Géza Tóth is at the Department of Mathematics, Massachusetts Institute of Technology, Cambridge, MA 02139, USA and at the A. Rényi Institute of Mathematics, Hungarian Academy of Sciences, Budapest, POB 127, 1364 Hungary. *geza@renyi.hu.*

Acknowledgments

Work on this paper has been supported by NSF grant DMS-99-70071. Géza Tóth was also supported by OTKA-T-020914 and OTKA-F-22234.

Expansive Motions and the Polytope of Pointed Pseudo-Triangulations

Günter Rote
Francisco Santos
Ileana Streinu

Abstract

We introduce the polytope of pointed pseudo-triangulations of a point set in the plane, defined as the polytope of infinitesimal expansive motions of the points subject to certain constraints on the increase of their distances. Its 1-skeleton is the graph whose vertices are the pointed pseudo-triangulations of the point set and whose edges are flips of interior pseudo-triangulation edges.

For points in convex position we obtain a new realization of the associahedron, i.e., a geometric representation of the set of triangulations of an n-gon, or of the set of binary trees on n vertices, or of many other combinatorial objects that are counted by the Catalan numbers. By considering the 1-dimensional version of the polytope of constrained expansive motions we obtain a second distinct realization of the associahedron as a perturbation of the positive cell in a Coxeter arrangement.

Our methods produce as a by-product a new proof that every simple polygon or polygonal arc in the plane has expansive motions, a key step in the proofs of the Carpenter's Rule Theorem by Connelly, Demaine and Rote (2000) and by Streinu (2000).

1 Introduction

Polytopes for combinatorial objects. Describing all instances of a combinatorial structure (e.g. trees or triangulations) as vertices of a polytope is often a step towards giving efficient optimization algorithms on those structures. It also leads to quick prototypes of enumeration algorithms using known vertex enumeration techniques and existing code [2, 9].

One particularly nice example is the associahedron, (see Figure 4 for an example): the vertices of this polytope correspond to Catalan structures. The Catalan structures refer to any of a great number of combinatorial objects which are counted by the Catalan numbers (see the extensive list in Stanley [24, ex. 6.19, p. 219]). Some of the most notable ones are the triangulations of a convex polygon, binary trees, the ways of evaluating a product of n factors when multiplication is not associative (hence the name associahedron), and monotone lattice paths that go from one corner of a square to the opposite corner without crossing the diagonal.

In this paper we describe a new polyhedron whose vertices correspond to *pointed pseudo-triangulations.*

Pseudo-triangulations. Pseudo-triangulations, as well as the closely related geodesic triangulations of simple polygons, have been used in Computational Geometry in applications such as visibility [18–20, 23], ray shooting [12], and kinetic data structures [1,14]. The *minimum* or *pointed pseudo-triangulations* introduced in Streinu [25] have applications to non-colliding motion planning of planar robot arms. They also have very nice combinatorial and rigidity theoretic properties. The polytope we define in this paper adds to the former, and is constructed exploiting the latter.

Expansive motions. An expansive motion on a set of points P is an infinitesimal motion of the points such that no distance between them decreases.

Expansive motions were instrumental in the first proof of the Carpenter's Rule Theorem by Connelly, Demaine and Rote [6]: Every simple polygon or polygonal arc in the plane can be unfolded into convex position without collisions. Streinu [25] built on this work, realizing the importance of pseudo-triangulations in connection with expansive motions and studying their rigidity properties. This paper provides a systematic study of expansive motions in one and two dimensions. The expansive motions of a set of n points in the plane form a polyhedral cone of dimension $2n-3$ (the *expansion cone*). As by-products of our approach we get a new proof of the existence of expansive motions for non-convex polygons and polygonal arcs (Theorem 4.3) and a characterization of the extreme rays of the expansion cone of a planar point set in general position, as equivalence classes of pointed pseudo-triangulations with one convex hull edge removed, modulo rigid subcomponents (Proposition 2.8).

Our tool is the introduction of *constrained expansions* as expansive motions with a special lower bound on the edge length increase. They form a polyhedron obtained by translation of the facets of the expansion cone. Our main result is the following (see a more precise statement as Theorem 3.1):

Theorem. *Let P be a set of n points in general position in the plane, b of them in the boundary of the convex hull. Then, there is a choice of constraints which produces as constrained expansions of P a simple polyhedron of dimension $2n-3$ with a unique maximal bounded face of dimension $2n-b-3$ whose vertices correspond to pointed pseudo-triangulations and edges correspond to flips between them.*

The flips mentioned in the statement are a certain neighborhood structure among pointed pseudo-triangulations (flips of interior edges). See the details in Section 2.

Two appearances of the associahedron. For points in convex position, pseudo-triangulations coincide with triangulations. We prove (Corollary 5.5) that, in this case, our construction gives a polytope affinely equivalent to the standard $(n-3)$-dimensional associahedron obtained as a secondary polytope of the point set [28, Section 9.2]. Perhaps surprisingly, this shows that the secondary polytope of n points in convex position in the plane (which lives in $\mathbb{R}^n$) can naturally be embedded as a face in a $(2n-3)$-dimensional unbounded polyhedron.

The associahedron appears again as the analog of our construction for points in one dimension (Section 5.3).

Rigidity. The connection of these results with rigidity theory is also worth mentioning. Pointed pseudo-triangulations are special instances of infinitesimally minimally rigid frameworks in dimension 2, whose combinatorial structure is well understood (see [13]). One-dimensional minimally rigid frameworks are trees, another well understood combinatorial structure. Adding the constraint of expansiveness is what leads to pointed pseudo-triangulations in 2d, and to the special non-crossing and alternating trees which appear in Section 5.3.

Future perspectives. It is our hope that the insight into one- and two-dimensional motions may eventually lead to generalizations to higher dimensions. There is no satisfactory definition of an analog of pseudo-triangulations in 3 dimensions. The 3-dimensional version of the robot arm motion planning problem, with potential applications to computational biology (protein folding), is much more challenging.

Overview. In Section 2 we give the preliminary definitions and results. Section 3 contains the main result, the construction of the polytope of pointed pseudo-triangulations (*ppt-polytope*). Section 4 applies the main result to get a new proof for the existence of expansive motions for non-convex polygons and polygonal arcs in the plane. In Section 5 we present an alternative construction of the ppt-polytope and two special cases leading to the associahedron: points in convex position and the polytope of constrained expansions in dimension 1. Section 6 attempts to put the results in 1 and 2 dimensions into a broader perspective, with the aim of extending the results to higher dimensions and to point sets which are not in general position. We conclude with some final comments in Section 7.

2 Preliminaries

Abbreviations and conventions. Throughout this paper we will assume general position for our point sets, i.e. we assume that no $d+1$ points in $\mathbb{R}^d$ lie in the same hyperplane (unless otherwise specified). We abbreviate "polytope of pointed pseudo-triangulations" as *ppt-polytope*, "one-degree-of-freedom mechanism" as *1DOF mechanism* and "pseudo-triangulation expansive mechanism" as *pte-mechanism.*

For an ordered sequence of $d+1$ points $q_0, \ldots, q_d \in \mathbb{R}^d$, $\det(q_0, \ldots, q_d)$ denotes the determinant of the $(d+1)\times(d+1)$ matrix with columns $(q_0, 1), \ldots,$ $(q_d, 1)$. Equivalently, $\det(q_0, \ldots, q_d)$ equals $d!$ times the Euclidean volume of the simplex with those $d+1$ vertices, with a sign depending on the orientation.

Pseudo-triangulations. A *pseudo-triangle* is a simple polygon with only three convex vertices (called *corners*) joined by three inward convex polygonal chains, see Figure 1a. In particular, every triangle is a pseudo-triangle. A *pseudo-triangulation* is a partitioning of the convex hull of a point set $P = \{p_1, \ldots, p_n\}$ into pseudo-triangles using P as vertex set.

Pseudo-triangulations are *graphs embedded on P*, i.e. graphs drawn in the plane on the vertex set P and with straight-line segments as edges. We will work with other graphs embedded in the plane. If edges intersect only at their end-points, as is the case for pseudo-triangulations, the graphs will be called *non-crossing* or *plane graphs.* A graph is *pointed at a vertex v* if there is (locally) an angle at v strictly larger than π and containing no edges. Under our general position assumption, convex-hull vertices are pointed for any embedded graph, as are vertices of degree at most two. A graph is called *pointed* if it is pointed at every vertex. Parts (b) and (c) of Figure 1 (including the broken edges) show two pointed pseudo-triangulations of a certain point set.

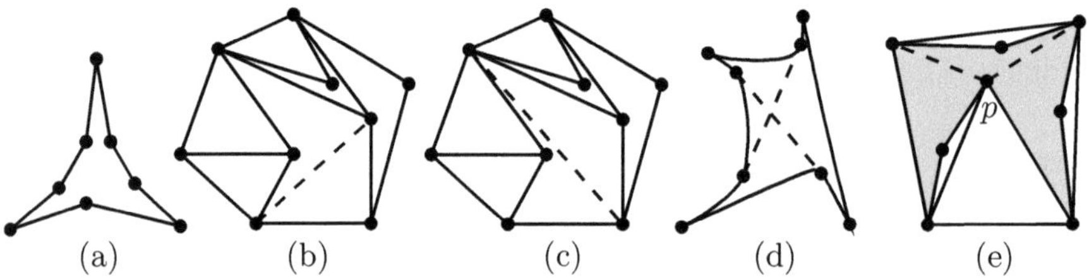

Fig. 1. (a) A pseudo-triangle. (b) A minimum, or pointed, pseudo-triangulation. (c) The broken edge in (b) is flipped, and gives another pointed pseudo-triangulation. (d) A schematic drawing of the flip operation. (e) The two edges involved in a flip may share a common vertex p.

The following properties of pseudo-triangulations were initially proved for the slightly different situation of pseudo-triangulations of *convex objects* by Pocchiola and Vegter [18, 19]. For completeness, we sketch the easy proofs (see also [3]).

Lemma 2.1. (Streinu [25]) *Let P be a set of n points in general position in the plane. Let G be a pointed and non-crossing graph on P.*

(a) *G has at most* $2n-3$ *edges, with equality if and only if it is a pseudo-triangulation.*

(b) *If G is not a pseudo-triangulation, then edges can be added to it keeping it non-crossing and pointed.*

Proof. (a) If the graph G is not connected, we can analyze the components separately. So let us assume that it is connected. Let e and f be the numbers of edges and bounded faces in G. Let a^+ and a^- denote the number of angles which are $> \pi$ and $< \pi$, respectively. (By our general position assumption, there are no angles equal to π.) Clearly, $a^+ + a^- = 2e$. Pointedness means $a^+ = n$ and, since any bounded face has at least three convex vertices, $a^- \geq 3f$ with equality if and only if G is a pseudo-triangulation. The equation $2e \geq n + 3f$, together with Euler's formula $e = n + f - 1$, implies $e \leq 2n - 3$ (and $f \leq n - 2$).

(b) The basic idea is that the addition of geodesic paths (i.e., paths which have shortest length among those sufficiently close to them) between convex vertices of a polygon keeps the graph pointed and non-crossing and, unless the polygon is a pseudo-triangle, there is always some of these geodesic paths going through its interior. □

Streinu [25] proved the following additional properties of pointed pseudo-triangulations, which we do not need for our results but which may interest the reader:

- Every pseudo-triangulation on n points has at least $2n - 3$ edges, with equality if and only if it is pointed. Hence, pointed pseudo-triangulations are the pseudo-triangulations with the minimum number of edges. For this reason they are called *minimum* pseudo-triangulations in [25]. In contrast with part (b) of Lemma 2.1, not every pseudo-triangulation contains a pointed one. An example of this is a regular pentagon with its central point, triangulated as a wheel. Hence, a *minimal* pseudo-triangulation is not always pointed.

- The graph of any pointed pseudo-triangulation has the Laman property: it has $2n - 3$ edges and the subgraph induced on any k vertices has at most $2k - 3$ edges. This property characterizes generically minimally rigid graphs in the plane ([15], see also [13]); that is, graphs which are minimally rigid in almost all their embeddings in the plane.

- All pointed pseudo-triangulations can be obtained starting with a triangle and adding vertices one by one and adding or adjusting edges, in much the same way as the Henneberg construction of generically minimally rigid graphs (cf. [13, page 113]), suitably modified to give pointed pseudo-triangulations in intermediate steps (see the details in [25]).

The other crucial properties of pointed pseudo-triangulations that we use are that all interior edges can be flipped in a natural way (part (a) of the following statement) and that the graph of flips between pointed pseudo-triangulations of any point set is connected. Both results were known to

Pocchiola and Vegter for pseudo-triangulations of convex objects (see [18, 19]). Parts (b) and (c) of Figure 1 show a flip between pointed pseudo-triangulations. An $O(n^2)$ bound on the diameter of the flip graph is proved in [3].

Lemma 2.2. (Flips between pointed pseudo-triangulations) *Let P be a point set in general position in the plane.*

(a) (**Definition of Flips**) *When an interior edge (not on the convex hull) is removed from a pointed pseudo-triangulation of P, there is a unique way to put back another edge to obtain a different pointed pseudo-triangulation.*

(b) (**Connectivity of the flip graph**) *The graph whose vertices are pointed pseudo-triangulations and whose edges correspond to flips of interior edges is connected.*

Proof. [3, 25] (a) When we remove an interior edge from a pointed pseudo-triangulation we get a planar and pointed graph with $2n-4$ edges. The same arguments of the proof of Lemma 2.1 imply now that $a^- = 3f + 1$. Hence, the new face created by the removal must be a pseudo-quadrilateral (that is, a simple polygon with exactly four convex vertices).

In any pseudo-quadrilateral there are exactly two ways of inserting an interior edge to divide it into pseudo-triangles, which can be obtained by the shortest paths between opposite convex vertices of the pseudo-quadrilateral (see the details in Lemma 2.1 of [26], and a schematic drawing in Figure 1d). One of these two is the edge we have removed, so only the other one remains. Note that the two interior edges of a pseudo-quadrilateral may be incident to the same vertex, see Figure 1e. This can only happen when the interior angle at this vertex is bigger than π.

(b) Let p be a convex hull vertex in P. Pointed pseudo-triangulations in which p is not incident to any interior edge are just pointed pseudo-triangulations of $P \setminus \{p\}$ together with the two tangent edges from p to the convex hull of the rest. By induction, we assume all those pointed pseudo-triangulations to be connected to each other. To show that all others are also connected to those, just observe that if a pointed pseudo-triangulation has an interior edge incident to p, then a flip on that edge inserts an edge not incident to p. (The case of Figure 1e cannot happen for a hull vertex p.) Hence the number of interior edges incident to p decreases. □

Infinitesimal rigidity. In this paper we work mostly with points in dimensions $d = 2$ and $d = 1$. Occasionally we will use superscripts to denote the components of the vectors $p_i = (p_i^1, \ldots, p_i^d)$.

An *infinitesimal motion* on a point set $P = \{p_1, \ldots, p_n\} \in \mathbb{R}^d$ is an assignment of a velocity vector $v_i = (v_i^1, \ldots, v_i^d)$ to each point p_i, $i = 1, \ldots, n$. The *trivial infinitesimal motions* are those which come from (infinitesimal) rigid transformations of the whole ambient space. In $\mathbb{R}^2$ these are the translations (for which all the v_i's are equal vectors) and rotations with a certain center p_0 (for which each v_i is perpendicular and proportional to the segment p_0p_i). Trivial motions form a linear subspace of dimension $\binom{d+1}{2}$ in the linear space $(\mathbb{R}^d)^n$ of all infinitesimal motions. Two infinitesimal motions whose difference is a trivial motion will be considered equivalent, leading to a reduced space of *non-trivial* infinitesimal motions of dimension $dn - \binom{d+1}{2}$. In particular, this is $n-1$ for $d=1$ and $2n-3$ for $d=2$. Rather than performing a formal quotient of vector spaces we will "tie the framework down" by fixing $\binom{d+1}{2}$ variables. E.g., for $d=1$ we can choose:

$$v_1 = 0 \tag{1}$$

and for $d=2$ (assuming w.l.o.g. that $p_2^2 \neq p_1^2$):

$$v_1^1 = v_1^2 = v_2^1 = 0 \tag{2}$$

Here, p_1 and p_2 can be any two vertices. A different choice of normalizing conditions only amounts to a linear transformation in the space of infinitesimal motions.

In rigidity theory, a graph $G = (P, E)$ embedded on P is customarily called a *framework* and denoted by $G(P)$. We will use the term framework when we want to emphasize its rigidity-theoretic properties (stresses, motions), but we will use the term graph when speaking about graph-theoretic properties, even if graph is embedded on a set P. For a given framework $G = (P, E)$, an infinitesimal motion such that $\langle p_i - p_j, v_i - v_j\rangle = 0$ for every edge $ij \in E$ is called a *flex* of G. This condition states that the length of the edge ij remains unchanged, to first order. The trivial motions are the flexes of the complete graph, provided that the vertices span the whole space $\mathbb{R}^d$. A framework is *infinitesimally rigid* if it has no non-trivial flexes. It is *infinitesimally flexible* or an *infinitesimal mechanism* otherwise.

Infinitesimal motions are to be distinguished from global motions, which describe paths for each point throughout some time interval. In this paper we are not concerned with global motions, nor their associated concept of rigidity, weaker than infinitesimal rigidity ([7, Theorem 4.3.1] or [13, page 6]). Let us also insist that we distinguish between *infinitesimal motions* (of the point set) and *flexes* (of the framework or embedded graph), while the terms flex and infinitesimal motion are sometimes equivalent in the rigidity theory literature.

The (*infinitesimal*) *rigidity map* $M_{G(P)}\colon (\mathbb{R}^d)^n \to \mathbb{R}^{E(G)}$ is a linear map associated with an embedded framework $G(P)$, $P \subset \mathbb{R}^d$. It sends each infinitesimal motion $(v_1, \dots, v_n) \in (\mathbb{R}^d)^n$ to the vector of infinitesimal edge increases $(\langle p_i - p_j, v_i - v_j\rangle)_{ij\in E}$. When no confusion arises, it will be simply denoted as M. The number $\langle p_i - p_j, v_i - v_j\rangle$ is called the *strain* on the edge ij in the engineering literature. As usual, the image of M is denoted by $\operatorname{Im} M = \{\, f \mid f = Mv \,\}$. The matrix of M is called the rigidity matrix. In this matrix, the row indexed by the edge $ij \in E$ has 0 entries everywhere except in the i-th and j-th group of d columns, where the entries are $p_i - p_j$ and $p_j - p_i$, respectively.

The kernel of M (after reducing $\mathbb{R}^{dn}$ to $\mathbb{R}^{dn-\binom{d+1}{2}}$ by forgetting trivial motions) is the space of flexes of $G(P)$. In particular, a framework is infinitesimally rigid if and only if the kernel of its associated rigidity map M is the subspace of trivial motions. In general, the dimension of the (reduced) space of flexes is the *degree of freedom* (DOF) of the framework. A 1DOF mechanism is a mechanism with one degree of freedom.

Finally, *expansive (infinitesimal) motions* $v_1, \ldots, v_n$ are those which simultaneously increase (perhaps not strictly) all distances: $\langle p_i - p_j, v_i - v_j \rangle \geq 0$ for every pair i, j of vertices. A mechanism is *expansive* if it has non-trivial expansive flexes.

The following results of Streinu [25] can be obtained as a corollary of our main result (see the proof after the statement of Theorem 3.1).

Proposition 2.3. (Rigidity of pointed pseudo-triangulations [25])

(a) *Pointed pseudo-triangulations are minimally infinitesimally rigid (and therefore rigid).*

(b) *The removal of a convex hull edge from a pointed pseudo-triangulation yields a 1DOF expansive mechanism (called a* pseudo-triangulation expansive mechanism *or shortly a* pte-mechanism*).* □

Part (a) is in accordance with the fact that the graph of any pointed pseudo-triangulation has the Laman property, and hence is generically rigid in the plane. It is a trivial consequence of (a) that the removal of an edge creates a (not necessarily expansive) 1DOF mechanism. The expansiveness of pte-mechanisms (part (b)) was proved in [25] using the Maxwell-Cremona correspondence between self-stresses and 3-d liftings of planar frameworks, a technique that was introduced in [6].

Self-stresses. A *self-stress* (or an *equilibrium stress*) on a framework $G(P)$ (see [27] or [6, Section 3.1]) is an assignment of scalars ω_{ij} to edges such that $\forall i \in P$, $\sum_{ij \in E} \omega_{ij}(p_i - p_j) = 0$. That is, the self-stresses are the row dependences of the rigidity matrix M. The proof of the following lemma is then straightforward.

Lemma 2.4. *Self-stresses form the orthogonal complement of the linear subspace* $\operatorname{Im} M \subset \mathbb{R}^{\binom{d}{2}}$. *In other words,* $(\omega_{ij})_{ij \in E}$ *is a self-stress if and only if for every infinitesimal motion* $(v_1, \ldots, v_n) \in (\mathbb{R}^d)^n$ *the following identity holds:*

$$\sum_{ij \in E} \omega_{ij} \langle p_i - p_j, v_i - v_j \rangle = 0 \qquad \square$$

As an example, the following result gives explicitly a stress for the complete graph on any affinely dependent point set:

Lemma 2.5. *Let* $\sum_{i=1}^{n} \alpha_i p_i = 0$, $\sum \alpha_i = 0$, *be an affine dependence on a point set* $P = \{p_1, \dots, p_n\}$. *Then,* $\omega_{ij} = \alpha_i \alpha_j$ *for every* i, j *defines a self-stress of the complete graph* G *on* P.

Proof. For any $p_i \in P$ we have:

$$\sum_{ij \in G} \omega_{ij}(p_i - p_j) = \sum_{j=1}^{n} \alpha_i \alpha_j (p_i - p_j) = \alpha_i p_i \sum_{j=1}^{n} \alpha_j - \alpha_i \sum_{j=1}^{n} \alpha_j p_j = 0. \qquad \square$$

Let us analyze here the case of $d+2$ points $P = \{p_1, \dots, p_{d+2}\}$ in general position in $\mathbb{R}^d$ (this is the first non-trivial case, because no self-stress can arise between affinely independent points). It can be easily checked that, under these assumptions, removing any single edge from the complete graph on P leaves a minimally infinitesimally rigid graph. This implies that the complete graph has a unique self-stress (up to a scalar factor). This self-stress is the one given in Lemma 2.5 for the unique affine dependence on P. The coefficients of this dependence can be written as:

$$\alpha_i = (-1)^i \det([p_1, \dots, p_{d+2}] \backslash \{p_i\}).$$

(Recall that $\det(q_0, \dots, q_d)$ is $d!$ times the signed volume of the simplex spanned by the $d+1$ points $q_0, \dots, q_d \in \mathbb{R}^d$.)

The special case $d = 2$, $n = 4$ will be extremely relevant to our purposes, and it will be convenient to renormalize the unique self-stress as follows:

Lemma 2.6. *The following gives a self-stresses for any four points* $p_1, \dots, p_4$ *in general position in the plane:*

$$\omega_{ij} := \frac{1}{\det(p_i, p_j, p_k) \det(p_i, p_j, p_l)} \tag{3}$$

where k *and* l *are the two indices other than* i *and* j.

Proof. Set $\alpha_i = (-1)^i \det([p_1, \dots, p_4] \backslash \{p_i\})$ in Lemma 2.5 and divide all the ω_{ij}'s of the resulting self-stress by the non-zero constant

$$-\det(p_1, p_2, p_3) \det(p_1, p_2, p_4) \det(p_1, p_3, p_4) \det(p_2, p_3, p_4). \qquad \square$$

The direct application of Lemma 2.5 would give as ω_{ij} a product of two determinants, rather than the inverse of such a product. The reason why we prefer the self-stress of Lemma 2.6 is because of the signs it produces. The reader can easily check, considering the two cases of four points in convex position and one point inside the triangle formed by the other three, that with the choice of Lemma 2.6 boundary edges always receive positive stress and interior edges negative stress. This uniformity is good for us because in both cases pointed pseudo-triangulations are the graphs obtained deleting from the complete graph any single interior edge.

The expansion cone. We are given a set of n points $P = (p_1, \ldots, p_n)$ in $\mathbb{R}^d$ that are to move with (unknown) velocities $v_i \in \mathbb{R}^d$, $i = 1, \ldots, n$. An *expansive* motion is a motion in which no inter-point distance decreases. This is described by the system of homogeneous linear inequalities:

$$\langle p_i - p_j, v_j - v_i \rangle \geq 0, \qquad \forall\, 1 \leq i < j \leq n \tag{4}$$

and hence defines a polyhedral cone.

The only motions in the intersection of all facets of the cone are the trivial motions. Thus, when we add normalizing equations like (1) or (2), we get a *pointed polyhedral cone* containing the origin as a vertex. We call it the cone of expansive motions or simply the *expansion cone* of P.

An extreme ray of the expansion cone is given by a maximal set of inequalities satisfied with equality by non-trivial motions. Each inequality corresponds to an edge of the point set, so that the ray corresponds to a graph embedded in our point set. The cardinality of this set of edges is at least the dimension of the cone minus 1, but may be much larger. Let's analyze the low dimensional cases.

For $d = 1$ the expansion cone is not very interesting. Let's assume that the points $p_i \in \mathbb{R}$ are labeled in increasing order $p_1 < p_2 < \cdots < p_n$. Then:

Proposition 2.7. *The expansion cone in one dimension has $n-1$ extreme rays corresponding to the motions where $p_1, \ldots, p_i$ remain stationary and the points $p_{i+1}, \ldots, p_n$ move away from them at uniform speed:*

$$0 = v_1 = v_2 = \cdots = v_i < v_{i+1} = \cdots = v_n \tag{5}$$

Proof. Note that the actual values of p_i are immaterial in this case. The expansion cone is given by the linear system $v_j \geq v_i$, $1 \leq i \leq j \leq n$ plus the extra condition $v_1 = 0$, and any maximal set of inequalities satisfied with equality and yet not trivial is obviously given by (5). □

The 2d case is more complex and requires additional terminology. 1DOF mechanisms may contain rigid subcomponents (r-components, cf. [13]): maximal sets of some k vertices spanning a Laman subgraph on $2k - 3$ edges. The r-components of pte-mechanisms are themselves pseudo-triangulations spanning convex subpolygons including all points in their interior. Adding edges to complete each r-component to a complete subgraph yields a *collapsed pte-mechanism* (see Figures 2 and 3).

Proposition 2.8. *In dimension 2, the extreme rays of the expansion cone correspond to the collapsed pte-mechanisms.* □

The proof will be given in Section 4.1, after we have determined the extreme rays of a perturbed version of the polytope.

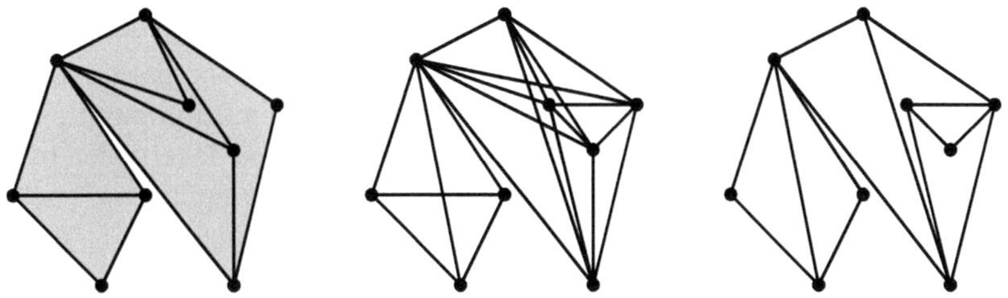

Fig. 2. A pte-mechanism with rigid sub-components (convex subpolygons) drawn shaded, the corresponding collapsed pte-mechanism, and another pte-mechanism that yields the same expansive motion.

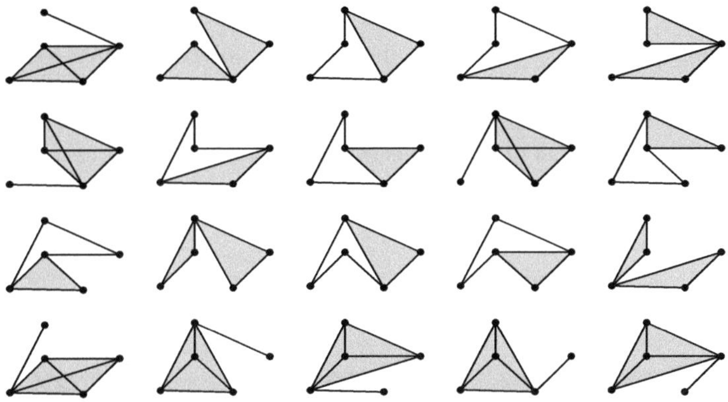

Fig. 3. The collapsed pte-mechanisms corresponding to the 20 extreme rays of the expansion cone for a planar point set of 5 points. The rigid sub-components (complete subgraphs) are shaded.

The polytope of constrained expansions. The construction we will give in section 3 can roughly be interpreted as separating the pseudo-triangulations contained in the same collapsed pte-mechanisms, to obtain a polyhedron whose vertices correspond to distinct pseudo-triangulations. The original expansion cone is highly degenerate: its extreme rays contain information about *all* the bars whose length is unchanged by a motion of a 1DOF expansive mechanism. We would like to perturb the constraints (4) to eliminate these degeneracies and recover pure pseudo-triangulations. We do so by giving up homogeneity, i.e., by translating the facets of the expansion cone. Our system will become:

$$\langle v_j - v_i, p_i - p_j \rangle \geq f_{ij}, \qquad \forall 1 \leq i < j \leq n \tag{6}$$

for some numbers f_{ij}. In some cases we will change these inequalities to equations for the edges on the convex hull of the given point set.

$$\langle v_j - v_i, p_i - p_j \rangle = f_{ij}, \qquad \text{for the convex hull edges } ij. \tag{7}$$

Section 3 proves our main result, Theorem 3.1: For any point set in general position in the plane and for some appropriate choices of the parameters f_{ij}, (6) defines a polyhedron whose vertices are in bijection with pointed pseudo-triangulations and all lie in a unique maximal bounded face given by (7). We call this face the "polytope of pointed pseudo-triangulations" or ppt-polytope.

A similar thing in 1d is done in Section 5.3, with the surprising outcome that the (unique) maximal bounded face of the polyhedron turns out to be an associahedron with vertices corresponding to *non-crossing alternating trees* (which are Catalan structures, as shown in [10]). The next paragraph prepares the ground for this result.

The associahedron. The associahedron is a polytope which has a vertex for every triangulation of a convex n-gon, and in which two vertices are connected by an edge of the polytope if the two triangulations are connected by an edge flip. Equivalently, various types of Catalan structures are reflected in the associahedron. Fig. 4 shows an example.

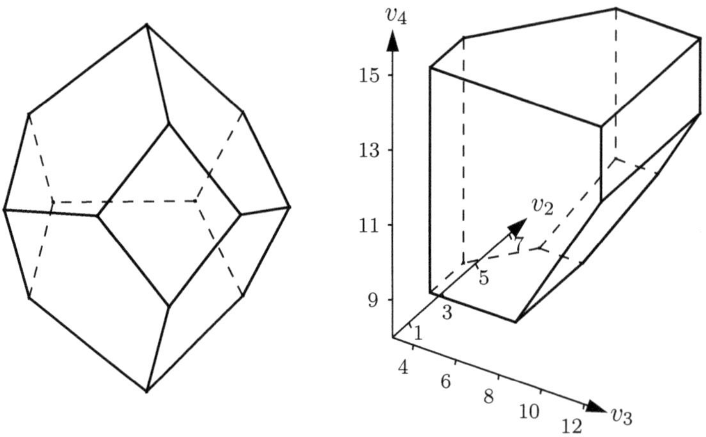

Fig. 4. The three-dimensional associahedron. The vertices represent all triangulations of a convex hexagon or all possible ways to insert parentheses into the product $a*b*c*d*e$. Left: a symmetric representation, the secondary polytope of a regular hexagon. Right: our representation, from Section 5.3. Both pictures are orthogonal projections.

There is an easy geometric realization of this polytope associated with each set P of n points *in convex position* in the plane, as a special case of a secondary polytope (Gel′fand, Zelevinskiĭ, and Kapranov [11], see also Ziegler [28, Section 9.2]). Every triangulation is represented by a vector $(a_1, \ldots, a_n)$ of n component with the entry a_i being simply the sum of the areas of all triangles of the triangulation that are incident to the i-th vertex. We will refer to this realization as the *classical realization* of the associahedron. It depends on the location of the vertices of the convex n-gon, but all polytopes that one gets in this way are combinatorially equivalent. Their face lattice is the poset of polygonal subdivisions of the n-gon or, in the terminology of the previous paragraphs, non-crossing and pointed graphs embedded in P and containing the n convex hull edges. But observe that the word "pointed" is superfluous for a graph with vertices in convex position. The order structure in this poset is just inclusion of edge sets (in reverse order since maximal graphs represent vertices).

Dantzig, Hoffman, and Hu [8, Section 2], and independently de Loera et al. [17] in a more general setting, have given other representations of the set of triangulations as the vertices of a 0-1-polytope in $\binom{n}{3}$ variables corresponding to the possible triangles of a triangulation (the *universal* polytope), or in $\binom{n}{2}$ variables corresponding to the possible edges of a triangulation. These realizations are in a sense most natural, but they have higher dimensions and have more adjacencies between vertices than the associahedron. Every classical associahedron, however, arises as a projection of the universal polytope. The first published realization of an associahedron is due to Lee [16], but it is not fully explicit. A few earlier and more complicated ad-hoc realizations that were never published are mentioned in Ziegler [28, Section 0.10].

In this paper the associahedron appears in two forms. First, we will show that for n points in convex position our polytope of pointed pseudo-triangulations is affinely equivalent to the secondary polytope of the configuration, which is a classical associahedron (Section 5.2). Second, as mentioned before, our construction adapted to a one-dimensional point configuration produces in a natural way an associahedron (Section 5.3). Notice that in dimension 1 the coordinates p_i can be eliminated from the constraints (6). Only the order of points along the line matters. One can also look at the whole *arrangement* of hyperplanes of the form

$$v_j - v_i = g_{ij}.$$

This arrangement, for various special values of g, has been the object of extensive combinatorial studies. For $g \equiv 0$ it is the classical *Coxeter* or *reflection* arrangement of type A_n. The case $g \equiv 1$ has been studied by Postnikov and Stanley [22]. The expansion cone of a 1d point set is the *positive cell* in the arrangement A_n, and our associahedron is a bounded face of the polyhedron obtained by translating the facets of this cell.

A different realization of the associahedron based in the root system of type A_n has recently appeared in [5]. It is interesting that these two new associahedra are not affinely equivalent to any classical associahedron obtained as a secondary polytope, or to one another. Also, that we are trying to get a simple polyhedron, in contrast to the above-mentioned choices of g which lead to highly degenerate arrangements.

3 The Main Result: the Polytope of Pointed Pseudo-Triangulations

In this section we prove our main result.

Theorem 3.1. *For every set $P = \{p_1, \ldots, p_n\}$ of $n \geq 3$ planar points in general position, there is a choice of f_{ij}'s for which equations* (6) *together with the normalizing equations* (2) *define a simple polyhedron $\bar{X}_f(P)$ of dimension $2n-3$ with the following properties*:

1. *The face poset of the polyhedron equals the opposite of the poset of pointed and non-crossing graphs on P, by the map sending each face to the set of edges whose corresponding equations* (6) *are satisfied with equality over that face. In particular:*
   (a) *Vertices of the polyhedron are in 1-to-1 correspondence with pointed pseudo-triangulations of P.*
   (b) *Bounded edges correspond to flips of interior edges in pseudo-triangulations, i.e., to pseudo-triangulations with one interior edge removed.*
   (c) *Extreme rays correspond to pseudo-triangulations with one convex hull edge removed.*
2. *The face $X_f(P)$ obtained by changing to equalities* (7) *those inequalities from* (6) *which correspond to convex hull edges of P is bounded* (*hence a polytope*) *and contains all vertices. In other words, it is the unique maximal bounded face, and its 1-skeleton is the graph of flips among pointed pseudo-triangulations.*

The proof is a consequence of lemmas proved throughout this section. Theorem 3.9 states that the choice $f_{ij} := \det(a, p_i, p_j) \det(b, p_i, p_j)$ produces the desired object, where a and b are any fixed points in the plane. In Section 5.1 we will derive a more canonical description of the polyhedron (and the polytope) in question.

It turns out that the polyhedron $\bar{X}_f(P)$ is the most convenient object for the proof. The properties of the polytope $X_f(P)$ (part 2 of the theorem) and the extreme ray description of the expansion cone (Proposition 2.8), which may be more interesting by themselves, are then easily derived.

Before going on, let us see that Theorem 3.1 implies Proposition 2.3. Observe that a framework is minimally infinitesimally rigid if and only if the hyperplanes $\langle p_i - p_j, v_i - v_j \rangle = 0$ corresponding to its edges ij meet transversally and at a single point, in the $(2n-3)$-dimensional space given by equations (2). Part 1.a of our theorem says that this happens for the $2n-3$ translated hyperplanes $\langle p_i - p_j, v_i - v_j \rangle = f_{ij}$ corresponding to any pointed pseudo-triangulation, hence giving part (a) of Proposition 2.3. An (infinitesimally) expansive 1DOF mechanism is one whose corresponding hyperplanes intersect in a line contained in the expansion cone. Part 1.c of the theorem says that this happens for a pointed pseudo-triangulation with one hull edge removed, giving part (b) of Proposition 2.3.

The polyhedron and the polytope of constrained expansions. The solution set $v \in (\mathbb{R}^2)^n$ of the system of inequalities (6) together with the normalizing equations (2) will be called the *polyhedron of constrained expansions* $\bar{X}_f(P)$ for the set of points P and perturbation parameters (constraints) f. We will frequently omit the point set P when it is clear from the context. A solution v may satisfy some of the inequalities in (6) with equality: the corresponding edges $E(v)$ of G are said to be *tight* for that solution. In the same way, for a face K of $\bar{X}_f$ we call *tight edges* of K and denote $E(K)$ the edges whose equations are satisfied with equality over K (equivalently, over a relative interior point of K). This is the correspondence that Theorem 3.1 refers to: the edges $E(K)$ of a face K form the pointed and non-crossing graph corresponding to that face.

When $f \equiv 0$, we just get the expansion cone $\bar{X}_0$ itself (in this sense, our notations $\bar{X}_0$ and $\bar{X}_f$ are consistent.) This cone equals the recession cone of $\bar{X}_f$, for any choice of f. (The recession cone of a polyhedron is the cone of vectors parallel to infinite rays contained in the polyhedron.) We will first establish a few properties of the expansion cone.

Lemma 3.2. (a) *The expansion cone $\bar{X}_0$ is a pointed polyhedral cone of full dimension $2n-3$ in the subspace defined by the three equations* (2). (*In this context, "pointed" means that the origin is a vertex of the cone.*)

(b) *Consider the set $E(v)$ of tight edges for any feasible point $v \in \bar{X}_0$. If $E(v)$ contains*

(i) *two crossing edges,*

(ii) *a set of edges incident to a common vertex with no angle larger than π (witnessing that $E(v)$ is not pointed at this vertex), or*

(iii) *a convex subpolygon,*

then $E(v)$ must contain the complete graph between the endpoints of all involved edges. In case (iii), *this complete graph also includes all points inside the convex subpolygon.*

Proof. (a) The dilation (scaling motion) $v_i := p_i$ satisfies all inequalities (4) strictly. By adding a suitable rigid motion, the three equations (2) can be satisfied, too, without changing the status of the inequalities (4), and so we get a relative interior point in the $(2n-3)$-dimensional subspace (2).

If the cone were not pointed, it would contain two opposite vectors v and $-v$. From this we would conclude that $\langle v_j - v_i, p_j - p_i \rangle = 0$ for all i, j, and hence v would be a flex of the complete graph on P. By the normalizing equations (2), v must then be 0.

(b) We first consider (iii), which is the most involved case. Let v be an expansive motion which preserves all edge lengths of some convex polygon. First we see that v preserves all distances between polygon vertices: indeed, if it preserves lengths of polygon edges but is not a trivial motion of the polygon then the angle at some polygon vertex p_i infinitesimally decreases, because the sum of angles remains constant. But decreasing the angle at p_i while preserving the lengths of the two incident edges implies that the distance between the two vertices adjacent to p_i in the polygon decreases. This is a contradiction.

By choosing p_1 and p_2 in (2) to be polygon vertices, the above implies that the polygon remains stationary under v. Now no interior point p_i can move with respect to the polygon, without decreasing the distance to some polygon vertex: If $v_i \neq 0$, there is at least one hull vertex p_j in the half-plane $\langle p_i - p_j, v_i \rangle < 0$. The edge ij will then violate condition (4).

Case (ii) is similar: If the edges incident to a vertex p_i do not move rigidly, at least one angle between two neighboring edges must decrease, and, this angle being less than π, this implies that the distance between the endpoints of these edges decreases, a contradiction.

For case (i), we apply Lemma 2.4 to our given four-point set in convex position, with the self-stress of Lemma 2.6, which is positive for the four hull edges and negative for the two diagonals. This implies that this four-point set can have no non-trivial expansive motion which is not strictly expansive on at least one of the two diagonals. □

As an immediate consequence of Lemma 3.2(a), we get:

Corollary 3.3. *$\bar{X}_f(P)$ is a $(2n-3)$-dimensional unbounded polyhedron with at least one vertex, for any choice of parameters f.* □

It is easy to derive part 2 of Theorem 3.1 from part 1. For every vertex or bounded edge of $\bar{X}_f(P)$, the set $E(v)$ contains all convex hull edges of P. On the contrary, for any unbounded edge (ray) of $\bar{X}_f(P)$, the set $E(v)$ misses some convex hull edge of P. Hence, by setting to equalities the inequalities corresponding to convex hull edges we get a face $X_f(P)$ of $\bar{X}_f(P)$ which contains all vertices and bounded edges of $\bar{X}_f(P)$, but no unbounded edge.

In order to prove part 1, we first need to check that indeed X_f is a bounded face, and hence a polytope which we call the *polytope of constrained expansions* or pce-polytope for the set of points P and perturbation parameters f.

Lemma 3.4. *For any choice of f, $X_f(P)$ is a bounded set.*

Proof. Suppose that $v_0 + tv$ is in X_f for all $t \geq 0$. Then we must have $v \in X_0$. Hence, it suffices to show that $X_0 = 0$, i.e. that the framework consisting of all convex hull edges has no non-trivial expansive flexes. This is an immediate consequence of Lemma 3.2b(iii). □

Reducing the problem to four points. We call a choice of the constants $f = (f_{ij}) \in \mathbb{R}^{\binom{n}{2}}$ *valid* if the corresponding polyhedron $\bar{X}_f$ of constrained expansions has the combinatorial structure claimed in Theorem 3.1.

Lemma 3.5. *A choice of $f \in \mathbb{R}^{\binom{n}{2}}$ is valid if and only if the graph $E(v)$ of tight edges corresponding to any feasible point $v \in \bar{X}_f(P)$ is non-crossing and pointed.*

Proof. Necessity is trivial, by definition of being valid. To see sufficiency note that, by Corollary 3.3, $\bar{X}_f$ has dimension $2n-3$. Thus, any vertex v of the polyhedron is incident to at least $2n-3$ faces $E(v)$. If $E(v)$ is non-crossing and pointed, Lemma 2.1 implies that $E(v)$ has *exactly* $2n-3$ incident faces and is a pointed pseudo-triangulation. In particular, the polyhedron is simple. Also, since the tight edges of faces incident to v are different subgraphs of $E(v)$, the poset of faces incident to the vertex v is the poset of all subgraphs of the pointed pseudo-triangulation $E(v)$.

It remains only to show that every pointed pseudo-triangulation actually appears as a vertex, for which we use a somewhat indirect argument, based on the fact that the flip graph is connected. This type of argument has also been used by Carl Lee for the case of the associahedron in [16], where it is attributed to Gil Kalai and Micha Perles.

Since the polytope is simple, every vertex v is incident to $2n-3$ edges of $\bar{X}_f$. The sets of tight edges corresponding to them are the $2n-3$ subgraphs of $E(v)$ obtained removing a single edge. We denote by K_{ij} the polyhedral edge corresponding to the removal of ij. By Lemma 3.4, if ij is interior then K_{ij} is bounded. Hence, it is incident to another vertex, which must correspond to a pointed pseudo-triangulation that completes $E(v) - \{ij\}$. By Lemma 2.2(a), this can only be the one obtained from $E(v)$ by a flip at ij. Together with the fact that the flip graph is connected (Lemma 2.2(b)) and that $\bar{X}_f$ has at least one vertex, this implies that all pointed pseudo-triangulations appear as vertices of $\bar{X}_f$, and hence that all pointed and non-crossing graphs appear as well.

Also, the extreme rays have the structure predicted in Theorem 3.1. For a convex hull edge ij, K_{ij} must be an unbounded edge because there is no other pointed pseudo-triangulation that contains $E(v) - \{ij\}$. □

We now conclude that valid perturbation vectors $f \in \mathbb{R}^{\binom{n}{2}}$ can be recognized by looking at 4-point subsets only.

Lemma 3.6. *A choice of $f \in \mathbb{R}^{\binom{n}{2}}$ is valid if and only if it is valid when restricted to every four points of P.*

Proof. By the previous Lemma, if f is not valid for P then there is a point v of $\bar{X}_f$ for which the graph $E(v)$ is either non-pointed or crossing. In either case, there is a subset of four points $P' \subseteq P$ on which the induced subgraph is non-pointed or crossing. Let v' and f' denote v and f restricted to P'.

Then, v' is in $\bar{X}_{f'}(P')$ and the graph $E(v')$ is crossing or not pointed, hence f' is not valid on P'. Contradiction. □

The case of four points.

Theorem 3.7. *A choice of perturbation parameters $f \in \mathbb{R}^{\binom{4}{2}}$ on four points $P = (p_1, p_2, p_3, p_4)$ forms a valid choice if and only if*

$$\sum_{1 \le i < j \le 4} \omega_{ij} f_{ij} > 0, \tag{8}$$

where the ω_{ij}'s are the unique self-stress on the four points, with signs chosen as in Lemma 2.6.

For a set of four points $P = (p_1, p_2, p_3, p_4)$, we denote by G_{ij} the graph on P whose only missing edge is ij. Recall that the choice of self-stress on four points has the property that G_{ij} is pointed and non-crossing (equivalently, ij is interior) if and only if ω_{ij} is negative.

Since $\bar{X}_f(P)$ is five-dimensional, for every vertex v the set $E(v)$ contains at least five edges. Therefore $E(v)$ is either the complete graph or one of the graphs G_{ij}. Theorem 3.7 is then a consequence of Lemma 3.5 and the following statement.

Lemma 3.8. *Let $R := \sum_{1 \le i < j \le 4} \omega_{ij} f_{ij}$. For every edge kl, the following properties are equivalent:*

1. *The graph G_{kl} appears as a vertex of $X_f(P)$.*

2. *R and ω_{kl} have opposite signs.*

Proof. The graph G_{kl} appears as a face if and only if the (unique, since G_{kl} is rigid) motion with edge length increase f_{ij} for every edge ij other than kl has edge length increase on kl greater than f_{kl}. In this case, by rigidity of G_{kl}, the face is actually a vertex. But, for this motion:

$$\begin{aligned} 0 &= \sum_{1 \le i < j \le 4} \omega_{ij} \langle p_j - p_i, v_j - v_i \rangle = \sum_{1 \le i < j \le 4} \omega_{ij} f_{ij} + \omega_{kl}(\langle p_k - p_l, v_k - v_l \rangle - f_{kl}) \\ &= R + \omega_{kl}(\langle p_k - p_l, v_k - v_l \rangle - f_{kl}). \end{aligned}$$

Hence, $\langle p_k - p_l, v_k - v_l \rangle > f_{kl}$ is equivalent to R and ω_{kl} having opposite sign. □

Observe that the previous lemma implicitly includes the statement that the complete graph appears as a vertex if and only if $R = 0$. The *only if* part of this is actually an easy consequence of Lemma 2.4. In this case X_f degenerates to a single point.

To complete the proof of Theorem 3.1 we still need to show that valid choices of perturbation parameters exist:

Theorem 3.9. *Let a and b be any two points in the plane. For any point set $P = \{p_1, \dots, p_n\}$ in general position in the plane, the following choice of parameters f is valid:*

$$f_{ij} = \det(a, p_i, p_j) \det(b, p_i, p_j) \tag{9}$$

Proof. This follows from the following Lemma, taking into account Lemma 3.6 and Theorem 3.7.

Lemma 3.10. *For any two points a and b and for any four points $p_1, \dots, p_4$ in general position in the plane, we have*

$$\sum_{1 \le i < j \le 4} \omega_{ij} f_{ij} = 1,$$

where f_{ij} is given by (9) and ω_{ij} are the self-stress of Lemma 2.6.

Proof. Let us consider the four points p_i as fixed and regard $R = \sum \omega_{ij} f_{ij}$ as a function of a and b.

$$R(a, b) = \sum_{1 \le i < j \le 4} \det(a, p_i, p_j) \det(b, p_i, p_j) \omega_{ij}.$$

For fixed b, $R(a, b)$ is clearly an affine function of a. We claim that $R(p_i, b) = 1$ for each of the four points $p_1, \dots, p_4$, which implies that $R(a, b)$ is constantly equal to 1. To prove the claim, without loss of generality we take $a = p_1$. Now, $R(p_1, b)$ is an affine function of b. By a similar argument as before, it suffices to show $R(p_1, b) = 1$ for the three affinely independent points $b = p_2, p_3, p_4$. Without loss of generality we look only at $R(p_1, p_2)$. Then, $f_{ij} = 0$ for every i, j except $f_{34} = \det(p_1, p_3, p_4) \det(p_2, p_3, p_4)$. Hence,

$$R(p_1, p_2) = \omega_{34} f_{34} = \frac{\det(p_1, p_3, p_4) \det(p_2, p_3, p_4)}{\det(p_3, p_4, p_1) \det(p_3, p_4, p_2)} = 1. \qquad \square$$

This proof is quite easy, but it does not provide much intuition why this choice of f works. The first valid choice that we found by heuristic considerations was the function

$$f'_{ij} = \tfrac{1}{2} \cdot \left(|p_i|^2 + |p_j|^2 + \langle p_i, p_j \rangle \right) \cdot |p_i - p_j|^2. \tag{10}$$

The intuition behind this is as follows. Looking back to Lemma 3.6, there are two cases of four points: in convex position and as a triangle with a point in the middle. In both cases we want to avoid the situation that all interior edges (inside the convex hull) are tight, while the hull edges expand at least at their prescribed rate f_{ij}. Thus, we want f_{ij} to be big for the "peripheral" edges and small for the "central" edges. (This goal is in accordance with Theorem 3.7, as our choice of ω_{ij} is positive on boundary edges and negative on interior ones.)

A function which has this property of being on average bigger on the border of a region than in the middle is the convex function $|x|^2$. When we integrate $|x|^2$ over the edge $p_i p_j$ and multiply the result by the edge length (because f_{ij} is expressed in terms of the derivative of the *squared* edge length), we get (10), up to a multiplicative constant.

The parameters f'_{ij} are valid. Indeed, it can be checked that $\sum \omega_{ij} f'_{ij} = 1$ holds for all 4-tuples of points: setting $a = b = 0$ in the definition (9) of f, the difference

$$f'_{ij} - f_{ij} =: g_{ij} = \left[|p_i|^2 |p_i|^2 + |p_j|^2 |p_j|^2 - (|p_i|^2 + |p_j|^2)\langle p_i, p_j \rangle \right] /2$$

satisfies $\sum_{i,j} \omega_{ij} g_{ij} = 0$, which follows from Lemma 2.4 with $v_i = |p_i|^2 p_i/2$.

Of course, the equation $\sum \omega_{ij} f'_{ij} = 1$ can be trivially checked by expanding the values of ω_{ij} and f'_{ij} with the help of a computer algebra package. Attempts to find a more classical proof have lead to the function f_{ij} of (9).

4 Applications of the Main Result

4.1 The Expansion Cone

As mentioned in the previous section, the expansion cone is the recession cone $\bar{X}_0$ of the pce-polyhedron $\bar{X}_f$, whose structure we know. The extreme rays of $\bar{X}_0$ are precisely the extreme rays of $\bar{X}_f$, shifted to start at 0, but parallel rays of $\bar{X}_f$ give rise to only one ray of $\bar{X}_0$, of course.

Studying when this happens will allow us to give now a rather easy proof of the characterization of the extreme rays of $\bar{X}_0$ (Proposition 2.8): We conclude from Theorem 3.1 that the extreme rays correspond to pointed pseudo-triangulations with one hull edge removed, i.e., pte-mechanisms. Any convex subpolygon in a pte-mechanism must be rigid in the mechanism, according to Lemma 3.2b(iii). This corresponds to the fact that every convex subpolygon of a pointed pseudo-triangulation contains a pointed pseudo-triangulation of that polygon and the enclosed points, and is therefore rigid. We still have to show that these r-components are the only subcomponents that move rigidly in the (unique) flex v on a pte-mechanism $G(P)$.

Lemma 4.1. *Let $P' \subset P$ be a maximal subset that moves (infinitesimally) rigidly by the unique flex v of the pte-mechanism $G(P)$.*

(a) *Then P' contains all points of P within its convex hull,*

(b) *G contains no edge which either crosses the boundary of the convex hull of P' or gives a non-pointed graph together with the boundary of P', and*

(c) *G contains all boundary edges of the convex hull of P'.*

Proof. (a) A subset $P' \subset P$ moves rigidly if and only if $E(v)$, considered in $\bar{X}_0$, contains the complete subgraph spanned by P'. Then part (a) follows from Lemma 3.2b(iii).

(b) An edge $ij \in G \subseteq E(v)$ in either of those two situations would, by Lemma 3.2b(i) or (ii), imply that the complete graph on $P' \cup ij$ is part of $E(v)$. and hence i and j are rigidly connected to P'. On the other hand, either of the two conditions implies that one of i or j is not in P', violating maximality of P' as a subset which moves rigidly.

(c) Assume that a hull edge ij of P' is missing in G. Let G' be the graph obtained adding ij to G. Since P' moves rigidly, G' is still a 1DOF mechanism. On the other hand, part (b) implies that G' is still a pointed and non-crossing graph. Since G had $2n-4$ edges, G' has $2n-3$ edges and, by Lemma 2.1, it is a pointed pseudo-triangulation. This is a contradiction, because pointed pseudo-triangulations are infinitesimally rigid (Proposition 2.3). □

If follows from the last statement that the rigidly moving components are precisely the convex subpolygons of the pte-mechanism, and two pte-mechanisms yield the same motion (extreme ray) if and only if they lead to the same collapsed pte-mechanism, thus concluding the proof of Proposition 2.8. □

4.2 Strictly Expansive Motions and Unfoldings of Polygons

Lemma 4.2. *Let $G(P)$ be a non-crossing and pointed framework in the plane. Then, $G(P)$ has a non-trivial expansive flex if and only if it does not contain all the convex hull edges. In this case, there is an expansive motion that is strictly expansive on all the convex hull edges not in G.*

Proof. If all the convex hull edges are in G, then Lemma 3.4 implies the statement: the face of $\bar{X}_f$ corresponding to G is bounded and, hence, it degenerates to the origin in $\bar{X}_0$. If a certain convex hull edge ij is not in G, then we extend G to a pointed pseudo-triangulation, according to Lemma 2.1. Removing ij yields a pte-mechanism, whose expansive motion is strictly expansive on ij. Adding all such motions for the different missing hull edges gives the stated motion. □

This immediately gives the following theorem.

Theorem 4.3. *Let $G(P)$ be a non-crossing non-convex plane polygon or a plane polygonal arc that does not lie on a straight line. Then there is an expansive motion that is strictly expansive on at least one edge.* □

This statement has been crucial to show that every simple polygon in the plane can be unfolded by a global motion into convex position and every polygonal arc can be straightened, without collisions [6, 25]. The proof given in those papers is based on several reduction steps between infinitesimal motions, self-stresses, and polyhedral terrains. The above new proof is completely independent, although not less indirect.

Actually, one can work a little harder in the proof of Lemma 4.2 and show that *any* edge $ij \notin G$ that is not contained inside a convex subpolygon of G can be chosen to be strictly expansive. (The proof constructs an appropriate pte-mechanism by a flipping argument, applied to the minimal convex subpolygon enclosing the chosen edge.) By adding several motions of $G(P)$ one can obtain an expansive motion that is strictly expansive on *all* eligible edges, and hence, in Theorem 4.3, there is a motion that is strictly expansive on all edges $ij \notin G$. This is actually the statement that was proved in [6], in a more general setting.

5 Other Constructions

In this section we present three related results: a different representation for the ppt-polytope that is less dependent on some seemingly arbitrary choice of parameters f, and the two constructions leading to the associahedron: 2-dimensional points in convex position and the one-dimensional expansion polytope.

5.1 A redefinition of the ppt-polyhedron

Let $P = \{p_1, \ldots, p_n\}$ be a fixed point configuration in general position. As before, with each possible choice of parameters $f = (f_{ij}) \in \mathbb{R}^{\binom{n}{2}}$ we associate the polyhedron $\bar{X}_f$ defined by the constraints (2) and (6), and the polytope X_f obtained setting to equalities the inequalities corresponding to convex hull edges. The case $f \equiv 0$ produces the expansion cone, with the polytope degenerating to a point. The choice $f_{ij} = \det(0, p_i, p_j)^2$, among others, produces our polytope of pointed pseudo-triangulations, according to Theorem 3.9. But the results of Section 3 imply that actually any other choice of f_{ij}'s would provide (combinatorially) the same polyhedron and polytope as long as it satisfies inequality (8) for every four points p_{i_1}, p_{i_2}, p_{i_3} and p_{i_4}, where the ω_{ij}'s are the self-stress on the four points with sign chosen as in Lemma 2.6.

In this section we give a new construction for this polyhedron, with the advantage that it does not depend on any choice of parameters. It has the disadvantage, however, that it involves more variables: one for each of the $\binom{n}{2}$ possible edges among the n points.

The basic idea is to study the image of the previously defined ppt-polytope under the rigidity map $M\colon \mathbb{R}^{2n} \to \mathbb{R}^{\binom{n}{2}}$ for the complete graph on P, using the variables $\delta_{ij} := \langle p_i - p_j, v_i - v_j \rangle$. The following lemma is a stronger version of Lemma 2.4.

Lemma 5.1. (Equivalence of parameters) *Let $\delta = (\delta_{ij})$ be a vector in $\mathbb{R}^{\binom{n}{2}}$. Then, the following properties are equivalent:*

(a) $\delta \in \operatorname{Im} M$

(b) *For any four points p_{i_1}, p_{i_2}, p_{i_3} and p_{i_4} in P one has*

$$\sum_{i<j\in\{i_1,i_2,i_3,i_4\}} \omega_{ij}\delta_{ij} = 0$$

where the ω_{ij}'s are a nonzero self-stress of the complete graph on those four points.

Proof. The implication (a)⇒(b) follows directly from one direction of Lemma 2.4. Lemma 2.4 also gives (b)⇒(a) for each quadruple: if (a) holds, then for each four points there is an infinitesimal motion $(a_{i_1}, a_{i_2}, a_{i_3}, a_{i_4})$ whose image by the rigidity map of the four points are the six relevant entries of δ. The motion for a quadruple is not unique, but any two choices differ only by a trivial motion of the quadruple.

To define a global motion $(a_1, \dots, a_n)$ of the whole configuration, let us start by setting $a_1 = (0,0)$ and $a_2 = (0,b)$, where b must be the unique number satisfying $\langle p_1 - p_2, (0,b)\rangle = \delta_{12}$. (We assume without loss of generality that p_1 and p_2 do not have the same y-coordinate.) The condition $\delta = M(a)$ on the edges $1i$ and $2i$ then uniquely defines a_i for every $i \neq 1,2$, because these two equations are linearly independent, the directions $p_i - p_1$ and $p_i - p_2$ being not parallel. To see that this global motion satisfies $\delta = M(a)$ also for any other edge ij ($i \neq 1,2$, $j \neq 1,2$), it is sufficient to consider the quadruple (p_1, p_2, p_i, p_j). By construction, the motion (a_1, a_2, a_i, a_j) satisfies the condition $\delta = M(a)$ for five of the six edges in the quadruple. Assumption (b) says that $\sum_{k,l\in\{1,2,i,j\}} \omega_{kl}\delta_{ij} = 0$ which, by Lemma 2.4, implies that δ restricted to (p_1, p_2, p_i, p_j) is the set of edge increases produced by some motion $v = (v_1, v_2, v_i, v_j)$. This motion can be normalized to $v_1 = (0,0)$ and $v_2 = (0,b)$ and then it must coincide with a by our uniqueness argument above. □

This, together that the observation that the kernel of M consists only of trivial motions immediately gives the following lemma.

Lemma 5.2. *For any $f \in \mathbb{R}^{\binom{n}{2}}$, the polyhedron $\bar{X}_f$ is linearly isomorphic to the one defined in $\binom{n}{2}$ variables $\delta_{ij} = \delta_{ji}$ by the following $\binom{n}{4}$ equations and $\binom{n}{2}$ inequalities.*

$$\begin{aligned} \textstyle\sum_{i<j\in\{i_1,i_2,i_3,i_4\}} \omega_{ij}\delta_{ij} = 0, &\qquad \forall\, i_1, i_2, i_3, i_4 \in \{1,\dots,n\}, \\ \delta_{ij} \geq f_{ij}, &\qquad \forall\, i,j \in \{1,\dots,n\}. \end{aligned}$$

Moreover, setting to equalities the inequalities corresponding to convex hull edges produces a polytope linearly isomorphic to X_f. □

By making the change of variables $d_{ij} = f_{ij} - \delta_{ij}$ and taking into account that $\sum \omega_{ij} f_{ij} = 1$ for any four points, for the valid choices f of Theorem 3.9, we conclude:

Theorem 5.3. *For any given point set $P = \{p_1, \ldots, p_n\}$ in the plane in general position, the following $\binom{n}{4}$ equations and $\binom{n}{2}$ inequalities define a simple polyhedron in $\mathbb{R}^{\binom{n}{2}}$ linearly isomorphic to the polyhedron $\bar{X}_f(P)$ of Theorem 3.1, with f_{ij} chosen as in Theorem 3.9.*

- $\sum \omega_{ij} d_{ij} = 1$ *for every quadruple, where the ω_{ij}'s of each equation are the self-stress on the corresponding quadruple stated in Lemma 2.6.*
- $d_{ij} \leq 0$ *for every variable.*

The maximal bounded face in the polyhedron is obtained setting to equalities the inequalities corresponding to convex hull edges. □

The $\binom{n}{4}$ equations are highly redundant. It follows from the proof of Lemma 5.1 that the $\binom{n-2}{2}$ quadruples involving two fixed vertices are sufficient. Subtracting this number from the number $\binom{n}{2}$ of variables actually gives the right dimension $2n - 3$ of the polyhedron.

5.2 Convex position and the associahedron

Suppose now that our n points $P = \{p_1, \ldots, p_n\}$ are in (ordered) convex position. In this subsection all indices are regarded modulo n. In Section 2 we noticed that the polytope of pointed pseudo-triangulations of P is combinatorially the same thing as the secondary polytope, which for a convex n-gon is an associahedron. We prove here that, in fact, the secondary polytope and the ppt-polytope are affinely isomorphic.

The first problem we encounter is that so far we have only facet descriptions for the ppt-polytope, while the secondary polytope is defined by the coordinates of its vertices. We recall that the secondary polytope lives in $\mathbb{R}^n$ and that the i-th coordinate of the vertex corresponding to a certain triangulation T equals the total area of all triangles of T incident to p_i. Denote this area as $\text{Area}_T(p_i)$. For convenience we will work with a normalized definition of area of a triangle with vertices p, q and r as being equal to $|\det(p, q, r)|$. We also assume our points to be ordered counter-clockwise, so that $\det(p_i, p_j, p_k)$ is positive if and only if i, j and k appear in this order in the cyclic ordering of $\{1, \ldots, n\}$. In this way

$$\text{Area}_T(p_i) := \sum_{l=1}^{t-1} \det(p_i, p_{j_l}, p_{j_{l+1}})$$

where $\{p_{j_1}, \ldots, p_{j_t}\}$ is the ordered sequence of vertices adjacent to p_i in T.

Our first task is to compute the coordinates of the vertices of the ppt-polytope. Notice that by definition the coordinates corresponding to edges of T are zero, since the inequalities corresponding to the edges of T are satisfied with equality. It will turn out that we do not need all the coordinates, but only those corresponding to *almost-convex-hull edges* $p_{i-1}p_{i+1}$.

Lemma 5.4. *Let T be a triangulation of P. Then, in the ppt-polytope of Theorem 5.3, the coordinate $d_{i-1,i+1}$ corresponding to an almost-convex-hull edge equals*

$$d_{i-1,i+1} = -\det(p_{i-1}, p_i, p_{i+1})\left(\mathrm{Area}_T(p_i) - \det(p_{i-1}, p_i, p_{i+1})\right).$$

Proof. Let $p_{j_1}, \ldots, p_{j_t}$ be the ordered sequence of vertices adjacent to p_i in T, with $p_{j_1} = p_{i+1}$ and $p_{j_t} = p_{i-1}$. We will prove by induction on $k = 2, \ldots, t$ that the coordinate $d_{j_1 j_k}$ of the ppt-polytope vertex corresponding to T equals

$$d_{j_1 j_k} = -\det(p_i, p_{j_1}, p_{j_k})\,\mathrm{Area}(p_{j_1}, \ldots, p_{j_k})$$

where $\mathrm{Area}(p_{j_1}, \ldots, p_{j_k})$ denotes the area of the polygon with vertices $p_{j_1}, \ldots,$ p_{j_k}, in that order. This reduces to the formula in the statement for $k = t$.

The base case $k = 2$ is trivial: since $j_1 j_2$ is an edge in the triangulation, we have $d_{j_1 j_2} = 0$. To compute $d_{j_1 j_k}$ for $k > 2$ we consider the quadruple $p_i, p_{j_1}, p_{j_{k-1}}, p_{j_k}$. The only non-zero d's on this quadruple are $d_{j_1 j_{k-1}}$ and $d_{j_1 j_k}$. (For $k = 3$, $d_{j_1 j_{k-1}}$ is also zero.) Hence, the equation $\sum \omega_{\alpha\beta} d_{\alpha\beta} = 1$ for this quadruple reduces to

$$d_{j_1 j_k}\omega_{j_1 j_k} + d_{j_1 j_{k-1}}\omega_{j_1 j_{k-1}} = 1.$$

From this we infer the stated value for $d_{j_1 j_k}$ from the known values of the other quantities:

$$\begin{aligned}
\omega_{j_1 j_k} &= \frac{1}{\det(p_i, p_{j_1}, p_{j_k})\det(p_{j_1}, p_{j_{k-1}}, p_{j_k})}, \\
\omega_{j_1 j_{k-1}} &= \frac{-1}{\det(p_i, p_{j_1}, p_{j_{k-1}})\det(p_{j_1}, p_{j_{k-1}}, p_{j_k})}, \quad \text{and} \\
d_{j_1 j_{k-1}} &= -\det(p_i, p_{j_1}, p_{j_{k-1}})\,\mathrm{Area}(p_{j_1}, \ldots, p_{j_{k-1}}).
\end{aligned}$$

(The last equation is the inductive hypothesis.) □

This Lemma immediately implies the following:

Corollary 5.5. (a) *The affine map*

$$a_i = -\frac{d_{i-1,i+1}}{\det(p_{i-1}, p_i, p_{i+1})} + \det(p_{i-1}, p_i, p_{i+1})$$

gives the coordinates $(a_1, \ldots, a_n)$ of a triangulation in the secondary polytope in terms of its coordinates $(d_{ij})_{ij \in \binom{n}{2}}$ in the ppt-polytope of Theorem 5.3.

(b) *The substitution $d_{ij} = f_{ij} - \langle p_i - p_j, v_i - v_j \rangle$ in the above formula gives an affine map $\mathbb{R}^{2n} \to \mathbb{R}^n$ sending the ppt-polytope of Theorem 3.1 to the secondary polytope, whenever the f_{ij}'s are a valid choice satisfying*

$$\sum_{1 \le i < j \le 4} \omega_{ij} f_{ij} = 1$$

for every quadruple, such as the choices in Theorem 3.9. □

Observe that Corollary 5.5 implies that, for points in convex position, we can consider the ppt-polytope of Theorem 5.3 as lying in the n-dimensional space given by the coordinates $d_{i-1,i+1}$. For the polyhedron, we additionally need the coordinates $d_{i,i+1}$. (These are zero on the polytope, but not on the polyhedron.)

5.3 The 1-Dimensional Case: the Associahedron Again

By considering 1-dimensional expansive motions, in this section we will recover the associahedron via a different route. The analogy of this construction to the 2-dimensional case will become even more apparent in Section 6.

The polytope of constrained expansions in dimension 1. In the 1-dimensional case we will rewrite constraints (6) in the form

$$v_j - v_i \geq g_{ij}, \qquad \forall i < j. \tag{11}$$

One set of inequalities is equivalent to the other under the change of constants $g_{ij}(p_j - p_i) = f_{ij}$, for any $i < j$. This reformulation explicitly shows that the solution set does not depend on the point set $P = \{p_1, \ldots, p_n\}$ that we choose. We denote this solution set $\bar{X}_g$, to mimic the notation of the 2-dimensional case.

It is easy to see that the polyhedron $\bar{X}_g$ is full-dimensional and it contains no lines if we add the normalization equation $v_1 = 0$. Hence, after normalization, it has dimension $n-1$ and contains some vertex. For any vertex v or for any feasible point $v \in \bar{X}_g$, we may look at the set $E(v)$ of tight inequalities at v:

$$E(v) := \{\, ij \mid 1 \leq i < j \leq n, v_j - v_i = g_{ij} \,\}$$

We regard $E(v)$ as the set of edges of a graph on the vertices $\{1, \ldots, n\}$.

One may get various polyhedra by choosing different numbers g_{ij} in (11). We choose them with the following properties.

$$g_{il} + g_{jk} > g_{ik} + g_{jl}, \qquad \forall 1 \leq i < j \leq k < l \leq n. \tag{12}$$

For $j = k$ we use this with the interpretation $g_{jj} = 0$, so we require

$$g_{il} > g_{ik} + g_{kl}, \qquad \forall 1 \leq i < k < l \leq n. \tag{13}$$

One way to satisfy these conditions is to select

$$g_{ij} := h(t_j - t_i), \quad \forall i < j \tag{14}$$

for an arbitrary strictly convex function h with $h(0) = 0$ and arbitrary real numbers $t_1 < \cdots < t_n$. The simplest choice is $h(t) = t^2$ and $t_i = i$, yielding $g_{ij} = (i - j)^2$.

Two edges ij and jk with $i < j < k$ are called *transitive edges*, and edges ik and jl with $i < j < k < l$ are called *crossing edges*.

Lemma 5.6. *If g satisfies* (12–13) *and $v \in \bar{X}_g$, then $E(v)$ cannot contain transitive or crossing edges.*

Proof. If we have two transitive edges $ij, jk \in E(v)$ this means that $v_j - v_i = g_{ij}$ and $v_k - v_j = g_{jk}$. This gives $v_k - v_i = g_{ij} + g_{jk} < g_{ik}$, by (13), and thus v cannot be in $\bar{X}_g$ because it violates (6). The other statement follows similarly. □

Non-crossing alternating trees. A graph without transitive edges is called an *alternating* or *intransitive* graph: every path in an alternating graph changes continually between up and down.

Lemma 5.7. *A graph on the vertex set $\{1, \ldots, n\}$ without transitive or crossing edges cannot contain a cycle.*

Proof. Assume that C is a cycle without transitive edges. Let i and m be the lowest and the highest-numbered vertex of a cycle C, and let ik be an edge of C incident to i, but different from im. The next vertex on the cycle after k must be between i and k; continuing the cycle, we must eventually reach m, so there must be an edge jl which jumps over k, and we have a pair ik, jl of crossing edges. □

Since the polyhedron is $(n-1)$-dimensional, the set $E(v)$ of a vertex v must contain at least $n-1$ edges. We have just seen that it is acyclic, and hence it must be a tree and contain exactly $n-1$ edges. So we get:

Proposition 5.8. *$\bar{X}_g$ is a simple polyhedron. The tight inequalities for each vertex correspond to non-crossing alternating trees.* □

We will see below that $\bar{X}_g$ contains in fact *all* non-crossing alternating trees as vertices.

A new realization of the associahedron. Let's look at the combinatorial properties of these trees. Alternating trees have been studied in combinatorics in several papers, see for example [21, 22] or [24, Exercise 5.41, pp. 90–92] and the references given there. Non-crossing alternating trees were studied only by Gelfand, Graev, and Postnikov, under the name of "standard trees". They proved the following fact [10, Theorem 6.4].

Proposition 5.9. *The non-crossing alternating trees non $n+1$ points are in one-to-one correspondence with the binary trees on n vertices, and hence their number is the n-th Catalan number $\binom{2n}{n}/(n+1)$.* □

The bijection given in [10] to prove this fact is that the vertices of the binary tree correspond to the edges of the alternating tree. It is easy to see that every non-crossing alternating tree must contain the edge $1n$. Removing this edge splits the tree into two parts; they correspond to the two subtrees of the root in the binary tree. The two parts are handled recursively. Even

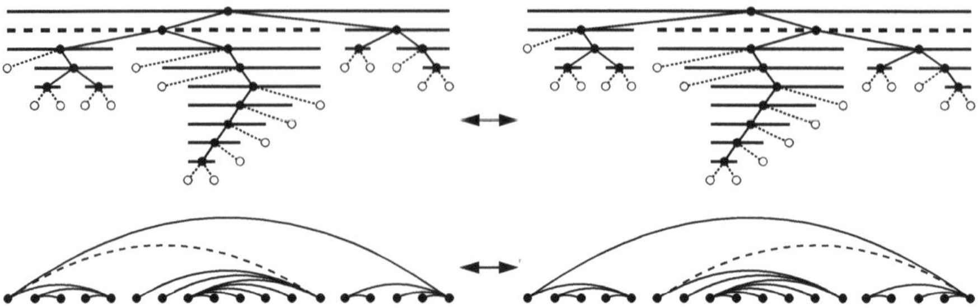

$((\,(a((bc)(de)))\ (f(g(((((hi)j)k)l)m)))\,)((no)(p(qr))))\ \longleftrightarrow\ ((a((bc)(de)))(\,(f(g(((((hi)j)k)l)m)))\ ((no)(p(qr)))\,))$

Fig. 5. The bijection between binary trees (up), non-crossing alternating trees (middle) and bracketings (bottom). The flipping operation in each case is shown, and the elements affected by the flip are highlighted.

simpler is the bijection to bracketings (ways to insert $n-1$ pairs of parentheses in a string of n letters). Just change edge ij by a parenthesis enclosing the i-th and j-th letter.

Figure 5 gives an example of these correspondences, including the correspondence of the respective flipping operations: If we remove any edge $e \neq 1n$ from a non-crossing alternating tree T, there is precisely one other non-crossing alternating tree T' which shares the edges $T - \{e\}$ with T. This exchange operation corresponds under the above bijection to a tree rotation of the binary tree, and to a single application of the associative law (remove a pair of interior parentheses and insert another one in the only possible way) in a bracketing.

Observe that still we don't know that *all* the non-crossing and alternating trees appear as vertices of X_g. But this is easy to prove: since X_g is simple of dimension $d-2$, its graph is regular of degree $d-2$. And it is a subgraph of the graph of rotations between binary trees, which is also $(d-2)$-regular and connected. Hence, the two graphs coincide.

If we now look at faces of $\bar{X}_g$, rather than vertices, the tight edges for each of them form a non-crossing alternating *forest*. Such a forest G is an expansive mechanism if and only if the edge $1n$ is not in G: If that edge is present, p_1 and p_n are fixed and any other p_i must approach one of the two. If the edge $1n$ is not present, then let i be the maximum index for which $1i$ is present. Then the motion $v_1 = \cdots = v_i = 0 < v_{i+1} = \cdots = v_n$ is an expansive flex of G.

Hence, $\bar{X}_g$ has a unique maximal bounded face, the facet given by the equation $v_n - v_1 = g_{1n}$ and corresponding to the edge $1n$ alone. This facet is then an $(n-2)$-dimensional simple polytope that we denote X_g. The $n-2$ neighbors of each vertex v correspond to the $n-2$ possible exchanges of edges different from $1n$ in the tree $E(v)$.

Theorem 5.10. *X_g is a simple $(n-2)$-dimensional polytope whose face poset is that of the associahedron. Vertices are in one-to-one correspondence with the non-crossing alternating trees on n vertices. Two vertices are adjacent if and only if the two non-crossing alternating trees differ in a single edge.*

$\bar{X}_g$ is an unbounded polyhedron with the same vertex set as X_g. The extreme rays correspond to the non-crossing alternating trees with the edge $1n$ removed.

Proof. Only the statement regarding the face poset remains to be proved. This can be proved in two ways: On the one hand, we already know that the graphs of X_g and of the $(d-2)$-associahedron coincide (the latter being the graph of rotations between binary trees). And simple polytopes with the same graph have also the same face poset. (This is a result of Blind and Mani; see [28, Section 3.4]). As a second proof, the correspondence between non-crossing alternating trees and bracketings trivially extends to a correspondence between non-crossing alternating forests containing $1n$ and "partial bracketings" in a string of n letters which include the parentheses enclosing the whole string. The poset of such things is the face poset of the associahedron (see [24], Proposition 6.2.1 and Exercise 6.33). In particular, the face-poset of X_g is a subposet of the face-poset of the associahedron, and two polytopes of the same dimension cannot have their face posets properly contained in one another. This is true in general by topological reasons, but specially obvious in our case since we know our polytopes to be simple and their vertices to correspond one to one. Each vertex is incident to exactly $\binom{d-2}{i}$ faces of dimension i in both polytopes, and two subsets of cardinality $\binom{d-2}{i}$ cannot be properly contained in one another. □

A result which is related to Theorem 5.10 was proved by Gelfand, Graev, and Postnikov [10, Theorem 6.3], in a setting dual to ours: there a triangulation of a certain polytope was constructed. The non-crossing alternating trees correspond to the *simplices* of the triangulation. It is shown explicitly that the simplices form a partition of the polytope. Certain numbers g_{ij} are then associated with the *vertices* of the polytope to show that the triangulation is a projection of the boundary of a higher-dimensional polytope. Incidentally, the numbers that were suggested for this purpose are $(i-j)^2$, which coincides with the simple proposal given above after (14), but the calculations are not given in the paper [10].

One easily sees that the conditions (12–13) on g are also necessary for the theorem to hold: If any of these conditions would hold as an equality or as an inequality in the opposite direction, the argument of Lemma 5.6 would work in the opposite direction, and certain non-crossing alternating trees would be excluded. Thus, (12–13) are a complete characterization of the "valid" parameter values g_{ij}.

Further Remarks. The result we presented in this section is surprising in two ways: first, that it produces such a well studied object as the associahedron; second, that it requires additional types of linear constraints that are not needed in dimension 2. Indeed, inequalities (13) in the 1-dimensional case are the exact analog of inequalities (8) in the 2-dimensional case, as we will see in Section 6. But the constraints (12) have no analog. This second aspect makes the task of producing 3d generalization of the constructions of this paper more challenging, as there does not seem to be a straightforward pattern for producing the linear constraints whose instantiations in 1d and 2d give the polytopes of expansive motions.

The conditions (12–13) leave a lot of freedom for the choice of the variables g_{ij}. We have an $\binom{n}{2}$-dimensional parameter space. This is in contrast to the classical representation, which has $2n$ parameters (the coordinates of n points in the plane). If we select the parameters to be integral, we obtain an associahedron which is an integral polytope (this is also true for the classical associahedron for a polygon with integer vertices.) But observe that, in fact, the associahedra obtained here are in a sense much more special than the classical associahedra obtained as secondary polytopes. They have $n-2$ pairs of parallel facets, given by the equations $v_i = v_1 + g_{1i}$ and $v_i = v_n - g_{in}$ (i.e., corresponding to the pairs of edges $1i$ and in). See the right picture of Figure 4, where $n = 5$. This is no surprise, since we have constructed our polytope by perturbing (a region of) a Coxeter arrangement, whose hyperplanes are in extremely non-general position.

One can check that classical associahedra have no parallel facets. Hence, are associahedra are not affinely equivalent to the classical ones. Recently, Chapoton, Fomin and Zelevinsky [5] have shown another construction of the associahedron having as facet normals the root system of type A_n (more generally, they show a similar construction for all root systems). Still, our associahedra are not affinely equivalent to theirs: Comparing the right part of our Figure 4 with their Figure 2 we see that both are 3-dimensional associahedra with three pairs of parallel facets but in their realization the three remaining facets share a vertex while in ours they do not. It is however conceivable that they are related by projective transformations.

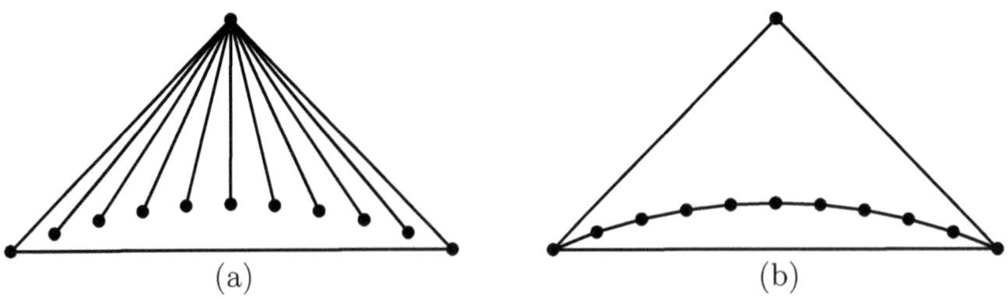

Fig. 6. (a) The non-crossing alternating trees appear as a face of a ppt-polytope. (b) Another associahedron in the same ppt-polytope.

The non-crossing alternating trees also appear directly as a face of the ppt-polytope. Consider the configuration of points shown in Figure 6, consisting of a convex chain $\{p_1, \ldots, p_n\}$ and an additional point p_0. The pseudotriangulations containing all the edges from p_0 to the other points (Figure 6a), correspond to the non-crossing alternating trees on $p_1, \ldots, p_n$, forming an associahedron of dimension $n-2$. Another associahedron (of dimension $n-3$) appears as a different face of this ppt-polytope: the face corresponding to the edges of the convex chain (Figure 6b). These two faces are disjoint.

Relations to optimization and the Monge matrices. Constraints of the form (11) are commonly found in project planning and critical path analysis. The variables v_i represent unknown start times of tasks, and the constraints (11) specify waiting conditions between different tasks. For example, the minimization of $v_n - v_1$ is a longest path problem in an acyclic network. The different vertices of X_g correspond to the different optimal solutions when we apply various linear objective functions.

Matrices $G = (g_{ij})$ with the property (12) are said to have the *Monge property*, if we set $g_{ii} = 0$ and $g_{ij} = \infty$ for $i > j$. The Monge property has received a great deal of attention in optimization because it arises often in applications and it characterizes special classes of optimization problems that can be solved efficiently, see [4] for a survey.

The dual linear program of the linear programming problem (11) (with a suitable objective function) is a minimum-cost flow problem on an acyclic network with edges ij for $1 \leq i < j \leq n$, and cost coefficients $-g_{ij}$. Network flow is actually one of the oldest areas in optimization in which the Monge property has been applied, and where it has been shown that optimal solutions can be obtained by a greedy algorithm in certain cases. The non-crossing alternating trees are just the different possible subgraphs of those edges which carry flow in an optimal solution.

6 Towards a General Framework, in Arbitrary Dimension

To make more explicit the analogy between our 1d and 2d constructions we'll rewrite the 1d construction back in terms of inequalities (6) instead of (11). As we mentioned at the beginning of Section 5.3, the way to do this is to substitute $g_{i,j} = f_{i,j}/(p_j - p_i)$ everywhere. In particular, the constraints (13) become

$$\frac{f_{ik}}{p_i - p_k} + \frac{f_{kl}}{p_k - p_l} + \frac{f_{il}}{p_l - p_i} > 0.$$

But $\omega_{ik} = \frac{1}{p_i - p_k}$, $\omega_{kl} = \frac{1}{p_k - p_l}$ and $\omega_{il} = \frac{1}{p_l - p_i}$ define a self-stress on any 1-dimensional point set $\{p_i, p_k, p_l\}$. Hence, inequalities (13) are the exact analog of inequalities (8) of the 2d case. The difference is that in 2d these inequalities are necessary and sufficient for a choice of parameters to be "valid", while in 1d we need the additional constraints (12), which do not follow from (13) as the following example shows:

$$g_{12} = g_{23} = g_{34} = 1, \qquad g_{13} = g_{24} = 2.2, \qquad g_{14} = 3.3.$$

Let's see what would be a generalization of the 1d and 2d cases to arbitrary dimension. For any finite point set $P = \{p_1, \ldots, p_n\}$ spanning $\mathbb{R}^d$ (in general position or not) there is a well-defined cone of expansive motions $\bar{X}_0$ of dimension $nd - \binom{d+1}{2}$, which is the positive region in the arrangement of hyperplanes defined by the constraints (4). Every choice of perturbation parameters $(f_{ij})_{i,j\in\{1,\ldots,n\}}$, via inequalities (6), produces a polyhedron of constrained expansions $\bar{X}_f$, whose faces are in correspondence with some graphs embedded on P. If the f_{ij}'s are sufficiently generic, this polyhedron is simple and:

- The face poset of $\bar{X}_f$ is isomorphic to a subposet of the graphs embedded on P, ordered by reverse inclusion.
- The maximal graphs in this subposet are some of the infinitesimally minimally rigid frameworks on P.
- Adjacent vertices on the polyhedron correspond to graphs differing only by one edge.

The main question is whether there is a clever choice of the f_{ij}'s or a sensible choice of a special family of minimally rigid frameworks to be the vertices of the polyhedron which admits a nice and simple geometric characterization, as non-crossing alternating trees in 1d or pointed pseudo-triangulations in 2d do. In particular, what should be the analog of pointed pseudo-triangulations on a point set in general position in 3d? One difficulty is that the combinatorics of minimally rigid graphs in dimension three is not fully understood. Perhaps a good case to start with would be points in convex and general position in dimension 3, for which the convex hull edges provide a very canonical minimally rigid framework to start with.

Another observation is that both in the 1d and 2d cases that we have treated here, the resulting polytope $\bar{X}_f$ has a unique maximal bounded face. This is related to the graph of flips between the special minimally rigid frameworks considered being connected and regular, which is an extra desirable property.

It would even be interesting to have a good generalization of the 2d case to point sets which are not in general position. If we use the 2d definition of f from equations (9) with many points on a common line, there are solutions in which all the inequalities are tight. In this way we get essentially the one-dimensional expansion cone of Proposition 2.7, when we project all vectors v_i on the direction of that line. (This is the same situation as when all relations (13) are satisfied as equations.) One way to get rid of this degeneracy is to "perturb the perturbations" by adding an infinitesimal component of the one-dimensional expansion parameters of Section 5.3: instead of f we use $f_{ij}^{(2)} + \varepsilon f_{ij}^{(1)}$, where $f^{(2)}$ is valid for two-dimensional point sets, $f_{ij}^{(1)}$ satisfy the restrictions required for the 1d perturbation parameters, and $\varepsilon > 0$ is sufficiently small. The situation would be like in Figure 6a when the points on the convex chain converge to being collinear. We have not further investigated this idea.

Another consideration which may help to treat degenerate point sets and arbitrary-dimensional ones is that the ideas in Section 5.1 generalize with not much difficulty under very weak general position assumptions. Let us first try to generalize Lemma 5.1. We need to decide what is the right analog of the statement in part (b). We propose the following. Remember that a *circuit* of a point set $P = \{p_1, \ldots, p_n\} \subset \mathbb{R}^d$ is a minimal affinely dependent subset. If P is in general position, its set of circuits $\mathcal{C}(P)$ is $\binom{P}{d+2}$. In general, any set of $d+2$ points spanning d dimensions contains a unique circuit, but this circuit can have less than $d+2$ elements. In any case, the complete graph on a circuit C has (up to a constant factor) a unique self stress $(\omega_{ij})_{i,j \in C}$ with no vanishing ω_{ij}, which is the one given by Lemma 2.5.

Let us consider the following linear maps:

$$\begin{array}{rccc} M\colon & \mathbb{R}^{dn} & \longrightarrow & \mathbb{R}^{\binom{n}{2}} \\ & (v_1, \ldots, v_n) & \longmapsto & \langle v_i - v_j, p_i - p_j \rangle, \\ \Delta\colon & \mathbb{R}^{\binom{n}{2}} & \longrightarrow & \mathbb{R}^{\mathcal{C}(P)} \\ & \{f_{ij}\}_{ij \in \binom{n}{2}} & \longmapsto & \Big\{ \sum\limits_{ij \in C} \omega_{ij} f_{ij} \Big\}_{C \in \mathcal{C}(P)}. \end{array}$$

M is just the rigidity map of the complete graph on P. The equivalence of (b) and (c) in Lemma 5.1 can be rephrased as

$$\ker \Delta = \operatorname{Im} M, \text{ for every planar point set in general position.}$$

But in fact the same is true in arbitrary dimension and under much weaker assumptions than general position:

Proposition 6.1. *For any point set $P \subset \mathbb{R}^d$, $\operatorname{Im} M \subseteq \ker \Delta$. If P has d affinely independent points whose spanned hyperplane contains no other point of P, then the reversed inclusion also holds.*

Proof. That $\operatorname{Im} M \subset \ker \Delta$ (i.e., $\Delta \circ M = 0$) is one direction of Lemma 2.4. To prove $\ker \Delta \subset \operatorname{Im} M$, let $f \in \mathbb{R}^{\binom{n}{2}}$ be such that $\sum_{ij \in C} \omega_{ij} f_{ij} = 0$ for every circuit C. We want to find a motion $a = (a_1, \ldots, a_n)$ for which $M(a) = f$. Without loss of generality assume that $p_1, \ldots, p_d$ are the claimed affinely independent points. Let us fix a motion $(a_1, \ldots, a_d)$ for the first d points satisfying the $\binom{d}{2}$ equations of the system $Ma = f$ concerning these points. This motion exists and is unique up to an arbitrary translational and rotational component, because the complete graph on any affinely independent point set is minimally infinitesimally rigid.

By assumption, $\{p_1, \ldots, p_d, p_i\}$ is an affine basis for every $i > d$, and hence the complete graph on it is again minimally infinitesimally rigid. Thus, there is a motion $(a'_1, \ldots, a'_d, a_i)$ satisfying the $\binom{d+1}{2}$ corresponding equations of the system $Ma = f$. Adding translations and rotations we can assume $a'_j = a_j$, $j = 1, \ldots, d$. So we have constructed a motion $(a_1, \ldots, a_n)$ which restricts to the one chosen for the d first points and which satisfies all the equations for pairs of points that include one of the first d.

It remains to show that the equations are also satisfied for the edges kl with $k, l > d$. We follow the same idea as in the proof of Lemma 5.1. Our assumption on the point set implies that the unique circuit C contained in $\{p_1, \ldots, p_d, p_k, p_l\}$ uses both p_k and p_l. To simplify notation, assume that this circuit is $\{p_1, \ldots, p_i, p_k, p_l\}$. By Lemma 2.4, there is a feasible motion $(a'_1, \ldots, a'_i, a'_k, a'_l)$. By translations and rotations we assume $a_j = a'_j$ for $j = 1, \ldots, i$. Observe now the value of $\langle v - u, p_{j_1} - p_{j_2} \rangle$, for any of the points in the circuit and for any vectors v and u, depends only on the projection of v and u to the affine subspace spanned by C. Since the complete graphs on $\{p_1, \ldots, p_i\}$, on $\{p_1, \ldots, p_i, p_k\}$ and on $\{p_1, \ldots, p_i, p_l\}$ are minimally infinitesimally rigid *when motions are restricted to that subspace*, we conclude that the projections of a_k and a_l to that affine subspace coincide with the projections of a'_k and a'_l. In particular, $\langle a_l - a_k, p_l - p_k \rangle = \langle a'_l - a'_k, p_l - p_k \rangle = f_{kl}$. □

Hence, Lemma 5.1 and its corollary, Lemma 5.2, hold in this generalized setting, with one equation per circuit instead of one equation per 4-tuple. The weakened general position assumption for d of the points holds for every planar point set, since, by Sylvester's theorem, any finite set of points in the plane, not all on a single line, has a line passing through only two of the points.

In dimension 3, however, the same is not true, and actually there are point sets for which $\operatorname{Im} M \neq \ker \Delta$. Consider the case of six points, three of them in one line and three in another, with the two lines being skew (not parallel and not crossing). These two sets of three points are the only two circuits in the point set. In particular, $\ker \Delta$ has at most codimension 2 in $\mathbb{R}^{15}$, i.e., it has dimension at least 13. On the other hand, $\operatorname{Im} M$ has at most the dimension of the reduced space of motions, $18 - 6 = 12$.

7 Final Comments

We describe some open questions and ideas for further research. The main questions related to this work are how to extend the constructions from dimensions 1 and 2 to 3 and higher, and how to treat subsets in special position in 2d. The expectation is that this would give a coherent definition for "pseudo-triangulations" in higher dimensions. Some ideas in this direction have been mentioned in Section 6.

Is our choice of f_{ij}'s in Section 3 essentially unique? The set of valid choices for a fixed point set is open in $\mathbb{R}^{\binom{n}{2}}$. But, what if we restrict our attention to choices for which, as in Theorem 3.9, each f_{ij} depends only on the points p_i and p_j, and not the rest of the configuration? Observe that if we want this, then Theorem 3.7 provides an infinite set of conditions on the infinite set of unknowns $\{f(p,q) : p,q \in \mathbb{R}^2\}$. It follows from Lemma 5.1 that adding to a valid choice $(f_{ij})_{i,j\in\{1,\dots,n\}}$ any vector $(\delta_{ij})_{i,j\in\{1,\dots,n\}}$ in the image of the rigidity map we still get a valid choice. And, of course, we can also scale any valid choice by a positive constant. This gives a half-space of valid choices of dimension $\binom{n}{2}+1$. Is this all of it?

It would also be interesting to see if there is a deeper reason behind Lemma 3.10. We have actually been able to extend the identity $\sum \omega_{ij} f_{ij} = 1$ to a more general class of planar graphs than just the complete graph on four vertices: to wheels (graphs of pyramids). A wheel is a cycle with an additional vertex attached to every vertex of the cycle. For a wheel embedded in the plane in general position, formulas that are quite similar to (3) in Lemma 2.6 define a self-stress ω_{ij} on its edges which make the identity $\sum \omega_{ij} f_{ij} = 1$ true. (Since the wheel is infinitesimally rigid and has $2n-2$ edges, the self-stress is unique up to a scalar factor.) We take this as a hint that the identity of Lemma 3.10 might be an instance of a more general phenomenon which we don't fully understand.

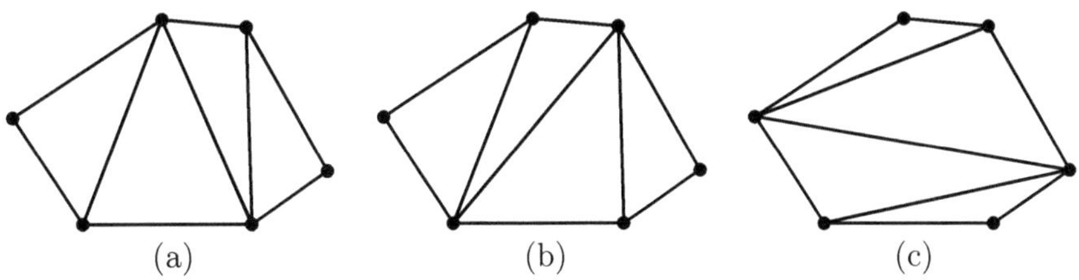

Fig. 7. (a) The Delaunay triangulation of a point set in convex position. (b-c) The triangulations minimizing and maximizing the objective function $(p_1,\dots,p_n)$ over the ppt-polytope, respectively. The triangulations in (b) and (c) are invariant under affine transformations, whereas (a) is not.

Another issue, with which we raised the introduction, is that having a representation of combinatorial structures as vertices of a polytope opens the way for selecting a particular structure, by optimizing some linear functional over the polytope. For example, the minimization of the objective function with coefficient vector $(|p_1|^2,\dots,|p_n|^2)$ over the secondary polytope gives the Delaunay triangulation. The opposite choice gives the furthest-site Delaunay triangulation. The most natural choice of objective function for the polytope of pointed pseudo-triangulations is $(p_1,\dots,p_n)$ or its opposite, i.e., minimize or maximize $\sum_i \langle p_i, v_i\rangle$ over all constrained expansions which are tight on convex hull edges. Even if, for points in convex position, our ppt-polytope is affinely isomorphic to the secondary polytope, this functional on the ppt-polytope does not, in general, give the Delaunay triangulation of those points, see Figure 7. In fact, the result on the ppt-polytope is invariant under affine transformations of the point set, while the Delaunay triangulation is not. The properties of the pointed pseudo-triangulations that are defined in this way await further studies.

Added in proof. The second author, together with David Orden, has extended the main construction of this paper to a simple polyhedron of dimension $3n - 3$ with a unique maximal bounded face whose vertices are *all* the pseudo-triangulations of the point set. Bounded edges correspond to either classical edge-flips or to the creation or destruction of pointedness at a vertex by the deletion or inclusion of a single edge. The face poset of this polyhedron is (essentially) the poset of all non-crossing graphs on the point set.

References

[1] Pankaj Agarwal, Julien Basch, Leonidas Guibas, John Hershberger, and Li Zhang, Deformable free space tilings for kinetic collision detection. in Bruce Randall Donald, Kevin Lynch and Daniela Rus (eds.), *Algorithmic and Computational Robotics: New Directions*, 4th Int. Workshop on Alg. Found. Robotics (WAFR), A K Peters, 2000, pp. 83–96.

[2] David Avis and Komei Fukuda, Reverse search for enumeration. *Discrete Appl. Math.* **65** (1996), 21–46.

[3] Hervé Brönnimann, Lutz Kettner, Michel Pocchiola, and Jack Snoeyink. Counting and enumerating pseudo-triangulations with the greedy flip algorithm. Manuscript, submitted for publication, 2001.

[4] Rainer E. Burkard, Bettina Klinz, and Rüdiger Rudolf, Perspectives of Monge properties in optimization. *Discrete Appl. Math.* **70** (1996), 95–161.

[5] Frederic Chapoton, Sergey Fomin, and Andrei Zelevinsky, Polytopal realizations of generalized associahedra, preprint `math.CO/0202004`, 25 pages, February 2002.

[6] Robert Connelly, Erik D. Demaine, and Günter Rote, Straightening polygonal arcs and convexifying polygonal cycles. *Discrete & Computational Geometry*, to appear. Preliminary version in *Proc. 41st Ann. Symp. on Found. of Computer Science (FOCS 2000)*, Redondo Beach, California; IEEE Computer Society Press, Washington, D.C., 2000, pp. 432–442; an extended version is available as technical report B 02-02, Freie Universität Berlin, Institut für Informatik, 2002.

[7] Robert Connelly and Walter Whiteley, Second-order rigidity and prestress tensegrity frameworks. *SIAM Journal on Discrete Mathematics* **9** (1996), 453–491.

[8] George B. Dantzig, Alan J. Hoffman, T. C. Hu, Triangulations (tilings) and certain block triangular matrices. *Math. Programming* **31** (1985), 1–14.

[9] Komei Fukuda, cdd code.
`http://www.ifor.math.ethz.ch/staff/fukuda/fukuda.html`

[10] Israel M. Gelfand, M. I. Graev, and Alexander Postnikov, Combinatorics of hypergeometric functions associated with positive roots. In: Arnold, V. I. et al. (ed.), *The Arnold-Gelfand Mathematical Seminars: Geometry and Singularity Theory*. Boston, Birkhäuser. pp. 205–221 (1997)

[11] Israel M. Gel′fand, Andrei V. Zelevinskiĭ, and M. M. Kapranov, Discriminants of polynomials in several variables and triangulations of Newton polyhedra. *Leningrad. Math. J.* **2** (1991), 449–505, translation from *Algebra Anal.* **2** (1990), No. 3, 1–62.

[12] Michael Goodrich and Roberto Tamassia, Dynamic ray shooting and shortest paths in planar subdivisions via balanced geodesic triangulations. *J. Algorithms* **23** (1997), 51–73.

[13] Jack Graver, Brigitte Servatius, and Hermann Servatius, *Combinatorial Rigidity*. AMS Graduate Studies in Mathematics vol. 2, 1993.

[14] David Kirkpatrick, Jack Snoeyink, and Bettina Speckmann, Kinetic collision detection for simple polygons. *International Journal of Computational Geometry and Applications* **12** (2002), 3–27. Extended abstract in *Proc. 16th Ann. Symposium on Computational Geometry*, pp. 322–330, 2000.

[15] G. Laman, On graphs and rigidity of plane skeletal structures. *J. Engineering Math.* **4** (1970), 331–340.

[16] Carl Lee, The associahedron and triangulations of the n-gon. *European J. Combinatorics* **10** (1989), 551–560.

[17] Jesús A. de Loera, Serkan Hoşten, Francisco Santos, and Bernd Sturmfels, The polytope of all triangulations of a point configuration. *Documenta Mathematica, J. DMV* **1** (1996), 103–119.

[18] Michel Pocchiola and Gert Vegter, Topologically sweeping visibility complexes via pseudo-triangulations. *Discr. & Comput. Geometry* **16** (1996), 419–453.

[19] Michel Pocchiola and Gert Vegter, The visibility complex. *Int. J. Comput. Geom. Appl.* **6** (1996), 279–308.

[20] Michel Pocchiola and Gert Vegter, Pseudo-triangulations: Theory and applications. In *Proc. 12th Ann. ACM Sympos. Comput. Geom.*, ACM Press, May 1996, pp. 291–300.

[21] Alexander Postnikov, Intransitive trees. *J. Combin. Theory*, Ser. A **79** (1997), 360–366.

[22] Alexander Postnikov and Richard P. Stanley, Deformations of Coxeter hyperplane arrangements. *J. Combin. Theory*, Ser. A **91** (2000), 544–597.

[23] B. Speckmann and C. D. Tóth, Vertex π-guards in simple polygons. In *Abstracts of the 18th European Workshop on Computational Geometry* (EuroCG), April 10–12, 2002, Warszawa, Poland, pp. 12–15. Full version submitted for publication.

[24] Richard Stanley, *Enumerative Combinatorics*. Vol. 2, Cambridge University Press, 1999.

[25] Ileana Streinu, A combinatorial approach to planar non-colliding robot arm motion planning, In *Proc. 41st Ann. Symp. on Found. of Computer Science (FOCS 2000)*, Redondo Beach, California, November 2000, pp. 443–453.

[26] Ileana Streinu, A combinatorial approach to planar non-colliding robot arm motion planning, preprint, 2002.

[27] Walter Whiteley, Rigidity and Scene Analysis, in: Jacob E. Goodman and Joseph O'Rourke (eds.) *Handbook of Discrete and Computational Geometry*, CRC Press, 1997, pp. 893–916.

[28] Günter Ziegler, *Lectures on Polytopes* (2nd ed.), Springer-Verlag, 1999.

About the Authors

Günter Rote is at the Institut für Informatik, Freie Universität Berlin, Takustraße 9, D-14195 Berlin, Germany, *rote@inf.fu-berlin.de*.

Francisco Santos is at the Departamento de Matemáticas, Estadística y Computación, Universidad de Cantabria, E-39005 Santander, Spain, *santos@matesco.unican.es*.

Ileana Streinu is at the Department of Computer Science, Smith College, Northampton, MA 01063, USA, *streinu@cs.smith.edu*.

Acknowledgments

We thank Ciprian Borcea for pointing out some relevant references in the mathematical literature. This work started at the Workshop on Pseudo-triangulations held at the McGill University Bellairs Institute in Barbados, January 2001, partially funded by NSF grant CCR-0104370, and continued during a visit of the third author to the Graduiertenkolleg *Combinatorics, Probability and Computing* at Freie Universität Berlin, supported by Deutsche Forschungsgemeinschaft, grant GRK 588/1. Work by Ileana Streinu was partially supported by NSF RUI Grants CCR-9731804 and CCR-0105507. Work by Francisco Santos was partially supported by grants PB97–0358 and BMF2001-1153 of the Spanish Dirección General de Enseñanza Superior.

Some Recent Quantitative and Algorithmic Results in Real Algebraic Geometry

Marie-Françoise Roy

Abstract

This paper offers a short survey of two topics. In the first section we describe a new bound on the Betti numbers of semi-algebraic sets whose proof rely on the Thom-Milnor bound for algebraic sets and the Mayer-Vietoris long exact sequence. The second section describes the best complexity results and practical implementations currently available for finding real solutions of systems of polynomial equations and inequalities. In both sections, the consideration of critical points will play a key role.

1 Bounding Betti numbers of semi-algebraic sets

According to Petrowsky/Oleinik/Thom/Milnor [18–20, 30], the sum of all Betti numbers of an algebraic set defined by polynomials of degree d in $\mathbb{R}^k$ does not exceed

$$d(2d-1)^{k-1}. \tag{1}$$

The exponential character of Thom-Milnor's bound (1) is unavoidable since the solution sets to the system of equations,

$$(X_1-1)(X_1-2)\cdots(X_1-d)=\cdots=(X_k-1)(X_k-2)\cdots(X_k-d)=0 \tag{2}$$

consists of d^k points.

The proof of Thom-Milnor's bound (1) relies on the basic results of Morse theory. Consider a smooth bounded hypersurface M. For a well chosen direction $(u_1,\ldots,u_k)\in\mathbb{R}^k$, consider the family of sweeping hyperplanes H_t defined by

$$u_1x_1+\ldots+u_kx_k-t=0,$$

and consider the half space L_t defined by $u_1x_1+\ldots+u_kx_k-t<0$. Changes in the topology of $M\cap L_t$ occur only for values of t where the hyperplane H_t is tangent to M at a critical point. At each critical point a handle is attached to $M\cap L_t$. The dimension of the handle depends on the signature of the Hessian quadratic form [17]. So the sum of the Betti numbers increases by 1 at most when a critical point is crossed by the sweeping hyperplane. The conditions for the sweeping hyperplane to be well chosen is that the number of critical points is finite, and the Hessian quadratic form at the each critical point is nondegenerate. The set of bad hyperplanes is of smaller dimension than the set of all hyperplanes. When the smooth bounded hypersurface is algebraic and defined by an equation of degree d, $H = 0$, a good sweeping hyperplane being given, there are at most $d(d-1)^{k-1}$ isolated critical points satisfying the equations

$$H = \frac{\partial H}{\partial X_1} = \ldots = \frac{\partial H}{\partial X_{k-1}} = 0,$$

according to Bezout theorem. Thus the sum of the Betti numbers for a bounded smooth algebraic hypersurface is at most $d(d-1)^{k-1}$. The Thom-Milnor bound for general algebraic sets (1) uses as a key ingredient this special case [7].

A transfer argument proves the same bound for a general real closed field R. Semi-algebraic geometry can be studied over a general real closed field , once adequate definitions for connected components, homology groups, etc, are given [7].

Betti numbers of semi-algebraic sets (i. e. sets defined by a boolean combination of polynomial inequalities) have also been considered. In particular, the number of nonempty sign conditions defined by a family of polynomials has been studied [1] as well as the number of connected components of all these nonempty sign conditions [21]. It is proved in [3] that the number of connected components of the nonempty sign conditions defined by s polynomials in k variables of degree at most d on an algebraic set of real dimension k' defined by an equation of degree at most d is

$$\sum_{j \le k'} \binom{O(s)}{j} O(d)^k. \tag{3}$$

This bound is separated into a combinatorial part (the dependence on s) which depends only on the real dimension k' of the algebraic set and an algebraic part (the dependence on d) which depends on the dimension k of the ambient space.

Let us give a sketch of the proof of this result. Let $\mathcal{P} = \{P_1, \ldots, P_s\} \subset \mathrm{R}[X_1, \ldots, X_k]$ be a set of s polynomials with degrees bounded by d, and $\mathrm{Z}(\mathcal{Q}, \mathrm{R}^k)$ of real dimension k', an algebraic set defined by a set $\mathcal{Q}$ of polynomial equations of degree at most d.

The main idea is to use the following proposition which says that in order to find points in connected components of a given semi-algebraic set it is enough to find points in connected components of some algebraic sets. We need the following definitions and results: the field of algebraic Puiseux series in δ with coefficients in R [31], $\mathrm{R}\langle\delta\rangle$, is a real closed field [7]. In this ordered field, the element δ is infinitesimal: positive and smaller than any positive element of R.

Proposition 1. *Let $C \subset \mathrm{R}^k$ be a non-empty semi-algebraically connected component of a basic closed semi-algebraic set defined by*

$$\sum_{Q \in \mathcal{Q}} Q^2 = P_1 = \cdots = P_\ell = 0, P_{\ell+1} > 0, \cdots, P_s > 0.$$

There exists an algebraic set W defined by equations

$$\sum_{Q \in \mathcal{Q}} Q^2 = P_1 = \cdots = P_\ell = 0, P_{i_1} = \delta \cdots P_{i_m} = \delta,$$

(with $\{i_1, \ldots, i_m\} \subset \{\ell+1, \ldots, s\}$) such that a semi-algebraically connected component C' of W is contained in the extension of C to $\mathrm{R}\langle\delta\rangle$ (i.e. the subset of $\mathrm{R}\langle\delta\rangle^k$ defined by the same polynomial inequalities than those defining C).

When the polynomials of $\mathcal{P}$ are in general position, there are no nonemtpy intersections of more than k' algebraic sets defined by polynomials of $\mathcal{P}$ over the zeroes of $\mathcal{Q}$, and the bound

$$\sum_{j \le k'} \binom{s}{j} 3^j d(2d-1)^{k-1}. \tag{4}$$

is obtained, using Proposition 1 and applying the Thom-Milnor bound (1) for algebraic sets. When the polynomials of $\mathcal{P}$ are not in general position, there may be too many nonempty algebraic sets to consider. However, it is intuitively clear that general position corresponds to the maximum number nonempty connected components, and a precise statement confirming this intuition can be proved. So, the key of the proof of (3) is to use a technique of perturbation to put the family of polynomials in general position. Perturbations by infinitesimals are used, and the number of polynomials is increased. We have now $\sum_{1 \le j \le k'} \binom{O(s)}{j} 3^j$ algebraic cases to consider, and apply the Thom-Milnor bound for algebraic sets in each case.

The bound (3) has been generalized and improved significantly recently as follows (see [5]).

If σ is a sign condition realizable over $\mathrm{Z}(\mathcal{Q}, \mathrm{R}^k)$, let $b_i(\sigma)$ denote the i-th Betti number of the realization of σ

$$R(\sigma) = \{x \in \mathrm{R}^k \mid \wedge_{Q \in \mathcal{Q}} Q(x) = 0 \wedge_{P \in \mathcal{P}} \mathrm{sign}(P(x)) = \sigma(P)\}$$

and

$$b_i(\mathcal{P}, \mathcal{Q}) = \sum_{\mathcal{R}(\sigma) \ne \emptyset} b_i(\sigma).$$

We write $b_i(d, k, k', s)$ for the maximum of $b_i(\mathcal{P}, \mathcal{Q})$ over algebraic sets, $\mathrm{Z}(\mathcal{Q}, \mathrm{R}^k)$ of real dimension k', defined by polynomial equations of degree at most d and over all $\mathcal{P}$ consisting of s polynomials in k variables, each of degree at most d.

Theorem 1.

$$b_i(d,k,k',s) \le \sum_{1\le j\le k'-i} \binom{s}{j} 4^j d(2d-1)^{k-1}. \tag{5}$$

As in (3), this bound is separated into a combinatorial part (the dependence on s) which depends only on the real dimension k' of $\mathrm{Z}(\mathcal{Q},\mathrm{R}^k)$ and an algebraic part (the dependence on d) which depends on the dimension k of the ambient space.

Note that, $b_0(\mathcal{P},\mathcal{Q})$ is the number of semi-algebraically connected components of basic semi-algebraic sets defined by $\mathcal{P}$ over $\mathrm{Z}(\mathcal{Q},\mathrm{R}^k)$, so that Theorem 1 when $i=0$ is nothing but a more precise version of (3).

Proposition 1 gives only information about the number of connected components, so it cannot be used when dealing with arbitrary Betti numbers. The technique of the proof of Theorem 1 is quite different from the proof of (3). The dimension argument using deformations and general position formerly used for the number of connected components is replaced by another dimension argument using algebraic topology techniques that we describe now.

The main ingredient of the proof of Theorem 1 is the following proposition. According to this proposition, the Betti numbers of a long intersection of sets is bounded in terms of the Betti numbers of short unions of the same sets. Short means here that only unions of less than $k'-i$ sets have to be considered for the i-th Betti number.

Proposition 2. *Let $S_1,\dots,S_s \subset \mathrm{R}^k$ be closed semi-algebraic sets, contained in a real algebraic set $\mathrm{Z}(\mathcal{Q},\mathrm{R}^k)$ of dimension k', and S their intersection.*

For $J \subset \{1,\dots,s\}$, let $T_J = \cup_{j\in J} S_j$, with the convention $T_\emptyset = \mathrm{Z}(\mathcal{Q},\mathrm{R}^k)$. Then, for $0\le i\le k'$

$$b_i(S) \le b_{k'}(T_\emptyset) + \sum_{1\le j\le k'-i} \; \sum_{J\subset\{1,\dots,s\},\#(J)=j} (b_{i+j-1}(T_J) + b_{k'}(T_\emptyset)).$$

The key tool for the proof of the proposition are the Mayer-Vietoris inequalities. They are immediate consequences of the Mayer-Vietoris long exact sequence which is a basic and fundamental tool in algebraic topology .

Proposition 3. *Let S_1, S_2 be two closed and bounded semi-algebraic sets with non-empty intersection. Then,*

$$b_i(S_1) + b_i(S_2) \le b_i(S_1\cup S_2) + b_{i-1}(S_1\cap S_2), \tag{6}$$

$$b_i(S_1\cup S_2) \le b_i(S_1) + b_i(S_2) + b_i(S_1\cap S_2), \tag{7}$$

Once estimates on long intersections are replaced by estimates on short unions, the following Lemmas are proved, also using Mayer-Vietoris inequalities. The first one relates short unions to short intersections.

Lemma 1. *Let $S_1, \ldots, S_j$, be closed semi-algebraic sets and T be their union. Then,*

$$b_i(T) \le \sum_{1 \le \ell \le j} \sum_{L \subset \{1,\ldots j\}, \#(L)=\ell} b_{i-\ell+1}(S_L),$$

where $S_L = \cap_{j \in L} S_j$.

Let

$$W_0 = \mathcal{R}((\bigvee_{1 \le i \le j} P_i^2(P_i^2 - \delta^2) = 0), \mathrm{Ext}(Z_r, \mathrm{R}\langle\delta\rangle)),$$

$$W_1 = \mathcal{R}((\bigvee_{1 \le i \le j} P_i^2(P_i^2 - \delta^2) \ge 0), \mathrm{Ext}(Z_r, \mathrm{R}\langle\delta\rangle)),$$

.

Lemma 2.

$$b_i(W_0) \le (4^j - 1)d(2d-1)^{k-1}.$$

Lemma 3.

$$b_i(W_1) \le (4^j - 1)d(2d-1)^{k-1} + b_i(\mathrm{Z}(\mathcal{Q}, \mathrm{R}\langle\delta\rangle^k)).$$

The proof of Theorem 1 follows easily from Proposition 2 and these Lemmas.

Similar techniques make it possible to bound also the Betti numbers for closed semi-algebraic sets. Let $\mathcal{P} \subset \mathrm{R}[X_1, \ldots, X_k]$ be finite. A $\mathcal{P}$-closed formula is a quantifier free $\mathcal{P}$-formula without negation, more precisely $\mathcal{P}$-closed formulae are constructed as follows

atoms $P = 0$, $P \ge 0$, $P \le 0$, $P \in \mathcal{P}$ are $\mathcal{P}$-closed formulae,

if Φ_1 and Φ_2 are $\mathcal{P}$-closed formulae, $\Phi_1 \wedge \Phi_2$ and $\Phi_1 \vee \Phi_2$ are $\mathcal{P}$-closed formulae.

The set of points satisfying a $\mathcal{P}$-closed formula is a closed semi-algebraic set.

Let $\mathcal{Q} \subset \mathrm{R}[X_1, \ldots, X_k]$ be finite. We consider a semi-algebraic set $S \subset \mathrm{R}^k$ defined as the intersection of $\mathrm{Z}(\mathcal{Q}, \mathrm{R}^k)$, with the realization of a $\mathcal{P}$-closed formula.

Let $\mathcal{P} = \{P_1, \ldots, P_s\} \subset \mathrm{R}[X_1, \ldots, X_k]$ be a set of s polynomials with degrees bounded by d, and $\mathrm{Z}(\mathcal{Q}, \mathrm{R}^k)$ of dimension k', an algebraic set defined by a set $\mathcal{Q}$ of polynomial equations of degree at most d. We write $\bar{b}(d, k, k', s)$ for the maximum of $b(S)$, the sum of the Betti numbers, over semi-algebraic sets $S \subset \mathrm{R}^k$ defined as the intersection of $\mathrm{Z}(\mathcal{Q}, \mathrm{R}^k)$ of dimension k', with the realization of a $\mathcal{P}$-closed formula.

Theorem 2.

$$\bar{b}(d, k, k', s) \le \sum_{0 \le i \le k'} \sum_{0 \le j \le k'-i} \binom{s}{j} 6^j d(2d-1)^{k-1}. \tag{8}$$

2 Deciding the emptyness of systems of equalities and inequalities

In all this section, polynomials are with coefficients in an ordered field K embedded in a real closed field R annd C = R[i] is the algebraic closure of R.

We consider the following problem: compute a point in every connected components of nonempty sign conditions of a set of polynomials $\mathcal{P}$ on the zero set of $\mathcal{Q}$.

2.1 Quasi optimal complexity results

A famous method for dealing with this question is Cylindrical Algebraic Decomposition [10, 11]. Since the Cylindrical Algebraic Decomposition is producing polynomials of degree doubly exponential in the number of variables [15], its practical use is limited to a very small number of variables.

On the other hand, the results of the previous section show that the size of the output is at least single exponential in the number of variables, precisely

$$\sum_{1\leq j\leq k'} \binom{s}{j} 4^j d(2d-1)^{k-1}.$$

according to Theorem 1. So there is hope for designing an algorithm with single exponential complexity, while single exponential complexity is unavoidable because of (2).

Starting from [14], many algorithms based on the critical point method, i.e. finding sample points of non-empty sign conditions as critical points of a well chosen function, have been described for this problem. Their complexity is single exponential in the number of variables (see for example [4, 8, 9, 16, 22, 23]).

In the most recent results , the algebraic part (dependence on the degree of polynomials) and the combinatorial part (dependence on the number of polynomials) are separated. Moreover the degrees of the polynomials in the output are independent of the number of polynomials.

For example, finding a point in every connected component of sign conditions defined by a list of s polynomials in k variables of degree d on an algebraic set of dimension k' can be done in time

$$\sum_{j\leq k'} \binom{s}{j} 4^j s d^{O(k)} \tag{9}$$

according to [4, 6] and the description of the points involves polynomials of degree $O(d)^k$.

The main ideas of this result are the following. It is possible to consider only equalities, using Proposition 1. Taking sums of squares does not increase too much the degree and makes the algebraic part of the problem independent of the number of polynomials. In order to avoid considering too many algebraic cases, infinitesimal deformations allow to go to general position. Techniques of approximating algebraic sets [28] are used in order to make deformations with a fixed number of infinitesimals without changing the dimension of the algebraic being deformed. The idea is to find $\binom{k'}{k}$ approximating algebraic sets of dimension k' defined by $k-k'$ equations covering $\mathrm{Z}(\mathcal{Q},\mathrm{R}^k)$. These algebraic sets are approximating in the sense that every point of $\mathrm{Z}(\mathcal{Q},\mathrm{R}^k)$ is infinitesimally close to one of these approximating algebraic sets. Then the algorithm performs the construction on these algebraic sets. It is possible to come back to the original algebraic set by an algebraic limit process where infinitesimals tend to 0. Finally, subroutines for algebraic sets give a point in every connected component by computing critical points and using polynomial system solving techniques [12,13,24,25].

2.2 Efficient algorithms and implementation

In spite of these complexity results, taking a small system of polynomial equalities and inequalities (written on a half page) and deciding if there exist solutions remains a computational challenge.

A software project using the critical point method is under development [26]. The first step of the project was to design efficient methods dealing with polynomial systems of equalities having a finite number of real solutions [12,13,24,25]. The second step consists in designing efficient methods finding a point in every connected component of an algebraic set [2,29]. The third step devoted to the case of general systems of inequalities is under study and has not been implemented yet.

Systems of equalities

We explain first why the algorithm sketched in the complexity section, though quasi optimal from a complexity point of view, does not provide an efficient method to find a point in every connected component of an algebraic set. One reason is that when we are using sums of squares to obtain one single equation we double the degree and introduce artificial singularities. Another reason is that several infinitesimals are used. (See [29]).

The efficient method we present now comes from [2,29]. Its avoids completely the use of infinitesimals.

The basic idea In order to motivate the general construction that follows, let us consider the case of a smooth algebraic hypersurface defined by a single equation

$$Z = \{X \in \mathrm{C}^k \mid H(X_1,\ldots,X_n) = 0\},$$

i.e. such that

$$\mathrm{Grad}_M(H) = \left[\frac{\partial H}{\partial X_1}(M),\ldots,\frac{\partial H}{\partial X_k}(M)\right]$$

never vanishes on the zeroes of H in C^k. A simple way to determine a point in every connected component of $Z \cap \mathrm{R}^k$ consists in computing a finite number of points of Z containing the local minima on $Z \cap \mathrm{R}^k$ of the distance function to a well chosen point. These points will be critical points of the distance function.

Suppose we are given a point $A \in \mathrm{R}^k$, which does not belong to the algebraic set. Let us consider the set of spheres of center A touching the corresponding algebraic set, i.e. those who have a tangent hyperplane which is also tangent to the algebraic hypersurface and the set of contact points. If this set is finite , it contains the local minima on $Z \cap \mathrm{R}^k$ of the distance function to A and is contained in the set of points of R^k which are solutions of the system:

$$\mathcal{C} = \{x \in \mathrm{C}^k \mid H(M) = 0, \mathrm{Grad}_M(H) // \overrightarrow{AM}\}.$$

It can happen, for example if $Z \cap \mathrm{R}^k$ contains a sphere of center A , that $\mathcal{C}$ has an infinite number of solutions. However, when A is well chosen, $\mathcal{C}$ has only a finite number of complex solutions [2, 29].

Suppose now that Z has a nonempty singular locus denoted by $\mathrm{Sing}(Z)$: these are the points M of C^k where H and $\mathrm{Grad}_M(H)$ are simultaneously zero. It is clear that every point in $M \in \mathrm{Sing}(Z)$ is a solution of $\mathcal{C}$. If $\mathrm{Sing}(Z)$ is finite, the system $\mathcal{C}$ still has, when A is well chosen, a finite number of complex solutions. On the other hand, if $\mathrm{Sing}(Z)$ is infinite, the system $\mathcal{C}$ has an infinite number of complex solutions for every choice of A. It is impossible to use the efficient methods of polynomial system solving which work only for a finite number of points. However, using that, under convenient hypotheses, the set of singular points of Z is of dimension smaller than the dimension of Z, it willl be possible to lower gradually the dimension. It is important to remark that even if we start from one single equation, we have to deal with several equations in the second step of the process.

Note that the notion of dimension we consider for this induction is not the real dimension of the set, but its dimension as an algebraic set. The two notions differ as it is posible to have an hypersurface Z of algebraic dimension $k-1$ whose real points are of dimension $< k-1$, completely included in the singular local of Z. Take for example the curve defined by $X^2 + Y^2$, its dimension is 1, the real dimension of its real zero set is 0, and the singular locus coincides with the set of real zeroes.

Singular locus Let Z be an algebraic subset of C^k of dimension d defined by equations $\{P_1, \ldots, P_s\}$, and $\{Q_1, \ldots, Q_r\}$ be generators of the radical ideal $I(Z)$ of polynomials vanishing on Z. A singular point of Z is a point $M \in Z$ such that the rank of the jacobian matrix associated to $\{Q_1, \ldots, Q_r\}$ is not greater than $k-d$ in M. We denote

$$\mathrm{Sing}(Z) = \{M \in Z \mid \mathrm{Rank}(\overrightarrow{\mathrm{Grad}}_M\ (Q_1), \ldots, \overrightarrow{\mathrm{Grad}}_M\ (Q_r)) < k-d\}.$$

The consideration of generators of the radical ideal $I(Z)$ is essential. Consider for example the polynomials $X^2 = 0$ and $X = 0$. They define the same algebraic sset $Z = \{0\} \subset \mathrm{C}$. However the derivative of the polynomial defining Z, is zero everywhere on Z in the first case and equal to 1 in the second case. The singular locus of Z, defined using the second equation, is empty and its dimension is smaller that the dimension of Z. This elementary example illustrates that in order to describe correctly the singular local, it is not possible to consider any set $\{P_1, \ldots, P_s\}$ defining Z, it is important to use polynomials $\{Q_1, \ldots, Q_r\}$ defining Z and generating the radical ideal $I(Z)$.

The general method The algorithm is based on the following result.

Theorem 3. *Let Z be an equidimensional algebraic set of dimension d and $S = \{Q_1, \ldots, Q_r\}$ a set of polynomials of $\mathrm{K}[X_1, \ldots, X_k]$ such that $I(Z) = \langle Q_1, \ldots, Q_r \rangle$.*

Given $A \in \mathrm{C}^k$, consider the algebraic set:

$$\mathcal{C}(Z, A) = \{M \in Z \mid \mathrm{Rank}(\overrightarrow{\mathrm{Grad}_M}\,(Q_1), \ldots, \overrightarrow{\mathrm{Grad}_M}\,(Q_r), \overrightarrow{AM}) \leq k - d\}.$$

If D is a big enough natural number, there exists at least a point A in $\{1 \ldots D\}^k$ such that :

- *$\mathcal{C}(Z, A)$ meets every connected component of $Z \bigcap \mathrm{R}^k$,*
- *$\mathcal{C}(Z, A) = \mathrm{Sing}(Z) \bigcup Z_0$,*

where Z_0 is a finite set of points of C^k. Moreover, $\dim(\mathcal{C}(Z, A)) < \dim(Z)$.

Note that we take as equations generators of $I(Z)$. Without this, we would not have the condition $\dim(\mathcal{C}(Z, A)) < \dim(Z)$, which will be essential for the termination of the algorithm. Not also that without the equidimensionnality assumption, we might miss connected components of lower dimension of the algebraic set.

The principle of the algorithm is simple: we decompose the algebraic set in equidimensional algebraic subsets and compute equations of the ideal of polynomials vanishing on these equidimensional algebraic subsets. For each of these sets Z_i, we chose A and consider the algebraic subset $\mathcal{C}(Z_i, A)$. We go on with this process until we get polynomial systems of dimension 0.

For $B \in \mathrm{C}^k$, $\mathcal{Q} = \{Q_1, \ldots Q_s\} \subset K[X_1, \ldots, X_n]^s$, and $d \in \mathrm{N}$, $0 \leq d < k$, define $\Delta_{B,d}(\mathcal{Q})$, to be the set of all minors of order $(k - d + 1, k - d + 1)$ of the matrix

$$\left[\left[\frac{\partial Q_i}{\partial X_j} \right]_{(i=1\ldots n, j=1\ldots s)} \middle| \overrightarrow{BM} \right]$$

Before writing down the algorithm computing a list of systems with a finite number of solutions whose zero set intersects every connected component of $\mathrm{Z}(S, \mathrm{R}^k)$ let us list the required subroutines:

- **EquiDimDec** : takes as input a system of polynomial equations S and outputs a finite number of polynomial systems $\{S_1, \ldots, S_n\}$ such that $\mathrm{Z}(S, \mathrm{C}^k) = \cup_{i=1,\ldots,n} \mathrm{Z}(S_i, \mathrm{C}^k)$, the ideal generated by S_i is radical and $\mathrm{Z}(S_i, \mathrm{C}^k)$ is equidimensional.

- **Dim** : takes as input a system of polynomial equations S and outputs the dimension of $\mathrm{Z}(S, \mathrm{C}^k)$.

- **Minors** : takes as input a finite system of polynomials $\mathcal{P}$, a natural number d and a point $A \in \mathrm{K}^k$ and returns $\Delta_{A,d}(\mathcal{Q})\rangle)$.

Algorithm 1. • **Input :** *A polynomial system S in $\mathrm{K}[X_1, \ldots, X_n]$.*

- **Output:** *A finite set of polynomial systems $\{S_1, \ldots, S_m\}$ with a finite number of complex solutions defining at least a point every connected component of $\mathrm{Z}(S, \mathrm{R}^k)$.*

- **Procedure :**
  - $T :=$ EquiDimDec(S), result $:= \emptyset$,
  - *Choose* $A \notin \mathrm{Z}(S, \mathrm{K}^k)$.
  - *while* $T \neq \emptyset$
    * *Remove U from T, define* $d = \dim(U)$,
    * *if* $d = 0$ *then* result $:=$ result $\cup\, S$,
    * *else*
      · *(*)* $V = \text{Minors}(U, d, A) \bigcup U$, $u = \text{Dim}(V)$
      · *if $u = d$ chose another point A and return to (*).*
      · $\mathrm{T} := \mathrm{T} \cup \text{EquiDimDec}(V)$
  - *return* result.

A drawback in the preceding algorithm is the number and size of the minors to compute. So the method implemented is an improvement of the preceding algorithm. Triangular systems of polynomials are used to identify one single minor to be computed [2, 29].

A good practical behaviour has been observed for this improved version [2, 29]. As expected the size of the output is quite reasonable compared to the output of Cylindrical Algebraic Decomposition. This algorithm has been succesfully applied to polynomial systems for which the computations of Cylindrical Algebraic Decomposition does not terminate [2, 29].

The complexity of the algorithm is not $d^{O(k)}$, contrarily to the algorithm sketched in the section on complexity. There are several reasons for that: the number of points A to consider, the complexity of the various subroutines, the induction process. However in all examples having been computed, the first choice of A was good.

Systems of inequalities

This paragraph describes work in progress, not implemented yet.

Using Proposition 1, in order to find a sample point in every connected components of a nonempty sign conditions of a set of polynomials $\mathcal{P}$ on the zero set of $\mathcal{Q}$, it is enough to consider all the subsets of $I \subset \mathcal{P}^\star = \{P, P-\delta, P+\delta\}$ such that the corresponding set of zeroes of $I \cup \{Q\}$ is nonempty, and to take a point at least in every connected component defined by these equations. It is thus possible to use several times the algorithm of the preceding section for systems of equalities.

There is a significant difference though since the coefficients now belong to $\mathrm{K}[\delta]$. A key point of the implementation will be an efficient arithmetic for infinitesimals. Note that compared to the theoretical algorithm with good theoretical complexity sketched above, there is only one infinitesimal involved.

Finally after the polynomial sytesm solving phase, it is necessary to study roots, in the field of Puiseux series $\mathrm{R}\langle\delta\rangle$, of univariate polynomials with coefficiens in $\mathrm{K}[\delta]$ and to determine the sign of other polynomials at these points. The real root counting and sign determination process based on subresultants [27] being quite slow, a method using Newton-Puiseux method and a variant of Uspensky's algorithm, which is based on Descartes's rule of signs, could be used.

The preceding algorithm does not have the complexity

$$\sum_{j\le k'} 4^j \binom{s}{j} s d^{O(k)}$$

of the algorithm sketched in the first part of the section. Besides the fact that the basic algebraic subtroutine ha s not complexity $d^{O(k)}$ as we have already seen, it may happen that $\mathcal{P}$ is not in general position on the zeroes of $\mathcal{Q}$ and that more than $\sum_{j\le k'} 4^j\binom{s}{j}$ subsets of $\mathcal{P}^\star$ have to be considered. However the description of the points involves polynomials of degree $d^{O(k)}$ so that the size of the output should be satisfactory.

Let us finally point out that, in the algorithms of real algebraic geometry, it remains a challenge to design algorithms having simultaneously a good theoretical (quasi optimal) complexity and a good practical behaviour in terms of efficient implementation.

References

[1] N. Alon *The number of polytopes, configurations and real matroids.* Mathematika, 33, 62-71 (1986).

[2] P. Aubry, F. Rouillier, M. Safey El Din *Real solving for real algebraic varieties defined by several equations.* (2000), submitted.

[3] S. Basu, R. Pollack, M.-F. Roy, *On the number of cells defined by a family of polynomials on a variety.* Mathematika, 43, 120-126 (1996).

[4] S. Basu, R. Pollack, M.-F. Roy, *On the Combinatorial and Algebraic Complexity of Quantifier Elimination.* Journal of the ACM, 43, 1002–1045 (1996).

[5] S. Basu, R. Pollack, M.-F. Roy, *On Bounding the Betti Numbers of the Connected Components Defined by a Set of Polynomials and of Closed Semi-algebraic Sets.* In preparation.

[6] S. Basu, R. Pollack, M.-F. Roy, *Algorithms in real algebraic geometry.* Book in preparation.

[7] J. Bochnak, M. Coste, M.-F. Roy, *Real algebraic geometry* (second edition, in english). Springer-Verlag.

[8] J. Canny, *The Complexity of Robot Motion Planning.* MIT Press (1987).

[9] J. Canny, *Some Algebraic and Geometric Computations in PSPACE.* Proc. Twentieth ACM Symp. on Theory of Computing, 460-467 (1988).

[10] G. E. Collins, *Quantifier elimination for real closed fields by cylindrical algebraic decomposition.* Lect. Notes in Comp. Sci., 33, 515-532. Springer-Verlag (1975).

[11] M. Coste,*Effective semi-algebraic geometry.* Lect. Notes in Comp. Sci., 391, 1-27. Springer-Verlag (1989).

[12] L. Gonzalez-Vega, F. Rouillier, M.-F. Roy, *Symbolic recipes for polynomial system solving.* In: Some tapas for computer algebra, A. Cohen ed. Springer, 34-65 (1999).

[13] L. Gonzalez-Vega, F. Rouillier, M.-F. Roy, G. Trujillo, *Symbolic recipes for real solutions.* In: Some tapas for computer algebra, A. Cohen ed. Springer, 121-167 (1999).

[14] D. Grigor'ev, N. Vorobjov, *Solving Systems of Polynomial Inequalities in Subexponential Time.* J. Symbolic Comput., 5, 37–64, (1988).

[15] J. Heintz, T. Recio, M.-F. Roy, *Algorithms in real algebraic geometry and applications to computational geometry.* Discrete and Computational Geometry: Papers from the DIMACS Special Year. DIMACS series in Discrete Mathematics and Theoretical Computer Science AMS-ACM, 6, 137-163 (1991).

[16] J. Heintz, M.-F. Roy, P. Solerno, *On the Complexity of Semi-Algebraic Sets.* Proc. IFIP 89, San Francisco, 293-298. North-Holland (1989).

[17] J. Milnor, *Morse Theory,* Annals of Mathematical Studies, Princeton University Press (1963).

[18] J. Milnor, *On the Betti numbers of real algebraic varieties.* Proc. AMS, 15, 275-280 (1964).

[19] O. A. Oleinick, *Estimates on the Betti numbers of real algebraic hypersurfaces.* Mat. Sb., 28 (70), 635-640 (1951).

[20] I. Petrowsky, *On the topology of real plane algebraic curves.* An. Math., 39 (1), 187-209, (1938).

[21] R. Pollack, M.-F. Roy, *On the number of cells defined by a set of polynomials.* C. R. Acad. Sci. Paris, 316, 573- 577 (1993).

[22] J. RENEGAR, *On the computational complexity and geometry of the first order theory of the reals.* J. of Symbolic Comput.,13 (3), 255-352 (1992).

[23] J. RENEGAR, *Recent progress on the complexity of the decision problem for the reals.* Discrete and Computational Geometry: Papers from the DIMACS Special Year. DIMACS series in Discrete Mathematics and Theoretical Computer Science, AMS-ACM., 6, 287-308 (1991).

[24] F. ROUILLIER, *Algorithmes efficaces pour l'étude des zéros réels des systèmes polynomiaux.* Thèse, Université de Rennes I (1996).

[25] F. ROUILLIER, *Solving zero-dimensional systems through the rational univariate representation.* Journal of Applicable Algebra in Engineering, Communication and Computing,9, 433–461 (1999).

[26] F. ROUILLIER, *Real Solving*, http://www.loria.fr/ rouillie .

[27] M.-F. ROY *Basic algorithms in real algebraic geometry: from Sturm theorem to the existential theory of reals,*, Lectures on Real Geometry in memoriam of Mario Raimondo, de Gruyter Expositions in Mathematics, 1-67 (1996).

[28] M.-F. ROY,N. VOROBJOV, *Computing the complexification of a semi-algebraic set,* Math. Zeitschrift 239, 131-142 (2002).

[29] M. SAFEY EL DIN *Résolution réelle des systèmes polynomiaux en dimension positive.* Thèse, Université Paris VI (2001).

[30] R. THOM, *Sur l'homologie des variétés réelles.* In: Differential and combinatorial topology, 255-265. Princeton University Press (1965).

[31] R. J. WALKER, *Algebraic Curves.* Princeton University Press (1950).

About Author

Marie-Françoise Roy is at Institut de Recherches Mathématiques de Rennes (UMR CNRS 6625), Campus de Beaulieu, 35042 Rennes Cedex, France. *mfroy@maths.univ-rennes1.fr.*

Acknowledgments

The author would like to thank Saugata Basu, Richard Pollack and Fabrice Rouillier for useful discussions.

A Discrete Isoperimetric Inequality and Its Application to Sphere Packings

Peter Scholl
Achill Schürmann
Jörg M. Wills

Abstract

We consider finite packings of equal spheres in Euclidean 3–space E^3. The convex hull of the sphere centers is the packing polytope. In the first part of the paper we prove a tight inequality between the surface area of the packing polytope and the number of sphere centers on its boundary, and investigate in particular the equality cases. The inequality follows from a more general inequality for cell complexes on packing polytopes.

In the second part we compare these results with densest sphere packings measured by parametric density up to 12 spheres. These results are obtained by some computer calculation and hence only most probably the densest packings. Independent of this restriction these packings have remarkable proberties: The packing polytopes for $n \neq 11$ and sufficiently large parameter are the deltahedra. And they coincide with the densest sphere packings in physics (clusters of atoms) and in coding theory, measured by minimum energy measures.

1 Introduction

What is the densest packing of finitely many equal spheres, if no container is given? The answer depends on the underlying density measure.

The problem plays an increasing role for the investigation of microclusters, which are more and more important in physics and nanotechnique. Microclusters can be modeled by finite sphere packings. In physics their density is measured by various minimum energy measures (cf. e.g. [BSW], [HJ], [WJ]). The complexity of the problem requires computer–aided methods, at least for more than 4 spheres.

In this paper we use a geometric approach for the density of a sphere packing. The convex hull of the sphere centers is called the packing polytope. If not explicitly mentioned, we always assume $\dim P = 3$. Otherwise we call P degenerate or simply a packing polygon. For packing polygons there is much literature (e.g. [FG], [Gr], [GWZ], [Schü1]).

In the first part of this paper we give a tight isoperimetric inequality between the surface area of the packing polytope and the number n of sphere centers on its boundary. This inequality is rather obvious for lattice packings. The extremal sphere packings with minimal surface area have some interesting properties. In particular they have at most 12 vertices, at most 20 facets and their vertex angles are either 1/12 or 2/12 or 3/12. Further for each n there are only finitely many packing polytopes with minimal surface area.

In the second part we compare these results for $n \leq 12$ with most probably densest sphere packings measured by parametric density and appropriate parameter. These packings have a nice geometric property: For $n \neq 11$ they are just the classical deltahedra. So only a few of them ($n = 4$ and 6) are lattice packings.

Further they coincide with the optimal microclusters predicted by energetic methods in physics. For $n \leq 10$ they also coincide with the optimal sphere clusters found by Sloane, Conway, Graham et al. [GS],[SHDC] with the second moment density measure.

The results underline the observation that minimum energy principles and minimum volume principles, i.e. maximum density generate most probably the same densest sphere packings.

2 An inequality for the surface area of sphere packings

Let B denote the unit ball in E^3 and $C = \{c_1, \ldots, c_n\}$ a discrete set such that $C + B$ is a finite packing. Then $P = \text{conv } C$ is a packing polytope with at most n vertices. We call $C + B$ a lattice packing if $C \subset L$ for some lattice L such that $L + B$ is a packing. The packing polytope P then is a lattice polytope.

We prove a general and tight inequality for the surface area $F(P)$. For this we recall the external angle at a vertex of a polytope, i.e. the portion of the unit sphere centered at the vertex, that is covered by the cone of outer normals of support planes at this vertex. E.g. the 4 external angles of a regular tetrahedron are 1/4 each. If P is degenerated to a segment, then the external angles at the two endpoints are 1/2 each and this is obviously the maximum value for $n \geq 2$.

Theorem 2.1. *a) Let P be a packing polytope with $n(P)$ boundary points and all external angles at the vertices $\leq 5/12$. Then*

$$\sqrt{12}(n(P) - 2) \ \leq F(P) \tag{1}$$

and the bound is tight for all $n \geq 4$.

b) Equality holds if and only if each facet of P permits a triangulation with equilateral triangles of edge length 2.

For lattice packings the result holds without the restriction $\leq 5/12$, and it follows easily from Ehrhart's formula and the reciprocity law (cf. [GW1]) or even simpler from Euler's formula (3) and the fact that each triangle of a packing lattice has area of at least $\sqrt{3}$. The difficulties come from nonlattice packings.

The tight assumption on the external angles is the crucial point of the theorem, because otherwise it is false: If P is a slightly "disturbed line" (i.e. $P+B$ sausage-like) then one can make $n(P)$ arbitrarily large and $F(P)$ arbitrarily small. The inequality (1) is a counterpart to the simple and well known inequality for the total contact or kissing number $K(P)$ of the n points on bd$P$, namely

$$K(P) \leq 3(n(P)-2) \tag{2}$$

which holds for the same equality cases as (1), but without the restriction $\leq 5/12$. (2) is a direct consequence of Euler's equation (3) and the fact that in a triangulation on bd$P$ the number of triangles is 2/3 of the number of edges. Theorem 2.1 will be derived from the more general theorem 3.1 on cell complexes. Note that packing polygons may not satisfy (1): E.g., if we consider arbitrary packing polygons with vertices from a hexagonal lattice, then (1) is satisfied with equality, if we count interior lattice points twice and the surface area on both sides of the polygon. Polygons with an edge longer than $\sqrt{12}$ can then locally be modified (cf. [Schü2]) to attain a polygon not satisfying (1).

The equality cases in (1), lattice or non-lattice, are of particular interest, as they have some nice and simple properties:

Theorem 2.2. *Equality in (1) implies the following:*

a) The external angles at the vertices are $i/12,\ i=1,2,3$.

b) P has at most 12 vertices and at most 20 facets.

c) For each n there are only finitely many packing polytopes.

d) If $C+B$ *is a lattice packing, then the lattice is the critical lattice* fcc.

The third theorem requires some definitions, which will be given in the next section. The proofs of the three theorems will be given in the last sections.

3 Boundary complexes

Our third result is rather technical. The definition and the proof are based on ideas from [FG, GWZ]. For this let us define the boundary complex of P as a collection $S=(S_0,S_1,S_2)$, where

1. $S_0 = C \cap \mathrm{bd}(P)$ is the set of vertices (or 0–faces) of S.

2. S_1 is a collection of (geodesic) segments $[uv]$ within bd$P$, called edges (or 1–faces) of S and measured with respect to the intrinsic metric.

The edges $[uv]$ are of positive length $|uv|$ and satisfy $[uv] \cap S_0 = \{u, v\}$. Moreover, the intersection of two edges is either empty or a vertex of S.

3. S_2 is a collection of triangles $[uvw]$ within bd$P$, called 2–faces. All triangles are chosen such that

$$[uvw] \cap S_0 = \{u, v, w\} \quad \text{and} \quad [uvw] \cap S_1 = \{[uv], [uw], [vw]\}.$$

Note that the definition allows edges $[uu]$, called loops in the sequel. As a consequence S_2 may contain 2–faces $[uuv]$ with only two edges, namely $[uu]$ and $[uv]$ twice.

An edge $[uv]$ of S is either an edge of two, one ore none 2–face. In the latter case we say it is a nonbounding edge; if it is an edge of only one 2–face $[uvw]$ and $w \notin \{u, v\}$, then it is a bounding edge. In this way we exclude the edges which bound a triangle together with a loop only. A loop might be a bounding edge.

The union supp$(S) = S_0 \cup S_1 \cup S_2$ is called the support of S. Of course, the support of S is triangulable. Moreover, it is well known that any polygon and hence any boundary of a packing polytope is triangulable. Within the finitely many boundary complexes of a packing polytope P, and all triangulations of their boundary respectively, there is one, not necessarily unique, minimizing

$$\sum_{[uv] \in S_1} |uv|,$$

a minimum boundary complex of P. We say a packing polytope is free of short loops if the minimum boundary complex contains no loop $[uu]$ of length $|uu| < 2$. Note that there is no boundary complex of P with a loop $[uu]$ of length $|uu| < 2$ if the packing polytope is free of short loops.

The area $F(S)$ of S is defined by

$$F(S) = \sum_{[uvw] \in S_2} |uvw|,$$

the perimeter $U(S)$ of S by

$$U(S) = \sum_{[uv] \in S_1 \text{ bounding}} |uv| + \sum_{[uv] \in S_1 \text{ nonbounding}} 2|uv|$$

and the Euler–Poincaré characteristic

$$\chi(S) = a_0(S) - a_1(S) + a_2(S), \tag{3}$$

where $a_i(S)$ denotes the number of elements (i–faces) in S_i. By our definition $a_0(S)$ depends only on P and not on the chosen boundary complex S. So we might as well write $a_0(P)$. Moreover $F(S), U(S)$ and $\chi(S)$ depend only on supp(S) and not on the chosen triangulation (cf. [GWZ]). So we may write $F(\text{supp}(S)), U(\text{supp}(S))$ and $\chi(\text{supp}(S))$ as well. In particular $F(\text{bd}P)$ is the surface area of P and $\chi(\text{bd}P) = 2$.

We show

Theorem 3.1. *a) Every boundary complex S of a packing polytope without short loops satisfies*

$$a_0(S) \leq \frac{1}{\sqrt{12}} F(S) + \frac{1}{4} U(S) + \chi(S). \tag{4}$$

b) Equality holds if and only if there is a boundary complex S' of supp(S) *such that all edges in S'_1 are of length* 2.

The result is closely related to the main result in [FG] (cf. also [GWZ], [Gr]). Note that in the equality case every 2–faces in S'_2 is of area $\sqrt{3}$.

4 Parametric Density

For a useful geometric density measure of finite sphere packings one considers all intrinsic volumes [S] (or quermassintegrals, or Minkowski functionals). The optimization is a difficult problem, if the number n of spheres is ≥ 4. So the general inequality for F in theorem 2.1 and the discussion of the equality cases in theorem 2.2 may help to solve this problem at least for $n \leq 12$. We now consider the appropriate geometric density measure.

Let V denote the volume and $\rho > 0$. Then

$$\delta(B, C, \rho) = nV(B)/V(P + \rho B)$$

is the parametric density of the packing $C + B$ with respect to ρ. The parameter ρ controls the influence of the boundary region of a packing (cf. [BHW]). Maximal δ is equivalent to minimal $V(P + \rho B)$. With Steiner's formula [S] and $V(B) = \frac{4}{3}\pi$ we have

$$V(P + \rho B) = V(P) + F(P)\rho + M(P)\rho^2 + \frac{4}{3}\pi\rho^3 \tag{5}$$

where F denotes the surface area and M the integral of mean curvature [S]. The optimality problem now is to minimize $V(P + \rho B)$ for a given $n \in \mathbb{N}$ and $\rho > 0$ among all packings $C + B$. Obviously the shape of a best packing depends essentially on ρ.

For $n = 1$ and 2 the solution is trivial and for $n = 3$ quite simple. For $n = 4$ and 5 it needs some investigation, but can be done without computer (cf. [BW]). For $n \geq 6$ the difficulties grow rapidly. The existence of a best packing for a given ρ is guaranteed by Blaschke's selection theorem [S].

From some general results (cf. [BHW], [SW], [GW2]) it is known that for each $n \geq 3$ there is a critical parameter $\rho_c(n)$ such that the best sphere packing for $\rho \leq \rho_c(n)$ is a degenerate linear packing (a sausage) and for $\rho > \rho_c$ is a cluster, i.e. $\dim C = 3$ (for $n \neq 3$). So except for $n = 3$ the center polytope of a best sphere packing is never planar. It is further known that $\rho_c(3) = 1{,}102\ldots, \rho_c(4) = 1{,}058\ldots$ and with computational methods [Sch1] that $1 < \rho_c(n) < 1{,}05$ for $4 < n < 56$.

In the next section we investigate the optimal packing polytopes with respect to parametric density and compare them with the optimal packings in physics [WJ] and coding theory [SHDC]. For this we use the results from section 2.

5 Deltahedra are good packing polytopes

As it is hard to compute the optimal sphere packings with respect to parametric density directly, we use the results from section 2 as follows: We determine all best packings with respect to F and hence also of K from $n = 4$ to 12, which leads to 19 packing polytopes, listed in table 1. For completeness we add $n = 3$, although it is not a packing polytope.

For these 19 cases we compute M and V, i.e. their normalizations M/n and V/n, listed in columns 2 and 3. In column 4 we classify their structure and distinguish between lattice packings, i.e. subsets of the critical lattice (fcc) and others, namely hexagonal close packings (hcp), packings with pentagonal (pent) or icosahedral symmetry (ico) or none of these (irr). Of course the structures for $n = 3$ and 4 are not only fcc. In column 5 we distinguish between deltahedra (Δ), where all facets are regular triangles, and other shapes. It turns out that this is the most interesting characterization, because the deltahedra are just the best packing polytopes computed in physics (P) for $n \neq 11$ and (for $n \leq 10$) in coding theory (C) in the last two columns. In columns 7 and 8 the intervals $\rho_{\min} \leq \rho \leq \rho_{\max}$ are given, for which the corresponding packing polytope is better than the linear packing (sausage) and the other packing polytope with same n of the list. This can easily be calculated. It is remarkable that only 3 of the 12 lattice packings, but 7 of the 8 nonlattice packings coincide with packings predicted in physics.

As it is difficult to understand the structure and shape of these 19 polytopes, they are shown in figures 1 and 2. In figure 1 we show just those which coincide with those found in physics. In figure 2 are the remaining 10 packing polytopes. In all cases the balls are shrunken and the $K(P)$ bonds between them (i.e. the edges of the packing polytopes) are shown as bars. This helps to understand the geometry. The conjecture is now that the packing polytopes with minimal F and maximal K are optimal with respect to parametric density:

Conjecture 5.1. *The packing polytopes in table 1 are optimal for the given intervals $\rho_{\min} \leq \rho \leq \rho_{\max}$ compared with all packing polytopes of same n.*

Table 1. Densest known packings for $3 \leq n \leq 12$

n	M/n	V/n	structure	Δ	$\rho_{\min}$	$\rho_{\max}$	P	C
3	3.142	0.000	fcc	Δ	1.103	∞	P	C
4	2.866	0.236	fcc	Δ	1.059	∞	P	C
5	2.701	0.377	hcp	Δ	1.048	∞	P	C
6	2.462	0.629	fcc	Δ	1.049	∞	P	C
7	2.402	0.673	fcc	–	–	–	–	–
7	2.355	0.689	pent	Δ	1.036	∞	P	C
8	2.356	0.707	fcc	–	1.043	1.108	–	–
8	2.356	0.707	fcc	–	1.043	1.108	–	–
8	2.232	0.859	irr	Δ	1.108	∞	P	C
9	2.321	0.733	fcc	–	1.041	1.160	–	–
9	2.113	1.013	irr	Δ	1.160	∞	P	C
10	2.293	0.754	fcc	–	–	–	–	–
10	2.105	0.943	fcc	–	1.037	1.497	–	–
10	2.016	1.143	irr	Δ	1.497	∞	P	C
11	2.099	0.943	fcc	–	1.036	1.314	–	–
11	1.949	1.203	irr	–	1.314	∞	P	–
12	2.094	0.943	fcc	–	–	–	–	–
12	2.094	0.943	fcc	–	–	–	–	–
12	1.911	1.100	hcp	–	1.028	2.027	–	–
12	1.824	1.454	ico	Δ	2.027	∞	P	–

This conjecture has been checked by computer for various classes of packing polytopes. These results support that the conjecture might be true. The computational methods are described briefly in the next section.

6 Computational Methods and Techniques

Computation of parametric density is more complicated than the Lennard–Jones potential or other density measures in physics [WJ] and also more complicated than the second moment used in coding theory [SHDC]. So the methods used there are not helpful here. On the other hand one has general results with methods from convex and discrete geometry, as e.g. theorems 2.1 and 2.2. Here we survey the part of experimental mathematics. The results on packing polytopes with minimal surface area F and maximal contact number K give a strong pre-selection of optimal packing polytopes with respect to parametric density; in particular if n is small, say $n \leq 12$.

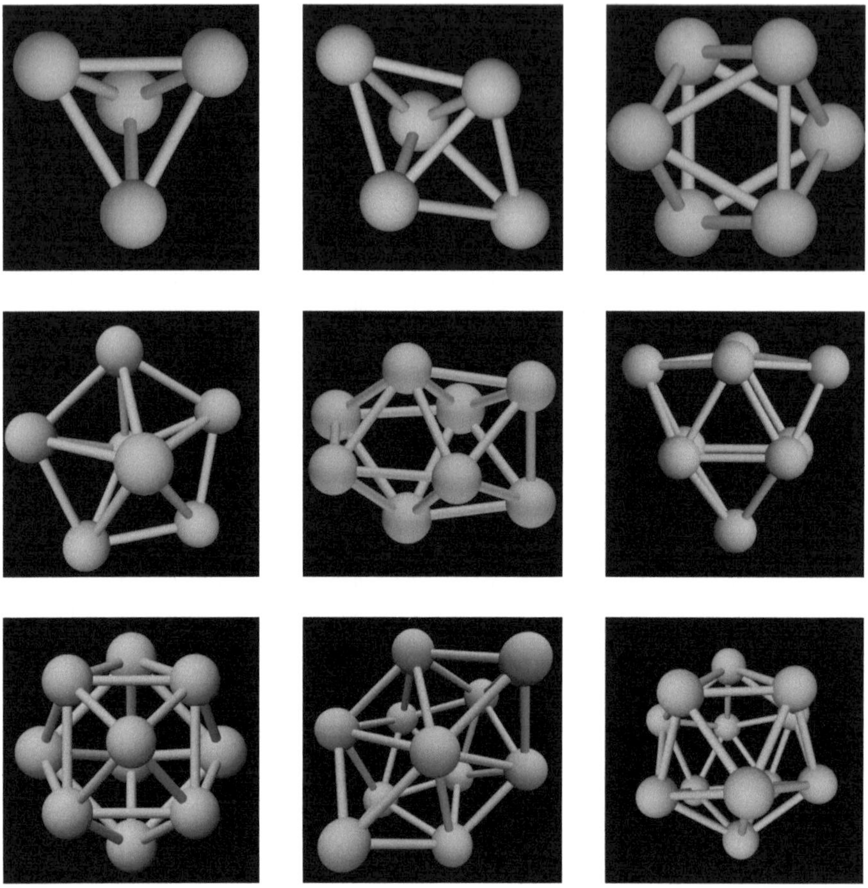

Fig. 1. Dense packings of 4 to 12 spheres with minimal surface area, minimal mean width and maximal contact number

With an appropriate computer program (cf. [Sch2]) some series of "good candidates" with small V and M are checked. Good candidates are subsets of densest infinite packings, i.e. of the densest lattice packing (fcc) and some densest nonlattice packing, e.g. hcp (hexagonal close packing) which keep V small. Further dense sphere packings with pentagonal or icosahedral symmetry, which are only a little less dense than the critical lattice, and highly symmetric, which keeps M small. In all cases the subsets with minimal surface area (as in theorem 2.1) were checked. For $n \leq 12$ this seems sufficient, as earlier investigations also suggest ([GaW], [SW]). For $n \geq 13$ investigations become more complicated, as the best packing polytopes contain interior sphere-centers and hence also squares (cuboctahedron) or larger triangles (icosahedron) on their surface.

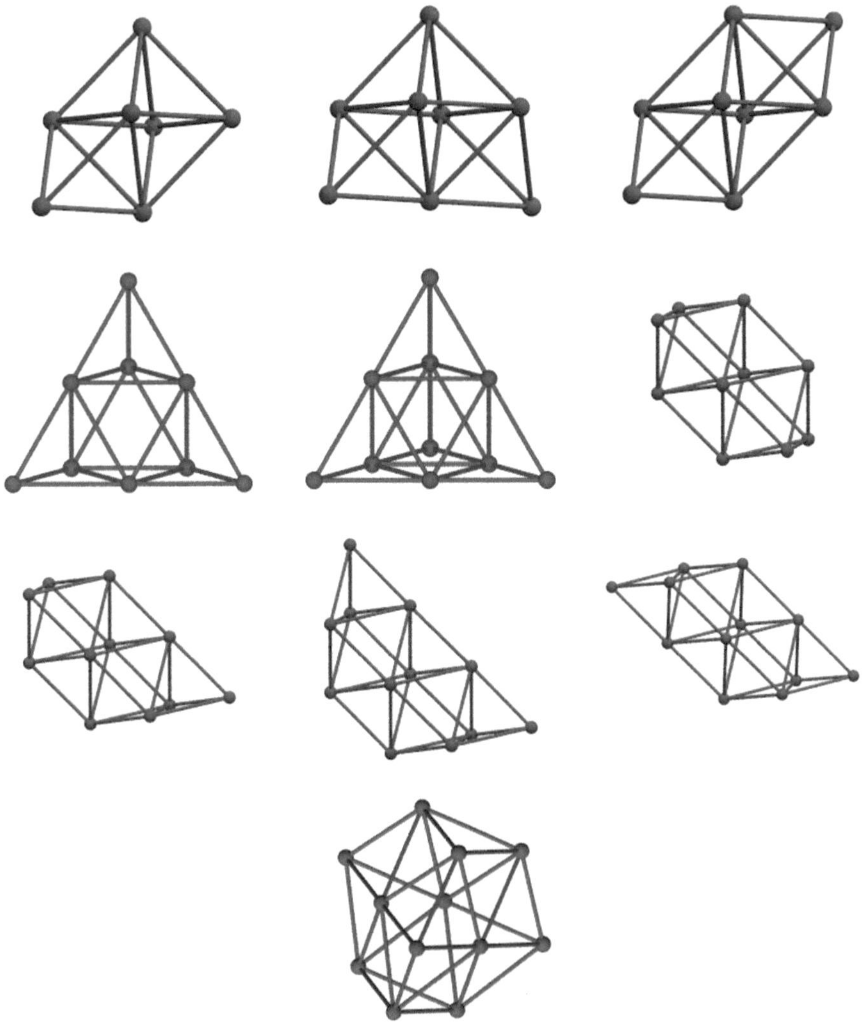

Fig. 2. Dense packings of 7 to 12 spheres with minimal surface area, maximal contact number but no minimal mean width

But for $n \leq 12$ the program indicates that all relevant candidates were checked. For details we refer to [Sch2].

7 Proof of theorem 2.1

For the proofs of theorems 2.1 and 2.2 we need a result on exterior angles (cf. [M]):

Lemma 7.1. *Let $\gamma(v, P)$ be the external angle of P at the vertex v and $\beta(v, F)$ the internal angle of the facet F of P at the vertex $v \in F$. Then*

$$2\gamma(v, P) + \sum_{v \in F} \beta(v, F) = 1. \tag{6}$$

Proof. The result is a special consequence of two general Euler–type relations. For a polyhedral cone $K \subset E^d$ with apex o and any pair of faces $F \subset G$ of K let $\beta(F, G)$ and $\gamma(F, G)$ be the internal and external angles of G at its face F, normalized so that the total angle is 1. Then (cf. [M], theorem 1 and 2) with the natural convention $\beta(F, F) = \gamma(F, F) = 1$ we have:

$$\sum_F \beta(o, F)\gamma(F, K) = 1$$
$$\sum_F (-1)^{\dim F} \beta(o, F)\gamma(F, K) = 0$$

where the sum extends over all faces F of K, including $\{o\}$ and K. If we add both identities, then all summands with odd $\dim F$ vanish. For $d = 3$ we get

$$2 \sum_{\dim F = 2} \beta(o, F)\gamma(F, K) + 2\gamma(o, K) = 1$$

Now $\gamma(F, K) = \frac{1}{2}$, if $\dim F = 2$. We replace K by P, and o by v and obtain (6).

□

As one referee pointed out, the lemma can also be proved with elementary spherical geometry and without McMullen's formulae. We now come to the proof of theorem 2.1: a) The inequality (1) is a special case of inequality (4) of theorem 3.1 with $\text{supp}(S) = \text{bd}P$, $U(S) = 0$ and $\chi(S) = 2$. We first show that the conditions in theorem 2.1 on the external angle is a consequence of the loop condition in theorem 3.1. For this we cut bd$P$ along a short loop and along the segment between the vertex v and the end w of the loop.

If we unfold this into the plane, we get a polygon P'. Let v' be the image of v and w', w'' the images of w after the cut. The edges $v'w'$ and $v'w''$ have length ≥ 2, and $\|w' - w''\| < 2$, as the polygonal line from w' to w'' is < 2. So the internal angle of P' at v' is $< \pi/3$. Hence the sum $\sum_{v \in F} \beta(v, F) < 1/6$ and with lemma 7.1 we get $\gamma(v, P) > 5/12$.

For the proof that (1) is tight for all n we construct two appropriate infinite series of lattice polytopes from the critical lattice fcc. In table 1 we have examples for $n \leq 12$, so it remains to show for $n \geq 13$. Let a, b, c be a base of the fcc lattice, with mutual angles $\pi/3$.

1. Let P_λ be the prism with base $\mathrm{conv}(o, a, b, a+b)$ and hight λc for some $\lambda \in \mathbb{N}$. If $\lambda = 1$, one has a fundamental cell of fcc. All lattice points of P are on its boundary and their cardinality is $4(\lambda+1)$, $\lambda \in \mathbb{N}$. So (1) holds with equality for $n \equiv 0 \mod 4$, if $n \geq 8$.

2. If one cuts off one or both acute vertices of a prism constructed in 1), one obtains polytopes with lattice point number $4(\lambda+1) - 1$ and $4(\lambda+1) - 2$, $\lambda \in \mathbb{N}$, whose surface still can be triangulated into regular triangles. So (1) holds with equality for $n \equiv 2 \mod 4$ and $n \equiv 3 \mod 4$, if $n \geq 6$.

3. Let Q_λ be the prism with base $\mathrm{conv}(o, 2a, 2b, 2(a+b))$ and hight λc, $\lambda \in \mathbb{N}$. Its lattice point number is $9(\lambda+1)$ and the lattice point number on the boundary is $8(\lambda+1)+2$, $\lambda \in \mathbb{N}$. If one cuts off one acute vertex, one obtains a polytope with triangulable surface and $8(\lambda+1)+1$ boundary points. If one cuts off one acute vertex and at the other acute vertex a regular tetrahedron with 4 lattice points, one obtains an appropriate polytope with $8(\lambda+1)-3 = 8\lambda+5$ boundary points. From these two cases follows that (1) holds with equality for $n \equiv 1 \mod 4$ and $n \geq 13$. This proves the equality cases in (1).

Proof of theorem 2.1 b): From theorem 3.1 b) follows that equality in (1) is equivalent to the existence of a triangulation of $\mathrm{bd}P$ such that all edges have length 2. If an edge of length 2 would cross (at least) one edge of P, then the endpoints of this edge would have Euclidean distance < 2 in contradiction to the fact that P is a packing polytope. The number of triangles in the equality case is $2(n(P)-2)$. So the number of edges is $3(n(P)-2)$.

8 Proof of theorem 2.2

a) If (1) holds with equality, then each facet of P permits a triangulation with equilateral triangles. Hence all facets have internal angles $\pi/3$ or $2\pi/3$ at the vertices, i.e. the normalized $\beta(v, F)$ are either $1/6$ or $2/6$.
From (6) we get

$$\gamma(v, P) = \frac{1}{2} - \sum_{v \in F} \frac{1}{2} \beta(v, F)$$

From $\dim P = 3$ follows that a least 3 facets meet at v. As each β is either $1/6$ or $2/6$, at most 5 can meet at v. So

$$\gamma(v, P) = i/12, \quad i = 1, 2, 3.$$

are the only possible values.

b) This is a simple consequence of a): As each external angle is $\geq 1/12$, there are at most 12 of them. Each vertex is at most 5–valent and each facet has at least 3 vertices. So there are at most 20 facets. The regular icosahedron shows that both bounds are attained.

c) For each n there are at most finitely many facets with at most n points and hence at most finitely many combinatorial possibilities to arrange the facets to the boundary of a convex polytope. From Cauchy's rigidity theorem (cf. [C]) for convex polytopes follows c).

d) Let $C+B$ be a lattice packing, i.e. $C \subset L$, L a lattice, $L+B$ a packing, and $P = \mathrm{conv}C$ a lattice polytope. Further let equality hold in (1). Then the vertices of P are at most 5–valent and the angles of the facets at the vertices are either $\pi/3$ or $2\pi/3$. We classify the vertices of P by ordered pairs (l, m), where l denotes the valence of the vertex and m the number of "large" angles $2\pi/3$ at the vertex.

The pairs (3,1), (4,2) and (5,1) lead to degenerate (flat) P. So only the following 5 pairs are possible: (3,0), (3,2), (4,0), (4,1) and (5,0), which we consider in this given order: In all cases let the vertex v be the origin.

1) (3,0): The 3 edges at v have pairwise angles of $\pi/3$. So they are a base of L , which is necessarily the critical lattice.

2) (3,2): Let $\angle ab = \pi/3$ and $\angle ac = \angle bc = 2\pi/3$. Then $a, b, -c$ are a base of L, which is again the critical lattice.

3) (4,0): Let a, b, c, d be the edges in cyclic order. They span a rhombic cone. Elementary symmetry arguments yield

$$\begin{aligned} a+c &= \lambda(b+d) \\ \text{and} \qquad b+d &= \mu(a+c) \end{aligned}$$

with some $\lambda, \mu \in \mathbb{N}$, hence $\lambda = \mu = 1$ and $a + c = b + d$. Observe that $c, d \notin \mathrm{aff}(a, b)$ and $a, b \notin \mathrm{aff}(c, d)$. So a, b, c, d span a quadratic cone and any 3 of them are a base of L. These bases are different from those in 1) and 2), but again they generate the critical lattice.

4) (4,1): Again let a, b, c, d the edges in cyclic order and let $\angle a, b = 2\pi/3$. So $\mathrm{dist}(a, b) = 2\sqrt{3}$. We assume that $v, a, b, c, d \in L$. Then v, a and b generate the planar hexagonal (critical) lattice $H \subset L$. Further $c, d \notin H' = \mathrm{aff}(v, a, b)$ and $\mathrm{dist}(c, d) = 2$. The line $l = \mathrm{aff}(c, d)$ contains infinitely many points of L at distance 2. Hence $l \cap H' = \emptyset$ and $\mathrm{dist}(c, H') = \mathrm{dist}(d, H')$. Now

$$\mathrm{dist}(a, d) = \mathrm{dist}(b, c) = \mathrm{dist}(v, c) = \mathrm{dist}(v, d) = 2$$

implies that $\mathrm{aff}(c, d)$ and $\mathrm{aff}(a, b)$ are parallel. But $\mathrm{dist}(a, b) = 2\sqrt{3} \neq \mathrm{dist}(c, d) = 2$, which contradicts the lattice property. So the vertex configuration (4,1) does not occur in a lattice packing.

5) (5,0): After the cases 1)-4) we can assume that all vertices are of type (5,0). This leads to the classical result that P is the regular icosahedron (with 12 vertices), which is for no lattice a packing polytope.

9 Proof of theorem 3.1

Proof of a) and b): Since S is free of short loops, it contains no 1–face $[uu]$ of length $|uu| < 2$. The proof follows mainly the proof in [FG] (see also [GWZ] for a generalization): We assume (4) is false. Then there exists a boundary complex S with a minimal number of simplices

$$a(S) = a_0(S) + a_1(S) + a_2(S),$$

for which (4) is false. If S consists only of vertices, then (4) is true. Hence S contains at least one edge.

If S_1 contains a nonbounding edge, then let S' be the boundary complex without it. Thus $a_0(S') = a_0(S)$, $a_1(S') = a_1(S) - 1$ and $a_2(S') = a_2(S)$, and therefore $a(S') < a(S)$, which implies that (4) holds for S'. Moreover $\chi(S) = \chi(S') - 1$, $F(S) = F(S')$, $U(S) \geq U(S') + 4$ and we find that (4) is also true for S.

Thus we may assume that S_1 does not contain nonbounding edges. Within S_1 there is an edge of maximal length, say $[uv]$. Further we may assume that our counterexample is chosen among all boundary complexes with the same support, such that $|uv|$ is minimal among all edges of maximal length. Now we consider three cases and show that $|uv|$ is not a boundary edge nor an inner edge where we call an edge an inner edge if it is neither a bounding nor a nonbounding edge.

1) Assume $[uv]$ is a bounding edge, possibly a loop, that bounds a 2–face $[uvw]$ of S. In this case $[uv]$ is the largest side of $[uvw]$ and by Lemma 1 in [FG] we know:

$$\frac{F([uvw])}{2\sqrt{3}} + |uv| \geq |uw| + |vw|,$$

with equality if and only if $[uvw]$ is a regular triangle of edge length 2. Now let S' be the boundary complex S without $[uv]$ and $[uvw]$. Then $a_0(S') = a_0(S)$, $\chi(S') = \chi(S)$, $F(S') = F(S) - F([uvw])$ and $U(S') = U(S) + |uw| + |vw| - |uv|$. Since S' satisfies (4) because of $a(S') < a(S)$, we find that S also satisfies (4). Hence $[uv]$ is not a bounding edge.

2) Assume $[uv]$ is an inner edge and bounds exactly one 2–face $[uuv]$. In this case $|uv|$ is at least as large as the loop $|uu|$. With Lemma 1 in [FG] we derive

$$\frac{F([uvw])}{2\sqrt{3}} \geq |uu|,$$

with equality if and only if $[uuv]$ is a regular triangle with edges of length 2. As in case 1, let S' be the boundary complex, given by S without $[uv]$ and $[uuv]$. Then $a_0(S') = a_0(S)$, $\chi(S') = \chi(S)$, $F(S') = F(S) - F([uuv])$ and $U(S') = U(S) + |uu|$. Thus S' satisfies (4) and therefore we find that S also satisfies (4).

3) Assume $[uv]$ is an inner edge and bounds exactly two 2–faces $[uvw]$ and $[uvx]$. By assumption $[uv]$ is the maximal side of both adjacent triangles

and therefore the quadrilateral $[uvwx]$ is convex (cf. [FG]). By Lemma 2 in [FG] we then know

$$\frac{2F([uvwx])}{\sqrt{3}} - U([uvwx]) + 4 \geq 0,$$

with equality if and only if $[uvw]$ and $[uvx]$ are regular triangles of edge length 2. Now let S' be the boundary complex, given by S without $[uv]$, $[uvw]$ and $[uvx]$. Then $a_0(S') = a_0(S)$, $\chi(S') = \chi(S) - 1$, $F(S') = F(S) - F([uvwx])$ and $U(S') = U(S) + U(S')$. Thus again because of $a(S') < a(S)$ we find that S' and consequently S satisfies (4). Taking both cases 2) and 3) into consideration, $[uv]$ is not an inner edge. This proves a).

Finally, going through the three cases and the case of nonbounding edges before, we find the assertion on the equality cases verified, which proves assertion b).

References

[BHW] U. Betke, M. Henk and J. M. Wills, Finite and infinite packings, J. reine angew. Math., 453 (1994), 165–191

[BSW] K. Böröczky jr., U. Schnell and J. M. Wills, Quasicrystals, Parametric Density and Wulff–Shape, in: *Directions in Math. Quasicrystals*, eds. M. Baake and R. V. Moody, AMS, CRM–Series, Montreal, 2000

[BW] K. Böröczky jr. and J. M. Wills, Finite sphere packings and critical radii, *Beitr. Algebra Geom.*, 38 (1997), 193–211

[C] R. Connelly, Rigidity, in: *Handbook of Convex Geometry*, eds. P.M. Gruber and J.M. Wills, North–Holland, Amsterdam, 1993

[CS] J. H. Conway and N. J. A. Sloane, *Sphere Packings, Lattices and Groups*, 3rd ed., Springer, New York, 1999

[DGH] M. Dyer, P. Gritzmann and A. Hufnagel, On the complexity of computing mixed volumes, *SIAM J. Comput.*, 27 (1998), 356–400

[FG] J.H. Folkman and R.L. Graham, A Packing Inequality for Compact Convex subsets of the Plane, *Canad. Math. Bull.*, 12 (1969), 745-752

[GaW] P. M. Gandini and J. M. Wills, On Finite Sphere–Packings, *Mathematica Pannonica*, 3/1 (1992), 19–29

[GWZ] R.L. Graham, H.S. Witsenhausen and H.J. Zassenhaus, On Tightest Packings in the Minkowski Plane, *Pacific J. Math.*, 41 (1972), 699-715

[GS] R. L. Graham and N. J. A. Sloane, Penny–Packing and two–dimensional codes, *Discrete Comp. Geom.*, 5 (1990), 1–11

[GW1] P. Gritzmann and J. M. Wills, Lattice points, in: *Handbook of Convex Geometry*, eds. P. M. Gruber and J. M. Wills, North–Holland, Amsterdam, 1993

[GW2] P. Gritzmann and J. M. Wills, Finite packing and covering, in: *Handbook of Convex Geometry*, eds. P. M. Gruber and J. M. Wills, North–Holland, Amsterdam, 1993

[Gr] H. Groemer, Über die Einlagerung von Kreisen in einen konvexen Bereich, *Math. Z.*, 73 (1960), 285–294

[G] B. Grünbaum, *Convex Polytopes*, Wiley & Sons, London, 1967

[HJ] J.E. Hearn and R.L. Johnston, Calcium and Strontium Clusters with Many-body Potentials, *J. Chem. Phys.*, 107 (1997), 4674–4687

[J] N. W. Johnson, Convex polyhedra with regular faces, *Canad. J. Math.*, 18 (1966), 169–200

[M] P. McMullen, Non-linear angle-sum relations for polyhedral cones and polytopes, *Math. Proc. Camb. Phil. Soc.*, 78 (1975), 247-261

[S] J. R. Sangwine-Yager, Mixed volumes, in: *Handbook of Convex Geometry*, eds. P. M. Gruber and J. M. Wills, North–Holland, Amsterdam, 1993

[SW] U. Schnell and J. M. Wills, Densest packings of more than three d–spheres are nonplanar, *Discrete Comp. Geom.*, 24 (2000), 539–549

[Sch1] P. Scholl, *Microcluster im fcc–Gitter*, Diplom–Thesis, University of Siegen, 1999

[Sch2] P. Scholl, *Sphere Packing Database*, `http://www.math.uni-siegen.de/wills/spdb/`, 2000–2001

[Schü1] A. Schürmann, *Plane Finite Packings*, Dissertation, University of Siegen, 2000; Shaker–Verlag, Aachen, 2000

[Schü2] A. Schürmann, On extremal finite packings, submitted

[SHDC] N. J, A. Sloane, R. H. Hardin, T. D. S. Duff and J. H. Conway, Minimal–Energy Clusters of Hard Spheres, *Discrete Comp. Geom.*, 14 (1995), 237–259

[WJ] N. T. Wilson and R. L. Johnston, Modelling gold clusters with an empirical many–body potential, *Europ. Phys. J. D*, 12 (2000), 161–169

[Z] V. A. Zalgaller, *Convex polyhedra with regular faces*, Sem. in Math., Steklov MI, Leningrad; Amer. Transl., New York, 1969

Acknowledgments

We want to thank the two referees for very careful reading and many helpful suggestions.
The first mentioned author was supported by the "Deutsche Forschungsgemeinschaft" (DFG) under grant Wi 1254/9.
The second author was partially supported by the annual prize of the "Industrie und Handelskammer" (IHK) for applied science in 2001.

Peter Scholl, Achill Schürmann and Jörg M. Wills
Department of Mathematics
University of Siegen
57068 Siegen
Germany
e–mail: {*peter.scholl, achill, joerg.wills*}*@math.uni-siegen.de*

On the Number of Maximal Regular Simplices Determined by n Points in $\mathbb{R}^d$

Zvi Schur
Micha A. Perles
Horst Martini
Yaakov S. Kupitz

1 Introduction

A set $V = \{x_1, \ldots, x_n\}$ of n distinct points in Euclidean d-space $\mathbb{R}^d$ determines $\binom{n}{2}$ distances $||x_j - x_i||$ $(1 \leq i < j \leq n)$. Some of these distances may be equal. Many questions concerning the distribution of these distances have been asked (and, at least partially, answered). E.g., what is the smallest possible number of distinct distances, as a function of d and n? How often can a particular distance (say, one) occur and, in particular, how often can the largest (resp., the smallest) distance occur?

In this paper we will consider mainly the *largest* distance, i.e., the diameter diam (V) of the set V. It will be convenient to associate with the set V its *diameter graph* $D(V)$. The vertices of $D(V)$ are the points of V, and the edges are the pairs $\{x, y\}$, where $x, y \in V$ and $||x - y|| =$ diam (V).

We denote by $e(V)$ the number of edges of $D(V)$, and define

$$e(d, n) := \max\{e(V) : V \subset \mathbb{R}^d,\ \#V = n\}\,.$$

In § 2 we give a short survey of known results about $e(d, n)$. Here we only mention two trivial facts and one basic known result.

$$e(d, n) = \binom{n}{2} \text{ iff } n \leq d + 1\,; \tag{1}$$

$$e(1, n) = 1 \text{ for } n \geq 2\,; \tag{2}$$

$$e(2, n) = n \text{ for } n \geq 3\,. \tag{3}$$

The late Zvi Schur (cf. the acknowledgement at the end of this article) started to investigate more delicate properties of the graph $D(V)$. For $2 \leq k \leq d+1$, a k-clique in $D(V)$ (i.e., a complete subgraph of $D(V)$ of order k) corresponds to a regular $(k-1)$-simplex of edge length diam (V) with vertices in V.

One could ask what is the maximum possible number of k-cliques in $D(V)$, as a function of d, n and k.

We denote by $e_k(V)$ the number of k-cliques in $D(V)$ and define

$$e_k(d,n) := \max\{e_k(V) : V \subset \mathbb{R}^d,\ \#V = n\}\,.$$

Clearly $e_1(V) = \#V = n,\ e_2(V) = e(V)$, and for $k \geq 2$ one has $e_k(V) \leq \binom{n}{k}$, with equality iff V is the set of vertices of a regular $(n-1)$-simplex. Note that if all distances between points of V are equal, then V is an affinely independent set in $\mathbb{R}^d$, and thus $n = \#V \leq d+1$. It follows that

$$e_1(d,n) = n\,,$$

$$e_2(d,n) = e(d,n)\,.$$

For $2 \leq k \leq d+1$ one has $e_k(d,n) \leq \binom{n}{k}$, with equality iff $n \leq d+1$.

For $k \geq d+2$, $e_k(d,n) = 0$. We will show later (Corollary 5.4) that $e_{d+1}(d,n) = 1$ for $n \geq d+1$.

The object of this paper is the following conjecture of Zvi Schur.

Conjecture 1.1. *For $n \geq d+1$, one has*

$$e_d(d,n) = n\,.$$

This assertion is trivial for $d = 1$. For $d = 2$ it reduces to the result quoted in (3) above on the number of occurrences of the diameter in a set of n points in the plane (see § 2 below).

The proof of the case $d = 3$ of this conjecture (see Theorem 7.1 below) forms the core of this paper.

A simple construction of a set V of n points in $\mathbb{R}^d$, $n \geq d+1$, with $e_d(V) = n$ is described in § 3 below. In view of this construction, Conjecture 1.1 reduces to the following statement.

Conjecture 1.1*. *If V is a set of n points in $\mathbb{R}^d$, $n \geq d+1$, then there are at most n regular $(d-1)$-simplices of edge length* diam (V) *with vertices in V.*

2 Results about $e(d,n)$

$$d = 1: \quad e(1,n) = 1 \quad \text{for } n \geq 2 \text{ (trivial)}\,,$$

$$d = 2: \quad e(2,n) = n \quad \text{for } n \geq 3\,.$$

This was probably first mentioned (and proved) in [6], Theorem 3. See [17], pp. 213-214, for a recent discussion of this matter and related references.

For a description of the extremal configurations, i.e., finite sets $V \subset \mathbb{R}^2$ with $e(V) = \#V$, see [21] or [12]. (These configurations are "skeletons" of Reuleaux polygons, see [15] and [14].)

The smallest distance among n points in $\mathbb{R}^2$ $(n \geq 2)$ is attained at most $\lfloor 3n - \sqrt{12n-3} \rfloor$ times. This was conjectured by O. Reutter [18] and proved by H. Harborth [10], see again [17], pp. 211-212, for a recent discussion.

The extremal configurations are described in [13]. These configurations also yield the maximum number of equilateral triangles whose edge length is the minimum distance among n points in $\mathbb{R}^2$.

The maximum possible number of occurrences of any particular distance within a set of n points in $\mathbb{R}^2$ grows faster than linear. For details, see [6] or [17], p. 144, Theorem 10.4.

$$d = 3: \quad e(3, n) = 2n - 2 \quad \text{for} \quad n \geq 4 .$$

This was conjectured by A. Vázsonyi (Weiszfeld) as mentioned in [6], p. 250, and proved independently by B. Grünbaum [9], A. Heppes [11] and S. Straszewicz [19], cf. [17], pp. 214-215.

Estimates of $e(d, n)$ for $d \geq 4$ depend on Construction 3.2, due to H. Lenz, as described in the next section.

For a discussion of the maximum number of equilateral triangles (of fixed or variable size) determined by n points in the plane (or in $\mathbb{R}^d$, $d \geq 2$) see [1].

3 Two Constructions

Here we describe two constructions of sets of n points of diameter λ in $\mathbb{R}^d$, $n \geq d+1$. The first construction yields sets with n regular $(d-1)$-simplices of edge length λ, and it is a special case of the "standard examples" described in § 4 below. The second construction beats the first with respect to the number of regular k-simplices of edge length λ for $1 \leq k \leq d-2$ (in particular, regarding the number of diameters $e(V)$) when n is sufficiently large and $d \geq 4$.

Construction 3.1. Let $\triangle^d$ $(d \geq 3)$ be a regular simplex inscribed in the unit sphere, centered at the origin, in $\mathbb{R}^d$. The vertices $p_0, p_1, \dots, p_d$ of $\triangle^d$ satisfy

$$\langle p_i, p_j \rangle = \begin{cases} 1 & \text{if} \quad i = j , \\ -\frac{1}{d} & \text{if} \quad i \neq j . \end{cases}$$

Thus $\sum_{i=0}^{d} p_i = o$, and $\langle p_i - p_j, p_i - p_j \rangle = ||p_i - p_j||^2 = 2(1 + \frac{1}{d}) \quad \text{for } i \neq j$.

The edge length λ of $\triangle^d$ is $\lambda_d = \sqrt{2(1 + \frac{1}{d})}$.

Denote by c the center of the $(d-2)$-face $[p_0, \dots, p_{d-2}]$ of $\triangle^d$, i.e.,

$$c = \frac{1}{d-1} \sum_{i=0}^{d-2} p_i \quad \left(= -\frac{1}{d-1}(p_{d-1} + p_d) \right) .$$

Denote by C the circle of radius $\sqrt{1 + \frac{2}{d-1}}$ centered at c that passes through p_{d-1} and p_d. Furthermore, denote by A the short (open) arc of C with endpoints p_{d-1} and p_d. Note that the vectors $p_i - c$ $(i = 0, 1, \dots, d-2)$ are perpendicular to the vectors $p_j - c$ $(j = d-1, d)$, and therefore the plane aff C is perpendicular to the $(d-2)$-flat aff $\{p_0, \dots, p_{d-2}\}$.

Choose $n-d-1$ distinct points $p_{d+1},\dots,p_{n-1}$ on A, and define $V = \{p_0,\dots,p_{n-1}\}$. Then diam $(V) = \lambda_d$ and, for $0 \le i < j < n$, $||p_i - p_j|| = \lambda_d$ iff either $0 \le i < j \le d$, or $0 \le i \le d-2$ and $d+1 \le j < n$.

The graph $D(V)$ has precisely n d-cliques: $d+1$ d-cliques correspond to the facets of $\triangle^d$, and the remaining ones are of the form $\{p_0, p_1, \dots, p_{d-2}, p_k\}$, $d+1 \le k < n$.

Construction 3.2 (H. Lenz, 1955)**.** This construction was first published by P. Erdős in [7]. Erdős attributes it to H. Lenz, and states that Lenz himself never published it. For a recent discussion see [17], p. 148 [1].

Assume $n \ge d \ge 2$ and $d = 2m$ or $d = 2m+1$. Let $e_1,\dots,e_d$ be an orthonormal basis of $\mathbb{R}^d$. For $1 \le i \le m$, denote by C_i the unit circle centered at the origin o that passes through e_{2i-1} and e_{2i}. Denote by A_i the short open arc of C_i with endpoints e_{2i} and e_{2i-1}. For $1 \le i \le m$, choose a set V_i' of $\lfloor \frac{n-d+i-1}{m} \rfloor$ points on A_i and define

$$V := \{e_1,\dots,e_d\} \cup \bigcup_{i=1}^{m} V_i' \,.$$

Then $\#V = n$, and the diameter of V is $\sqrt{2}$. To describe the diameter graph $D(V)$, we choose the following notation:

$$m' = \begin{cases} m & \text{if} \quad d = 2m\,, \\ m+1 & \text{if} \quad d = 2m+1\,, \end{cases}$$

$$V_i = V \cap C_i = \{e_{2i-1}, e_{2i}\} \cup V_i' \ \text{ for } i = 1,\dots,m, \ \ V_{m+1} = \{e_d\} \text{ if } d = 2m+1\,.$$

Then V is the *disjoint union* of the sets V_i, $1 \le i \le m'$, and the distance between two points $x, y \in V$ is $\sqrt{2}$ iff either x and y belong to different sets V_i, V_j, or $\{x, y\} = \{e_{2i-1}, e_{2i}\}$ for some $1 \le i \le m$. In other words, if $d = 2m$ is even, then $D(V)$ is a balanced m-partite graph on n vertices plus m extra edges. If $d = 2m+1$ is odd, then $D(V)$ is a balanced m-partite graph on $n-1$ vertices plus m extra edges and an extra universal vertex ($= e_d$).

A rather simple calculation (see [17], p. 135, Ex. 9.5, or [16], p. 196) leads to explicit expressions for $e(V)$:

If $d = 2m$ is even and $n = am + r$, $a \ge 2$, $0 \le r \le m$, then

$$e(V) = \frac{1}{2}(1 - \frac{1}{m})n^2 - \frac{r(m-r)}{2m} + m$$

($= e(T_m(n)) + m$, where $e(T_m(n))$ is the number of edges of the balanced m-partite graph – the Turán graph – $T_m(n)$.) Note that $0 \le \frac{r(m-r)}{2m} \le \frac{m}{8}$.

If $d = 2m+1$ is odd, then

$$e(V) = e(T_m(n-1)) + m + n - 1\,.$$

[1]The reference in [17] to Lenz (1955) is erroneous and should be replaced by Erdős (1960), which is [7] in our list of references.

This example shows that, unlike the case $d = 3$ where $e(3,n) = 2n - 2$ is linear in n, if $d \geq 4$ then $e(d,n)$ is quadratic in n. In fact, Erdős proved (in [7]) that the leading term $\frac{1}{2}(1 - \frac{1}{m})n^2$ in the preceding expression for $e(V)$ is asymptotically the right order of magnitude of $e(d,n)$ for $n \longrightarrow \infty$ (d fixed), i.e., $e(d,n) \sim \frac{1}{2}(1 - \frac{1}{m})n^2$ as $n \longrightarrow \infty$.

The asymptotic behaviour of the error term, $e(d,n) - \frac{1}{2}(1 - \frac{1}{m})n^2$, is also known: for $d \geq 4$ even it is $\Theta(n)$ (i.e., bounded between $c_1 n$ and $c_2 n$, for some constants $0 < c_1 < c_2 < \infty$, see again [7]), and for $d \geq 5$ odd it is $\Theta(n^{4/3})$, see [8].

Remarks.

1) From Construction 3.2 we can easily extract some lower bounds on $e_k(d,n)$. For $2 \leq k \leq m$ $(= \lfloor \frac{d}{2} \rfloor)$ we find that

$$\left(\binom{n}{k} \geq \right) \; e_k(d,n) \geq c_{d,k} \cdot n^k \,,$$

and

$$\left(\binom{n}{d-k} \geq \right) \; e_{d-k}(d,n) \geq c_{d,d-k} \cdot n^k \,,$$

where $c_{d,k}$ and $c_{d,d-k}$ are suitable positive constants. (For n large, we can take both $c_{d,k}$ and $c_{d,d-k}$ close to $\binom{m}{k}/m^k$.)

Thus we know the exact rate of growth of $e_k(d,n)$ (d, k fixed, $d \geq 4$, $n \to \infty$) when $k \leq m$. When $m < k \leq d$, we encounter a gap between the exponent k in the upper bound and the exponent $d - k$ in the lower bound. This gap has been narrowed in a few cases. Thus

$$cn^2 \leq e_3(4,n) \leq c' n^{65/23} \,.$$

(See Remark 3 below for the lower bound, and [3] for the upper bound.) Also $e_3(5,n) \leq c'' n^{26/9}$, see [1].

2) If we do not insist that the repeated distance ($\sqrt{2}$) be the diameter of V, then we can improve the construction slightly. Instead of choosing V_i from one quarter-circle of C_i draw $\lceil \frac{1}{4} \# V_i \rceil$ squares inscribed in C_i and choose V_i from their vertices, omitting vertices from one square if $\# V = n$ is not divisible by 4. This will increase the number of occurrences of the distance $\sqrt{2}$ by almost $n - m$.

Moreover, if $n \equiv 0 (\text{mod } 4)$ and $d = 2m$, then the increase will be exactly $n - m$ and the repeated distance ($\sqrt{2}$) will occur exactly $e(T_m(n)) + n$ times.

This construction has been used by P. Brass to determine the maximum number of occurrences of unit distances in a set of n points in $\mathbb{R}^4$ *exactly*, see [4] and [20].

3) The Lenz construction can be improved in dimension 4 by using unequal circles: Given a number $n \geq 5$, choose k to be the smallest odd integer $\geq \frac{n}{2}$. Thus $k \geq 3$. Let P be a regular k-gon of diameter 1 in the e_1, e_2 plane, centered at the origin, and let C_1 be the circumcircle of P. The radius r_1 of

C_1 satisfies

$$r_1 = \frac{1}{2\cos\left(\frac{\pi}{2k}\right)} \le \frac{1}{\sqrt{3}} \, .$$

Define $r_2 = \sqrt{1-r_1^2}$ $\left(\ge \sqrt{\frac{2}{3}} > \frac{1}{2}\right)$, and let C_2 be a circle of radius r_2 in the e_3, e_4 plane which is centered at the origin. Draw a chord of length 1 in C_2, and let A_2 be the short arc of C_2 determined by that chord. Let V consist of the k vertices of P, the two endpoints of A_2 and $n-k-2$ additional points on A_2. V has diameter 1 and the diameter graph $D(V)$ is a complete bipartite graph $K_{k,n-k}$, with an additional hamiltonian circuit on one side, and one additional edge on the other side. This graph has exactly $k(n-k+1)$ triangles (one more if $k=3$). This number is approximately $\frac{n^2}{4}$. Thus

$$e_3(4,n) = \Omega(n^2) \, .$$

Agarwal and Sharir [2] recently showed that $e_3(4,n) = O(n^{2+\varepsilon})$ for all $\varepsilon > 0$. (Their result is even more general.)

4 The standard examples

In this section we describe, for all $d \ge 3$ and $n > d$, certain sets $V \subset \mathbb{R}^d$ with $e_d(V) = \#V = n$, i.e., sets V of n points in $\mathbb{R}^d$ such that the number of regular $(d-1)$-simplices of edge length diam(V) with vertices in V is exactly n (or, in other words, such that the number of d-cliques in the diameter graph $D(V)$ is exactly n).

Our sets V will consist of the $d+1$ vertices of a regular d-simplex $\triangle^d$, plus $n-d$ additional points that lie on certain circular arcs that connect pairs of vertices of $\triangle^d$. The sets described in Construction 3.1 are just a special case of these sets. The sets described here will be called *standard examples.*

Later, in Theorem 7.1, we will show that for every set V of n points in $\mathbb{R}^3$ $(n \ge 4)$ one has $e_3(V) \le n$, where equality holds iff V is similar to a standard example, i.e., is obtained from a standard example by dilatation and a rigid motion.

Define $\lambda_d = \sqrt{2(1+\frac{1}{d})}$. Let $\triangle^d \subset \mathbb{R}^d$ be a regular d-simplex of edge length λ_d, with vertices $p_0, p_1, \ldots, p_d$. We assume that the barycenter of the vertices of $\triangle^d$ is the origin, i.e.,

$$\sum_{i=0}^{d} p_i = o \, . \tag{4}$$

For $i, j \in \{0, 1, \ldots, d\}$ we have

$$||p_i - p_j||^2 = \begin{cases} 0 & \text{if} \quad i = j \, , \\ \lambda_d^2 & \text{if} \quad i \ne j \, . \end{cases}$$

Thus

$$||p_i - p_j||^2 = \langle p_i - p_j, p_i - p_j\rangle = ||p_i||^2 - 2\langle p_i, p_j\rangle + ||p_j||^2 = (1 - \delta_{ij})\lambda_d^2 \,.$$

Summation over j, for i fixed, yields

$$(d+1)\,||p_i||^2 - 2\,\langle p_i, \sum_{j=0}^{d} p_j\rangle + \sum_{j=0}^{d} ||p_j||^2 = d\,\lambda_d^2$$

or, in view of (4),

$$(d+1)\,||p_i||^2 + \sum_{j=0}^{d} ||p_j||^2 = 2(d+1)\,.$$

Since this holds for all i, the norm $||p_i||$ does not depend on i and, in fact,

$$||p_i|| = 1 \text{ for } i = 0, 1, \ldots, d \text{ and } \langle p_i, p_j\rangle = -\frac{1}{d} \text{ for } 0 \leq i < j \leq d\,.$$

Note that if $x \neq o$, then for some $j \in \{0, 1, \ldots, d\}$ we have $\langle x, p_j\rangle \leq 0$, and therefore

$$||x - p_j||^2 = \langle x - p_j, x - p_j\rangle = ||x||^2 - 2\langle x, p_j\rangle + ||p_j||^2 \geq ||x||^2 + 1 > 1\,.$$

In other words, we have

Observation 4.1. *All the vertices of $\triangle^d$ lie on the unit sphere (= the boundary of the ball of radius 1 centered at o), and no other ball of radius 1 includes all vertices of $\triangle^d$.*

Remark: This is, of course, a special case of the general result from convexity on the unicity of the circumscribed sphere. But we could not locate this result in the literature, and in our case it follows easily from the calculations above.

For $i \neq j$, denote by c_{ij} the center of the $(d-2)$-face of $\triangle^d$ complementary to the edge $[p_i, p_j]$. Thus

$$c_{ij} = \frac{1}{d-1} \sum \{p_k : k \in \{0, 1, \ldots, d\} \setminus \{i, j\}\} = -\frac{1}{d-1}(p_i + p_j)\,. \tag{5}$$

One can easily verify that

$$||p_i - c_{ij}|| = ||p_j - c_{ij}|| = \sqrt{\frac{d+1}{d-1}} \quad \left(= \sqrt{1 + \frac{2}{d-1}}\right),$$

and that the vectors $p_i - c_{ij}$ and $p_j - c_{ij}$ are orthogonal to $p_k - c_{ij}$ for all $k \in \{0, 1, \ldots, d\} \setminus \{i, j\}$.

Denote by C_{ij} the circle centered at c_{ij} that passes through p_i and p_j, and by A_{ij} the short open arc of C_{ij} with endpoints p_i and p_j.

If $k \neq i, j$, then the vector $p_k - c_{ij}$ is perpendicular to any linear combination of $p_i - c_{ij}$ and $p_j - c_{ij}$. It follows, by the Pythagorean Theorem, that p_k is equidistant from all points of the circle C_{ij}, i.e., $||p_k - x|| = ||p_k - p_i|| = \lambda_d$ for all $x \in C_{ij}$. If $x \in A_{ij}$, then the distances from x to the endpoints p_i and p_j are smaller than $||p_i - p_j|| (= \lambda_d)$.

It follows that if the set V consists of the $d+1$ vertices $p_0, p_1, \dots, p_d$ of $\triangle^d$ and $n-d-1$ additional points on one of the arcs A_{ij}, as in Construction 3.1, then diam $(V) = \lambda_d$, and the only $(d-1)$-simplices of edge length λ_d with vertices in V are the $d+1$ facets of $\triangle^d$ and the convex hulls of the $n-d-1$ sets

$$\{x\} \cup \{p_k : k \in \{0, 1, \dots, d\} \setminus \{i, j\}\} ,$$

where $x \in V \cap A_{ij}$, i.e., $e_{d-1}(V) = n = \#V$.

What happens if we add points on different arcs A_{ij} to the vertex set of $\triangle^d$? The key to the answer lies in the following lemma.

Lemma 4.2. *Suppose* $x \in A_{ij}$, $y \in A_{kl}$ $(i, j, k, l \in \{0, 1, \dots, d\}, i \neq j,\ k \neq l)$. *Then*

$$\text{(a)}\ ||x - y|| < \lambda_d \quad \text{if} \quad \{i, j\} \cap \{k, l\} \neq \emptyset ,$$

$$\text{(b)}\ ||x - y|| > \lambda_d \quad \text{if} \quad \{i, j\} \cap \{k, l\} = \emptyset .$$

The proof of this lemma is quite technical and is deferred to the end of this section.

From Lemma 4.2 we conclude that if V consists of the $d+1$ vertices $p_0, p_1, \dots, p_d$ of $\triangle^d$ and $n-d-1$ additional points chosen from $\bigcup\{A_{ij} : 0 \leq i < j \leq d\}$, then the number of regular $(d-1)$-simplices of edge length λ_d with vertices in V is exactly n. But diam (V) equals λ_d iff every two arcs A_{ij} that carry points of V have an endpoint in common. This condition can be met in two ways: either all arcs A_{ij} that carry points of V have a common endpoint p_i, or there are just three arcs that form a “triangle” with vertices A_{ij} , A_{jk} , A_{ik}. This yields the definition of a standard example.

Definition 4.3. A *standard example* of n points in $\mathbb{R}^d$ $(n > d)$ consists of the $d+1$ vertices $p_0, p_1, \dots, p_d$ of $\triangle^d$ and $n-d-1$ additional points chosen either from $\bigcup_{j=1}^{d} A_{0j}$, or from $A_{01} \cup A_{02} \cup A_{12}$.

Proof of Lemma 4.2:

(a) If $\{i, j\} = \{k, l\}$, then both points x, y lie on the same arc A_{ij}. Since A_{ij} is an open short circular arc, the distance $||x - y||$ is strictly smaller than the distance between the endpoints p_i and p_j. But $||p_i - p_j|| = \lambda_d$.

If $\#(\{i, j\} \cap \{k, l\}) = 1$, then we may assume that i, j, k are distinct and $l = i$, i.e., $x \in A_{ij}$, $y \in A_{ik}$. Fix x and regard $y = y(t)$ as a point that moves on A_{ik} at a constant speed from p_i to p_k. Since $||x - p_i|| < \lambda_d = ||x - p_k||$, it suffices to show that the distance $||x - y(t)||$ increases monotonically as $y(t)$ moves on A_{ik} from p_i to p_k. We will show that $\frac{d}{dt}\left(||x - y(t)||^2\right) > 0$.

Note that

$$x - y(t) = (x - c_{ik}) - (y(t) - c_{ik}) ,$$

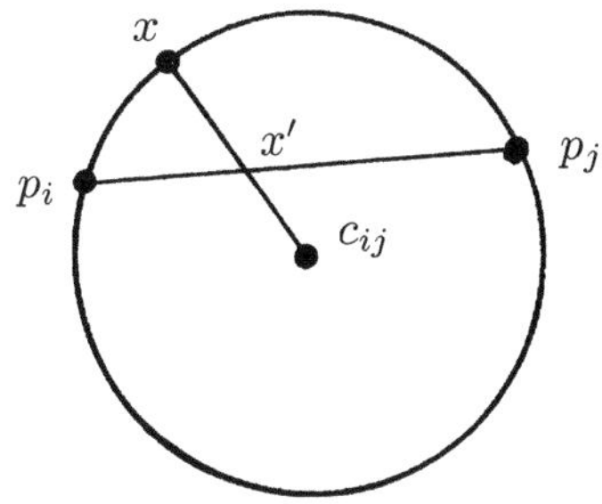

Fig. 1.

and thus

$$||x - y(t)||^2 = ||x - c_{ik}||^2 + ||y(t) - c_{ik}||^2 - 2\langle x - c_{ik}, y(t) - c_{ik}\rangle .$$

Here $||y(t) - c_{ik}||^2$ is constant, since $y(t)$ moves on A_{ik}. It follows that

$$\frac{d}{dt}(||x - y(t)||^2) = -2\langle x - c_{ik}, y'(t)\rangle .$$

In order to prove that the derivative is positive, it suffices to show that

$$\langle x, y'(t)\rangle \ < \ \langle c_{ik}, y'(t)\rangle . \tag{6}$$

Looking at Fig. 1, we see that the point x can be written as an affine combination of p_i, p_j and c_{ij}, where the coefficients of p_i, p_j are positive, and the coefficient of c_{ij} is negative, i.e.,

$$x = \alpha p_i + \beta p_j - \gamma c_{ij}\,, \quad \alpha > 0, \beta > 0, \gamma > 0, \quad \alpha + \beta + \gamma = 1 \tag{7}$$

(α, β, γ are the barycentric coordinates of x with respect to the affine basis p_i, p_j, c_{ij} of the plane aff C_{ij}).

To get (6), it suffices to show that:

$$\langle p_i, y'(t)\rangle < \langle c_{ik}, y'(t)\rangle\,, \tag{8}$$

$$\langle p_j, y'(t)\rangle = \langle c_{ik}, y'(t)\rangle\,, \tag{9}$$

$$\langle c_{ij}, y'(t)\rangle > \langle c_{ik}, y'(t)\rangle\,. \tag{10}$$

(Combining (8), (9) and (10) with the coefficients α, β, γ and using (7), we get (6).)

Now we proceed to prove these three statements one by one.

Since $y(t)$ moves on the circular arc A_{ik} at a constant speed from p_i to p_k, the vector $y'(t)$ is tangent to A_{ik} at $y(t)$, pointing towards p_k (away from p_i), and is perpendicular to the radius vector $y(t) - c_{ik}$, i.e.,

$$\langle p_i, y'(t)\rangle \ < \ \langle y(t), y'(t)\rangle = \langle c_{ik}, y'(t)\rangle \ < \ \langle p_k, y'(t)\rangle\,, \tag{11}$$

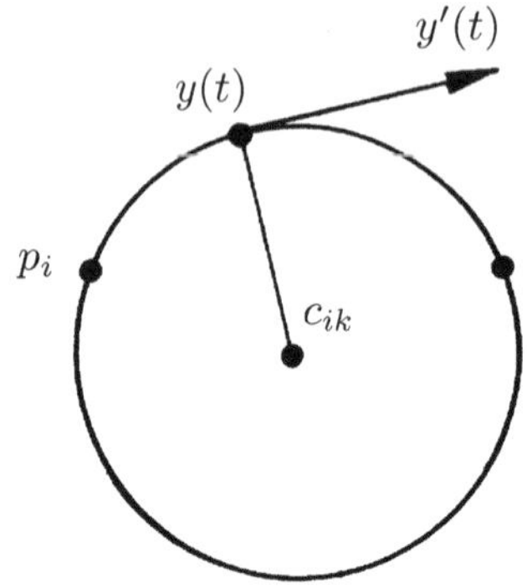

Fig. 2.

see Fig. 2.

This proves (8). Moreover, since the vector $y'(t)$ lies in the plane aff C_{ik}, it is a linear combination of $p_i - c_{ik}$ and $p_k - c_{ik}$. Therefore $y'(t)$ is perpendicular to $p_\nu - c_{ik}$ for all $\nu \in \{0, 1, \ldots, d\} \setminus \{i, k\}$, in particular for $\nu = j$. This means that $\langle p_j, y'(t)\rangle = \langle c_{ik}, y'(t)\rangle$, which proves (9).

As noted earlier (see (5)),

$$c_{ij} = \frac{-1}{d-1}\,(p_i + p_j) = c_{ik} + \frac{1}{d-1}(p_k - p_j)$$

and

$$\langle p_j, y'(t)\rangle \overset{(9)}{=} \langle c_{ik}, y'(t)\rangle \overset{(11)}{<} \langle p_k, y'(t)\rangle\,.$$

It follows that $\langle c_{ij}, y'(t)\rangle > \langle c_{ik}, y'(t)\rangle$, which proves (10). □

(b) Assume i, j, k, l are all different. Fix x on A_{ij}, and regard y as a moving point on A_{kl}. Denote by m_{kl} the midpoint of the arc A_{kl} and assume, without loss of generality, that y lies between p_k and m_{kl} (possibly $y = m_{kl}$). Note that $||x - p_k|| = \lambda_d$. We assume that $y = y(t)$ moves on A_{kl} at a constant speed from p_k to m_{kl} and propose to show that the distance $||x - y(t)||$ is a strictly increasing function of t. It suffices to show that $\frac{d}{dt}(||x - y(t)||^2) > 0$ for $y(t)$ strictly between p_k and m_{kl}.

As in part (a) we have

$$||x - y(t)||^2 = ||x - c_{kl}||^2 + ||y(t) - c_{kl}||^2 - 2\,\langle x - c_{kl}, y(t) - c_{kl}\rangle\,,$$

and therefore

$$\frac{d}{dt}\left(||x - y(t)||^2\right) = -2\,\langle x - c_{kl}, y'(t)\rangle\,.$$

For the derivative to be positive we must have

$$\langle x, y'(t)\rangle < \langle c_{kl}, y'(t)\rangle\,. \tag{12}$$

Exactly as in part (a) (see (7)), it suffices to show that

$$\langle p_i, y'(t)\rangle = \langle c_{kl}, y'(t)\rangle\,, \tag{13}$$

$$\langle p_j, y'(t)\rangle = \langle c_{kl}, y'(t)\rangle\,, \tag{14}$$

$$\langle c_{ij}, y'(t)\rangle > \langle c_{kl}, y'(t)\rangle\,. \tag{15}$$

The vector $y'(t)$ is tangent to the circular arc A_{kl} at the point $y(t)$, and points towards m_{kl} and p_l, away from p_k. Therefore

$$\langle p_k, y'(t)\rangle \;<\; \langle y(t), y'(t)\rangle = \langle c_{kl}, y'(t)\rangle \;<\; \langle m_{kl}, y'(t)\rangle\,. \tag{16}$$

Since the midpoint $\frac{1}{2}(p_k + p_l)$ of the chord $[p_k, p_l]$ of A_{kl} is an interior point of the radius $[c_{kl}, m_{kl}]$, we find that

$$\langle c_{kl}, y'(t)\rangle \;<\; \langle \frac{1}{2}(p_k + p_l), y'(t)\rangle\,. \tag{17}$$

As in part (a), we find that the vector $y'(t)$ is perpendicular to $p_\nu - c_{kl}$ for all $\nu \in \{0, \ldots, d\} \setminus \{k, l\}$, in particular for $\nu = i$ and $\nu = j$. Thus

$$\langle \frac{1}{2}(p_i + p_j), y'(t)\rangle = \langle p_i, y'(t)\rangle = \langle p_j, y'(t)\rangle = \langle c_{kl}, y'(t)\rangle\,. \tag{18}$$

This proves (13) and (14).

From (17) and (18) we conclude

$$\langle p_k + p_l, y'(t)\rangle \;>\; \langle p_i + p_j, y'(t)\rangle\,.$$

But

$$c_{ij} = -\frac{1}{d-1}(p_i + p_j) \;\text{ and }\; c_{kl} = -\frac{1}{d-1}(p_k + p_l)\,.$$

Hence $\langle c_{ij}, y'(t)\rangle \;>\; \langle c_{kl}, y'(t)\rangle$, which proves (15). □

5 A proof that $e_{d+1}(d, n) = 1$

In this paragraph we prove

Theorem 5.1. *A subset V of $\mathbb{R}^d$ with diameter $\lambda > 0$ contains the vertices of at most one regular d-simplex of edge length λ.*

Using the notation introduced in the Introduction, we just show that $e_{d+1}(d, n) = 1$ for all $n \geq d + 1$; see Corollary 5.4 below.

Proof: Assume $\triangle_1^d$ and $\triangle_2^d$ are two regular d-simplices of edge length λ with vertices in V. Since nothing changes when we dilate or translate V, we shall assume, without loss of generality, that $\lambda = \lambda_d = \sqrt{2(1 + \frac{1}{d})}$, and that $\triangle_1^d$ is the regular simplex $\triangle^d$ of edge length λ_d with vertices $p_0, \ldots, p_d$, as described in the previous paragraph.

Thus $\langle p_i, p_j\rangle = -\frac{1}{d}$ for $i \neq j$ and $= 1$ for $i = j$ (see § 4, before Observation 4.1), and $\sum_{i=0}^{d} p_i = o$.

Definition 5.2. The set of all points $z \in \mathbb{R}^d$ that satisfy $||z - p_i|| \leq \lambda_d$ for $i = 0, \ldots, d$ is called the *Reuleaux simplex* $R(\triangle^d)$ associated with $\triangle^d$.

Remark 5.3. The set $R(\triangle^d)$ is the intersection of the balls of radius λ_d centered at the vertices of $\triangle^d$, i.e.,

$$R(\triangle^d) = \bigcap_{i=0}^{d} B(p_i, \lambda_d)\,,$$

where $B(p, \lambda)$ $(= p + \lambda B(0, 1))$ is the closed ball of radius λ centered at p. Our $R(\triangle^d)$ is, of course, a special case of the "Kugelpolyeder" introduced by A. Heppes in [11] to prove Vázsonyi's Conjecture (see [17], p. 215, for a modern presentation of the proof due to Heppes). In dimension 2, $R(\triangle^d)$ is the well known Reuleaux triangle, which is a set of constant width $\lambda_2 = \sqrt{3}$, but for $d \geq 3$, the diameter of $R(\triangle^d)$ is larger than λ_d, as follows from Lemma 4.2 (b).

In fact, the closures of the arcs A_{ij}, $0 \leq i < j \leq d$, can be viewed as the "edges" of $R(\triangle^d)$, and Lemma 4.2 (b) says that the distance between two points lying in the relative interiors of disjoint "edges" is $> \lambda_d$. Hence diam $(R(\triangle^d)) > \lambda_d$. (This implies that $R(\triangle^d)$ $(d \geq 3)$ is not of constant width, since the width of $R(\triangle^d)$, even of $B(p_0, \lambda_d) \cap B(p_1, \lambda_d)$, in the direction of $p_1 - p_0$ is exactly λ_d.)

Back to the proof of Theorem 5.1. From Definition 5.2 it follows that $V \subset R(\triangle^d)$, since $\{p_0, p_1, \ldots, p_d\} \subset V$ and diam $V = \lambda_d$. In particular, the vertices of $\triangle_2^d$ are in $R(\triangle^d)$.

We are going to show that $R(\triangle^d)$ is inscribed in the unit sphere, same as $\triangle^d$, and touches the unit sphere only at the points $p_0, p_1, \ldots, p_d$. If this is true, then the second simplex $\triangle_2^d$ is also inscribed in the unit sphere. But by Observation 4.1, if a regular d-simplex of edge length λ_d lies within a ball of radius 1 in $\mathbb{R}^d$, then all its vertices lie on the boundary of that ball. Therefore the vertices of $\triangle_2^d$ lie in the intersection of $R(\triangle^d)$ with the unit sphere, and thus coincide with the vertices $p_0, \ldots, p_d$ of $\triangle^d$. Hence $\triangle_2^d = \triangle^d$.

It remains to show that if $z \in R(\triangle^d)$, then $||z|| \leq 1$, and equality holds iff $z \in \{p_0, p_1, \ldots, p_d\}$. The point z can be expressed as a linear combination of the p_i's: $z = \sum_{i=0}^{d} \zeta_i p_i$. Since $\sum_{i=0}^{d} p_i = o$, we also have $z = \sum(\zeta_i - \rho)p_i$, for any $\rho \in \mathbb{R}$. By choosing $\rho = \min\{\zeta_i : 0 \leq i \leq d\}$, we obtain a representation where all coefficients are non-negative, and at least one coefficient is zero.

Assume therefore, without loss of generality, that $z = \sum_{i=0}^{d} \zeta_i p_i$, where $\zeta_i \geq 0$ for $i = 1, 2, \ldots, d$ and $\zeta_0 = 0$. Define $\zeta = \sum_{i=1}^{d} \zeta_i$. If $\zeta < 1$, then

$z \in \text{int}(\triangle^d)$, hence $||z|| < 1$. If $\zeta = 1$, then z lies on the facet $[p_1, \ldots, p_d]$ of $\triangle^d$. Hence $||z|| \leq 1$, with equality iff $z = p_i$ for some $1 \leq i \leq d$.

There remains the case $\zeta > 1$. Note that

$$\langle z, p_0 \rangle = \langle \sum_{i=1}^{d} \zeta_i p_i \,, p_0 \rangle = \sum_{i=1}^{d} \zeta_i \langle p_i, p_0 \rangle = -\frac{1}{d} \sum_{i=1}^{d} \zeta_i = -\frac{\zeta}{d} \,.$$

Since $z \in R(\triangle^d)$, we have $||z - p_0|| \leq \lambda_d$. Thus

$$\begin{aligned} 2 + \tfrac{2}{d} &= \lambda_d^2 \geq ||z - p_0||^2 = \langle z - p_0, z - p_0 \rangle = \langle z, z \rangle - 2\langle z, p_0 \rangle + \langle p_0, p_0 \rangle \\ &= ||z||^2 + \tfrac{2}{d}\zeta + 1 > ||z||^2 + \tfrac{2}{d} + 1 \,. \end{aligned}$$

Therefore $||z||^2 < 1$, as required. □

Corollary 5.4. $e_{d+1}(d, n) = 1$.

Remark 5.5. The Reuleaux simplex $R(\triangle^d)$ is a source of many interesting questions. E.g., we do not even know the exact diameter of $R(\triangle^d)$.

6 Some Lemmata

In this section we state and prove four elementary geometric lemmata concerning distances in $\mathbb{R}^3$. These lemmata will be used in the proof of the case $d = 3$ of Schur's Conjecture (Conjecture 1.1).

Lemma 6.1. *Let H be an open half-plane in $\mathbb{R}^3$ with boundary line L. Direct L, and denote by P_α $(0 \leq \alpha \leq \pi)$ the rotation of $\mathbb{R}^3$ by α radians about the axis L. The half-planes H and $P_\alpha H$ bound a wedge of dihedral angle α. Given any two points $a, b \in H$, the distance between a and $P_\alpha b$ is an increasing function of α for $0 \leq \alpha \leq \pi$.* (In other words: *if a and b are dots on facing pages of a closed book, then the distance between a and b increases as we open the book.*)

Proof: We may assume, without loss of generality, that L is the y-axis , that H is the half-plane $x > 0$ of the (x, y)-plane $z = 0$, and that the origin is the point of L nearest to b. By an appropriate change of scale we can make the distance from b to L equal to 1. Thus $b = (1, 0, 0)$, $a = (x_0, y_0, 0)$ with $x_0 > 0$, and $P_\alpha b = (\cos\alpha, 0, \sin\alpha)$. Thus

$$||P_\alpha b - a||^2 = (\cos\alpha - x_0)^2 + y_0^2 + \sin^2\alpha = x_0^2 + y_0^2 + 1 - 2x_0 \cos\alpha \,,$$

which is an increasing function of α for $0 \leq \alpha \leq \pi$. □

In the sequel S_k stands for a regular k-simplex $(1 \leq k \leq 3)$.

Lemma 6.2. *Suppose S_1 is a closed line segment of length λ, and S_2 is an equilateral triangle of edge length λ, both in $\mathbb{R}^3$. If*

$$\text{diam}\,(S_1 \cup S_2) = \lambda \,, \tag{19}$$

then either

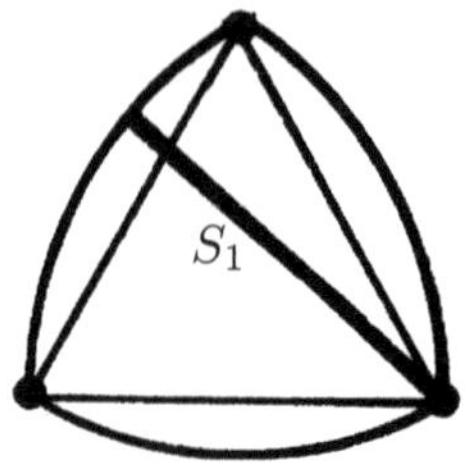

Fig. 3.

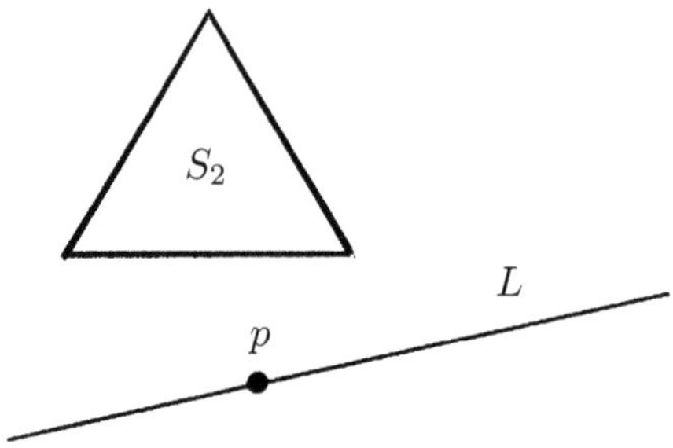

Fig. 4.

(a) *S_1 and S_2 share a vertex, or*

(b) *S_1 crosses* aff S_2*, i.e., the line* aff S_1 *meets the plane* aff S_2 *in a unique point, and this point lies in the relative interior of S_1.*

Proof: Assume S_1 does not cross aff S_2. There are three possible cases:

(i) $S_1 \subset$ aff S_2;

(ii) S_1 meets aff S_2 in a unique point, which is a vertex of S_1;

(iii) S_1 lies entirely in one of the two open half-spaces bounded by aff S_2.

In case (i), S_1 must be a diameter of the Reuleaux triangle that circumscribes S_2, and therefore one of the endpoints of S_1 must be a vertex of S_2 (Fig. 3).

In case (ii), let p be the endpoint of S_1 that lies in aff S_2, and let q be the other endpoint of S_1. The point p may be either a vertex of S_2, as required in the Lemma, or a non-extreme point of S_2, or it may lie outside S_2. If p is a non-extreme point of S_2, then p is the midpoint of a segment $[a, b] \subset S_2$. But then one of the distances $||q - a||$, $||q - b||$ is larger than λ, contrary to (19). If p lies outside S_2, we draw through p a line $L \subset$ aff S_2 that does not meet S_2 (Fig. 4).

Denote by H the open half-plane bounded by L that includes S_2, and by H' the open half-plane bounded by L that includes $S_1 \setminus \{p\}$. Denote by α the angle between H and H', $0 < \alpha < \pi$.

If we rotate H' around the axis L back to H, then q reaches a point q_1 in H. Since we have diam $(S_1 \cup S_2) = \lambda$, the distance from q (and also from p) to any vertex of S_2 is at most λ, and therefore, by the preceding Lemma, the distance from q_1 to any vertex of S_2 is smaller than λ. We conclude that p belongs to the Reuleaux triangle that circumscribes S_2, and q_1 lies in the interior of this Reuleaux triangle. Thus $||p - q_1|| < \lambda$. But $||p - q_1|| = ||p - q|| = \lambda$, a contradiction.

Note that in the cases considered so far (namely (i) and (ii)), we found that S_1 and S_2 share a vertex p. The other vertex q of S_1 is therefore at a distance λ from a vertex of S_2. Now we pass to case (iii):

S_1 lies in an open half-space bounded by aff S_2. Move S_1 towards S_2 in a direction perpendicular to aff S_2, until it touches aff S_2. Denote the translate of S_1 thus obtained by S_1'. The applied translation strictly decreases the distance between any point of S_1 and any fixed point of S_2. S_2 and S_1' satisfy the conditions of case (i) or (ii), but the distance between any vertex of S_1' and any vertex of S_2 is smaller than λ, contrary to our observation in the preceding paragraph. This shows that case (iii) is impossible, and completes the proof of Lemma 6.2. □

Lemma 6.3. *Suppose S_2 and S_2' are two distinct equilateral triangles of edge length λ in $\mathbb{R}^3$, and* diam $(S_2 \cup S_2') = \lambda$. *Then either*

(1) *S_2 and S_2' share an edge, or*

(2) *S_2 and S_2' have a unique common vertex p.*

In the second case, the edge of S_2 opposite p crosses aff S_2' *in the relative interior of* $\text{cone}_p(S_2')$, *and vice versa: the edge of S_2' opposite p crosses* aff S_2 *in* relint $\text{cone}_p(S_2)$.

Remark: The claim that S_2 and S_2' share at least one vertex is a special case of a result of V.L. Dol'nikov [5]: In the diameter graph of a finite set $V \subset \mathbb{R}^3$, every two odd circuits have at least one vertex in common.

Proof: By Theorem 5.1

$$\text{aff}\,(S_2 \cup S_2') = \mathbb{R}^3 .$$

If S_2 and S_2' have no vertex in common, then, by Lemma 6.2, each edge of S_2' crosses aff S_2, which is impossible.

For the rest of the proof, we may assume that S_2 and S_2' share exactly one vertex. Denote the common vertex by p, and the remaining vertices by q, r for S_2 and by q', r' for S_2'.

By Lemma 6.2, the edge $[q, r]$ of S_2 crosses aff S_2' in a point z, $z \in$ relint $[q, r]$. The point z belongs to the Reuleaux triangle that circumscribes S_2'.

If z is not in relint $\text{cone}_p(S_2')$ ($=$ relint $\text{cone}_p(q', r')$), then z lies in one of the two shaded caps in Fig. 5.

Assume, e.g., that z lies either on or beyond the edge $[p, q']$ of S_2'. But then aff $(S_2) \cap$ aff (S_2') is the line aff (p, z), and the edge $[q', r']$ of S_2' does not cross aff S_2, contrary to Lemma 6.2. □

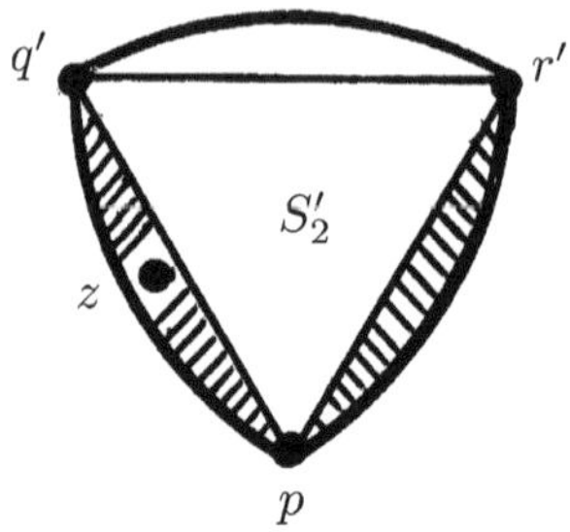

Fig. 5.

Lemma 6.4. *Suppose S_2 is an equilateral triangle and S_3 is a regular tetrahedron, both of edge length λ and both in $\mathbb{R}^3$. If* diam $(S_2 \cup S_3) = \lambda$, *then either*

(a) *S_2 is a facet of S_3, or*

(b) *S_2 and S_3 share an edge pq, and the edge of S_3 opposite pq crosses S_2.*

Proof: Assume S_2 is *not* a facet of S_3. By Lemma 6.3, S_2 shares (at least) a vertex with every facet of S_3. It follows that S_2 shares with S_3 exactly two vertices, say p and q. Denote the remaining vertices of S_3 by r, s, and the remaining vertex of S_2 by t, and denote by o the midpoint of pq. Then t lies on the circle C with center o and radius $\frac{\sqrt{3}}{2} \cdot \lambda$ that passes through r and s. Note that $||r - s|| = \lambda$.

If a point $c \in C$ lies outside the short circular arc $\hat{rs}$ of C, then the distance from c to r or to s is larger than λ. It follows that $t \in$ relint $\hat{rs}$, and therefore the edge $[r, s]$ of S_3 crosses the triangle S_2 $(= [p, q, t])$. $\square$

7 Main result: $e_3(3, n) = n$.

Let us now return to the case $d = 3$ of the standard examples described in § 4.

Let $\triangle^3 \subset \mathbb{R}^3$ be a regular tetrahedron of edge length λ, with vertices p_0, p_1, p_2, p_3. For $i, j \in \{0, 1, 2, 3\}$, $i \neq j$, define c_{ij} to be the midpoint of the edge $[p_k, p_l]$ of $\triangle^3$ opposite $[p_i, p_j]$ ($\{i, j, k, l\} = \{0, 1, 2, 3\}$).

Denote by C_{ij} the circle (of radius $\frac{\sqrt{3}}{2}\lambda$) with center c_{ij} that passes through p_i and p_j, and let A_{ij} be the open short circular arc of C_{ij} with endpoints p_i, p_j.

If V is a set of n points ($n \geq 4$) that consists of $\{p_0, p_1, p_2, p_3\}$ and $n - 4$ additional points of $\bigcup\{A_{ij} : 0 \leq i < j \leq 3\}$, then there are exactly n equilateral triangles of edge length λ with vertices in V.

First, there are the four facets of $\triangle^3$. Every additional point $v \in V \cap A_{ij}$ contributes the triangle $[p_k, p_l, v]$, $\{i, j, k, l\} = \{0, 1, 2, 3\}$, and nothing else; p_k and p_l are the only points of V whose distance from v is λ (see Lemma 4.2).

If we want the diameter of V to be λ, we must refrain from using opposite arcs A_{ij} and A_{kl} ($\{i,j,k,l\} = \{0,1,2,3\}$). Thus we may assume that

$$V \setminus \text{ vert } \triangle^3 \subset A_{01} \cup A_{02} \cup A_{03} \text{ or that } V \setminus \text{ vert } \triangle^3 \subset A_{12} \cup A_{23} \cup A_{31}\,.$$

Now we are ready to formulate (and prove) the main result of this paper.

Theorem 7.1.

(a) *If V is a set of n points in $\mathbb{R}^3$ ($n \geq 4$), then there are at most n equilateral triangles of edge length* diam (V) *with vertices in V.*

(b) *The maximum number n is attained iff V is one of the standard examples described above, i.e.,*

$$\text{vert } \triangle^3 \subset V \subset \text{ vert } \triangle^3 \cup A_{01} \cup A_{02} \cup A_{03}$$

or

$$\text{vert } \triangle^3 \subset V \subset \text{ vert } \triangle^3 \cup A_{12} \cup A_{23} \cup A_{31}\,.$$

Proof: Assume diam $(V) = \lambda$; we proceed by induction on n. The initial case $n = 4$ is obvious.

In the induction step $n - 1 \longrightarrow n$ ($n \geq 5$) we distinguish a number of cases that together cover all possibilities.

Case I: There is a point $p \in V$ that belongs to at most one triangle. (Here, and in the sequel, "triangle" stands for "equilateral triangle of edge length λ with vertices in V".)

Define $V' = V \setminus \{p\}$. If diam $(V') < \lambda$, then there is not even one triangle in V. Assume, therefore, that diam $V' = \lambda$. By the induction hypothesis, there are at most $n-1$ triangles in V'. Putting p back, we add at most one triangle. In order to obtain n triangles on V, we must have $n-1$ triangles on V', and one triangle that uses p. By the induction hypothesis, V' must consist of the four vertices p_0, p_1, p_2, p_3 of a regular tetrahedron $\triangle^3$ of edge length λ, plus $n-5$ points on the arcs A_{ij}, $0 \leq i < j \leq 3$, never using two opposite arcs. By Lemma 6.4, the (unique) triangle that uses p must share an edge $[p_i, p_j]$ with $\triangle^3$, and p must lie on the arc A_{ij}. Since

$$\text{diam } (V) = \text{ diam } (V') = \text{ diam } (\triangle^3) = \lambda\,,$$

the opposite arc A_{kl} ($\{i,j,k,l\} = \{0,1,2,3\}$) must be free of points of V' (Lemma 4.2).

Case II: There are three triangles with a common edge $[a,b]$. The tips q_1, q_2, q_3 of these triangles lie on a circle C of radius $\frac{\sqrt{3}}{2}\lambda$ with center $c = \frac{1}{2}(a+b)$, perpendicular to $[a,b]$. All the angles $\sphericalangle q_i c q_j$ are $\leq 2\sin^{-1}\left(\frac{\sqrt{3}}{3}\right) \sim 70^o 32'$.

It follows that q_1, q_2 and q_3 lie on a short arc of C. Assume q_2 lies between q_1 and q_3 on that arc. We will show that the only triangle that uses q_2 is $[q_2, a, b]$. Assume, on the contrary, that there is another triangle $T = [q_2, x, y]$. This triangle does not use q_1, nor q_3, since $||q_i - q_2|| < ||q_3 - q_1|| \leq \lambda$ for

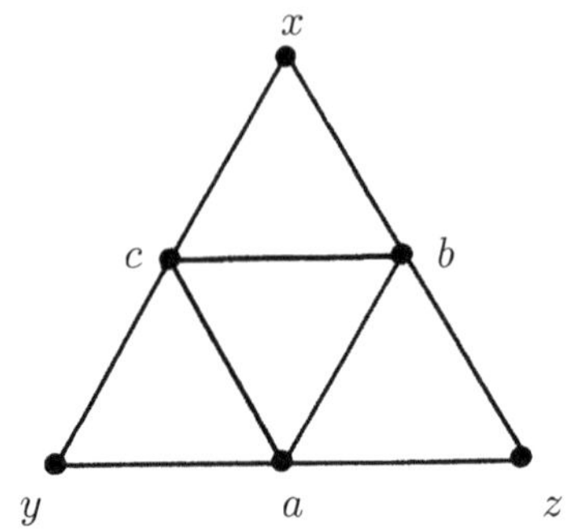

Fig. 6.

$i = 1, 3$. But T does share a vertex with $[a, b, q_1]$ and with $[a, b, q_3]$, by Lemma 6.3.

Assume, without loss of generality, that $T = [q_2, a, y]$ with $y \neq b$. If $y \in$ aff $\{q_2, a, b\}$, then $||y - b|| = \sqrt{3} \cdot \lambda > \lambda$, which is impossible. Assume therefore, without loss of generality, that y lies in the open half space of $\mathbb{R}^3$ bounded by aff $\{q_2, a, b\}$ that contains q_3 (and misses q_1).

But then the two triangles $[a, b, q_1]$ and $T = [q_2, a, y]$ share a unique vertex a, and the two cones $\text{cone}_a(b, q_1)$ and $\text{cone}_a(q_2, y)$ have only the apex a in common. This contradicts the second part of Lemma 6.3.

Thus we have shown that q_2 belongs to only one triangle, and we are back in Case I.

Case III: There is a regular tetrahedron $\triangle^3$ of edge length λ with vertices in V. If the only triangles in V are the facets of $\triangle^3$, then there are only four triangles, and $4 < n$. If there is another triangle T, then, by Lemma 6.4, T shares an edge with $\triangle^3$. This edge belongs to three triangles, and we are back in Case II.

Case IV: There are four triangles $[a, b, c]$, $[x, b, c]$, $[a, y, c]$, $[a, b, z]$ with vertices in V (see Fig. 6).

If, say, $x = y$, then $[a, b, c, x]$ is a regular tetrahedron, and we are back in Case III. Assume, therefore, that $x \neq y \neq z \neq x$.

If there are no other triangles, then we are done ($4 < 6 \leq n$). If there is another triangle T, then T shares a vertex with each of the triangles listed above, by Lemma 6.3. If T shares an edge with $[a, b, c]$, then we are back in Case II. If T contains only one vertex of $[a, b, c]$, say a, then it must contain also x, but then $||x - a|| = \lambda$, $[a, b, c, x]$ is a regular tetrahedron and we are back in Case III.

Case V: All triangles in V have a common vertex.

There are at most $n - 1$ segments $[p, x]$ of length λ with $x \in V$. Every triangle uses two of these segments. If the number of triangles is $\leq n - 1$, we are done. Otherwise we find a segment $[p, x]$ that is an edge of three triangles, which leads us back to Case II.

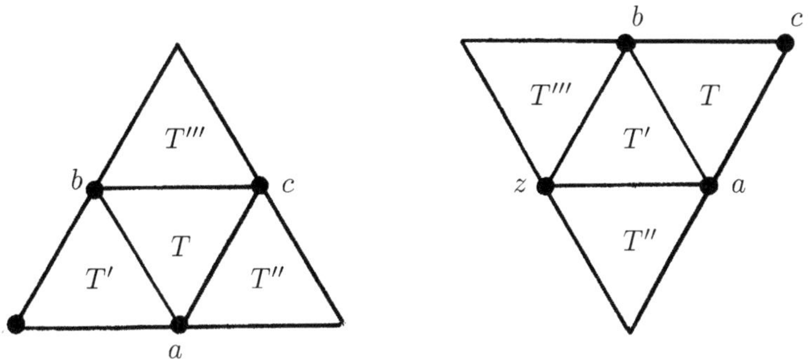

Fig. 7.

Case VI: There are two (distinct) triangles $T = [a, b, c]$ and $T' = [a, b, z]$ with common edge $[a, b]$.

Let T'' be another triangle. By Lemma 6.3, T'' shares a vertex with T, and also with T'. Thus, if neither a nor b are in T'', then both c and z are vertices of T'', hence $||c - z|| = \lambda$, and $[a, b, c, z]$ is a regular tetrahedron in V. This leads us back to Case III.

Assume, therefore, that every other triangle T'' contains either a or b; but not both. (Otherwise we return to Case II.) If all triangles T'' use a (or all use b), then we are back in Case V. Thus we may assume that one triangle T'' uses a, but not b, and another triangle T''' uses b, but not a. By Lemma 6.3, T'' and T''' share a vertex y.

If $c \neq y \neq z$, then the triangle $[a, b, y]$ takes us back to Case II, and if $y = c$ or $y = z$, then T, T', T'' and T''' take us back to Case IV, see Fig. 7 (left hand side: $y = c$; right hand side: $y = z$).

Now we come to the last case.

Case VII: Two distinct triangles never share an edge, or (in view of Lemma 6.3): Every two distinct triangles in V have a unique vertex in common.

Suppose T_1 and T_2 are triangles with a common vertex a. If all other triangles use a as well, then we are back in Case V. If not, then there is another triangle T_3 that does not contain a. If b is the common vertex of T_3 and T_1, and c is the common vertex of T_3 and T_2, then $T_4 = [a, b, c]$ is an equilateral triangle. T_4 shares the edge $[a, b]$ with T_1, $[a, c]$ with T_2 and $[b, c]$ with T_3, contrary to the defining hypothesis of Case VII. □

8 Concluding Remarks

1) Each of the extremal sets V in Theorem 7.1 (b) has exactly $2n - 2$ diametrical segments. I.e., these sets are also extremal with respect to the Grünbaum-Heppes-Straszewicz theorem ($e(3, n) = 2n - 2$).

2) Let V be a set of n points in $\mathbb{R}^3$. By Theorem 7.1, $e_3(V) = n$ implies $e_4(V) = 1$. If $e_4(V) = 0$, then the maximum possible value of $e_3(V)$ is $n-1$ provided $n \geq 6$. The sets $V \subset \mathbb{R}^3$ with $\#V = n$, $e_3(V) = n-1$ and $e_4(V) = 0$ can be described explicitly. They all consist of a central point z and $n-1$ points on a sphere of radius diam (V) centered at z, and they all satisfy $e(V) = 2n-2$, same as the extremal sets of Theorem 7.1.

3) Zvi Schur's manuscript (see the acknowledgements below) contains a lengthy treatment of the four-dimensional case. He even claims to have proved the case $d = 4$ of Conjecture 1.1. Unfortunately, so far we have not been able to verify the details. We may come back to these matters in a subsequent paper.

4) We wish to thank the referees for their valuable comments and, in particular, for rounding out the background material in paragraphs 2 and 3.

References

[1] B. M. Ábrego, S. Fernández-Merchant: On the maximum number of equilateral triangles I. *Discrete Comput. Geom.* **23** (2000), 129-135.

[2] P. K. Agarwal, M. Sharir: On the number of congruent simplices in a point set. In: 'SCG 01', *17th ACM Symp. Comput. Geom.* (2001), 1-9.

[3] T. Akutsu, H. Tamaki, T. Tokuyama: Distribution of distances and triangles in a point set and algorithms for computing the largest common point sets. *Discrete Comput. Geom.* **20** (1998), 307-331, ACM Symposium on Computational Geometry (Nice, 1997).

[4] P. Brass: On the maximum number of unit distances among n points in dimension four. In: *Intuitive Geometry*, I. Bárány et al., eds., Bolyai Soc. Math. Studies **4** (1997), 277-290.

[5] V. L. Dol'nikov: Some properties of graphs of diameters. *Discrete Comput. Geom.* **24** (2000), 293-299.

[6] P. Erdős: On sets of distances of n points. *Amer. Math. Monthly* **53** (1946), 248-250.

[7] P. Erdős: On sets of distances of n points in Euclidean space. *Magyar Tudom. Akad. Matem. Kut. Int. Közl.* (*Publ. Math. Inst. Hung. Acad. Sci.*) **5** (1960), 165-169.

[8] P. Erdős, J. Pach: Variations on the theme of repeated distances. *Combinatorica* **10** (1990), 261-269.

[9] B. Grünbaum: A proof of Vázsonyi's conjecture. *Bull. Research Council Israel*, Section A 6 (1956), 77-78.

[10] H. Harborth: Solution to Problem 664A. *Elem. Math.* **29** (1974), 14-15.

[11] A. Heppes: Beweis einer Vermutung von A. Vázsonyi. *Acta Math. Acad. Sci. Hungar.* **7** (1956), 463-466.

[12] Y. S. Kupitz: *Extremal Problems in Combinatorial Geometry.* Aarhus University, Lecture Notes Series **53** (1979).

[13] Y. S. KUPITZ: On the maximal number of appearences of the minimal distance among n points in the plane. *Colloquia Math. Soc. János Bolyai* **63** (Intuitive Geometry, Szeged 1991), North-Holland, 1994, 217-244.

[14] Y. S. KUPITZ, H. MARTINI: On the isoperimetric inequalities for Reuleaux polygons. *Journal of Geometry* **68** (2000), 171-191.

[15] Y. S. KUPITZ, H. MARTINI, B. WEGNER: A linear-time construction of Reuleaux polygons. *Beiträge Algebra Geom.* **37** (1996), 415-427.

[16] Y. S. KUPITZ, M. A. PERLES: Extremal theory for convex matchings in convex geometric graphs. *Discrete Comput. Geom.* **15** (1996), 195-220.

[17] J. PACH, P. K. AGARWAL: *Combinatorial Geometry.* John Wiley & Sons, 1995.

[18] O. REUTTER: Problem 664A. *Elem. Math.* **27** (1972), p. 19.

[19] S. STRASZEWICZ: Sur un problème géométrique de P. Erdős. *Bull. Acad. Polon. Sci. Cl.* III, **5** (1957), 39-40, IV-V.

[20] P. VAN WAMELEN: The maximum number of unit distances among n points in dimension four. *Beiträge Algebra Geom.* **40** (1990), 475-477.

[21] D. R. WOODALL: Thrackles and Deadlock. In: Combinatorics, Proc. Conf. Comb. Math. (D. Welsh, ed.), Academic Press, London, 1971, 335-347.

About Authors

Yaakov S. Kupitz is at the Institute of Mathematics, The Hebrew University of Jerusalem, Jerusalem, ISRAEL *kupitz@math.huji.ac.il.*

Horst Martini is at the Faculty of Mathematics, University of Chemnitz, 09107 Chemnitz, GERMANY *horst.martini@mathematik.tu-chemnitz.de.*

Micha A. Perles is at the Institute of Mathematics, The Hebrew University of Jerusalem, Jerusalem, ISRAEL *perles@math.huji.ac.il.*

Acknowledgments

The work on this paper was inspired by a manuscript of Zvi Schur, who was born in Jerusalem in 1923. About 1942 he started to study mathematics at the Hebrew University. In 1947 he obtained his Master's degree (Thesis advisor: Prof. Michael Fekete; Topic: "Linear transformations of infinite sequences"). From 1953 to 1959 he worked as a teaching assistant at the Mathematics Institute of the Hebrew University, and did research towards a Ph.D. under the guidance of Prof. H. Hanani from the Technion in Haifa. Later he moved to Ramat Gan (near Tel-Aviv). There he taught mathematics in high school until his retirement, and died in 1996. He left a widow, Esther, two daughters and a son. The fourth author was a student of Zvi Schur in high school. The three living authors would like to thank Mrs. Schur, who made her husband's manuscript available to them.

Balanced Lines, Halving Triangles, and the Generalized Lower Bound Theorem

Micha Sharir
Emo Welzl

Abstract

A recent result by Pach and Pinchasi on so-called balanced lines of a finite two-colored point set in the plane is related to other facts on halving triangles in 3-space and to a special case of the Generalized Lower Bound Theorem for convex polytopes.

1 Introduction

The following three facts are related to each other.

Fact A Let R and B be two disjoint finite planar sets, so that $|R \cup B| = 2n$ is even and $R \cup B$ is in general position (i.e., no three points are collinear). Points in R and B are referred to as 'red' and 'blue,' respectively. A line ℓ is *balanced* (w.r.t. (R, B)) if ℓ passes through a red point and a blue point, and on both sides of ℓ, the number of red points minus the number of blue points is the same.

The number of balanced lines is at least $\min\{|R|, |B|\}$.

If R and B can be separated by a line (but also in other configurations), this number is attained.

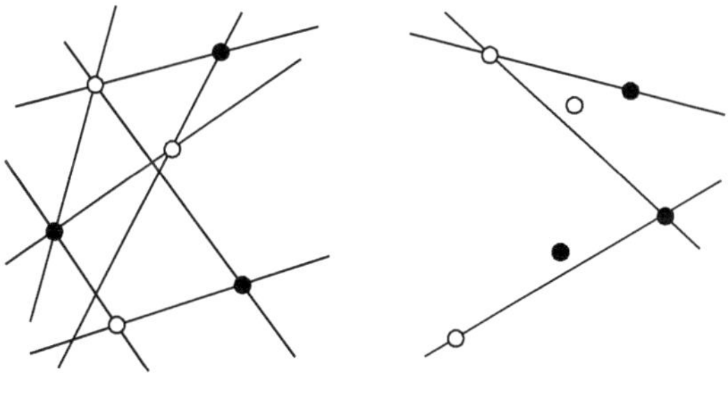

Fig. 1. Balanced Lines.

Fact B $n \in \mathbf{N}$. Let Q be a set of $2n+1$ points in 3-space in general position (i.e., no four points are coplanar). A *halving triangle of* Q is a triangle spanned by three points in Q such that the plane containing the three points equipartitions the remaining points of Q.

The number of halving triangles is at least n^2.

If Q is in convex position (but also in other configurations), this number is attained.

Fact C $d \in \mathbf{N}$, even[1]. Let $\mathcal{P}$ be a convex polytope which is the bounded intersection of $d+4$ halfspaces in general position in d-space, i.e., no $d+1$ bounding hyperplanes meet in a common point. (Therefore, either $\mathcal{P}$ is empty, or it is a simple convex d-polytope with at most $d+4$ facets. All vertices are incident to d edges. Our set-up is chosen in this way, in order to have a clean relation to the other statements.) Let its edges be oriented according to a generic linear function (edges are directed from smaller to larger value; 'generic' means that the function evaluates to distinct values at the vertices of $\mathcal{P}$).

The number of vertices with $d/2-1$ *outgoing edges is at most the number of vertices with* $d/2$ *outgoing edges.*

If $\mathcal{P}$ is empty (but also for other polytopes), this is tight.

(In fact, for any d (whether odd or even), and for all $1 \leq j \leq \lceil d/2 \rceil$, the number of vertices with $j-1$ outgoing edges is at most the number of vertices with j outgoing edges. And for d odd, and $j = \lceil d/2 \rceil$, these numbers are even equal. But that will not be relevant in our context.)

(A) has been proved by J. Pach and R. Pinchasi [7], answering a question of G. Baloglou's. (The statement in [7] is restricted to the case $|R| = |B| = n$. Then a balanced line must have the same number of red and blue points on each side, and there are at least n such balanced lines. But see Remark 2.1 below.) (C) is a very special case of the Generalized Lower Bound Theorem (GLBT) for simple polytopes, which—in turn—is part of the necessity part of the g-Theorem proved by R. P. Stanley [8] (thereby answering a conjecture by P. McMullen, who later provided also an alternative proof [6]); cf. also [10]. It was recently shown that (B) and (C) can be derived from each other [9]. In Section 2 we present a simple proof of the equivalence (A⇔B). That is, (A)–(C) are equivalent to each other.[2] In Section 3, we give an alternative proof of the equivalence (A⇔C). Clearly, that is already implied by (A⇔B⇔C), but we include here an argument for this specific setting for the sake of completeness.

On one hand, this means that the result of [7] admits a proof that is considerably simpler than their original proof, via the GLBT. On the other

[1] For reference to previous and forthcoming facts: $d+4=m=2n$.

[2] Of course, true statements are always equivalent; we mean that these facts can be derived from each other in a fashion that is significantly simpler compared to the proofs of the individual statements.

hand, Pach and Pinchasi's proof has merits of its own, because (i) no purely combinatorial proof of the GLBT (such as that in [7]) has been previously known (not even for the special case (C) equivalent to the balanced line problem), and (ii) that proof is based on allowable sequences in the dual, and thus (A) applies also for oriented matroids.

2 Balanced Lines and Halving Triangles

We first transform the balanced lines problem (A) to yet another problem (D) involving halving triangles in three dimensions, which appears to be new.

Assume that the points of $R \cup B$ (as in (A)) lie in the plane $z = 1$. Map these points onto the unit sphere centered at the origin O by: $R \ni r \mapsto r^* := r/\|r\|$, and $B \ni b \mapsto b^* := -b/\|b\|$. Let S_0 denote the resulting set of projected points, and put $S = S_0 \cup \{O\}$. By a small perturbation of $R \cup B$ that does not change the combinatorial type of this set, we may assume that S is in general position.

Observe the following properties, whose proofs are straightforward:

(i) The xy-plane, $\pi_0 : z = 0$, separates S_0 into sets of cardinalities $|R|$ and $|B|$.

(ii) For $r \in R$ and $b \in B$, the line passing through r and b is a balanced line iff the triangle Or^*b^* is a halving triangle of S. In particular, this establishes a correspondence between the balanced lines in $R \cup B$ and those halving triangles of S that are incident to O and are crossed by π_0 (i.e., π_0 intersects their relative interior).

(iii) The point O is an extreme point of S if and only if R and B are separated by a line.

Moreover, we can apply a reverse transformation as follows. Let Q be any set of $2n + 1$ points in 3-space in general position. Let $q_0 \in Q$ be a fixed point, and let π_0 be a plane of Q that passes through q_0 and through no other point of Q. Let π be a plane parallel to π_0. Map each point $q \in Q \setminus \{q_0\}$ to the point of intersection of π with the line that passes through q and q_0. Denote by R (resp. B) the subset of points on π that are images of points of Q that lie in the side of π_0 that contains (resp. does not contain) π.

(iv) A triangle $q_0q_1q_2$, for $q_1, q_2 \in Q$, is a halving triangle crossed by π_0 if and only if the line that passes through the images of q_1 and q_2 is a balanced line w.r.t. (R, B).

These properties imply the equivalence (A⇔D) of the result of Pach and Pinchasi and the following assertion (D).

Fact D $n \in \mathbf{N}$. Let Q be a set of $2n + 1$ points in 3-space in general position. Let $q_0 \in Q$ be a fixed point, and let π_0 be a plane of Q that passes through q_0 and through no other point of Q, and separates $Q \setminus \{q_0\}$ into two sets of cardinalities k and $2n - k$.

There are at least $\min\{k, 2n-k\}$ *halving triangles of* Q *that are incident to* q_0 *and are crossed by* π_0.

If q_0 is an extreme point of Q (but also in other situations), this number is attained.

Let us first show that, indeed, *for* q_0 *extreme, the number of halving triangles of* Q *that are incident to* q_0 *and are crossed by* π_0 *equals* $\min\{k, 2n-k\}$. Project $Q_0 = Q \setminus \{q_0\}$ centrally from q_0 onto a plane parallel to a supporting plane of Q at q_0; denote the projected set by Q_0^*. The plane π_0 projects to a line λ that separates Q_0^* into sets of cardinalities k and $2n-k$. It is then easy to check that, for points $q_1, q_2 \in Q_0$, the triangle $q_0q_1q_2$ is a halving triangle of Q crossed by π_0 if and only if the segment $q_1^*q_2^*$, connecting the images q_1^*, q_2^* of q_1, q_2, is a halving edge[3] of Q_0^* that is crossed by the line λ. By Lovász' Lemma [3, 5], the number of such edges is exactly $\min\{k, 2n-k\}$.

We proceed to a *proof of implication (D* $\Rightarrow$ *B)*. Suppose (D) holds. Consider a set Q of $2n+1$ points. Let π_q, for $q \in Q$, be pairwise parallel planes such that $\pi_q \cap Q = \{q\}$ for each $q \in Q$. Every halving triangle Δ of Q is crossed by exactly one of these planes which is also incident to a vertex of Δ (a plane crosses a triangle if it contains one of the three vertices, and separates the other two). Hence, there are at least

$$\sum_{i=1}^{2n+1} \min\{i-1, 2n+1-i\} = n^2$$

halving triangles, which implies (B). (By the preceding argument, equality is attained when Q is in convex position.)

Finally, let us provide the *proof of implication (B* $\Rightarrow$ *D)*. Suppose that assertion (D) is false. Thus there exist a set Q of $2n+1$ points, a parameter $0 \le k \le 2n$, a point $q_0 \in Q$ and a plane π_0 passing through q_0 and partitioning $Q \setminus \{q_0\}$ into two sets of cardinalities k and $2n-k$, such that the number c of halving triangles of Q incident to q_0 and crossed by π_0 is strictly smaller than $\min\{k, 2n-k\}$. First, we project $Q_0 = Q \setminus \{q_0\}$ from q_0 onto a sphere centered at q_0; let Q_0' denote the resulting set of projected points, and $Q' = Q_0' \cup \{q_0\}$. In this way, the collection of halving triangles incident to q_0 did not change, nor did the number of points on either side of π_0. Therefore Q', q_0 and π_0 still provide a configuration contradicting (D). Now let π_q, for $q \in Q_0'$, be planes parallel to π_0 with $\pi_q \ni q$ for each q. If necessary, rotate π_0 slightly about q_0 so that $\pi_q \cap Q' = \{q\}$ for each $q \in Q_0'$. As in the previous argument, every halving triangle of Q' is crossed by exactly one of the planes in $\{\pi_0\} \cup \{\pi_q \mid q \in Q_0'\}$ (which is also incident to a vertex of the triangle). Since all points apart from q_0 are extreme in Q', the number of halving

[3] An edge whose containing line equipartitions $Q_0^* \setminus \{q_1^*, q_2^*\}$.

triangles of Q' is exactly

$$\underbrace{c-\min\{k,2n-k\}}_{<0}+\underbrace{\sum_{i=1}^{2n+1}\min\{i-1,2n+1-i\}}_{=n^2}<n^2\ . \qquad (1)$$

The equivalence (B⇔D), and thus (B⇔A) is established.

Remark 2.1. Consider $Q_0' \cup \{q_0\}$ as in the argument just given. Let π' be another plane through q_0 that partitions Q_0' into sets of cardinalities k' and $2n-k'$, and let c' be the number of halving triangles incident to q_0 and crossed by π' (this is also the number of such halving triangles in the original Q). Since the left-hand side of (1) is equal to the number of halving triangles of $Q_0' \cup \{q_0\}$, it follows that

$$c-\min\{k,2n-k\}=c'-\min\{k',2n-k'\}\ .$$

Hence, if there were a configuration contradicting (D), then there would also be one with a plane π_0 that equipartitions $Q \setminus \{q_0\}$, and, thus, if there were a configuration contradicting (A), then there would also be one with $|R|=|B|$. That is, the 'special case' of (A) treated in [7] immediately entails the more general formulation in (A).

3 Balanced Lines and the GLBT

We want to exhibit a more direct relation between (A) and (C). We will not do so with (C) itself, though, but replace it by the following assertion (E), which is known to be equivalent to (C) by the Gale transform [9].

Fact E $m \in \mathbf{N}$, even. Let S be a set of m points in 3-space, and let ρ be a directed ray pointing at its apex x, such that $S \cup \{x\}$ is in general position, and ρ is disjoint from S and from all segments connecting points in S. An oriented triangle spanned by three points in S is called a *j-triangle of S*, if there are exactly j points of S on its positive side.[4] We say that ρ *enters a j-triangle Δ of S*, if it intersects Δ from the positive side to the negative side of it (i.e., x is on the negative side of Δ). If ρ crosses Δ from the negative to the positive side, then we say that ρ *leaves* Δ. Let $g_j(x,S)$ be the number of j-triangles entered by ρ minus the number of j-triangles left by ρ.

$$g_{m/2-2}(x,S)\geq 0\ .$$

If x is extreme in $S \cup \{x\}$ (but also in other situations), equality holds.

In fact, as already suggested by the above notation, it can be shown that $g_j(x,S)$ is a function of x that is independent of the choice of the ray

[4] The orientation of the triangle declares one side of the plane it spans as the positive side. Obviously, the opposite orientation of a j-triangle is an $(m-3-j)$-triangle.

ρ pointing at it. From this it immediately follows that if x is extreme in $S \cup \{x\}$, then $g_j(x, S) = 0$ for all j. Using the Gale transform (see [9]), one can show that if x is not extreme in $S \cup \{x\}$, then there is a simple polytope $\mathcal{P}$ in $\mathbf{R}^{m-4}$ with at most m facets, such that the so-called g-vector of $\mathcal{P}$ is exactly the vector $(g_j(x, S))_{j=0}^{\lfloor (m-4)/2 \rfloor}$; see [9] for details (if x is extreme, $\mathcal{P}$ is the empty polytope). The nonnegativity of this vector is the GLBT.[5] We refer to [9] for the equivalence (E⇔C); see also [4].

We have prepared the ground for a *proof of equivalence (D⇔E).* Assume the set-up of statement (D); recall that Q has $2n+1$ points. Put $Q_0 = Q \setminus \{q_0\}$, and let $H(\tilde{q})$, for any $\tilde{q} \in \pi_0$, denote the number of halving triangles of $Q_0 \cup \{\tilde{q}\}$ that are incident to $\tilde{q}$ and are crossed by π_0. We draw a line ℓ in π_0 passing through q_0, move a point $\tilde{q}$ along ℓ from infinity to q_0, and keep track of the changes in $H(\tilde{q})$ during this motion (see [2] for related results obtained through this continuous motion paradigm, and [1] for a thorough treatment of the combinatorial changes occurring in such a motion).

Initially, $\tilde{q}$ is an extreme point of $Q_0 \cup \{\tilde{q}\}$ and so $H(\tilde{q}) = \min\{k, 2n-k\}$. As $\tilde{q}$ moves along ℓ, $H(\tilde{q})$ changes only when $\tilde{q}$ becomes coplanar with three points $a, b, c \in Q_0$, so that the plane passing through these four points bounds two open halfspaces, one of which contains $n-1$ points of Q_0 and the other $n-2$. Three cases may arise, as illustrated in Figure 2.

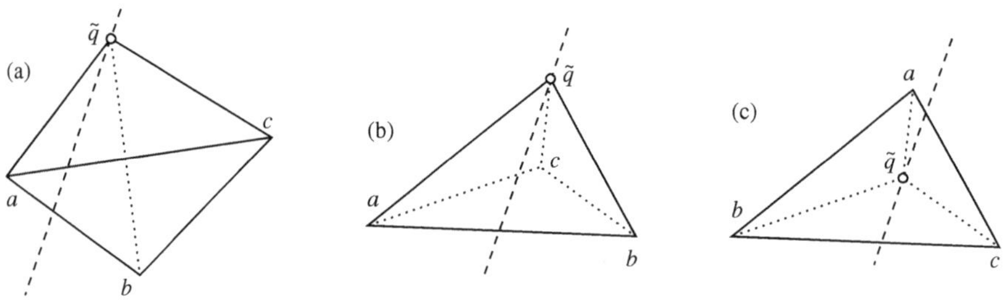

Fig. 2. Three cases of coplanarity: (a) The four points are in convex position. (b) The four points are not in convex position but $\tilde{q}$ is an extreme point of the quadruple. (c) The four points are not in convex position and $\tilde{q}$ is the middle point.

(a) The four points $a, b, c, \tilde{q}$ are in convex position, say in this counterclockwise order, see Figure 2(a).

(b) The four points are not in convex position but $\tilde{q}$ is an extreme point of the quadruple, and, say, c lies in the interior of $\tilde{q}ab$, see Figure 2(b).

(c) The four points are not in convex position and $\tilde{q}$ is the non-extreme ('middle') point, see Figure 2(c).

[5] And the characterization of all possible g-vectors is the g-Theorem.

Each case is further divided into two subcases, depending on whether $\tilde{q}$ reaches the plane of abc from the side containing $n-2$ points of Q_0 (subcase (i)), or from the side containing $n-1$ points of Q_0 (subcase (ii)). Let δ denote the line of intersection of π_0 and the plane of abc (drawn as a dashed line in Figure 2).

In case (a.i), the triangles abc, $\tilde{q}ac$ were halving triangles of $Q_0 \cup \{\tilde{q}\}$ before $\tilde{q}$ reached the plane of abc, and the triangles $\tilde{q}ab$, $\tilde{q}bc$ are halving triangles after $\tilde{q}$ leaves that plane. If δ does not cross the quadrangle $abc\tilde{q}$ (at the time of coplanarity) then π_0 does not cross the triangle $\tilde{q}ac$ before $\tilde{q}$ reaches the plane of abc, and does not cross $\tilde{q}ab$, qac after $\tilde{q}$ leaves that plane. Hence $H(\tilde{q})$ does not change in this case. On the other hand, if δ crosses $abc\tilde{q}$ then π_0 crosses the triangle $\tilde{q}ac$ before $\tilde{q}$ reaches the plane of abc, and crosses exactly one of the triangles $\tilde{q}ab$, $\tilde{q}ac$ afterwards. Hence $H(\tilde{q})$ does not change in this case either. Case (a.ii) is treated in a fully symmetric manner, and $H(\tilde{q})$ does not change in this subcase as well. In cases (b.i) and (b.ii) the local behavior at $\tilde{q}$ is the same as in the corresponding subcases (a.i) and (a.ii), so $H(\tilde{q})$ does not change in these cases either.

In case (c.i), the triangle abc was a halving triangle of $Q \cup \{\tilde{q}\}$ before $\tilde{q}$ reached the plane of abc, and the triangles $\tilde{q}ab$, $\tilde{q}bc$ and $\tilde{q}ac$ are halving triangles after $\tilde{q}$ leaves that plane. The line δ always crosses exactly two of these three triangles, which means that $H(\tilde{q})$ *increases* by 2 in this subcase. By a symmetric reasoning, $H(\tilde{q})$ *decreases* by 2 in subcase (c.ii). In each of these subcases, abc spans, depending on its orientation, an $(n-2)$-triangle and an $(n-1)$-triangle of Q_0. In case (c.i), $\tilde{q}$ *enters* the $(n-2)$-triangle[6] spanned by abc, or more precisely, the ray on which $\tilde{q}$ moves to q_0 enters this $(n-2)$-triangle. In case (c.ii), $\tilde{q}$ *leaves* the $(n-2)$-triangle spanned by abc.

We have shown

$$H(q_0) = \min\{k, 2n-k\} + 2g_{n-2}(q_0, Q_0) .$$

So $H(q_0) \geq \min\{k, 2n-k\}$ iff $g_{n-2}(q_0, Q_0) \geq 0$. The latter is the assertion of (E). This completes the proof.

Remark 3.1. This implication does not hold if we consider the number of j-triangles of Q, for $j \leq n-2$, that are incident to q_0 and are crossed by π_0. In this case, $H(\tilde{q})$ changes by $+2$ when $\tilde{q}$ enters a $(j-1)$-triangle of Q_0 or when $\tilde{q}$ leaves a j-triangle of Q_0, and $H(\tilde{q})$ changes by -2 when $\tilde{q}$ leaves a $(j-1)$-triangle of Q_0 or when $\tilde{q}$ enters a j-triangle of Q_0. In this case we have

$$H(q_0) = 2\min\{j+1, n-2-j, k, 2n-k\} + 2\left(g_j(q_0, Q_0) - g_{j-1}(q_0, Q_0)\right),$$

which does not lead to the same implication as in the preceding proof.

[6]That is, $\tilde{q}$ approaches the plane of abc from the side that contains $n-2$ points of Q_0.

4 Discussion

The purpose of this paper is to show the relation between the balanced lines problem (A) of [7], some problems involving halving triangles in 3-space, and the Generalized Lower Bound Theorem. This sheds some extra light on the result in [7] and explains the difficulty in obtaining a purely combinatorial proof of (A), as experienced in [7]. It highlights the additional merit of the proof of [7], in providing, implicitly, the first purely combinatorial proof of the special case of the Generalized Lower Bound Theorem described in (C). In doing so, we also obtained the property (D), which seems to be new. Several interesting challenges remain.

- Can one obtain a direct and simpler proof of the balanced lines result (A)?
- Can one obtain a purely combinatorial proof of the Generalized Lower Bound Theorem, beyond the special case established (indirectly) here?

References

[1] Artur Andrzejak and Emo Welzl, In between k-sets, j-facets, and i-faces: (i,j)-partitions, *Discrete Comput. Geom.* **29** (2003), 105–131.

[2] Artur Andrzejak, Boris Aronov, Sariel Har-Peled, Raimund Seidel, and Emo Welzl, Results on k-sets and j-facets via continuous motion, *in* "Proc 14th Ann. ACM Symp. on Comput. Geom." (1998), 192–199.

[3] Paul Erdős, László Lovász, A. Simmons, and Ernst G. Straus, Dissection graphs of planar point sets, *in* A Survey of Combinatorial Theory (J. N. Srivastava et al., Eds.), North Holland (1973), 139–149.

[4] Carl W. Lee, Winding numbers and the generalized lower-bound conjecture, *DIMACS Series in Discrete Mathematics and Theoretical Computer Science* **6** (1991), 209–219.

[5] László Lovász, On the number of halving lines, *Ann. Universitatis Scientarium Budapest, Eötvös, Sectio Mathematica* **14** (1971), 107–108.

[6] Peter McMullen, On simple polytopes, *Inventiones Math.* **113** (1993), 419–444.

[7] János Pach and Rom Pinchasi, On the number of balanced lines, *Discrete Comput. Geom.* **25** (2001), 611–628.

[8] Richard P. Stanley, The number of faces of simplicial polytopes and spheres, *in* "Discrete Geometry and Convexity", (J. E. Goodman, E. Lutwak, J. Malkevitch, and R. Pollack, Eds.), *Annals New York Academy of Sciences* **440** (1985), 212–223.

[9] Emo Welzl, Entering and leaving j-facets, *Discrete Comput. Geom.* **25** (2001), 351–364.

[10] Günter M. Ziegler, *Lectures on Polytopes*, Graduate Texts in Mathematics, Springer-Verlag (1995).

About Authors

Micha Sharir is at the School of Mathematical Sciences, Tel Aviv University, Tel Aviv 69978, Israel and at the Courant Institute of Mathematical Sciences New York University, New York, NY 10012, USA; *sharir@math.tau.ac.il*.

Emo Welzl is at the Institute of Theoretical Computer Science, ETH Zürich, CH-8092 Zürich, Switzerland; *emo@inf.ethz.ch*.

Acknowledgments

The authors thank Boris Aronov, János Pach, Rom Pinchasi, and Günter Ziegler for valuable discussions and remarks. A preliminary version of this paper has appeared in Proc. 17th Ann. ACM Symp. on Comput. Geom. (2001) 315-318.

Quantizing Using Lattice Intersections

N. J. A. Sloane
B. Beferull-Lozano

Abstract

The usual quantizer based on an n-dimensional lattice Λ maps a point $x \in \mathbb{R}^n$ to a closest lattice point. Suppose Λ is the intersection of lattices $\Lambda_1, \ldots, \Lambda_r$. Then one may instead combine the information obtained by simultaneously quantizing x with respect to each of the Λ_i. This corresponds to decomposing $\mathbb{R}^n$ into a honeycomb of cells which are the intersections of the Voronoi cells for the Λ_i, and identifying the cell to which x belongs. This paper shows how to write several standard lattices (the face-centered and body-centered cubic lattices, the root lattices D_4, E_6^*, E_8, the Coxeter-Todd, Barnes-Wall and Leech lattices, etc.) in a canonical way as intersections of a small number of simpler, decomposable, lattices. The cells of the honeycombs are given explicitly and the mean squared quantizing error calculated in the cases when the intersection lattice is the face-centered or body-centered cubic lattice or the lattice D_4.

1 Introduction

An n-dimensional lattice Λ determines a partition of $\mathbb{R}^n$ into cells which are congruent copies of the Voronoi cell containing the origin. A vector quantizer based on Λ ([11], [16], [18]) is a map $Q_\Lambda : \mathbb{R}^n \to \mathbb{R}^n$ which maps $x \in \mathbb{R}^n$ to the closest lattice point, i.e. to the lattice point at the center of the Voronoi cell containing x. (In the case of a tie, one of the closest lattice points is chosen at random.)

Suppose Λ is the intersection of n-dimensional lattices $\Lambda_1, \ldots, \Lambda_r$. A recent paper [4] discusses (among other things) the possibility of replacing the quantizer Q_Λ by a "multiple description quantizer" which simultaneously quantizes with respect to each of the Λ_i, i.e. computes

$$(Q_{\Lambda_1}(x), Q_{\Lambda_2}(x), \ldots, Q_{\Lambda_r}(x)) . \tag{1}$$

This gives rise to a different partition of $\mathbb{R}^n$: the cells are now the intersections of the Voronoi cells of the individual Λ_i, and (1) specifies the cell to which x belongs.

There are several reasons for investigating quantizers of this type.

(i) If the Λ_i are simpler lattices than Λ then (1) may be easier to compute than $Q_\Lambda(x)$.

(ii) Perhaps this approach will lead to new insights on the so-far intractable problem of finding good lattice quantizers in high dimensions (cf. [11, Chap. 2]). Even in 24 dimensions the best lattice quantizer presently known, the Leech lattice, is quite complicated to analyze and to implement — its Voronoi cell has 16969680 faces and over 10^{21} vertices ([11, Chaps. 21, 22, 23, 25], [3], [30], [31]).

(iii) The individual $Q_{\Lambda_i}(x)$ could be communicated over separate channels; in the event of one or more channels failing a reasonably good approximation to x will still be obtained. Other multiple description quantizers have recently been studied in, for example, [13], [29].

(iv) Initially it seemed possible that this approach might lead to quantizers with a lower mean squared error that can be obtained from lattice quantizers. The investigations reported here now make this unlikely (but see the remarks in the final section).

However, our focus here is not on applications but on the theoretical aspects of these quantizers, and in particular on two questions: which lattices have a "nice" description as the intersection of simpler lattices, and what are the associated "honeycombs," or decompositions of space into cells?

Reference [4] gives one very appealing example, shown here in Figure 1. The familiar planar hexagonal lattice A_2 (large circles) can be obtained as the intersection of three rectangular lattices, all rotations of each other by 120°: these are the lattices generated respectively by the two vectors OA, the two vectors OB and the two vectors OC. If we use generator matrices to specify these lattices (the rows span the lattices) then the three rectangular lattices Λ_1, Λ_2, Λ_3 have generator matrices

$$\begin{bmatrix} \sqrt{3} & 0 \\ 0 & 1 \end{bmatrix}, \quad \begin{bmatrix} \frac{\sqrt{3}}{2} & \frac{1}{2} \\ \frac{-\sqrt{3}}{2} & \frac{3}{2} \end{bmatrix}, \quad \begin{bmatrix} \frac{-\sqrt{3}}{2} & \frac{1}{2} \\ \frac{\sqrt{3}}{2} & \frac{3}{2} \end{bmatrix} \tag{2}$$

and their intersection has generator matrix

$$\begin{bmatrix} \sqrt{3} & 1 \\ 0 & 2 \end{bmatrix}, \tag{3}$$

which is indeed a copy of the A_2 lattice.

In this case the honeycomb contains four kinds of cells, as shown by the heavy (solid) lines in Fig. 2. For example, if the point x being quantized is close to $(0,0)$, then $Q_{\Lambda_1}(x) = Q_{\Lambda_2}(x) = Q_{\Lambda_3}(x) = (0,0)$, and the cell containing x is the intersection of the Voronoi cells containing $(0,0)$ of the three Λ_i. This is the horizontally shaded hexagon in Fig. 2. Just to the North of this hexagon the cell is the intersection of the Voronoi cell for Λ_1 that contains $(0,1)$ with the Voronoi cells at $(0,0)$ for Λ_2 and Λ_3: this is the small cross-hatched equilateral triangle. There are two further cells that

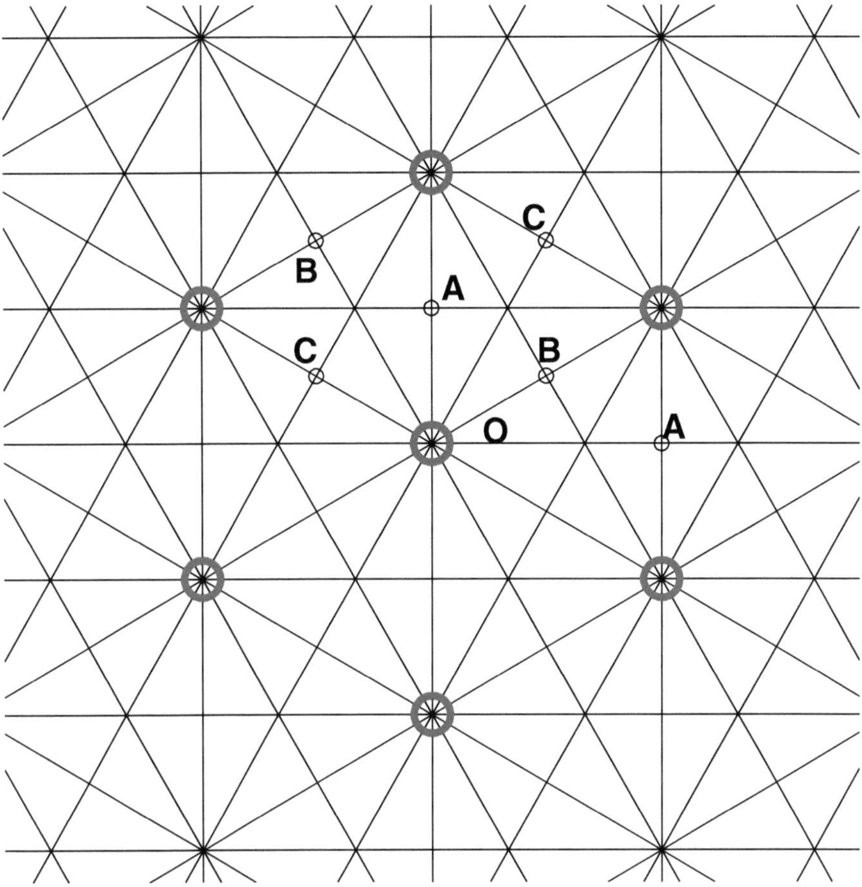

Fig. 1. The hexagonal lattice (heavy circles) as the intersection of three rectangular lattices (spanned by the vectors OA, OB and OC resp.).

are obtained in a similar manner: the diagonally shaded isosceles triangle and the larger vertically shaded equilateral triangle. This example will be discussed further in Section 4.

The present paper was prompted by the question: how can Figures 1 and 2 be generalized to higher dimensions?

Section 2 gives some terminology and defines the normalized mean squared error G that will be used when comparing quantizers. Section 3 describes some general constructions for writing a lattice as an intersection and gives a number of examples, including the body-centered cubic (bcc) and face-centered cubic (fcc) lattices D_3^* and D_3, the root lattices D_4, E_6^*, E_8, the Coxeter-Todd lattice K_{12}, the Barnes-Wall lattices BW_n and the Leech lattice Λ_{24}. We focused attention on these lattices because A_2, D_3^*, D_4, E_6^*,

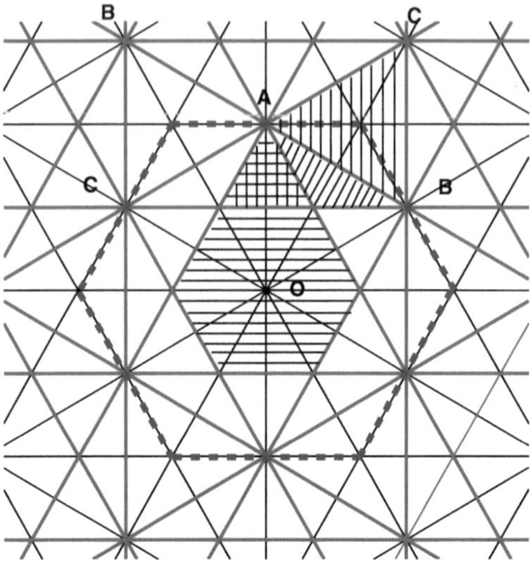

Fig. 2. Honeycomb associated with Voronoi cells in Fig. 1.

E_8, K_{12}, BW_{16} and Λ_{24} are the best quantizers currently known[1] in their dimensions [11]. In fact A_2 is optimal among all two-dimensional quantizers [14], and D_3^* is optimal among three-dimensional lattice quantizers [2].

In Section 4 we reexamine the quantizer of Figures 1 and 2. Then Sections 5, 6 and 7 determine the honeycombs and mean squared errors for the bcc, fcc and D_4 lattices, respectively. The final section contains some conclusions and comments.

Table 1 summarizes the main decompositions mentioned in this paper.

We repeat that we do not propose that the quantizers described in this paper will have practical applications, nor that they are simpler than the usual lattice quantizers based on the intersection lattices. Our interest is in the geometry of the honeycombs associated with the quantizers. These do not seem to have been studied before[2] — they are not mentioned for example in Okabe, Boots and Sugihara [23], nor in Wells [33], even in the chapter on *Interpenetrating three-dimensional nets.* There has been considerable interest among physicists in recent years in soap froths, both in connection with the study of colloids, dendritic polymers, etc., and in the Kelvin problem of finding the minimal-area honeycomb in $\mathbb{R}^3$ ([26], [32], [36]). So it may be of interest to see a class of honeycombs that have a purely geometrical construction.

We end this section by describing how the cells of the honeycombs in Sections 5–7 were found. Many cells could be found by elementary geomet-

[1] Assuming always that the variable to be quantized is uniformly distributed over a large ball in $\mathbb{R}^n$.

[2] Of course there is an extensive literature dealing with the Voronoi and Delaunay cell decompositions associated with various lattices — see for example [7], [8], [10], [11], [12].

Table 1. Summary of decompositions described in this paper.

Lattice	Copies	Component lattice	Sections
A_2	3	rectangular	1, 3, 4
A_3^* (bcc)	3	"rectangular"	3, 5
A_3 (fcc)	4	"prismatic"	3, 6
D_4	3	$\mathbb{Z}^4$	3, 7
E_6^*	4	A_2^3	3
E_8	15	$\mathbb{Z}^8$	3
E_8	10	A_2^4	3
E_8	5	D_4^2	3
K_{12}	21	A_2^6	3
Leech	4095	$\mathbb{Z}^{24}$	3
BW_n	$\prod_{j=1}^{m-1}(2^j - 1)$	$\mathbb{Z}^n$	3

rical reasoning. But in complicated cases we carried out some or all of the following steps.

(i) To find the cell containing a point $x \in \mathbb{R}^3$ or $\mathbb{R}^4$ we first quantized x using each of the lattices Λ_i in turn. For each Λ_i, we determined the Voronoi cell C_i containing x, or, more precisely, the equations to the hyperplanes bounding C_i.

(ii) Linear programming (in MATLAB) was then used to determine the vertices of the cell containing x. We let w range over a set of 130 points on a sphere centered at x (taken from the tables of spherical codes in [19]), and, for each w, we maximized the inner product (w, z) subject to the constraints that z lie in the polytope formed by the intersection of *all* the hyperplanes found in (i). Any such solution z is a vertex of the cell, and since the w's are essentially random, the 130 solutions should include all the vertices of the cell.

(iii) The convex hull program QHULL [1], [15] was used to find the convex hull of these vertices.

At this point we have candidates for all the cells in the honeycomb. Since it is theoretically possible (although unlikely) that step (ii) might have failed to find all the vertices of a cell, we now verified by hand that the cells fitted together to form a proper tiling of the space.

(iv) The XGobi program [28] for displaying multi-dimensional data was used to help visualize the cells and their neighbors.

To compute the volumes and second moments of the cells we decomposed the cells into simplices and used the formulae in [7] and [11, Chap. 21].

2 Notation

We assume the n-dimensional lattice Λ is the intersection of n-dimensional lattices $\Lambda_1, \ldots, \Lambda_r$. The intersections of the Voronoi cells for the Λ_i partition $\mathbb{R}^n$ into a tesselation or *honeycomb*. Let $\mathcal{P}_1, \ldots, \mathcal{P}_k$ be representatives for the different polytopes or *cells* that appear in the honeycomb.

Let p_i $(i = 1, \ldots, k)$ be the probability that a randomly chosen point in $\mathbb{R}^n$ (uniformly distributed over a very large ball, say) belongs to $\mathcal{P}_i$, and let $N_i = p_i V / V_i$, where V_i is the volume of $\mathcal{P}_i$ and $V = \sqrt{\det \Lambda}$ is the volume of a fundamental region or Voronoi cell for Λ. Let $\mathcal{V}$ be the particular Voronoi cell for Λ that contains the origin. Then the honeycomb is periodic with "tile" equal to the part lying in $\mathcal{V}$, and there are N_1 cells of type $\mathcal{P}_1$ per copy of $\mathcal{V}$, N_2 cells of type $\mathcal{P}_2$, etc. Also

$$V = N_1 V_1 + \cdots + N_k V_k \ . \tag{4}$$

We assume that a point that falls into a cell of type $\mathcal{P}_i$ is quantized as the centroid c_i of that cell, in which case the mean squared error is

$$U_i = \int_{\mathcal{P}_i} \|x - c_i\|^2 dx \ .$$

The mean squared error for the full quantizer is then

$$U = \sum_{i=1}^{k} N_i U_i = V \sum_{i=1}^{k} \frac{p_i U_i}{V_i} \ . \tag{5}$$

We now derive the expression that we will use as a measure of the normalized mean squared error of this quantizer. Suppose we are quantizing a random variable $X \in \mathbb{R}^n$, with differential entropy per dimension $h(X)$, and whose support contains a large number of points of Λ.

A general theorem of Zador about vector quantizers ([34], [35], [18, Eq. (20)]) implies that in this "high-rate" case the average mean squared error per dimension $U/(nV)$ can be approximated by

$$\frac{U}{nV} \approx G \, 2^{2(h(X)-R)} \ , \tag{6}$$

where R bits/symbol is the quantizing rate and G depends on the positions of the quantizing points but is independent of X.

Since G does not depend on the distribution of X, we may choose any convenient distribution in order to calculate G, and we assume that X is uniformly distributed over a large region of $\mathbb{R}^n$, or, equivalently, that X is uniformly distributed over V. Then

$$h(X) = \frac{1}{n} \log_2 V, \quad \text{so} \quad 2^{2h(X)} = V^{2/n} \ . \tag{7}$$

To calculate R, observe that we need $h(p_1, \ldots, p_k) = -\sum_{i=1}^{k} p_i \log_2 p_i$ bits to specify the type of cell to which the quantized point belongs, and a further $\sum_{i=1}^{k} p_i \log_2 N_i = \sum_{i=1}^{k} p_i \log_2(p_i V/V_i)$ bits to specify the particular one of the N_i cells of that type. This requires a total of $\log_2 V - \sum_{i=1}^{k} p_i \log_2 V_i$ bits, and then R is this quantity divided by n, so that

$$2^{-2R} = V^{-2/n} \prod_{k=1}^{k} V_i^{2p_i/n} . \tag{8}$$

From (5)–(8) we obtain

$$G = \frac{\sum_{i=1}^{k} \frac{p_i U_i}{V_i}}{n\left(\prod_{i=1}^{k} V_i^{p_i}\right)^{\frac{2}{n}}} , \tag{9}$$

the normalized mean squared error per dimension, which we take as our figure of merit for a quantizer. The numerator of (9) is equal to U/V (see (5)). Note that when there is only one kind of cell (9) reduces to the familiar formula

$$G = \frac{U}{nV^{1+\frac{2}{n}}} \tag{10}$$

for a lattice quantizer ([7], [11], [16]).

Incidentally, a different expression from (9) for the figure of merit was used in a recent paper of Kashyap and Neuhoff [20]. By defining the rate of the quantizer in a different way they end up with a different expression in the denominator. However, we believe our formula gives a fairer comparison. This point is discussed in more detail in [27].

Further terminology. Let Λ be a lattice in $\mathbb{R}^n$. The dual lattice will be denoted by Λ^*. The *norm* of a vector $x \in \mathbb{R}^n$ is its squared length (x, x). A *similarity* σ is a linear map from $\mathbb{R}^n$ to $\mathbb{R}^n$ such that there is a real number N with $(\sigma x, \sigma y) = N(x, y)$ for $x, y \in \mathbb{R}^n$. If Λ and M are similar lattices we write $\Lambda \cong M$. A lattice Λ is said to be N-*modular* if $\Lambda \cong \Lambda^*$ under a similarity that multiplies norms by N (cf. [25]). For example, the root lattices $\mathbb{Z}^n$ and E_8 are 1-modular, A_2 is 3-modular, and D_4 is 2-modular. We write $\langle\, \Lambda_1, \Lambda_2, \ldots \,\rangle$ for the lattice generated by the union of the lattices $\Lambda_1, \Lambda_2, \ldots$. Two lattices or polytopes are *congruent* if one can be mapped to the other by an element of $SO(n)$.

3 Writing a lattice as an intersection

Let Λ be a lattice in $\mathbb{R}^n$. We wish to write Λ as an intersection

$$\Lambda = \Lambda_1 \cap \Lambda_2 \cap \cdots \cap \Lambda_r \qquad (11)$$

where r is small and the Λ_i are pairwise congruent and as "simple" as possible. Ideally we would like each Λ_i to be a direct sum of congruent copies of a fixed low-dimensional lattice K such as $\mathbb{Z}$, A_2 or D_4, but this is not always possible. In the example shown in Fig. 1, for instance, the Λ_i are rectangular rather than square lattices. If this is not possible, we ask that the Λ_i be decomposable into a direct sum of as many congruent low-dimensional sublattices as possible.

If we were going to investigate the honeycombs associated with higher-dimensional intersections such as those for E_6^* or E_8, we would impose an additional formal requirement that the Λ_i form an orbit under some subgroup of the automorphism group $Aut(\Lambda)$, in order to guarantee that the honeycomb be symmetric. However, for the low-dimensional examples studied in Sections 5–7, we were able to achieve this symmetry by using the natural decompositions, without introducing the machinery of group theory.

This section describes some general methods for finding intersections. We will make use of the following version of de Morgan's law:

Lemma 1. *If $\Lambda_1, \ldots, \Lambda_r$ are lattices in $\mathbb{R}^n$ then*

$$\Lambda_1 \cap \cdots \cap \Lambda_r \;=\; \langle\, \Lambda_1^*, \ldots, \Lambda_r^* \,\rangle^* \,.$$

Proof. See for example [24, Section 82F]. ∎

Method 1: Partitioning the minimal vectors for Λ^*. For the first method, suppose the dual lattice Λ^* is generated by its minimal vectors (i.e. the vectors of minimal nonzero norm), and let K be one of $\mathbb{Z}$, A_2 or D_4. Let K have dimension κ and be N-modular. Then we attempt to partition the minimal vectors of Λ^* into a number of copies of the minimal vectors of a suitably rescaled version of $K^{n/\kappa}$.

If this can be done, let Λ_i^* be the lattice generated by the ith part of the partition. The hypotheses guarantee that $\Lambda_i^* \cong \Lambda_i \cong K^{n/\kappa}$, and, by the Lemma,

$$\Lambda = \Lambda_1 \cap \cdots \cap \Lambda_r \ ,$$

where r = number of minimal vectors of Λ^* divided by number of minimal vectors of K.

Whether this partitioning is possible is an interesting question in its own right. For example, can the 240 minimal vectors of E_8 be partitioned into 15 copies of the minimal vectors of (a scaled version of) $\mathbb{Z}^8$, i.e. into 15 coordinate frames[3] or into 10 copies of the minimal vectors of A_2^4? Partial answers are given below.

We now give a number of examples, beginning with the case when $K = \mathbb{Z}$.

$\mathbf{D_4}$. The lattice D_4 may be taken to have generator matrix

$$\begin{bmatrix} 2 & 0 & 0 & 0 \\ 0 & 2 & 0 & 0 \\ 0 & 0 & 2 & 0 \\ 1 & 1 & 1 & 1 \end{bmatrix} \tag{12}$$

and then the 24 minimal vectors consist of eight of the form $(\pm 2, 0, 0, 0)$ and 16 of the form $(\pm 1, \pm 1, \pm 1, \pm 1)$. D_4 is 2-modular and is generated by its minimal vectors. The minimal vectors may be partitioned into three coordinate frames, consisting of ± 1 times the rows of each of the matrices

$$\begin{bmatrix} 2 & 0 & 0 & 0 \\ 0 & 2 & 0 & 0 \\ 0 & 0 & 2 & 0 \\ 0 & 0 & 0 & 2 \end{bmatrix}, \quad \begin{bmatrix} +1 & +1 & +1 & +1 \\ +1 & -1 & +1 & -1 \\ +1 & -1 & -1 & +1 \\ +1 & +1 & -1 & -1 \end{bmatrix}, \quad \begin{bmatrix} -1 & +1 & +1 & +1 \\ -1 & -1 & +1 & -1 \\ -1 & -1 & -1 & +1 \\ -1 & +1 & -1 & -1 \end{bmatrix}. \tag{13}$$

After applying the Lemma and rescaling, we conclude that if Λ_1, Λ_2, Λ_3 ($\cong \mathbb{Z}^4$) have the generator matrices given in (13), then

$$\Lambda_1 \cap \Lambda_2 \cap \Lambda_3 = \Lambda\,, \tag{14}$$

where Λ has generator matrix

$$\begin{bmatrix} 4 & 0 & 0 & 0 \\ 2 & 2 & 0 & 0 \\ 2 & 0 & 2 & 0 \\ 2 & 0 & 0 & 2 \end{bmatrix}, \tag{15}$$

and is another version of D_4 on the scale at which its minimal norm is 8. Equation (14) may also be verified directly, without appealing to the Lemma. The group generated by the second matrix in (13) and diag$\{-1, +1, +1, +1\}$ is a symmetric group S_3 permuting the Λ_i; it is also a subgroup of $Aut(\Lambda)$. The honeycomb for this example is studied in Section 7.

Barnes-Wall lattices. The preceding example can be generalized using orthogonal spreads. Let BW_n ($n = 2^m$, $m = 1, 2, \ldots$) denote the n-dimensional Barnes-Wall lattice ([11], [21], [22]). In particular, $BW_2 \cong \mathbb{Z}^2$, $BW_4 \cong D_4$, $BW_8 \cong E_8$. For $n \neq 8$, BW_n is 2-modular, while as already mentioned BW_8 is 1-modular.

[3] A *coordinate frame* in $\mathbb{R}^n$ is a set of $2n$ vectors $\pm v_1, \ldots, \pm v_n$ with $(v_i, v_i) =$ a constant, $(v_i, v_j) = 0$ if $i \neq j$.

It is known that the minimal vectors of BW_n may be partitioned into $\prod_{j=1}^{m-1}(2^j-1)$ coordinate frames, which are transitively permuted by symmetries of $Aut(BW_n)$. This is a consequence of the existence of an orthogonal spread in the orthogonal vector space $\Omega^+(2m,2)$ of maximal Witt index ([5], [6], [21]).

It follows that BW_n can be written as the intersection of $\prod_{j=1}^{m-1}(2^j-1)$ copies of $\mathbb{Z}^n$. In particular, E_8 is the intersection of 15 copies of $\mathbb{Z}^8$. An explicit method for constructing such an intersection for E_8 is given below.

Intersections of smaller numbers of lattices are possible, although they are less symmetric and therefore less satisfactory. For example in (14) it is also true that $\Lambda_1 \cap \Lambda_2 = \Lambda \cong D_4$. Similarly, E_8 is (up to a similarity) the intersection of $2\mathbb{Z}^8$ and the lattice (similar to $2\mathbb{Z}^8$) with generator matrix

$$\begin{bmatrix} 1 & 1 & 1 & 1 & 0 & 0 & 0 & 0 \\ 0 & 0 & 0 & 0 & 1 & 1 & 1 & 1 \\ 1 & -1 & 0 & 0 & 1 & -1 & 0 & 0 \\ 0 & 0 & 1 & -1 & 0 & 0 & 1 & -1 \\ 1 & 0 & -1 & 0 & -1 & 0 & 1 & 0 \\ 0 & 1 & 0 & -1 & 0 & -1 & 0 & 1 \\ 1 & 0 & 0 & -1 & 0 & 1 & -1 & 0 \\ 0 & 1 & -1 & 0 & 1 & 0 & 0 & -1 \end{bmatrix}$$

But this representation of E_8 is in no way canonical, and the resulting honeycomb is not interesting.

Eisenstein and Hurwitzian lattices. Smaller intersections which *are* canonical can be obtained if we change K from $\mathbb{Z}$ to A_2 or D_4. For example, the minimal vectors of E_8 can be partitioned into 10 copies of the minimal vectors of A_2^4. As in [11], let $\mathcal{E} = \{a + b\omega : a, b \in \mathbb{Z}\}$, $\omega = e^{2\pi i/3}$, denote the ring of Eisenstein integers. The six units in $\mathcal{E}$ are ± 1, $\pm\omega$, $\pm\bar{\omega}$. When regarded as a two-dimensional real lattice $\mathcal{E}$ is similar to A_2. As an $\mathcal{E}$-module, E_8 has generator matrix

$$\begin{bmatrix} \theta & 0 & 0 & 0 \\ 0 & \theta & 0 & 0 \\ 1 & 1 & 1 & 0 \\ 0 & 1 & -1 & 1 \end{bmatrix}, \quad \text{where} \quad \theta = \omega - \bar{\omega}\,. \tag{16}$$

Inner products are computed using the Hermitian inner product $(u, v) = \sum u_i \bar{v}_i$. See [11, Chapters 2 and 7] for further details. The minimal vectors consist of 24 of the form $(u\theta, 0, 0, 0)$, where u is a unit in $\mathcal{E}$, and $8 \times 3^3 = 216$ which are congruent mod θ to one of the eight nonzero codewords of the tetracode [11, Chap. 3].

A partition of these 240 vectors into 10 copies of the minimal vectors of A_2^4 was found by graph coloring. A graph was constructed with the 40 projectively distinct vectors as nodes and with edges corresponding to pairs of non-orthogonal vectors. A coloring with 10 colors was then found with the help of a program supplied by David Johnson. The ten copies of $\mathcal{E}^4 \cong A_2^4$ are shown in Table 2. Only one from each complex conjugate pair is shown.

Table 2. Decomposition of minimal vectors of E_8 into ten copies of $\mathcal{E}^4 \cong A_2^4$. Each row generates a copy of $\mathcal{E}^4 \cong A_2^4$. The complex conjugates of the last four rows have been omitted.

θ	0	0	0	0	θ	0	0	0	0	θ	0	0	0	0	θ
0	1	-1	1	1	0	-1	-1	1	-1	0	1	1	1	1	0
0	1	-1	ω	1	0	$-\omega$	$-\bar{\omega}$	1	$-\omega$	0	$\bar{\omega}$	1	ω	ω	0
0	1	$-\omega$	ω	1	0	$-\omega$	$-\omega$	1	-1	0	ω	1	1	ω	0
0	1	$-\omega$	$\bar{\omega}$	1	0	$-\bar{\omega}$	-1	1	$-\omega$	0	1	1	ω	$\bar{\omega}$	0
0	1	$-\omega$	1	1	0	-1	$-\bar{\omega}$	1	$-\bar{\omega}$	0	$\bar{\omega}$	1	$\bar{\omega}$	1	0

By applying the Lemma we obtain a representation of E_8 as an intersection of 10 copies of A_2^4. In fact (since $\mathcal{E}$ is itself 3-modular) we may omit the final step of taking the duals of the lattices in Table 2. Let $\Lambda_1, \ldots, \Lambda_{10}$ be the ten versions of A_2^4 generated by the rows of Table 2 and their complex conjugates. Then their intersection is easily seen to be the version of E_8 with generator matrix θ times (16). This decomposition is probably not unique, and it would be nice to know which version has the largest symmetry group.

Again just two lattices suffice: E_8 is also the intersection of the first two lattices in Table 2.

We may also write E_8 as the intersection of five copies of D_4^2. For this we regard E_8 as a 2-dimensional module over the ring $\mathbb{H} \cong D_4$ of Hurwitzian quaternions [11, p. 55]. The five copies of D_4^2 have generator matrices

$$\begin{bmatrix} 1+i & 0 \\ 0 & 1+i \end{bmatrix}, \begin{bmatrix} 1 & 1 \\ 1 & -1 \end{bmatrix}, \begin{bmatrix} 1 & i \\ 1 & -i \end{bmatrix}, \begin{bmatrix} 1 & j \\ 1 & -j \end{bmatrix}, \begin{bmatrix} 1 & k \\ 1 & -k \end{bmatrix}. \tag{17}$$

E_6^* may be written as the intersection of four copies of A_2^3, with generator matrices

$$\begin{bmatrix} \theta & 0 & 0 \\ 0 & \theta & 0 \\ 0 & 0 & \theta \end{bmatrix}, \begin{bmatrix} 1 & 1 & 1 \\ 1 & \omega & \bar{\omega} \\ 1 & \bar{\omega} & \omega \end{bmatrix}, \begin{bmatrix} 1 & 1 & \omega \\ 1 & \omega & 1 \\ 1 & \bar{\omega} & \bar{\omega} \end{bmatrix}, \begin{bmatrix} 1 & 1 & \bar{\omega} \\ 1 & \omega & \omega \\ 1 & \bar{\omega} & 1 \end{bmatrix}. \tag{18}$$

This was found by partitioning the 72 minimal vectors of E_6 (by hand) into four copies of the minimal vectors of A_2^3 and using the Lemma.

Method 2: Congruence bases and norm-doubling maps. The second method is based on the observation that several well-known lattices Λ have the property that for some prime π, the vectors in some of the classes of $\Lambda/\pi\Lambda$ can be partitioned into coordinate frames. For example, Conway's proof of the uniqueness of the Leech lattice Λ_{24} [11, Chap. 12] considers the classes of $\Lambda_{24}/2\Lambda_{24}$. A consequence of the numerical identity

$$\begin{aligned}\frac{n_0}{1}+\frac{n_4}{2}+\frac{n_6}{2}+\frac{n_8}{48} &= \frac{1}{1}+\frac{196560}{2}+\frac{16773120}{2}+\frac{398034000}{48} \\ &= 16777216 = 2^{24}\,, \end{aligned} \tag{19}$$

where n_j is the number of vectors in Λ_{24} of norm j, is that, for the classes of $\Lambda_{24}/2\Lambda_{24}$ in which the minimal norm is 8, the minimal vectors in the class form a coordinate frame or *congruence base*. A similar property holds for the D_4, E_8, K_{12} and other lattices (cf. [9]). This gives a representation of Λ_{24} as an intersection of $398034000/48 = 8292375$ copies of $\mathbb{Z}^{24}$. However, the following argument, due to J. H. Conway (personal communication), shows that if the lattice has a suitable norm-doubling map (cf. [11, p. 239], [9]) then we can also partition the minimal vectors into coordinate frames and obtain a smaller intersection.

Suppose a lattice $\Lambda \subseteq \mathbb{R}^n$ has the structure of a free module over a ring J with inner product $(\ ,\ )$ (cf. [11, Chap. 2]). In the present application J will be either $\mathbb{Z}$ or $\mathcal{E}$. Let $a = n/\dim_{\mathbb{R}} J$. Consider the classes of $\Lambda/2\Lambda$. Suppose there is an integer m with the property that each class either contains no vectors of norm $2m$, or else all the vectors of norm $2m$ in the class can be partitioned into sets of $2a$ vectors $\pm v_1, \pm v_2, \ldots, \pm v_a$ where $(v_i, v_j) = 0$ if $i \neq j$.

Suppose in addition there is a *norm-doubling* map T, a similarity from Λ into Λ such that $(Tu, Tu) = 2(u, u)$ for $u \in \Lambda$, with the extra property that $2\Lambda \subset T\Lambda$. Then we may conclude that the vectors of norm m in Λ may also be partitioned into sets of $2a$ mutually orthogonal vectors.

To see this, let $u \in \Lambda$ have norm m. Then $v = Tu$ has norm $2m$, and by the hypotheses is part of a coordinate frame $\pm v_1, \ldots \pm v_a$, where $v_i = v + 2w_i$, say, with $w_1 = 0$. We can write $2w_i = Tw_i'$ for some w_i', so $v_i = T(u + w_i')$. Since T is a similarity, the set $\pm(u + w_i')$ is a coordinate frame containing u.

Examples

(i) $\Lambda = D_4$ or E_8, $J = \mathbb{Z}$, $m = 2$, $T =$ direct sum of respectively 2 or 4 copies of $\begin{bmatrix} 1 & 1 \\ 1 & -1 \end{bmatrix}$, with $T^2 = 2I$. The classes of $D_4/2D_4$ and $E_8/2E_8$ and the associated congruence bases are given in [11, Chap. 6]. The analogue of (19) for E_8 reads

$$1 + \frac{240}{2} + \frac{2160}{16} = 2^8\,.$$

We obtain decompositions of the minimal vectors of D_4 into three coordinate frames, as already seen in (13), and of the minimal vectors of E_8 into 15 coordinate frames as also discussed above. To get an explicit decomposition in the latter case, note that a coordinate frame of norm 4 vectors has the form $Tu + 2w_i = T(u + Tw_i)$. So the coordinate frame of norm 2 vectors consists of the vectors of minimal norm in the translate $u+T\Lambda$. For E_8 these consist of seven sets of the form shown on the left in (20) and eight of the form shown on the right:

$$\begin{bmatrix} + & 0 & + & 0 & 0 & 0 & 0 & 0 \\ + & 0 & - & 0 & 0 & 0 & 0 & 0 \\ 0 & + & 0 & + & 0 & 0 & 0 & 0 \\ 0 & + & 0 & - & 0 & 0 & 0 & 0 \\ 0 & 0 & 0 & 0 & + & 0 & + & 0 \\ 0 & 0 & 0 & 0 & + & 0 & - & 0 \\ 0 & 0 & 0 & 0 & 0 & + & 0 & + \\ 0 & 0 & 0 & 0 & 0 & + & 0 & - \end{bmatrix}, \quad \frac{1}{2}\begin{bmatrix} + & + & + & + & + & + & + & + \\ + & + & + & + & - & - & - & - \\ + & + & - & - & + & + & - & - \\ + & + & - & - & - & - & + & + \\ + & - & + & - & + & - & + & - \\ + & - & + & - & - & + & - & + \\ + & - & - & + & + & - & - & + \\ + & - & - & + & - & + & + & - \end{bmatrix}. \tag{20}$$

Then E_8 is also the intersection of the 15 copies of $\mathbb{Z}^8$ having these generator matrices.

(ii) $\Lambda =$ Leech lattice Λ_{24}, $J=\mathbb{Z}$, $m=4$, $T=(I+i)$, where $i \in Aut(\Lambda_{24})$ is given in [11, Fig. 6.7], and satisfies $i^2=-I$, with $(I+i)(I-i)=T(I-i)=2I$. The coordinates frames of norm 4 vectors consist of the 48 vectors of minimal norm in the translates $u+(I-i)\Lambda$ where $(u,u)=4$. We obtain a decomposition of the minimal vectors of Λ_{24} into 4095 coordinate frames, and a representation of Λ_{24} as the intersection of 4095 copies of $\mathbb{Z}^{24}$.

We do not know if it is possible to write the Leech lattice as the intersection of 2730 copies of A_2^{12}. Since 196560/96 is not an integer, there is no analogous decomposition as an intersection of copies of D_4^6.

(iii) $\Lambda =$ Coxeter-Todd lattice K_{12}, a 6-dimensional $\mathcal{E}$-module, $J=\mathcal{E}$, $m=6$, $T=$ the map given in [9, Eq. (43)], with $T^2+T+2=0$. The classes of $K_{12}/2K_{12}$ are given in [9], and the analogue of (19) reads

$$1+\frac{756}{2}+\frac{4032}{1}+\frac{20412}{12}=2^{12}\,.$$

The coordinate frames of norm 6 vectors consist of the 12 minimal vectors in the translates $u+(T+I)K_{12}$, $(u,u)=6$. By combining these coordinate frames in sets of three, by taking the union of the sets $\alpha\{\pm v_1,\ldots,\pm v_{12}\}$ with $\alpha=1,\omega$ and $\bar{\omega}$, we obtain a decomposition of the minimal vectors of K_{12} into 21 copies of the minimal vectors of A_2^6, and, via the Lemma, a representation of K_{12} as the intersection of 21 copies of A_2^6. An explicit decomposition, not shown here, was found by the graph coloring method mentioned earlier.

Method 3: First principles. If the above methods fail, as they do for the bcc and fcc lattices, we can always fall back on a direct attack from first principles. The following method handles the hexagonal, bcc and fcc lattices in a unified manner. We list the vectors of small norms in the lattice, and look for a partition of some subset of these vectors which produces a small number of congruent, decomposable lattices whose intersection is similar to the original lattice.

For the hexagonal lattice, which we take to be generated by $(0,1)$ and $(-\frac{\sqrt{3}}{2}, \frac{1}{2})$, there are six vectors of norm 1, namely $(0, \pm 1)$, $(\pm\frac{\sqrt{3}}{2}, \pm\frac{1}{2})$, and six of norm 3, namely $(\pm\sqrt{3}, 0)$, $(\pm\frac{\sqrt{3}}{2}, \pm\frac{3}{2})$. Then (2) is obtained by partitioning these 12 vectors into three sets of size 4.

For the bcc lattice generated by $(1,0,0)$, $(0,1,0)$, $(\frac{1}{2}, \frac{1}{2}, \frac{1}{2})$, the vectors of small norms are the following:

shape	norm	number
0 0 0	0	1
$\frac{1}{2}$ $\frac{1}{2}$ $\frac{1}{2}$	$\frac{3}{4}$	8
1 0 0	1	6
1 1 0	2	12

We take the 18 vectors of norms 1 and 2 and partition them into three sets of size 6. The resulting lattices have generator matrices

$$\begin{bmatrix} 1 & 0 & 0 \\ 0 & 1 & 1 \\ 0 & 1 & -1 \end{bmatrix}, \quad \begin{bmatrix} 0 & 1 & 0 \\ 1 & 0 & 1 \\ 1 & 0 & -1 \end{bmatrix}, \quad \begin{bmatrix} 0 & 0 & 1 \\ 1 & 1 & 0 \\ 1 & -1 & 0 \end{bmatrix} \tag{21}$$

and their intersection has generator matrix

$$\begin{bmatrix} 2 & 0 & 0 \\ 0 & 2 & 0 \\ 1 & 1 & 1 \end{bmatrix}, \tag{22}$$

which is indeed another version of the bcc lattice. This decomposition of the bcc lattice is the simplest we have found, and will be discussed further in Section 5.

For the fcc lattice generated by $(1,1,0)$, $(1,-1,0)$, $(0,1,-1)$, the vectors of small norms are:

shape	norm	number
0 0 0	0	1
1 1 0	2	12
2 0 0	4	6
2 1 1	6	24
2 2 0	8	12
3 1 0	10	24
2 2 2	12	8

The simplest intersection we have found is formed by taking the 32 vectors of norms 6 and 12 and partitioning them into four sets of size 8. The resulting lattices have generator matrices

$$\begin{bmatrix} 2 & 1 & 1 \\ 1 & 2 & -1 \\ -2 & 2 & 2 \end{bmatrix}, \quad \begin{bmatrix} 1 & 2 & 1 \\ -1 & 1 & 2 \\ 2 & -2 & 2 \end{bmatrix}, \quad \begin{bmatrix} 1 & 1 & 2 \\ 2 & -1 & 1 \\ 2 & 2 & -2 \end{bmatrix}, \quad \begin{bmatrix} 2 & -1 & -1 \\ -1 & 2 & -1 \\ 2 & 2 & 2 \end{bmatrix} \tag{23}$$

and their intersection is the fcc lattice with generator matrix $(3,3,0)$, $(3,-3,0)$, $(0,3,-3)$. We will return to this example in Section 6.

Method 4: Using intersections of codes. A fourth method, which however has not yet led to any interesting examples, is to reduce the problem to the analogous question for codes. Let $\Lambda(C)$ denote the lattice obtained by applying Construction A to a binary linear code C ([11, Chap. 5]). If $C_1, \ldots, C_r$ are codes of length n whose intersection is a code C, then

$$\Lambda(C) = \Lambda(C_1) \cap \cdots \cap \Lambda(C_r) .$$

This can be generalized to nonbinary codes [11, Chaps. 7, 8], in particular to the case where the C_i are nonbinary codes whose intersection is binary.

4 The hexagonal lattice as an intersection of three lattices

We now begin our study of the honeycombs formed by some of the intersections described in Section 3. Once the cells have been found it is generally straightforward to compute their volumes and second moments using the techniques presented in [7] or [11, Chap. 21].

The hexagonal lattice is the intersection of the three rectangular lattices given by (2), as in Fig. 1. The honeycomb is shown in Fig. 2. We now compute the mean squared error for this quantizer. This analysis differs from that given in [4], which used a different set of representation points for the quantizer.

There are four types of cells. The origin is contained in a hexagon $\mathcal{P}_1$ (horizontally shaded in Fig. 2) of edge length $1/\sqrt{3}$, area $V_1 = \sqrt{3}/2$ and second moment $U_1 = 5\sqrt{3}/72$. The second type of cell is a small equilateral triangle $\mathcal{P}_2$ (cross-hatched), with $V_2 = \sqrt{3}/12$, $U_2 = \sqrt{3}/432$. $\mathcal{P}_3$ is an isosceles triangle (diagonally shaded), with $V_3 = \sqrt{3}/12$, $U_3 = 5\sqrt{3}/1296$. The fourth type, $\mathcal{P}_4$, is a larger equilateral triangle (vertically shaded), with $V_4 = \sqrt{3}/4$, $U_4 = \sqrt{3}/48$.

The incidences between the different types of cells are shown in Fig. 3. Here (and in similar diagrams in later sections) a circle containing i refers to a cell of type $\mathcal{P}_i$, and an edge

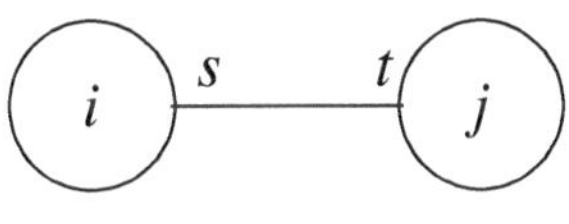

indicates that $\mathcal{P}_i$ and $\mathcal{P}_j$ share a common face of the maximal dimension (here 1), and the edge labels indicate that each $\mathcal{P}_i$ is adjacent to s cells of type $\mathcal{P}_j$ and each $\mathcal{P}_j$ to t cells of type $\mathcal{P}_i$.

The Voronoi cell $\mathcal{V}$ for the intersection lattice is enclosed by the broken lines in Fig. 2. In the notation of Section 2, $\mathcal{V}$ contains $N_1 = 1$ copy of $\mathcal{P}_1$,

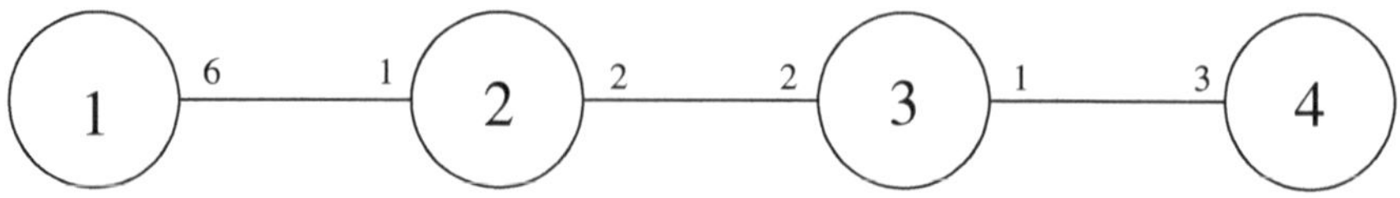

Fig. 3. Incidences between cells in honeycomb for hexagonal lattice.

$N_2 = 6$ copies of $\mathcal{P}_2$, $N_3 = 6$ copies of $\mathcal{P}_3$ and $N_4 = 6 \times \frac{1}{3} = 2$ copies of $\mathcal{P}_4$, and the volume equation (4) reads

$$2\sqrt{3} = \frac{\sqrt{3}}{2} + \frac{\sqrt{3}}{2} + \frac{\sqrt{3}}{2} + \frac{\sqrt{3}}{2} \,. \tag{24}$$

Thus the probabilities $p_1, \ldots, p_4$ of a randomly chosen point in the plane belonging to a cell of each type are all equal to 1/4. From (9), the mean squared error for this quantizer is $G = 2^{7/4}/27 = 0.1246\ldots$ This value (and corresponding values of G found in the next three Sections) is considerably worse than the value $0.080188\ldots$ for the hexagonal lattice itself.

5 The bcc lattice as an intersection of three lattices

The bcc lattice is the intersection of the three "rectangular" lattices $\Lambda_1, \Lambda_2, \Lambda_3$ defined in (21). Each Λ_i is congruent to $\mathbb{Z} \times \sqrt{2}\mathbb{Z} \times \sqrt{2}\mathbb{Z}$, and has as Voronoi cell a brick with square cross-section. There is an obvious symmetric group S_3 that permutes the Λ_i.

Again the honeycomb contains four types of cells. The origin is contained in the intersection of the Voronoi cells at 0 for the three Λ_i. This is a cube, $\mathcal{P}_1$, with vertices $(\pm 1/2, \pm 1/2, \pm 1/2)$, volume $V_1 = 1$, and second moment $U_1 = 1/4$. Across each square face of $\mathcal{P}_1$ is a square pyramid $\mathcal{P}_2$, such as that with base $(1/2, \pm 1/2, \pm 1/2)$, apex $(1,0,0)$, $V_2 = 1/6$, $U_2 = 11/600$. Across each triangular face of $\mathcal{P}_2$ is a tetrahedron $\mathcal{P}_3$, such as that with vertices $(1/2, \pm 1/2, 1/2)$, $(1,0,0)$, $(1,0,1/2)$, $V_3 = 1/24$, $U_3 = 1/512$. Finally, across the other three faces of $\mathcal{P}_3$ we reach a fourth type of cell, $\mathcal{P}_4$, a quarter-octahedron, which occurs in two orientations, one having vertices such as

$$\left(\frac{1}{2}, 0, \frac{1}{2}\right), \quad \left(\frac{1}{2}, 0, 1\right), \quad \left(1, 0, \frac{1}{2}\right), \quad (1,0,1), \quad \left(\frac{1}{2}, \pm\frac{1}{2}, \frac{1}{2}\right), \tag{25}$$

the other having vertices such as

$$\left(1, \frac{1}{2}, \frac{1}{2}\right), \quad \left(1, 0, \frac{1}{2}\right), \quad \left(1, \frac{1}{2}, 0\right), \quad (1,0,0), \left(\frac{1}{2}, \frac{1}{2}, \frac{1}{2}\right), \left(\frac{3}{2}, \frac{1}{2} \frac{1}{2}\right) \,. \tag{26}$$

These may be described as quarters of squat octahedra. E.g., (25) is a quarter of the octahedron with vertices $(\pm 1, 0, \pm 1)$, $(1/2, \pm 1/2, 1/2)$. For $\mathcal{P}_4$ we have $V_4 = 1/12$, $U_4 = 1/192$.

No further types of cell appear: every face of either version of $\mathcal{P}_4$ leads to a $\mathcal{P}_3$. The incidence diagram is shown in Fig. 4.

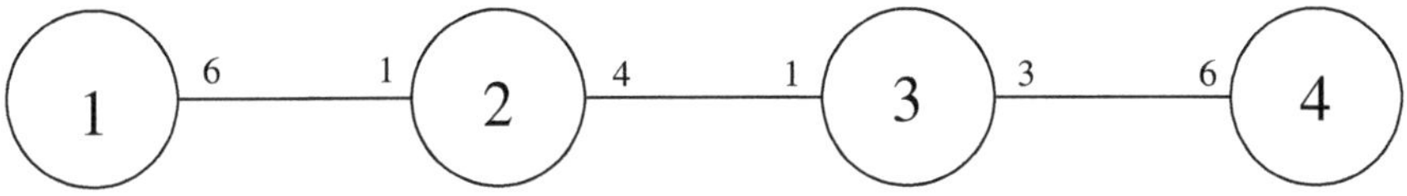

Fig. 4. Incidences between cells for honeycomb for bcc lattice.

The Voronoi cell $\mathcal{V}$ for the intersection lattice is a truncated octahedron with 24 vertices $(0, \pm 1/2, \pm 1)$. This contains $N_1 = 1$ copy of $\mathcal{P}_1$, $N_2 = 6$ copies of $\mathcal{P}_2$ and $N_3 = 24$ copies of $\mathcal{P}_3$. The cells of type $\mathcal{P}_4$ partially overlap $\mathcal{V}$. There are 12 of type (25), intersecting $\mathcal{V}$ in a tetrahedron such as that with vertices $(1/2, \pm 1/2, 1/2)$, $(1/2, 0, 1)$, $(1, 0, 1/2)$, with volume $1/24$. There are also 24 of type (26), intersecting $\mathcal{V}$ in a tetrahedron such as $(1/2, 1/2, 1/2)$, $(1, 0, 0)$, $(1, 0, 1/2)$, $(1, 1/2, 0)$, with volume $1/48$. The volume equation (4) reads

$$\begin{aligned} 4 &= 1 \times 1 + 6 \times \frac{1}{6} + 24 \times \frac{1}{24} + 12 \times \frac{1}{24} + 24 \times \frac{1}{48} \\ &= 1 + 1 + 1 + 1 \, , \end{aligned} \tag{27}$$

so again the probabilities p_i of a randomly chosen point belonging to a cell of given type are all equal to $1/4$. The mean squared error is

$$G = \frac{751\sqrt{3}}{9600} = 0.1355\ldots$$

6 The fcc lattice as an intersection of four lattices

The fcc lattice is the intersection of the four lattices $\Lambda_1, \ldots, \Lambda_4$ defined in (23). Each of these has Gram matrix equivalent to

$$\begin{bmatrix} 6 & -3 & 0 \\ -3 & 6 & 0 \\ 0 & 0 & 12 \end{bmatrix}$$

and is a direct sum $\sqrt{3}A_2 \oplus \sqrt{12}\mathbb{Z}$, with Voronoi cell a hexagonal prism. The Λ_i look more symmetrical if they are written in the coordinates used to describe the root lattice A_3 ($\cong D_3$), that is, using four coordinates that add to 0. Then Λ_1 has generator matrix

$$\begin{bmatrix} 0 & 2 & -1 & -1 \\ 0 & -1 & 2 & -1 \\ 3 & -1 & -1 & -1 \end{bmatrix}$$

and the others are given by cyclic shifts of these columns. This shows that there is a symmetric group S_4 permuting the Λ_i. However, the three-dimensional coordinates given in (23) are more convenient for computations.

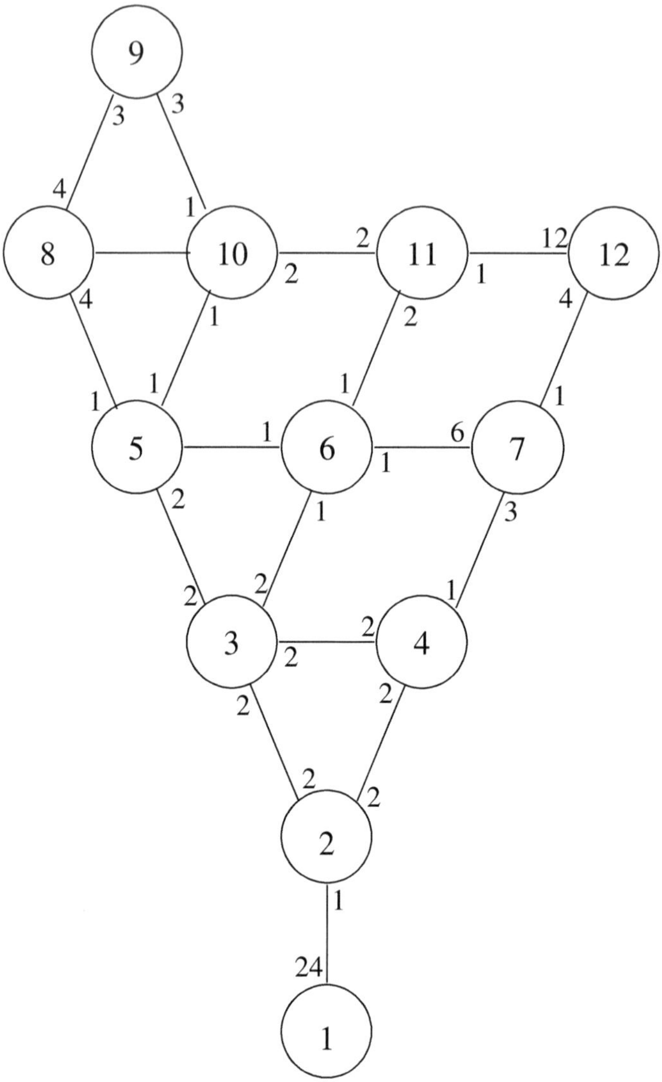

Fig. 5. Incidences among cells of fcc honeycomb.

This honeycomb is the most complicated we have analyzed and we shall give only a brief description. There are twelve types of cells, $\mathcal{P}_1, \ldots, \mathcal{P}_{12}$, whose parameters are summarized in Table 3 and whose incidences are shown in Fig. 5. Figures 6 and 7 shows cross-sections through the honeycomb along the planes $z = 0$ and $z = 0.35$.

The following is a brief description of the cells, including coordinates for one cell of each type.

Table 3. The twelve types of cells in the fcc honeycomb, showing numbers of vertices, edges, faces (v, e, f), the number per fcc cell (N_i), and their volumes, probabilities and second moments(V_i, p_i, U_i).

i	v	e	f	N_i	V_i	p_i	U_i
1	26	48	24	1	9	1/6	1449/160
2	5	8	5	24	1/8	1/18	41/3200
3	6	12	8	24	1/4	1/9	9/320
4	5	8	5	24	1/8	1/18	41/3200
5	5	9	6	24	1/10	2/45	427/50000
6	4	6	4	48	1/40	1/45	443/320000
7	10	18	10	8	27/40	1/10	41013/160000
8	6	12	8	6	1	1/9	47/180
9	5	9	6	8	3/8	1/18	189/3200
10	4	6	4	24	1/40	1/90	1897/1280000
11	5	8	5	24	3/40	1/30	2319/400000
12	22	36	16	2	63/10	7/30	213597/40000

$\mathcal{P}_1$. Obtained from cube by pushing in corners and pulling out edge midpoints. Vertices: all cyclic shifts and sign changes of $(3/2, 0, 0)$, $(1, 1, 0)$, $(3/4, 3/4, 3/4)$.

$\mathcal{P}_2$. Pyramid with kite-shaped base. Base: $(3/2, 0, 0)$, $(1, 1, 0)$, $(1, 0, 1)$, $(3/4, 3/4, 3/4)$, apex: $(3/2, 1/2, 1/2)$.

$\mathcal{P}_3$. Irregular octahedron with vertices $(3/2, 0, 0)$, $(3/2, 3/2, 0)$, $(1, 1, 0)$, $(2, 1, 0)$, $(3/2, 1/2, \pm 1/2)$.

$\mathcal{P}_4$. Congruent to $\mathcal{P}_2$. Vertices: $(3/2, 0, 0)$, $(3/4, 3/4, 3/4)$, $(3/2, 1/2, 1/2)$, $(1/2, 3/2, 1/2)$, $(1, 1, 0)$.

$\mathcal{P}_5$. Irregular polyhedron with six faces. Vertices: $(3/2, 0, 0)$, $(2, 1, 0)$, $(2, 0, 1)$, $(9/5, 3/5, 3/5)$, $(3/2, 1/2, 1/2)$.

$\mathcal{P}_6$. Tetrahedron: $(2, 1, 0)$, $(3/2, 3/2, 0)$, $(3/2, 1/2, 1/2)$, $(9/5, 3/5, 3/5)$.

$\mathcal{P}_7$. "Flying saucer": hexagonal base (cyclic shifts of $(3/2, 3/2, 0)$ and $(9/5, 3/5, 3/5)$) with four vertices above it (cyclic shifts of $(3/2, 1/2, 1/2)$ and $(3/4, 3/4, 3/4)$ at apex).

$\mathcal{P}_8$. Irregular octahedron with vertices $(2, \pm 1, 0)$, $(2, 0, \pm 1)$, $(3/2, 0, 0)$, $(3, 0, 0)$.

$\mathcal{P}_9$. Regular tetrahedron (vertices $(2, 1, 0)$, $(2, 0, 1)$, $(3, 0, 0)$, $(3, 1, 1)$) with triangular cap (apex $(9/4, 3/4, 3/4)$) on one face.

$\mathcal{P}_{10}$. Irregular tetrahedron with vertices $(2, 1, 0)$, $(2, 0, 1)$, $(9/4, 3/4, 3/4)$, $(9/5, 3/5, 3/5)$.

$\mathcal{P}_{11}$. Another pyramid with kite-shaped base. Base: $(3/2, 3/2, 0)$, $(9/4, 3/4, 3/4)$, $(9/5, 3/5, 3/5)$, $(12/5, 6/5, 3/5)$, apex: $(2, 1, 0)$.

$\mathcal{P}_{12}$. See Fig. 8. Has 26 vertices, four hexagonal and 12 kite-shaped faces. Vertices are all permutations of $(3/2, 3/2, 0)$, $(3/2, 3/2, 3)$, $(9/4, 3/4, 3/4)$, $(9/4, 9/4, 9/4)$, $(9/5, 3/5, 3/5)$, $(12/5, 9/5, 9/5)$, $(12/5, 6/5, 3/5)$.

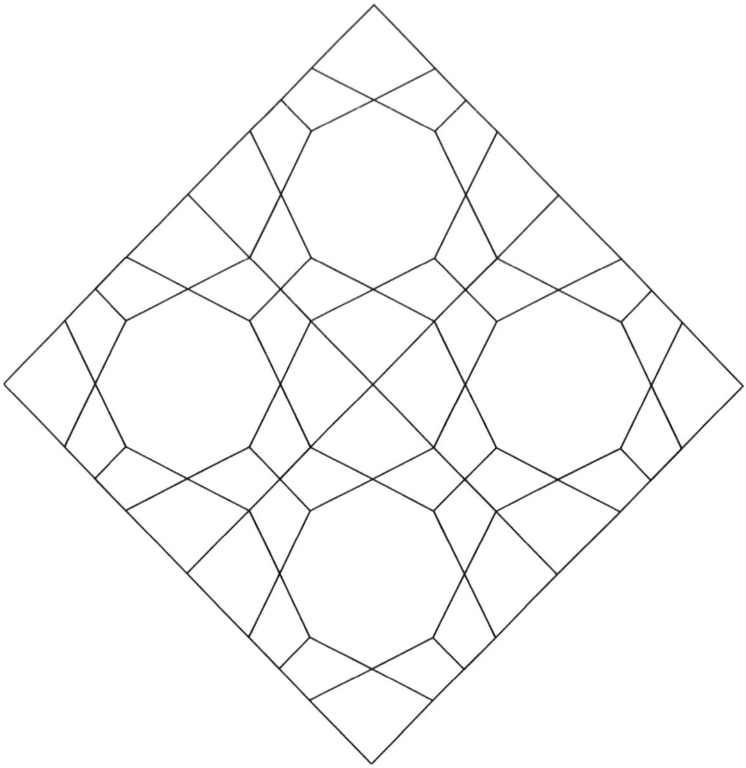

Fig. 6. Cross-section of fcc honeycomb along plane $z = 0$ ($-3 \le x \le 9$, $-6 \le y \le 6$), with origin at center of octagon on left. Only three cells are visible: $\mathcal{P}_1$ (octagon), $\mathcal{P}_2$ (small kite), $\mathcal{P}_9$ (large dart).

If we decompose $\mathbb{R}^3$ into Voronoi cells for the intersection lattice Λ (24), just three of the twelve types of cells are cut by the boundary walls. Cells of type $\mathcal{P}_9$ are cut into three equal pieces, cells of type $\mathcal{P}_{11}$ are cut in half, and cells of type $\mathcal{P}_{12}$ are cut into four equal ice-cream cone shaped pieces. The base of each cone is at the center of $\mathcal{P}_{12}$, and the top contains one of the hexagons and parts of the neighboring faces.

The mean squared error is

$$G = \frac{12269777}{816480000}\, 3^{28/135}\, 5^{8/27}\, 7^{38/45} = 0.1572\ldots .$$

7 The D_4 lattice as an intersection of three cubic lattices

This is the nicest example we have found. The three cubic lattices Λ_1, Λ_2, Λ_3 are defined in (12) and their intersection $\Lambda \cong D_4$ in (13). There are just four types of cells, $\mathcal{P}_1$, $\mathcal{P}_2$, $\mathcal{P}_3$, $\mathcal{P}_4$, whose properties are summarized in Table

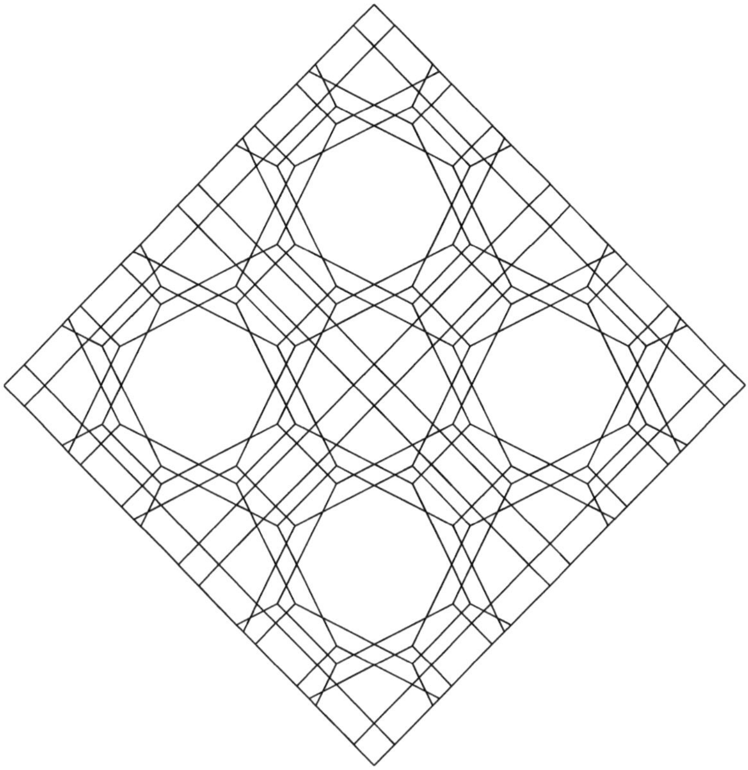

Fig. 7. Cross-section through fcc honeycomb along plane $z = 0.35$. Cross-sections of all 12 types of cells can be seen.

4 and whose intersections are shown in Fig. 9. We use coordinates (a, b, c, d) for points in $\mathbb{R}^4$.

$\mathcal{P}_1$ is the 4-dimensional regular polytope known as a 24-cell ([11], [12]). The Voronoi cells for Λ_1, Λ_2, Λ_3 at the origin are all cubes, whose intersection is bounded by the hyperplanes

$$|a| \leq 1,\ |b| \leq 1,\ |c| \leq 1,\ |d| \leq 1,\ |a| + |b| + |c| + |d| \leq 1,$$

which is the 24-cell with vertices of the form $(\pm 1, \pm 1, 0, 0)$.

Across each of the 24 octahedral faces of $\mathcal{P}_1$ we reach an octahedral-based pyramid $\mathcal{P}_2$, such as that with vertices $(1, \pm 1, 0, 0)$, $(1, 0, \pm 1, 0)$, $(1, 0, 0, \pm 1)$ and $(2, 0, 0, 0)$ (the apex).

There are eight other faces of $\mathcal{P}_2$, regular tetrahedra; these lead to copies of $\mathcal{P}_3$, which is an irregular simplex such as that with vertices $(1, 1, 0, 0)$, $(1, 0, 1, 0)$, $(1, 0, 0, 1)$, $(2, 0, 0, 0)$, $(1, 1, 1, 1)$.

Finally, three of the five faces of each $\mathcal{P}_3$ lead to cells of the fourth type, $\mathcal{P}_4$. This can best be described as the product of two skew equilateral triangles of

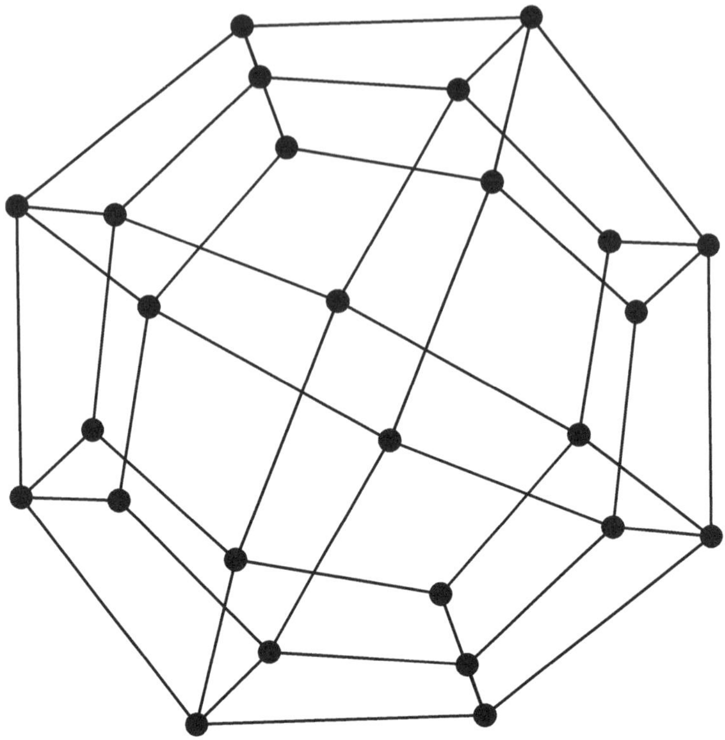

Fig. 8. 26-vertex cell $\mathcal{P}_{12}$.

different sizes (just as a tetrahedron in three dimensions is the product of two skew line segments). Take an equilateral triangle with vertices $p = (2, 1, 1, 0)$, $q = (1, 1, 0, 0)$, $r = (1, 0, 1, 0)$ and another with vertices $P = (2, 0, 0, 0)$, $Q = (1, 1, 1, 1)$, $R = (1, 1, 1, -1)$. Then $\mathcal{P}_4$ is their convex hull. There are nine tetrahedral faces given by the convex hull of an edge of the first triangle and an edge of the second triangle.

If we decompose $\mathbb{R}^4$ into Voronoi cells for the intersection lattice, only cells of type $\mathcal{P}_4$ are cut by the boundary walls. Each $\mathcal{P}_4$ is divided into three equal pieces, a typical piece being an "ice-cream cone" whose center is at the center of $\mathcal{P}_4$ and whose three-dimensional face is the convex hull of the second triangle (P, Q, R) and any edge of the first triangle.

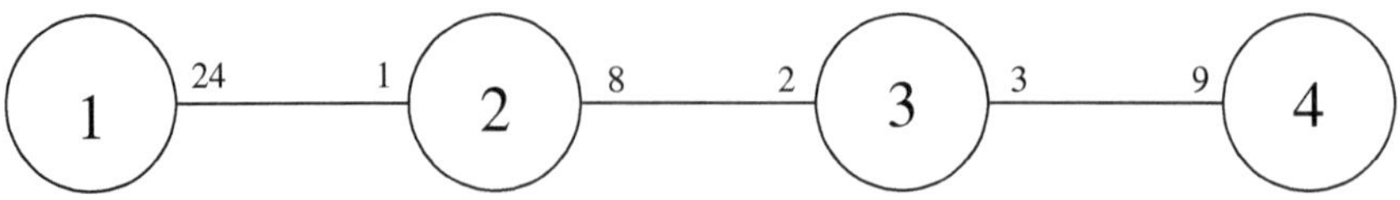

Fig. 9. Incidences among cells of D_4 honeycomb.

Table 4. Cells in D_4 honeycomb, showing numbers of vertices and 3-dimensional faces (v, f), the number per D_4 cell (N_i) and their volumes, probabilities and second moments (V_i, p_i, U_i).

i	v	f	N_i	V_i	p_i	U_i
1	24	24	1	8	1/4	104/15
2	7	9	24	1/3	1/4	8/105
3	5	5	96	1/12	1/4	11/900
4	6	9	32	1/4	1/4	1/20

The volume equation (4) then reads

$$32 = 1 \times 8 + 24 \times \frac{1}{3} + 96 \times \frac{1}{12} + 96 \times \frac{1}{4} \times \frac{1}{3}.$$

Again a random point is equally likely to fall into a cell of any of the four types. The mean squared error is

$$G = \frac{757}{8400} \, 2^{1/8} \, 3^{1/4} = 0.1293\ldots .$$

8 Conclusions and comments

In each of the honeycombs of Sections 4, 5 and 7 just four types of cells occurred. This is easily explained in the case of the D_4 honeycomb: there are three equivalent lattices Λ_1, Λ_2, Λ_3, and the associated quantizers are essentially making binary decisions about the location of a point with respect to the intersection lattice. So the space is divided up into regions that can be labeled 000, 001, 011 and 111.

This argument doesn't quite apply to the A_2 or bcc honeycombs, since there the individual lattices themselves are not fully symmetric (rectangular rather square in the A_2 case, for instance). So it is fortuitous that only four cells occur.

In contrast, the fcc honeycomb of Section 6 shows that the number of cells can increase rapidly in less fortunate cases with more component lattices. It would be interesting to see how complicated are the E_6^* and E_8 honeycombs mentioned in Section 3.

As for practical applications, it is possible that better quantizers could be obtained by amalgamating less symmetrical cells. For example, in Fig. 2, the diagonally and vertically shaded triangles could be amalgamated to give a honeycomb made up of regular hexagons and equilateral triangles with the same edge length as the hexagons. The new honeycomb will have larger absolute error but a smaller normalized error G. We did not investigate this possibility for the higher-dimensional examples.

Another topic for future research is the design of quantizers by taking a simple initial lattice Λ_1 and combining it with several other lattices which are both rotations *and translations* of Λ_1.

References

[1] C. B. Barber, D. P. Dobkin and H. T. Huhdanpaa, The Quickhull algorithm for convex hulls, *ACM Trans. Mathematical Software*, **22** (1996), 469–483.

[2] E. S. Barnes and N. J. A. Sloane, The optimal lattice quantizer in three dimensions, *SIAM J. Algebraic Methods*, **4** (1983), 30–41.

[3] Y. Be'ery and J. Snyders, Fast decoding of the Leech lattice, *IEEE J. Select. Areas Comm.*, **7** (1989), 959–967.

[4] B. Beferull-Lozano and A. Ortega, Efficient quantization for overcomplete expansions in $\mathbb{R}^N$, *IEEE Trans. Inform. Theory*, **49** (2003), 129–150.

[5] A. R. Calderbank, P. J. Cameron, W. M. Kantor and J. J. Seidel, $\mathbb{Z}_4$-Kerdock codes, orthogonal spreads, and extremal Euclidean line-sets, *Proc. London Math. Soc.*, **75** (1997), 436–480.

[6] A. R. Calderbank, R. H. Hardin, E. M. Rains, P. W. Shor and N. J. A. Sloane, A group-theoretic framework for the construction of packings in Grassmannian spaces, *J. Algebraic Combin.*, **9** (1999), 129–140.

[7] J. H. Conway and N. J. A. Sloane, Voronoi regions of lattices, second moments of polytopes, and quantization, *IEEE Trans. Inform. Theory*, **28** (1982), 211-226.

[8] J. H. Conway and N. J. A. Sloane, Fast quantizing and decoding algorithms for lattice quantizers and codes, *IEEE Trans. Inform. Theory*, **28** (1982), 227–232.

[9] J. H. Conway and N. J. A. Sloane, The Coxeter-Todd lattice, the Mitchell group, and related sphere packings, *Math. Proc. Camb. Phil. Soc.*, **93** (1983), 421–440.

[10] J. H. Conway and N. J. A. Sloane, The cell structures of certain lattices, in *Miscellanea mathematica*, P. Hilton, F. Hirzebruch and R. Remmert, editors, Springer-Verlag, NY, 1991, pp. 71–107.

[11] J. H. Conway and N. J. A. Sloane, *Sphere Packings, Lattices and Groups*, 3rd ed., Springer-Verlag, New York, 1998.

[12] H. S. M. Coxeter, *Regular Polytopes*, Dover, NY, 3rd ed., 1973.

[13] S. N. Diggavi, N. J. A. Sloane and V. A. Vaishampayan, Asymmetric multiple description lattice vector quantizers, *IEEE Trans. Information Theory*, **48** (2002), 174–191.

[14] L. Fejes Tóth, Sur la représentation d'une population infinie par une nombre fini d'éléments, *Acta Mathematica Academiae Scientiarum Hungaricae*, **10** (1959), 299–304.

[15] Geometry Center, *Home page for Qhull*, published electronically at http://www.geom.umn.edu/software/qhull.

[16] A. Gersho and R. M. Gray, *Vector Quantization and Signal Compression*, Kluwer, Boston, 1992.

[17] R. M. Gray, *Source Coding Theory*, Kluwer, Boston, 1990.

[18] R. M. Gray and D. L. Neuhoff, Quantization, *IEEE Trans. Inform. Theory*, **44** (1998), 2325–2383.

[19] R. H. Hardin, N. J. A. Sloane and Warren D. Smith, Tables of Spherical Codes, available electronically from http://www.research.att.com/~njas/packings/.

[20] N. Kashyap and D. L. Neuhoff, On quantization with the Weaire-Phelan partition, IEEE Trans. Inform. Theory, **47** (2001), 2538–2543.

[21] G. Nebe, E. M. Rains and N. J. A. Sloane, The Invariants of the Clifford Groups, *Designs, Codes and Cryptography*, **24** (2001), 99–121.

[22] G. Nebe, E. M. Rains and N. J. A. Sloane, A simple construction for the Barnes-Wall lattices, in *Codes, Graphs, and Systems: A Celebraton of the Life and Career of G. David Forney, Jr. on the Occasion of his Sixtieth Birthday*, ed. R. E. Blahut and R. Koetter, Kluwer, 2002, pp. 333–342.

[23] A. Okabe, B. Boots and K. Sugihara, *Spatial Tesselations: Concepts and Applications of Voronoi Diagrams*, Wiley, NY, 1992.

[24] O. T. O'Meara, *Introduction to Quadratic Forms*, Springer-Verlag, NY, 1971.

[25] H.-G. Quebbemann, Modular lattices in Euclidean spaces, *J. Number Theory*, **54** (1995), 190–202.

[26] N. Rivier, Kelvin's conjecture on minimal froths and the counter-example of Weaire and Phelan, *Phi. Mag. Lett.* **69** (1997), 297–303.

[27] N. J. A. Sloane and V. A. Vaishampayan, A Zador-Like formula for quantizers based on periodic tilings, *IEEE Trans. Inform. Theory*, **48** 2002, 3138–3140.

[28] D. F. Swayne, D. Cook and A. Buja, XGobi: interactive dynamic graphics in the X window system with a link to S, in *ASA Proceedings Section on Statistical Graphics*, Amer. Stat. Assoc., Alexandria, VA, 1991, pp. 1–8.

[29] V. A. Vaishampayan, N. J. A. Sloane and S. D. Servetto, Multiple description vector quantization with lattice codebooks: design and analysis, *IEEE Trans. Inform. Theory*, **47** (2001), 1718–1734.

[30] A. Vardy, Even more efficient bounded-distance decoding of the hexacode, the Golay code, and the Leech lattice, *IEEE Trans. Inform. Theory*, **41** (1995), 1495–1499.

[31] A. Vardy and Y. Be'ery, Maximum-likelihood decoding of the Leech lattice, *IEEE Trans. Inform. Theory*, **39** (1993), 1435–1444.

[32] D. Weaire and R. Phelan, A counter-example to Kelvin's conjecture on minimal surfaces, *Phil. Mag. Lett.*, **69** (1994), 107–110.

[33] A. F. Wells, *Three-Dimensional Nets and Polyhedra*, Wiley, NY, 1977.

[34] P. L. Zador, *Development and Evaluation of Procedures for Quantizing Multivariate Distributions*, Ph.D. Dissertation, Stanford Univ., 1963.

[35] P. L. Zador, Asymptotic quantization error of continuous signals and their quantization dimension, *IEEE Trans. Inform. Theory*, **28** (1982), 139–148.

[36] P. Ziherl and R. D. Kamien, Maximizing entropy by minimizing area: towards a new principle of self-organization, arXiv: cond-mat/0103171, 2001.

About Authors

N. J. A. Sloane is at the Information Sciences Research Center, AT&T Shannon Lab, Florham Park, New Jersey 07932-0971; *njas@research.att.com*.

B. Beferull-Lozano is at the Swiss Federal Institute of Technology – Lausanne, EPFL-IC-LCAV, CH-1015 Lausanne, Switzerland; *baltasar.beferull@epfl.ch*.

Acknowledgments

We thank John Conway, Suhas Diggavi, Eric Rains and Vinay Vaishampayan for helpful discussions. Deborah Swayne provided valuable help in running XGobi, and David Johnson kindly supplied the graph coloring program that was used in Section 3.

This work was carried out during B. Beferull-Lozano's visit to AT&T Labs in the summer of 2001.

Note on a Generalization of Roth's Theorem

József Solymosi

Abstract

We give a simple proof that for sufficiently large N, every subset of $[N]^2$ of size at least δN^2 contains three points of the form $\{(a,b),(a+d,b),(a,b+d)\}$.

In this note we give a simple proof for a theorem of Ajtai and Szemerédi [1]. In their proof Ajtai and Szemerédi used and iterated Szemerédi's theorem about long arithmetic progressions in dense sets of integers [8]. A more general theorem of Fürstenberg and Katznelson also implies Theorem 1, but does not give bound on N as it uses ergodic theory [2]. After improving the bound in Szemerédi's theorem, Gowers asked for a quantitative proof of Theorem 1 [3, 4].

Theorem 1 *(Ajtai-Szemerédi) For any real number $\delta > 0$ there is a natural number N_0 such that for $N > N_0$ every subset of $[N]^2$ of size at least δN^2 contains a triple of the form $\{(a,b),(a+d,b),(a,b+d)\}$ for some integer $d \neq 0$.*

The key of the proof is a lemma of Ruzsa and Szemerédi [7]. A subgraph of a graph G is a *matching* if every vertex has degree one. A matching M is an *induced matching* if there are no other edges of G between the vertices of M.

Lemma 2 *(Ruzsa-Szemerédi) If G_n is the union of n induced matchings, then $e(G_n) = o(n^2)$.*

The lemma, with a simple proof deduced from Szemerédi's Regularity Lemma, can be also found in a survey paper of Komlós and Simonovits [5].

Proof of Theorem 1: Let S be a subset of the grid $[N]^2$ of size at least δN^2. We refer to a point of the grid with its coordinates, which are pairs $(i,j); i,j \in \{1,2,\ldots,N\}$. Let us define a bipartite graph $G(A,B)$ with vertex sets $A = \{v_1,\ldots,v_N\}$ and $B = \{w_1,\ldots,w_N\}$. Two vertices v_i and w_j are connected by an edge iff $(i,j) \in S$ (see Fig. 1).

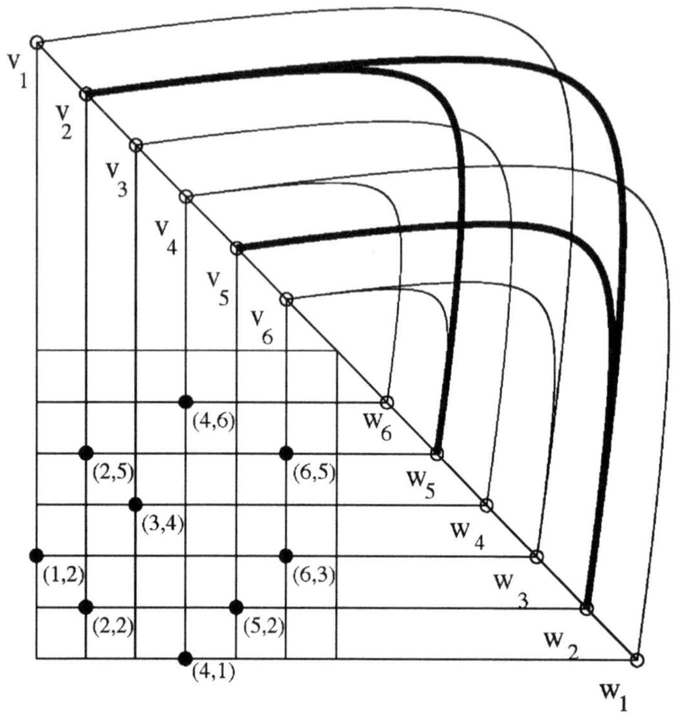

Fig. 1. Converting points into edges

Let us partition the edges of G according to their length, $(v_i, w_j) \sim (v_l, w_m)$ iff $i+j = l+m$. Every partition class is a matching, so we can apply Lemma 2 to G. If N is large enough, then at least one matching is not induced. A triple of edges $(v_i, w_m), (v_i, w_j), (v_l, w_m)$ such that $(v_i, w_j) \sim (v_l, w_m)$ guarantees a triple in S, $\{(a, b), (a + d, b), (a, b + d)\}$ (see bold edges in Fig.1). □

The only known proof of Lemma 2 uses Szemerédi's Regularity Lemma [8], so while the proof is quantitative, it gives a tower-type bound on $N_0 = N_0(\delta^{-1})$. It would be very important to find another, maybe analytical proof for Lemma 2 to get a better bound.

References

[1] M. Ajtai and E. Szemerédi, Sets of Lattice Points That Form No Squares. *Studia Scientiarum Mathematicarum Hungarica.* **9** (1974), 9–11.

[2] H. Fürstenberg and Y. Katznelson, A density version of the Hales-Jewett theorem. *J.d'Analyse Math.***57** (1991), 64–119.

[3] W.T. Gowers, A new proof of Szemerédi's theorem. *GAFA, Geom. Funct. Anal.* **11** (2001) 465–588.

[4] W.T. Gowers, Rough structure and classification. *GAFA, Geom. Funct. Anal.* Special Volume - GAFA2000 "Visions in Mathematics", Tel Aviv, 1999. Part I, 79–117.

[5] J. Komlós and M. Simonovits, Szemerédi's Regularity Lemma and its applications in graph theory. in: Combinatorics, Paul Erdős is eighty, Vol. 2 (Keszthely, 1993), 295–352, Bolyai Soc. Math. Stud., 2, János Bolyai Math. Soc., Budapest, 1996.

[6] K.F. Roth On certain sets of integers, *J.London Math. Soc.* **28** (1953), 245–252.

[7] I.Z. Ruzsa and E. Szemerédi, Triple systems with no six points carrying three triangles. in: Colloquia Mathematica Societatis János Bolyai, 18. Combinatorics, Keszthely (Hungary), 1976, 939–945.

[8] E. Szemerédi, On sets of integers containing no k elements in arithmetical progression. *Acta Arithmetica* **27** (1975), 199–245.

About Author

József Solymosi is at the Department of Mathematics, University of California, San Diego, La Jolla, CA 92093-0112, USA. *solymosi@math.ucsd.edu.*

Acknowledgments

I thank Timothy Gowers and Tibor Szabó for useful discussions. Work on this paper has been supported by the Berlin-Zürich European Graduate Program "Combinatorics, Geometry, and Computation" and by the Computer and Automation Research Institute, Hungarian Academy of Sciences.

Arrangements, Equivariant Maps and Partitions of Measures by k-Fans

Siniša T. Vrećica
Rade T. Živaljević

Abstract

We study topological and combinatorial structures that arise in the problem of finding α-partitions of m spherical measures by k-fans. An elegant construction of I. Bárány and J. Matoušek, [3], [4], shows that this problem can be reduced to the question whether there exists a G-equivariant map $f : V_2(\mathbb{R}^3) \to V_{\mathcal{P}} \setminus \cup\mathcal{A}_{\mathcal{P}}$ where G is a subgroup of a dihedral group $\mathbb{D}_{2n}$ while the target space $V_{\mathcal{P}} \setminus \cup\mathcal{A}_{\mathcal{P}}$ is the complement of a G-invariant, linear subspace arrangement $\mathcal{A}_{\mathcal{P}}$. We demonstrate that in many cases the relevant topological obstructions for the existence of these equivariant maps can be computed by a variety of geometric, combinatorial and algebraic ideas.

1 Preliminaries

The problem of simultaneous partitions of measures by k-fans has attracted a lot of attention in the last few years, see [3], [4] and [23]. For original motivation and related developments the reader is referred to [1], [5], [12], [13], [16], [18]. Aside from its natural appearance and relevance for discrete and computational geometry, this question serves also as a very interesting test problem for equivariant topological methods.

A k-fan is a point x in an Euclidean plane (Euclidean sphere), called the center of the fan, together with k semilines (great semicircles) $l_1, \ldots, l_k$ emanating from it. It is always assumed that semilines or semicircles in a k-fan $\mathfrak{p} = (x; l_1, \ldots, l_k)$ are enumerated counter clockwise, which is in agreement with the standard orientation on the ambient 2-manifold. Let $\mu_1, \mu_2, \ldots, \mu_m$ be Borel probability measures in the plane or on a sphere. We assume in this paper that all measures μ_j are *proper* in the sense that $\mu_j([a, b]) = 0$ for any line segment (circular arc) $[a, b]$ and that $\mu_j(U) > 0$ for each nonempty open set $U \subset \mathbb{R}^2$ ($U \subset S^2$). All the results can be extended by a standard limit argument to more general measures, including the counting measures of finite sets, see [3], [20], [21] for similar constructions and related examples.

We focus our attention mainly on the spherical case. This is justified by the fact that a partition result valid for a 2-sphere $K \subset \mathbb{R}^3$ with the center at

$a \in \mathbb{R}^3$, implies a similar result for the projective plane $\mathbb{P}^2(a) := \{l \mid \dim(l) = 1 \text{ and } a \in l\}$ of all lines concurrent with a. By taking a 2-plane D such that $a \notin D$, in other words by choosing an affine chart for the projective plane $\mathbb{P}^2(a)$, one obtains the corresponding result for the Euclidean plane. Note (see [3] for details and more precise formulations) that in the Euclidean case the center of the fan or even some of the semilines may lay at the line at infinity and thus not be visible in the ambient Euclidean plane.

We denote by σ_i the open angular sector between l_i and l_{i+1}, $i = 1, \ldots, n$. As it turns out, sometimes it may be more convenient to use the notation $\mathfrak{p} = (x; \sigma_1, \sigma_2, \ldots, \sigma_k)$ for the associated k-fan. Here and throughout the paper, we adopt the circular order $1 \prec 2 \prec \ldots \prec n \prec 1$ of indices and their addition "modulo n", so for example σ_n is the open angular sector between l_n and l_1. Let $(\alpha_1, \alpha_2, \ldots, \alpha_k)$ be a vector of positive real numbers where $\alpha_1 + \alpha_2 + \ldots + \alpha_k = 1$. Following [3], and taking into account our simplifying assumptions that we deal only with *proper* measures, we say that a k-fan $(x; l_1, \ldots, l_k) = (x; \sigma_1, \sigma_2, \ldots, \sigma_k)$ is an α-partition for the collection $\{\mu_j\}_{j=1}^m$ of measures if $\mu_j(\sigma_i) = \alpha_i$ for all $i = 1, \ldots, k$ and $j = 1, \ldots, m$.

Definition 1.1. A vector $\alpha \in \mathbb{R}^k$ is called admissible, or more precisely, (m,k)-*admissible* if for any collection of m *proper* measures $\mu_1, \ldots, \mu_m$ there exists a simultaneous α-partition. The collection of all (m,k)-admissible vectors is denoted by $\mathcal{A}_{m,k}$.

Problem 1.2. ([3]) Find an explicit characterization of the set $\mathcal{A}_{m,k}$. In other words the problem is to determine for what combinations of m, k and $\alpha \in \mathbb{R}^k$ one can guarantee that for any collection of measures $\mathcal{M} = \{\mu_1, \mu_2, \ldots, \mu_m\}$, there exists an α-partition for $\mathcal{M}$.

According to the original paper [3], see also Sections 2 and 3 of this paper, the following problem naturally arises in connection with Problem 1.2.

Problem 1.3. Given a subgroup G of the dihedral group $\mathbb{D}_{2n}$, the configuration space $X_{\mathcal{P}}$ and the test arrangement $\mathcal{A}_{\mathcal{P}}$, associated with the Problem 1.2 (defined in Section 2), find necessary and sufficient conditions for α and integers k and m such that there does *not* exist a G-equivariant map

$$f : X_{\mathcal{P}} \to V_{\mathcal{P}} \setminus \cup \mathcal{A}_{\mathcal{P}}. \tag{1}$$

The obstructions to the existence of equivariant maps (1) are identified as elements in the group N_G of G-coinvariants where $N := H_2(V_{\mathcal{P}} \setminus \cup \mathcal{A}_{\mathcal{P}}; \mathbb{Z})$ is the second homology group of the complement. Recall ([7], section II.2) that the group of coinvariants of a G-module M is the quotient group $M_G := M/D$ where D is a subgroup of M generated by all elements of the form $gm - m$, $(g \in G, m \in M)$. This shows why an aspect of the general α-partition problem is the following problem about the arrangements of subspaces:

Problem 1.4. Let $\mathcal{A}_{\mathcal{P}}$ be the G-invariant, linear subspace arrangement which appears in Problems 1.2 and 1.3. Determine the G-module structure on the group $H_p(V \setminus \mathcal{A}, \mathbb{Z})$.

The analysis given in [3] shows that the most interesting cases for the problem of the existence of α-partitions are $(m,k) = (3,2), (2,3), (2,4)$. Perhaps the most intriguing is the case $(m,k) = (2,4)$ where the measures are partitioned in the largest possible number of conical sectors. Bárány and Matoušek proved in [3] that $(\frac{2}{5}, \frac{1}{5}, \frac{1}{5}, \frac{1}{5}) \in \mathcal{A}_{2,4}$. Subsequently in [4] they proved that $(\frac{1}{4}, \frac{1}{4}, \frac{1}{4}, \frac{1}{4})$ is also $(2,4)$-admissible. R. Živaljević proved in [23] that $\mathcal{A}_{3,2} = \{(p,q) \mid p+q = 1,\, p,q > 0\}$.

1.1 This paper

In this paper we give new proofs of results of Bárány and Matoušek, see Theorems 4.1 and 5.1. The proof of $(\frac{1}{4}, \frac{1}{4}, \frac{1}{4}, \frac{1}{4}) \in \mathcal{A}_{2,4}$ is particularly short and transparent and serves as a good illustration of the general principle that one should use, whenever possible, the full group of symmetries in a problem of the existence of equivariant maps. We show, Theorem 4.2, that for a "generic" vector $\alpha = (\alpha_1, \alpha_2, \alpha_3, \alpha_4)$ there *does* exist a $\mathbb{Z}_n$-equivariant map $f : V_2(\mathbb{R}^3) \to V_{\mathcal{P}} \setminus \cup\mathcal{A}(\alpha)$. This is in agreement with [3] where it was shown by example that such a map exists in the case $\alpha = (\frac{1}{n}, \frac{1}{n}, \frac{1}{n}, 1 - \frac{3}{n})$ for n sufficiently large.

The combinatorial/topological technique developed here allows in principle to check if there exist G-equivariant maps from spheres of appropriate dimension to complements of arbitrary, G-invariant, linear subspace arrangements. This is obviously of some independent interest and we hope that this technique will find other interesting geometric and combinatorial applications in the future.

Finally we prove (Theorem 6.2) an equipartition result which is a common generalization of the fact that $(\frac{1}{3}, \frac{1}{3}, \frac{1}{3}) \in \mathcal{A}_{2,3}$, proved independently in [5] and [16] (see also [3] and [18]), and the partition result $B(n, n-2)$ from [17]. This theorem can be viewed as a step in the direction of studying higher dimensional α-partitions by k-fans.

2 Configuration spaces and test maps

I. Bárány and J. Matoušek introduced in [3] a very nice idea which permitted them to reduce the α-partition problem to the problem of the existence of equivariant maps. In this reduction they utilize the well known *configuration space/test map* method, see review papers [2], [6], [14], [20], [22] for more information and references to the original papers where these ideas were developed. Recall that the essence of the method is the construction of a *configuration* space $X_{\mathcal{P}}$, the test space $V_{\mathcal{P}}$ and a test map $f : X_{\mathcal{P}} \to V_{\mathcal{P}}$, associated to the problem $\mathcal{P}$, [22].

Given two Borel, probability measures μ and ν on S^2, one of them, say μ, is used in the definition of the configuration or *candidate space* $X_{\mathcal{P}} = X_\mu$, while the other measure ν is used for the definition of the appropriate *test map* $F_\nu : X_\mu \to W_n \subset R^n$. Here is a brief review of this construction. As before, we denote by $(x; l_1, \ldots, l_n)$ a spherical n-fan, where x is a point in S^2 and $l_1, \ldots, l_n$ is a sequence of great semicircles emanating from x which is ordered counterclockwise when viewed from the point $2x$. Let $\sigma_i, i = 1, \ldots, n$ be the open angular sector on S^2 between l_i and l_{i+1} $(l_{n+1} := l_1)$. Then the candidate space X_μ is defined by

$$X_\mu := \{(x; l_1, \ldots, l_n) \mid (\forall i = 1, \ldots, n)\ \mu(\sigma_i) = \tfrac{1}{n}\}. \tag{2}$$

Note that $(x; l_1, \ldots, l_n) \in X_\mu$ is uniquely determined by the pair (x, l_1) or equivalently the pair (x, y), where y is the unit tangent vector to l_1 at x. Hence, the space X_μ is homeomorphic to the Stiefel manifold $V_2(\mathbb{R}^3)$ of all orthonormal 2-frames in $\mathbb{R}^3$. By completing a 2-frame to a positively oriented 3-frame in $\mathbb{R}^3$, one obtains the first of the following well known homeomorphisms, $V_2(\mathbb{R}^3) \cong SO(3) \cong RP^3$. Let $\mathbb{R}^n$ be an Euclidean space with a preferred orthonormal basis $e_1, e_2, \ldots, e_n$ and the associated coordinate functions $x_1, x_2, \ldots, x_n$. Let W_n be the hyperplane in $\mathbb{R}^n$ defined by $W_n := \{x \in \mathbb{R}^n \mid x_1 + x_2 + \ldots + x_n = 0\}$ and suppose that the α-vector has the form $\alpha = (\frac{a_1}{n}, \frac{a_2}{n}, \frac{a_3}{n}, \frac{a_4}{n})$ where $a_1 + a_2 + a_3 + a_4 = n$ and a_i are positive integers.

Then the test map $F_\nu : X_\mu \to W_n \subset \mathbb{R}^n$ and the fundamental test linear subspace $L = L(\alpha) \subset W_n$ are defined as follows,

$$F_\nu(x, y) = F_\nu(x; l_1, \ldots, l_n) = (\nu(\sigma_1) - \tfrac{1}{n}, \nu(\sigma_2) - \tfrac{1}{n}, \ldots, \nu(\sigma_n) - \tfrac{1}{n}) \tag{3}$$

$$L = L(\alpha) := \{x \in \mathbb{R}^n \mid z_1(x) = z_2(x) = z_3(x) = z_4(x) = 0\}, \text{ where} \tag{4}$$

$$\begin{aligned} z_1(x) &= x_1 + x_2 + \ldots + x_{a_1}, & z_2(x) &= x_{a_1+1} + \ldots + x_{a_1+a_2}, \\ z_3(x) &= x_{a_1+a_2+1} + \ldots + x_{a_1+a_2+a_3}, & z_4(x) &= x_{a_1+a_2+a_3+1} + \ldots + x_n. \end{aligned} \tag{5}$$

Let $\mathcal{A} = \mathcal{A}(\alpha)$ be the smallest $\mathbb{Z}_n$-invariant, linear subspace arrangement in $\mathbb{R}^n$ which contains $L(\alpha)$. Hence, $\mathcal{A}(\alpha)$ is precisely the arrangement which consists of all subspaces of the form $L_g(\alpha) := g(L(\alpha))$, $g \in \mathbb{Z}_n$, together with their intersections. Note that $D = D(\mathcal{A}) := \cup\mathcal{A}(\alpha) \subset W_n$. Occasionally it may be convenient to view $\mathcal{A}(\alpha)$ as an arrangement both in W_n and $\mathbb{R}^n$ but the primary choice is the space W_n, being the natural target space for the test map F_ν. The cyclic group $\mathbb{Z}_n = \{1, \omega, \ldots, \omega^{n-1}\}$ acts both on the candidate space X_μ and the test space W_n by the action $\omega(x; l_1, \ldots, l_n) := (x; l_2, \ldots, l_n, l_1)$ and $\omega(x_1, \ldots, x_n) := (x_2, \ldots, x_n, x_1)$ for $(x; l_1, \ldots, l_n) \in X_\mu$ and $(x_1, \ldots, x_n) \in W_n$. The test map F_ν is clearly $\mathbb{Z}_n$-equivariant. As a $\mathbb{Z}_n$-space, X_μ is homeomorphic to the manifold $V_2(\mathbb{R}^3) \cong SO(3)$ with the action defined by $\omega(x, y) := (x, R_x(\frac{2\pi}{n})(y))$, where $R_x(\theta) : \mathbb{R}^3 \to \mathbb{R}^3$ is the rotation around the axes determined by x, through the angle θ.

Proposition 2.1. ([3]) *Let $\alpha = (\frac{a_1}{n}, \frac{a_2}{n}, \frac{a_3}{n}, \frac{a_4}{n}) \in \mathbb{R}^4$ be a vector such that both n and a_i are positive integers and $a_1 + a_2 + a_3 + a_4 = n$. Let us suppose that there does not exist a $\mathbb{Z}_n$-equivariant map $F : V_2(\mathbb{R}^3) \to M(\alpha)$, where $M(\alpha) := W_n \setminus \cup\mathcal{A}(\alpha)$. Then, for any two measures μ and ν on S^2, there always exists a 4-fan which simultaneously α-partitions both μ and ν.*

The following proposition was also established in [3]. Its proof relies on the fact that if n is an odd integer, then there always exists a $\mathbb{Z}_n$-equivariant map $f : S^3 \to V_2(\mathbb{R}^3)$.

Proposition 2.2. *Suppose that n is an odd integer and let us assume that there does not exist a $\mathbb{Z}_n$-equivariant map $F : S^3 \to M(\alpha)$. Then $\alpha \in \mathcal{A}_{2,4}$, i.e. for any two measures μ and ν on S^2, there always exists a 4-fan which simultaneously α-partitions both μ and ν.*

3 $\mathbb{Z}_n$-invariant arrangements

An immediate consequence of Proposition 2.1 is that the topological structure of the complement $M(\alpha) := W_n \setminus \cup\mathcal{A}(\alpha)$ of the arrangement $\mathcal{A}(\alpha)$ is apparently of great importance for the α-partition problem. In this section we collect all the relevant facts about $\mathcal{A}(\alpha)$ and linear subspace arrangements in general.

3.1 Generalities about arrangements

Recall that the area of arrangements of subspaces is a well studied subject with many interesting applications and connections with other fields, [15]. One of the main research themes has been the dependence of the topology of the complement $M := V \setminus D(\mathcal{A})$ and the compactified link $\hat{D}(\mathcal{A}) := D(\mathcal{A}) \cup \{\infty\}$, $D := \cup\mathcal{A}$, on the combinatorics of the *intersection poset* P. Recall that the *intersection poset* $P = P_{\mathcal{A}}$ is an abstract poset that records the (reversed) containment relation in $\mathcal{A}$, $(P_{\mathcal{A}}, \leq) \cong (\mathcal{A}, \supseteq)$. As usual, [6], $\Delta(Q)$ is the order complex of Q, i.e. the simplicial complex of all chains in Q while $P_{<p} := \{x \in P \mid x < p\}$ is the (open) lower cone of p in P.

The following formula of M. Goresky and R. MacPherson, [10], is of fundamental importance:

$$\tilde{H}_i(M;\mathbb{Z}) \cong \bigoplus_{p\in P} \tilde{H}^{n-2-i-d(p)}(\Delta(P_{<p});\mathbb{Z}). \tag{6}$$

It reveals that the homology of the complement $M = \mathbb{R}^n \setminus D(\mathcal{A})$ is completely determined by the intersection poset of the arrangement and the associated dimension function $d : P \to N$, $d(p) := \dim(p)$. A closely related formula of G. Ziegler and R. Živaljević, [19],

$$\hat{D} \simeq \bigvee_{p\in P} \Delta(P_{<p}) * S^{d(p)}, \tag{7}$$

says that the homotopy type of the one-point compactification $\hat{D} := D \cup \{+\infty\}$ of $D = D(\mathcal{A}) = \cup\mathcal{A}$, splits into a wedge of *joins* of combinatorially defined spaces and spheres. Both formulas (6) and (7) can be used in the analysis of the homology of the complement $M = W_n \setminus \cup\mathcal{A}(\alpha)$. Keeping in mind the intended applications, see Section 4, we will be mainly interested in the $\mathbb{Z}_n$-module structure on the group $H_2(M;\mathbb{Z})$. It turns out that this module is quite sensitive to the arithmetic of the vector $\alpha = (\frac{a_1}{n}, \frac{a_2}{n}, \frac{a_3}{n}, \frac{a_4}{n})$.

3.2 Circular 4-partitions

In light of the cyclic $\mathbb{Z}_n$-action on $\mathbb{R}^n$, we assume that vectors $e_1, e_2, \ldots, e_n$ are put in a circular order $e_1 \prec e_2 \prec \ldots \prec e_n \prec e_1$. Similarly, the indices $1, 2, \ldots, n$ are also put in a circular order with the addition defined modulo n. Consequently, for $i, j \in [n] := \{1, \ldots, n\}$, the circular interval $I = [i,j]_\prec = [i,j] = \{i, i+1, \ldots, j\}$ is well defined both for $i \leq j$ and $i \geq j$. Given a (circular) interval $I = [i,j]_\prec$, let $e_I := e_i + e_{i+1} + \ldots + e_j$ be the associated "characteristic" vector. A circular 4-partition $\mathcal{J} = \{I_1, I_2, I_3, I_4\}$, $I_j = [a_j, b_j], j = 1, \ldots, 4$, is a disjoint collection of intervals such that $\cup\, \mathcal{J} = \cup_{j=1}^4 I_j = [n]$. We always assume that $a_1 \prec b_1 \prec a_2 \prec b_2 \prec \ldots \prec a_4 \prec b_4$. The group $\mathbb{Z}_n$ acts on "everything in sight" and for example, $g(\mathcal{J}) := \{g(I_j)\}_{j=1}^4 = \{[a_j + k, b_j + k]_\prec\}_{j=1}^4$ is also a circular 4-partition of $[n]$, where $g = \omega^k \in \mathbb{Z}_n$.

The following question is intimately related to the problem of describing the group $H_2(W_n \setminus \cup\mathcal{A}(\alpha);\mathbb{Z})$. It is answered by Proposition 3.1. Here and elsewhere we denote by $\mathcal{L}(K)$ the linear subspace of $\mathbb{R}^n$ generated by $K \subset \mathbb{R}^n$.

Question: Describe all circular 4-partitions $\mathcal{J} = \{I_1, I_2, I_3, I_4\}$ such that for some $g \in \mathbb{Z}_n \setminus \{1\}$,

$$\dim \mathcal{L}\{e_I \mid I \in \mathcal{J} \cup g(\mathcal{J})\} = \dim \mathcal{L}\{e_{I_1}, \ldots, e_{I_4}, e_{g(I_1)}, \ldots, e_{g(I_4)}\} \leq 5. \quad (8)$$

An α-vector $\alpha(\mathcal{J})$ of a circular 4-partition $\mathcal{J}$ is defined by $\alpha(\mathcal{J}) := (\frac{|I_1|}{n}, \frac{|I_2|}{n}, \frac{|I_3|}{n}, \frac{|I_4|}{n})$ where $|I|$ is the cardinality of I. Sometimes it is convenient to identify the indices $1, \ldots, n$ with the vertices $v_1, \ldots, v_n$ of the regular polygon P_n inscribed in a unit circle. Then a circular 4-partition $\mathcal{J} = \{I_j\}_{j=1}^4$ defines a circular 4-partition $\mathcal{J}' = \{I_j'\}_{j=1}^4$ of the circle S^1. Then the α-vector records the lengths of consecutive arcs (intervals) in the circular partition $\mathcal{J}'$. We silently identify α-vectors which differ only by a cyclic permutation, i.e. $\alpha(\mathcal{J}) \in \mathbb{R}^4/\mathbb{Z}_4$.

Proposition 3.1. *The following equivalences give a characterization of circular 4-partitions with small dimension of the associated linear span.*

$$\exists_{g\neq 1}\ \dim \mathcal{L}\{e_I \mid I \in \mathcal{J} \cup g(\mathcal{J})\} = 4 \iff \exists_{p,q}\ \alpha(\mathcal{J}) = (p, q, p, q) \quad (9)$$

$$\exists_{g\neq 1}\ \dim \mathcal{L}\{e_I \mid I \in \mathcal{J} \cup g(\mathcal{J})\} = 5 \iff \begin{array}{ll} \alpha(\mathcal{J}) = (\frac{1}{3}, \frac{1}{3}, p, q) & \text{or} \\ \alpha(\mathcal{J}) = (p, q, p, p+q) & \text{or} \\ \alpha(\mathcal{J}) = (p, q, q, q) & (p \neq q). \end{array} \quad (10)$$

Proof: The left implications "$\Longleftarrow$" in both equivalences (9) and (10) are obvious. Let us assume that $\dim \mathcal{L}\{e_I \mid I \in \mathcal{J} \cup g(\mathcal{J})\} \leq 5$ and try to see how this constraint affects the corresponding 4-partition $\mathcal{J}$. The following Claim does not require a proof.

> **Claim:** Let $\{J_1, J_2, \ldots, J_m\}$ be a collection of disjoint circular intervals in $[n]$ such that $\cup_{j=1}^m J_j$ is also an interval in $[n]$. Let $J \subset \cup_{j=1}^m J_j$ be an interval. Then $\dim \mathcal{L}\{e_{J_1}, \ldots, e_{J_m}\} = \dim \mathcal{L}\{e_{J_1}, \ldots, e_{J_m}, e_J\} = m$ implies that $J = \cup_{i \in K} J_i$ for some $K \subset [m]$.

Let $|I_1| = \min\{|I_j|\}_{j=1}^4$, i.e. we assume that I_1 is a circular interval in $\mathcal{J}$ of minimum length. Depending on the position of $g(I_1)$ relative to $\mathcal{J} = \{I_1, I_2, I_3, I_4\}$, we discuss the following three cases,

(1) $g(I_1) \in \mathcal{J}$ and $I_1 \cup g(I_1)$ is an interval in $[n]$,

(2) $g(I_1) \in \mathcal{J}$ and $I_1 \cup g(I_1)$ is *not* an interval in $[n]$,

(3) $g(I_1) \notin \mathcal{J}$.

In the first case it is easily checked that $\alpha(\mathcal{J}) = (p, q, q, q)$ where $q = |I_1|$. In the second case $\alpha(\mathcal{J})$ must be either of the form (p, q, p, q) or of the form $(p, q, p, p + q)$. Finally if $g(I_1) \notin \mathcal{J}$ we apply the Claim on intervals $\{J_i\}_{i=1}^k$, where $k = 4$ or $k = 5$, which arise when we remove $g(I_1)$ from $[n]$ and from individual elements of the 4-partition $\mathcal{J} = \{I_1, I_2, I_3, I_4\}$. A conclusion is that for some rearrangement I, J, K, L of intervals in $\mathcal{J}$, either $|I| = |J|$ or $|I| = |J| + |K|$ or $|I| = |J| + |K| + |L|$. The last possibility is ruled out. The first is already covered by cases (1) and (2). Finally, in the case $|I| = |J| + |K|$, a simple analysis shows that $\alpha(\mathcal{J})$ must be either of the form $(\frac{1}{3}, \frac{1}{3}, p, q)$ or of the form $(p, q, p, p + q)$.

3.3 When $\mathcal{A}(\alpha)$ can be replaced by $\mathcal{A}(\beta)$?

Two arrangements $\mathcal{A}$ and $\mathcal{B}$ in $\mathbb{R}^n$ are *isomorphic* if there exists a non-singular linear map $C : \mathbb{R}^n \to \mathbb{R}^n$ such that $K \in \mathcal{A} \Leftrightarrow C(K) \in \mathcal{B}$. Suppose that $G \subset GL(n, \mathbb{R})$ is a group of symmetries of both $\mathcal{A}$ and $\mathcal{B}$. If $C : \mathbb{R}^n \to \mathbb{R}^n$ is G-equivariant, i.e. if $Cg = gC$ for each $g \in G$, we say that $\mathcal{A}$ and $\mathcal{B}$ are G-isomorphic. Finally, $\mathcal{A}$ and $\mathcal{B}$ are *weakly* G-isomorphic or wG-isomorphic if there exists an automorphism $\theta : G \to G$ of the group G, such that $C(g \cdot x) = \theta(g) \cdot C(x)$.

If the arrangements $\mathcal{A}$ and $\mathcal{B}$ are G-isomorphic then both the associated compactified links $\hat{D}(\mathcal{A})$, $\hat{D}(\mathcal{B})$ and the complements $M(\mathcal{A})$, $M(\mathcal{B})$ are homeomorphic as G-spaces. Similarly, if $\mathcal{A}$ and $\mathcal{B}$ are wG-isomorphic, then we have a similar conclusion, if we modify the action of G on $\hat{D}(\mathcal{B})$ and $M(\mathcal{B})$ by the automorphism θ, i.e. define a new action on the ambient space of the arrangement $\mathcal{B}$ by $g \cdot_\theta x := \theta(g) \cdot x$.

The importance of these notions is based on the following observation.

Proposition 3.2. *Suppose that $\mathcal{A}(\alpha)$ and $\mathcal{A}(\beta)$ are the arrangements in $\mathbb{R}^n$ associated to vectors $\alpha = (\frac{a_1}{n}, \frac{a_2}{n}, \frac{a_3}{n}, \frac{a_4}{n})$ and $\beta = (\frac{b_1}{n}, \frac{b_2}{n}, \frac{b_3}{n}, \frac{b_4}{n})$. Suppose that $\mathcal{A}(\alpha)$ and $\mathcal{A}(\beta)$ are $w\mathbb{Z}_n$-isomorphic. Then for any free $\mathbb{Z}_n$-action on the 3-sphere S^3, there exists a $\mathbb{Z}_n$-equivariant map $F : S^3 \to M(\mathcal{A}(\alpha))$ if and only if there exists an equivariant map $G : S^3 \to M(\mathcal{A}(\beta))$.*

Proof: By assumption there exists an automorphism $\theta : \mathbb{Z}_n \to \mathbb{Z}_n$ and a linear isomorphism $C : \mathbb{R}^n \to \mathbb{R}^n$, such that for each $v \in \mathbb{R}^n$ and $g \in \mathbb{Z}_n$, $C(g \cdot v) = \theta(g) \cdot C(v)$. In other words, there exists an integer p relatively prime to n, such that $C(\omega(v)) = \omega^p(C(v))$. From here one deduces that there exists a homeomorphism $C' : M(\mathcal{A}(\alpha)) \to M(\mathcal{A}(\beta))$ which is $(\mathbb{Z}_n, \theta)$-equivariant in the sense that $C'(\omega(x)) = \omega^p(C'(x))$ for each $x \in M(\mathcal{A}(\alpha))$. Note that for any two free, $\mathbb{Z}_n$-actions ρ_1, ρ_2 on S^3, there exists a $\mathbb{Z}_n$-equivariant map $f : (S^3, \rho_1) \to (S^3, \rho_2)$. This follows from the fact that S^3 is a 3-dimensional, 2-connected CW-complex, and an elementary obstruction theory argument, see e.g. [22] for similar constructions. Given a free $\mathbb{Z}_n$-action on S^3, suppose that $F : S^3 \to M(\mathcal{A}(\alpha))$ is a $\mathbb{Z}_n$-equivariant map. Then $C' \circ F : S^3 \to M(\mathcal{A}(\beta))$ is $(\mathbb{Z}_n, \theta)$-equivariant in the sense that $C' \circ F(\omega(x)) = \omega^p(C' \circ F(x))$ for each $x \in S^3$. Define a new free, $\mathbb{Z}_n$-action on S^3 by $(\omega, x) \mapsto \omega^p(x)$. Since any two actions on S^3 are "equivalent", there exists an equivariant map $H : S^3 \to S^3$ such that $H(\omega^p(x)) = \omega(H(x))$ for each $x \in S^3$. It follows that $G := C' \circ F \circ H : S^3 \to M(\mathcal{A}(\beta))$ is a map which satisfies the condition $G(\omega^p(x)) = \omega^p(G(x))$, i.e. G is a $\mathbb{Z}_n$-equivariant map from S^3 to $M(\mathcal{A}(\beta))$.

Proposition 3.3. *Let $\alpha = (1 - \frac{3}{n}, \frac{1}{n}, \frac{1}{n}, \frac{1}{n})$ and $\beta = (\frac{p}{n}, \frac{q}{n}, \frac{q}{n}, \frac{q}{n})$, where q is an integer, relatively prime to n. Then the associated arrangements $\mathcal{A}(\alpha)$ and $\mathcal{A}(\beta)$ are $w\mathbb{Z}_n$-isomorphic.*

Proof: Let $\mathcal{J} = \{I_1, I_2, I_3, I_4\}$ be a circular 4-partition of $[n]$ such that $|I_1| = |I_2| = |I_3| = q$ and $|I_4| = p$. Define $c_1 = e_{I_1} := \sum_{j \in I_1} e_j$ and let $c_k := \omega^{qk}(c_1) = e_{\omega^{qk}(I_1)}$. Note that for example, $c_2 = e_{I_2}, c_3 = e_{I_3}$. We immediately observe that the arrangement $\mathcal{A}(\beta)$ has the same description, relative the bases $c_1, c_2, \ldots, c_n$ as the arrangement $\mathcal{A}(\alpha)$ relative the basis $e_1, e_2, \ldots, e_n$. In other words the linear map $C : \mathbb{R}^n \to \mathbb{R}^n$, defined by $C(e_i) = c_i$ is an isomorphism of arrangements $\mathcal{A}(\alpha)$ and $\mathcal{A}(\beta)$. The cyclic permutation of vectors $c_1, c_2, \ldots, c_n$ is easily identified as the map $\omega^q : \mathbb{R}^n \to \mathbb{R}^n$. On the other hand, $\omega^q(v) = C \circ \omega \circ C^{-1}(v)$ and the result follows.

3.4 $H_2(M(\mathcal{A}(\alpha)))$ as a $\mathbb{Z}_n$-module

The fundamental group $\pi_1(M(\mathcal{A}(\alpha)))$ of the complement $M(\mathcal{A}(\alpha)) = W_n \setminus \cup\mathcal{A}(\alpha)$ is trivial. This is a consequence of the fact that each of the subspaces $K \in \mathcal{A}(\alpha)$ has the codimension in W_n greater or equal to 3. From here, or alternatively from formulas (6) and (7), it follows that $H_1(M(\mathcal{A}(\alpha)); \mathbb{Z}) = 0$. Hence, the first nontrivial group is $H_2(M(\mathcal{A}(\alpha)); \mathbb{Z})$, which is precisely the group relevant for the α-partition problem.

Let us collect here some useful facts about $H_2(M(\mathcal{A}(\alpha)))$ which will be used later in the computation of relevant obstruction classes. More generally, let $L_1, L_2, \ldots, L_k$ be a collection of a-dimensional, linear subspaces in an $(a+b)$-dimensional, Euclidean space V. Let $\mathcal{A}$ be the associated arrangement with the associated intersection poset $P_{\mathcal{A}}$, compactified union (link) $\hat{D}(\mathcal{A}) = \cup\,\mathcal{A} \cup \{+\infty\} \subset V \cup \{+\infty\} \cong S^{a+b}$, and the complement $M(\mathcal{A}) = V \setminus \cup\,\mathcal{A}$. For a given $x \in L_i \setminus \cup_{j\neq i} L_j$, let $D_\epsilon(x) = x + D_\epsilon$ be a small disc around x, where $D_\epsilon = \{y \in L_i^\perp \mid \langle y, y\rangle \leq \epsilon\}$. "Small" means that ϵ is chosen so that $D_\epsilon(x) \cap (\cup_{j\neq i} L_j) = \emptyset$. We assume that $D_\epsilon(x)$ is oriented, typically by the orientation inherited from some orientations on V and L_i which are prescribed in advance. The fundamental class of the pair $(D_\epsilon(x), \partial D_\epsilon(x))$ determines a homology class in $H_b(V, M(\mathcal{A}); \mathbb{Z})$, which we denote by $[x]$ and call the class *near the point* x. Note that by the Excision axiom, $[x]$ does not depend on ϵ. Moreover, by the Homotopy axiom, $[x]$ does not change if x is moved inside a connected component of $L_i \setminus \cup_{j\neq i} L_j$. Similarly, in light of the isomorphism $H_b(V, M(\mathcal{A})) \to H_{b-1}(M(\mathcal{A}))$, the class $[\![x]\!] := \partial[x]$, which is also often called the *class near the point* x, has all these invariance properties as well.

Let us show that the class $[x]$ is always nontrivial. By formula (7), the homotopy type of the one-point compactification $\hat{D}(\mathcal{A}) = \cup\,\mathcal{A} \cup \{+\infty\}$ has the decomposition of the form,

$$\hat{D}(\mathcal{A}) \simeq \hat{L}_1 \vee \hat{L}_2 \vee \ldots \vee \hat{L}_k \vee \ldots \tag{11}$$

where the displayed factors correspond to elements $p \in P(\mathcal{A})$ of minimum dimension $d(p) = a$. Let $c(\hat{L}_i) \in H^b(V, M(\mathcal{A}); \mathbb{Z})$ be the cohomology class which is Poincaré-Alexander dual of the fundamental homology class $[\hat{L}_i] \in H_a(\hat{D}(\mathcal{A}); \mathbb{Z})$ associated with the sphere $\hat{L}_i$ in $\hat{V}$. Then $c(\hat{L}_i)([N]) \in \mathbb{Z}$ is essentially the intersection number $[\hat{l}_i] \cap [N]$, whenever this number is correctly defined, for example if N is a manifold and the intersection $\hat{L}_i \cap N$ is transversal. From here we see that $c(\hat{L}_i)([x]) = 1$, hence $[x]$ must be nontrivial. All this also applies to classes $[\![x]\!] = \partial[x] \in H_{b-1}(M(\mathcal{A}))$. We have already made an observation that $[x]$ does not depend on x if this point is moved within the same connected component of $L_i \setminus \cup_{j\neq i} L_j$. As a corollary we obtain the following result, needed in the proof of Theorem 4.2.

Lemma 3.4. *If the codimension of $L_{ij} := L_i \cap L_j$ in L_i is greater than* 1 *for each $j \neq i$, then $[x]$ is a well defined class in $H_b(V, M(\mathcal{A}))$ which does not depend on x whatsoever.*

What if $\mathrm{codim}_{L_i}(L_i \cap L_j) = 1$ for some $j \neq i$? Then L_{ij}, being a hyperplane in both L_i and L_j, decomposes these spaces into the union of closed halfspaces, $L_i = L_i^1 \cup L_i^2$ and $L_j = L_j^1 \cup L_j^2$ respectively. By gluing the halfspaces L_i^1 and L_j^1 along the common boundary L_{ij} and by adding the infinite point $+\infty$, one obtains a sphere $K \subset \hat{V}$. Recall that the decomposition (7) involves a choice of generic points in all elements of the arrangement $\mathcal{A}$, see the proof of Theorem 2.2. and Section 3 in [19]. The points can be chosen in halfspaces L_i^1 and L_j^1, which implies that the sphere K appears in decomposition (11) as the factor $\hat{L}_{ij} * \Delta(P_{<p_{ij}}) \cong S^{a-1} * S^0 \cong S^a$, where $p_{ij} \in P$ corresponds to $L_{ij} \in \mathcal{A}$. This guarantees that the fundamental class $[K]$ of K is nontrivial in $H_a(\hat{D}(\mathcal{A}))$. Let $c(K) \in H^b(V, M(\mathcal{A}); \mathbb{Z})$ be the cohomology class, Poincaré-Alexander dual to $[K]$. Then $c(K)([x_1]) = \pm 1$ if x_1 is in the interior of the halfspace L_i^1 but $c(K)([x_2]) = 0$ if x_2 belongs to the interior of the complementary halfspace L_i^2. Hence, $[x_1] \neq [x_2]$.

4 Main theorems about $\mathbb{Z}_n$-equivariant maps

Theorem 4.1. ([3]) *There does not exist a $\mathbb{Z}_n$-equivariant map $f : S^3 \to M(\mathcal{A}(\alpha))$ where $\alpha = (\frac{1}{5}, \frac{1}{5}, \frac{1}{5}, \frac{2}{5})$ and as a consequence $(\frac{1}{5}, \frac{1}{5}, \frac{1}{5}, \frac{2}{5}) \in \mathcal{A}_{2,4}$.*

Theorem 4.2. *Suppose that neither the vector $\alpha = (\frac{a_1}{n}, \frac{a_2}{n}, \frac{a_3}{n}, \frac{a_4}{n})$ nor any of its cyclic permutations is equal to a vector of the form (r, s, s, s), $(\frac{1}{3}, \frac{1}{3}, r, s)$, or $(r, s, r, r+s)$ for some positive rational numbers r and s. Then there does exist a $\mathbb{Z}_n$-equivariant map*

$$F : S^3 \to M(\mathcal{A}(\alpha))$$

where $M(\mathcal{A}(\alpha)) = W_n \setminus \cup\mathcal{A}(\alpha)$ is the complement of the arrangement $\mathcal{A}(\alpha)$.

Proofs of T. 4.1 and T. 4.2: The proof of Theorem 4.1 given in [3] was based on Dold's theorem, [9]. Our proofs of both Theorems 4.1 and 4.2 use elements of equivariant obstruction theory as presented in [8], see also [22], (II) Sections 4.1–4.3. For convenience we often use the following abbreviations

$$P = a_1, \quad Q = a_1 + a_2, \quad R = a_1 + a_2 + a_3, \quad S = a_1 + a_2 + a_3 + a_4, \tag{12}$$

where of course $S = n$. Also, as in Section 3.2, the indices range over all integers with the assumption that $e_p = e_q$ if p and q are congruent modulo n.

We start with an observation that the existence of an equivariant map $f : S^3 \to Y$, where Y is an arbitrary $\mathbb{Z}_n$-space, does not depend on a particular choice of a free $\mathbb{Z}_n$-action on S^3, a fact which was already used in the proof of Proposition 3.2. Indeed, for any two $(k-1)$-connected, k-dimensional, free $\mathbb{Z}_n$-complexes K_1 and K_2, for $k \geq 2$, there always exists a $\mathbb{Z}_n$-equivariant map $g : K_1 \to K_2$, cf. [22], (I) Section 3.2. This allows us to use any free action on S^3 and our favorite choice is described as follows. Multiplication by $\omega = \exp 2\pi i/n$ defines a $\mathbb{Z}_n$-action on both the unit circle $U = \{z \in \mathbb{C} \mid |z| = 1\}$ and the regular polygon P_n with vertices $v_i = \omega^{i-1}$, $i = 1, 2, \ldots, n$ inscribed in U. Let $S^3 = S^1 * S^1$ be the simplicial complex obtained as a join of two copies of $S^1 \cong P_n$, with the induced action of $\mathbb{Z}_n$, $\omega(tx + (1-t)y) =$

$t\omega(x) + (1-t)\omega(y)$. For convenience, let us denote the vertices of the second copy of P_n by $w_1, w_2, \ldots, w_n$, so a typical 3-simplex in S^3 is of the form $[v_i, v_{i+1}; w_j, w_{j+1}] = [v_i, v_{i+1}] * [w_j, w_{j+1}]$. According to Section 3, the space $M = W_n \setminus \cup\mathcal{A}(\alpha)$ is 1-connected, hence it is 2-simple in the sense that $\pi_1(M)$ acts trivially on $\pi_2(M)$, [8], Section II.3, and by Hurewicz theorem $\pi_2(M) \cong [S^2, M] \cong H_2(M;\mathbb{Z})$. As a consequence, the relevant part of the obstruction exact sequence, [8], [22], has the following form,

$$[S^3, M]_{\mathbb{Z}_n} \xrightarrow{\theta} \mathrm{Im}\left\{[S^3_{(2)}, M]_{\mathbb{Z}_n} \longrightarrow [S^3_{(1)}, M]_{\mathbb{Z}_n}\right\} \xrightarrow{\tau} H^3_{\mathbb{Z}_n}(S^3, A) \tag{13}$$

where $S^3_{(1)}$ and $S^3_{(2)}$ are respectively the 1- and 2-skeleton of $S^3 = P_n * P_n$ and $A = H_2(M;\mathbb{Z})$ is viewed as a $\mathbb{Z}_n$-module. Since $[S^3_{(1)}, M]_{\mathbb{Z}_n} = \{*\}$ is one-element set and $[S^3_{(2)}, M]_{\mathbb{Z}_n}$ is nonempty, the sequence (13) reduces to,

$$[S^3, M]_{\mathbb{Z}_n} \longrightarrow \{*\} \xrightarrow{\tau} H^3_{\mathbb{Z}_n}(S^3, A). \tag{14}$$

The exactness of (14) means that the set $[S^3, M]_{\mathbb{Z}_n}$ is nonempty if and only if a special element $\tau(*) \in H^3_{\mathbb{Z}_n}(S^3, A)$ is equal to zero. Note that $\tau(*)$ depends only on M, for example it does not depend on the chosen CW-structure on S^3. The class $\tau(*)$ is evaluated by the following standard procedure, cf. [22], (II) Sections 4.1-4.3. Let $h : S^3 \to W_n$ be an arbitrary $\mathbb{Z}_n$-equivariant, simplicial map which is in general position in the following sense. For any simplex σ in S^3 and any linear space U in the arrangement $\mathcal{A}(\alpha)$, if the intersection $h(\sigma) \cap U$ is nonempty, then $\dim(\sigma) = \dim(h(\sigma)) = 3$, $\dim(U) = n-3$ and $h(\sigma) \cap U$ is a single point in the interior of $h(\sigma)$. It is not difficult to check that a typical map is in general position and that every equivariant, simplicial map can be put in general position by a small perturbation. For each general position, $\mathbb{Z}_n$-equivariant map $h : S^3 \to W_n$, there is an associated obstruction cocycle $c(h) \in C^3_{\mathbb{Z}_n}(S^3, A) = \mathrm{Hom}_{\mathbb{Z}_n}(C_3(S^3), A)$, $[c(h)] = \tau(*)$. Given an oriented 3-simplex σ in S^3, then $c(h)(\sigma) \in H_2(M;\mathbb{Z})$ is described as the image $h_*([\partial(\sigma)])$ of the fundamental class of $\partial(\sigma) \cong S^2$ by the map $h_* : H_2(\partial(\sigma);\mathbb{Z}) \to H_2(M;\mathbb{Z})$. Here is a very convenient choice for a general position map $h : S^3 \to W_n$. Similar "test" maps have already been used in [22] for the computation of obstruction elements $\tau(*)$. Let $e_1, e_2, \ldots, e_n$ be the standard orthonormal basis in $\mathbb{R}^n$ and let $x_1, x_2, \ldots, x_n$ be the associated dual linear functions. The points $\{u_i\}_{i=1}^n$, where $u_i := e_i - \frac{1}{n}\sum_{i=1}^n e_i$, are vertices of a regular simplex in W_n. Note that they are cyclically permuted by the ambient action of $\mathbb{Z}_n$ on W_n. Let $h_1, h_2 : P_n \to W_n$ be simplicial, $\mathbb{Z}_n$-equivariant maps such that $h_1(v_k) = u_k$, $h_2(w_k) = u_k$ and let $h = h_1 * h_2 : S^3 \to W_n$, $h(tx + (1-t)y) := th_1(x) + (1-t)h_2(y)$, be an associated simplicial map. It is not difficult to see that h is a map in general position. Indeed, it is sufficient to test for which simplices $\sigma = \sigma_1 * \sigma_2 \subset P_n * P_n = S^3$ the image $h(\sigma)$ intersect the subspace $L = L(\alpha)$ defined by (4) and (5). It is easily checked that there are exactly 4 such simplices. Indeed, by inspection we see that only the simplices,

$$\begin{aligned} \theta_1 &= [v_P, v_{P+1}; w_R, w_{R+1}], \quad \theta_2 = [v_Q, v_{Q+1}; w_S, w_{S+1}], \\ \theta_3 &= [v_R, v_{R+1}; w_P, w_{P+1}], \quad \theta_4 = [v_S, v_{S+1}; w_Q, w_{Q+1}], \end{aligned} \tag{15}$$

have the desired property. Moreover, the map h restricted on these simplices is 1 – 1 and a complete list of the points in the preimage $h^{-1}(L)$ together with simplices where they belong is the following,

$$\begin{array}{lllll} x_1 &=& \frac{a_1}{n}v_P + \frac{a_2}{n}v_{P+1} + \frac{a_3}{n}w_R + \frac{a_4}{n}w_{R+1} &\in& \theta_1 \cap h^{-1}(L) \\ x_2 &=& \frac{a_2}{n}v_Q + \frac{a_3}{n}v_{Q+1} + \frac{a_4}{n}w_S + \frac{a_1}{n}w_{S+1} &\in& \theta_2 \cap h^{-1}(L) \\ x_3 &=& \frac{a_3}{n}v_R + \frac{a_4}{n}v_{R+1} + \frac{a_1}{n}w_P + \frac{a_2}{n}w_{P+1} &\in& \theta_3 \cap h^{-1}(L) \\ x_4 &=& \frac{a_4}{n}v_S + \frac{a_1}{n}v_{S+1} + \frac{a_2}{n}w_Q + \frac{a_3}{n}v_{Q+1} &\in& \theta_4 \cap h^{-1}(L) \end{array} \tag{16}$$

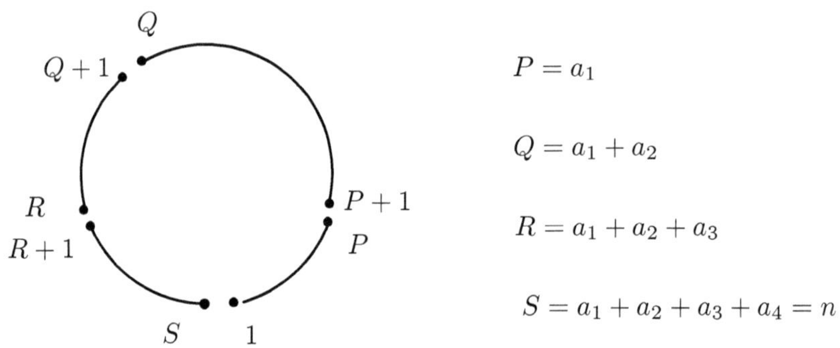

Fig. 1.

Note that if $y_i := h(x_i)$ for $i = 1, \ldots, 4$ then,

$$\begin{array}{lllllll} y_1 &=& y_3 &=& \frac{a_1}{n}u_P + \frac{a_2}{n}u_{P+1} + \frac{a_3}{n}u_R + \frac{a_4}{n}u_{R+1} &\in& L \\ y_2 &=& y_4 &=& \frac{a_2}{n}u_Q + \frac{a_3}{n}u_{Q+1} + \frac{a_4}{n}u_S + \frac{a_1}{n}u_{S+1} &\in& L \end{array} \tag{17}$$

which, in light of the fact that $u_i = e_i - e$ where $e = \frac{1}{n}(e_1 + \ldots + e_n)$, can also be expressed as

$$\begin{array}{lllll} y_1 &=& y_3 &=& \frac{a_1}{n}e_P + \frac{a_2}{n}e_{P+1} + \frac{a_3}{n}e_R + \frac{a_4}{n}e_{R+1} - e \\ y_2 &=& y_4 &=& \frac{a_2}{n}e_Q + \frac{a_3}{n}e_{Q+1} + \frac{a_4}{n}e_S + \frac{a_1}{n}e_{S+1} - e. \end{array} \tag{18}$$

All the simplices θ_i are oriented coherently with the orientation on $S^3 = P_n * P_n$ inherited from the positively (counterclockwise) oriented polygons P_n. Let us find now an explicit form of the obstruction cocycle $c(h) \in C^3_{\mathbb{Z}_n}(S^3, A) := \mathrm{Hom}_{\mathbb{Z}_n}(C_3(S^3), H_2(M;\mathbb{Z}))$. By definition of the obstruction cocycle, $c(h)(\theta)$ is nonzero only if θ is one of the simplices $\omega^k(\theta_i)$, $i = 1, \ldots, 4$ and for example $c(h)(\theta_i) = h_*([\partial\theta_i]) \in H_2(M;\mathbb{Z})$. Following the terminology of Section 3.4, we see that $c(h)(\theta_i) = \epsilon_i[y_i]$ where $\epsilon_i \in \{-1, +1\}$, $y_i := h(x_i) \in L$ and $[y_i] = [h(x_i)] \in H_2(M;\mathbb{Z})$ is the generating class of M *near* *the point* $y_i \in L$. The sign ϵ_i is just the intersection number of the oriented simplex $h(\theta_i)$ with the oriented subspace L. Recall that L is defined as the linear subspace in $\mathbb{R}^n$ orthogonal to the vectors $e_{I_1} = e_1 + \ldots + e_{a_1}, e_{I_2} = e_{a_1+1} + \ldots + e_{a_1+a_2}, e_{I_3} = e_{a_1+e_2+1} + \ldots + e_{a_1+a_2+a_3}, e_{I_4} = e_{a_1+a_2+a_3+1} + \ldots + e_n$. The orientation on L is chosen so that a frame in L is positive iff together with the frame $[e_{I_1}, e_{I_2}, e_{I_3}, e_{I_4}]$ in the orthogonal complement $L^{\perp}$, it has the same orientation as the standard positive frame $[e_1, \ldots, e_n]$ in $\mathbb{R}^n$. The orientation of the target space W_n is chosen so that together with the orthogonal vector $e_{[n]} = e_1 + \ldots + e_n$ it defines the positive orientation on $\mathbb{R}^n$. It is not difficult to check that the intersection number of an oriented simplex $\sigma = \text{conv}\{z_1, z_2, z_3, z_4\} \subset W_n$ with L is equal to the sign of the following Gramm matrix,

$$A = \{\langle Z_i, e_{I_j} \rangle\}_{i,j=1}^4 \tag{19}$$

where $Z_i := z_i + \frac{1}{n}\sum_{i=1}^n e_i$. From here we easily find out that,

$$\begin{array}{ll} \epsilon_1 = \text{sgn}(1234) = +1 & \epsilon_2 = \text{sgn}(2341) = -1 \\ \epsilon_3 = \text{sgn}(3412) = +1 & \epsilon_4 = \text{sgn}(4123) = -1. \end{array} \tag{20}$$

We are almost ready to evaluate the element $[c(h)] = \tau(*) \in H^3_{\mathbb{Z}_n}(S^3, A)$. What remains to be done is to calculate, or better to describe in transparent geometric terms, the nontrivial elements of the cohomology group $H^3_{\mathbb{Z}_n}(S^3, A)$. For this it is convenient to replace the simplicial complex structure $P_n * P_n$ on S^3 and the associated simplicial chain complex $\{C_i(S^3, \mathbb{Z})\}_{i=0}^3$ by a much more economical $\mathbb{Z}_n$-invariant, cell complex structure on S^3 and the associated cellular chain complex $\{D_i(S^3, \mathbb{Z})\}_{i=0}^3$. The obstruction class $c(h)$, computed relative to the new cell complex structure is denoted by $c'(h)$. The new cell complex has n cells $\{\omega^j d_i\}_{j=0}^{n-1}$ in each dimension $i = 0, 1, 2, 3$ where,

$$d_0 = w_0 \qquad d_1 = [w_0, w_1] \qquad d_2 = v_0 * P_n \qquad d_3 = [v_0, v_1] * P_n. \tag{21}$$

The associated cellular chain complex has the usual form,

$$0 \;\rightarrow\; \mathbb{Z}(\mathbb{Z}_n)d_3 \;\overset{\omega-1}{\rightarrow}\; \mathbb{Z}(\mathbb{Z}_n)d_2 \;\overset{N}{\rightarrow}\; \mathbb{Z}(\mathbb{Z}_n)d_1 \;\overset{\omega-1}{\rightarrow}\; \mathbb{Z}(\mathbb{Z}_n)d_0 \;\rightarrow\; 0 \tag{22}$$

where $N = 1+\omega+\ldots+\omega^{n-1} \in \mathbb{Z}(\mathbb{Z}_n)$. On applying the functor $\text{Hom}_{\mathbb{Z}_n}(\cdot, A)$, where $A = H_2(M;, \mathbb{Z})$ is viewed as a $\mathbb{Z}_n$-module, one obtains the sequence,

$$0 \;\longleftarrow A \;\overset{\omega-1}{\longleftarrow}\; A \;\overset{N}{\longleftarrow}\; A \;\overset{\omega-1}{\longleftarrow}\; A \;\longleftarrow\; 0. \tag{23}$$

Hence,

$$H^3_{\mathbb{Z}_n}(S^3, A) \cong A/\text{Im}(\omega - 1). \tag{24}$$

Note that the cell $d_3 = [v_0, v_1] * P_n$ contains precisely four simplices of the form $\omega^k(\theta_i)$, namely the simplices $\omega^{-P}\theta_1, \omega^{-Q}\theta_2, \omega^{-R}\theta_3$ and $\omega^{-S}\theta_4$. It follows that the value of the obstruction cocycle $c'(h) : \mathbb{Z}(\mathbb{Z}_n) \rightarrow A$ at d_3 is

$$c_1 := c'(h)(d_3) = \omega^{-P}[\![y_1]\!] - \omega^{-Q}[\![y_2]\!] + \omega^{-R}[\![y_3]\!] - \omega^{-S}[\![y_4]\!] \in A, \tag{25}$$

where $[\![y_i]\!] = [\![h(x_i)]\!]$ are the classes in $H_2(M(\mathcal{A}(\alpha)))$, associated to points y_i, described in Section 3.4 (equations (16) and (17)). It follows that $\tau(*) = [c'(h)] = c_1 + \mathrm{Im}(\omega - 1) \in A/\mathrm{Im}(\omega - 1)$. Alternatively, $\tau(*)$ can be represented by $c_2 := [\![y_1]\!] - [\![y_2]\!] + [\![y_3]\!] - [\![y_4]\!] = 2[\![y_1]\!] - 2[\![y_2]\!]$ since obviously $c_1 - c_2 \in \mathrm{Im}(\omega - 1)$, $y_1 = y_3$ and $y_2 = y_4$. Consequently, the question of the existence of a $\mathbb{Z}_n$-equivariant map $f : S^3 \to M(\mathcal{A}(\alpha))$ is reduced to the question whether $c_2 \in \mathrm{Im}(\omega - 1)$.

Proof of T. 4.2 (cont.): By Proposition 3.1, if $\alpha = (\frac{a_1}{n}, \frac{a_2}{n}, \frac{a_3}{n}, \frac{a_4}{n}) \in \mathbb{R}^4$ is not of the form $(r, s, s, s), (\frac{1}{3}, \frac{1}{3}, r, s)$ or $(r, s, r, r + s)$ for some positive rational numbers r and s, then $\mathcal{A}(\alpha)$ is an arrangement which satisfies the conditions of Lemma 3.4. Hence both the class $[y] \in H_3(W_n, M(\mathcal{A}(\alpha)))$ and $[\![y]\!] \in H_2(M(\mathcal{A}(\alpha))$ do not depend on the position of $y \in L$. Consequently, $c_2 := [\![y_1]\!] - [\![y_2]\!] + [\![y_3]\!] - [\![y_4]\!] = 2[\![y_1]\!] - 2[\![y_2]\!] = 0$, the obstruction $\tau(*)$ vanishes and the proof of the theorem is completed.

Proof of T. 4.1 (cont.): Let us show that $c_2 = 2[\![y_1]\!] - 2[\![y_2]\!]$ is a nonzero element in the group $A/(\omega - 1)A$ where $A = H_2(M(\alpha); \mathbb{Z})$ and $\alpha = (\frac{1}{5}, \frac{1}{5}, \frac{1}{5}, \frac{2}{5})$. In this case W_5 is a 4-dimensional space and $\mathcal{A}(\alpha) = \{L_0, \ldots, L_4\}$ is an arrangement of 5 lines where $L_i := \omega^i(L)$. The line $L = L_0$ is determined in $\mathbb{R}^5$ by equations $x_1 = x_2 = x_3 = x_4 + x_5 = 0$. It follows from (18) that $y_1 = \frac{1}{5}e_4 - \frac{1}{5}e_5 = -y_2$. Note that $[\![y_1]\!] - [\![y_2]\!]$ is, by results of Section 3.4, a nonzero element in A. Actually, since $W_5 \setminus \cup \mathcal{A}(\alpha)$ has the same $\mathbb{Z}_5$-homotopy type as the punctured sphere $\mathbb{S} := S^3 \setminus \cup_{i=0}^4 L_i$ it is not difficult to see that A as a $\mathbb{Z}_5$-module fits in the following exact sequence

$$0 \longrightarrow \mathbb{Z}[\mathbb{Z}_5]t \xrightarrow{\mu} \mathbb{Z}[\mathbb{Z}_5]a \oplus \mathbb{Z}[\mathbb{Z}_5]b \xrightarrow{\nu} A \longrightarrow 0 \tag{26}$$

where $\mu(t) = (1 + \omega + \ldots + \omega^4)(a - b)$ and $\nu(a) = [\![y_1]\!]$, $\nu(b) = [\![y_2]\!]$. The associated sequence of coinvariants has the following form

$$\mathbb{Z}t \xrightarrow{\hat{\mu}} \mathbb{Z}a \oplus \mathbb{Z}b \xrightarrow{\hat{\nu}} A/(\omega - 1)A \longrightarrow 0 \tag{27}$$

where $\hat{\mu}(t) = 5(a - b)$. It follows that $2(a - b)$ is a nonzero element in $\mathrm{Coker}(\hat{\mu})$ i.e. $2([\![y_1]\!] - [\![y_2]\!])$ is a nonzero element in $A/(\omega - 1)A$. This completes the proof of Theorem 4.1.

Remark 4.3. A natural question to ask is what about the remaining cases. In other words what happens if $\alpha = (\frac{a_1}{n}, \frac{a_2}{n}, \frac{a_3}{n}, \frac{a_4}{n})$ is, up to a cyclic permutation of coordinates, of the form $(\frac{p}{n}, \frac{q}{n}, \frac{q}{n}, \frac{q}{n}), (\frac{1}{3}, \frac{1}{3}, r, s)$ or of the form $(r, s, r, r + s)$ for some positive rational numbers r and s. Another natural question is whether one can extend the analysis of equivariant maps presented in this paper by using the complete group of symmetries $\mathbb{D}_{2n}$ of the configuration space X_μ, not just its subgroup $\mathbb{Z}_n$.

5 The equipartition case $\alpha = (\frac{1}{4}, \frac{1}{4}, \frac{1}{4}, \frac{1}{4})$

The equipartition case, $\alpha = (\frac{1}{4}, \frac{1}{4}, \frac{1}{4}, \frac{1}{4})$, deserves a special attention. First of all this is the most symmetric case. Indeed, the natural group of symmetries for this problem is the dihedral group $\mathbb{D}_8$ of order 8. In addition to the cyclic group $\mathbb{Z}_n = \{1, \omega, \omega^2, \omega^3\}$, there is an involution θ defined on the candidate space $X_\mu \cong V_2(\mathbb{R}^3)$ and the test space W_4 by $\theta(x, y) = (-x, y)$ for $(x, y) \in V_2(\mathbb{R}^3)$ and $\theta(t_1, t_2, t_3, t_4) = (t_4, t_3, t_2, t_1)$ for $(t_1, t_2, t_3, t_4) \in W_4$ respectively. The test map F_ν defined by (3) is still equivariant with respect to the larger group $\mathbb{D}_8$. Note that the arrangement $\mathcal{A}(\alpha)$ degenerates in the case $n = 4$ to a single point $0 \in W_4$ so, along the lines of Proposition 2.1, the equipartition problem is reduced to the question of the existence of a $\mathbb{D}_8$-equivariant map $F : V_2(\mathbb{R}^3) \to W_4 \setminus \{0\}$.

The equipartition case $\alpha = (\frac{1}{4}, \frac{1}{4}, \frac{1}{4}, \frac{1}{4})$ was originally established by Bárány and Matoušek in [4] by a topological argument which used in an essential way the fact that the test map $F_\nu : X_\mu \to W_4$ has special additional properties. They showed that the equivariance with respect to a smaller group $\mathbb{Z}_2 \oplus \mathbb{Z}_2 := \langle \omega^2, \theta \rangle \subset \mathbb{D}_8$ is not sufficient to guarantee that an equivariant map $F : V_2(\mathbb{R}^3) \to W_4 \setminus \{0\}$ does not exist. However, if $F = F_\nu$ for some measure ν then they proved by a direct argument that such a map cannot exist. Therefore one might get an impression from the original approach and the proof of Theorems 4.1 and 4.2 that the case $\alpha = (\frac{1}{4}, \frac{1}{4}, \frac{1}{4}, \frac{1}{4})$ is not amenable to direct equivariant methods. Our objective in this section is to show that this is not the case. Moreover, we use this opportunity to illustrate the so called *singularity method* which introduces ideas from the theory of bordisms in their rudimentary form. Note that we could also rely on the obstruction theory approach, used in the proof of Theorems 4.1 and 4.2.

Theorem 5.1. *There does not exist a $\mathbb{D}_8$-equivariant map*

$$F : V_2(\mathbb{R}^3) \to W_4 \setminus \{0\}.$$

Proof: We will need basic facts about the *transversality* of maps and manifolds, see [11] Chapter II. Note that a basic observation is that a map can be improved to a map transverse to a given submanifold by small local perturbations. This means that all transversality results still hold for G-equivariant smooth maps where G is a finite group acting freely on the base manifold.

For any map $f : M \to W_4$, the zero set $Z(f)$ is defined by $Z(f) = \{y \in M \mid f(y) = 0\}$. If M is a smooth, compact, 3-dimensional manifold, a generic smooth map is *transverse* to $0 \in W_4$ in the sense that $df : T_x(M) \to T_0(W_4)$ is an isomorphism for each $x \in Z(f)$. In this case $Z(f)$ is a finite set and the cardinality $|Z(f)|$ of the zero set does not change by a small perturbation of the map f.

Let us suppose that a $\mathbb{D}_8$-equivariant map $F : V_2(\mathbb{R}^3) \to W_4 \setminus \{0\}$ does exist, in other words that $Z(F) = \emptyset$. This property is preserved by a small perturbation of the map so, without a loss of generality it can be assumed that

F is a smooth map. Since W_4 is an affine space, any two $\mathbb{D}_8$-equivariant maps $F_0, F_1 : V_2(\mathbb{R}^3) \to W_4$ are equivariantly homotopic, for example one can use the linear homotopy $\alpha : V_2(\mathbb{R}^3) \times I \to W_4$, $\alpha(x,t) = (1-t)F_0(x) + tF_1(x)$. By a small perturbation, α can be made transverse to $0 \in W_4$. The transversality condition implies that the zero set $Z(\alpha)$ is a 1-dimensional submanifold with boundary of $V_2(\mathbb{R}^3) \times I$. We could have assumed from the beginning that α is "constant" in the vicinity of bases, i.e. that $\alpha(x,t) = \alpha(x,0)$ for $t < \epsilon$ and $\alpha(x,t) = \alpha(x,1)$ for $1-\epsilon < t$. This means that $Z(\alpha)$ is "orthogonal" to bases $V_2(\mathbb{R}^3) \times \{i\}$, $i = 0, 1$. The boundary $\partial Z(\alpha)$ has a natural decomposition $\partial Z(\alpha) = Z(F_0) \cup Z(F_1)$. By passing to the orbit space $M := V_2(\mathbb{R}^3)/\mathbb{D}_8$, we finally obtain that there is a bordism between the 0-dimensional manifolds $Z(F_0)/\mathbb{D}_8$ and $Z(F_1)/\mathbb{D}_8$, see Figure 2. It immediately follows that

$$|Z(F_0)/\mathbb{D}_8| = |Z(F_1)/\mathbb{D}_8| \pmod 2. \tag{28}$$

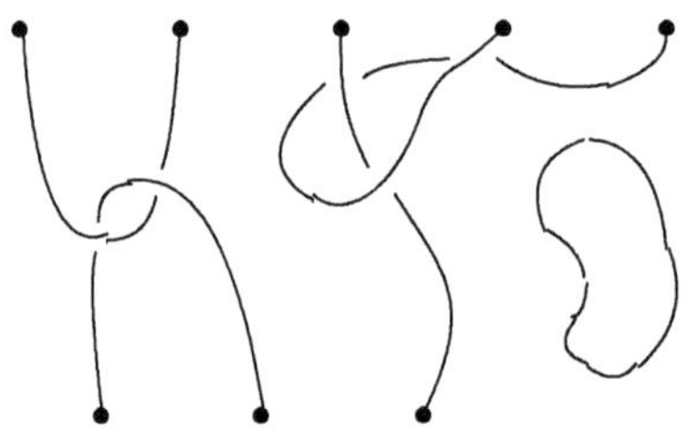

Fig. 2.

This means that it is sufficient to construct a smooth $\mathbb{D}_8$-equivariant map $F_1 : V_2(\mathbb{R}^3) \to W_4$ such that $|Z(F_1)/\mathbb{D}_8| = 1$. Indeed, a smooth, $\mathbb{D}_8$-equivariant homotopy between a map F_0 with $Z(F_0) = \emptyset$ and F_1 with $|Z(F_1)/\mathbb{D}_8| = 1$ would contradict (28).

Claim: There exists a $\mathbb{D}_8$-equivariant map $F_1 : V_2(\mathbb{R}^3) \to W_4$, transverse to $0 \in W_4$, such that

$$|Z(F_1)/\mathbb{D}_8| = 1.$$

Proof of the Claim: Let μ be the uniform probability measure on S^2. Consequently, a 4-fan $(x; l_1, l_2, l_3, l_4)$ in the configuration space X_μ is characterized by the property that $C_1 := l_1 \cup l_3$ and $C_2 := l_2 \cup l_4$ are two mutually orthogonal great circles in S^2. The corresponding $\mathbb{D}_8$-orbits are clearly identified as an unordered pairs $\{C_1, C_2\}$ of mutually orthogonal great circles in S^2, or equivalently pairs $\{P_1, P_2\}$ of orthogonal 2-planes which contain the origin $0 \in \mathbb{R}^3$. Let K be the circle in $\mathbb{R}^3$, described as the intersection of the plane $z = 1$ with the cylinder $(x-1)^2 + y^2 = 1$. Let ν be the measure on S^2 defined by

$$\nu(U) := m(\mathrm{Cone}(U) \cap K) \tag{29}$$

where $\text{Cone}(U) = \cup_{\lambda \geq 0} \lambda U$ is the cone with vertex 0 generated by U and m is the uniform measure (the arc length) on the circle K.

Let us show that the test map $F_1 = F_\nu$ has the desired property, i.e. that there exists precisely one μ-equipartition $\{P_1, P_2\}$ which is a ν-equipartition as well. Indeed, an easy geometric argument shows that the only pair of planes with this property is $P_1 = \{y = 0\}$ and the plane P_2 spanned by vectors $v_2 = (1,1,1)$ and $v_4 = (1,-1,1)$. The plane P_1 can be also described as the plane spanned by vectors $v_1 = (0,0,1)$ and $v_3 = (2,0,1)$.

What remains to be checked is that the map $F_1 = F_\nu : V_2(\mathbb{R}^3) \to W_4$ is transverse to 0. In other words we have to show that for each (μ,ν)-equipartition $e = (x; l_1, l_2, l_3, l_4) \in Z(F_1)$, there is an open neighborhood U of e such that F_1 is a homeomorphism from U to some neighborhood of $0 \in W_4$. Since $V_2(\mathbb{R}^3)$ is a covering space of $V_2(\mathbb{R}^3)/\mathbb{D}_8$, we can describe such a neighborhood by describing a neighborhood in the latter space. A neighborhood N_1 of the equipartition $\{P_1, P_2\}$ is informally described as the collection of all ordered pairs $\{Q_1, Q_2\}$ of 2-planes such that $Q_1 \perp Q_2$ and $Q_1 \in V(P_1), Q_2 \in V(P_2)$, where $V(P_i)$ is a small neighborhood of P_i. A neighborhood N_2 of $0 \in W_4$ is described as the collection of all $(t_1, t_2, t_3, t_4) \in W_4$ such that, aside from the condition $t_1 + t_2 + t_3 + t_4 = 0$, there is a condition $|t_i| < \epsilon_i$. We observe now that both N_1 and N_2 can be, by appropriate choices, put in a one-to-one correspondence with the open set N_3 of all ordered triples $(x,y,z) \in K^3$, where $x \in V(v_1), y \in V(v_2), z \in V(v_4)$ are the points on the circle K which belong to the respective neighborhoods of v_1, v_2 and v_4. For example note that, given such a triple (x,y,z), the plane Q_2 is spanned by vectors v_1 and v_4 while Q_1 is the unique plane orthogonal to Q_2 which contains the vector v_1.

This finishes the proof of the Claim and the proof of Theorem 5.1 is now complete.

6 Higher dimensions

A natural problem is to study k-fan partitions of measures on higher dimensional spheres. As it was shown in [18] (Theorems 3.3 and 3.4), the higher generality actually yields some information about lower dimensional cases. The following result generalizes Theorem 2.2 from [17] and illustrates the use of cohomological index theory in equivariant problems. As in Section 1 we assume that all measures are *proper* in the sense that $\mu(C) = 0$ for each great sphere $C \subset S^{n-1}$ and that $\mu(U) > 0$ for each open set $U \subset S^{n-1}$.

Definition 6.1. Let μ be a probability measure on the sphere $S^{n-1} \subset \mathbb{R}^n$. Let $L \subset \mathbb{R}^n$ be a 2-plane, $L^\perp$ its orthogonal complement and $S(L)$ the associated great circle in S^{n-1}. Recall that an orthogonal decomposition $\mathbb{R}^n \cong U \oplus U^\perp$ yields a well known *join* decomposition $S^{n-1} \cong S(U) * S(U^\perp)$, [22]. Suppose that $S(L)$ is oriented and dissected into k arcs $a_1, \ldots, a_k$ and that the arcs are ordered in the positive direction determined by the orientation. Let $\sigma_i := a_i * S(L^\perp) \subset S^{n-1} = S(L) * S(L^\perp)$ and let $\mathfrak{q} = (\sigma_1, \dots, \sigma_k)$ be the associated (ordered) dissection of the sphere S^{n-1}. This dissection will be referred to as a k-fan on the sphere S^{n-1}.

Theorem 6.2. *Suppose that k is an odd prime and m is an integer such that $2n-4 \geq (k-1)m$. Then for any collection $\mu_0, \mu_1, \dots, \mu_m$ of $m+1$ (proper) probability measures on S^{n-1} there exists a k-fan $\mathfrak{q} = (\sigma_1, \dots, \sigma_k)$ on the sphere S^{n-1} such that $\mu_i(\sigma_j) = 1/k$ for each $i = 0, 1, \dots, m$ and each $j = 1, \dots, k$.*

Proof: As in Section 2, one of the measures (say μ_0) is used for the construction of the candidate space $X_{\mu_0} := \{\mathfrak{q} = (\sigma_1, \dots, \sigma_k) \mid \mu_0(\sigma_1) = \dots = \mu_0(\sigma_k) = 1/k\}$. Let $W_k := \{x \in \mathbb{R}^k \mid x_1 + \dots + x_k = 0\}$. Define the test map $f_i : X_{\mu_0} \to W_k$ for the measure μ_i ($i \geq 1$) by the formula

$$f_i(\mathfrak{q}(x,y)) = (\mu_i(\sigma_1), \dots, \mu_i(\sigma_m)) - (1/k, \dots, 1/k).$$

The total test map $f : X_{\mu_0} \to W_k^{\oplus m}$ is defined by $f(\mathfrak{q}) = (f_1(\mathfrak{q}), \dots, f_m(\mathfrak{q}))$. The group $\mathbb{Z}_k$ acts on both X_{μ_0} and $W_k^{\oplus m}$ and the test map f is clearly $\mathbb{Z}_k$-equivariant. The group $\mathbb{Z}_k$ can be identified with a subgroup of $SO(2) \cong S^1$. Both groups $SO(2)$ and $\mathbb{Z}_k$ act on $V_{n,2}$ by rotating the pair (x,y) in the plane $L = \text{span}\{x,y\}$. Note that X_{μ_0} and $V_{n,2}$ are homeomorphic as $\mathbb{Z}_k$-spaces. This can be shown by observing first that $V_{n,2}$ is $\mathbb{Z}_k$-homeomorphic to the space X_ν of all regular k-fans where ν is the uniform probability measure on S^{n-1}. On the other hand a $\mathbb{Z}_k$-equivariant homeomorphism $X_{\mu_0} \to X_\nu$ is obtained by iterating the process of taking (twice) the midpoints of all arcs a_i. Since $\mathfrak{q}$ is a solution of the problem if and only if $f(\mathfrak{q}) = 0$ the proof of the theorem will be complete once we establish the following result.

Theorem 6.3. *Suppose that $2n-4 \geq (k-1)m$ where m and n are positive integers and k is an odd prime. Then there does not exist a $\mathbb{Z}_k$-equivariant map*

$$f : V_{n,2} \to S(W_k^{\oplus m})$$

where $S(V)$ is the unit sphere in V.

Proof: Suppose that a $\mathbb{Z}_k$-equivariant map $f : V_{n,2} \to S(W_k^{\oplus m})$ does exist. Following the usual approach of the *ideal valued index theory*, [22] (II), one has a commutative diagram

$$\begin{array}{ccc} H^*(V_{n,2}/\mathbb{Z}_k;\mathbb{Z}) & \xleftarrow{f_1} & H^*(S(W_k^{\oplus m})/\mathbb{Z}_k;\mathbb{Z}) \\ \alpha\uparrow & & \uparrow\beta \\ H^*(B\mathbb{Z}_k;\mathbb{Z}) & \xleftarrow[\cong]{} & H^*(B\mathbb{Z}_k;\mathbb{Z}) \end{array} \tag{30}$$

Since $\dim(S(W_k^{\oplus m})/\mathbb{Z}_k) = (k-1)m - 1$ we conclude that $\beta(z) = 0$ for each element $z \in H^*(B\mathbb{Z}_k;\mathbb{Z})$ of degree $\geq (k-1)m$. Since $\alpha(z) = f_1 \circ$

$\beta(z) = 0$, in order to reach a contradiction it is sufficient to find an element $z \in H^{2n-4}(B\mathbb{Z}_k;\mathbb{Z})$ such that $\alpha(z) \neq 0$. Let $G^+_{n,2} \cong V_{n,2}/SO(2)$ be the Grassmann manifold of all oriented 2-planes in $\mathbb{R}^n$. Since $\mathbb{Z}_k \subset SO(2)$, there is a commutative diagram:

$$\begin{array}{ccccc} S^1 & \longrightarrow & V_{n,2} & \longrightarrow & G^+_{n,2} \\ {\scriptstyle \times k}\downarrow & & \downarrow & & \downarrow \\ S^1 & \longrightarrow & V_{n,2}/\mathbb{Z}_k & \xrightarrow{\pi} & G^+_{n,2} \\ & & \downarrow & & \downarrow \\ & & B\mathbb{Z}_k & \xrightarrow{\tau} & BSO(2) \end{array} \tag{31}$$

It was shown in [17] that the Euler class e of the first row of (31) is nonzero. Moreover $e = 2g$ where g is a generator of the group $H^{2n-4}(G^+_{n,2};\mathbb{Z})$. A comparison of spectral sequences for the (horizontal) S^1-fibrations in (31) shows that $\pi^*(e) \neq 0$. The reader can easily reconstruct the details following the proof for the case $k = 3$ given in [17], pages 28-29. The lower right corner of (31) yields in cohomology the following diagram:

$$\begin{array}{ccc} H^*(V_{n,2}/\mathbb{Z}_k;\mathbb{Z}) & \xleftarrow{\pi^*} & H^*(G^+_{n,2};\mathbb{Z}) \\ {\scriptstyle \alpha}\uparrow & & \uparrow{\scriptstyle \gamma} \\ H^*(B\mathbb{Z}_k;\mathbb{Z}) & \xleftarrow[\tau^*]{} & H^*(BSO(2);\mathbb{Z}) \end{array} \tag{32}$$

Since e is a characteristic class, $e = \gamma(e')$ for some $e' \in H^*(BSO(2);\mathbb{Z})$. It follows that $z := \pi^*(e) = \alpha(\tau^*(e'))$. This completes the proof of the theorem since $z \neq 0$ and $\deg(z) = 2n - 4 \geq (k-1)m$.

References

[1] J. Akiyama, A. Kaneko, M. Kano, G. Nakamura, E. Rivera-Campo, S. Tokunaga, and J. Urrutia. Radial perfect partitions of convex sets in the plane. In *Discrete and Computational Geometry* (J. Akiyama et al. eds.), *Lect. Notes Comput. Sci. 1763, 1–13.* Springer, Berlin 2000.

[2] I. Bárány. Geometric and combinatorial applications of Borsuk's theorem. *New trends in Discrete and Computational Geometry*, János Pach, ed., Algorithms and Combinatorics 10, Springer-Verlag, Berlin, 1993.

[3] I. Bárány, J. Matoušek. Simultaneous partitions of measures by k-fans. *Discrete Comp. Geometry*, 25 (2001), 317–334.

[4] I. Bárány, J. Matoušek. Equipartitions of two measures by a 4-fan. (preprint).

[5] S. Bespamyatnikh, D. Kirkpatrick, and J. Snoeyink. Generalizing ham sandwich cuts to equitable subdivisions. *Discrete Comput. Geom.*, 24 (2000), 605–622.

[6] A. Björner, Topological methods. In R. Graham, M. Grötschel, and L. Lovász, editors, *Handbook of Combinatorics.* North-Holland, Amsterdam, 1995.

[7] K.S. Brown. *Cohomology of groups.* Springer-Verlag, New York, Berlin, 1982.

[8] T. tom Dieck, *Transformation groups.* de Gruyter Studies in Math. 8, Berlin, 1987.

[9] A. Dold, Simple proofs of some Borsuk-Ulam results. *Contemporary Math.* 19 (1983), 65–69.

[10] M. Goresky and R. MacPherson. *Stratified Morse Theory.* Volume 14 of *Ergebnisse der Mathematik und ihrer Grenzgebiete*, Springer, Berlin, Heidelberg, New York, 1988.

[11] M. Golubitsky, V. Guillemin. *Stable Mappings and Their Singularities.* Graduate Texts in Mathematics 14, Springer–Verlag 1973.

[12] H. Ito, H. Uehara, and M. Yokoyama. 2-dimension ham-sandwich theorem for partitioning into three convex pieces. In *Discrete and Computational Geometry* (J. Akiyama et. al eds.), *Lect. Notes Comput. Sci. 1763*, 129–157. Springer, Berlin 2000.

[13] A. Kaneko, M. Kano. Balanced partitions of two sets of points in the plane. *Comput. Geom. Theor. Appl.*, 13(4) (1999), 253–261.

[14] J. Matoušek. *By the Borsuk-Ulam Theorem; Lectures on Topological Methods in Combinatorics and Geometry* (in preparation).

[15] P. Orlik, H. Terao. *Arrangements of Hyperplanes.* Grundlehren der mathematischen Wissenschaften 300, Springer-Verlag 1992.

[16] T. Sakai. Radial partitions of point sets in R^2. Manuscript, Tokoha Gakuen University, 1998.

[17] S. Vrećica, R. Živaljević. The ham sandwich theorem revisited. *Israel J. Math.* 78 (1992), 21–32.

[18] S. Vrećica, R. Živaljević. Conical equipartitions of mass distributions. *Discrete Comput. Geom.*, 225 (2001), 335–350.

[19] G.M. Ziegler and R.T. Živaljević. Homotopy types of subspace arrangements via diagrams of spaces. *Math. Ann.*, 295 (1993), 527–548.

[20] R.T. Živaljević. Topological methods. in *CRC Handbook of Discrete and Computational Geometry*, J.E. Goodman, J. O'Rourke, eds. CRC Press, Boca Raton 1997.

[21] R.T. Živaljević. The Tverberg–Vrećica problem and the combinatorial geometry on vector bundles. *Israel J. Math.* 111 (1999), 53–76.

[22] R.T. Živaljević. User's guide to equivariant methods in combinatorics I and II. *Publ. Inst. Math. Belgrade*, 59(73) (1996), 114–130, and 64(78) (1998), 107–132.

[23] R.T. Živaljević. Combinatorics and topology of partitions of spherical measures by 2 and 3 fans. (arXiv:math.CO/0203028 Mar 2002).

Qualitative Infinite Version of Erdős' Problem About Empty Polygons

Tudor Zamfirescu

The well-known problem of Erdős-Szekeres in Discrete Geometry is about realizations of convex polygons P with a given number of vertices in a finite set $F \subset \mathbb{R}^2$, which means that $V(P)$, the vertex set of P, is a subset of F [5], [6]. See the very good survey [11].

This problem admits an obvious generalization to higher dimensions, observed already in [5], and further studied in [3], [12], [10], [14], [8], [9].

Another possible generalization consists in dropping the convexity assumption.

A *polyhedron* is a finite union P of $(d-1)$-dimensional convex polytopes in $\mathbb{R}^d$, such that

1) the intersection of any two of them is either empty or a face of each of them,

2) any point of P has a neighbourhood in P homeomorphic to $\mathbb{R}^{d-1}$,

3) the complement of P in $\mathbb{R}^d$ has precisely two components, one of which is bounded and called *interior* of P.

The facial structure of P determines through its inclusions the combinatorial class of P. The above $(d-1)$-dimensional polytopes are special faces of P called *facets*; if they are simplices then P is *simplicial.*

It is possible to sharpen the higher-dimensional Erdős-Szekeres problem by looking for realizations in a given combinatorial class. Notice that combinatorially distinct polytopes and a fortiori combinatorially distinct polyhedra may have the same number of vertices.

Let $\mathcal{P}$ be the space of all polyhedra and $\mathcal{K}$ the space of all compact sets in $\mathbb{R}^d$. With the Pompeiu-Hausdorff distance ρ, $\mathcal{K}$ is complete. If $K, K' \in \mathcal{K}$ and $\rho(K, K') < \varepsilon$, we say that K' is an *ε-neighbour* of K.

A compact set is called *typical* if it has all typical properties; and a property **P** is *typical* if the set of all compact sets not enjoying **P** is of first Baire category in $\mathcal{K}$.

We say that a polyhedron P is *realizable* in some set $S \in \mathbb{R}^d$ if there is a polyhedron with vertices in S combinatorially equivalent to P.

More specially, we say that P is *realizable with empty interior* in S if there is a polyhedron with vertices in S combinatorially equivalent to P, the interior of which does not meet S.

A variant of the Erdős-Szekeres problem, due to Erdős, deals with polygons realizable with empty interior in finite subsets of the plane [4]. Generalizations to higher dimensions were studied by Valtr [14], Bisztriczky and Soltan [1], Bisztriczky and Harborth [2].

In the planar case, and similarly in higher dimensions, if n is large enough, there are arbitrarily large finite sets, and of course infinite sets, in which no convex n-gon is realizable with empty interior. Concretely, in the planar case, this is so for $n \geq 7$. And in 3 dimensions, for $n \geq 22$.

Let us remain for a moment in the plane, and keep thinking of the Erdős problem. It seems difficult to pick up empty polygons realized in an arbitrary compact set. But typical compact sets are nowhere dense; this might offer a slight chance...

A polyhedron will be called *combinatorially convex* if it is combinatorially equivalent to the boundary of a convex polytope.

For $M \subset \mathbb{R}^d$ and $x \in \mathbb{R}^d$, let

$$\Delta_M(x) = \{\| y - x\|^{-1}(y - x) : y \in M \setminus \{x\}\}.$$

We investigate now the realization of polytopes in a typical compact set.

Two results, one by Wieacker and the other by Gruber, will be especially useful.

Lemma 1 [15]. *A typical compact set contains no affinely dependent sets.*

Lemma 2 [7]. *Let $K \in \mathcal{K}$. A typical compact set contains, for any $\varepsilon > 0$ a subset homothetic to a compact ε-neighbour of K.*

These lemmas immediately imply the following theorem.

Theorem 1. *Every simplicial polyhedron, and no other polyhedron, is realizable in a typical compact set.*

Proof. Let P be a simplicial polyhedron. A small perturbation of P leaves P in the same combinatorial class $\mathcal{C}$, i.e. if $\eta > 0$ is small enough, $P' \in \mathcal{P}$, card$V(P)$ =card$V(P')$ and $\rho(P, P') < \eta$, then $P' \in \mathcal{C}$. Let $\varepsilon > 0$. By Lemma 2, a typical compact set contains a set homothetic to a finite set F satisfying $\rho(F, V(P)) < \varepsilon$. For $\eta > 0$, we can choose, for each vertex v of $V(P)$, a point $f(v)$ in F at distance smaller than η from v. For $\eta < \varepsilon$ small enough, f is injective. The set $f(V(P)) \subset F$ is an η-neighbour of $V(P)$, and there is a polyhedron P' with $V(P') = f(V(P))$, itself an η-neighbour of P, and therefore in $\mathcal{C}$ for η small enough. This yields the realization.

The realization of nonsimplicial polyhedra in a typical compact set would contradict Lemma 1.

Corollary 1. *Among the points of a typical compact set we can find, for every $n \geq d+1$, the vertices of a simplicial convex polytope with n vertices.*

This was the pendant of the Erdős-Szekeres situation. Now we turn to the empty-polyhedron problem, pendant of the Erdős problem. Then the requirement is that their interior does not meet the typical compact set.

We obtain here a somewhat surprising result.

Theorem 2. *Every combinatorially convex simplicial polyhedron and no other polyhedron is realizable with empty interior in a typical compact set.*

The proof of Theorem 2 will use the following lemma.

Lemma 3 [16]. *If K is a typical compact set, $x \in K$ and N is a neighbourhood of x, then $\Delta_{K\cap N}(x)$ is dense in some hemisphere of $\mathbb{S}^{d-1}$.*

Proof of Theorem 2. Let the combinatorially convex simplicial polyhedron P and the boundary P^* of a simplicial convex polytope belong to the same combinatorial class $\mathcal{C}$. It is enough to show that the set of all $K \in \mathcal{K}$ in which P is not realizable is nowhere dense.

Indeed, let $\mathcal{O}$ be open in $\mathcal{K}$.

We find a finite set $F \in \mathcal{O}$. Let P' be a small homothetic copy of P^* such that $F \cap P' = \emptyset$ and $F \cup V(P')$ still belongs to $\mathcal{O}$.

Now, for $\varepsilon > 0$ very small, any compact set C in an ε-neighbourhood of $F \cup V(P')$ is still in $\mathcal{O}$ and contains card$V(P)$ suitable points close to those in $V(P')$, such that the boundary of their convex hull Q belongs to $\mathcal{C}$ and int$Q$ includes no further points of C. This can be seen in the following way.

First, we choose ε so that each polytope P'' with card$V(P'')$ =card$V(P)$ and

$$\rho(V(P''), V(P')) < \varepsilon$$

belongs to $\mathcal{C}$. This implies that, for each $x \in V(P')$, $B(x,\varepsilon) \cap P'(x,\varepsilon) = \emptyset$, where $B(x,\varepsilon)$ is the open ball of radius ε around x and

$$P'(x,\varepsilon) = \operatorname{conv} \bigcup_{y \in V(P') \setminus \{x\}} B(y,\varepsilon).$$

Then, we choose from C, in each ball $B(x,\varepsilon)$ with $x \in V(P')$, a point closest to some hyperplane separating $C \cap B(x,\varepsilon)$ from $P'(x,\varepsilon)$.

Hence P is realizable with empty interior.

Why can no simplicial polyhedron which is not combinatorially convex be realized with empty interior?

Consider a realization of such a polyhedron P. Then, at some vertex v of P, for any neighbourhood N of v, $\Delta_{P\cap N}(v)$ is not contained in any hemisphere of $\mathbb{S}^{d-1}$. This together with Lemma 3 implies that points of

K must lie in the interior of P; thus the realization cannot be with empty interior.

The following corollary is a direct strengthening of Corollary 1.

Corollary 2. *Among the points of a typical compact set K we can find, for every $n \geq d+1$, the vertices of a simplicial convex polytope with n vertices, which does not contain any further points of K in its interior.*

References

[1] T. Bisztriczky, V. Soltan, Some Erdős-Szekeres type results about points in space, *Monatsh. Math.* **118** (1994) 33-40.

[2] T. Bisztriczky, H. Harborth, On empty convex polytopes, *J. Geometry* **52** (1995) 25-29.

[3] L. Danzer, B. Grünbaum, V. Klee, Helly's theorem and its relatives, *Convexity (Seattle, 1961)* 101-179, Proc. Pure Math. Vol VII A.M.S., Providence, R.I., 1963.

[4] P. Erdős, Some more problems on elementary geometry, *Austr. Math. Soc. Gaz.* **5** (1978) 52-54.

[5] P. Erdős, G. Szekeres, A combinatorial problem in geometry, *Compositio Math.* **2** (1935) 463-470.

[6] P. Erdős, G. Szekeres, On some extremum problems in elementary geometry, *Ann. Univ. Sci. Budapest. Eötvös Sect. Math.* **3-4** (1961) 53-62.

[7] P. Gruber, Your picture is everywhere, *Rend. Sem. Mat. Messina II* **14**, 1 (1991) 123-128.

[8] S. Johnson, A new proof of the Erdős-Szekeres convex k-gon result, *J. Comb. Theory A* **42** (1986) 318-319.

[9] G. Károlyi, Ramsey-remainder for convex sets and the Erdős-Szekeres theorem, *Discr. Appl. Math.*, to appear.

[10] G. Károlyi, P. Valtr, Sets in $\mathbb{R}^d$ without large convex subsets, to appear.

[11] W. Morris, V. Soltan, The Erdős-Szekeres problem on points in convex position – a survey, *Bull. Amer. Math. Soc.* **37**, 4, (2000) 437-458.

[12] T. S. Motzkin, Cooperative classes of finite sets in one and more dimensions, *J. Comb. Theory* **3** (1967) 244-251.

[13] P. Valtr, Sets in $\mathbb{R}^d$ with no large empty convex subsets, *Discrete Math.* **108** (1992) 115-124.

[14] P. Valtr, *Several results related to the Erdős-Szekeres theorem,* Dissertation, Charles University, Prague, 1996.

[15] J. A. Wieacker, The convex hull of a typical compact set, *Math. Ann.* **282** (1988) 637-644.

[16] T. Zamfirescu, The strange aspect of most compacta, to appear.

About the Author

Tudor Zamfirescu is at Fachbereich Mathematik, Universität Dortmund, 44221 Dortmund, Germany; *tudor.zamfirescu@mathematik.uni-dortmund.de*.

MIX
Papier aus verantwortungsvollen Quellen
Paper from responsible sources
FSC® C105338

If you have any concerns about our products,
you can contact us on
ProductSafety@springernature.com

In case Publisher is established outside the EU,
the EU authorized representative is:
Springer Nature Customer Service Center GmbH
Europaplatz 3, 69115 Heidelberg, Germany

Printed by Libri Plureos GmbH
in Hamburg, Germany

Algorithms and Combinatorics 25

Springer-Verlag Berlin Heidelberg GmbH

Boris Aronov • Saugata Basu
János Pach • Micha Sharir

Editors

Discrete and Computational Geometry

The Goodman-Pollack Festschrift

With 287 Figures

Springer

Editors

Boris Aronov

Department of Computer
and Information Science
Polytechnic University
Six MetroTech Center
Brooklyn, NY 11201, USA
e-mail: aronov@poly.edu

Saugata Basu

School of Mathematics
Georgia Institute of Technology
Atlanta, GA 30332-0160, USA
e-mail: saugata@math.gatech.edu

János Pach

City College
and Courant Institute
251 Mercer St.
New York, NY 10012, USA
e-mail: pach@cims.nyu.edu

Micha Sharir

School of Computer Science
Tel Aviv University
Tel Aviv 69978, Israel
e-mail: michas@tau.ac.il

Cataloging-in-Publication Data applied for

Bibliographic information published by Die Deutsche Bibliothek

Die Deutsche Bibliothek lists this publication in the Deutsche Nationalbibliografie; detailed bibliographic data is available in the Internet at <http://dnb.ddb.de>.

Mathematics Subject Classification (2000):
52-XX, 68-XX, 05-XX, 14-XX, 57-XX, 60-02

ISSN 0937-5511
ISBN 978-3-642-62442-1 ISBN 978-3-642-55566-4 (eBook)
DOI 10.1007/978-3-642-55566-4

http://www.springer.de

Originally published by Springer-Verlag Berlin Heidelberg New York in 2003
Softcover reprint of the hardcover 1st edition 2003

Typeset by the authors. Edited and reformatted by Kurt Mattes, Heidelberg, using a Springer LaTeX macro package
Cover design: *design & production* GmbH, Heidelberg

Printed on acid-free paper 46/3142/LK - 5 4 3 2 1 0

Preface

Eli Goodman and Ricky Pollack were born at about the same time and brought up in the same city, within the same social milieu. They had many musical, literary, and scientific interests in common that drew them to the same events and locations. Their orbits were bound to intersect.

Precisely where and when it happened, they disagree. Was it at a concert in the late fifties? In a group of bridge enthusiasts? Or perhaps a little later, at the Heights campus of New York University, where each of them taught? In fact, Eli and Ricky disagree about a great many little things and details: this seems to be their working method, the secret of their relationship – carefully testing commonly accepted views and concepts, no matter how self-evident they might appear, by exposing them to each other's critical eyes and words.

Both of them received a first-class mathematical education. Ricky's thesis was in number theory, under the supervision of Harold Shapiro at New York University, while Eli became an algebraic geometer under the direction of Heisuke Hironaka and Steven Kleiman at Columbia University. Both of them published remarkable results in their respective research areas and had a taste of scientific success. In the mid 1970s, apparently on divergent paths, both of them took sabbaticals, and soon after each decided to try something new.

Ricky spent a semester at McGill University, collaborating with a group of excellent combinatorists. In particular, he learned about and was fascinated by the Erdős–Szekeres conjecture, which states that in every set of more than 2^{n-2} points in general position in the plane one can find n points that form the vertex set of a convex n-gon. He spent two months trying to come up with a proof. Unbeknownst to him, Eli had been working in topological graph theory and had recently turned his attention to some problems in combinatorial geometry. Shortly after Ricky's return from Montreal, he bumped into Eli at Courant Institute, at a seminar talk on graph algorithms. "What are you doing here?" – they asked each other in surprise. A few weeks later Eli called Ricky: – "Have you ever heard of the Erdős–Szekeres problem? I have some ideas ..."

They started meeting once or twice a week and, based on Eli's preliminary ideas, they soon developed a natural combinatorial encoding of a planar point configuration. This later became known as an "allowable sequence," and turned out to be an important tool in combinatorial geometry that led to spectacular breakthroughs such as Ungar's "book" proof of Scott's conjecture, according to which every set of n noncollinear points in the plane determines at least $2\lfloor n/2 \rfloor$ distinct directions. Dozens of joint papers evolved from this first fruitful collaboration, some of them developing theoretical aspects of the material, others solving problems that had seemed insurmountable by other methods. To their surprise, Eli and Ricky found themselves drawn more to combinatorial geometry than to their original fields. They loved this newly-found playground, and made it their own. They made a permanent move and never looked back. Gradually their joint work spread out to encompass various related topics in discrete geometry, and later included additional authors as well. Of their many ground-breaking discoveries we mention two. (1) They revitalized geometric transversal theory by showing that, using the notion of "order types," one can establish many exciting Helly-type theorems in d-space for the existence of a k-dimensional flat intersecting all members of a family of convex bodies, for $0 < k < d$. (2) They were among the first mathematicians to use real algebraic geometry to enumerate order types, simplicial polytopes, and other geometric objects. Their surprising results found numerous applications in ray shooting, in computer graphics, and elsewhere.

In 1984, after it had become clear that their work had implications not only in the well-established discipline of discrete geometry but also in the then nascent field of computational geometry, the opportunity arose of founding a journal devoted to both fields. Three publishers were interested, but Eli and Ricky decided on Springer-Verlag (whose New York office was then under the leadership of Walter Kaufmann-Bühler) as the best of the choices, and the journal *Discrete & Computational Geometry* was born the following year. Its first issue appeared in 1986, and the journal has thrived since then, having grown from under 400 pages a year at the start to nearly 1300 pages currently. At the same time the field has grown enormously. It is hard to overestimate the role Eli, Ricky, and their journal have played in these developments. The marriage of the two fields proved to be fruitful and long-lasting. DCG is now the leading journal in a fast growing area along the borderline of pure mathematics and computer science. Eli and Ricky co-organized the first Special Year at DIMACS (1989–90) and several major conferences dedicated to the subject (Santa Cruz 1986, Mount Holyoke 1996, Ascona 1999, and Oberwolfach 2000), as well as a number of special sessions at meetings of the American Mathematical Society. The investment of seemingly infinite resources on their part has paid off: they have built bridges between remote groups of researchers working on similar problems as well as bridges between various subdisciplines. In this sense, they are perhaps the two most conspicuous "founding fathers" of the field. As such, their names have become inseparable.

The idea of editing a collection of papers dedicated to Eli and Ricky was proposed by N. Prabhu, a former student of Ricky's, to whom we are grateful. We were soon overwhelmed by an avalanche of excellent manuscripts submitted by leading researchers in discrete and computational geometry. They contributed state-of-the-art results covering virtually every aspect of the field. In the selection process, we followed the simple rule that each paper should be refereed according to the high standards adopted by Eli's and Ricky's journal. It is most appropriate that the publisher of this volume is Springer-Verlag, under whose auspices the journal is produced. We are very grateful for their enthusiastic support.

Each editor of this volume regards himself as a student, a close friend, a collaborator, and a co-author of Eli and Ricky – all in one. They have shaped our lives, our mathematical thinking, our views of the world. We wish that their golden age last forever to the benefit of our community.

Boris Aronov
Saugata Basu
János Pach
Micha Sharir

Contents

On the Complexity of Many Faces in Arrangements of Pseudo-Segments and Circles

Pankaj K. Agarwal
Boris Aronov
Micha Sharir

Abstract

We obtain improved bounds on the complexity of many faces in an arrangement of pseudo-segments, circles, or unit circles. The bounds are worst-case optimal for unit circles; they are also worst-case optimal for the case of pseudo-segments except when the number of faces is very small, in which case our upper bound is a polylogarithmic factor away from the best-known lower bound. For general circles, the bounds nearly coincide with the best-known bounds for the number of incidences between points and circles.

1 Introduction

Problem statement and motivation. The *arrangement* $\mathcal{A}(\Gamma)$ of a finite collection Γ of curves or surfaces in $\mathbb{R}^d$ is the decomposition of the space into relatively open connected cells of dimensions $0, \ldots, d$ induced by Γ, where each cell is a maximal connected set of points lying in the intersection of a fixed subset of Γ and avoiding all other elements of Γ. The *combinatorial complexity* (or *complexity* for short) of a cell ϕ in $\mathcal{A}(\Gamma)$, denoted as $|\phi|$, is the number of faces of $\mathcal{A}(\Gamma)$ of all dimensions that lie on the boundary of ϕ. Besides being interesting in their own right, due to the rich geometric, combinatorial, algebraic, and topological structure that they possess, arrangements also lie at the heart of numerous geometric problems arising in a wide range of applications, including robotics, computer graphics, and molecular modeling. The study of arrangements of lines and hyperplanes has a long, rich history, but most of the work until the 1980s dealt with the combinatorial structure of the entire arrangement or of a single cell in the arrangement (which, in this case, is a convex polyhedron); see [17] for a summary of early work. Motivated by problems in computational and combinatorial geometry, various combinatorial and algorithmic issues involving substructures of arrangements of hyperplanes and hypersurfaces have received considerable attention, mostly during the last two decades; see [5] for a recent survey.

This paper studies the so-called *many-faces* problem for arrangements of pseudo-segments or circles in the plane. (A set of arcs is called a family of *pseudo-segments* if every pair of arcs intersect in at most one point.) More precisely, given a set Γ of n arcs and a set P of m points in the plane, none lying on any arc in Γ, let $K(P,\Gamma)$ be the combined combinatorial complexity of the cells of $\mathcal{A}(\Gamma)$ that contain at least one point of P. We wish to obtain an upper bound for the maximum value of $K(P,\Gamma)$, as a function of n and m, for the cases where Γ is a set of pseudo-segments or a set of circles. The study of the complexity of many faces, and the accompanying algorithmic problem of computing many faces, in planar arrangements (as studied, e.g., in [3, 16]) has several motivations: (i) It arises in a variety of problems involving 3-dimensional arrangements [8, 19]. (ii) It is closely related to the classical problem in combinatorial geometry of bounding the number of *incidences* between points and curves, as studied in numerous papers, including [14, 23, 26–28]. Informally, in both cases we have points and curves; in the case of incidences, the points lie on the curves and an incidence is a pair (p,γ), where point p lies on curve γ. In the case of many faces, the points lie "in between" the curves, and we are essentially interested in "extended incidences," involving pairs (p,γ), where point p can reach curve γ without crossing any other curve (i.e., γ appears on the boundary of the face containing p). The incidence problem for points and curves has attracted considerable attention in combinatorial and computational geometry; see the papers cited above. The problem of many faces is typically harder than the (already quite hard) corresponding incidence problem. (iii) The many-faces problem is the "loosest" (i.e., least restricted) of all problems that study substructures in arrangements. It poses the greatest challenge because there is less structure to exploit. Tackling this problem has led to the derivation of various tools, such as the Combination Lemma [20, 25], which are interesting in their own right, and have many algorithmic applications; see, e.g., [2] for one such recent application.

Previous results. An early paper by Canham [11] initiated the study of the many-faces problem for line arrangements. After a number of intermediate results, tight bounds on the complexity of many faces in line and pseudo-line arrangements were obtained by Clarkson *et al.* [14], using an approach based on random sampling. This work and a series of subsequent papers proved near-optimal or nontrivial bounds on the complexity of many faces in arrangements of line segments, circles, and other classes of curves in the plane, and in arrangements of hyperplanes in higher dimensions; see [5] and the references therein. Aronov *et al.* [7] showed, using a fairly involved analysis, that the complexity of m distinct faces in an arrangement of n segments in the plane is $O(m^{2/3}n^{2/3} + n\alpha(n) + n\log m)$, which is optimal in the worst case except for a small range of m near the value $n^{1/2}$. Unlike the case of lines, their proof does not immediately extend to the case of pseudo-segments. In fact, some key properties of segments that are used in the proof do not hold for arbitrary collections of pseudo-segments, but do hold if we assume that the pseudo-segments are *extendible*; see below for details and further discussion. The best-known bound on the complexity of m distinct faces in an arrangement of n circles in the plane is $O(m^{3/5}n^{4/5}4^{\alpha(n)/5} + n)$. If all circles are congruent, then the bound is $O(m^{2/3}n^{2/3}\alpha^{1/3}(n) + n)$; here $\alpha(n)$ is the extremely slowly growing inverse of the Ackermann's function [25]. These bounds were obtained in [14].

As mentioned above, the many-faces problem is closely related to the incidence problem, which, given a set Γ of curves and a set P of points in the plane, asks for bounding the number of pairs $(p, \gamma) \in P \times \Gamma$ such that $p \in \gamma$. For example, the tight bounds on the maximum number of incidences between points and lines (or segments, or pseudo-lines) are asymptotically the same as the maximum complexity of m distinct faces in an arrangement of n lines, viz., $\Theta(m^{2/3}n^{2/3} + m + n)$ [14]. (Note, though, that the best-known bound for the complexity of many faces in an arrangement of *line segments*, mentioned above, is slightly weaker [7].) The same was true for arrangements of circles (except for the tiny $4^{\alpha(n)/5}$ factor in the leading term) until recently, when Aronov and Sharir [9] obtained an improved bound of $O(m^{2/3}n^{2/3} + m^{6/11+3\varepsilon}n^{9/11-\varepsilon} + m + n)$, for any $\varepsilon > 0$, on the number of incidences between points and circles. This bound has been slightly strengthened by Agarwal *et al.* [4] to $O(m^{2/3}n^{2/3} + m^{6/11}n^{9/11}\kappa(m^3/n) + m + n)$, where $\kappa(r) = (\log r)^{O(\alpha^2(r))}$.[1] Aronov and Sharir raised the question whether a similar bound can be obtained for the complexity of many faces in circle arrangements, which, after the cases of lines and segments, is one of the natural next problem instances to be tackled.

Our results: The case of extendible and general pseudo-segments. A set S of x-monotone Jordan curves (resp. arcs) is called a family of *pseudo-lines* (resp. *pseudo-segments*) if any two of them intersect in at most one point and they cross each other at that point. A family S of pseudo-segments is called *extendible* if there exists a family Γ of pseudo-lines, such that each $s \in S$ is contained in some $\gamma \in \Gamma$. See a recent paper of Chan [12], where extendible pseudo-segments are discussed. In particular, not every family of pseudo-segments is a family of extendible pseudo-segments; the simplest demonstration of this fact is depicted in Figure 1. Chan has shown that a family of n x-monotone pseudo-segments can be transformed into a family of $O(n \log n)$ extendible pseudo-segments by cutting each of the given pseudo-segments into at most $O(\log n)$ pieces, in a "segment-tree" fashion.

We prove that the complexity of m distinct faces in an arrangement of n extendible pseudo-segments with X intersecting pairs is $O(m^{2/3}X^{1/3} + n \log n)$. The best lower bound, which is constructed using straight segments, is $\Omega(m^{2/3}X^{1/3} + n\alpha(n))$ [7]. Hence, our bound is worst-case tight when the first term dominates, and is otherwise within a logarithmic factor of the lower

[1] Hereafter, slightly abusing the notation, we will write $\kappa(r)$ for *any* function of the form $(\log r)^{O(\alpha^2(r))}$, irrespective of the value of the implied constant. Hence κ may refer to different functions in different bounds, all having this same general form.

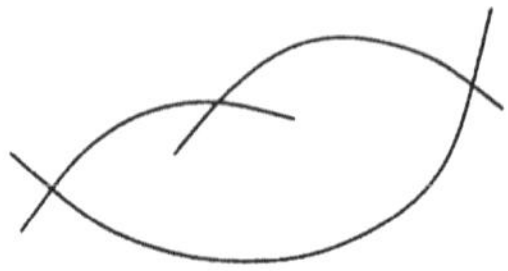

Fig. 1. Three pseudo-segments that do not form an extendible family.

bound. Thus, since $X = O(n^2)$, the bound is $O(m^{2/3}n^{2/3} + n\log n)$, which is worst-case optimal for $m = \Omega(n^{1/2}\log^{3/2} n)$.

A closer inspection of the argument in [7] shows that, with some obvious modifications, it also applies to the case of extendible pseudo-segments, thus yielding the bound $O(m^{2/3}X^{1/3} + n\alpha(n) + n\log m)$, which beats our bound only when $\log m = o(\log n)$ and $\log m = \omega(\alpha(n))$. Nevertheless, our proof is simpler than that in [7].

By Chan's result, this bound implies an upper bound of $O(m^{2/3}X^{1/3} + n\log^2 n)$ for the complexity of m faces in an arrangement of arbitrary x-monotone pseudo-segments; this bound also holds when the pseudo-segments are not x-monotone, but each of them has only $O(1)$ locally x-extremal points. Again, this is worst-case optimal, unless m is small. For example, substituting $X = O(n^2)$, the bound becomes $O(m^{2/3}n^{2/3} + n\log^2 n)$, which is worst-case optimal for $m = \Omega(n^{1/2}\log^3 n)$.

The analysis of the cases of extendible and general pseudo-segments is important for two independent reasons. First, we obtain nontrivial bounds (which are worst-case optimal or near-optimal) for these cases. In doing so, we obtain a proof that is much simpler than the one given in [7] and, of course, applies also to the case of segments. As mentioned, our bounds are known to be worst-case optimal, unless the value of m is small (about $n^{1/2}$ or smaller). The bound in [7] (modified for the case of extendible pseudo-segments) is slightly better, but these two bounds differ only in a narrow range of m and only by at most a logarithmic factor. Second, the result for extendible pseudo-segments is used as a major tool in our derivation of the bounds for the case of circles.

The case of circles. Next, we answer the question raised by Aronov and Sharir affirmatively: Let C be a set of circles in the plane and P a set of points, not lying on any circle. As defined earlier, we will use $K(P,C)$ to denote the combined combinatorial complexity of the faces of $\mathcal{A}(C)$ that contain at least one point of P. Set $K(m,n) = \max K(P,C)$, with the maximum taken over all families C of n circles and all families P of m points. We prove that

$$K(m,n) = O\Big(m^{2/3}n^{2/3} + m^{6/11}n^{9/11}\kappa(m^3/n) + n\log n\Big),$$

where $\kappa(r)$ is, as above, $(\log r)^{O(\alpha^2(r))}$.

Let $K'(m,n)$ denote the maximum value of $K(P,C)$ with the added assumption that all pairs of circles intersect. In this case, following the analysis by Agarwal *et al.* [4], we obtain the following improved bound:

$$K'(m,n) = O\left(m^{2/3}n^{2/3} + m^{1/2}n^{5/6}\log^{1/2}(m^3/n) + n\log n\right).$$

If not all pairs of circles intersect, we obtain a bound that depends on X, the number of intersecting pairs of circles. Let $K(m,n,X) = \max K(P,C)$, with the maximum taken over all families P of m points and C of n circles with X intersecting pairs. We show that

$$K(m,n,X) = O\left(m^{2/3}X^{1/3} + m^{6/11}X^{4/11}n^{1/11}\kappa(m^3X^2/n^5) + n\log n\right).$$

These three bounds are nearly the same as the new corresponding bounds for incidences, given in [4, 9], apart from polylogarithmic factors. Note that the bound of $O(m^{3/5}n^{4/5}4^{\alpha(n)/5} + n)$, obtained by Clarkson *et al.* [14], is slightly better than the ones stated here for $m \leq (n^{1/3}/4^{\alpha(n)/3})\log^{5/3} n$. For example, $K(m,n) = O(n)$ if $m \leq n^{1/3}/4^{\alpha(n)/3}$.

Face-curve incidences. Our general technique is similar to the one used in [9], i.e., we first prove a weaker bound, which is almost optimal for large values of m, by cutting the circles of C into extendible pseudo-segments and using the bound for extendible pseudo-segments that we derive separately. Next, to handle small values of m, we use a partitioning scheme in the "dual space," decompose the problem into many subproblems, bound the complexity for each subproblem using the weaker bound, and estimate the overall complexity as we merge the subproblems.

However, several new ideas are needed to carry out each of these steps. First, we introduce the notion of *face-curve incidences* between the given collection of arcs and a set of marked faces in its arrangement, where a face-curve incidence is a pair (f,γ), where f is a marked face and γ appears along ∂f. Thus, even if a curve appears many times along a face boundary, we count it only once in this new measure. We show that it suffices to bound the number of those face-curve incidences in order to bound the complexity of the given marked faces. The advantage of incidences is that, with a careful extension of their definition to arrangements of subsets of the given set of curves, this measure is *additive* with respect to both the number of face-marking points and the number of curves. This makes it considerably easier to partition the set of arcs into various subsets, bound the complexity of marked faces in each subarrangement, and then merge (i.e., add the face-curve incidence counts for) the subarrangements. Once a bound on the number of face-curve incidences is obtained, it can be converted to a bound on the actual complexity of these faces (see Lemmas 2.1, 2.2, and 2.3).

Previous techniques (e.g., that of [7]) have faced the same problem of merging subarrangements into the whole arrangement, and solved it using *combination lemmas*, which provide relations between the complexity of the marked faces in the subarrangements and the complexity of the marked faces in the whole arrangement. These combination lemmas (presented, e.g., in [20, 25]) are more involved, and generally yield weaker bounds when the partition into subarrangements consists of many recursive levels, as is the case in the analysis presented in this paper.

The case of unit circles. Finally, for the case in which all circles in C are congruent (the case of "unit circles"), we show that the complexity of m distinct faces in an arrangement of n congruent circles with X intersecting pairs, is $O(m^{2/3}X^{1/3}+n)$. This bound is asymptotically tight in the worst case, in contrast with the same asymptotic upper bound for the case of incidences [14, 26, 27], which is far away from the best-known, near-linear lower bound. Note that the improvement here is rather marginal—we only remove the factor $\alpha(n)^{1/3}$ from the leading term, appearing in the previous bound of [14].

The paper is organized as follows. In Section 2, we introduce the notion of face-curve incidences and establish several general properties of this measure, including a relationship between the number of face-curve incidences and the actual complexity of the corresponding faces. Next, we establish in Section 3 complexity bounds for the case of extendible (and general) pseudo-segments. Section 4 derives the bounds for general circles, and Section 5 establishes an optimal bound for congruent circles.

2 Incidences between Curves and Faces

Let Γ be a set of n Jordan arcs in the plane, each pair of which intersect in at most s points. For a point p not lying on any arc in Γ, let f_p denote the face of $\mathcal{A}(\Gamma)$ that contains p. Let P be a finite set of points in which no point lies on any arc in Γ. For a subset $G \subseteq \Gamma$, we define $I_\Gamma(P,G)$ to be the number of pairs $(p,\gamma) \in P \times G$ such that an arc of γ appears on ∂f_p. Note that f_p is defined as a face of the entire arrangement $\mathcal{A}(\Gamma)$ rather than a face of $\mathcal{A}(G)$; it is in fact a subset of the face of $\mathcal{A}(G)$ that contains p. Note also that a pair (p,γ) is counted only once, even if γ contains more than one edge of ∂f_p. If the set Γ is obvious from the context, we will simply use $I(P,G)$ to denote $I_\Gamma(P,G)$; in particular we will write $I(P,\Gamma)$ for $I_\Gamma(P,\Gamma)$.

Lemma 2.1. *Let Γ be a set of Jordan arcs in the plane such that every pair of arcs intersect in at most s points. Let P be a set of points so that none of them lies on any arc of Γ and so that no face of $\mathcal{A}(\Gamma)$ contains more than one point of P. Then*

$$I_\Gamma(P,\Gamma) \le K(P,\Gamma) = O(\lambda_t(I_\Gamma(P,\Gamma))),$$

where $t=s$ if every arc in Γ is either an unbounded curve that separates the plane or a closed Jordan curve, and $t=s+2$ otherwise; $\lambda_t(n)$ is the maximum length of an (n,t)-Davenport–Schinzel sequence.

Proof. Let n_p be the number of arcs of Γ that appear on the boundary of f_p, for a point $p \in P$. Then $I_\Gamma(P,\Gamma) = \sum_{p \in P} n_p$, and this is clearly a lower bound for $K(P,\Gamma)$. By a result of Guibas *et al.* [18], the complexity of f_p is $O(\lambda_t(n_p))$, where $t = s$ if every arc in Γ is either an unbounded curve separating the plane or a closed curve, and $t = s+2$ otherwise. Since each face of $\mathcal{A}(\Gamma)$ contains at most one point of P, $K(P,\Gamma) = \sum_{p \in P} O(\lambda_t(n_p)) = O(\lambda_t(I_\Gamma(P,\Gamma)))$. □

The quantity $K(P,\Gamma) - I(P,\Gamma)$ is closely related to the notion of *excess* introduced by Aronov and Sharir [8]. Specifically, the excess of a face ϕ is the number of edges bounding ϕ minus the number of distinct arcs of Γ that appear on the boundary of ϕ. A result of Sharir [24] implies the following:

Lemma 2.2. *Let Γ be a set of n line segments in the plane, and let P be a set of points, none lying on any segment, so that no face of $\mathcal{A}(\Gamma)$ contains more than one point of P. Then*

$$K(P,\Gamma) = I(P,\Gamma) + O(n \log\log n).$$

A close inspection of the proof given in [24] shows that it also holds for extendible pseudo-segments:

Lemma 2.3. *Let Γ be a set of n extendible pseudo-segments in the plane, and let P be a set of points, none lying on any pseudo-segment, so that no face of $\mathcal{A}(\Gamma)$ contains more than one point of P. Then*

$$K(P,\Gamma) = I(P,\Gamma) + O(n \log\log n).$$

The following lemma will be crucial in proving the bounds on the complexity of many faces.

Lemma 2.4. *Let $G \subseteq \Gamma$ be a subset of g arcs, and let P be a set of m points, none lying on any arc, so that no face of $\mathcal{A}(\Gamma)$ contains more than one point of P. Then*

$$I_\Gamma(P,G) \leq 2m + 2g + K(P,G).$$

Note the difference between this lemma, which deals with the case where G is a subset of Γ, and Lemma 2.1, which deals only with the case $G = \Gamma$. The difference lies in the fact that a face in $\mathcal{A}(G)$ may contain many points of P, and each of its edges may appear on the boundary of many marked faces of $\mathcal{A}(\Gamma)$. Lemma 2.4 shows that the number of these additional multiple occurrences of edges is bounded by $2m + 2g$. (The lemma also holds for $G = \Gamma$, but then the bound in Lemma 2.1 is better.)

Proof. Let F be the set of faces of $\mathcal{A}(G)$ that contain points of P. Let f be a face in F that contains $m_f > 0$ points of P, say, $p_1, \ldots, p_{m_f}$. Refer to Figure 2. The corresponding faces f_{p_j} of $\mathcal{A}(\Gamma)$, for $j = 1, \ldots, m_f$, are pairwise-disjoint connected regions within f, since each face of $\mathcal{A}(\Gamma)$ is assumed to contain at most one point of P. Suppose ∂f has ξ_f connected components. For each connected component, we choose a point q_j, for $1 \le j \le \xi_f$, that lies in the complement of f bounded by that component. We decompose each connected component of ∂f into maximal connected portions, so that each portion overlaps with the boundary of a single face f_{p_i} of $\mathcal{A}(\Gamma)$; such a portion might appear on ∂f_{p_i} in many disconnected pieces; see ∂f_{p_2} in Figure 2. Let $\gamma_1, \ldots, \gamma_{h_f}$ denote the resulting partition of ∂f. Then the points of P lying in f contribute at most $h_f + |f|$ to $I(P, G)$, where $|f|$ is the number of edges in ∂f. Hence,

$$I(P,G) \le \sum_{f \in F} (h_f + |f|) = K(P,G) + \sum_{f \in F} h_f .$$

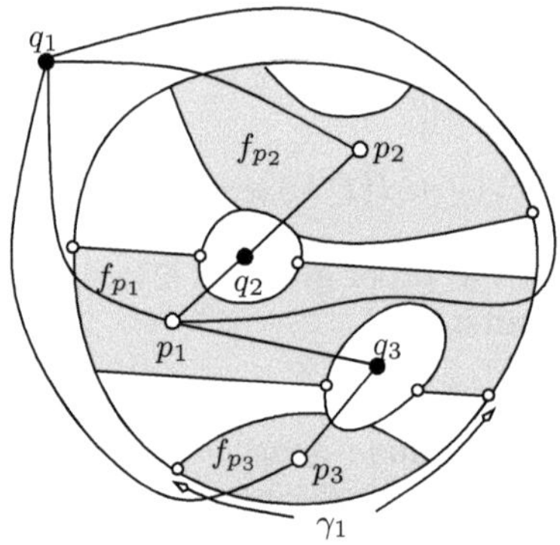

Fig. 2. Construction of the bipartite graph to bound $I(P,G)$ within a single face of $\mathcal{A}(G)$; small white circles along ∂f denote the partition of ∂f into $\gamma_1, \gamma_2, \ldots$.

In order to bound h_f, we construct a planar bipartite graph whose vertices are the points p_j, for $j = 1, \ldots, m_f$, on one side, and the points q_j, for $j = 1, \ldots, \xi_f$, on the other side. For each γ_l, if γ_l is a portion of the jth connected component of ∂f and overlaps with ∂f_{p_i}, we connect p_i to q_j by an edge; we draw the edge as an arc passing through γ_l; see Figure 2. This can easily be done so that these edge drawings are pairwise disjoint except at their endpoints. The resulting graph is planar and has no faces of degree two, although there may be multiple edges between a pair of vertices, such as p_1 and q_1 in Figure 2. Hence, the number h_f of edges in the graph is at most $2(m_f + \xi_f) - 4$.

The points of P are partitioned among the faces of F, so $\sum_{f \in F} m_f = m$. Moreover, $\sum_{f \in F} (\xi_f - 1) \le |G| = g$. Indeed, $\xi_f - 1$ is the total number of "islands" (inner boundary components) inside the face f, and an arc of G cannot belong to more than one island. This completes the proof of the lemma. □

A useful property of $I_\Gamma(P, G)$, which justifies its introduction, is given in the following lemma; its proof is immediate from the definition.

Lemma 2.5. *$I_\Gamma(\cdot,\cdot)$ is additive in both variables: If $P = P_1 \dot\cup P_2$, where P_1 and P_2 are disjoint subsets of marking points, so that no face of $\mathcal{A}(\Gamma)$ contains more than one point of P, and if $G = G_1 \dot\cup G_2$, where G_1 and G_2 are disjoint subsets of G, then*

$$I_\Gamma(P_1 \dot\cup P_2, G) = I_\Gamma(P_1, G) + I_\Gamma(P_2, G), \text{ and}$$
$$I_\Gamma(P, G_1 \dot\cup G_2) = I_\Gamma(P, G_1) + I_\Gamma(P, G_2).$$

3 The Case of Pseudo-Segments

A collection Γ of bounded Jordan arcs (resp., unbounded Jordan curves, each separating the plane) in the plane is called a family of *pseudo-segments* (resp., *pseudo-lines*) if every pair of them intersect in at most one point (resp., in exactly one point), where they cross each other. A collection Γ of n x-*monotone* pseudo-segments is called a family of *extendible pseudo-segments* if there exists a family Γ_0 of x-*monotone* pseudo-lines, so that each arc in Γ is contained in some pseudo-line of Γ_0. The basic properties of extendible pseudo-segments are mentioned in the introduction, and presented in more detail in [12].

The main result of this section gives a bound on the complexity of m distinct faces in an arrangement of n extendible pseudo-segments. By combining this bound with the machinery in [12], we also obtain an upper bound on the complexity of many faces in an arrangement of arbitrary x-monotone pseudo-segments. This bound also holds for non-x-monotone pseudo-segments, provided that each of them has only $O(1)$ locally x-extremal points. As far as we know, this is the first study of these cases. Besides being interesting in its own right, the case of extendible pseudo-segments will be used as a main tool in our derivation of the bounds for the case of general circles, presented in the next section.

A weaker bound for extendible pseudo-segments. Let Γ be a set of n x-monotone extendible pseudo-segments, and let P be a set of m points in the plane so that no point lies on any pseudo-segment or on any vertical line passing through an endpoint of a pseudo-segment.

Lemma 3.1. *The maximum complexity of m distinct faces in an arrangement of n extendible pseudo-segments in the plane is $O(m^{2/3}n^{2/3} + n^{4/3})$.*

Proof. For each $p \in P$ for which f_p is not x-monotone, partition f_p into x-monotone subfaces by erecting vertical segments up and down from each pseudo-segment endpoint that lies on ∂f_p, until they meet another pseudo-segment (or extend all the way to $\pm\infty$). The number of resulting subfaces is at most $m + 4n$. Let $P_0 \supseteq P$ be a new set of marking points, one in each of the new subfaces. We apply Székely's technique [27] to bound the complexity of these $O(m+n)$ x-monotone subfaces, using the same approach as in [15]. Namely, we define a graph G with the set P_0 of marking points as vertices. Two points $p, p' \in P_0$ are connected by an edge if there exists a pseudo-segment $s \in \Gamma$ that appears on the top boundaries of these two faces (resp., on their bottom boundaries), so that s does not appear along the top (resp., bottom) boundary of any other marked subface between these two appearances. An illustration of a portion of such a graph is given in Figure 3. As shown in [15], one can draw the edges of G as arcs in the plane,

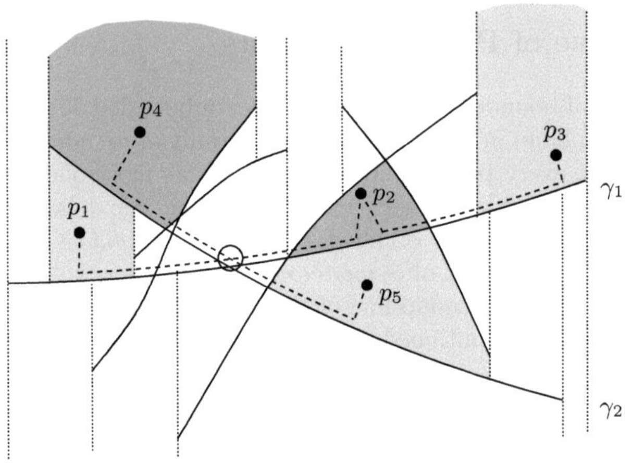

Fig. 3. Three edges of the graph G. The edge (p_1, p_2) connecting p_1 and p_2 along the upper side of the arc γ_1 and the edge (p_4, p_5) connecting p_4 to p_5 along the upper side of the arc γ_2 cross at an intersection point of γ_1 and γ_2.

so that they intersect only at points of intersection between the curves of Γ. The graph G may have multiple edges connecting the same pair of points. However, since the pseudo-segments are extendible and the subfaces are x-monotone, the edge multiplicity in the resulting graph is at most four. This is shown as follows.

Define, as in [12], a relation on Γ, so that for $s, s' \in \Gamma$, $s \prec_\Gamma s'$ if s and s' intersect and, s lies below s' slightly to the left of their intersection point. As noted in [12], Γ is a collection of extendible pseudo-segments if and only if $\prec_\Gamma$ is a partial order. Let L be a family of pseudo-lines, so that each $s \in \Gamma$ is contained in some $\tilde{s} \in L$. Define a total order on L so that, for $\gamma, \gamma' \in L$, $\gamma \prec_L \gamma'$ if γ lies below γ' to the left of their intersection point. By construction, $\prec_L$ is a linear extension of $\prec_\Gamma$.

Claim. *Let f be an x-monotone subface, as constructed above, and let $s \in \Gamma$ be a pseudo-segment appearing on the top (resp., bottom) boundary of f. Then f lies fully below (resp., above) the pseudo-line $\tilde{s} \in L$ containing s.*

Proof. Suppose to the contrary that the top portion of ∂f, which is a connected x-monotone curve, crosses $\tilde{s}$, say to the right of s (clearly, the boundary cannot cross s itself). Consider the leftmost such crossing. Let $t \in \Gamma$

be the pseudo-segment along which the crossing takes place, and let $\tilde{t}$ be the pseudo-line in L containing t. By definition, we have $\tilde{t} \prec_L \tilde{s}$. On the other hand, follow the top boundary of f from s to the right, and let $s = s_1, s_2, \ldots, s_j = t$ be the sequence of pseudo-segments that we encounter between s and t. See Figure 4. By definition, we have $s_i \prec_\Gamma s_{i+1}$ and thus

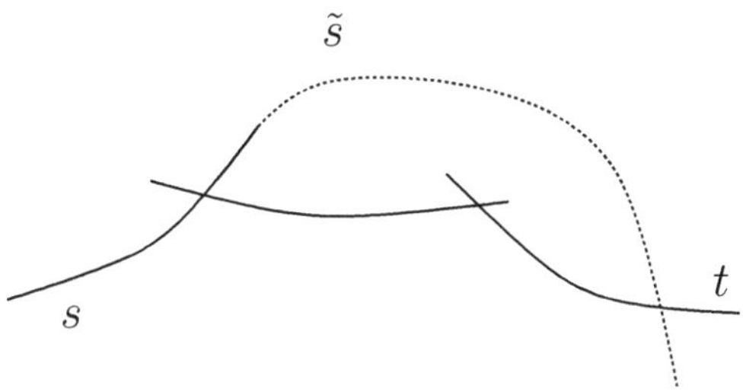

Fig. 4. Illustration to the claim in the proof of Lemma 3.1.

$\tilde{s}_i \prec_L \tilde{s}_{i+1}$, for each $i = 1, \ldots, j-1$. Therefore $\tilde{s} \prec_L \tilde{t}$, a contradiction that establishes the asserted claim. The cases where $\tilde{s}$ meets ∂f to the left of s, or where s appears along the bottom boundary of f, are handled in a fully symmetric manner. □

Now let f and f' be two (x-monotone sub-)faces that are connected by at least five edges in G. Then there exist three distinct pseudo-segments, s_1, s_2, and s_3 that appear, say, along the top boundaries of both f and f'. Let E denote the lower envelope of the three corresponding pseudo-lines $\tilde{s}_1, \tilde{s}_2$, and $\tilde{s}_3$. The above claim implies that f and f' lie fully below E, and each of them touches E at three distinct points. Since E consists of three connected arcs, each contained in a different pseudo-line, it follows easily that this configuration yields an impossible planar drawing of $K_{3,3}$. See Figure 5 for an illustration.

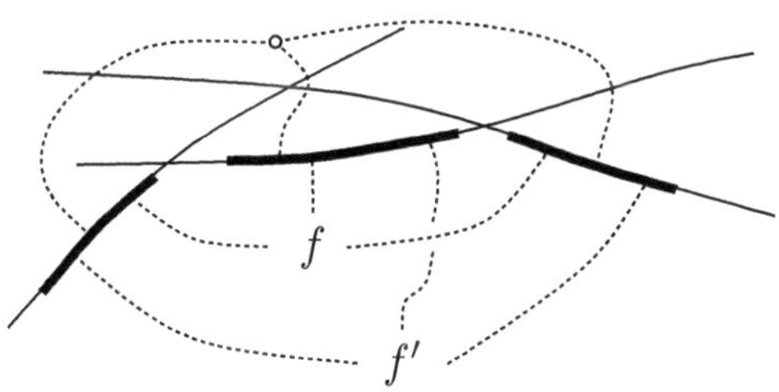

Fig. 5. Impossible drawing of $K_{3,3}$ when 3 distinct pseudo-segments bound f and f' on their top sides.

Hence the edge multiplicity of G is at most 4, and the lemma now follows exactly as in [15, 27], using the crossing lemma of Leighton and of Ajtai et al. (see [22]). □

Next, we obtain an improved bound on $K(P,\Gamma)$ using a decomposition in dual space. It suffices to obtain a bound on $I(P,\Gamma)$ since, by Lemma 2.2,

$$K(P,\Gamma) = I(P,\Gamma) + O(n \log\log n).$$

Cuttings. Although the following discussion applies to (and is presented for) any dimension d, we only need it for $d = 2$ (in this section) and $d = 3$ (when treating the case of circles).

Let H be a set of m hyperplanes in $\mathbb{R}^d$, and let S be a set of n points in $\mathbb{R}^d$. For a simplex Δ, we use $H_\Delta \subseteq H$ to denote the set of hyperplanes that cross (i.e., meet the interior of) Δ, and S_Δ to denote $S \cap \Delta$. Set $m_\Delta = |H_\Delta|$ and $n_\Delta = |S_\Delta|$. Let k_Δ be the number of vertices of $\mathcal{A}(H)$ that lie inside Δ.

Let $1 \le r \le m$ be a parameter and Δ a simplex. A simplicial subdivision Ξ of Δ is called a $(1/r)$-*cutting* of H (with respect to Δ) if at most m/r hyperplanes of H cross any simplex of Ξ. We will use Chazelle's *hierarchical cuttings* [13] to construct a $(1/r)$-cutting Ξ of H. In this approach, one chooses a sufficiently large constant r_0 and sets $\nu = \lceil \log_{r_0} r \rceil$. One then constructs a sequence of cuttings $\Xi_0, \Xi_1, \ldots \Xi_\nu = \Xi$, where Ξ_i is a $(1/r_0^i)$-cutting of H. The initial cutting Ξ_0 is simply Δ itself. The cutting Ξ_i is obtained from Ξ_{i-1} by computing, for each $\tau \in \Xi_{i-1}$, a $(1/r_0)$-cutting Ξ_i^τ of H_τ within τ. It is shown in [13] that $|\Xi_i| \le c r_0^{di}$, for some constant $c > 0$ that only depends on d. Hence, $|\Xi| = O(r^d)$. The reason for using hierarchical cuttings is that they yield a better recurrence for $I(P,\Gamma)$, in the sense that the overhead terms that appear in the recurrence have better dependence on r than what can be achieved with the usual one-step method for constructing cuttings.

Chazelle's technique is presented only for the case of hyperplanes. However, in the planar case it also applies to other families of curves. In particular, it holds for families of pseudo-lines. The only technical difference is that, instead of simplices (i.e., triangles), one needs to use vertical pseudo-trapezoids (see, e.g., [1] for details).

A stronger bound for extendible pseudo-segments. We now return to our analysis of pseudo-segments. The bound of Lemma 3.1 is worst-case tight when $m \ge n$, so we only need to consider the case $m < n$. Let Γ_0 be the set of pseudo-lines containing the pseudo-segments of Γ. For simplicity, we may assume that no two pseudo-segments lie on the same pseudo-line. We apply the recent duality transform of Agarwal and Sharir [6], which maps Γ_0 into a set Γ_0^* of n dual points, and maps P to a set P^* of m dual x-monotone pseudo-lines, so that the above/below relationships between points and pseudo-lines are preserved.

We fix a parameter $r \ge 1$, to be determined later, and construct a hierarchical $(1/r)$-cutting Ξ of $\mathcal{A}(P^*)$, as just described. Ξ consists of $O(r^2)$ pseudo-trapezoids, and it is constructed in $\lceil \log_{r_0} r \rceil$ phases, for some constant $r_0 > 1$. Let Ξ_i be the i-th layer of the cutting; we have $|\Xi_i| \le c r_0^{2i}$. If a pseudo-trapezoid Δ of the final cutting Ξ contains more than n/r^2 points of Γ_0^*, then we split it further into subtrapezoids, each of which contains at most n/r^2 points of Γ_0^*. Let $\tilde{\Xi}$ denote the resulting $(1/r)$-cutting. Using the notation introduced above, $|\tilde{\Xi}| = O(r^2)$, $n_\Delta = |(\Gamma_0^*)_\Delta| \leq n/r^2$, and $m_\Delta = |P_\Delta^*| \leq m/r$, for every pseudo-trapezoid $\Delta \in \tilde{\Xi}$. By choosing the marking points of P generically, and by exploiting the flexibility available in drawing the dual family P^* of pseudo-lines (see [6]), we may assume that all the points of Γ_0^* lie in the interiors of the cells of each of the cuttings in the hierarchy. Thus $\sum_{\Delta \in \tilde{\Xi}} n_\Delta = n$.

Lemma 3.2. *Let τ be a pseudo-trapezoid in one of the cuttings Ξ_i. Let $P' \subseteq P$ be a subset of the marking points. Then $I(P' \setminus P_\tau, \Gamma_\tau) = O(|P'| \log^* |P'| + |\Gamma_\tau|)$.*

Proof. By definition, for any point $p \in P' \setminus P_\tau$, the dual pseudo-line p^* does not cross the pseudo-trapezoid τ, and therefore passes below all the points of $(\Gamma_0^*)_\tau$, or above all these points. In primal space, p lies below all the pseudo-lines in $(\Gamma_0)_\tau$ or above all these pseudo-lines. In particular, every such p lies in the unbounded face φ of $\mathcal{A}(\Gamma_\tau)$—since the pseudo-segments are bounded, by definition, $\mathcal{A}(\Gamma_\tau)$ has a single unbounded face.

Let γ be a connected component of $\partial\varphi$, and let n_γ be the number of pseudo-segments of Γ_τ that appear on γ. Partition γ into maximal connected portions (referred to as *blocks*), each overlapping the boundary of a single face f_p (in the entire $\mathcal{A}(\Gamma)$), for some $p \in P' \setminus P_\tau$; let m_γ be the number of blocks into which γ is partitioned. We showed in the proof of Lemma 2.4

$$\sum_\gamma n_\gamma \leq n_\tau = |\Gamma_\tau| \quad \text{and} \quad \sum_\gamma m_\gamma \leq 2(|P'| + \xi_\tau) - 4, \tag{1}$$

where ξ_τ is the number of connected components of $\partial\varphi$. As a matter of fact, the proof of Lemma 2.4 shows that this serves as an upper bound on the number of blocks δ along any subset of components, with ξ_τ replaced by the size of that subset.

Fix a connected component γ of φ, and let $m_\gamma \geq 3$ denote the number of blocks δ into which γ has been partitioned. (Components with $m_\gamma \leq 2$ will be handled separately.) Enumerate these blocks as $\delta_1, \ldots, \delta_{m_\gamma}$ in their circular counterclockwise order along γ. Encode each block δ_i as a (linear, rather than circular) sequence of the pseudo-segments that appear along δ_i in order, but (i) use different symbols for the two different sides of each pseudo-segment, and also use, if necessary, two different symbols for a side of a pseudo-segment, to account for the possible "wrap-around" of that side when the circular sequence is being linearized (see [25, Section 5.2] for details), and (ii) record only one appearance of each of these symbols in a block, even if it appears there several times. Let σ_γ denote the concatenation of these "block-sequences." By construction, the length of σ_γ is an upper bound on the contribution of γ to $I(P' \setminus P_\tau, \Gamma_\tau)$, and the sum of these lengths, over all components γ, is an upper bound for the full quantity $I(P' \setminus P_\tau, \Gamma_\tau)$. If the last symbol of a block is the same as the first symbol of the next block, we delete one of them; at most m_γ symbols are deleted. The resulting sequence is a Davenport-Schinzel sequence of order three, composed of at most $4n_\gamma$ symbols, and consisting of m_γ blocks, each composed of distinct symbols. The analysis of Davenport-Schinzel sequences of order three, as presented in [25, Section 2.2], implies that the length of σ_γ is $O(km_\gamma \alpha_k(m_\gamma) + kn_\gamma)$, for any integer k, where α_k is the inverse of the k-th Ackermann's function. Choosing $k = 3$, we obtain $|\sigma_\gamma| = O(m_\gamma \log^* m_\gamma + n_\gamma)$.

We sum this bound over all connected components γ for which $m_\gamma \geq 3$. Let t denote the number of such components. Then they contribute at least $3t$ to the left-hand side of the second inequality of (1), implying that $3t \leq 2|P'| + 2t - 4$, or $t \leq 2|P'| - 4$. Hence,

$$\sum_{m_\gamma \geq 3} m_\gamma \leq 2(|P'| + t) - 4 = O(|P'|).$$

For components γ with $m_\gamma \leq 2$, the total length of their associated sequences σ_γ is at most $8 \sum_\gamma n_\gamma$. Hence, using (1), the total length of all the sequences σ_γ is

$$O\Bigg(\sum_{m_\gamma \geq 3} (m_\gamma \log^* m_\gamma + n_\gamma) + \sum_{m_\gamma \leq 2} n_\gamma\Bigg) = O(|P'| \log^* |P'| + n_\tau),$$

and this is easily seen to imply the lemma. □

By Lemma 2.5,

$$I(P, \Gamma) = \sum_{\Delta \in \Xi} I(P, \Gamma_\Delta) = \sum_{\Delta \in \Xi} \big(I(P_\Delta, \Gamma_\Delta) + I(P \setminus P_\Delta, \Gamma_\Delta)\big). \qquad (2)$$

Instead of bounding the right-hand side directly, we use a recursive approach, based on the hierarchy of cuttings $\Xi_0, \Xi_1, \ldots, \Xi_\nu = \Xi$ that underlies the construction of Ξ.

$$\begin{aligned}
I(P, \Gamma) &\leq \sum_{\Delta \in \Xi_1} \big(I(P_\Delta, \Gamma_\Delta) + I(P \setminus P_\Delta, \Gamma_\Delta)\big) \\
&\leq \sum_{\Delta \in \Xi_1} \sum_{\tau \in \Xi_2^\Delta} \big(I(P_\tau, \Gamma_\tau) + I(P_\Delta \setminus P_\tau, \Gamma_\tau)\big) + \sum_{\Delta \in \Xi_1} a(m \log^* m + n_\Delta) \\
&\qquad\qquad\qquad\qquad\qquad\qquad\qquad\qquad \text{(by Lemma 3.2)} \\
&\leq \sum_{\tau \in \Xi_2} \big(I(P_\tau, \Gamma_\tau) + a(n_\tau + (m/r_0) \log^* m)\big) + a(n + cr_0^2 m \log^* m) \\
&\cdots \\
&\leq \sum_{\tau \in \Xi_i} I(P_\tau, \Gamma_\tau) + ian + a'm \log^* m \sum_{j=0}^{i-1} r_0^j,
\end{aligned}$$

for all $i = 1, \ldots, \lceil \log_{r_0} r \rceil$, where a is the constant in the bound of Lemma 3.2, c is the constant in the bound for the size of the cutting, and $a' = acr_0^2$. We thus obtain the following.

$$\begin{aligned} I(P,\Gamma) &\le \sum_{\tau\in\Xi} I(P_\tau,\Gamma_\tau) + O(n\log r + mr\log^* m) \\ &\le \sum_{\tau\in\tilde{\Xi}} I(P_\tau,\Gamma_\tau) + O(n\log r + mr\log^* m) \\ &\le \sum_{\tau\in\tilde{\Xi}} (K(P_\tau,\Gamma_\tau) + 2m_\tau + 2n_\tau) + O(n\log r + mr\log^* m), \end{aligned} \qquad (3)$$

where the last inequality follows from Lemma 2.4. Substituting the value of $K(P_\tau,\Gamma_\tau)$ from Lemma 3.1, and using the fact that $m_\tau \le m/r$ and $n_\tau \le n/r^2$, for each τ, and that $|\tilde{\Xi}| = O(r^2)$, we obtain

$$\begin{aligned} I(P,\Gamma) &= O(n\log r + mr\log^* m) + \sum_{\tau\in\tilde{\Xi}} O\left(m_\tau^{2/3}n_\tau^{2/3} + n_\tau^{4/3}\right) \\ &= O\left(n\log r + mr\log^* m + m^{2/3}n^{2/3} + \frac{n^{4/3}}{r^{2/3}}\right). \end{aligned}$$

Choosing $r = \lceil n/m \rceil$ and using Lemma 2.2, we have

$$K(P,\Gamma) = O\left(m^{2/3}n^{2/3} + n(\log(n/m) + \log^* n + \log\log n)\right).$$

We note that the near-linear terms dominate only when m is smaller than, or is very close to $n^{1/2}$. For such values of m, the first near-linear term is $O(n\log n)$ and thus dominates all the others. Hence we obtain the following bound, which coincides with the one in [7] for all but very small values of m.

Theorem 3.3. *The maximum complexity of m distinct faces in an arrangement of n extendible pseudo-segments in the plane is $O(m^{2/3}n^{2/3} + n\log n)$.*

We next refine Theorem 3.3, to obtain a bound that depends on the number X of intersections between the pseudo-segments of Γ. This is done using the following, fairly standard approach. Put $s = \lceil n^2/X \rceil$, and construct a $(1/s)$-cutting of $\mathcal{A}(\Gamma)$ that consists of $O(s+s^2X/n^2) = O(s)$ vertical pseudo-trapezoids, each crossed by at most n/s pseudo-segments [10]. We apply Theorem 3.3 to bound the complexity of the marked faces within each cell, add up the resulting complexity bounds, and also add the complexity of the zones of the cell boundaries to account for faces not confined to a single cell (as in [14]). The overall complexity of the zones is $O(s) \cdot O(\frac{n}{s}\alpha(\frac{n}{s})) = O(n\alpha(n))$ [25]. This leads to the following result.

Theorem 3.4. *The maximum complexity of m distinct faces in an arrangement of n extendible pseudo-segments in the plane with X intersecting pairs is $O(m^{2/3}X^{1/3} + n\log n)$.*

The case of arbitrary pseudo-segments. We next extend the analysis to the case of arbitrary x-monotone pseudo-segments. This is an easy consequence of Chan's analysis [12]. Namely, we cut the n given pseudo-segments into $O(n \log n)$ subarcs, which constitute a family of extendible pseudo-segments, and then apply Theorem 3.4 to the new collection, observing that the cuts do not change X. We thus obtain:

Theorem 3.5. *The maximum complexity of m distinct faces in an arrangement of n x-monotone pseudo-segments in the plane, with X intersecting pairs, is $O(m^{2/3}X^{1/3} + n \log^2 n)$.*

Remark. As already noted, the same bound also holds for collections of pseudo-segments that are not x-monotone, provided that each of them has only $O(1)$ locally x-extremal points. By cutting each pseudo-segment at its x-extremal points, we obtain a family of $O(n)$ x-monotone pseudo-segments, and can then apply Theorem 3.5 to the new collection.

4 The Case of Circles

In this section, we derive an improved bound on the complexity of many faces in an arrangement of circles in the plane. Let C be a set of n circles in the plane with X intersecting pairs, and let P be a set of m points, none of which lies on any of the circles. By Lemma 2.1,

$$K(P,C) = O(I(P,C)). \tag{4}$$

We first prove a weak bound on $K(P,C)$ by cutting the circles into pseudo-segments and using the results of the preceding section, and then derive an improved bound by decomposing the problem into subproblems using cuttings in dual space, similar to the approach used for pseudo-segments.

A weaker bound. Agarwal *et al.* [4] showed that any family of n circles in the plane can be cut into $O(n^{3/2}\kappa(n))$ x-monotone pseudo-segments, where $\kappa(n) = (\log n)^{O(\alpha^2(n))}$. Using the result of Chan [12], mentioned above, we can decompose each of the resulting pseudo-segments into $O(\log n)$ subarcs, that collectively constitute a family Γ of extendible pseudo-segments. Then $I_C(P,C) \le I_\Gamma(P,\Gamma)$. Using Theorem 3.4, the inequality (4), and the fact that $X = O(n^2)$ in the worst case, we obtain the following lemma (where the two logarithmic factors, one incurred by cutting the arcs further into extendible pseudo-segments, and one appearing in the bound of Theorem 3.4, are both subsumed by the factor $\kappa(n)$).

Lemma 4.1. *The maximum complexity of m distinct faces in an arrangement of n circles in the plane is $O(m^{2/3}n^{2/3} + n^{3/2}\kappa(n))$.*

If every pair of circles in C intersect, then a recent result by Agarwal *et al.* [4] shows that C can be cut into $O(n^{4/3})$ pseudo-segments, and thus into $O(n^{4/3}\log n)$ extendible pseudo-segments, which implies the following bound.

Lemma 4.2. *The maximum complexity of m distinct faces in an arrangement of n pairwise-intersecting circles in the plane is*

$$O(m^{2/3}n^{2/3} + n^{4/3}\log^2 n).$$

In Lemma 4.1, the term $n^{3/2}\kappa(n)$ becomes dominant when m is smaller than roughly $n^{5/4}$. In order to obtain an improved bound for small values of m, we (i) choose a parameter r, depending on n and m, (ii) partition C into $O(r^3)$ subsets, each of size at most n/r^3, so that the points of P lie in at most m/r distinct faces of the arrangement of each subset, in addition to faces in the common exterior or in the common interior of the circles in the subset, (iii) use Lemma 4.1 to bound the complexity of the faces in question in each subarrangement, and (iv) analyze the cost of overlaying all the subarrangements. Although this technique is similar in spirit to an analogous approach used in [9] for the case of incidences, it becomes considerably more involved when analyzing the complexity of many faces.

Decomposing into subproblems. Using hierarchical cuttings in dual space, we decompose the problem of estimating $I(P,C)$ into subproblems, each involving appropriate subsets of P and C. We use the standard lifting transformation, as in [9], to map circles to points, and points to planes, in $\mathbb{R}^3$: A circle γ of radius ρ and center (a,b) in the plane is mapped to the point $\gamma^* = (a, b, a^2+b^2-\rho^2) \in \mathbb{R}^3$, and a point $p=(\xi,\eta)$ in the plane is mapped to the plane $p^* : \; z = 2\xi x + 2\eta y - (\xi^2+\eta^2)$ in $\mathbb{R}^3$. As is easily verified, a point p lies on (resp., inside, outside) a circle γ if and only if the dual plane p^* contains (resp., passes above, below) the dual point γ^*. Let P^* denote the set of planes dual to the points of P, and let C^* denote the set of points dual to the circles of C. No three planes of P^* pass through a common line, as all planes of P^* are tangent to the paraboloid $\Pi : \; z = x^2+y^2$.

We apply the hierarchical cutting procedure, reviewed in the preceding section, to P^* and C^* in the dual 3-dimensional space, with respect to a sufficiently large simplex that contains C^* and all vertices of $\mathcal{A}(P^*)$, with a value of r that will be fixed later. Let Ξ denote the resulting hierarchical $(1/r)$-cutting. If $|C^*_\Delta| > n/r^3$ for any simplex $\Delta \in \Xi$, then we split it further into subsimplices, each containing at most n/r^3 points of P^*. This step creates at most r^3 new simplices. Let $\tilde{\Xi}$ denote the resulting cutting. The size of $\tilde{\Xi}$ is also $O(r^3)$. For each simplex $\Delta \in \tilde{\Xi}$, we have $m_\Delta = |P^*_\Delta| \le m/r$ and $n_\Delta = |C^*_\Delta| \le n/r^3$. Finally, for a simplex $\Delta \in \tilde{\Xi}$, let C_Δ be the subset of circles in C that are dual to the points of C^*_Δ, and let P_Δ denote the set of points of P dual to the planes of P^*_Δ. Since no point of P lies on any circle and we can choose them generically, we may assume that all points of C^* lie in the interiors of the simplices of the cutting. We thus have $\sum_\Delta n_\Delta = n$.

We define similar quantities for the simplices of intermediate cuttings in the hierarchy.

Obtaining the improved bound. We will follow the notation introduced above for computing a $(1/r)$-cutting. We first prove the following lemma.

Lemma 4.3. *Let Δ be a simplex in one of the cuttings Ξ_i. Let $P' \subseteq P$ be a subset of the marking points. Then $I(P' \setminus P_\Delta, C_\Delta) \leq a(|P'| + n_\Delta)$, for an absolute constant $a \geq 1$.*

Proof. For any point $p \in P' \setminus P_\Delta$, the dual plane p^* does not cross the simplex Δ. If p^* lies below (resp., above) Δ, and therefore below (resp., above) all points of C^*_Δ, then p lies in the common exterior (resp., common interior) of the circles in C_Δ. Since the complexity of the common exterior or common interior of n_Δ circles in the plane is $O(n_\Delta)$ [21], we obtain that $K(P' \setminus P_\Delta, C_\Delta) = O(n_\Delta)$. The claim now follows from Lemma 2.4. □

We proceed now in a manner similar to the case of pseudo-segments. Applying (2) to $\Xi_1, \Xi_2, \ldots$ in succession, and noticing that $\sum_{\Delta \in \Xi_i} n_\Delta = n$ for each i, we have

$$
\begin{aligned}
I(P,C) &\leq \sum_{\Delta \in \Xi_1} \left(I(P_\Delta, C_\Delta) + I(P \setminus P_\Delta, C_\Delta) \right) \\
&\leq \sum_{\Delta \in \Xi_1} I(P_\Delta, C_\Delta) + \sum_{\Delta \in \Xi_1} a(m + n_\Delta) \quad \text{(by Lemma 4.3)} \\
&\leq \sum_{\Delta \in \Xi_1} \left(\sum_{\tau \in \Xi_2^\Delta} I(P_\tau, C_\tau) + I(P_\Delta \setminus P_\tau, C_\tau) \right) + a(n + c r_0^3 m),
\end{aligned}
$$

where c is the constant of proportionality in the bound for the size of the cutting Ξ_i. Setting $a' = acr_0^3$ and using Lemma 4.3 again to bound $I(P_\Delta \setminus P_\tau, C_\tau)$, we obtain

$$
\begin{aligned}
I(P,C) &\leq \sum_{\tau \in \Xi_2} \left(I(P_\tau, C_\tau) + a\left(n_\tau + \frac{m}{r_0} \right) \right) + an + a'm \\
&\leq \sum_{\tau \in \Xi_2} I(P_\tau, C_\tau) + 2an + a'm(1 + r_0^2),
\end{aligned}
$$

because $\sum_\tau n_\tau = n$ and $|\Xi_2| \leq cr_0^6$. Continuing in this manner and recalling that for any simplex $\tau \in \Xi_j$, $m_\tau \leq m/r_0^j$, and that $|\Xi_j| \leq cr_0^{3j}$, we obtain

$$
I(P,C) \leq \sum_{\tau \in \Xi_i} I(P_\tau, C_\tau) + ian + a'm \sum_{j=0}^{i-1} r_0^{2j}.
$$

We may apply the last stage $i = \lceil \log r \rceil$ with $\tilde{\Xi}$ replacing Ξ, since $\tilde{\Xi}$ satisfies the same inequalities as does Ξ, with c replaced by $c+1$. We thus obtain

$$\begin{aligned} I(P,C) &\le \sum_{\tau \in \tilde{\Xi}} I(P_\tau, C_\tau) + O(n \log r + mr^2) \\ &= \sum_{\tau \in \tilde{\Xi}} O(K(P_\tau, C_\tau) + m_\tau + n_\tau) + O(n \log r + mr^2). \end{aligned} \tag{5}$$

Substituting the bound from Lemma 4.1 in (5) and using the inequalities $m_\tau \le m/r$, $n_\tau \le n/r^3$, and $|\Xi| = O(r^3)$, we obtain

$$\begin{aligned} I(P,C) &= \sum_{\tau \in \tilde{\Xi}} O(m_\tau^{2/3} n_\tau^{2/3} + n_\tau^{3/2} \kappa(n_\tau)) + O(mr^2 + n \log r) \\ &= O(m^{2/3} n^{2/3} r^{1/3} + (n/r)^{3/2} \kappa(n/r^3) + mr^2 + n \log r). \end{aligned}$$

Choose $r = \left\lceil n^{5/11} \kappa^{6/11}(m^3/n)/m^{4/11} \right\rceil$. Note that $1 \le r \le m$ when $n^{1/3} \le m \le cn^{5/4}\kappa^{3/2}(n)$, for an appropriate constant c. If $m < n^{1/3}$ then $K(m,n) = O(n)$, as follows, e.g., from [14]. For $m > n^{1/3}$, the term mr^2 is dominated by $(n/r)^{3/2}$. If $m = \Omega(n^{5/4}\kappa^{3/2}(n))$, we use the bound of Lemma 4.1, to conclude that in this case $K(m,n) = O(m^{2/3}n^{2/3})$. Using this and substituting the value of r for the case when $1 \le r \le m$, we have, as in [4,9],

$$I(P,C) = O(m^{2/3}n^{2/3} + m^{6/11}n^{9/11}\kappa^{2/11}(m^3/n) + n \log n).$$

In accordance with our practice, we rewrite $\kappa^{2/11}(\cdot)$ as $\kappa(\cdot)$, since both functions have a similar asymptotic form, with different constants in the exponent. Using (4), we obtain the following main result of the paper.

Theorem 4.4. *The maximum complexity of m distinct faces in an arrangement of n arbitrary circles in the plane is*

$$O(m^{2/3}n^{2/3} + m^{6/11}n^{9/11}\kappa(m^3/n) + n \log n),$$

where $\kappa(r) = (\log r)^{O(\alpha^2(r))}$.

We can extend Theorem 4.4 to obtain an upper bound for $K(m,n,X)$, which takes into account the number X of intersecting pairs of circles in C. This is done exactly as in the case of pseudo-segments. That is, put $s = \lceil n^2/X \rceil$, and construct a $(1/s)$-cutting of $\mathcal{A}(C)$ that consists of $O(s + s^2X/n^2) = O(s)$ cells, each crossed by at most n/s circles [10]. By decomposing further the cells of the cutting, if needed, we may also assume that each cell contains at most m/s points of P, while the number of cells remains $O(s)$. Apply Theorem 4.4 to bound the complexity of the marked faces within each cell, add up the resulting complexity bounds, and also add the complexity of the zones of the cell boundaries to account for faces not confined to a single cell (as in [14]). The complexity of the zones is

$$O(s) \cdot O\left(\lambda_4\left(\frac{n}{s}\right)\right) = O(s) \cdot O\left(\frac{n}{s} \cdot 2^{\alpha(n/s)}\right) = O(n \cdot 2^{\alpha(n)}).$$

This leads to the following result.

Theorem 4.5. *The maximum complexity of m distinct faces in an arrangement of n arbitrary circles in $\mathbb{R}^2$ with X intersecting pairs is* $O(m^{2/3}X^{1/3} + m^{6/11}X^{4/11}n^{1/11}\kappa(m^3X^2/n^5) + n\log(X/n))$.

The case of pairwise intersecting circles can be handled in a similar manner, using Lemma 4.2 to substitute the value of $K(P_\tau, C_\tau)$ in (5). Omitting the straightforward details, we obtain:

Theorem 4.6. *The maximum complexity of m distinct faces in an arrangement of n arbitrary pairwise-intersecting circles in the plane is*

$$O(m^{2/3}n^{2/3} + m^{1/2}n^{5/6}\log^{1/2}(m^3/n) + n\log n).$$

Remark. Note that the bound of $O(m^{3/5}n^{4/5}4^{\alpha(n)/5} + n)$, obtained by Clarkson *et al.* [14] on the complexity of m distinct faces in an arrangement of n circles in $\mathbb{R}^2$, is slightly better than the ones stated in the above three theorems, for $m \le (n^{1/3}/4^{\alpha(n)/3})\log^{5/3} n$. For example, as mentioned above, $K(m,n) = O(n)$ if $m \le n^{1/3}/4^{\alpha(n)/3}$.

5 The Case of Unit Circles

In this section we prove the following worst-case optimal bound on the maximum complexity of many faces in an arrangement of unit circles in the plane.

Theorem 5.1. *The maximum combinatorial complexity of m distinct faces in an arrangement of n unit circles in the plane with X intersecting pairs is* $\Theta(m^{2/3}X^{1/3} + n)$.

Proof. Let C be a collection of n unit circles in the plane and P a collection of m points marking (lying in the interior of) distinct faces in $\mathcal{A}(C)$. We aim to bound the total complexity of the marked faces. By Lemma 2.1, it suffices to bound the number $I = I(P,C)$ of incidences between the marked faces and the circles.

Note that $m = O(X+n)$, as the total number of faces in the arrangement is at most $2X + n + 1$, since the rightmost vertex of every bounded face is either one of the at most $2X$ arrangement vertices or one of the n rightmost points of the circles, and each point can be used only once in this manner. In the remainder of the proof we assume, without loss of generality, that the union of the circles of C is connected, so $X = \Omega(n)$ and $m = O(X)$. The analysis can easily be extended to the case in which the union is disconnected.

The analysis begins in a manner similar to that for the case of a line arrangement, as presented in [15], and its variant used in Section 3. For each circle $\gamma \in C$, we distinguish between faces touching γ "from the inside" and those that touch γ "from the outside." We construct two separate (multi-) graphs G_- and G_+ to encode the two types of face-circle incidences. The graphs are drawn as prescribed in [15], and briefly reviewed in Section 3.

More precisely, the graph G_- has P as its set of vertices. For each face-circle incidence along the "inside" of a circle $c \in C$ of $\mathcal{A}(C)$, fix a point of c on the face boundary to represent the incidence. Two consecutive representative points are connected along c and each of them is connected to the point marking the face it is incident to. The graph G_+ is constructed similarly, and encodes the "outer" incidences. The total number $|G_-| + |G_+|$ of edges in the two graphs is exactly I, by definition.

The analysis of Clarkson *et al.* [14] implies that the multiplicity of any edge of G_- is at most two. Actually, a stronger property holds: It is impossible for two distinct faces to touch three distinct unit circles on their interior sides (the argument is essentially the same as the one illustrated in Figure 5). Hence, arguing as in [15, 27], $|G_-|$ is $O(m^{2/3}X_-^{1/3} + m)$, where X_- is the number of edge crossings in G_-. Since, by construction, an edge crossing in G_- is also an intersection point of a pair of circles in C, and no two edge crossings in G_- can use the same intersection point of the same pair of circles, it follows that $X_- \leq 2X$ and $|G_-| = O(m^{2/3}X^{1/3} + m) = O(m^{2/3}X^{1/3})$ (the latter estimate follows from the fact that $m = O(X)$).

Handling the graph G_+ is somewhat more involved. It is shown in [14] that G_+ can be manipulated as follows. We first disregard the faces of the arrangement that lie outside all circles of C, if any of them are marked, because they can contribute at most $6n - 12$ (for $n \geq 3$) to $K(P, C)$ [21]. Each remaining marked face is enclosed by at least one circle of C and thus has diameter at most 2. We overlay the arrangement of the circles of C with the unit grid. Each circle meets the grid lines at most 8 times, so the total number of circle arcs that are part of the drawing of G_+ and are met by the grid lines is at most $8n$; we remove the edges corresponding to these arcs from G_+. It can now be shown (adapting the analysis given in [14]) that in what remains of G_+, the edge multiplicities are all bounded by a constant. Hence we can apply an analysis similar to that above to conclude that $|G_+|$, and thus also the overall face-circle incidence count, are $O(m^{2/3}X^{1/3} + m + n) = O(m^{2/3}X^{1/3} + n)$ (the latter estimate follows, as above, from the fact that $m = O(X)$). This completes the proof of the upper bound.

To see that the bound is tight in the worst case, consider an arrangement of n lines which has m faces whose combined complexity is $\Theta(m^{2/3}n^{2/3} + n)$ (see [22] for details). We can then "bend" the lines slightly into large but congruent circles without changing the combinatorial structure of any face. This shows that the bound is worst-case tight when $X = \Theta(n^2)$. For smaller values of X, put $k = \lceil n^2/X \rceil$, and take k copies of the preceding construction, placed far away from each other, each involving $\lfloor n/k \rfloor$ circles and $\lfloor m/k \rfloor$ faces, of combined complexity (within a single copy)

$$\Theta\left(\left(\frac{m}{k}\right)^{2/3}\left(\frac{n}{k}\right)^{2/3} + \frac{n}{k}\right).$$

Together, we have at most n congruent circles and at most m faces in their arrangement. The number of intersecting pairs is at most $k \cdot (n/k)^2 \leq X$, and the overall complexity of all the marked faces is

$$k \cdot \Theta\left(\left(\frac{m}{k}\right)^{2/3}\left(\frac{n}{k}\right)^{2/3} + \frac{n}{k}\right) = \Theta\left(\frac{m^{2/3}n^{2/3}}{k^{1/3}} + n\right) = \Theta(m^{2/3}X^{1/3} + n).$$

□

References

[1] P.K. Agarwal, T. Hagerup, R. Ray, M. Sharir, M. Smid and E. Welzl, Translating a planar object to maximize point containment, *Proc. 10th European Sympos. Algorithms*, 2002, 42–53.

[2] P. Agarwal, R. Klein, C. Knauer, S. langerman, P. Morin, M. Sharir and M. Soss, Computing the maximum detour and spanning ratio of 2 and 3-dimensional paths, trees and cycles, in preparation.

[3] P.K. Agarwal, J. Matoušek and O. Schwarzkopf, Computing many faces in arrangements of lines and segments, *SIAM J. Comput.* 27 (1998), 491–505.

[4] P.K. Agarwal, E. Nevo, J. Pach, R. Pinchasi, M. Sharir and S. Smorodinsky, Lenses in arrangements of pseudocircles and their applications, submitted for publication. A preliminary version appeared in *Proc. 18th ACM Symp. Comput. Geom.* (2002), 123–132.

[5] P.K. Agarwal and M. Sharir, Arrangements and their applications, in: *Handbook of Computational Geometry* (J.-R. Sack and J. Urrutia, eds.), North-Holland, Amsterdam, 2000, pp. 49–119.

[6] P.K. Agarwal and M. Sharir, Pseudo-line arrangements: Duality, algorithms and applications, *Proc. 13th ACM-SIAM Sympos. Discrete Algorithms*, 2002, pp. 800-809.

[7] B. Aronov, H. Edelsbrunner, L. Guibas and M. Sharir, Improved bounds on the number of edges of many faces in arrangements of line segments, *Combinatorica* 12 (1992), 261–274.

[8] B. Aronov and M. Sharir, Triangles in space, or: Building (and analyzing) castles in the air, *Combinatorica* 10 (1990), 137–173.

[9] B. Aronov and M. Sharir, Cutting circles into pseudo-segments and improved bounds on incidences, *Discrete Comput. Geom.* 28 (2002), 475–490.

[10] M. de Berg and O. Schwarzkopf, Cuttings and applications, *Intl. J. Comput. Geom. Appl.* 5 (1995), 343–355.

[11] R.J. Canham, A theorem on arrangements of lines in the plane, *Israel J. Math.* 7 (1969), 393–397.

[12] T.M. Chan, On levels in arrangements of curves, *Discrete Comput. Geom.* 29 (2003), 375–393.

[13] B. Chazelle, Cutting hyperplanes for divide-and-conquer, *Discrete Comput. Geom.* 9 (1993), 145–158.

[14] K. Clarkson, H. Edelsbrunner, L. Guibas, M. Sharir, and E. Welzl, Combinatorial complexity bounds for arrangements of curves and spheres, *Discrete Comput. Geom.* 5 (1990), 99–160.

[15] T. Dey and J. Pach, Extremal problems for geometric hypergraphs, *Discrete Comput. Geom.* 19 (1998), 473–484.

[16] H. Edelsbrunner, L. Guibas and M. Sharir, The complexity and construction of many faces in arrangements of lines and of segments, *Discrete Comput. Geom.* 5 (1990), 161–196.

[17] B. Grünbaum, Arrangements of hyperplanes, *Congr. Numer.* 3 (1971), 41–106.

[18] L. J. Guibas, M. Sharir, and S. Sifrony, On the general motion planning problem with two degrees of freedom, *Discrete Comput. Geom.*, 4 (1989), 491–521.

[19] D. Halperin and M. Sharir, Improved combinatorial bounds and efficient techniques for certain motion planning problems with three degrees of freedom, *Comput. Geom. Theory Appl.* 1(5) (1992), 269–303.

[20] S. Har-Peled, *The Complexity of Many Cells in the Overlay of Many Arrangements*, M.Sc. Thesis, Dept. Computer Science, Tel Aviv University, 1995. See also: Multicolor combination lemmas, *Comput. Geom. Theory Appls.* 12 (1999), 155–176.

[21] K. Kedem, R. Livne, J. Pach and M. Sharir, On the union of Jordan regions and collision-free translational motion amidst polygonal obstacles, *Discrete Comput. Geom.* 1 (1986), 59–71.

[22] J. Pach and P.K. Agarwal, *Combinatorial Geometry*, Wiley-Interscience, New York, 1995.

[23] J. Pach and M. Sharir, On the number of incidences between points and curves. *Combinatorics, Probability and Computing* 7 (1998), 121–127.

[24] M. Sharir, Excess in arrangements of segments, *Inform. Process. Lett.* 58 (1996), 245–247.

[25] M. Sharir and P.K. Agarwal, *Davenport Schinzel Sequences and Their Geometric Applications*, Cambridge University Press, New York, 1995.

[26] J. Spencer, E. Szemerédi and W. T. Trotter, Unit distances in the Euclidean plane, in *Graph Theory and Combinatorics* (B. Bollobás, Ed.), Academic Press, New York, NY, 1984, pp. 293–303.

[27] L. Székely, Crossing numbers and hard Erdős problems in discrete geometry, *Combinatorics, Probability and Computing* 6 (1997), 353–358.

[28] E. Szemerédi and W. Trotter, Jr., Extremal problems in discrete geometry, *Combinatorica* 3 (1983), 381–392.

About Authors

Pankaj K. Agarwal is at the Department of Computer Science, Duke University, Durham, NC 27708-0129, USA; *pankaj@cs.duke.edu*.

Boris Aronov is at the Department of Computer and Information Science, Polytechnic University, Brooklyn, NY 11201-3840, USA; *aronov@poly.edu*.

Micha Sharir is at the School of Computer Science, Tel Aviv University, Tel-Aviv 69978, Israel, and Courant Institute of Mathematical Sciences, New York University, New York, NY 10012, USA; *sharir@cs.tau.ac.il*.

Acknowledgments

Work on this paper has been supported by a joint grant from the U.S.-Israeli Binational Science Foundation. Work by Pankaj Agarwal was also supported by Army Research Office MURI grant DAAH04-96-1-0013, by a Sloan fellowship, by NSF grants ITR-333-1050, EIA-98-70724, EIA-99-72879, CCR-97-32787, and CCR-00-86013. Work by Boris Aronov was also supported by NSF Grants CCR-99-72568 and ITR CCR-00-81964. Work by Micha Sharir was also supported by NSF Grants CCR-97-32101 and CCR-00-98246, by a grant from the Israel Science Fund (for a Center of Excellence in Geometric Computing), and by the Hermann Minkowski–MINERVA Center for Geometry at Tel Aviv University.

Polyhedral Cones of Magic Cubes and Squares

Maya Ahmed
Jesús De Loera
Raymond Hemmecke

Abstract

Using computational algebraic geometry techniques and Hilbert bases of polyhedral cones we derive explicit formulas and generating functions for the number of magic squares and magic cubes.

Magic cubes and squares are very popular combinatorial objects (see [2, 19, 21] and their references). A *magic square* is a square matrix whose entries are nonnegative integers and whose row sums, column sums, and main diagonal sums add up to the same integer number s. We will call s the *magic sum* of the square. In the literature there have been many variations on the definition of magic squares. For example, one popular variation of our definition adds the restriction of using the integers $1, \ldots, n^2$ as entries (such magic squares are commonly called *natural* or *pure* and a large part of the literature consists of procedures for constructing such examples, see [2, 19, 21]), but in this article the entries of the squares will be arbitrary nonnegative integers. We will consider other kinds of restrictions instead:

Semi-magic squares is the case when only the row and column sums are considered. This apparent simplification has in fact a very rich theory and several open questions remain (see [9, 14, 26] and references within. Semi-magic squares are called magic squares in these references). *Pandiagonal magic squares* are magic squares with the additional property that any broken-line diagonal sum adds up to the same integer (see Figure 1).

There are analogous definitions in higher dimensions. A *semi-magic hypercube* is a d-dimensional $n \times n \times \cdots \times n$ array of n^d non-negative integers, which sum up to the same number s for any line parallel to some axis. A *magic hypercube* is a semi-magic cube that has the additional property that the sums of all the main diagonals, the 2^{d-1} copies of the diagonal $x_{1,1,\ldots,1}, x_{2,2,\ldots,2}, \ldots, x_{n,n,\ldots,n}$ under the symmetries of the d-cube, are also equal to the magic sum. For example, in a $2 \times 2 \times 2$ cube there are 4 diagonals with sums $x_{1,1,1}+x_{2,2,2} = x_{2,1,1}+x_{1,2,2} = x_{1,1,2}+x_{2,2,1} = x_{1,2,1}+x_{2,1,2}$. We can see a magic $3 \times 3 \times 3$ cube in Figure 2 (the number 14 is at the central

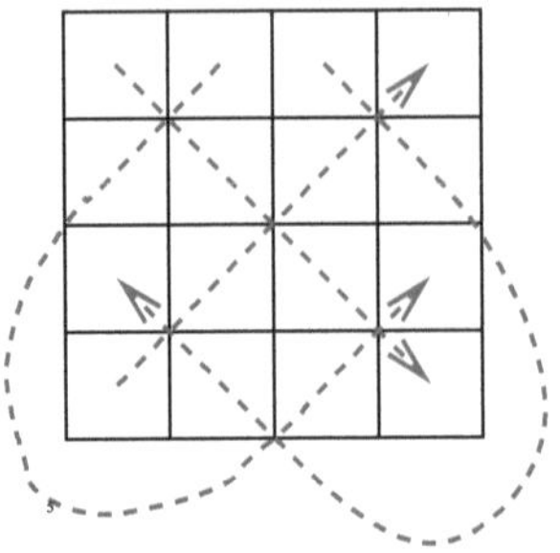

12	0	5	7
0	12	7	5
7	5	0	12
5	7	12	0

Fig. 1. Four broken diagonals of a square and a pandiagonal magic square.

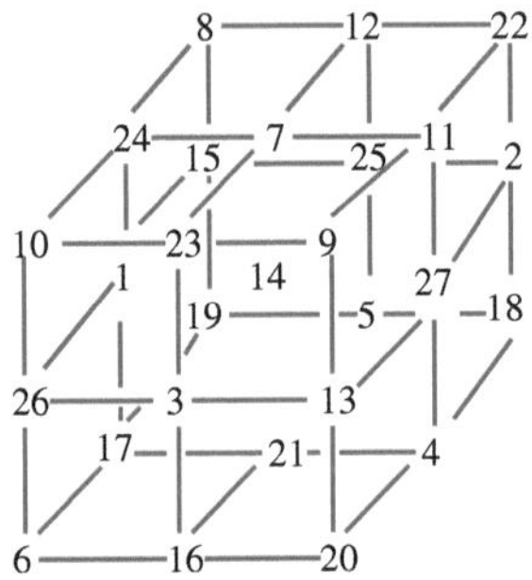

Fig. 2. A magic cube.

(2, 2, 2) position). From now on, when referring to any of these structures, we will use the terminology *magic arrays.*

Two fundamental problems about magic arrays are **(1)** enumerating such arrays and **(2)** generating particular elements. In this paper we address these two issues from a discrete-geometric perspective. The work of Ehrhart and Stanley [14, 15, 25, 26] when applied to the study of semi-magic squares showed that many enumerative and structural properties of magic arrays can actually be formulated in terms of polyhedral cones. The conditions of constant magic sum can be written in terms of a system $\{x|Ax = 0, x \geq 0\}$, where the vector x has as many entries as there are cells in the array (labeled $x_{i_1,i_2,\ldots,i_d}$), and a matrix A with entries $0, 1$ or -1 forces the different possible sums to be equal.

The purpose of this note is to study the convex polyhedral cones defined by magic squares, pandiagonal magic squares, semi-magic hypercubes, and magic hypercubes. In particular we study the Hilbert bases and extreme rays of these cones. We have used computational polyhedral geometry and commutative algebra techniques to derive explicit counting formulas for the four families of magic arrays we defined. Similar derivations had been done earlier for semi-magic squares [27, §4]. The interested reader can download the complete extreme ray information and Hilbert bases from `www.math.ucdavis.edu/~deloera/RESEARCH/magic.html`

Hilbert bases for these cones of magic arrays are special finite sets of nonnegative integer arrays that generate every other nonnegative integer array as a linear nonnegative integer combination of them. Most of our arguments will actually use minimal Hilbert bases which are smallest possible and unique [24]. Due to their size and complexity, our calculations of Hilbert bases and extreme rays were done with the help of a computer. We explain later on our algorithmic methods.

Having a Hilbert basis allows the generation of *any* magic array in the family, and makes trivial the construction of unlimited numbers of such objects or simply to list *all* magic arrays of fixed small size. Another benefit is that a Hilbert basis can be used to compute generating functions for the number of magic arrays from the computation of Hilbert series of the associated affine semigroup ring. We carry on these calculations using Gröbner bases methods. Finally minimal integer vectors along extreme rays of a cone are in fact also members of the Hilbert basis.

It is well-known from the work of Ehrhart [13] that for any rational pointed cone, if its lattice points receive a grading (e.g. by total sum of the entries, or in this case magic sum), then the function that counts the lattice points of fixed graded value is a *quasipolynomial.* A function $f : \mathbb{N} \to \mathbb{C}$ is quasipolynomial if there is an integer $N > 0$ and polynomials $f_0, \ldots, f_{N-1}$ such that $f(m) = f_i(m)$ if $m \equiv i \ (mod\ N)$. The integer N, which is not unique, will be called a *quasi-period* of f. If it is the smallest quasi-period it will be called the *period of the quasipolynomial.* It is a natural question to investigate when the quasipolynomial is actually a polynomial, i.e when the period is one. We study this question for the four families of magic arrays. It is known that for the cone of semi-magic squares the quasi-polynomial is actually a polynomial (i.e. period is one). This follows from a well-known result of Ehrhart that assures that, for integral polytopes, the function that counts lattice points inside their integral dilations is a polynomial. We prove that the same argument does not work for other magic arrays. This will involve studying the extreme rays of the various cones. In general, determining exactly the period is a delicate issue as seen from Example 4.6.27 [27].

Consider the convex hull P of real nonnegative arrays (of given size) all whose mandated sums equal 1. We call the polytope P the *polytope of stochastic magic arrays.* For example, the stochastic semi-magic squares are the well-known bistochastic matrices ($n \times n$ matrices whose row and column sums are one) and P is the famous Birkhoff-von Neumann polytope [9, 26]. It is easy to see that the polytope P can be written as $P = \{x \in R^d : x \geq 0, \text{ and } Bx = 1\}$ where the matrix B has $\{0,1\}$ entries. B has as many rows as axial sums (row, column, diagonals, etc), and the columns of B correspond to the entries of the magic array.

In [6], Bona presented a proof that the counting function of semi-magic $3\times3\times3$ cubes is a quasi-polynomial of non-trivial period. In our first theorem we extend this by actually computing an explicit generating function and quasipolynomial formulas for the number of semi-magic $3 \times 3 \times 3$ cubes.

Theorem 0.1. *Denote by $SH_n^d(s)$ the number of semi-magic d-dimensional hypercubes with n^d entries. We have the following results*

1. *From the Hilbert bases for the cones of $3 \times 3 \times 3$ semi-magic cubes we obtain the generating function.*
   $\sum_{s=0}^{\infty} SH_3^3(s)t^s = \frac{t^8+5t^7+67t^6+130t^5+242t^4+130t^3+67t^2+5t+1}{(1-t)^9(1+t)^2}$ $(= 1 + 12t + 132t^2 + 847t^3 + 3921t^4 + 14286t^5 + 43687t^6 + 116757t^7 + \ldots)$.
   In other words, $SH_3^3(s) =$

$$\begin{cases} \frac{9}{2240}s^8 + \frac{27}{560}s^7 + \frac{87}{320}s^6 + \frac{297}{320}s^5 + \frac{1341}{640}s^4 + \frac{513}{160}s^3 + \frac{3653}{1120}s^2 + \frac{627}{280}s + 1 & \text{if } 2|s, \\ \frac{9}{2240}s^8 + \frac{27}{560}s^7 + \frac{87}{320}s^6 + \frac{297}{320}s^5 + \frac{1341}{640}s^4 + \frac{513}{160}s^3 + \frac{3653}{1120}s^2 + \frac{4071}{2240}s + \frac{47}{128} & \text{otherwise.} \end{cases}$$

2. *The number of vertices of the polytope of stochastic semi-magic $n \times n \times n$ cubes is bounded below by $(n!)^{2n}/n^{n^2}$. The polytopes of stochastic $3 \times 3 \times 3 \times 3$ semi-magic $3 \times 3 \times 3$ cubes and $3 \times 3 \times 3 \times 3$ hypercubes are not integral.*

We also computed an explicit generating function for the number of $3 \times 3 \times 3$ magic cubes.

Theorem 0.2. *Let $MC_n(s)$ denote the number of $n \times n \times n$ magic cubes. Then, $MC_n(s)$ is a quasipolynomial of degree $(n-1)^3 - 4$ for $n \geq 3, n \neq 4$. For $n = 4$ it has degree $(4-1)^3 - 3 = 24$. For $n = 3$, using the minimal Hilbert basis for the cones of $3 \times 3 \times 3$ magic cubes, we computed $\sum_{s=0}^{\infty} MC_3(s)t^s = \frac{t^{12}+14\,t^9+36\,t^6+14\,t^3+1}{(1-t^3)^5}$ $(= 1 + 19\,t^3 + 121\,t^6 + 439\,t^9 + 1171\,t^{12} + 2581\,t^{15} + 4999\,t^{18} + \ldots)$. Thus, in terms of a quasipolynomial formula we have:*

$$MC_3(s) = \begin{cases} \frac{11}{324}s^4 + \frac{11}{54}s^3 + \frac{25}{36}s^2 + \frac{7}{6}s + 1 & \text{if } 3|s, \\ 0 & \text{otherwise.} \end{cases}$$

The polytope of stochastic $3 \times 3 \times 3 \times 3$ magic hypercubes is not integral.

Our next contribution is to continue the enumerative analysis done in [4]. These authors wrote down formulas for the number of magic squares of orders 3 and 4. We have corrected a minor mistake in the 4×4 formula of [4, page 8] (the 3×3 case has been known since 1915 [20]), we find further values for order 5 magic squares and we give evidence supporting one of their conjectures [4, page 9].

Theorem 0.3. *If $M_n(s)$ denotes the number of $n \times n$ magic squares of magic sum s, then , from the minimal Hilbert bases for the cones of 4×4 and 5×5 magic squares, we obtain*
$\sum_{s=0}^{\infty} M_4(s)t^s = \frac{t^8+4t^7+18t^6+36t^5+50t^4+36t^3+18t^2+4t+1}{(1-t)^4(1-t^2)^4}$ $(= 1 + 8t + 48t^2 + 200t^3 + 675t^4 + 1904t^5 + 4736t^6 + 10608t^7 + 21925t^8 + \ldots)$,

specifically we obtain that $M_4(s) =$

$$\begin{cases} \frac{1}{480}\,s^7 + \frac{7}{240}\,s^6 + \frac{89}{480}\,s^5 + \frac{11}{16}\,s^4 + \frac{49}{30}\,s^3 + \frac{38}{15}\,s^2 + \frac{71}{30}\,s + 1 & \text{if } 2|s, \\ \frac{1}{480}\,s^7 + \frac{7}{240}\,s^6 + \frac{89}{480}\,s^5 + \frac{11}{16}\,s^4 + \frac{779}{480}\,s^3 + \frac{593}{240}\,s^2 + \frac{1051}{480}\,s + \frac{13}{16} & \text{otherwise.} \end{cases}$$

We also know the values of $M_5(s)$ *for* $s \leq 6$. *The polytope of stochastic magic squares is not integral for* $n > 2$.

Finally, we continue the work started in [1, 16] for the study of pandiagonal magic squares. Here we investigate their Hilbert bases, as an application we recomputed the formulas of Halleck (see [16, Chapters 8,10]. The integrality of the polytope of panstochastic magic squares was fully solved in [1].

Theorem 0.4. *Let* $MP_n(s)$ *denote the number of* $n \times n$ *pandiagonal magic squares of magic sum* s, *then from the Hilbert bases for the cones of* 4×4 *and* 5×5 *pandiagonal magic squares we obtain*

$$MP_4(s) = \begin{cases} \frac{1}{48}(s^2 + 4s + 12)(s + 2)^2 & \text{if } 2|s, \\ 0 & \text{otherwise.} \end{cases}$$

$$MP_5(s) = \frac{1}{8064}(s + 4)(s + 3)(s + 2)(s + 1)(s^2 + 5s + 8)(s^2 + 5s + 42).$$

Here is the plan for the paper: In Section 1 we review the notion of (minimal) Hilbert bases and how we computed them. We show how to use a Hilbert basis to compute a generating function that counts the number of nonnegative integer arrays of given magic sum. In that section we recall some basic facts about polyhedral cones, Ehrhart polynomials, and commutative semigroup rings (see [11, 28]). Finally, in Section 2, we discuss the specific details for the four theorems above, each appearing in a separate subsection. We close this introduction remarking that the algebraic-geometric techniques used here are not the only useful computational tools. In fact, there has been a surge of interest on such techniques with good practical results (see [3, 12, 29]).

1 Hilbert bases for counting and element generation

Let A be an integer $d \times n$ matrix, we study pointed cones of the form $C = \{x | Ax = 0, x \geq 0\}$. A cone is *pointed*, if it does not contain any linear subspace besides the origin. It is well-known that pointed cones admit also a representation as the set of all possible nonnegative real linear combinations of finitely many vectors, the so called *extreme rays* of the cone (see page 232 of [24]). As an example we consider the cone of 3×3 magic matrices. This cone is defined by the system of equations

$$x_{11}+x_{12}+x_{13}=x_{21}+x_{22}+x_{23}=x_{31}+x_{32}+x_{33}$$
$$x_{11}+x_{12}+x_{13}=x_{11}+x_{21}+x_{31}=x_{12}+x_{22}+x_{32}=x_{13}+x_{23}+x_{33}$$
$$x_{11}+x_{12}+x_{13}=x_{11}+x_{22}+x_{33}=x_{31}+x_{22}+x_{13},$$

and the inequalities $x_{ij} \geq 0$. In our example for 3×3 magic squares the cone C has dimension 3, it is a cone based on a quadrilateral, thus it has 4 rays (see Figure 3). It is easy to see that all other cones that we will treat for magic arrays are also solutions of a system $Ax = 0, x \geq 0$, where A is a matrix with $0, 1, -1$ entries. For a given cone C we are interested in $S_C = C \cap \mathbb{Z}^n$, the *semigroup of the cone C*.

An element v of S_C is called *irreducible* if a decomposition $v = v_1 + v_2$ for $v_1, v_2 \in S_C$ implies that $v_1 = 0$ or $v_2 = 0$. A *Hilbert basis* for C is a finite set of vectors $HB(C)$ in S_C such that every other element of S_C is a positive integer combination of elements in $HB(C)$. A *minimal Hilbert basis* $HB(C)$ is inclusion minimal with respect to all other Hilbert bases of C. As a consequence all elements of the minimal Hilbert basis $HB(C)$ are irreducible and $HB(C)$ is unique.

1	0	2
2	1	0
0	2	1

2	0	1
0	1	2
1	2	0

0	2	1
2	1	0
1	0	2

1	2	0
0	1	2
2	0	1

1	1	1
1	1	1
1	1	1

Fig. 3. The Hilbert basis for the cone of 3×3 magic squares. The top four squares are the rays of the cone.

A natural question is then, *how can we compute the minimal Hilbert basis of a cone C?* Several research communities have developed algorithms for computing Hilbert bases having different applications in mind: integer programming and optimization [17], commutative algebra [7, 23, 28], and constraint programming [10, 22]. In our calculations of minimal Hilbert bases we used extensively the novel project-and-lift algorithm presented in [18] and implemented in `4ti2` by R. Hemmecke. On the other hand we were able to corroborate independently most of our results using a different algorithm, the cone decomposition algorithm, implemented in `NORMALIZ` by Bruns and Koch [7]. Similar ideas were also discussed in [27]. Now we present brief descriptions of these two methods.

Hemmecke's algorithm for computing the Hilbert basis H of a pointed rational cone C expressed as $\{z : Az = 0, z \in \mathbb{R}^n_+\}$ proceeds as follows: Let $\pi_j : \mathbb{R}^n \to \mathbb{R}^j$ be the projection onto the first j coordinates.

Let $K_j := \{\pi_j^n(v) : v \in \ker_{\mathbb{Z}}(A)\}$, $K_j^+ := K_j \cap (\mathbb{R}_+^{j-1} \times \mathbb{R}_+)$, and $K_j^- := K_j \cap (\mathbb{R}_+^{j-1} \times \mathbb{R}_-)$, where $\ker_{\mathbb{Z}}(A) = \{z : Az = 0, z \in \mathbb{Z}^n\}$ denotes the integral kernel (or null space) of A. Observe that K_j^+ and K_j^- are semi-groups under vector addition. Let H_j^+ and H_j^- denote the unique inclusion minimal generating sets of the semi-groups $(K_j^+, +)$ and $(K_j^-, +)$. Clearly, $H = H_n^+$, since $K_n^+ = C$.

The idea of the project-and-lift algorithm is to start with H_1^+, which is easy to compute, and to compute $H_{j+1}^+ \cup H_{j+1}^-$ from H_j^+. This last step is done by a completion procedure (similar to s-pair reduction in Buchberger's algorithm [11]) and is based on the fact that for any vector $v \in \ker_{\mathbb{Z}}(A)$ with $\pi_{j+1}(v) \in H_{j+1}^+ \cup H_{j+1}^-$, the vector $\pi_j(v)$ can be written as a non-negative integer linear combination of elements in H_j^+. Since many unnecessary vectors are already thrown away when H_{j+1}^+ is extracted from $H_{j+1}^+ \cup H_{j+1}^-$, intermediate results are kept comparably small and larger problems can be solved.

The cone-decomposition algorithm, used in `NORMALIZ` triangulates the cone C into finitely simplicial cones. A cone is simplicial if it is spanned by exactly n linearly independent vectors $v_1, \ldots, v_n$. There are many possible triangulations, and any of these can be used. For each simplicial cone consider the parallepiped $\Pi = \{\lambda_1 v_1 + \cdots + \lambda_n v_n \in \mathbb{Z}^n | \lambda_i \in [0,1), \}$. It is easy to see that the finite set of points $G_i = \Pi \cap \mathbb{Z}^n$ generates the semigroup. The computation of G_i can be done via direct enumeration and the knowledge that $|G_i|$ is the same as the number of cosets of the quotient of $\mathbb{Z}^n$ by the Abelian group generated by the cone generators.

This way, each simplicial cone σ_i in the triangulation of C provides us with a set of generators G_i. From the union $G = \cup G_i = \{w_1, \ldots, w_m\}$, which obviously generates $C \cap \mathbb{Z}^n$, we need to find a subset $H \subset G$ whose elements are irreducible and still generate $C \cap \mathbb{Z}^n$. The subset H is constructed recursively, starting from the empty set, in the k-th step we check if $w_k - h \in C$ for some $h \in H$. If yes, delete w_k from the list and go to the next iteration; otherwise remove all those h in H which satisfy $h - w_k \in C$ and add w_k to H before passing to the next step. Clearly, since we have the inequality representation of the cone, it is easy to decide whether a vector belongs to the cone or not.

With any d-dimensional rational pointed polyhedral cone $C = \{Ax = 0, x \geq 0\}$ and a field k we associate a *semigroup ring*, $R_C = k[y^a : a \in S_C]$, where there is one monomial $y_1^{a_1} y_2^{a_2} \ldots y_d^{a_d}$ in the ring for each element $a = (a_1, \ldots, a_d)$ of the semigroup S_C. By the definition of a Hilbert basis we know that every element of S_C can be written as a finite linear combination $\sum \mu_i h_i$ where the μ_i are nonnegative integers. Thus R_C is in fact a finitely generated k-algebra, with one generator per element of a Hilbert bases. Therefore R_C can be written as the quotient $k[x_1, x_2, \ldots, x_N]/I_C$: Once we have the Hilbert basis $H = \{h_1, \ldots, h_N\}$ for the cone C, I_C is simply the kernel of the polynomial map $\phi : k[x_1, x_2, \ldots, x_N] \longrightarrow k[y_1, y_2, \ldots y_d]$, where $\phi(x_i) = y^{h_i}$

and for $h_i = (a_1, a_2, \ldots, a_d)$ we set $y^{h_i} = y_1^{a_1} y_2^{a_2} \ldots y_d^{a_d}$. There are standard techniques for computing this kind of kernel (see [28] and references within).

It is important to observe that, for our cones of magic arrays, we can give a natural grading to R_C. A magic array can be thought of as a monomial on the ring and its degree will be its magic sum. For example, all the elements of the Hilbert bases of 3×3 magic squares are elements of degree 3. Once we have a graded k-algebra we can talk about its decomposition into the direct sum of its graded components $R_C = \bigoplus R_C(i)$, where each $R_C(i)$ collects all elements of degree i and it is a k-vector space (where $R_C(0) = k$). The function $H(R_C, i) = dim_k(R_C(i))$ is the *Hilbert function* of R_C. Similarly one can construct the *Hilbert-Poincaré series* of R_C, $H_{R_C}(t) = \sum_{i=0}^{\infty} H(R_C, i)t^i$.

Lemma 1.1. *Let C be a pointed rational cone, with Hilbert basis $\{h_1, \ldots, h_N\}$. Let the degree of a variable x_i in the ring $k[x_1, \ldots, x_N]$ be the magic sum of the its corresponding Hilbert basis element h_i. Let R_C be the (graded) semigroup ring obtained from the minimal Hilbert basis of a cone C of magic arrays. Then the number of distinct magic arrays of magic constant s equals the value of the Hilbert function $H(R_C, s)$.*

Proof: By the definition of a Hilbert basis we have that every magic array in the cone C can be written as a linear integer combination of the elements of the Hilbert basis. The elements of $HB(C) = \{h_1, h_2, \ldots, h_N\}$ are not affinely independent therefore there are different combinations that produce the same magic array. We have some dependencies of the form $\sum a_i h_i = \sum a_j h_j$ where the sums run over some subsets of $\{1, \ldots, N\}$. We consider such identities as giving a single magic array. The dependencies are precisely the elements of the toric ideal I_C, that give $R_C = k[x_1, x_2, \ldots, x_N]/I_C$. Every such dependence is a linear combination of generators of any Gröbner basis of the ideal I_C. Thus, if we encode a magic array X as a monomial in variables $x_1, \ldots, x_N$ whose exponents are the coefficients of the corresponding Hilbert basis elements that add to X, we are counting the equivalence classes modulo I_C. These are called *standard monomials.* Finally, it is known that the number of standard monomials of graded degree i equals the dimension of $R_C(i)$ as a k-vector space [11, Chapter 9]. □

It is known that the Hilbert-Poincaré series of R_C can be expressed as a rational function of the form $H_{R_C}(t) = \frac{p(t)}{\Pi_{i=1}^{r}(1-t^{\delta_i})}$. where δ_i can be read from the rays of the cone C; they correspond to the denominators of the vertices of the polytope of stochastic arrays whose dilations give the cone C (see Theorem 4.6.25 [27] and Theorem 2.3 in [26]). To compute the Hilbert-Poincaré series we relied on the computer algebra package `CoCoA` [8], that has implementations for different algorithms of Hilbert series computations [5]. The basic idea comes from the theory of Gröbner bases (see [11, §9]). It is known that the initial ideal of I_C with respect to any monomial order gives a monomial ideal J and the Hilbert functions of $k[x_1, x_2, \ldots, x_N]/I_C$ and $k[x_1, x_2, \ldots, x_N]/J$ are equal. Computing the Hilbert function of the monomial ideal J is a combinatorial problem which can be solved by an inclusion-exclusion type procedure [5] that eliminates variables at each iteration.

We illustrate the above algebraic techniques calculating a formula for the number of 3×3 magic squares, where x_5 corresponds to the matrix with all entries one, at the bottom of Figure 3, and the other 4 variables x_1, x_2, x_3, x_4 correspond to the magic squares on top of Figure 3, as they appear from left to right. The ideal I_C given by the kernel of the map is generated by the two relations $x_1x_4 - x_5^2$, $x_2x_3 - x_1x_4$. The first relation means, for example, that the sum of magic square 1 with magic square 4 is the same as twice the magic square 5. The CoCoA commands that compute the Hilbert-Poincaré series is

```
L:=[3,3,3,3,3];
Use S::=Q[x[1..5]],Weights(L);
I:=Ideal(x[1]*x[4]-x[5]^2,x[1]*x[4]-x[2]*x[3]);
Poincare(S/I);
---  Non-simplified HilbertPoincare' Series  ---
(1-2x[1]^6+x[1]^12)/((1-x[1]^3)(1-x[1]^3)(1-x[1]^3)(1-x[1]^3)(1-x[1]^3))
```

Note that to carry out the computation it is necessary to specify a weight for the variables. In our case the weights are simply the magic sums of the array. It is known that from a rational representation like this one can directly recover a quasipolynomial (see [27, §4]).

$$M_3(s) = \begin{cases} \frac{2}{9}s^2 + \frac{2}{3}s + 1 & \text{if } 3|s, \\ 0 & \text{otherwise.} \end{cases}$$

We have seen already that magic arrays are nonnegative integer solutions of a system $Ax = 0, x \geq 0$, where A is a matrix with $\{0, 1, -1\}$ entries. This system defines a pointed rational polyhedral cone C. One can set up the cone C as the union of all (real-valued) dilations of the polytope of stochastic magic arrays $P = \{x \in R^d : x \geq 0, \text{ and } Bx = 1\}$. For a positive number n we denote $E(P, n)$ the number of lattice points in the dilation $nP = \{nx | x \in P\} = \{x \in R^d : x \geq 0, \text{ and } Bx = n \cdot 1\}$. Note that when n is integer, $E(P, n)$ for the P polytope of stochastic magic arrays counts the number of integral magic arrays of magic sum n. If we let n take real values, then the union of the different dilations of P as n changes is the pointed polyhedral cone C. This is easy to see since any magic square of magic sum λ satisfies the equations $Ax = 0, x \geq 0$, thus all dilations are contained in the cone C. On the other hand, any solution x to the system of inequalities that defines C is a magic square of real valued magic sum λ. Dividing all entries of the array x by λ we obtained a magic array that satisfies the system $Bx = 1, x \geq 0$, thus the cone C is contained in the union of all dilations of P. It can be verified that the rays of the cone C are given by all scalar multiples of vertices of P. For our purposes the main result is a theorem of Ehrhart:

Lemma 1.2 ([13]). *For a rational k-polytope P embedded in $\mathbb{R}^d$, in particular the polytope of stochastic magic arrays, the counting function $E(P,n)$ is a quasipolynomial in n whose degree equals k and whose period is less than or equal to the least common multiple of the denominators of the vertices of P.*

For example, for 3×3 magic squares the vertices of the polytope of stochastic magic squares are obtained by dividing the first 4 magic squares in Figure 3 by 3. In this case the periodicity of the function is exactly three. Although in all our computations the period of the quasipolynomial turned out to be equal to the least common multiple of the denominators of the vertices of P, this is not true in general (see Example 4.6.27 in [27]).

2 Families of Magic Arrays.

2.1 Semi-magic Hypercubes: Theorem 0.1

We consider first the $3\times3\times3$ semi-magic cube. Bona [6] had already observed that a Hilbert basis must contain only elements of magic constant one and two. Here we provide the 12 Hilbert basis elements of magic constant 1. There are 54 of magic constant 2, which we are not listing here, but can be downloaded from `www.math.ucdavis.edu/~deloera/RESEARCH/magic.html`

$$
\begin{array}{c}
(0,0,1,1,0,0,0,1,0,1,0,0,0,1,0,0,0,1,0,1,0,0,0,1,1,0,0)\\
(0,0,1,0,1,0,1,0,0,0,1,0,1,0,0,0,0,1,1,0,0,0,0,1,0,1,0)\\
(1,0,0,0,0,1,0,1,0,0,1,0,1,0,0,0,0,1,0,0,1,0,1,0,1,0,0)\\
(0,0,1,1,0,0,0,1,0,0,1,0,0,0,1,1,0,0,1,0,0,0,1,0,0,0,1)\\
(1,0,0,0,1,0,0,0,1,0,0,1,1,0,0,0,1,0,0,1,0,0,0,1,1,0,0)\\
(0,0,1,0,1,0,1,0,0,1,0,0,0,0,1,0,1,0,0,1,0,1,0,0,0,0,1)\\
(0,1,0,1,0,0,0,0,1,0,0,1,0,1,0,1,0,0,1,0,0,0,0,1,0,1,0)\\
(0,1,0,1,0,0,0,0,1,1,0,0,0,0,1,0,1,0,0,0,1,0,1,0,1,0,0)\\
(1,0,0,0,1,0,0,0,1,0,1,0,0,0,1,1,0,0,0,0,1,1,0,0,0,1,0)\\
(1,0,0,0,0,1,0,1,0,0,0,1,0,1,0,1,0,0,0,1,0,1,0,0,0,0,1)\\
(0,1,0,0,0,1,1,0,0,1,0,0,0,1,0,0,0,1,0,0,1,1,0,0,0,1,0)\\
(0,1,0,0,0,1,1,0,0,0,0,1,1,0,0,0,1,0,1,0,0,0,1,0,0,0,1)
\end{array}
$$

From the Hilbert basis and using `CoCoA` to compute the number of magic cubes we obtain the stated rational generating function. Now we claim the number of vertices of the polytope of stochastic semi-magic $n\times n\times n$ cubes is bounded below by $(n!)^{2n}/n^{n^2}$. This follows from a bijection between integral stochastic semi-magic cubes and $n\times n$ latin squares: Each 2-dimensional layer or slice of the integral stochastic cubes are permutation matrices (by Birkhoff-Von Neumann theorem), the different slices or layers cannot have overlapping entries else that would violate the fact that along a line the sum of the entries equals one. Thus make the permutation coming from the first slice be the first row of the latin square, the second slice permutation gives the second row of the latin square, etc. From well-known bounds for latin squares we obtain the lower bound.

The polytope of stochastic semi-magic $3\times 3\times 3$ cubes is actually not equal to the convex hull of integral semi-magic cubes. This follows because the 54 elements of degree two in the Hilbert basis, when appropriately normalized, give rational stochastic matrices that are all vertices. In other words, the Birkhoff-von Neumann theorem [24, page 108] about stochastic semi-magic matrices is false for $3\times 3\times 3$ stochastic semi-magic cubes. Finally, the polytope of 4-dimensional semi-magic hypercubes has a non-integral vertex (each row is a 3-cube worth of values):

$$\begin{aligned}1/3*(&0,2,1,2,1,0,1,0,2,1,1,1,0,1,2,2,1,0,2,0,1,1,1,1,0,2,1,\\&2,0,1,0,1,2,1,2,0,1,2,0,1,1,1,1,0,2,0,1,2,2,1,0,1,1,1,\\&1,1,1,1,1,1,1,1,1,1,0,2,2,1,0,0,2,1,1,2,0,0,1,2,2,0,1)\end{aligned}$$

2.2 Magic Hypercubes: Theorem 0.2

The function that counts magic cubes is a quasipolynomial whose degree is the same as the dimension of the cone of magic cubes minus one. For small values (e.g $n=3,4$) we can directly compute this. We present an argument for its value for $n>4$:

Lemma 2.1. *Let B be the $(3n^2+4)\times n^3$ matrix with $0,1$ entries determining axial and diagonal sums. In this way we see that $n\times n\times n$ magic cubes of magic sum s are the integer solutions of $Bx=(s,s,\dots,s)^T, x\geq 0$. For $n>4$ the kernel of the matrix B has dimension $(n-1)^3-4$.*

Proof. It is known that for semi-magic cubes the dimension is $(n-1)^3$ [4], which means that the rank of the submatrix B' of B without the 4 rows that state diagonal sums is $n^3-(n-1)^3$. It remains to be shown that the addition of the 4 sum constraints on the main diagonals to the defining equations of the $n\times n\times n$ semi-magic cube increases the rank of the defining matrix B by exactly 4.

Let us denote the n^3 entries of the cube by $x_{1,1,1},\dots,x_{n,n,n}$ and consider the $(n-1)\times(n-1)\times(n-1)$ sub-cube with entries $x_{1,1,1},\dots,x_{n-1,n-1,n-1}$. For a semi-magic cube we have complete freedom to choose these $(n-1)^3$ entries. The remaining entries of the $n\times n\times n$ magic cube become known via the semi-magic cube equations, and all entries together form a semi-magic cube. For example: $x_{n,1,1}=-\sum_{i=1}^{n-1}x_{i,1,1}$, $x_{1,n,n}=\sum_{i=1}^{n-1}\sum_{j=1}^{n-1}x_{i,j,1}$, $x_{n,n,n}=-\sum_{i=1}^{n-1}\sum_{j=1}^{n-1}\sum_{k=1}^{n-1}x_{i,j,k}$.

However, for the magic cube, 4 more conditions have to be satisfied along the main diagonals. Employing the above semi-magic cube equations, we can rewrite these 4 equations for the main diagonals such that they involve only the variables $x_{1,1,1},\dots,x_{n-1,n-1,n-1}$. Thus, as we will see, the complete freedom of choosing values for the variables $x_{1,1,1},\dots,x_{n-1,n-1,n-1}$ is restricted by 4 independent equations. Therefore the dimension of the kernel of B is reduced by 4.

Let us consider the 3 equations in $x_{1,1,1}, \ldots, x_{n-1,n-1,n-1}$ corresponding to the main diagonals $x_{1,1,n}, \ldots, x_{n,n,1}$, $x_{1,n,1}, \ldots, x_{n,1,n}$, and $x_{n,1,1}, \ldots, x_{1,n,n}$. They are linearly independent, since the variables $x_{n-1,n-1,1}$, $x_{n-1,1,n-1}$, and $x_{1,n-1,n-1}$ appear in exactly one of these equations. The equation corresponding to the diagonal $x_{1,1,1}, \ldots, x_{n,n,n}$ is linearly independent from the other 3, because, when rewritten in terms of only variables of the form $x_{i,j,k}$ with $1 \leq i, j, k < n$, it contains the variable $x_{2,2,3}$, which for $n > 4$ does not lie on a main diagonal and is therefore not involved in one of the other 3 equations. This completes the proof. □

We consider now the $3 \times 3 \times 3$ magic cubes. There are 19 elements in the Hilbert basis and all of them have magic sum value of 3. This already indicates that there is a quasipolynomial counting formula since there are no elements of magic sum not divisible by 3.

$$(2,1,0,1,0,2,0,2,1,0,2,1,2,1,0,1,0,2,1,0,2,0,2,1,2,1,0)$$
$$(1,1,1,1,0,2,1,2,0,0,2,1,2,1,0,1,0,2,2,0,1,0,2,1,1,1,1)$$
$$(1,2,0,1,0,2,1,1,1,1,0,2,2,1,0,0,2,1,1,1,1,0,2,1,2,0,1)$$
$$(2,1,0,1,1,1,0,1,2,1,1,1,1,1,1,1,1,1,0,1,2,1,1,1,2,1,0)$$
$$(2,0,1,0,2,1,1,1,1,0,2,1,2,1,0,1,0,2,1,1,1,1,0,2,1,2,0)$$
$$(2,1,0,0,2,1,1,0,2,1,0,2,2,1,0,0,2,1,0,2,1,1,0,2,2,1,0)$$
$$(1,1,1,2,0,1,0,2,1,1,2,0,0,1,2,2,0,1,1,0,2,1,2,0,1,1,1)$$
$$(0,1,2,2,0,1,1,2,0,1,2,0,0,1,2,2,0,1,2,0,1,1,2,0,0,1,2)$$
$$(1,2,0,2,0,1,0,1,2,2,0,1,0,1,2,1,2,0,0,1,2,1,2,0,2,0,1)$$
$$(0,2,1,1,0,2,2,1,0,1,0,2,2,1,0,0,2,1,2,1,0,0,2,1,1,0,2)$$
$$(1,1,1,1,1,1,1,1,1,1,1,1,1,1,1,1,1,1,1,1,1,1,1,1,1,1,1)$$
$$(2,0,1,1,2,0,0,1,2,1,2,0,0,1,2,2,0,1,0,1,2,2,0,1,1,2,0)$$
$$(1,0,2,0,2,1,2,1,0,0,2,1,2,1,0,1,0,2,2,1,0,1,0,2,0,2,1)$$
$$(0,2,1,2,0,1,1,1,1,2,0,1,0,1,2,1,2,0,1,1,1,1,2,0,1,0,2)$$
$$(1,1,1,0,2,1,2,0,1,1,0,2,2,1,0,0,2,1,1,2,0,1,0,2,1,1,1)$$
$$(0,1,2,1,1,1,2,1,0,1,1,1,1,1,1,1,1,1,2,1,0,1,1,1,0,1,2)$$
$$(1,0,2,1,2,0,1,1,1,1,2,0,0,1,2,2,0,1,1,1,1,2,0,1,0,2,1)$$
$$(1,1,1,1,2,0,1,0,2,2,0,1,0,1,2,1,2,0,0,2,1,2,0,1,1,1,1)$$
$$(0,1,2,1,2,0,2,0,1,2,0,1,0,1,2,1,2,0,1,2,0,2,0,1,0,1,2)$$

From this information, and using CoCoA, we can derive the desired formula for the count that appears in Theorem 0.2. Finally we include below an extreme ray for the cone of magic $3 \times 3 \times 3 \times 3$ hypercubes. Dividing its entries by 15 we get a rational vertex of the polytope of stochastic magic $3 \times 3 \times 3 \times 3$ hypercubes.

8	7	0	0	8	7	7	0	8	4	4	7	5	2	8	6	9	0	3	4	8	10	5	0	2	6	7
4	4	7	5	2	8	6	9	0	1	10	4	8	5	2	6	0	9	10	1	4	2	8	5	3	6	6
3	4	8	10	5	0	2	6	7	10	1	4	2	8	5	3	6	6	2	10	3	3	2	10	10	3	2

2.3 Magic Squares: Theorem 0.3

4×4 magic squares: Our calculations using 4ti2 show that there are 20 elements in the Hilbert basis for the cone $C_{M4\times4}$ of 4×4 magic squares. The 8 elements of magic sum one (not 7 as reported in [4]) and the 12 elements of magic sum 2 are listed below. To save space we present the squares as vectors $(x_{11}, \ldots, x_{14}, x_{21}, \ldots, x_{24}, x_{31}, \ldots, x_{34}, x_{14}, \ldots, x_{44})$.

$$\begin{array}{ll}
(0,0,1,0,0,1,0,0,0,0,0,1,1,0,0,0) & (0,0,1,0,1,0,0,0,0,1,0,0,0,0,0,1) \\
(1,0,0,0,0,0,1,0,0,0,0,1,0,1,0,0) & (0,0,0,1,1,0,0,0,0,0,1,0,0,1,0,0) \\
(0,1,0,0,0,0,0,1,0,0,1,0,1,0,0,0) & (1,0,0,0,0,0,0,1,0,1,0,0,0,0,1,0) \\
(0,0,0,1,0,1,0,0,1,0,0,0,0,0,1,0) & (0,1,0,0,0,0,1,0,1,0,0,0,0,0,0,1)
\end{array}$$

These 8 permutation matrices are exactly all the magic squares of magic sum 1. The rest of the minimal Hilbert basis consists of magic sum 2 magic squares:

$$\begin{array}{ll}
(1,0,1,0,0,0,0,2,0,1,1,0,1,1,0,0) & (0,0,2,0,0,1,0,1,1,1,0,0,1,0,0,1) \\
(0,0,1,1,0,1,0,1,1,0,1,0,1,1,0,0) & (1,1,0,0,0,1,1,0,0,0,0,2,1,0,1,0) \\
(1,0,0,1,1,1,0,0,0,1,0,1,0,0,2,0) & (0,1,0,1,1,1,0,0,0,0,1,1,1,0,1,0) \\
(1,0,0,1,0,0,1,1,1,0,1,0,0,2,0,0) & (0,2,0,0,1,0,1,0,0,0,1,1,1,0,0,1) \\
(1,1,0,0,1,0,1,0,0,1,0,1,0,0,1,1) & (0,1,0,1,2,0,0,0,0,1,1,0,0,0,1,1) \\
(1,0,1,0,0,0,1,1,1,1,0,0,0,1,0,1) & (0,0,1,1,0,1,1,0,2,0,0,0,0,1,0,1)
\end{array}$$

Using CoCoA's Hilbert series computation we obtain the generating function stated in Theorem 0.3.

5×5 **magic squares** The 5×5 magic squares are the first challenging case. We were unable so far to recover the Hilbert series for this case. By using the fact that the Hilbert basis is a generating set we can easily compute several values of the Hilbert function, i.e. the numbers of magic squares for small values of the magic sum. Using the generators we consider all possible sums of them with small coefficients, making sure that repeated squares are only counted once. The values below allow us to prove that there is no polynomial formula that fits those values via interpolation. We use the Ehrhart-Macdonald reciprocity laws [26] that give us other 6 values of the function, which together with known roots allow for interpolation. There is no solution for the resulting linear system.

magic sum	total number magic squares
1	20
2	449
3	6792
4	67,063
5	484,419
6	2,750,715

The following table lists the number of elements in the Hilbert basis. All the elements for all the Hilbert bases we have computed can be obtained at `www.math.ucdavis.edu/~deloera/RESEARCH/magic.html`

magic sum	number of HB elements
1	20
2	240
3	1392
4	1584
5	1192
6	160
7	224
9	16
	4828

Finally we prove the rest of Theorem 0.3. We construct integral extreme ray vectors that, when its entries are divided by 1/2, give a fractional vertex of the polytope of stochastic magic squares: Let $n \geq 6$ and let P_{n-2} be an $(n-2) \times (n-2)$ permutation matrix that does not contain a non-zero entry on its two main diagonals. Let R_n be the $n \times n$ matrix that is constructed as follows:

- $R_{n,i,j} = 2 * P_{n-2,i-1,j-1}$ for $i = 2, \ldots, n-1$, $j = 2, \ldots, n-1$,
- $R_{n,1,j} = R_{n,n,j} = 0$ for $j = 2, \ldots, n-1$,
- $R_{n,i,1} = R_{n,i,n} = 0$ for $i = 2, \ldots, n-1$,
- $R_{n,1,1} = R_{n,n,1} = R_{n,1,n} = R_{n,n,n} = 1$.

Since $n-2 \geq 4$, there exists a permutation matrix P_{n-2} with no non-zero entries on its main diagonals. Thus, R_n is well-defined.

Lemma 2.2. *By construction, R_n is a magic square of size n with magic constant 2, in addition, for $n \geq 6$, R_n is an extremal ray of the cone of $n \times n$ magic squares.*

Proof: Suppose that R_n is not an extremal ray of the magic square cone. Therefore, there exists a non-zero magic square $\bar{R}_n$ with magic constant $s > 0$ whose support is strictly contained in the support of R_n. Since every row and column must have at least one non-zero entry, $\bar{R}_n$ must have a zero in one of the corners, that is without loss of generality, $\bar{R}_{n,1,1} = 0$. Since $s = \sum_{i=1}^{n} \bar{R}_{n,i,1} = \sum_{j=1}^{n} \bar{R}_{n,n,j}$, we obtain $s = \bar{R}_{n,n,1} = \bar{R}_{n,n,1} + \bar{R}_{n,n,n}$. Thus, $\bar{R}_{n,n,n} = 0$. But this contradicts $0 < s = \sum_{i=1}^{n} \bar{R}_{n,i,i} = 0$. Therefore, $\bar{R}_n$ does not exist, implying that R_n is an extremal ray. □

2.4 Pandiagonal Magic Squares: Theorem 0.4

Let us denote by $MP_n(s)$ the number of $n \times n$ pandiagonal magic squares with magic sum s. As in the case of magic squares the function $MP_n(s)$ is a quasipolynomial in s of degree equal to the dimension of the cone plus one.

Halleck [16] computed the dimension of the cone to be $(n-2)^2$ for odd n and $(n-2)^2+1$ for even n (degree of the quasipolynomial $MP_n(s)$ is one less than these). For the 4×4 pandiagonal magic squares a fast calculation corroborates that there are 8, magic-sum-2, generators. In his investigations, Halleck [16] identified a much larger generating set.

$$
\begin{array}{ll}
(1,0,0,1,0,1,1,0,1,0,0,1,0,1,1,0) & (1,0,1,0,0,1,0,1,0,1,0,1,1,0,1,0) \\
(0,0,1,1,1,1,0,0,0,0,1,1,1,1,0,0) & (1,0,1,0,1,0,1,0,0,1,0,1,0,1,0,1) \\
(0,1,0,1,0,1,0,1,1,0,1,0,1,0,1,0) & (0,1,1,0,1,0,0,1,0,1,1,0,1,0,0,1) \\
(1,1,0,0,0,0,1,1,1,1,0,0,0,0,1,1) & (0,1,0,1,1,0,1,0,1,0,1,0,0,1,0,1)
\end{array}
$$

From the Hilbert basis we can calculate the formula stated in Theorem 0.4 using `CoCoA`. Finally we verify that the 5×5 pandiagonal magic squares have indeed a polynomial counting formula. This case requires in fact no calculations thanks to earlier work by [1] who proved that for $n=5$ the only pandiagonal rays are precisely the pandiagonal permutation matrices. It is easy to see that only 10 of the 120 permutation matrices of order 5 are pandiagonal:

$$
\begin{array}{l}
(0,0,0,1,0,1,0,0,0,0,0,1,0,0,0,0,0,1,0,1,0,0,0) \\
(1,0,0,0,0,0,0,1,0,0,1,0,0,0,0,0,0,1,0,0,1,0,0) \\
(0,0,1,0,0,1,0,0,0,0,0,0,1,0,0,1,0,0,0,0,0,0,1) \\
(0,0,0,0,1,0,1,0,0,0,0,0,1,0,1,0,0,0,0,0,1,0,0) \\
(0,0,1,0,0,0,0,0,1,0,1,0,0,0,0,0,1,0,1,0,0,0,0) \\
(0,0,0,0,1,0,0,1,0,0,1,0,0,0,0,0,0,1,0,0,1,0,0,0) \\
(0,0,0,1,0,0,1,0,0,0,0,0,0,1,0,0,1,0,0,1,0,0,0,0) \\
(1,0,0,0,0,0,1,0,0,0,0,0,1,0,1,0,0,0,0,0,1,0) \\
(0,1,0,0,0,0,0,0,1,0,0,1,0,0,1,0,0,0,0,0,0,0,1,0) \\
(0,1,0,0,0,0,0,0,1,0,1,0,0,0,0,0,1,0,0,0,0,0,0,1)
\end{array}
$$

Once more a simple `CoCoA` calculation shows that the counting function equals indeed a polynomial, $\frac{1}{8064}(s+4)(s+3)(s+2)(s+1)(s^2+5s+8)(s^2+5s+42)$, as claimed in the Theorem.

References

[1] Alvis, D. and Kinyon, M. *Birkhoff's theorem for Panstochastic matrices*, Amer. Math. Monthly, (2001), Vol 108, no.1, 28-37.

[2] Andrews, W.S. *Magic squares and cubes*, second edition, Dover Publications, Inc., New York, N.Y. 1960.

[3] Beck, M. and Pixton D. The Ehrhart polynomial of the Birkhoff Polytope e-print: arXiv:math.CO/0202267, (2002)

[4] Beck, M., Cohen, M., Cuomo, J., and Gribelyuk, P. *The number of "magic squares" and hypercubes*, e-print: arXiv:math.CO/0201013, (2002).

[5] Bigatti, A.M *Computation of Hilbert-Poincaré Series*, J. Pure Appl. Algebra, 119/3, (1997), 237–253.

[6] Bona, M., *Sur l'enumeration des cubes magiques*, C.R.Acad.Sci.Paris Ser.I Math. 316 (1993), no.7, 633-636.

[7] Bruns, W. and Koch, R. `NORMALIZ`, *computing normalizations of affine semigroups*, Available via anonymous ftp from ftp//ftp.mathematik.uni-onabrueck.de/pub/osm/kommalg/software/

[8] Capani, A., Niesi, G., and Robbiano, L., `CoCoA`, *a system for doing Computations in Commutative Algebra*, Available via anonymous ftp from `cocoa.dima.unige.it`, (2000).

[9] Chan, C.S. and Robbins, D.P. *On the volume of the polytope of doubly stochastic matrices*, Experimental Math. Vol 8, No. 3, (1999), 291-300.

[10] Contejean, E. and Devie, H. *Resolution de systemes lineaires d'equations diophantienes* C. R. Acad. Sci. Paris, 313: 115-120, 1991. Serie I

[11] Cox, D., Little, J., and O'Shea, D. *Ideals, varieties, and Algorithms*, Springer Verlag, Undergraduate Text, 2nd Edition, 1997.

[12] De Loera, J.A. and Sturmfels B. *Algebraic unimodular counting* to appear in Mathematical Programming.

[13] Ehrhart, E. *Sur un probléme de géométrie diophantienne linéaire II*, J. Reine Angew. Math. 227 (1967), 25-49.

[14] Ehrhart, E. *Figures magiques et methode des polyedres* J.Reine Angew.Math. 299/300(1978), 51-63.

[15] Ehrhart, E. *Sur les carrés magiques*, C.R. Acad. Sci. Paris 227 A, (1973), 575-577.

[16] Halleck, E.Q. *Magic squares subclasses as linear Diophantine systems*, Ph.D. dissertation, Univ. of California San Diego, 2000, 187 pages.

[17] Henk, M. and Weismantel, R. *On Hilbert bases of polyhedral cones*, Results in Mathematics, 32, (1997), 298-303.

[18] Hemmecke, R. *On the Computation of Hilbert Bases of Cones*, in: "Mathematical Software, ICMS 2002", A.M. Cohen, X.-S. Gao, N. Takayama, eds., World Scientific, 2002. Software implementation `4ti2` available from http://www.4ti2.de.

[19] Gardner, M. *Martin Gardner's New mathematical Diversions from Scientific American*, Simon and Schuster, New York 1966. pp 162-172

[20] MacMahon, P.A. *Combinatorial Analysis*, Chelsea, 1960volumes I and II reprint of 1917 edition.

[21] Pasles, P.C. *The lost squares of Dr. Franklin*, Amer. Math. Monthly, 108, (2001), no. 6, 489-511.

[22] Pottier, L. *Bornes et algorithme de calcul des générateurs des solutions de systémes diophantiens linéaires*, C. R. Acad. Sci. Paris, 311, (1990) no. 12, 813-816.

[23] Pottier, L. *Minimal solutions of linear Diophantine systems: bounds and algorithms*, In Rewriting techniques and applications (Como, 1991), 162–173, Lecture Notes in Comput. Sci., 488, Springer, Berlin, 1991.

[24] Schrijver, A. *Theory of Linear and Integer Programming*. Wiley-Interscience, 1986.

[25] Stanley, R.P. *Linear homogeneous diophantine equations and magic labellings of graphs*, Duke Math J. 40 (1973), 607-632.

[26] Stanley, R.P. *Combinatorics and commutative algebra*, Progress in Mathematics, 41, Birkhaüser Boston, MA, 1983.

[27] Stanley, R.P. *Enumerative Combinatorics*, Volume I, Cambridge, 1997.

[28] Sturmfels, B. *Gröbner bases and convex polytopes*, university lecture series, vol. 8, AMS, Providence RI, (1996).

[29] Vergne M. and Baldoni-Silva W. *Residues formulae for volumes and Ehrhart polynomials of convex polytopes.* manuscript 81 pages. available at math.ArXiv, CO/0103097.

About Authors

Maya Ahmed, Jesús De Loera, and Raymond Hemmecke are at the Dept. of Mathematics, University of California, Davis, CA, USA, {ahmed, deloera, ramon}@math.ucdavis.edu.
This research was supported by NSF Grant DMS-0073815.

Congruent Dudeney Dissections of Triangles and Convex Quadrilaterals – All Hinge Points Interior to the Sides of the Polygons

Jin Akiyama
Gisaku Nakamura

Abstract

Let α and β be polygons with the same area. A *Dudeney dissection* of α to β is a partition of α into parts which can be reassembled to produce β in the following way. Hinge the parts of α like a chain along the perimeter of α, then fix one of the parts to form β with the perimeter of α going into its interior and with its perimeter consisting of the dissection lines in the interior of α, without turning the pieces over. In this paper we discuss a special type of Dudeney dissection of triangles and convex quadrilaterals in which α is congruent to β and call it a *congruent Dudeney dissection.* In particular, we consider the case where all hinge points are interior to the sides of the polygon α and β. For this case, we determine all triangles and convex quadrilaterals which have congruent Dudeney dissections.

1 Introduction

A geometric dissection is a cutting of a geometric figure into a finite number of pieces which can be rearranged to form another figure. Many beautiful and important results on dissections have been discovered in the last two millennia [7, 11]. Among many results on planar dissections, the following result obtained independently by Wallace [12], Bolyai [4] and Gerwien [9] is important: *An arbitrary polygon can be transformed to any other polygon of the same area by partitioning into a finite number of pieces and reassembling the pieces in some suitable way.*

Henry E. Dudeney introduced in [6] a partition of an equilateral triangle α into parts that can be reassembled in some way, without turning the pieces over, to form a square β of the same area (Figure 1.1). An examination of Dudeney's method of partition motivated us to introduce the notion of Dudeney dissection of a polygon.

Definition 1.1. Let α and β be convex polygons with the same area. A *Dudeney dissection of α to β* is a partition of α into finite number of parts which can be reassembled to produce β as follows. Hinge the parts of α on

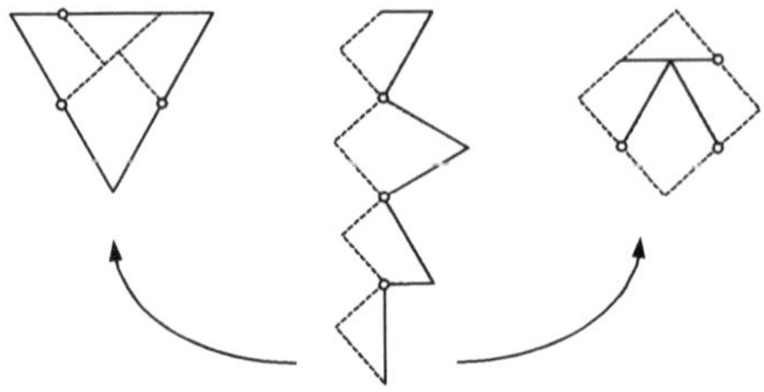

Fig. 1.1.

points interior to the sides of α, then fix one of the parts and rotate the remaining parts about the fixed part to form β in such a way that:

- All of the perimeter of α is in the interior of β.
- The perimeter of β consists of the dissection lines in the interior of α.
- The pieces of α are never turned over.

Dudeney dissections will be denoted simply by the notation DD and the dissected and hinged version of β is called a *Dudeney partner of the dissected and hinged version of* α. When no ambiguity occurs we will simply say that β is a Dudeney partner of α. Throughout this paper β will denote a Dudeney partner of α. In a related paper [1], we discussed procedures for obtaining Dudeney dissections of convex quadrilaterals to convex quadrilaterals, quadrilaterals to parallelograms, triangles to parallelograms, parallel hexagons to trapezoids, parallel hexagons to triangles, and trapezoidal pentagons to trapezoids. Frederickson's book [7, 8] surveys results involving a more general procedure called fully hinged dissections. Some of our results are related to those in [1, 2] but are obtained by different methods.

In this paper, we consider a special case of Dudeney dissection in which polygon β is congruent to α itself. That is, we *turn* α *inside-out.* We refer to such a dissection as a congruent Dudeney dissection of polygon α and denote it simply by CDD of α (Figure 1.2). We determine necessary and sufficient

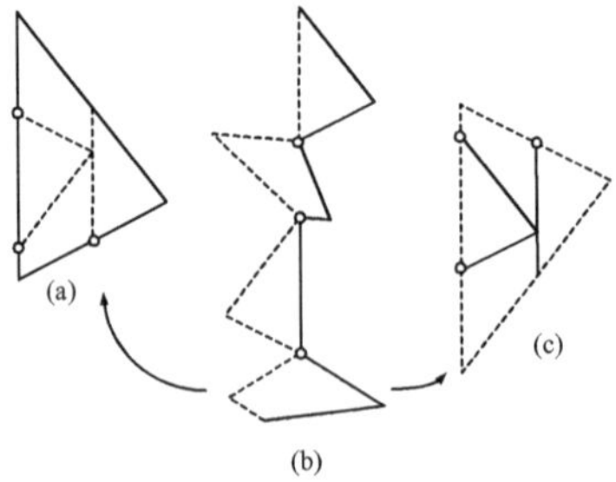

Fig. 1.2.

conditions under which such dissections exist. The discussion is confined only to convex polygons.

2 Simple Observations on Dudeney Dissection of Polygons

A congruent Dudeney dissection is a Dudeney dissection, so we begin by considering conditions under which a polygon α has a Dudeney dissection to some polygon β.

Throughout this paper, the sides of polygon α will be drawn using solid lines while the lines of the dissection will be drawn using dotted lines (Figure 2.1(a)). We refer to the dotted lines as dissection lines and the resulting parts of the polygon as *components of the dissection.*

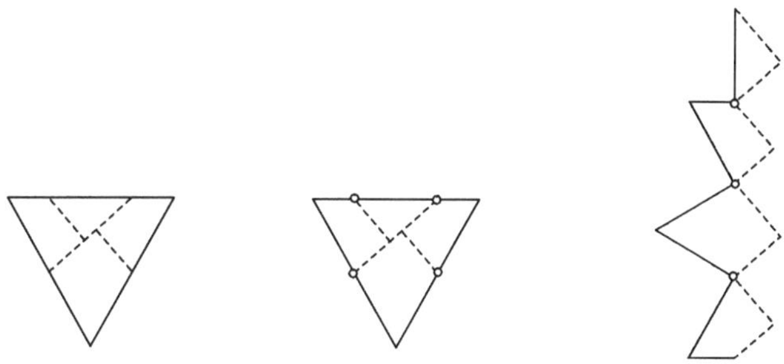

a dissection of α (a)

a hinged polygon of α (b)

a string of components of α (c)

Fig. 2.1.

Consider a dissection of α and attach hinges to all the points on the perimeter of α from which at least one dissection line emanates. Call the resulting polygon a *hinged dissection of* α. The hinge points are denoted by small circles (Figure 2.1(b)). Throughout this paper we consider Dudeney dissections of polygons in which all its hinges are located at *interior points* of the sides of the polygon. Hinge points can and will be suppressed from a hinged polygon, as appropriate, to obtain a *string of components of* α (Figure 2.1(c)).

Suppose β is a Dudeney partner of α. The following results are immediate consequences of the fact that both α and β are convex.

Proposition 2.1. *In a DD, every component of a polygon is convex.*

Proposition 2.2. *In a DD, every component is bounded by both sides of* α *and dissection lines.*

3 A relation between Dudeney dissections and tilings

Let β be a Dudeney partner of a polygon α. Since every vertex-angle of α is less than π, when α is transformed to β, each vertex of α meets at least two other vertices of α at an interior point of β, or at least one other vertex at an interior point on a dissection line or a side of β (Figure 3.1). Therefore the average size of the vertex-angles of α is less than or equal to $\frac{2}{3}\pi$.

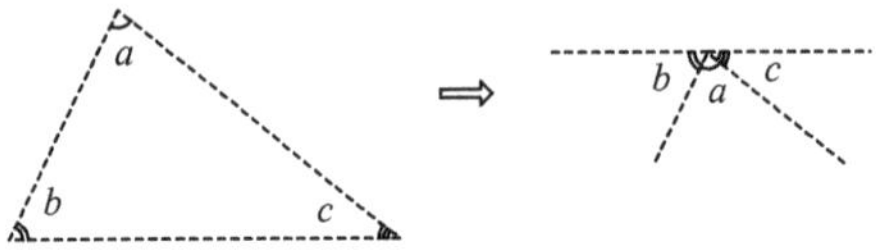

Fig. 3.1.

Theorem 3.1. *Let α be a polygon that has a DD. Then α is either a triangle, a quadrilateral, a pentagon or a hexagon.*

Proof. Suppose that α is an n-gon with a DD. The average size of the vertex angles of α is equal to $\frac{n-2}{n}\pi$. But $\frac{n-2}{n}\pi \leq \frac{2}{3}\pi$, only when $n \leq 6$. □

Let us give a few definitions which connect Dudeney dissections to tilings of the plane. Denote by α' the polygon obtained from a polygon α by rotating it by 180° around a hinge point on a side of α. Call this rotation a *half-turn* (Figure 3.2(a)). It is clear that the hinge point which is the center of rotation remains always on the perimeter of α'. A *concatenation* of α and α' is a polygon obtained by putting α and α' together along a common side (Figure 3.2(b)). Some of the simplest periodic tilings of the plane are generated by multiple translations of a polygon along two linearly independent vectors, say X and Y. The group of transformations on the plane generated by two independent translations is denoted by P_1, and periodic tilings generated by P_1-groups are also called *P_1-tilings.* A polygon α for which a P_1-tiling of the plane exists is called a *P_1-tiler.* Some other periodic tilings of the plane are generated by transformation groups generated by a P_1-group and additional transformations of the plane. One such transformation group is generated by a P_1-group and a half-turn, say H. Such a transformation group is called a *P_2-group.* Periodic tilings generated by P_2-groups will be called *P_2-tilings.* It can be shown that a plane tiling is a P_2-tiling if it has a P_1-tiler which is congruent to a concatenation of some convex polygon α and an half-turn α' of α. A convex polygon α is said to satisfy *condition* P_2 if some concatenation of α and a half-turn α' of it is a P_1-tiler. A polygon α satisfying condition P_2 is called a *P_2-tiler.* It is also known in the theory of crystallography [5] that there are 17 generating groups for periodic tilings, two of which are P_1 and P_2.

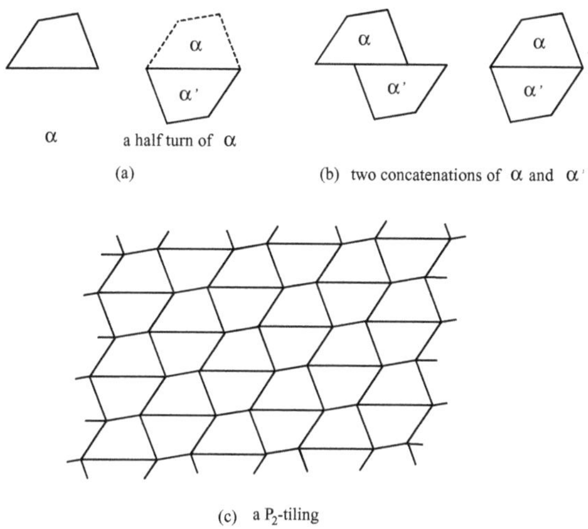

Fig. 3.2.

Let β be a Dudeney partner of α. We discuss the relationship between Dudeney dissections of α to β and two tilings of the plane, one by α and another by β. Consider a specific hexagon α which has a Dudeney dissection to β. Suppose that α can be dissected into four parts α_1, α_2, α_3 and α_4 along the dotted lines as shown in (Figure 3.3(a)). Since β is a Dudeney partner of α, β shares a common part α_1 of α, if we fix α_1 and rotate the remaining parts α_2, α_3 and α_4 about α_1 (Figure 3.3(b)). Note that the three points S, T and U as shown in Figure 3.3(b) are collinear. Fix a part α_4 of α, then rotate the remaining parts about α_4. This results in a hexagon β' congruent to β (bounded by dotted lines) on the right hand side of α, sharing α_4 in common (Figure 3.3(c)). Note that the straight segment SU in Figure 3.3(c) concides with the one in Figure 3.3(b). This implies that β's, and β' in Figure 3.3(b) and in Figure 3.3(c), respectively, do not overlap and there are no gaps between them (Figure 3.3(d)). Note that the three points S, V and W in Figure 3.3(d) are collinear. Fix a part α_3 of α and rotate the remaining parts about α_3, then we again obtain a polygon congruent to β (bounded by dotted lines) just under α_3, sharing α_3 in common and with no overlapping or gaps (Figure 3.3(e)). Finally fix a part α_2 of α and rotate the remaining parts about α_2. By repeatedly iterating this process we obtain a part of a plane tiling by β, (drawn with dotted lines (Figure 3.3(f)).

In a similar way, considering α instead of β, we can generate a sector of a plane tiling of the plane by α. Moreover, it follows from the tiling method that tilings by both α and β are P_2-tilings. It is easy to prove that the operation mentioned above can be applied to an arbitrary α and any of its Dudeney partners β, since at every stage of this procedure every part of α has a side which faces the exterior region of α and the same is true for every part of β. This example leads to the following result.

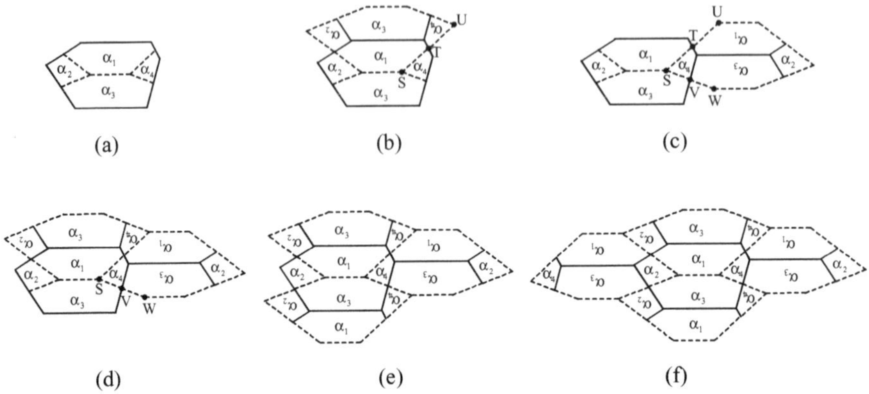

Fig. 3.3.

Theorem 3.2. *Let α be a polygon that has a Dudeney dissection. Then α satisfies the condition P_2.*

Proof. Let us prove that a convex polygon α having a Dudeney dissection can tile the plane. For this purpose, take an arbitrary hinge point lying in the interior of a side of the polygon α, and call the point H as in Figure 3.4. A dissecting line (indicated by a dotted line in Figure 3.4) emanates from H toward the interior of α, and we take an arbitrary point P different from H on this line segment. Call the two components of the dissection separated by this line of dissection u and v, respectively. Take points S and T on the side of α containing H in such a way that S lies on a side of u, and T on a side of v.

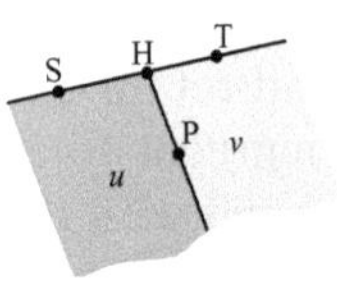

Fig. 3.4.

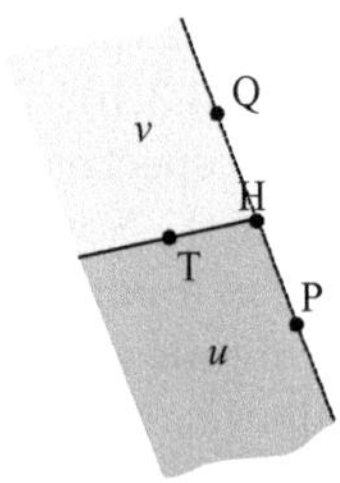

Fig. 3.5.

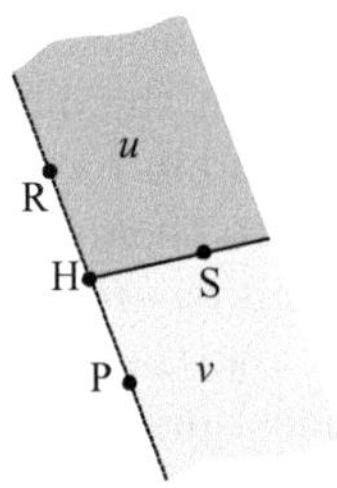

Fig. 3.6.

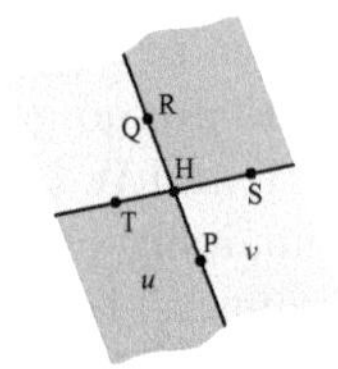

Fig. 3.7.

If we perform the procedure of Dudeney dissection by fixing the component u and rotating the remaining parts, the component v will end up exactly above u with no overlap and no gap in between, as indicated in Figure 3.5. Also, if we fix v instead and rotate the remaining parts, the component u will end up above v with no overlap and no gap in between as in Figure 3.6. The point Q in Figure 3.5 indicates where the point P would be after the rotation of the component v, and the point R in Figure 3.6 indicates where

P would be after the rotation of u. It is clear that the points P, H and Q in Figure 3.5 and the points P, H and R respectively in Figure 3.6 are collinear, and therefore if we attach the figures in Figure 3.5 and Figure 3.6 in such a way that points P and H in both figures come together, respectively, then we end up with the figure in Figure 3.7, where two components u and v are pieced together with no overlap and no gap in between. In this way we see that the configuration of Figure 3.7 is obtained by performing half-turns on both u and v simultaneously (but in opposite directions). One can apply the same procedure to any of the hinge points lying in the interior of the sides of the convex polygon α. Thus, applying this procedure to every one of the hinge points of the dissection, we end up with the situation where two pairs of adjacent dissection components (such as u and v) come together along a side of the polygon, with one pair lying within the polygon α and the other pair lying outside, in such a way that the pairs are half-turns of each other. One can furthermore draw convex polygons congruent to α, and containing copies of the dissection components which lie outside of α, in such a way that the original polygon α will be surrounded without overlapping and gaps by a finite number of polygons each of which is congruent to α'. If we repeat the process with each of the adjacent polygons, we can see that the plane can be tiled by congruent copies of α. Furthermore, since half-turns of the dissection components can be achieved at each hinge point of the dissection of α, it is clear that the tiling thus obtained is a P_2-tiling. □

According to results from tiling theory [5, 10], the only convex polygons which tile the plane are triangles, quadrilaterals, special kinds of pentagons and three different types of hexagons. This fact coincides with the statement in Theorem 3.1. In order to determine all polygons which have congruent Dudeney dissections, we proceed by discussing four cases separately: triangles, quadrilaterals, pentagons and hexagons. The latter two are discussed in [3].

4 Dudeney Dissections and Congruent Dudeney Dissections of Triangles

It is well known that every triangle satisfies condition P_2, i.e., every triangle is a P_2-tiler.

Let β be a Dudeney partner of α. In order to find a Dudeney dissection of α to β, we superimpose a plane tiling using β on another one using α (Figure 4.1). We call this method *superimposition of tilings*. Superimposition of tilings is a powerful tool for finding a dissection of α to β when α, β are given. However, since there are infinitely many ways of superimposition, care must be taken to decide how to superimpose tilings to obtain Dudeney's dissections. We refer to the resulting procedure as the *dissection structure method*. In the following proofs we combine the two methods to determine the appropriate dissections.

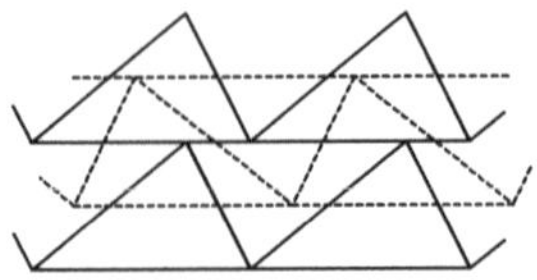

Fig. 4.1.

By a *forest of dissection lines* of a polygon α we mean a collection of dissection lines of α (Figure 4.2(a)). Forests of dissection lines which have the same adjacency pattern may yield different dissections (Figure 4.2(b)). So, we define a *dissection structure* of polygon α to be the geometrical configuration arising from a forest of dissection lines embedded in the polygon.

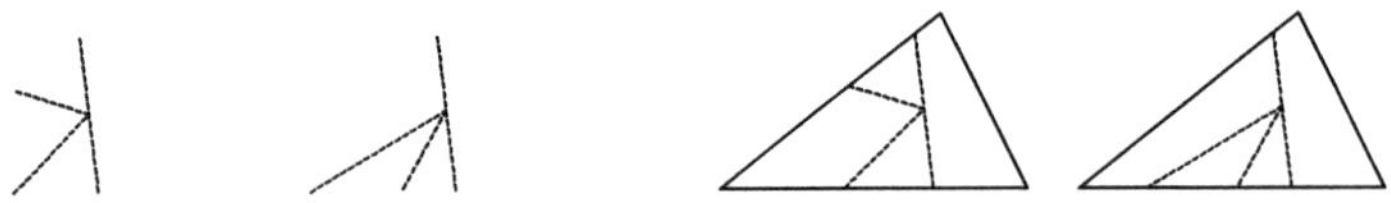

(a) forests of dissection lines (b) dissection structures of α

Fig. 4.2.

Proposition 4.1. *Suppose that a triangle α has a Dudeney partner β. Then every dissection structure of α contains the forest of dissection lines illustrated in Figure 4.3.*

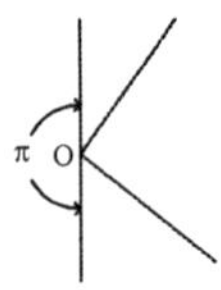

Fig. 4.3.

Proof. This follows at once from the fact that the sides of β must be made up of dissection lines of α and the three vertices of β must meet at an interior point O of α (Figure 4.3). □

Theorem 4.2. *Every triangle has a Dudeney dissection to another triangle.*

Proof. The proof is by construction. Suppose that α has a hinged dissection that reassembles to β. Since α contains the forest of dissection lines illustrated in Figure 4.3, α has exactly 4 hinge points on its sides. We divide the proof into four cases depending on how many hinge points lie on each side of α.

Case 1. One side contains 2 hinge points and each of the other two sides contains one hinge point (Figure 4.4(a)).

Case 2. Each of two sides contains 2 hinge points (Figure 4.4(b)).

Case 3. One side contains 3 hinge points and another side contains one (Figure 4.4(c)).

Case 4. One side contains 4 hinge points.

Since Case 4 cannot happen, we have only three cases to consider.

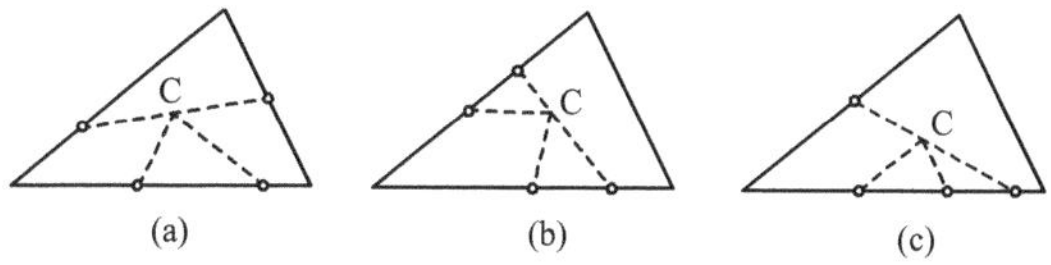

Fig. 4.4.

In all cases, α is dissected into 4 parts, but each hinged polygon corresponding to Case 2 or Case 3 has a component containing 2 vertices, say A, B of α. Then A and B are mapped to different interior points of β. It follows from the observation at the beginning of Section 3 that we map at least two vertices of α to each of these points, which is impossible. So we need to consider only Case 1 (Figure 4.4(a)).

We use the notation shown in Figure 4.5.

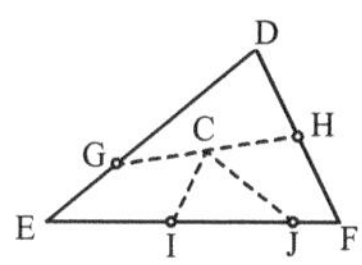

Fig. 4.5.

In order to obtain a Dudeney partner β of α, it is necessary for G and H to be the midpoints of DE and DF, respectively. Therefore EF is parallel to GH. Since component ICGE rotates about I toward IJ and component JFHC rotates about J toward JI, we have IJ = EI + JF, resulting in the dissection illustrated in Figure 4.6(a). The string of components which transforms α into β is shown in Figure 4.6.

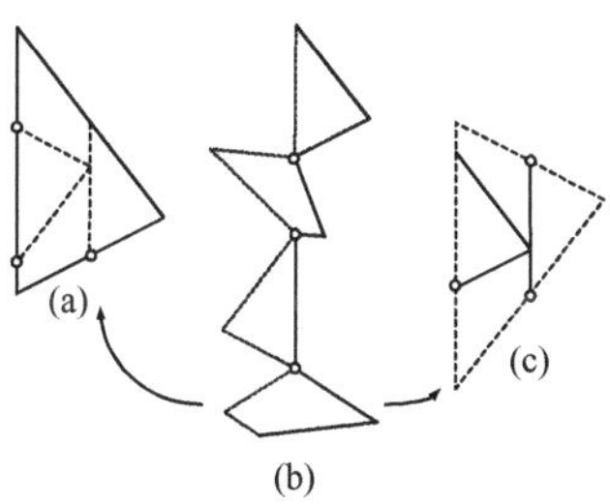

Fig. 4.6.

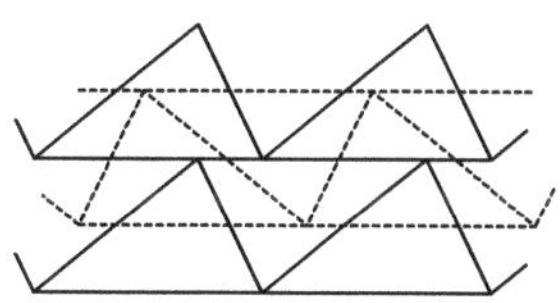

Fig. 4.7.

Figure 4.7 illustrates the tiling using α superimposed on the tiling using β. □

Theorem 4.3. *Every triangle has a congruent Dudeney dissection.*

Proof. The proof is by construction. Let α be a triangle, then by Theorem 4.2, α must have a Dudeney dissection to another triangle β. It also follows from Theorem 4.2 that α has the dissection structure illustrated in Figure 4.5.

To make β congruent to α, choose an arbitrary point C in the interior of the segment joining the midpoints G and H of ED and DF.

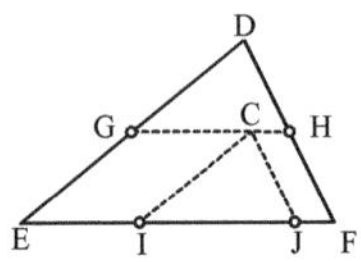

Fig. 4.8.

Draw the segment CI parallel to GE, and CJ parallel to HF (Figure 4.8). Dissect α into four parts along the segments GH, CI and CJ, and create a string of components, which define a congruent Dudeney dissection of α (Figure 4.9).

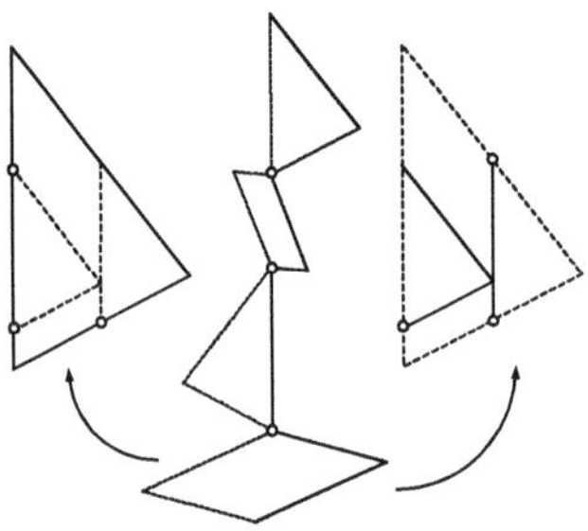

Fig. 4.9.

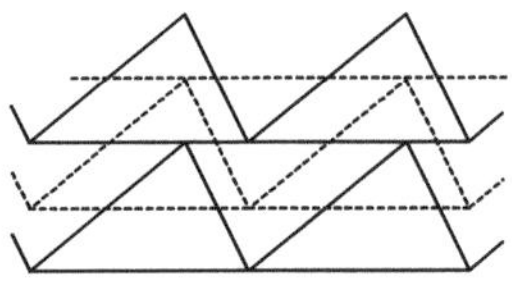

Fig. 4.10.

Figure 4.10 illustrates the tiling using α (solid lines) superimposed on the tiling of α using dotted lines, where the tiling represented by solid lines is shifted upward by half of the height of the tiling represented by dotted lines. □

5 Dudeney Dissections and Congruent Dudeney Dissections of Convex Quadrilaterals

It is well known in the theory of tiling that every quadrilateral satisfies the condition P_2, i.e. every quadrilateral is a P_2-tiler (Figure 5.1).

In the proofs that we present in this section, we again combine the two methods, *superimposition of tilings* and the *dissection structure method* in order to determine the appropriate dissections.

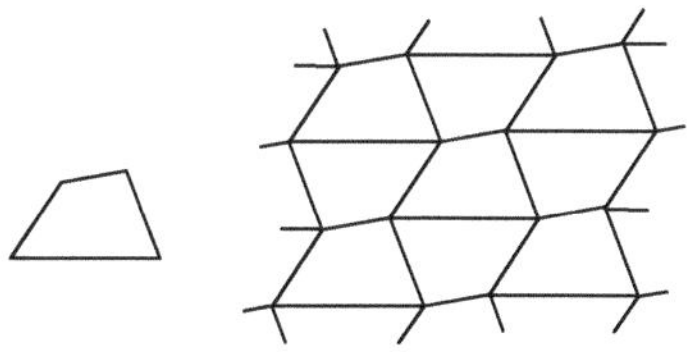

Fig. 5.1.

Proposition 5.1. *Let α be a quadrilateral which has a Dudeney dissection to another quadrilateral β. Then its dissection structure contains one of the forests of dissection lines illustrated in Figure 5.2(a), (b), (c), (d), (e) and possibly other independent dissection lines.*

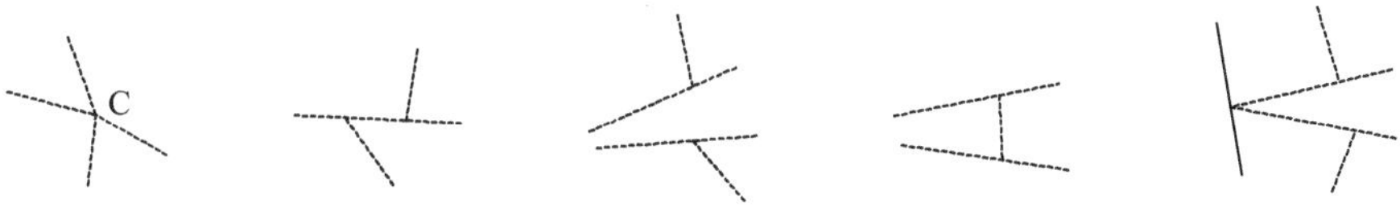

(a) F_1 (b) F_2 (c) F_3 (d) F_4 (e) F_5

Fig. 5.2. Forests of dissection lines

Proof. There are two possible ways in which the four vertices of β meet inside α. The four vertices of β can meet at one interior or boundary point C of α covering 360°about C (Figure 5.2(a)) or each of two pairs of adjacent vertices of β meet at two different interior or boundary points of α, each pair covering 180° about the specified points, respectively (Figure 5.2(b), (c), (d), (e)). □

Denote by F_1, F_2, F_3, F_4 and F_5 the forest of dissection lines in Figure 5.2(a), (b), (c), (d) and (e), respectively.

Proposition 5.2. *Let α be a quadrilateral which has a Dudeney dissection to another quadrilateral, and let each side of α have at least one hinge point. Then there are eight essentially different dissection structures for α, which are shown in Figure 5.3.*

Proof. If the forest of dissection lines of α is isomorphic to F_1, F_2, F_4 in Figure 5.2(a), (b), (d) the dissection has four hinge points. Five cases arise depending on how these four points are distributed along the sides of α.

4a: Each of the four sides of α contains one hinge point.

4b: Each of two sides of α contains one hinge point and another side contains two hinge points.

4c: Each of two sides of α contains two hinge points.

4d: A side of α contains three hinge points and another side contains one hinge point.

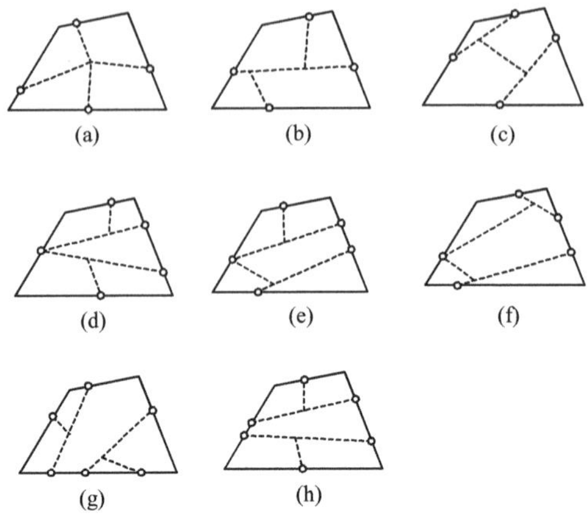

Fig. 5.3.

4e: A side of α contains four hinge points.

If the forest of dissection lines of α is isomorphic to F_5 in Fig. 5.2(e), where two dissection lines meet at a point on a side of α, there are five hinge points. The following cases arise:

5a: Each of three sides of α contains one hinge point, and the other side contains two hinge points.

5b: Each of two sides of α contains one hinge point and another side contains three hinge points.

5c: Each of two sides of α contains two hinge points and another side contains one hinge point.

5d: A side of α contains four hinge points and another side contains one hinge point.

5e: A side of α contains three hinge points and another side contains two hinge points.

5f: A side of α contains five hinge points.

If the forest of dissection lines of α is isomorphic to F_3 in Figure 5.2 (c), there will be six hinge points. When six hinge points are on the sides of α, we have the following nine cases:

6a: Each of three sides of α contains one hinge point and the other side contains three hinge points.

6b: Each of two sides of α contains one hinge point and each of the other two sides contains two hinge points.

6c: Each of two sides contains one hinge point, and another side contains four hinge points.

6d: Each of three sides of α contains one, two, and three hinge points, respectively.

6e: Each of three sides of α contains two hinge points.

6f: A side of α contains five hinge points and another side contains one hinge point.

6g: A side of α contains four hinge points and another side contains two hinge points.

6h: Each of two sides of α contains three hinge points.

6i: A side of α contains six hinge points.

Among these 20 cases, only 4a, 5a, 6a and 6b satisfy the condition that every side of α has a hinge point. Corresponding to these four cases, we exhaust all possible dissection structures. These are eight dissection structures altogether, which are illustrated in Figure 5.3. □

Refer to the dissection structures shown in Figure 5.3 (a), (b), (c), ..., (h) as A, B, ..., H, respectively.

Remark 5.3. The dissection structures corresponding to the rest cases are considered in [13].

Proposition 5.4. *There is a way of dissecting every quadrilateral α so that it has a Dudeney dissection with dissection structure A (shown in Figure 5.3(a)).*

Proof. The proof is by construction. Choose any interior point C of α and dissect α along each segment connecting C with the midpoint of each side of α (Figure 5.4). Place hinges at three of these points. The resulting string of components transforms to another quadrilateral β as illustrated in Figure 5.5. Figure 5.6 shows how two tilings, one using α drawn with solid lines and another using β drawn with dotted lines, overlap. □

Fig. 5.4.

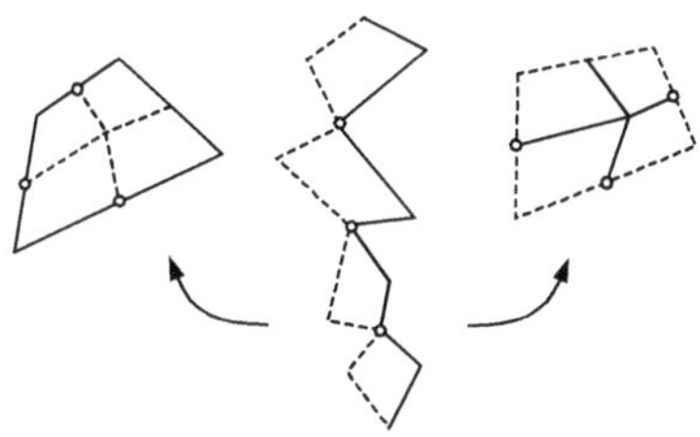

Fig. 5.5.

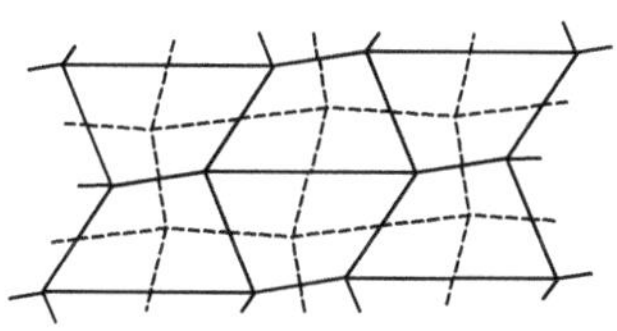

Fig. 5.6.

Theorem 5.5. *Every quadrilateral α has a congruent Dudeney dissection.*

Proof. The proof is by the superimposition method. Let T and S be a plane tilings using α as in Figure 5.7 where T is drawn using solid lines, and

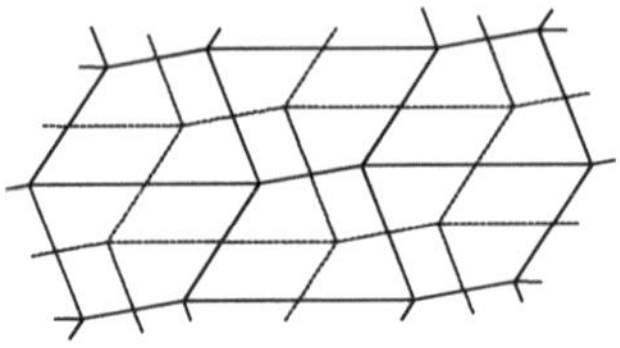

Fig. 5.7.

Fig. 5.8.

S with dotted lines. Let S and T be such that each solid side bisects a dotted side and each dotted side bisects a solid side (Figure 5.7).

This configuration gives two different congruent Dudeney dissections of α (Figure 5.8(a), (b)). Note that in each dissection two of the components are parallelograms. Figure 5.9 shows how the string of components obtained transform to α. □

Proposition 5.6. *Every quadrilateral α has a Dudeney dissection with dissection structure B (in Figure 5.3(b)).*

Proof. For α to have a Dudeney dissection, it is necessary that each of the four hinge points P, Q, R, S lies on the midpoint of a side of α (Figure 5.10). Let T and U be two distinct points which are arbitrarily chosen on the interior of the segment QS (Figure 5.10). Dissect α along the segment QS, PU and RT.

The string of components obtained by suppressing one hinge point transforms to another quadrilateral as illustrated in Figure 5.11. □

Proposition 5.7. *A quadrilateral with dissection structure B does not have a congruent Dudeney dissection.*

Proof. Let α be a quadrilateral with dissection structure B. Then its Dudeney partner β is a trapezoid since two pairs of adjacent vertex angles of β cover 180°. For the dissection to be a congruent Dudeney dissection, α must also be a trapezoid. Let $\alpha =$ CDEF and let G, H, I, J be the midpoints of FC, CD, DE and EF, respectively (Figure 5.12). For α to be congruent to β, it is necessary that the dissection line emanating from G is parallel to FJ, and the dissection line emanating from I is parallel to DH. However, these two dissection lines meet at the same point on HJ which gives dissection structure A. Hence α cannot have a dissection structure which is B. □

Proposition 5.8. *Any convex quadrilateral has a Dudeney dissection using dissection structure C.*

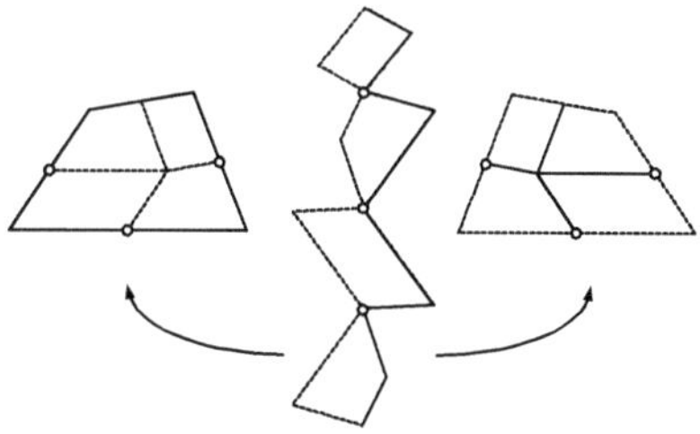

Fig. 5.9.

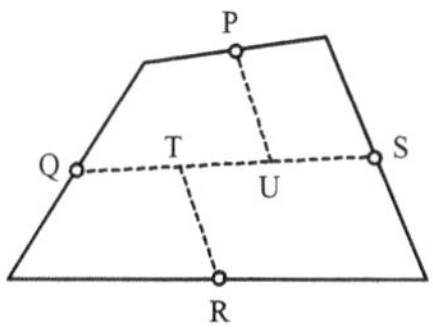

Fig. 5.10.

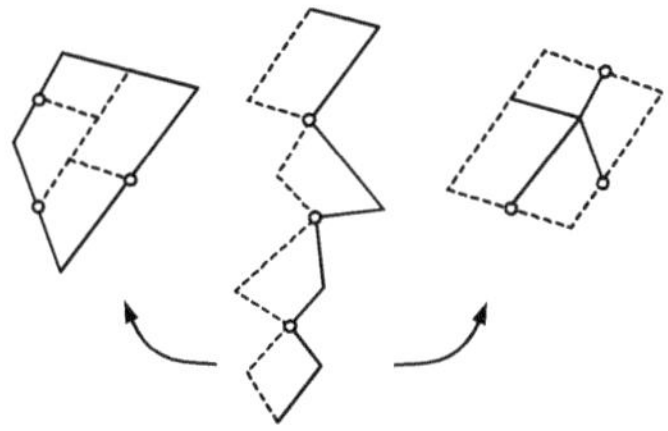

Fig. 5.11.

Proof. For α to have a Dudeney dissection, it is necessary that each of the four hinge points E, F, G, H lies on the midpoint of a side AB, BC, CD, DA of α. Choose points I and J arbitrarily on the segments HE, FG, respectively (Figure 5.13). Dissect α along the segments HE, GF and IJ. Figure 5.14 illustrates a string of components of α that transforms to α and to another quadrilateral β. Figure 5.15 shows how two tilings, one using α and the other using β can be superimposed to obtain the dissections. □

Proposition 5.9. *A quadrilateral α has a congruent Dudeney dissection with dissection structure C if and only if $\alpha = ABCD$ is a parallelogram with $BD = BC$.*

Proof. Let E, F, G, H be the midpoints of AB, BC, CD, DA, respectively (Figure 5.16). For α to be congruent to a parallelogram β, we must have BD = BC. Then the distance between the pair of parallel lines AD and BC is equal to the distance between EH and FG. Choose I and J such that ∠JIH = ∠BAD. Dissect α along the segments HE, GF and IJ. A congruent Dudeney dissection of α with dissection structure C is obtained. The above

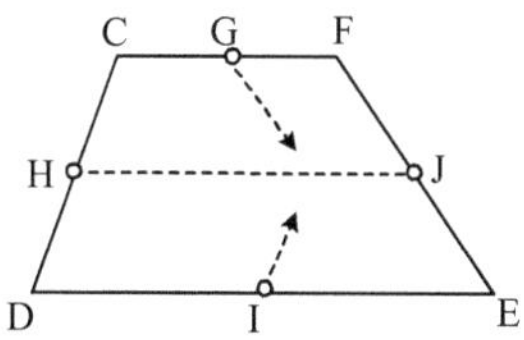

Fig. 5.12.

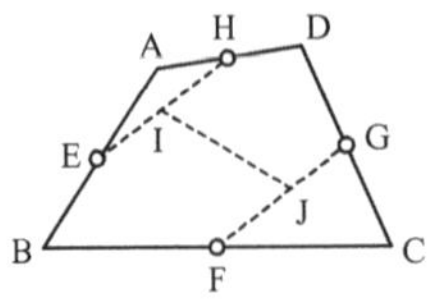

Fig. 5.13.

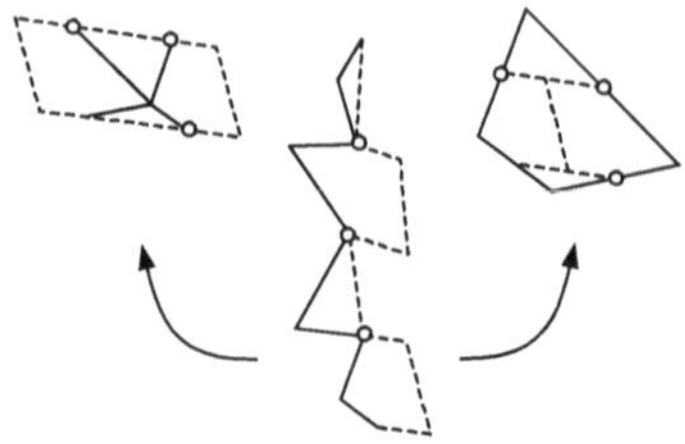

Fig. 5.14.

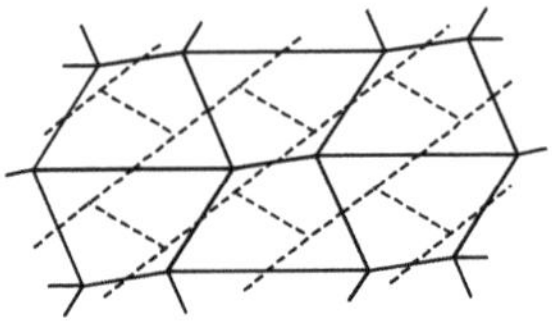

Fig. 5.15.

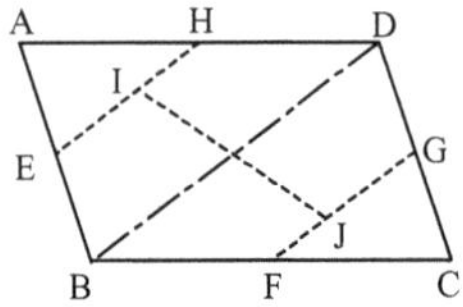

Fig. 5.16.

dissection results in a string of components which transforms to two congruent parallelograms α and β (Figure 5.17). Figure 5.18 illustrates how tilings using α and β can be superimposed to generate the dissections. □

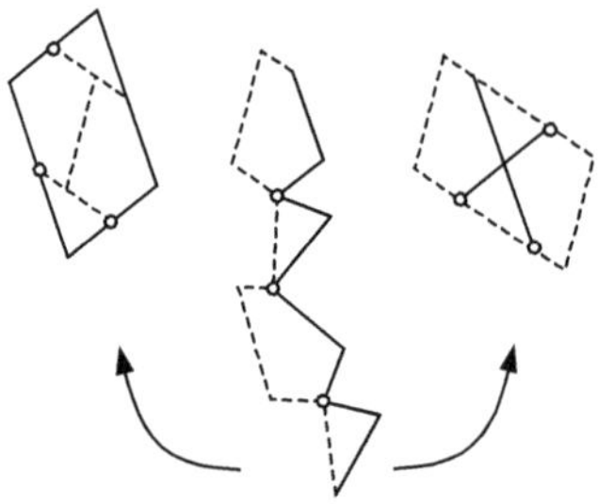

Fig. 5.17.

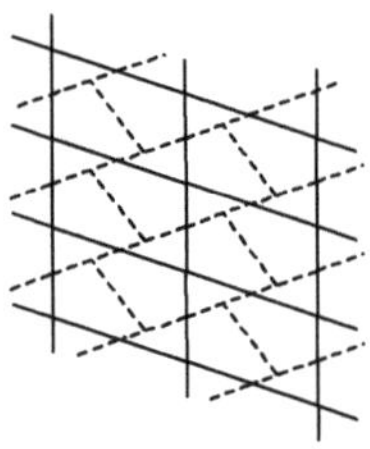

Fig. 5.18.

Refer to a vertex at which three or more dissection lines meet as a *cutpoint* of the dissection structure. In Figure 5.19, cutpoints are represented by black circles.

Proposition 5.10. *No Dudeney dissections of quadrilaterals with dissection structures D, E or F exist.*

Proof. Suppose first that such Dudeney dissection of α exists, and let β be a Dudeney partner of α. We consider two cases depending on whether the cutpoints of the dissection structure are on a side of α (Case 2) or not (Case 1).

Case 1. If α has a Dudeney dissection with structure D, E, or F, there are at least five angles (denoted by dots in Figure 5.20) which must correspond

Fig. 5.19.

to vertices of β. Since these angles correspond to different vertices of β, it has five vertices, and thus β cannot be a quadrilateral.

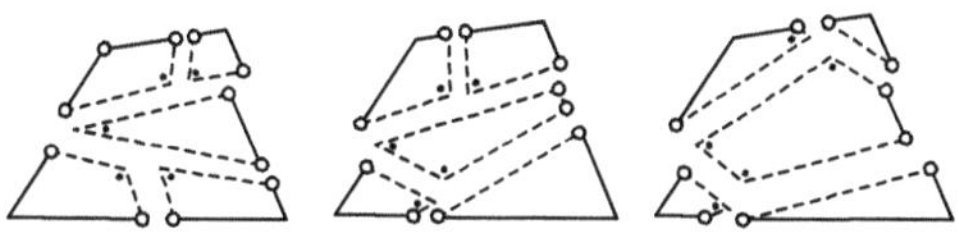

Fig. 5.20.

Case 2. It is possible that some of the cutpoints of the dissection structure lie on a side of α (Figure 5.21(a), (b), (c) and (d)).

Suppose that α has a Dudeney dissection. Then a tiling by α would also give a tiling by β by theorem 3.2. Figure 5.22 illustrates the tiling by α with the dissection structure shown in Figure 5.21(a). No matter how the dissection lines of α are adjusted, in the tiling by β there is a quadrilateral (bounded by dotted lines) which has a smaller area than α, which is a contradiction. Similar contradictions result if we assume that α has a Dudeney dissection with the dissection structure as shown in Figure 5.21(b), (c), and (d). □

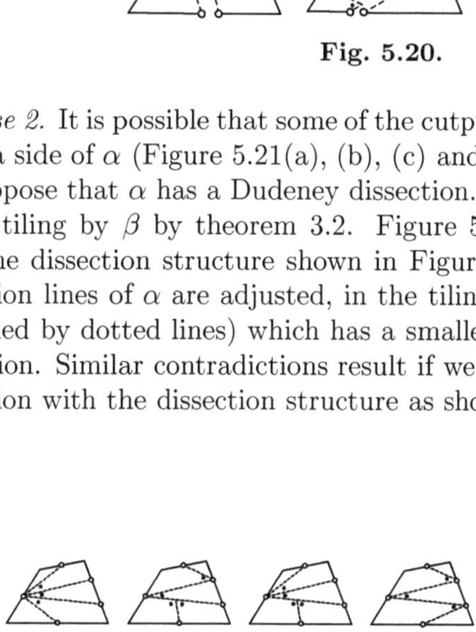

(a) (b) (c) (d)

Fig. 5.21.

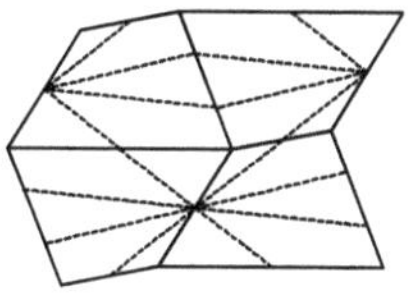

Fig. 5.22.

Proposition 5.11. *A quadrilateral α, with dissection structure G, does not have a Dudeney dissection to another quadrilateral β.*

Proof. Let α be a quadrilateral with dissection structure G, and β its corresponding Dudeney partner. Adjust the location of the hinge points of α so as to obtain a concave hexagon β (bounded by dotted lines in Figure 5.23). Five components of α coincide exactly with five components of β. Hence α has a Dudeney dissection to a hexagon β.

We try to reconfigure β to a quadrilateral by moving the hinge points of α. If point S is moved to the location M, then point R moves to location L automatically (Figure 5.24).

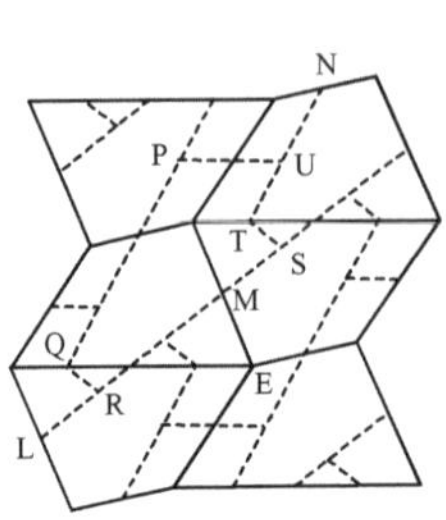

Fig. 5.23.

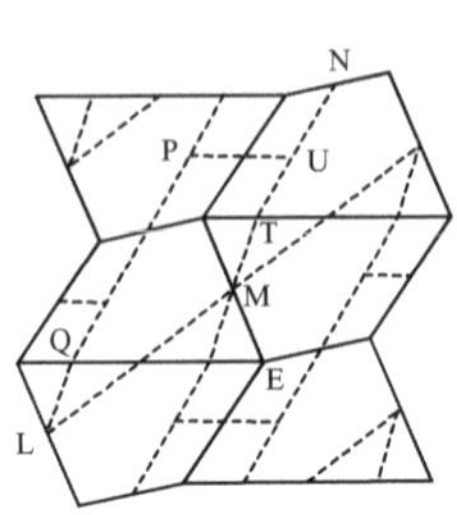

Fig. 5.24.

If the three points M, T and U are collinear so are P, Q and L, since Figure 5.24 represents a tiling of the plane by α. The three points M, T and U are collinear if and only if CD is parallel to FG (Figure 5.25). Thus we restrict the discussion only to those quadrilaterals which satisfy this condition. The tiling by α shown in Figure 5.25 gives a Dudeney dissection of α to another quadrilateral (Figure 5.26). There would, however, be a hinge point on a vertex of β (represented by black circles in Figure 5.26), which we exclude by definition. □

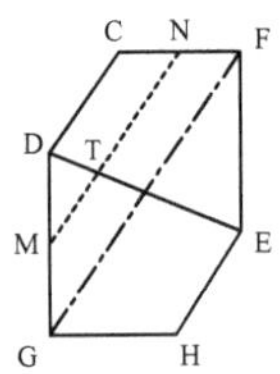

Fig. 5.25.

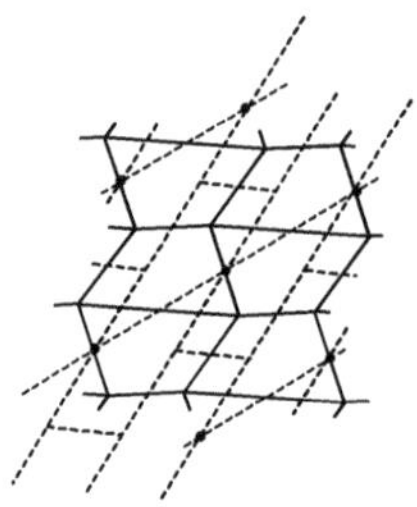

Fig. 5.26.

Proposition 5.12. *No quadrilaterals, except parallelograms, with dissection structure H have Dudeney dissections to other quadrilaterals.*

Proof. Let α be a quadrilateral with dissection structure H. We consider a plane tiling by α. No matter how the location of hinge points are adjusted, no tiling with another polygon can be obtained since there are only the strip domains (represented by the shaded portion in Figure 5.27). This implies that general quadrilaterals with dissection structure H have no Dudeney dissections to other polygons. If α is a parallelogram, however, it follows from the tiling shown in Figure 5.28 that α has a Dudeney dissection to another parallelogram (Figure 5.29). □

Proposition 5.13. *Let $n \geq 2$ be an arbitrary integer, and let α be a parallelogram of size $1 \times n$. Then α has a congruent Dudeney dissection.*

Proof. Let α be a parallelogram of size 1×2 dissected as shown in Figure 5.30, where E and F are the midpoints of AB and CD, respectively, and

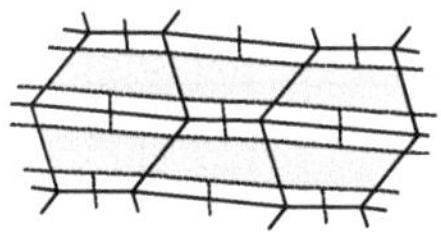

Fig. 5.27.

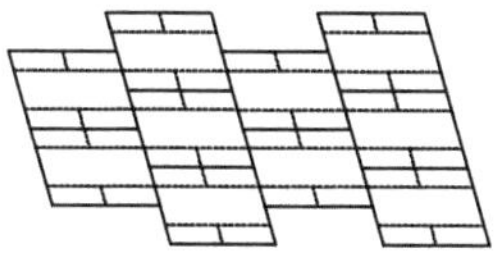

Fig. 5.28.

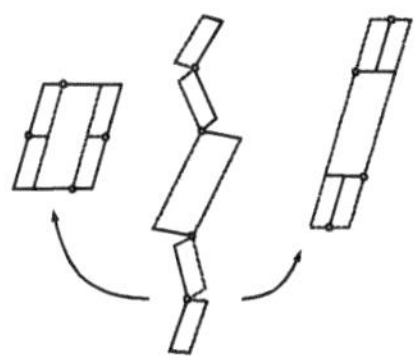

Fig. 5.29.

$AG = ID = \frac{1}{4}AD$, $BH = JC = \frac{1}{4}BC$. Then the Dudeney partner of α is its reflection (Figure 5.31). In particular, if α is a 1×2 rectangle, then we obtain a congruent Dudeney dissection of the rectangle (Figure 5.32).

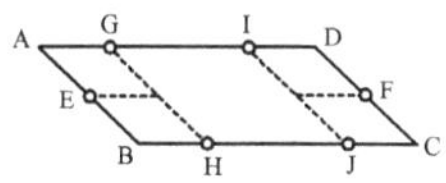

Fig. 5.30.

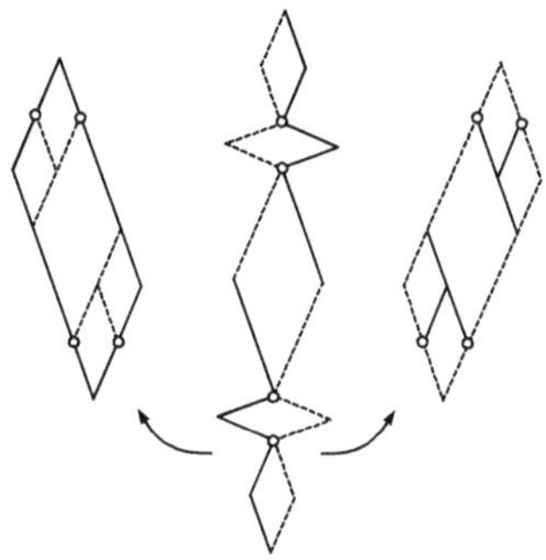

Fig. 5.31.

If α is a rectangle of size 1×4, dissect it as shown in Figure 5.33. Then the string of components of α has two additional parts which are two unit squares (Figure 5.33(b)). This is a congruent Dudeney dissection of α. It is straightforward to see that this construction generalizes to rectangles of size $1 \times n$ for every integer $n \geq 2$. □

Figure 5.34 illustrates the tiling of plane with a rectangle of size 1×4.

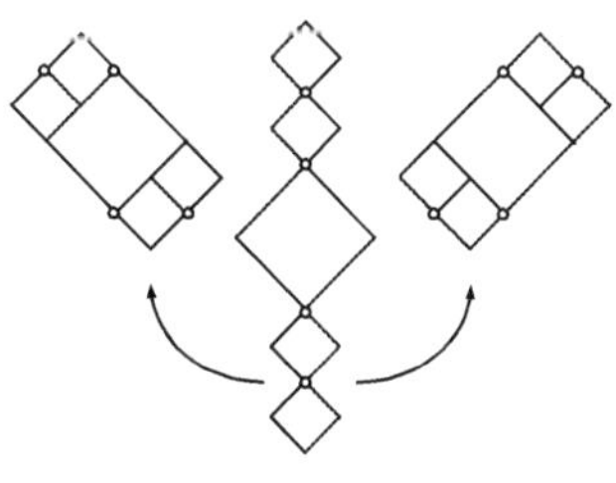

Fig. 5.32.

Remarks: In the last section of the paper, we checked all possible dissection structures of convex quadrilaterals, and determined which of them produce Dudeney dissections. Since we aimed to find congruent Dudeney

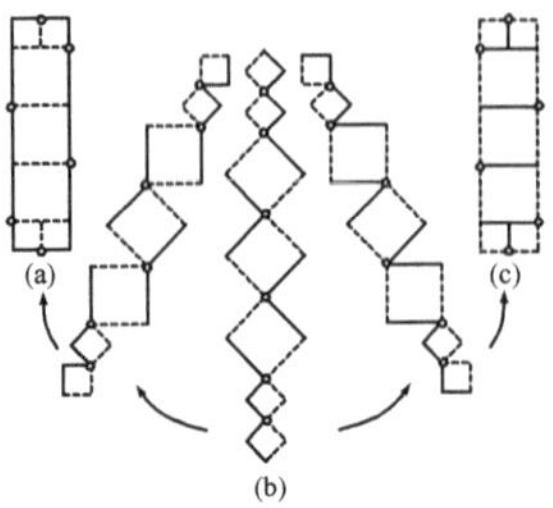

Fig. 5.33.

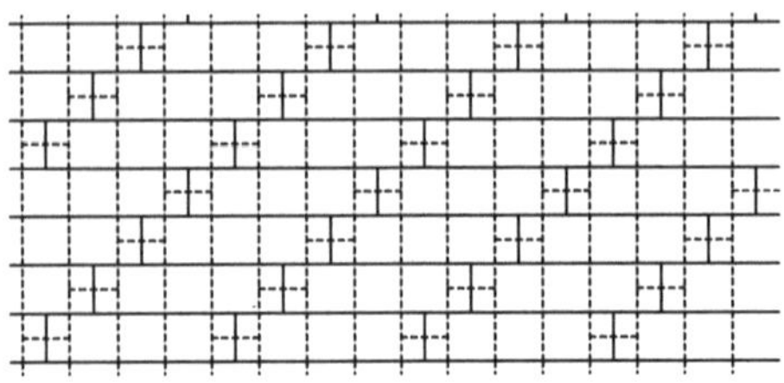

Fig. 5.34.

dissections which are applicable to most quadrilaterals, we dealt with arbitrary quadrilaterals or trapezoids, i.e, those whose sides have different lengths and which have different angles at their vertices. There are, however, many quadrilaterals with special properties such as isosceles trapezoids, parallelograms, rhombuses, rectangles and squares, which may have Dudeney dissections not arising from our constructions due to their own pecularities. For example, Figure 5.35 shows a congruent Dudeney dissection of a rectangle of size 1×2, which was was not obtained by the methods used in this section.

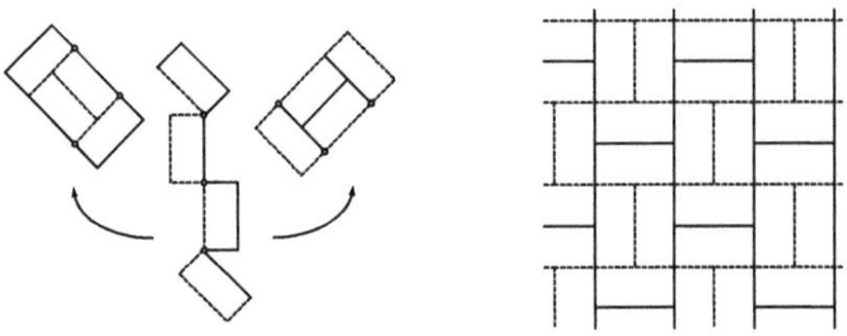

(a) (b)

Fig. 5.35.

References

[1] J. Akiyama and G. Nakamura, Dudeney dissection of polygons, *Discrete and Computational Geometry*, Japanese Conference, JCDCG 1998 (J. Akiyama et. al. (eds.)), Lecture Notes in Computer Science, Vol. 1763, pp.14-29, Springer-Verlag, 2000.

[2] J. Akiyama and G. Nakamura, Dudeney dissections of polygons and polyhedrons –A Survey, *Discrete and Computational Geometry*, Japanese Conference, JCDCG 2000, (J. Akiyama et al. (eds.)), Lecture Notes in Computer Science, Vol. 2098, pp.1-30, Springer-Verlag, 2001.

[3] J. Akiyama and G. Nakamura, Determination of all convex polygons which are chameleons – Congruent Dudeney dissections of polygons –, IEICE Trans. Fundamentals, Vol. E86-A, No. 5, pp. 978–986, 2003.

[4] F. Bolyai, *Tentamen juventutem*, Typis Collegii Reformatorum per Josephum et Simeonem Kali (in Hungarian), 1832.

[5] H.S.M. Coxeter, *Introduction to Geometry*, Wiley & Sons, 1965.

[6] H.E. Dudeney, *The Canterbury Puzzles and Other Curious Problems*, W. Heinemann, 1907.

[7] G.N. Frederickson, *Dissections: Plane & Fancy*, Cambridge University Press, 1997.

[8] G.N. Frederickson, *Hinged Dissections: Swinging & Twisting*, Cambridge University Press, 2002.

[9] P. Gerwien, Zerschneidung jeder beliebigen Anzahl von Gleichen geradlinigen Figuren in dieselben Stücke, *Journal für die reine und angewandte Mathematik (Crelle's Journal)* **10**, pp. 228-234 and Taf. III, 1833.

[10] D.A. Klarner, *The Mathematical Gardner*, Wadsworth International, 1981.

[11] H. Lindgren, *Geometric Dissections*, D. Van Nostrand Company, 1964.

[12] W. Wallace, (Ed.) *Elements of Geometry* (8th ed.), Bell & Bradfute, first six books of Euclid, with a supplement by John Playfair, 1831.

[13] J. Akiyama and G. Nakamura, Congruent Dudeney dissections of quadrilaterals (in Japanese), Tech. Report, RIED, Tokai Univ., pp. 1–45, 1999.

About Authors

Jin Akiyama and Gisaku Nakamura are at the Research Institute of Educational Development, Tokai University, 2-28-4 Tomigaya, Shibuya-ku, Tokyo 151–0063 Japan; *fwjb5117@mb.infoweb.ne.jp*

Acknowledgments

The authors would like to thank referees, Yuji Ito, Jorge Urrutia and Toshinori Sakai for their very careful and helpful suggestions, and the editors for their helpful suggestions and nice contribution.

Computing the Hausdorff Distance of Geometric Patterns and Shapes

Helmut Alt
Peter Braß
Michael Godau
Christian Knauer
Carola Wenk

Abstract

A very natural distance measure for comparing shapes and patterns is the Hausdorff distance. In this article we develop algorithms for computing the Hausdorff distance in a very general case in which geometric objects are represented by finite collections of k-dimensional simplices in d-dimensional space. The algorithms are polynomial in the size of the input, assuming d is a constant. In addition, we present more efficient algorithms for special cases like sets of points, or line segments, or triangulated surfaces in three dimensions.

1 Introduction

In application areas like computer vision or pattern recognition, it is often necessary to compare shapes and patterns and to have a numerical value describing their similarity or dissimilarity. Mostly, these geometric objects are compact subsets of $\mathbb{R}^2$ or $\mathbb{R}^3$, or of higher dimensional space in some cases. The most natural distance measure for such objects P and Q apparently is the *Hausdorff distance* where for each point on one object we consider the closest point on the other one and then maximize over all these values. More formally, the Hausdorff distance between P and Q is defined as

$$\delta(P,Q) = \max(\tilde{\delta}(P,Q), \tilde{\delta}(Q,P))$$

where

$$\tilde{\delta}(A,B) = \max_{x \in A} \min_{y \in B} ||x-y||$$

is the *directed Hausdorff distance* from A to B. We assume throughout this paper that the underlying metric $||x-y||$ is the Euclidean metric, but many of our considerations are valid for other commonly used metrics, as well. The directed Hausdorff distance is interesting on its own because $\tilde{\delta}(P,Q)$ is a measure of similarity between P and some *part* of Q.

First results on computing the Hausdorff distance between two convex polygons in $\mathbb{R}^2$ were obtained in [8] and for two finite sets of points or line segments in [6]. Other previous research is concerned with *matching* shapes under certain allowable motions minimizing the Hausdorff distance or *simpli-*

fying shapes within a certain tolerance with respect to the Hausdorff distance. For a survey see [7].

In many cases patterns and shapes are modeled as finite sets of points, finite sets of line segments representing for example, curves or two-dimensional shapes, or triangulated surfaces in three dimensions. Here, we unify all these examples by considering two finite sets P and Q consisting of n and m k-dimensional simplices in d-dimensional space. The Hausdorff distance is defined by considering P and Q as the union of the sets of all points lying in the individual simplices. We give algorithms for this general situation in section 3.

Besides giving algorithms for the general setting we will, in section 4, consider special cases such as point patterns in arbitrary dimensions and sets of triangles in three dimensions, finding more efficient algorithms for these instances.

2 Points and Line Segments in Two Dimensions

For two finite sets of points P and Q in arbitrary dimension the straightforward algorithm, calculating all pairwise distances and maximizing, has running time $O(nm)$. An improved algorithm will be given in Section 4.1. In two dimensions we obtain an asymptotically faster algorithm by first constructing the Voronoi diagram $V(P)$ of P. Then we perform a line sweep on $V(P)$ and Q finding for each point in Q its nearest neighbor in P. The maximum over all distances of these pairs gives the directed Hausdorff distance $\tilde{\delta}(Q,P)$. $\tilde{\delta}(P,Q)$ can be determined analogously and altogether we have an algorithm of running time $O((n+m)\log(n+m))$.

A generalization of this approach to sets of line segments in two dimensions can be found in [6]. For the sake of completeness we present the main ideas of this algorithm here. We assume that the line segments are pairwise *noncrossing*, which means that any pair of segments either does not intersect or has an intersection point that does not lie in the interior of both segments. For example, simple polygons or polygonal chains fulfill this condition. The algorithm is based on the following lemma.

Lemma 2.1. *Let P and Q be finite sets of line segments in $\mathbb{R}^2$. Then the directed Hausdorff distance $\tilde{\delta}(P,Q)$ is attained either at an endpoint of a line segment in P or at an intersection point of a line segment in P with an edge of the Voronoi diagram $V(Q)$.*

This follows from the observation that when moving along a line segment e in P within a Voronoi cell of a line segment f in Q the distance to f is bitonic, i.e., first monotone decreasing and then monotone increasing. Consequently, the distance is maximal either at an endpoint of e or at a point where e leaves the Voronoi cell of f.

Lemma 2.1 gives $O(nm)$ candidate points where the Hausdorff distance can be attained. Their number can be reduced to linear by the following observation:

If we move along a Voronoi edge s bounding the Voronoi cell of some line segment f in Q, then the distance to f is again a bitonic function. Consequently, if several line segments of P are intersecting s, then the maximum distance of an intersection point to f is attained at one of the two extreme intersection points on s. Therefore, for each Voronoi edge of Q we have to consider only the leftmost and rightmost intersection points.

These linearly many candidates are then determined by two line sweeps over P and $V(Q)$. The first sweep, which is performed from left to right and in which each Voronoi edge of Q is deleted as soon as its leftmost intersection point with an edge in P is found, determines all leftmost intersection points of the edges of $V(Q)$ with P. By performing a second sweep from right to left all rightmost intersection points are found. In an analogous manner the directed Hausdorff distance from Q to P is found.

Summarizing, we have

Theorem 2.2. *[6] The Hausdorff distance between two sets $P, Q \subset \mathbb{R}^2$ consisting of n and m points or pairwise not properly intersecting line segments can be found in time $O((n+m)\log(n+m))$*

3 The General Case

In this section we consider the general case of computing the Hausdorff distance between two families P and Q of k-dimensional simplices in $\mathbb{R}^d$, where k, d are constants. Most of the ideas and results of this section appeared originally in the Ph.D. dissertation of Michael Godau [13].

In order to determine the directed Hausdorff distance $\tilde{\delta}(P,Q)$ we assign to each simplex T in Q a distance function d_T: $\mathbb{R}^d \to \mathbb{R}$
$d_T(x) = \min_{y \in T} ||x-y||^2$ for all $x \in \mathbb{R}^d$. (For reasons of simplicity we work with the square of the Euclidean distance rather than the Euclidean distance itself.) Then the squared distance of any $x \in \mathbb{R}^d$ to Q is given by $g(x) = \min\{d_T(x) | T \text{ simplex of } Q\}$, i.e., by the *lower envelope* of the m functions d_T. We obtain the directed Hausdorff distance $\tilde{\delta}(P,Q)$ as the square root of the *maximum of the lower envelope* g over all points in P.

In order to find this maximum, we consider each simplex S of P separately and decompose each function d_T into algebraic pieces. The following lemma, which is a generalization of the so-called *0-dimensionality lemma* in [13] and might be of independent interest, gives a necessary condition for the maximum of the lower envelope of a set of convex functions on a compact convex set.

Lemma 3.1. *Let K be a convex, compact subset of $\mathbb{R}^d$, F a set of m continuous convex functions from K to $\mathbb{R}$, and $g: K \to \mathbb{R}$ their lower envelope. Assume that $c = \max_{x \in K} g(x)$ is not attained at the boundary of K, and let $a \in \text{Int}(K)$ be the smallest point with respect to the lexicographic order with $g(a) = c$. Then $m \geq d+1$ and:*

a) $f(a) = c$ *for at least* $d+1$ *functions* $f \in F$.

b) If all functions in F *are twice continuously differentiable then there is a neighborhood* U *of* a *so that these* $d+1$ *functions coincide in no point in* U *other than* a.

Proof. The maximum of g must be attained somewhere in K because of the compactness of K. So if it does not occur at the boundary it must occur in the interior. Since $\{x \in K | g(x) = c\}$ is a compact subset of $\mathrm{Int}(K)$ its minimal element a with respect to the lexicographic order exists. Let $P_a \in \mathbb{R}^{d+1}$ be the point (a, c).

Let $M = \{f \in F | f(a) = c\}$, then $M \neq \emptyset$ and $f(a) > c$ for all $f \notin M$. Consider the *graphs* of the functions $f \in F$, i.e., the sets $G_f = \{(x, f(x)) | x \in K\}$. Because of the convexity of the functions in F, there exists for each $f \in M$ a hyperplane H_f that is tangent to G_f in P_a. More precisely, if H_f^+ denotes the upper halfspace determined by H_f, then $P_a \in H_f$ and $G_f \subset H_f^+$, i.e., $H_f(x) \leq f(x)$ for all $x \in K$, where in the latter inequality we identified H_f with the affine function whose graph is H_f.

We now claim

Claim 1. $\bigcap_{f \in M} H_f = \{P_a\}$.

Proof of Claim 1. Assume otherwise. Then there exists a straight line $l \subset \bigcap_{f \in M} H_f$ with $P_a \in l$. l lies below all $G_f, f \in F$. Consider the projection l' of l to K and consider l as an affine function from l' to $\mathbb{R}$. Then $a \in l'$, $l(a) = f(a) = c$ for all $f \in M$, and $l(x) \leq f(x)$ for all $x \in l', f \in F$, i.e., $l(x) \leq g(x)$ for all $x \in l'$. Consider some convex neighborhood $V \subset K$ of a and the line segment $s = V \cap l'$. Now, there are two possibilities:

1. If l is a constant function then $l(x) = l(a) = c$ for all $x \in s$. Since $l(x) \leq g(x)$ and c is the maximal value of g, $g(x) = c$ for all $x \in s$. Since a lies in the interior of s this contradicts the lexicographic minimality of a.

2. If l is not constant then $l(x) > c$ for all $x \in s$ on one side of a. Since $l(x) \leq g(x)$ this contradicts the maximality of c.

So in any case, we get a contradiction which proves Claim 1. □

Claim 1 states that the hyperplanes $H_f, f \in M$ intersect in one point. Therefore, it must be $|M| \geq d+1$ which proves part a) of the lemma.

In order to prove part b) select a subset $M' \subset M$ of $d+1$ elements with $\bigcap_{f \in M'} H_f = \{P_a\}$. Then, considering each H_f as a function from $\mathbb{R}^d$ to $\mathbb{R}$, we have

Claim 2. *There is a constant* $\alpha > 0$ *such that for all* $x \in \mathbb{R}^d$ *there are functions* $f_1, f_2 \in M'$ *with* $H_{f_1}(x) - H_{f_2}(x) \geq \alpha \|x - a\|$.

Proof of Claim 2. Consider the sphere $C = \{x \in \mathbb{R}^d | \, \|x - a\| = 1\}$ and let $\alpha = \min_{x \in C} \max_{f_1, f_2 \in M'} H_{f_1}(x) - H_{f_2}(x)$ which exists because of the compactness of C. Also $\alpha > 0$ since $\bigcap_{f \in M'} H_f = \{P_a\}$. For arbitrary $x \in \mathbb{R}^d, x \neq a$ let x' be the point on C on the straight line through a and x, i.e., $x' = a + (x - a)/\|x - a\|$. Then there are $f_1, f_2 \in M'$ with $h(x') = H_{f_1}(x') - H_{f_2}(x') \geq \alpha$. Since h is an affine function with $h(a) = 0$ we have $h(x) = h(x')\|x - a\| \geq \alpha\|x - a\|$, which proves claim 2. □

On the other hand, we have

Claim 3. *There is a constant $\beta > 0$ such that for all $x \in K$*

$$H_f(x) \leq f(x) \leq H_f(x) + \beta\|x - a\|^2.$$

Proof of Claim 3. The first inequality was explained before, the second one is obtained by the Taylor-expansion of f about a, since the second derivatives of f are bounded in the compact set K. □

To finish the proof of Lemma 3.1 consider the functions f_1, f_2 from claim 2 for $x = a + \varepsilon$, where $\varepsilon \in \mathbb{R}^d$ is sufficiently short such that $a + \varepsilon \in K$. Then, using claims 2 and 3 we get:

$$\begin{aligned} f_1(a+\varepsilon) - f_2(a+\varepsilon) &\geq H_{f_1}(a+\varepsilon) - H_{f_2}(a+\varepsilon) - \beta\|\varepsilon\|^2 \\ &\geq \alpha\|\varepsilon\| - \beta\|\varepsilon\|^2 \end{aligned}$$

The latter expression is greater then 0 for sufficiently small $\|\varepsilon\| > 0$ which shows that f_1 and f_2 do not coincide in some neighborhood of a except in a itself. This finishes the proof of Lemma 3.1 □

Lemma 3.1 can be applied to the problem of computing the Hausdorff-distance between P and Q in order to obtain a finite number of candidates where the directed Hausdorff distance $\tilde{\delta}(P, Q)$ can occur.

First let us have a closer look at the n distance functions determined by the simplices T in Q. For a face T' of T let $A_{T'}$ be the affine space spanned by T' and $f_{T'} : \mathbb{R}^d \to \mathbb{R}$ the function that assigns to $x \in \mathbb{R}^d$ its squared distance to $A_{T'}$. Each $f_{T'}$ is a quadratic and, hence, a convex and twice continuously differentiable function. Also, for the restriction of the distance function d_T to $V_{T'}$ where $V_{T'}$ is the Voronoi cell of T' within the Voronoi diagram of all sites (i.e., all lower-dimensional simplices) of T (see Figure 1) we have $d_T|V_{T'} = f_{T'}|V_{T'}$, so d_T is a convex, piecewise quadratic function with $2^k - 1$ pieces determined by all Voronoi cells.

Lemma 3.2. *Let S be a k-dimensional simplex in P. Then the directed Hausdorff-distance $\tilde{\delta}(S, Q)$ occurs at some point a on some face S' of S of dimension $k' \leq k$ where $k' + 1$ distance functions $f_{T'}(x)$ for faces T' of simplices in Q have the same value. Furthermore the point a is isolated in the sense that in S' there is a neighborhood U of a so that for no point in U other than a these $k' + 1$ functions have the same value.*

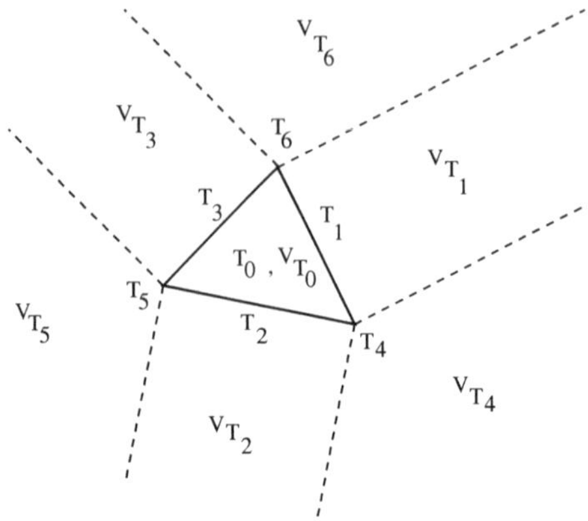

Fig. 1. Voronoi cells of a triangle $T_0 \subset \mathbb{R}^2$ and its faces.

Proof. Let S' be a face of S of minimal dimension k' that contains a point a where the directed Hausdorff distance $\tilde{\delta}(S,Q)$ is attained, i.e., $\tilde{\delta}(S,Q) = \min_{x \in Q} \|a - x\|$. Observe that a cannot lie at the boundary of S'.

Let us identify the affine span $A_{S'}$ with $\mathbb{R}^{k'}$, consider S' as a compact subset of $\mathbb{R}^{k'}$, and consider all distance functions d_T and $f_{T'}$ as restricted to $A_{S'}$, i.e., $\mathbb{R}^{k'}$. For $x \in A_{S'}$, the distance $\tilde{\delta}(\{x\}, Q)$ of x to Q is given by the lower envelope $g(x)$ of the functions $d_T(x)$ for all simplices T of Q.

Assume that $a \in S'$ is the point where g attains its maximum that is minimal with respect to the lexicographic order. Since all d_T are convex, we can apply Lemma 3.1 a) to this setting and, therefore, $k'+1$ distance functions d_T for simplices T in Q have the same value $c = g(a)$ in a.

Therefore, also $k'+1$ distance functions $f_{T'}$ for faces T' of simplices in Q have the value c in a. Let $F = \{f_{T'} | T' \text{ face of a simplex in } Q, f_{T'}(a) = c\}$. Then, we have:

Claim 4. *There is a neighborhood V of a so that the lower envelope h of the functions in F restricted to V has a maximum in a and a is the lexicographically smallest point with that property in V.*

Proof of Claim 4. Consider the complement of F, i.e., the set $\bar{F} = \{f_{T'} | T'$ face of a simplex in $Q, f_{T'}(a) \neq g(a)\}$. Since all functions involved are continuous, there must be a neighborhood V of a so that $f(x) \neq g(x)$ for all $x \in V$ and all $f \in \bar{F}$. On the other hand, for any $x \in \mathbb{R}^{k'}$ there is a face T' of a simplex in Q with $g(x) = f_{T'}(x)$. Consequently, for any $x \in V$ there is an $f \in F$ with $g(x) = f(x)$. Therefore in V, the lower envelope of the functions in F does not exceed g, i.e., $h(x) \leq g(x)$ for all $x \in V$. On the other hand, $h(a) = g(a)$ so a is a point in V where the lower envelope h attains its maximum, because g does. There cannot be any point b in V lexicographically smaller than a with this property, since we would have $g(b) \geq h(b) = h(a) = g(a)$, i.e., $g(b) = g(a)$ which would contradict the lexicographic minimality of a. This proves claim 4. □

Consider a convex compact subset $K \subset V$ that contains a in its interior and apply Lemma 3.1 b) to the functions in F restricted to K. This proves Lemma 3.2. □

Finally, in order to compute $\tilde{\delta}(P,Q)$, we consider each face of every simplex in P separately, i.e., we have a set of $n(2^k - 1)$ simplices S, each of dimension at most k.

Lemma 3.2 gives a straightforward algorithm for computing $\tilde{\delta}(S,Q)$:

1. For each selection $T_1, \ldots, T_{k'+1}$ of faces of simplices of Q, where k' is the dimension of S:

   1a. Solve the system of equations consisting of k' quadratic equations

   $$f_{T_1}(x) = f_{T_2}(x), \ldots, f_{T_{k'}}(x) = f_{T_{k'+1}}(x)$$

   and $d - k'$ linear equations defining the affine span A_S.

   1b. For each isolated solution x of this system test whether

   a) $x \in S$,
   b) $x \in V_{T_i}$ for $i = 1, \ldots, d+1$, and
   c) $f_T(x) \geq f_{T_1}(x)$ for all faces T of simplices in Q with $x \in V_T$.

2. For all x which passed the test in step 1b consider $f_{T_1}(x)$ and return the maximum of all these values as $\tilde{\delta}(S,Q)$.

This procedure is carried out for each face S of each simplex in P, and the maximum value obtained is $\tilde{\delta}(P,Q)$. Likewise, $\tilde{\delta}(Q,P)$ and thus, $\delta(P,Q)$, can be determined.

For the analysis of the running time we observe that there are $n(2^k - 1) = O(n)$ faces of simplices in P. For each of them, at most $\binom{m(2^k-1)}{k+1} = O(m^{k+1})$ selections of faces are made in step 1. For each selection, we spend constant time in step 1a and time $O(m)$ in step 1b. Hence, we spend $O(nm^{k+2})$ for computing $\tilde{\delta}(P,Q)$ and $O(nm^{k+2} + n^{k+2}m)$ for computing $\delta(P,Q)$.

This result is summarized in the following theorem where, as in the theorems thereafter, we only give the running time for computing the directed Hausdorff distance. If this running time is $T(n,m)$ then the Hausdorff distance itself can be computed in time $O(T(n,m) + T(m,n))$.

Theorem 3.3. *Given two sets $P, Q \subseteq \mathbb{R}^d$ of n and m k-dimensional simplices, we can compute $\tilde{\delta}(P,Q)$ in $O(nm^{k+2})$ time.*

Besides this straightforward approach, we also can use results obtained in the theory of lower envelopes. Essentially, Lemma 3.2 says that the maximum of the lower envelope is attained at a *vertex* of the lower envelope of the functions restricted to some face of the polytope S. Fortunately, vertices of lower envelopes can be determined efficiently. In fact, Agarwal et al.

[1, 16] give a Las Vegas algorithm for computing the vertices of the lower envelope of m partially defined k-variate algebraic functions of degree two in $O(m^{k+\varepsilon})$ time for $k > 1$ and in $O(m\alpha(m)\log m)$ time for $k = 1$, where $\alpha(m)$ is a functional inverse of the Ackermann function. So we can proceed as follows: For each face F of S we compute the vertices of the lower envelope of the distance functions $\{d_T \mid T \text{ is a simplex of } Q\}$ restricted to F with the algorithm of Agarwal et al. Since we have to do this for all $O(n)$ faces of simplices in P, we obtain

Theorem 3.4. *Given two sets $P, Q \subseteq \mathbb{R}^d$ of n and m k-dimensional simplices ($k \geq 1$), we can compute $\tilde{\delta}(P,Q)$ in $O(nm^{k+\varepsilon})$ randomized expected time, for any $\varepsilon > 0$, if $k > 1$, and in $O(nm\alpha(m)\log m)$ randomized expected time, if $k = 1$.*

4 Efficient Algorithms for Special Problems

4.1 Point Patterns

The Voronoi based approach of section 2 for finite sets of points does not yield efficient algorithms in higher dimensions, because the complexity of constructing the Voronoi cells is too large. However, Agarwal et al. [2] give a Las Vegas algorithm for computing for each point $p \in P$ in an n–point set P one closest neighbor in an m–point set Q in $O((n+m)\log(n+m) + (nm)^{1+\varepsilon-1/(1+\lceil d/2 \rceil)})$ expected time, for any $\varepsilon > 0$.

With the help of more sophisticated techniques, like, e.g., efficient point-location data structures and hierarchical cuttings (which did not exist at the time [2] was published), it is possible to improve the algorithm slightly and get:

Theorem 4.1. *The Hausdorff distance between two sets $P, Q \subset \mathbb{R}^d$ consisting of n and m points can be found in $O((n+m+(nm)^{1-1/(1+\lceil d/2 \rceil)})\log(n+m))$ randomized expected time.*

4.2 Triangles in Three Dimensions

Surfaces in three dimensions are in many cases modeled as a triangular mesh. We give an algorithm for computing the Hausdorff distance not only for triangulated surfaces but, more generally, for sets of triangles in $\mathbb{R}^3$. For the remainder, P and Q are sets of n and m triangles. Theorem 3.4 gives a bound of $O(nm^{2+\varepsilon})$ for computing the directed Hausdorff distance $\tilde{\delta}(P,Q)$. By using a line-sweep algorithm on the affine spans of the triangles involved and parametric search, we can obtain an $O(nm^2 \log^{O(1)}(mn))$–time algorithm. Here, we present a third approach, which is asymptotically faster than cubic in the input size. However we need to require that within one set the triangles do not intersect properly, i.e., any two distinct triangles in the set do not intersect in their relative interiors. Let us assume, that $\delta > 0$ is fixed and let P^0 denote the set of vertices of P.

We have that $\tilde{\delta}(P,Q) \leq \delta$ iff for each point of P there is a point of Q within distance at most δ. Therefore it is reasonable to look at the set of all points of distance at most δ to Q; in the following $\mathsf{nh}_\delta(Q) = \{x \in \mathbb{R}^3 \mid \mathrm{d}(x,Q) \leq \delta\}$ denotes the *δ-neighborhood* of Q, and $\mathsf{bd}_\delta(Q) = \{x \in \mathbb{R}^3 \mid \mathrm{d}(x,Q) = \delta\}$ denotes the *boundary of the δ-neighborhood* of Q. Our results are based on the following simple observation: The directed Hausdorff distance from P to Q is at most δ iff all vertices of P are contained in the δ-neighborhood of Q, and none of the triangles in P intersects the boundary $\mathsf{bd}_\delta(Q)$. More formally:

Lemma 4.2. *Let P and Q be compact sets in $\mathbb{R}^3$, and $\delta > 0$. Then*

$$\tilde{\delta}(P,Q) < \delta \iff P^0 \subset \mathsf{nh}_\delta(Q) \text{ and } P \cap \mathsf{bd}_\delta(Q) = \emptyset.$$

So we are left with the task of verifying whether $P^0 \subset \mathsf{nh}_\delta(Q)$ ('inclusion property'), and $P \cap \mathsf{bd}_\delta(Q) = \emptyset$ ('intersection property').

The inclusion property can easily be checked in $O(nm)$ steps by computing the distance of each vertex in P^0 to each triangle in Q in $O(1)$ time. This method also identifies the triangles $\Delta \in P$ that contain a vertex outside of $\mathsf{nh}_\delta(Q)$.

Lemma 4.3. *We can decide whether $P^0 \subset \mathsf{nh}_\delta(Q)$ in $O(nm)$ time.*

In the following we describe an efficient algorithm to verify the intersection property.

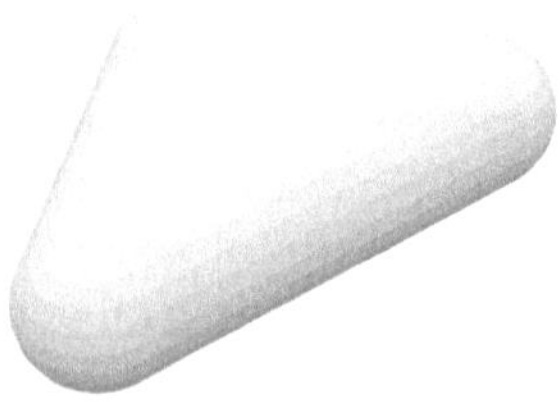

For a triangle Δ, the set $\mathsf{nh}_\delta(\Delta)$ (called a *kreplach* in [5]) is the convex hull of three copies of a δ-ball centered at the vertices of Δ; it is the (non-disjoint) union of three balls of radius δ around the vertices of Δ, three cylinders of radius δ around the edges of Δ, and a triangular prism of height 2δ containing Δ in its center. As usual, we mean by the complexity of $\mathsf{bd}_\delta(Q)$ its number of vertices, edges, and 2-faces. By a result of Agarwal and Sharir [4, 5], the boundary $\mathsf{bd}_\delta(Q)$ has complexity $O(m^{2+\varepsilon})$, and can be computed in $O(m^{2+\varepsilon})$ randomized expected time for any $\varepsilon > 0$; the algorithm computes a description of $\mathsf{bd}_\delta(Q)$ where each 2-face is partitioned into semialgebraic[1] surface patches of constant description complexity. Each of these surface patches is contained in one spherical, cylindrical, or triangular portion of $\mathsf{bd}_\delta(\Delta)$ for some $\Delta \in Q$ (the same Δ that contains the corresponding 2-face), and is bounded by at most four arcs. Each arc in turn is part of the intersection of the portion of the boundary that contains the patch with either a plane, or a δ-sphere,

[1]A set $S \subseteq \mathbb{R}^d$ is called *semialgebraic* if it satisfies a *polynomial expression.* A polynomial expression is any finite Boolean combination of *atomic polynomial expressions*, which are of the form $P(\mathbf{x}) \leq 0$, where $P \in \mathbb{R}[x_1, \ldots, x_d]$ is a d-variate polynomial.

or a δ-cylinder. A polynomial expression defining a patch is formed by the conjunction of five atomic expressions of degree at most two: one polynomial equation describing the portion of $\mathsf{bd}_\delta(\Delta)$ that contains the patch (i.e., a cylinder, a sphere, or a plane), and at most four polynomial inequalities defining the arcs (again these are equations describing a cylinder, a sphere, or a plane).

In order to verify the intersection property we need a method to detect intersections between the triangles in P and the surface patches of bd_δ. We apply a standard approach suggested in [10] and [9], and transform this problem to a semialgebraic point-location problem.

Lemma 4.4. *Let Ω be a set of k semialgebraic sets of constant description complexity in $\mathbb{R}^3$. For any $\varepsilon > 0$ we can build a data structure of size $O(k^{14+\varepsilon})$ in $O(k^{14+\varepsilon})$ randomized expected time, such that for any query triangle Δ we can decide in $O(k^\varepsilon)$ time whether Δ intersects Ω.*

Proof. Let $\Delta(p_1, p_2, p_3; x)$ be a polynomial expression that defines a triangle Δ depending on its three vertices p_1, p_2, and p_3, i.e., $\Delta = \{x \in \mathbb{R}^3 \mid \Delta(p_1, p_2, p_3; x) \text{ holds}\}$; we can form Δ as the conjunction of three linear inequalities, and one linear equation. Let $\Gamma(q; x)$ be a polynomial expression that defines a set $\Gamma \in \Omega$, depending on a sequence of real parameters q, i.e., $\Gamma = \{x \in \mathbb{R}^3 \mid \Gamma(q; x) \text{ holds}\}$. For some fixed Γ, consider the set $C_\Gamma = \{(p_1, p_2, p_3) \in \mathbb{R}^9 \mid (\exists x : \Delta(p_1, p_2, p_3; x) \wedge \Gamma(q; x)) \text{ holds}\}$. If we look at $\mathbb{R}^9$ as the configuration space of the set of all triangles in 3-space, then C_Γ is the set of (the parameters of) all triangles that intersect Γ. By quantifier elimination [12] we can find a polynomial expression $C_\Gamma(q; p_1, p_2, p_3)$ that defines C_Γ; therefore this set is semialgebraic, too.

Let F denote the set of $O(k)$ many polynomials that appear in the atomic polynomial expressions forming the expressions C_Γ. With an algorithm from [9, 14] we can compute a point-location data structure of size $O(k^{14+\varepsilon})$ in $O(k^{14+\varepsilon})$ time for the varieties defined by F. Since the signs of all polynomials in F, and therefore the validity of each polynomial expression C_Γ is constant for each cell of the decomposition of $\mathbb{R}^9$ induced by these varieties, the claim follows. □

Lemma 4.5. *For any $\varepsilon > 0$ we can decide whether $P \cap \mathsf{bd}_\delta(Q) = \emptyset$ in $O(nm^\varepsilon + m^{2+\varepsilon} n^{13/14+\varepsilon})$ randomized expected time.*

Proof. In a first step we compute a description of $\mathsf{bd}_\delta(Q)$ with the algorithm of Agarwal/Sharir. This can be done in $O(m^{2+\varepsilon})$ time, and yields a set of $O(m^{2+\varepsilon})$ semialgebraic surface patches of constant description complexity that partition the boundary of $\mathsf{nh}_\delta(Q)$. Now we distinguish two cases:

$m^{28} \leq n$: We run the algorithm of Lemma 4.4 to build a data structure of size $O(m^{28+\varepsilon})$ in $O(m^{28+\varepsilon})$ time that supports triangle intersection queries to $\mathsf{bd}_\delta(Q)$ in $O(m^\varepsilon)$ time, and then we query this data structure with all triangles in P to test for intersections in $O(nm^\varepsilon)$ steps. The total time spent is $O(m^{28+\varepsilon} + nm^\varepsilon) = O(nm^\varepsilon)$.

$n \le m^{28}$: We partition $\mathsf{bd}_\delta(Q)$ into $g = \lceil m^{2+\varepsilon}/n^{1/14} \rceil$ groups of $k = n^{1/14} \le m^2$ surface patches each. For each group, we run the algorithm of Lemma 4.4 to build a data structure of size $O(k^{14+\varepsilon})$ in $O(k^{14+\varepsilon})$ time that supports triangle intersection queries in $O(k^\varepsilon)$ time, and then we query this data structure with all triangles in P to test for intersections in $O(nk^\varepsilon)$ steps. The total time spent is $O(g(k^{14+\varepsilon} + nk^\varepsilon)) = O(gn^{1+\varepsilon/14}) = O(m^{2+\varepsilon}n^{13/14+\varepsilon})$.

This algorithm can also determine the triangles $\Delta \in P$ that intersect $\mathsf{bd}_\delta(Q)$. □

Putting Lemma 4.3 and Lemma 4.5 together, we obtain

Lemma 4.6. *For any $\varepsilon > 0$ we can compute the set $X = \{\Delta \in P \mid \tilde{\delta}(\Delta, Q) > \delta\}$ in $O(nm + m^{2+\varepsilon}n^{13/14+\varepsilon})$ randomized expected time.*

Using the well-known technique by Clarkson and Shor, c.f. [11], we can easily turn the algorithm for the decision problem into a randomized procedure that actually computes the minimal distance, and obtain

Theorem 4.7. *Given two sets $P, Q \subseteq \mathbb{R}^3$ of n and m triangles with the property that no two triangles in Q intersect in their interior, we can compute $\tilde{\delta}(P, Q)$ in $O(nm \log n + m^{2+\varepsilon}n^{13/14+\varepsilon})$ randomized expected time, for any $\varepsilon > 0$.*

Proof. We follow a strategy similar to that proposed in [3]. Initially we set $\delta = 0$ and $X = P$. Then we repeat the following steps until X becomes empty: Choose a random triangle $\Delta \in X$, and compute $\delta' = \tilde{\delta}(\Delta, Q)$ in $O(m^{2+\varepsilon})$ time according to Theorem 3.4. Set δ to $\max(\delta, \delta')$. Now compute the set $X' = \{\Delta \in X \mid \tilde{\delta}(\Delta, Q) > \delta\}$ in $O(nm + m^{2+\varepsilon}n^{13/14+\varepsilon})$ time with the algorithm from Lemma 4.6. Finally set X to X'.

Obviously the last value of δ will be $\tilde{\delta}(P, Q)$. As is shown in [11], the expected number of iterations is $O(\log n)$, and therefore the expected time to compute $\tilde{\delta}(P, Q)$ with this algorithm is $O(nm \log n + m^{2+\varepsilon}n^{13/14+\varepsilon})$. □

References

[1] P. K. Agarwal, B. Aronov, and M. Sharir Computing envelopes in four dimensions with applications *SIAM J. Comput.*, 26:1714–1732, 1997

[2] P. K. Agarwal, J. Matoušek, and S. Suri Farthest neighbors, maximum spanning trees and related problems in higher dimensions *Comput. Geom. Theory Appl.*, 1(4):189–201, 1992

[3] P. K. Agarwal and M. Sharir Efficient randomized algorithms for some geometric optimization problems *Discrete Comput. Geom.*, 16:317–337, 1996

[4] P. K. Agarwal and M. Sharir Motion planning of a ball amid polyhedral obstacles in three dimensions In *Proc. 10th ACM-SIAM Sympos. Discrete Algorithms*, pages 21–30, 1999

[5] P. K. Agarwal and M. Sharir Pipes, cigars, and kreplach: The union of Minkowski sums in three dimensions In *Proc. 15th Annu. ACM Sympos. Comput. Geom.*, pages 143–153, 1999

[6] H. Alt, B. Behrends, and J. Blömer Approximate matching of polygonal shapes *Ann. Math. Artif. Intell.*, 13:251–266, 1995

[7] H. Alt and L. J. Guibas Discrete geometric shapes: Matching, interpolation, and approximation In J.-R. Sack and J. Urrutia, editors, *Handbook of Computational Geometry*, pages 121–153. Elsevier Science Publishers B.V. North-Holland, Amsterdam, 2000

[8] M. J. Atallah A linear time algorithm for the Hausdorff distance between convex polygons *Inform. Process. Lett.*, 17:207–209, 1983

[9] B. Chazelle, H. Edelsbrunner, L. J. Guibas, and M. Sharir A singly-exponential stratification scheme for real semi-algebraic varieties and its applications *Theoret. Comput. Sci.*, 84:77–105, 1991

[10] B. Chazelle and M. Sharir An algorithm for generalized point location and its application *J. Symbolic Comput.*, 10:281–309, 1990

[11] K. L. Clarkson and P. W. Shor Applications of random sampling in computational geometry, II *Discrete Comput. Geom.*, 4:387–421, 1989

[12] G. E. Collins Quantifier elimination for real closed fields by cylindrical algebraic decomposition In *Proc. 2nd GI Conference on Automata Theory and Formal Languages*, volume 33 of *Lecture Notes Comput. Sci.*, pages 134–183 Springer-Verlag, 1975

[13] M. Godau *On the complexity of measuring the similarity between geometric objects in higher dimensions* PhD thesis, Department of Computer Science, Freie Universitt Berlin, 1999

[14] V. Koltun Almost tight upper bounds for vertical decompositions in four dimensions In *Proc. 42nd Annu. IEEE Sympos. Found. Comput. Sci.*, 2001

[15] J. Matoušek and O. Schwarzkopf On ray shooting in convex polytopes *Discrete Comput. Geom.*, 10(2):215–232, 1993

[16] M. Sharir and P. K. Agarwal *Davenport-Schinzel Sequences and Their Geometric Applications* Cambridge University Press, New York, 1995

About Authors

The authors are at the Institute for Computer Science, Free University of Berlin, Takustr. 9, 14195 Berlin, Germany;
{*alt,brass,godau,knauer,wenk*}*@inf.fu-berlin.de.*

Acknowledgments

Part of this research was funded by the Deutsche Forschungsgemeinschaft (DFG) under grants Al 253/4-3 and Br 1465/5-2 (Heisenberg scholarship).

A Sum of Squares Theorem for Visibility Complexes and Applications

Pierre Angelier
Michel Pocchiola

Abstract

We present a new method to implement in constant amortized time the flip operation of the so-called Greedy Flip Algorithm, an optimal algorithm to compute the visibility complex of a collection of pairwise disjoint bounded convex sets of constant complexity (disks). The method uses simple data structures and only the left-turn predicate for disks; it relies, among other things, on a sum of squares like theorem for visibility complexes stated and proved in this paper. (The sum of squares theorem for a simple arrangement of lines states that the average value of the square of the number of vertices of a face of the arrangement is bounded by a constant.)

1 Introduction

This paper is concerned with visibility among convex obstacles (disks for short) in the euclidean plane [5, 34]. A better understanding of this is intimately involved in the practical concerns of computer graphics and motion planning in robotics. An abstract object related to this problem is the visibility complex, a 2-dimensional regular cell complex whose underlying topological space is the quotient space of the space of rays modulo the visibility relation: two rays are said mutually visible if they are supported by the same (straight) line and if the line-segment joining the origins of the two rays lies in free space, the complement of the obstacles [43]. The one-skeleton of the visibility complex is (roughly speaking) homeomorphic to the tangent visibility graph of the collection of obstacles; in particular there is a two-to-one correspondence between the set of vertices of the visibility complex and the set of free undirected bitangents of the collection of obstacles. Applications of visibility graphs and visibility complexes are found in path planning [13, 27, 33], network optimization [31, 32], global visibility [11, 14, 41, 47], and pattern recognition [21]; visibility complexes of simple polygons have been elegantly characterized in the framework of oriented matroid theory [1, 7] as a class of bipartite graphs [36, 37, 49]. The related notions of pseudotriangles and pseudotriangulations, which have played a key role in the design of efficient

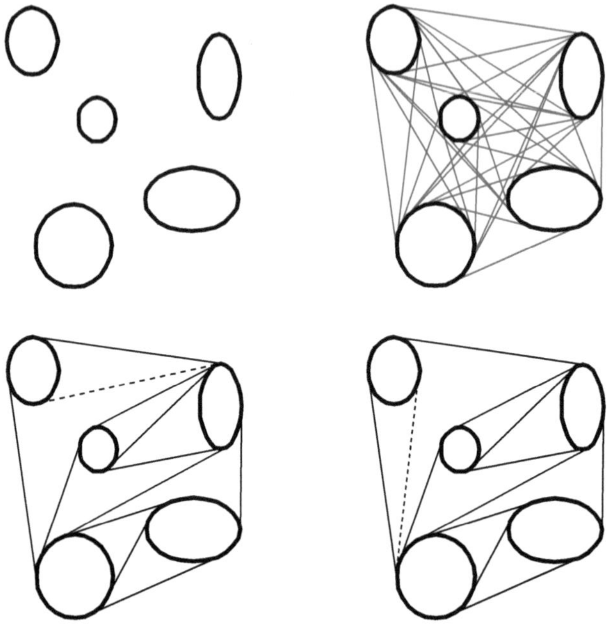

Fig. 1. A collection of five convex obstacles, its set of free undirected bitangents and two of its pseudotriangulations related by a flip operation. (A pseudotriangulation is a maximal set of pairwise disjoint undirected bitangents. Two pseudotriangulations are adjacent or related by a flip operation if their symmetric difference is a pair, that is, a set having exactly two elements.)

tangent visibility graphs or visibility complexes algorithms [42, 43], have applications to ray-shooting [9,19,41], tangent visibility graphs counting [39–41], covering and separation [44], collision detection [2, 23–25], stretchability and realizability of pseudoline arrangements [18, 39, 41], and, last but not least, planning expansive motions of polygonal chains [12, 35, 50]. A key feature of pseudotriangulations that is largely exploited in the applications mentioned above is the existence of a well-defined flip operation on every non-hull bitangent of a pseudotriangulation, e.g., in Figure 1 the two pseudotriangulations are related by a flip operation performed on the dashed bitangents; properties of the flip graph of pseudotriangulations of a given point set are reported in [6, 8, 45, 48]. A challenging question is to extend these notions or applications or both to 3D environments.

In this paper we focus on the design of efficient algorithms for tangent visibility graphs (or, equivalently, visibility complexes). We describe an optimal time and linear working space tangent visibility graph algorithm whose primitive predicate is the *left-turn predicate for disks*: the left-turn relation $\chi_1(u, v)$ on ordered pairs of (directed) bitangents asserts that u and v touch a same directed disk and that the determinant of the 2×2 matrix whose column vectors are the directions of u and v taken in this order is positive. For point

Table 1. Optimal time and linear space tangent visibility graph or visibility complex algorithms for obstacles with constant complexity.

Source	Obstacles	Data structures	Predicate
Edelsbrunner-Guibas [16]	points	Simple	left-turn
Pocchiola-Vegter [42]	disks	Splittable queues	χ_2, χ_3
Rivière [46]	line segments	Simple	left-turn
This paper	disks	Simple	left-turn

obstacles the left-turn relation coincides with the so-called counterclockwise relation pqr on triplets of points, which asserts that the points p, q, r are encountered in cyclic order $p, q, r, p, q \ldots$ when traversing counterclockwise the circle through p, q and r.

Our starting point is the so-called greedy flip algorithm (GFA), an optimal time and linear working space output sensitive algorithm to compute the visibility complex of obstacles of constant complexity like line segments, circles, ellipses, etc., [42]. (For a complete bibliography on the algorithms for visibility graphs we refer to [5,32,42].) The GFA realizes a topological sweep of the visibility complex guided by some partial order $\prec$ on the set of free bitangents. This partial order is defined as the transitive closure of a subset of χ_1 relations: $u \prec v$ if there exists a sequence $u_1 = u, u_2, \ldots, u_r = v$ of bitangents such that $\chi_1(u_i, u_{i+1})$ and the (counterclockwise) interval of directions[1] from the direction of u_i to the direction of u_{i+1} avoids, say, the horizontal direction ($i = 1, 2 \ldots, r-1$). The sweep structure S, some sub-complex of the visibility complex, is associated with a pseudotriangulation $\mathrm{G}(S)$ via the bijective operator sink() that carries a 2-cell of the visibility complex to its *sink*, that is, the maximum bitangent of its boundary with respect to $\prec$. (See Figure 2 (A).) The sweep of a bitangent v of the visibility complex boils down to some local change in the sweep structure S plus the flip of v in the pseudotriangulation $\mathrm{G}(S)$, that is, the computation of the bitangent $\varphi(v)$ that leaves the pseudotriangle $\mathrm{R}(v)$ of $\mathrm{G}(S)$ incident upon v and locally to the right of v and enters the pseudotriangle $\mathrm{L}(v)$ of $\mathrm{G}(S)$ incident upon v and locally to the left of v. (In the case where v is a hull-bitangent $\mathrm{R}(v)$ or $\mathrm{L}(v)$ is not defined and $\varphi(v)$ is obtained from v by reversing its direction.) It is known that $\mathrm{R}(v)$, $\mathrm{L}(v)$, and $\varphi(v)$ depend only on v and not on both v and S; more precisely, considered as the $\prec$-increasing sequences of bitangents that

[1] A *direction* is an element of $\mathbb{S}^1 = \{(x, y) \in \mathbb{R}^2 \mid x^2 + y^2 = 1\}$. For a and $b \in \mathbb{S}^1$, the closed interval $[a, b]$ from a to b is defined to be $[a, b] = \exp([u, v])$ where $\exp : x \in \mathbb{R} \to (\cos x, \sin x) \in \mathbb{S}^1$ is the universal covering map of the 1-sphere, $a = \exp(u), b = \exp(v)$ with $u \leq v < u + 2\pi$. $\mathbb{S}^1$ will be also considered as an interval of $\mathbb{S}^1$. The horizontal and vertical directions are $\exp 0$ and $\exp \pi/2$.

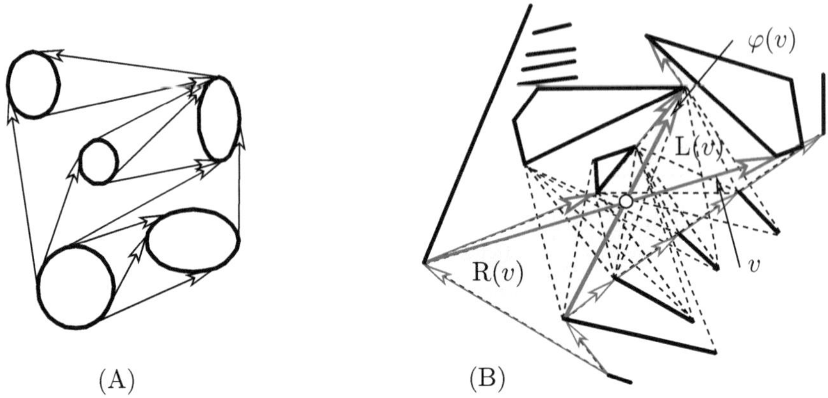

Fig. 2. (A) The greedy pseudotriangulation $G(S_0) = \text{sink}(S_0)$ associated with the initial sweep structure S_0, that is, the set of 2-cells of the visibility complex that contain the rays with horizontal direction. (B) The bitangent $\varphi(v)$ is the minimal bitangent that crosses v from its right side to its left side. (The bitangents crossing v are shown dashed.) The bitangent $\varphi(v)$ is also the bitangent that leaves the right greedy pseudotriangle $R(v)$ and enters the left greedy pseudotriangle $L(v)$ associated with v.

appear in their boundaries, the pseudotriangles $R(v)$ and $L(v)$ are the lexicographic smallest pseudotriangles incident upon v and locally to the right and to the left of v. The pseudotriangles $R(v)$ and $L(v)$ are called the *right and left greedy pseudotriangles* associated with v. Equivalently the bitangent $\varphi(v)$ can be defined as the minimal bitangent crossing v from its right side to its left side or as the sink of the 2-face of the visibility complex whose source (that is, the minimum bitangent of its boundary with respect to $\prec$) is v. (See Figure 2 (B).)

Besides linked structures to represent the incidence relations in the sweep structure and in its associated pseudotriangulation the GFA uses auxiliary data structures to implement efficiently the flip operation. Two parts of the boundary of each pseudotriangle of the associated pseudotriangulation, called AWAKE and ASLEEP in [42], are represented by splittable queues, a data structure for ordered lists that supports the enqueue and dequeue operations either at the head or at the tail of the list, and the split operation at an atom preceded by a search of the atom; the implementation of the splittable queues should guarantee a constant amortized time complexity per operation on a sequence of operations (this can be done, for example, with red-black trees with parent pointers or with randomized search trees). In [42] it is shown that the flip operation requires a constant number of enqueue and split operations, and thus is constant amortized. The dequeue and enqueue operations are guided by the evaluation of the predicate $\chi_2(u, v)$ which states that the determinant of the 2×2 matrix whose column vectors are the directions of u and v taken in this order is positive, while the split operations are guided by the evaluation of the predicate $\chi_3(u,v)$ which states that the triplet of points (tail of u, head of u, head of v) is counterclockwise. Some comments on the bit-complexity of the predicates χ_1, χ_2 and χ_3 are in order at this point. Although the bit-complexities of the predicates χ_2 and χ_3 for polygonal obstacles are identical (they boil down to the counterclockwise predicate on vertices and difference of vertices of the input polygons) the situation is different in general. To fix the ideas the evaluation of the afore mentioned predicates for circles given by center and radius requires the evaluation of the sign of expressions of the following type

$$A_1 + A_2\sqrt{B_2} + A_3\sqrt{B_3} + A_4\sqrt{B_2B_3}$$

where the A_i, B_i are polynomials in the inputs whose degrees depend on the predicate. (Explicit formulas are given in [3].) By a repeated application of the principle that if $a > 0$ and $b < 0$ then the expressions $a + b\sqrt{c}$ and $a^2 - b^2c$ have the same sign one can show that the evaluation of χ_3 requires the evaluation of the sign of an irreducible polynomial of degree at most 16 while χ_2 (or χ_1) requires the evaluation of an irreducible polynomial of degree at most 6. Therefore the bit-complexities of χ_2 and χ_3 are $6b + O(1)$ and $16b + O(1)$ where b is the bit-complexity of the inputs, that is, the bit-complexities of the radius and coordinates of the center of the circle.

Since the result of the flip operation depends only on the χ_1 relations and not on the χ_2 nor χ_3 relations or since the bit-complexity of χ_1 is smaller than the bit-complexity of χ_3 it is justifiable to ask whether the flip operation can be efficiently implemented using only the left-turn predicate for disks. A related question is whether the flip operation can be implemented without having recourse to the splittable queue data structure. We answer affirmatively both questions by playing on a sequence of properties that involve both the partial order $\prec$ and its dual, that is, the partial order $\prec_*$ defined on the same set of bitangents by $u \prec_* v$ if $v \prec u$. Any operator Φ defined with respect to $\prec$ (e.g., $\varphi, \mathrm{G}, \mathrm{R}, \mathrm{L}$) has a dual version defined with respect to $\prec_*$. This dual version is denoted Φ_*. In particular we introduce in the picture,

1. the *dual pseudotriangulation* $\mathrm{G}_*(S)$, image of S under the operator sour() that carries a 2-cell of the visibility complex to its *source*, that is, the minimum bitangent of its boundary with respect to $\prec$ or the maximum bitangent of its boundary with respect to $\prec_*$. Equivalently $\mathrm{G}_*(S)$ can be defined as the image of $\mathrm{G}(S)$ under the inverse or dual version φ_* of the operator φ (see Figure 3 (A)).

2. the operator φ_{R} that associates with a non-hull bitangent v the maximal bitangent crossing v and leaving $\mathrm{R}(v)$. Equivalently $\varphi_{\mathrm{R}}(v)$ can be defined as the bitangent crossing v from its right side to its left side whose crossing point with v is the closest to the tail of v, or as the bitangent that leaves the right greedy pseudotriangle associated with v

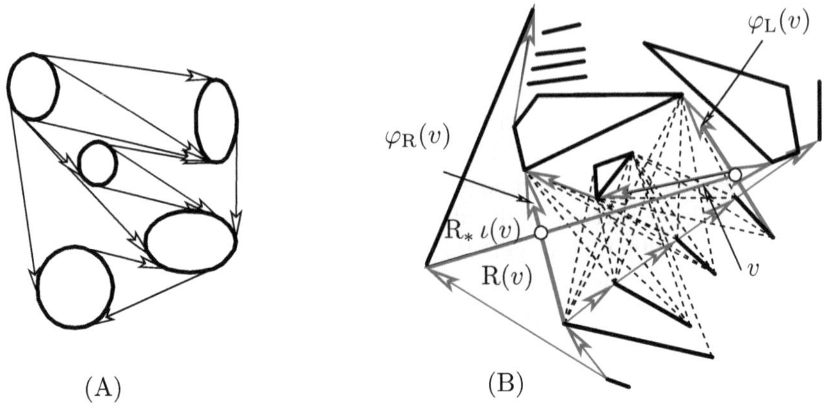

(A) (B)

Fig. 3. (A) The dual greedy pseudotriangulation $\mathrm{G}_*(S_0) = \varphi_*(G(S_0)) = \mathrm{sour}(S_0)$ associated with the initial sweep structure S_0, that is, the set of 2-cells of the visibility complex that contain the rays with horizontal direction. (B) The bitangent $\varphi_{\rm R}(v)$ is the maximal bitangent crossing v and leaving $\mathrm{R}(v)$. The bitangent $\varphi_{\rm R}(v)$ is also the bitangent that leaves $\mathrm{R}(v)$ and enters the dual version $\mathrm{R}_*(\iota(v))$ of the right greedy pseudotriangle associated with the bitangent $\iota(v)$ obtained from v by reversing its direction.

and enters the dual version of the right greedy pseudotriangle associated with the bitangent $\iota(v)$ obtained from v by reversing its direction. We denote by $\varphi_{\rm L}$ the left version of $\varphi_{\rm R}$, that is, the maximal bitangent crossing v and entering $\mathrm{L}(b)$. (See Figure 3 (B).)

Thus the main result of the paper can be stated as follows.

Theorem 1. *The flip operation of the greedy flip algorithm can be implemented in constant amortized time using only the left-turn predicate for disks. Furthermore the only necessary data structures are linked structures to represent the incidence relations in the sweep structure S and its associated pseudotriangulations $\mathrm{G}(S)$ and $\mathrm{G}_*(S)$, a list to store the minimal bitangents of $\mathrm{G}(S)$, and for each minimal bitangent v pointers to the arcs of $\mathrm{G}(S)$ that $\varphi_{\rm R}(v)$ and $\varphi_{\rm L}(v)$ leave and pointers to the arcs of $\mathrm{G}_*(S)$ that $\varphi_{\rm R}(v)$ and $\varphi_{\rm L}(v)$ enter.* □

In the forthcoming paper [38] the problem of computing the initial sweep structure S_0 and its associated pseudotriangulations $\mathrm{G}(S_0)$ and $\mathrm{G}_*(S_0)$, using only the left-turn predicate, is shown to be linear-time reducible to the problem of sorting, according to the vertical direction, the points of the disks with horizontal direction. Thus the GFA can be implemented efficiently using only the left-turn predicate modulo some initial sorting. Computing a suitable initial sweep structure to start the GFA using only the left-turn predicate remains an open problem. The complexity analysis of our flip method is based on a theorem similar to the sum of squares theorem for a simple

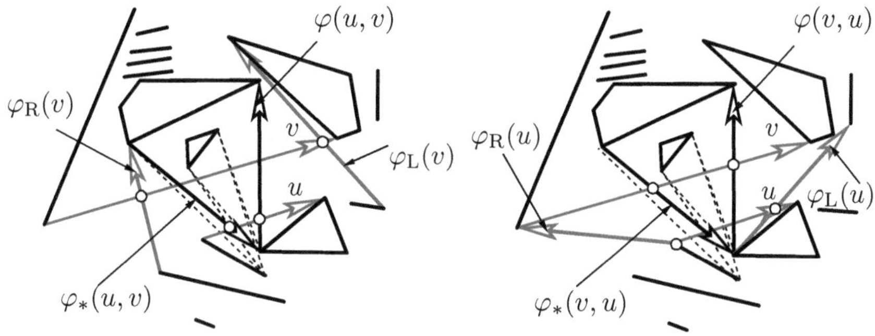

Fig. 4. A pair (u, v) such that $\varphi(u, v) = \varphi(v, u)$ and $\varphi_*(u, v) = \varphi_*(v, u)$.

arrangement of lines in the plane. This last theorem states that the average value of the square of the number of vertices of a face of the arrangement is a constant. (This is a well-known consequence of the linear bound on the complexity of the so-called zone of a line in an arrangement of lines; see [10, 17], [13, Chap. 8], [15, Chap. 5], and [4, 22] for an higher dimensional analogue.) The statement of our sum of squares theorem requires the introduction of a certain binary operator $\varphi(u, v)$ on the set of bitangents which is defined as follows. For every subset κ of pairwise disjoint free undirected bitangents let $\mathcal{B}_\kappa$ be the set of free bitangents that intersect transversally no bitangent of κ, let $\prec_\kappa$ be the restriction of $\prec$ to $\mathcal{B}_\kappa$, and for every bitangent v of $\mathcal{B}_\kappa$ let $\varphi^\kappa(v)$ be the minimal bitangent of $\mathcal{B}_\kappa$ crossing v from its right side to its left side. If no bitangent crossing v crosses a bitangent of κ then $\varphi^\kappa(v) = \varphi(v)$. In particular $\varphi^\emptyset = \varphi$. We will see that the operator φ^κ is well-defined, one-to-one, onto, and that its inverse coincides with its dual version φ_*^κ. Then we set

$$\varphi(u, v) = \varphi^{\kappa(v)}(u), \tag{1}$$

where $\kappa(v)$ is the undirected version of the set $\{\varphi_{\mathrm{R}}(v), \varphi_{\mathrm{L}}(v)\}$, and we denote by $\varphi_*(u, v) = \varphi_*^{\kappa(v)}(u)$ its dual version. (See Figure 4.) We are now ready to state our sum of squares theorem for visibility complexes.

Theorem 2 (Sum of Squares Theorem). *The number of pairs of free bitangents (u, v) such that $\varphi(u, v) = \varphi(v, u)$ and $\varphi_*(u, v) = \varphi_*(v, u)$ is bounded by a constant times the number of free bitangents.* □

As we shall see the set of pairs of bitangents that satisfy the condition of the sum of squares theorem is closely related to the multiset of bitangents visited during the course of the GFA. For point obstacles, our theorem can be restated — without the φ and φ_* binary operators to make it easier to grasp — as follows : the number of 4-uplets $xyzt$ of points of a set P of n points in general position such that (1-) x, y, z, t are in convex position and appear in this order when traversing clockwise their convex hull, and (2-) no point of P

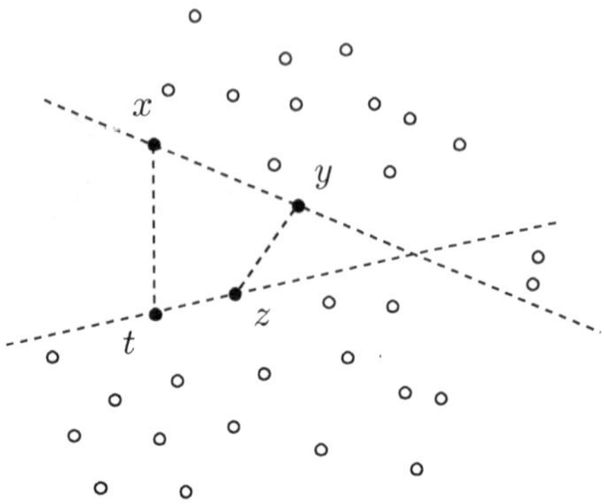

Fig. 5. A 4-uplet $xyzt$ of points of a finite set of points P that satisfies the condition of the sum of squares theorem for points : the points x, y, z, t are in convex position and the dashed region contains no point of P.

lies in the intersection of the right half-plane of the line going from x to y and the right half-plane of the line going from z to t, is a $O(n^2)$. (See Figure 5.) It is no hard to see that this bound implies the following weak form of the sum of squares theorem for simple arrangement of lines: the average value of the product of the numbers of vertices of the right and left boundaries of a face of an arrangement of lines in general position is a constant. Finally we mention that this bound is essentially the bound used by Edelsbrunner and Guibas to analyze the complexity of their topological sweep method [16].

The paper is organized as follows. In Section 2 we extend the theory behind the GFA developed in [42] to the case where the collection of obstacles is augmented with a set κ of pairwise disjoint undirected bitangents : we generalize the flip operation to some configurations of possibly overlapping pseudotriangles (Theorem 3), we introduce the visibility complex $\mathbb{V}_\kappa$ associated with κ and a finite collection of operators φ_α^κ defined on the set of cells of $\mathbb{V}_\kappa$ related to their incidence relations, and we give a characterization of the chains of the 1-skeleton of $\mathbb{V}_\kappa$ that are boundaries of 2-cells of the visibility complex (Theorem 4); then we show that the operator φ^κ is well-defined, coincides with one of the φ_α^κ's, and satisfies the so-called minimum element, right-to-left and flip properties (Theorems 5 and 6) which were stated in [42] in the case where κ is the emptyset. Finally we describe an optimal flip method that uses simple data structures and only the χ_2 predicate. We do not claim originality for the method but for the proof of its optimality, which is based on the sum of squares theorem. Section 3 is devoted to the study of the dual order $\prec_*$, the operator φ_{R}, and to the proof of the sum of squares theorem. As a first step we isolate a key property of the left-turn predicate for disks (Theorem 9). Then we successively

1. relate the bitangents of the greedy pseudotriangles to the orbits of the φ_α^κ's (Theorems 10 and 11) and establish a duality relation between the

greedy pseudotriangles defined with respect to the partial order $\prec$ and the ones defined with respect to the dual partial order $\prec_*$ (Theorem 12);

2. give several characterizations of the operator φ_{R} (Theorems 13 and 14), study the equation $\varphi(v) = \varphi_{\mathrm{R}}(v)$ (Theorem 15), and explain how $\varphi_{\mathrm{R}}(v)$ can be efficiently computed in the dynamic setting of the GFA using only the left-turn predicate (Theorems 16 and 17);

3. prove the sum of squares theorem and state sufficient conditions, used in our flip method, for a pair of bitangents to satisfy the conditions of the sum of squares theorem (Theorems 19 and 20).

In the fourth section we describe the optimal flip method based on the left-turn predicate χ_1 and analyze its complexity, thus proving Theorem 1. We have implemented both methods for polygons and circles using the C++ library **CGAL** (binaries are available at the url http://www.di.ens.fr/~gecoal). Experimental results are discussed and reported in [3]. In the fifth and last section we mention further research suggested by our results.

2 The Theory behind the Greedy Flip Algorithm Revisited and Extended

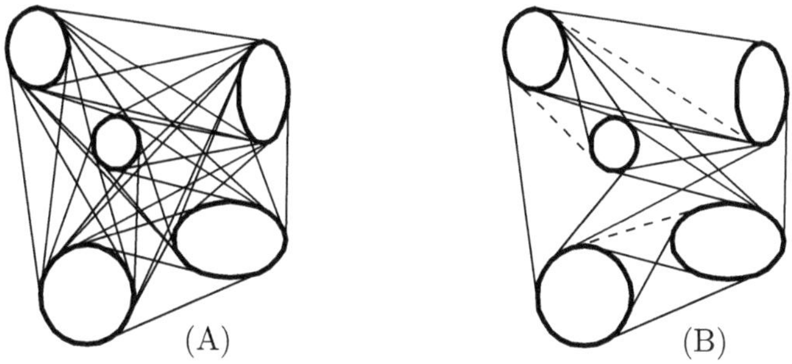

Fig. 6. (A) A collection of five obstacles with its set of free undirected bitangents. (B) A collection of five obstacles with its set of free undirected bitangents intersecting transversally none of the three dashed bitangents.

2.1 Bitangents and pseudotriangulations

Throughout the paper we consider a finite family $o_1, o_2, \ldots, o_n$ of $n \geq 2$ pairwise disjoint bounded closed 2-dimensional convex subsets of the Euclidean plane $\mathbb{R}^2$, disks or obstacles for short; for the ease of the exposition we restrict our attention to the case of *smooth disks in general position*, that is, there is a well-defined undirected tangent line through each boundary point and there is no line tangent to three disks[2]. An *undirected bitangent* is a closed line-segment tangent to two disks at its endpoints and whose interior intersects none of the two disks, and a *directed bitangent* or simply a *bitangent* is an undirected bitangent with a "direction" assigned to it and indicated in figures by an arrow (a formal definition is given in the next subsection). A *free bitangent* is a bitangent whose undirected version is included in free space $\mathbb{F}$, that is, the complement in the plane of the interiors of the disks. We emphasize that, thanks to our smoothness and generic position assumptions, two free undirected bitangents are disjoint or intersect transversally at an interior point. For every subset κ of pairwise disjoint undirected bitangents we introduce the subset $\mathcal{B}_\kappa$ of free bitangents whose undirected versions intersect transversally no bitangent of κ. By definition the oriented versions of the bitangents of κ and the hull-bitangents of the collection of disks are elements of $\mathcal{B}_\kappa$. A *pseudotriangulation* of the o_i's is a maximal subset of pairwise disjoint free undirected bitangents. It is known that a pseudotriangulation contains $3n-3$ bitangents, that subdivide free space into $2n-2$ pseudotriangles (recall that a pseudotriangle τ is a subset of the plane bounded by three convex arcs pairwise tangent at their endpoints such that τ is contained in the triangle formed by these endpoints as in the figure below) plus an unbounded face, and that every pair of adjacent (along a bitangent) and non-overlapping pseudotriangles share exactly two common undirected tangent lines which are the supporting lines of two free crossing undirected bitangents (the diagonals of the pseudoquadrangle union of the two adjacent pseudotriangles). This last property remains valid under the weaker assumption that the adjacent pseudotriangles are non-overlapping only locally around the common bitangent.

Theorem 3. *Two pseudotriangles defined on the same set of disks, adjacent along a common undirected bitangent v, and non overlapping locally around v share exactly two common tangent lines. One of these lines is the supporting line of the bitangent v and the other line is the supporting line of a free bitangent crossing v and touching at its endpoints the boundaries of the pseudotriangles.*

Proof. Let τ and τ' be the two pseudotriangles adjacent along v and let z_0 be the undirected supporting line of v. Assume wlog that the horizontal tangents to τ and τ' are distinct. Let u_0 be the direction of the interval $[\exp 0, \exp \pi]$ parallel to v. The proof is based on the following three observations:

(1-) the dual curve of a pseudotriangle, that is, the map z_τ that associates with the direction u the tangent to $\partial\tau$ with direction u is well-defined and antipodal, that is, $z_\tau(-u) = -z_\tau(-u)$;

[2]However we allow line segments in the boundaries of the disks. Degenerate configurations of possibly non-smooth disks are treated using standard symbolic perturbation techniques: smoothness is obtained by replacing the obstacles by their Minkowski sums with a small enough circle and degeneracies are removed by inflating slightly the disks along suitable directions. See also Figure 17.

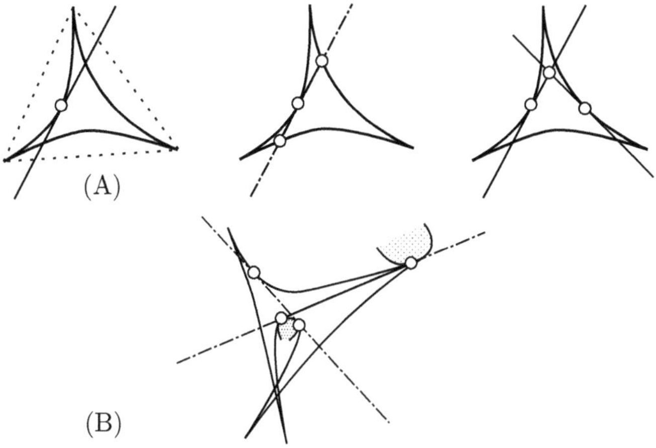

Fig. 7. (A) A triple of pseudotriangles. (B) Two adjacent pseudotriangles non overlapping locally around their common bitangent share two common tangent lines.

(2-) for $u \in \mathbb{S}^1$ the intersection of τ and $z_\tau(u)$ is a line segment;

(3-) for $u \neq v \in \mathbb{S}^1$ the intersection point of $z_\tau(u)$ and $z_\tau(v)$ belongs to τ.

Let $\tilde{z}_\tau(u)$ be the signed distance from the origin of the plane to the line $z_\tau(u)$ and let $\Delta(u) = \tilde{z}(u) - \tilde{z}'(u)$. Since τ and τ' are adjacent along v but non overlapping locally around v the continuous function $\Delta(u)$ vanishes at u_0 but does not change its sign in the vicinity of u_0. According to the intermediate value theorem the continuous function $\Delta(u) = \tilde{z}(u) - \tilde{z}'(u)$ should vanish at a direction $u_1 \neq u_0$ on the interval $(\exp 0, \exp \pi)$ since $\Delta(\exp 0) \times \Delta(\exp \pi) = -(\Delta(\exp 0))^2 < 0$ and since $\Delta(u)$ does not change its sign in the vicinity of u_0. Let z_1 be the tangent to τ and τ' with direction u_1. Now we prove that the line z_1 is the unique (up to reorientation) tangent $\neq z_0$ to both τ and τ', and is the supporting line of a free bitangent touching at its endpoints the pseudotriangles τ and τ'. Let s and s' be the intersections of z_0 with τ and τ'. Since τ and τ' are non overlapping locally around v the intersection of the line segments s and s' is v. The lines z_0 and z_1 are crossing at a point that belongs to both s and s', and consequently belongs to the interior of the bitangent v. It follows that z_1 is the unique common tangent $\neq z_0$ to τ and τ'. Let A be a touching point of z_1 and τ, let A' be a touching point of z_1 and τ', and let B be the crossing point of z_0 and z_1. Since τ and τ' are non overlapping around v the points A, B and A' appear in this order along the line l, and consequently $A \neq A'$. It follows that the line segment $[A, A']$ contains a free undirected bitangent crossing v and touching at its endpoints the pseudotriangles τ and τ'. □

2.2 Visibility complexes

The visibility complex $\mathbb{V}_\kappa$ is a 2-dimensional regular cell complex[3] equipped with a (kind of) branched covering map[4] $\varpi_\kappa : \mathbb{V}_\kappa \to \mathbb{D}$ onto the set of directed lines in the plane. $\mathbb{V}_\kappa$ and ϖ_κ are obtained from the space of *rays* $\mathbb{R}^2 \times \mathbb{S}^1$ and the map that associates with the ray $(p, u) \in \mathbb{R}^2 \times \mathbb{S}^1$ the line through p and direction u using standard topological operations like cutting along a curve, taking quotient space and quotient map, factorization, and compactification. (We refer to [20, 29, 30, 51] for background material on topology.)

As a first step we must start thinking directed lines and directed disks as subsets of the space of rays $\mathbb{R}^2 \times \mathbb{S}^1$. Thus we consider a (directed) line as the set of rays (p, u) where u is the direction of the line and where p ranges over its undirected version and we consider a bitangent as a subset of its supporting line. Similarly we consider a *counterclockwise directed disk* or *positive cycle* as the set of rays (p, u) where p ranges over the boundary of its underlying disk and where u is the direction of the tangent line at p to the disk that contains the disk in its left half-plane, and a *clockwise directed disk* or *negative cycle* as the set of ordered pairs (p, u) where p ranges over the boundary of its underlying disk and where u is the direction of the tangent line at p to the disk that contains the disk in its right half-plane. A cycle and a line *touch* each other or are *tangent* to each other if their intersection is nonempty. Applying this vocabulary, we see that there is a well-defined tangent line going from a first cycle to a second cycle, that is, the point of touching of the line with the first cycle comes before the point of touching with the second cycle as we travel in the direction of the line. This line is the supporting line of a well-defined bitangent which is said to leave the first cycle and to enter the second cycle. A bitangent leaving a positive cycle and entering a negative cycle is said to be a *left-right* bitangent; similarly we introduce *left-left, right-right*, and *right-left* bitangents. A *left-xx* bitangent is a left-right or left-left bitangent; similarly we introduce the *right-xx, xx-left*,

[3] A (finite regular) cell complex is a topological space X and a partition of X into a finite number of 'cells' $\sigma_i \subset X$ (that is, the σ_i's are nonempty, pairwise-disjoint and $X = \bigcup \sigma_i$) such that (1-) $(\overline{\sigma}_i, \overline{\sigma}_i \setminus \sigma_i) \sim (\mathbb{B}^n, \mathbb{S}^{n-1})$, for some $n = n(i)$, (2-) each $\overline{\sigma}_i \setminus \sigma_i$ is a union of σ_j's. Here $\overline{\sigma}_i$ denotes the closure of σ_i and $\sim$ denotes homeomorphism, $\mathbb{B}^n = \{x \in \mathbb{R}^n \mid \langle x, x\rangle \leq 1\}$ is the unit ball of $\mathbb{R}^n$, and $\mathbb{S}^{n-1} = \{x \in \mathbb{R}^n \mid \langle x, x\rangle = 1\}$ is the unit sphere of $\mathbb{R}^n$ ($\langle x, y\rangle$ is the scalar product $x_1y_1 + \cdots + x_ny_n$ of the points $x = (x_1, \ldots, x_n)$ and $y = (y_1, \ldots, y_n)$ of $\mathbb{R}^n$). The union X^n of the σ_i's with $n(i) \leq n$ is called the n-skeleton. The connected components of $X^n \setminus X^{n-1}$ are called n-cells (0-cells and 1-cells are called vertices and edges). The dimension of the complex is the integer n such that $X = X^n$. A complex is said to be *canonical* if its n-cells are the connected components of the set of points of X^n that have a neighborhood homeomorphic to $\mathbb{R}^n$. Finally we use the term *graph* as synonymous with 1-dimensional cell complex.

[4] A (continuous) map $p : X \to Y$ between topological spaces is called a *covering map* if every point of Y has a neighborhood V such that every connected component of $p^{-1}(V)$ is mapped homeomorphically onto V by p. A map $p : X \to Y$ between topological spaces is called a *branched covering map* if there exists a subspace B (the space of branched points) of Y of codimension one — for this paper — such that p restricted to $X \setminus p^{-1}(B)$ is a covering map onto $Y \setminus B$.

and *xx-right* bitangents. The positive cycle associated with the disk o_i is denoted c_{+i}, its negative cycle is denoted c_{-i}, and the line and the bitangent going from c_i to c_j are denoted z_{ij} and v_{ij}.

As a second step we introduce the subset $\mathcal{V}$ of the space of rays defined as the union of the bitangents of $\mathcal{B}$, the cycles associated with the disks, and the *cycles associated with the endpoints of the bitangents of* $\mathcal{B}$, that is, the $\{x\} \times \mathbb{S}^1$ where x ranges over the endpoints of the undirected versions of the bitangents of $\mathcal{B}$. The connected components of $\mathcal{V} \setminus \mathcal{B}$ are termed *primitive arcs*, and the concatenation of adjacent primitive arcs is termed a *composite arc*. The set of (primitive and composite) arcs is denoted $\mathcal{A}$. A *closed chain* of $\mathcal{V}$ is a simple curve $\gamma : [0,1] \to \mathcal{V}$ non-decreasing with respect to the space of directions, that is, $\gamma(t) = (p(t), u(t))$ where $u(t) = \exp\theta(t)$ with $\theta : [0,1] \to \mathbb{R}$ non-decreasing. $\gamma(0)$ and $\gamma(1)$ are called the *tail* and the *head* of the chain γ. An *open chain* is the restriction of a closed chain to the open interval $(0,1)$, and a *semi-open* chain is the restriction of a closed chain to the semi-open interval $[0,1)$ or $(0,1]$. The *atoms* of a chain γ are the bitangents and the (maximal) arcs included in its range $[\gamma]$. We will only work with chains whose ranges are the union of their atoms, and we will consider two chains γ and γ' as equivalent if $\gamma'(t) = \gamma(h(t))$ where $h : [0,1] \to [0,1]$ is increasing. Thus bitangents are closed chains, arcs are open chains, the concatenation of chains is a well-defined operation, and a chain is the concatenation of its atoms. The source and the sink of a chain are the first and last atoms of its closed version; in particular the source and the sink of a chain are bitangents. A chain is termed *convex* if its projection on the plane (the first factor of the space of rays) is a convex curve. A bitangent and a chain touch each other or are tangent to each other if their intersection is nonempty and if the bitangent is not an atom of the chain.

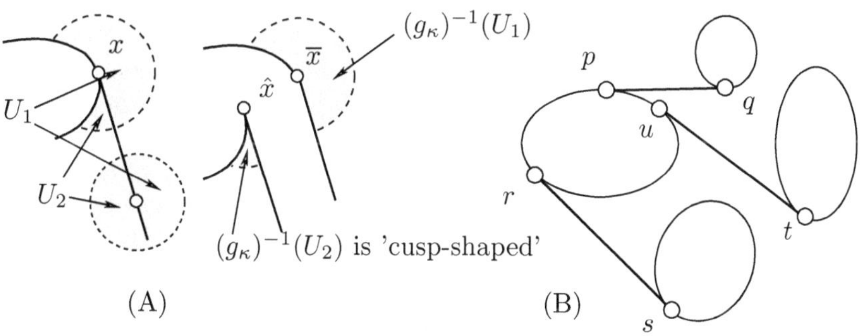

Fig. 8. (A) The surface $\mathbb{F}_\kappa$ is obtained by cutting $\mathbb{F}$ along the bitangents of κ. (B) Here κ is a collection of three bitangents and the boundary of $\mathbb{F}_\kappa$ is composed of a unique (closed) curve. The pre-images under g_κ of the endpoints p, q, r, s, t, u of the bitangents of κ are encountered in cyclic order $\hat{q}, \overline{p}, \overline{r}, \overline{s}, \hat{s}, \hat{r}, \hat{u}, \overline{t}, \hat{t}, \overline{u}, \hat{p}, \overline{q}, \hat{q}, \overline{p}, \cdots$ when traversing counterclockwise the boundary of $\mathbb{F}_\kappa$. The six cusps $\hat{p}, \hat{r}$, etc., decompose the boundary into six 6 convex curves $\hat{q}\overline{p}\,\overline{r}\,\overline{s}\hat{s}$, $\hat{s}\hat{r}$, etc.

As a third step we introduce the surface with boundary $\mathbb{F}_\kappa$ obtained by cutting $\mathbb{F}$ along the bitangents of κ and we write $g_\kappa : \mathbb{F}_\kappa \to \mathbb{F}$ for its associated projection. (Formally $\mathbb{F}_\kappa$ is defined as a gluing : let $\tau \supseteq \kappa$ be a pseudotriangulation that contains κ; pick the union $\mathbb{F}_\tau$ of disjoint copies of the closures of the faces (pseudotriangles + unbounded face) of the induced subdivision of free space and glue the copies of the faces that are adjacent along a bitangent of $\tau \setminus \kappa$ to obtain $\mathbb{F}_\kappa$. The map $g_\kappa : \mathbb{F}_\kappa \to \mathbb{F}$ is the factor of the natural projection $f : \mathbb{F}_\tau \to \mathbb{F}$ through the quotient map $q : \mathbb{F}_\tau \to \mathbb{F}_\kappa$.) For every $v \in \kappa$ and every $x \in v$ the pre-image under g_κ of a path connected closed neighborhood U of x small enough that $U \setminus v$ has two connected components U_1 and U_2 is the union of disjoint copies of the topological closures in $\mathbb{F}$ of U_1 and U_2. For x endpoint of a bitangent of κ we denote by $\hat{x}$ and $\overline{x}$ the two elements of its pre-image under g_κ where $\hat{x}$ denotes the element whose neighborhoods are 'cusp-shaped' as indicated in Figure 8 (A). The set of $\hat{x}$'s is denoted J_κ and its elements are called the cusps of $\mathbb{F}_\kappa$. The cusps of $\mathbb{F}_\kappa$ subdivide its boundary into a set of curves whose projections under g_κ are convex curves in the plane composed of bitangents of κ and connected pieces of the boundaries of the disks. The number of such curves is twice the number of bitangents of κ plus the number of disks that are disjoint from $\bigcup \kappa$. These curves are termed the *views* associated with κ. (See Figure 8 (B).)

We come now to the definition of $\mathbb{V}_\kappa$. The space $\mathbb{V}_\kappa$ is a quotient space of the product space $\mathbb{F}_\kappa \times \mathbb{S}^1$ modulo a certain equivalence relation R_κ. To define this equivalence relation we proceed in two steps. Firstly, we identify the two pre-images of every ray of the union of the directed versions of the bitangents of κ under the direct product $g_\kappa \times 1_{\mathbb{S}^1}$ of g_κ and the identity map $1_{\mathbb{S}^1}$ of $\mathbb{S}^1$. The resulting space is denoted $\mathcal{U}_\kappa$. We denote by $\omega : \mathcal{U}_\kappa \to \mathbb{D}$ the map that associates with the pair (p, u) the line through $g_\kappa(p)$ and direction u, and we introduce the subset $\mathcal{V}_\kappa$ of $\mathcal{U}_\kappa$ defined as the union of the bitangents[5] of $\mathcal{B}_\kappa$, the cycles associated with the disks, and the *cycles associated with the cusps*, that is, the $\{x\} \times \mathbb{S}^1$ where x ranges over J_κ. The connected components of $\mathcal{V}_\kappa \setminus \bigcup \mathcal{B}_\kappa$ are called the *primitive arcs* of $\mathcal{V}_\kappa$ and their set is denoted $\mathcal{A}_\kappa$. Now comes the second step of the identification procedure. The space $\mathbb{V}_\kappa$ is obtained from the space $\mathcal{U}_\kappa$ by identification of the elements of $\mathcal{U}_\kappa$ that are in the same connected component of the pre-image of a line under ω and the quotient map is denoted $q_\kappa : \mathcal{U}_\kappa \to \mathbb{V}_\kappa$. The *backward view* $\mathrm{Bv}(r)$ and the *forward view* $\mathrm{Fv}(r)$ of $r \in \mathbb{V}_\kappa$ are the views that contain the origins of the tail and the head of r. We write

$$\mathcal{U}_\kappa \xrightarrow{q_\kappa} \mathbb{V}_\kappa \xrightarrow{\varpi_\kappa} \mathbb{D}$$

[5]We have implicitly identified a bitangent of $\mathcal{B}_\kappa$ with its pre-image in $\mathcal{U}_\kappa$ under the factor map of the direct product $g_\kappa \times 1_{\mathbb{S}^1}$ through the quotient map $p : \mathbb{F}_\kappa \times \mathbb{S}^1 \to \mathcal{U}_\kappa$; a similar identification is assume for the cycles. Thus $\mathcal{V}_\kappa$ is identified with a subset of $\mathcal{V}$, and $\mathcal{A}_\kappa$ is identified with a subset of $\mathcal{A}$.

for the factorization of ω through the quotient map q_κ. And finally, for each cycle c of $\mathcal{V}_\kappa$ we introduce the curve $\gamma_c : \mathbb{S}^1 \to \mathbb{V}_\kappa$ that associates with the direction u the equivalence class of the points of c with direction u, and we set $z_c = \varpi_\kappa \circ \gamma_c$. A curve γ_c is called a *cycle* of $\mathbb{V}_\kappa$. The space $\mathbb{V}_\kappa$ and the map ϖ_κ satisfy the properties (P1), (P2) and (P3) that follow.

(P1). $\mathbb{V}_\kappa = \bigcup \sigma_i$ is a 2-dimensional canonical cell complex whose set of vertices is $q_\kappa(\mathcal{B}_\kappa)$ and whose set of edges is $q_\kappa(\mathcal{A}_\kappa)$. A *closed chain* of $\mathbb{V}_\kappa$ is a simple curve $\gamma : [a, b] \subset \mathbb{S}^1 \to \mathbb{V}_\kappa^1$ parametrized by its set of directions, that is, the direction of $\gamma(u)$ is u. $\gamma(a)$ and $\gamma(b)$ are called the *source* and the *sink* of the chain. We introduce also the notions of *open* and *semi-open* chains. The *atoms* of a chain γ are the vertices and the edges of $\mathbb{V}_\kappa$ included in its range $[\gamma]$. We will only work with chains whose ranges are the union of their atoms. Thus vertices are closed chains, edges are open chains, the concatenation of chains is a well-defined operation, and a chain is the concatenation of its atoms. The operator ρ_κ that associates with a chain γ of $\mathcal{V}_\kappa$ the chain $q_\kappa \circ \gamma$ of $\mathbb{V}_\kappa$ is well-defined, one-to-one and onto. The inverse under ρ_κ of the chain γ of $\mathbb{V}_\kappa$ is termed the envelope of γ and is denoted $\varepsilon(\gamma)$. In particular the envelope of a vertex is a bitangent and the envelope of an edge is a primitive arc. The vertices of γ whose envelopes are not atoms of $\varepsilon(\gamma)$ are said to touch the chain γ. A chain is said to be *convex* if its envelope is convex. By a slight abuse of language a vertex of $\mathbb{V}_\kappa$ will be termed a bitangent of $\mathbb{V}_\kappa$.

(P2). The closure of a 2-cell $\overline{\sigma}_i$ is mapped homeomorphically by ϖ_κ onto the union of the closures of 2-cells of the arrangement of the z_c's. To describe the boundary of a 2-cell it is convenient to use the global parametrization of the space $\mathbb{D}$ of directed lines in the plane $\mathbb{R}^2$ given by the homeomorphism that associates with the directed line z the ordered pair whose first component is the direction of the line and whose second component $\tilde{z}$ is the signed distance from the origin of the plane to the line, that is, the common value of the scalar products of the points of the line and the direction orthogonal to the direction of the line and pointing in its left half-plane. With this parametrization in mind the closure of a bounded 2-cell σ_i of $\mathbb{V}_\kappa$ is mapped homeomorphically under ϖ_κ onto a domain of type

$$[z, z'] = \{(u, v) \mid u \in I, \tilde{z}(u) \leq v \leq \tilde{z}'(u)\}$$

where z and z' are images under ϖ_κ of chains γ and γ' of $\mathbb{V}_\kappa$ defined on the same interval $I = [a, b]$ and with the same endpoints, that is, $\gamma(a) = \gamma'(a)$ and $\gamma(b) = \gamma(b')$ where $I = [a, b]$. The chains γ and γ' are termed the *right boundary* and the *left boundary* of σ_i and are denoted $\mathrm{rc}(\sigma_i)$ and $\mathrm{lc}(\sigma_i)$, respectively. The *source* and the *sink* of σ_i are defined to be the common source of γ and γ' and the common sink of γ and γ'. They are denoted $\mathrm{sour}(\sigma_i)$ and $\mathrm{sink}(\sigma_i)$, respectively. (See Figure 9 and Figure 10.) There are also two unbounded 2-cells which are homeomorphic to the domain

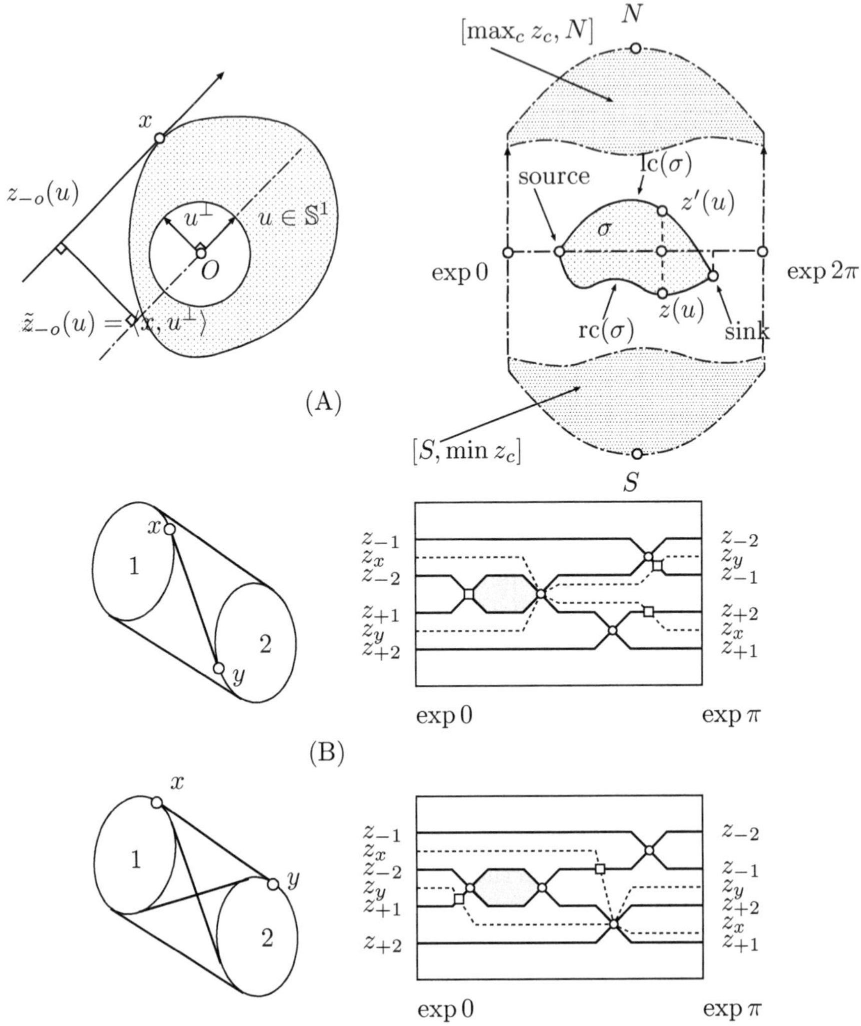

Fig. 9. (A) The space of directed lines. (B) The restriction to the domain $\exp([0,\pi]) \times \mathbb{R}$ of the arrangement of the z_c's for two disks associated with a 'separating' bitangent or with a 'non-separating' bitangent: each curve z_c is stretched and bended so that it stays roughly parallel to the equator of the space of lines. The crossing points of the curves z_c are marked with a small square or a small disk depending on whether they are supporting lines of unfree or free bitangents, and the dashed region represents one of the two connected components of the set of lines that separate the two disks.

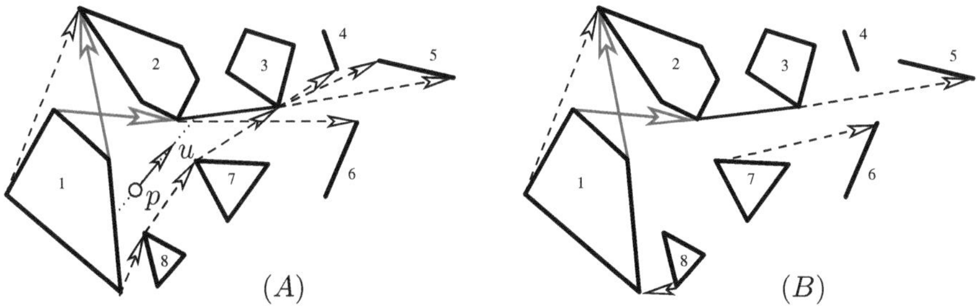

Fig. 10. Here $\kappa = \{v_{2,3}\}$. (A). The source and the sink of the 2-face that contains the equivalence class of the ray (p,u) are the bitangents $v_{-1,2}$ and $v_{1,-2}$, respectively. The sequences of bitangents of the chains $\mathrm{rc}(\sigma)$ and $\mathrm{lc}(\sigma)$ are $v_{-1,2}, v_{2,-6}, v_{2,3}, v_{3,5}, v_{3,-5}, v_{3,4}, v_{-7,3}, v_{-8,-7}, v_{1,-8}, v_{1,-2}$ and $v_{-1,2}, v_{-1,-2}, v_{1,-2}$, respectively. (B) $\varphi_{\mathrm{righ}}(v_{1,-2}) = v_{-7,-6}$, $\varphi_{\mathrm{left}}(v_{1,-2}) = v_{-1,-2}$, $\varphi_{\mathrm{forw}}(v_{1,-2}) = v_{3,5}$, $\varphi_{\mathrm{back}}(v_{1,-2}) = v_{-8,-1}$

$[S, \min_c\{z_c\}]$ of lines that contain the disks in their left half-planes and the domain $[\max_c\{z_c\}, N]$ of lines that contain the disks in their right half-planes. We write γ_+ and γ_- for the chains pre-images of the curves $\min_c\{z_c\}$ and $\max_c\{z_c\}$ under ϖ_κ.

(P3). The map ϖ_κ is locally modeled by a finite family $p_i : T_i \to S_i$ of (kind of) branched covering maps indexed by the integers of the interval $[-6, +6]$. We refer to Figures 11–15. The S_i's are defined to be $\mathbb{R}^2 = \{(x,y)\}$, except S_1 and S_{-1} which are defined to be the half-planes $\{x + y \leq 0\}$ and $\{x + y \geq 0\}$ of $\mathbb{R}^2$. $T_0 = S_0$,$T_{-1} = S_{-1}$, $T_1 = S_1$, and the other T_i's are gluings : we glue along some of their boundaries 2-dimensional horizontal cones or complements of cones $A_i^j \subset \{z = h_{ij}\}$ of $\mathbb{R}^3$ $(j = 1, 2, \ldots, j_i)$ whose vertical projections in the plane T_0 are unions of cones of the family of cones generated by two of the 12 vectors $\exp(\pi/4 + k\pi/6)$ $(k \in \mathbb{N})$. The cone A_i^j is labeled j in the figures. The gluing is done along the half-lines that are generated by the same vector of the family $\{\exp(\pi/4 + k\pi/2)\}$ (labeled α in the Figures 11–15). Finally the p_i's are defined to be the restrictions to the T_i's and S_i's of the projection $(x,y,z) \in \mathbb{R}^3 \mapsto (x,y) \in \mathbb{R}^2$. To every $B \in S_i$ we assign the sequence $\mu(B)$ of indexes $j_1 j_2, \ldots$ of the cones A_i^j pierced by the vertical line through B directed along the increasing values of z. The values of $\mu()$ on each connected components of the complement in S_i of the projection under p_i of the 1-skeleton of T_i are indicated in Figures 11–15. The exact values of the h_{ij} are not relevant for our purpose: only the values of $\mu()$ are. By 'locally modeled' we mean that every point x of $\mathbb{V}_\kappa$ has a neighborhood W such that the restriction $\varpi_{\kappa|_W}$ of ϖ_κ to W and $\varpi_\kappa(W)$ is equivalent to one of the p_i's, that is,

(1-) there are an index $i = i(x)$ and two homeomorphisms $\alpha : T_i \to W$ and $\beta : S_i \to \varpi_\kappa(W)$ such that $p_i \circ \alpha = \beta \circ \varpi_{\kappa|_W}$, and $\beta(x,y) = (u(x), v(y)) \in \mathbb{S}^1 \times \mathbb{R}$ where u and v are increasing functions, and

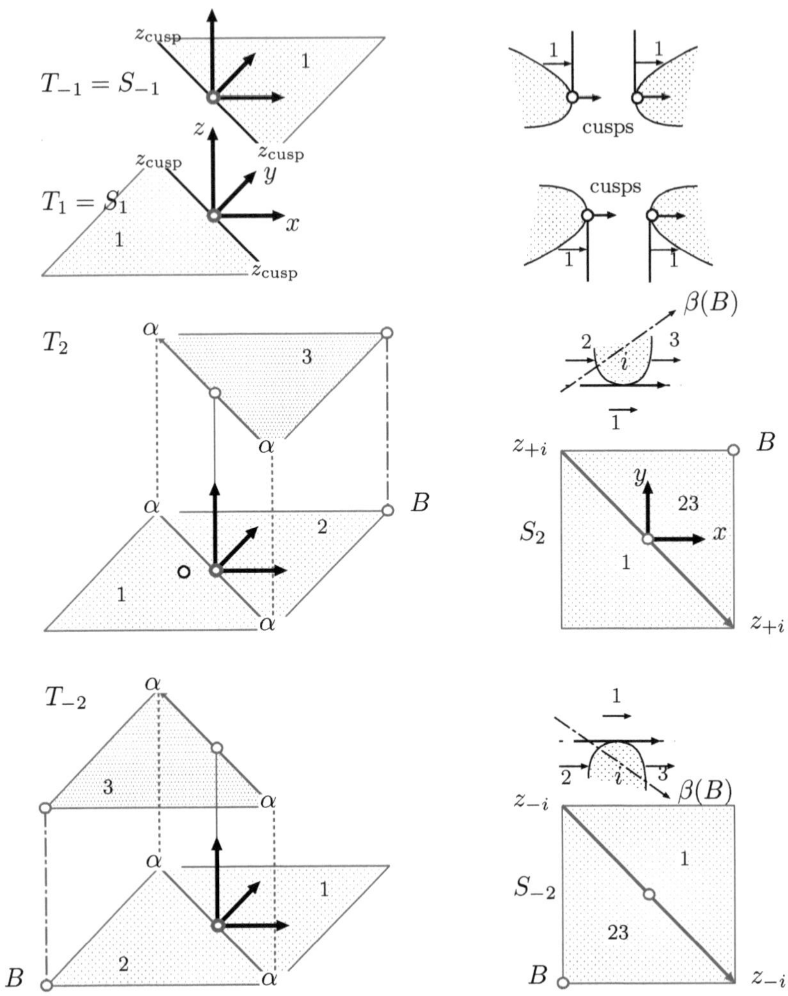

Fig. 11. The spaces $T_{\pm 1}$ and $T_{\pm 2}$ are models for the neighborhoods of a point on an edge. We will use the more handy notations $\sigma_{\text{back}}, \sigma_{\text{forw}}$ for the faces σ_2, σ_3, and σ_{righ} or σ_{left} for σ_1 depending on whether the edge is positive or negative.

(2-) for every $\ell \in V = \varpi_\kappa(W)$ the order of the points $x_1, x_2, \ldots, x_k$ of the pre-image of ℓ under $\varpi_{\kappa|_W}$ induced by the order of their images s_i under $g_\kappa \times 1_{\mathbb{S}^1}$ along the line ℓ corresponds to the order of the $\alpha^{-1}(x_i)$'s along the vertical line through $\beta^{-1}(\ell)$.
Of course the function $x \mapsto i(x)$ is constant on each cell σ of $\mathbb{V}_\kappa$. Thus a 0- or 1-cell e whose points have neighborhoods modeled by p_i is incident to j_i 2-cells σ_j that correspond to the j_i cones A_i^j of T_i ($j = 1, 2, \ldots, j_i$). To denote these 2-cells we will use the more handy notations σ_α where α belongs

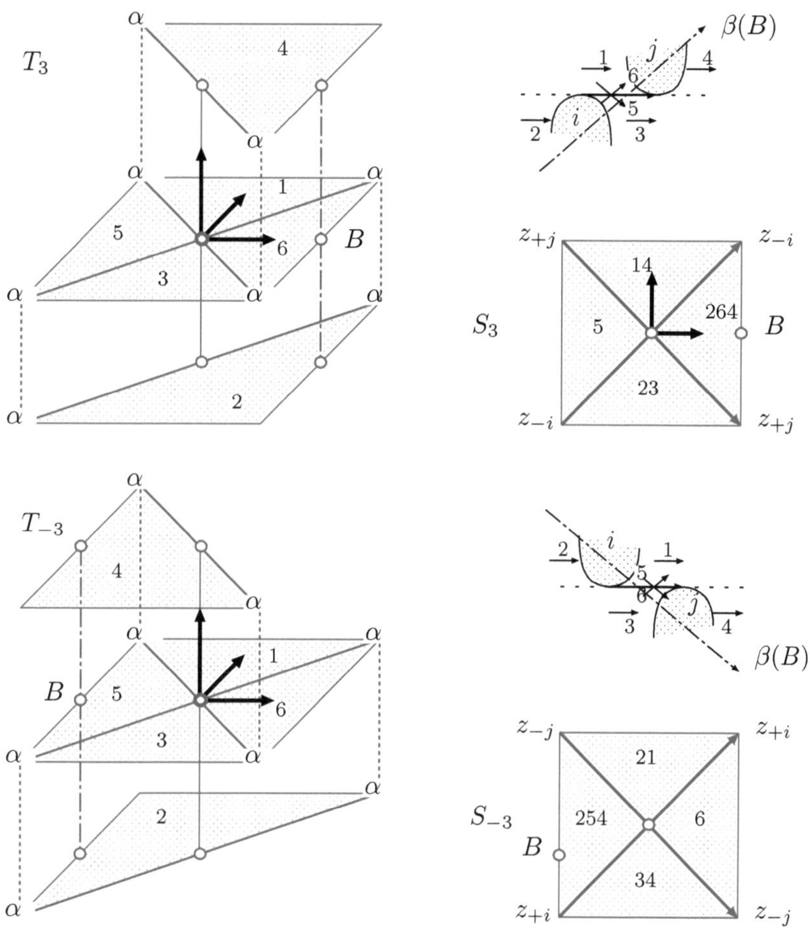

Fig. 12. The spaces T_3 and T_{-3} are models for the neighborhoods of the vertices not in κ of type right-left and left-right. We will use the more handy notations $\sigma_{\text{left}}, \sigma_{\text{back}}, \sigma_{\text{righ}}, \sigma_{\text{forw}}, \sigma_{\text{sink}}$ and σ_{sour} for the faces $\sigma_1, \sigma_2, \sigma_3, \sigma_4, \sigma_5$ and σ_6.

to the alphabet

$$\Lambda = \{\text{sour}, \text{sink}, \text{righ}, \text{left}, \text{back}, \text{forw}, \text{sour}_h, \text{sink}_h, \text{sour}_t, \text{sink}_t\}$$

as explained in the captions of the Figures 11–15. For e a vertex or an edge and $\alpha \in \Lambda$ the sink of the cell $\sigma_\alpha(e)$ (if defined) is denoted $\varphi^\kappa_\alpha(e)$ or $\varphi_\alpha(e;\kappa)$. The sink of the cell(s) with source v is denoted $\varphi^\kappa_{\text{sink}}(v)$ or $\varphi_{\text{sink}}(v;\kappa)$. (Note that if $v \in \kappa$ then $\varphi^\kappa_{\text{sink}}(v) = \iota(v)$.)

We close this subsection with a characterization of chains which are boundary of a 2-face of the visibility complex. The proof is left to the reader.

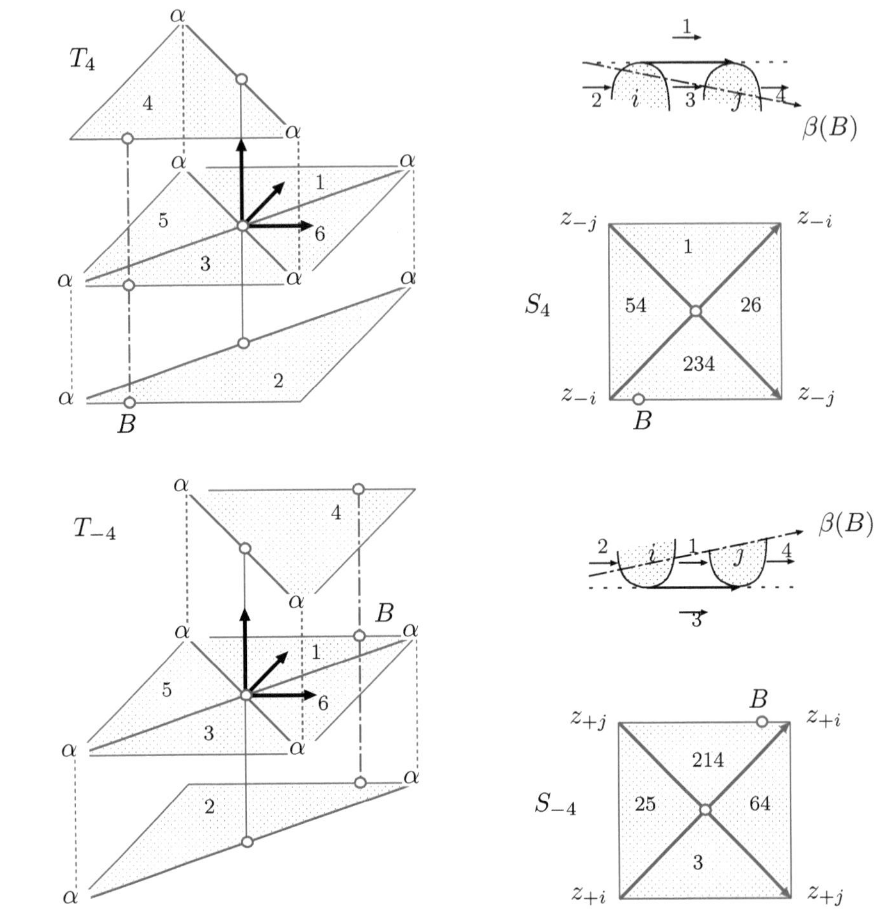

Fig. 13. The spaces T_4, and T_{-4} are models for the neighborhoods of the vertices not in κ of type right-right and left-left. We will use the more handy notations $\sigma_{\text{left}}, \sigma_{\text{back}}, \sigma_{\text{righ}}, \sigma_{\text{forw}}, \sigma_{\text{sink}}$ and σ_{sour} for the faces $\sigma_1, \sigma_2, \sigma_3, \sigma_4, \sigma_5$ and σ_6.

Theorem 4 (Face Description). *Let σ be a 2-face of $\mathbb{V}_\kappa$ whose source is not a bitangent of κ nor a hull-bitangent. Then its right boundary $\mathrm{rc}(\sigma)$ is the concatenation of three non empty convex subchains*

$$\mathrm{rc}(\sigma) = \mathrm{rc}_1(\sigma)\, \mathrm{rc}_2(\sigma)\, \mathrm{rc}_3(\sigma) \tag{2}$$

whose atoms $a \neq \mathrm{sour}(\sigma), \mathrm{sink}(\sigma)$ are characterized by $\varphi^\kappa_{\text{back}}(a)$, $\varphi^\kappa_{\text{left}}(a)$ and $\varphi^\kappa_{\text{forw}}(a) = \mathrm{sink}(\sigma)$, respectively. In particular the envelopes Σ_i of the chains $\mathrm{rc}_i(\sigma)$ satisfy the three following properties

(1-) the bitangents of Σ_1 but its source and the bitangents of Σ_3 but its sink are bitangents of (the oriented version of) κ;

(2-) no free bitangent enters Σ_1 nor leaves Σ_3;

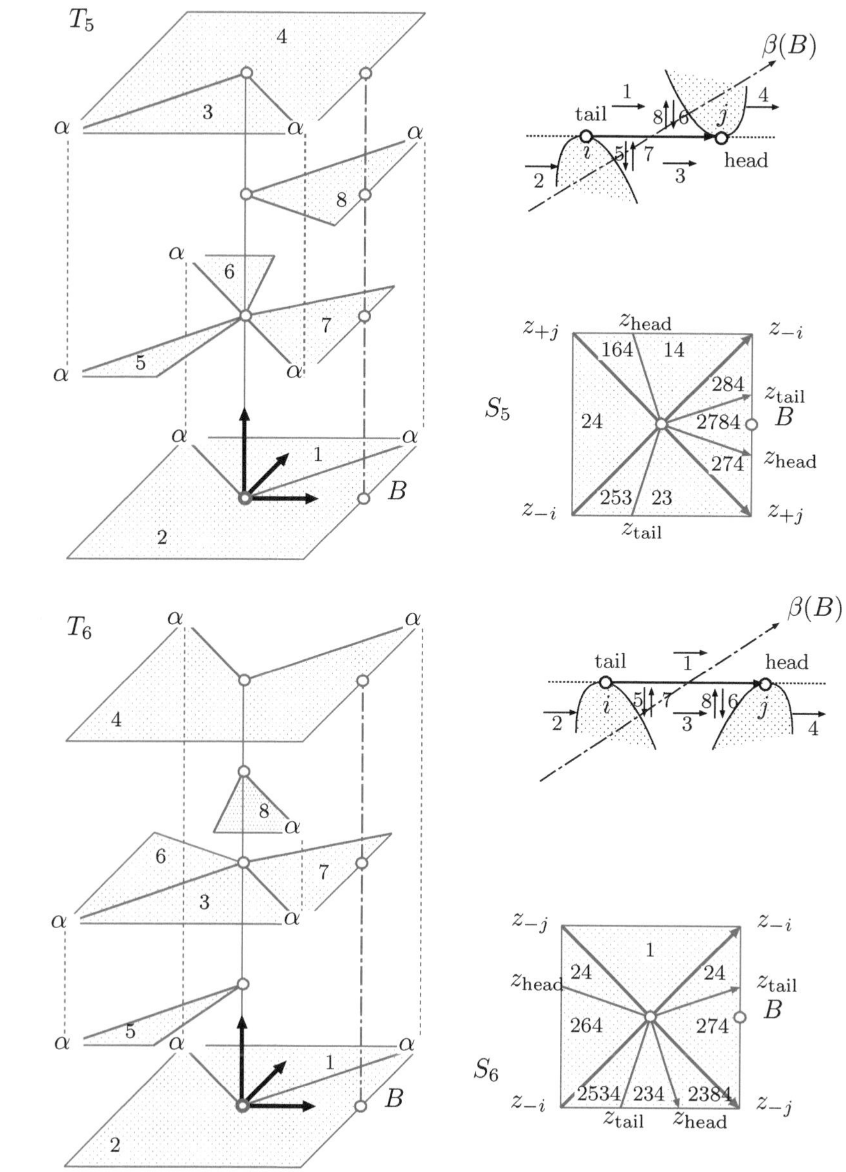

Fig. 14. The spaces T_5 and T_6 are models for the neighborhoods of the vertices of κ of type right-left and right-right. We will use the more handy notations $\sigma_{\rm left}, \sigma_{\rm back}, \sigma_{\rm righ}, \sigma_{\rm forw}$, for the faces $\sigma_1, \sigma_2, \sigma_3, \sigma_4$ and $\sigma_{{\rm sink}_t}, \sigma_{{\rm sink}_h}$, $\sigma_{{\rm sour}_t}$, and $\sigma_{{\rm sour}_h}$ for the faces $\sigma_5, \sigma_6, \sigma_7$, and σ_8.

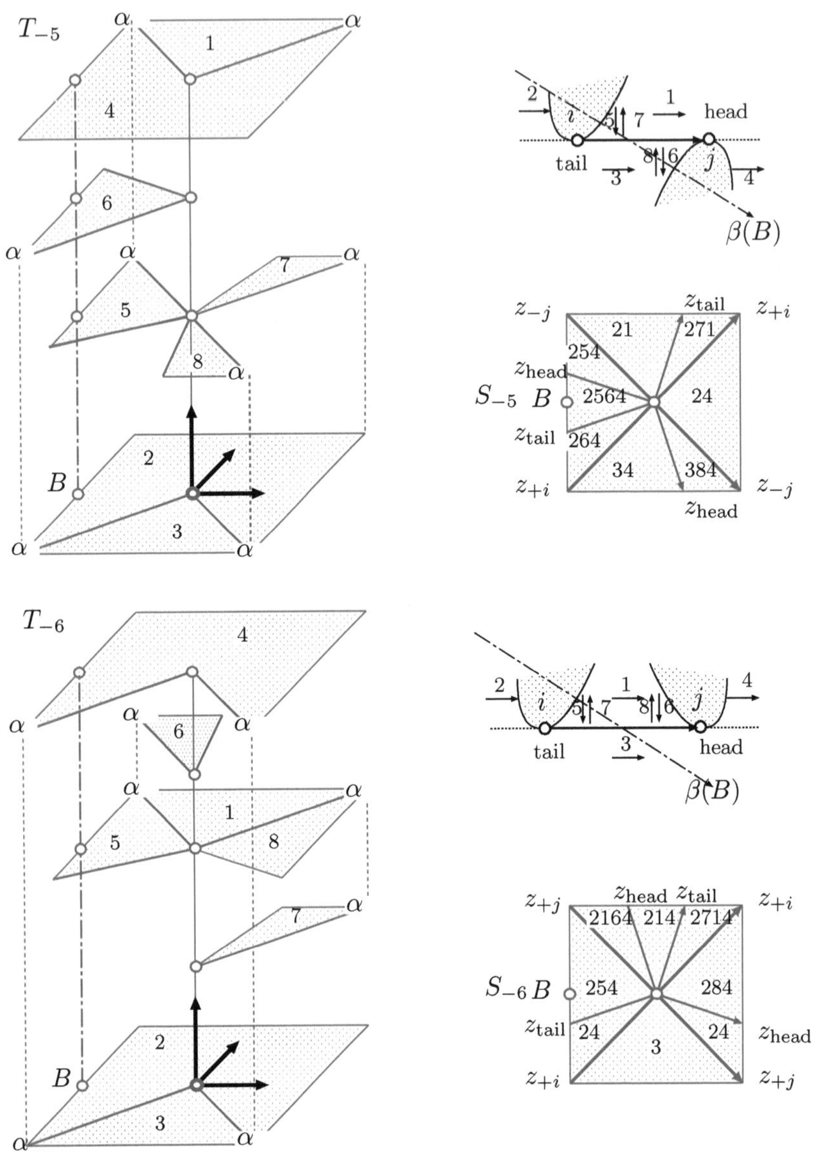

Fig. 15. The spaces T_{-5} and T_{-6} are models for the neighborhoods of the vertices of κ of type left-right and left-left. We will use the more handy notations $\sigma_{\text{left}}, \sigma_{\text{back}}, \sigma_{\text{righ}}, \sigma_{\text{forw}}$, for the faces $\sigma_1, \sigma_2, \sigma_3, \sigma_4$ and $\sigma_{\text{sink}_t}, \sigma_{\text{sink}_h}$, σ_{sour_t}, and σ_{sour_h} for the faces σ_5, σ_6, σ_7, and σ_8.

(3-) no free bitangent enters nor leaves Σ_2.
Conversely any closed chain Σ *of bitangents and arcs that is the concatenation of three convex subchains* $\Sigma_1\Sigma_2\Sigma_3$ *that satisfy the three properties above is the envelope of the right boundary of a face whose source and sink are the source and the sink of* Σ. □

Remark 1. *A similar theorem can be stated in the case where the source* v *of* σ *is a bitangent of* κ *or is a hull-bitangent. For example if* $v \in \kappa$ *and if* σ *is the 2-face associated with the tail of* v *(that is, the face* $\sigma_7 = \sigma_{\text{sour}_t}$ *in the notations of Figures 14 and 15) its left boundary is reduced to the chain* $ve_{\text{tail}}(v)\iota(v)$ *where* $e_{\text{tail}}(v)$ *is the edge supported by the cycle associated with the cusp corresponding to the tail of* v *whose range of directions is* $(u, -u)$ *where* u *is the direction of* v, *and its right boundary* $\text{rc}(\sigma)$ *is decomposed into four convex subchains* $\text{rc}_1(v)\,\text{rc}_2(v)\,\text{rc}_3(v)\,\text{rc}_4(v)$ *where* $\text{rc}_4(v) = \iota(v)$ *is the sink of* σ. *The chains* rc_1, rc_2, *and* rc_3 *satisfy the three properties mentioned in the above theorem. We will use freely this extension without more details.*

2.3 The minimum element, right-to-left, flip and minimal elements properties

For technical but good reasons it is now preferable to work in the cover spaces $\mathcal{W}, \mathcal{W}_\kappa$, and $\mathbb{W}_\kappa$ of $\mathcal{V}, \mathcal{V}_\kappa$, and $\mathbb{V}_\kappa$ induced by the covering maps $p_\kappa : \mathbb{F}_\kappa \times \mathbb{R} \to \mathbb{F}_\kappa \times \mathbb{S}^1$ that associate with $x = (o(x), \theta(x))$ the ray with origin o and direction $(\cos\theta, \sin\theta)$; the real $\theta(x)$ is called the angle of x and ι denotes now the operation that increases the angle of a ray by π. So $\mathcal{W} = p_\emptyset^{-1}(\mathcal{V}), \mathcal{W}_\kappa = p_\kappa^{-1}(\mathcal{V}_\kappa)$, and $\mathbb{W}_\kappa$ is the quotient space of $\mathbb{F}_\kappa \times \mathbb{R}$ by the equivalence relation R^1_κ defined by : $(x, x') \in R^1_\kappa$ if $\theta(x) = \theta(x')$ and $(p(x), p(x')) \in R_\kappa$. Elements of $\mathbb{F}_\kappa \times \mathbb{R}, \mathbb{W}_\kappa$ are still called rays, X^0_κ, X^1_κ and X^2_κ denote the sets of 0-,1- and 2-cells of $\mathbb{W}_\kappa$ (we will use the same symbols for the operators corresponding to the φ^κ_α's and ρ_κ's), and $\prec_\kappa$ denotes the partial order defined on the 1-skeleton of $\mathbb{W}_\kappa$ as the transitive closure of the relation defined on pair of rays tangent to a same directed disk by $x \prec_\kappa x'$ if $\theta(x) < \theta(x')$. It is not hard to see that two crossing bitangents of X^0_κ are comparable with respect to $\prec_\kappa$. The design and correctness of the GFA is based on the three following theorems stated and proved in [42] in the case where κ is the empty set; up to some minor modifications the proofs are also valid in the case $\kappa \neq \emptyset$ and are not reported here.

Theorem 5 (Minimum Element Property). *The operator* $\varphi^\kappa : X^0_\kappa \to X^0_\kappa$ *that associates with* $v \in X^0_\kappa$ *the minimum element of the set of* $u \in X^0_\kappa$ *such that* u *crosses* v *and* $v \prec_\kappa u$ *is well-defined, one-to-one, and onto. Furthermore if* v *is a non-hull bitangent then* $\varphi^\kappa(v) = \varphi^\kappa_{\text{sink}}(v)$ *and if* v *is a hull bitangent then* $\varphi^\kappa(v) = \iota(v)$.

Recall that a filter I of a poset $(P, \preceq)$ is a subset of P such that if $x \in I$ and $x \preceq y$, then $y \in I$. The sets $I(\theta)$ of bitangents with angles $\geq \theta$ are examples of filters of the poset $(X^0_\kappa, \prec_\kappa)$. Each filter I of $(X^0_\kappa, \prec_\kappa)$ is associated with a greedy pseudotriangulation $\mathrm{G}(I) \subset I$ and a planar sub-complex $\mathrm{S}(I)$ of $\mathbb{W}_\kappa$ — the sweep structure S mentioned in the introduction. The pseudotriangulation

$$\mathrm{G}(I) = \{v_1, v_2, \ldots, v_{3n-3}\} \subset I \tag{3}$$

is defined as follows: (1-) v_1 is a minimal element of I, and (2-) v_{i+1} is a minimal element of the subset of bitangents of I crossing none of the bitangents $v_1, v_2, \ldots, v_i$; since crossing bitangents are comparable, $\mathrm{G}(I)$ is well-defined and is a supset of $\min_{\prec_\kappa} I$. The planar sub-complex $\mathrm{S}(I)$ is defined as the set of 1- and 2-cells σ of $\mathbb{W}_\kappa$ such that $\mathrm{sink}(\sigma) \in I$ but not $\mathrm{sour}(\sigma)$, that is,

$$\mathrm{S}(I) = \{\, \sigma \in X_\kappa^{\geq 1} \mid \mathrm{sink}(\sigma) \in I,\ \mathrm{sour}(\sigma) \notin I \,\}; \tag{4}$$

$\mathrm{S}(I)$ contains an edge e (and its incident faces) included in γ_c per cycle c, and for each face $\sigma \in \mathrm{S}(I)$ an edge $\sigma^+ \in \mathrm{lc}(\sigma)$ and an edge $\sigma^- \in \mathrm{rc}(\sigma)$; for σ a face in $\mathrm{S}(I)$ the bitangent $\mathrm{sink}(\sigma) \in I$ is termed right-minimal if $\mathrm{sink}(\sigma) = \mathrm{sink}(\sigma^-)$; similarly $\mathrm{sink}(\sigma) \in I$ is termed left-minimal if $\mathrm{sink}(\sigma) = \mathrm{sink}(\sigma^+)$.

Theorem 6 (Flip, Right-to-left and Minimal Elements Properties). *Let I be a filter of X_κ^0, let $v \in \mathrm{G}(I)$, and let $u \in I$ crossing v. Then*

(1-) $\mathrm{G}(I) = I \setminus \varphi^\kappa(I) \subset \mathrm{sink}(\mathrm{S}(I))$;

(2-) $v \prec_\kappa u$;

(3-) v is a minimal element of I iff. v is a right-minimal and left-minimal element of I. □

For $v \in \mathrm{G}(I)$, let T be the pseudotriangle of the unoriented version of $\mathrm{G}(I)$ incident upon the unoriented version $\overline{v}$ of v and lying locally around $\overline{v}$ in the right half-plane of the supporting line of v, and let u be the minimal bitangent of $\mathrm{G}(I)$ whose unoriented version lies in the boundary of T. We denote by $\mathrm{R}(v, I)$ the unique (semi-open) chain of $\mathcal{W}$ whose tail is the tail of u and whose head is the head of $\iota(u)$ whose projection in the plane is the boundary of T. Note that since $\mathrm{G}(I)$ is a pseudotriangulation $\iota(u)$ is the unique bitangent of $p_\kappa^{-1}(\mathcal{B}_{\mathrm{G}(I)})$ touching the chain $\mathrm{R}(v, I)$. By a slight abuse of language $\mathrm{R}(v, I)$ is called a pseudotriangle of $\mathrm{G}(I)$. According to the Flip Property, if v is right-minimal in I then $\varphi^\kappa(v)$ leaves $\mathrm{R}(v, I)$; in that case $\mathrm{R}(v, I)$ is independent of I and is simply denoted $\mathrm{R}(v; \kappa)$; walking along the chain $\mathrm{R}(v; \kappa)$, starting from the tail of v, we find successively the convex subchains $\mathrm{R}^1(v; \kappa), \mathrm{R}^2(v; \kappa), \mathrm{R}^3(v; \kappa)$, and $\mathrm{R}^4(v; \kappa)$ of $\mathrm{R}(v; \kappa)$. We introduce the left version $\mathrm{L}(v, I) := \iota^{-1}\, \mathrm{R}\, \iota(v, I)$ of $\mathrm{R}(v, I)$ and we omit the mention of κ when $\kappa = \emptyset$.

Remark 2. *An interesting consequence of the Flip Property is that the set of subsets of pairwise non-crossing undirected non-hull bitangents ordered by inclusion is an abstract polytope [8, 28]. This abstract polytope is realizable in some $\mathbb{R}^d$ as a perturbation of the cone of expansive motions of some point set in the plane [48].*

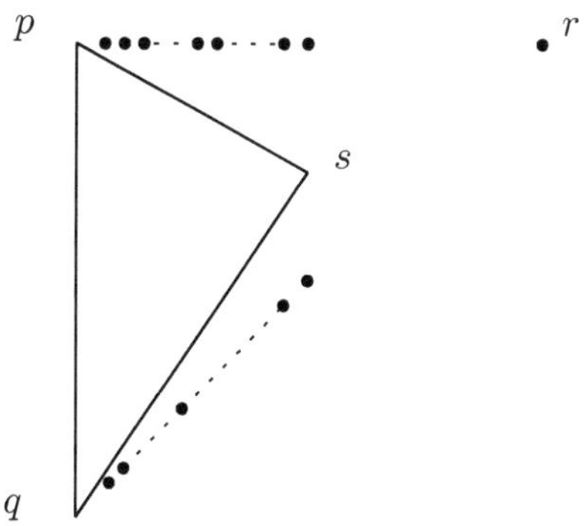

Fig. 16. In the configuration $\{(i,0),(i,i-n)\}_{1\leq i\leq n/2-1}$ augmented with the point $r=(n,0)$ and the triangle joining the three points $p=(0,0), q=(0,-n)$ and $s=(n/2-1,-n/4-1/2)$ we have $\sum_v |\mathrm{R}'(v)| = \Omega(n^2)$ but $|\mathcal{B}| = O(n)$.

2.4 The Greedy Flip Algorithm (GFA)

We assume now that $\kappa = \emptyset$. The GFA maintains a representation of the sweep structure $\mathrm{S}(I)$ and a representation of (the pseudotriangles of) $\mathrm{G}(I)$ when the filter I of $(X^0, \prec)$ describes a maximal chain of filters of the interval $[I(0), I(\pi)]$, — recall that $I(\theta)$ is the filter of bitangents with angle $\geq \theta$. We use a linked structure to represent the incidence relations in $\mathrm{S}(I)$ and $\mathrm{G}(I)$. Furthermore an atom of $\mathrm{S}(I)$ points (via the sink() operator) to the corresponding bitangent of $\mathrm{G}(I)$. Finally we have access to the linked structure via the set of 'minimal' faces of $\mathrm{S}(I)$, that is, the faces whose corresponding bitangents in $\mathrm{G}(I)$ are minimal in the filter I. It should be clear that the update of the data structure when a minimal bitangent v of I is deleted from I can be done in constant time provided we have access to the two arcs of $\mathrm{G}(I)$ touched by $\varphi(v)$. We explain now how these two arcs can be computed in constant amortized time using only the predicate χ_2. A crucial consequence of the right-to-left property is that a bitangent touching $\mathrm{R}(v)$ can only touch some specific subchain(s) of $\mathrm{R}(v)$; in particular $\varphi(v)$ leaves either the chain $\mathrm{R}^2(v)$ or the first atom of $\mathrm{R}^3(v)$ (which in that case should be an arc). This motivates the introduction of the chains $\mathrm{R}^{\mathrm{lea}}(v) = \mathrm{R}^2(v)\,\mathrm{aR}^3(v)$ and $\mathrm{R}^{\mathrm{ent}}(v) = \mathrm{R}^3(v)\,\mathrm{R}^4(v)$ —where $\mathrm{aR}^3(v)$ is the chain reduced to the first atom of $\mathrm{R}^3(v)$ or the empty chain depending on whether this atom is an arc or not.

Theorem 7. *We claim that*

(1-) $\mathrm{R}^4(v)$ *is an arc;*

(2-) any bitangent that enters $\mathrm{R}(v)$ *enters* $\mathrm{R}^{\mathrm{ent}}(v)$*;*

(3-) any bitangent that leaves $\mathrm{R}(v)$ *leaves* $\mathrm{R}^{\mathrm{lea}}(v)$ *or* $\mathrm{R}^4(v)$*.*

In particular there is no bitangent touching the chain $\mathrm{R}^1(v)$ *and the bitangent* $\varphi(v)$ *leaves the chain* $\mathrm{R}^{\mathrm{lea}}(v)$*.* □

Since $\varphi(v)$ leaves $\mathrm{R}^{\mathrm{lea}}(v)$ and enters $\mathrm{L}^{\mathrm{ent}}(v)$ — the left version of $\mathrm{R}^{\mathrm{lea}}(v)$ that is formally defined as the chain $\iota^{-1}\,\mathrm{R}^{\mathrm{lea}}\,\iota(v)$ — a natural idea is to merge according to the χ_2 relations the increasing sequences of bitangents of the chains $\mathrm{R}^{\mathrm{lea}}(v)$ and $\mathrm{L}^{\mathrm{ent}}(v)$ until we detect the bitangent $\varphi(v)$. This leads to the following piece of code where the chains are considered as lists of arcs, where $\mathrm{bit}(r, l)$ is the bitangent going from the supporting cycle of the arc r to the supporting cycle of the arc l, and where r^+ and l^+ are the sinks of r and l.

Function FLIP2(v : BITANGENT) : PAIR OF ARCS
/* returns the arcs of $\mathrm{G}(I)$ that $\varphi(v)$ leaves and enters */
1 $C \leftarrow \mathrm{R}^{\mathrm{lea}}(v)$; $C' \leftarrow \mathrm{L}^{\mathrm{ent}}(v)$;
2 χ_2-WALK(C, C');
3 **return** $(\mathrm{Top}(C), \mathrm{Top}(C'))$

Procedure χ_2-WALK(C, C': LIST OF ARCS)
0 **while** $C \neq \emptyset$ and $C' \neq \emptyset$
1 $r \leftarrow \mathrm{Top}(C)$; $l \leftarrow \mathrm{Top}(C')$; $\varphi \leftarrow \mathrm{bit}(r, l)$;
2 **if** (φ leaves r and enters l) **exit**;
3 **if** $\chi_2(r^+, l^+)$ $C \leftarrow \mathrm{Pop}(C)$ **else** $C' \leftarrow \mathrm{Pop}(C')$;

Let $\mathrm{R}'(v)$ be the set of bitangents of the chain $\mathrm{R}^{\mathrm{lea}}(v)$ that precede $\varphi(v)$, that follow the minimal bitangent of the chain $\mathrm{R}^{\mathrm{lea}}(v)$, and that do not appear in the boundary of $\sigma_{\mathrm{sour}}(v)$. Since a bitangent is incident to a constant (6) number of faces, the amortized cost of the function FLIP2 on the whole set $\mathcal{B}$ of free bitangents is proportional to the average value of the size of $\mathrm{R}'(v)$. As illustrated in Figure 16 this average value is not constant. However, thanks to the sum of squares theorem to be proved in the next section, a better bound is obtained if we restrict our attention to the set of bitangents whose images by φ leave a clockwise directed disk.

Theorem 8. *We claim that* $\sum_{v \in \varphi^{-1}(\mathcal{B}')} |\mathrm{R}'(v)| = O(|\mathcal{B}|)$, *where* $\mathcal{B}'$ *is the set of free bitangents that leave a clockwise directed disk.* □

Thanks to this bound it is now not hard to modify FLIP2 to obtain an optimal algorithm : instead of running χ_2-WALK on the pair $(\mathrm{R}^{\mathrm{lea}}, \mathrm{L}^{\mathrm{ent}})$ we run it in tandem or synchronously on the four pairs $(\mathrm{R}^2, \mathrm{L}^2)$, $(\mathrm{aR}^3, \mathrm{L}^2)$, $(\mathrm{R}^2, \mathrm{aL}^3)$ and $(\mathrm{aR}^3, \mathrm{aL}^3)$.

Remark 3. *The weaker bound* $\sum_{v \in \mathcal{B}} |\mathrm{R}'(v)| = O(n^2)$ *also follows from our analysis. Thus the procedure* FLIP2 *is worst-case optimal, that is, optimal when the visibility complex has quadratic complexity.*

Remark 4. *Since the lists* $\mathrm{aR}^3(v)$ *and* $\mathrm{aL}^3(v)$ *have zero or one element the* χ_2 *test at line 3 of the procedure* χ_2-WALK *is useless when running the procedure on the three pairs* $(\mathrm{aR}^3, \mathrm{L}^2)$, $(\mathrm{R}^2, \mathrm{aL}^3)$ *and* $(\mathrm{aR}^3, \mathrm{aL}^3)$. *In these three cases the procedure* FLIP2 *is easily modified to use only the* χ_1 *predicate. This observation will be used in the design of our flip method based only on* χ_1.

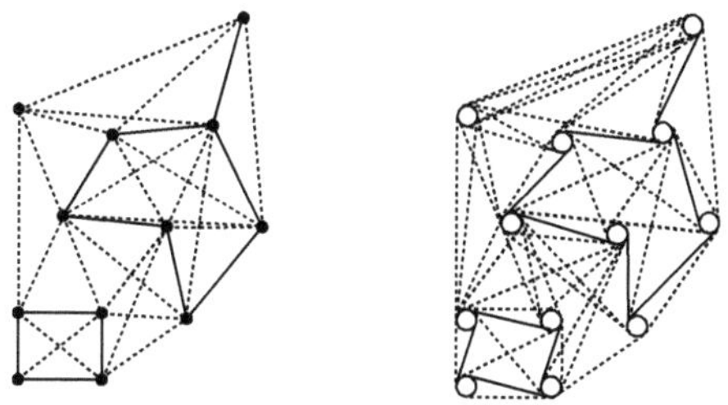

Fig. 17. This figure suggests how visibility graphs of polygons with holes fit into the theory. These are visibility graphs that arise as limits of tangent visibility graphs of circles and right-left bitangents.

Remark 5. *The above considerations are also valid in the case κ nonempty provided we define $\mathrm{R}^{\mathrm{lea}}(v;\kappa)$ to be the chain $\mathrm{R}^2(v;\kappa)$ augmented with the maximal prefix subchain of $\mathrm{R}^3(v;\kappa)$ that contain only bitangents of κ, and $\mathrm{R}'(v;\kappa)$ to be the set of bitangents of $\mathrm{R}^{2_1}(v;\kappa)$ that are neither in the boundary of $\sigma_{\mathrm{sour}}(v)$ nor in κ. Thus the amortized cost of the modified version of the function* FLIP2, *as explain above, is bounded by the maximal size of a view induced by κ. Thus the set $\mathcal{B}_\kappa$ is computable in optimal time $O(\#\mathcal{B}_\kappa + n\log n)$ under the assumption that the views induced by κ have constant complexities. In particular the classical visibility graph of a polygon with holes with a total of n vertices is computable in optimal $O(k + n\log n)$ time and linear space — here k is the size of the visibility graph. (See Figure 17.)*

3 The Sum of Squares Theorem

3.1 The left-turn or counterclockwise predicate for disks

Before going further in our analysis of the greedy flip algorithm we isolate a simple but crucial property of the left-turn predicate on a collection of 3 disks. This property will be repeatedly used in the next subsection to prove the correction of the flip method based only on the left-turn predicate, and its recognition is a first step toward a comprehensive study of the left-turn predicate in the spirit of the Knuth monograph on axioms and hulls [26]. A sequence of bitangents $t_1 t_2 \dots t_p$ such that $\chi_1(t_i, t_{i+1})$ $(i = 1, \dots, p-1)$ and $\chi_1(t_p, \iota(t_1))$ (equivalent to $\chi_1(t_1, t_p)$) is called a χ_1-sequence. Recall that v_{ij} denotes the bitangent going from the cycle c_i to the cycle c_j, and observe that $\iota(v_{ij}) = v_{-j-i}$. We start with a simple geometric lemma whose proof is left to the reader (cf. Figure 18).

Lemma 1. *Let c_i and c_j be two cycles and let $l(B)$ be the half-line that enters c_i at point B. We denote by r_0 and r_1 the half-lines that enter c_j and whose directions are the direction of $l(B)$ and its opposite. Let E and F be the tails of the bitangents v_{-ij} and v_{ij}. For X a point in the plane we denote by r_X the half-line through X that enters c_j and for X on the boundary of*

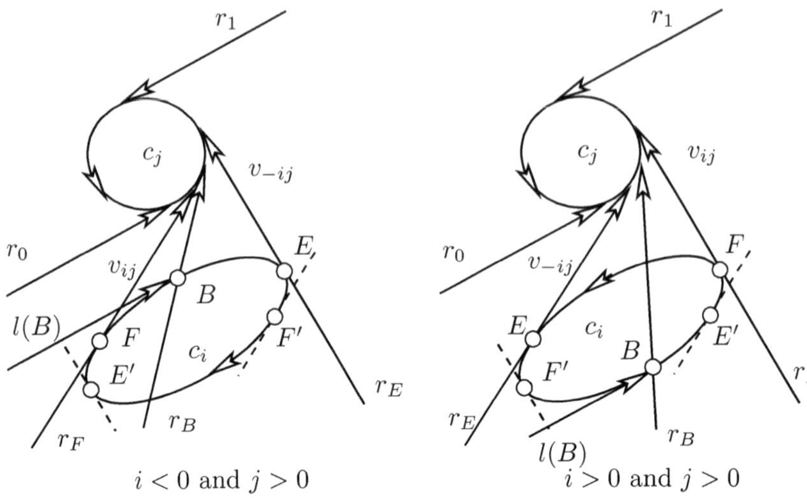

Fig. 18. Illustrate Lemma 1.

the underlying disk of c_i we denote by X' the touching point of the tangent to c_i whose direction is the opposite of the direction of the tangent to c_i through X. We claim that the half-lines r_0, r_B and r_1 enter in this order the cycle c_j (when traversing the boundary of o_j counterclockwise) iff. B belongs to the (counterclockwise) arc EF iff. the half-line $l(B)$ lies on the left side of the half-line r_B. Furthermore if c_i is a negative (resp. positive) cycle then

1. *the half-line r_X pierces the disk o_i iff. r_F, r_X, r_E (resp. r_E, r_X, r_F) enter in this order the cycle c_j;*
2. *the half-line r_0 pierces o_i iff. B belongs to the arc FE' (resp. $E'F$);*
3. *the half-line r_1 pierces o_i iff. B belongs to the arc $F'E$ (resp. EF').*

In particular if $B \in EF$ and c_i is a negative cycle then r_0 and r_1 do not pierce o_i. □

Theorem 9. *Consider three cycles c_i, c_j, and c_k whose underlying disks are distinct. Assume that the bitangents v_{ki}, v_{kj} and v_{-ij} are free and that $\chi_1(v_{ki}, v_{kj})$. We claim that the three following sequences*
(1-) $v_{ki}v_{kj}v_{-ij}$;
(2-) $v_{ki}v_{ij}v_{-ij}$ under the assumption that c_i is a negative cycle;
(3-) $v_{kj}v_{-ij}v_{-i,-j}$ under the assumption that c_j is a positive cycle;
are χ_1-sequences.

Proof. Since an axial symmetry reverses the χ_1-relations and the signs of the disks Claim (1-) is valid for the triple (i, j, k) iff. it is valid for the triple $(-j, -i, -k)$. Therefore it is sufficient to consider the case $k > 0$. We use the

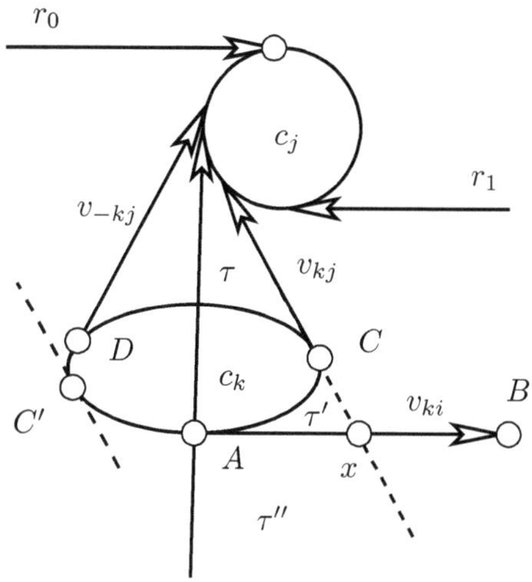

Fig. 19. Illustrate the proof of Theorem 9.

notations introduced in the statement of Lemma 1 and/or Figure 18 plus the notations of Figure 19, — τ is the region swept by an half-line that enters c_j and rotates from r_A to r_C; this region is decomposed into three subregions τ, τ', τ'' as indicated in Figure 19 —, and we write that $r < r' < r''$ to state that the half-lines r, r', and r'' enter in this circular order the cycle c_j when traversing the boundary of its underlying disk counterclockwise. Claim (1) is therefore equivalent to say that $r_0 < r_C < r_E < r_1$ which follows easily from the following observations.

1. $r_0 < r_A < r_C < r_1$. Indeed since $\chi_1(v_{ki}, v_{kj})$ the point A belongs to the arc $C'C$ and consequently $A \notin DC'$ (here we use $k > 0$). Therefore according to Lemma 1 applied to the cycles c_k, c_j and c_A the half-lines r_0, r_A, and r_1 enter in this order the cycle c_j and r_1 does not pierce o_k.

2. $r_A < r_B < r_1$. Indeed r_A is the supporting half-line of the bitangent $v_{lj}(= v_{Aj})$ where l is the negative cycle associated with $[AB]$ and r_1 does not cross o_l. Here we use the second claim of Lemma 1 applied to the cycles c_k, c_j and c_A.

3. $B \in EF$. Since $r_0 < r_B < r_1$ (from observations 1. and 2.) and according to Lemma 1.

4. $E \notin \tau \cup \tau' \cup \tau''$ or, equivalently $r_A < r_C < r_E$. Because v_{Ej} is free and $[EB] \subset o_i$ which is disjoint from o_j, o_k and v_{kj}.

5. if $i < 0$ then $r_0 < r_B < r_E < r_1$. Indeed since $B \in EF$ (observation 3.) then, according to Lemma 1, the half-lines r_0 and r_1 do not pierce o_i.

6. if $i > 0$ then $r_C < r_E < r_B$. Indeed if it is not the case then r_C crosses o_i and consequently since v_{kj} is free there is a point of o_i in τ' and, consequently, $E \in \tau'$: impossible.
   Similarly $r_A < r_C < r_B$ (otherwise $E \in \tau'$).

We turn now to the proof of Claim 2. According to Claim 1 the point B belongs to the arc EE' and consequently does not belong to the arc EF'. It follows according to Lemma 1 that r_1 does not pierce o_i and consequently that v_{ki} enters the arc EF. This proves Claim 2. The proof of Claim 3 is similar. □

3.2 Dual order

The dual partial order $\prec_*$ is defined on the same set of bitangents X^0 by $u \prec_* v$ if and only if $v \prec u$. Any operator $\Phi : X \to \mathcal{P}(X)$ defined with respect to the partial order $\prec$ has a dual version defined with respect to the dual partial order $\prec_*$; this dual version is denoted Φ_* and satisfies similar properties. In particular $\varphi_*(v)$ is the $\prec_*$-minimal bitangent in the set of bitangents $u \in X^0$ that cross v and that $\prec_*$-succeed v (Minimum Element Property), and $G_*(J) = J \setminus \varphi_*(J)$ where $G_*(J)$ is the greedy (with respect to $\prec_*$) pseudotriangulation associated with the filter J of $(X^0, \prec_*)$ (Flip Property), and $\mathrm{R}_*(v)$ and $\mathrm{L}_*(v)$ are the dual greedy pseudotriangles associated with the bitangent v. One can easily check that $\varphi_* = \varphi^{-1}$ and that $\varphi_* G(I) = G_*(I_*)$, — here I is a filter of $(X^0, \prec)$ and I_* is its "dual filter", that is, the filter of $(X^0, \prec_*)$ defined by $I_* = X^0 \setminus I$.

Let I be a filter with $u \in \min_\prec I$ and let e be a primitive arc of $\mathcal{W}$ with source u and sink v. According to the flip property the pseudotriangle $\mathrm{R}(v)$ or $\mathrm{L}(v)$ coincides with $\mathrm{R}(\varphi(u), I \setminus u)$ or with $\mathrm{L}(\varphi(u), I \setminus u)$ which in turn can be expressed as a concatenation of the chains $\mathrm{R}^i(u)$, $\mathrm{R}^{i_1}(u)$, $\mathrm{R}^{i_2}(u)$, $\iota(e)$, and their left versions — here $\mathrm{R}^{i_1}, \mathrm{R}^{i_2}$ are the prefix and suffix subchains (possibly reduced to the empty chains[6]) of the chain $\mathrm{R}^i(v)$ cut at (the tail of) $\varphi(v)$ for $i = 2, 3$. The explicit decomposition depends on the *type* of the primitive arc (or edge) under consideration : A *positive* primitive arc (that is, an arc supported by a positive cycle or a counterclockwise directed disk) with source u, with sink v, and supported by the cycle c_i is said to be a *0-cusp* arc if u enters c_i and v leaves c_i; a *sink-cusp* arc if both u and v enter c_i; a *source-cusp* arc if both u and v leave c_i (source-cusp arcs and sink-cusp arcs are also termed *1-cusp* arcs); a *2-cusp* arc if u leaves c_i and v enters c_i (and similar terms are introduce for negative arcs). See Table 2. Similar relations hold in the dual setting (cf. Table 3) and for negative arcs. We use now these relations to prove by induction three theorems on the structure of the greedy pseudotriangles. The first theorem describes the greedy pseudotriangles in terms of the orbits of the φ_α's and the operator $\rho = \rho_\emptyset$.

[6]If $\varphi(v)$ does not leave the chain $\mathrm{R}^i(v)$ we set $\mathrm{R}^{i_1}(v) = \mathrm{R}^i(v)$ and $\mathrm{R}^{i_2}(v) = \emptyset$ $(i = 2, 3)$.

Table 2. Relations between greedy pseudotriangles associated with the source u and the sink v of a positive arc e.

0-cusp arc $v = \varphi_{\text{forw}}(u)$ $u = \varphi_{\text{back}_*}(v)$ 	$\mathrm{R}(v) = \mathrm{R}(\varphi(u), I \setminus u)$ $\mathrm{R}^1(u) = ue\,\mathrm{R}^1(v)$ $\mathrm{R}^2(v) = \mathrm{R}^{2_1}(u)$ $\mathrm{R}^3(v) = \mathrm{R}^{3_1}(u)\varphi(u)\,\mathrm{L}^{2_2}(u)$ $\mathrm{R}^4(v) = \mathrm{L}^{3_2}(u)\iota(e)$
sink cusp arc $v = \varphi_{\text{righ}}(u)$ $u = \varphi_{\text{left}_*}(v)$ 	$\mathrm{L}(v) = \mathrm{R}(\varphi(u), I \setminus u)$ $\mathrm{L}^1(v) = \mathrm{R}^{2_1}(u)$ $\mathrm{L}^2(v) = \mathrm{R}^{3_1}(u)\varphi(u)\,\mathrm{L}^{2_2}(u)$ $\mathrm{L}^3(v) = \mathrm{L}^{3_2}(u)\iota(e)$ is an arc $\mathrm{L}^4(v) = \emptyset, \qquad \mathrm{R}^1(u) = ue$
source cusp arc $v = \varphi_{\text{left}}(u)$ $u = \varphi_{\text{righ}_*}(v)$ 	$\mathrm{R}(v) = \mathrm{L}(\varphi(u), I \setminus u)$, $u \notin \gamma_+$ $\mathrm{R}^1(v) = \mathrm{L}^{2_1}(u), \mathrm{L}^1(u) = u$ $\mathrm{R}^2(v) = \mathrm{L}^{3_1}(u)\varphi(u)\,\mathrm{R}^{2_2}(u)$ $\mathrm{R}^3(v) = \mathrm{R}^{3_2}(u)$ $\mathrm{R}^4(v) = \mathrm{R}^4(u)\iota(e)$
2-cusp arc $v = \varphi(u)$ $u = \varphi_*(v)$ 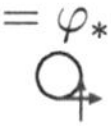	$\mathrm{L}(v) = \mathrm{L}(\varphi(u))$ $\mathrm{L}^1(v) = v\,\mathrm{R}^{2_2}(u)$ $\mathrm{L}^2(v) = \mathrm{R}^{3_2}(u)$ $\mathrm{L}^3(v) = \mathrm{R}^4(u)\iota(e)$ $\mathrm{L}^4(v) = \emptyset$

Theorem 10 (Orbits Theorem). *Let v be a free bitangent not on γ_+ and let $e_{j0}v_{j1}e_{j1}, \ldots, v_{jk_j}e_{jk_j}$ with $k_j \geq 0$ be the chain $\mathrm{R}^j(v)$ where v_{jk} stands for a bitangent and e_{jk} for an arc possibly empty ($j = 1, 2, 3, 4$). We claim (under the assumption that v_{31} or v_{21} or both are well-defined) that $v_{31} = \varphi_{\text{back}}(v), v_{21} = \varphi_{\text{righ}}(v)$, and $\rho(e_{20}v_{21}) = \mathrm{lc}_3(\sigma_{\text{righ}}(v))$. Furthermore $v_{1,i+1} = \varphi_{\text{forw}}(v_{1i})$, $v_{2,i+1} = \varphi_{\text{back}}(v_{2i})$, and $v_{3,i+1} = \varphi_{\text{forw}}(v_{3i})$.*

Proof. We first show that $v_{31} = \varphi_{\text{back}}(v)$, under the assumption that v_{31} is well-defined, that is, v is a left-xx bitangent. Let $v_1 = \varphi_{\text{back}_*}(v)$, $\sigma := \sigma_{\text{sour}}(v_1)$, and let $v_1\rho(e_1)v_2\rho(e_2)\ldots$ be the chain $\mathrm{rc}_1(\sigma)$. Note that v appears in this chain. The arc e_1 is a positive 0-cusp arc, and the arc e_i ($i \geq 2$) is a positive source-cusp arc. Therefore if $v_1 \notin \gamma_+$ then, according to Table 2,

Table 3. Relations between dual greedy pseudotriangles associated with the source u and the sink v of a positive arc e.

0-cusp arc $v = \varphi_{\text{forw}}(u)$ $u = \varphi_{\text{back}_*}(v)$ 	$\mathrm{R}_*(u) = \mathrm{R}_*(\varphi_*(v), I_* \setminus v)$ $\mathrm{R}^1_*(v) = ve\,\mathrm{R}^1_*(u)$ $\mathrm{R}^2_*(u) = \mathrm{R}^{2_1}_*(v)$ $\mathrm{R}^3_*(u) = \mathrm{R}^{3_1}_*(v)\varphi_*(v)\,\mathrm{L}^{2_2}_*(v)$ $\mathrm{R}^4_*(u) = \mathrm{L}^{3_2}_*(v)\iota(e)$
sink cusp arc $v = \varphi_{\text{righ}}(u)$ $u = \varphi_{\text{left}_*}(v)$ 	$\mathrm{R}_*(u) = \mathrm{L}_*(\varphi_*(v), I_* \setminus v),\ v \notin \gamma_+$ $\mathrm{R}^1_*(u) = \mathrm{L}^{2_1}_*(v)$ $\mathrm{R}^2_*(u) = \mathrm{L}^{3_1}_*(v)\varphi_*(v)\,\mathrm{R}^{2_2}_*(v)$ $\mathrm{R}^3_*(u) = \mathrm{R}^{3_2}_*(v)\iota(e)$ is an arc $\mathrm{R}^4_*(u) = \emptyset, \qquad \mathrm{L}^1_*(v) = v$
source cusp arc $v = \varphi_{\text{left}}(u)$ $u = \varphi_{\text{righ}_*}(v)$ 	$\mathrm{L}_*(u) = \mathrm{R}_*(\varphi_*(v), I_* \setminus v)$ $\mathrm{L}^1_*(u) = \mathrm{R}^{2_1}_*(v), \mathrm{R}^1_*(v) = ve$ $\mathrm{L}^2_*(u) = \mathrm{R}^{3_1}_*(v)\varphi_*(v)\,\mathrm{L}^{2_2}_*(v)$ $\mathrm{L}^3_*(u) = \mathrm{L}^{3_2}_*(v)$ $\mathrm{L}^4_*(u) = \mathrm{L}^4_*(v)\iota_*(e)$
2-cusp arc $v = \varphi(u)$ $u = \varphi_*(v)$ 	$\mathrm{L}_*(u) = \mathrm{L}_*(\varphi_*(v))$ $\mathrm{L}^1_*(u) = u\,\mathrm{R}^{2_2}_*(v)$ $\mathrm{L}^2_*(u) = \mathrm{R}^{3_2}_*(v)$ $\mathrm{L}^3_*(u) = \mathrm{R}^4_*(v)\iota_*(e),$ $\mathrm{L}^4_*(u) = \emptyset$

$\mathrm{R}^3(v_2) = \mathrm{R}^{3_1}(v_1)\varphi(v_1)\,\mathrm{L}^{2_2}(v_1)$ and $\mathrm{R}^3(v_{i+1}) = \mathrm{R}^{3_2}(v_i)$ for $i \geq 2$. It follows that $\min_\prec \mathrm{R}^3(v_i) = \varphi(v_1)(= \varphi_{\text{back}}(v))$ for $i \geq 2$, and in particular $v_{31} := \min_\prec \mathrm{R}^3(v) = \varphi_{\text{back}}(v)$. Now if $v_1 \in \gamma_+$ then v_2 is the successor of v_1 on γ_+, $\mathrm{R}^3(v_3) = \iota(v_2)$, and $\mathrm{R}^3(v_{i+1}) = \mathrm{R}^{3_2}(v_i)$ for $i \geq 3$. Since $\varphi_{\text{back}}(v) = \iota(v_2)$ the result follows. Now we show that $v_{21} = \varphi_{\text{righ}}(v)$. Let $v' = \min_\prec \mathrm{R}^2(v)$ and let e be the first atom of $\mathrm{R}^2(v)$ if this atom is an arc. According to the right-to-left property there is no bitangent touching the chain $\mathrm{R}^1(v)$ and no bitangent entering the chain $\mathrm{R}^2(v) = ev' \cdots$. Therefore the chain $\mathrm{R}^1(v)ev'$ (which contains one cusp point) is a suffix subchain of the left boundary of the face whose sink is v'; therefore $v' = \varphi_{\text{righ}}(v)$. Finally the last part of the theorem is just a reformulation of Theorem 7 in terms of the φ_α's. □

The second theorem links $\mathrm{R}(v)$ and $\mathrm{R}(u)$ where u is the source of the 2-cell $\sigma_{\text{righ}}(v)$. The tail of the bitangent $\varphi(v)$ cuts the chain $\mathrm{R}^{\text{lea}}(v)$ into two subchains, denoted $\mathrm{R}^{\text{lea}_1}(v)$ and $\mathrm{R}^{\text{lea}_2}(v)$ where $\mathrm{R}^{\text{lea}_1}(v)$ comes before $\mathrm{R}^{\text{lea}_2}(v)$ when walking along $\mathrm{R}^{\text{lea}}(v)$.

Theorem 11 (Orbits Theorem Continued). *Let σ be a face of the visibility complex $\mathbb{V}$ with source u and sink u' not on γ_+. Let v be a bitangent of the chain $\mathrm{lc}_2(\sigma)$ and let κ be the unoriented version of the set of bitangents of the chain $\mathrm{lc}_2(\sigma)$ smaller than v. We claim that $\mathrm{R}^{\text{lea}}(v)$ is a prefix subchain of $\mathrm{R}^{\text{lea}}(v;\kappa)$, and that*

$$\mathrm{R}^{\text{lea}}(v;\kappa) = eu'\,\mathrm{R}^{\text{lea}_2}(u)\iota(e') \tag{5}$$

where e and e' are arcs characterized by $\rho(ue') = \mathrm{lc}_1(\sigma)$ and $\rho(eu') = \mathrm{lc}_3(\sigma)$, respectively.

Proof. Let $u_1u_2\ldots v$ be the increasing sequence of bitangents of the chain $\mathrm{lc}_2(\sigma)$ smaller than v. Since the edge with source u_i and sink u_{i+1} is a positive 0-cusp edge, the chain $\mathrm{R}^{\text{lea}}(u_{i+1})$ is a prefix subchain of $\mathrm{R}^{\text{lea}}(u_i)$ (see Table 2). It is therefore sufficient to show the Lemma in the case $v = u_1$. Let now $v_0\rho(e_0)v_1\rho(e_1)\ldots v_p\rho(e_p)$ be the chain $\mathrm{lc}_1(\sigma)$ where $v_0 = u$ and where e_i is the primitive arc with source v_i and sink v_{i+1} where $v_{p+1} = u_1$. We successively treat the cases $p \neq 0$ and $p = 0$. For $p \neq 0$ the primitive arc e_0 is a negative 0-cusp arc, the primitive arcs $e_1, e_2, \ldots, e_{p-1}$ are negative source-cusp arcs, and the arc e_p is a negative 2-cusp arc. Therefore, according to the negative version of Table 2, $\mathrm{L}^3(v_1) = \mathrm{L}^{3_1}(v_0)\varphi(v_0)\,\mathrm{R}^{2_2}(v_0)$ and $\mathrm{L}^4(v_1) = \mathrm{R}^{3_2}(v_0)\iota(e_0)$, or, after concatenation,

$$\mathrm{L}^3(v_1)\,\mathrm{L}^4(v_1) = \mathrm{L}^{3_1}(v_0)\varphi(v_0)\,\mathrm{R}^{\text{lea}_2}(v_0)\iota(e_0);$$

similarly

$$\begin{aligned}\mathrm{L}^3(v_{i+1})\,\mathrm{L}^4(v_{i+1}) &= \mathrm{L}^{3_2}(v_i)\,\mathrm{L}^4(v_i)\iota(e_i)\\ &= (\mathrm{L}^{3_1}(v_i))^{-1}\,\mathrm{L}^3(v_i)\,\mathrm{L}^4(v_i)\iota(e_i), \qquad (i = 1,2,\ldots,p-1),\end{aligned}$$

(here we use the notation $\gamma_1^{-1}\gamma$ for γ_2 when $\gamma = \gamma_1\gamma_2$) and

$$\begin{aligned}\mathrm{R}^{\text{lea}}(u_1) &= \mathrm{L}^{3_2}(v_p)\,\mathrm{L}^4(v_p)\iota(e_p)\\ &= (\mathrm{L}^{3_1}(v_p))^{-1}\,\mathrm{L}^3(v_p)\,\mathrm{L}^4(v_p)\iota(e_p);\end{aligned}$$

from which we deduce that

$$\begin{aligned}\mathrm{R}^{\text{lea}}(u_1) &= (\mathrm{L}^{3_1}(v_p))^{-1}\ldots(\mathrm{L}^{3_1}(v_i))^{-1}\ldots(\mathrm{L}^{3_1}(v_1))^{-1}\\ &\quad\ \mathrm{L}^{3_1}(v_0)\varphi(v_0)\,\mathrm{R}^{\text{lea}_2}(v_0)\iota(e_0)\iota(e_1)\ldots\iota(e_p)\\ &= e''\varphi(v_0)\,\mathrm{R}^{\text{lea}_2}(v_0)\iota(e')\end{aligned}$$

where e'' is a suffix subarc of the arc $\mathrm{L}^{3_1}(v_0)$ and where e' is the concatenation of the arcs $e_0, \ldots, e_p$, that is, the envelope of $\mathrm{lc}_1(\sigma)$. This proves that $\mathrm{R}^2(u_1)$

is not reduced to an arc and consequently according to the Orbits Theorem $e'' = e$. This finishes the proof in the case $p \neq 0$. If $p = 0$, then the arc e_0 with source u and sink u_1 is a positive source-cusp arc or a negative sink-cusp arc. In the first case e' is the empty arc and, according to Table 2, $\mathrm{R}^2(u_1) = \mathrm{L}^{3_1}(v_0)\varphi(v_0)\,\mathrm{R}^{2_2}(v_0)$ and $\mathrm{R}^3(u_1) = \mathrm{R}^{3_2}(v_0)$. In the second case $e_0 = e'$, $\mathrm{R}^2(u_1) = \mathrm{L}^{3_1}(v_0)\varphi(v_0)\,\mathrm{R}^{2_2}(v_0)$, and $\mathrm{R}^3(u_1) = \mathrm{R}^{3_2}(v_0)\iota(e_0)$. In both cases

$$\mathrm{R}^{\mathrm{lea}}(u_1) = \mathrm{L}^{3_1}(v_0)\varphi(v_0)\,\mathrm{R}^{\mathrm{lea}_2}(v_0)\iota(e')$$

and we conclude as in the case $p \neq 0$. □

The third and last theorem of this subsection provides a duality relation between the greedy pseudotriangles. This duality relation plays a crucial role in the proof of the sum of squares theorem and is used to prove that some pseudotriangles are adjacent, and consequently share a common bitangent (cf. Theorem 3).

Theorem 12 (Duality Theorem). $u \in \mathrm{R}^{2_1}(v)$ *iff* $v \in \mathrm{L}_*^{2_1}(u)$.

Proof. Let $u \in [v, \iota(v)]$. We claim that
(0-) if u' and v are not comparable then $\varphi(u') \notin \mathrm{R}(v)$;
(1-) if u belongs to $\mathrm{R}(v)$ then $\varphi_*(u) \prec v$;
(2-) if u touchs $\mathrm{R}(v)$ then $v \preceq \varphi_*(u)$;
(3-) if u belongs to $\mathrm{R}^1(v)$ then v belongs to $\mathrm{R}_*^1(u)$;
(4-) if u leaves or belongs to $\mathrm{R}^{\mathrm{lea}}(v)$ then v belongs to or leaves $\mathrm{L}_*^{\mathrm{lea}}(u)$;
(5-) if u belongs to or enters $\mathrm{R}^{\mathrm{ent}}(v)$ then v belongs to or leaves $\mathrm{R}_*^{\mathrm{lea}}(u)$;
(6-) if u leaves $\mathrm{R}^4(v)$ then v leaves $\mathrm{L}_*^4(u)$;
from which our theorem follows easily.

Proof of Claim (0). Let J be the filter generated by u' and v. Since u' and v are non comparable they are both minimal elements of J. Since u' is a minimal element of J the bitangent $\varphi(u')$ is not a bitangent of $\mathrm{G}(J)$; since v is a minimal element of J the bitangents of the pseudotriangle $\mathrm{R}(v) = \mathrm{R}(v, J)$ are bitangents of $\mathrm{G}(J)$. It follows that $\varphi(u') \notin \mathrm{R}(v)$.

Proof of Claim (1). According to Claim (0) the bitangents v and $\varphi_*(u)$ are comparable; since $u \in \mathrm{R}(v)$ it follows that $\varphi_*(u)$ does not belong to the filter of bitangents greater than v. Consequently $\varphi_*(u) \prec v$.

Proof of Claim (2). If u touchs $\mathrm{R}(v)$ then u pierces a bitangent v' of $\mathrm{R}(v)$ from right-to-left; therefore, according to the right-to-left property, $v' \preceq \varphi_*(u')$, and consequently $v \preceq \varphi_*(u')$ since $v \preceq v'$.

Proof of Claims (3,4,5,6). They are proved by induction of the linearly ordered set S of bitangents u that belong to or touch the chain $\mathrm{R}(b)$. The case $u = v$ is trivial. Suppose the result is true for a bitangent u, and let u'

be the successor of u in S. The bitangents u and u' are the source and the sink of an primitive arc e whose type, thanks to Theorem 7, is restricted.

If u' belongs to $\mathrm{R}^1(v)$ then so does u and the arc e is a 0-cusp positive arc. By induction $v \in \mathrm{R}^1_*(u)$ and, according to Table 2, $\mathrm{R}^1_*(u') = u' e\, \mathrm{R}^1_*(u)$. Thus $v \in \mathrm{R}^1_*(u')$ which proves the result for u'.

Suppose now that u' belongs to or leaves $\mathrm{R}^{\mathrm{lea}}(v)$. The arc e can be a:

(i) negative source-cusp or positive sink-cusp arc in which case u is the last bitangent of $\mathrm{R}^1(v)$ and u' is the first bitangent of $\mathrm{L}^2(v)$; by induction $v \in \mathrm{R}^1_*(u)$ and, according to Table 2, $\mathrm{R}^1_*(u) = \mathrm{L}^{2_1}_*(u')$ in both cases. Therefore $v \in \mathrm{L}^{2_1}_*(u') \subseteq \mathrm{L}^{\mathrm{lea}}_*(u')$.

(ii) negative 2-cusp arc in which case u is the last bitangent of $\mathrm{R}^1(v)$ and u' leaves the first atom of $\mathrm{L}^2(v)$; by induction $v \in \mathrm{R}^1_*(u)$ and, according to Table 2, $\mathrm{R}^1_*(u) = u\, \mathrm{L}^{2_2}_*(u')$. Therefore $v = u$, from which we deduce that v leaves $\mathrm{L}^{\mathrm{lea}}_*(u')$ since $u = \varphi_*(u')$, or $v \in \mathrm{L}^{2_2}_*(u')$ which is a subchain of $\mathrm{L}^{\mathrm{lea}}_*(u')$.

(iii) positive source-cusp or a negative sink-cusp arc in which case u belongs to or leaves $\mathrm{R}^{\mathrm{lea}}(v)$, $u' \notin \mathrm{R}(v)$ (and thus $v \preceq \varphi_*(u')$ according to Claim 3); by induction v belongs to or leaves $\mathrm{L}^{\mathrm{lea}}_*(u)$ and, according to Table 2, $\mathrm{L}^2_*(u) = \mathrm{R}^{3_1}_*(u')\varphi_*(u')\, \mathrm{L}^{2_2}_*(u')$, $\mathrm{L}^3_*(u) = \mathrm{L}^{3_2}_*(u')\iota(e)$ (positive) or $\mathrm{L}^3_*(u) = \mathrm{L}^{3_2}_*(u')$ (negative). Since $v \preceq \varphi_*(u')$, v does not belong to nor leaves $\mathrm{R}^{3_1}_*(u')$. Therefore $v = \varphi_*(u')$ or $v \in \mathrm{L}^{2_2}_*(u')$ or v leaves $\mathrm{L}^{2_2}_*(u')$ or v leaves $\mathrm{L}^{3_2}_*(u')$. In all cases v leaves or belongs to $\mathrm{L}^{\mathrm{lea}}_*(u')$.

(iv) negative 0-cusp arc in which case u belongs to or leaves $\mathrm{R}^2(v)$, $u' \in \mathrm{R}^2(v)$; by induction v belongs to or leaves $\mathrm{L}^{\mathrm{lea}}_*(u)$ and, according to Table 2, $\mathrm{L}^2_*(u) = \mathrm{L}^{2_1}_*(u')$ and $\mathrm{L}^3_*(u) = \mathrm{L}^{3_1}_*(u')\varphi_*(u')\, \mathrm{R}^{2_2}(u')$; and Claim (3) gives $\varphi_*(u') \prec v$ from which we deduce that v does not leave nor belongs to $\mathrm{R}^{2_2}_*(u')$. It remains that $v = \varphi_*(u')$ or v leaves or belongs to $\mathrm{L}^{2_1}_*(u')$ or v leaves $\mathrm{L}^{3_1}_*(u')$. In all case v leaves or belongs to $\mathrm{L}^{\mathrm{lea}}_*(u')$.

(v) positive 0-cusp arc in which case u enters $\mathrm{R}^{\mathrm{lea}}(v)$ and u' leaves $\mathrm{R}^3(v)$; by induction v belongs to or leaves $\mathrm{R}^{\mathrm{lea}}_*(u)$ and, according to Table 2, $\mathrm{R}^3_*(u) = \mathrm{R}^{3_1}_*(u')\varphi_*(u')\, \mathrm{L}^{2_2}_*(u')$ and $\mathrm{R}^4_*(u) = \mathrm{L}^{3_2}_*(u')\iota(e)$; Claim (3) gives $\varphi_*(u') \preceq_* v$ from which we deduce that b does not leave nor belongs to $\mathrm{R}^{2_2}_*(u')$. It remains $v = \varphi_*(u')$ or b leaves or belongs to $\mathrm{L}^{2_2}_*(u')$ or leaves $\mathrm{L}^{3_2}_*(u')$ which is an arc. In all cases v leaves or belongs to $\mathrm{L}^{\mathrm{lea}}_*(u')$.

We assume now that u' enters or belongs to $\mathrm{R}^{\mathrm{ent}}(v)$. We have to show that v belongs to or leaves $\mathrm{R}^{\mathrm{lea}}_*(u') = \mathrm{R}^3_*(u')\, \mathrm{R}^4_*(u')$. The arc e can be a:

(i) positive 2-cusp arc; in that case u' enters the first atom of $\mathrm{R}^3(v)$, and u leaves the first atom of $\mathrm{R}^3(v)$ or is the last bitangent of $\mathrm{R}^2(v)$ (in both case u belongs to or leaves $\mathrm{R}^{\mathrm{lea}}(v)$); by induction v belongs to or leaves $\mathrm{L}^{\mathrm{lea}}_*(u)$ and, according to Table 2, $\mathrm{L}^2_*(u) = \mathrm{R}^{3_2}_*(u')$ and $\mathrm{L}^3_*(u) = \mathrm{R}^4_*(u')\iota(e)$. Therefore v leaves or belongs to $\mathrm{R}^{\mathrm{lea}}_*(u')$.

(ii) positive sink-cusp; in that case u enters or belongs to $\mathrm{R}^{\mathrm{ent}}(v)$ and $u' \notin \mathrm{R}(v)$; by induction v leaves or belongs to $\mathrm{R}^{\mathrm{lea}}_*(u)$ and, according to Table 2, $\mathrm{R}^3_*(u) = \mathrm{R}^{3_2}_*(u')$ and $\mathrm{R}^4_*(u) = \mathrm{R}^4(u')\iota(e)$. Therefore v leaves or belongs to $\mathrm{R}^{\mathrm{lea}}_*(u')$.

(iii) positive 0-cusp arc; in that case u enters or belongs to $\mathrm{R}^3(v)$ and u' belongs to $\mathrm{R}^3(v)$; by induction v leaves or belongs to $\mathrm{R}_*^{\mathrm{lea}}(u)$ and, according to Table 2, $\mathrm{R}_*^3(u) = \mathrm{R}_*^{3_1}(u')\varphi_*(u')\,\mathrm{L}^{2_2}(u')$ and $\mathrm{R}_*^4(u) = \mathrm{L}^{3_2}(u')\iota(e)$. Since $u' \in \mathrm{R}(v)$ one has $v \prec_* \varphi_*(u')$ therefore v does not leave nor belongs to $\mathrm{L}^{2_2}(u')$; $\mathrm{R}^3(v)$ is nonempty and therefore v is left-xx and consequently v does not leave $\mathrm{R}_*^4(u)$. Therefore $v = \varphi_*(u')$ or v leaves or belongs to $\mathrm{R}_*^{3_1}(u')$. Thus v leaves or belongs to $\mathrm{R}_*^{\mathrm{lea}}(u')$.

(iv) negative 0-cusp arc; in that case u leaves $\mathrm{R}^4(v)$ and $u' \notin \mathrm{R}(v)$; by induction v leaves $\mathrm{L}_*^4(u)$ and, according to Table 2, $\mathrm{L}_*^4(u) = \mathrm{R}_*^{3_2}(u')\iota(e)$. Therefore v leaves $\mathrm{R}_*^{\mathrm{lea}}(u')$.

(v) negative sink-cusp; in that case u enters or belongs to $\mathrm{R}^{\mathrm{ent}}(v)$ and $u' \notin \mathrm{R}(v)$. By induction v leaves or belongs to $\mathrm{R}_*^{\mathrm{ent}}(u)$ and, according to Table 2, $\mathrm{R}_*^3(u) = \mathrm{R}_*^{3_2}(u')\iota(e)$ et $\mathrm{R}_*^4(u) = \emptyset$. Therefore v leaves or belongs to $\mathrm{R}_*^{3_2}(u')$ and consequently v leaves or belongs to $\mathrm{R}_*^{\mathrm{lea}}(u')$.

Assume now that u' leaves $\mathrm{R}^4(v)$. The arc e can be a:

(i) negative sink-cusp arc; in that case u leaves $\mathrm{R}^4(v)$; by induction v leaves $\mathrm{L}_*^4(u)$ and, according to Table 2, $\mathrm{L}_*^4(u) = \mathrm{L}_*^4(u')\iota(e)$. Therefore v leaves $\mathrm{L}_*^4(u')$.

(ii) negative 2-cusp arc; in that case u enters or belongs to $\mathrm{R}^{\mathrm{ent}}(v)$. By induction v leaves or belongs to $\mathrm{R}_*^{\mathrm{ent}}(u)$ and, according to Table 2, $\mathrm{R}_*^3(u) = \mathrm{L}_*^4(u')\iota(e)$ and $\mathrm{R}_*^4(u) = \emptyset$. Therefore v leaves $\mathrm{L}_*^4(u')$. □

Table 4. Relations between greedy pseudotriangles associated with the source v and the sink v' of a positive primitive arc e in the case where v is a bitangent of κ and, consequently, $\phi(v;\kappa) = \iota(v)$.

0-cusp arc  v is an obstacle	$\mathrm{R}^1(v;\kappa) = ve\,\mathrm{R}^1(v';\kappa)$ $\mathrm{R}^2(v';\kappa) = \mathrm{R}^2(v;\kappa)$ $\mathrm{R}^3(v';\kappa) = \mathrm{R}^3(v;\kappa)$ $\mathrm{R}^4(v';\kappa) = \mathrm{R}^4(v;\kappa)\iota(v)\iota(e)$
sink cusp arc  v is an obstacle	$\mathrm{L}^1(v';\kappa) = \mathrm{R}^2(v;\kappa)$ $\mathrm{L}^2(v';\kappa) = \mathrm{R}^3(v;\kappa)$ $\mathrm{L}^3(v';\kappa) = \mathrm{R}^4(v;\kappa)\iota(v)\iota(e)$ $\mathrm{L}^4(v';\kappa) = \emptyset, \mathrm{R}^1(v;\kappa) = ve$
source cusp arc  v is an obstacle	$\mathrm{R}^1(v';\kappa) = \mathrm{L}^2(v;\kappa), \mathrm{L}^1(v;\kappa) = v$ $\mathrm{R}^2(v';\kappa) = \mathrm{L}^3(v;\kappa)$ $\mathrm{R}^3(v';\kappa) = \mathrm{L}^4(v;\kappa)\iota(v)$ $\mathrm{R}^4(v';\kappa) = \iota(e)$

Remark 6. *Table 2 is also valid in the case κ non-empty under the assumption that $v \notin \kappa$ and can be easily completed in case $v \in \kappa$ as indicated in Table 4. Therefore very similar theorems can be proved in the case κ non-empty; they are freely used at few places in the sequel without more details.*

3.3 The unary operator φ_{R}

This subsection is devoted to the study of the operator φ_{R} that associates with a non-hull bitangent v of X^0 the maximal bitangent crossing v and leaving $\mathrm{R}(v)$. The left version $\varphi_{\mathrm{L}}(v)$ of $\varphi_{\mathrm{R}}(v)$ is defined similarly as the maximal bitangent crossing v and entering $\mathrm{L}(v)$. This operator provides not only the clue to the formulation of the sum of squares theorem but also to the flip method based on the left-turn predicate. We start by two characterizations of φ_{R} before proving that $\varphi_{\mathrm{R}}(v)$ can be computed efficiently in the dynamic setting of the GFA using only the left-turn predicate.

Theorem 13. *Let v be a non-hull bitangent of X^0. We claim that $\varphi_{\mathrm{R}}(v)$ is the bitangent of the interval $[v, \iota(v)]$ crossing v whose crossing point with v is the closest to the tail of v, and that $\varphi_{\mathrm{R}}(v)$ is the bitangent leaving $\mathrm{R}(v)$ and entering $\iota\,\mathrm{L}_*(v)$.*

Proof. Let v, u, u' be three bitangents such that u' crosses u from its right side to its left side and v crosses both u' and the supporting half-line u^+ of u with origin the tail of u from left to right as illustrated in the figure below.

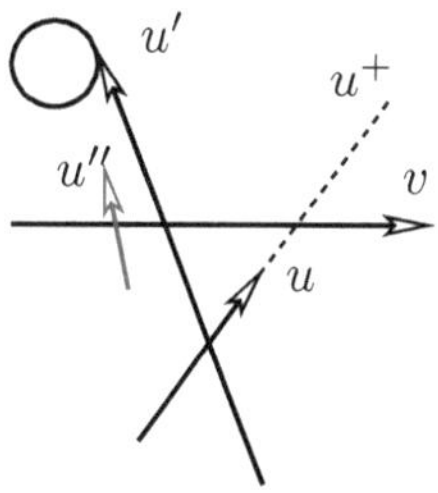

Then we claim that there exists a bitangent u'' crossing v whose crossing point with v is closer to the tail of v than the crossing points of u' and u^+ with v. Indeed let κ be a maximal subset of pairwise non crossing bitangents that cross neither u nor u'; the bitangents of κ subdivide free space into pseudotriangles and one pseudoquadrangle τ whose diagonals are u and u'; walking in clockwise order around the boundary of τ starting at the tail of u and ending at the head of u' we found three convex chains C_1, C_2, C_3 in this order where C_1 is the empty chain if u leaves a positive cycle and where C_3 is the empty chain if u' enters a positive cycle; clearly v intersects the chain C_2 and, consequently, intersects a bitangent u'' in a point which is closer to the tail of v than the crossing points of u and u' with v. Therefore the bitangent v' of the interval $[v, \iota(v)]$ crossing v whose crossing point with v is the closest to the tail of v is well-defined and crosses no bitangent crossing v. We show that v' leaves $\mathrm{R}^{\mathrm{lea}}(v)$ (and consequently is the maximal one). Let u' crossing v from its right side to its left side, that does not leave $\mathrm{R}^{\mathrm{lea}}(v)$ and whose crossing point with v is closer to the tail of v than the crossing point of $\varphi(v)$ with v. We show that u' is not v'. If u' crosses $\varphi(v)$, we are done since v' does not cross $\varphi(v)$. So assume now that u' does not cross $\varphi(v)$. In that case $\varphi(v)$ leaves the chain $\mathrm{R}^2(v)$ and u' crosses a bitangent $u \in \mathrm{R}^{2_2}(v)$ from its right side to its left side (here we use the greedyness of $\mathrm{R}(v)$). Moreover, as the angle of u is such that $\Theta(\varphi(v)) < \Theta(u) < \Theta(u')$, v intersects the half-line u^+. Therefore, according to the claim at the beginning of the proof, u' is not v'. We now prove the last part of the theorem. The left dual version of the first part states that $\varphi_{\mathrm{L}_*}(v)$, the largest (for $\prec_*$) bitangent that leaves $\mathrm{L}_*^{\mathrm{lea}}(v)$, is the bitangent crossing v whose crossing point with v is the closest to the tail of v. Therefore $\varphi_{\mathrm{L}_*}(v) = \iota_* \varphi_{\mathrm{R}}(v)$, and consequently $\varphi_{\mathrm{R}}(v)$ enters $\iota\,\mathrm{L}_*(v)$. Uniqueness follows Theorem 3. □

Let $\rho(e)$ be a positive 0-cusp edge with source v and sink v_R. In the next theorem we show that $\varphi_{\mathrm{R}}(v_R)$ belongs to the chain $\mathrm{rc}_2(v)$ and that $\varphi_{\mathrm{R}}(v_R)$ splits the chain $\mathrm{rc}(v)$ into two subchains that are subchains of $\mathrm{R}(v)$ and $\iota\,\mathrm{R}_*(v)$, respectively. This result is at the core of the flip method using the left-turn predicate.

Theorem 14 (Bridge Theorem). *Let $\rho(e)$ be a positive 0-cusp edge with source v and sink v_R, let $v_R S(v)$ be the prefix (for $\prec$) subchain of $\mathrm{R}_*\,\iota(v_R)$ cut at the head of $\varphi_{\mathrm{R}}(v_R)$, let $T(v)$ be the suffix subchain of $\mathrm{R}^{\mathrm{lea}}(v_R)$ cut at the tail of $\varphi_{\mathrm{R}}(v_R)$, and let $E = veS(v)\varphi_{\mathrm{R}}(v_R)T(v)\varphi(v)$. We claim that $\mathrm{R}^{\mathrm{lea}}(v_R)$ is a prefix subchain of $\mathrm{R}^{\mathrm{lea}}(v)$, that $\mathrm{rc}(v) = \rho(E)$, and that the chain $S(v)$ is a suffix subchain of $\iota\,\mathrm{R}_*(v)$.*

Proof. The first part of the Theorem follows immediately from Table 2. There is exactly one cusp point on the chain $veS(v)\varphi_{\mathrm{R}}(v_R)$, namely the head of $\varphi_{\mathrm{R}}(v_R)$ (if this bitangent is xx-left) or the second cusp point of $\mathrm{R}_*\,\iota(v_R)$. Similarly the chain $\varphi_{\mathrm{R}}(v_R)T(v)\varphi(v)$ also has only one cusp point. Let $E = E_1E_2E_3$ be the decomposition of E by these two cusp points. Since the source and sink of E are respectively equal to v and $\varphi(v)$, to prove $\rho(E) = \mathrm{rc}(v)$ it is, according to Theorem 4, sufficient to prove that:

(1) there is no bitangent entering E_1;
(2) there is no bitangent touching E_2;
(3) there is no bitangent leaving E_3.

These three claims follow immediately from the fact that $\varphi_{\mathrm{R}}(v_R)$ is the maximal bitangent entering $\mathrm{R}_*\,\iota(v_R)$ and the maximal bitangent leaving $\mathrm{R}^{\mathrm{lea}}(v_R)$ (Theorem 13). We now prove the last part of the Theorem. Let I be a filter such that v is a minimal element of I. Then v_R belongs to $\mathrm{G}(I)$ and consequently $\varphi_*(v_R)$ belongs to $\mathrm{G}_*(I_*)$. The chain $S(v)$ is a subchain of $\iota\,\mathrm{R}_*^{\mathrm{lea}_2}(v_R)$; we show that $\mathrm{R}_*^{\mathrm{lea}_2}(v_R)$ is a subchain of $\mathrm{R}_*(\varphi_*(v_R), I_*)$. To prove this result we pick a bitangent u leaving $\mathrm{R}_*^{\mathrm{lea}_2}(v_R)$ (if it exists) and we show that $u \notin \mathrm{G}_*(I_*)$. It should be clear that the bitangent u crosses a bitangent

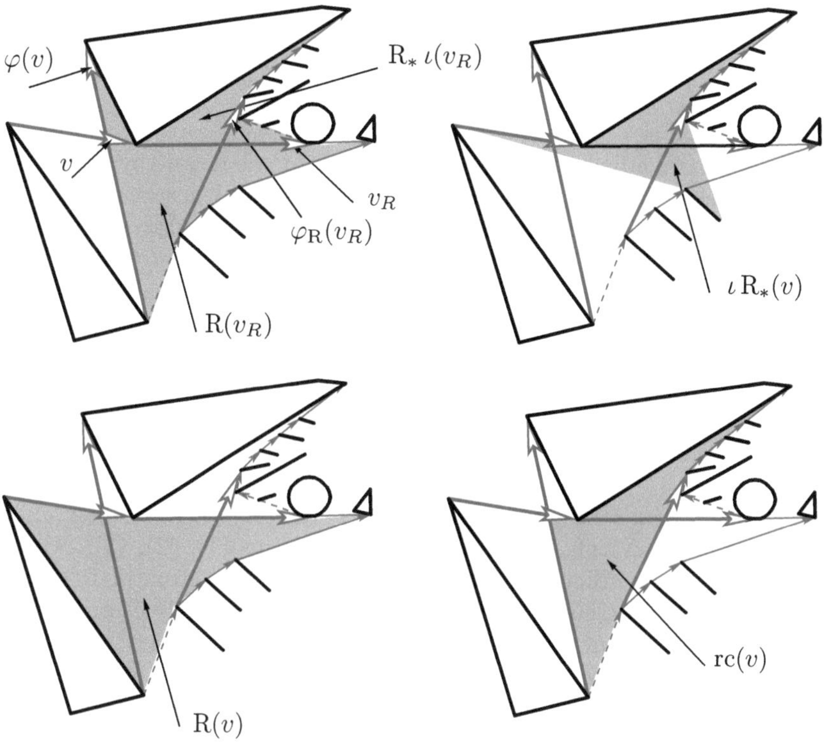

Fig. 20. The bitangent $\varphi_\mathrm{R}(v_R)$ is an atom of the envelope of $\mathrm{rc}_2(v)$ and is a *bridge* between $\mathrm{R}^{\mathrm{lea}}(v)$ and $\iota \mathrm{R}_*(v)$.

$u' \in \mathrm{R}_*^{2_1}(v_R)$. We prove that $u' \in \mathrm{G}_*(I_*)$ thus proving $u \notin \mathrm{G}_*(I_*)$. According to Table 2, $\mathrm{R}_*^2(v) = \mathrm{R}_*^{2_1}(v_R)$, and consequently $u' \in \mathrm{R}_*^2(v)$ from which we deduce that $u' \prec v$. Since v is a minimal element of I, we have $u' \notin I$. Furthermore, since u' belongs to $\mathrm{R}_*(v)$ we have $v \prec \varphi(u')$ and consequently $\varphi(u') \in I$. This achieves to prove that u' belongs to $\mathrm{G}_*(I_*)$. □

The case where $\varphi(v) = \varphi_\mathrm{L}(v)$ (or $\varphi(v) = \varphi_\mathrm{R}(v)$) requires a special treatment in our new flip method. In the next theorem we identify this situation.

Theorem 15 (Equation $\varphi_\mathrm{L}(v) = \varphi(v)$). *The five following assertions are equivalent:*

1. $\varphi(v) = \varphi_\mathrm{L}(v)$.
2. $\varphi(v)$ *does not cross* $\varphi_{\mathrm{righ}_*}(v)$.
3. $\varphi_\mathrm{L}(v)$ *does not cross* $\varphi_{\mathrm{righ}}(v)$.
4. $\varphi(v)$ *leaves the first atom of* $\mathrm{R}^2(v)$ *or leaves the first atom of* $\iota \mathrm{R}_*^2(v)$

5. $\varphi_{\mathrm{L}}(v)$ *leaves the first atom of* $\mathrm{R}^2(v)$ *or leaves the first atom of* $\iota\,\mathrm{R}^2_*(v)$*.*

Furthermore if $\varphi_{\mathrm{L}}(v)$ *leaves the first atom of* $\iota\,\mathrm{R}^2_*(v)$ *then so does* $\varphi_{\mathrm{R}}(v)$ *and* $\varphi_{\mathrm{R}}(v) = \iota\varphi_*(v)$*.*

Proof. Given two adjacent pseudotriangles τ and τ' we denote by $\Phi(\tau,\tau')$ the bitangent that leaves τ and enters τ' (cf. Theorem 3). Let $\sigma = \sigma_{\mathrm{righ}}(v)$, $u = \varphi_{\mathrm{righ}}(v)$, $u' = \varphi_{\mathrm{righ}_*}(v)$, $eu = \epsilon(\mathrm{lc}_3(\sigma))$, $u'e' = \epsilon(\mathrm{lc}_1(\sigma))$, $\Delta = \mathrm{R}(v;\kappa)$, and $\Delta_* = \mathrm{R}_*(v;\kappa_*)$ where κ and κ_* are the sets of bitangents of $\mathrm{lc}_3(\sigma)$ respectively smaller and greater than v. According to Theorem 11 the chain $\mathrm{R}^{\mathrm{lea}}(v)$ is a prefix subchain of Δ^{lea} and $\Delta^{\mathrm{lea}} = eu\,\mathrm{R}^{\mathrm{lea}_2}(u')\iota(e')$, from which we deduce that (1-) $\varphi(v) = \Phi(\Delta, \mathrm{L}(v))$ and $\varphi_{\mathrm{R}}(v) = \Phi(\Delta, \iota\,\mathrm{L}_*(v))$, (2-) $\varphi(v)$ and u are non crossing, and (3-) $\varphi(v)$ and u' are non crossing iff. (Claim 4') $\varphi(v)$ leaves e or $\iota(e')$. Similarly according to the dual version of Theorem 11 the chain $\mathrm{R}^{\mathrm{ent}}_*(v)$ is a prefix subchain of Δ^{ent}_* and $\Delta^{\mathrm{ent}}_* = e'u'\,\mathrm{R}^{\mathrm{ent}_2}_*(u)\iota_*(e)$ from which we deduce that (1-) $\varphi_{\mathrm{L}}(v)$ and u' are non crossing, (2-) $\varphi_{\mathrm{L}}(v)$ and u are non crossing iff. (Claim 5') $\varphi_{\mathrm{L}}(v)$ leaves e or $\iota(e')$, and (3-) $\varphi_{\mathrm{L}}(v) = \Phi(\iota\Delta_*, \mathrm{L}(v))$ and $\iota\varphi_*(v) = \Phi(\iota\Delta_*, \iota\,\mathrm{L}_*(v))$. It follows that Claims (1), (2), (3), (4') and (5') are equivalent. Since e and $\iota(e')$ contain the first atom of $\mathrm{R}^2(v)$ and $\iota\,\mathrm{R}^2_*(v)$ respectively, Claim (4) implies Claim (4'). Conversely Claim (4') and Claim (1) imply Claim (4). We prove now the last part of the Theorem. If $\varphi(v)$ leaves $\iota(e')$ so does $\varphi_{\mathrm{R}}(v)$ since $\varphi(v) \preceq \varphi_{\mathrm{R}}(v)$. Therefore $\varphi_{\mathrm{R}}(v)$ leaves $\iota\Delta_*$ and the identity $\iota\varphi_*(v) = \Phi(\iota\Delta_*, \iota\,\mathrm{L}_*(v))$ implies $\varphi_{\mathrm{R}}(v) = \iota\varphi_*(v)$. □

Now we show that $\varphi_{\mathrm{R}}(v)$ can be computed efficiently with the χ_1 predicate in the dynamic setting of the GFA. We first examine the case where v is the sink of a 1- or 2-cusp edge.

Theorem 16. *Let* $\rho(e)$ *be a positive sink cusp edge or a negative source cusp edge with source* v *and sink* v_R*. Then* $\varphi_{\mathrm{L}}(v_R)$ *coincides either with* $\varphi_{\mathrm{L}}(v)$ *or with* $\iota\varphi_*(v_R)$ *depending on whether* $\varphi_{\mathrm{L}}(v) \neq \varphi(v)$ *or not. Let* $\rho(e)$ *be a negative 2-cusp edge with source* v *and sink* v_R *then* $\varphi_{\mathrm{R}}(v_R) = \iota(v)$*.*

Proof. Suppose first that $\rho(e)$ is a positive sink cusp edge or a negative source cusp edge. In both cases, according to Table 2,

$$\mathrm{L}^{\mathrm{ent}}(v_R) = \mathrm{R}^{3_1}(v)\varphi(v)\,\mathrm{L}^{\mathrm{ent}_2}(v)[\iota(e)].$$

where in the above formula the arc $\iota(e)$ is added when e is positive. If $\varphi(v) \neq \varphi_{\mathrm{L}}(v)$ then $\varphi_{\mathrm{L}}(v)$ is the maximal bitangent entering $\mathrm{L}^{\mathrm{ent}_2}(v)$, and therefore is the maximal bitangent entering $\mathrm{L}^{\mathrm{ent}}(v_R)$. Theorem 13 yields $\varphi_{\mathrm{L}}(v_R) = \varphi_{\mathrm{L}}(v)$. Assume now that $\varphi_{\mathrm{L}}(v) = \varphi(v)$. Then there are no bitangents entering $\mathrm{L}^{\mathrm{ent}_2}(v)$ and $\varphi_{\mathrm{L}}(v_R)$ enters $\mathrm{R}^{3_1}(v)$ which is the first atom of $\mathrm{L}^2(v_R)$. Since $\varphi_{\mathrm{L}}(v_R) = \iota\varphi_{\mathrm{R}_*}(v_R)$, the bitangent $\varphi_{\mathrm{R}_*}(v_R)$ leaves the first atom of $\iota_*\,\mathrm{L}^2(v_R)$ and consequently, according to left dual version of Theorem 15, $\varphi_{\mathrm{R}_*}(v_R) = \varphi_*(v_R)$, that is, $\varphi_{\mathrm{L}}(v_R) = \iota\varphi_*(v_R)$. Suppose now that $\rho(e)$ is a negative two cusp edge. According to Table 2 the arc $\mathrm{R}^3(v_R)$ is equal to $\mathrm{R}^4(v)\iota(e)$. Since $\iota(v)$ is the sink of the chain $\mathrm{R}^4(v)$ and since there are no bitangents leaving $\iota(e)$, $\iota(v)$ is the maximal bitangent leaving $\mathrm{R}^{\mathrm{lea}}(v)$. We conclude using Theorem 13. □

We now examine the case where v is the sink of a positive 0-cusp edge.

Theorem 17. *Let $\rho(e)$ be a positive 0-cusp edge with source v and sink v_R and let I be a filter with $v \in \min_\prec I$. Then $\varphi_{\mathrm{R}}(v_R)$ is computable as the bitangent that leaves $\mathrm{R}^{\mathrm{lea}}(v)$ and enters $\iota\,\mathrm{R}_*^{\mathrm{lea}}(v)$ using only the predicate χ_1 in time $|\,\mathrm{rc}(v)|+|\,\mathrm{R}^{2_1}(v)|$ if $\varphi_{\mathrm{R}}(v_R)$ leaves a negative cycle (that is, a clockwise directed disk), and time $|\,\mathrm{rc}(v)|$ otherwise.*

Proof. We use the following function.

Function PHIR(v, v_R : BITANGENT) : PAIR OF ARCS
Comments/* returns the arcs of $\mathrm{R}(v)$ and $\mathrm{R}_*\,\iota(v_R)$
that $\varphi_{\mathrm{R}}(v_R)$ leaves and enters, respectively */
$C \leftarrow \mathrm{R}^{\mathrm{lea}}(v)$; $C' \leftarrow \iota\,\mathrm{R}_*^{\mathrm{lea}}(v)$ in reverse order;
χ_1-WALK(C, C');
return (Top(C), Top(C'))

Procedure χ_1-WALK(C, C' : LIST OF ARCS)
while $C \neq \emptyset$ and $C' \neq \emptyset$
  $r \leftarrow \mathrm{Top}(C)$; $l \leftarrow \mathrm{Top}(C')$; $u \leftarrow \mathrm{bit}(r, l)$;
  if $\chi_1(r^+, u)$ $C \leftarrow \mathrm{Pop}(C)$
  elseif $\chi_1(l^+, u)$ $C' \leftarrow \mathrm{Pop}(C')$
  else exit;

Its complexity is clearly the sum of the sizes of the chains $\mathrm{R}^{2_1}(v)$ and $S(v)$ where $S(v)$ is the chain introduced in Theorem 14. This shows that the complexity of χ_1-WALK is at most $|\,\mathrm{rc}(v)| + |\,\mathrm{R}^{2_1}(v)|$. Using a tandem walk (as explained in the previous section) we can reduce this complexity to $|\,\mathrm{rc}(v)|$ in case where $\varphi_{\mathrm{R}}(v_R)$ leaves a positive cycle. It remains to prove the correction of the method. Call r_g and l_g the arcs of $\mathrm{R}^{\mathrm{lea}}(v)$ and $\mathrm{R}_*^{\mathrm{ent}}\,\iota(v_R)$ that $\varphi := \varphi_{\mathrm{R}}(v_R)$ leaves and enters, respectively. The correction of χ_1-WALK is a straightforward consequence of the following four claims:

1. If $l \prec l_g$ and $r = r_g$ then $\chi_1(u, r_g^+)$ and $\chi_1(l^+, u)$.

2. If $l = l_g$ and $r \prec r_g$ then $\chi_1(r^+, u)$.

3. If $l \prec l_g, r \prec r_g$ and $\chi_1(u, r^+)$ then $\chi_1(l^+, u)$.

4. $\chi_1(\varphi, l_g^+)$ and $\chi_1(\varphi, r_g^+)$.

Let c_i and c_j be the supporting cycles of the arcs $r \preceq r_g$ and $l \preceq l_g$. We denote by p the touching point of r^+ with c_i and by q the touching point of l^+ with c_j. The direction of the line or bitangent v is denoted $\vec{v}$.

Proof of Claim 1. We first prove the relation $\chi_1(u, r_g^+)$. Let x be the intersection point of the supporting lines of l^+ and r_g^+. We observe that (1) $\chi_2(l^+, r_g^+)$ holds, (2) v_{xj} is free and $\vec{l^+} = \vec{v}_{xj}$, that is, x, q appear in this order along the supporting line of l^+, and (3) either (Case 1) v_{xi} is free and $\vec{r_g^+} = \vec{v}_{xi}$ with c_i positive or (Case 2) v_{ix} is free and $\vec{r_g^+} = \vec{v}_{ix}$. In the first case one has $\chi_1(v_{xj}, v_{xi})$. Therefore, according to Theorem 9 Claim (3-) applied to the cycles c_j, c_i and c_x, the sequence $v_{ij}v_{xi}v_{-ji}$ is a χ_1-sequence and consequently $\chi_1(v_{ij}, v_{xi})$, that is, $\chi_1(u, r_g^+)$. In the second case one has $\chi_1(v_{xj}, v_{ix})$ or, equivalently, $\chi_1(v_{x-i}, v_{xj})$. Therefore, according to Theorem 9 Claim (1-) applied to the cycles c_{-i}, c_j and c_x, the sequence $v_{x-i}v_{xj}v_{ij}$ is a χ_1-sequence and consequently $\chi_1(v_{ij}, v_{ix})$, that is, $\chi_1(u, r_g^+)$. We turn now to the the proof of the relation $\chi_1(l^+, u)$. We redefine x to be the intersection point of the supporting lines of l^+ and φ and we observe that (1-) v_{xj} is free and $\vec{l^+} = \vec{v}_{xj}$, (2-) v_{ix} is free and $\vec{\varphi} = \vec{v}_{ix}$, and (3-) $\chi_1(v_{xj}, v_{ix})$ since $v_R \prec l^+ \prec \varphi \prec \iota(v_R)$. Therefore, according to Theorem 9 Claim (1-) applied to the cycles c_{-i}, c_j and c_x, the sequence $v_{x-i}v_{xj}v_{ij}$ is a χ_1-sequence and consequently $\chi_1(v_{xj}, v_{ij})$, that is, $\chi_1(l^+, u)$.

Proof of Claim 2. Let x be the intersection point of the supporting lines of φ and r^+. We observe that (1-) $\chi_2(r^+, \varphi)$ holds, (2-) v_{xj} is free and $\vec{\varphi} = \vec{v}_{xj}$, (3-) c_i is a negative cycle, and (4-) either (Case 1) v_{xi} is free and $\vec{r^+} = \vec{v}_{xi}$ or (Case 2) v_{ix} is free and $\vec{r^+} = \vec{v}_{ix}$. In the first case one has $\chi_1(v_{xi}, v_{xj})$. Therefore, according to Theorem 9 Claim (2-), the sequence $v_{xi}v_{ij}v_{-ij}$ is a χ_1-sequence and consequently $\chi_1(v_{xi}, v_{ij})$, that is, $\chi_1(r^+, u)$. In the second case $\chi_1(v_{xj}, v_{x-i})$; therefore, according to Theorem 9 Claim (3-) applied to the cycles c_j, c_{-i} and c_x, the sequence $v_{-ij}v_{x-i}v_{-j-i}$ is a χ_1-sequence and consequently $\chi_1(v_{x-i}, v_{-j-i})$, that is, $\chi_1(r^+, u)$.

Proof of Claim 3. Let y (resp. x) be the intersection point of the supporting lines of l^+ (resp. r^+) and φ. We observe that (1-) c_i is a negative cycle, (2-) x and y appear in this order along the supporting line of φ and the line segment $[xy]$ is free, (3-) $\chi_2(l^+, \varphi)$ and $\chi_2(r^+, \varphi)$ hold, (4-) v_{yj} is free and $\vec{l^+} = \vec{v}_{yj}$, that is, y, q appear in this order along the supporting line of l^+, and (5-) either (Case 1) v_{xi} is free and $\vec{r^+} = \vec{v}_{xi}$ or (Case 2) v_{ix} is free and $\vec{r^+} = \vec{v}_{ix}$. The supporting line of φ splits the convex-hull of x and c_i into two negative cycles denoted $c_{i'}$ and $o_{i''}$ where $o_{i'}$ lies on the right of the supporting line of φ (note that $o_{i'}$ and $o_{i''}$ might be reduced to the point x.) Let α and β be the points of touching of the bitangents v_{xi} and v_{ix} with the disk o_i. We note that $u(= v_{ij})$ leaves the cycle $c_{i'}$, $c_{i''}$, or the arc $\beta\alpha$. (cf. figure below.) We have $\vec{\varphi} = \vec{v}_{uy}$ and $\chi_1(v_{yj}, v_{uy})$. Therefore, according to Theorem 9 Claim (1-) applied to the cycles $c_{-i'}, c_j$ and c_y, the sequence $v_{y,-i'}v_{yj}v_{i'j}$ is a χ_1-sequence and consequently

$$\chi_1(v_{yj}, v_{i'j}). \tag{6}$$

We have $\vec{\varphi} = \vec{v}_{-i''y}$ and $\chi_1(v_{yj}, v_{-i''y})$, that is, $\chi_1(v_{yi''}, v_{yj})$. Therefore, according to Theorem 9 Claim (2-) applied to the cycles $c_{i''}, c_j$ and c_y, the sequence $v_{yi''}v_{i''j}v_{-i''y}$ is a χ_1-sequence and consequently $\chi_1(v_{yi''}, v_{i''j})$ which can be rewritten as $\chi_2(v_{i''j}, \varphi)$. This last relation proves that $v_{i''j}$ leaves the arc γx of $c_{i''}$ where γ is the point of tangency of the tangent line to $c_{i''}$ with direction $\vec{\varphi}$. Since by hypothesis $\chi_1(u, r^+)$ the bitangent u leaves the arc pp' of the disk o_i where p' is the tangency point of the supporting line to c_i with direction $-\vec{r^+}$. We distinguish now the two cases:

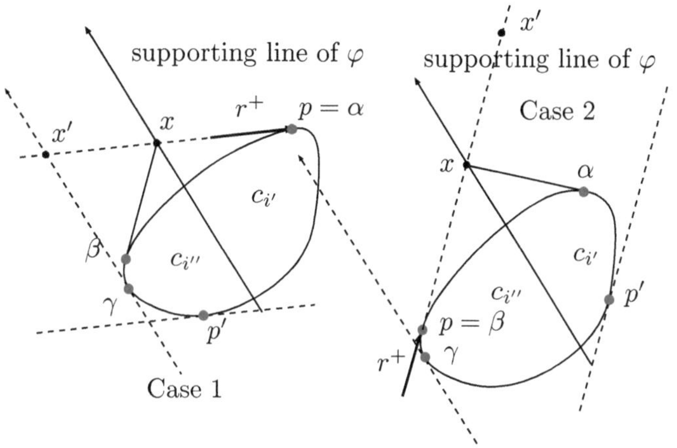

Case 1. In that case $p = \alpha$ and the points p, p', γ, x appear in this order along the boundary of o_i. Since the arcs pp' and γx are disjoint it follows that $v_{ij} = v_{i'j}$ and we are done thanks to the Equation (6).

Case 2. In that case $p = \beta$, γ appear on the boundary of o_i, and the points $\beta, \alpha, p', \gamma$ appears in this order along the boundary of o_i. Therefore $v_{ij} = v_{i'j}$ and we are done thanks to the Equation (6) or v_{ij} leaves the arc $\beta\alpha$. In this latter case we apply Theorem 9 to the cycles c_i, c_j and c_k where c_k is the line segment $[x, y]$ oriented counterclockwise.

Proof of Claim 4. This is clear since $v_R \preceq r_g^- \prec \varphi \prec r_g^+ \prec \iota(v_R)$, and $v_R \preceq l_g^- \prec \varphi \prec l_g^+ \preceq \iota(v_R)$. □

3.4 Proof of the sum of squares theorem

We now turn to the proof of the sum of squares theorem. The proof is split into several lemmas. As a first step we reformulate the duality theorem in terms more suitable for our subsequent analysis. Recall that $\mathrm{R}'(v)$ denotes the set of bitangents $u \neq \varphi_{\mathrm{righ}}(v)$ of the chain $\mathrm{R}^{2_1}(v)$ not in $\sigma_{\mathrm{sour}}(v)$ (cf. Theorem 8) and let $r(v)$ be the set of bitangents of the chain $\mathrm{rc}_2(\sigma_{\mathrm{righ}}(v))$. We denote by $Z_R(v)$ the union of sets $\mathrm{R}'(v), r(v)$, and $\mathrm{R}'_*(v)$, and by $Z_L(v)$ its left version, that is, the union of the sets $\mathrm{L}'(v), l(v)$, and $\mathrm{L}'_*(v)$ where $l(v)$ is the left version of $r(v)$, that is, the set of bitangents of the chain $\mathrm{lc}_2(\sigma_{\mathrm{left}}(v))$.

We combine the duality theorem and the obvious observation that $u \in r(v)$ iff. $v \in l(u)$ to obtain the following theorem.

Theorem 18 (Duality Theorem bis). *$u \in Z_R(v)$ iff. $v \in Z_L(u)$.* □

The second step of the proof of the sum of squares theorem is to show, among other more technical things, that if u and v are non crossing and $\varphi(u,v) = \varphi(v,u)$ then $u \in Z_R(v)$ or $u \in Z_L(v)$. To this end we are going to show that the bitangents of $Z_R(v)$ are bitangents of the right boundary of some face of some visibility complex. Let $\kappa(v)$ be the (unoriented version of the) 2-set $\{\varphi_{\mathrm{R}}(v), \varphi_{\mathrm{L}}(v)\}$. We assume that $\varphi_{\mathrm{R}}(v) \neq \varphi_{\mathrm{L}}(v)$ and we introduce the 2-face $\sigma(v)$ of $\mathbb{V}_{\kappa(v)}$ defined by the line segment $[x_1, x_2]$ directed from x_1 to x_2 where x_1 and x_2 are the crossing points of the unoriented version of the bitangent v with the unoriented versions of the bitangents $\varphi_{\mathrm{R}}(v)$ and $\varphi_{\mathrm{L}}(v)$, respectively. We study the right boundary of $\sigma(v)$. If $\varphi_{\mathrm{R}}(v)$ and $\varphi_{\mathrm{L}}(v)$ leave the same cycle then the right boundary of the face $\sigma(v)$ is reduced to the curve γ_p where p is the tail of $\varphi_{\mathrm{R}}(v)$ or the tail of $\varphi_{\mathrm{L}}(v)$ depending on whether $\varphi_{\mathrm{R}}(v)$ leaves a negative or positive cycle. So assume now that $\varphi_{\mathrm{R}}(v)$ and $\varphi_{\mathrm{L}}(v)$ do not leave the same cycle. In particular this implies that the minimal bitangent of $\mathrm{R}^2(v)$ is well-defined and coincides with $\varphi_{\mathbf{righ}}(v)$. Let E be the subchain of $\mathrm{R}^{\mathrm{lea}}(v)$ with source $\varphi_{\mathbf{righ}}(v)$ and sink $\varphi(v)$ when $\varphi_{\mathbf{righ}}(v) \prec \varphi(v)$, that is, $\varphi(v) \neq \varphi_{\mathrm{L}}(v)$ otherwise E is not defined. We define a chain C as follows:

1. if $\varphi_{\mathrm{L}}(v)$ and $\varphi_{\mathrm{R}}(v)$ enter the same disk, then $\varphi(v) = \varphi_{\mathrm{R}}(v)$, $\varphi_*(v) = \iota_*\varphi_{\mathrm{L}}(v)$, and we set

$$C = \begin{cases} E\epsilon & \text{if } \varphi_{\mathrm{R}}(v) \text{ and } \varphi_{\mathrm{L}}(v) \text{ enter in this order the first atom of } \mathrm{L}^2(v) \\ E & \text{if } \varphi_{\mathrm{R}}(v) \text{ and } \varphi_{\mathrm{L}}(v) \text{ enter in this order the first atom of } \iota\,\mathrm{L}^2_*(v) \end{cases}$$

   where ϵ is the subarc of $\mathrm{L}^2(v)$ with source $\varphi_{\mathrm{R}}(v)$ and sink $\varphi_{\mathrm{L}}(v)$.

2. if $\varphi_{\mathrm{R}}(v)$ and $\varphi_{\mathrm{L}}(v)$ do not enter the same disk, then the minimal bitangent of $\mathrm{L}^2(v)$ is well-defined, coincides with $\varphi_{\mathbf{left}}(v)$, and we set

$$C = \begin{cases} \emptyset & \text{if } \varphi(v) = \varphi_{\mathrm{L}}(v) \text{ (Case 1)} \\ E\epsilon & \text{if } \varphi(v) = \varphi_{\mathrm{R}}(v) \text{ (Case 2)} \\ E & \text{otherwise (Case 3)} \end{cases}$$

   where ϵ is the subarc of $\mathrm{L}^2(v)$ with source $\varphi(v)$ and sink $\varphi_{\mathbf{left}}(v)$. In Case 1, $\varphi_{\mathrm{L}}(v)$ leaves the first atom of $\mathrm{R}^2(v)$, and in Case 2, $\varphi_{\mathrm{R}}(v)$ enters the first atom of $\mathrm{L}^2(v)$ (but not $\varphi_{\mathrm{L}}(v)$).

Let now D be the envelope of the right boundary of the face $\sigma := \sigma_{\mathbf{righ}}(v)$ and let C_* be the dual version of the chain C. Since the sink $u = \varphi_{\mathbf{righ}}(v)$ of

D is the source of C and since the source u' of D is the sink of C_* we can introduce the concatenation $\Sigma(v)$ of the chains $\overline{C}_*$ (that is, the chain C_* in reverse order), D, and C.

Lemma 2. *$\varphi_{\mathrm{R}}(v)$ crosses no bitangent of the chain $\Sigma(v)$.* □

Proof. Since $\varphi_{\mathrm{R}}(v)$ leaves $\mathrm{R}^{\mathrm{lea}}(v)$ it is clear that $\varphi_{\mathrm{R}}(v)$ crosses no bitangent of C (and similarly no bitangent of C_*). We show now that $\varphi_{\mathrm{R}}(v)$ crosses no bitangent of D. According to Theorem 11 the bitangent $\varphi_{\mathrm{R}}(v)$ leaves $e, \iota(e')$ or $\mathrm{R}^{\mathrm{lea}_2}(u')$. In the first two cases $\varphi_{\mathrm{R}}(v)$ is a bitangent of $\mathrm{lc}(\sigma)$, and consequently crosses no bitangent of $\mathrm{rc}(\sigma)$. We now examine the third case, that is, $\varphi_{\mathrm{R}}(v)$ leaves the chain $\mathrm{R}^{\mathrm{lea}_2}(u')$. Let e'' be the edge with source u' that u' enters. Assume first that e'' is a 1-cusp edge; in that case the chain $\mathrm{rc}_2(\sigma)$ is a subchain of $\mathrm{R}^2(u')$ and we are done. Assume now that e'' is a (positive) 0-cusp edge and let u'_R be the sink of e''; in that case, according to the Bridge Theorem, $\varphi_{\mathrm{R}}(v)$ leaves the pseudotriangle on the left of $\varphi_{\mathrm{R}}(u'_R)$ in the quadrangle obtained by merging $\mathrm{R}(u')$ and $\mathrm{R}_*\,\iota(u'_R)$; since this pseudotriangle contains the envelope of $\mathrm{rc}(\sigma)$ (Bridge Theorem) we are done. □

Lemma 3. *The chain $\Sigma(v)$ is the envelope of the right chain of the 2-face $\sigma(v)$.*

Proof. We only prove the lemma in the case where nor $\varphi(v)$ nor $\iota\varphi_*(v)$ coincides with $\varphi_{\mathrm{L}}(v)$ or $\varphi_{\mathrm{R}}(v)$. The other cases can be treated similarly. In that case $\Sigma(v)$ contains exactly two cusps that split $\Sigma(v)$ into the three convex chains

$$\Sigma_1(v) = \overline{e'\,\mathrm{R}_*^{3_1}(v)\varphi_*(v)}, \tag{7}$$
$$\Sigma_2(v) = \overline{(e_{20_*}(v)u')^{-1}\,\mathrm{R}_*^{2_1}(v)}r(v)(e_{20}(v)u)^{-1}\,\mathrm{R}^{2_1}(v) \tag{8}$$
$$\Sigma_3(v) = e\,\mathrm{R}^{3_1}(v)\varphi(v) \tag{9}$$

where $u'e'$, $r(v)$ and eu are the envelopes of the chains $\mathrm{rc}_1(\sigma)$, $\mathrm{rc}_2(\sigma)$ and $\mathrm{rc}_3(\sigma)$; therefore
(1-) there is no bitangent of $\mathbb{V}^0_{\kappa(v)}$ entering or leaving $\Sigma_2(v)$
(2-) there is no bitangent leaving $\Sigma_3(v)$ and $\Sigma_3(v)$ is an arc concatenated with a bitangent of $\mathbb{V}^0_{\kappa(v)} \setminus \kappa(v)$.
(3-) there is no bitangent entering $\Sigma_1(v)$ and $\Sigma_1(v)$ is the concatenation of an arc with a bitangent of $\mathbb{V}^0_{\kappa(v)} \setminus \kappa(v)$.
Therefore, according to Theorem 4 and since the bitangents of the chain $\Sigma(v)$ are in $\mathbb{V}^0_{\kappa(v)}$, the chain $\Sigma(v)$ is the envelope of the right boundary of the face of $\mathbb{V}_{\kappa(v)}$ with sink $\varphi(v)$ and source $\varphi_*(v)$. This proves in particular that $\varphi(v)$ and $\varphi_*(v)$ are crossing. We show now that $\varphi(v)$ is the sink of the face $\sigma(v)$. Let x be the intersection point of $\varphi(v)$ with v and let $r(\theta)$ be the ray in $\mathbb{V}_\kappa$ with origin x and angle θ. We observe that the backward view of the ray $r(\theta)$ does not change when θ ranges over the semi-open interval

$[\Theta(v), \Theta(\varphi(v)))$; consequently $\varphi(v)$ belongs to the boundary of the face $\sigma(v)$ and consequently, since $\varphi(v)$ and v are crossing, $\varphi(v)$ is the sink of the face $\sigma(v)$. A dual argument shows that $\varphi_*(v)$ is the source of $\sigma(v)$. □

It should be clear at this point that the bitangents of $Z_R(v)$ are bitangents of the chain $\Sigma_2(v)$. The next step is to describe the greedy pseudotriangles $\mathrm{R}(u,v) = \mathrm{R}(u;\kappa(v))$ and $\mathrm{L}(u,v) = \mathrm{L}(u;\kappa(v))$ of the bitangents u of the chain $\Sigma_2(v)$ and to evaluate $\varphi(u,v)$. This is the purpose of the three following lemmas.

Lemma 4. *Let $u \in \Sigma_2(v)$. Then $\mathrm{R}(u,v) = \mathrm{R}(u)$ and $\varphi(u,v) = \varphi(u; \{\varphi_{\mathrm{L}}(v)\})$.*

Proof. Let C_1 and C_2 be the backward and forward views of the rays of the face $\sigma_{\mathrm{righ}}(u)$ of $\mathbb{V}_{\kappa(v)}$, let D_1 and D_2 be the backward and forward views of the rays of the face $\sigma_{\mathrm{left}}(u)$ of $\mathbb{V}_{\kappa(v)}$, let C_3 be the backward view of the rays of $\sigma_{\mathrm{back}}(u)$ in case u is a left-xx bitangent, and let C_4 be the forward view of the rays of $\sigma_{\mathrm{forw}}(u)$ in case u is a xx-left bitangent. One can check that these chains are distinct, and consequently, since $\varphi_{\mathrm{R}}(v)$ belongs to D_1 and since $\varphi_{\mathrm{L}}(v)$ belongs to D_2, the chains C_1, C_2, C_3, C_4 do not contain $\varphi_{\mathrm{R}}(v)$ nor $\varphi_{\mathrm{L}}(v)$. This proves that the bitangents of $\mathrm{R}(u)$ belong to $X^0_{\kappa(v)}$, and consequently that $\mathrm{R}(u,v) = \mathrm{R}(u)$. We now prove that $\varphi(u,v) = \varphi(u; \{\varphi_{\mathrm{L}}(v)\})$. Since $\varphi_{\mathrm{L}}(v) \in \kappa(v)$ one has $\varphi(u; \{\varphi_{\mathrm{L}}(v)\}) \preceq \varphi(u,v)$ (Theorem 5). Let w be a bitangent that leaves $\mathrm{R}(u)$ and pierces u. If $w \prec \varphi(u,v)$ then w pierces the forward view of the rays in the face $\sigma(v)$, that is, pierces $\varphi_{\mathrm{L}}(v)$; therefore $\varphi(u; \{\varphi_{\mathrm{L}}(v)\}) \not\preceq \varphi(u,v)$. This proves that $\varphi(u,v) = \varphi(u; \{\varphi_{\mathrm{L}}(v)\})$. □

Lemma 5. *Let $u \in \Sigma_2(v)$, let*

$$C(v) = v_1 e_1 v_2 e_2 \dots e_p v_{p+1} = C'(v) v_{p+1} \tag{10}$$

with $p \geq 0$ be the subchain of $\mathrm{L}^{\mathrm{ent}}(v)$ with source $v_1 = \varphi(v)$ and sink $v_{p+1} = \varphi_{\mathrm{L}}(v)$, and let e_0 be the first atom of the chain $\Sigma_3(v)$ if this atom is an arc. Assume that u is a xx-right bitangent. Then $\varphi_{\mathrm{L}}(v)$ and $\varphi_{\mathrm{forw}}(u,v) = \varphi^{\kappa(v)}_{\mathrm{forw}}(u)$ are comparable and one of the three following cases holds

1. *$\varphi_{\mathrm{forw}}(u,v) \prec \varphi_{\mathrm{L}}(v)$ and $\mathrm{L}^{\mathrm{ent}}(u,v)$ is a prefix subchain of $e_0 C'(v)$.*
2. *$\varphi_{\mathrm{L}}(v) = \varphi_{\mathrm{forw}}(u,v)$ and $\mathrm{L}^{\mathrm{ent}}(u,v) = e_0 C(v)$.*
3. *$\varphi_{\mathrm{L}}(v) \prec \varphi_{\mathrm{forw}}(u,v)$ and $\mathrm{L}^{\mathrm{ent}}(u,v) = e_0 C(v) e_{p+1}$ where e_{p+1} is the suffix subarc of the first atom of $\mathrm{L}^3(u)$ with source $\varphi_{\mathrm{L}}(v)$ and sink $\varphi_{\mathrm{forw}}(u) (= \varphi_{\mathrm{forw}}(u,v))$.*

Furthermore if u is an xx-left bitangent then $\mathrm{L}^{\mathrm{ent}}(u,v) = e_0 C(v) e_{p+1}$ where e_{p+1} is the suffix subarc of the first atom of $\mathrm{L}^3(u)$ with source $\varphi_{\mathrm{L}}(v)$ and sink $\iota(u)$.

Proof. According to the Orbit Theorem (1-) $\mathrm{L}^1(u,v)$ is the suffix subchain of $\Sigma_2(v)$ with source u, and (2-) the minimal bitangent of $\mathrm{L}^2(u,v)$ is the minimal bitangent of $\Sigma_3(v)$, that is, $\varphi(v)$ (cf. definition of $\Sigma(v)$) or $\mathrm{L}^2(u,v)$ is reduced to an arc. Now according to the Orbit Theorem $v_{i+1} = \varphi_{\mathrm{forw}}(v_i)$ for $i < p$; since for $t \in \mathrm{L}^2(v)$ with $v_1 \prec t$ one has $t \in X^0_{\kappa(v)}$ it follows that $\varphi_{\mathrm{forw}}(v_i, v) = v_{i+1}$ for $i < p$ and either v_p is xx-right or v_p is xx-left and $\varphi_{\mathrm{forw}}(v_p, v) = v_{p+1}$. This proves our lemma. □

Lemma 6. *Let $u \in \Sigma_2(v)$. Then the bitangents $\varphi(u,v), \varphi(v)$ and $\varphi_{\mathrm{L}}(v)$ are comparable and distinct, and one of the three following cases holds:*

1. $\varphi(u,v) \prec \varphi(v)$, $\varphi(u,v) = \varphi_{\mathrm{R}}(u)$ *is xx-left, and* $\varphi(v)$ *is left-xx*
2. $\varphi_{\mathrm{L}}(v) \prec \varphi(u,v)$, $\varphi(u,v) = \varphi(u)$ *is xx-right, and* $\varphi_{\mathrm{L}}(v)$ *is right-xx*
3. $\varphi(v) \prec \varphi(u,v) \prec \varphi_{\mathrm{L}}(v)$ *and if* $u \in Z_R(v)$ *then* $\varphi(u,v) = \varphi(v,u)$.

Proof. According to Lemma 4, $\mathrm{R}(u,v) = \mathrm{R}(u)$. Assume first that $\varphi(u,v) \prec \varphi(v)$. In that case $\varphi(u,v)$ enters the (positive) arc e_0. It follows that $\varphi(u,v)$ is the maximal bitangent of X^0 leaving $\mathrm{R}^{\mathrm{lea}}(u,v)$. Since $\mathrm{R}(u,v) = \mathrm{R}(u)$, $\varphi(u,v)$ is the maximal bitangent of X^0 leaving $\mathrm{R}(u)$, and consequently $\varphi(u,v) = \varphi_{\mathrm{R}}(u)$. Assume now that $\varphi_{\mathrm{L}}(v) \prec \varphi(u,v)$. In that case $\varphi(u,v)$ enters the (negative) arc e_{p+1} which belongs to $\mathrm{L}^3(u)$. Since $\mathrm{R}(u,v) = \mathrm{R}(u)$, the bitangent $\varphi(u,v)$ leaves $\mathrm{R}(u)$ and enters $\mathrm{L}(u)$, that is, $\varphi(u,v) = \varphi(u)$. Finally assume that $\varphi(v) \prec \varphi(u,v) \prec \varphi_{\mathrm{L}}(v)$ and that $u \in Z_R(v)$. According to Lemma 5, the bitangent $\varphi(u,v)$ enters $\mathrm{L}^{\mathrm{ent}}(v)$ and crosses v. According to the duality theorem $v \in Z_L(u)$, and consequently v belongs to $X^0_{\kappa(u)}$. The bitangent $\varphi(v,u)$ is therefore well-defined. We prove successively that $\varphi(v,u) \preceq \varphi(u,v)$ and $\varphi(u,v) \preceq \varphi(v,u)$. Since v belongs to $Z_L(u)$, v belongs to $\Sigma_2(u)$, and consequently, thanks to the left version of Lemma 4, $\varphi(v,u) = \varphi(v; \{\varphi_{\mathrm{R}}(u)\})$. The bitangent $\varphi(u,v)$ crosses v but $\varphi_{\mathrm{R}}(u)$ doesn't, therefore $\varphi(u,v) \neq \varphi_{\mathrm{R}}(u)$. Since the bitangents $\varphi(u,v)$ and $\varphi_{\mathrm{R}}(u)$ leave $\mathrm{R}(u)$ they do not cross and the minimum element property in $X^0_{\{\varphi_{\mathrm{R}}(u)\}}$ yields $\varphi(v; \{\varphi_{\mathrm{R}}(u)\}) \preceq \varphi(u,v)$ from which we deduce $\varphi(v,u) \preceq \varphi(u,v)$. Since $\varphi(u,v)$ and $\varphi(v,u)$ both enter $\mathrm{L}(u,v)$ the opposite inequality will follow from the fact that $\varphi(u,v)$ crosses no bitangent of $\mathrm{R}^{2_1}(v,u)$. The knowledge of this chain is provided by the left version of Lemma 5 which we can apply because $\varphi_{\mathrm{R}}(u)$ and $\varphi_{\mathrm{L}}(u)$ do not enter the same disk. This last claim is a consequence of $\varphi_{\mathrm{R}}(u) \neq \varphi(u)$ which in turn is a consequence of the existence of at least two bitangents leaving $\mathrm{R}(u)$ and crossing u, namely $\varphi(u,v)$ and $\varphi_{\mathrm{R}}(u)$. A bitangent $t \in \mathrm{R}^{2_1}(v,u)$ is therefore equal to $\varphi(u)$ or belongs to $\mathrm{R}^{2_2}(u)$. Since $\varphi(u,v)$ leaves $\mathrm{R}(u)$, $\varphi(u,v)$ crosses no such bitangent t. □

We are now ready to prove what we announced in the introduction of this subsection.

Lemma 7. *Let u, v non crossing with $\varphi(u,v) = \varphi(v,u)$. Then*

1. $\varphi_{\mathrm{L}}(v) \neq \varphi_{\mathrm{R}}(v)$ *and* $\varphi(u,v)$ *pierces* u *and* v.

2. $u \in Z_R(v)$ *or* $u \in Z_L(v)$ *(the left version of* Z_R*)*

3. If $u \in Z_R(v)$ *then* $\varphi(v) \neq \varphi_{\mathrm{L}}(v)$.

4. $v \prec \varphi(u)$ *and* $u \prec \varphi(v)$.

Proof. We prove successively Claims 1, 2, 3, and 4.

Proof of Claim 1. The bitangent $\varphi(u,v)$ crosses u and $\varphi(v,u)$ crosses v so the second part of the claim is clear. The existence of $\varphi(u,v)$ implies that u does not intersect $\varphi_{\mathrm{R}}(v)$ and therefore $\varphi(u,v) \neq \varphi_{\mathrm{R}}(v)$. Since $\varphi(u,v) = \varphi(v,u)$ it follows that $\varphi(v,u) \neq \varphi_{\mathrm{R}}(v)$ and therefore there are at least two bitangents intersecting v. This proves that $\varphi_{\mathrm{L}}(v) \neq \varphi_{\mathrm{R}}(v)$.

Proof of Claim 2. Without loss of generality we can assume that $\varphi(u,v)$ crosses u and v in that order. We prove successively that u belongs to $\Sigma_2(v)$ and $u \in Z_R(v)$. Let x be the crossing point of v with $\varphi(u,v)$ and let y_1, y and y_2 denote the crossing points of u with $\varphi_{\mathrm{R}}(u), \varphi(u,v)$ and $\varphi_{\mathrm{L}}(u)$, respectively.

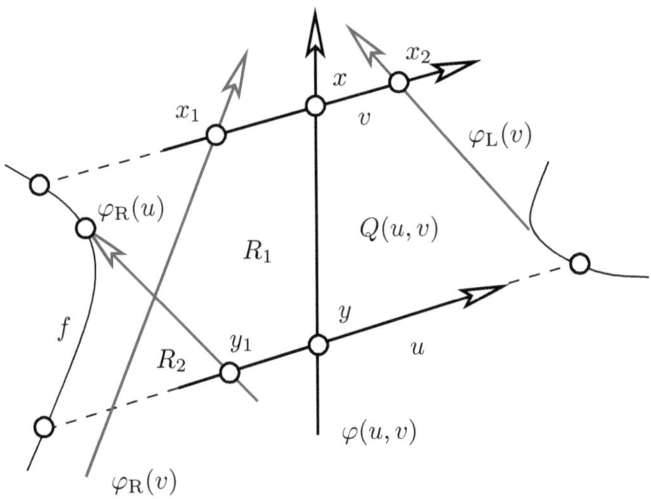

With the help of the face $\sigma_{\mathrm{sour}}(v,u) \in \mathbb{V}_{\kappa(u)}$ with source v and sink $\varphi(v,u)$ we define a pseudoquadrangle $Q(v,u)$ as follows. Let s and s' be the sub-segments of $(g_{\kappa(u)} \times 1_{\mathbb{S}^1})(v)$ and $(g_{\kappa(u)} \times 1_{\mathbb{S}^1})(\varphi(v,u))$ with head x and let e be the sub-chain of the backward view of $\sigma_{\mathrm{sour}}(v,u)$ defined by the tails of s and s'. The interior of the pseudotriangle $\delta := ss'e$ is a subset of the projection in the plane of the face $\sigma_{\mathrm{sour}}(v,u)$ therefore it is a subset of free space. Since the segment $[y_1, y]$ crosses s' from its left side to its right side, it follows that y_1 belongs to e or the segment $[y_1, y]$ crosses s. We show that the latter case does not hold. Indeed otherwise $[y_1, y]$ would cross $s \setminus [x_1, x]$ (because u and v do not cross) and consequently $\varphi_{\mathrm{R}}(v)$ would cross one of the two segments $[y, x]$ or $[y_1, y]$. The two cases are impossible because $[y, x] \subset \varphi(u, v)$, $[y_1, y] \subset u$ and $\varphi_{\mathrm{R}}(v)$ crosses neither of the bitangents $\varphi(u, v)$ or u. This achieves to prove $y_1 \in e$. We define $Q(v, u)$ to be the pseudoquadrangle defined by the three segments s, $[y, x]$, $[y_1, y]$ and the subchain of e defined by y_1 and the tail point of s. Since $Q(v, u) \subset \delta$, the interior of $Q(v, u)$ is a subset of free space. The bitangent $\varphi_{\mathrm{R}}(v)$ crosses s but does not cross $\varphi(v, u)$ nor $[y_1, y]$, therefore $\varphi_{\mathrm{R}}(v)$ leaves or crosses the chain e. In both cases $\varphi_{\mathrm{R}}(v)$ partitions $Q(v, u)$ in two regions; let R_1 be the one containing y. Let t be the sub-segment of $(g_\emptyset \times 1_{\mathbb{S}^1})(u)$ with head y_1, t' be the sub-segment of $(g_\emptyset \times 1_{\mathbb{S}^1})(\varphi_{\mathrm{R}}(u))$ with tail y_1 and let f be the sub-arc of the left backward view of u in $\mathbb{V}$ defined by the tail and head points of t and t' respectively. The pseudotriangle $\tau := tft'$ is a subset of free space. If $\varphi_{\mathrm{R}}(v)$ leaves e, we define R_2 as being equal to τ. Otherwise $\varphi_{\mathrm{R}}(v)$ crosses $\varphi_{\mathrm{R}}(u)$ and therefore partitions the interior of τ in two regions and we define R_2 to be the one containing y_1. Merging the two adjacent (along a sub-segment of $\varphi_{\mathrm{R}}(u)$) regions R_1 and R_2 yields a pseudoquadrangle Q.

The construction operated above to build $Q(v, u)$ using the face $\sigma_{\mathrm{sour}}(v, u)$ and its backward view can be performed on the face $\sigma_{\mathrm{sour}}(u, v)$ and its forward view to yield the interior free pseudoquadrangle $Q(u, v)$. Merging the two pseudoquadrangles $Q(v, u)$ and $Q(u, v)$ adjacent along the segment $[y, x]$, defines an interior free pseudoquadrangle having $[x_1, x_2]$ and $(g_{\kappa(v)} \times 1_{\mathbb{S}^1})(u)$ as two sides. This proves that u belongs to $\Sigma_2(v)$. To prove that u belongs to $Z_R(v)$ we use the description of the chain $\Sigma_2(v)$ given in Lemma 3 and prove that $u \notin \mathrm{rc}_2(\sigma_{\mathrm{sour}}(v))$. This last conclusion together with its dual version will allow us to conclude. Let t be a bitangent of $\mathrm{rc}_2(\sigma_{\mathrm{sour}}(v))$ with $t \neq \varphi_{\mathrm{righ}}(v)$, we prove that $t \neq u$. Note that $v = \varphi_{\mathrm{left}_*}(t)$. Moreover, according to the Duality Theorem, v belongs to $\mathrm{L}_*^{21}(t)$ and therefore $\varphi_*(t)$ does not leave the the first atom of $\mathrm{L}_*^2(t)$. This last conclusion together with the left dual version of Theorem 15 prove that $\varphi_{\mathrm{L}}(t)$ crosses v (first case) or $\varphi_*(t)$ leaves the first atom of $\iota_* \mathrm{L}^2(t)$ (second case). In the first case $t \neq u$ because $\varphi_{\mathrm{L}}(u)$ does not cross v. In the second case assume $t \in X^0_{\kappa(v)}$ otherwise it is clear that $t \neq u$. The bitangent $\iota\varphi_*(t)$ is smaller than $\varphi_{\mathrm{left}}(t)$ $(= \varphi(v))$ and is the maximal bitangent crossing t and greater than t. Consequently $\varphi(t, v) \preceq \iota\varphi_*(t)$ and therefore $\varphi(t, v) \prec \varphi(v)$. This last inequality implies that $\varphi(t, v)$ does not cross v and achieves to prove $t \neq u$.

Proof of Claim 3. It is sufficient to show that there are at least two bitangents entering $\mathrm{L}(v)$ and crossing v. The first of these two bitangents is $\varphi_{\mathrm{L}}(v)$ and does not cross u. Since $v \in Z_L(u)$, v belongs to $\Sigma_2(u)$, and, according to the left version of Lemma 4, $\mathrm{L}(v, u) = \mathrm{L}(v)$. Consequently the bitangent $\varphi(v, u)$ crosses u and v and enters $\mathrm{L}(v)$. $\varphi(u, v)$ is the second bitangent we were looking for.

Proof of Claim 4. Assume $u \in Z_R(v)$. The preceding claim states that $\varphi_{\mathrm{L}}(v) \neq \varphi(v)$ therefore the sink of the chain Σ introduced in Lemma 3 is equal to $\varphi(v)$. This proves $u \prec \varphi(v)$. The left version of this result yields $u \prec \varphi(v)$ for $u \in Z_L(v)$. To prove $v \prec \varphi(u)$ simply swap the roles of u and v. □

So far we have identified the set of pairs (u, v) such that $\varphi(u, v) = \varphi(v, u)$. In the next lemma we study the equation $\varphi(u, v) = \varphi(u', v')$. Let $\mathcal{G}$ denote the set of pairs $(u, v) \in X^0 \times X^0$ such that $\varphi(u, v) = \varphi(v, u)$ and $\varphi_*(u, v) = \varphi_*(v, u)$. We partition $\mathcal{G}$ into two subsets $\mathcal{G}_R$ and $\mathcal{G}_L$ where $\mathcal{G}_R$, $\mathcal{G}_L$ contain the pairs $(u, v) \in \mathcal{G}$ such that $u \in Z_R(v)$, $u \in Z_L(v)$, respectively.

Lemma 8. *Let $(u, v), (u', v') \in \mathcal{G}_R$ where v, v' are xx-left bitangents and where u, u' are xx-right bitangents. Assume that $\varphi(u, v) = \varphi(u', v')$ and that $v' \not\preceq v$. Then $v \prec v'$ and $\iota(v) \prec \varphi_{\mathrm{forw}}(u', v')$.*

Proof. We first prove that

$$\varphi_{\mathrm{L}}(v) \prec \iota(v) \prec \varphi_{\mathrm{L}}(v') \prec \iota(v'). \tag{11}$$

Let $t_0 = \varphi(v, u)$. Since t_0 enters $\mathrm{L}(v, u)$ and since $\mathrm{L}(v, u) = \mathrm{L}(v)$ (cf. Lemma 6) we can introduce the suffix subchain $C(v)$ of $\mathrm{L}(v)$ with source t_0. Let $t_0, t_1, \ldots$ be the sequence of bitangents defined by $t_{i+1} = \varphi_{\mathrm{forw}}(t_i)$ for $i \geq 0$. According to the Orbits Theorem

$$C(v) = t_0 \epsilon_0 t_1 \epsilon_1 \ldots t_p \epsilon$$

where p is the first index such that t_p is an xx-right bitangent and where ϵ is the arc with source t_p and sink $\iota(v)$ (here we use the fact that v is an xx-left bitangent). Similarly we introduce the suffix subchain $C(v')$ of $\mathrm{L}(v')$ with source $t'_0 = \varphi(v', u')$ and write

$$C(v') = t'_0 \epsilon'_0 t'_1 \epsilon'_1 \ldots t'_{p'} \epsilon'$$

where $t'_{i+1} = \varphi_{\mathrm{forw}}(t'_i)$ and where ϵ' is the arc with source $t'_{p'}$ and sink $\iota(v')$. Since by hypothesis $t_0 = t'_0$ and $v' \not\preceq b$ it follows that

$$C(v') = C(v)\iota(e)$$

where e is an arc with source v and sink v'. Since $\varphi_{\mathrm{L}}(v)$ and $\varphi_{\mathrm{L}}(v')$ are the maximal bitangents entering $C(v)$ and $C(v')$, respectively, it follows that $\varphi_{\mathrm{L}}(v) = \varphi_{\mathrm{L}}(v')$ or $\varphi_{\mathrm{L}}(v')$ enters $\iota(e)$ which proves Equation (11). It remains to prove that $\varphi_{\mathrm{L}}(v) \neq \varphi_{\mathrm{L}}(v')$. If $\varphi_{\mathrm{L}}(v) = \varphi_{\mathrm{L}}(v')$ then according to Lemma 6 one has $\varphi(u; \{\varphi_{\mathrm{L}}(v)\}) = \varphi(u'; \{\varphi_{\mathrm{L}}(v)\})$ and consequently $u = u'$ since $\varphi^{\{\varphi_{\mathrm{L}}(v)\}}$ is one-to-one. But $u = u'$ implies that $\varphi(v, u) = \varphi(v', u')$ and consequently $v = v'$ since $\varphi^{\kappa(u)}$ is one-to-one. We also note that, according to our analysis, $\varphi(u, v), \varphi_{\mathrm{L}}(v), \varphi_{\mathrm{L}}(v')$ are pairwise non crossing. We prove now that $\iota(v) \prec \varphi_{\mathrm{forw}}(u', v')$. According to Lemma 7 Claim 4 the bitangent

$\varphi_{\text{forw}}(u', v')$ enters $\mathrm{L}(v')$ or $\varphi_{\mathrm{L}}(v') \prec \varphi_{\text{forw}}(u', v')$. In the latter case we are done since we have seen that $\iota(v) \prec \varphi_{\mathrm{L}}(v')$. Assume now that $\varphi_{\text{forw}}(u', v')$ enters $\mathrm{L}(v')$, and consequently enters $C(v')$. We claim that

$$\varphi_{\mathrm{L}}(v) \prec \varphi_{\text{forw}}(u', v') \tag{12}$$

from which we deduce that $\varphi_{\text{forw}}(u', v')$ enters $\iota(e)$, and consequently that $\iota(v) \prec \varphi_{\text{forw}}(u', v')$. We prove successively that

(0-) $\varphi_{\mathrm{L}}(v')$ and u are non crossing;

(1-) $\varphi_{\mathrm{L}}(v)$ and u' are crossing;

(2-) $\varphi_{\mathrm{L}}(v) \prec \iota(u')$;

(3-) $\varphi_{\mathrm{L}}(u', v') \prec \varphi_{\text{forw}}(u', v')$;

(4-) $\varphi_{\mathrm{L}}(u', v') = \iota\varphi_*(u'; T)$ where T is the set of bitangents of the chain $\mathrm{L}^2(u', v')$;

(5-) $\varphi_{\mathrm{L}}(v) \preceq \iota\varphi_*(u'; T)$;

from which Equation (12) follows. Claim (0-) is a direct application of the minimum element property and the observation that $\varphi_*(u) \prec \iota_*\varphi_{\mathrm{L}}(v') \prec \varphi(u; \{\varphi_{\mathrm{L}}(v)\})$ — the first inequality follows Equation (11) and Lemma 7 Claim 4 applied to φ_*, and the second inequality follows from the fact that $\varphi_{\mathrm{L}}(v')$ and $\varphi(u, v)$ $(= \varphi(u; \{\varphi_{\mathrm{L}}(v)\}))$ enter $\mathrm{L}^{\text{ent}}(v')$ with $\varphi(u, v) \prec \varphi_{\mathrm{L}}(v')$. To prove Claim (1-) we observe that if $\varphi_{\mathrm{L}}(v)$ and u' are non crossing we can write that $\varphi(u; G) = \varphi(u'; G)$ where $G = \{\varphi_{\mathrm{L}}(v), \varphi_{\mathrm{L}}(v')\}$ and consequently that $u = u'$ since φ^G is one-to-one. Claim (2-) is a consequence of $\varphi_{\mathrm{L}}(v) \prec \varphi_{\mathrm{L}}(v')$, $\varphi_{\mathrm{L}}(v') \preceq \iota\varphi_*(v')$ and $\iota\varphi_*(v') \prec \iota(u')$ (Lemma 7 Claim 4 applied to φ_*). To prove Claim (3-) we observe that $\varphi_{\mathrm{L}}(u', v')$ enters $\mathrm{L}^{\text{ent}}(u', v')$ and that $\varphi_{\text{forw}}(u', v')$ is the sink of $\mathrm{L}^{\text{ent}}(u', v')$ (see the Orbits Theorem with u' is an xx-right bitangent). Claim (4-) is obvious and Claim (5-) is the affiliation of Claims (1-) and (2-), $\varphi_{\mathrm{L}}(v) \in X_T^0$ and the minimum element property in X_T^0. □

Proof of the Sum of Squares Theorem. Let now I be a filter of $(X_0, \prec)$ and let $\mathcal{G}(I), \mathcal{G}_R(I), \mathcal{G}_L(I)$ be the sets of pairs $(u, v) \in \mathcal{G}, \mathcal{G}_R, \mathcal{G}_L$ respectively, such that $v \in I \setminus \iota(I)$. The sum of squares Theorem can be restated as $|\mathcal{G}(I)| = O(|\mathcal{B}|)$ and it is clearly sufficient to prove that $|\mathcal{G}_R(I)| = O(|\mathcal{B}|)$. Let $\mathcal{G}_R^+(I)$ be the set of $(u, v) \in \mathcal{G}_R(I)$ such that $\varphi(u, v) \notin \iota(I)$, and let $\mathcal{G}_R^-(I)$ be the set of $(u, v) \in \mathcal{G}_R(I)$ such that $\varphi_*(u, v) \in I$. We claim that

(1-)$\mathcal{G}_R(I) = \mathcal{G}_R^+(I) \cup \mathcal{G}_R^-(I)$;

(2-)$\mathcal{G}_R^-(I) = (\mathcal{G}_R^+)_*(J)$ where $J = X^0 \setminus \iota(I)$;

(3-)$|\mathcal{G}_R^+(I)| = O(|\mathcal{B}|)$;

from which the theorem follows easily. Claim (1-) follows $\varphi(u, v) \preceq \iota\varphi_*(u, v)$. Claim (2-) follows $(\mathcal{G}_R^+)_*(J) = \{(u, v) \in (\mathcal{G}_R)_*(J) \mid \varphi_*(u, v) \in I\}$ and $(\mathcal{G}_R)_*(J) = \mathcal{G}_R(I)$. To prove Claim (3-) we argue as follows. Since $Z_R(v)$ contains at most one xx-left bitangent the number of pairs (u, v) of $\mathcal{G}_R(I)$ such that u is xx-left is a $O(|\mathcal{B}|)$. Similar arguments allow us to restrict our attention to the case where v is xx-left and u is xx-right. Let $E(v)$ denote the set of bitangents u such that $(u, v) \in \mathcal{G}_R^+(I)$. It is sufficient to prove that if the pairs (u, v) and (u', v') satisfy the assumptions of Lemma 8 then $u' = \min_{\prec} E(v')$. Let u'' be a bitangent of $Z_R(v')$ with $u'' \prec u'$ (otherwise we are done). Since two consecutive bitangents t, t' of $Z_R(v')$ are related by the relation $t' = \varphi_{\text{forw}_*}(t, v')$ and since, according to the Orbits Theorem, $\varphi(t, v') \prec \varphi_{\text{forw}}(t, v')$, we have $\varphi(u', v') \prec \varphi_{\text{forw}}(u', v') \preceq \varphi(u'', v')$, from which we deduce, using Lemma 8, that $\iota(v) \prec \varphi(u'', v')$, and consequently $u'' \notin E(v')$. This proves that $u' = \min_{\prec} E(v')$. □

3.5 Pairs of bitangents that satisfy the conditions of the Sum of Squares Theorem

We identify now a set of pairs (u, v) that satisfy the conditions of the sum of squares theorem. As we shall see in the next section this set is closely related to the multiset of bitangents visited during the course of the flip method based on the left-turn predicate. We denote by v_R the successor of v in the pseudotriangle $\mathrm{R}(v)$.

Theorem 19. *Let (u, v) be a pair of bitangents that satisfies one of the three following conditions:*
(C1-) $u \in \mathrm{R}'(v)$ and $\varphi_{\mathrm{R}}(v)$ is a right-xx bitangent;
(C2-) $u \in r(v)$, $\varphi_{\mathrm{R}}(v)$ and $\varphi_{\mathrm{R}_*}(u)$ are right-xx bitangents, and $\varphi_{\text{righ}}(v)$ is an xx-left bitangent (or $\varphi_{\text{left}}(u)$ is a right-xx bitangent);
(C3-) $v \in \mathrm{L}'(u)$ and $\varphi_{\mathrm{L}}(u)$ is an xx-left bitangent.
Then $(u, v) \in \mathcal{G}_R$.

Proof. Assume that the pair (u, v) satisfies (C1-). We claim that
(1-) $\varphi_{\mathrm{R}}(v)$ and $\varphi(v)$ are right-xx;
(2-) $\varphi_{\mathrm{L}_*}(v) \neq \varphi_*(v)$ and $\varphi_{\mathrm{L}}(v) \neq \varphi(v)$;
(3-) $\varphi(u) \preceq \varphi_{\mathrm{L}}(u) \prec \varphi_{\mathrm{L}}(v) \preceq \pi\varphi_*(v)$;
from which we deduce, according to Lemma 6 and its dual version, that $\varphi(u, v) = \varphi(v, u)$ and $\varphi_*(u, v) = \varphi_*(v, u)$, that is, $(u, v) \in \mathcal{G}_R$. Claim (1-) is clear since by assumption $\varphi_{\mathrm{R}}(v)$ is right-xx. Claim (2-) is consequence of Claim (1-), $u \in \mathrm{R}'(v)$ and Theorem 15. To prove Claim (3-) we use the following Lemma.

Lemma 9. *Let $t' = \varphi_{\text{back}}(t)$ where t' enters the negative cycle that t leaves. Then $\varphi_{\mathrm{L}}(t') \prec \varphi(t)$ (because $\varphi(t) = \min_{\prec} \mathrm{L}^3(t')$ and $\varphi_{\mathrm{L}}(t')$ enters $\mathrm{L}^{\text{ent}}(t')$).*

According to this lemma $\varphi_{\mathrm{L}}(u) \prec \varphi(\varphi_{\text{righ}}(v))$. Furthermore, since $\varphi_{\mathrm{L}}(v)$ crosses $\varphi_{\text{righ}}(v)$, $\varphi(\varphi_{\text{righ}}(v)) \preceq \varphi_{\mathrm{L}}(v)$ (Minimum Element Property). Therefore $\varphi_{\mathrm{L}}(u) \prec \varphi_{\mathrm{L}}(v)$. Since $\varphi(u) \preceq \varphi_{\mathrm{L}}(u)$ is clear we are done.

Assume now that u verifies (C2-). We claim that
(1-) $\varphi_{\mathrm{R}}(v)$ and $\varphi(v)$ are right-xx;
(2-) $\varphi_{\mathrm{R}_*}(u)$ is right-xx and $\varphi(u)$ is xx-left;
(3-) $\varphi_{\mathrm{L}_*}(v) \neq \varphi_*(v)$ and $\varphi_{\mathrm{L}}(v) \neq \varphi(v)$;
from which we deduce, according to Lemma 6, that $\varphi(u, v) = \varphi(v, u)$ and $\varphi_*(u, v) = \varphi_*(v, u)$, that is, $(u, v) \in \mathcal{G}_R$.

The proof under (C3-) is similar using the left version of Lemma 6 and its dual version. □

Theorem 20. *Let $Y_R(v)$ be the set of bitangents u such that one of three following conditions is satisfied*

(C1-) $u \in \mathrm{R}'(v)$ and $\varphi_{\mathrm{R}}(v_R)$ is a right-xx bitangent;

(C2-) $u \in r(v)$, $u_L \in \mathrm{L}^1(u)$, $v_R \in \mathrm{R}^1(v)$ and $\varphi_{\mathrm{R}}(v_R)$ is a right-xx bitangent, $\varphi_{\mathrm{L}}(u_L)$ is an xx-left bitangent and $\varphi_{\mathrm{righ}}(v)$ is a right-left bitangent;

(C3-) $u \in \mathrm{R}'_*(v)$ and $\varphi_{\mathrm{L}}(u_L)$ is an xx-left bitangent.

Let u and v be such that $u \in Y_R(v)$ (iff. $v \in Y_L(u)$). Assume that $u \neq \min_\prec Y_R(v)$ and $v \neq \min_\prec Y_L(u)$. Then $(u,v) \in \mathcal{G}$.

Proof. Let (u,v) be a pair of bitangents that satisfy the assumptions of the theorem. We claim that

(1-) $u \in \mathrm{R}'(v)$ or $\varphi_{\mathrm{L}}(u)$ is xx-left;

(2-) $v \in \mathrm{L}'(u)$ or $\varphi_{\mathrm{R}}(v)$ is right-xx;

from which we deduce — since the two conditions $u \in \mathrm{R}'(v)$ and $v \in \mathrm{L}'(u)$ (which is equivalent to $u \in \mathrm{R}'_*(v)$ by the duality theorem) cannot occur simultaneously — that the pair (u,v) satisfies the assumptions of Theorem 19 and consequently that $(u,v) \in \mathcal{G}$. We now prove Claim (1-) (Claim (2-) is the left version of Claim (1)). Assume that $Y_R(v) \setminus \mathrm{R}'(v)$ contains at least two bitangents – otherwise we are done – and let $c = \min_\prec Y_R(v) \setminus \mathrm{R}'(v) = \min_\prec Y_R(v)$ and c_L be its successor in $\mathrm{L}^1(c)$. Let $\rho(e)$ be an edge of $\mathrm{rc}_2(\sigma_{\mathrm{righ}}(v))$ or an edge of $\rho(\mathrm{R}^2_*(v))$ and let t and t' be its source and sink respectively. Assume that $\varphi_{\mathrm{L}}(t)$ is xx-left; we prove that $\varphi_{\mathrm{L}}(t')$ is also xx-left. The arc e is either a negative 0-cusp arc (case (i)) or a negative source cusp arc (case (ii)). In case (i), Table 2 tells us that $\varphi_{\mathrm{L}}(t')$ enters $\mathrm{L}^{\mathrm{ent}}(t)$ and consequently $\varphi_{\mathrm{L}}(t')$ is xx-left. In case (ii), Table 2 tells us that $\varphi_{\mathrm{L}}(t')$ enters the first atom of $\mathrm{L}^2(t')$ or enters $\mathrm{L}^{\mathrm{ent}}(t)$, consequently $\varphi_{\mathrm{L}}(t')$ is xx-left. Since c belongs to $Y_R(v)$ the bitangent $\varphi_{\mathrm{L}}(c_L)$ is an xx-left bitangent. It follows that all the bitangents $\varphi_{\mathrm{L}}(u)$ are xx-left bitangents for $u \in Y_R(v) \setminus \mathrm{R}'(v)$ and $u \neq \min_\prec Y_R(v)$. □

We are now in position to justify Theorem 8. Assume that $\mathrm{R}'(v)$ is not empty and that $\varphi(v)$ is right-xx. Then $\varphi_{\mathrm{R}}(v_R)$ is right-xx. It follows, according to the above theorem, that the set of pairs (u,v) such that $u \in \mathrm{R}'(v)$ and $\varphi(v)$ right-xx is a subset of $\mathcal{G}$, and consequently, thanks to the sum of squares theorem, its size is a big-O of the size of $\mathcal{B}$.

4 Optimal flip using the left-turn predicate

We maintain as described previously not only $\mathrm{G}(I)$ and $\mathrm{S}(I)$ but also the dual pseudotriangulation $\mathrm{G}_*(I_*)$; furthermore for each bitangent $t \in \mathrm{G}(I)$ which is right-minimal (resp. left-minimal) we maintain a pointer to the arc of $\mathrm{G}(I)$ that $\varphi_{\mathrm{R}}(t)$ leaves (resp. $\varphi_{\mathrm{L}}(t)$ enters) and a pointer to the arc of $\mathrm{G}_*(I_*)$ that $\varphi_{\mathrm{R}}(t)$ enters (resp. $\varphi_{\mathrm{L}}(t)$ leaves). Let v be minimal in the

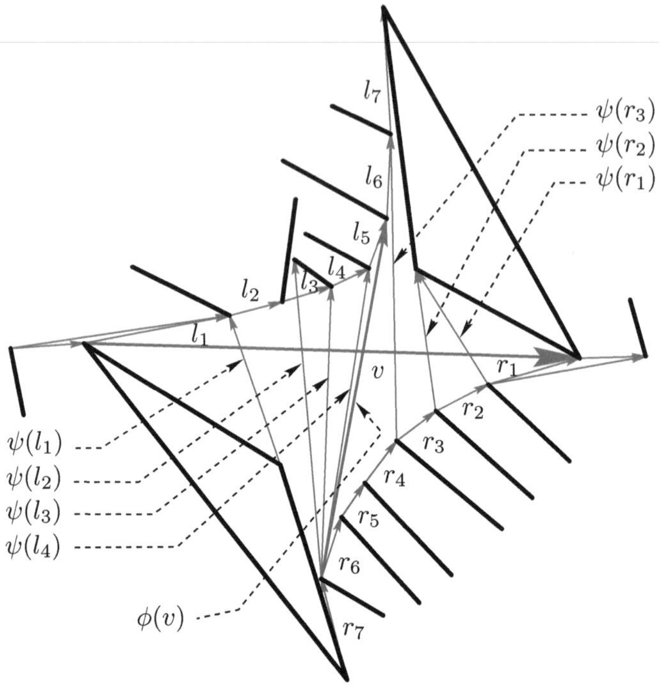

Fig. 21. Flipping v using only χ_1. Here $r_u = r_3$, $l_5 = l_q$, $r_6 = r_p$ and $\psi(r_6) = \psi(l_5) = \varphi(v)$.

filter I, and let v_R, v_L be its successors in $\mathrm{R}(v), \mathrm{L}(v)$, respectively. Thanks to Theorem 15 we can detect in constant time if $\varphi(v)$ coincides with $\varphi_{\mathrm{R}}(v)$ or $\varphi_{\mathrm{L}}(v)$; in that case we are done. We assume now that $\varphi_{\mathrm{R}}(v)$ and $\varphi_{\mathrm{L}}(v)$ are both $\neq \varphi(v)$, and consequently $\varphi(v)$ does not leave the first atom of $\mathrm{R}^2(v)$ nor enter the first atom of $\mathrm{L}^2(v)$. If v and v_R are separated by a cusp then v_R is left minimal in $I' = I \setminus v$ and $\varphi_{\mathrm{L}}(v_R) = \varphi_{\mathrm{L}}(v)$ (cf. Theorem 16). If v and v_R are not separated by a cusp then v_R is right-minimal and we compute $\varphi_{\mathrm{R}}(v_R)$, as explained in the previous section, to restore the invariant of our algorithm (cf. Theorem 17). A similar computation is done with v_L. It remains to compute $\varphi(v)$. We set $r_1 = v_R$ if v and v_R are separated by a cusp otherwise we set $r_1 = \varphi_{\mathrm{R}}(v_R)$. If r_1 is a left-xx bitangent then $\varphi(v)$ is also a left-xx bitangent and therefore leaves the first atom (which is an arc) of $\mathrm{R}^3(v)$. Similarly if l_1 is an xx-right bitangent then $\varphi(v)$ is an xx-right bitangent and enters the first atom (which is an arc) of $\mathrm{L}^3(v)$. In both cases the method FLIP2 is easily adapted to work only with the χ_1 predicate (cf. Remark 4). Assume now that r_1 and l_1 are right-xx and xx-left bitangents, respectively. Let $l_2 \prec \ldots \prec l_{\max}$ be the increasing sequence of bitangents of $\mathrm{L}^2(v)$ greater than l_1. We denote by l_q the last element of this sequence which is smaller than $\varphi(v)$. Note that according to the bridge theorem $l_1, \ldots, l_q$ are atoms of $\mathrm{lc}(v)$. We introduce similar notations for $\mathrm{R}^2(v)$, that is, $r_1, r_2, \ldots, r_p, \ldots, r_{\max}$. According to the duality theorem the pseudotriangles $\mathrm{R}(v)$ and $\mathrm{R}_*(l_i)$ are incident to the same bitangent (v) and they don't locally overlap $(i \leq q)$. It follows (cf. Theorem 3) that there is a unique bitangent $\psi(l_i)$ that leaves $\mathrm{R}(v)$ and enters $\iota\,\mathrm{R}_*(l_i)$. The operators ψ, φ_{R}, and φ are related as described in the following theorem.

Theorem 21. *There is an index q' with $1 \leq q' \leq q$ such that $\psi(l_i) = \varphi_{\mathrm{R}}(l_{i+1}) \neq \varphi(v)$ for $1 \leq i < q'$ and $\psi(l_i) = \varphi(v)$ for $q' \leq i \leq q$. Moreover*

1. *For $1 \leq i < q'$ the bitangent $\psi(l_i)$ enters the chain $\iota\,\mathrm{R}_*^{\mathrm{lea}}(\varphi_*(l_{i+1}), I'_*)$ where $I' = I \setminus v$.*

2. *For $q' \leq i < q$ the bitangent $\iota\varphi_*(l_{i+1})$ leaves the first atom of $\mathrm{L}^3(v)$ and the bitangent $\psi(l_i)$ enters the first atom of $\mathrm{L}^3(v)$.*

3. *$\psi(l_q)$ enters the arc l_q^+ which is a subarc of the last atom of the chain $\iota\,\mathrm{R}_*^{\mathrm{lea}}(\varphi_*(l_{q+1}), I_*)$.*

Proof. According to Theorem 11 and since l_i is an atom of the chain $\mathrm{lc}_2(v)$ the chain $\mathrm{R}^{\mathrm{lea}}(l_i)$ is a prefix subchain of the chain $e\varphi(v)\,\mathrm{R}^{\mathrm{lea}_2}(v)e'$ where e is a suffix subarc of $\mathrm{L}^{3_1}(v)$. Moreover $\varphi_{\mathrm{R}}(l_{i+1}) \prec \varphi(l_i) \preceq \varphi_{\mathrm{R}}(l_i)$. If v_L belongs to $\mathrm{L}^2(v)$ the arc e' is empty or is a primitive arc; otherwise $\iota(v_L)$ leaves e' and intersects l_1. It follows that $\varphi_{\mathrm{R}}(l_i) \prec \iota(v_L)$ for $i \geq 2$, and therefore $\varphi_{\mathrm{R}}(l_i)$ does not leave e' for $2 \leq i \leq q$. If $\varphi_{\mathrm{R}}(l_i)$ leaves $\mathrm{R}^{\mathrm{lea}}(v)$ for all the indices $2 \leq i \leq q$ we set $q' = q$, otherwise let $q' < q$ be such that $\varphi_{\mathrm{R}}(l_i)$ leaves $\mathrm{R}^{\mathrm{lea}}(v)$ for $2 \leq i \leq q'$ and $\varphi_{\mathrm{R}}(l_i)$ leaves e for $q' < i \leq q$.

Let $1 \leq i < q'$. The bitangent $\varphi_{\mathrm{R}}(l_{i+1})$ leaves $\mathrm{R}^{\mathrm{lea}_2}(v)$ and therefore crosses v. To prove that $\varphi_{\mathrm{R}}(l_{i+1}) = \psi(l_i)$ it remains to prove that $\varphi_{\mathrm{R}}(l_{i+1})$ enters $\iota\,\mathrm{R}_*^{\mathrm{lea}}(l_i)$. According to the dual version of Theorem 15 and since $v = \varphi_{\mathrm{righ}_*}(l_{i+1})$ and $\varphi_{\mathrm{R}}(l_{i+j})$ crosses v, $\varphi_{\mathrm{R}}(l_{i+1}) \neq \iota\varphi_*(l_{i+1})$ and consequently $\varphi_{\mathrm{R}}(l_{i+1})$ enters $\iota\,\mathrm{L}_*^{\mathrm{lea}_2}(l_{i+1})$. The arc e_i with source l_i and sink l_{i+1} is a positive 0-cusp primitive arc, so according to Table 2

$$\mathrm{R}_*^{\mathrm{lea}}(l_i) = \mathrm{R}_*^{3_1}(l_i)\varphi_*(l_{i+1})\,\mathrm{L}_*^{\mathrm{lea}_2}(l_{i+1})\iota(e_i).$$

This equality proves that $\varphi_{\mathrm{R}}(l_{i+1})$ enters $\iota\,\mathrm{R}_*^{\mathrm{lea}}(l_i)$. To complete the proof of Claim 1 it remains to show that $\varphi_{\mathrm{R}}(l_{i+1})$ enters $\iota\,\mathrm{R}_*^{\mathrm{lea}}(\varphi_*(l_{i+1}), I'_*)$. This will be a consequence of the fact that $C(v) = \varphi_*(l_{i+1})\,\mathrm{L}_*^{\mathrm{lea}_2}(l_{i+1})$ is a chain of $G_*(I'_*)$. Since $l_{i+1} \in I'$, we have $\varphi_*(l_{i+1}) \in G_*(I'_*)$ and the orbits theorem tells us that the bitangents of the chain $C(v)$ also belong to $G_*(I'_*)$. We now show that there is no bitangent of $G_*(I'_*)$ leaving the chain $C(v)$. Since $v \in G_*(I'_*)$ and $v = \varphi_{\mathrm{righ}_*}(l_{i+1})$, the bitangents of $\mathrm{R}_*^2(l_{i+1})$ are bitangents of $G_*(I'_*)$. Since a bitangent leaving $C(v)$ crosses a bitangent of $\mathrm{R}_*^2(l_{i+1})$, $G_*(I'_*)$ contains no bitangent leaving $C(v)$. This achieves to prove that $C(v)$ is a chain of $G_*(I'_*)$; the result follows.

Assume $q' \neq q$ and let $q' \leq i < q$. The bitangent $\varphi_{\mathrm{R}}(l_{i+1})$ leaves the arc e. Since e is a suffix subarc of $\mathrm{L}^{3_1}(v)$, $\varphi_{\mathrm{R}}(l_{i+1})$ leaves $\mathrm{L}^{3_1}(v)$ and $\varphi(v)$ enters the first atom of $\mathrm{L}^3(v)$. Moreover, $\varphi_{\mathrm{R}}(l_{i+1})$ leaving the first atom of $\mathrm{R}^2(l_{i+1})$, the dual version of Theorem 15 states that $\varphi_{\mathrm{R}}(l_{i+1})$ $(= \iota\varphi_{\mathrm{L}_*}(l_{i+1}))$ is equal to $\iota\varphi_*(l_{i+1})$. To finish the proof of Claim 2 it remains to prove that $\varphi(v)$ enters $\iota\,\mathrm{L}_*^{\mathrm{lea}}(l_i)$. Since $\varphi_*(l_{i+1})$ $(= \min_{\prec_*} \mathrm{R}_*^3(l_i)$ by the orbits theorem) is xx-left, $\mathrm{R}_*^2(l_i)$ is not reduced to an arc and therefore contains v $(= \varphi_{\mathrm{righ}_*}(l_i))$. Let $u = \max_{\prec_*} \mathrm{R}_*^2(l_i)$; the bitangent u is xx-left and the orbits theorem allows us to claim that $\varphi(v) \prec \varphi(u) \prec \iota(u)$. Since $\iota\varphi_*(l_{i+1})$ $(= \varphi_{\mathrm{R}}(l_{i+1}))$ leaves $\mathrm{L}^{3_1}(v)$ we have $\iota\varphi_*(l_{i+1}) \prec \varphi(v)$; it follows that $\varphi(v)$ enters the subarc of $\iota\,\mathrm{R}_*^{3_1}(l_i)$ whose source and sink for $\prec_*$ are $\iota(u)$ and $\iota\varphi_*(l_{i+1})$ respectively. Consequently $\varphi(v)$ enters $\iota\,\mathrm{R}_*^{\mathrm{lea}}(l_i)$.

We now prove Claim 3. It is clear that $\varphi(v)$ enters l_q^+. As $l_{q+1} = \varphi_{\mathrm{forw}}(l_q)$ (orbits theorem) the arc l_q^+ is a subarc of $\iota\,\mathrm{R}_*^4(l_q)$ thus $\varphi(v)$ enters $\iota\,\mathrm{R}_*^{\mathrm{lea}}(l_q)$ and $\varphi(v) = \psi(l_q)$. We now prove that $\mathrm{L}_*^{\mathrm{lea}_2}(l_{q+1})l_q^+$ is a chain of the pseudotriangle $G_*(I_*)$. The result will follow because the source (for $\prec_*$) of this chain is $\varphi_*(l_{q+1})$. We already know, thanks to the orbits theorem, that the bitangents of this chain belong to $G_*(I_*)$ because $\varphi_*(l_{q+1}) \in G_*(I_*)$. We must show that there is no bitangent of $G_*(I_*)$ leaving it. To this end we exhibit a chain $C(v)$ composed of bitangents of $G_*(I_*)$ intersected by any bitangent leaving $\mathrm{L}_*^{\mathrm{lea}_2}(l_{q+1})l_q^+$. The source (for $\prec_*$) of this chain will be $\varphi_*(l_q)$; we show that this bitangent enters $\mathrm{R}_*^{\mathrm{ent}}(l_{q+1})$. This is a consequence of the three simple following remarks:

(i) $v \prec_* \varphi_*(l_q) \prec_* \varphi_*(l_{q+1})$ (orbits theorem);
(ii) $v = \min_{\prec_*} \mathrm{R}_*^2(l_q)$ (orbits theorem);
(iii) v enters $\mathrm{R}_*^{\mathrm{ent}}(l_{q+1})$ (duality theorem).

The chain $C(v)$ which we were looking for is the suffix (for $\prec_*$) subchain of $\mathrm{R}_*^{\mathrm{ent}}(l_{q+1})$ with source $\varphi_*(l_q)$. □

We come back to the computation of $\varphi(v)$. A first idea is to compute successively the bitangents $\psi(l_i)$ for $i = 1, 2, \ldots, q'$ with the procedure PhiR. Unfortunately the repeated use of this procedure is in general to costly. To cut the budget we first compute the bitangents $\psi(r_1), \psi(r_2), \ldots, \psi(r_u)$ — with the left version PhiL of the procedure PhiR — where u is the first index such either $\psi(r_u) = \varphi(v)$ or $\psi(r_u)$ is an xx-left bitangent. In this later case we can now safely compute the bitangents $\psi(l_1), \psi(l_2), \ldots, \psi(l_q)$ with the function PhiRNew(v, l_{i+1}) which differs from PhiR only in the initialization of the variables C and C':

Function PhiRNew$(v, l$: Bit.): Pair of Arcs
$C \leftarrow [r_u, \ldots, r_{\max}]$; $C' \leftarrow \iota\,\mathrm{R}_*^{\mathrm{lea}}(\varphi_*(l), I_*)c$;
χ_1-Walk(C, C');
return (Top(C), Top(C'));

here c is a suffix subarc of the first atom of $\mathrm{L}^3(v)$ added to the last atom of $\iota\,\mathrm{R}_*^{\mathrm{lea}}(\varphi_*(b'), I_*)$ in the case where this last atom is itself a subarc of the first atom of $\mathrm{L}^3(v)$ (added to treat case (2-) of Theorem 21). We call this procedure FLIP1. We show now that its complexity is constant amortized.

Let $1 \le i < q'$; we have $\psi(l_i) = \varphi_{\mathrm{R}}(l_{i+1})$. Let $r_u, \ldots, r_m$ be the sequence of bitangents of C visited during the procedure PHIRNEW(v, l_{i+1}). As $\psi(l_i)$ leaves $\mathrm{R}^{\mathrm{lea}}(v)$ and crosses v it should be clear that $p \le m$. According to Theorem 17, the number of bitangents of the chain C' visited by the procedure is $1 + m - u + |\operatorname{rc}(l_i)|$. When $\psi(l_i)$ is a left-xx bitangent the procedure PHIRNEW(v, l_{i+1}) can be too costly. Running tandem walks as in Theorem 17 allows us to compute $\psi(l_i)$ in time $|\operatorname{rc}(l_i)|$ when $\psi(l_i)$ is a left-xx bitangent. Assume now that $\psi(l_i)$ is a right-xx bitangent. We show that the bitangents $r_u, \ldots, r_m$ all belong to $Y_R(l_i)$ except maybe for r_p. This will prove that the complexity of PHIRNEW(v, l_{i+1}) is at most $1 + |Y_R(l_i)| + |\operatorname{rc}(l_i)|$ when $\psi(l_i)$ is a right-xx bitangent. If $u \le p$ then the bitangents $r_u, \ldots, r_p$ belong to $\operatorname{rc}(v)$ whereas l_i belongs to $\operatorname{lc}(v)$. According to the definition of $r(l_i)$, this implies that the bitangents $r_u, \ldots, r_p$ belong to $r(l_i)$. If $p \ne u$, let j be an index such $u \le j \le p - 1$; as $\varphi_{\mathrm{R}}(l_{i+1})$ $(= \psi(l_i))$ is a right-xx bitangent, we have $r_j \in Y_R(l_i)$ by definition of $Y_R(l_i)$. Assume that $p \ne m$ otherwise we are done. According to Theorem 10 we have $r_{j+1} = \varphi_{\mathrm{back}}(r_j)$ for $p \le j \le m - 1$. Furthermore $r_p = \varphi_{\mathrm{righ}}(l_i)$ which implies that r_p is the minimal bitangent of $\mathrm{R}^2(l_i)$ (cf. Theorem 10). Using the fact that the bitangents of $\mathrm{R}^2(l_i)$ are in the orbit of r_p under the operator φ_{back} (cf. Theorem 10), we obtain $r_j \in \mathrm{R}'(l_i)$ for $p + 1 \le j \le m$. As $\varphi_{\mathrm{R}}(l_{i+1})$ is a right-xx bitangent, the bitangents $r_{p+1}, \ldots, r_m$ belong to $Y_R(l_i)$. The complexity of the procedure FLIP1 is therefore at most the sum of the following quantities:

(1-) $|Y_R(v)| + |\operatorname{rc}(v)|$ to compute $\varphi_{\mathrm{R}}(v_R)$ with the procedure PHIR(v, v_R);

(2-) $|Y_L(v)| + |\operatorname{lc}(v)|$ to compute $\varphi_{\mathrm{L}}(v_L)$ with the procedure PHIL(v, v_L);

(3-) $\sum_{1 \le i \le u-1} |\operatorname{lc}(r_i)|$ to compute the xx-right $\psi(r_i)$'s with the procedure PHILNEW(v, r_{i+1});

(4-) $|\operatorname{lc}(r_u)| + q + |\mathrm{L}'(r_u)|$ to compute the bitangent $\psi(r_u)$ with the procedure PHILNEW(v, r_{u+1});

(5-) $\sum_{1 \le i \le q'-1} 1 + |Y_R(l_i)| + |\operatorname{rc}(l_i)|$ to compute the $\psi(l_i)$'s with the procedure PHIRNEW(v, l_{i+1}) $(i = 1, \ldots, q' - 1)$;

(6-) $p - u + 1$ to compute $\psi(l_{q'}) = \varphi(v)$.

For a bitangent t, there is at most one bitangent v such that $t \in \operatorname{rc}(v)$. The same holds for $\operatorname{lc}(v)$. Moreover the sum $\sum_v |\operatorname{rc}(v)|$ is a $O(|\mathcal{B}|)$. Theorem 20 connects $Y_R(v)$ to $\mathcal{G}$ so with the help of Theorem 2 we can claim that $|Y_R(v)|$ is amortized constant. It follows that the cost of FLIP1 is amortized constant. This closes the proof of Theorem 1.

5 Conclusion and further research

We have presented a new method to compute in optimal output sensitive time and linear space the tangent visibility graph or visibility complex of a collection of disks of constant complexity. The method is built on top of the greedy flip algorithm of the second author of this paper and G. Vegter [42].

This is the first optimal algorithm for general disks (circles, ellipses, etc) that uses only simple data structures and only the counterclockwise or left-turn predicate for disks (modulo the result reported in [38]), once the disks are sorted with respect to some canonical direction. We believe that tuning a combinatorial model of the left-turn predicate on triplets of disks in the plane—as oriented matroids [7] or Knuth CCC-systems [26] are for the left-turn predicate on triplets of points in the plane— is an interesting line of research for the future from which our presentation might benefit as well as the implementation of our algorithms.

References

[1] J. Abello and K. Kumar. Visibility graphs and oriented matroids. In R. Tamassia and I. G. Tollis, editors, *Graph Drawing (Proc. GD '94)*, volume 894 of *Lecture Notes Comput. Sci.*, pages 147–158. Springer-Verlag, 1995.

[2] P. K. Agarwal, J. Basch, L. J. Guibas, J. Hershberger, and L. Zhang. Deformable free space tiling for kinetic collision detection. In B. Donald, K. Lynch, and D. Rus, editors, *Algorithmic and Computational Robotics: New Directions (Proc. 4th Workshop Algorithmic Found. Robotics)*, pages 83–96. A. K. Peters, 2001.

[3] P. Angelier. *Algorithmique des graphes de visibilité*. PhD thesis, Ecole Normale Supérieure (Paris), February 2002.

[4] B. Aronov, J. Matoušek, and M. Sharir. On the sum of squares of cell complexities in hyperplane arrangements. *J. Combin. Theory Ser. A*, 65:311–321, 1994.

[5] T. Asano, S. K. Ghosh, and T. C. Shermer. Visibility in the plane. In J.-R. Sack and J. Urrutia, editors, *Handbook of Computational Geometry*, pages 829–876. Elsevier Science Publishers B.V. North-Holland, Amsterdam, 2000.

[6] S. Bespamyatnikh. An efficient algorithm for enumeration of triangulations. 10th Annual Fall Workshop on Computational Geometry. `http://www.cs.ubc.ca/~besp/cgw.ps.gz`, 2000.

[7] A. Björner, M. Las Vergnas, N. White, B. Sturmfels, and G. M. Ziegler. *Oriented Matroids*. Cambridge University Press, Cambridge, 1993.

[8] H. Brönnimann, L. Kettner, M. Pocchiola, and J. Snoeyink. Counting and enumerating pseudotriangulations with the greedy flip algorithm. Submitted for publication, 2001.

[9] B. Chazelle, H. Edelsbrunner, M. Grigni, L. J. Guibas, J. Hershberger, M. Sharir, and J. Snoeyink. Ray shooting in polygons using geodesic triangulations. *Algorithmica*, 12:54–68, 1994.

[10] B. Chazelle, L. J. Guibas, and D. T. Lee. The power of geometric duality. *BIT*, 25:76–90, 1985.

[11] F. Cho and D. Forsyth. Interactive ray tracing with the visibility complex. *Computers and Graphics (Special Issue on Visibility - Techniques and Applications)*, 23(5):703–717, 1999.

[12] R. Connelly, E. D. Demaine, and G. Rote. Straightening polygonal arcs and convexifying polygonal cycles. In *Proc. 41th Annu. IEEE Sympos. Found. Comput. Sci.*, pages 432–442, 2000.

[13] M. de Berg, M. van Kreveld, M. Overmars, and O. Schwarzkopf. *Computational Geometry: Algorithms and Applications.* Springer-Verlag, Berlin, Germany, 2nd edition, 2000.

[14] F. Durand, G. Drettakis, and C. Puech. Fast and accurate hierarchical radiosity using global visibility. *ACM Transactions on Graphics*, 18(2):128–170, 1999.

[15] H. Edelsbrunner. *Algorithms in Combinatorial Geometry*, volume 10 of *EATCS Monographs on Theoretical Computer Science.* Springer-Verlag, Heidelberg, West Germany, 1987.

[16] H. Edelsbrunner and L. J. Guibas. Topologically sweeping an arrangement. *J. Comput. Syst. Sci.*, 38:165–194, 1989. Corrigendum in 42 (1991), 249–251.

[17] H. Edelsbrunner, J. O'Rourke, and R. Seidel. Constructing arrangements of lines and hyperplanes with applications. *SIAM J. Comput.*, 15:341–363, 1986.

[18] J. E. Goodman. Pseudoline arrangements. In J. E. Goodman and J. O'Rourke, editors, *Handbook of Discrete and Computational Geometry*, chapter 5, pages 83–110. CRC Press LLC, Boca Raton, FL, 1997.

[19] M. T. Goodrich and R. Tamassia. Dynamic ray shooting and shortest paths in planar subdivisions via balanced geodesic triangulations. *J. Algorithms*, 23:51–73, 1997.

[20] J. L. Gross and T. W. Tucker. *Topological Graph Theory.* John Wiley & Sons, 1987.

[21] M. Hagedoorn, M. Overmars, and R. C. Veltkamp. A robust affine invariant similarity measure based on visibility. In *Abstracts 16th European Workshop Comput. Geom.*, pages 112–116. Ben-Gurion University of the Negev, 2000.

[22] D. Halperin. Arrangements. In J. E. Goodman and J. O'Rourke, editors, *Handbook of Discrete and Computational Geometry*, chapter 21, pages 389–412. CRC Press LLC, Boca Raton, FL, 1997.

[23] D. Kirkpatrick, J. Snoeyink, and B. Speckmann. Kinetic collision detection for simple polygons. In *Proc. 16th Annu. ACM Sympos. Comput. Geom.*, pages 322–329, 2000.

[24] D. Kirkpatrick and B. Speckmann. Separation sensitive kinetic separation for convex polygons. In *Proc. Japan Conf Disc. Comp. Geom.*, number 2098 in Lecture Notes Comput. Sci., pages 222–236. Springer Verlag, 2001.

[25] D. Kirkpatrick and B. Speckmann. Kinetic maintenance of context-sensitive hierarchical representations of disjoint simple polygons. In *Proc. 18th Annu. ACM Sympos. Comput. Geom.*, pages 179–188, 2002.

[26] D. E. Knuth. *Axioms and Hulls*, volume 606 of *Lecture Notes Comput. Sci.* Springer-Verlag, Heidelberg, Germany, 1992.

[27] J.-C. Latombe. *Robot Motion Planning.* Kluwer Academic Publishers, Boston, 1991.

[28] P. MacMullen. Modern developments in regular polytopes. In T. Bisztriczky, P. McMullen, R. Schneider, and A. I. Weiss, editors, *Polytopes: Abstract, Convex and Computational*, pages 97–124. Kluwer Academic Publischers, 1994.

[29] W. S. Massey. *A Basic Course in Algebraic Topology.* Springer-Verlag, 1991.

[30] G. McCarty. *Topology : An Introduction with Application to Topological Groups.* Dover, 1988.

[31] J. S. B. Mitchell. Shortest paths and networks. In J. E. Goodman and J. O'Rourke, editors, *Handbook of Discrete and Computational Geometry*, chapter 24, pages 445–466. CRC Press LLC, Boca Raton, FL, 1997.

[32] J. S. B. Mitchell. Geometric shortest paths and network optimization. In J.-R. Sack and J. Urrutia, editors, *Handbook of Computational Geometry*, pages 633–701. Elsevier Science Publishers B.V. North-Holland, Amsterdam, 2000.

[33] N. Nilsson. A mobile automaton: An application of artificial intelligence techniques. In *Proc. IJCAI*, pages 509–520, 1969.

[34] J. O'Rourke. Visibility. In J. E. Goodman and J. O'Rourke, editors, *Handbook of Discrete and Computational Geometry*, chapter 25, pages 467–480. CRC Press LLC, Boca Raton, FL, 1997.

[35] J. O'Rourke. Computational geometry column 39. *SIGACT News*, 31(3):47–49, 2000.

[36] J. O'Rourke and I. Streinu. Vertex-edge pseudo-visibility graphs: Characterization and recognition. In *Proc. 13th Annu. ACM Sympos. Comput. Geom.*, pages 119–128, 1997.

[37] J. O'Rourke and I. Streinu. The vertex–edge visibility graph of a polygon. *Comput. Geom. Theory Appl.*, 10:105–120, 1998.

[38] M. Pocchiola. Computing pseudo-triangulations efficiently. In preparation, 2002.

[39] M. Pocchiola and G. Vegter. Order types and visibility types of configurations of disjoint convex plane sets (extended abstract). Technical Report 94-4, Labo. Inf. Ens, Jan. 1994.

[40] M. Pocchiola and G. Vegter. Minimal tangent visibility graphs. *Comput. Geom. Theory Appl.*, 6:303–314, 1996.

[41] M. Pocchiola and G. Vegter. Pseudo-triangulations: Theory and applications. In *Proc. 12th Annu. ACM Sympos. Comput. Geom.*, pages 291–300, 1996.

[42] M. Pocchiola and G. Vegter. Topologically sweeping visibility complexes via pseudo-triangulations. *Discrete Comput. Geom.*, 16:419–453, Dec. 1996.

[43] M. Pocchiola and G. Vegter. The visibility complex. *Internat. J. Comput. Geom. Appl.*, 6(3):279–308, 1996.

[44] M. Pocchiola and G. Vegter. On polygonal covers. In B. Chazelle, J. Goodman, and R. Pollack, editors, *Advances in Discrete and Computational Geometry*, volume 223 of *Contemporary Mathematics*, pages 257–268. AMS, Providence, 1999.

[45] D. Randall, G. Rote, F. Santos, and J. Snoeyink. Counting triangulations and pseudo-triangulations of wheels. In *Proc. 13th CCCG*, Univ. of Waterloo, Ont, 2001.

[46] S. Rivière. *Calculs de Visibilité dans un Environnement 2D.* PhD thesis, Université Joseph Fourier, Grenoble, France, 1997.

[47] S. Rivière, R. Orti, F. Durand, and C. Puech. Using the visibility complex for radiosity computation. In *ACM Workshop Appl. Comput. Geom.*, May 1996.

[48] G. Rote, F. Santos, and I. Streinu. Expansive motions and the polytope of pointed pseudo-triangulations. Manuscript, 2001.

[49] I. Streinu. Stretchability of star-like pseudo-visibility graphs. In *Proc. 15th Annu. ACM Sympos. Comput. Geom.*, pages 274–280, 1999.

[50] I. Streinu. A combinatorial approach to planar non-colliding robot arm motion planning. In *Proc. 41th Annu. IEEE Sympos. Found. Comput. Sci.*, pages 443–453, 2000.

[51] W. Thurston. *Three-Dimensional Geometry and Topology, Volume 1.* Princeton University Press, New Jersey, 1997.

About Authors

Pierre Angelier and Michel Pocchiola are at the Department of Computer Science, Ecole Normale Supérieure, UMR 8548 (CNRS), Paris, France;
Pierre.Angelier@ens.fr,Michel.Pocchiola@ens.fr

Acknowledgements

A preliminary version of this work appeared in the *Proceedings of the 17th Annual ACM Symposium on Computational Geometry, Medford, pages 302-311, June 2001.* Work on this paper has been partially supported by the Actions de Recherche Coopérative *Geometrica* and *Visibility 3D* of INRIA (FRANCE) and by the IST Programme of the EU as a Shared-cost RTD (FET Open) Project under Contract No IST-2000-26473 (ECG - Effective Computational Geometry for Curves and Surfaces).

On the Reflexivity of Point Sets

Esther M. Arkin
Sándor P. Fekete
Ferran Hurtado
Joseph S. B. Mitchell
Marc Noy
Vera Sacristán
Saurabh Sethia

Abstract

We introduce a new measure for planar point sets S that captures a combinatorial distance that S is from being a convex set: The *reflexivity* $\rho(S)$ of S is given by the smallest number of reflex vertices in a simple polygonalization of S. We prove combinatorial bounds on the reflexivity of point sets and study some closely related quantities, including the *convex cover* number $\kappa_c(S)$ of a planar point set, which is the smallest number of convex chains that cover S, and the *convex partition* number $\kappa_p(S)$, which is given by the smallest number of convex chains with pairwise-disjoint convex hulls that cover S.

1 Introduction

In this paper, we study a fundamental combinatorial property of a discrete set, S, of points in the plane: What is the minimum number, $\rho(S)$, of *reflex vertices* among all of the *simple polygonalizations* of S? A *polygonalization* of S is a closed tour on S whose straight-line embedding in the plane defines a connected cycle without crossings, i.e., a simple polygon. A vertex of a simple polygon is *reflex* if it has interior angle greater than π. We refer to $\rho(S)$ as the *reflexivity* of S. We let $\rho(n)$ denote the maximum possible value of $\rho(S)$ for a set S of n points.

In general, there are many different polygonalizations of a point set S. There is always at least one: simply connect the points in angular order about some point interior to the convex hull of S (e.g., the center of mass suffices). A set S has precisely one polygonalization if and only if it is in convex position; in general, though, a point set has numerous polygonalizations. Studying the set of polygonalizations (e.g., counting them, enumerating them, or generating a random element) is a challenging and active area of investigation in computational geometry [4, 5, 9, 17, 19, 34].

The reflexivity $\rho(S)$ quantifies, in a combinatorial sense, the degree to which the set of points S is in convex position. See Figure 1 for an example. We remark that there are other notions of combinatorial "distance" from convexity of a point set S, e.g., the minimum number of points to delete from

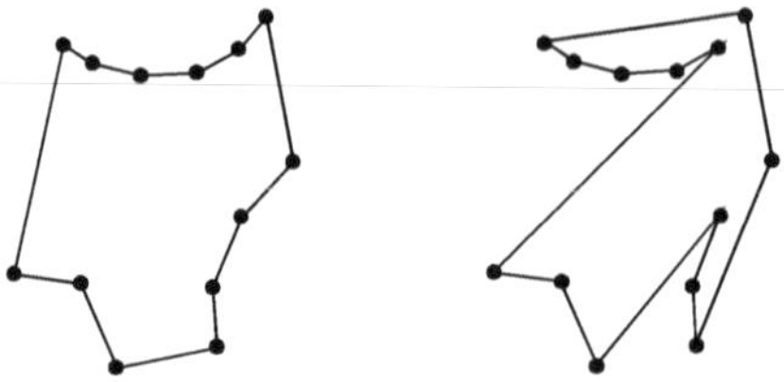

Fig. 1. Two polygonalizations of a point set, one (left) using 7 reflex vertices and one (right) using only 3 reflex vertices.

S in order that the remaining point set is in convex position, the number of convex layers, or the minimum number of changes in the orientation of triples of points of S in order to transform S into convex position.

We have conducted a formal study of reflexivity, both in terms of its combinatorial properties and in terms of an algorithmic analysis of the complexity of computing it, exactly or approximately. Some of our attention is focussed on the closely related *convex cover number* of S, which gives the minimum number of convex chains (subsets of S in convex position) that are required to cover all points of S. For this question, we distinguish between two cases: The *convex cover number*, $\kappa_c(S)$, is the smallest number of convex chains required to cover S; the *convex partition number*, $\kappa_p(S)$, is the smallest number of convex chains with pairwise-disjoint convex hulls required to cover S. Note that nested chains are feasible for a convex cover but not for a convex partition.

Motivation. In addition to the fundamental nature of the questions and problems we address, we are also motivated to study reflexivity for several other reasons:

(1) An application motivating our original investigation is that of meshes of low stabbing number and their use in performing ray shooting efficiently. If a point set S has low reflexivity or a low convex partition number, then it has a triangulation of low stabbing number, which may be much lower than the general $O(\sqrt{n})$ upper bound guaranteed to exist ([1, 20, 33]). For example, if the reflexivity is $O(1)$, then S has a triangulation with stabbing number $O(\log n)$.

(2) Classifying point sets by their reflexivity may give us some structure for dealing with the famously difficult question of counting and exploring the set of all polygonalizations of S. See [19, 34] for some references to this problem.

(3) There are several applications in computational geometry in which the number of reflex vertices of a polygon can play an important role in the complexity of algorithms. If one or more polygons are *given* to us, there are many problems for which more efficient algorithms can be written with complexity in terms of "r" (the number of reflex vertices), instead of "n" (the total number of vertices), taking advantage of the possibility that we may have $r \ll n$

for some practical instances (see, e.g., [21,25]). The number of reflex vertices also plays an important role in convex decomposition problems for polygons; see Keil [26] for a recent survey, and see Agarwal, Flato, and Halperin [2] for applications of convex decompositions to computing Minkowski sums of polygons.

(4) Reflexivity is intimately related to the issue of convex cover numbers, which has roots in the classical work of Erdős and Szekeres [15,16], and has been studied more recently by Urabe et al. [23,24,30,31].

(5) Our problems are related to some problems in curve (surface) reconstruction, where the goal is to obtain a "good" polygonalization of a set of sample points (see, e.g., [7,12,13]).

Related Work. The study of convex chains in finite planar point sets is the topic of classical papers by Erdős and Szekeres [15,16], who showed that any point set of size n has a convex subset of size $t = \Omega(\log n)$. This is closely related to the convex cover number κ_c, since it implies an asymptotically tight bound on $\kappa_c(n)$, the worst-case value for sets of size n. There are still a number of open problems related to the exact relationship between t and n; see, for example, [28] for recent developments.

Other issues have been considered, such as the existence and computation ([14]) of large "empty" convex subsets (i.e., with no points of S interior to their hull); this is related to the convex partition number, $\kappa_p(S)$. It was shown by Horton [22] that there are sets with no empty convex chain larger than 6; this implies that $\kappa_p(n) \geq n/6$.

Tighter worst-case bounds on $\kappa_p(n)$ were given by Urabe [30, 31], who shows that $\lceil (n-1)/4 \rceil \leq \kappa_p(n) \leq \lceil 2n/7 \rceil$ and that $\kappa_c(n) = \Theta(n/\log n)$ (with the upper and lower bounds having a gap of roughly a factor of 2). (Urabe [32] also studies the convex partitioning problem in $\Re^3$, where, in particular, the upper bound on $\kappa_p(n)$ is shown to be $\lceil 2n/9 \rceil$.) Most recently, Hosono and Urabe [24] have obtained improved bounds on the size of a partition of a set of points into disjoint convex quadrilaterals, which has the consequence of improving the upper bound on $\kappa_p(n)$: $\kappa_p(n) \leq \lceil 5n/18 \rceil$ and $\kappa_p(n) \leq (3n+1)/11$ for $n = 11 \cdot 2^{k-1} - 4$ $(k \geq 1)$. The remaining gaps in the constants between upper and lower bounds for $\kappa_c(n)$ and $\kappa_p(n)$ (as well as the gap that our bounds exhibit for reflexivity in terms of n) all point to the apparently common difficulty of these combinatorial problems on convexity.

For a given set of points, we are interested in polygonalizations of the points that are "as convex as possible". This has been studied in the context of TSP (traveling salesperson problem) tours of a point set S, where convexity of S implies (trivially) the optimality of a convex tour. Convexity of a tour can be characterized by two conditions. If we drop the global condition (i.e., no crossing edges), but keep the local condition (i.e., no reflex vertices), we get "pseudo-convex" tours. In [18] it was shown that any set with $|S| \geq 5$ has such a pseudo-convex tour. It is natural to require the global condition of simplicity instead, and minimize the number of local violations – i.e., the number of reflex vertices. This kind of problem is similar to that of minimizing the total amount of turning in a tour, as studied by Aggarwal et al. [3].

The number of polygonalizations on n points is, in general, exponential in n; García et al. [19] prove a lower bound of $\Omega(4.64^n)$.

Another related problem is studied by Hosono et al. [23]: Compute a polygonalization P of a point set S such that the interior of P can be decomposed into a minimum number ($f(S)$) of empty convex polygons. They prove that $\lfloor (n-1)/4 \rfloor \leq f(n) \leq \lfloor (3n-2)/5 \rfloor$, where $f(n)$ is the maximum possible value of $f(S)$ for sets S of n points. The authors conjecture that $f(n)$ grows like $n/2$. For reflexivity $\rho(n)$, we show that $\lfloor n/4 \rfloor \leq \rho(n) \leq \lceil n/2 \rceil$ and conjecture that $\rho(n)$ grows like $n/4$, which, if true, would imply that $f(n)$ grows like $n/2$.

We mention one final related problem. A *convex decomposition* of a point set S is a convex planar polygonal subdivision of the convex hull of S whose vertices are S. Let $g(S)$ denote the minimum number of faces in a convex decomposition of S, and let $g(n)$ denote the maximum value of $g(S)$ over all n-point sets S. It has been conjectured ([29]) that $g(n) = n + c$ for some constant c, and it is known that $g(n) \leq 3n/2$ ([29]) and that $n + 2 \leq g(n)$ ([6]).

Summary of Main Results. In this paper, we prove the following results on reflexivity:

- Tight bounds on the worst-case value of $\rho(S)$ in terms of n_I, the number of points of S interior to the convex hull of S; in particular, we show that $\rho(S) \leq \lceil n_I/2 \rceil$ and that this upper bound can be achieved by a class of examples.
- Upper and lower bounds on $\rho(S)$ in terms of $n = |S|$; in particular, we show that $\lfloor n/4 \rfloor \leq \rho(n) \leq \lceil n/2 \rceil$.
- Upper and lower bounds on "Steiner reflexivity", which is defined with respect to the class of polygonalizations that allow Steiner vertices (not from the input set S).

In Section 6 we study a closely related problem – that of determining the "inflectionality" of S, defined to be the minimum number of inflection edges (joining a convex to a reflex vertex) in any polygonalization of S. We give an $O(n \log n)$ time algorithm to determine an inflectionality-minimizing polygonalization, which we show will never need more than 2 inflection edges.

Additional Results. We also summarize additional results we have obtained in this line of research; in the interest of preserving space here, the proofs of the following results appear in the extended paper [8]:

- In the case in which S has two layers, we show that $\rho(S) \leq \lceil n/4 \rceil$, and this bound is tight.

- We prove that it is NP-complete to compute the convex cover number ($\kappa_c(S)$) or the convex partition number ($\kappa_p(S)$), for a given point set S.
- We give polynomial-time approximation algorithms, having approximation factor $O(\log n)$, for the problems of computing convex cover number, convex partition number, or Steiner reflexivity of S.
- We give efficient exact algorithms to test if $\rho(S) = 1$ or $\rho(S) = 2$.

2 Preliminaries

Throughout this paper, S will be a set of n points in the plane $\Re^2$. A polygonalization, P, of S is a simple polygon whose vertex set is S. Let $\mathcal{P}$ be the set of all polygonalizations of S. Note that $\mathcal{P}$ is not empty, since any point set S having $n \geq 3$ points has at least one polygonalization (e.g., the star-shaped polygonalization obtained by sorting points of S angularly about a point interior to the convex hull of S).

Each vertex of a simple polygon P is either *reflex* or *convex*, according to whether the interior angle at the vertex is greater than π or less than or equal to π, respectively. We let $r(P)$ (resp., $c(P)$) denote the number of reflex (resp., convex) vertices of P. We define the *reflexivity* of a planar point set S to be $\rho(S) = \min_{P \in \mathcal{P}} r(P)$. Similarly, the *convexivity* of a planar point set S is defined to be $\chi(S) = \max_{P \in \mathcal{P}} c(P)$. Note that $\chi(S) = n - \rho(S)$. We let $\rho(n) = \max_{|S|=n} \rho(S)$.

We let CH(S) denote the *convex hull* of S. The point set S is partitioned into (convex) *layers*, $S_1, S_2, \ldots$, where the first layer is given by the set S_1 of points of S on the boundary of CH(S), and the ith layer, S_i ($i \geq 2$) is given by the set of points of S on the boundary of CH$(S \setminus (S_1 \cup \cdots \cup S_{i-1}))$. We say that S has k *layers* or *onion depth* k if $S_k \neq \emptyset$, while $S_{k+1} = \emptyset$. We say that S is in *convex position* (or forms a *convex chain*) if it has one layer (i.e., $S = S_1$).

A *Steiner point* is a point not in the set S that may be added to S in order to improve some structure of S. We define the *Steiner reflexivity* $\rho'(S)$ to be the minimum number of reflex vertices of any simple polygon with vertex set $V \supset S$. We let $\rho'(n) = \max_{|S|=n} \rho'(S)$. A *convex cover* of S is a set of subsets of S whose union covers S, such that each subset is a convex chain (a set in convex position). A *convex partition* of S is a partition of S into subsets each of which is in convex position, such that the convex hulls of the subsets are pairwise disjoint. We define the *convex cover number*, $\kappa_c(S)$, to be the minimum number of subsets in a convex cover of S. We similarly define the *convex partition number*, $\kappa_p(S)$. We denote by $\kappa_c(n)$ and $\kappa_p(n)$ the worst-case values for sets of size n.

Finally, we state a basic property of polygonalizations of point sets.

Lemma 2.1. *In any polygonalization of S, the points of S that are vertices of the convex hull of S are convex vertices of the polygonalization, and they*

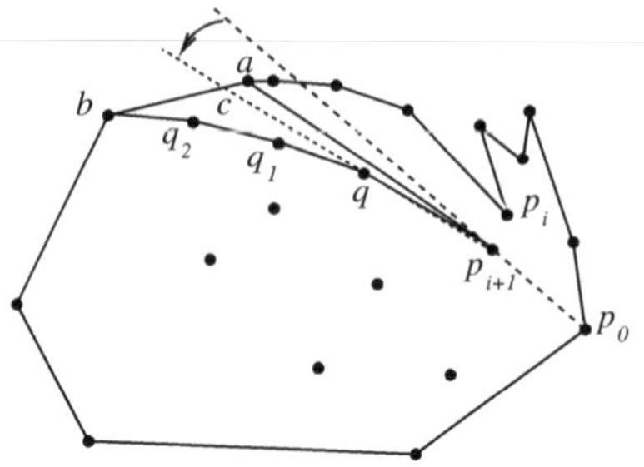

Fig. 2. Computing a polygonalization with at most $\lceil n_I/2 \rceil$ reflex vertices.

occur in the polygonalization in the same order in which they occur along the convex hull.

Proof. Any polygonalization P of S must lie within the convex hull of S, since edges of the polygonalization are convex combinations of points of S. Thus, if $p \in S$ is a vertex of CH(S), then the local neighborhood of P at p lies within a convex cone, so p must be a convex vertex of P.

Consider a clockwise traversal of P and let p and q be two vertices of CH(S) occurring consecutively along P. Then p and q must also appear consecutively along a clockwise traversal of the boundary of CH(S), since the subchain of P linking p to q partitions CH(S) into a region to its left (which is outside the polygon P) and a region to its right (which must contain all points of S not in the subchain). □

3 Combinatorial Bounds

3.1 Reflexivity

One of our main combinatorial results establishes an upper bound on the reflexivity of S that is worst-case tight in terms of the number n_I of points *interior* to the convex hull, CH(S), of S. Since, by Lemma 2.1, the points of S that are vertices of CH(S) are required to be convex vertices in any (non-Steiner) polygonalization of S, the bound in terms of n_I seems to be quite natural.

Theorem 3.1. *Let S be a set of n points in the plane, n_I of which are interior to the convex hull CH(S). Then $\rho(S) \leq \lceil n_I/2 \rceil$.*

Proof. We describe a polygonalization in which at most half of the interior points are reflex. We begin with the polygonalization of the convex hull vertices that is given by the convex polygon bounding the hull. We then iteratively incorporate interior points of S into the polygonalization. Fix a point p_0 that lies on the convex hull of S. At a generic step of the algorithm, the following invariants hold: (1) our polygonalization consists of a simple

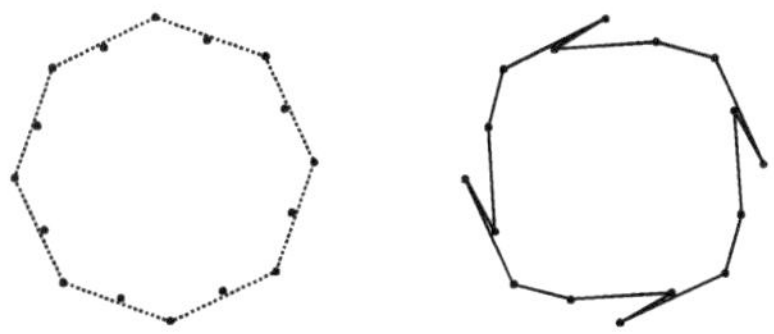

Fig. 3. Left: The configuration of points, $S_0(n)$, which has reflexivity $\rho(S_0(n)) \geq \lceil n_I/2 \rceil$. Right: A polygonalization having $\lceil n_I/2 \rceil$ reflex vertices.

polygon, P, whose vertices form a subset of S; and (2) all points $S' \subset S$ that are not vertices of P lie interior to P; in fact, the points S' all lie within the subpolygon, Q, to the left of the diagonal p_0p_i, where p_i is a vertex of P such that the subchain of ∂P from p_i to p_0 (counter-clockwise) together with the diagonal p_0p_i forms a convex polygon (Q). If S' is empty, then P is a polygonalization of S and we are done; thus, assume that $S' \neq \emptyset$. Define p_{i+1} to be the first point of S' that is encountered when sweeping the ray $\overrightarrow{p_0p_i}$ counter-clockwise about its endpoint p_0. Then we sweep the subray with endpoint p_{i+1} further counter-clockwise, about p_{i+1}, until we encounter another point, q, of S'. (If $|S'| = 1$, we can readily incorporate p_{i+1} into the polygonalization, increasing the number of reflex vertices by one.) Now the ray $\overrightarrow{p_{i+1}q}$ intersects the boundary of P at some point $c \in ab$ on the boundary of Q.

As a next step, we modify P to include interior points p_{i+1} and q (and possibly others as well) by replacing the edge ab with the chain $(a, p_{i+1}, q, q_1, \ldots, q_k, b)$, where the points q_i are interior points that occur along the chain we obtain by "pulling taut" the chain (q, c, b). In this "gift wrapping" fashion, we continue to rotate rays counter-clockwise about each interior point q_i that is hit until we encounter b. This results in incorporating at least two new interior points (of S') into the polygonalization P, while creating only one new reflex vertex (at p_{i+1}). It is easy to check that the invariants (1) and (2) hold after this step. □

In fact, the upper bound of Theorem 3.1, $\rho(S) \leq \lceil n_I/2 \rceil$, is *tight* in the worst case, as we now argue based on the special configuration of points, $S = S_0(n)$, in Figure 3. The set $S_0(n)$ is defined for any integer $n \geq 6$, as follows: $\lceil n/2 \rceil$ points are placed in convex position (e.g., forming a regular $\lceil n/2 \rceil$-gon), forming the convex hull CH(S), and the remaining $n_I = \lfloor n/2 \rfloor$ interior points are also placed in convex position, each one placed "just inside" CH(S), near the midpoint of an edge of CH(S). The resulting configuration $S_0(n)$ has two layers in its convex hull.

Lemma 3.2. *For any* $n \geq 6$, $\rho(S_0(n)) \geq \lceil n_I/2 \rceil \geq \lfloor n/4 \rfloor$.

Proof. Let $(x_1, x_2, \ldots, x_{\lceil n/2 \rceil})$ denote the points of $S_0(n)$ on the convex hull, in clockwise order, and let $(v_1, v_2, \ldots, v_{\lfloor n/2 \rfloor})$ denote the remaining points of $S_0(n)$, with v_i just inside the convex hull edge (x_i, x_{i+1}). We define $x_{\lceil n/2 \rceil + 1} = x_1$.

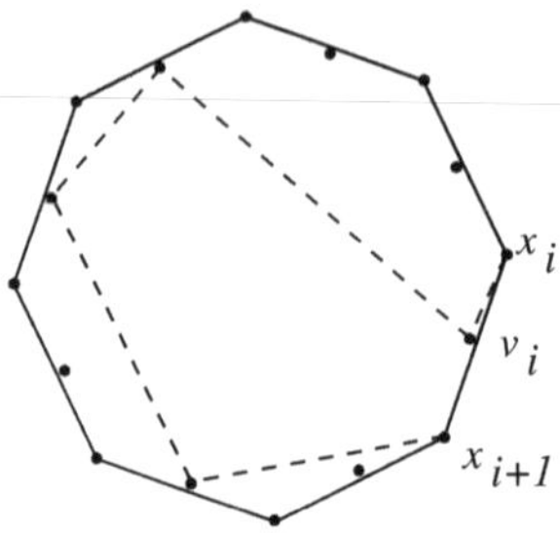

Fig. 4. Proof of the lower bound: $\rho(S_0(n)) \geq \lceil n_I/2 \rceil \geq \lfloor n/4 \rfloor$.

Consider any polygonalization, P, of $S_0(n)$. From Lemma 2.1 we know that the points x_i are convex vertices of P, occurring in the order $x_1, x_2, \ldots, x_{\lceil n/2 \rceil}$ around the boundary of P. Consider the subchain, γ_i, of ∂P that goes from x_i to x_{i+1}, clockwise around ∂P. Let m_i denote the number of points v_j, interior to the convex hull of $S_0(n)$, that appear along γ_i.

If $m_i = 0$, $\gamma_i = x_i x_{i+1}$. If $m_i = 1$, then $\gamma_i = x_i v_i x_{i+1}$ and v_i is a reflex vertex of P; to see this, note that v_i lies interior to the triangle determined by x_i, x_{i+1}, and any v_j with $j \neq i$. If $m_i > 1$, then we claim that (a) v_i must be a vertex of the chain γ_i, (b) v_i is a convex vertex of P, and (c) any other point v_j, $j \neq i$, that is a vertex of γ_i must be a reflex vertex of P. This claim follows from the fact that the points x_i, x_{i+1}, and any nonempty subset of $\{v_j : j \neq i\}$ are in convex position, with the point v_i interior to the convex hull. Refer to Figure 4, where the subchain γ_i is shown dashed.

Thus, the number of reflex vertices of P occurring along γ_i is in any case at least $\lceil m_i/2 \rceil$, and we have

$$\begin{aligned} \rho(S_0(n)) &\geq \sum \lceil m_i/2 \rceil \\ &\geq \left\lceil \sum (m_i/2) \right\rceil = \lceil n_I/2 \rceil \geq \lfloor n/4 \rfloor . \end{aligned}$$

□

Since $n_I \leq n$, the corollary below is immediate from Theorem 3.1 and Lemma 3.2. The gap in the bounds for $\rho(n)$, between $\lfloor n/4 \rfloor$ and $\lceil n/2 \rceil$, remains an intriguing open problem. While our combinatorial bounds are worst-case tight in terms of n_I (the number of points of S whose convexity/reflexivity is not forced by the convex hull of S), they are not worst-case tight in terms of n.

Corollary 3.3. $\lfloor n/4 \rfloor \leq \rho(n) \leq \lceil n/2 \rceil$.

Based on experience with a software tool developed by A. Dumitrescu that computes, in exponential time, the reflexivity of user-specified or randomly generated point sets, as well as the proven behavior of $\rho(n)$ for small values of n (see Section 3.5), we make the following conjecture:

Conjecture 3.4. $\rho(n) = \lfloor n/4 \rfloor$.

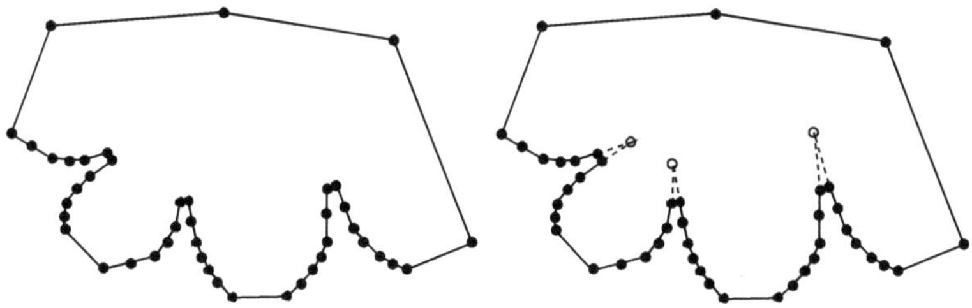

Fig. 5. Left: A point set S having reflexivity $\rho(S) = r$. Right: The reflexivity of S when Steiner points *are* permitted is substantially reduced from the no-Steiner case: $\rho'(S) = r/2$.

3.2 Steiner Points

If we allow Steiner points in the polygonalizations of S, the reflexivity of S may go down substantially, as the example in Figure 5 shows. In fact, the illustrated class of examples shows that the use of Steiner points may allow the reflexivity to go down by a factor of two. The Steiner reflexivity, $\rho'(S)$, of S is the minimum number of reflex vertices of any simple polygon with vertex set $V \supset S$. We conjecture that $\rho'(S) \geq \rho(S)/2$ for any set S, which would imply that this class of examples (essentially) maximizes the ratio $\rho(S)/\rho'(S)$.

Conjecture 3.5. *For any set S of points in the plane, $\rho'(S) \geq \rho(S)/2$.*

We have seen (Corollary 3.3) that $\lfloor n/4 \rfloor \leq \rho(n) \leq \lceil n/2 \rceil$. We now show that allowing Steiner points in the polygonalization allows us to prove a smaller upper bound, while still being able to prove roughly the same lower bound:

Theorem 3.6.

$$\left\lceil \frac{n-1}{4} \right\rceil - 1 \leq \rho'(n) \leq \left\lceil \frac{n}{3} \right\rceil .$$

Proof. For the upper bound, we give a specific method of constructing a polygonalization (with Steiner points) of a set S of n points. Sort the points S by their x-coordinates and group them into consecutive triples. Let p_{n+1} denote a (Steiner) point with a very large positive y-coordinate and let p_0 denote a (Steiner) point with a very negative y-coordinate. Each triple, together with either point p_{n+1} or point p_0, forms a convex quadrilateral. Then, we can polygonalize S using one reflex (Steiner) point per triple, as shown in Figure 6, placed very close to p_{n+1} or p_0 accordingly. This polygonalization has at most $\lceil n/3 \rceil$ reflex points.

For the lower bound, we consider the configuration of n points, S, used in Urabe [30] to prove that $\kappa_p(n) \geq \lceil (n-1)/4 \rceil$. For this set S of n points, let P be a Steiner polygonalization having r reflex vertices. Then the simple polygon P can be partitioned into $r + 1$ (pairwise-disjoint) convex pieces;

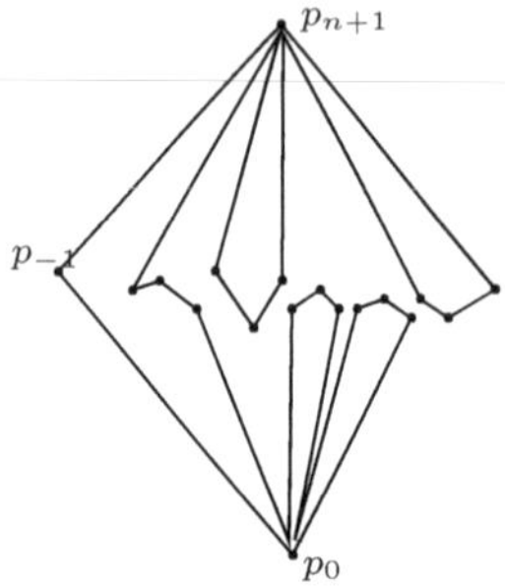

Fig. 6. Polygonalization of n points using only $\lceil n/3 \rceil$ reflex (Steiner) points.

this is a simple observation of Chazelle [10] (see Theorem 2.5.1 of [27]). The points S occur as a subset of the vertices of these pieces; thus, the partitioning also decomposes S into at most $r+1$ subsets, each in convex position. Since $\kappa_p(n) \geq \lceil (n-1)/4 \rceil$, we get that $r \geq \lceil (n-1)/4 \rceil - 1$. □

3.3 Two-Layer Point Sets

Let S be a point set that has two (convex) layers. It is clear from our repeated use of the example in Figure 3 that this is a natural case that is a likely candidate for worst-case behavior. With a very careful analysis of this case, we are able to obtain tight combinatorial bounds on the worst-case reflexivity in terms of n. The proof of the following theorem is quite technical and long, so it is deferred to the extended paper [8].

Theorem 3.7. *Let S be a set of n points having two layers. Then $\rho(S) \leq \lceil n/4 \rceil$, and this bound is tight in the worst case.*

Remark. Using a variant of the polygonalization given in the proof of Theorem 3.7, it is possible to show that a two-layer point set S in fact has a polygonalization with at most $\lceil n/3 \rceil$ reflex vertices such that none of the edges in the polygonalization pass through the interior of the convex hull of the second layer. (The polygonalization giving upper bound of $\lceil n/4 \rceil$ requires edges that pass through the interior of the convex hull of the second layer.) This observation may be useful in attempts to reduce the worst-case upper bound ($\rho(n) \leq \lceil n/2 \rceil$) for more general point sets S.

3.4 Convex Cover/Partition Numbers

As a consequence of the Erdős-Szekeres theorem [15, 16], Urabe has given bounds on the convex cover number of a set of n points:

$$\frac{n}{\log_2 n + 2} < \kappa_c(n) < \frac{2n}{\log_2 n - \log_2 e} .$$

Urabe [30] and Hosono and Urabe [24] have obtained bounds as well on the convex partition number of an n-point set:

$$\left\lceil \frac{n-1}{4} \right\rceil \leq \kappa_p(n) \leq \left\lceil \frac{5n}{18} \right\rceil .$$

While it is trivially true that $\kappa_c(S) \leq \kappa_p(S)$ the ratio $\kappa_p(S)/\kappa_c(S)$ for a set S may be as large as $\Theta(n)$; the set $S = S_0(n)$ (Figure 3) has $\kappa_c(S) = 2$, but $\kappa_p(S) \geq n/4$.

The fact that $\kappa_p(S) \leq \rho(S) + 1$ follows easily by iteratively adding $\rho(S)$ segments to an optimal polygonalization P, bisecting each reflex angle. The result is a partitioning of P into $\rho(S)+1$ convex pieces. Thus, we can obtain a convex partitioning of S by associating a subset of S with each convex piece of P, assigning each point of S to the subset associated with any one of the convex pieces that has the point on its boundary.

We believe that the relationship between reflexivity ($\rho(S)$) and convex partition number ($\kappa_p(S)$) goes the other way as well: A small convex partition number should imply a small reflexivity. In particular, we have invested considerable effort in trying to prove the following conjecture:

Conjecture 3.8. $\rho(S) = O(\kappa_p(S))$.

The reflexivity can be as large as *twice* the convex cover number ($\rho(S) = 2\kappa_p(S)$), as illustrated in the example of Figure 7; however, this is the worst class of examples we have found so far.

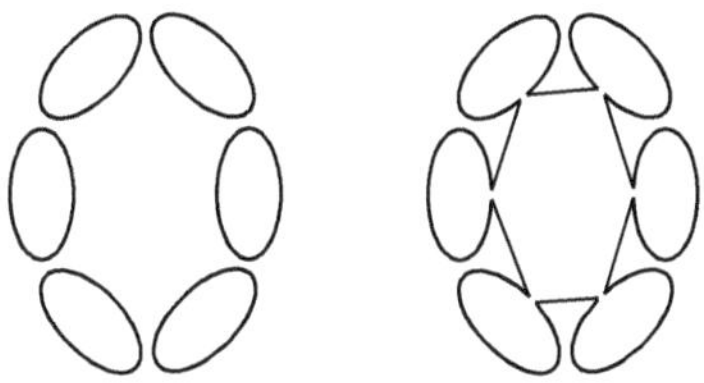

Fig. 7. An example with $\rho(S) = 2\kappa_p(S)$. Each thick oval shape represents a numerous subset of points of S in convex position.

Turning briefly to Steiner reflexivity, it is not hard to see that $\rho'(S) = O(\kappa_p(S))$ (see [8]). Thus, a proof of Conjecture 3.8 would follow from the validity of Conjecture 3.5.

3.5 Small Point Sets

It is natural to consider the exact values of $\rho(n)$, $\kappa_c(n)$, and $\kappa_p(n)$ for small values of n. Table 1 below shows some of these values, which we obtained through (sometimes tedious) case analysis. Aichholzer and Krasser [5] have recently applied their software that enumerates point sets of size n of all distinct order types to verify our results computationally; in addition, they

Table 1. Worst-case values of ρ, κ_c, κ_p for small values of n.

n	$\rho(n)$	$\kappa_c(n)$	$\kappa_p(n)$
≤ 3	0	1	1
4	1	2	2
5	1	2	2
6	2	2	2
7	2	2	2
8	2	2	3
9	3	3	3
10	3	–	–

have obtained the result that $\rho(10) = 3$. (Experiments are currently under way for $n = 11$; values of $n \geq 12$ seem to be intractable for enumeration.)

4 Complexity

In the extended paper [8], we prove lower bounds on the complexity of computing the convex cover number, $\kappa_c(S)$, and the convex partition number, $\kappa_p(S)$. The proof for the convex cover number uses a reduction of the problem 1-in-3 SAT and is inspired by the hardness proof for the ANGULAR METRIC TSP given in [3]. The proof for the convex partition number uses a reduction from PLANAR 3 SAT.

Theorem 4.1. *It is NP-complete to decide whether for a planar point set S the convex cover number $\kappa_c(S)$ or the convex partition number $\kappa_p(S)$ is below some threshold k.*

So far, the complexity status of determining the reflexivity of a point set remains open. However, the apparently close relationship between convex cover/partition numbers and reflexivity leads us to believe the following:

Conjecture 4.2. *It is NP-complete to determine the reflexivity $\rho(S)$ of a point set.*

5 Algorithms

We have obtained a number of algorithmic results on computing, exactly or approximately, reflexivity and convex cover/partition numbers. We summarize the results here, but we defer the proofs of the theorems to the extended paper [8].

Theorem 5.1. *Given a set S of n points in the plane, in $O(n \log n)$ time one can compute a polygonalization of S having at least $\chi(S)/2$ convex vertices, where $\chi(S) = n - \rho(S)$ is the convexity of S.*

Theorem 5.2. *Given a set S of n points in the plane, the convex cover number $\kappa_c(S)$, the convex partition number $\kappa_p(S)$, and the Steiner reflexivity $\rho'(S)$ can each be computed approximately, within a factor $O(\log n)$, in polynomial time.*

For small values of r, we have devised particularly efficient algorithms that check if $\rho(S) \leq r$ and, if so, produce a witness polygonalization having at most r vertices. Of course, the case $r = 0$ is trivial, since that is equivalent to testing if S lies in convex position (which is readily done in $O(n \log n)$ time, which is worst-case optimal). One can obtain an $n^{O(r)}$ algorithm by enumerating over all combinatorially distinct (with respect to S) convex subdivisions of CH(S) into $O(r)$ convex faces, testing that the subsets of S within each face are in convex position, and then checking all possible ways to order these $O(r)$ convex chains to form a circuit that may form a simple polygon. With a careful analysis of the cases $r = 1, 2$, we show that testing if $\rho(S) = 1$ can be done in time $O(n \log n)$, which we prove is worst-case optimal, and that testing if $\rho(S) = 2$ can be done in $O(n^3 \log n)$ time. See [8] for details.

6 Inflectionality of Point Sets

Consider a clockwise traversal of a polygonalization, P, of S. Then, convex (resp., reflex) vertices of P correspond to *right* (resp., *left*) turns. In computing the reflexivity of S we desire a polygonalization that minimizes the number of left turns. In this section we consider the related problem in which we want to minimize the number of *changes* between left-turning and right-turning during a traversal that starts (and ends) at a point interior to an edge of P. We define the minimum number of such transitions between left and right turns to be the *inflectionality*, $\phi(S)$, of S, where the minimum is taken over all polygonalizations of S. Clearly, $\phi(S)$ must be an even integer; it is zero if and only if S is in convex position. Somewhat surprisingly, it turns out that $\phi(S)$ can only take on the values 0 or 2:

Theorem 6.1. *For any finite set S of n points in the plane, $\phi(S) \in \{0, 2\}$, with $\phi(S) = 0$ precisely when S is in convex position. In $O(n \log n)$ time, one can determine $\phi(S)$ as well as a polygonalization that achieves inflectionality $\phi(S)$.*

Proof. If S is in convex position, then trivially $\phi(S) = 0$. Thus, assume that S is not in convex position. Then $\phi(S) \neq 0$, so $\phi(S) \geq 2$. We claim that $\phi(S) = 2$. For simplicity, we assume that S is in general position.

Consider the ℓ nested convex polygons, $C_1, C_2, \ldots, C_\ell$, whose boundaries constitute the ℓ layers (the "onion") of the set S; these can be computed in time $O(n \log n)$ [11].

We construct a "spiral" polygonalization of S based on taking one edge, ab, of C_1, and replacing it with a pair of right-turning chains from a to $p \in S \cap C_\ell$ and from b to p. The two chains exactly cover the points of S on layers $C_2, \ldots, C_\ell$. A constructive proof that such a polygonalization exists is based on the following claim:

Claim 6.2. *For any* $1 \leq m < \ell$ *and any pair,* $a, b \in S$*, of vertices of* C_m*, there exist two purely right-turning chains,* $\gamma_a = (a, u_1, u_2, \ldots, u_i, p)$ *and* $\gamma_b = (b, v_1, v_2, \ldots, v_j, p)$*, such that the points of* S *interior to* C_m *are precisely the set* $\{u_1, u_2, \ldots, u_i, p, v_i, v_2, \ldots, v_j\}$.

Proof of Claim. We prove the claim by induction on m. If $m = \ell - 1$, the claim follows easily, by a case analysis as illustrated in Figure 8.

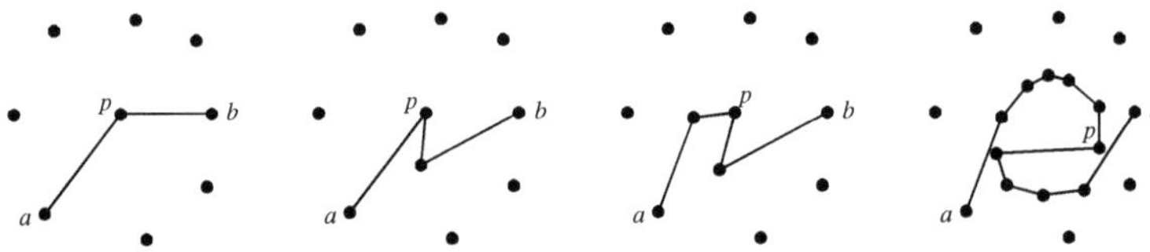

Fig. 8. Simple case in the inductive proof: $m = \ell - 1$. There are four subcases, left to right: (i) C_ℓ is a single point; (ii) C_ℓ is a line segment determined by two points of S; (iii) C_ℓ is a triangle determined by three points of S; or (iv) C_ℓ is a convex polygon whose boundary contains four or more points of S.

Assume that the claim holds for $m \geq k + 1$ and consider the case $m = k$. If C_{k+1} is either a single vertex or a line segment (which can only happen if $k + 1 = \ell$), the claim trivially follows; thus, we assume that C_{k+1} has at least three vertices. We let u_1 be the vertex of C_{k+1} that is a *left tangent* vertex with respect to a (meaning that C_{k+1} lies in the closed halfplane to the right of the oriented line au_1); we let v be the left tangent vertex of C_{k+1} with respect to b. Refer to Figure 9. If $v = u_1$, we define v_1 to be the vertex of C_{k+1} that is the counter-clockwise neighbor of u_1; otherwise, we let $v_1 = v$. Let a' be the counter-clockwise neighbor of v_1. Let b' be the counter-clockwise neighbor of u_1. (Thus, b' may be the same point as v_1.) By the induction hypothesis, we know that there exist right-turning chains, $\gamma_{a'}$ and $\gamma_{b'}$, starting from the points a' and b', spiraling inwards to a point p interior to C_{k+1}. Then we construct γ_a to be the chain from a to u_1, around the boundary of C_{k+1} clockwise to a', and then along the chain $\gamma_{a'}$. Similarly, we construct γ_b to be the chain from b to v_1, around the boundary of C_{k+1} clockwise to b', and then along the chain $\gamma_{b'}$. □

The proof of the above claim is constructive; the required chains are readily obtained in $O(n \log n)$ time, given the convex layers. This concludes the proof of the theorem. □

7 Open Problems

There are a number of interesting open problems that our work suggests. First, there are the four specific conjectures mentioned throughout the paper;

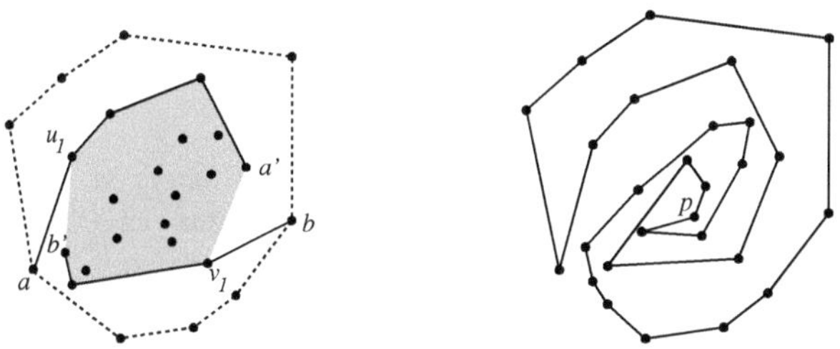

Fig. 9. Left: Constructing the spiraling chains γ_a and γ_b. Right: An example of the resulting spiral polygonalization.

these represent to us the most outstanding open questions raised by our work. In addition, we mention two other areas of future study:

1. Instead of minimizing the number of reflex vertices, can we compute a polygonalization of S that minimizes the sum of the *turn angles* at reflex vertices? (The turn angle at a reflex vertex having interior angle $\theta > \pi$ is defined to be $\theta - \pi$.) This question was posed to us by Ulrik Brandes. It may capture a notion of goodness of a polygonalization that is useful for curve reconstruction. The problem differs from the angular metric TSP ([3]) in that the only turn angles contributing to the objective function are those of reflex vertices.

2. What can be said about the generalization of the reflexivity problem to polyhedral surfaces in three dimensions? This may be of particular interest in the context of surface reconstruction.

References

[1] P. K. Agarwal. Ray shooting and other applications of spanning trees with low stabbing number. *SIAM J. Comput.*, 21:540–570, 1992.

[2] P. K. Agarwal, E. Flato, and D. Halperin. Polygon decomposition for efficient construction of Minkowski sums. *Comput. Geom. Theory Appl.*, 21:39–61, 2002.

[3] A. Aggarwal, D. Coppersmith, S. Khanna, R. Motwani, and B. Schieber. The angular-metric traveling salesman problem. In *Proc. 8th ACM-SIAM Sympos. Discrete Algorithms*, pages 221–229, 1997.

[4] O. Aichholzer, F. Aurenhammer, and H. Krasser. Enumerating order types for small point sets with applications. In *Proc. 17th Annu. ACM Sympos. Comput. Geom.*, 2001, pp. 11–18.

[5] O. Aichholzer and H. Krasser. The point set order type data base: a collection of applications and results. In *Proc. 13th Canad. Conf. Comput. Geom.*, Waterloo, Canada, 2001, pp. 17–20.

[6] O. Aichholzer and H. Krasser. Personal communication, 2001.

[7] N. Amenta, M. Bern, and D. Eppstein. The crust and the β-skeleton: Combinatorial curve reconstruction. *Graphical Models and Image Processing*, 60:125–135, 1998.

[8] E. M. Arkin, S. P. Fekete, F. Hurtado, J. S. B. Mitchell, M. Noy, V. Sacristán, and S. Sethia. On the reflexivity of point sets. `arXiv:cs.CG/0210003`, 2002.

[9] T. Auer and M. Held. Heuristics for the generation of random polygons. In *Proc. 8th Canad. Conf. Comput. Geom.*, pages 38–43, 1996.

[10] B. Chazelle. *Computational geometry and convexity.* Ph.D. thesis, Dept. Comput. Sci., Yale Univ., New Haven, CT, 1979. Carnegie-Mellon Univ. Report CS-80-150.

[11] B. Chazelle. On the convex layers of a planar set. *IEEE Trans. Inform. Theory*, IT-31(4):509–517, July 1985.

[12] T. K. Dey and P. Kumar. A simple provable algorithm for curve reconstruction. In *Proc. 10th ACM-SIAM Sympos. Discrete Algorithms*, pages 893–894, Jan. 1999.

[13] T. K. Dey, K. Mehlhorn, and E. A. Ramos. Curve reconstruction: Connecting dots with good reason. In *Proc. 15th Annu. ACM Sympos. Comput. Geom.*, pages 197–206, 1999.

[14] D. P. Dobkin, H. Edelsbrunner, and M. H. Overmars. Searching for empty convex polygons. *Algorithmica*, 5:561–571, 1990.

[15] P. Erdős and G. Szekeres. A combinatorial problem in geometry. *Compositio Math.*, 2:463–470, 1935.

[16] P. Erdős and G. Szekeres. On some extremum problem in elementary geometry. *Ann. Univ. Sci. Budapest*, 3-4:53–62, 1960.

[17] J. Erickson. Generating random simple polygons. `http://compgeom.cs.uiuc.edu/~jeffe/open/randompoly.html`

[18] S. P. Fekete and G. J. Woeginger. Angle-restricted tours in the plane. *Comput. Geom. Theory Appl.*, 8(4):195–218, 1997.

[19] A. García, M. Noy, and J. Tejel. Lower bounds for the number of crossing-free subgraphs of K_n. In *Proc. 7th Canad. Conf. Comput. Geom.*, pages 97–102, 1995.

[20] J. Hershberger and S. Suri. A pedestrian approach to ray shooting: Shoot a ray, take a walk. *J. Algorithms*, 18:403–431, 1995.

[21] S. Hertel and K. Mehlhorn. Fast triangulation of the plane with respect to simple polygons. *Inform. Control*, 64:52–76, 1985.

[22] J. Horton. Sets with no empty convex 7-gons. *Canad. Math. Bull.*, 26:482–484, 1983.

[23] K. Hosono, D. Rappaport, and M. Urabe. On convex decompositions of points. In *Proc. Japanese Conf. on Discr. Comp. Geom. (2000)*, Vol 2098 of *Lecture Notes Comput. Sci.*, pages 149–155. Springer-Verlag, 2001.

[24] K. Hosono and M. Urabe. On the number of disjoint convex quadrilaterals for a planar point set. *Comp. Geom. Theory Appl.*, 20:97–104, 2001.

[25] F. Hurtado and M. Noy. Triangulations, visibility graph and reflex vertices of a simple polygon. *Comput. Geom. Theory Appl.*, 6:355–369, 1996.

[26] J. M. Keil. Polygon decomposition. In J.-R. Sack and J. Urrutia, editors, *Handbook of Computational Geometry*, pages 491–518. Elsevier Science Publishers B.V. North-Holland, Amsterdam, 2000.

[27] J. O'Rourke. *Computational Geometry in C.* Cambridge University Press, 2nd edition, 1998.

[28] J. Pach (ed.). *Discrete and Computational Geometry*, 19, Special issue dedicated to Paul Erdös, 1998.

[29] E. Rivera-Campo and J. Urrutia. Personal communication, 2001.

[30] M. Urabe. On a partition into convex polygons. *Discrete Appl. Math.*, 64:179–191, 1996.

[31] M. Urabe. On a partition of point sets into convex polygons. In *Proc. 9th Canad. Conf. Comput. Geom.*, pages 21–24, 1997.

[32] M. Urabe. Partitioning point sets into disjoint convex polytopes. *Comput. Geom. Theory Appl.*, 13:173–178, 1999.

[33] E. Welzl. Geometric graphs with small stabbing numbers: Combinatorics and applications. In *Proc. 9th Internat. Conf. Fund. Comput. Theory*, Lecture Notes Comput. Sci., Springer-Verlag, 1993.

[34] C. Zhu, G. Sundaram, J. Snoeyink, and J. S. B. Mitchell. Generating random polygons with given vertices. *Comput. Geom. Theory Appl.*, 6:277–290, 1996.

About Authors

Esther M. Arkin and Joseph S. B. Mitchell are at the Dept. of Applied Mathematics and Statistics, Stony Brook University, Stony Brook, NY 11794-3600, USA; {*estie,jsbm*}*@ams.sunysb.edu.*

Sándor P. Fekete is at the Department of Mathematical Optimization, TU Braunschweig, Pockelsstr. 14, D-38106 Braunschweig, Germany; *sandor.fekete@tu-bs.de.*

Ferran Hurtado, Marc Noy, and Vera Sacristán are at the Dept. de Matemàtica Aplicada II, Universitat Politècnica de Catalunya, Pau Gargallo, 5, E-08028, Barcelona, Spain;
{*Ferran.Hurtado,Marc.Noy,Vera.Sacristan*}*@upc.es.*

Saurabh Sethia is at the Department of Computer Science, Oregon State University, Corvallis, OR 97331, USA. *saurabh@cs.orst.edu.* (This work was conducted while S. Sethia was at the Dept. of Computer Science, Stony Brook University.)

Acknowledgments

We thank Adrian Dumitrescu for valuable input on this work, including a software tool for calculating reflexivity of point sets. We thank Oswin Aichholzer for applying his software to search all combinatorially distinct small point sets. This collaborative research between the Universitat Politècnica de Catalunya and Stony Brook University was made possible by a grant from the Joint Commission USA-Spain for Scientific and Technological Cooperation Project 98191. E. Arkin acknowledges additional support from the National Science Foundation (CCR-9732221, CCR-0098172). S. Fekete acknowledges travel support by the Hermann-Minkowski-Minerva Center for Geometry at Tel Aviv University. F. Hurtado, M. Noy, and V. Sacristán acknowledge support from CUR Gen. Cat. 1999SGR00356, CUR Gen. Cat. 2001SGR00224, MCYT-FEDER BFM2002-0557, and Proyecto DGES-MEC PB98-0933. J. Mitchell acknowledges support from NSF (CCR-9732221, CCR-0098172) and NASA Ames Research Center (NAG2-1325).

Geometric Permutations of Large Families of Translates

A. Asinowski
A. Holmsen
M. Katchalski
H. Tverberg

Abstract

Let F be a finite family of disjoint translates of a compact convex set K in $\mathbf{R}^2$, and let l be an ordered line meeting each of the sets. Then l induces in the obvious way a total order on F. It is known that, up to reversals, at most three different orders can be induced on a given F as l varies. It is also known that the families are of six different types, according to the number of orders and their interrelations. In this paper we study these types closely, focusing on their relations to the given set K, and on what happens as $|F| \to \infty$.

1 Introduction

Let F be a non-empty family of disjoint convex sets in the plane. Then a *transversal* of F is a line meeting all sets of F. If the line is oriented one gets a certain ordering of F, which is reversed if the orientation is changed. Such a pair of orderings is called a *geometric permutation* of F (abbreviated: a GP of F). We describe a GP by listing the sets using one of the two orders. In 1957 Hadwiger [7] proved that if F is finite, then F admits a transversal if and only if F has an ordering, $\prec$, satisfying the following condition: whenever $X \prec Y \prec Z$, $\{X, Y, Z\}$ admits the GP (XYZ). This result shows that once we know, for each triple of sets from F, which GPs it admits, we know whether F admits a transversal or not. No further knowledge of F is needed.

Using Hadwiger's theorem it is for instance easy to see that if each triple from F admits exactly one GP, while each quadruple of F has a transversal, then F has a transversal. One longstanding conjecture by Grünbaum [6] stated that if F consists of at least five pairwise disjoint translates of a given compact convex set K, and each quintuple from F has a transversal, then F has one. This conjecture was first proved in a weaker form (5 replaced by 128) by Katchalski [10] and then, as stated, by Tverberg [13]. In both proofs GPs played an important role.

Quite apart from their role as a tool, the GPs have turned out to be interesting objects of study in their own right. A natural question to ask is: Given $|F|$, how many different GPs can F have? Edelsbrunner and Sharir

[3] showed that the number of GPs is at most $2|F|-2$, for $|F| \geq 4$, and Katchalski, Lewis, and Zaks [9] gave an example of a family admitting this number of GPs. In the present paper we study only what we shall call K-families. A *K-family* F consists of finitely many disjoint translates of a compact convex set K. We require K to be an *oval*, i.e. to have interior points, in order to avoid trivial exceptional cases in some of our statements.

In the case of a K-family F, the upper bound $2|F|-2$ can be replaced by 3, for $|F| \geq 3$. A refined version of this statement was proved by Asinowski, Holmsen, and Katchalski in [1]:

Theorem 0. *The set of GPs of a given K-family, if non-empty, must be of one of the following* Types.

1. $\{(W)\}$
2. $\{(WABW'), (WBAW')\}$ *(For this Type we assume that W and W' are not both empty)*
3. $\{(WABCW'), (WBCAW')\}$
4. $\{(WABCXW'), (WBXACW')\}$
5. $\{(WABCW'), (WACBW'), (WBACW')\}$
6. $\{(WABXCW'), (WBAXCW'), (WACBXW')\}$

(To clarify the notation: If F is of Type 4, say, then F has two GPs, which can be written as $(Y_1Y_2\dots Y_mABCXY'_1\dots Y'_n)$ and $(Y_1Y_2\dots Y_mBXACY'_1\dots Y'_n)$, where $m \geq 0, n \geq 0$.) Note that if F is of Type 3, with $|F| = 3$, then F can also be considered to be of Type 2. In all other cases, F has a unique Type. If F is of Type x and cardinality n we sometimes say that F is of Type x_n.

The main purpose of the present paper is to refine Theorem 0 in various ways. Firstly we want to see how it is affected if we specialize K and $|F|$. Secondly, we study what happens when $|F| \to \infty$. Apart from their intrinsic value, these results may also make the GPs a better tool for studies of transversal problems for K-families. For many of these problems deal with a special K, or a special class of K's.

For a simple example, let K be a circular disc. Then the reader will easily construct a K-family of Type 5, with $|F| = 3$. A little experimentation will indicate that none exists for $|F| = 4$, and we shall in fact give a rigorous proof of that. If we instead let K be a semicircular disc, then figure 1 shows a K-family of Type 5 and size 4 and it will be clear that one can be found for any size.

The reader should be aware that this paper deals with a limited area within the larger field of transversals of, and in, all dimensions, as well as GPs and their higher-dimensional analogues. For more information consult i.e. [4], [5], [11].

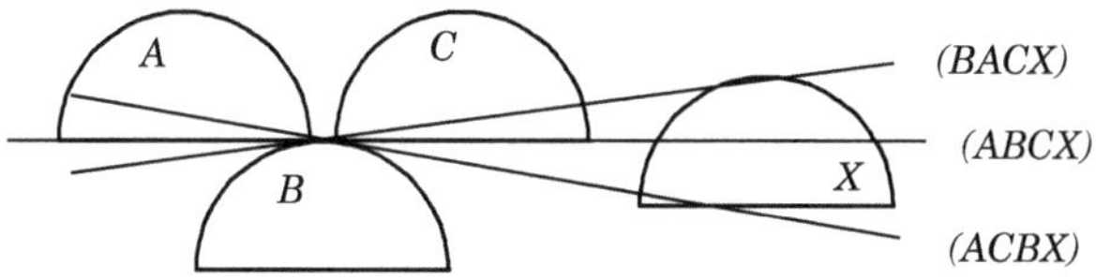

Fig. 1. A K-family of Type 5 and size 4 (i.e. Type 5_4). The two non-horizontal lines touch two semidiscs, the horizontal one touches three.

The material presented here originates in the Master's theses of Asinowski [2] and Holmsen [8], who were supervised by, respectively, Katchalski and Tverberg. All the four authors participated in the revisions and expansions which resulted in the present paper.

2 Types 1 and 2

We start by introducing some basic definitions. Let K be a given oval, and x a given Type, and assume that there are arbitrarily large K-families of Type x. We then say that K is *x-good.* Let furthermore, for an x-good K, $F_1, F_2, \ldots$ be K-families, where each family F_n is of Type x and has at least n members. We shall then say that the sequence $\{F_n\}$ is *x-good.* (Note that, when $m \neq n$, F_m bears no relation to F_n, except that both are K-families).

Let now $\{F_n\}$ be an x-good K-sequence, and let $\{t_n\}$ be a sequence of lines, where t_n, of slope $s_n \in \mathbf{R} \cup +\infty$, is a transversal for F_n. If $\{s_n\}$ converges to a "number" in $\mathbf{R} \cup \{+\infty, -\infty\}$, we say that the corresponding direction is x-good for K. Note that if *one* sequence $\{t_n\}$ shows x-goodness of K, then *any* sequence $\{t'_n\}$, where t'_n is a transversal for F_n, does the same, and gives the same x-good direction. For an easy argument shows that $\lim_{n\to\infty} s'_n = \lim_{n\to\infty} s_n$. (Since K has positive, finite area and finite diameter, there is a positive constant c_K such that some two sets of F_n are at least $c_k\sqrt{n}$ apart, but then for some other positive constant c'_K, the angle between any lines meeting these two sets is at most $c'_K/\sqrt{n}$. This shows the equality of the two limits.)

It is clear that any oval K admits Type 1_n for all n, and that every direction is 1-good for K. For Type 2 the situation is somewhat different.

Here, and also later, the concept of an *extreme* tangent will be important, so we take a close look at it. For a given oval K most of its boundary points p will be smooth, i.e. K has exactly one tangent at p which is then declared extreme. The set of non-smooth boundary points is at most countable. If p is non-smooth, the tangents at p fill a double cone with apex at p. Then the boundary of that cone is the union of two tangents declared to be extreme at p. A little thought shows that we could have defined an extreme tangent at p as a tangent which is the limit of secants through p, to the curve bdry(K). This will be useful later on. We now show the following theorem.

Theorem 1. *Let K be an oval.*

a) K admits Type 2_n for all $n \geq 3$.

b) A direction is 2-good for K if and only if K has an extreme tangent in that direction.

Note that a square has only two 2-good directions, because of *b)*, while for a circular disc all directions are 2-good, also by *b)*.

Proof of Theorem 1. For *a)* we choose a smooth boundary point p of K (we remind the reader that there are at most countably many non-smooth boundary points). Assume that p is not contained in a segment of the boundary of K. The case where p is contained in a boundary segment is treated similarly, and is left to the reader. It is no serious restriction to assume that K is in the halfplane $y \leq 0$, that p is the origin, and the x-axis the unique tangent. Consider a line l through p, with a small positive slope. Then l is not a tangent to K, but becomes one, l', when lifted by the right amount. We now place a translate K^* of K in the unique position where it touches the negative x-axis and the part of l' in the halfplane $y \geq 0$ (see figure 2). It is now clear that if the slope of l is small enough (dependent on K and n), we can place $n-2$ further translates, touching the x-axis from above such that the x-axis induces $(K^*KK_1 \ldots K_{n-2})$ while l' induces $(KK^*K_1 \ldots K_{n-2})$ and no third GP exists.

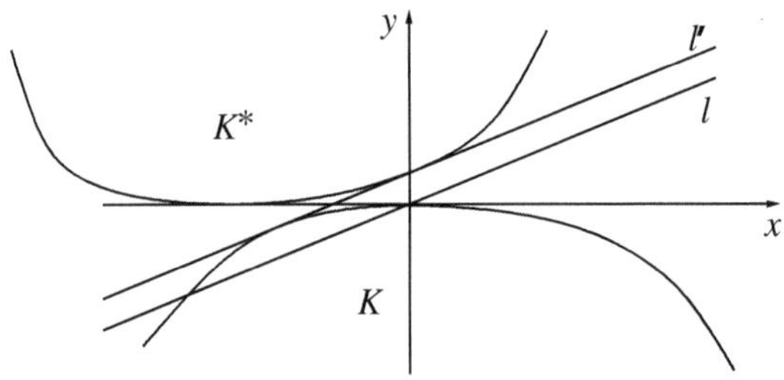

Fig. 2. The x-axis and l' are common separating tangents for K and K'

In order to prove the sufficiency part of *b)* we note that in the construction just given we did not use the smoothness at p to the full extent. If K has, say, a horizontal extreme tangent t at some point p, then if t is not a limit of ascending secant lines through p, it must be a limit of descending ones. In the latter case we may assume after reflection in the y-axis that K is located as in the proof of *a)*, with the x-axis being the limit of ascending secant lines through p. Then the argument above is still valid.

For the proof of the necessity part of *b)* let, say, the horizontal direction be 2-good for K and for the 2-good K-sequence $\{F_n\}$. We have to prove that K has a horizontal extreme tangent, and so we assume that this is not the case.

In a segment in bdry(K) every point, except possibly the endpoints, is smooth. Our assumption thus implies that the two horizontal tangents to K meet K in single points p^+ (at the top) and p^- (at the bottom). By non-extremeness K has two tangents, one of slope $\epsilon > 0$ and one of slope $-\epsilon$ at p^+, and also at p^-.

Consider now the K-family $\{F_n\}$, with GPs $\pi_{1,n} = (W_n A_n B_n W'_n)$ and $\pi_{2,n} = (W_n B_n A_n W'_n)$. Let t_n be one of the two tangents to A_n and B_n which separate A_n from B_n. Then t_n is a transversal for F_n. For let C belong to F_n without meeting t_n. One of the closed halfplanes defined by t_n, say the one containing A_n, has C in its interior. Then conv$(A_n \cup C)$ does not meet B_n, hence the GP $(A_n B_n C)$ does not exist. But then one of $\pi_{1,n}$ and $\pi_{2,n}$ does not exist either.

With t_n being a transversal of F_n, we know from the second paragraph of this section that we can choose n such that the slope of t_n is less than ϵ in absolute value. Let, say, A_n be in the lower one, and B_n in the upper one of the two closed halfplanes bounded by t_n, and let p_n^+ be the top point of A_n. Then A_n is contained in the cone which has its apex in p_n^+ and its left (right) boundary ray of slope ϵ $(-\epsilon)$. Hence t_n can only meet A_n, which it does, in p_n^+. Similarly, for large n, the other separating tangent t'_n also contains p_n^+, and so $t_n \cap t'_n \in A_n$. By symmetry, $t_n \cap t'_n \in B_n$, which leads to the final contradiction $A_n \cap B_n \neq \emptyset$. □

3 On parallelograms

Parallelograms behave in a nice and simple way, as expressed by the following.

Theorem 2. *Let P be a parallelogram.*

a) P admits Type x_n for every $n \geq x$, $x = 1, 2, 3, 4$.

b) P does not admit Type 5 or 6.

c) If an oval K does not admit Type 5_3 then K is a parallelogram.

d) For $2 \leq x \leq 4$ P has just two x-good directions.

e) If an oval K has just two 2-good directions, then K is a parallelogram.

Proof of Theorem 2. After an affine transformation, which does not affect transversal properties and GPs, we may assume P to be a square. We first prove *a)*, *b)*, and *d)*.

For *b)*, note that any two parallel rectangles can be strictly separated by a horizontal or vertical line (or both). Therefore all ascending transversals define the same GP, and similarly for all descending transversals, so a family of disjoint parallel rectangles can have at most 2 GPs, thus Types 5 and 6 are impossible.

Here *a)* is already known for the cases $x = 1$ and $x = 2$. The remaining two cases are dealt with by the following construction.

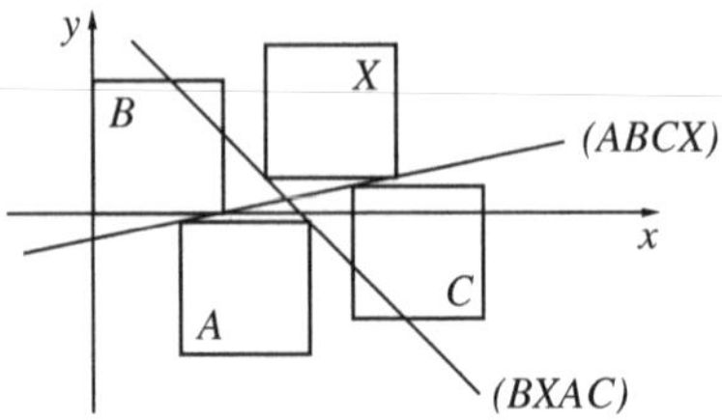

Fig. 3. The two induced GPs

Let ϵ be a number in $(0, 1/3)$ and let A, B, C, and X be squares, with axis-parallel sides of length 3 and SW (= South West = lower left) corners in $(2, -3-\epsilon)$, $(0, 0)$, $(6, -3+3\epsilon)$, and $(4, 4\epsilon)$, respectively. Then the upper non-separating common tangent to A and C has slope ϵ and induces $(ABCX)$, while the separating common tangent to X and A has slope -5ϵ and induces $(BXAC)$. By *b)*, there cannot be a third GP. See figure 3 for an illustration.

One can now place, for any $n \geq 4$, $n-4$ further squares $S_1, S_2, \ldots, S_{n-4}$ with the SW corner of S_i in $(4i+10, -1)$. Then the family $\{A, B, C, X, S_1, \ldots, S_{n-4}\}$ is clearly of Type 4_n provided ϵ has been chosen small enough. And the family $\{A, B, X, S_1, \ldots, S_{n-4}\}$ is of Type $3_{(n-1)}$, as the GP (BAX) does not exist.

Clearly *d)* follows from Theorem 1.

Here we prove *c)* only partly, for centrally symmetric K. The general case will be dealt with in section 5. Let K be centrally symmetric, but not a parallelogram. We start with the case when K has a segment in its boundary. We can assume, after an affine transformation, that K has its center in $(0, 1)$, while the x-axis intersects K in the segment $[(-1, 0), (1, 0)]$. This segment is also the top of $-K$ (the reflection of K in the origin). Because of the central symmetry, $-K$ is also a translate of K, in fact $-K = K + (0, -2)$. The two vertical tangents to K, given by $x = \pm a$, $a > 0$, are also tangents to $-K$. Here $a > 1$, as K is not a square. If the right tangent intersects K and $-K$ in single points (a, b) and $(a, b-2)$ we get, for a sufficiently small $\epsilon > 0$, the three desired GPs for $K + (0, \epsilon)$, $-K$ and $K + (2a, 2b-2)$. See figure 4, below, for an illustration. If, however, the intersection mentioned is a segment, we choose an almost vertical tangent t to K, with a negative slope, which meets K in a single point p close to, or on the segment. Now push $-K$ rightwards until it touches t and place a third translate so that it is separated by t from K and (the displaced) $-K$, while it touches t at the intersection of t and the x-axis. Then again the three desired GPs exist, after K is lifted a little.

If K is strictly convex (i.e the boundary of K contains no proper segment), then K and $-K$ can be assumed to meet in the smooth point $(0, 0)$ with the x-axis as the common tangent, and $(0, 1)$ as the center of K. Let p and $q = (q_1, q_2)$ be the leftmost and rightmost points of K, and translate K so that p moves to $(q_1, 0)$. Then $K + (0, \epsilon)$, $-K$, and the translated K will have the desired GPs for all small $\epsilon > 0$.

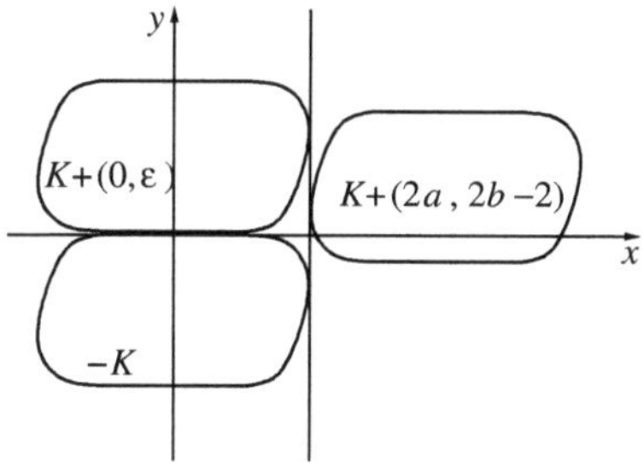

Fig. 4. K is not a parallelogram $\Rightarrow$ K admits Type 5_3.

We finally prove *e)*. We have to find three 2-good directions, for any given oval K, not a parallelogram. Now bdry(K) is either a polygon, or infinitely many non parallel segments, or it contains an arc with no segments. Each of these three cases are easily dealt with using Theorem 1 *b)*. □

4 On ellipses

By an *ellipse* we mean an oval with boundary an ellipse in the usual sense. Ellipses, like parallelograms, have an easily describable behavior with regard to GPs, as seen from the following.

Theorem 3. *Let E be an ellipse.*

a) *E is 1-good and 2-good for all directions.*

b) *E is not x-good for $x > 2$.*

c) *E does not admit the Types 4 and 6.*

d) *E admits Type 3_n or 5_n only for $n = 3$.*

Remark. Statement *b)* was also proved by Smorodinsky, Mitchell, and Sharir in [12].

Proof of Theorem 3. Clearly *a)* holds (and is a consequence of Theorem 1). Obviously *b)* will follow from *c)* and *d)*, while corollary 1 in section 7 will prove *c)*. As to *d)* we first observe that if F_n is a K-n-family of Type 3_n, with $n \geq 4$, W_n or W_n' must be non-empty. If W_n is non-empty then F_n has a subfamily $\{A_n, B_n, C_n, Y_n\}$ which allows (at least) the GPs $(Y_n A_n B_n C_n)$ and $(Y_n B_n C_n A_n)$. If W_n' is non-empty, there will be GPs $(A_n B_n C_n Y_n')$ and $(B_n C_n A_n Y_n')$ or, differently expressed, $(Y_n' A_n C_n B_n)$ and $(Y_n' C_n B_n A_n)$. If F_n is of Type 5_n, we can ignore the first GP $(W_n A_n B_n C_n W_n')$ and apply the same reasoning. Thus *d)* will follow from the lemma below. □

Lemma 1. *When K is a circular disc, a K-family can not have both of the GPs $(ABCD)$ and $(ACDB)$.*

Proof of Lemma 1. We prove the lemma by contradiction, so assume we have $(ABCD)$ and $(ACDB)$. Now we may assume that $(ABCD)$ is a *critical* GP, i.e. there is a unique line inducing $(ABCD)$. This assumption is justified by a shrinking argument often used in the study of transversal problems. To shrink an oval A we first chose as shrinking center a point a_0 in A. The shrinking process will then consist in reducing A continously, by letting a parameter λ vary from 1 downwards and letting for every λ A_λ be the set $\{a_0+\lambda(a-a_0), a \in A\} = \lambda A+(1-\lambda)a_0$. If F is a K-family $\{K+v_i, i \in I\}$ and the sets $K+v_i$ are shrunk simultaneously with centers k_i+v_i, where $k_i \in K$ we get for every λ a λK-family $F_\lambda = \{\lambda(K+v_i)+(1-\lambda)(k_i+v_i), i \in I\}$ $= \{\lambda K + v_i + (1-\lambda)k_i, i \in I\}$.

In the present case we first shrink the discs around their centers until one of the two GPs of F_λ becomes critical. Thus after renaming the discs we may assume one of the GPs of $\{A, B, C, D\}$ to be critical. If the critical GP is $(ACDB)$ we shrink further, but now with centers chosen on the transversal which induces $(ACDB)$. At some stage also the other GP becomes critical, and so our assumption is justified.

Let l be the unique line that determines $(ABCD)$. We may assume that l is horizontal, and that we meet the sets in the order $ABCD$ when we traverse l from left to right.

Now the reader will easily verify that a GP of a family is critical exactly when it induces a critical GP for some subfamily of size three. From now on we can clearly assume that two sets of our family lie below l. In our case (BCD) cannot be a critical GP. For then there will exist a vertical line that separates D strictly from A and B, so that we cannot have (ADB), and hence not $(ACDB)$.

We now make an observation, more general than needed here, but we will also use it later. Let a K-family F, where K need not be a disc, be given. Then any directed transversal t meets F first in a translate, say A, which is not in the convex hull of the others (which is the same as saying that if one picks a point of A, that point is not in the convex hull of the corresponding points of the other translates). For assume the falsity of the statement on t. By Carathéodory's theorem this means that, say, $A \subset \mathrm{conv}(B \cup C)$ or $A \subset \mathrm{conv}(B \cup C \cup D)$ where B, C, D are other members of F. In the first case t must induce BAC so it cannot meet A first. If $A \subset \mathrm{conv}(B \cup C \cup D)$, then t must enter, say, $\mathrm{conv}(B \cup C)$ before it can get to A, and then leave A before it has left $\mathrm{conv}(B \cup C)$. But clearly the part of t before A, together with A, separates B from C in $\mathrm{conv}(B \cup C)$, so that t cannot meet both B and C after A.

By the observation just made it follows that if (ACD) is critical then the centers of the discs must form a convex quadrilateral with cyclic order $ABCD$, but this violates the fact that we have $(ACDB)$, as shown in section 5 of [13].

Now consider the case where (ABD) is the critical GP. The convex hull of the centers must be a triangle, or else we end up with a violation of the cyclic order, as before. Let m be the vertical line which is tangent to the right side of B. Clearly D cannot lie to the left of m, and if D lies to the right of m, D would not meet $\mathrm{conv}(A \cup B)$. Thus m must meet D. We obtain a contradiction as follows. Move D away from A in the horizontal direction, until it becomes tangent to m. Note that while doing this, the disjointness of the discs will be preserved. Now B and D lie in opposite quadrants defined by l and m, and $C \cap l$ lies between $B \cap l$ and $D \cap l$. It is easy to see that this contradicts the disjointness of the discs.

Finally, consider the case where (ABC) is the critical GP. The convex hull of the centers must then be a quadrilateral. We also see that the cyclic order of the centers must be $ABDC$. We now move D away from A, B, and C, in the direction of l. This will cause (ADB) (and thus $(ACDB)$) to become a critical GP. But the situation we have described cannot occur. To see this, assume without loss of generality that the discs are of diameter 1 and let a, b, c, d denote the centers of A, B, C, D, respectively. Further let (x_z, y_z) denote the coordinates of the center of the disc Z. Assume that a and c lie on the x-axis, with c at the origin and $x_a < -1$. Since (ABC) is a critical GP it follows that b lies on the line $y = 1$. Thus, d must lie in the first quadrant, below the line $y = 1$, and by the disjointness of the discs d must lie outside the closed unit circle centered at the origin. Let m be the line through a and b. Since (ADB) is a critical GP, the distance from d to m must equal 1, and the distance from c to m must be less than 1. We therefore have $x_a < -1 < x_b < 0$. Let m' be the line parallel to m which goes through the point d. Let b' be the orthogonal projection of b on the line m'. The orthogonal projections of the centers on the line m' must have the order $acdb$. This implies that d must lie below b' on the line m'. The situation is illustrated in figure 5.

Figure 5 indicates that the point b' has distance less than 1 from c. In order to see this, we find that

$$b' = \left(x_b + \frac{1}{\sqrt{1 + (x_a - x_b)^2}}, 1 + \frac{x_a - x_b}{\sqrt{1 + (x_a - x_b)^2}} \right)$$

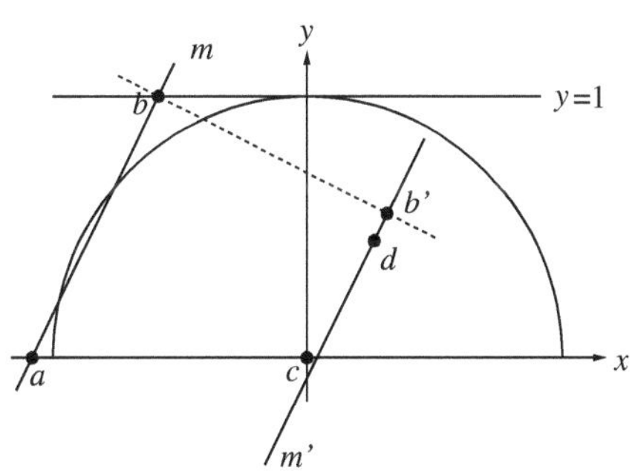

Fig. 5. The centers of the discs.

and thus

$$|cb'|^2 = x_b^2 + 2 + \frac{2x_a}{\sqrt{x_a^2 + x_b^2 - 2x_a x_b + 1}}$$

For a fixed slope of m, $|cb'|$ increases when m is moved to the right. Thus we only have to consider the limit cases when $b = (0,1)$ or $a = (-1,0)$. Easy calculations show that the distance 1 is assumed only for $a = (-1,0)$, $b = (-1,1)$. Thus b' belongs to the open unit disc centered at the origin, and so does d. But this contradicts the disjointness of the discs C and D. □

5 On Minkowski symmetrization

This classical symmetrization consists in associating with a convex body K another, centrally symmetric one K', defined by $K' = \frac{1}{2}K - \frac{1}{2}K$. It has turned out to be useful also in transversal problems for translates (cf. [13], for instance). Let $F = \{K + v_i, i \in I\}$ be a family of the type studied in this paper. Then $F' = \{K' + v_i, i \in I\}$ is easily seen to be a family of the same type. Assume now that F has, say, a horizontal transversal t, and that K has height 1, i.e. the minimal horizontal strip containing K has width 1. Then some horizontal strip of width 1 contains all the v_i. Conversely, if a horizontal strip of width 1 covers all the v_i, F has a horizontal transversal. Now, as is easily seen, F' also has height 1. This means in view of the preceding arguments, that F' has a horizontal transversal if and only if F has one. The same applies to every direction, of course.

What is important to note here, as we are concerned with GPs, is that when a GP is induced by, say, a horizontal transversal of F, then any horizontal transversal of F' induces the corresponding GP of F' (recall here that all parallel transversals induce the same GP on a given F). For F can be transformed continuously into F' via the families $F_s = \{((1-s)K + sK') + v_i, i \in I\}$, $s \in [0,1]$. Straightforward calculations show that F_s is a family of the type under study and that $(1-s)K + sK'$ has the same height as K. By continuity it follows that F_0 $(= F)$ has the same "horizontal" GPs as F_1 $(= F')$. The same holds, of course, for any direction. An example of this is illustrated in figure 6.

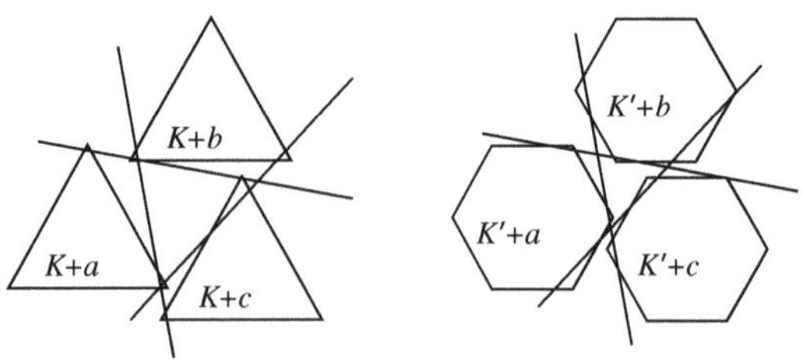

Fig. 6. $\{K + v, v \in \{a,b,c\}\}$ and $\{K' + v, v \in \{a,b,c\}\}$ have the same GPs in the same directions.

We can now complete the proof of Theorem 2*c)*. We proved that if the oval K is centrally symmetric and does not admit Type 5_3, then K is a parallelogram. If K is not centrally symmetric and does not admit Type 5_3, we know from the considerations above that the symmetrization K' does not admit Type 5_3, and thus is a parallelogram. Hence K' and, accordingly, K have only two 2-good directions. But then, by Theorem 2*e)*, K is a parallelogram. We shall make more use of symmetry later.

6 Type 3

Every oval admits Type 2_3 and therefore Type 3_3. But by Theorem 3 an ellipse does not admit Type 3_n for any $n \geq 4$. Now, in view of section 5, we can say even more; If K is affinely equivalent to a set of constant width (i.e. all its orthogonal projections have the same length), K does not admit Type 3_n for $n \geq 4$. For if K is a set of constant width then its symmetrization K' is known to be a disc, which does not admit Type 3_n.

We shall now deduce a necessary condition for a centrally symmetric K to be 3-good for, say, the horizontal direction, and then see how it can be sharpened to become both necessary and sufficient.

Let the 3-good K-sequence $\{F_n\}$ be given, so that we have, for $n = 1, 2, \ldots$, $(W_n A_n B_n C_n W'_n)$ and $(W_n B_n C_n A_n W'_n)$. Let the good direction for $\{F_n\}$ be horizontal. Consider an F_n. If we translate each member of it by the same vector v_n, the new family F'_n is still a K-family. If we do this for every n (v_n may depend on n) the sequence $\{F'_n\}$ is a 3-good sequence, with the horizontal direction being good. This means that if we choose some fixed translate A of K, then we may assume that A belongs to every F_n and that the two GPs of F_n are $(W_n A B_n C_n W'_n)$ and $(W_n B_n C_n A W'_n)$.

We now ignore the C_n's. From Theorem 1 and parts of its proof we find that the common separating tangents t_n and t'_n to A and B_n both converge to, say, the upper horizontal tangent to A. (Note: the assumed falsity of Theorem 1 does not enter into those parts we use here). We assume this tangent to be the x-axis.

Let a_n (a'_n) be the point on $t_n \cap A$ ($t'_n \cap A$) nearest to $t_n \cap t'_n$. Then the distance from $t_n \cap t'_n$ to $\mathrm{conv}(\{a_n, a'_n\}) \subset A$ converges to 0, because t_n and t'_n converge to the x-axis. Therefore $\{t_n \cap t'_n\}$ has an accumulation point in A, so that after passing to a subsequence of $\{F_n\}$, we may assume $\{t_n \cap t'_n\}$ to converge to a point p in A. Considering the similarly defined points b_n and b'_n, we find that p is an accumulation point for a sequence $\{b''_n\}$, with $b''_n \in B_n$. Thus, after passing once more to a subsequence, we can assume that $\{B_n\}$ converges to a set B. B is of course a translate of K and $A \cap B$ is either the point p or a segment containing p. We can assume p to be the origin.

If we now treat the "new" F_n in the same way as the original one, but ignore B_n this time, we can add the further assumption that $\{C_n\}$ converges to C, a translate of K, and that $A \cap C$ is a point or a segment in a horizontal tangent to A. That tangent can not be the lower one, as for large n the almost horizontal transversals to F_n would then be impossible. Thus the tangent is the x-axis, and so C is a horizontal translate of B.

We now draw an important conclusion: The top (and hence the bottom, by central symmetry) of K must be flat. For if it is a single point, the bottom point of B must coincide with the top point of A, since B meets A. But the same holds for C vs. A, and thus B and C must be equal, which contradicts, for large n, $B_n \cap C_n = \emptyset$.

Let the length of the top segment of K be 1, say. Assume that B is to the left of C. The intersection of B and C need not be empty, but it is clear that it can be at most 1-dimensional. This again implies that a horizontal segment in B (C) can meet C (B) in at most one point. Consider now B_1 and C_1, obtained by translating B to the left and C to the right until they both meet A in just one point. Then $C_1 = B_1 + (2,0)$. Since the longest horizontal chord of B_1 meets C_1 in at most one point, it follows that its length is at most 2. Here equality can only happen if $B = B_1$ and $C = C_1$. It is easy to see which are the longest horizontal chords of K: By convexity and central symmetry they form a parallelogram (possibly degenerate), centrally symmetric about the symmetry center of K. In particular the horizontal central chord is a maximal one, and so we have: If K is 3-good for a given direction, then its two boundary segments in that direction must be at least half as long as the corresponding central chord.

The necessary condition just found, is almost sufficient, in the following sense: If a centrally symmetric K has, say, a horizontal segment in its boundary, with the horizontal central chord of length strictly less than twice the length of the segment, then K is 3-good. For then we can place two translates B and C on top of a third, A, so that B and C are disjoint, but both have 1-dimensional intersection with A. With $B_n = B + (0, n^{-1})$, $C_n = C + (0, n^{-1})$ we have for all large n the almost horizontal transversals inducing AB_nC_n and B_nC_nA which demonstrate 3-goodness of K (cf. the sufficiency part of Theorem 1).

If however the length of the parallel central chord is exactly twice that of the segment, an additional condition enters: At least one of the endpoints of the segment must be smooth.

In order to see the sufficiency of this condition we proceed as in the preceding paragraph, but now B and C touch and they both meet A in just one point. If, say, the left endpoint of the top segment of K is smooth, we saw in the proof of Theorem 1 how to get a B^* slightly further to the left than B, such that the x-axis induces B^*A, while a small-sloped tangent to A and B^* induces (AB^*). As B^* is further to the left than B, we can move C a little to the left to get C^*, disjoint from B^*, but having 1-dimensional intersection with A. Lifting C^* a little we get C^{**}, disjoint also from A, and now it is clear that the GPs (AB^*C^{**}) and $(B^*C^{**}A)$ exist and are induced by almost horizontal transversals. Note that A is disjoint from $\text{conv}(B^* \cup C^{**})$ so the GP B^*AC^{**} does not exist. Thus the construction just shows 3-goodness of

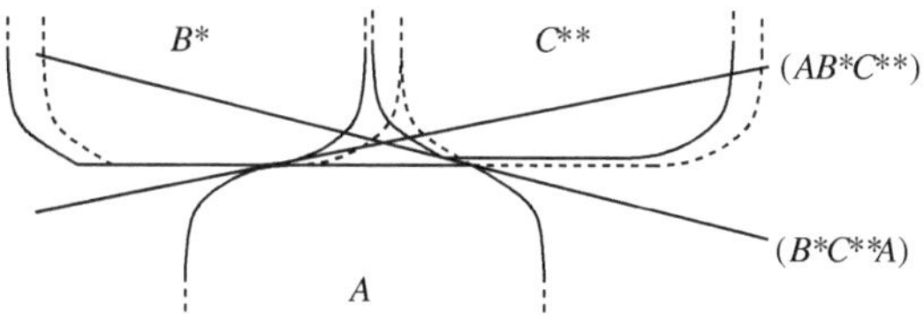

Fig. 7. The lines induce the two GPs of Type 3.

K. See figure 7 for an illustration. We note, for later use, that the same K is also 5-good. In fact, we could have created a B^{**} by translating B^* very little downwards along the small-sloped tangent mentioned above before we lifted C^*, and thus obtained also $(B^{**}AC^{**})$.

It remains to check that if the base and top of K have both endpoints non-smooth, and has length half that of the horizontal central chord, then the horizontal direction is not 3-good for K.

In this situation B and C intersect, either in just the common end, e, of their horizontal central chords, or in a non-horizontal segment having e as its midpoint. Now, for any 3-good $\{F_n\}$, with A_n constant, $(= A)$, and B_n converging to B, C_n to C, B_n has to contain e for large n. For consider the non-horizontal extreme tangent to A at the origin, which we after an affine transformation may assume to be the y-axis. Thus A lies in the fourth quadrant with the axes as extreme tangents at the origin. Then B_n cannot lie in the closed left half-plane for large n, which forces it to meet the open right half-plane, and thus to meet the open first quadrant. Thus B_n is obtained by translating B a distance $o(1)$ upwards and rightwards.

Now, because of central symmetry, $B = -A$ and it is then clear that the point e is in the second open quadrant, or on the positive y-axis, and in the latter case the segment from e to the origin is of course contained in B. In the former case all the tangents to B at e has negative slope. In both cases B contains a rectangle with its NE corner at e, which means that for large n, B_n will contain e.

By symmetry (not central!) we get a contradiction: Also C_n contains e for large n.

Some definitions are needed to formulate Theorem 4. Let a *side* of an oval K be the intersection of K and a tangent to K, of positive length. If K is centrally symmetric and the length of a side is at least half that of the parallel central chord, the side is said to be *long*. If the length exceeds half the chord length, then the side is *very long*. Then we have

Theorem 4. *Let K be a centrally symmetric oval.*

a) *K admits Type 3.*

b) *K is 3-good for a direction if and only if the two corresponding sides are either very long, or long (but not very long) with at least one smooth endpoint.*

c) K has at most two 3-good directions.

Here *a)* and *b)* have just been proved, while *c)* will follow from the following technical lemma:

Lemma 2. *A centrally symmetric K has at most three pairs of long sides. If it has three pairs, it is an affine image of a hexagon H_a defined by $0 \le x \le a$, $0 \le y \le a$, $|y - x| \le 1$, with $a \ge 2$. H_a has at most one pair of very long sides.*

Lemma 2 will imply *c)*, in view of *b)*.

Proof of Lemma 2. Consider first the case when K is a hexagon, with all sides long. After an affine transformation we can assume its vertices to be $(0,1)$, $(0,0)$, $(1,0)$, (a,b), $(a,b+1)$, and $(a-1,b+1)$, with $0 < b \le a-1 > 0$. If K has only two very long sides we choose the transformation such that $(0,0)$ is not on any of them. The line through $(0,0)$, parallel to the side $[(1,0),(a,b)]$, meets the side $[(a,b),(a,b+1)]$ in the point $(a, ab/(a-1))$. Now the chord $[(0,0),(a,ab/(a-1))]$ is not longer than the parallel central chord, which is at most twice as long as the parallel side $[(1,0),(a,b)]$. Thus $a - 0 \le 2(a-1)$, i.e. $a \ge 2$.

Consider now the horizontal chord $[(0,b/(a-1)),(2,b/(a-1))]$. One of its endpoints must be a vertex, for if not, horizontal chords slightly above it will have lengths > 2, which is a contradiction. If $(0,b/(a-1))$ is a vertex, b=a-1 and the lemma holds, as $K = H_a$ and the horizontal and vertical sides are not very long. If $(2,b/(a-1))$ is a vertex, then $a = 2$. If now $b = 1$, then $K = H_2$, which is an affine image of the regular hexagon, for which the lemma clearly holds. If $a = 2$, but $b < 1$, all sides are long, only the vertical ones are very long, and so we have not chosen the right affine transformation.

In the case when K is not a hexagon we assume the center of symmetry to be in the origin. If the lemma does not hold chose six long sides ρ, σ, τ, $-\rho$, $-\sigma$, $-\tau$, with the stated order corresponding to the one along bdry(K). Let $K_+ = \mathrm{conv}(\rho \cup \sigma \cup \tau)$, $K_- = -K_+$. As K is not a hexagon we may assume that $\rho \cap (-\tau) = (-\rho) \cap \tau = \emptyset$. K_+ is a four-, five-, or six-sided polygon, with only one side α that meets ρ and τ but not σ. We translate K_+ so that the midpoint of that side goes to $(0,0)$. After a similar treatment of K^-, the union of the two translates will be a centrally symmetric oval K_1. The sides $\pm\rho_1$, $\pm\sigma_1$, and $\pm\tau_1$ (the translates of $\pm\rho$, $\pm\sigma$, $\pm\tau$) are long sides of K_1. $\pm\rho_1$ and $\pm\tau_1$ are in fact very long, while $\pm\sigma_1$ are very long unless $\pm\sigma_1$ are parallel to α. K_1 is a $2p$-gon with at least four very long sides and $p \le 5$. After $p-3$ further applications of the procedure just described, we get a hexagon with at least four very long sides. But that has been proved impossible. □

7 Type 4

Any oval K admits Types 1, 2, and 3, but it may not admit Type 4. One property, useful for the discussion of which sets admit Type 4 is given by

Lemma 3. *If an oval K admits Type 4, then there are four translates of K, and two transversals, demonstrating this, which are tangents to all four translates.*

Proof of Lemma 3. We first note a consequence of the general observation made during the proof of Lemma 1: If K admits Type 4, so that we have translates of K admitting $(ABCX)$ and $(BXAC)$, then the translates are convexly independent, as each translate is at an end of one GP. Thus we can speak of their cyclic order.

We now shrink A, B, C, and X towards their circumcenters until one of the GPs becomes critical (cf. the proof of Lemma 1). Let a, b, c, and x be points on the critical transversal, in A, B, C, and X, respectively, and then shrink further with a, b, c, and x as centers of contraction, until the other GP also becomes critical. Now observe that the set of GPs allow the name changes $A \to B \to X \to C \to A$; $A \leftrightarrow X$, $B \leftrightarrow C$; $A \to C \to X \to B \to A$. This fact allows the assumption that $(ABCX)$ is induced by a transversal which touches, and separates, X and C, while it either touches A separating it from C, or touches B, separating it from C. Here only the second case can occur, for in the first case it is clear that the cyclic order of the translates is $ABCX$. This is however incompatible with $(BXAC)$, as shown in [13], section 5.

The argument just given shows, by symmetry, that for $(BXAC)$, either (XAC) or (BXA) must be the critical induced GP. In the second case the relabeling $A \to B \to X \to C \to A$ shows that we can assume that we are in the first case. Thus the full GPs are $(ABCX)$ and $(BXAC)$, while the critical shorter ones are (BCX) and (XAC).

We can now assume that $(ABCX)$ is induced by the x-axis and $(BXAC)$ by the y-axis, with C being in the first quadrant. Then X must be in the fourth quadrant, while A touches the positive y-axis from the left and B touches the negative x-axis from below. Let A^* be the upwards translate of A, touching the x-axis, and B^* the leftwards translate of B, touching the y-axis. Clearly $A^* \cap X = B^* \cap X = B^* \cap C = \emptyset$. The couple$(A^*, B^*)$ is a translate of (C, X) and so $A^* \cap B^* = \emptyset$. (A^*, C) is a translate of (B^*, X) and so $A^* \cap C = \emptyset$. The x-axis induces A^*B^*CX, the y-axis B^*XA^*C, and so Lemma 3 is proved. □

Corollary 1. *An ellipse does not admit Types 4 and 6.*

Proof of Corollary 1. It suffices to prove that a circle does not admit Type 4. This follows from Lemma 3, for when two lines touch the circles A, B, C, and X, and one of them induces $ABCX$, the other will induce $ACBX$. □

It seems natural to ask the following question: Which Ks do not admit Type 4 ? We know, using Corollary 1, that a K affinely equivalent to a set of constant width does not admit Type 4, as the Minkowski symmetrization of the latter is a disc. But there are other examples, too, for instance all regular n-gons with $n \not\equiv 0 \pmod 4$.

As to 4-goodness, we have the following theorem.

Theorem 5. *A given centrally symmetric oval K is 4-good if and only if it is 3-good.*

Proof of Theorem 5. We start with the "only if" part. Let K be 4-good, and let $\{F_n\}$ be a 4-good sequence for K. Thus we have GPs $(W_n A_n B_n C_n X_n W_n')$ and $(W_n B_n X_n A_n C_n W_n')$, $n = 1, 2, \ldots$. If we ignore the sets B_n we get a sequence $\{F_n^*\}$ where, for every n, F_n^* is either of Type 3 or admits a further GP. But when we deduced the necessary condition for K being of Type 3, we did not use fully the fact that the n'th K-family was of Type 3.: We just used the fact that there was for every n, a K-family having two GPs of the Type used to define Type 2, and did not care about a possible third one. Thus our $\{F_n\}$ shows that K satisfies the necessary condition for K being of Type 3. But that condition was also sufficient, and thus K is 3-good if it is 4-good.

Let now K be 3-good. By Theorem 4, K has one or two pairs of long sides corresponding to good directions. Consider such a pair, assumed to be horizontal. Let K^+, K^- denote, respectively, the top and bottom segment of K, and use a similar notation for translates of K. We place four translates of K: A, B, C, and X, such that B^-, A^+, X^-, and C^+ are, say, on the x-axis, with say, $B^- \cap A^+ = (-1, 0)$, $A^+ \cap X^- = (0, 0)$, $X^- \cap C^+ = (1, 0)$. We now have to move the translates a little, in order to make them disjoint, and having the desired GPs induced by almost horizontal transversals.

The easier case is when K^+ is very long. After small displacements of B, X, and C, each of $B \cap A$, $A \cap X$, and $X \cap C$ is a small horizontal segment. We then lift X and C by the same small amount. Now the line through the midpoints of $A \cap B$ and $X \cap C$ induces $(\text{int}(A)\ \text{int}(B)\ \text{int}(C)\ \text{int}(X))$, while the falling, separating, common tangent for X and A induces $(\text{int}(B)\ XA\ \text{int}(C))$. (Note that the concept of GPs does not require the sets to be compact translates of one another). After slight moves of B and C the sets A, B, C, and X are disjoint and the two lines mentioned induce the desired GPs and have small slopes.

When K^+ is not very long, $X \cap B$ and $A \cap C$ are non-empty (but at most 1-dimensional) so the procedure for construction of the desired GPs has to be modified slightly. Now K^+ is assumed to have at least one smooth endpoint. As the Type of any K-family is clearly invariant under an affine transformation of the plane, we can assume that the right endpoint of K^+ (and thus the left endpoint of K^-) is smooth. We start by moving A and X equal small distances, A to the left, X to the right. Then we get $A \cap X = B \cap X = A \cap C = \emptyset$ and $\dim(B \cap A) = \dim(X \cap C) = 1$. If X and C are lifted a little, the descending separating common tangent to A and X will induce the GP $(\text{int}(B)\ XA\ \text{int}(C))$ (here we use the smoothness condition), and the line through the midpoints of $B \cap A$ and $X \cap C$ will induce $(\text{int}(A)\ \text{int}(B)\ \text{int}(C)\ \text{int}(X))$ as before. After slight moves of B and C we again have disjointness and the desired GPs.

It is easily checked in both cases that the four sets have no other GP, and it is then clear how to construct, in both cases, a 4-good sequence for K. □

8 Type 5

Here we have the following theorem.

Theorem 6. *Let the oval K be centrally symmetric.*

a) K admits Type 5_3 if and only if it is not a parallelogram.

b) K is 5-good if and only if it has a long side with at least one smooth endpoint.

Proof of Theorem 6. Here *a)* follows from Theorem 2*b)* and *c)*. For *b)* we start with the necessity and assume K to be 5-good for a given direction. Let K^+ (defined as in section 7) have both endpoints non-smooth, and let A, B, C be the limit translates obtained as in the proof of Theorem 4, although B here plays the role of A there, so in F_n we have A_n, B, and C_n. Now Lemma 1 requires the slope of the separating common tangents for B and A_n, B and C_n to converge to zero. This requires (remember the central symmetry) both A_n and C_n to be above the x-axis for all large n, but then the GP A_nBC_n does not exist.

For the sufficiency, let K have a long side with a smooth endpoint. During the proof of Theorem 4 we noted in the case when K^+ is long, but not very long, that K is then, because of the smooth endpoint, 5-good. The argument given there also works when K^+ is very long (in that case we gave a simpler argument for just 3-goodness). □

9 Type 6

This Type is interesting as a K which admits Type 6 can easily be seen to admit the other Types. Our theorem about 6-goodness is

Theorem 7. *Let the oval K be centrally symmetric.*

a) If K is 6-good, it has a long side with at least one smooth endpoint.

b) If K has a very long side with at least one smooth endpoint, or a long side with both endpoints smooth, then K is 6-good.

Proof of Theorem 7. A K-n-family of Type 6 becomes a K-$(n-1)$-family of Type 5 by omission of X. Therefore *a)* follows from Theorem 6 *b)*. For the proof of *b)* it is convenient to interchange A and B, and C and X, in the definition of Type 6. Thus the proof of *b)* will consist in finding almost horizontal lines inducing GPs $(BACX)$, $(ABCX)$, and $(BXAC)$ for appropriately chosen translates A, B, C, and X.

We start out as in the proof of Theorem 5. In the first case we can assume that the left endpoint of K^+ is smooth. We first translate X and C by $(-\epsilon, \epsilon^2)$ for some positive ϵ. If ϵ is small enough, the falling, separating, common tangent for X and A will induce the GP (int B)XA(int C). Consider t_ϵ, the upper tangent to A from $(1-\epsilon, \epsilon^2)$ $(= X \cap C)$. The smoothness condition implies that we may assume (possibly after choosing a smaller ϵ) that $t_\epsilon \cap A$ is a segment or a smooth point. We may also assume, after a slight move of B, that $t_\epsilon \cap B = t_\epsilon \cap A$. Now smoothness implies that t_ϵ induces the GP A(int C)(int X). This means that if we move B slightly away from A it is possible, by wriggling t_ϵ a little, to get GPs AB(int C)(int X) and BA(int C)(int X). Thus we have the desired GPs after moving X a little away from C.

If K^+ is long, but not very long, we start by translating X and C by $(\epsilon, 0)$, where $\epsilon > 0$. This is to free C from A, and X from B and A. Now translate X and C by $(0, \epsilon^2)$. If ϵ is small enough the falling, separating, common tangent for X and A will (now by the smoothness condition) induce (int(B) XA int(C)). Just as above we may assume $t_\epsilon \cap A$ to be a smooth point or a segment, while $t_\epsilon \cap A = t_\epsilon \cap B$, where t_ϵ is the upper tangent to A from $(1+\epsilon, \epsilon^2)$. Now X and C are moved slightly against each other so that $X \cap C$ becomes a segment met by t_ϵ in an interior point, and then one can again move X away from C, A away from B, and wriggle t_ϵ to get the desired GPs. □

As a corollary to *b)* of Theorem 6 and *a)* of Theorem 7 we get a nice result about two important classes of ovals.

Theorem 8. *Assume that the oval K is either strictly convex, or a polygon. Then every sufficiently large K-family has at most two geometric permutations*

Proof of Theorem 8. If K is strictly convex (i.e. K has no segment in its boundary) then the same applies to K', the Minkowski symmetrization of K, as will follow from the next section. Thus K' does not have a long side, and so is neither 5-good nor 6-good. But, as shown in section 5, to every K-family of Type x_n there corresponds a K'-family of Type x_n (and conversely). Thus K is neither 5-good nor 6-good, which implies the statement in the theorem.

If K is a polygon, it is well known that K' is also a polygon, so that none of its sides has a smooth endpoint. Thus the theorem holds for K' and hence, as above, for K. □

Remark. We do not know whether, in Theorem 7*b)*, the smoothness at both ends of K^+ is really necessary, or whether a different kind of condition has to be introduced for the case when K^+ is not very long.

10 More on symmetrization

Theorems 4-7 were all formulated for a centrally symmetric K, for the sake of simplicity. But they have their counterparts for a general K, which can be found once one is aware of how certain properties of K relate to the same properties of K'. Some of these properties were discussed in Section 5, and the remaining ones are dealt with below.

To start with let K intersect its horizontal tangents t^+ and t^- in segments of length l^+ and l^-, one or both of which may be degenerate. The corresponding segments of $K-K$ are determined by the former and one finds their lengths to be $l^+ + l^-$, so that for K' their lengths are $(l^+ + l^-)/2$. In particular if those of K' are degenerate those of K are, too. Similar statements hold for all directions.

As was observed in section 6, in a centrally symmetric oval a central chord is at least as long as any parallel chord. Applying this to K', and recalling the construction of K' we find that for any given direction K and K' have the same length of maximal chords in that direction. This fact makes it easy to find the counterparts of those theorems where long and very long segments enter, once we have realized certain connections between smoothness of K and smoothness of K'.

Consider a K for which the upper horizontal tangent t^+ intersects K in the segment $[p^+, q^+]$ with p^+ to the left, and assume that K is smooth at p^+. Then we know that the horizontal direction is 2-good for K, and that t^+ is the limit of tangents with positive slopes having 2-good directions for K. From section 5 we then know that the horizontal direction is 2-good for K' and a limit for 2-good ascending directions for K'. From central symmetry it now follows that the left endpoint of the top segment of K' is smooth, and so is its reflection in the center of K', which is of course the right endpoint of the base segment. Thus smoothness transfers from K to K'.

The reasoning above can be reversed. It will then show that if K' has a smooth endpoint at, say, the left end of its top boundary segment, then K will have either a top boundary segment with smooth left endpoint, a base boundary segment with smooth right endpoint, or (if not any of these) a horizontal tangent which intersects K in one point and is the limit of ascending tangents to K.

References

[1] A. Asinowski, A. Holmsen, and M. Katchalski, The triples of geometric permutations for families of disjoint translates, *Discrete Math.* **241** (2001), 23–32.

[2] Andrei Asinowski, Geometric permutations for planar families of disjoint translates of a convex set, Master's thesis, Technion – Israel Institute of Technology, December 1999.

[3] Herbert Edelsbrunner and Micha Sharir, The maximum number of ways to stab n convex nonintersecting sets in the plane is $2n-2$, *Discrete Comput. Geom.* **5** (1990), 35–42.

[4] J. E. Goodman and R. Pollack, Hadwiger's transversal theorem in higher dimensions, *J. Amer. Math. Soc.* **1** (1988), 301–309.

[5] J. E. Goodman, R. Pollack, and R. Wenger. Geometric transversal theory, in *New trends in Discrete and Computational Geometry*, J. Pach, Ed., vol. 10 of Algorithms and Combinatorics. Springer-Verlag, Heidelberg, 1993, pp. 163–198.

[6] B. Grünbaum, On common transversals, *Arch. Math.* **9** (1958), 465–469.

[7] H. Hadwiger, Über Eibereiche mit gemeinsamer Treffgeraden, *Portugalie Math.* **16** (1957), 23–29.

[8] A. Holmsen, Geometriske transversaler for disjunkte familier av translater i planet, Master's thesis, University of Bergen, January 2000.

[9] M. Katchalski, T. Lewis, and J. Zaks, Geometric permutations for convex sets, *Discrete Math.* **54** (1985), 271–284.

[10] Meir Katchalski, A conjecture of Grünbaum on common transversals, *Math. Scand.* **59** (1986), 192–198.

[11] R. Pollack and R. Wenger, Necessary and sufficient conditions for hyperplane transversals, *Combinatorica* **10** (1990), 307–311.

[12] S. Smorodinsky, J. S. B. Mitchell, and M. Sharir, Sharp bounds on geometric permutations of pairwise disjoint balls in $\mathbf{R}^d$, *Discrete Comput. Geom.* **23** (2000), 247–259.

[13] H. Tverberg, Proof of Grünbaum's conjecture on common transversals for translates, *Discrete Comput. Geom.* **4** (1989), 191–203.

About Authors

Andrei Asinowski and Meir Katchalski are at the Faculty of Mathematics, Technion – Israel Institute of Technology, Haifa 3200, Israel.

Andreas Holmsen and Helge Tverberg are at the Departement of Mathematics, University of Bergen, Johs. Brunsgt. 12, 5008 Bergen, Norway.

Email addresses:
andrei@techunix.technion.ac.il (A. Asinowski),
andreash@mi.uib.no (A. Holmsen),
meirk@techunix.technion.ac.il (M. Katchalski),
tverberg@mi.uib.no (H. Tverberg).

Integer Points in Rotating Convex Bodies

Imre Bárány
Jiří Matoušek

Abstract

Let K be a planar convex body symmetric about the origin. We define $P(K)$ as the probability of $\tau K \cap \mathbf{Z}^2 \neq \{0\}$, where $\tau \in SO(2)$ is a random rotation around the origin and $\mathbf{Z}^2$ denotes the integer lattice, and we let $P(v) = \inf\{P(K) : \text{vol}(K) = v\}$. By Minkowski's theorem, $P(v) = 1$ for $v > 4$, and $P(v) = 0$ for $v < \pi$. We describe the behavior of $P(v)$ in the intervals $[\pi, \pi + \varepsilon_0]$ and $[4 - \varepsilon_0, 4]$ for a small positive constant ε_0.

1 Introduction and results

Let K be a bounded open convex body in the plane symmetric about 0. Let $\text{vol}(K)$ denote its area. Let τ be a rotation around 0 by a random angle (uniformly distributed in $[0, 2\pi)$). We put

$$P(K) = \text{Prob}\left[\mathbf{Z}^2 \cap \tau K \neq \{0\}\right],$$

and for $v \geq 0$, we define

$$P(v) = \inf\{P(K) : \text{vol}(K) = v\}.$$

We have $P(\pi) = 0$, as is witnessed by the open unit disc, and Minkowski's classical theorem ([Min96]) implies that $P(v) = 1$ for any $v > 4$. Our goal is to understand the behavior of the function $P(v)$ on the interval $[\pi, 4]$, in particular, near the endpoints of this interval. We will show, for instance, that for $v = 4 - \varepsilon$, $P(v)$ behaves like $1 - const \cdot \sqrt{\varepsilon}$. This is the interval where the condition of Minkowski's theorem just fails to hold by a small ε. So our result says that in this case the conclusion of Minkowski's theorem fails to hold with probability at most $O(\sqrt{\varepsilon})$.

Areas close to π. The following theorem describes the behavior of $P(v)$ in the range $\pi \leq v \leq \pi + \varepsilon_0$ for a suitable positive constant $\varepsilon_0 > 0$. For a given $\varepsilon \in (0, \varepsilon_0)$, let K_ε denote the open convex body indicated in Fig. 1. This K_ε is very close to the unit disc. It is symmetric under rotation by $\frac{\pi}{2}$, and the segments AB, BC, CD, and DE all have equal length, which is set

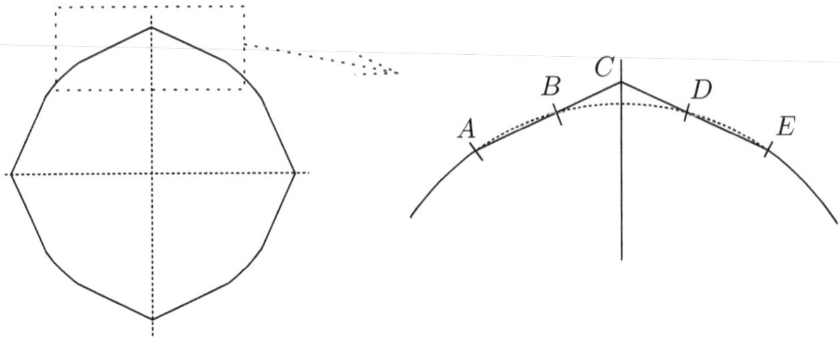

Fig. 1. The body K_ε.

in such a way that $\mathrm{vol}(K_\varepsilon) = \pi + \varepsilon$. An elementary computation shows that $\alpha \sim \sqrt[3]{3\varepsilon}$ as $\varepsilon \to 0$, where α is the length of the arc BD and $f \sim g$ means $\lim \frac{f}{g} = 1$. It follows that

$$P(K_\varepsilon) \sim \tfrac{2}{\pi}\sqrt[3]{3\varepsilon} = (0.918\ldots)\varepsilon^{1/3}.$$

Theorem 1.1 *There is a positive constant ε_0 such that for $\varepsilon \in [0, \varepsilon_0]$,*

$$P(\pi + \varepsilon) = P(K_\varepsilon).$$

Moreover, if $\mathrm{vol}(K) = \pi + \varepsilon$ *and* $P(K) = P(K_\varepsilon)$ *then K is a rotated copy of K_ε, i.e. K_ε is the unique extremal body.*

Areas close to 4. For areas close to 4, we do not characterize the extremal bodies, but most likely they are simply squares. Let Q denote the square $[-1,1]^2$; then the square $Q' = \sqrt{1 - \frac{\varepsilon}{4}}Q$ has area $4 - \varepsilon$ and

$$P(Q') = 1 - \tfrac{4}{\pi}\arccos\sqrt{1 - \tfrac{\varepsilon}{4}} \approx 1 - \tfrac{2}{\pi}\sqrt{\varepsilon}.$$

We believe that Q' is the extremal body for small enough ε. The order of magnitude of $1 - P(4 - \varepsilon)$ is given by the next result.

Theorem 1.2 *There exist constants $\varepsilon_0 > 0$ and $c_1, c_2 > 0$ such that for all $\varepsilon \in (0, \varepsilon_0]$, we have*

$$1 - c_1\sqrt{\varepsilon} \le P(4 - \varepsilon) \le 1 - c_2\sqrt{\varepsilon}.$$

The proofs of both theorems use elementary geometric arguments. The lower bound in Theorem 1.2 is shown by Q'. For the upper bound we have to use something like Minkowski's theorem, and indeed in one of the stages of the proof we argue similarly to the usual proof of Minkowski's theorem, see e.g. [PA95] or [Gru95].

Both proofs have analogous global structure: we treat separately thin bodies K, showing that $P(K)$ is much larger for them than the optimal value.

In these arguments, we estimate the contribution to $P(K)$ of an individual lattice point and then we sum up over all lattice points.

In the second part of each proof, we consider bodies with not too large circumradius. For Theorem 1.1, having established the existence of a body attaining the infimum, we proceed by local modifications of the boundary that increase the volume and do not increase $P(K)$. This is somewhat similar in spirit to proofs of isoperimetric inequalities by symmetrizations, for example. Since the extremal body is not completely simple, several types of modifications are needed.

For Theorem 1.2 the modification strategy does not quite work for bodies that are close to the square because there are too many cases to consider with too much computation. This is why we couldn't characterize the extremal bodies. So we have a different strategy for bodies that are not thin: we show that such a K can be closely approximated by a "tile". These tiles are parallelograms or hexagons and they form nice lattice packings, although they do not necessarily fill the whole plane. Then we analyze several possible shapes of the tiles and show that in every case $P(K) \geq 1 - O(\sqrt{\varepsilon})$.

Unfortunately, the proofs are quite lengthy and contain many tedious details. It would be nice to find shorter proofs. Also, determining the extremal body for areas close to 4 seems feasible but at the price of a detailed, lengthy, and exhausting case analysis.

Another interesting aspect would be investigating the stability of the inequalities in the above theorems. The question is similar to Bonnesen's inequality (see [Bon29] or the survey by Groemer [Gro95]). Assume K is 0-symmetric, has area $\pi + \varepsilon$ and $P(K) < const \cdot \varepsilon^{\frac{1}{3}}$. Does it follow that K is close to the unit disc? Or more quantitatively: how close is then K to the unit disc?

A further open problem is extending the results to higher dimensions. As the planar case indicates, characterization of extremal bodies may be quite difficult or even out of reach, and even determining the correct orders of magnitude of $P(K)$ near the critical values ($\mathrm{vol}(B^d)$ and 2^d in dimension d) seems challenging at present. It looks likely that for small enough ε

$$P(\mathrm{vol}(B^d) + \varepsilon) = O\left(\varepsilon^{\frac{d(d-1)}{2(d+1)}}\right) \text{ and } P(2^d - \varepsilon) = 1 - O\left(\varepsilon^{\frac{d(d-1)}{4}}\right);$$

these inequalities are suggested by examples similar to K_ε and Q'.

2 Notation and preliminaries

In the sequel, by a *body*, we mean a symmetric open convex body in the plane.

Let S^1 denote the unit circle. We identify the set of all rotations around the origin with S^1. For a body K and $z \in \mathbf{Z}^2$, let $T_z(K) = \{\tau \in S^1 : z \in \tau K\}$ and $T(K) = \bigcup_{z \in \mathbf{Z}^2 \setminus \{0\}} T_z(K)$; thus, $P(K) = \mu(T(K))$, where μ is the one-dimensional Lebesgue measure on S^1 scaled so that $\mu(S^1) = 1$.

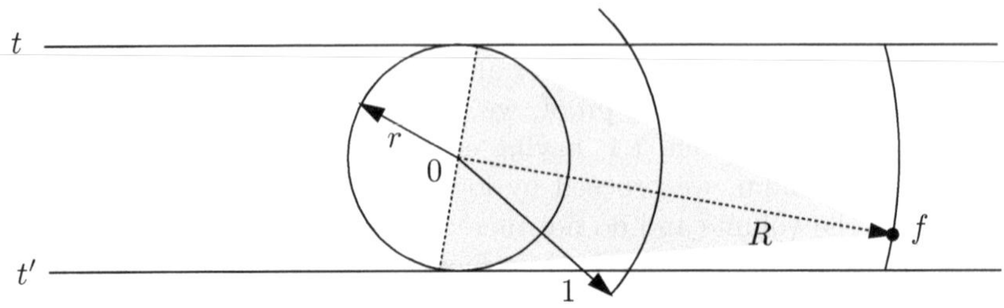

Fig. 2. The inscribed and circumscribed circles.

The following notation concerning various parameters of the body K will be used. Let R denote the radius of the smallest enclosing circle of K, and let r be the radius of the largest circle inscribed in (the closure of) K (see Fig. 2). Let f be a point of the boundary ∂K at distance R from 0. The largest inscribed circle touches ∂K at some two diametrically opposite points, and it has tangents t and t' common with K at these points. In the figure, K is thus contained in the horizontal strip between t and t', and we have

$$2rR \leq \mathrm{vol}(K) \leq 4rR;$$

the second inequality is obvious from the containment and the first one follows by considering the gray triangle, which must be contained in K.

Minimal bodies. Call a body K *minimal* if $P(K) = P(\mathrm{vol}(K))$. For some of the subsequent arguments, we will need the existence of minimal bodies. We prove it by a compactness argument, but such an argument needs a bound on the circumradius of the considered bodies. Later on, we will show that only bodies with a limited circumradius can come close to being minimal. With such a bound on the circumradius, we will be able to use the following lemma.

Lemma 2.1 *For $v \in (\pi, 4)$ and $R \geq 1$, let $\mathcal{K}(v,R)$ denote the set of all symmetric open convex bodies of area v and circumradius at most R. Whenever $\mathcal{K}(v,R) \neq \emptyset$, there exists a $K_{min} \in \mathcal{K}(v,R)$ such that $P(K_{min}) = \inf_{K \in \mathcal{K}(v,R)} P(K)$.*

Proof. We note that if the circumradius is bounded then the inradius r is bounded away from 0.

Write $\mathcal{K} = \mathcal{K}(v,R)$, and equip $\mathcal{K}$ with the Hausdorff metric h: $h(K,K') = \min(h_1(K,K'), h_1(K',K))$, where $h_1(K,K') = \sup_{x\in K} \inf_{y \in K'} |x-y|$. This $\mathcal{K}$ is a compact metric space (use the Arzelà–Ascoli theorem; this is a similar argument as in the Blaschke selection theorem).

Let $P = \inf_{K\in\mathcal{K}} P(K)$. There is a sequence $\{K_n\}_{n=1}^{\infty}$ of bodies in $\mathcal{K}$ with $P(K_n) \to P$, and by compactness, we may assume that it converges in the Hausdorff metric. We show that the limit body, which we denote by K_{min}, has the required property.

Let $\tau \in T(K_{min})$. Let z be an integer point with $z \in \tau K_{min}$. Since K_{min} is open, a neighborhood of z is contained in τK_{min} and we also have $z \in \tau K_n$ for all sufficiently large n. Consequently, we get

$$T(K_{min}) \subseteq \bigcup_{n=1}^{\infty} \bigcap_{i=n}^{\infty} T(K_i).$$

Write $A_n = \bigcap_{i=n}^{\infty} T(K_i)$. Since $\mu(T(K_n)) \to P$, we obtain $\mu(A_n) \leq P$. We have $A_1 \subseteq A_2 \subseteq \cdots$, and by the σ-additivity of μ, we obtain $P(K_{min}) = \mu(\bigcup_{n=1}^{\infty} A_n) \leq P$. □

3 Proof of Theorem 1.1

We consider bodies K with $\mathrm{vol}(K) \in (\pi, \pi + \varepsilon_0]$ for a small constant $\varepsilon_0 > 0$. We divide the proof into two parts. In the first part, we show that bodies K with large circumradius satisfy $P(K) \geq c > 0$ for an absolute constant c. Therefore, we can disregard them, since $P(K_\varepsilon) \to 0$ as $\varepsilon \to 0$ for the extremal body K_ε.

3.1 The thin case

Here we consider bodies with $R \geq \sqrt{2}$. For this case, we prove the following more precise result:

Proposition 3.1 *If $R \geq \sqrt{2}$ and $0 < \mathrm{vol}(K) < 4$, then $P(K) \geq c \cdot \mathrm{vol}(K)$ for a suitable absolute constant $c > 0$. (In fact, the result holds for any fixed $R > 1$ with a suitable $c = c(R)$.)*

Proof. Let

$$Z = \{z \in \mathbf{Z}^2 \setminus \{0\} : |z| \leq \tfrac{1}{\sqrt{2}} R, \gcd(z_1, z_2) = 1\}$$

denote the set of all primitive integer vectors of length at most $\frac{1}{\sqrt{2}}R$. For each $z \in Z$, let u_z be the arc on the unit circle centered at $\frac{z}{|z|}$ and of length $\ell_z = \frac{r}{2|z|}$.

We claim that for all rotations τ that bring the point $\frac{f}{|f|}$ into the arc u_z, we have $z \in \tau K$; see Fig. 3. To this end, it suffices to show that u_z is contained in the thick arc on the unit circle in the figure. Simple calculation shows that the length of the thick arc is actually at least $2\frac{r(R-|z|)}{R \cdot |z|} \geq \frac{r}{2|z|}$ (as $|z| \leq \frac{1}{\sqrt{2}}R$).

Next, we claim that all the arcs u_z, $z \in Z$, are pairwise disjoint. Assume that $z, w \in Z$ are consecutive in the angular order around the origin; we need to show that the angle $\alpha = \angle z0w$ is at least $\frac{1}{2}(\ell_z + \ell_w)$. Since z and w are consecutive primitive vectors, the triangle $z0w$ contains no integer

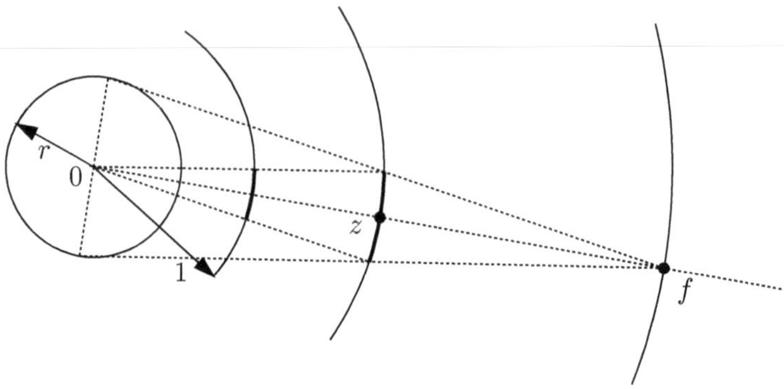

Fig. 3. Rotations with $z \in \tau K$.

point and so it has area $\frac{1}{2}$ (Pick's theorem); at the same time, the area is $\frac{1}{2}|z| \cdot |w| \cdot \sin\alpha \leq \frac{1}{2}|z| \cdot |w| \cdot \alpha$. So $\alpha \geq \frac{1}{|z|\cdot|w|}$. The desired inequality $\alpha > \frac{1}{2}(\ell_z + \ell_w)$ reduces to $r(|z|+|w|) < 4$, which holds in view of $|z|, |w| < R$ and $2rR \leq \mathrm{vol}(K) < 4$. The disjointness of the arcs u_z is proved.

We now calculate

$$P(K) \geq \sum_{z \in Z} \frac{r}{2|z|} \geq c_1 r R \geq c\,\mathrm{vol}(K)$$

for a sufficiently small $c > 0$. This proves Proposition 3.1. □

3.2 The round case

Proposition 3.1 shows that for area $v \leq \pi + \varepsilon_0$, all bodies close to being minimal have circumradius $\leq \sqrt{2}$. By Lemma 2.1, we know that minimal bodies exist for these v. So we assume that K is a minimal body, necessarily of circumradius at most $\sqrt{2}$, and we derive a series of claims about the shape of K. We already know that $P(K) \leq P(K_\varepsilon) = O(\varepsilon^{1/3})$. Our basic tool for establishing further properties of a minimal K is the following observation.

Lemma 3.2 *Let $K, \tilde{K}$ be bodies of circumradius at most $\sqrt{2}$ and such that $P(\tilde{K}) \leq P(K) < 1$ and $\mathrm{vol}(\tilde{K}) > \mathrm{vol}(K)$. Then K is not minimal.*

Proof. Let $L = q\tilde{K}$ where $q < 1$ is a scaling factor such that $\mathrm{vol}(L) = \mathrm{vol}(K)$; it suffices to check that $P(L) < P(\tilde{K})$.

Consider the set $T_{e_1}(\tilde{K})$ of all rotations τ bringing the point $e_1 = (0,1)$ into $\tau\tilde{K}$. This set is a union of open intervals, and it is easy to check that $T_{e_1}(L) \subseteq T_{e_1}(\tilde{K})$ and, moreover, for each interval $\tilde{I}$ of $T_{e_1}(\tilde{K})$, sufficiently small parts of $\tilde{I}$ near the endpoints (indicated by thick arcs in Fig. 4) are disjoint with $T_{e_1}(L)$. Since $T(\tilde{K})$ is not the whole S^1 by the assumption $P(\tilde{K}) \leq P(K) < 1$, the inclusion $T(L) \subset T(\tilde{K})$ is proper, and we have $P(L) < P(\tilde{K})$. □

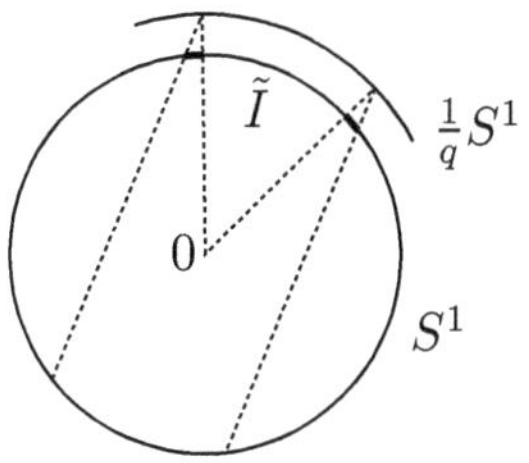

Fig. 4. Shrinking makes $P(\tilde{K})$ smaller.

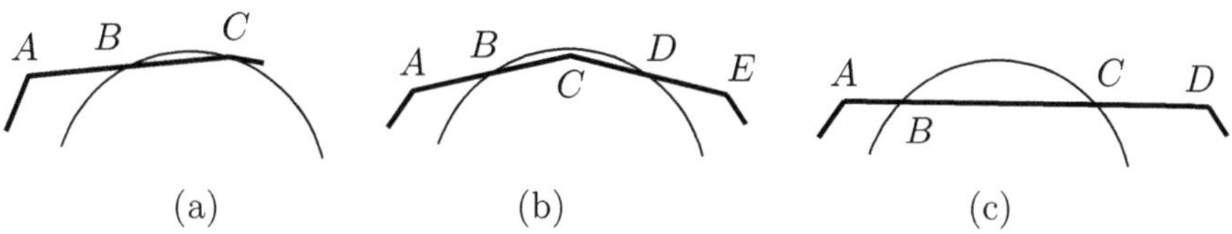

Fig. 5. The cases in condition (P3).

Note that the assumption $P(\tilde{K}) \leq P(K)$ is satisfied if $\tilde{K} \cap S^1 \subseteq K \cap S^1$ or, more generally, if $\tilde{K} \cap S^1 \subseteq (K \cap S^1) \cup E$ for a set E of measure 0.

Next, we introduce some terminology. The boundary of K is partitioned into

- *circle arcs*, which are the connected components of $S^1 \cap \partial K$,
- *inside arcs*, which are the connected components of the part of ∂K lying strictly inside S^1, and
- *outside arcs*, which are the connected components of the part of ∂K lying strictly outside of S^1.

Note that the inside and outside arcs are open intervals in ∂K, and there are at most countably many of them.

Lemma 3.3 *A minimal body K, $\pi < \mathrm{vol}(K) \leq \pi + \varepsilon_0$, has the following properties:*

(P1) Each outside arc of K consists of two segments.

(P2) Each inside arc of K consists of one or two segments.

(P3) Each segment of an inside arc continues by a collinear segment of an outside arc in at least one direction; the possible cases are shown in Fig. 5. In case (a), we have $|AB| = |BC|$, in case (b) $|AB| = |BC|$ and $|CD| = |DE|$, and in case (c), $|AB| + |CD| \geq |BC|$.

Proof. If there is an outside arc, with endpoints x and y (as in Fig. 6), that is not made of two segments, we can produce a body $\tilde{K}$ by replacing

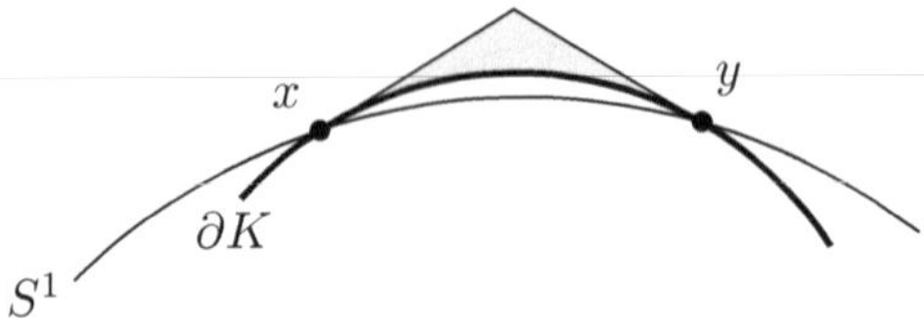

Fig. 6. Replacing an outside arc.

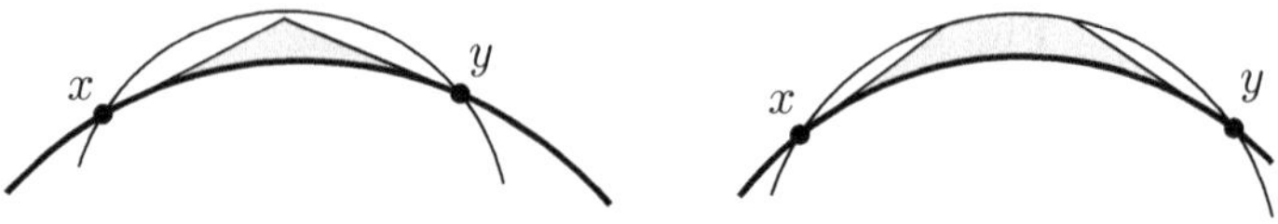

Fig. 7. Transforming an inside arc.

the arc by two segments tangent to ∂K at x and at y as in the figure. The circumradius of $\tilde{K}$ is below $\sqrt{2}$, because the considered outside arc is short (of length $O(\varepsilon_0^{1/3})$) and K contains the inscribed circle of radius at least $\frac{1}{3}$. Of course, the arc symmetric to the considered arc xy is transformed in the same way, to make $\tilde{K}$ symmetric; the modifications in all the subsequent steps are made symmetrically as well (without mentioning this explicitly). By Lemma 3.2, we get that K is not minimal unless it has property (P1).

To get property (P2), we proceed similarly. An inside arc with endpoints x and y as in Fig. 7 is replaced by two segments tangent to ∂K at x and at y; we obtain either a single inside arc (as in the left figure) or two inside arcs and a new circle arc (as in the right figure). Note that this works even if, say, the tangent segment to ∂K at x is tangent to S^1; in such case, we let the boundary of the modified body continue with an arc of S^1.

As for (P3), suppose that a segment $s = xy$ of an inside arc does not extend to either side within ∂K, as in one of the cases in Fig. 8. Then there are tangents to ∂K at x and at y making a positive angle with s (we already know that outside arcs consist of two segments!), and we can enlarge K to $\tilde{K}$ as indicated.

Thus, ∂K looks like one of the cases in Fig. 5 near each inside arc, and it remains to check the claimed equalities and inequalities concerning the lengths of the segments. For example, in case (a), if $|AB| < |BC|$, a suitable body $\tilde{K}$ can be produced by adding the gray triangle and removing the dashed triangle in Fig. 9(a). Appropriate modifications of K in cases (b) and (c) are indicated in figures (b) and (c). This proves Lemma 3.3. □

Fig. 8. An inside arc that does not extend.

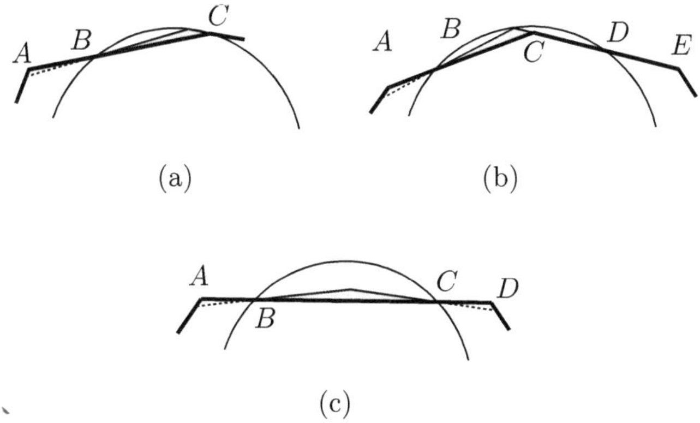

Fig. 9. Proving the inequalities in (P3).

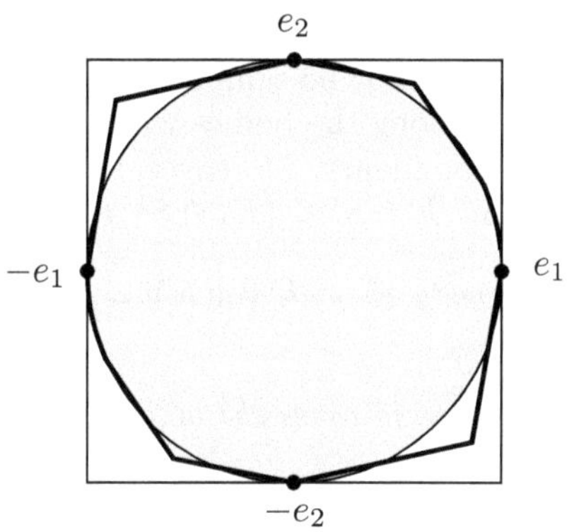

Fig. 10. Placing K into the square.

Lemma 3.4 *A minimal body K can be rotated in such a way that it is contained in the square $[-1,1]^2$ and the sides of the square are tangent to ∂K at their midpoints $\pm e_1$ and $\pm e_2$ (Fig. 10).*

Proof. By the equalities and inequalities in condition (P3), we see that the angular length of each inside arc is at most a constant multiple of the angular lengths of the adjacent outside arcs. The total angular length of the outside arcs is at most $P(K) = O(\varepsilon_0^{1/3})$. If $C = \partial K \cap S^1$ denotes the union of the circle arcs, we thus have $\mu(C) > \frac{3}{4}$, and so there is a rotation of K such that C is dense in some neighborhood of each of the four integer points on S^1. For such a rotation, the tangents to S^1 at $\pm e_1$, $\pm e_2$ are tangents to ∂K as well. □

From now on, we assume that the minimal body K is rotated as in Lemma 3.4. Let s_{++}, s_{+-}, s_{--}, s_{-+} denote the arcs of S^1 in the first, second, third, fourth quadrant, respectively, and similarly, let $k_{++}, \ldots, k_{-+}$

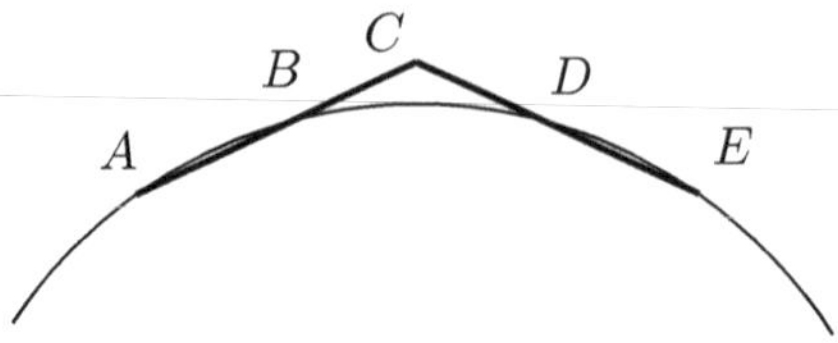

Fig. 11. The only possible type of outside arcs.

be the arcs of ∂K. Let K_1 be the body bounded by k_{++}, s_{+-}, k_{--}, and s_{-+} (indicated by gray shading in Fig. 10), and let K_2 be bounded by the complementary arcs s_{++}, k_{+-}, s_{--}, and k_{-+}.

Let $\tau_{\pi/2}$ denote the counterclockwise rotation by $\frac{\pi}{2}$. Let $\varepsilon_i = \mathrm{vol}(K_i) - \pi$, $i = 1, 2$. We have $\varepsilon = \mathrm{vol}(K) - \pi = \varepsilon_1 + \varepsilon_2$ and $P(K) \geq \max(P(K_1), P(K_2))$, with equality only if $K_1 \cap S^1 \subseteq \tau_{\pi/2}(K_2 \cap S^1)$ or $K_2 \cap S^1 \subseteq \tau_{\pi/2}(K_1 \cap S^1)$. If $\varepsilon_1 > \varepsilon_2$, let $\tilde{K} = K_1 \cup \tau_{\pi/2} K_1$. Then $\mathrm{vol}(\tilde{K}) = 2\varepsilon_1 + \pi > \mathrm{vol}(K)$ and $P(\tilde{K}) = P(K_1) \leq P(K)$, and so K cannot be minimal. Therefore $\varepsilon_1 = \varepsilon_2 = \frac{\varepsilon}{2}$ and K_1 minimizes $\mu(K_1 \cap S^1)$ among the bodies with area $\pi + \frac{\varepsilon}{2}$ containing the arcs s_{+-} and s_{-+} in their boundary. Moreover, after possibly interchanging K_1 and K_2, we may assume $K \cap S^1 \subseteq (K_1 \cap S^1) \cup \tau_{\pi/2}(K_1 \cap S^1)$.

Lemma 3.5 *A minimal body K with* $\mathrm{vol}(K) = \pi + \varepsilon \leq \pi + \varepsilon_0$ *has the following properties:*

(P4) No inside arc of K is as in cases (b) or (c) in Fig. 5.

(P5) Each of the two segments of an outside arc continues by a collinear segment of an inside arc.

(P6) Each of the two segments of each outside arc continues by a single-segment inside arc as in Fig. 11, where $|AB| = |BC| = |CD| = |DE|$.

Proof. Concerning (P4), we consider the situation as in case (b), and we suppose that $|0E| \geq |0A|$ (if not, we proceed symmetrically). We choose the point $D' \in S^1$ as in Fig. 12 so that the arcs BD' and DF on S^1 have the same lengths, and we let the polygonal line $E'C'F$ be a mirror reflection of the polygonal line ECB. First we construct a new body $\tilde{K}_1$ from K_1 by replacing the portion $ABCDEF$ of ∂K_1 by $AE'D'C'F$. The angle GAE' is convex since $|E'0| \geq |A0|$. We have $\mu(\tilde{K}_1 \cap S^1) = \mu(K_1 \cap S^1)$ and $\mathrm{vol}(\tilde{K}_1)$ exceeds $\mathrm{vol}(K_1)$ by the area of the triangle ABE'. Then we put $\tilde{K} = \tilde{K}_1 \cup \tau_{\pi/2}\tilde{K}_1$; we have $\mathrm{vol}(\tilde{K}) > \mathrm{vol}(K)$ and $P(\tilde{K}) \leq P(K)$. Hence K was not minimal by Lemma 3.2. Case (c) can be excluded by an analogous construction, and so (P4) holds.

Next, we consider a segment of an outside arc with endpoint $x \in S^1$ and we check (P5). We distinguish two cases. First, suppose that there are points of $S^1 \setminus K$ on the left of x and arbitrarily close to it, as in Fig. 13 left. Then we form a new body $\tilde{K}_1$ as in the left figure, by intersecting K_1 with the

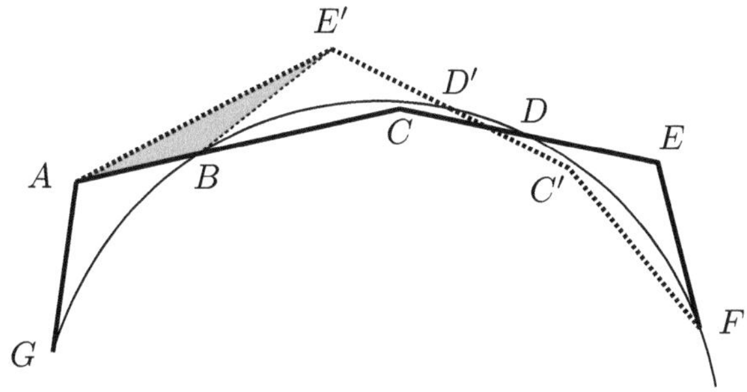

Fig. 12. Excluding case (b).

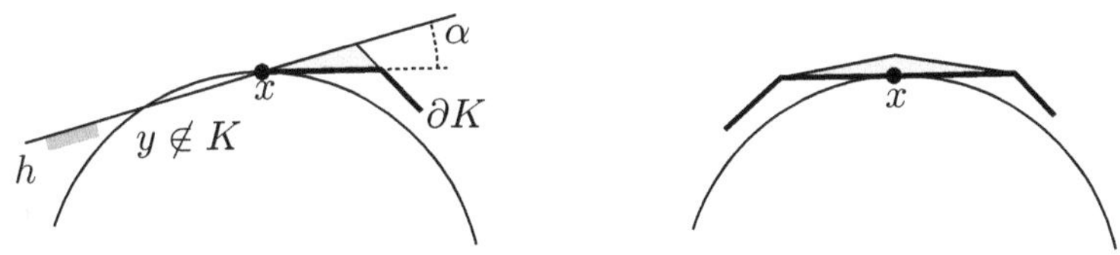

Fig. 13. The proof of property (P5).

(open) halfplane h and adding the gray triangle. If the angle α is sufficiently small, we have $\mathrm{vol}(\tilde{K}_1) > \mathrm{vol}(K_1)$ and $\tilde{K}_1 \cap S^1 \subseteq K_1 \cap S^1$. The argument is finished as the one above for the condition (P4).

Next, the situation can be as in the right figure, with segments of outside arcs of ∂K_1 emanating from x in both directions. Then $\tilde{K}_1$ can be produced by adding the indicated triangle. In this way, $\tilde{K}_1 \cap S^1$ has the extra point x, but Lemma 3.2 still applies.

We have already shown that each outside arc BCD continues with inside arcs on both sides, each formed by a single segment, so the situation is as in Fig. 11, with $|AB| = |BC|$ and $|CD| = |DE|$; this already implies $|BC| = |CD|$ as well. □

In order to prove Theorem 1.1, it now suffices to verify that K has only one outside arc in the first quadrant. Call the portion of ∂K consisting of an outside arc and the two adjacent inside arcs, i.e. from A to E in Fig. 11, a *dent*. Suppose that there are at least two dents. The dents can be moved and interchanged along ∂K_1 without changing the area or $P(K_1)$. Thus, we may suppose that the considered two dents are adjacent as in Fig. 14. But then the construction as in Fig. 12 used above for proving (P4) can be applied to the portion from B to D' and we get non-minimality of K. Theorem 1.1 is proved. □

4 Proof of Theorem 1.2

The proof consists of three parts. First (Section 4.1) we treat bodies with relatively large circumradius $R \geq 3.1$ (thin bodies). Next (Section 4.2), we

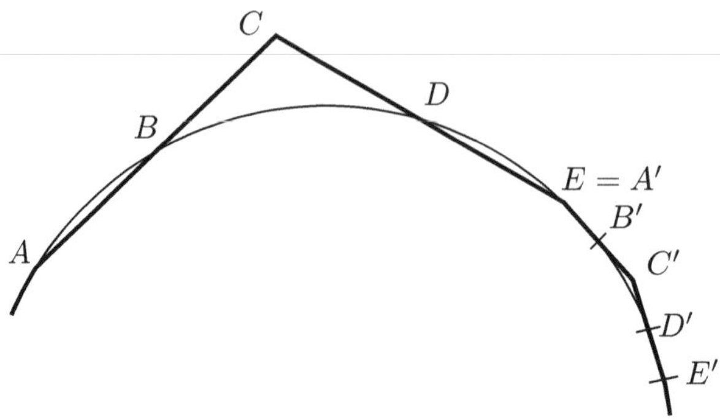

Fig. 14. Two adjacent dents.

deal with bodies that have circumradius below 3.1 but are not very close to the square. For both these cases, we show that, for $\mathrm{vol}(K) = 4 - \varepsilon$, $P(K) \geq 1 - O(\varepsilon)$ and so these bodies cannot be minimal. In the last part (Section 4.3), we consider bodies with small Hausdorff distance to the square and use tiles and their floating bodies to finish the proof.

4.1 Thin bodies

This part is devoted to the proof of the following:

Proposition 4.1 *Let K be a body with* $\mathrm{vol}(K) = 4 - \varepsilon$ *and circumradius* $R \geq 3.1$ *and let* $\varepsilon < \varepsilon_1$ *for a sufficiently small* $\varepsilon_1 > 0$. *Then* $P(K) \geq 1 - c\varepsilon$ *for a positive constant* $c > 0$. *Thus, such K cannot be minimal.*

For the time being we only assume $R > 2.2$, since some of the auxiliary results obtained in the proof will be useful later. Note that this implies $r < 2/R < 0.91$.

Let $\mathrm{vol}(K) = 4 - \varepsilon$ and suppose that K has been rotated so that $K \cap \mathbf{Z}^2 = \{0\}$. Let h be a vector of length $\frac{\varepsilon}{R}$ orthogonal to $0f$, where, as before, f is a point on ∂K where the circumradius is attained. Set

$$K^+ = K + [-h, h].$$

Then $\mathrm{vol}(K^+) \geq 4 - \varepsilon + 4\varepsilon = 4 + 3\varepsilon > 4$, and so there is at least one nonzero integer point in $K^+ \setminus K$. Call a point $z \in K^+ \setminus K$ *moderate* if there is a line ℓ passing through z and avoiding K that intersects the line $0f$ at angle at most $\frac{\pi}{4}$ (Fig. 15).

Lemma 4.2 *If* $R \geq 2.2$ *then there is at least one moderate integer point in* $K^+ \setminus K$.

Proof. As was noted above, $K^+ \setminus K$ contains some nonzero integer point z, as well as $-z$. Choose the notation so that z is no farther from f than

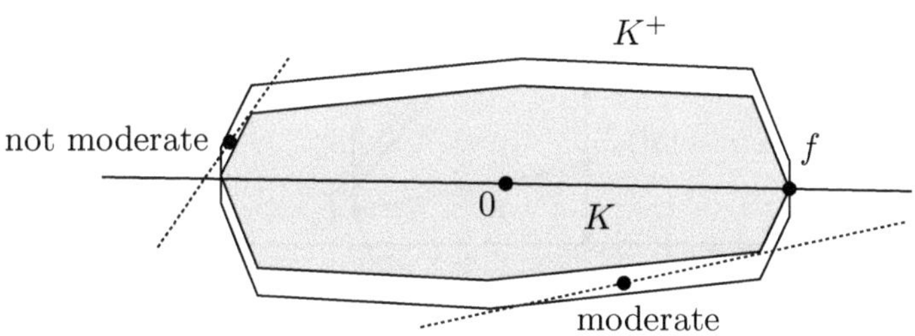

Fig. 15. The body K^+ and the definition of a moderate point.

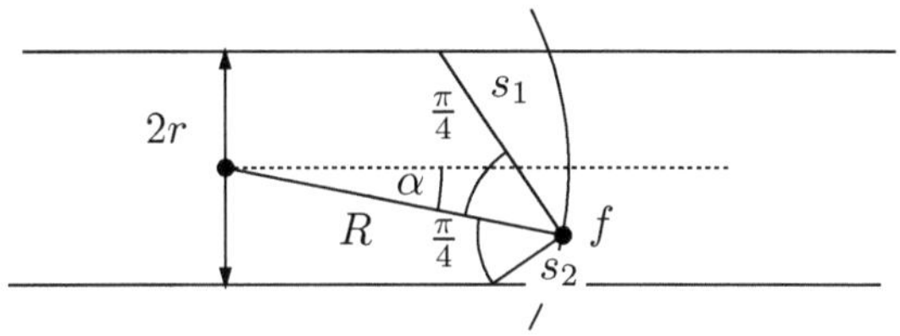

Fig. 16. Estimating s_1 and s_2.

$-z$. Assume that z is not moderate. Choose a line ℓ separating z from K, let ℓ^+ be the closed halfplane determined by ℓ and not containing 0, and let $K^\pm = K^+ \setminus (\ell^+ \cup (-\ell^+))$ (so we cut off by ℓ and by its symmetric copy). We show first that $\mathrm{vol}(K^\pm) > 4$. As $\mathrm{vol}(K^+) > 4 + 3\varepsilon$ it suffices to show that $\mathrm{vol}(K^+ \cap \ell^+) < \varepsilon$.

Let α be the angle between the ray $0f$ and the parallel strip of width $2r$ enclosing K. Then $\sin\alpha \le \frac{r}{R}$ (see Fig. 16). We may assume, by symmetry, that f is below the middle line of the strip. We estimate first the lengths, s_1 and s_2, of the segments indicated in Fig. 16.

Using $R > 2.2$ and $r < \frac{2}{R}$, it is easy to see that s_2 is the longest when $\alpha = 0$, and then it is equal to $\sqrt{2}r$. So $s_2 \le \sqrt{2}r$. For s_1, we calculate

$$s_1 = \frac{r + R\sin\alpha}{\sin(\alpha + \frac{\pi}{4})} \le R \cdot \frac{\frac{1}{2} + \sin\alpha}{\sin(\alpha + \frac{\pi}{4})}.$$

The right-hand side is maximized for the maximum possible α, i.e. the one with $\sin\alpha = \frac{r}{R} \le \frac{1}{2}$, in which case it yields $s_1 < 1.04R$.

The intersection of ℓ with K^+ is a segment whose length $s \le \max(s_1 + |h|, s_2 + |h|) \le 1.1R$ (see Fig. 17). The cut-off area of $K^+ \cap \ell^+$ is at most s times the width of the strip between ℓ and $\ell + h$, which is at most $|h| \cos\frac{\pi}{4}$. Thus

$$\mathrm{vol}(K^+ \cap \ell^+) \le s|h|\cos\frac{\pi}{4} \le 1.1R \cdot \frac{\varepsilon}{R} \cdot \frac{1}{\sqrt{2}} < \varepsilon.$$

We have shown that $\mathrm{vol}(K^\pm) > 4$. Consequently $K^\pm$ contains another nonzero integer point w. We again choose the sign in such a way that w is closer to f than $-w$.

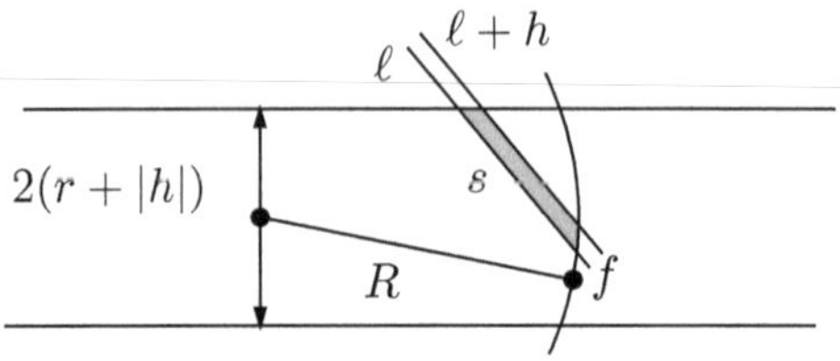

Fig. 17. Bounding the portion cut off by ℓ.

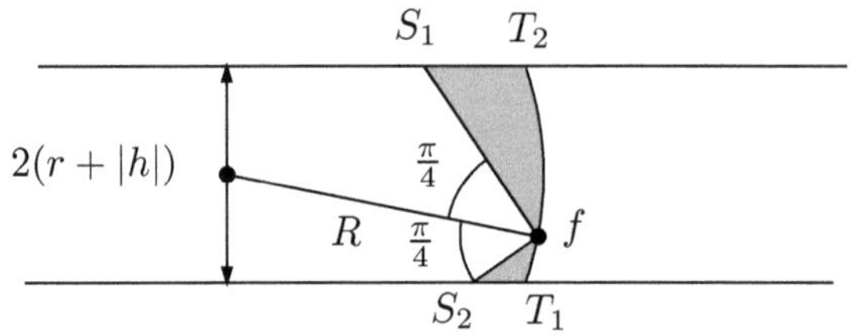

Fig. 18. Bounding the distance of z and w.

If w is moderate we are done, so we assume that w is not moderate. It follows that both z and w lie in the region indicated in Fig. 18 whose diameter is either S_1T_1 or S_2T_2. The segment S_2T_2 is the longest when $\alpha = 0$, and then $|S_2T_2| < \sqrt{5}(r + |h|) < \sqrt{5}$. As for S_1T_1, we have $|S_1T_1| \leq (1 + O(h)) \cdot \sqrt{4r^2 + |S_1T_2|^2}$ and $|S_1T_2| \leq (r + R \sin\alpha)\cot(\alpha + \frac{\pi}{4})$. Using $\sin\alpha \leq \frac{r}{R}$, we obtain $|S_1T_2| \leq 2r$ and it follows that $|S_1T_1| < \sqrt{8}$ (in fact, more detailed calculations reveal that it is even at most $\sqrt{5}$, which eliminates Case 3 below). Consequently $|z - w| < \sqrt{8}$.

We have to consider the following cases: $z - w$ equals $\pm e_1$ or $\pm e_2$ (Case 1), or $\pm e_1 \pm e_2$ (Case 2), or $\pm 2e_1 \pm e_2$ or $\pm e_1 \pm 2e_2$ (Case 3), or $\pm 2e_1$ or $\pm 2e_2$ (Case 4).

Since the convex quadrilateral $zw(-z)(-w)$ is contained in K^+ and it consists of four triangles of the same area as the lattice triangle $0zw$, it follows that the area of $0zw$ is at most 1. Therefore the area of $0zw$ equals either 1 or $\frac{1}{2}$. The area of $0zw$ is 1 if it contains no lattice point in its interior and has exactly one lattice point on its boundary (apart from its vertices). So the area of $0zw$ is 1 if and only if Case 4 occurs.

Case 1 We can assume, by symmetry, that $z = w + e_1$ with w having non-negative x-component. As the area of $0zw$ is $\frac{1}{2}$, the y-component of z and w is ± 1, see Fig. 19. Since $\angle 0fw \leq \frac{\pi}{4}$, f lies in the union of the two discs indicated in the figure. At the same time, f is outside the disc of radius $|z| - |h|$ centered at 0. If the x-component of w is 0, then R is certainly below 2.2, and if it at least 1, then the point f is so far away from the line $0w$ that the area of the triangle $0fw$ is larger than 1.

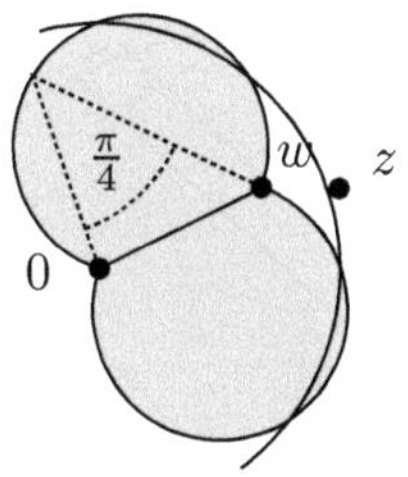

Fig. 19. The case $z = w + e_1$.

Case 2 We assume, by symmetry again, that $z - w = e_1 + e_2$, and $w = (n, n-1)$ for some nonnegative integer n. Again, f is in the double disc and is outside the disc of radius $|z| - |h|$, and the argument is finished in the same way as before.

Case 3 We assume $z - w = 2e_1 + e_2$ and $w = (2n+1, n)$ for some integer n. Again, f is in the double disc and is outside the disc of radius $|z| - |h|$ and the argument is finished similarly.

Case 4 Assume $z = w + 2e_1$, and $w = (n, 1)$ with an integer $n \geq -1$. As the area of $0zw$ is 1, K^+ is very close to the parallelogram with vertices $\pm z, \pm w$. Consequently, f is very close to z or w. But it is also in the double disc over the segment $0w$ which is impossible unless $n = -1$. In this case, however, $R \approx \sqrt{2}$, contradicting to $R > 2.2$.

Lemma 4.2 is proved. □

Next, we prove that a given z cannot serve as a moderate integer point in $\tau(K^+ \setminus K)$ for too many rotations τ.

Lemma 4.3 *Let K be such that $R \geq 3.1$, let $z \in \mathbf{Z}^2 \setminus \{0\}$, and let $J(z) = \{\tau \in S^1 : z \in \tau(K^+ \setminus K)$ is moderate$\}$. Then the measure of $J(z)$ is at most $O(\frac{\varepsilon}{R \cdot |z|})$, where* $\text{vol}(K) = 4 - \varepsilon$.

Proposition 4.1 is an immediate consequence of Lemmas 4.2 and 4.3. Indeed, we obtain

$$1 - P(K) \leq \sum_{z \in \mathbf{Z}^2 \setminus \{0\}, |z| \leq R} O\left(\frac{\varepsilon}{R \cdot |z|}\right) = O(\varepsilon)$$

as claimed by Proposition 4.1.

Proof of Lemma 4.3. We fix $z \in \mathbf{Z}^2 \setminus \{0\}$. First rotate K into the position where z lies on the segment $0f$, and fix this position as the starting one: $\tau = 0$. Then continue rotating in the positive (counterclockwise) direction until z first becomes moderate. This happens after a rotation by an angle τ_0; see Fig. 20. By symmetry, it suffices to bound the measure of the set $J^+(z) = \{\tau \in [0, \frac{\pi}{2}] : z$ is moderate for $\tau K\}$.

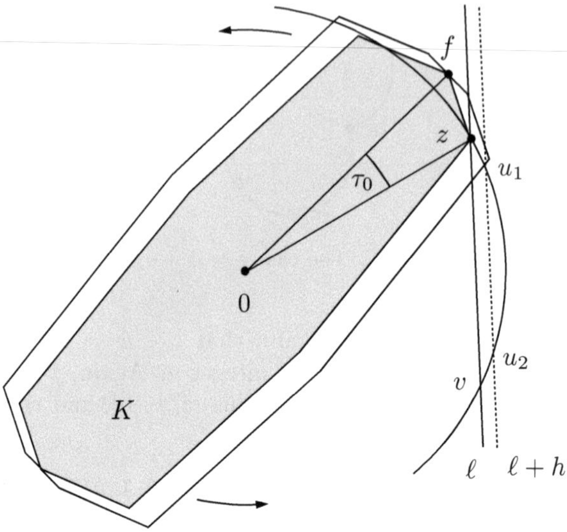

Fig. 20. The proof of Lemma 4.3.

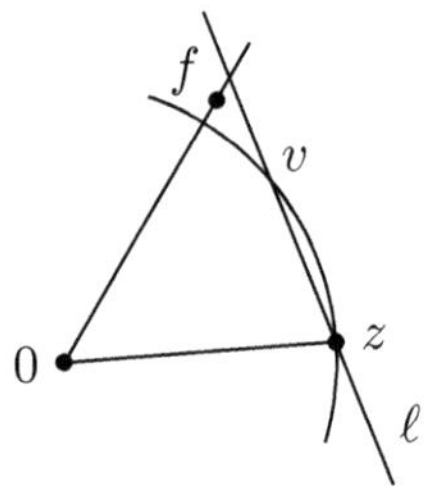

Fig. 21. The argument for (C1).

Let $\ell = \ell(z)$ be a line passing through z and witnessing that z is moderate for K. This ℓ intersects the circle $|z|S^1$ at points z and v. We will establish the following two conditions:

(C1) $|z - f| \leq |v - f|$.

(C2) The translated line $\ell + h$ intersects the circle $|z|S^1$.

We verify (C1). The angle $\angle z0f$ is smaller than $\frac{\pi}{2}$, so we may assume that $\angle v0f \leq \frac{\pi}{2}$ too. Since ℓ intersects the ray $0f$ beyond f, both z and v are on the same side of f. If v were closer to f than z, then $\angle vz0 < \frac{\pi}{2}$ (Fig. 21), and since we know that $\angle 0fz \leq \frac{\pi}{4}$, we would get $\angle z0f > \frac{\pi}{4}$. Then the triangle $0zf$ would have area $\frac{1}{2}|z| \cdot |f| \sin \angle z0f \geq \frac{1}{2} \cdot 3.1 \cdot \frac{1}{\sqrt{2}} > 1.09$. At the same time, this triangle is almost contained in K: by moving z by at most $|h| = \frac{\varepsilon}{R}$ we obtain a point $z' \in K$. The areas of the triangles $z0f$ and $z'0f$ differ by at most $\frac{1}{2} \cdot \frac{\varepsilon}{R} \cdot |f| < \frac{\varepsilon}{2}$, and so we would get $\text{vol}(K) > 4$.

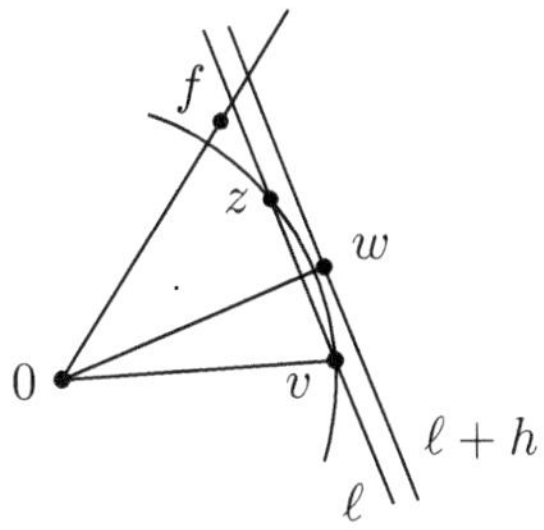

Fig. 22. Bounding the angle $w0z$.

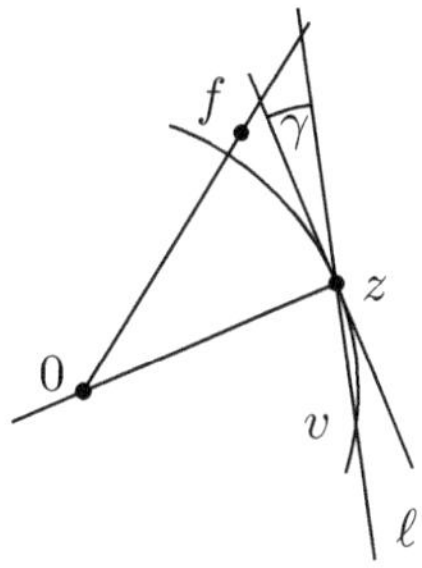

Fig. 23. Bounding the angle γ.

The proof of (C2) is similar. Assume that $\ell + h$ does not meet the circle $|z|S^1$ and let $w \in \ell + h$ be nearest to 0. We show that $\angle 0zv$ must be very close to $\frac{\pi}{2}$; if it is the case, then the previous argument applies, showing that the area of the triangle $0zf$ would be too large. By a simple calculation (see Fig. 22), we find that $\sin \angle w0z \leq \sqrt{\varepsilon}$, and it follows that $\angle 0zv \geq \frac{\pi}{2} - \sqrt{\varepsilon}$. This proves (C2).

Let u_1 and u_2 be the intersections of $\ell + h$ with the circle $|z|S^1$, where u_1 is closer to z and u_2 is closer to v. We now claim that all the rotations $\tau \in J^+(z)$ satisfy $\tau_0 \leq \tau \leq \tau_0 + \angle z0u_1$. Indeed, $\tau_0 \leq \tau$ follows from the definition of τ_0, and for $\tau \in (\tau_0 + \angle z0u_1, \tau_0 + \angle z0u_2)$, z is too far from τK to be in $\tau(K^+ \setminus K)$. Assume now that $\tau \in [\tau_0 + \angle z0u_2, \frac{\pi}{2})$. Rotate z clockwise by angle $\tau - \tau_0$ and denote the rotated point by w. If τ were in $J^+(z)$, i.e., if z were moderate for τK, then $w \in K^+ \setminus K$, and the area of the triangle $f0w$ is too large again since $\angle f0w \geq \angle f0u_2 \geq \frac{\pi}{4}$.

Let γ denote the angle made by ℓ with the circle $|z|S^1$ at the intersection point z. We want to show that γ is bounded away from 0; see Fig. 23. This is again similar to the proof of (C1). The angle $\angle z0f$ is at least $\frac{\pi}{4} - \gamma$, and as we have calculated before, if this angle is very near to $\frac{\pi}{4}$ then the triangle $z0f$ has too large area. So we may assume $\gamma > \gamma_0$ for a positive constant γ_0.

By the above, we can estimate the measure of $J^+(z)$ by the angle $z0u_1$, and here simple estimates show that this angle is bounded by $O(\frac{\varepsilon}{|z| \cdot R\gamma})$; cf. Fig. 24. This concludes the proof of Lemma 4.3. □

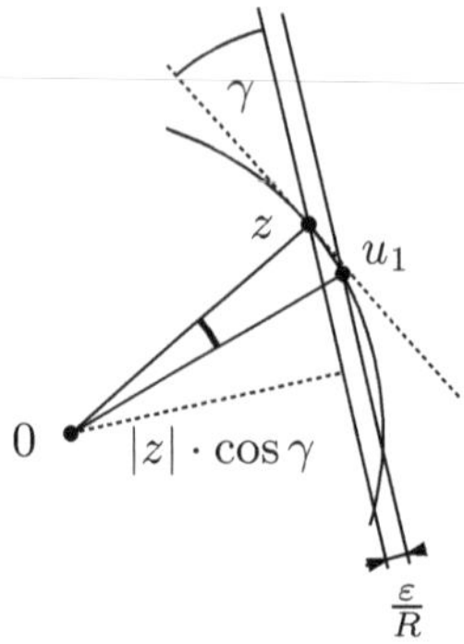

Fig. 24. Bounding the angle $z0u_1$.

Note that, in the proof above, the condition $R > 3.1$ is only used in the form $|z| \cdot |f| > 3.1$. This also holds when $|z| \geq \sqrt{2}$ and $|f| = R > 2.2$. We will need this observation later in the "thick" case:

Lemma 4.4 *Let K be such that $R \geq 2.2$ and $\mathrm{vol}(K) = 4-\varepsilon$. Assume $z \in \mathbf{Z}^2$ with $|z| \geq \sqrt{2}$. Then the measure of $J(z)$ is at most $O(\frac{\varepsilon}{R\cdot|z|})$.*

4.2 Fat bodies and tiles

Proposition 4.5 *For any $\Delta > 0$ there exist $c > 0$ and $\varepsilon_1 > 0$ such that if K is a body with $\mathrm{vol}(K) = 4 - \varepsilon$ and $R \leq 3.1$, where $\varepsilon \in (0, \varepsilon_1]$, such that all rotated copies τK have Hausdorff distance at least Δ from the square Q, then $P(K) \geq 1 - c\varepsilon$. Consequently, such K cannot be minimal if ε is sufficiently small in terms of Δ.*

For the proof, we adopt the following strategy. We note that if K has a lattice-point-free position and area close to 4, then $\frac{1}{2}K$ almost tiles the plane. We then discuss the possible shapes of such almost-tiles.

We define a *basic parallelogram tile* T_4 as is indicated in Fig. 25 on the left, where $0 \leq \alpha_1 \leq \alpha_2 \leq \frac{\pi}{4}$. The tiles of $T_4 + 2\mathbf{Z}^2$ have disjoint interiors, but they do not have to fill the plane completely: one of the omitted areas is indicated in the picture as a shaded parallelogram. Similarly, we define a *basic hexagonal tile* T_6 according to Fig. 25 right, with $0 \leq \beta_1, \beta_2 \leq \frac{\pi}{4}$ and $0 \leq \beta_0 \leq \frac{\pi}{2}$. A *parallelogram tile* is an image of a basic parallelogram tile under a lattice-preserving linear transform A, and similarly for a *hexagonal tile*. By a *tile* we mean a parallelogram tile or a hexagonal tile.

We recall that for a convex body $C \subset \mathbf{R}^2$, the *ε-floating body of C*, denoted by $C_{\varepsilon\text{-float}}$, is the intersection of all (closed) halfplanes h such that $\mathrm{vol}(C \cap h) \geq \mathrm{vol}(C) - \varepsilon$. We note that if C is a convex polygon with all angles bounded from below by some constant then all points of the boundary of C are at distance $O(\sqrt{\varepsilon})$ from $C_{\varepsilon\text{-float}}$.

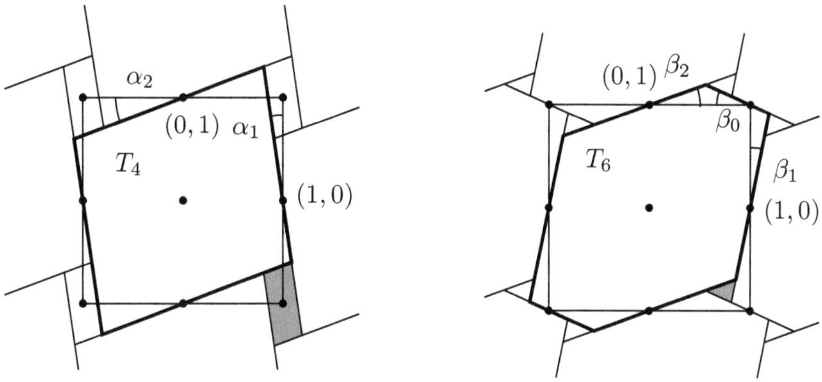

Fig. 25. A basic parallelogram tile and a basic hexagonal tile.

Lemma 4.6 *Let K be a body with* $\mathrm{vol}(K) = 4 - \varepsilon$ *and* $K \cap \mathbf{Z}^2 = \{0\}$*, where* $\varepsilon > 0$ *is sufficiently small. Then there is a tile T with* $T_{\varepsilon\text{-float}} \subseteq K \subseteq T$.

Proof. This is similar to the usual proof of Minkowski's theorem (see e.g. [Gru95] or [PA95]). Let $K_0 = \frac{1}{2}K$ and consider the system of bodies $K_0 + \mathbf{Z}^2$; since $K \cap \mathbf{Z}^2 = \{0\}$ these bodies are pairwise disjoint. We consider the smallest $t_1 \geq 1$ such that the boundaries of the bodies in $t_1 K_0 + \mathbf{Z}^2$ first touch; say that $t_1 K_0$ touches $t_1 K_0 + z_1$. Let $K_1 = t_1 K_0$. By the symmetry and convexity of K_1, $\frac{1}{2}z_1 \in \partial K_1$. Let ℓ_1 be a line through $\frac{1}{2}z_1$ separating K_1 and $K_1 + z_1$; then K_1 is contained in the strip between ℓ_1 and $-\ell_1$. Next, enlarge K_1 to $K_1(t_2) = K_1 + [-t_2 h, t_2 h]$, where h is a unit vector parallel to ℓ_1 and $t_2 \geq 0$ is the smallest such that $K_1(t_2)$ touches some $K_1(t_2) + z_2$, $z_1 \neq z_2$. If several choices of z_2 are possible we take the one such that $\frac{1}{2}z_2$ is encountered first along the boundary of the body when going from $\frac{1}{2}z_1$. Let $K_2 = K_1(t_2)$. Again $\frac{1}{2}z_2 \in \partial K_2$, and we let ℓ_2 be a line through $\frac{1}{2}z_2$ separating K_2 from $K_2 + z_2$. The body K_2 is now contained in the parallelogram P delimited by the lines $\pm\ell_1$ and $\pm\ell_2$, as in Fig. 26. If the parallelograms in $P + \mathbf{Z}^2$ do not overlap then $T = 2P$ is the desired (parallelogram) tile, as we will check later. If they do overlap then we enlarge K_2 further; in this third stage, we use the parameterization $K_2(t_3) = (1 - t_3)K_2 + t_3 P$, $t_3 \in [0, 1]$. We consider the smallest t_3 such that $K_3 = K_2(t_3)$ touches some $K_2(t_3) + z_3$, $z_3 \neq z_1, z_2$. We separate K_3 and $K_3 + z_3$ by a line ℓ_3. The lines $\pm\ell_3$ cut off P into a hexagon H, and we claim that $T = 2H$ is our tile. Indeed, if the hexagons in $H + \mathbf{Z}^2$ overlapped, then we could still enlarge K_3 so that it touches some $K_3 + z_4$, but this is impossible by a planarity argument (connect the centers of the bodies by edges through the contact points).

Since the triangle $0z_1z_2$ contains no lattice points distinct from the vertices, it has area $\frac{1}{2}$ and there is a lattice-preserving linear transform A_1 sending z_1 to e_1 and z_2 to e_2. If the parallelogram $2A_1(P)$ contains no integer point distinct from 0 in its interior then it is a basic parallelogram tile ac-

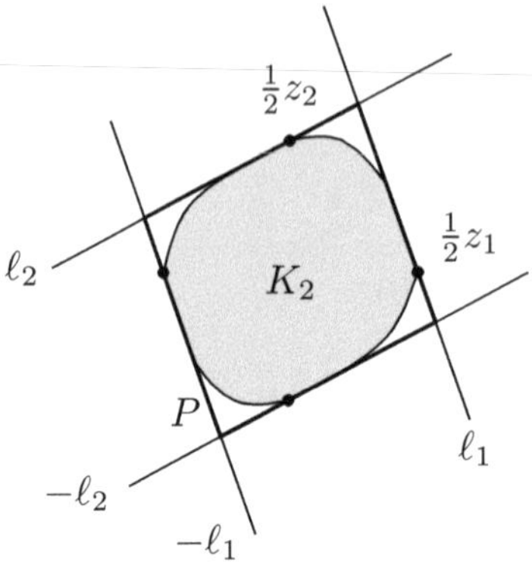

Fig. 26. Proof of Lemma 4.6.

cording to our definition. If it does contain such an integer point then it has to contain $(1,1)$ or $(-1,1)$; say that it contains $(1,1)$. If this point is cut off by a halfplane avoiding e_1 and e_2 (and $(-1,-1)$ is cut off by the symmetric halfplane) then it is easy to check that no other integer point can remain. Thus, in this case, $T = 2H$ is a hexagonal tile.

The claim $T_{\varepsilon\text{-float}} \subseteq K$ follows from $\text{vol}(K) = 4 - \varepsilon$ and $\text{vol}(T) \leq 4$. $\square$

More properties of the tiles. If T had a very sharp angle then its ε-floating body would have a very large circumradius. Since we assume that K has circumradius at most 3.1, we see that T does not have any very sharp angles and ∂T is nowhere farther than $O(\sqrt{\varepsilon})$ from $T_{\varepsilon\text{-float}}$ and thus also from K. Moreover, if a point $x \in \partial T$ is at distance at least $\delta > 0$ (independent of ε) away from the vertices of T then the distance of x to K is at most $C_\delta \varepsilon$, where C_δ is a suitable function of δ, again by simple properties of the floating body. These properties can be checked by hand (as the tiles are very simple convex polygons), or found in [Lei86].

Let A be the lattice-preserving linear map transforming a basic tile into the tile T; then Ae_1 and Ae_2 are (primitive) integer points on ∂T. The first few possible lengths of primitive integer vectors are 1, $\sqrt{2}$, $\sqrt{5}$, $\sqrt{10}$,.... We will discuss the possible values of Ae_1 and of Ae_2 and the circumradii of K in the various cases.

Since each tile has circumradius at least $\sqrt{2}$, we always have $R \geq \sqrt{2} - O(\sqrt{\varepsilon})$. Further, if $|Ae_1| > \sqrt{2}$ or $|Ae_2| > \sqrt{2}$ then $R \geq \sqrt{5} - O(\sqrt{\varepsilon})$, and if $|Ae_1| > \sqrt{5}$ or $|Ae_2| > \sqrt{5}$ then $R \geq \sqrt{10} - O(\sqrt{\varepsilon})$. The latter case is excluded by the assumption $R < 3.1$.

Thus, we know that $|Ae_1|, |Ae_2| \leq \sqrt{5}$. Since the triangle $0(Ae_1)(Ae_2)$ has area $\frac{1}{2}$, we find that necessarily $|Ae_1| = 1$ or $|Ae_2| = 1$. By symmetry, we may thus assume $Ae_2 = e_2$. Let us first consider the case where T is a parallelogram tile. By discussing the possible values of Ae_1, we arrive at the cases (a)–(e) in Fig. 27 for the parallelogram tiles.

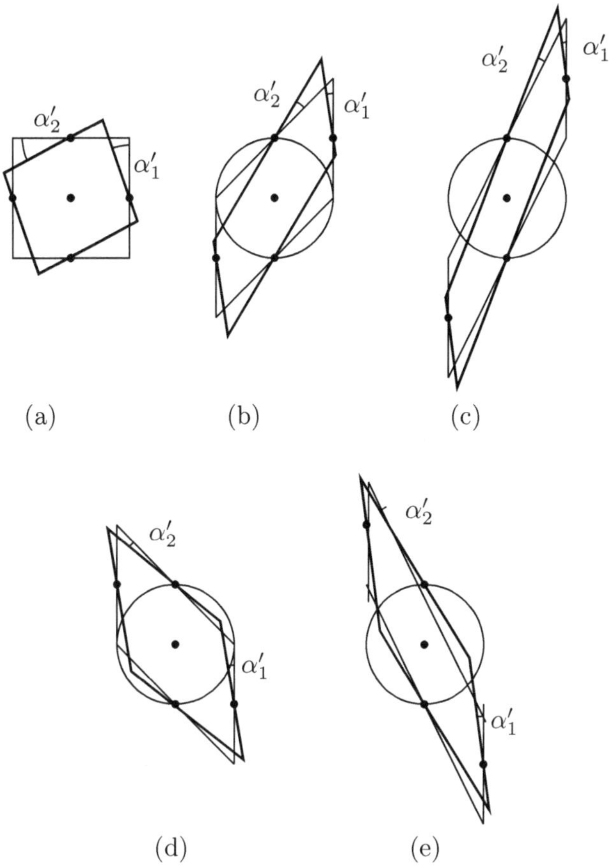

Fig. 27. The possible cases of parallelogram tiles.

For these parallelogram tiles, we have $\min(\alpha_1', \alpha_2') = O(\sqrt{\varepsilon})$. This is because the area left out in the tiling by T (shaded in Fig. 25) is at most ε, as $\mathrm{vol}(T) \geq \mathrm{vol}(K) = 4 - \varepsilon$.

A similar analysis yields the two possibilities (f) and (g) for hexagonal tiles shown in Fig. 28.

A case analysis. In the case (c), the tile has a too large radius.

In the cases (b), (e), and (g), we have $R \geq \sqrt{5} - O(\sqrt{\varepsilon})$ and so Lemma 4.2 applies, showing that a moderate lattice point can be made responsible for each lattice-point-free position of K, and further by Lemma 4.4, the contribution of the lattice points different from $\pm e_1, \pm e_2$ is only $O(\varepsilon)$. Moreover, troubles with $\pm e_1$ or $\pm e_2$ can occur only if the circle S^1 intersects ∂T at very small angle γ or very close to a vertex. (Here and in the sequel, by "very small" we mean "below a constant that can be made arbitrarily small by choosing ε_1 and Δ sufficiently small", but the constant must not depend on ε.) In all the cases (b), (e), (g) we thus need not worry about the inter-

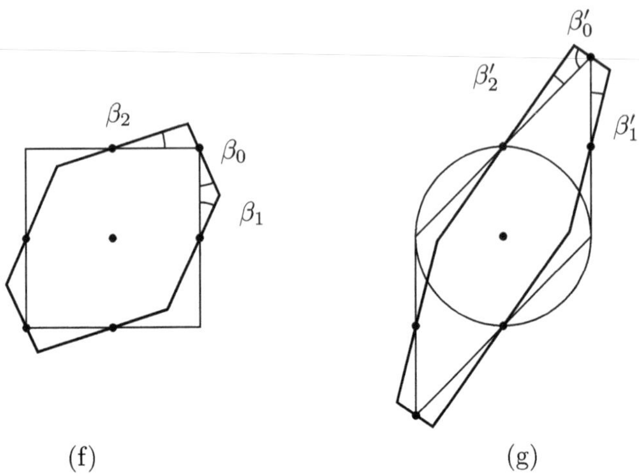

Fig. 28. The possible hexagonal tiles.

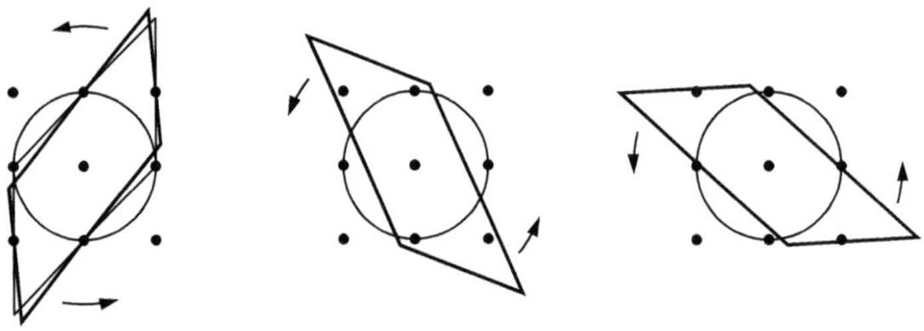

Fig. 29. Rotating a tile close to T_0.

sections of ∂T with S^1 at $\pm e_2$. The other intersections of ∂T with S^1 are fine unless the following conditions hold:

- In the case (b), both the angles α_1' and α_2' are very small and so T is close to $T_0 = \operatorname{conv}\{\pm e_1, \pm(1,2)\}$.
- In the case (e), the two vertices of T closer to the origin are very close to S^1. Since $\alpha_1' = O(\sqrt{\varepsilon})$ or $\alpha_2' = O(\sqrt{\varepsilon})$, this is possible only if T is very close to a reflected copy of T_0.
- In the case (g), the two vertices of T closest to the origin are very close to S^1, and then both β_1' and β_2' are very small and T is again close to a reflected copy of T_0.

Consider the case (b) when T is very close to T_0. Suppose that T (drawn thick) is positioned as in Fig. 29 on the left and starts rotating, together with K, counterclockwise. After angle interval $O(\varepsilon)$, e_2 enters K, and before it leaves K, $(-1,1)$ is already in (middle picture). After $(-1,1)$ leaves, there is an interval of length only $O(\varepsilon)$ after which e_1 enters. So $1 - P(K) = O(\varepsilon)$.

Similar arguments work for (e) and (g) and they are omitted.

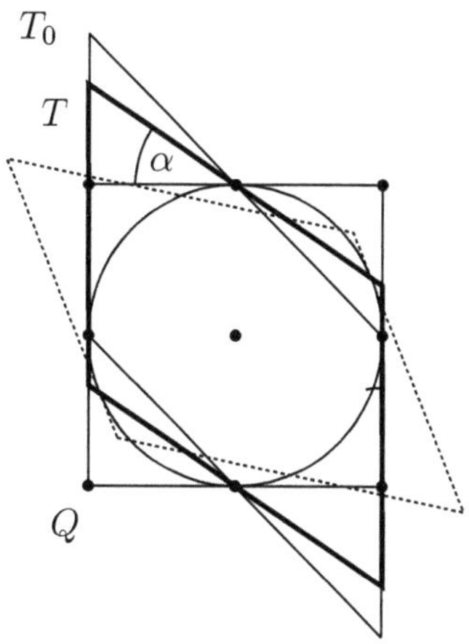

Fig. 30. The cases (a) and (d1).

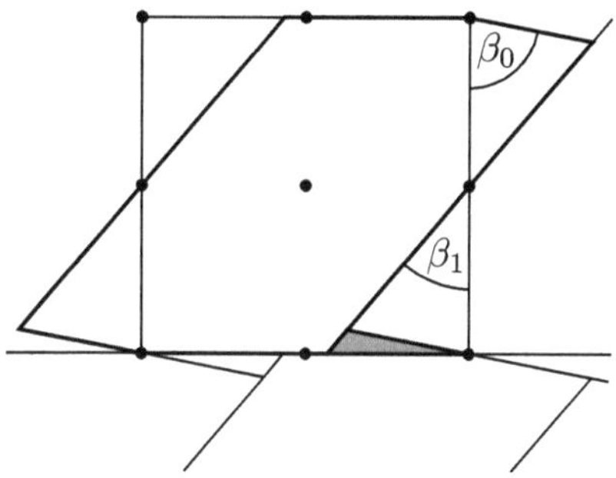

Fig. 31. The case (f).

We divide the case (d) into two subcases: (d1) when α'_1 is $O(\sqrt{\varepsilon})$, and (d2) when $\alpha'_2 = O(\sqrt{\varepsilon})$ (note that these cases are not symmetric).

In the cases (d1), and (a), we know that two sides of T are nearly vertical and so T is very close to a parallelogram "interpolating" between T_0 and the square Q; see Fig. 30. If T starts rotating from the position drawn thick in either direction, then e_2 for clockwise direction and $(-1, 1)$ for counterclockwise direction enter K after angular interval $O(\varepsilon)$ *unless* the angle α is very small, and in that case T is very close to Q. For counterclockwise rotation, when $(-1, 1)$ is going to leave K, as in the dashed position in the drawing, $-e_1$ has already entered. We see that $1 - P(K) = O(\varepsilon)$ unless T is very close to Q. A similar analysis shows that $1 - P(K) = O(\varepsilon)$ always in the case (d2).

Finally, in the case (f), we clearly have $P(K) = O(\varepsilon)$ if neither β_1 nor β_2 are very small. Thus, suppose that β_2 is very small as in Fig. 31. Now since the area of the left-out part (shaded) must be at most 2ε, either β_1 is very small, in which case T is almost the square Q, or β_0 is very close to $\frac{\pi}{2}$, and then we are in the situation already discussed for the cases (a) and (d1). This concludes the proof of Proposition 4.5. □

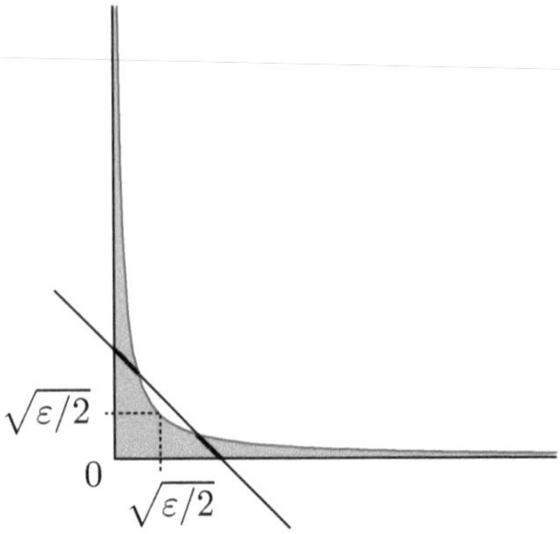

Fig. 32. The floating body of a quadrant.

4.3 Bodies very close to the square

We want to prove that $1 - P(4-\varepsilon) = O(\sqrt{\varepsilon})$ for all $\varepsilon \in [0, \varepsilon_0]$. Let K be a body at Hausdorff distance at most Δ from the square Q. By Lemma 4.6, there is a tile T with $T_{\varepsilon\text{-float}} \subseteq K \subseteq T$. Further let T^+ denote the tile T enlarged by a factor of $1 + \frac{\varepsilon}{4}$, so that $\mathrm{vol}(T^+) > 4$. Then we have $P(T^+) = 1$ by Minkowski's theorem and we can bound

$$\begin{aligned} 1 - P(K) &= (1 - P(T^+)) + (P(T^+) - P(K)) \le P(T^+) - P(T_{\varepsilon\text{-float}}) \\ &\le \mu(S^1 \cap (T^+ \setminus T_{\varepsilon\text{-float}})) + \mu(\sqrt{2}\, S^1 \cap (T^+ \setminus T_{\varepsilon\text{-float}})), \end{aligned}$$

where μ is the uniform probability measure on S^1.

The types of tiles coming into consideration as T are (a), (d), and (f). The treatment of the type (f) will be slightly different.

For all these types, the intersection of $T^+ \setminus T_{\varepsilon\text{-float}}$ with S^1 is far from the corners of T^+. Near each of these intersections, the set $T^+ \setminus T_{\varepsilon\text{-float}}$ can be covered by a strip of width 2ε, say, and so $\mu(S^1 \cap (T^+ \setminus T_{\varepsilon\text{-float}}))$ is at most four times the length of the intersection of S^1 with a parallel strip of width 2ε, which is $O(\sqrt{\varepsilon})$.

We note that if T is a hexagonal tile of the type (f), then a quarter-circle of S^1 is completely contained in T. This quarter-circle alone guarantees $P(T) = 1$, and by passing from T to $T_{\varepsilon\text{-float}}$ we lose a portion of at most $O(\sqrt{\varepsilon})$ from this quarter-circle. Therefore, $1 - P(K) = O(\sqrt{\varepsilon})$ follows without considering the second circle $\sqrt{2}\, S^1$ at all.

For the tile types (a) and (d), it remains to bound the length of the arcs of the second circle $\sqrt{2}\, S^1$ inside $T^+ \setminus T_{\varepsilon\text{-float}}$. We note that since T is a parallelogram tile close to the square, it has a corner with angle close to $\frac{\pi}{2}$ near $(-1, -1)$, and the second circle intersects such corner (if at all) at an angle close to $\frac{\pi}{4}$. Also, the ε-floating body of T near the second circle is almost like the ε-floating body of a full quadrant (see Fig. 32), which is bounded by the hyperbola $xy = \frac{\varepsilon}{2}$. Then $\mu(\sqrt{2}\, S^1 \cap (T^+ \setminus T_{\varepsilon\text{-float}}))$ is at most twice (say) the length of the two thick segments in Fig. 32. The latter is easily seen to be always less than $2\sqrt{\varepsilon}$. □

References

[Bon29] T. Bonnesen. *Problémes des isopérimétries et des isépihanes.* Gauthier-Villars, Paris, 1929.

[Gro95] H. Groemer. Stability of geometric inequalities. In *Handbook of Convex Geometry*, North-Holland 1995, pages 125–150.

[Gru95] P. M. Gruber. Geometry of numbers. In *Handbook of Convex Geometry*, North-Holland 1995, pages 739–764.

[Lei86] K. Leichtweiss. Zur Affinoberfläche konvexer Körper. *Manuscripta Math.*, 56:429–464, 1986.

[Min96] H. Minkowski. *Geometrie der Zahlen.* Teubner, Leipzig, 1896.

[PA95] J. Pach, P. Agarwal. *Combinatorial geometry.* Wiley, New York, 1995.

About Authors

Imre Bárány is affiliated with Rényi Institute of Mathematics, Hungarian Academy of Sciences, POBox 127, 1364 Budapest, Hungary, and with Department of Mathematics, University College London, Gower Street, London WC1E 6BT, England.

Jiří Matoušek is affiliated with Department of Applied Mathematics of the Charles University and with Institute for Theoretical Computer Science (ITI), Malostranské nám. 25, 118 00 Praha 1, Czech Republic.

Acknowledgments

Work by Imre Bárány was supported by Hungarian National Foundation Grants T 029255, T 016391, and T 020914. Work by Jiří Matoušek was supported by Charles University grants No. 158/99 and 159/99.

Complex Matroids
Phirotopes and Their Realizations in Rank 2

Alexander Below
Vanessa Krummeck
Jürgen Richter-Gebert

Abstract

The motivation for this article comes from the desire to link two seemingly incompatible worlds: oriented matroids and dynamic geometry. Oriented matroids have proven to be a perfect tool for dealing with sidedness information in geometric configurations (for instance for the computation of convex hulls). Dynamic geometry deals with elementary geometric constructions in which moving certain free elements controls the motion of constructively dependent elements. In this field the introduction of complex coordinates has turned out to be a "key technology" for achieving a consistent continuous movement of the dependent elements. The additional freedom of an ambient complex space makes it possible to bypass disturbing singularities. Unfortunatly, complex coordinates seem to make it impossible to use oriented matroids which are heavily based on real numbers.

In this paper we introduce a generalization of oriented matroid chirotopes for complex point configurations, the *phirotopes* (short for *phase chirotopes*). Unlike chirotopes, already in rank 2 these phirotopes exhibit interesting behavior full of geometric meaning. Realizability questions in rank 2 lead to a necessary and sufficent algebraic condition. As a first application, we show that a certain class of incidence theorems (Miquel's theorem among others) already hold on the phirotope level. Finally, somewhat surprisingly, we find that every phirotope of rank 2 naturally encodes a chirotope of rank 4.

Chirotopes and oriented matroids (see [1,9]) have been a very important topic of diverse investigations during the last 25 years. They can be considered as a combinatorial abstraction and generalization of the behavior of the determinants of the $d \times d$ submatrices of a real $d \times n$ matrix. Chirotopes are functions $\chi\colon \{1,\ldots,n\}^d \to \{-1,0,+1\}$ that associate to each d-tuple of indices the sign of an (abstract) determinant function. Hence they are purely combinatorial objects. However, not all such sign functions are chirotopes. They have to satisfy two combinatorial axioms: First they should be *alternating* and second they are not allowed to obviously violate the Grassmann-Plücker relations (which are certain fundamental polynomial identities shared by the values of the $d \times d$ subdeterminants of any $d \times n$ matrix). Hence chirotopes are sign functions that have a reasonable chance to come from the subdeterminants of a real $d \times n$ matrix. In fact for $d = 2$ all chirotopes are *realizable* in this sense and for $d = 3$ non-realizable chirotopes exist only for $n \geq 9$.

Chirotopes are very geometric objects. If we consider the columns of a matrix as vectors $X_1, \dots, X_n$ in $\mathbb{R}^d$ the corresponding chirotope encodes a lot of information on the relative position of these vectors. However, the chirotope of the matrix is insensitive to multiplying a vector with a positive scalar. Thus the chirotope describes the relative position information of the rays $R_i := \{rX_i | r \in \mathbb{R}^+\}$. We can also compactify the situation by intersecting the rays with an affine hyperplane H of $\mathbb{R}^d$ that does not pass through the origin (w.l.o.g. we may assume that this hyperplane is not parallel to any of the rays). If a ray R_i does not intersect H then $-R_i$ does. Thus we can consider our vector configuration as (signed) homogeneous coordinates of an oriented affine point configuration in H. Each point of the configuration in H is either equipped with a positive or a negative sign depending on whether R_i or $-R_i$ intersects H. Many geometric properties of the affine configuration can be read off from the chirotope: convex hulls, Radon partitions, hyperplane separations, etc. Another way to think of a realizable chirotope comes from viewing the vectors $X_1, \dots, X_n$ in $\mathbb{R}^d$ as normals for (oriented) linear hyperplanes. These hyperplanes separate $\mathbb{R}^d$ into a cell complex of polyhedral cones. The chirotope encodes exactly the combinatorial structure of this (signed) cell complex. Obviously, this information is insensitive to multiplication of the X_i by positive scalars.

Reorientation of a point, i.e. replacing X_i by $-X_i$, has a very controllable effect on the corresponding chirotope. We simply have to reverse all signs $\chi(\lambda_1, \dots, \lambda_d)$ whenever $i \in \{\lambda_1, \dots, \lambda_d\}$. If we consider the minimal equivalence classes of chirotopes that are stable under reorientation (i.e. the set $[\chi]$ of all chirotopes that are obtained from χ by a sequence of reorientations) we get the *reorientation classes.* They are invariant under projective transformations of the affine point configuration.

The aim of this paper is to transfer this setup from real to complex vector spaces. Since determinants of complex-valued matrices are complex numbers, the notion of *sign* has to be suitably modified. Instead of signs we consider phases: The *phase* of a non-zero complex number z is $\omega(z) := z/|z|$. Furthermore, we set $\omega(0) = 0$. Thus the image space of the complex phase-function is $S^1 \cup \{0\}$, the complex unit circle and zero. Now we introduce a new concept called *phirotopes.* The word "phirotope" is short for **ph**ase-c**hirotope**. A phirotope is an alternating function $\varphi\colon \{1, \dots, n\}^d \to S^1 \cup \{0\}$ that does not obviously violate the Grassmann-Plücker relations (this will be made precise in Section 1).

In the terminology of Dress and Wenzel [3–5] phirotopes are cryptomorphic to matroids with coefficients on the fuzzy ring $\mathbb{C}//\mathbb{R}^+$. Although for matroids with coefficients there is a general theory that supplies basic facts about Grassmann-Plücker maps, duality and Tutte groups, phirotopes deserve an extensive investigation on their own right. They are the natural *geometric* complexification of chirotopes. Here "geometric" means that a phirotope φ of a complex vector configuration $X = (X_1, \ldots, X_n)$ admits a canonically defined reorientation class $[\varphi]$ that is a projective invariant of the complex point configuration with complex homogeneous coordinates X. Thus in the realizable case a phirotope encapsulates information on the relative positions of points in the complex projective space $\mathbb{CP}^{d-1}$. Since the geometry of $\mathbb{CP}^{d-1}$ is very closely related to the geometry of configurations of circles, phirotopes are closely related to abstract configurations of circles. This effect already arises in the rank-2 case. This article focuses on these rank-2 effects. As one of the main results we will show that most rank-2 phirotopes are non-realizable and we will derive a non-obvious syzygy on five elements that ensures the realizability of a rank-2 configuration (Section 3). This sharply contrasts the chirotope case: Rank-2 chirotopes are *always* realizable. From this syzygy one can derive an interesting theorem in elementary geometry on angles between five points in the plane that was not known before to the best of our knowledge (Section 4). Furthermore realizable phirotopes admit an interestingly rigid behavior: If not all points are on a circle, then a phirotope determines the corresponding point configuration up to Moebius transformations (Section 3).

We will also study the counterpart of *oriented points* in the complex affine setup. Here the orientation of a point is not either "+" or "−"; it can be an arbitrary phase on S^1. We will briefly outline the relation of phirotopes to a structure that one could call "oriented complex projective geometry". If we consider the space of all complex oriented points on $\mathbb{CP}^d$ each point of $\mathbb{CP}^d$ is associated to a S^1 fiber. The entire space of oriented points is topologically a S^{2d-1}. This is exactly the Hopf fibration (Section 2).

We want to close this introduction with two remarks. The first remark is on alternative ways of complexifying chirotopes. In their papers [2] and [11] Björner and Ziegler propose a sign function with image space $\{0, +, -, i, j\}$. On $\mathbb{R}$ this function behaves like the usual sign function. The image of the upper complex half-plane is i, the image of the lower complex half-plane is j. Modeling their sign function to operate consistently on determinants they arrive at a combinatorial characterization of what they call "complex matroids". This theory is purely combinatorial since the image space is finite (in contrast to our theory: there are already uncountably many rank-2 phirotopes). Many concepts of oriented matroid theory have been transferred by Ziegler and Björner to the setup of complex matroids. From our viewpoint however this theory has a big disadvantage for possible applications we have in mind (see below): it does not admit a nice reorientation theory. In particular, it is not possible to model the effect of a reorientation of a point on the level of complex matroids. Therefore this structure of complex matroids is not a projective theory in the sense that it carries information on a specific homogenization of a projective configuration that cannot be factored out on the combinatorial level.

The second remark is about the possible applications that led us to the study of this specific structure of phirotopes. In the area of *dynamic geometry* one studies real configurations of elementary geometry (say a ruler and compass construction). But instead of viewing this configuration as a static picture one is interested to move the base points of a construction and watch the effect on dependent elements. One of the fundamental problems in this area is to model and control the behavior of the dependent elements in a reasonable and preferably continuous manner. A big breakthrough in the field of dynamic geometry was the insight that one can get a nicely closed theory if one embeds the real setup into a complexified ambient space. By this singularities that were unavoidable in the real setup could be easily resolved [6–8]. However, by going to the complexified setup one looses control on the orientations of points and lines in the real setup and it becomes difficult to consistently talk about objects like rays, segments, circular arcs, etc. One of the intended aims of the research connected to phirotopes is to find a nice synthesis of complexified dynamic geometry (that can very elegantly deal with configurations and dependencies under motion) and oriented matroid theory (that can very nicely deal with everything concerning orientations). Whether such a synthesis is possible and how it would finally look like is still an open and challenging question.

1 Phirotopes

In this section we introduce the idea of complex matroids in terms of *phirotopes*. We start with phirotopes of complex vector configurations. Their definition is closely related to the definition of chirotopes on real vector configurations. Similarly to chirotopes we establish a set of axioms based on the Grassmann-Plücker relations in such a way that chirotopes turn out to be a special case of phirotopes. We define the general concept of realizability of phirotopes and consider reorientations.

Let us start by recalling the definition of chirotopes of real vector configurations: Let $V = (V_1, \ldots, V_n)$ be a set of n real vectors in $\mathbb{R}^d$ that linearly span $\mathbb{R}^d$ and let $E = \{1, \ldots, n\}$ be its finite index set. Then the rank-d chirotope χ of this vector configuration is defined on all d-tuples of E by the function

$$\begin{array}{rlcl} \chi: & E^d & \longrightarrow & \{-1, 0, +1\} \\ & (\lambda_1, \ldots, \lambda_d) & \longmapsto & \operatorname{sign}(\det(V_{\lambda_1}, \ldots, V_{\lambda_d})). \end{array}$$

Chirotopes of real vector configurations evaluate the signs of certain real determinants. If we just replace the real vectors by complex vectors we get a complex determinant. Since the notion of sign does not make sense for complex numbers it is generalized by the *phase*:

Definition 1.1 (PHASE ω OF A COMPLEX NUMBER)
For any complex number $z = re^{i\alpha} \in \mathbb{C}$ with $r \in \mathbb{R}_0^+$, $\alpha \in \mathbb{R}$, we define its

phase $\omega(z) \in S^1 \cup \{0\}$ *by*

$$\omega(z) = \begin{cases} 0 & \textit{if } r = 0 \\ e^{i\alpha} & \textit{if } r \neq 0. \end{cases}$$

We replace the *sign* of the real determinant in the definition of chirotopes of real vector configurations by the *phase* of a complex determinant in order to define phirotopes of complex vector configurations:

Definition 1.2 (PHIROTOPE φ OF A COMPLEX VECTOR CONFIGURATION) *Let $Z = (Z_1, \ldots, Z_n)$ be a set of n complex vectors in $\mathbb{C}^d$ that linearly span $\mathbb{C}^d$ and let $E = \{1, \ldots, n\}$ be its finite index set. Then the* rank-d phirotope *φ_Z of this vector configuration is defined on all d-tuples of E by the function*

$$\begin{array}{rlcl} \varphi_Z : & E^d & \longrightarrow & S^1 \cup \{0\} \\ & (\lambda_1, \ldots, \lambda_d) & \longmapsto & \omega(\det(Z_{\lambda_1}, \ldots, Z_{\lambda_d})). \end{array}$$

The image space of a phirotope is no longer a discrete set. It is the union of the unit circle S^1 and $\{0\}$. In particular, it is uncountable. This is a decisive point where our approach differs from the complex matroids introduced by Günter Ziegler [2]. His system is still discrete operating on the plus and minus signs of the real and imaginary parts of complex numbers. Our construction can be found in a more general setup in the papers of Dress and Wenzel [4, 5]. According to their definition we look at matroids with coefficients in the fuzzy ring $\mathbb{C}//\mathbb{R}^+$.

We follow the leitmotif of chirotopes by giving an "abstract" axiomatization of phirotopes. It should meet two demands: First, phirotopes of complex vector configurations should be phirotopes. Second, phirotopes with a real image (i.e. an image in $\{-1, 0, +1\} \subset \{S^1 \cup \{0\}\}$) should be chirotopes.

A chirotope is an alternating function that does not obviously contradict the Grassmann-Plücker relations. The Grassmann-Plücker relations state the following: Given a (real or complex) vector configuration $X = (X_1, \ldots, X_n)$ of n vectors in rank d with index set $E = \{1, \ldots, n\}$, for any two subsets $\tau = \{\tau_1, \ldots, \tau_{d-1}\}$ and $\mu = \{\mu_1, \ldots, \mu_{d+1}\}$ of E with $d-1$ elements and $d+1$ elements, respectively, the following equation always holds:

$$\sum_{k=1}^{d+1} (-1)^k \det(X_{\tau_1}, \ldots, X_{\tau_{d-1}}, X_{\mu_k}) \cdot \det(X_{\mu_1}, \ldots, \widehat{X_{\mu_k}}, \ldots, X_{\mu_{d+1}}) = 0$$

where $\widehat{X_{\mu_k}}$ invokes the usual notion for omitting this vector. On the axiomatic level chirotopes are not determinants themselves. However, they model the behavior of determinants. The following axiom ensures that the sign maps cannot contradict the Grassmann-Plücker relations locally: For any rank-d chirotope χ and

$$s_k = (-1)^k \chi(\tau_1, \ldots, \tau_{d-1}, \mu_k) \cdot \chi(\mu_1, \ldots, \widehat{\mu_k}, \ldots, \mu_{d+1})$$

for $k = 1, \ldots, d+1$ there have to exist $r_1, \ldots, r_{d+1} \in \mathbb{R}^+$ such that $\sum_{k=1}^{d+1} r_k s_k = 0$.

The $r_k s_k$ play the role of the products of the determinants in the Grassmann-Plücker relations. In order to sum up to zero their signs (expressed by the s_k) either have to include at least one "+" *and* one "−" or they all have to be zero.

In the context of phirotopes we have to consider the Grassmann-Plücker relations for complex vector configurations. We also have to require that any phirotope φ is an alternating function, i.e. interchanging two entries flips the sign of the function.

Definition 1.3 (Phirotope (General Definition))
Let $E = \{1, \ldots, n\}$ be a finite index set. A function $\varphi : E^d \longrightarrow S^1 \cup \{0\}$ on all d-tuples of E is called a rank-d *phirotope if*

(a) the function φ is alternating, and

(b) for any two subsets $\tau = \{\tau_1, \ldots, \tau_{d-1}\}$ and $\mu = \{\mu_1, \ldots, \mu_{d+1}\}$ of E and

$$\omega_k = (-1)^k \varphi(\tau_1, \ldots, \tau_{d-1}, \mu_k) \cdot \varphi(\mu_1, \ldots, \widehat{\mu_k}, \ldots, \mu_{d+1})$$

for $k = 1, \ldots, d+1$ there exist $r_1, \ldots, r_{d+1} \in \mathbb{R}^+$ such that $\sum_{k=1}^{d+1} r_k \omega_k = 0$.

We call φ uniform *if its image is contained in S^1.*

Note that by mapping the phases of a phirotope to their absolute values "1" or "0" we get the basis function of a matroid (as for chirotopes). In the sequel however we will only consider uniform phirotopes.

Remark: (Rank-2 Phirotope)
In case of a rank-2 phirotope on at least four points condition (b) reads as follows: For any four indices $\lambda_1, \lambda_2, \lambda_3, \lambda_4 \in E$ and

$$\begin{aligned} \omega_1 &= + \ \varphi(\lambda_1, \lambda_2) \cdot \varphi(\lambda_3, \lambda_4) \\ \omega_2 &= - \ \varphi(\lambda_1, \lambda_3) \cdot \varphi(\lambda_2, \lambda_4) \\ \omega_3 &= + \ \varphi(\lambda_1, \lambda_4) \cdot \varphi(\lambda_2, \lambda_3), \end{aligned}$$

there exist $r_1, r_2, r_3 \in \mathbb{R}^+$ such that $\sum_{k=1}^{3} r_k \omega_k = 0$.

Let us have a look at the geometric meaning of this particular condition. The three complex numbers $r_k \omega_k$ have to sum up to zero. If we picture them as vectors in $\mathbb{R}^2$ there are possible and impossible configurations:

(a) Possible configurations:
(a.1) all vectors are zero; (a.2) one vector is zero, the other two

vectors point towards opposite directions; (a.3) two vectors point towards the same, the third vector towards the opposite direction; (a.4) zero lies in the interior of the convex hull of the three vectors

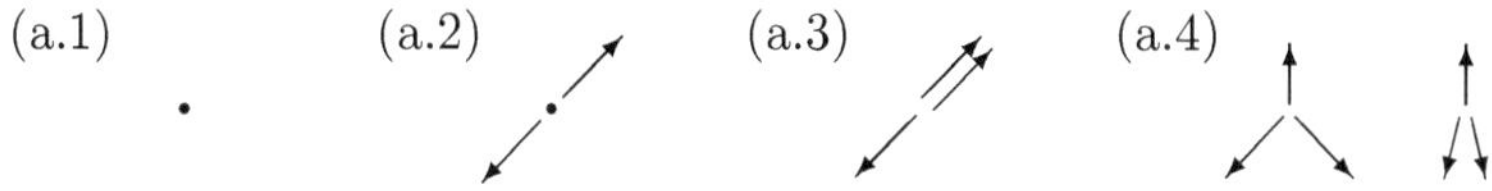

(b) Impossible configurations:
(b.1) one vector is zero, the other two vectors do not point towards opposite directions; (b.2) all three vectors point towards the same direction; (b.3) the convex hull C of the three vectors is 2-dimensional, and zero does not lie in the interior of C, i.e. either zero lies on the boundary of the convex hull or on the outside

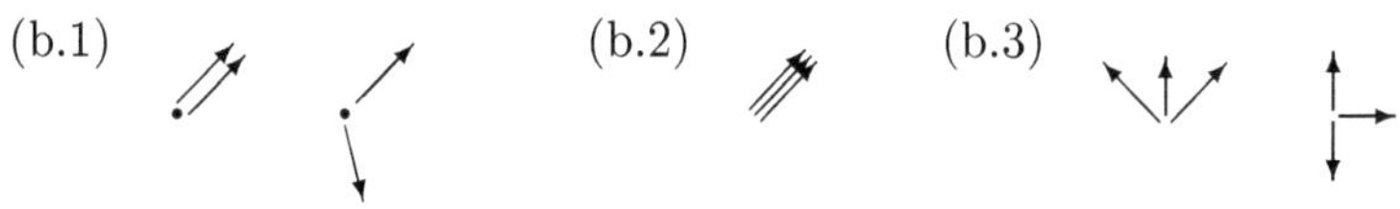

It is straightforward to prove that phirotopes of complex vector configurations are general phirotopes. Also real-valued phirotopes are clearly chirotopes.

Similarly to chirotopes we now define realizability of a general phirotope:

Definition 1.4 (Realizability of a Phirotope)
A rank-d phirotope $\varphi : E^d \longrightarrow S^1 \cup \{0\}$ with index set $E = \{1, \ldots, n\}$ is called realizable *if there exists a vector configuration $Z = (Z_1, \ldots, Z_n) \in \mathbb{C}^{d \times n}$ with $\varphi_Z = \varphi$.*

The notion of reorientation is very important in the theory of chirotopes: Given a sign vector $\rho = (\rho_1, \ldots, \rho_n) \in \{-1, +1\}^n$ the reorientation χ^ρ of the chirotope χ is defined as

$$\chi^\rho(\lambda_1, \ldots, \lambda_d) = \rho_{\lambda_1} \cdots \rho_{\lambda_d} \cdot \chi(\lambda_1, \ldots, \lambda_d).$$

Realizability of chirotopes is invariant under reorientations since determinants operate linearly on the columns of matrices. We can define reorientations for phirotopes in the complex setup by considering phase vectors instead of sign vectors:

Definition 1.5 (Reorientation of a phirotope)
A reorientation *of a rank-d phirotope φ with index set $E = \{1, \ldots, n\}$ and with phase vector $\rho = (\rho_1, \ldots, \rho_n) \in (S^1)^n$ is the map*

$$\begin{array}{rccl} \varphi^\rho = \varphi^{(\rho_1, \ldots, \rho_n)} : & E^d & \longrightarrow & S^1 \cup \{0\} \\ & (\lambda_1, \ldots, \lambda_d) & \longmapsto & \rho_{\lambda_1} \cdots \rho_{\lambda_d} \cdot \varphi(\lambda_1, \ldots, \lambda_d). \end{array}$$

The preservation of realizability is guaranteed by the following lemma:

Lemma 1.6 (Reorientations Maintain Realizability)
If a phirotope φ is realizable then every reorientation is also realizable.

Proof: If $Z = (Z_1, \dots Z_n)$ is a realization of the phirotope φ and φ^ρ a reorientation with phase vector $\rho = (\rho_1, \dots, \rho_n)$ then $(\rho_1 Z_1, \dots, \rho_n Z_n)$ is a realization of φ^ρ because

$$\begin{aligned}\omega(\det(\rho_{\lambda_1} Z_{\lambda_1}, \dots, \rho_{\lambda_d} Z_{\lambda_d})) &= \rho_{\lambda_1} \cdots \rho_{\lambda_d} \cdot \omega(\det(Z_{\lambda_1}, \dots, Z_{\lambda_d})) \\ &= \rho_{\lambda_1} \cdots \rho_{\lambda_d} \cdot \varphi(\lambda_1, \dots, \lambda_d) \\ &= \varphi^\rho(\lambda_1, \dots, \lambda_d). \qquad \square\end{aligned}$$

At this point we have defined phirotopes both of complex vector configurations and in general in such a way that phirotopes embed chirotopes as a natural substructure. We have defined realizability and reorientations of phirotopes from a purely algebraic point of view. In the next section we will focus on the geometrical meaning of it.

2 Homogeneous Coordinates

The aim of this section is to get a more geometric picture of phirotopes and their realizations in the complex setup. We use homogeneous coordinates to interpret linear vector configurations as affine point configurations and present a way to view them graphically (in particular for rank $d = 2$). We consider projective transformations and cross ratios on the phases. We also define *chirotopal phirotopes.*

2.1 An Affine Picture

If we are given a real vector configuration of n nonzero vectors in $\mathbb{R}^d$ we can interpret these vectors as the homogeneous coordinates of a projective point configuration in the real projective space $\mathbb{RP}^{d-1} = (\mathbb{R}^d - \{0\}) / (\mathbb{R} - \{0\})$. We get the projective picture in the usual way by treating all vectors as lines through the origin. If we distinguish between the two rays of each line we end up with an *oriented* projective picture in the sense of Stolfi [10]: For each vector $X \in \mathbb{R}^d$ we get a positive and a negative ray $R^+ = \{rX | r \in \mathbb{R}^+\}$ and $R^- = \{-rX | r \in \mathbb{R}^+\}$, respectively. This means we are now in the *oriented* projective space $\mathbb{RP}^{d-1}_\pm = (\mathbb{R}^d - \{0\}) / \mathbb{R}^+$ which we can also think of as a double covering of the real projective space or alternatively as a fibration over the real projective space with typical fiber $S^0 = \{-1, +1\}$. Intersecting the rays with an appropriate $(d-1)$-dimensional hyperplane we obtain an affine point configuration with positive and negative points.

We usually intersect with the hyperplane consisting of all vectors with last coordinate 1. In order to get from a real affine point configuration to a linear vector configuration we use the same hyperplane which means that we identify $\mathbb{R}^{d-1}$ with $\mathbb{R}^{d-1} \times \{1\} \subset \mathbb{R}^d$.

Therefore whenever we start with an affine point configuration in $\mathbb{R}^{d-1}$ we can calculate the associated chirotope by all possible $d \times d$ determinants of the resulting homogenized vectors. Actually the chirotope can be read off from the orientations of the simplices spanned by all d-tuples of the (affine) points. Vice versa, if we start with a realizable chirotope we will find a realization whose vectors have to be treated as (signed) homogeneous coordinates of some affine point configuration. The last coordinate determines the sign of the associated point.

To draw a picture based on this notion we need *two* hyperplanes (connected by the points at infinity according to the double covering of the real projective space $\mathbb{RP}^{d-1}$). Alternatively, we can just use *one* hyperplane with the points marked positive or negative.

Example: Given a chirotope in rank 2, we can get a realization as a linear vector configuration:

$$\left.\begin{array}{l}\chi(AB) = - \\ \chi(AC) = + \\ \chi(BC) = -\end{array}\right\} \rightsquigarrow A = \begin{pmatrix} 1 \\ 1 \end{pmatrix} B = \begin{pmatrix} 3 \\ 1 \end{pmatrix} C = \begin{pmatrix} -2 \\ -1 \end{pmatrix} = - \begin{pmatrix} 2 \\ 1 \end{pmatrix}$$

We can draw a picture with either two hyperplanes, that is lines in $\mathbb{R}^2$ or with just one line in $\mathbb{R}^2$ with positive and negative points:

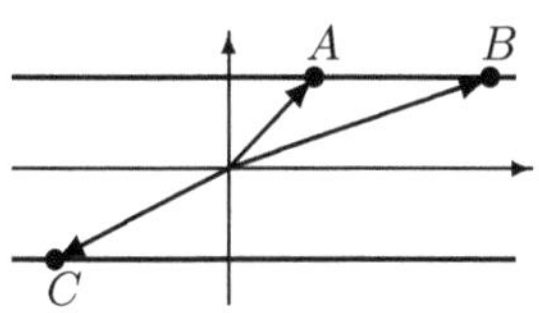

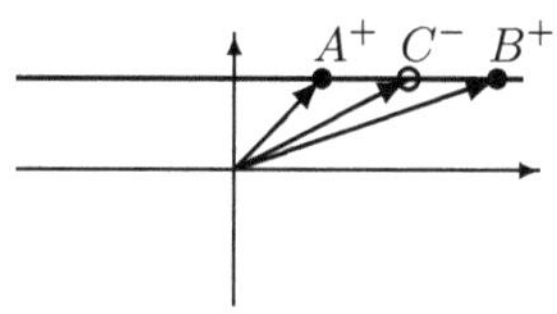

Let us now switch to the complex setup. Using the hyperplane consisting of all vectors with *complex* last coordinate 1 we can homogenize an affine complex point configuration in $\mathbb{C}^{d-1}$ by adding a complex 1 as an additional entry. We get a complex vector configuration in $\mathbb{C}^d$ from which we can calculate the associated phirotope by all $d \times d$ determinants of the resulting vectors.

The homogenization actually lifts the affine point configuration into the complex *projective* space $\mathbb{CP}^{d-1} = (\mathbb{C}^d - \{0\}) / (\mathbb{C} - \{0\})$. Multiplication with any complex scalar (that is some positive real scalar *and* a phase) still represents the same affine point. In order to dehomogenize we rewrite any complex vector $Z \in \mathbb{C}^d$ with a non-zero d-th coordinate in the following way

$$Z = \begin{pmatrix} z_1 \\ \vdots \\ z_{d-1} \\ z_d \end{pmatrix} = |z_d| \cdot \omega(z_d) \begin{pmatrix} \frac{z_1}{z_d} \\ \vdots \\ \frac{z_{d-1}}{z_d} \\ 1 \end{pmatrix} = r_Z \cdot \omega_Z \begin{pmatrix} z \\ 1 \end{pmatrix}$$

with $r_Z = |z_d| \in \mathbb{R}^+$, $\omega_Z = \omega(z_d) \in S^1$ and $z \in \mathbb{C}^{d-1}$. Then $z \in \mathbb{C}^{d-1}$ encodes the coordinates of the represented affine point.

If we start with a realizable phirotope we want to interpret the vectors of a realization again as representatives of an affine point configuration. In order to remain consistent with the given phirotope we once again have to distinguish between the classic projective space $\mathbb{CP}^{d-1}$ and its oriented version $\mathbb{CP}^{d-1}_{S^1} = \mathbb{C}^d - \{0\}/\,\mathbb{R}^+$ where we can only neglect positive real scalars. The orientation is given by the phase of the last entry of each vector. These different phases play the role of the different signs in the real case. This pertains to the fact that multiplying vectors with positive real scalars does not change the phirotope whereas multiplying with a phase corresponds to reorientation as defined in Section 1. In fact, the oriented complex space $\mathbb{CP}^{d-1}_{S^1}$ is a fibration over the complex projective space $\mathbb{CP}^{d-1}$ with fiber S^1, namely the Hopf fibration. Our space $\mathbb{CP}^{d-1}_{S^1}$ is in fact isomorphic to S^{2d-1} as one can easily see: A vector in $\mathbb{CP}^{d-1}$ has d complex entries $z = (z_1, \dots, z_d)$; one gets a representative of z in $\mathbb{CP}^{d-1}_{S^1}$ by scaling down with a positive factor $r \in \mathbb{R}^+$ such that $|rz_1| + |rz_2| + \dots + |rz_d| = 1$. For $d = 2$ we get the well known (smallest) Hopf fibration $S^3 \to S^2$ with typical fiber S^1, since $\mathbb{CP}^1_{S^1} \sim S^3$ and $\mathbb{CP}^1 \sim S^2$. The fibers S_1 correspond to the orientations. Let us define the phase of a complex vector $Z \in \mathbb{C}^d$ and its affine representative:

Definition 2.1 (PHASE AND AFFINE REPRESENTATIVE OF A COMPLEX VECTOR)
Given a vector $Z \in \mathbb{C}^d$ in (oriented) homogeneous coordinates with d-th entry $z_d \neq 0$ given as

$$Z = r_Z \omega_Z \begin{pmatrix} z \\ 1 \end{pmatrix} \quad \text{with } r_Z \in \mathbb{R}^+, \omega_Z \in S^1, z \in \mathbb{C}^{d-1}.$$

We call ω_Z its phase *and z its* affine representative.

If $z_d = 0$, the phase of Z is defined to be the phase $\omega_Z = \omega(z_k)$ of the last non-zero entry z_k $(k < d)$ and Z is said to represent the point at infinity in direction $(\frac{z_1}{z_k}, \dots, \frac{z_{k-1}}{z_k}, 1, 0, \dots, 0) \in \mathbb{C}^{d-1}$.

Now we can consider each vector $Z \in \mathbb{C}^d$ of a realization as its affine representative $z \in \mathbb{C}^{d-1}$ together with its phase information ω_Z. Let us therefore speak of the *point* $Z \in \mathbb{CP}^{d-1}_{S^1}$.

2.2 Point Configurations in $\mathbb{CP}^1_{S^1}$

According to the above definition the complex determinant of any two points $A, B \in \mathbb{CP}^1_{S^1}$ with affine representatives a and b and phases ω_A and ω_B is given by

$$\det(A, B) = r_A \cdot r_B \cdot \omega_A \cdot \omega_B \cdot (a - b) \in \mathbb{C}.$$

Notation 2.2 *If we are dealing with point configurations and realizations we are interested rather in the points of the realization than in their indices. In order not to get lost in subscripts let us simply write $\varphi(A, B)$ instead of*

$\varphi(\lambda_1, \lambda_2)$ *where* $\lambda_1, \lambda_2 \in E = \{1, \ldots, n\}$ *are the indices of the points* $A = Z_{\lambda_1}$ *and* $B = Z_{\lambda_2}$.

Let us also abbreviate $\det(A, B)$ *by square brackets* $\det(A, B) = [A, B]$.

With this notation the phirotope of a complex vector configuration Z is given by

$$\varphi_Z(A, B) = \omega([A, B]) = \omega_A \cdot \omega_B \cdot \omega(a - b) \in S^1 \cup \{0\}$$

for all pairs of indices in E. This means that the value of a phirotope on two points is the product of three complex numbers: the phase of the first point, the phase of the second point and the phase of the difference of their affine representatives. In rank 2 we can draw the points in $\mathbb{CP}^1_{S^1}$ very nicely by first drawing the affine representative and second, adding an arrow to each point of the affine point configuration that indicates the direction of its phase.

Example: Consider the complex point configuration Z of three points in complex rank 2:

$$A = \begin{pmatrix} 1+i \\ 1 \end{pmatrix}, \quad B = \begin{pmatrix} 2+6i \\ 2i \end{pmatrix} = 2e^{i\frac{\pi}{2}} \begin{pmatrix} 3-i \\ 1 \end{pmatrix},$$

$$C = \begin{pmatrix} 2i \\ -1-i \end{pmatrix} = \sqrt{2}e^{i\frac{5}{4}\pi} \begin{pmatrix} -1-i \\ 1 \end{pmatrix}.$$

Therefore we get the affine representatives $a = 1+i$, $b = 3-i$, $c = -1-i$ with phases $\omega_A = 1$, $\omega_B = e^{i\frac{\pi}{2}}$, $\omega_C = e^{i\frac{5}{4}\pi}$.

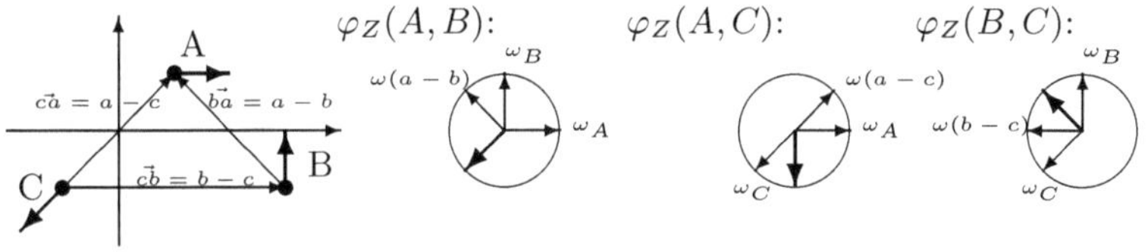

2.3 Projective Transformations

Linear transformations on $\mathbb{C}^d$ do not only preserve the equivalence classes of the complex projective space $\mathbb{CP}^{d-1} = (\mathbb{C}^d - \{0\})/(\mathbb{C} - \{0\})$, but also of the oriented complex projective space $\mathbb{CP}^{d-1}_{S^1} = (\mathbb{C}^d - \{0\})/\mathbb{R}^+$. The induced transformations on (either of) these quotient spaces are the projective transformations. They act on the phirotope of a complex vector configuration as multiplication with the phase of their determinants. Only projective transformations with positive real determinant leave the phirotope of a vector configuration unchanged. However, one can check that all other projective transformations give rise to a reorientation of the phirotope and, as the next lemma shows, that in a realization of a uniform phirotope the first $d+1$ points can be (almost) arbitrarily chosen.

Lemma 2.3 (Freedom of Choice for the First $d+1$ Points) *Given a realizable uniform rank-d phirotope φ on $n \geq d+1$ points, for any choice of affine representatives for the first $d+1$ points (in general position) we can find a realization of φ. In other words, choose any $d+1$ points $z_1, \ldots, z_{d+1} \in \mathbb{C}^{d-1}$ in general position. Then there is a realization $Z = (Z_1, \ldots, Z_{d+1}, Z_{d+2}, \ldots, Z_n)$ of φ such that*

$$Z_k = r_{Z_k} \omega_{Z_k} \begin{pmatrix} z_k \\ 1 \end{pmatrix}$$

for some $r_{Z_k} \in \mathbb{R}^+$ and phases $\omega_{Z_k} \in S^1$ for $k = 1, \ldots, d+1$.

Proof: Let $Y = (Y_1, \ldots, Y_n)$ be a realization of φ. By the following argument we will find a projective transformation with matrix M which leaves the phirotope invariant and maps $Y_1, \ldots, Y_{d+1}$ to $\lambda_1 \binom{z_1}{1}, \ldots, \lambda_{d+1} \binom{z_{d+1}}{1}$ for some $\lambda_1, \ldots, \lambda_{d+1}$ (the λ_k stand for the products $r_k \omega_k$):

The equations $M(Y_k) = \lambda_k \binom{z_k}{1}$ give rise to $(d+1) \cdot d = d^2 + d$ linear equations in the d^2 entries of the matrix M and the $d+1$ unknowns λ_k. The Y_k are in general position since φ is uniform. Hence one can show that these equations are actually linearly independent. Hence there is a solution for M and the λ_k *up to a complex scalar multiple.* Notice that for any d vectors $X_1, \ldots, X_d \in \mathbb{C}^d$ we have $\det(M(X_1), \ldots, M(X_d)) = \det(M) \cdot \det(X_1, \ldots, X_d)$. Therefore the phases of $\det(M(X_1), \ldots, M(X_d))$ and $\det(X_1, \ldots, X_d)$ differ exactly in $\omega(\det(M))$. Since $Y_1, \ldots, Y_{d+1}$ and $z_1, \ldots, z_{d+1}$ are in general position the matrix M has full rank. Therefore we can adjust the complex scalar multiple mentioned above such that the determinant of M has phase 1. This matrix has the desired properties. □

It is also not hard to prove that the phases ω_{Z_k} are determined by the affine representatives up to multiplication with dth roots of unity. (Below we will see this for the case $d = 2$.)

2.4 Cross Ratios

The cross ratio of four vectors $A, B, C, D \in \mathbb{C}^2$ is defined as

$$cr(A, B|C, D) = \frac{[A, C][B, D]}{[A, D][B, C]}.$$

It is invariant under the multiplication of the vectors with non-zero complex scalars. Therefore it makes sense to define it for points in $\mathbb{CP}^1_{S^1}$ and for points in $\mathbb{CP}^1$. Furthermore, it is invariant under projective transformations of $\mathbb{CP}^1$, i.e. Moebius transformations. There is a lot of geometric meaning to cross ratios. If the points can be associated with finite affine representatives $a, b, c, d \in \mathbb{C}$, then the cross ratio amounts to

$$cr(A, B|C, D) = cr(a, b|c, d) = \frac{(a-c)(b-d)}{(a-d)(b-c)}.$$

Since the phases cancel out, the cross ratio is an invariant of the affine points regardless of the phases. It is not hard to show that if the cross ratio is real, then the points a, b, c, d lie on a circle, namely c and d on the same (opposite) side(s) of the line through a and b if and only if the cross ratio is positive (negative). Also, if the cross ratio is complex and its imaginary part is positive, then if the points a, b and c are oriented positively (negatively) as points on the euclidean plane, then point d lies inside (outside) of the circle through a, b, and c. Vice versa if the imaginary part of the cross ratio is negative. All this is information which is encoded already in the phase of the cross ratio.

For a phirotope φ we define the (abstract) *cross ratio phase*:

$$cr_\varphi(A, B|C, D) = \frac{\varphi(A, C)\varphi(B, D)}{\varphi(A, D)\varphi(B, C)}.$$

For realizable phirotopes obviously $cr_\varphi(A, B|C, D) = \omega(cr(A, B|C, D))$. Even if no realizability information is known about a phirotope, there are some interesting properties of these cross ratio phases:

Lemma 2.4 (Properties of Cross Ratio Phases)
Let φ be a uniform phirotope in rank $d = 2$. Then

1. *For all permutations $\pi \in S_4(A, B, C, D)$*

   $$cr_\varphi(A, B|C, D) \in \mathbb{R} \quad \text{if and only if} \quad cr_\varphi(\pi(A), \pi(B)|\pi(C), \pi(D)) \in \mathbb{R}.$$

   Also, if $cr_\varphi(A, B|C, D) \notin \mathbb{R}$, then for all permutations $\pi \in S_4(A, B, C, D)$

   $$\operatorname{sign}(\operatorname{Im}(cr_\varphi(A, B|C, D))) = \operatorname{sign}(\pi)\cdot\operatorname{sign}(\operatorname{Im}(cr_\varphi(\pi(A), \pi(B)|\pi(C), \pi(D)))).$$

2. *If $cr_\varphi(A, B|C, D) \notin \mathbb{R}$, then $cr(A, B|C, D)$ is determined by φ for all realizations of φ.*

3. *All cross ratio phases of φ are real if and only if there is a reorientation of φ which is a chirotope, i.e. such that all reoriented phirotope values are in $\{-1, +1\}$.*

4. *Out of the set of cross ratio phases on five points $\{cr_\varphi(A, B|C, D)$, $cr_\varphi(A, B|C, E)$, $cr_\varphi(A, B|D, E)$, $cr_\varphi(A, C|D, E)$, $cr_\varphi(B, C|D, E)\}$ there is either no real value, one real value, or they are all real.*

5. *If $cr_\varphi(A, B|C, D) \notin \mathbb{R}$, then no phirotope value containing two points out of $\{A, B, C, D\}$ can be "flipped", i.e. for any choice of two elements $K, L \in \{A, B, C, D\}$ the alternating function φ' defined as $\varphi'(X, Y) = \varphi(X, Y)$ for all $\{X, Y\} \neq \{K, L\}$ and $\varphi'(K, L) = -\varphi(K, L)$ is not a phirotope.*

Proof:

1. It is easy to see that $cr_\varphi(B,A|C,D) = cr_\varphi(A,B|D,C) = 1/cr_\varphi(A,B|C,D)$. For these two permutations the two assertions obviously hold.

   From the Grassmann-Plücker relations we know that there have to exist $r_1, r_2, r_3 \in \mathbb{R}^+$ such that

   $$+r_1\varphi(A,B)\varphi(C,D) - r_2\varphi(A,C)\varphi(B,D) + r_3\varphi(A,D)\varphi(B,C) = 0,$$

   or equivalently

   $$-r_1 cr_\varphi(A,C|B,D) - r_2 cr_\varphi(A,B|C,D) + r_3 = 0. \qquad (1)$$

   It is now clearly visible that $cr_\varphi(A,C|B,D)$ is real if and only if $cr_\varphi(A,B|C,D)$ is. Also their imaginary parts (if any) have opposite signs. These three permutations are generators for the permutation group on four elements, so the relations hold for all permutations.

2. Equation (1) gives two real linear constraints on the $r_k \in \mathbb{R}^+$ (one for the real part and one for the imaginary part of the equation). Since they are independent and the Grassmann-Plücker relations guarantee the existence of at least one solution, there is only this one solution up to multiplication with a positive real scalar.

   For each realization of φ the Grassmann-Plücker relations

   $$\begin{aligned} [A,B][C,D] - [A,C][B,D] + [A,D][B,C] &= 0 \quad \text{are equivalent to} \\ -cr(A,C|B,D) - cr(A,B|C,D) + 1 &= 0. \end{aligned}$$

   The phases of these cross ratios are exactly the cross ratio phases defined by the phirotope. So setting the r_k to the absolute values $r_1 = |cr(A,C|B,D)|$, $r_2 = |cr(A,B|C,D)|$, and $r_3 = 1$ is a solution to equation (1). Since this is the only solution with $r_3 = 1$, the cross ratio $cr(A,B|C,D) = r_2 cr_\varphi(A,B|C,D)$ is uniquely determined.

3. We have to find complex numbers ρ_K on the unit circle such that for all K, L we have $\rho_K\rho_L\varphi(K,L) \in \mathbb{R}$. We specify three points A, B, C and choose $\rho_A = \sqrt{\frac{\varphi(B,C)}{\varphi(A,B)\varphi(A,C)}}$ (any of the two roots) and $\rho_K = \frac{1}{\rho_A\varphi(A,K)}$. Notice now that this solves the equations

   $$\begin{aligned} \varphi^\rho(A,K) &= \rho_A\rho_K\varphi(A,K) &= 1 \quad &\text{for } K \neq A \text{ and} \\ \varphi^\rho(B,C) &= \rho_B\rho_C\varphi(B,C) &= 1. & \end{aligned}$$

   But also all other reoriented phirotope values are now real: Since $cr_\varphi = cr_{\varphi^\rho}$, and therefore $cr_{\varphi^\rho}(A,B|C,K) = \frac{\varphi^\rho(A,C)\varphi^\rho(B,K)}{\varphi^\rho(A,K)\varphi^\rho(B,C)} \in \mathbb{R}$, we have $\varphi^\rho(B,K) \in \mathbb{R}$. Finally, since $cr_{\varphi^\rho}(A,K|B,L) = \frac{\varphi^\rho(A,B)\varphi^\rho(K,L)}{\varphi^\rho(A,L)\varphi^\rho(K,B)} \in \mathbb{R}$, also $\varphi^\rho(K,L) \in \mathbb{R}$ for all other K, L.

4. Note that if two of the cross ratio phases are real, w.l.o.g. $cr_\varphi(A,B|C,E)$ and $cr_\varphi(A,B|C,D) \in \mathbb{R}$, then also $cr_\varphi(A,B|D,E) = cr_\varphi(A,B|C,E)/cr_\varphi(A,B|C,D) \in \mathbb{R}$. It is easy to see that then all of the cross ratio phases made up of four of the five points are real.

5. The reason is again a Grassmann-Plücker relation. We have seen above that there is only one solution of equation (1) if the cross ratio is not real. Flipping one phirotope value would force the corresponding r_k to flip as well. This number would now be negative, meaning that there is no positive solution. Hence the Grassmann-Plücker relations would be violated.

□

Item 3 of this lemma motivates the following definition:

Definition 2.5 (CHIROTOPAL/NON-CHIROTOPAL PHIROTOPE)
We call a phirotope chirotopal *if all its cross ratio phases are real. On the other hand, if some cross ratio phases have non-zero imaginary part, we call the phirotope* non-chirotopal.

3 Realizations of Uniform Rank-2 Phirotopes

In this section we concentrate on the realization of a given uniform phirotope in rank 2. Phirotopes on three and four points are always realizable. Phirotopes on five points are not realizable in general. However, we come up with an interesting additional condition that guarantees realizability. This condition also ensures realizability of phirotopes on more than five points.

3.1 Realizations on Three and Four Points

Lemma 3.1 *A uniform rank-2 phirotope φ on three points is always realizable. In fact, if we choose affine representatives $a, b, c \in \mathbb{C}$ for the points A, B, and C then there are exactly two sets of phases $\omega_A, \omega_B, \omega_C$ that realize φ.*

Proof: We have to find $\omega_A, \omega_B, \omega_C \in S^1$ such that

$$\varphi(A,B) = \omega_A \cdot \omega_B \cdot \omega(a-b) \tag{2}$$
$$\varphi(A,C) = \omega_A \cdot \omega_C \cdot \omega(a-c) \tag{3}$$
$$\varphi(B,C) = \omega_B \cdot \omega_C \cdot \omega(b-c). \tag{4}$$

If we multiply equation (2) with equation (3) and divide by equation (4) we get

$$\omega_A^2 = \frac{\varphi(A,B)}{\omega(a-b)} \cdot \frac{\varphi(A,C)}{\omega(a-c)} \cdot \frac{\omega(b-c)}{\varphi(B,C)}$$

which gives the two solutions for ω_A and accordingly the values for ω_B and ω_C by the other two equations. □

We get a similar result for realizations of phirotopes on four points:

Lemma 3.2 *Uniform rank-2 phirotopes on four points are always realizable. Moreover, in a non-chirotopal phirotope, if the realizations of the first three points are known, the fourth point is determined as well.*

Proof: Chirotopal phirotopes are realizable since rank-2 chirotopes are realizable.

For non-chirotopal phirotopes let us first show the uniqueness of the realization of the fourth point: By Lemma 2.4.2 the cross ratio of the realization $\gamma = cr(A,B|C,D) = \frac{[A,C][B,D]}{[A,D][B,C]}$ is determined by the phirotope. Therefore if the realizations of A, B, and C are known then because of

$$[[A,C]B - \gamma[B,C]A, D] = 0$$

the realization of D is also known up to its phase. But the phase is determined by any phirotope value containing D, e.g. by $\varphi(A,D)$.

Let us turn to the existence of the realization. By Lemma 3.1 we can find a realization of the first three points A, B, and C. But now a point D can be computed by the above method (first get the actual cross ratio value, then compute D itself). It is now easy to check that this D is indeed a realization (i.e. that $\varphi(K,D) = \omega([K,D])$ for $K = A,B,C$). □

Explicitly, if we are given a rank-2 phirotope φ on four points we are actually given the six values

$$\varphi(A,B),\ \varphi(A,C),\ \varphi(B,C),\ \varphi(A,D),\ \varphi(B,D),\ \varphi(C,D) \in S^1.$$

We have to find four vectors

$$A = r_A\omega_A\begin{pmatrix} a \\ 1 \end{pmatrix},\quad B = r_B\omega_B\begin{pmatrix} b \\ 1 \end{pmatrix},$$
$$C = r_C\omega_C\begin{pmatrix} c \\ 1 \end{pmatrix},\quad D = r_D\omega_D\begin{pmatrix} d \\ 1 \end{pmatrix} \in \mathbb{CP}^1_{S^1}$$

such that besides equations (2), (3), and (4) the following equations hold:

$$\varphi(A,D) = \omega_A \cdot \omega_D \cdot \omega(a-d) \tag{5}$$
$$\varphi(B,D) = \omega_B \cdot \omega_D \cdot \omega(b-d) \tag{6}$$
$$\varphi(C,D) = \omega_C \cdot \omega_D \cdot \omega(c-d). \tag{7}$$

As in the proof of Lemma 3.1, we choose $a,b,c \in \mathbb{C}$ arbitrarily, but distinct, and get two solutions for the phase ω_A. Choosing one of these solutions determines the phases ω_B and $\omega_C \in S^1$ and therefore the first three points $A,B,C \in \mathbb{CP}^1_{S^1}$.

Since we use (oriented) homogeneous coordinates the phases in the cross ratio used in the proof of Lemma 3.2 cancel out and we get the following formula for the affine representative d of point D (recall that $\gamma = cr[A,B|C,D]$):

$$d = \frac{(a-c)b - (b-c)a\gamma}{(a-c) - (b-c)\gamma}.$$

Geometrically the situation is as follows: If we divide equation (5) by equation (6) we get

$$\omega\left(\frac{a-d}{b-d}\right)=\frac{\varphi(C,D)}{\varphi(B,D)}\cdot\frac{\omega_B}{\omega_A}$$

which tells us that the complex point d lies on a circle $\mathcal{C}_1$ through the complex points a and b with angle $\sphericalangle(\vec{da},\vec{db})=\omega\left(\frac{a-d}{b-d}\right)=\frac{\varphi(C,D)}{\varphi(B,D)}\cdot\frac{\omega_B}{\omega_A}$ at its circumference.

If we do the same calculation for equations (5) and (7) and also for equations (6) and (7) we get

$$\omega\left(\frac{a-d}{c-d}\right)=\frac{\varphi(C,D)}{\varphi(A,D)}\cdot\frac{\omega_C}{\omega_B}\quad\text{and}\quad\omega\left(\frac{b-d}{c-d}\right)=\frac{\varphi(B,D)}{\varphi(A,D)}\cdot\frac{\omega_C}{\omega_B}$$

which defines us similar circles $\mathcal{C}_2$ (for a, c and d) and $\mathcal{C}_3$ (for b, c and d).

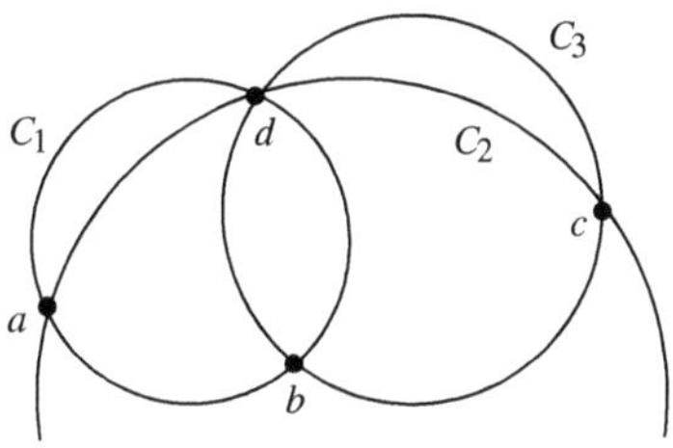

Circles $\mathcal{C}_1, \mathcal{C}_2, \mathcal{C}_3$

Interestingly, it is ultimately the Grassmann-Plücker relations which guarantee that these three circles meet in a point (they were used to show Lemma 2.4.2).

3.2 Realizations on Five Points

Not all phirotopes in rank 2 on five points are realizable. Even so, we can give a simple (algebraic) condition on the phirotope which is equivalent to realizability.

Notation 3.3 *Since we will often need the* squares *of the phirotope values we introduce a new notation:* $[\![K,L]\!]:=(\varphi(K,L))^2$.

Lemma 3.4 (ALGEBRAIC CONDITION ON THE REALIZABILITY OF A FIVE-POINT PHIROTOPE)
Let φ be a uniform non-chirotopal rank-2 phirotope on five points $\{A,B,C,D,E\}$. It is realizable if and only if for the squares of the phirotope values the following algebraic relation holds:

$$\sum_{\substack{\pi\in S_4(A,B,C,D)\\ \pi(A)<\pi(D)}}\operatorname{sign}(\pi)[\![E,\pi(A)]\!][\![\pi(A),\pi(B)]\!][\![\pi(B),\pi(C)]\!][\![\pi(C),\pi(D)]\!][\![\pi(D),E]\!]=0. \tag{8}$$

We call this the five-point condition. *If it holds the realization is unique up to Moebius transformations with real determinant.*

For a chirotopal phirotope the condition above is trivial (the square phases are all 1), also every rank-2 chirotope is realizable, but the realization is by no means unique.

Proof: Since the phirotope is not chirotopal, w.l.o.g. $cr_\varphi(A,B|C,D) \notin \mathbb{R}$, $cr_\varphi(A,B|C,E) \notin \mathbb{R}$, and $cr_\varphi(A,B|D,E) \notin \mathbb{R}$ (see Lemma 2.4.4). First we show that the algebraic condition is necessary. Let φ be realized by a vector configuration (A,B,C,D,E).

By Corollary 2.3 about the freedom to choose the positions of three points in the realization, assume $E = \omega_E \cdot (1,0)$ the point at infinity. By a reorientation of the phirotope, we can assume that all $\omega_K = 1$. Let $a,\dots,d \in \mathbb{C}$ denote the corresponding (finite) affine points, i.e. $A = (a,1)$ etc. In particular, $E = (1,0)$. (A reorientation leaves the five-point condition unchanged.)

Let $s(K,L)$ denote the slope of the line through two affine points k and l. More precisely,

$$s(K,L) = \frac{\mathrm{Im}(k-l)}{\mathrm{Re}(k-l)}.$$

The slopes satisfy a well known quadrilateral set relation:

$$\begin{aligned} 0 &= (s(A,B),s(C,D) \mid s(A,C),s(B,D) \mid s(A,D),s(B,C)) \\ &:= \quad (s(A,B)-s(B,D))\,(s(A,C)-s(B,C))\,(s(A,D)-s(C,D)) \\ &\quad -(s(A,B)-s(B,C))\,(s(A,C)-s(C,D))\,(s(A,D)-s(B,D)). \end{aligned}$$

We have enough freedom in the choice of the first three points of our realization to force that none of the slopes is infinite.

The slopes are linked very closely to the square phases: The slope $s(c)$ of a complex number c on the unit circle is given by

$$s(c) = \frac{\mathrm{Im}(c)}{\mathrm{Re}(c)} = \frac{c-\overline{c}}{i(c+\overline{c})} = \frac{c-1/c}{i(c+1/c)} = \frac{c^2-1}{ic^2+i}.$$

Here $\overline{c}$ represents the complex conjugate of c which for points on the unit circle equals $1/c$, and i the imaginary unit. It follows that for any complex numbers c and d on the unit circle, we have

$$\begin{aligned} s(c)-s(d) &= \det\begin{pmatrix} s(c) & s(d) \\ 1 & 1 \end{pmatrix} \\ &= \frac{1}{(ic^2+i)(id^2+i)} \quad \det\begin{pmatrix} c^2-1 & d^2-1 \\ ic^2+i & id^2+i \end{pmatrix} \\ &= \frac{1}{(ic^2+i)(id^2+i)} \quad \det\begin{pmatrix} 1 & -1 \\ i & i \end{pmatrix} \ \det\begin{pmatrix} c^2 & d^2 \\ 1 & 1 \end{pmatrix} \\ &= \frac{2i}{(ic^2+i)(id^2+i)} \quad (c^2-d^2). \end{aligned}$$

In the quadrilateral set equation above we substitute the slope differences with the differences of square phases; after canceling out the common factors we get an equivalent formula:

$$\begin{aligned} 0 &= ([\![A,B]\!] - [\![B,D]\!])([\![A,C]\!] - [\![B,C]\!])([\![A,D]\!] - [\![C,D]\!]) \\ &\quad -([\![A,B]\!] - [\![B,C]\!])([\![A,C]\!] - [\![C,D]\!])([\![A,D]\!] - [\![B,D]\!]) \\ &= +[\![A,B]\!][\![B,C]\!][\![C,D]\!] - [\![A,B]\!][\![B,D]\!][\![D,C]\!] - [\![A,C]\!][\![C,B]\!][\![B,D]\!] \\ &\quad +[\![A,C]\!][\![C,D]\!][\![D,B]\!] + [\![A,D]\!][\![D,B]\!][\![B,C]\!] - [\![A,D]\!][\![D,C]\!][\![C,B]\!] \\ &\quad -[\![B,A]\!][\![A,C]\!][\![C,D]\!] + [\![B,A]\!][\![A,D]\!][\![D,C]\!] + [\![B,C]\!][\![C,A]\!][\![A,D]\!] \\ &\quad -[\![B,D]\!][\![D,A]\!][\![A,C]\!] + [\![C,A]\!][\![A,B]\!][\![B,D]\!] - [\![C,B]\!][\![B,A]\!][\![A,D]\!]. \end{aligned}$$

The last equality is valid since $[\![K,L]\!] = [\![L,K]\!]$. Now each summand is of the form

$$\operatorname{sign}(\pi)[\![\pi(A),\pi(B)]\!][\![\pi(B),\pi(C)]\!][\![\pi(C),\pi(D)]\!]$$

for some permutation $\pi \in S_4(A,B,C,D)$, and these are all permutations with $\pi(A) < \pi(D)$.

We chose $E = (1,0)$, hence $[\![E,K]\!] = 1$ for all $K \neq E$. Hence we can multiply each summand by the two terms $[\![E,\pi(A)]\!][\![\pi(D),E]\!]$ and obtain the condition (8) on the square phases.

We will show now that this condition is also sufficient for realizability. We realize the three points A, B, and E with the only restriction that $E = (1,0)$. As in the previous section we get a unique solution for C from the cross ratio phase $cr_\varphi(A,B|C,E)$ (which was assumed not to be real). Equally, we get a unique solution for D from $cr_\varphi(A,B|D,E) \notin \mathbb{R}$.

We obtain a vector configuration which conforms to all phirotope values but one: All that is left to show is $\varphi(C,D) = \omega([C,D])$. But notice that both values satisfy the five-point condition (8): the left hand side by assumption and the right hand side since the validity of the equation is necessary for each point configuration. Since all phases were assumed to be 1 and $E = (1,0)$, this equality reads

$$(m_1 - m_2)[\![C,D]\!] = m_1[\![A,C]\!] - m_2[\![A,D]\!]$$

where $m_1 = ([\![A,B]\!] - [\![B,D]\!])([\![A,C]\!] - [\![B,C]\!])$ and $m_2 = ([\![A,B]\!] - [\![B,C]\!])([\![A,D]\!] - [\![B,D]\!])$.

This equality is linear in the square phase $[\![C,D]\!]$. (If it is degenerate, i.e. $m_1 = m_2$, then necessarily some square phases must be equal. This we can avoid by a slightly perturbing reorientation.) Therefore at least $\varphi(C,D)^2 = \omega([C,D])^2$. Now since $cr_\varphi(A,B|C,D)$ is not real, Lemma 2.4.5 says the sign of a single phirotope value cannot be flipped. So we really have $\varphi(C,D) = \omega([C,D])$ and the phirotope is realized. □

We encapsulate the sufficiency proof into a lemma of its own:

Lemma 3.5 (SUFFICIENCY OF FIVE-POINT CONDITION FOR REALIZABILITY)
Given a non-chirotopal five-point rank-2 phirotope φ which satisfies the five-point condition (8), if the phirotope of a vector configuration Z equals φ in all but one values, then $\varphi = \varphi_Z$.

The formula (8) consists of 12 summands and is equivalent to the following formula on 24 summands where we sum over *all* permuations in S_4:

$$0 = \sum_{\pi \in S_4(A,B,C,D)} \operatorname{sign}(\pi)[\![E,\pi(A)]\!][\![\pi(A),\pi(B)]\!][\![\pi(B),\pi(C)]\!][\![\pi(C),\pi(D)]\!][\![\pi(D),E]\!].$$

This equivalence holds since each permutation π and its opposite permutation π' (defined by $\pi'(A) = \pi(D)$, $\pi'(B) = \pi(C)$, $\pi'(C) = \pi(B)$, $\pi'(D) = \pi(A)$) have the same sign.

3.3 Realizations on More Than Five Points

Theorem 3.6 (REALIZABILITY AND RIGIDITY OF PHIROTOPES WITH MANY POINTS)
Let φ be a non-chirotopal uniform rank-2 chirotope on $n \geq 5$ points. It is realizable if and only if the five-point condition (8) is satisfied for all five-point subsets. In this case the realization is unique up to Moebius transformation with real determinant.

Notice that all chirotopal rank-2 phirotopes are realizable since all rank-2 chirotopes are realizable, but that this realization is by no means unique. So this remarkable rigidity in the realization is really due to the complex setup.

Proof: We proceed by induction on the number n of points. For $n = 5$ the statement follows from Lemma 3.4. Assume the statement is true for all uniform non-chirotopal rank-2 phirotopes on $n - 1$ points and let φ be such a phirotope on n points.

If φ is non-chirotopal then there is one cross ratio phase $cr_\varphi(A, B|C, D) \notin \mathbb{R}$. Let K be an element not in $\{A, B, C, D\}$. The phirotope restricted on the elements except K is still non-chirotopal, hence realizable – even uniquely up to Moebius transformations with real determinant. Fix a realization. By Lemma 2.4.4 w.l.o.g. also $cr_\varphi(A, B|C, K) \notin \mathbb{R}$. Hence there is a unique location for K conforming to the phirotope values $\varphi(A, K)$, $\varphi(B, K)$, and $\varphi(C, K)$.

For all five-point sets $\{A, B, C, K, Y\}$ (with $Y \notin \{A, B, C, K\}$) the five-point condition is satisfied. Hence by Lemma 3.5 also the phirotope value $\varphi(Y, K)$ conforms with the realization. □

4 Relations to elementary geometry

In Section 2 we have learned that four points A, B, C, D in the plane are cocircular if and only if $cr_\varphi(A, B|C, D) \in \mathbb{R}$. This immediately relates phirotopes to elementary geometric considerations. We now will explore some of these relations.

4.1 A formula about angles

The five-point formula that we derived in Lemma 3.4 of the last section may seem like a very abstract invariant-theoretic syzygy. Nevertheless, it has some implications that can be expressed on an (almost) elementary geometric level. In what follows let $\sphericalangle A|BC$ be the counterclockwise angle under which a point A sees the points B and C (in this order). Consider the following problem that can be expressed in terms of elementary geometry:

Problem 4.1 *Let A, B, C, D, E be five points in the usual euclidean plane. Let $\alpha, \beta, \gamma, \delta, \epsilon$ be the following five (cyclically shifted) differences of angles:*

$$\begin{aligned}
\alpha &= \sphericalangle A|CD - \sphericalangle B|CD,\\
\beta &= \sphericalangle B|DE - \sphericalangle C|DE,\\
\gamma &= \sphericalangle C|EA - \sphericalangle D|EA,\\
\delta &= \sphericalangle D|AB - \sphericalangle E|AB,\\
\epsilon &= \sphericalangle E|BC - \sphericalangle A|BC.
\end{aligned}$$

If you know the numbers $\alpha, \beta, \gamma, \delta$, what is the value of ϵ?

Let us first make a rough "degree-of-freedom count" that provides evidence that there is indeed a relation between these five numbers. Each of these numbers is an invariant under Moebius transformations of the plane (as we will see below). Thus we can choose points A, B, C arbitrarily (as long as they are distinct) and use the remaining four degrees of freedom (of placing D and E) in order to "adjust" the numbers $\alpha, \beta, \gamma, \delta$. These numbers already determine the configuration up to Moebius transformation. After this one can directly calculate the number ϵ.

But what is the relation between these numbers? At first sight one is tempted to derive the relation by using angle-sum-in-triangle conditions, however this does not lead to any result (try it!). In fact, the true relation is a kind of reformulation of our five-point Formula (8) in Lemma 3.4. In order to see this, identify the plane with $\mathbb{C}$ and observe that we can express the angle differences as phases of cross ratios.

$$\begin{aligned}
\mathbf{a} &:= e^{i\alpha} = cr_\varphi(A, B|C, D),\\
\mathbf{b} &:= e^{i\beta} = cr_\varphi(B, C|D, E),\\
\mathbf{c} &:= e^{i\gamma} = cr_\varphi(C, D|E, A),\\
\mathbf{d} &:= e^{i\delta} = cr_\varphi(D, E|A, B),\\
\mathbf{e} &:= e^{i\epsilon} = cr_\varphi(E, A|B, C).
\end{aligned}$$

In particular this implies that the angle differences are invariant under projective transformations in $\mathbb{CP}^1$ which are exactly the Moebius transformations. We are now going to rewrite the five-point formula of brackets

$$\begin{array}{ll} +[\![A,B]\!][\![B,C]\!][\![C,D]\!][\![D,E]\!][\![E,A]\!] - [\![A,B]\!][\![B,C]\!][\![C,E]\!][\![E,D]\!][\![D,A]\!] & \\ -[\![A,B]\!][\![B,D]\!][\![D,C]\!][\![C,E]\!][\![E,A]\!] + [\![A,B]\!][\![B,D]\!][\![D,E]\!][\![E,C]\!][\![C,A]\!] & \\ +[\![A,B]\!][\![B,E]\!][\![E,C]\!][\![C,D]\!][\![D,A]\!] - [\![A,B]\!][\![B,E]\!][\![E,D]\!][\![D,C]\!][\![C,A]\!] & \\ -[\![A,C]\!][\![C,B]\!][\![B,D]\!][\![D,E]\!][\![E,A]\!] + [\![A,C]\!][\![C,B]\!][\![B,E]\!][\![E,D]\!][\![D,A]\!] & \\ +[\![A,C]\!][\![C,D]\!][\![D,B]\!][\![B,E]\!][\![E,A]\!] - [\![A,C]\!][\![C,E]\!][\![E,B]\!][\![B,D]\!][\![D,A]\!] & \\ +[\![A,D]\!][\![D,B]\!][\![B,C]\!][\![C,A]\!][\![E,A]\!] - [\![A,D]\!][\![D,C]\!][\![C,B]\!][\![B,E]\!][\![E,A]\!] & = 0 \end{array}$$

into a corresponding formula of cross ratios. By abuse of notation we identified the points with their homogeneous coordinates. Since this identity is invariant to multiplication of any of the points by a scalar factor we can assume w.l.o.g. that $[A,D] = [B,E] = [C,A] = [D,B] = [E,C] = 1$. We then get $\frac{1}{\mathbf{a}} = \frac{\varphi(A,D)\varphi(B,C)}{\varphi(A,C)\varphi(B,D)} = \varphi(B,C)$ and similarly $\frac{1}{\mathbf{b}} = \varphi(C,D)$, $\frac{1}{\mathbf{c}} = \varphi(D,E)$, $\frac{1}{\mathbf{d}} = \varphi(E,A)$, $\frac{1}{\mathbf{e}} = \varphi(A,B)$. Substituting these replacements into the formula and multiplying by $\mathbf{a}^2\mathbf{b}^2\mathbf{c}^2\mathbf{d}^2\mathbf{e}^2$ we obtain the desired formula:

$$\begin{array}{ll} -1 & \\ +\mathbf{a}^2\mathbf{c}^2 + \mathbf{c}^2\mathbf{e}^2 + \mathbf{e}^2\mathbf{b}^2 + \mathbf{b}^2\mathbf{d}^2 + \mathbf{d}^2\mathbf{a}^2 & \\ -\mathbf{a}^2\mathbf{c}^2\mathbf{e}^2 - \mathbf{c}^2\mathbf{e}^2\mathbf{b}^2 - \mathbf{e}^2\mathbf{b}^2\mathbf{d}^2 - \mathbf{b}^2\mathbf{d}^2\mathbf{a}^2 - \mathbf{d}^2\mathbf{a}^2\mathbf{c}^2 & \\ +\mathbf{a}^2\mathbf{b}^2\mathbf{c}^2\mathbf{d}^2\mathbf{e}^2 & = 0. \end{array}$$

4.2 Incidence Theorems

As a second relation to elementary geometry we note that there are several incidence theorems about circles and lines that hold already on the level of phirotopes. For this we say that four points in a phirotope φ are *cocircular* if $cr_\varphi(A,B|C,D) \in \mathbb{R}$, regardless whether the phirotope is realizable or not. In the realizable case this agrees with the usual notion of cocircularity (we have to consider lines as circles of infinite radius).

Theorem 4.2 (Miquel's Theorem for phirotopes) *Any uniform rank-2 phirotope has the following property: If the quadruples* $(1,2,3,4)$, $(1,2,5,6)$, $(2,3,6,7)$, $(3,4,7,8)$, $(1,4,5,8)$ *are cocircular, then so is* $(5,6,7,8)$.

Proof: The hypotheses of the theorem and Lemma 2.4.1 imply that

$$\frac{\varphi(1,2)\varphi(3,4)}{\varphi(1,4)\varphi(3,2)} \in \mathbb{R},\ \frac{\varphi(1,5)\varphi(6,2)}{\varphi(1,2)\varphi(6,5)} \in \mathbb{R},\ \frac{\varphi(2,3)\varphi(7,6)}{\varphi(2,6)\varphi(7,3)} \in \mathbb{R},$$

$$\frac{\varphi(3,7)\varphi(8,4)}{\varphi(3,4)\varphi(8,7)} \in \mathbb{R},\ \frac{\varphi(4,1)\varphi(5,8)}{\varphi(4,8)\varphi(5,1)} \in \mathbb{R}.$$

Multiplying these expressions and canceling out common terms we are left with

$$\frac{\varphi(5,8)\varphi(7,6)}{\varphi(5,6)\varphi(7,8)},$$

which has to be real as well. Hence $5, 6, 7, 8$ are cocircular. □

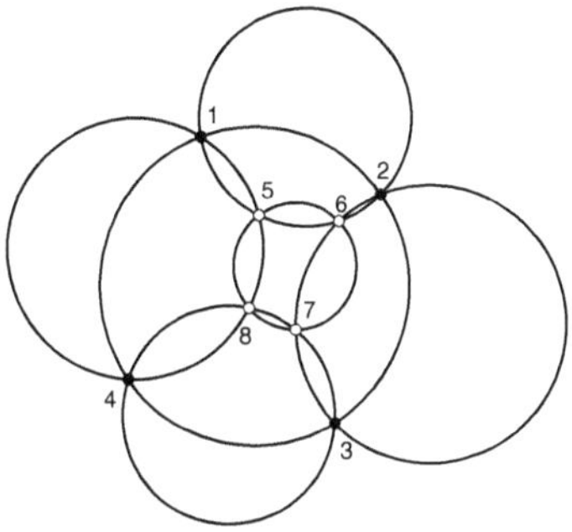

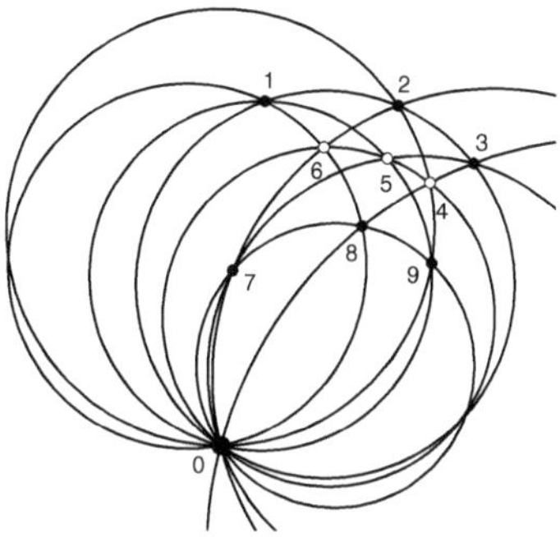

Miquel's theorem (left) does hold for phirotopes. However the circular Pappos' theorem (right) does not.

One might be tempted to conjecture that all incidence theorems on circles and points automatically hold for rank-2 phirotopes. However, this is not the case as the following example shows. Consider the set $\mathcal{C}$ of circles in $\mathbb{CP}^1$ through a given point 0 of the plane. Through any pair of points in $\mathbb{CP}^1 \setminus \{0\}$ there is exactly one circle in $\mathcal{C}$ passing through both of them. Conversely, any non-tangent pair of circles in $\mathcal{C}$ has one point other than 0 in common. The incidence structure of $\mathcal{C}$ considered as lines and of $\mathbb{CP}^1 \setminus \{0\}$ taken as points is isomorphic to a real affine plane (as can be easily checked by a Moebius transformation taking 0 to the point at infinity). In particular, this implies that any pure incidence theorem in the plane has its counterpart in $\mathcal{C}$. The last figure shows on the right a circular version of Pappos' theorem: "If for nine points in $\mathbb{CP}^1$ the quadruples $(0, 1, 2, 3)$, $(0, 1, 5, 9)$, $(0, 1, 6, 8)$, $(0, 2, 4, 9)$, $(0, 2, 6, 7)$, $(0, 3, 4, 8)$, $(0, 3, 5, 7)$ and $(0, 7, 8, 9)$ are cocircular, then so is $(0, 4, 5, 6)$."

Theorem 4.3 (NON-PAPPOS FOR PHIROTOPES)
The circular Pappos' theorem does not hold for phirotopes.

Proof: Consider the realization of the circular Pappos' configuration from the picture and its corresponding phirotope φ_{pap}. In this picture, except of the hypotheses and the conclusion no additional cocircularity relations hold. We have to show that there is a phirotope that satisfies all hypotheses of the theorem but not the conclusion. (This phirotope cannot be realizable.) We get this by a perturbation of φ_{pap}. For sufficiently small $\varepsilon > 0$ we set

$$\varphi(\lambda) := \begin{cases} \varphi_{pap}(\lambda) + \varepsilon & \text{for } \lambda = (4, 5) \\ \varphi_{pap}(\lambda) + \sigma\varepsilon & \text{for } \lambda = (5, 6) \\ \varphi_{pap}(\lambda) & \text{otherwise} \end{cases}$$

with $\sigma \in \{+1, -1\}$ chosen according to conditions specified below. The only Grassmann-Plücker relation that includes $\varphi(4,5)$ and $\varphi(5,6)$ and for which all three terms define collinear directions is $[4,5][6,0] - [4,6][5,0] + [4,0][5,6]$ (for all other such Grassmann-Plücker relations the corresponding points are not cocircular.) For small ε the only Grassmann-Plücker relation that may be violated by φ is this one. However, the cyclic order $(4,5,6,0)$ along these four points implies that the complex numbers $[4,5][6,0]$ and $[4,0][5,6]$ point in one direction and that $[4,6][5,0]$ points in the opposite direction. Hence by suitable choice of σ we can ensure that also this Grassmann-Plücker relation remains valid. Therefore φ is a phirotope that satisfies all hypotheses of the circular Pappos' theorem, but violates its conclusion. □

It is an interesting question whether there is a specific structure that makes some incidence theorems true on the level of phirotopes and others not? So far we cannot give a definite answer to this question. What made the perturbation in the previous proof work is that none of the vertex pairs 45, 46, 56 occur in another cocircularity quadruple. By the same proof we can conclude that no pure incidence theorem of $\mathbb{RP}^2$ can hold in the circular version for arbitrary phirotopes.

However, at least we can provide a large class of incidence theorems in addition to Miquel's theorem that are true on the level of phirotopes. They share the same proof structure as the proof of Miquel's theorem. We only sketch their construction and the proof of their validity.

The structure of our class of theorems is based on orientable cubic 2-manifolds. These are cell-decompositions of an orientable 2-manifold, for

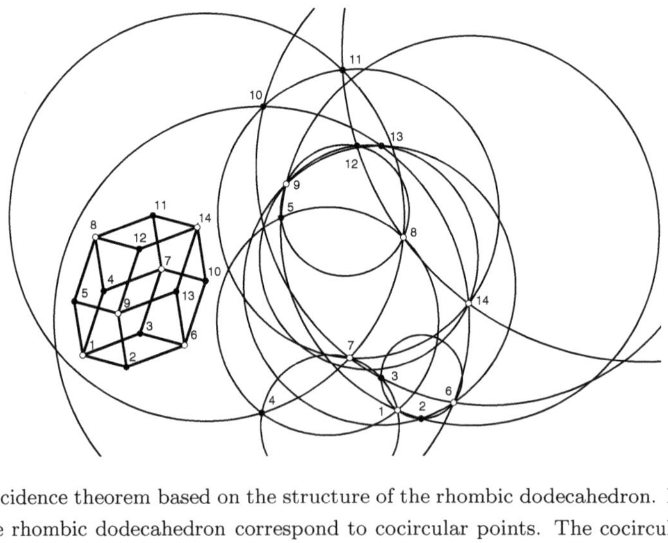

An incidence theorem based on the structure of the rhombic dodecahedron. Faces of the rhombic dodecahedron correspond to cocircular points. The cocircularity of the last four points is satisfied automatically.

which each cell topologically corresponds to a 4-gon and for which the intersection of two cells is either the empty set, a point or an edge. Concrete examples are the cube and the rhombic dodecahedron. However, also non-spherical manifolds are admitted in our construction. Furthermore we require that the vertex/edge graph of such a manifold is bipartite (this is in particular the case for all cubical polytopes, i.e. the spherical case).

Theorem 4.4 *Let $M \subset \{1,\ldots,n\}^4$ be the set of vertex quadruples of an orientable cubical manifold on n points with bipartite edge graph. Let $m \in M$ and let φ be a phirotope on n points for which for all quadruples $m' \in M \setminus \{m\}$ the corresponding points are cocircular. Then the points of m are also cocircular in φ.*

Proof: The proof is a generalization of our proof of Miquel's theorem. Let the vertices of the manifold bipartitely be colored by black labels $B_1,\ldots,B_{n_B}$ and white labels $W_1,\ldots,W_{n_W}$. Along each 4-gon of M we have two black and two white points in alternating order. Furthermore, let the manifold be equipped with an orientation which induces an orientation on each 4-gon in M. We assume that the quadruples representing 4-gons in M are ordered with respect to the orientation and that they start with black points. For a particular 4-gon $q_1 = (B,W,B',W') \in M$ we consider the cross ratio $cr_\varphi(B,B'|W,W') = \frac{\varphi(B,W)\varphi(B',W')}{\varphi(B,W')\varphi(B',W)}$. This cross ratio being real expresses the cocircularity of the corresponding quadruple of points in φ. This cross ratio is invariant on whether that quadruple starts with B or with B'. This is where we use the phirotope axioms (by Lemma 2.4.1). Now consider a quadruple $q_2 = (B,W'',B'',W) \in M$ that is adjacent to q_1 along the edge (BW). If it starts with B the letter W has to be the last entry, due to our orientation condition. Hence in the corresponding cross ratio $\frac{\varphi(B,W'')\varphi(B'',W)}{\varphi(B,W)\varphi(B'',W'')}$ the entry corresponding to (BW) occurs in the denominator. In a product it cancels out with the corresponding term of q_1. Thus forming the product of all cross ratios corresponding to quadruples of $M \setminus m$ everything cancels out except of the terms corresponding to edges of m. The remaining terms form exactly the cross ratio corresponding to m. Hence cocircularity of the quadruples of $M \setminus m$ implies the cocircularity of the points in m. □

5 Second Proof of the 5-Point Formula

In Section 3 we have proved the 5-point formula by relating it to quadrilateral-set relations of real arrangements of lines. In this section we want to present an alternative proof that exhibits an interesting generalization of this formula.

Notation 5.1 *We use the shorthand expressions*

$$\begin{aligned}\zeta(A,B,C,D,E) &:= [\![A,B]\!]^2[\![B,C]\!]^2[\![C,D]\!]^2[\![D,E]\!]^2[\![E,A]\!]^2\\ \pi(A,B,C,D,E) &:= (\pi(A),\pi(B),\pi(C),\pi(D),\pi(E))\end{aligned}$$

with $\pi \in S_5$ a permutation of five letters.

With this notation we can write the five-point formula (8) as

$$\sum_{\pi \in S_5} \sigma(\pi)\zeta(\pi(A,B,C,D,E)) = 0.$$

Here the left side of the formula differs by a factor of 10 from the original formula since every cycle is counted 10 times by ten different permutations. It is important to observe that a cyclical shift of the five letters, as well as reversing the order does not alter the sign of the permutation. Hence all ten permutations that describe a cycle have the same sign. Now observe that we have $[\![X,Y]\!]^2 = [X,Y]/\overline{[X,Y]} = [X,Y]/[\overline{X},\overline{Y}]$. If we set

$$\zeta(A,B,C,D,E|A',B',C',D',E') := \frac{[A,B][B,C][C,D][D,E][E,A]}{[A',B'][B',C'][C',D'][D',E'][E',A']}$$

then the five-point formula (8) reads

$$\sum_{\pi \in S_5} \sigma(\pi)\zeta(\pi(A,B,C,D,E)|\pi(\overline{A},\overline{B},\overline{C},\overline{D},\overline{E})) = 0.$$

It is an amazing fact that this formula does even hold for arbitrary five points A',B',C',D',E' instead of the complex conjugates. We will prove the following generalization of Lemma 3.4.

Theorem 5.2 *For arbitrary ten vectors $A,\dots,E,A',\dots,E' \in \mathbb{C}^2$ we have*

$$\sum_{\pi \in S_5} \sigma(\pi)\zeta(\pi(A,B,C,D,E)|\pi(A',B',C',D',E')) = 0.$$

Proof: We can multiply this formula by the common denominator $[A',B'][A',C']\cdots[D',E']$ and get a multihomogeneous bracket polynomial. This polynomial is quadratic in each of the ten letters. Thus the polynomial being zero is invariant under linear transformations of $\mathbb{C}^2$ and scalar multiplication of each point. Hence we may assume w.l.o.g. that each point X is of the form $(x,1)$. The bracket polynomial then becomes a polynomial in ten complex variables $a,\dots,e,a',\dots,e'$. Since this polynomial is quadratic in each variable it is sufficient to prove it for the special cases $a,b,c,d,e \in \{0,1,2\}$. For most of these 3^5 cases the formula will degenerate immediately such that it vanishes trivially. In particular, if three of the variables are equal then every summand vanishes since it contains a bracket of two identical points. It remains to prove the formula for the case were two pairs of variables are equal. By the symmetry of the formula it suffices to study the case $a=b$, $c=d$. Since the three points of the test set can be interchanged arbitrarily by a suitable projective transformation we only have to check the case

$a = b = 0,\ c = d = 1,\ e = 2$. Substituting these values into the formula all but four terms disappear and we are left with

$$\begin{array}{r}2[A',B'][B',E'][E',C'][C',D'][D',A']\\ -2[A',B'][B',D'][D',C'][C',E'][E',A']\\ -2[A',B'][B',C'][C',D'][D',E'][E',A']\\ +2[A',B'][B',E'][E',D'][D',C'][C',A'].\end{array}$$

The term $2[A',B'][C',D']$ appears as a common factor in all terms. Vanishing of either $[A',B']$ or $[C',D']$ forces the whole expression to vanish. Thus we can w.l.o.g. divide by $2[A',B'][C',D']$. We are left with

$$\begin{array}{rcl}[B',E'][E',C'][D',A']+[B',D'][C',E'][E',A'] & & \\ -[B',C'][D',E'][E',A']-[B',E'][E',D'][C',A'] & = & \\ [B',E']([E',C'][D',A']-[E',D'][C',A']+[E',A'][C',D']) & & \\ +[E',A']([B',D'][C',E']-[B',C'][D',E']+[B',E'][D',C']) & = & 0.\end{array}$$

The last equality holds, since the terms in parentheses are Grassmann-Plücker relations. This proves the claim. □

6 Is there an n-point formula?

The last section showed that the 5-point formula from Section 3 can be considered as a special case of a 12-summand bracket syzygy on ten points. In fact we strongly conjecture that this syzygy is just one example of a class of rank-2 syzygies on $2n$ points. We now formulate this version of the syzygy

$$\sum_{\pi\in S_5}\sigma(\pi)\zeta(\pi(A,B,C,D,E|A',B',C',D',E'))=0$$

that performs the summation over all cycles on n letters instead of just 5 letters. We define

$$\zeta(A_1,\ldots,A_n|A_1',\ldots,A_n') := \frac{[A_1,A_2][A_2,A_3]\cdots[A_{n-1},A_n][A_n,A_1]}{[A_1',A_2'][A_2',A_3']\cdots[A_{n-1}',A_n'][A_n',A_1']}.$$

For a permutation $\pi\in S_n$ and a sequence of points $(A_1,\ldots,A_n)$ we set $\pi(A_1,\ldots,A_n) := \pi(A_{\pi(1)},\ldots,A_{\pi(n)})$.

Conjecture 6.1 *Let $A_1,\ldots,A_n,A_1',\ldots,A_n'\in K^2$ be $2n$ points in a 2-dimensional vector space over a commutative field K. Then the following formula holds*

$$\sum_{(1,\pi_2,\ldots,\pi_n)\in S_n}\sigma(\pi)\zeta(\pi(A_1,,\ldots,A_n)|\pi(A_1',,\ldots,A_n'))=0.$$

By multiplying with a common denominator we again obtain a multihomogeneous bracket polynomial. In the conjecture we did on purpose not sum over *all* permutations but only over those that leave the first element invariant, since otherwise it would vanish trivially for all even n. Thus in the above conjecture each cycle contributes twice to the summation, once literally and once in reversed order. This in turn makes the conjecture trivially true for all n with $n = 3 \mod 4$ or with $n = 0 \mod 4$, since in these cases a cycle and its reversed copy have opposite signs. All other cases are non-trivial. The case $n = 5$ is exactly our five-point formula. That case $n = 6$ was algebraically checked by the computer algebra system Mathematica (so it could be considered as proven, modulo mistakes in the CAS). For larger cases the syzygy is so large that it is almost impossible to check it symbolically. For instance, for $n = 9$ the summation ranges over 20160 terms; each of them expands into 2^{36} monomials.

The cases $n \in \{9, 10, 13, 14\}$ were checked on several numerical randomly generated examples. No counterexamples were found, so it is extremely likely that the formula also holds in these cases. However, a general proof still seems to be out of reach.

7 A Connection To Rank-4 Chirotopes

There are two big differences of our complexification of oriented matroids to the approach of Björner and Ziegler [2]. First, phirotopes are, in contrast to complex matroids in the sense of [2], continuous objects, i.e. there is a continuum of different phirotopes for fixed parameters n and d. Second, the phirotopes admit a natural concept of reorientation classes. This implies that they are good structures for describing geometric invariants in $\mathbb{CP}^d$.

From a geometric point of view the second difference is a great advantage of our approach, while the first difference could be considered as a kind of disadvantage. A large portion of the oriented matroid literature is based on the fact that the set of oriented matroids (for fixed n and d) is finite. For instance, all the considerations about *extension spaces*, and the *space of all oriented matroids* take advantage of the fact that these spaces can be represented by a nice finite partially ordered set. In the approach of Björner and Ziegler many of these concepts and even some theorems carry over to the complex setup. In our approach this seems to be impossible at first sight.

In what follows we will investigate how one can define *combinatorial* invariants for reorientation classes based on the notion of phirotopes. These combinatorial structures are still "geometric" in the above sense. However, they have all prerequisites to make the usual combinatorial investigations. The crucial point is that the cross ratios cr_φ are invariant under reorientations of φ. Thus any combinatorial stratification of the image space of cr_φ defines a combinatorial invariant of projective configurations in $\mathbb{CP}^d$. Following the spirit of this paper, we will elaborate this setup for $d = 2$. Among the different possibilities for a stratification of $\mathbb{C} \cup \{\infty\}$ we will consider the following one:

$$s(z) := \begin{cases} 0 & \text{if } z \in \mathbb{R} \cup \infty \\ \text{sign}(\text{Im}(z)) & \text{otherwise .} \end{cases}$$

For given φ on index set E we now study the map

$$\begin{array}{rrcl} \chi_\varphi\colon & E^4 & \to & \{1,-1,0\} \\ & (A,B,C,D) & \mapsto & s(cr_\varphi(A,B|C,D)). \end{array}$$

Remark: It may seem that this sign function is relatively coarse and one should prefer the finer stratification

$$s'(z) := \begin{cases} \text{sign}(z) & \text{if } z \in \mathbb{R} \\ i \cdot \text{sign}(\text{Im}(z)) & \text{if } z \in \mathbb{C} \setminus \mathbb{R} \\ \infty & \text{if } z = \infty. \end{cases}$$

which takes values in $\{0,1,-1,i,-i,\infty\}$. However, it is not difficult to prove that for non-chirotopal phirotopes the stratification on the cross ratios defined by s' can be reconstructed by the values of the stratification using s.

We are now heading for the (slightly surprising) fact that χ_φ is again an ordinary chirotope. For this we define a function $g : \mathbb{C}^2 \to \mathbb{R}^4$ which maps

$$A = r_A\omega_A \begin{pmatrix} a \\ 1 \end{pmatrix} \mapsto \begin{pmatrix} \text{Re}(a) \\ \text{Im}(a) \\ |a|^2 \\ 1 \end{pmatrix} \quad \text{and} \quad A = r_A\omega_A \begin{pmatrix} 1 \\ 0 \end{pmatrix} \mapsto \begin{pmatrix} 0 \\ 0 \\ 1 \\ 0 \end{pmatrix}.$$

It is not hard to see that the induced function $g : \mathbb{CP}^1_{S^1} \to \mathbb{RP}^3_\pm$ is continuous. The following lemma shows how the signs of the imaginary part of the cross ratio phases of a complex rank-2 point configuration is linked to signs of determinants of the induced real rank-4 point configuration.

Lemma 7.1 (2-PHIROTOPE / 4-CHIROTOPE CONNECTION)
For any points $A,B,C,D \in \mathbb{C}^2$

$$\text{sign}(\text{Im}(cr(A,B|C,D))) = \text{sign}(\det(g(A),g(B),g(C),g(D))).$$

Proof: If all vectors A,B,C,D have finite corresponding affine points, i.e. $A = r_A\omega_A\binom{a}{1}$ etc. then by simple algebraic transformations

$$\text{Im}(cr(A,B|C,D)) \cdot |a-d|^2 \cdot |b-c|^2 = \det(g(A),g(B),g(C),g(D)).$$

If one point is infinite, w.l.o.g. $D = r_A\omega_A\binom{1}{0}$ then also easily

$$\text{Im}(cr(A,B|C,D)) \cdot |b-c|^2 = \det(g(A),g(B),g(C),g(D)).$$

□

Theorem 7.2 (Rank-2 Phirotope Gives Rank-4 Chirotope) *For any uniform rank-2 phirotope φ on n points the function $\chi_\varphi(A, B, C, D)$ defines a chirotope in rank 4 on n points.*

Proof: If φ is chirotopal then χ_φ is 0 for all four-tuples and the theorem is trivially true. Assume now that φ is non-chirotal.

The function χ_φ is by Lemma 2.4.1 clearly alternating.

The absolute value of χ_φ is the basis function of a matroid. For this we have to show that the Steinitz exchange axiom holds: for any two bases B_1 and B_2 (in our case quadruples of points with cross ratio non-real) and for any element $e \in B_1 - B_2$ there is an element $f \in B_2 - B_1$ such that $B_1 - \{e\} \cup \{f\}$ is a basis. This is true since if $B_2 = \{A, B, C, D\}$ and $E \notin B_2$, then by Lemma 2.4.4 either $\{A, B, C, E\}$ or $\{A, B, D, E\}$ is a basis.

Finally, we have to show that the 3-term Grassmann-Plücker relations hold for χ_φ, i.e. that for any A, B, C, D, E_1, E_2 there are numbers $r_1, r_2, r_3 \in \mathbb{R}^+$ such that

$$\begin{aligned} &+ r_1\, \chi_\varphi(A, B, E_1, E_2)\, \chi_\varphi(C, D, E_1, E_2) \\ &- r_2\, \chi_\varphi(A, C, E_1, E_2)\, \chi_\varphi(B, D, E_1, E_2) \\ &+ r_3\, \chi_\varphi(A, D, E_1, E_2)\, \chi_\varphi(B, C, E_1, E_2) \quad = \quad 0. \end{aligned}$$

In order to show that these numbers exist we will find a partial realization of the phirotope φ, i.e. a configuration of vectors in $\{A', B', C', D', E_1', E_2'\}$ in $\mathbb{C}^2$ such that

$$\varphi(K, L) = \omega(\det(K', L'))$$

for all $K \in \{A, B, C, D\}$ and $L \in \{E_1, E_2\}$.

Choose A', E_1', E_2' in general position. Then if $cr_\varphi(A, B|E_1, E_2) \notin \mathbb{R}$ by Lemma 2.4.2 the cross ratio is determined and there is a unique vector B' conforming with the phirotope. If $cr_\varphi(A, B|E_1, E_2) \in \mathbb{R}$, choose any cross ratio with the right phase and realize B' as above. Find realizations C' and D' analoguously.

Notice that the phirotope values $\varphi(K, L)$ with $K \in \{A, B, C, D\}$ and $L \in \{E_1, E_2\}$ are exactly those which occur in the formula for the cross ratios $cr_\varphi(A, B|E_1, E_2) = \frac{\varphi(A,E_1)\varphi(B,E_2)}{\varphi(A,E_2)\varphi(B,E_1)}$ etc. Hence the cross ratio phases occurring in the Grassmann-Plücker relations conform with the phases of the cross ratios of this vector configuration.

The image of this vector configuration under the function g defined above gives rise to a configuration of vectors in $\mathbb{R}^4$. Now $\text{sign}(\det(A', B', E_1', E_2')) = \text{sign}(\text{Im}(cr_\varphi(A, B|E_1, E_2))) = \chi_\varphi(A, B, E_1, E_2)$. The Grassmann-Plücker relations hold for signs of determinants, hence they also hold for χ_φ. □

We conclude our article with an open problem and a conjecture that may give a partial answer to the problem.

Problem 7.3 *Clearly not all 4-chirotopes are of the form χ_φ for some phirotope φ. Find necessary conditions that characterize the possible 4-chirotopes of that kind.*

Conjecture 7.4 *All χ_φ are (combinatorially) convex, i.e. each χ_φ defines a matroid polytope (see [1] for a definition).*

References

[1] A. Björner, M. Las Vergnas, B. Sturmfels, N. White, G. M. Ziegler: *Oriented Matroids*, Cambridge University Press 1993, second ed. 1999

[2] A. Björner, G. M. Ziegler: *Combinatorial stratifications of complex arrangements*, J. Amer. Math. Soc. **5** (1992), 105-149

[3] A. W. M. Dress: *Duality Theory for Finite and Infinite Matroids with Coefficients*, Adv. in Math. **59** (1986), 97–123

[4] A. W. M. Dress, W. Wenzel: *Endliche Matroide mit Koeffizienten* Bayreuth. Math. Schr. **26** (1988), 37–98

[5] A. W. M. Dress, W. Wenzel: *Grassmann-Plücker Relations and Matroids with Coefficients*, Adv. in Math. **86** (1991), 68–110

[6] U. Kortenkamp: *Foundations of dynamic geometry*, PhD-thesis, ETH Zürich, 1999, `http://www.inf.fu-berlin.de/~kortenka/Papers/diss.pdf`.

[7] U. Kortenkamp, J. Richter-Gebert:, *Grundlagen Dynamischer Geometrie*, in: "Zeichnung - Figur - Zugfigur" (H.-J. Elschenbroich, Th. Gawlick, H.-W. Henn, Hrsg.) pp. 123-144, Verlag Franzbecker, Hildesheim 2001

[8] U. Kortenkamp, J. Richter-Gebert:, *Complexity Issues in Dynamic geometry*, Festschrift of the Smale Fest, to appear

[9] J. Richter-Gebert, G. M. Ziegler: *Oriented Matroids*, CRC Handbook on "Discrete and Computational Geometry" (J.E. Goodman, J. O'Rourke, eds.) pp. 111–132, CRC Press, Boca Raton, New York 1997

[10] J. Stolfi: *Oriented Projective Geometry*, Academic Press 1991

[11] G. M. Ziegler: *What is a complex matroid ?* Special issue on "Oriented Matroids" (eds. J. Richter-Gebert, G. M. Ziegler), Discrete Comput. Geom. **10** (1993), 313–348

Covering the Sphere by Equal Spherical Balls

Károly Böröczky, Jr.
Gergely Wintsche

Abstract

We show that for any acute φ, there exists a covering of S^d by spherical balls of radius φ such that no point is covered more than $400d \ln d$ times. It follows that the density is of order at most $d \ln d$, and even at most $d \ln \ln d$ if the number of balls is polynomial in d. If the number of equal spherical balls is $d+3$ then we determine the optimal arrangement.

At the end, we described how our and other peoples results yield estimates for the largest origin centred Euclidean ball contained in the convex hull of N points chosen from the sphere.

1 Introduction

Packings of equal balls on the sphere have been investigated rather intensively since the middle of the 20th century (see the books J.H. Conway and N.J.A. Sloane [8], and Ch. Zong [20], and the survey G. Fejes Tóth and W. Kuperberg [13]), and the optimal arrangement is known in many cases. On the other hand, covering S^d by equal spherical balls seems to be technically harder, only the case $d = 2$ is well–understood (see the book L. Fejes Tóth [14]).

Theorem 1.1. *For any $d \geq 3$ and $0 < \varphi < \frac{\pi}{2}$, S^d can be covered by spherical balls of radius φ in a way that no point of S^d is covered more than $400\, d \ln d$ times.*

The order of the bound in Theorem 1.1 is close to be optimal because there always exists a point that is covered by $d+1$ spheres. The proof of Theorem 1.1 is based on an idea due to P. Erdős and C.A. Rogers [12] who verified the analogous statement in the Euclidean case.

Next we consider the density of coverings. Let $\varphi < \frac{\pi}{2}$, and hence at least $d+2$ spherical balls of radius φ are needed in order to cover S^d. If $\varphi \geq \arccos \sqrt{\frac{1}{d+1}}$ then the $(d+1)$–dimensional regular crosspolytope provides a thin covering; namely,

$$S^d \text{ can be covered by } 2d+2 \text{ spherical balls of radius } \varphi \geq \arccos\sqrt{\tfrac{1}{d+1}}, \text{ and hence the density of the covering is less than } d+1. \tag{1}$$

Therefore in order to obtain coverings with low density, we may confine our attention to the case $\varphi \leq \arccos\sqrt{\frac{1}{d+1}}$. We deduce using Theorem 1.1 that

Corollary 1.2. *For any* $0 < \varphi \leq \arccos\frac{1}{\sqrt{d+1}}$, S^d *can be covered by*

$$c\cos\varphi \cdot \frac{1}{\sin^d\varphi} \cdot d^{\frac{3}{2}} \ln(1 + d\cos^2\varphi)$$

spherical balls of radius φ, *and the density of the arrangement is at most*

$$c \cdot d\ln(1 + d\cos^2\varphi)$$

where c *is an absolute constant.*

Here the improvement $\ln(1 + d\cos^2\varphi)$ compared to $\ln d$ (that is essential only if φ is rather large) uses an idea of M. Kochol [16]. We note that for $\varphi < \arccos\sqrt{\frac{1}{d+1}}$, C.A. Rogers [18] verified the existence of a covering of S^d by spherical balls of radius φ whose density is at most

$$\left(1 + c \cdot \tfrac{\ln\ln d}{\ln d}\right) \cdot d \cdot \left(\ln d + \ln\tfrac{1}{\sin\varphi}\right)$$

for some absolute constant c. His proof is rather technical, and uses coverings of S^d by Euclidean $(d+1)$–dimensional caps. We will sketch an argument at the end of Section 5 that yields the upper bound

$$\left(1 + c \cdot \tfrac{\ln\ln d}{\ln d}\right) \cdot d\ln d \tag{2}$$

for the density. The optimality of Corollary 1.2 is discussed in Section 6.

Now we turn to coverings of S^d by small number of equal spherical balls. It is natural to guess that the optimal covering of $2d+2$ equal spherical balls is determined by the $(d+1)$–dimensional regular crosspolytope. More generally, we believe

Conjecture 1.3. *For* $d \geq 3$ *and* $d+2 \leq m \leq 2d+2$, *let* m *equal spherical balls of minimal radius cover* S^d. *Then the convex hull of the centres in* $\mathbb{E}^{d+1}$ *is the direct sum of mutually orthogonal* $\lceil\frac{d+1}{m-d-1}\rceil$ *and* $\lfloor\frac{d+1}{m-d-1}\rfloor$*–dimensional regular simplices of circumradius one, and the total number of these simplices is* $m-d-1$.

The case $d = 2$ has been actually handled earlier by L. Fejes Tóth if $m = 6$ (see [14]), and by K. Schütte if $m = 5$ (see [19]). For $m = d+2$,

Conjecture 1.3 is common knowledge (and can be verified say similarly to the proof of Theorem 1.4). The only other case known is when $d = 3$ and $m = 8$ (see L. Dalla, D.G. Larman, P. Mani–Levitska, Ch. Zong [10]). We now consider the case of $d + 3$ balls:

Theorem 1.4. *If $d + 3$ equal spherical balls with minimal possible radius cover S^d then the convex hull of the centres in $\mathbb{E}^{d+1}$ is the convex hull of a $\lfloor \frac{d+1}{2} \rfloor$ and a $\lceil \frac{d+1}{2} \rceil$-dimensional regular simplex of circumradius one.*

The paper is organized as follows: Sections 2 and 3 are preparations for the proof of Theorem 1.1 in Section 4. Corollary 1.2 is verified in Section 5, and Section 6 exhibits some examples about the optimality of Corollary 1.2. After that, we prove Theorem 1.4 in Section 7. Finally, Section 8 collects estimates about the largest origin centred ball contained of polytopes of N vertices inscribed into the unit ball. This problem is equivalent to determining the Banach–Mazur distance of ellipsoids from the polytopes of at most N vertices.

2 Covering a symmetric space

P. Erdős and C.A.Rogers [12] proved essentially that $\mathbb{E}^d$ can be covered by equal spheres in a way that no point of $\mathbb{E}^d$ is contained in more than $c \cdot d \ln d$ points for some absolute constant c. Below, we prove Lemma 2.1 that formulates the core of their argument. We choose to formulate Lemma 2.1 in terms of a symmetric space X because we need it both for some torus (when the group G is just the additive group on X), and for S^{d-1} (when the group G is the orthogonal group); and in addition, the proof is the same for say tori any for any symmetric space of finite volume.

Lemma 2.1. *Let G be a locally compact group such that the Haar measure μ is a probability measure, and let X be a factor of G equipped with the resulting measure $\tilde{\mu}$. We throw $N \geq 4$ copies of a measurable subset Θ of X uniformly and independently with respect to μ, and for any copies $\Theta_1, \ldots, \Theta_N$, we write X_i to denote the family of points of X that are contained in exactly i out of $\Theta_1, \ldots, \Theta_N$. If $\mu(\Theta) < 1$ then*

(i) $\mathbb{E}(\tilde{\mu}(X_0)) < e^{-N\tilde{\mu}(\Theta)}$;

(ii) $\mathbb{E}\left(\sum_{i=k}^{\infty} \tilde{\mu}(X_i)\right) < e^{-N\tilde{\mu}(\Theta)}$ *if* $k > 4N\tilde{\mu}(\Theta)$.

Proof: (i) simply follows by the relations $\mathbb{E}(\tilde{\mu}(X_0)) = (1 - \tilde{\mu}(\Theta))^N$ and $1 - t < e^{-t}$.

The second statement does require some argument. We may readily assume that $k \leq N$, and hence $\mu(\Theta) < \frac{1}{4}$. If $k = N$ then $\mathbb{E}(\tilde{\mu}(X_N)) = \tilde{\mu}(\Theta)^N$ and $\mu(\Theta) < \frac{1}{4}$ yield (ii). Therefore let $k < N$.

When integrating with respect to $\tilde{\mu}$ or μ, we simply write dx or dg for the variables $x \in X$ and $g \in G$, respectively, and we write χ_H to denote the indicator function of a set H. In addition, let $\Theta_i = g_i\theta$ for the ith random variable $g_i \in G$. The fundamental observation is that if $1 \le n \le N$ then

$$\chi_{X_n}(x) = \sum_{m=n}^{N} \frac{m!(-1)^{m-n}}{n!(m-n)!} \cdot \sum_{1\le i_1<\ldots i_m\le N} \prod_{j=1}^{m} \chi_{\Theta_{i_j}}(x).$$

This formula can be verified by considering the cases when x is covered exactly by n, by less than n or more than n sets out of $\Theta_1, \ldots, \Theta_N$. Since

$$\mathbb{E}(\tilde{\mu}(X_n)) = \int_G \cdots \int_G \left(\int_X \chi_{X_n}(x)\, dx \right) dg_1 \ldots dg_N;$$

$$\int_G \cdots \int_G \int_X \prod_{j=1}^{m} \chi_{\Theta_{i_j}}(x)\, dx dg_1 \ldots dg_N = \tilde{\mu}(\Theta)^m,$$

we conclude that

$$\begin{aligned} \mathbb{E}(\tilde{\mu}(X_n) &= \sum_{m=n}^{N} \frac{m!(-1)^{m-n}}{n!(m-n)!} \cdot \frac{N!}{m!(N-n)!} \cdot \tilde{\mu}(\Theta)^m \\ &= \frac{N!}{n!(N-n)!} \cdot \tilde{\mu}(\Theta)^n (1-\tilde{\mu}(\Theta))^{N-n}. \end{aligned}$$

Adding these expressions for $n \ge k$ results in the formula

$$\begin{aligned} \mathbb{E}\left(\sum_{n=k}^{\infty} \tilde{\mu}(X_n) \right) &= \frac{N!}{k!(N-k)!} \cdot \tilde{\mu}(\Theta)^k (1-\tilde{\mu}(\Theta))^{N-k} \times \\ &\quad \times \sum_{m=0}^{N-k} \frac{k!(N-k)!}{(k+m)!(N-k-m)!} \cdot \frac{\tilde{\mu}(\Theta)^m}{(1-\tilde{\mu}(\Theta))^m}. \end{aligned}$$

The quotient of the $(m+1)^{th}$ and the mth terms in the sum is

$$\frac{N-k-m}{k+m+1} \cdot \frac{\tilde{\mu}(\Theta)}{1-\tilde{\mu}(\Theta)} \le \frac{N-k}{k+1} \cdot \frac{\tilde{\mu}(\Theta)}{1-\tilde{\mu}(\Theta)} < 1.$$

We use the corresponding geometric series to estimate the sum behind the summation symbol, and hence

$$\mathbb{E}\left(\sum_{i=k}^{\infty} \tilde{\mu}(X_i) \right) \le \frac{N!}{k!(N-k)!} \cdot \tilde{\mu}(\Theta)^k (1-\tilde{\mu}(\Theta))^{N-k} \frac{(k+1)(1-\tilde{\mu}(\Theta))}{(k+1)-(N+1)\tilde{\mu}(\Theta)}.$$

Here $k > 4N\tilde{\mu}(\Theta)$ yields that

$$\frac{(k+1)(1-\tilde{\mu}(\Theta))}{(k+1)-(N+1)\tilde{\mu}(\Theta)} = 1 + \frac{(N-k)\tilde{\mu}(\Theta)}{(k+1)-(N+1)\tilde{\mu}(\Theta)} < \frac{4}{3}.$$

We recall that $t! = \Gamma(t+1)$ for a non–negative integer t, and

$$\left(\frac{t}{e}\right)^t \sqrt{2\pi t} < \Gamma(t+1) < \left(\frac{t}{e}\right)^t \sqrt{2\pi(t+1)} \tag{3}$$

holds according to the Stirling formula (see say E. Artin [1]). Therefore we deduce by $1 - t < e^{-t}$ that

$$\mathbb{E}\left(\sum_{i=k}^{\infty} \tilde{\mu}(X_i)\right) \le \left(\frac{eN\tilde{\mu}(\Theta)}{k(1-\tilde{\mu}(\Theta))}\right)^k \cdot e^{-N\tilde{\mu}(\Theta)} < e^{-N\tilde{\mu}(\Theta)}.$$

Q.E.D.

3 The volume of spherical balls

We need some basic estimates for the volume of spherical balls. We write $B(\varphi, x)$ to denote the spherical ball of radius φ and of centre x, and $|\cdot|$ to denote the volume (Hausdorff d–measure) on S^d. In addition, the volume of the unit Euclidean d–ball B^d is denoted by κ_d, and we note that $|S^d| = (d+1)\kappa_{d+1}$ (see say L. Fejes Tóth [14]).

Lemma 3.1. *Let $0 < \varphi < \frac{\pi}{2}$, and let $x \in S^d$.*

(i) $|B(\varphi, x)| > \kappa_d \varphi \sin^{d-1}\varphi > \kappa_d \sin^d \varphi$;

(ii) $|B(\varphi, t\,x)| < t^d \cdot |B(\varphi, x)|$ *if* $1 < t < \frac{\pi}{2\varphi}$;

(iii) $\dfrac{\kappa_d}{3\cos\varphi} \cdot \sin^d \varphi < |B(\varphi, x)| < \dfrac{\kappa_d}{\cos\varphi} \cdot \sin^d \varphi$ *if* $\varphi \le \arccos \frac{1}{\sqrt{d+1}}$.

Proof: The boundary of $B(\varphi, x)$ in S^d is the boundary of a Euclidean d–ball of radius $\sin\varphi$, as well. Therefore using polar coordinates on S^d shows that

$$|B(\varphi, x)| = d\,\kappa_d \int_0^{\varphi} \sin^{d-1} t\, dt. \tag{4}$$

We deduce that

$$\begin{aligned} \tfrac{\partial}{\partial\varphi} |B(\varphi, x)| &= d\,\kappa_d \cdot \sin^{d-1}(\varphi) > \kappa_d \cdot (\sin^{d-1}\varphi + \varphi \cdot (d-1)\sin^{d-2}\varphi\cos\varphi) \\ &= \tfrac{\partial}{\partial\varphi} (\kappa_d \varphi \cdot \sin^{d-1}\varphi), \end{aligned}$$

which in turn yields (i). Now (ii) follows by

$$\tfrac{\partial}{\partial t} |B(t\varphi, x)| = d\kappa_d\varphi \cdot \sin^{d-1}(t\varphi) < d\kappa_d\varphi \cdot t^{d-1}\sin^{d-1}\varphi < \tfrac{\partial}{\partial t} (t^d \cdot |B(\varphi, x)|).$$

In order to verify (iii), we rewrite (4) in the form

$$|B(\varphi, x)| = d\,\kappa_d \cdot \int_0^{\sin\varphi} \frac{1}{\sqrt{1-r^2}} \cdot r^{d-1} dr, \tag{5}$$

thus the upper bound is a consequence of $\frac{1}{\sqrt{1-r^2}} \leq \frac{1}{\cos\varphi}$ for $r \leq \sin\varphi$. On the other hand, if $(1-\frac{1}{d})\sin\varphi \leq r \leq \sin\varphi$ then $\tan\varphi \leq \sqrt{d}$ yields

$$\frac{\sqrt{1-r^2}}{\cos\varphi} \leq \frac{\sqrt{1-(1-\frac{1}{d})^2\sin^2\varphi}}{\cos\varphi} < \sqrt{1+\tfrac{2}{d}\tan^2\varphi} \leq \sqrt{3}.$$

Since $(1-\frac{1}{d})^d < \frac{1}{e}$, we conclude that

$$|B(\varphi,x)| > d\,\kappa_d \int_{(1-\frac{1}{d})\sin\varphi}^{\sin\varphi} \frac{1}{\sqrt{1-r^2}}\cdot r^{d-1}dr > \kappa_d\left(1-\frac{1}{e}\right)\cdot\frac{1}{\sqrt{3}\cos\varphi}\cdot\sin^d\varphi.$$

Therefore $\frac{e-1}{e\sqrt{3}} > \frac{1}{3}$ yields (iii). Q.E.D.

Now the probability measure of $B(\varphi,x)$ is

$$\Omega(\varphi) = \frac{|B(\varphi,x)|}{|S^d|}.$$

We deduce by the estimates $\sqrt{\frac{d+2}{2\pi}} > \frac{\kappa_d}{\kappa_{d+1}} > \sqrt{\frac{d+1}{2\pi}}$ (see U. Betke, P. Gritzmann and J. M. Wills [4]) and by elementary calculations that

Corollary 3.2. *Let* $0 < \varphi < \frac{\pi}{2}$.

(i) $\Omega(\varphi) > \dfrac{1}{\sqrt{2\pi(d+1)}}\cdot\sin^d\varphi;$

(ii) $\Omega(t\varphi) < t^d\cdot\Omega(\varphi)$ *if* $1 < t < \frac{\pi}{2\varphi}$;

(iii) *if* $\varphi \leq \arccos\frac{1}{\sqrt{d+1}}$ *then*
$$\frac{1}{3\sqrt{2\pi(d+1)}}\cdot\frac{1}{\cos\varphi}\cdot\sin^d\varphi < \Omega(\varphi) < \frac{1}{\sqrt{2\pi d}}\cdot\frac{1}{\cos\varphi}\cdot\sin^d\varphi;$$

(iv) $\dfrac{1}{3e\sqrt{2\pi}} < \Omega(\varphi) < \dfrac{1}{2}$ *if* $\varphi \geq \arccos\frac{1}{\sqrt{d+1}}$.

4 The proof of Theorem 1.1

In order to verify Theorem 1.1, first we consider the case $\varphi \geq \frac{\pi}{48d}$. The condition $\varphi \geq \frac{\pi}{48d}$ is needed in order to ensure that the probability measure of a spherical ball of radius φ is not too small when we apply Lemma 2.1. After that we project suitable Euclidean coverings stereographically onto S^d if $\varphi \leq \frac{\pi}{48d}$. In the Euclidean case, any ball B can be embedded into a torus of volume at most $2^d|B|$ according to the Minkowski–Hlawka Theorem (see say P.M. Gruber and C.G. Lekkerkerker [15]), therefore Lemma 2.1 yields the bound of right order. The argument will use the constant

$$\eta = \frac{1}{2d}.$$

Case I. $\varphi \geq \frac{\pi}{48d}$

We place spherical balls $B((1-\eta)\varphi, x_1), \ldots, B((1-\eta)\varphi, x_N)$ on S^d independently and with uniform distribution according to the probability measure μ where N will be fixed later. Let

$$k = \lceil 4N\Omega((1+\eta)\varphi) \rceil .$$

We write Y_k to denote the family of points that are covered by at least k–times by the N spherical balls $B((1+\eta)\varphi, x_1), \ldots, B((1+\eta)\varphi, x_N)$, and X_0 to denote the family of points of S^d that are not covered by the N spherical balls $B((1-\eta)\varphi, x_1), \ldots, B((1-\eta)\varphi, x_N)$. According to Lemma 3.1, we may choose $x_1, \ldots, x_N$ in a way that

$$\mu(Y_k) + \mu(X_0) < e^{-N\Omega((1+\eta)\varphi)} + e^{-N\Omega((1-\eta)\varphi)}.$$

We claim that if $N\Omega((1-\eta)\varphi) \geq 8d \ln d$ then

$$\mu(Y_k) + \mu(X_0) < \Omega(\eta\varphi), \tag{6}$$

which inequality is readily the consequence of the esimate

$$e^{-N\Omega((1-\eta)\varphi)} < \tfrac{1}{2}\,\Omega(\eta\varphi).$$

It follows by Corollary 3.2 that

$$\ln \frac{2}{\Omega(\eta\varphi)} < \ln \frac{2}{\eta^d \Omega(\varphi)} < \ln 2 + d\ln(2d) + \ln\sqrt{2\pi(d+1)} + d\ln \tfrac{1}{\sin\varphi} < 8d\ln d,$$

which in turn yields the claim (6). Therefore we choose $N = \left\lceil \frac{8d\ln d}{\Omega((1-\eta)\varphi)} \right\rceil$, and hence say $k < 400\, d\ln d$ holds according to $\frac{(1+\eta)^d}{(1-\eta)^d} < e^2$ and Corollary 3.2.

Let us verify that $B(\varphi, x_1), \ldots, B(\varphi, x_N)$ form a suitable cover of S^d. If $x \in S^d$ is not covered by $B(\varphi, x_1), \ldots, B(\varphi, x_N)$ then $B(\eta\varphi, x)$ is disjoint from $B((1-\eta)\varphi, x_1), \ldots, B((1-\eta)\varphi, x_N)$, contradicting (6). On the other hand, if some y is contained in at least k spherical balls out of $B(\varphi, x_1), \ldots, B(\varphi, x_N)$ then Y_k contains $B(\eta\varphi, y)$, which is absurd, as well. This completes the proof if $\varphi \geq \frac{\pi}{48d}$.

Case II. $\varphi \leq \frac{\pi}{48d}$

First we provide a covering in Euclidean space. We also write $B(x, r)$ to denote the ball of centre x and radius r in $\mathbb{E}^d$. It will be always clear from the context when we mean spherical, and when Euclidean balls.

Proposition 4.1. *For any $r > 0$, there exists a (periodic) covering of $\mathbb{E}^d$ by balls $\{B(x_i, (1-\eta)\, r)\}$ such that no point of $\mathbb{E}^d$ is contained in more than $200\, d\ln d$ out of the balls $\{B(x_i, (1+\eta)\, r)\}$.*

Proof: According to the Minkowski–Hlawka Theorem (see say P.M. Gruber and C.G. Lekkerkerker [15]), there exists a lattice Λ such that the balls of radius $(1+2\eta)\,r$ centred at the points of Λ form a packing, and $\det\Lambda \le 2^d(1+2\eta)^d\kappa_d\cdot r^d$. Therefore the torus $T=\mathbb{E}^d/\Lambda$ contains embedded Euclidean balls of radius $(1+2\eta)\,r$, and having a suitable cover of T yields the required cover of $\mathbb{E}^d$.

We write μ to denote the invariant probability measure on T, and hence

$$\mu(\eta r\,B^d)\ge \frac{\eta^d}{2^d(1+2\eta)^d}>\frac{1}{e\,4^d d^d}.$$

We place balls $B((1-2\eta)\varphi,x_1),\ldots,B((1-2\eta)\varphi,x_N)$ on T independently and with uniform distribution according to the probability measure μ where N will be fixed later. Let

$$k=\lceil 4N\mu((1+2\eta)r\,B^d)\rceil.$$

We write Y_k to denote the family of points that are covered by at least k–times by the N balls $B((1+2\eta)\varphi,x_1),\ldots,B((1+2\eta)\varphi,x_N)$, and X_0 to denote the family of points of S^d that are not covered by the N balls $B((1-2\eta)\varphi,x_1),\ldots,B((1-2\eta)\varphi,x_N)$. According to Lemma 3.1, we may choose $x_1,\ldots,x_N$ in a way that

$$\mu(Y_k)+\mu(X_0)<e^{-N\mu((1+2\eta)r\,B^d)}+e^{-N\mu((1-2\eta)r\,B^d)}.$$

We claim that if $N\mu((1-2\eta)r\,B^d)\ge 3d\ln d$ then

$$\mu(Y_k)+\mu(X_0)<\mu(\eta r\,B^d). \tag{7}$$

Now $d\ge 3$ yields that

$$\ln\frac{2}{\mu(\eta r\,B^d)}<\ln(2e)+d\ln d+d\ln 4<3d\ln d,$$

and hence $e^{-N\mu((1-2\eta)r\,B^d)}<\frac{1}{2}\cdot\mu(\eta r\,B^d)$. In turn, we conclude the claim (7).

Since $\frac{(1+2\eta)^d}{(1-2\eta)^d}<e^{\frac{5}{2}}$ follows by $2\eta<\frac{1}{2}$, we may choose N in a way that (7) holds and $k<200$. Therefore the proof of Proposition 4.1 can be completed by the arguments given in Case I. Q.E.D.

Using stereographic projection and coverings in Euclidean d-space, we now construct suitable coverings of S^d by spherical balls of radius $\varphi\le\frac{\pi}{48d}$.

Let p and q be a pair of opposite points of S^d, and we think about p as the north pole, and about q as the south pole. First we construct the covering for the southern hemisphere $\{x\in S^d:\langle x,q\rangle\ge 0\}$.

For $i=1,\ldots,4d$, we define

$$\Theta_i=\left\{x\in S^d\ :\cos\tfrac{(i-1)\pi}{8d}\le\langle x,q\rangle\le\cos\tfrac{i\cdot\pi}{8d}\right\}.$$

We identify the tangent hyperplane to S^d at q with $\mathbb{E}^d$. For any object X contained in $S^d\backslash p$, we write $\tilde{X}$ to denote the stereographic projection of X into $\mathbb{E}^d$ from p. In addition, if $x \in S^d\backslash p$ then $\alpha(x)$ denotes the angle of x and q. We observe that if $\gamma : [0,1] \to S^d\backslash p$ is a continuously differentiable curve then the length of $\tilde{\gamma}$ is

$$\int_0^1 |\tilde{\gamma}'(t)|\, dt = \int_0^1 \left(1 + \tan^2 \tfrac{\alpha(\gamma(t))}{2}\right) \cdot |\gamma'(t)|\, dt. \tag{8}$$

We start with the spherical ball Θ_1. According to Proposition 4.1, there exists a covering of $\widetilde{\Theta}_1$ by Euclidean balls $\{B(\tilde{x}_j, (1-\eta)\,\varphi)\}$ such that no point of $\mathbb{E}^d$ is contained in more than $200\, d \ln d$ out of the balls $\{B(\tilde{x}_j, (1+\eta)\,\varphi)\}$. We may assume that each of the balls $B(\tilde{x}_j, (1-\eta)\,\varphi)$ intersect $\widetilde{\Theta}_1$. It follows by (8) that corresponding family $\{B(x_j, \varphi)\}$ of spherical balls covers Θ_1. Since $\tan^2 \frac{\alpha(x)}{2} < \eta$ if $\alpha(x) \leq \frac{\pi}{8d} + 2\varphi$, we conclude that no point of S^d is contained in more than $200\, d \ln d$ out of the balls $\{B(x_j, \varphi)\}$.

Next let $2 \leq i \leq 4d$. We define $\psi_i = \frac{(2i-1)\pi}{16}$, what is $\alpha(x)$ if x is contained in the "median" level of Ω_i. Using Proposition 4.1 as above, we cover $\widetilde{\Theta}_i$ by Euclidean balls $\{B(\tilde{y}_j, (1-\eta)\, r_i)\}$ such that no point of $\mathbb{E}^d$ is contained in more than $200\, d \ln d$ out of the balls $\{B(\tilde{x}_j, (1+\eta)\, r_i)\}$ where

$$r_i = \left(1 + \tan^2 \frac{\psi_i}{2}\right) \cdot \varphi.$$

We may assume that each of the balls $B(\tilde{y}_j, (1-\eta)\, r_i)$ intersect $\widetilde{\Theta}_i$. Now if x is contained in a spherical ball of radius φ that intersects Ω_i then $|\alpha(x) - \psi_i| \leq \frac{5\pi}{24d}$. On the other hand,

$$\begin{aligned} \frac{\left|(1+\tan^2 \frac{\psi_i}{2}) - (1+\tan^2 \frac{\alpha(x)}{2})\right|}{1+\tan^2 \frac{\psi_i}{2}} \quad &\leq \quad (\tan \tfrac{\psi_i}{2} + \tan \tfrac{\alpha(x)}{2}) \times \\ &\times \frac{1+\tan \frac{\psi_i}{2} \tan \frac{\alpha(x)}{2}}{1+\tan^2 \frac{\psi_i}{2}} \cdot \tan \tfrac{5\pi}{48d} \; < \; \eta, \end{aligned}$$

where we used the fact that $\psi_i < \frac{\pi}{2}$. It follows that the family of spherical balls $\{B(y_j, \varphi)\}$ cover Ω_i, but no point of S^d is covered by more than $200\, d \ln d$ times.

Finally, the regions Ω_i were defined wide enough that even after repeating the same procedure for the northern hemisphere, no point of S^d is covered by more than $400\, d \ln d$ times by the spherical balls of radius φ. Q.E.D.

5 The proof of Corollary 1.2

In this section, $c_1, c_2, \ldots$ always denote some positive absolute constants. Theorem 1.1 yields right away that there exists a covering of S^d by spherical balls of radius φ of density at most $c_1 \cdot d \ln d$ for any φ. If $\varphi \le \arccos \frac{1}{d+1}$ then Lemma 3.1 yields that the number of spherical balls is at most

$$c_2 \cos\varphi \cdot \frac{1}{\sin^d \varphi} \cdot d^{\frac{3}{2}} \ln d. \tag{9}$$

Therefore we may assume that

$$d > 15 \quad \text{and} \quad \frac{1}{\sqrt{d+1}} \le \cos\varphi \le \frac{1}{4^{\frac{1}{4}}(d+1)^{\frac{3}{8}}}.$$

Let $m = \lceil 3(d+1)^2 \cos^4 \varphi \rceil$ and $k = \left\lceil \frac{d+1}{m} \right\rceil$. The idea is first to take pairwise orthogonal linear subspaces $L_1, \ldots, L_k$ that span $\mathbb{E}^{d+1}$, and satisfy $\dim L_i = m$ for $i = 1, \ldots, \left\lfloor \frac{d+1}{m} \right\rfloor$, and then to cover each $L_i \cap S^d$ using Theorem 1.1.

We define the acute angle ψ by the relation $\cos\psi = \sqrt{k} \cdot \cos\varphi$. According to Theorem 1.1, $L_i \cap S^d$ can be covered by N_i spherical $(m-1)$–balls of radius ψ where

$$N_i \le c_3 \cos\psi \cdot \frac{1}{\sin^{\frac{m-1}{2}} \psi} \cdot m^{\frac{3}{2}} \ln m.$$

Since for any $u \in S^d$, the length of the orthogonal projection of u into some L_i is at least $\frac{1}{\sqrt{k}}$, we obtain a covering of S^d by $n = \sum_1^k N_i$ spherical balls of radius φ where

$$n \le c_3 \left\lceil \frac{d+1}{m} \right\rceil^{\frac{3}{2}} \cdot \cos\varphi \cdot \frac{1}{\left(1 - \left\lceil \frac{d+1}{m} \right\rceil \cos^2\varphi\right)^{\frac{m-1}{2}}} \cdot m^{\frac{3}{2}} \ln m.$$

We observe that

$$m \le 4(d+1)^2 \cos^4 \varphi \le \sqrt{d+1}, \text{ and hence } \left\lceil \frac{d+1}{m} \right\rceil \le \frac{5}{4} \cdot \frac{d+1}{m}, \tag{10}$$

which in turn yields that $\left\lceil \frac{d+1}{m} \right\rceil \cos^2 \varphi < \frac{1}{2}$. On the other hand, $1-t > e^{-t-t^2}$ holds for $0 < t < \frac{1}{2}$, therefore

$$n < c_4 \, d^{\frac{3}{2}} \cdot \cos\varphi \cdot \frac{e^{\frac{m-1}{2} \cdot \left\lceil \frac{d+1}{m} \right\rceil^2 \cos^4 \varphi}}{e^{-\frac{m-1}{2} \cdot \left\lceil \frac{d+1}{m} \right\rceil \cos^2 \varphi}} \cdot \ln 4(d+1)^2 \cos^4 \varphi.$$

Here $\frac{m-1}{2} \left\lceil \frac{d+1}{m} \right\rceil^2 \cos^4 \varphi \le \frac{25}{96}$ by (10). In addition, we claim that

$$e^{-\frac{m-1}{2} \cdot \left\lceil \frac{d+1}{m} \right\rceil \cos^2 \varphi} > \sin^d \varphi. \tag{11}$$

Since $\left\lceil \frac{d+1}{m} \right\rceil < \frac{d+1}{m} + 1$ and $m^2 \le d+1$ yield that $\frac{m-1}{2} \cdot \left\lceil \frac{d+1}{m} \right\rceil < d$, we deduce (11) by $e^{-dt} > (1-t)^d$. In turn, we conclude Corollary 1.2. Q.E.D.

Remark 5.1. *If $\varphi < \frac{\pi}{2}$ then there exists a covering by spherical balls of radius φ whose density is at most*

$$\left(1 + c \cdot \tfrac{\ln \ln d}{\ln d}\right) \cdot d \ln d. \tag{12}$$

This estimate is a slight improvement of Corollary 1.2 if say $\varphi < \arccos \frac{1}{\sqrt[4]{d}}$ and d is large.

We sketch the proof that is based on an idea of C.A. Rogers [17]. Let $\eta = \frac{1}{d \ln d}$. We first distribute N centres $x_1, \ldots, x_N$ on S^d independently and with uniform distribution according to the probability measure on S^d for suitable N. According to Lemma 2.1, we may choose the centres in a way that the probability measure of the part of S^d not covered by balls $B((1-\eta)\,\varphi, x_i)$, $i = 1, \ldots, N$, is at most $e^{-N\Omega((1-\eta)\,\varphi)}$. Next we place as many non–overlapping copies of spherical balls of radius $\eta\,\varphi$ into the uncovered part as possible, say the spherical balls $B(\eta\,\varphi, y_i)$, $i = 1, \ldots, M$. Since $1 - t < e^{-t}$ holds for positive $t < 1$, we deduce that

$$M \leq \frac{(1 - \Omega((1-\eta)\,\varphi))^N}{\Omega(\eta\,\varphi)} < \frac{e^{-N \cdot \Omega((1-\eta)\,\varphi)}}{\Omega(\eta\,\varphi)}.$$

Now for any $z \in S^d$, the spherical ball $B(\eta\,\varphi, z)$ intersects either some $B((1-\eta)\,\varphi, x_i)$ or some $B(\eta\,\varphi, y_j)$, and hence we obtain a covering of S^d by $n = N + M$ spherical balls of radius φ where

$$n \leq N + \frac{e^{-N \cdot \Omega((1-\eta)\,\varphi)}}{\Omega(\eta\,\varphi)}.$$

Derivation shows that the optimal N is

$$N = \left\lfloor \frac{1}{\Omega((1-\eta)\,\varphi)} \cdot \ln \frac{\Omega((1-\eta)\,\varphi)}{\Omega(\eta\,\varphi)} \right\rfloor.$$

Since readily $\Omega((1-\eta)\,\varphi) < \frac{1}{2}$, we deduce that the density of the covering by n spherical balls of radius φ is

$$n \cdot \Omega(\varphi) < \frac{\Omega(\varphi)}{\Omega((1-\eta)\,\varphi)} \cdot \left(\ln \frac{\Omega((1-\eta)\,\varphi)}{\Omega(\eta\,\varphi)} + \sqrt{e}\right).$$

Finally we apply the estimate of Corollary 3.2 (ii). Since $\eta = \frac{1}{d \ln d}$, it follows that

$$n \cdot \Omega(\varphi) < \frac{1}{(1-\eta)^d} \cdot \left(\ln \frac{1}{\eta^d} + \sqrt{e}\right) < (1 + \tfrac{2}{\ln d}) \cdot (d \ln d + d \ln \ln d + \sqrt{e}).$$

In turn, we conclude (12).

6 Examples

Let us discuss the optimality of the estimates in Corollary 1.2 where we always assume that d is large:

Example 6.1. *If* $\gamma \geq 1$ *and* $\arccos \frac{\gamma}{\sqrt{d+1}} \leq \varphi < \frac{\pi}{2}$ *then some* $e^{c\gamma^2} \cdot d$ *spherical balls of radius* φ *cover* S^d *where* c *is a absolute constant; namely, the number of balls needed is linear in* d.

We note that at least $d+2$ spherical balls are needed to cover S^d.

Example 6.2. *If* $\frac{1}{\sqrt{d+1}} \leq \cos\varphi \leq \frac{\ln d}{\sqrt{d+1}}$ *then there exists a covering by spherical balls of radius* φ *with density at most* $c\,d \ln\ln d$ *where* c *is a absolute constant.*

For example, this case includes when the number of balls in the optimal covering is polynomial in d; namely, at most d^{α} for an absolute constant $\alpha \geq 2$. More precisely, $d^{\ln d+2}$ balls are sufficient.

Example 6.3. *If* $\varphi \leq \arcsin\sqrt{\frac{1}{d+1}}$ *then the density of any finite covering of* S^d *by spherical balls of radius* φ *is at least* $c\,d$ *where* c *is a positive absolute constant.* Therefore the upper bound in Corollary 1.2 is close to be optimal up to a factor of constant times $\ln d$.

We verify the lower bound for the density on S^d by comparing it to the Euclidean simplex bound. For a spherical d–simplex T of circumradius at most φ, the $d+1$ spherical balls of radius φ centred at the vertices cover T. We define the *associated density* to be the ratio of the sum of the volumes of the sectors of the $d+1$ balls, lying in the simplex, to the volume of the simplex. For some a finite covering of S^d by spherical balls of radius φ, we write P to denote the convex hull (in $\mathbb{E}^{d+1}$) of the centres, and we consider a triangulation of ∂P using the same set of vertices. The radial projections of these Euclidean simplices on S^d results in a spherical triangulation of S^d, which triangulation is known as *Delone triangulation.*

Let T be a spherical d–simplex of the Delone triangulation. Since the original system of spherical balls cover S^d, we deduce that the spherical circumradius of T is at most φ. Now the Euclidean d–simplex T' whose vertices coincide with the vertices of T, is of circumradius at most $\sin\varphi$. In addition, the Euclidean d–ball in aff T' centred at a vertex v of T' and of radius $\sin\varphi$ satisfies that its radial projection onto S^d is contained in the spherical ball centred at v and of radius φ. According to the simplex bound by H.S.M. Coxeter, L. Few and C.A. Rogers [9], the associated density for T' (which can be defined analogously) is minimal for the regular simplex, and hence the associated density for T' is at least $c'\,d$ for a positive absolute constant c'. Since $\cos\varphi \geq \sqrt{1-\frac{1}{d+1}}$, we conclude that the associated density for the spherical simplex T is at least $c\,d$. The same estimate for each Delone d–simplex yields the density bound $c\,d$ for the covering of S^d.

7 Covering S^d by d+3 balls

The proof of Theorem 1.4 use Euclidean tools, and depends on the following observation:

Lemma 7.1. *Some family of spherical balls of radius $\varphi < \frac{\pi}{2}$ cover S^d if and only if the convex hull of the centres in $\mathbb{E}^{d+1}$ contains a ball centred at the origin of radius* $\cos\varphi$.

Proof: We write $B(r, x)$ to denote the spherical ball of radius r and of centre x. First we assume that spherical balls of radius $\varphi < \frac{\pi}{2}$ cover S^d, and write P to denote the convex hull of the centres. The condition $\varphi < \frac{\pi}{2}$ yields that P contains the origin in its interior. If F is a facet of P then aff F cuts off a spherical ball $B(\psi, y)$ with $\psi < \frac{\pi}{2}$. Now y is a contained in a spherical ball of radius φ around a vertex of P, and hence $\psi \leq \varphi$. Therefore the Euclidean distance of aff F from the origin is at least $\cos\varphi$.

On the other hand, let P be a polytope whose vertices lie on S^d, and we assume that P contains the Euclidean ball centred at the origin of radius $\cos\varphi$. For any $y \in S^d$, P intersects the convex hull (in $\mathbb{E}^{d+1}$) of $B(\varphi, y)$, and hence $B(\varphi, y)$ contains a vertex v of P. Therefore $y \in B(\varphi, v)$. Q.E.D.

In order to verify Theorem 1.4, we seek the polytope inside the unit ball with the largest inscribed ball regardless of the place of the centre of the inscribed ball, and show that the centre of the inscribed ball is actually the origin for the optimal polytope. Therefore let $P \subset B^{d+1}$ be a polytope with at most $d+3$ vertices such that $r(P)$ is maximal, and let B be an inscribed ball of P.

If P has only $d+2$ vertices then it is a $(d+1)$–simplex. We choose a vertex v, and a point $u \in S^d$ such that the segment vu intersect the interior of P. Then the convex hull P' of P and u has $d+3$ vertices, and each facet of P' that touches B contains v. Therefore B can be enlarged from v in a way that the resulting ball is still contained in P', what is a contradiction.

Therefore P has actually $d+3$ vertices. According to the Radon Theorem, P is the convex hull of a k–dimensional simplex S and an m–dimensional simplex T such that $k+m = d+1$ holds for $k, m \geq 1$, and $S \cap T \neq \emptyset$. It follows that S and T intersect in one point that we denote by q, and q is not a vertex of P.

Next we suppose that a vertex v of P is contained in the interior of B^{d+1}, say v is a vertex of T. We choose a v' outside of P such that the ray $v'v$ intersects the interior of P, and for any facet F of P that does not contain v, the point v' lies in the open halfspace determined by aff F that intersects P. Now we define P' to be the convex hull of P and v', and observe that P' has $d+3$ vertices, as well. We choose a vertex w of T different from v. Since any facet of P contains either v or w, we deduce that any facet of P' contains either v' or w. On the other hand, no facet of P' containing v' touches B, thus B can be enlarged from w in a way that the resulting ball is still contained in P'. This contradiction yields that each vertex of P lies on S^d.

Finally, we suppose that there exist an edge vw of say S and a vertex u of P such that $d(u,v) \neq d(u,w)$. Let H be the hyperplane that bisects the edge vw perpendicularly, and we write X' to denote the image of any object X by the Steiner symmetrization with respect to H (see say T. Bonnesen and W. Fenchel [5]). We observe that $P' \subset B^{d+1}$, and P' is a polytope with $d+3$ vertices. Since $B' \subset P'$, we deduce that P' is an optimal polytope contained in B^{d+1}, as well. On the other hand, the vertex u' of P' lies in the interior of B^{d+1}, what is absurd. Therefore S and T are regular k and m–simplices, respectively, whose affine hulls are orthogonal, and meet at the common circumcentre of S and T.

We deduce that P has a unique inscribed ball that is centred at the origin and touches each facet. Any facet C of P is the convex hull of a $(k-1)$–face F of S and an $(m-1)$–face G of T. We observe that the distance of F and G from the origin is $\frac{1}{k}$ and $\frac{1}{m}$, respectively, and $F \subset (\mathrm{lin}\, G)^{\perp}$ and $G \subset (\mathrm{lin}\, F)^{\perp}$. Denoting the centroids of F and G by p and q, respectively, the distance of the origin from C is the distance of the origin from the segment pq. Since the segments op and oq are orthogonal, and of length $\frac{1}{k}$ and $\frac{1}{m}$, respectively, we deduce that $r(P) = \frac{1}{k^2+m^2}$. The condition $k+m = d+1$ yields that the sum k^2+m^2 is minimal if k and m are as close to $\frac{d+1}{2}$ as possible. Therefore the dimensions of S and T are $\lfloor \frac{d+1}{2} \rfloor$ and $\lceil \frac{d+1}{2} \rceil$ in this or in the reverse order. In turn, we conclude Theorem 1.4 by Lemma 7.1. Q.E.D.

We note that similar applications of the Radon Theorem occur in various context, say I. Bárány [2] applies it to the minimal ratio of the maximum distance to the minimal distance among $d+2$ points in $\mathbb{E}^d$.

8 Inradius of the inscribed polytopes

We have proved various estimates on the density of coverings by equal spherical balls. We would like to summarize these and some related results from another perspective; namely, we consider the largest origin centred Euclidean d–ball contained in the convex hull of a given number of points on S^{d-1}. More precisely, the covering problem on S^{d-1} can be formulated as follows: What is the minimal radius $\varphi(N)$ of N equal spherical balls that cover S^{d-1}? According to Lemma 7.1, $\cos\varphi(N)$ is the maximal radius of the Euclidean balls centred at the origin that are contained in the convex hull of N points chosen from S^{d-1}. Equivalently, $\frac{1}{\cos\varphi(N)}$ is the Banach–Mazur distance of ellipsoids from polytopes of at most N vertices.

If $N \geq \sqrt{2}^d$ then Corollary 1.2 and the trivial estimate $N \cdot \Omega(\varphi(N)) \geq 1$ yield

$$1 - \cos\varphi(N) = (1 + e(N,d)) \cdot \tfrac{1}{2} N^{\frac{-2}{d-1}}, \quad \text{where } |e(N,d)| < N^{\frac{-2}{d-1}}. \tag{13}$$

Therefore let $d+1 \le N \le \sqrt{2}^d$. Concerning lower bounds,

$$\cos\varphi(N) > \begin{cases} \left(1 - \frac{\ln N}{d} - o(1)\right) \cdot \sqrt{\frac{2\ln\frac{N}{d}}{d}} & \text{if } d^2 \le N \le \sqrt{2}^d; \\ c \cdot \sqrt{\frac{\ln\frac{N}{d}}{d}} & \text{if } d+1 \le N \le d^2, \end{cases} \tag{14}$$

where $o(1)$ denotes a function of d that tends to zero as d tends to infinity, and c is a positive absolute constant. Here if $d+1 \le N \le 2d$ then the arrangements described in Conjecture 1.3 provide the lower bound, while the case $2d < N \le \sqrt{2}^d$ is a consequence of Corollary 1.2 and the estimates

$$e^{\frac{d}{2}\cdot\cos^2\varphi(N)} < \sin^{-d}\varphi(N) < e^{\frac{d}{2}\cdot\cos^2\varphi(N)\cdot(1+\cos^2\varphi(N))}.$$

Let us turn to upper bounds when $d+1 \le N \le \sqrt{2}^d$. The trivial estimate $N \cdot \Omega(\varphi(N)) \ge 1$ yields that

$$\cos\varphi(N) < \sqrt{\frac{2\ln N}{d}}. \tag{15}$$

According to the lower bounds above, this estimate is a asymptotically optimal if $N \ge d^{\ln d}$, say, and good up to a multiplicative constant $\sqrt{2}$ if $N \ge d^2$. On the other hand, a better upper bound is needed if N is close to be linear in d. It is known that if $2d \le N \le d^2$ then

$$\cos\varphi(N) < c \cdot \sqrt{\frac{\ln\frac{N}{d}}{d}} \tag{16}$$

for some absolute constant c. It is still open whether $\cos\varphi(N) < c \cdot \sqrt{\frac{\ln\frac{N}{d}}{d}}$ if $d+2 < N < 2d$. A very satisfying way to prove this estimate would be by verifying Conjecture 1.3.

Next we discuss the two known arguments leading to (16). Both of them give an upper bound for the maximal relative volume $V(d,N)$ of a polytope with N vertices that is contained in the unit ball B^d. We note that readily

$$\cos\varphi(N) < \sqrt[d]{V(d,N)}.$$

Now B. Carl [6] (see also B. Carl and A. Pajor [7]) verified (16) using deep methods from the local theory of Banach spaces. Independently, I. Bárány and Z. Füredi [3] obtained the same estimate. Here we show how their elementary argument yields (16) because (16) is not stated explicitly in the paper, and even the experts seem to be unaware the fact that the proof of I. Bárány and Z. Füredi [3] leads to (16).

Theorem 3 of I. Bárány and Z. Füredi [3] states that

$$V(d,N) \le \binom{N}{k+1} \cdot \frac{\sqrt{k+1}}{k!}\left(1+\frac{1}{k}\right)^{\frac{k}{2}} \cdot \frac{\kappa_{d-k}}{\kappa_d}\left(\frac{d-k}{d\cdot k}\right)^{\frac{d-k}{2}} \tag{17}$$

holds for any $k = 1, \ldots, d-1$. This theorem is a consequence of the (elementary) fact that if S is a d simplex of circumradius one and $x \in S$ then S has a k–face F such that the closest point y of aff F to x lies in F and $d(x,y) \le \sqrt{\frac{d-k}{d\cdot k}}$. Since $\kappa_d = \frac{\pi^{\frac{d}{2}}}{\Gamma(\frac{d}{2}+1)}$ (see say L. Fejes Tóth [14]), the Stirling formula (3) and (17) yield that

$$V(d,N) < c_0^k \cdot N \cdot \frac{N^k d^{\frac{k}{2}}}{k^{\frac{d+3k}{2}}}$$

for some absolute constant c_0. In particular, if $\alpha = \frac{d}{k}$ then

$$\sqrt[d]{V(d,N)} < c_0^{\frac{1}{\alpha}} \cdot N^{\frac{1}{d}} \cdot \frac{N^{\frac{1}{\alpha}} \cdot d^{\frac{1}{2\alpha}}}{\sqrt{k} \cdot k^{\frac{3}{2\alpha}}} = c_0^{\frac{1}{\alpha}} \cdot N^{\frac{1}{d}} \cdot e^{\frac{\ln \frac{N}{d}}{\alpha}} \cdot \alpha^{\frac{3}{2\alpha}} \cdot \frac{1}{\sqrt{k}}.$$

Therefore choosing $k = \left\lfloor \frac{d}{2\ln \frac{N}{d}} \right\rfloor$ (as it is implicitly suggested by I. Bárány and Z. Füredi [3]) yields (16).

Finally, we note that estimating $V(d,N)$ does not give sufficient information about $\cos\varphi(N)$ if N is close to $d+1$ because the volume of the regular simplex inscribed into the unit balls larger than $\frac{1}{2^d}$ times the volume of the inscribed regular crosspolytope.

Remark 8.1. We would like to point out that if $d^2 \le N < 1.05^d$ then the argument above gives

$$\sqrt[d]{V(d,N)} < \left(1 + \frac{4\ln N}{d} + c_1 \cdot \frac{\ln\ln N}{\ln N}\right) \cdot \sqrt{\frac{2e\ln\frac{N}{d}}{d}}$$

for some absolute constant c_1. Here we stop at 1.05^d because say for $1.05^d < N < \sqrt{2}^d$, the bound $V(d,N) < \frac{N}{2^d}$ (see Gy. Elekes [11]) slowly takes over. Comparing to (15), we see that the quotient of the upper and the lower bound for $\cos\varphi(N)$ is at most asymptotically $\sqrt{e}$.

References

[1] E. Artin: The gamma function. Holt, Rinehart and Winston, 1964.

[2] I. Bárány: The densest $(n+2)$ set in $\mathbb{R}^n$. In: Intuitive Geometry (1991), Coll. Math. Soc. János Bolyai, No 63, North Holland, 1997.

[3] I. Bárány, Z. Füredi: Approximation of the sphere by polytopes having few vertices. *Proc. Amer. Math. Soc.*, 102 (1988), 651–659.

[4] U. Betke, P. Gritzmann, J. M. Wills: Slices of L. Fejes Tóth's sausage conjecture. *Mathematika*, 29 (1982), 194–201.

[5] T. Bonnesen, W. Fenchel. Theory of Convex bodies. BCS Assoc., Moscow (Idaho), 1987. Translated from German: Theorie der konvexen Körper, Springer, 1934.

[6] B. Carl: Inequalities of Bernstein–Jackson–type and the degree of compactness of operators in Banach spaces. *Ann. Inst. Fourier (Grenoble)*, 35 (1985), 79–118.

[7] B. Carl, A. Pajor: Gelfand numbers of operators with values in Hilbert space. *Invent. Math.*, 94 (1988), 479–504.

[8] J.H. Conway, N.J.A. Sloane: Sphere packings, Lattices and Groups. *Springer–Verlag*, 1989.

[9] H.S.M. Coxeter, L. Few, C.A. Rogers: Covering space with equal spheres. *Mathematika*, 6 (1959), 147–157.

[10] L. Dalla, D.G. Larman, P. Mani–Levitska, Ch. Zong: The blocking numbers of convex bodies. *Discrete Comput. Geom.*, 24 (2000), 267–277.

[11] Gy. Elekes: A geometric inequality and the complexity of computing the volume. *Disc. Comp. Geom.*, 1 (1996), 289–292.

[12] P. Erdős, C.A. Rogers: Covering space with convex bodies. *Acta Arith.*, 7 (1962), 281–285.

[13] G. Fejes Tóth and W. Kuperberg: Packing and covering. *In: Handbook of Convex Geometry*, P.M. Gruber and J.M. Wills (eds), North Holland, 1993, 799-860.

[14] L. Fejes Tóth: Regular Figures. *Pergamon Press*, 1964.

[15] P.M. Gruber, C.G. Lekkerkerker: Geometry of Numbers. North Holland, 1987.

[16] M. Kochol: Constructive approximation of a ball by polytopes. *Math. Slovaca*, 44 (1994), 99–105.

[17] C.A. Rogers. A note on coverings. *Mathematika*, 4 (1957), 1–6.

[18] C.A. Rogers: Covering a sphere with spheres. *Mathematika*, 10 (1963), 157–164.

[19] K. Schütte: Überdeckungen der Kugel mit höchstens acht Kreisen. *Math. Ann.*, 129 (1955), 181–186.

[20] Ch. Zong: Sphere packings. *Springer–Verlag*, 1999.

About Authors

Károly Böröczky, Jr. is at the Rényi Institute, Budapest, PO. Box. 127, H–1364 Hungary. *e-mail:* carlos@renyi.hu

Gergely Wintsche is at the Lóránd Eötvös University, Teacher Training Faculty, Budapest, Pázmány P. sétány 1/c, 1117 Hungary.
e-mail: wgerg@ludens.elte.hu

Acknowledgments

We would like to thank I. Bárány whose remarks improved the presentation, and who brought our attention to the paper P. Erdős and C.A. Rogers [12]. Work on this paper was supported by OTKA grants T 31 984 and T 30 012.

Lower Bounds for High Dimensional Nearest Neighbor Search and Related Problems

Allan Borodin
Rafail Ostrovsky
Yuval Rabani

Abstract

In spite of extensive and continuing research, for various geometric search problems (such as nearest neighbor search), the best algorithms known have performance that degrades exponentially in the dimension. This phenomenon is sometimes called the *curse of dimensionality*. Recent results [37,38,40] show that in some sense it is possible to avoid the curse of dimensionality for the approximate nearest neighbor search problem. But must the exact nearest neighbor search problem suffer this curse? We provide some evidence in support of the curse. Specifically we investigate the exact nearest neighbor search problem and the related problem of exact partial match within the asymmetric communication model first used by Miltersen [43] to study data structure problems. We derive non-trivial asymptotic lower bounds for the exact problem that stand in contrast to known algorithms for approximate nearest neighbor search.

1 Introduction

Background. One of the most intriguing problems concerning search structures in computational geometry is the following: Given n points in a d-dimensional Euclidean space (the *database*), pre-process the points so that queries of the form *"find in the database the closest point to location x"* can be answered quickly. More generally, we can define this *nearest neighbor search* (NNS) problem in any vector space, and with any metric (or even with a non-metric distance function). Recently, theoretical research into this problem gained some momentum, inspired in part by applications to multimedia information retrieval and data mining [10,20,26,27,34,47,49,50].

Trivially, one solution to the problem is to store the raw data, and in response to a query, to compute the distance from the query to each of the n points. Typically, as in the Euclidean case, this would take $O(nd)$ storage and $O(nd)$ search time. On the other extreme, if the set of possible queries is finite (e.g., in the Hamming cube), one can store the answer to each possible query in a dictionary keyed by the query. In the Hamming cube (as a typical case), this requires $\exp(d)$ storage, but merely $O(d)$ search time. Thus, the challenge is to find a solution enjoying the good aspects of both trivial solutions; that is, storage polynomial (preferably linear) in nd (the size of the data set), and search time polynomial (again preferably linear) in d (the size of the query). This problem is non-trivial in the range[1] $\log n \ll d \ll n^{\kappa}$, for all $\kappa > 0$. (In non-discrete spaces, such as the Euclidean case, even smaller values of d are challenging and for small d we might rephrase the challenge to allow search time polynomial, preferably linear, in $d + \log n$.)

To date, no such solution is known for arbitrary n and d in any reasonable setting (such as Euclidean space, or the Hamming cube). The common "wisdom" among researchers is that simultaneously getting poly(nd) storage and poly(d) search time is impossible. Moreover, it has been conjectured that either storage or search time must grow exponentially in d (at least for certain values of n). This conjecture is known as *the curse of dimensionality* [19]. Consequently, much of the present research emphasizes some restriction of the problem, such as considering "typical inputs", or approximate solutions.

Our results. This paper aims at providing some more persuasive evidence for the curse of dimensionality in a combinatorial setting. We examine NNS in the context of the cell probe model [52]. In the cell probe model, the database is stored in a data structure consisting of m memory cells, each containing b bits. A query is answered by probing in sequence t cells (the address of each cell may be a function of the query and of the contents of previously probed cells). We exploit the connection between asymmetric communication complexity and the cell probe model [44] to derive tradeoffs between the size m of the data structure and the number $t \cdot b$ of bits retrieved.

Specifically, consider the d-dimensional Hamming cube $C_d = \{0,1\}^d$. We analyze the communication game between Alice, who gets a query $q \in C_d$, and Bob, who gets a database $D \in C_d^n$ (think of D as a set of n points in C_d). They have to output one iff the minimum Hamming distance from q to a point in D is at most λ, and otherwise they have to output zero. The threshold $\lambda \in \{0, 1, 2, \ldots, d\}$ is a fixed parameter of the communication game. Notice that the function that Alice and Bob have to compute is a decision version of NNS. We refer to it as the λ-*neighbor problem.* Clearly, the NNS problem is at least as hard as the λ-neighbor problem. We show that if there is a randomized two-sided error protocol[2] to compute the λ-neighbor problem where Alice sends a bits and Bob sends b bits, then either $a = \Omega(\log n \log d)$ or $b = \Omega(n^{1-\varepsilon})$, for every $\varepsilon > 0$. We derive these bounds using the general *richness* technique developed in [46]. In fact, using this richness technique we can prove much stronger lower bounds for randomized one-sided error protocols (and hence for deterministic protocols) for the *complement* of the λ-neighbor problem. Namely, for every ϵ, there is a δ such that either Alice

[1] We use $f(n) \ll g(n)$ to denote that f is asymptotically smaller than g.

[2] For definiteness we say that a two-sided error protocol returns the correct answer with probability at least 2/3. A one-sided error protocol for a decision problem never incorrectly answers YES (=1) and incorrectly answers NO (=0) with probability at most 1/3.

sends $(1-\epsilon)d$ bits or Bob sends at least δnd bits. However, as we will show, a "direct application" of this method to the λ-neighbor problem appears to be impossible.[3] Our main conceptual contribution then is a way to restrict the instances of the λ-neighbor problem to a subset with a nicely structured communication matrix so as to be able to apply the richness technique to the λ-neighbor problem.

We achieve the restriction by considering a different well studied search problem of significant importance in its own right. In the *exact partial match problem*, the database consists of n points in the Hamming cube C_d. A query has 0-1 values assigned to some of the coordinates. The other coordinates are *don't cares*. In reply, we must check if the query matches one of the points in the database, comparing positions with assigned 0-1 values only. (Notice that the implicit distance function here is not a metric.) We show an easy reduction from the exact partial match problem to the λ-neighbor problem, for some value of λ. The reduction produces restricted instances of the λ-neighbor problem. We then proceed to show lower bounds for the communication complexity of exact partial match. We have to further restrict the instances to allow for the application of the richness technique.

The bounds on communication complexity imply lower bounds on exact partial match and on NNS in the Hamming cube. By a simple reduction, we also get similar results for instances of points in ℓ_p^d ($\mathbb{R}^d$ with distances measured by the L_p norm), for every $1 \le p < \infty$. We show that if a nearest neighbor query is answered in a constant number of probes t, then we either need super-polynomial storage ($2^{\Omega(\log n \log d)}$ cells), or else we need to retrieve nearly-linear $\Omega(n^{1-\varepsilon})$ bits from the database. These lower bounds stand in contrast with recent positive results on approximate search (see below). Alternatively, if we restrict ourselves to cells of size poly(d), then a polynomial number of cells is unattainable unless the number of rounds $t = \Omega(\log d)$. This improves upon the $\Omega(\sqrt{\log d})$ randomized two-sided error lower bound claimed in [46] for the "notoriously difficult" partial match problem. (Our definition of the exact partial match problem differs from the partial match problem defined in [46]. However, a d-dimensional instance of our problem can be easily embedded into a $2d$ dimensional instance of their problem so that our $\Omega(\log d)$ lower bound does apply to their problem as well.) We note in passing that deriving strong lower bound tradeoffs in the unrestricted cell probe model is a well-recognized and fundamental open problem (see below).

[3] As discussed in the related work mentioned below, the known upper bounds for an approximate version of the λ-neighbor problem use randomization with (depending on the implementation) 1-sided or 2-sided error. A 1-sided error lower bound for the non λ-neighbor problem does not imply a randomized lower bound for the λ-neighbor problem (although it does imply a *deterministic* lower bound). Hence our lower bound for the non λ-neighbor problem does not serve to distinguish between the exact and approximate versions of the nearest neighbor problem. Of course, a 2-sided error lower bound for non λ-neighbor implies the same bound for λ-neighbor.

Related work. There is an extensive body of research concerning nearest neighbor problems for small dimensional (e.g. 2 and 3) Euclidean space (see for example the text by de Berg *et al* [13]). Dobkin and Lipton's seminal paper [21] marks the beginning of work on the Euclidean case of arbitrary dimension. They achieve a discretization of the problem, so that (super-) exponential storage can be used to answer queries quickly. Dobkin and Lipton use $O\left(n^{2^{d+1}}\right)$ storage to allow $O(2^d \log n)$ search time. Clarkson [18] improves the storage requirement to $O\left(n^{(1+\delta)\lceil d/2\rceil}\right)$, paying $d^{O(d)} \log n$ search time. Improvements by Yao and Yao [53], Matoušek [41], and Agarwal and Matoušek [1] still give exponential in d storage and search time. Finally, Meiser [42], gives the best result to date (in terms of search time) — $O(d^5 \log n)$ search time using $O\left(n^{d+1+\delta}\right)$ storage.[4]

In the *approximate nearest neighbor search* (approximate NNS) problem, a query is answered by finding a database point whose distance from the query is within a factor of $(1+\varepsilon)$ of the distance to the closest database point. Usually, the parameter ε is fixed in the pre-processing phase. (To avoid confusion, we refer to the version of the problem requiring an exact answer as the *exact* NNS problem.) For approximate NNS in Euclidean space, Arya and Mount [6] give $O(1/\varepsilon)^d \log n$ search time using $O(1/\varepsilon)^d n$ storage. Clarkson [19] improves the dependence on ε to $(1/\varepsilon)^{(d-1)/2}$. Arya, Mount, Netanyahu, Silverman, and Wu [7] give $O((d/\varepsilon)^d \log n)$ search time using $O(n \log n)$ storage (the pre-processing does not depend on ε). In comparison with exact search, these results have better storage requirements. Still, their search time is better than the trivial $O(nd)$ for small d only (as are the results for exact search excluding Meiser's).

Kleinberg [38] gives $O(n + d\log^3 n)/\varepsilon^2$ search time using $nd \log^{O(1)} n/\varepsilon^2$ storage, thus providing asymptotic improvement over the trivial search time for all (non constant) d using polynomial storage. Another algorithm of Kleinberg (in the same paper) gives $d^2 \log^{O(1)} n/\varepsilon^2$ search time using $O(n \log d/\varepsilon^2)^{2d}$ storage. This is better than Meiser's exact search time, using similar storage. Indyk and Motwani [37] give improved bounds of $d \log^{O(1)} n$ search time using $O(1/\varepsilon)^d n \log^{O(1)} n$ storage. Their result extends to any L_p norm. They also show that for $1 \le p \le 2$, polynomial $(nd)^{O(1/\varepsilon^2)}$ storage can be used to answer correctly *most* queries in time $d \log^{O(1)} n/\varepsilon^2$. Independent of [37], Kushilevitz, Ostrovsky, and Rabani [40] give $d^2 \log^{O(1)} n/\varepsilon^2$ search time (for *all* queries) using $(nd)^{O(1/\varepsilon^2)}$ storage. Their result holds for the L_1 norm

[4]Meiser considers the more general problem of point location in arrangements of hyperplanes and obtains the storage bound $O\left(n^{d+\delta}\right)$ where n is the number of hyperplanes. The nearest neighbor problem with n data points is easily transformed into this general problem by considering the $\binom{n}{2}$ hyperplanes constructed by bisecting each pair of data points. To obtain the stated space bound, the given problem is transformed to a point location query among n hyperplanes in $(d+1)$-space which is obtained by replacing each point $(a_1, a_2, \ldots, a_d)a_d$ by the hyperplane $x_{d+1} = (x_1 - a_1)a_1 + \ldots + (x_d - a_d)$ and replacing the query point $(q_1, \ldots, q_d)$ by $(q_1, \ldots, q_d, \infty)$. The hyperplane directly below the transformed query point corresponds to the original query point's nearest neighbor.

too (in particular for the cube, with somewhat better bounds). With the exception of the first algorithm of Indyk and Motwani [37] (which is still exponential in d for small ε), all of these approximate NNS algorithms are randomized algorithms. Of related interest are results on approximate NNS for a large ε by Bern [14] and Chan [16] (for Euclidean space), and by Indyk [36] (for the L_∞ norm).

Thus, approximate NNS does not suffer from the curse of dimensionality,[5] at least not from the point of view of randomized algorithms and asymptotic bounds for fixed ε. In fact, most of the results mentioned above can be stated in terms of the cell probe model, using a small number of probes. For example, a λ-neighbor version of the algorithm of Kushilevitz et al. [40] for the cube can be implemented as a randomized two-sided error one round cell probe algorithm with $n^{O(1/\varepsilon^2)}$ cells, each containing one bit.[6] Using several such structures (for different distances) an approximate NNS implementation takes $O(\log\log d)$ rounds (using binary search on a geometric progression of λ's). Therefore, the lower bounds in this paper provide some evidence that in high dimension, exact NNS is indeed far more difficult than approximate NNS.

It is easy to see that the exact full match problem (equivalent to 0-neighbor) also does not suffer the curse of dimensionality. Indeed, one can pre-process the database by using a perfect hashing function thereby permitting a query search to be performed in one table lookup. Alternatively, one can arrange the database in a d-depth, n-leaf tree and then perform the query search by a simple search of the tree in time $O(d)$. We do not know of any analogous results for the partial match problem.

However, to the best of our knowledge, lower bounds for exact NNS (or the partial match query problem) in high dimensions do not seem sufficiently "convincing" to justify the curse of dimensionality conjecture. That is, either the models with respect to which lower bounds have been established seem quite restricted or the bounds are quite weak. One nice example of a well structured model (for both dynamic and static data structure problems) is Fredman's [28–30] semi-group model. The model is designed for searching problems (e.g., range queries) in which a semi-group value is associated with each data point and one wants to retrieve the semi-group sum of all data points in some specified set (e.g., satisfying a partial match or more generally satisfying a range query). The static model allows pre-processing of sums of arbitrary subsets of database points and Fredman derives strong lower bound tradeoffs between memory (the number of pre-computed subsets) and search time[7] (the number of semi-group additions on these pre-computed subsets).

[5] However, as one of the referees has pointed out, since the stated complexity bounds are now exponential in $1/\epsilon^2$, we might say that the curse of dimensionality has been replaced by a "curse of exactness".

[6] Using d bits per cell, it is possible to derive a one-sided error implementation.

[7] Associated with each possible query is a straight line program whose operations are either the semi-group addition $v_i = v_j + v_k$ or the scalar multiplication of a semi-group value $v_i = c \cdot v_j = v_j + v_j + \ldots + v_j$ by a positive integer c.

While this model can be used for decision problems (by letting the semigroup be the Boolean values under the operation of logical OR) the model is clearly restrictive in the limited way information can be obtained from the data structure.

In the setting of real Euclidean space (i.e., $\mathbb{R}^d$ with the usual L_2 metric), an appropriate model is the algebraic computation tree (or the closely related algebraic decision tree). Simply stated for real valued inputs, one can compute and test the signs of polynomials of these inputs. The complexity of such an algebraic computation tree is usually taken as the number of multiplications and tests. The algebraic computation model was first introduced by Ben-Or [12] (having been preceded by the algebraic decision tree model where one only counts tests but then restricts the degree of the polynomials allowed). Following a substantial chain of papers, Grigoriev and Karpinski [32] prove a randomized lower bound to determine membership in a polyhedron, the lower bound being (roughly speaking) the logarithm of the number of faces. Since there are instances of n data points $x_1, \ldots, x_n$ in $\mathbb{R}^d$ which give rise to a Voronoi diagram with $\Theta(n^{\lceil d/2 \rceil})$ faces, we can apply the polyhedron lower bound to derive an $\Omega(d \log n)$ randomized algebraic computation tree lower bound for deciding if a given query $q \in \mathbb{R}^d$ is closest to a given data point x_i. Computation tree models do not reflect storage usage and indeed this lower bound holds independent of the storage allowed and hence the bound provides a somewhat matching counter-part to the $d^5 \log n$ upper bound of Meiser. In the context of our paper, we note that the model does not allow modular operations (thereby precluding certain hashing functions), and the analysis used for the Grigoriev and Karpinski lower bound is not applicable for the case of finite domains such as the Hamming cube.

A model which does capture hashing and more combinatorial settings is introduced in Rivest [48] and further developed in Dolev, Harari and Parnas [22] and Dolev, Harari, Linial, Nisan and Parnas [23]. Rivest studies the *all* partial match problem and Dolev *et al.* study the *all* λ-neighbor problem where for each problem all database points satisfying the query must be found. In this model, each database point is hashed to a bucket and interesting tradeoffs are established between the number of buckets containing database points satisfying the query and the maximum size of a bucket. Another lower bound is by Indyk [36]. He shows a lower bound for approximate NNS under the L_∞ norm in the *indexing model* of Hellerstein, Koutsoupias, and Papadimitriou [35]. The indexing model tries to capture the cost of using external memory devices for large data sets, and appears to be computationally more restricted than the cell probe model. Indyk shows that the *superset query* problem of Hellerstein et al. reduces to $(1+\varepsilon)$-approximate NNS under the L_∞ norm, for any $\varepsilon < 1$. The lower bound that Hellerstein et al. give for superset query is weak, unless the storage redundancy is quite small. This does not seem to pose a serious theoretical restriction on a solution, though it may address important considerations in practice.

The cell probe model was formulated by Yao [52]. It is considered as the most general data structure model for proving lower bounds. Ajtai [2] and Xiao [51] obtain further lower bounds in this model. Miltersen [43–45] pioneered the connection between the cell probe model and asymmetric communication complexity. Miltersen, Nisan, Safra, and Wigderson [46] provide general methods for proving lower bounds for asymmetric communication complexity, including the richness technique used here. For more information on communication complexity, see the book by Kushilevitz and Nisan [39]. As Miltersen et al. observe, communication complexity cannot prove strong lower bounds for the cell probe model without restrictions (say on the number of rounds). Furthermore, they observe that lower bounds for the general cell probe model imply time-space tradeoffs for branching programs, one of the notoriously difficult problems in computational complexity. For the best lower bound on branching programs to date, and additional references, see Beame, Saks, and Thathachar [9], and the two recent papers of Ajtai [3,4].

As mentioned above, Rivest [48] analyses hashing based algorithms for partial match. (See Bentley and Sedgewick [11] for a more recent historical account.) Rivest conjectures that *any* $O(nd)$-sized data structure would require $\Omega(n^{1-s/d})$ time to search, where s is the number of exposed coordinates.[8] Also as mentioned above, using a *round elimination* technique, Miltersen et al. [46] claim a lower bound of $\Omega(\sqrt{\log d})$ on the number of probes required to find a partial match in the cell probe model with $(nd)^{O(1)}$ cells, each containing poly(d) bits. We improve this bound to $\Omega(\log d)$. Moreover, for the corresponding communication problem our lower bounds are on the total number of bits transmitted by either side, irrespective of the number of rounds.

Independent of (and complimentary to) our work, Chakrabarti, Chazelle, Gum and Lvov [15] have established an $\Omega(\log\log d/\log\log\log d)$ deterministic lower bound in the cell probe model for finding an approximate nearest neighbor. The approximation factor here can be as large as $2^{\lfloor(\log d)^{1-\varepsilon}\rfloor}$ for any $\varepsilon > 0$.

Subsequent to our conference publication, Barkol and Rabani [8] established a significantly improved lower bound for the λ-neighbor problem. They are able to derive a lower bound for *two-sided error* randomized asymmetric communication complexity protocols that is quite comparable to our one-sided error bound in Theorem 2.2. Namely, again assuming some necessary reasonable bounds on the range of the dimension d, they show:

Let $\lambda = \frac{d}{2} - \frac{2}{\ln 2}\sqrt{d\log n}$. For every $0 < \varepsilon < 1$ there exist $\delta > 0$, such that in every two-sided error protocol for the λ-neighbor problem in C_d, when n is sufficiently large then either the query side sends at least εd bits, or the database side sends at least n^δ bits.

---

[8] The Rivest conjecture was stated without any conditions on s but it seems reasonable to believe that the conjecture assumes that s is not very small (e.g. s is a constant) or very large (e.g. $d - c$, c a constant).

2 A Communication Complexity Lower Bound for the non λ-neighbor problem

We use the Miltersen et al. [46] *richness technique* to analyze the asymmetric communication game between Alice, who gets a query $q \in C_d$, and Bob, who gets a database $D \in C_d^n$.[9] For the *non λ-neighbor problem*, the players have to output one iff there does *not* exist a data base point $x \in D$ whose Hamming distance to the query q is at most λ. An $[a, b]$ protocol for a communication problem is a protocol in which Alice sends at most a bits and Bob sends at most b bits.

For the sake of completeness, we review the Miltersen et al. [46] richness technique (for one-sided error protocols). We associate a communication matrix M_f with any communication problem f. Namely we index the rows by the possible inputs for Alice and the columns by the possible inputs for Bob and the x, y entry of M_f is the value $f(x, y)$. A communication problem f is $[u, v]$-rich iff its communication matrix M_f has at least v columns each containing at least u ones. The richness technique is captured by the following *richness lemma:*

Lemma 2.1 (Miltersen et al. [46]). *Let f be $[u, v]$-rich. If f has a randomized one-sided error $[a, b]$ protocol, then M_f contains a submatrix of dimension at least $u/2^{a+2} \times v/2^{a+b+2}$ containing only 1-entries.*

The richness technique then is to show that a given communication problem is sufficiently rich yet does not contain large submatrices containing only 1-entries. We now apply this technique to the non λ-neighbour problem. We wish to derive asymptotic tradeoffs for "large" dimensions. Thus we consider an infinite sequence of problems for all sufficienlty large values of n, and dimension $d = d(n)$. Here we assume $\log n \ll d$.

Theorem 2.2. *For every $0 < \varepsilon < 1$ there exist δ and ν, such that for every $n > \nu$ there exists λ for which in every protocol for non λ-neighbor problem in C_d, either the query side sends at least $(1 - \varepsilon)d$ bits, or the database side sends at least δnd bits.*

Let $B_d(\lambda)$ denote the Hamming ball of radius λ around an arbitrary vector in C_d and let $B_d(A, \lambda)$ denote the set of cube points at distance at most λ from a point in $A \subseteq C_d$. In the proof of this theorem, we use the following standard inequalities (see, for example, [5]) :

Fact 2.3 (Entropy bound). *For $0 < p < \frac{1}{2}$, $|B_d(pd)| \leq 2^{(H(p)+o(1))d}$, where $H(p) = -p \log p - (1 - p) \log(1 - p)$ is the entropy function.*

Fact 2.4 (Harper's isoperimetric inequality [33]). *For $A \subseteq C_d$, let $r > 0$ be such that $|A| \geq |B_d(r)|$. Then, for every $\lambda > 0$, $|B_d(A, \lambda)| \geq |B_d(r + \lambda)|$.*

[9] That is, in order to simplify the analysis, we allow repetitions in the database. The results that follow would not change in any significant way if a database was defined to be n distinct vectors.

Fact 2.5 (Chernoff's bound). *For every* $a > 0$, $|B_d(d/2-a)| \le e^{-a^2/d} . 2^d$.

Proof of Theorem 2.2. We apply the richness technique. Take $\lambda = \frac{d}{2} - \sqrt{d\ln(2n)}$. (The hardest case seems to be to distinguish between a distance of at most $\frac{d}{2}$ and a distance of at least $\frac{d}{2}+1$.) Using Claim 2.5, $n|B_d(\lambda)| \le 2^{d-1}$. Thus, for every database at most half the queries are within a distance of λ of one of the points in the database. Therefore, for our choice of λ, non λ-neighbor is $[2^{d-1}, 2^{nd}]$-rich. We will show that for every $\varepsilon > 0$, there exists δ such that every $2^{\varepsilon d} \times 2^{(1-\delta)nd}$ submatrix of the communication matrix of the non λ-neighbor problem contains a zero entry.

Consider a set Q of queries of cardinality $2^{\varepsilon d}$. Let λ_Q be the largest integer such that $|B_d(\lambda_Q)| \le |Q|$. By Fact 2.3, there exists a constant $p = p(\varepsilon)$ such that $\lambda_Q \ge pd$. Let ξ be a constant such that $0 < \xi < p - \sqrt{\ln(2n)/d}$. By our assumption that $\log n \ll d$, we can choose ξ arbitarily close to p when n is sufficiently large. Then,

$$\begin{aligned}
|B_d(Q,\lambda)| &\ge |B_d(pd+\lambda)| \\
&= \left|B_d\left(\left(\frac{1}{2}+p\right)d - \sqrt{d\ln(2n)}\right)\right| \\
&\ge \left|B_d\left((\frac{1}{2}+\xi)d\right)\right| \\
&> 2^d - \left|B_d\left((\frac{1}{2}-\xi)d\right)\right| \\
&\ge 2^d - e^{-\xi^2 d}2^d \\
&= 2^d - 2^{(1-\xi^2)d},
\end{aligned}$$

where the first inequality follows from Fact 2.4, and the fourth inequality follows from Fact 2.5. Consider a probability distribution over C_d, with all points equally likely. From the final inequality above, the probability a random point from this distribution does not fall in $B_d(Q,\lambda)$ is less than $\frac{2^{(1-\xi^2)d}}{2^d} = 2^{-\xi^2 d}$. Consider now a distribution over C_d^n, with all databases equally likely. This distribution is equivalent to taking n independent samples from the uniform distribution over C_d. Thus, the probablility that none of the points of a random n point database fall in $B_d(Q,\lambda)$ is less than $2^{-\xi^2 nd}$. Put $\delta = \xi^2$, and the theorem follows. □

3 A Communication Complexity Lower Bound for the partial match problem

In the exact partial match problem, the database consists of n vectors v^1, v^2, $\ldots$, v^n in C_d, and a query is a vector in $\tilde{C}_d = \{0,1,*\}^d$. A query q matches a vector $v \in C_d$ iff for all $j \in \{1,2,\ldots,d\}$, either $q_j = *$ or $q_j = v_j$. A query q matches the database iff there exists $i \in \{1,2,\ldots,n\}$ such that q matches

v^i. We say that the coordinates j for which $q_j \neq *$ are *exposed* in q. We also refer to any j such that $q_j = *$ as a *don't care coordinate*, or simply as a *don't care*. For simplicity, we assume that n is a power of 2 (thus $\log n$ is an integer). We shall see that lower bounds for the partial match problem imply corresponding lower bounds for the exact NNS problem. (As far as we know, lower bounds for exact NNS cannot be used to imply lower bounds for the partial match problem.)

We wish to derive asymptotic tradeoffs for the partial match problem for "large" dimensions. Thus we again consider an infinite sequence of problems for all sufficiently large values of n, each of dimension $d = d(n)$. We assume that $\log n \ll d \leq n^{\kappa}$ for a positive $\kappa < 1$.[10]

The set of possible databases again includes all 2nd choices in C_d^n. The set of possible queries is restricted to

$$Q_{n,d} = \{q; |\{i; q_i \neq *\}| = \log n + 1\}.$$

In words, the queries are restricted to have exactly $\log n + 1$ exposed coordinates. Notice that the number of possible queries is exactly

$$2n\binom{d}{\log n + 1} = 2^{O(\log n \log d)}.$$

Our main result is the following theorem:

Theorem 3.1. *Let $0 < \varepsilon < 1 - \kappa$ be fixed. Suppose there is an $[a, b]$ (deterministic or randomized, two-sided error) communication protocol for* NPM. *Then, either $a = \Omega(\log n \log d)$, or $b = \Omega\left(n^{1-\varepsilon}\right)$.*

We first present the proof for one-sided error protocols. We then extend the proof to handle two-sided error protocols. The latter is somewhat more involved, yet it builds on the ideas for the one-sided error case. Throughout this section ε refers to the ε as stated in the theorem.

Lemma 3.2. NPM *is $\left[\frac{R}{5}, \frac{C}{6}\right]$-rich where $R = 2n\binom{d}{\log n+1}$ is the number of rows in M_{NPM} and $C = 2^{nd}$ is the number of columns in M_{NPM}.*

Proof. As in the previous section, consider a probability distribution over C_d, with all points equally likely. For $q \in Q_{n,d}$, the probability that q matches a random vector from this distribution is $\frac{1}{2n}$. As before, consider now a distribution over C_d^n, with all databases equally likely. The probability that q does not match a random database from this distribution is

$$\left(1 - \frac{1}{2n}\right)^n \geq e^{-1}.$$

If, however, less than $\frac{1}{6}$ of the columns (databases) contain at least a fraction of $\frac{1}{5}$ ones entries, then the fraction of ones entries in the communication matrix does not exceed $\frac{1}{3}$, a contradiction. □

[10]The assumption that the dimension $d \leq n^{\kappa}$ is only used for the case of 2-sided errors.

We say that two queries $q, q' \in Q_{n,d}$ are *consistent* if they agree on all coordinates which are exposed in both vectors. We say that q and q' are *ε-neighbors* iff they are consistent and the number of coordinates which are exposed in both of them is at least $\varepsilon \log n$. The notion of ε-neighbors is useful because of the following lemma.

Lemma 3.3. *If $q, q' \in Q_{n,d}$ are not ε-neighbors, then the fraction of vectors $v \in C_d$ such that both q and q' match v is at most $\frac{1}{4n^{2-\varepsilon}}$.*

Proof. If q and q' are not consistent, then by definition there is no vector that they both match. Otherwise, all of the coordinates that are exposed in both queries must have the same value in both. Furthermore, if both q and q' match a vector v, then for every j which is exposed in either q or q', v_j must equal the exposed bit. The total number of different coordinates exposed in either q or q' is at least $(2-\varepsilon)\log n + 2$. The lemma follows from computing the fraction of vectors with these bits fixed. □

Lemma 3.4. *For every $\delta > 0$ there exists ν such that for all $n > \nu$, for $d = d(n)$ the following holds. For every $q \in Q_{n,d}$, the number of its ε-neighbors is less than*

$$\binom{d}{\log n + 1}^{1-\varepsilon+\delta}.$$

Proof. The number N of neighbors of q is given by

$$\begin{aligned}
&\sum_{j=0}^{(1-\varepsilon)\log n+1} \binom{\log n + 1}{j}\binom{d - \log n - 1}{j} 2^j \\
&\le 2n^{1-\varepsilon}\binom{d - \log n - 1}{(1-\varepsilon)\log n + 1} \sum_{j=0}^{\log n+1}\binom{\log n + 1}{j} \\
&= 4n^{2-\varepsilon}\binom{d - \log n - 1}{(1-\varepsilon)\log n + 1} \\
&< 4n^{2-\varepsilon}\binom{d}{(1-\varepsilon)\log n + 1}.
\end{aligned}$$

Now,

$$\begin{aligned}
\frac{\binom{d}{\log n+1}}{\binom{d}{(1-\varepsilon)\log n+1}} &= \frac{((1-\varepsilon)\log n + 1)!\,(d - (1-\varepsilon)\log n - 1)!}{(\log n + 1)!\,(d - \log n - 1)!} \\
&\ge \left(\frac{d - \log n}{\log n + 1}\right)^{\varepsilon \log n}
\end{aligned}$$

$$
\begin{aligned}
&= \left(1 - \frac{\log n}{d}\right)^{\varepsilon \log n} \left(\frac{d}{\log n + 1}\right)^{\varepsilon \log n} \\
&\geq n^{-\varepsilon} \left(\frac{d}{\log n + 1}\right)^{\varepsilon \log n} \\
&\geq n^{-\varepsilon} \left(\frac{ed}{\log n + 1}\right)^{-\varepsilon} n^{-\varepsilon \log e} \left(\frac{d}{\log n + 1}\right)^{\varepsilon}.
\end{aligned}
$$

Therefore,

$$
N < 4n^{2+\varepsilon \log e} \left(\frac{ed}{\log n + 1}\right)^{\varepsilon} \left(\frac{d}{\log n + 1}\right)^{1-\varepsilon}.
$$

Let $\delta > 0$. As $d \gg \log n$, for n sufficiently large,

$$
4n^{2+\varepsilon \log e} \left(\frac{ed}{\log n + 1}\right)^{\varepsilon} \leq \left(\frac{d}{\log n + 1}\right)^{\delta}.
$$

□

We call a set $I \subseteq Q_{n,d}$ *ε-independent* iff for every $q, q' \in I$ such that $q \neq q'$, q and q' are not ε-neighbors. We conclude from the above lemma

Lemma 3.5. *For every δ such that $0 < \delta < \varepsilon/2$, there is ν such that for every $n > \nu$, every $R \subseteq Q_{n,d}$ of cardinality at least $\left(\frac{d}{\log n+1}\right)^{1-\delta}$ contains an ε-independent subset ind(R) of cardinality $n^{1-\varepsilon}$.*

Proof. Let n be sufficiently large so that Lemma 3.4 holds and $\left(\frac{d}{\log n+1}\right)^{\varepsilon-2\delta} \geq n^{1-\varepsilon}$. Consider the graph whose nodes are the elements of R, and a pair of nodes is an edge iff its endpoints are ε-neighbors. By Lemma 3.4, the maximum degree in this graph is less than $\left(\frac{d}{\log n+1}\right)^{1-\varepsilon+\delta}$. Therefore, it has an ε-independent set of size at least $\left(\frac{d}{\log n+1}\right)^{\varepsilon-2\delta} \geq n^{1-\varepsilon}$. □

For $q \in Q_{n,d}$ we denote $\mathcal{D}(q) = \{D \in C_d^n; q \text{ does not match } D\}$. For $R \subseteq Q_{n,d}$ we abuse notation and denote $\mathcal{D}(R) = \cap_{q \in R} \mathcal{D}(q)$.

Lemma 3.6. *For every δ such that $0 < \delta < \varepsilon/2$, there is ν such that for every $n > \nu$, if $R \subseteq Q_{n,d}$ has cardinality at least $\left(\frac{d}{\log n+1}\right)^{1-\delta}$, then $\mathcal{D}(R)$ has cardinality less than $2^{nd-n^{1-\varepsilon}/4}$.*

Proof. We examine the subset ind(R) of cardinality $n^{1-\varepsilon}$ from Lemma 3.5. Consider a distribution over C_d, with all points equally likely. The probability that any $q \in \text{ind}(R)$ matches a random point from this distribution is exactly $\frac{1}{2n}$. Let $q, q' \in \text{ind}(R)$, $q \neq q'$. By Lemma 3.3, the probability that both q and q' match a random point from this distribution is at most $\frac{1}{4n^{2-\varepsilon}}$. Therefore, by the inclusion-exclusion principle, the probability that a random point from the distribution is matched by at least one point in $\mathrm{ind}(R)$ is at least

$$n^{1-\varepsilon} \cdot \frac{1}{2n} - \binom{n^{1-\varepsilon}}{2} \cdot \frac{1}{4n^{2-\varepsilon}} \geq \frac{1}{2n^\varepsilon} - \frac{1}{8n^\varepsilon} = \frac{3}{8n^\varepsilon}.$$

Now, consider a distribution over C_d^n with all databases equally likely. The probability that none of the points in $\mathrm{ind}(R)$ match a random database from this distribution is at most $(1-3/8n^\varepsilon)^n \leq e^{-n^{1-\varepsilon}/4}$. □

Lemma 3.6 implies the following

Corollary 3.7. *For every δ such that $0 < \delta < \varepsilon/2$, there is ν such that for every $n > \nu$, the communication matrix of* NPM *does not contain a $\binom{d}{\log n+1}^{1-\delta} \times 2^{nd-n^{1-\varepsilon}/4}$ 1-monochromatic rectangle.*

Proof. Otherwise, we have a set $R \subseteq Q_{n,d}$ with $|R| \geq \binom{d}{\log n+1}^{1-\delta}$, and $|\mathcal{D}(R)| \geq 2^{nd-n^{1-\varepsilon}/4}$, in contradiction with Lemma 3.6. □

We now can conclude the proof of Theorem 3.1 for one-sided errors by applying the richness Lemma 2.1.

It remains to show how to extend this proof to the case of two-sided error. Miltersen *et al.* [46] prove a second form of the richness lemma which makes it possible to prove lower bounds for randomized algorithms having two-sided error. Essentially, instead of showing that every sufficiently large submatrix is not 1-monochromatic, we now need to show that every sufficiently large submatrix has a constant fraction of zeros. We now indicate how to apply this form of the richness lemma to establish Theorem 3.1 for two-sided error protocols.

Lemma 3.8. *For every δ such that $0 < \delta < \varepsilon/2$, there is ν such that for every $n > \nu$, every $R \subseteq Q_{n,d}$ of cardinality at least $\binom{d}{\log n+1}^{1-\delta}$ can be partitioned into sets $I_0, I_1, I_2, \ldots, I_f$ such that the following hold:*

1. *I_0 contains at most half of R; and,*
2. *for $j = 1, 2, \ldots, f$, I_j is an ε-independent set with $|I_j| = 2n^{1-\varepsilon}$, and*
3. *$f \ll 2^{n^\kappa}$. (Recall $\kappa < 1-\varepsilon$ and $d \leq n^\kappa$.)*

Proof. As long as at least half of R remains, repeatedly apply Lemma 3.5 to pick a set I_j with the desired properties, then remove it from R. (To be more precise, we have to slightly modify Lemma 3.5 to make each I_j have size $2n^{1-\varepsilon}$ assuming $R \subseteq Q_{n,d}$ is of cardinality at least $\frac{1}{2}\binom{d}{\log n+1}^{1-\delta}$.) Now f is trivially smaller than the total number of queries. Thus, $f \leq 2n\binom{d}{\log n+1} \ll 2^{n^\kappa}$, for sufficiently large n (as we assume that $d \leq n^\kappa$). □

We want most databases to match many points in the sets I_j, $j > 0$. For $I \subseteq Q_{n,d}$, we denote by $\mathcal{D}(I)$ the set of databases that match less than $\frac{1}{100}$ of the queries in I.[11] We have

Lemma 3.9. *For every $\varepsilon > 0$, there is ν such that for every $n > \nu$ the following holds: For every ε-independent $I \subseteq Q_{n,d}$ such that $|I| = 2n^{1-\varepsilon}$, we have $\mathcal{D}(I) \leq 2^{nd-n^{1-\varepsilon}/50}$.*

Proof. Consider a database chosen at random, all databases equally likely. We think of the database as being chosen in sequence, one point at a time, each point chosen independently of the others from a uniform distribution over C_d. Let $x_1, x_2, \ldots, x_n$ be the database points. Let $M_0, M_1, M_2, \ldots, M_{n^{1-\varepsilon}}$ be the following subsets of I: $M_0 = \emptyset$. M_i includes all of M_{i-1}, and if any query q in $I \setminus M_{i-1}$ matches one of the database points $x_{(i-1)n^{\varepsilon}+1}, \ldots, x_{in^{\varepsilon}}$, then M_i also contains one (arbitrarily chosen) such matching query q; thus, $|M_{i-1}| \leq |M_i| \leq |M_{i-1}| + 1$. Let $X_i = |M_i|$.

We show that for all i, $0 \leq i < n^{1-\varepsilon}$, $\Pr[X_{i+1} > X_i] \geq 1 - e^{-1/4}$. For all i, $|I \setminus M_i| \geq n^{1-\varepsilon}$. Therefore, by the proof of Lemma 3.6, the probability that a random database point is matched by one of the queries in $I \setminus M_i$ is at least $3/8n^{\varepsilon}$. For n^{ε} random points we get that the probability that none of these points are matched by a query in $I \setminus M_i$ is $(1 - 3/8n^{\varepsilon})^{n^{\varepsilon}} \leq e^{-1/4}$, for sufficiently large n.

Now, define $Y_0, Y_1, \ldots, Y_{n^{1-\varepsilon}}$: Y_0 is the initial expected size of $M_{n^{1-\varepsilon}}$. Y_i is the same expectation after choosing $x_1, \ldots, x_{in^{\varepsilon}}$. Notice that all these expectations are random variables (depending on the choice of $x_1, \ldots, x_{in^{\varepsilon}}$ with Y_0 being a constant). Further notice that $Y_{n^{1-\varepsilon}} = |M_{n^{1-\varepsilon}}|$. By definition, $Y_i = E[Y_{i+1} \mid Y_i]$. Also, $|Y_i - Y_{i+1}| \leq 1$. Therefore, the sequence $Y_0, Y_1, \ldots, Y_{n^{1-\varepsilon}}$ is a martingale. By Azuma's inequality, $\Pr[Y_{n^{1-\varepsilon}} < Y_0 - \lambda n^{(1-\varepsilon)/2}] < e^{-\lambda^2/2}$. Now, by the linearity of expectation, $Y_0 \geq \left(1 - e^{-1/4}\right) n^{1-\varepsilon}$. Set $\lambda = \frac{1}{5} n^{(1-\varepsilon)/2}$. We get, $\Pr[Y_{n^{1-\varepsilon}} < \left(1 - e^{-1/4} - 1/5\right) n^{1-\varepsilon}] < e^{-n^{1-\varepsilon}/50}$. Finally, notice that $1 - e^{-1/4} - \frac{1}{5} > \frac{1}{50}$ so that

$$\left(1 - e^{-1/4} - 1/5\right) n^{1-\varepsilon} > \frac{1}{100} |I|.$$

□

Theorem 3.10. *For every δ such that $0 < \delta < \varepsilon/2$, there exists ν such that for all $n > \nu$ the following holds: In the communication matrix of* NPM, *in every rectangle $R \times D$ with $|R| \geq \left(\frac{d}{\log n + 1}\right)^{1-\delta}$ and $|D| \geq 2^{nd-n^{1-\varepsilon}/100}$, at least a fraction of $\frac{1}{400}$ of the entries are zeros.*

Proof. Let n be sufficiently large, and let $R \times D$ be a rectangle satisfying the conditions of the theorem. By Lemma 3.8, at least half the queries in R can be partitioned into disjoint independent subsets $I_1, I_2, \ldots, I_f$,

---

[11] $\frac{1}{100}$ is a somewhat arbitrary constant.

$f \ll 2^{n^\kappa}$, for all $\kappa > 0$ and for n sufficiently large. For any j, $1 \le j \le f$, the number of databases that match less than $\frac{1}{100}$ of the queries in I_j is at most $2^{nd-n^{1-\varepsilon}/50}$, by Lemma 3.9. Therefore, the number of databases that match less than $\frac{1}{100}$ of the queries in any of the sets I_j is less than $2^{nd-n^{1-\varepsilon}/50+n^\kappa} \le 2^{nd-(n^{1-\varepsilon}/100)-1}$, for n sufficiently large and $0 < \kappa < 1-\varepsilon$. Thus, if we take all $2^{nd-n^{1-\varepsilon}/100}$ databases in D, at least half of them match at least $\frac{1}{100}$ of the queries in every set I_j. The theorem follows because the number of queries in these sets is at least half the total number of queries in R □

4 Consequences

Miltersen [44] shows that asymmetric communication complexity lower bounds can be used to derive lower bounds for the cell probe model. Specifically, if there is a deterministic (respectively, randomized) cell probe model solution to a "data structure" problem with parameters m (the number of cells), b (the maximum cell size) and t (the number of probes of the data structure), then there is a deterministic (respectively, randomized) asymmetric communication protocol for this problem with $2t$ rounds of communication[12] in which Alice sends $\log m$ bits in *each* of her messages and Bob sends b bits in *each* of his messages. That is, Alice (respectively, Bob) sends a total of at most $t \log m$ bits (respectively, tb bits). Using Theorem 2.2 and the connection to the cell probe model, we get the following lower bound for the λ-neighbor problem in the cell probe model.

Theorem 4.1. *Any one-sided randomized algorithm for the non λ-neighbor problem (and hence deterministic algorithm for the λ-neighbor problem or its complement) that makes makes t probes, either uses $2^{\Omega(d/t)}$ cells, or uses cells of size $\Omega(nd/t)$.*

We point out two extremes of Theorem 4.1:

- If the cell size is $d^{O(1)}$ and the number of cells is polynomial (in n) then the algorithm must make $\Omega(d/\log n)$ probes.
- If the algorithm answers a query in a constant number of probes, then either it uses $2^{\Omega(d)}$ cells, or requires the processing of a cell containing $\Omega(nd)$ bits.

In the same way, using Theorem 3.1 we obtain the following lower bound for the partial match problem.

Theorem 4.2. *Any randomized (two-sided error) cell probe algorithm for the exact partial match problem that makes t probes, either uses $2^{\Omega(\log n \log d/t)}$ cells, or uses cells of size $\Omega\left(n^{1-\varepsilon}/t\right)$.*

[12] In the asymmetric communication model, a message passed by either Alice or Bob is considered a round.

And now we get the following two extremes of Theorem 4.2:

- If the cell size is $d^{O(1)}$ and the number of cells is polynomial (in n) then the algorithm must make $\Omega(\log d)$ probes.
- If the algorithm answers a query in a constant number of probes, then either it uses $2^{\Omega(\log n \log d)}$ cells, or requires the processing of a cell containing $\Omega\left(n^{1-\varepsilon}\right)$ bits.

Lower bounds corresponding to Theorems 3.1 and 4.2 can be obtained for the λ-neighbor problem by applying the following reduction:

Theorem 4.3. *Let $Q(d, \lambda)$ denote the set of points in $\tilde{C}_d$ with exactly λ don't cares. Then, there exist functions $\varphi_A : Q(d, \lambda) \to C_{2d}$ and $\varphi_B : C_d^n \to C_{2d}^n$ with the following property: If $(q, D) \in Q(d, \lambda) \times C_d^n$, then q matches D iff for $(q', D') = (\varphi_A(q), \varphi_B(D))$, q' is a λ-neighbor of D'. Furthermore, both φ_A and φ_B can be computed efficiently (in linear time).*

Proof. For $x \in \tilde{C}_d$, define $y = \varphi_A(x) \in C_{2d}$ as follows. For $i \in \{1, 3, 5, \ldots, 2d-1\}$,

$$y_i y_{i+1} = \begin{cases} x_{\lceil i/2 \rceil} x_{\lceil i/2 \rceil} & x_{\lceil i/2 \rceil} \neq * \\ 01 & x_{\lceil i/2 \rceil} = * \end{cases}$$

Now, define φ_B by applying the transformation φ_A to each of the n points in D. Consider any point $x \in D$, and its image $x' \in D'$. Each don't care in q produces one mismatch between q' and x', regardless of the value of the corresponding coordinate in x. If q matches x, no additional mismatches are produced between q' and x'. Otherwise, there are at least two additional mismatches. □

Now, consider the r-neighbor problem in ℓ_p^d, $1 \leq p < \infty$, for $0 < r \in \mathbb{R}$. Analogous to its definition for the cube, this problem requires deciding whether or not the minimum distance between a query point and a database of n points is at most r. We have

Theorem 4.4. *For every $p \in \mathbb{R}$, $1 \leq p < \infty$, for every $\lambda \in \{0, 1, 2, \ldots, d\}$, there exists $r \in \mathbb{R}$, $r > 0$, such that exact partial match with queries in $Q(d, \lambda)$ reduces to the r-neighbor problem in ℓ_p^{2d}.*

Proof. For $p = 1$, the theorem follows from Theorem 4.3, as the points of C_{2d} are a subset of $\mathbb{R}^{2d}$, and the Hamming distance is equivalent to the L_1 distance for these points. For $p > 1$, the theorem follows from a monotonicity property of the L_p norm on C_{2d} (viewed as a subset of $\mathbb{R}^{2d}$): If $w, x, y, z \in C_{2d}$, then $\|w - x\|_1 < \|y - z\|_1$ iff $\|w - x\|_p < \|y - z\|_p$, where $\|\cdot\|_p$ denotes the L_p norm. (Notice that this monotonicity property does not hold for the L_∞ norm.) □

Finally, we mention some implications to other geometric search problems: First, exact NNS in Euclidean space is a special case of point location in an arrangement of hyperplanes (with $\binom{n}{2}$ hyperplanes, defining the Voronoi diagram). Therefore, our results imply lower bounds for point location.

Next, consider the cube C_d. Notice that the reduction in Theorem 4.3 proves a somewhat stronger claim than mentioned. It shows that exact partial match reduces to the problem of determining whether or not there is a database point at distance *precisely* λ from the query. So, we get a lower bound for this problem as well (for the Hamming cube). Now, consider the cube as a subset of $\mathbb{R}^d$ (for simplicity, we'll use the vectors $\{\pm 1\}^d$ here). The set of cube points at Hamming distance exactly λ from a cube point v lies on the hyperplane $x \cdot v = d - 2\lambda$. Therefore, we get a lower bound for the problem of determining whether or not a query point v lies on one of a collection of n hyperplanes (one hyperplane for each of the n database points). [13] As Chazelle points out [17], this problem can be viewed as a multi-dimensional generalization of the dictionary problem. The dictionary problem can be stated as follows: In the one-dimensional real line, we have a database of hyperplanes (zero-dimensional flats, i.e. points, in this case), queries are points, and the answer to a query is whether or not it is contained in the database. The problem can be solved in $O(1)$ probes per query via hashing. Our lower bounds show that a similar result for the multi-dimensional generalization is impossible. Erickson [24] proves lower bounds on the related *Hopcroft's problem* of deciding for a set of points and a set of hyperplanes whether or not there is a point that lies on one of the hyperplanes. This is not considered as a data structure problem, and the bounds are on the computation time as a function of the number n of points and the number m of hyperplanes. In a subsequent paper, Erickson [25] proves strong time space lower bounds in a structured model (called *partition graphs*) for an online version of Hopcroft's problem where the data base consists of points and the queries are hyperplanes.

Last, consider a geometric interpretation of exact partial match. The database points are vectors in C_d (viewed either as vectors in Z_2^d, or as vectors in $\mathbb{R}^d$). The query is an affine subspace of C_d (in either view), defined by the linear equations $x_i = q_i$ for all i such that $q_i \neq *$. For definiteness, assume that this subspace is given by an orthogonal basis plus a shift — this representation can be computed easily from a partial match query. Thus we have lower bounds for the problem of determining whether or not a query affine subspace contains at least one database point. (Miltersen et al. give rather strong lower bounds for the SPAN problem of determining whether or not a given linear subspace contains a given point.)

[13] By duality, we can interchange the role played by points (which become the data base) and hyperplanes (which become the queries)

5 Limitations of the Method

In this section we explain why a direct application of the richness technique does not appear to provide stronger lower bounds.

First, we consider why we cannot apply the richness technique directly to the the exact NNS problem, or, more precisely, the λ-neighbor decision problem.[14] (As noted before, the hardest case seems to be to distinguish between a distance of at most $\frac{d}{2}$ and a distance of at least $\frac{d}{2}+1$.) We again let $B_d(\lambda)$ denote the Hamming ball of radius λ (say around the all-zeros vector $0^d \in C_d$).

Claim 5.1. *For every $\lambda \in \{0,1,\ldots,d\}$, the communication matrix for the λ-neighbor problem contains a 1-monochromatic rectangle of size $|B_d(\lambda)| \times 2^{nd-d}$.*

Before we prove this claim, we point out its consequence. In the λ-neighbor problem, each database is close enough to at most $n\,|B_d(\lambda)|$ queries. Thus, the problem cannot be richer than $\left[n\,|B_d(\lambda)|, 2^{nd}\right]$. Therefore, the best lower bound we can hope to prove this way is the very weak conclusion: Either the query side sends $\Omega(\log n)$ bits or the database side sends $\Omega(d)$ bits.

Proof of Claim 5.1. Take all the queries in $B_d(\lambda)$. If 0^d is contained in the database then it produces a value of 1 with all these queries. If we pick a database at random, all databases equally likely, the database contains 0^d with probability more than 2^{-d}. □

We now consider the natural idea of restricting exact partial match to instances with fewer don't cares, in an attempt to prove better lower bounds. The hardest case seems to be when queries have exactly $\frac{d}{2}$ don't cares. In this case NPM is extremely rich. Almost all entries in the communication matrix are one. However, perhaps not surprisingly, we have the following:[15]

Claim 5.2. *The communication matrix for* NPM *restricted to queries with exactly $\frac{d}{2}$ don't cares contains a 1-monochromatic rectangle of size*

$$n^{-2}\binom{d}{d/2}2^{d/2} \times e^{-1}2^{nd}.$$

The consequence here is obvious. The total number of possible queries is $\binom{d}{d/2}2^{d/2}$. Thus, the best lower bound we can prove by the richness technique is the rather pathetic "either the query sends $\Omega(\log n)$ bits or the database sends $\Omega(1)$ bits".

---

[14] By "directly" we mean that we do not restrict the set of queries.

[15] Using similar arguments, one can show that considering the complement function does not help in this case.

Proof of Claim 5.2. Take the set of queries to be all possible queries with $\frac{d}{2}$ don't cares, and the first k bits fixed as zeros (k to be determined shortly). The number of such queries is

$$\binom{d-k}{d/2}2^{d/2-k} \geq \binom{d}{d/2}2^{d/2-2k}.$$

The number of cube points matched by at least one query is exactly 2^{d-k}. Therefore, the number of databases that are not matched by any query is

$$(2^d - 2^{d-k})^n = (1-2^{-k})^n 2^{nd} \geq e^{-n/2^k}2^{nd}.$$

Now, take $k = \log n$. □

Returning to the case of $\log n + 1$ exposed bits, is it possible to improve upon the proven bounds? If all possible queries are enumerated in some predefined order, the database can store the answer to all possible queries and the query player can then simply send the index of the query using $O(\log n \log d)$ bits (and the database player responds with the correct answer using one bit). Hence the bound on the query player is optimal. Finally, we ask if we can improve our lower bound on the database side to $\Omega(n)$? The following claim shows that our analysis cannot be improved significantly.

Claim 5.3. *For every integer c, there is $\nu > 0$ such that for every $n \geq \nu$, the communication matrix of* NPM *restricted to queries from $Q_{n,d}$ contains a 1-monochromatic rectangle of size $2^{-c(\log d+1)}|Q_{n,d}| \times 2^{nd-n\log e/2^c}$.*

Proof. We may assume that n is sufficiently large so that $\log n \gg c$. Take all queries in $Q_{n,d}$ with the first c bits fixed as zeros. The number of such queries is

$$\frac{n}{2^{c-1}}\binom{d-c}{\log n+1-c} \geq \frac{2n}{2^{c(\log d+1)}}\binom{d}{\log n+1}.$$

The number of cube points matched by at least one of these queries is 2^{d-c}. Therefore, the number of databases not matched by any of these queries is

$$(1-2^{-c})^n 2^{nd} \geq e^{-n/2^c}2^{nd}.$$

□

References

[1] P.K. Agarwal and J. Matoušek. Ray shooting and parametric search. *SICOMP*, 22:794-806, 1993

[2] M. Ajtai. A lower bound for finding predecessors in Yao's cell probe model. *Combinatorica*, 8:235–247, 1988.

[3] M. Ajtai. Determinism versus non-determinism for linear-time RAMs. In *Proc. of 31st STOC*, pp. 632-641, 1999.

[4] M. Ajtai. A non-linear time lower bound for Boolean branching programs. In *Proc. of 40th FOCS*, pp. 60-70, 1999

[5] N. Alon and J.H. Spencer. *The Probabilistic Method.* Wiley, 1992.

[6] S. Arya and D. Mount. Approximate nearest neighbor queries in fixed dimensions. In *Proc. of 4th SODA*, pp. 271–280, 1993.

[7] S. Arya, D. Mount, N. Netanyahu, R. Silverman, and A. Wu. An optimal algorithm for approximate nearest neighbor searching in fixed dimensions. *JACM*, 45(6):891-923, 1998.

[8] O. Barkol and Y. Rabani. Tighter bounds for nearest neighbor search and related problems in the cell probe model. In *Proc. of the 32nd STOC*, pp. 388-396, 2000.

[9] P. Beame, M. Saks, and J.S. Thathachar. Time-space tradeoffs for branching programs. In *Proc. of 39th FOCS*, pp. 254–263, 1998.

[10] J.S. Beis and D.G. Lowe. Shape indexing using approximate nearest-neighbor search in high-dimensional spaces. In *Proc. IEEE Conf. Comp. Vision Patt. Recog.*, pages 1000–1006, 1997.

[11] J.L. Bentley and R. Sedgewick. Fast algorithms for sorting and searching strings. In *Proc. of 8th SODA*, pp. 360–369, 1997.

[12] M. Ben-Or. Lower bounds for algebraic computation trees. In *Proc. of 15th STOC*, pp. 80-86, 1983.

[13] M. de Berg, M. van Kreveld, M. Overmars, O. Schwarzkopf. *Computational Geometry, Algorithms and Applications.* Springer, 1997.

[14] M. Bern. Approximate closest-point queries in high dimensions. *Information Processing Lett.*, 45:95–99, 1993.

[15] A. Chakrabarti, B. Chazelle, B. Gum, A. Lvov. A good neighbor is hard to find. In *Proc. of 31st STOC*, pp 305-311, 1999.

[16] T.M. Chan. Approximate nearest neighbor queries revisited. *Discrete and Computational Geometry*, 20:359-373, 1998.

[17] B. Chazelle. Private communication.

[18] K. Clarkson. A randomized algorithm for closest-point queries. *SIAM J. Computing*, 17:830-847, 1988.

[19] K. Clarkson. An algorithm for approximate closest-point queries. In *Proc. of 10th SCG*, pp. 160–164, 1994.

[20] S. Deerwester, S.T. Dumais, G.W. Furnas, T.K. Landauer, and R. Harshman. Indexing by latent semantic analysis. *J. Amer. Soc. Info. Sci.*, 41(6):391–407, 1990.

[21] D. Dobkin and R. Lipton. Multidimensional search problems. *SIAM J. Computing*, 5:181-186, 1976.

[22] D. Dolev, Y. Harari, and M. Parnas. Finding the neighborhood of a query in a dictionary. In *Proc. of 2nd ISTCS*, 1993.

[23] D. Dolev, Y. Harari, N. Linial, N. Nisan, and M. Parnas. Neighborhood preserving hashing and approximate queries. In *Proc. of 5th SODA*, pp. 251–259, 1994.

[24] J. Erickson. New lower bounds for Hopcroft's problem. *Discrete Comput. Geom.*, 16:389–418, 1996.

[25] J. Erickson. Space-time tradeoffs for emptiness queries. *SIAM J. Computing*, 29(6):1968-1996, 2000.

[26] R. Fagin. Fuzzy queries in multimedia database systems. In *Proc. of PODS*, pp 1-10, 1998.

[27] M. Flickner, H. Sawhney, W. Niblack, J. Ashley, Q. Huang, B. Dom, M. Gorkani, J. Hafner, D. Lee, D. Petkovic, D. Steele, and P.Yanker. Query by image and video content: the QBIC system. *IEEE Computer*, 28:23–32, 1995.

[28] M.L. Fredman. A lower bound on the complexity of orthogonal range queries. *Journal of the ACM*, 28:696-705, 1981

[29] M.L. Fredman. Lower bounds on the complexity of some optimal data structures. *SIAM J. Computing*, 10:1-10, 1981

[30] M.L. Fredman and D.J. Volper. The complexity of partial match retrieval in a dynamic setting. *Journal of Algorithms*, 3:68-78, 1982.

[31] D. Grigoriev. Randomized complexity lower bounds for arrangements and polyhedra. *Discrete and Computational Geometry*, 21:329-344, 1999.

[32] D. Grigoriev and M. Karpinski. Randomized $\Omega(n^2)$ lower bound for knapsack. *Proc. of 29th STOC*, pp. 76-85, 1997.

[33] L. Harper. Optimal numberings and isoperimetric problems on graphs. *Journal of Combinatorial Theory* 1:385–394, 1966.

[34] T. Hastie and R. Tibshirani. Discriminant adaptive nearest neighbor classification and regression. *Advances in Neural Information Processing Systems*, 8:409-415, 1996.

[35] J. Hellerstein, E. Koutsoupias, and C.H. Papadimitriou. On the analysis of indexing schemes. In *Proc. of PODS*, 1997.

[36] P. Indyk. On approximate nearest neighbors in non-Euclidean spaces. In *Proc. of 39th FOCS*, 1998.

[37] P. Indyk and R. Motwani. Approximate nearest neighbors: Towards removing the curse of dimensionality. In *Proc. of 30th STOC*, 1998.

[38] J. Kleinberg. Two algorithms for nearest-neighbor search in high dimensions. In *Proc. of 29th STOC*, pp. 599–608, 1997.

[39] E. Kushilevitz and N. Nisan. *Communication Complexity*. Cambridge University Press, 1997.

[40] E. Kushilevitz, R. Ostrovsky, and Y. Rabani. Efficient search for approximate neares neighbor in high dimensional spaces. *SICOMP*, 30(2): 457-474, 2000.

[41] J. Matoušek. Reporting points in halfspaces. *Computational Geometry and Applications*, 2:169-186, 1992.

[42] S. Meiser. Point location in arrangements of hyperplanes. *Information and Computation*, 106(2):286–303, 1993.

[43] P.B. Miltersen. The bit probe complexity measure revisited. In *Proc. of 10th STACS*, pp. 662–671, 1993.

[44] P.B. Miltersen. Lower bounds for union-split-find related problems on random access machines. In *Proc. of of 26th STOC*, pp. 625–634, 1994.

[45] P.B. Miltersen. On the cell probe complexity of polynomial evaluation. *Theoretical Computer Science*, 143:167–174, 1995.

[46] P.B. Miltersen, N. Nisan, S. Safra, and A. Widgerson. On data structures and asymmetric communication complexity. *JCSS*, 57(5): 37-49, 1998.

[47] A. Pentland, R.W. Picard, and S. Sclaroff. Photobook: tools for content-based manipulation of image databases. In *Proc. SPIE Conf. on Storage and Retrieval of Image and Video Databases II*, 1994.

[48] R. Rivest. Partial-match retrieval algorithms. *SIAM J. Computing*, 5:19–50, 1976.

[49] G. Salton. *Automatic Text Processing.* Addison-Wesley, 1989.

[50] A.W.M. Smeulders and R. Jain (eds). *Proc. 1st Workshop on Image Databases and Multi-Media Search*, 1996.

[51] B. Xiao. New Bounds in Cell Probe Model. Ph.D. thesis, UC San Diego, 1992.

[52] A.C. Yao. Should tables be sorted? *J. Assoc. Comput. Mach.*, 28:615–628, 1981.

[53] A.C. Yao and F.F. Yao. A general approach to d-dimension geometric queries. In *Proc. of 17th STOC*, pp. 163–168, 1985.

About Authors

Allan Borodin is at the Department of Computer Science, University of Toronto, Toronto M5S 3G4, Canada. *borodin@cs.toronto.edu.*

Rafail Ostrovsky is at the Math Sciences Research Center, Telcordia Technologies, 445 South Street, Morristown, NJ 07960-6438, USA. *rafail@research.telcordia.com.*

Yuval Rabani is at the Computer Science Department, Technion — Israel Institute of Technology, Haifa 32000, Israel. *rabani@cs.technion.ac.il.*

Acknowledgments

Part of this work was done while the first and third authors were visiting Telcordia Technologies and while the first and second authors were visiting the Technion. Work at the Technion supported by BSF grant 96-00402, by Ministry of Science contract number 9480198, and by a grant from the Fund for the Promotion of Research at the Technion.

The authors would like to thank Bernard Chazelle for many helpful comments. We are also indebted to the referees for their very constructive suggestions. In particular, we thank one of the referees for the clarification of Meiser's [42] space bound and for the reference to Erickson [25].

A Turán-type Extremal Theory of Convex Geometric Graphs

Peter Brass
Gyula Károlyi
Pavel Valtr

Abstract

We study Turán-type extremal questions for graphs with an additional cyclic ordering of the vertices, i.e. for convex geometric graphs. If a suitably defined chromatic number of the excluded subgraph is bigger than two then the results on convex geometric graphs resemble very much the classical results from the Turán theory. On the other hand, in the bipartite case we show some surprising differences, in particular for trees and forests. For example, the Turán function of some convex geometric forests is of the order $\Theta(n \log n)$, a growth rate that does not occur in the graph Turán theory. We also obtain still another proof of Füredi's $O(n \log n)$ bound on the number of unit distances in a convex n-gon, together with a lower bound showing the limits of this model. The exact growth of the Turán function for several infinite classes of convex geometric graphs is also determined.

1 Introduction

Turán-type problems ask for the maximum size of some structures of given order that do not contain a given substructure. They can be considered for most kinds of discrete structures. However, the situation is well-understood almost only in the classical case of abstract graphs. Already the seemingly direct generalization to three-uniform hypergraphs turns out to be difficult. For example, Turán's conjecture on the maximum number of 3-element subsets of an n-element set such that for no 4-element set all its 3-element-subsets are selected is still open. Also the case of directed graphs is not completely understood. This is surprising, since the graph results have been very useful in proving a number of bounds for extremal problems in graph theory and also in combinatorial geometry and number theory. Proofs of some extremal results outside graph theory have a similar combinatorial flavor. It seems that Turán-type theorems for various structures may provide appropriate combinatorial models for a large variety of extremal problems.

In this paper we study Turán-type questions for 'convex geometric graphs', which are the combinatorial structures underlying systems of sides and diagonals of a convex polygon, if we abstract from the position of the vertices and look only at their cyclic order. Turán-type results for convex geometric graphs were already obtained in various contexts by Kupitz, Pach, Perles and others [8, 20, 22, 23, 27]. Convex geometric graphs were considered also in a number of Ramsey-type questions [3, 18–20, 30].

A *convex geometric graph* (shortly *cgg*) $D = (V, E)$ is a graph with a cyclic ordering on the vertices. It is usually represented as a graph drawn in the plane so that V is the set of vertices of a convex polygon and E is a set of straight-line segments connecting some of the pairs of points of V. We assume the cyclic ordering then always to be the clockwise ordering of the vertex set. The underlying (abstract) graph corresponding to D will be denoted by $G(D)$, a *cgg* is determined by this underlying graph $G(D)$ and the (clockwise) cyclic ordering of its vertex set. We distinguish *cggs* only up to isomorphism of this combinatorial structure. For example, all *cggs* D with $G(D) = K_n$ are isomorphic. Fig. 1 shows all the four non-isomorphic *cggs* D with $G(D) = P_3$.

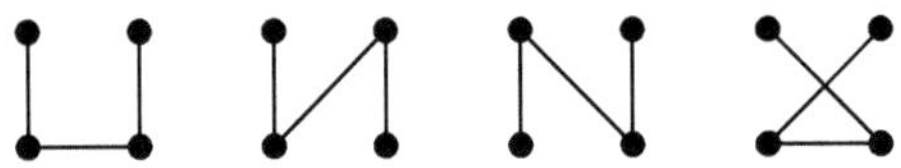

Figure 1

Any subset of the vertex set $V(D)$ of a *cgg* D which is contiguous with respect to the cyclic ordering of $V(D)$ will be called an *interval*. Any reference to the *first* or *last vertex* of an interval will be assumed with respect to the clockwise order.

Let D, D' be two *cggs*. We say that D' is a *sub-cgg* of D, if it can be obtained from D by removing some vertices and edges. We say that D *contains* D', if D' is isomorphic to a sub-*cgg* of D. Otherwise we say that D is D'*-free*.

For a *cgg* D, we investigate the maximum number $\mathrm{ex}_{\circlearrowright}(n, D)$ of edges in a D-free *cgg* on n vertices. Of course, $\mathrm{ex}_{\circlearrowright}(n, D) \geq \mathrm{ex}_{\mathrm{graph}}(n, G(D))$ holds for any *cgg* D, where $\mathrm{ex}_{\mathrm{graph}}(n, G)$ denotes the Turán function of G, i.e. it is the maximum number of edges in a G-free graph on n vertices. However, $\mathrm{ex}_{\circlearrowright}(n, D)$ depends very much on the order of the vertices of D and may be very different from $\mathrm{ex}_{\mathrm{graph}}(n, G(D))$. For example, results described in the next two sections show that if D is the first *cgg* in Fig. 1 then $\mathrm{ex}_{\circlearrowright}(n, D)$ is quadratic in n, and it is linear in n for the other three *cggs* in Fig. 1. Another example is the pair of the *cggs* with the underlying graph C_4, for which $\mathrm{ex}_{\circlearrowright}(n, \square) = \Theta(n^2)$ and $\mathrm{ex}_{\circlearrowright}(n, \bowtie) = \Theta(n^{3/2})$.

The theory of convex geometric graphs is quite distinct from the theory of (general) geometric graphs, although analogous questions have been studied in both models. The convexity assumption on the vertices removes many geometric difficulties and makes the object a purely combinatorial structure. It also guarantees that the questions for maximum edge numbers without some specified substructure do have answers. There is a unique maximal object (the complete cyclically ordered graph) in this theory, whereas there are many distinct geometric graphs of K_n. So for classical geometric graphs even the existence of Turán- and Ramsey-numbers for a given excluded sub-configuration is not obvious, and indeed is an open problem in many cases.

The paper is organized as follows. Section 2 is devoted to the analogues of classical results in Turán theory for convex geometric graphs. Exact results or estimates on the Turán function for some convex geometric trees and forests are given in Section 3. Further results and remarks are in Section 4. Section 5 contains the proofs of all numbered theorems (Theorems 1–9).

2 Classical extremal results in graph theory and their *cgg*-analogues

Let $\chi(G)$ denote the chromatic number of a graph G, and let $\mathrm{ex}_{\mathrm{graph}}(n, G)$ denote the Turán function of G. Let $T_{n,r}$ denote the Turán graph, i.e. the complete r-partite graph on n vertices with vertex classes of cardinality $\lfloor \frac{n}{r} \rfloor$ or $\lceil \frac{n}{r} \rceil$. Here are two classical results in Turán theory (see [31] for a survey):

Theorem (Turán). *Let $n \geq r \geq 3$. Then*

$$\mathrm{ex}_{\mathrm{graph}}(n, K_r) = e(T_{n,r-1}) = \tfrac{r-2}{r-1}\tbinom{n}{2} + O(n),$$

and the graph $T_{n,r-1}$ is the unique K_r-free graph on n vertices with $\mathrm{ex}_{\mathrm{graph}}(n, K_r)$ edges.

Theorem (Erdős-Stone-Simonovits). *Let G be a graph with $n \geq \chi(G) \geq 3$. Then*

$$\mathrm{ex}_{\mathrm{graph}}(n, K_{\chi(G)}) \leq \mathrm{ex}_{\mathrm{graph}}(n, G) \leq \mathrm{ex}_{\mathrm{graph}}(n, K_{\chi(G)}) + o(n^2).$$

Thus, if $\chi(G) > 2$ then $\mathrm{ex}_{\mathrm{graph}}(n, G)$ is quadratic in n and it is essentially determined by its chromatic number $\chi(G)$. If the excluded subgraph G is bipartite ($\chi(G) = 2$), the situation is much more difficult. The function $\mathrm{ex}_{\mathrm{graph}}(n, G)$ is then sub-quadratic by the Kővári-Sós-Turán Theorem.

Theorem (Kővári-Sós-Turán). *Let $r \leq s$. Then*

$$\mathrm{ex}_{\mathrm{graph}}(n, K_{r,s}) = O(n^{2-\frac{1}{r}}).$$

If G is bipartite, then either $\mathrm{ex}_{\mathrm{graph}}(n, G) = O(n)$ (if G is a forest) or $\Omega(n^{1+\varepsilon}) << \mathrm{ex}_{\mathrm{graph}}(n, G) << O(n^{2-\varepsilon'})$ for $\varepsilon, \varepsilon' > 0$ (otherwise): The $O(n)$-bound for forests is a folklore (e.g. see [31], [10, p. 165, Ex. 6], or [5, p. 137, Ex. 24]), a standard probabilistic argument [13] shows that if G contains a cycle of length $2r$ then $\mathrm{ex}_{\mathrm{graph}}(n, G) = \Omega(n^{1+\frac{1}{2r}})$ (this was improved in [24] to $\mathrm{ex}_{\mathrm{graph}}(n, G) = \Omega(n^{1+2/(3r+O(1))})$), and the $O(n^{2-\varepsilon'})$-bound follows from the Kővári-Sós-Turán Theorem.

The above three theorems have counterparts for *cggs*. Let D_r^{complete} be the unique *cgg* whose underlying graph is K_r. Let $D_{n,r}^{\text{Turan}}$ be a common notation for the *cggs* on n vertices such that $G(D_{n,r}^{\text{Turan}}) = T_{n,r}$ and that each of the r vertex classes of $T_{n,r}$ corresponds to an interval in $D_{n,r}^{\text{Turan}}$ (note that $D_{n,r}^{\text{Turan}}$ may not be unique: if there are several classes of size $\lfloor \frac{n}{r} \rfloor$ and several classes of size $\lceil \frac{n}{r} \rceil$, then non-isomorphic *cggs* may be obtained by different orderings of the classes). Here is an (extended) analogue of the Turán Theorem for *cggs*:

Proposition 1 (folklore). *Let $n \geq r \geq 3$, and let D be any cgg on r vertices such that any two consecutive vertices are connected by an edge. Then*

$$\mathrm{ex}_{\circlearrowleft}(n, D) = \mathrm{ex}_{\mathrm{graph}}(n, K_r) \quad \left(= \tfrac{r-2}{r-1}\tbinom{n}{2} + O(n)\right).$$

Moreover, if D is a circle, then a D-free cgg on n vertices has $\mathrm{ex}_{\circlearrowleft}(n, D)$ edges if and only if it is one of the cggs $D_{n,r-1}^{\text{Turan}}$.

Proof : Every $D_{n,r-1}^{\text{Turan}}$ is D-free, which gives the lower bound. The upper bound and the last sentence in the proposition follow directly from the Turán theorem. □

Thus the *cgg* shown in Figure 2 is not only the largest 15-vertex *cgg* without , but also the largest 15-vertex drawing without  or without .

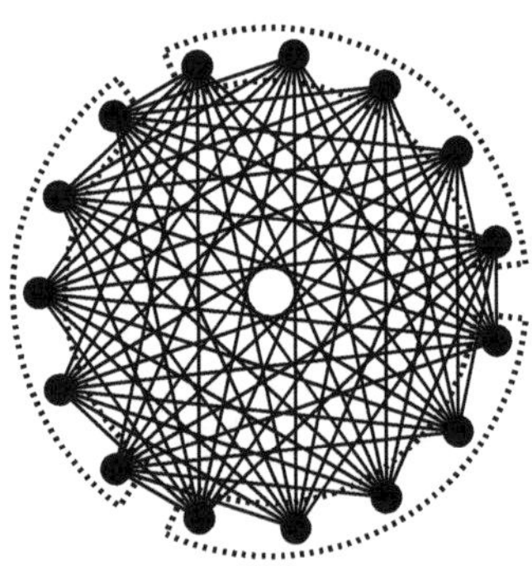

Figure 2

Before stating *cgg*-analogues of the other two classical results, we give a definition of the chromatic number of a *cgg*.

Definition. Let D be a *cgg*. The *(cyclic) chromatic number* $\chi_{\circlearrowleft}(D)$ of D is the minimum number of colors in a vertex coloring of D such that vertices joined by an edge have distinct colors, and each color class is an interval. If $\chi_{\circlearrowleft}(D) \leq 2$ then we say that D is *(cyclic-)bipartite*.

For example, $\chi_{\circlearrowleft}($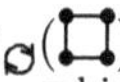$) = 4$ and $\chi_{\circlearrowleft}($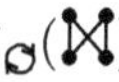$) = 2$. Clearly, $\chi_{\circlearrowleft}(D) \geq \chi(G(D))$ for any *cgg* D. Let $D_{r,s}^{\text{bipartite}}$ denote the *cgg* satisfying $G(D_{r,s}^{\text{bipartite}}) = K_{r,s}$ and $\chi_{\circlearrowleft}(D_{r,s}^{\text{bipartite}}) = 2$ (e.g. $D_{3,3}^{\text{bipartite}} =$ ). Here are *cgg*-analogues of the Erdős-Stone-Simonovits Theorem and the Kővári-Sós-Turán Theorem:

Theorem 1. *Let D be a cgg with $n \geq \chi_{\circlearrowleft}(D) \geq 3$. Then*

$$\text{ex}_{\circlearrowleft}(n, K_{\chi_{\circlearrowleft}(D)}) \leq \text{ex}_{\circlearrowleft}(n, D) \leq \text{ex}_{\circlearrowleft}(n, K_{\chi_{\circlearrowleft}(D)}) + o(n^2).$$

Theorem 2. *Let $r \leq s$. Then*

$$\text{ex}_{\circlearrowleft}(n, D_{r,s}^{\text{bipartite}}) = O(n^{2-\frac{1}{r}}).$$

So $\text{ex}_{\circlearrowleft}(n, D) = \left(1 - \frac{1}{\chi_{\circlearrowleft}(D) - 1}\right)\binom{n}{2} + o(n^2)$. As in the graph case, $\text{ex}_{\circlearrowleft}(n, D)$ is essentially determined by the chromatic number for all D with $\chi_{\circlearrowleft}(D) \geq 3$. If $\chi_{\circlearrowleft}(D) = 2$, then D is a sub-*cgg* of some complete bipartite *cgg* $D_{r,s}^{\text{bipartite}}$, and a non-trivial upper bound on $\text{ex}_{\circlearrowleft}(n, D)$ is given by Theorem 2.

Since there are bipartite graphs with matching lower bounds, Theorem 2 gives the correct order of magnitude for the corresponding cyclic-bipartite *cggs*. In particular, $\text{ex}_{\circlearrowleft}(n,$ $) = \Theta(n^{3/2})$, $\text{ex}_{\circlearrowleft}(n,$ $) = \Theta(n^{5/3})$, and $\text{ex}_{\circlearrowleft}(n, D_{r,s}^{\text{bipartite}}) = \Theta(n^{2-\frac{1}{r}})$ for $s > (r-1)!$ (the graph lower bounds $\text{ex}_{\text{graph}}(n, K_{r,s}) = \Omega(n^{2-\frac{1}{r}})$ are shown for $r = s = 2$ in [6,12,16], for $r = s = 3$ in [2,6], and for $s > (r-1)!$ in [2,21]).

Theorems 1 and 2 show that $\text{ex}_{\circlearrowleft}(n, D)$ is sub-quadratic if and only if D is cyclic-bipartite ($\chi_{\circlearrowleft}(D) = 2$). If D contains a cycle then $\text{ex}_{\circlearrowleft}(n, D) \geq \text{ex}_{\text{graph}}(n, G(D)) \geq \Omega(n^{1+\varepsilon})$ for some $\varepsilon = \varepsilon(G(D)) > 0$. Thus, $\text{ex}_{\circlearrowleft}(n, D)$ may be linear or nearly linear only for cyclic-bipartite forests D. This case will be especially studied in the next section.

In the non-bipartite case ($\chi_{\circlearrowleft}(D) > 2$), Theorem 1 gives asymptotic bounds on $\text{ex}_{\circlearrowleft}(n, D)$ but exact values of $\text{ex}_{\circlearrowleft}(n, D)$ are known only in cases given by Proposition 1 and by the Kupitz Perles Theorem now following after a few definitions.

Let $D_k^{\overset{\text{conv}}{\text{match}}}$ denote the *cgg* of a convex matching on $2k$ vertices, i.e. k disjoint edges $v_{2i-1}v_{2i}$ in convex position ($D_3^{\overset{\text{conv}}{\text{match}}} =$ ). Let $D_{n,k}^{\text{Kupitz-Perles}}$ denote any *cgg* obtained from any $D_{n,k}^{\text{Turan}}$ by adding all missing edges from the first vertex in each block (again, $D_{n,k}^{\text{Kupitz-Perles}}$ is possibly not unique). Clearly, $\chi_{\circlearrowleft}(D_k^{\overset{\text{conv}}{\text{match}}}) = k$, the *cgg* $D_{n,k-1}^{\text{Kupitz-Perles}}$ is $D_k^{\overset{\text{conv}}{\text{match}}}$-free and has $e(D_{n,k-1}^{\text{Kupitz-Perles}}) = e(T_{n,k-1}) + n - (k-1)$ edges.

Theorem (Kupitz-Perles [23]). *Let $n \geq 2k \geq 6$. Then*

$$\text{ex}_{\circlearrowleft}(n, D_k^{\overset{\text{conv}}{\text{match}}}) = e(T_{n,k-1}) + n - (k-1).$$

If $n > 2k$ and $n \neq 7$, then the graphs $D_{n,k-1}^{\text{Kupitz-Perles}}$ are the only extremal graphs in the Kupitz-Perles Theorem (see [23] for details). Figure 3 shows a $D_{15,3}^{\text{Kupitz-Perles}}$ which is the unique maximal *cgg* on 15 vertices that does not contain D_4^{match} (conv) = .

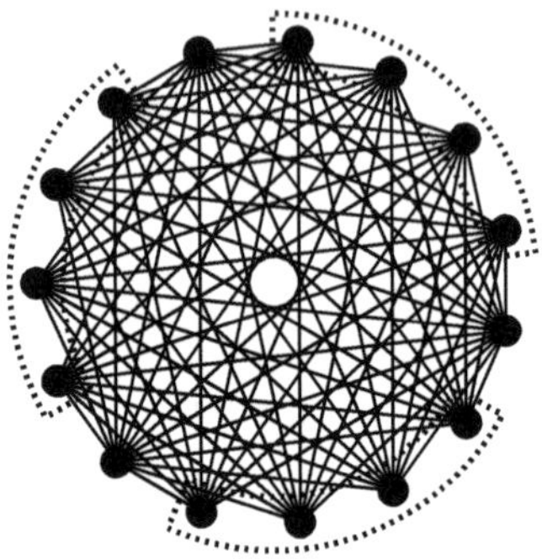

Figure 3

3 Excluded Trees and Forests

The maximum edge number of an outerplanar graph on n vertices is $2n-3$, this can be written as ex(n, ) $= 2n-3$. In fact we have

Theorem 3. *For $n \geq 4$,*

ex(n, ) $= 2n - 3$.

So we get a folded path as soon as we have a crossing, which differs from the situation for general geometric graphs, where the order of the Turán function of the folded path is known to be $\Theta(n \log n)$ [29].

But if we look for two folded paths, something surprising happens, depending on the mutual orientation:

Theorem 4.

ex(n, ) $= \Theta(n \log n)$.

Theorem 5.

ex(n, ) $= \Theta(n \log n)$.

Theorem 6. *For any $k \geq 1$,*

ex(n, 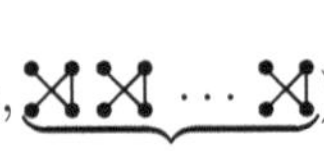) $= \Theta(n)$.

k times 

Theorem 5 gives a new proof of the following theorem:

Theorem (Füredi [15]). *The number of unit distances between the vertices of a convex n-gon is at most $O(n \log n)$.*

Proof: According to Theorem 5, it suffices to show that 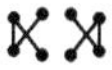 is not realizable with unit distances among points in convex position. In any such realization, the interior angles in each of the outer four corners would be less than $\frac{\pi}{2}$, since they would be base angles of isosceles triangles; but a convex polygon cannot have four vertices with interior angles less than $\frac{\pi}{2}$. □

It is not known if the bound $O(n \log n)$ in Füredi's theorem can be replaced by $O(n)$. See Paragraph 2 in Section 4 for further details.
Theorems 3–6 are also interesting for the following reasons:

- *An unusual growth rate:* In the graph Turán theory all forests have a linear Turán function. The growth rate $\Theta(n \log n)$ does not occur for any graph.

- *Excluding several disjoint copies:* In the graph Turán theory excluding several disjoint copies of a graph does not change the growth rate: soon after the first copy of the forbidden subgraph occur many further copies. But two copies of the *cgg*  with a given vertex order behave essentially different than one.

- *Differences caused by distinct vertex orderings:* We have already noticed that *cggs* with the same underlying graph may have different growth of their Turán functions $\mathrm{ex}_{\circlearrowright}$ because they may have different cyclic chromatic numbers. Theorem 4, 5, and 6 show that also *cggs* with the same underlying graph and the same cyclic chromatic number may have different growth rate of their Turán functions $\mathrm{ex}_{\circlearrowright}$ (Theorem 6 gives $\mathrm{ex}_{\circlearrowright}(n,$ $) = \Theta(n)$).

  Moreover, although  and  have Turán functions of the same order, the constructions of the extremal *cggs* and the proofs of the upper bounds are completely different. Figure 4 shows left a *cgg* without  and right a *cgg* without .

The theorems above show that there is a very subtle distinction between *cggs* D with $\mathrm{ex}_{\circlearrowright}(n, D) = \Theta(n)$ and such with $\mathrm{ex}_{\circlearrowright}(n, D) = \Theta(n \log n)$. We find the following problems most interesting on this structure:

Problem 1. *For which cggs D is $\mathrm{ex}_{\circlearrowright}(n, D) = O(n)$?*

Problem 2. *Does $\mathrm{ex}_{\circlearrowright}(n, D) = O(n\,\mathrm{polylog}(n))$ hold for all cyclic-bipartite forests D?*

The exact Turán functions have already been obtained for several infinite classes of bipartite *cggs*. Capoyleas and Pach determined them for the *cgg* $D_k^{\substack{\text{cross}\\ \text{match}}}$ consisting of k pairwise crossing edges ($D_4^{\substack{\text{cross}\\ \text{match}}} =$ ).

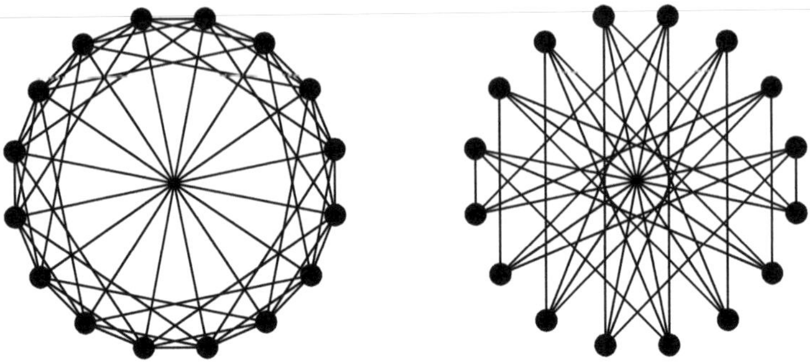

Figure 4

Theorem (Capoyleas-Pach [8, 27]). *Let* $n \geq 2k-1$. *Then*

$$\mathrm{ex}_{\circlearrowleft}(n, D_{k}^{\substack{\text{cross}\\ \text{match}}}) = 2(k-1)n - \tfrac{1}{2}(2k-1)(2k-2)\,.$$

Also the Turán function of the 'parallel' matching $D_{k}^{\substack{\text{par}\\ \text{match}}}$ of k edges is known ($D_{4}^{\substack{\text{par}\\ \text{match}}} =$ ):

Theorem (Kupitz [22]). *Let* $n \geq 2k$. *Then*

$$\mathrm{ex}_{\circlearrowleft}(n, D_{k}^{\substack{\text{par}\\ \text{match}}}) = (k-1)n.$$

The upper bound follows from the fact that the edges of K_n can be partitioned into n classes of 'index-parallel' edges, each of which contains at most $k-1$ edges of any $D_{k}^{\substack{\text{par}\\ \text{match}}}$-free *cgg*. Figure 5 shows left a *cgg* without  and right a *cgg* without .

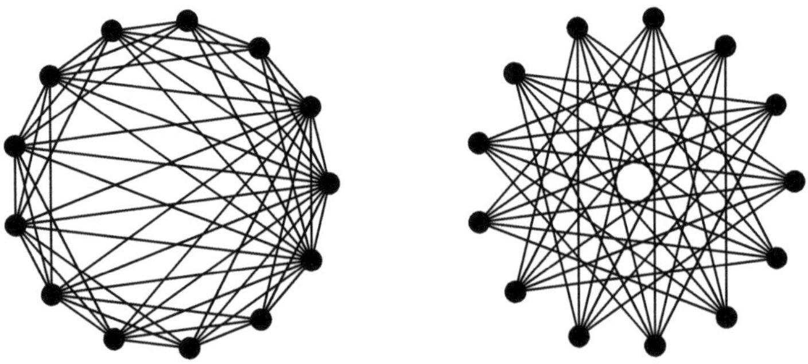

Figure 5

A *caterpillar* is a tree in which all vertices of degree > 1 lie on a single path. A beautiful result of Perles (quoted in a slightly different version in [25],

p. 292) shows that all bipartite crossing-free *cggs* of a caterpillar with v vertices (e.g. ) have the same Turán function:

Theorem (Perles). *Let F be a cyclic-bipartite crossing-free cgg of any caterpillar with v vertices. Then*

$$\mathrm{ex}_{\circlearrowright}(n, F) = \left\lfloor \tfrac{1}{2}(v-2)n \right\rfloor .$$

Here 'caterpillar' could also be replaced by 'tree', since the only trees that allow a bipartite crossing-free *cgg* are caterpillars. A slight extension of Perles' unpublished proof gives a common generalization of his result and the above result of Kupitz. The common generalization shows that we get the same Turán function, depending only on the vertex number, for a large class of excluded *cggs* F, even containing isolated vertices, as long as the first and last edge between the vertex classes in the left-to-right order belong to F (e.g.  or ):

Theorem 7. *Let F be a cyclic-bipartite, crossing-free cgg with v vertices that contains the two edges connecting consecutive vertices belonging to different vertex classes of F. Then*

$$\mathrm{ex}_{\circlearrowright}(n, F) = \left\lfloor \tfrac{1}{2}(v-2)n \right\rfloor .$$

So, again quite different from the graph situation, isolated vertices may change the growth of the Turán function.

As an example of a cyclic-bipartite *cgg* with crossings, for which we also can state the exact values of $\mathrm{ex}_{\circlearrowright}$ we give

Theorem 8. *For $n \geq 5$,*

$\mathrm{ex}_{\circlearrowright}(n,$  $) = \left\lfloor \tfrac{5}{2}n - 4 \right\rfloor .$

Here the extremal *cggs* consist of copies of K_4, glued together, as in Figure 6, with one triangle for odd n. These can be glued together along outer edges in any tree-like fashion (not only as a path as in Figure 6), and indeed these are the only extremal *cggs* for this problem. The order of the Turán function for general geometric graphs is the same $\Theta(n)$ [28].

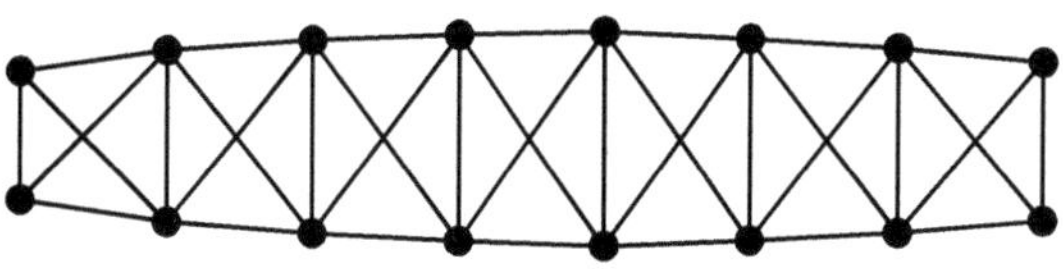

Figure 6

4 Further Results and Remarks

1. Ramsey-type problems. One can also ask Ramsey-type questions for excluded sub-*cggs* D: what is the smallest $n = R(D;k)$ such that whenever the edges of D_n^{complete} are colored with k colors, a monochromatic sub-*cgg* D occurs? In fact, there have been more papers on the Ramsey-aspects than on the Turán-aspects of *cggs*. For bipartite *cggs* these aspects are related in the usual way: if $n < R(D;k)$ we have $\text{ex}_{\circlearrowleft}(n, D) \geq \binom{n}{2}/k$, since the complete *cgg* can be partitioned into k sub-*cggs* that are D-free. So if $\text{ex}_{\circlearrowleft}(n, D) = O(n^{2-\varepsilon})$ (i.e. $\chi_{\circlearrowleft}(D) = 2$) we obtain a polynomial upper bound on $R(D;k)$:

$$R(D;k) \leq \min\left\{n \mid \text{ex}_{\circlearrowleft}(n,D) < \binom{n}{2}/k\right\} = O(k^{1/\varepsilon}) .$$

By this argument the two-color Ramsey number of a non-crossing bipartite path was obtained in [20] from a Turán theorem which was a special case of Theorem 7 above.

We note that also for $\chi_{\circlearrowleft}(D) \geq 3$ there is a connection of Turán- and Ramsey-type theorems by the chromatic number $\chi_{\circlearrowleft}(D)$. If D is 'geometrically connected' (connected as a subset of the plane, such as ) and $\chi_{\circlearrowleft}(D) \geq 3$, then $R(D;k) > (v(D)-1)(\chi_{\circlearrowleft}(D)-1)^{k-1}$ ($v(D)$ denotes the number of vertices of D). To construct such a k-coloring, take groups of $v(D)-1$ consecutive vertices and color all edges in each group by the first color, then color all edges between a set of $\chi_{\circlearrowleft}(D)-1$ groups by the second color, and continue the same way with the further colors. At least in some cases this coloring gives a good lower bound: Harborth e.a. [3, 18, 30] proved that this is indeed the extremal coloring for convex paths. Also it should be noted that in Harborth and Bialostocki's list [3] of the Ramsey numbers for all *cggs* with four vertices the *cgg* Ramsey number is much bigger than the graph Ramsey number just in those cases when $\chi_{\circlearrowleft}(D)$ is bigger than $\chi_{\text{graph}}(G(D))$. Thus $\chi_{\circlearrowleft}$ seems to be a relevant parameter in the study of *cggs*.

2. Unit distances in point sets in convex position. Although there is a short self-contained proof of Füredi's theorem in a very recent paper [7], our proof using Theorem 5 seems also interesting, since Theorem 5 also gives a lower bound of $\Omega(n \log n)$ in this *cgg*-model. This shows the limits of this model, and the type of configurations one would have to exclude as unit distance graphs to further improve the upper bound on the number of unit distances. The conjectured correct order is $O(n)$, the best general lower bound is $2n-7$ [11]. It should also be noticed that for centrally symmetric polygons this problem has been essentially solved [1], with $2n-3$ upper and $2n - O(\sqrt{n})$ lower bounds; but the *cggs* constructed for the lower bound of Theorem 5 are centrally symmetric, so in the *cgg* model central symmetry does not make any difference, whereas in the geometric problem it does.

It could of course be that  is just a bad choice for a non-realizable subgraph; another *cgg* which is not realizable as unit distance graph in a convex polygon is ; but the same construction as in Theorem 5 gives also $\mathrm{ex}_{\circlearrowleft}(n,$ $) = \Omega(n \log n)$. Other excluded bipartite configurations were also studied by Fishburn and Reeds in a similar model [14]. Still the *cgg* model is better than a graph model, since the only graphs with a linear Turán function are trees, and all trees are realizable as graphs with unit distances in convex polygons.

3. Relation between the Turán functions of a cgg and its underlying graph. The Turán-function $\mathrm{ex}_{\mathrm{graph}}(n, G(D))$ of the underlying graph $G(D)$ is a lower bound for the Turán-function $\mathrm{ex}_{\circlearrowleft}(n, D)$ of the *cgg* D; but it is not clear whether each graph G does have a *cgg* D with G as underlying graph for which these functions are (at least asymptotically) the same. For non-bipartite graphs G this follows from Theorem 1, since there is always a cyclic order of the vertices that does not increase the chromatic number. But for bipartite graphs, even for trees, it is open: does each graph G has a *cgg* D (a cyclic order of its vertices) that is as 'unavoidable' as G? For the path P_3 of length three (e.g.) the possible *cggs* are shown in Figure 1, and we have $\mathrm{ex}_{\mathrm{graph}}(n, P_3) = n - O(1)$, and $\mathrm{ex}_{\circlearrowleft}(n,$ $) = \frac{1}{3}n^2 + O(n)$, $\mathrm{ex}_{\circlearrowleft}(n,$ 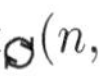$) = \mathrm{ex}_{\circlearrowleft}(n,$ $) = n$, and $\mathrm{ex}_{\circlearrowleft}(n,$ $) = 2n - 3$.

4. Adding some edges does not change the Turán function. Also we know of two simple constructions that preserve the order of magnitude of the Turán function (see Figure 7):

Theorem 9. *Let D, D' be two cggs such that D' differs from D only by the addition of a vertex w of degree one between two consecutive vertices u, v and of an edge incident to w ($D' \setminus \{w\} = D$, $|E(D')| = |E(D)| + 1$).*

(1) If uv and wv are edges of D', then $\mathrm{ex}_{\circlearrowleft}(n, D') \leq \mathrm{ex}_{\circlearrowleft}(n, D) + n$.

(2) If u, v, w have a common neighbor in D', then $\mathrm{ex}_{\circlearrowleft}(n, D') \leq 5\,\mathrm{ex}_{\circlearrowleft}(n, D)$.

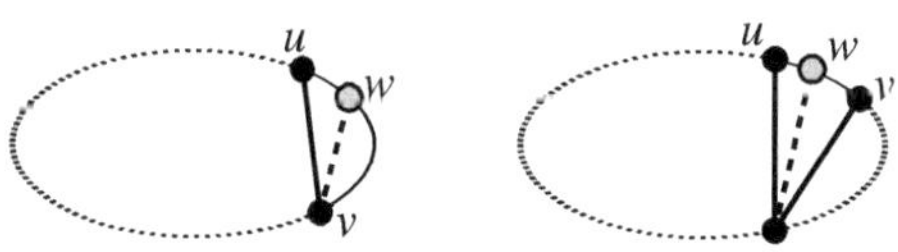

Figure 7

Although these constructions help to give an upper bound in some cases, the question for a linear upper bound is still open for most bipartite forest *cggs*.

5. Simple structure of cyclic chromatic number. Although the results presented in this paper indicate that the cyclic chromatic number plays a similar role for *cggs* as the chromatic number does for graphs, it is really a much simpler parameter.

The cyclic chromatic number is much easier to determine than the graph chromatic number: whereas it is NP-hard even to give a constant-factor approximation of the chromatic number of graphs, the cyclic chromatic number of *cggs* can be determined easily in polynomial time, by a simple greedy algorithm. One just chooses a starting vertex for the first color class, and extends each color class as far as possible before starting a new color class; repeating this for each vertex as starting vertex will find the cyclic coloring with minimum number of colors.

Similarly, the structure of critically k-chromatic *cggs* (whose chromatic number decreases if any edge is deleted) seems much simpler than that of critically k-chromatic graphs. A *cgg* on $(k-1)l+1$ vertices with only all length-l diagonals $(v_i v_{i+l})$ as edges is a critically k-chromatic *cgg*, and much simpler than large critically k-chromatic graphs.

6. Related structures. Besides the *cggs* studied in this paper, several other graph-like structures with some order relation on the vertex set seem natural objects, and would allow a similar excluded-substructure extremal theory. This is the case for bipartite graphs with a linear ordering on both vertex classes; these structures appeared already in literature as 0-1-matrices with forbidden submatrices [4, 15, 17]. Even more natural are graphs with a linear ordering on the vertices [9]. The extremal theories of these objects are related to each other, thus Lemma 4 of this paper is really an extremal theorem on bipartite ordered graphs, and Lemma 3 is an extremal theorem on linear ordered graphs. We plan to describe these extremal theories and their relations in another paper.

5 Proofs of the Results

Proof of Theorem 1 : The lower bound follows from $\chi_{\circlearrowleft}(D^{\text{Turan}}_{n,r-1}) = r-1$, so $D^{\text{Turan}}_{n,r-1}$ cannot contain any sub-*cgg* D with $\chi_{\circlearrowleft}(D) = r$. We only have to show the upper bound of the theorem: that for sufficiently big $n > n_0(k,m,\epsilon)$ each *cgg* D_n with n vertices and at least $e(D^{\text{Turan}}_{n,k-1}) + \epsilon n^2$ edges contains a sub-*cgg* $D^{\text{Turan}}_{mk,k}$, which in turn contains all *cgg*'s of chromatic number at most k and at most m vertices. By the Erdős-Stone Theorem the graph $G(D_n)$ will contain (for $n > n_0^{\text{Erdős-Stone}}(k, mk^3, \epsilon)$) a subgraph $T_{mk^3,k}$. Let $\hat{D}$ be the sub-*cgg* of D_n induced by this subgraph, so $G(\hat{D}) = T_{mk^3,k}$. We partition the mk^3 vertices of $\hat{D}$ in k boundary intervals of mk^2 consecutive vertices, and construct a bipartite graph with the k vertex-classes of $G(\hat{D})$ and these k intervals of $\hat{D}$ as vertices, joining a vertex-class to an interval if that interval contains at least m vertices of that class. Now by Hall's Theorem this bipartite graph contains a matching: for otherwise there is a subset of i classes such that only at most $i-1$ of the intervals contain each at least m vertices of at least one of the i classes. The remaining $k-i+1$ intervals contain then of each of the i classes at most $m-1$ in each interval, and of each of the $k-i$ classes together at most mk^2 (i.e. all). But

$$(k-i+1)i(m-1)+(k-i)mk^2 < (k-i+1)mk^2,$$

so there are not enough vertices to fill the the $(k-i+1)$ intervals each containing mk^2 vertices. Thus Hall's condition is satisfied, and there is a matching from the classes to the intervals. So we can select from each interval m vertices of its matched class, and get a sub-*cgg* $D^{\text{Turan}}_{mk,k}$. This completes the proof of the theorem. □

An alternative proof, replacing Hall's Theorem by a greedy selection of the vertex classes, was given by Pach [26].

Proof of Theorem 2 : Let D be a *cgg* on n vertices that does not contain $D^{\text{bipartite}}_{r,s}$ as sub-*cgg*. As in the Kővári-Sós-Turán-proof for the corresponding graph-result, we count the sub-*cggs* $D^{\text{bipartite}}_{1,r}$ (r-armed stars) twice: once by taking each vertex as center and selecting r neighbors, and once by taking each r-tuple of vertices adding a center. Each r-tuple defines r intervals of vertices, and each of these intervals may contain only at most $s-1$ possible centers of r-armed stars without generating a sub-*cgg* $D^{\text{bipartite}}_{s,r}$. This gives

$$\sum_{v\in V(D)} \binom{\deg(v)}{r} \le (s-1)r\binom{n}{r},$$

and since $\binom{x}{r}$ is a convex function of x, we get by application of Jensen's inequality the claimed bound on the number of edges $e=\frac{1}{2}\sum_{v\in V(D)}\deg(v)$. □

Proof of Theorem 3: Suppose there is a counterexample D of minimal vertex number. Let $v_iv_jv_k$ be a path of length two in D for which the number of vertices on the boundary between v_i and v_j on the arc that contains v_k is minimal. Let A be the set of vertices on the long arc between v_i and v_j ($v_k \notin A$), B the set of vertices between v_i and v_k, and C the set of vertices between v_k and v_j (so $V(D)=A\cup B\cup C\cup\{v_i,v_j,v_k\}$), as in Figure 8. Since the minimal degree in the minimum counterexample D is at least three, v_k has at least two further neighbors. If one of them is in A, we have a folded path . If one of the neighbors is in C, or two of the neighbors are in $B\cup\{v_i\}$, the path $v_iv_jv_k$ was not minimal. So there is no counterexample.

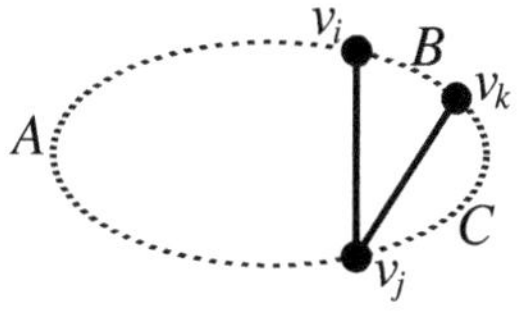

Figure 8

Proof of Theorem 4 : A -free *cgg* of n vertices and $\Omega(n \log n)$ edges can be constructed by taking all diagonals of length 2^i for $1 \leq 2^i < \frac{1}{2}n$. Suppose this *cgg* contains a sub-*cgg* . The two uncrossed (middle) edges cut off two disjoint intervals in the vertex set, each of which contains one . One of these intervals has length at most $\frac{1}{2}n$. So the middle edge of that is a diagonal of length 2^i, and the other two edges are shorter diagonals, of length at most 2^{i-1}. So they are not long enough to intersect.

The proof of an upper bound $\mathrm{ex}(n, \;) < 3n \log n$ proceeds in the other direction: we partition the edge set of a given -free n-vertex *cgg* D_n in $\log n$ sub-*cggs* $D_{n,i}$, $i = 1, \ldots, \log n$. The sub-*cgg* $D_{n,i}$ contains all edges of D_n whose length is between 2^{i-1} and $2^i - 1$. Let $\hat{D}_{n,i}$ be the sub-*cgg* obtained from $D_{n,i}$ by removing from each vertex the first and last of the incident edges. Suppose there are two non-intersecting edges in $\hat{D}_{n,i}$. Then $D_{n,i}$ contains a , where the uncrossed middle edges are the two non-intersecting edges in $\hat{D}_{n,i}$, and the outer edges are intersecting by the length restriction on edges of $D_{n,i}$. Thus

$$e(D_{n,i}) \leq 2n + e(\hat{D}_{n,i}) \leq 2n + \mathrm{ex}(n, \;) = 3n,$$

and

$$e(D_n) = \sum_{i=1}^{\log n} e(D_{n,i}) \leq 3n \log n.$$

Proof of Theorem 5 : To show that $\mathrm{ex}(n, \;) > cn \log n$ we use the construction indicated in Figure 9, which gives a sequence of bipartite *cggs* not containing with 2^k vertices. For that we pack two copies of the *cgg* with 2^{k-1} vertices (with bipartition $V_1 \cup V_2, V_1' \cup V_2'$) diagonally across each other, so the vertex classes are in the order V_1, V_2', V_2, V_1', and join V_1 and V_2' by 2^{k-1} parallel edges. This gives a *cgg* with 2^k vertices and $k2^{k-1}$ edges without as sub-*cgg*.

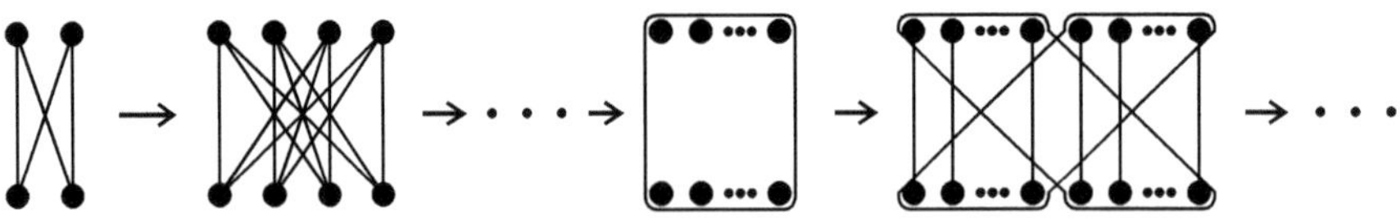

Figure 9

It remains to prove the upper bound. For this we reverse the construction process and use induction. Let D_n be a *cgg* with $n = 2^\ell$ vertices which is -free. We call a partition $V(D_n) = A \cup B \cup C \cup D$ an orthogonal partition if it is a partition of the vertex set in four consecutive sets A, B, C, D of consecutive vertices, $|A \cup C| = |B \cup D| = \frac{n}{2}$, and $|A| = |B|$, $|C| = |D|$. Let $e(X, Y)$ be the number of edges in D_n between X and Y. We call such an orthogonal partition balanced if

$$\left|e(A,B) - e(C,D)\right| \le \tfrac{3}{2}n \quad\text{and}\quad \left|e(A,D) - e(B,C)\right| \le \tfrac{3}{2}n,$$

so the numbers of edges between the opposite pairs of bipartite sub-*cggs* induced by our partition do not differ too much from each other.

Lemma 1. *There is always a balanced orthogonal partition.*

Let $A \cup B \cup C \cup D$ be such a balanced partition. Since

$$e(D_n) = e(A \cup C) + e(B \cup D) + e(A,B) + e(B,C) + e(C,D) + e(D,A)$$

$\le 2\,\mathrm{ex}_{\circlearrowleft}(\frac{n}{2},$  $) + e(A,B) + e(B,C) + e(C,D) + e(D,A),$

we have to bound the numbers of edges in the bipartite sub-*cggs* generated by our partition. For this we use

Lemma 2. *If X, Y are two consecutive classes of our partition and $e(X, Y) > n$ then the sub-cgg induced by $X \cup Y$ contains a folded path*  *with the fold oriented toward the other two classes of the partition.*

Thus if $e(A, B) > n$ and $e(C, D) > n$, each contains half of our forbidden configuration, together resulting in a sub-*cgg* . So if $e(A, B) > n$, then we have $e(C, D) \le n$, and by our assumption of a balanced partition we get $e(A, B) \le \frac{5}{2}n$. Thus we get $e(A, B) + e(C, D) \le \frac{7}{2}n$. The same argument holds for the other pair of bipartite sub-*cgg*'s, so $e(A, D) + e(B, C) \le \frac{7}{2}n$, and together

$\mathrm{ex}_{\circlearrowleft}(n,$ 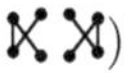 $) \le 2\,\mathrm{ex}_{\circlearrowleft}(\frac{n}{2},$  $) + 7n.$

This implies the upper bound $\mathrm{ex}_{\circlearrowleft}(n,$  $) = O(n \log n)$ claimed in the theorem. □

Proof of Lemma 1. Starting with an arbitrary orthogonal partition (A, B, C, D) we will construct a sequence of orthogonal partitions (A_i, B_i, C_i, D_i) and show that among them there is one with the claimed property. To construct new orthogonal partitions from old, we use the following moves (see Figure 10):

- $+$-translation: the first element of A is moved to D, and the last element of B is moved to C;
- $-$-translation: the last element of D is moved to A, and the first element of C is moved to B;

- rotation: the last element of A is moved to B, the last element of B is moved to C, the last element of C is moved to D, and the last element of D is moved to A.

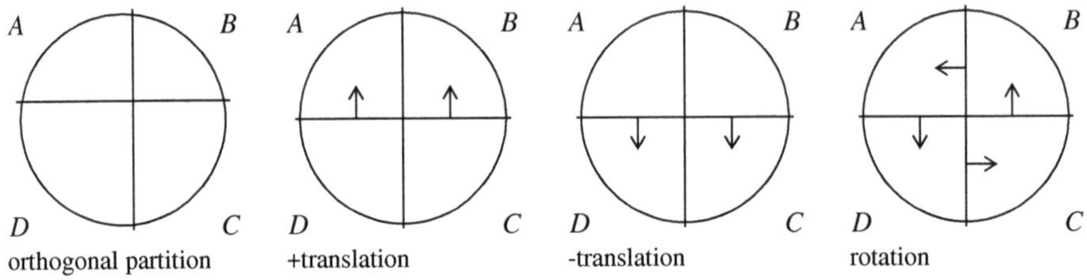

Figure 10

Each of these moves produces again an orthogonal partition. We construct now a sequence of orthogonal partitions from our arbitrary starting partition by the following rules:

- if $e(A,B) > e(C,D) + \frac{n}{2}$ we use a +-translation,
- if $|e(A,B) - e(C,D)| \leq \frac{n}{2}$ we use a rotation,
- if $e(A,B) < e(C,D) - \frac{n}{2}$ we use a −-translation.

Since each +-translation does not increase $e(A,B)$ and does not decrease $e(C,D)$, it does not increase the difference $e(A,B) - e(C,D)$. But since only one point of each A and B is removed, and these points are incident to at most $n-1$ edges, a +-translation decreases $e(A,B) - e(C,D)$ by less than n. So a +-translation can only be followed by another +-translation, or a rotation, never by a −-translation. And since a sufficiently long sequence of +-translations will result in $e(A,B) = 0$, each sequence of +-translations really ends with a rotation. The same argument holds for −-translations.

Starting now with an arbitrary orthogonal partition, we perform a sequence of translations until we reach for the first time $|e(A,B)-e(C,D)| \leq \frac{n}{2}$. Denote that partition by (A_1, B_1, C_1, D_1), and number the following partitions accordingly. Now in this sequence each translation move decreases the absolute value of the difference $|e(A,B)-e(C,D)|$, and a rotation move only happens if it is at most $\frac{n}{2}$. This rotation move possibly increases it again, but only by at most n. Thus for each orthogonal partition (A_i, B_i, C_i, D_i) in our sequence we have $|e(A_i, B_i) - e(C_i, D_i)| \leq \frac{3}{2}n$. Each rotation move rotates the bipartition $(A \cup D, B \cup C)$ one step further, so after $\frac{n}{2}$ rotation moves we have some (A_k, B_k, C_k, D_k) with $A_1 \cup D_1 = B_k \cup C_k$ and $B_1 \cup C_1 = A_k \cup D_k$. This does not necessarily mean that $A_1 = C_k$, $B_1 = D_k$, $C_1 = A_k$ and $D_1 = B_k$, but we can get one partition from the other by performing some further translation moves which do not destroy the property $|e(A_i, B_i) - e(C_i, D_i)| \leq \frac{3}{2}n$. Thus we have a closed sequence

$(A_i, B_i, C_i, D_i)_{i=1}^m$, in which at the end A and C, B and D have changed places, and for which always $|e(A_i, B_i) - e(C_i, D_i)| \leq \frac{3}{2}n$ holds.

We now need an index for which also $e(A_i, D_i) - e(B_i, C_i)$ has small absolute value. In each step this difference changes by at most $2n$, and the sequence of difference values changes the sign somewhere, since $e(A_m, D_m) - e(B_m, C_m) = -\Big(e(A_1, D_1) - e(B_1, C_1)\Big)$. Thus there is an index μ (before or after the sign change) with $|e(A_\mu, D_\mu) - e(B_\mu, C_\mu)| \leq n$. This is the orthogonal partition required by the lemma. □

Proof of Lemma 2. Suppose $e(X, Y) > n$, and remove recursively from X and Y all those points that have only at most one incident edge in that bipartite sub-*cgg*. Let x be the last point of X before the start of Y. This point has at least two neighbors y, y^* in Y, where we may assume that y is between x and y^*. y has a second neighbor x^* in X. Since x is the last point of X, the points occur in the order x^*, x, y, y^*. This is the claimed folded path . □

Proof of Theorem 6 : In the following we prove a slightly stronger statement:

Lemma 3. *There are constants $c_{k,l}$ such that each cgg D with n vertices and more than $c_{k,l}(n-1)$ edges contains, for each vertex v, either a sub-cgg*  *with all crossings oriented towards v or a sub-cgg* 
k l
with all crossings oriented away from v.

Figure 11 illustrates the two situations. This directly implies the theorem. □

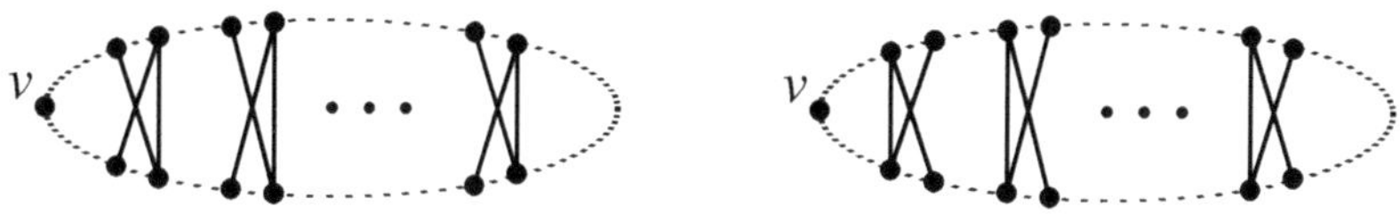

Figure 11

Proof of Lemma 3: The proof is by induction on k, for fixed l. We first observe that $c_{1,l} \leq 3$ for each l, for if D_n is a *cgg* with n vertices and more than $3(n-1)$ edges, and v is that marked vertex, then we can construct a sub-*cgg* $D_n^*(v)$ by removing all edges incident to v, and for each vertex $w \neq v$ the first edge from w to a point following v, and the last edge from w to a point preceding v. By this we remove at most $3(n-1)$ edges. If the remaining *cgg* $D_n^*(v)$ still contains an edge $\{w_1, w_2\}$, then this forms together with the removed edges from w_1, w_2 a  with its fold oriented towards v.

Consider now $k \geq 2$, l fixed, and let $D_{n,k}$ be a given *cgg* with a fixed vertex v, that does not contain a sub-*cgg*  with all crossings oriented
k

towards v or a sub-*cgg*  l with all crossings oriented away from v.

Let $v, v_1, \ldots, v_{n-1}$ denote the vertices of $D_{n,k}$ in the cyclic order. Construct the same way as in the case $k = 1$ a sub-*cgg* $D^*_{n,k}$ by removing all edges incident to v, and for each vertex $w \neq v$ the first and the last edge incident to w to vertices following resp. preceding v (again removing a total of at most $3(n-1)$ edges).

To bound the number of edges in $D^*_{n,k}$, we construct a system of intervals covering the vertex set. In $D^*_{n,k}$ the vertex v and some of the vertices following and preceding v will have no incident edges. Let x_1 be the first vertex following v that still has an incident edge, and y_1 its last neighbour preceding v; then x_1, y_1 defines an interval $\{x_1 \ldots y_1\}$ (not containing v) in the vertex-set of $D^*_{n,k}$. Construct further intervals $\{x_{i+1} \ldots y_{i+1}\}$ by the following rules

- if only isolated vertices are left between y_i and v, then stop.
- if there is a vertex between y_i and v that has a neighbor in the union of the previous intervals (between v and y_i), then let y_{i+1} be the last such vertex preceding v, and x_{i+1} the last such neighbor of y_{i+1} (in the union of the previous intervals).
- if there is a non-isolated vertex left between y_i and v, but no such vertex has a neighbor in the union of the previous intervals, then let x_{i+1} be the first non-isolated vertex following y_i, and y_{i+1} its last neighbor.

The system of intervals $\{x_i \ldots y_i\}$ constructed by this process has the following properties (see Figure 12):

- each vertex is contained in at most two of these intervals,
- each edge of $D^*_{n,k}$ is either contained in one interval, or goes between two consecutive intervals,
- each interval is separated from v by an edge $x_i y_i$, which forms in $D_{n,k}$, together with the edges removed in the construction of $D^*_{n,k}$, a  with its fold oriented towards v.

Now if there are a_i vertices between x_i and y_i, then the the number of edges within the interval $\{x_i \ldots y_i\}$ is at most $c_{k-1,l}a_i + 2a_i + 1$, for the sub-*cgg* between x_i and y_i may not contain either a  $k-1$ with all crossings oriented towards v or a sub-*cgg*  l with all crossings oriented away from v, and there are at most $2a_i + 1$ additional edges in the interval, incident to x_i or y_i. Taking the sum over all these intervals, and using that each vertex belongs to at most two intervals, we obtain an upper bound of at most $2(c_{k-1,l}+2)(n-1)$ for the number of edges of $D^*_{n,k}$ that are contained by some interval $\{x_i \dots y_i\}$.

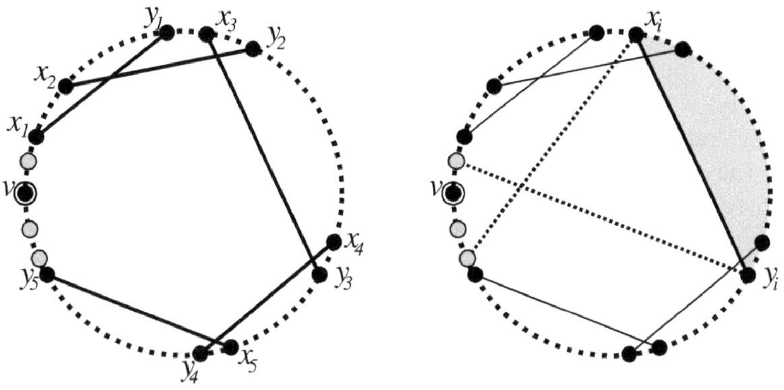

Figure 12

It remains to bound the number of edges in $D^*_{n,k}$ that join two distinct such intervals (which then are consecutive). Here we need another similar lemma.

Lemma 4. *There are constants $c^{\text{bip}}_{k,l}$ such that if W_1, W_2 are disjoint vertex intervals of a cgg D, with a total of n vertices in $W_1 \cup W_2$, and there are more than $c^{\text{bip}}_{k,l}(n-1)$ edges between W_1 and W_2, then for each vertex v following W_2 and preceding W_1 there is either a sub-cgg*  k *with all crossings oriented towards v or a sub-cgg*  l *with all crossings oriented away from v.*

Taking again the sum over all intervals $x_i y_i$ (this time over the pairs of consecutive intervals, separated at some arbitrary point of their overlap) and using again that each vertex occurs in at most two intervals, we get another bound of $2c^{\text{bip}}_{k,l}(n-1)$ for the number of those edges of $D^*_{n,k}$ that are not contained in a single interval. Together this shows that $D^*_{n,k}$ indeed can not have more than $(2c_{k-1,l} + 2c^{\text{bip}}_{k,l} + 4)(n-1)$ edges. Thus the claim of Lemma 3 is true with $c_{k,l} \le 2c_{k-1,l} + 2c^{\text{bip}}_{k,l} + 7$. It only remains to prove Lemma 4.

Proof of Lemma 4: The proof of this lemma is by induction on k and l. Let D be a *cgg* with disjoint vertex intervals W_1, W_2 and a marked vertex v, as in the lemma. The clockwise cyclic order on D induces a linear order on the vertices of W_1 and W_2 (see Figure 13), all references to 'first' and 'last' vertices in the following will be with respect to that order. All references to edges concern only edges between W_1 and W_2.

We first note that $c^{\text{bip}}_{1,l} \le 1$; for if we remove from each vertex in W_1 the last incident edge, and from each vertex of W_2 the first incident edge (a total of at most $|W_1|+|W_2|-1=n-1$ edges), then any remaining edge will form, together with the removed edges at its end vertices, a sub-*cgg*  with its crossing oriented towards v. With a symmetrical proof we find that $c^{\mathrm{bip}}_{k,1} \leq 1$.

Let now $k,l \geq 2$, and D, W_1, W_2 be given as in the lemma. Construct a new *cgg* D^* by removing from each vertex in W_1 and W_2 the first and last incident edge (a total of at most $2(n-1)$ edges). Some of the vertices will then have no incident edges between W_1 and W_2; discard those vertices, and denote the sets of the remaining vertices by W_1' and W_2', respectively. We will now cover (as in the previous lemma) W_1' and W_2' by a sequence of interval pairs, such that each remaining edge is contained in some interval pair, and we can apply the induction for each interval pair. For this end we assume that D does not contain either of the two desired sub-configurations. We will argue then that D^* cannot have more than $(c^{\mathrm{bip}}_{k-1,l}+c^{\mathrm{bip}}_{k,l-1}+1)(|W_1'|+|W_2'|-1)$ edges, which in turn proves the statement of the lemma with $c^{\mathrm{bip}}_{k,l} \leq c^{\mathrm{bip}}_{k-1,l}+c^{\mathrm{bip}}_{k,l-1}+3$.

Let x_1 be the first vertex in W_2', y_1 the first vertex in W_1', $Y_0=\{y_1\}$. Define a sequence of points and intervals by letting x_2 be the last neighbor of y_1 in W_2', and X_1 the interval $\{x_1 \ldots x_2\}$. Then let y_2 be the last vertex in W_1' that has a neighbor in the interval X_1, and Y_1 the interval $\{y_1 \ldots y_2\}$. Define now inductively

- x_j as the last vertex in W_2' that has a neighbor in Y_{j-2},
- X_{j-1} as the interval $\{x_{j-1} \ldots x_j\}$,
- y_j as the last vertex in W_1' that has a neighbor in X_{j-1}, and
- Y_{j-1} as the interval $\{y_{j-1} \ldots y_j\}$.

Stop this procedure if $y_j = y_{j-1}$ for some $j \geq 2$ or $x_j = x_{j-1}$ for some $j \geq 3$. Define $W_1^{(1)} = Y_1 \cup \ldots \cup Y_t$ and $W_2^{(1)} = X_1 \cup \ldots \cup X_s$, where in the first case $t=j-2$, $s=j-1$, whereas $t=s=j-2$ in the second case. Then every edge of D^* starting at $W_1^{(1)}$ ends in $W_2^{(1)}$, and vice versa. Moreover, every edge starting at Y_i ends in $X_i \cup X_{i+1}$ and every edge starting at X_i ends in $Y_{i-1} \cup Y_i$. If we repeat this procedure with the vertex sets $W_1' \setminus W_1^{(1)}$ and $W_2' \setminus W_2^{(1)}$, and so on, eventually we obtain partitions $W_1' = W_1^{(1)} \cup \ldots \cup W_1^{(r)}$ and $W_2' = W_1^{(1)} \cup \ldots \cup W_1^{(r)}$ such that each edge between W_1 and W_2 is between $W_1^{(m)}$ and $W_2^{(m)}$ for some $1 \leq m \leq r$. Thus, we only have to bound the number of edges of D^* between $W_1^{(m)}$ and $W_2^{(m)}$, for $1 \leq m \leq r$. Next we show how to do it for $m=1$.

For $1 \leq i$, write $X_i^+ = X_i \setminus \{x_i\}$, $X_i^- = X_i \setminus \{x_{i+1}\}$, $Y_i^+ = Y_i \setminus \{y_i\}$, and $Y_i^- = Y_i \setminus \{y_{i+1}\}$. These sets are not empty with the possible exception of X_1^- and X_1^+. From what we said above, each edge between $W_1^{(1)}$ and $W_2^{(1)}$ connects either Y_i^+ to X_i^- or Y_i^- to X_{i+1}^+ or y_i to X_i^+ or Y_i^+ to x_{i+1} for some i, or perhaps y_1 to x_1.

For each i the edges between Y_i^+ and X_i^- are all to the right of the edge from x_{i+1} to its neighbor in Y_{i-1}^+. This edge, together with its first and last neighbors (removed in the beginning of the proof, in the construction of D^*) forms a  with its crossing oriented towards v. So the bipartite sub-*cgg* induced by Y_i^+ and X_i^- may contain at most $c_{k-1,l}^{\text{bip}}(|Y_i^+| + |X_i^-| - 1)$ edges. Taking the sum over all possible values of i, we find at most $c_{k-1,l}^{\text{bip}}(|W_1^{(1)}| + |W_2^{(2)}| - 1)$ edges contributed by all the pairs Y_i^+ and X_i^-. A similar argument shows that there cannot be more than $c_{k,l-1}^{\text{bip}}(|W_1^{(1)}| + |W_2^{(2)}| - 1)$ edges connecting the possible pairs Y_i^- and X_{i+1}^+, since the edges between Y_i^- and X_{i+1}^+ are all to the left of the edges from y_{i+1} to its neighbour in X_i^-.

Note that there are at most $|W_2^{(1)}| - 1$ edges which connect vertices y_i to vertices in their respective sets X_i^+. Similarly, there are at most $|W_1^{(1)}| - 1$ edges which connect vertices x_{i+1} to vertices in their respective sets Y_i^+. Adding 1 for the possible edge y_1x_1 we conclude that there cannot be more than $(c_{k-1,l}^{\text{bip}} + c_{k,l-1}^{\text{bip}} + 1)(|W_1^{(1)}| + |W_2^{(2)}| - 1)$ edges of D^* between $W_1^{(1)}$ and $W_2^{(1)}$. The same way one can obtain similar estimates for the number of edges between $W_1^{(m)}$ and $W_2^{(m)}$. Given that the sets $W_i^{(m)}$ form a partition of W_i', we find that D^* cannot have more than $(c_{k-1,l}^{\text{bip}} + c_{k,l-1}^{\text{bip}} + 1)(|W_1'| + |W_2'| - 1)$ edges, which proves the claim of the lemma, and completes the proof of Theorem 6. □

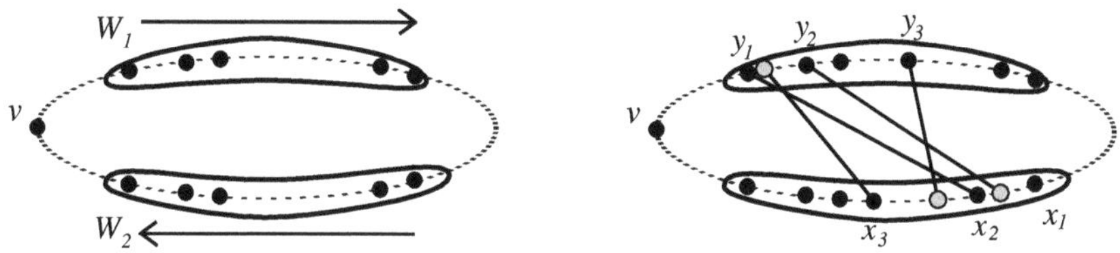

Figure 13

Proof of Theorem 7 : First we prove the upper bound $\mathrm{ex}_{\circlearrowleft}(n, F) \le \lfloor \frac{1}{2}(v-2)n \rfloor$. If F is not connected, we can add edges so that the obtained *cgg* F' is a crossing-free cyclic-bipartite tree with $v - 1$ edges. If a *cgg* contains F' then it certainly contains also F. Since the required upper bound is $\lfloor \frac{1}{2}(v-2)n \rfloor$ both for F and F', we may assume that F is a tree. We further follow Perles' (unpublished) proof of the upper bound in his Theorem.

Let $e_1, \dots, e_{v-1}$ be the edges of F listed from left to right. For $i = 1, \dots, v-1$, let $e_i = x_iy_i$, where x_i is the upper vertex of e_i. Further, let F_i be the smallest *cgg* with edges $e_1, \dots, e_i$. In particular, $F_{v-1} = F$.

Let G be a *cgg* with n vertices and e edges. Let E_i be the set of all ordered pairs (oriented edges) (α, β) for which there is a sub-*cgg* X of G isomorphic to F_i such that $\alpha \in V(X)$ corresponds to x_i and $\beta \in V(X)$ corresponds to y_i. We will prove by induction on i that $|E_i| \geq 2e - (i-1)n$ for each $i = 1 \ldots, v-1$.

E_1 is the set of all ordered pairs of neighboring vertices in G. Thus, $|E_1| = 2e$. For $i > 1$, E_i is obtained from E_{i-1} as follows. If $x_i = x_{i-1}$, then E_i is obtained from E_{i-1} by removing, for each vertex $\alpha \in V(G)$, the last edge $(\alpha, \beta) \in E_{i-1}$ starting in α. If $y_i = y_{i-1}$, then E_i is obtained from E_{i-1} by removing, for each vertex $\beta \in V(G)$, the first edge $(\alpha, \beta) \in E_{i-1}$ ending in β. It follows that $|E_i| \geq |E_{i-1}| - n$. Therefore, by induction, $|E_i| \geq 2e - (i-1)n$ for each i. In particular, if $e > \frac{1}{2}(v-2)n$ then $|E_{v-1}| > 0$. This gives the required upper bound $\mathrm{ex}_{\circlearrowleft}(n, F) \leq \lfloor \frac{1}{2}(v-2)n \rfloor$ (for F connected).

It remains to find *cgg*'s with n vertices giving the lower bound $\mathrm{ex}_{\circlearrowleft}(n, F) \geq \lfloor \frac{1}{2}(v-2)n \rfloor$. We give two constructions, a more general one that works only for connected F, and a special case that works for all F included in the theorem. In both cases the *cgg* giving the lower bound is independent of the structure of F, and depends only on its number of vertices v.

For the case of a connected F and vn even we use the following construction: Let A be any $v-2$-element subset of the cyclic group Z_n that is symmetric (i.e. $A = -A$), and $D_{n,A}$ be the *cgg* of the Cayley-graph over Z_n with A as generators, taken with the natural cyclic ordering. Then $D_{n,A}$ has the required number of edges and does not contain any bipartite crossing-free caterpillar on v vertices as sub-*cgg*. For any such sub-*cgg* D_{sub} induces a bipartition on its vertices $V(D_{\text{sub}})$; in this bipartition we label for each $x \in V(D_{\text{sub}})$ the edges in $D_{n,A}$ around x with $1 \ldots v-2$

- clockwise if x belongs to the 'bottom' class of the bipartition,
- counterclockwise if x belongs to the 'top' class of the bipartition.

In this labeling each edge of D_{sub} receives the same label from its top and bottom vertex. Consider now the $v-1$ edges of the embedded caterpillar D_{sub}, in their natural left-to-right order as used in the upper bound of the theorem. Then the labels increase strictly monotone along the this edge sequence. Since there are only $v-2$ distinct labels, there cannot be such an edge sequence of $v-1$ edges, so there is no v-vertex bipartite crossing-free caterpillar contained in $D_{n.A}$.

To adapt this construction for the case n and v both odd, we use the construction for n an $v+1$ with the additional restriction that A ($|A| = v-1$) contains the two middle diagonals $\lfloor \frac{n}{2} \rfloor$ and $\lceil \frac{n}{2} \rceil$. They form a cycle C_n in $D_{n,A}$, of which we remove a set of $\lceil \frac{n}{2} \rceil$ edges such that in each vertex at least one of the incident edges is removed. Denote this *cgg* by $D^*_{n,A}$. If we use now the same numbering scheme for $D_{n,A}$ and look at the induced numbers on $D^*_{n,A}$, we find that the edges are labeled by 1 to $v-1$, but in each vertex at least one of the labels $\lfloor \frac{v-2}{2} \rfloor$ and $\lceil \frac{v-2}{2} \rceil$ is missing. Thus we have now enough distinct labels for an increasing sequence of $v-1$ edge labels, but only if all the labels occur. But this can only happen if the labels $\lfloor \frac{v-2}{2} \rfloor$ and $\lceil \frac{v-2}{2} \rceil$ both occur in the sequence, where they have to occur consecutively, sharing a vertex. But this does not happen by the construction of the subset we removed.

Figure 14 shows three such *cggs* with 21 vertices and maximum edge number that do not contain a crossing-free bipartite tree of 8 vertices (e.g.), but the second and third contain copies of the crossing-free bipartite forest), and many copies of).

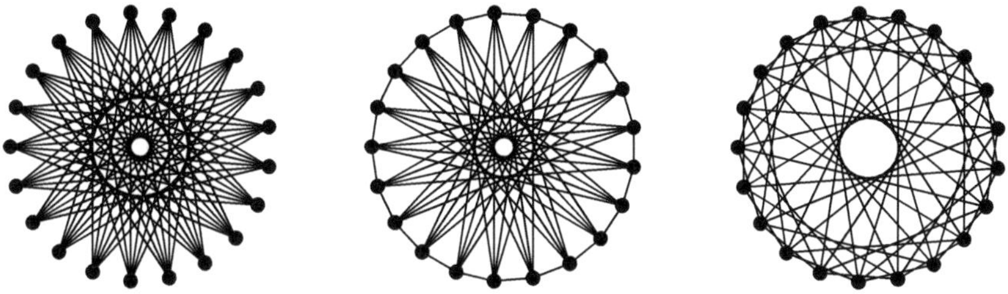

Figure 14

If the excluded forest F is not connected, we still get an extremal *cgg* $D_{n,v}$ with the correct edge number if we use only the $v-2$ middle diagonals from each vertex for v and n of distinct parity, or the $v-1$ middle edges, omitting the rightmost edge for $\lceil \frac{n}{2} \rceil$ consecutive vertices, for v and n of the same parity. To see that this *cgg* does not contain F, we look at the images e, e' in $D_{n,v}$ of the first and last edge of F. These edges do not intersect, and the sub-*cgg* D_{sub} bounded by these edges contains F, so it must contain at least v vertices. But in the graph $D_{n,v}$ any subgraph bounded by two non-crossing edges contains at most $v-1$ vertices, as can be seen by rotating one of the bounding edges until it shares an endpoint with the other. This proves the claim of Theorem 7. □

Proof of Theorem 8 : The proof is by induction on the vertex-number. The claim is obviously true for $n = 3, 4$. Let $n \geq 5$ and D be a *cgg* on n vertices that does not contain . Let $v_i v_j$ be an edge of D such that $i < j \leq i + \frac{n}{2}$, i and j are not consecutive ($j \neq i+1$), and all other edges in $v_i, v_{i+1}, \ldots, v_j$ (the part cut of by $v_i v_j$) are only joining consecutive vertices. Then the sub-*cgg* on $\{v_i, v_{i+1}, \ldots, v_j\}$ has $j - i + 1$ vertices and at most as many edges. Also, since each edge from $\{v_{i+1}, \ldots, v_{j-1}\}$ to $\{v_{j+1}, \ldots, v_{i-1}\}$ crosses $v_i v_j$, and D is -free, each point in $\{v_{i+1}, \ldots, v_{j-1}\}$ has at most one neighbor in $\{v_{j+1}, \ldots, v_{i-1}\}$. Thus $e(D) \leq e(D \setminus \{v_{i+1}, \ldots, v_{j-1}\}) + 2(j-i) - 1$. Let $k = j - i - 1$ (so there are $k \geq 1$ vertices between v_i and v_j), then we obtain, using the inductive hypothesis,

$$e(D) \leq \mathrm{ex}(n, \cdot) \leq \mathrm{ex}(n-k, \cdot) + 2k + 1 \leq \left\lfloor \tfrac{5}{2}(n-k) - 4 \right\rfloor + 2k + 1.$$

For $k \geq 2$ this proves the claim of the theorem, it remains only the case $k = 1$, i.e. $j = i + 2$, v_i, v_{i+1}, v_{i+2} are all joined by edges in D, and there is one further edge going from v_{i+1} across $v_i v_{i+2}$ to some point v_m.

Suppose now that $v_i v_m$ and $v_{i+2} v_m$ are also edges of D. Then D does not contain any edge from $\{v_{i+3}, \dots, v_{m-1}\}$ to $\{v_{m+1}, \dots, v_{i-1}\}$. Thus we can contract v_i, v_{i+1}, v_{i+2} to a single vertex v_i^{new} without creating a sub-*cgg* , if there was not one before. This contraction removes two vertices and five edges, so we can use the inductive hypothesis for the rest to obtain the claim of the theorem.

Suppose finally that at least one of $v_i v_m$ and $v_{i+2} v_m$ (we may assume $v_i v_m$) is not an edge of D; then we remove the edge $v_{i+1} v_m$ and add a new edge $v_i v_m$. This again does not create a , if there was not one before. This does not change the numbers of vertices and edges, but afterwards v_{i+1} is a point of degree two, which we can remove (losing one vertex and two edges) and apply the inductive hypothesis on the rest. This completes the proof of Theorem 8.

Proof of Theorem 9 :
(1) If we remove from a D'-free *cgg* for every vertex the last edge starting at that vertex, we obtain a D-free *cgg*. This process removed at most n edges, and proves claim (1).
(2) Consider a D'-free *cgg* on the vertex set V, $|V| = n$. To prove claim (2), we construct a subset E' of the edge set E with $|E'| \geq |E|/5$, such that at each vertex, between any two edges incident to that vertex in E', there is another incident edge in E that does not belong to the subset E'. This subset then gives a D-free *cgg*, since any sub-*cgg* of (V, E'), isomorphic to D, would extend to a sub-*cgg* of (V, E), isomorphic to D', which does not exist by assumption. To construct E', we define a graph G with vertex set E, in which each edge $e \in E$ is joined by a G-edge to each edge that shares and endpoint with e and immediately precedes or follows e in the cyclic ordering. This graph has a maximum degree at most four, so it contains an independent set whose size at least one fifth of the order of G, which is the claimed edge set. This proves claim (2).

References

[1] B.M. Ábrego and S. Fernández-Merchant: The unit distance problem for centrally symmetric convex polygons, *Discr. Comput. Geom.* **28** (2002), 467–473.

[2] N. Alon, L. Rónyai, and T. Szabó: Norm-graphs: Variations and applications, *J. Comb. Theory, Ser. B* **76** (1999), 280–290.

[3] A. Bialostocki and H. Harborth: Ramsey colorings for diagonals of convex polygons, *Abhandlungen der Braunschw. Wiss. Ges.* **47** (1996), 159–163.

[4] D. Bienstock and E. Győri: An extremal problem on sparse 0-1-matrices, *SIAM J. Discr. Math.* **4** (1991), 17–27.

[5] B. Bollobás: *Modern Graph Theory*, Springer, 1998.

[6] W. G. Brown: On graphs that do not contain a Thomsen graph, *Can. Math. Bull.* **9** (1966), 281–285.

[7] P. Braß and J. Pach: The maximum number of times the same distance can occur among the vertices of a convex n-gon is $O(n \log n)$, *Journal of Combinatorial Theory* Ser. A **94** (2001) 178–179.

[8] V. Capoyleas and J. Pach: A Turán-type theorem on chords of convex polygons, *J. Combin. Theory, Ser. B* **56** (1992), 9–15.

[9] J. Czipszer, P. Erdős, and A. Hajnal: Some extremal problems on infinite graphs, *Publ. Math. Inst. Hung. Acad. Sci., Ser. A* **7** (1962), 441-457.

[10] R. Diestel: *Graph Theory*, Springer, 1997.

[11] H. Edelsbrunner and P. Hajnal: A lower bound on the number of unit distances between the vertices of a convex polygon, *J. Combin. Theory, Ser. A* **56** (1991), 312–316.

[12] P. Erdős, A. Rényi, and V. T. Sós: On a problem of graph theory, *Studia Sci. Math. Hungar.* **1** (1966), 215–235.

[13] P. Erdős and H. Sachs: Reguläre Graphen gegebener Taillenweite mit minimaler Knotenzahl, *Wiss. Martin-Luther-Univ. Halle-Wittenberg, Math.-Naturwiss. Reihe* **12** (1963), 251–258.

[14] P.C. Fishburn and J.A. Reeds: Unit distances between the vertices of a convex polygon, *Comput. Geom. — Th. Appl.* **2** (1992), 81–91.

[15] Z. Füredi: The maximum number of unit distances in a convex n-gon, *J. Combin. Theory, Ser. A* **55** (1990), 316–320.

[16] Z. Füredi: Turán type problems, in: *Surveys in Combinatorics*, 253–300, Cambridge Univ. Press, Cambridge, 1991.

[17] Z. Füredi and P. Hajnal: Davenport-Schinzel theory of matrices, *Discr. Math.* **103** (1992), 233–251.

[18] H. Harborth and H. Lefmann: Coloring arcs of convex sets, *Discr. Math.* **220** (2000), 107–117.

[19] Gy. Károlyi, J. Pach, and G. Tóth: Ramsey-type results for geometric graphs I, *Discr. Comput. Geom.* **18** (1997), 247–255.

[20] Gy. Károlyi, J. Pach, G. Tóth, and P. Valtr: Ramsey-type results for geometric graphs II, *Discr. Comput. Geom.* **20** (1998), 375–388.

[21] J. Kollár, L. Rónyai, and T. Szabó: Norm-graphs and bipartite Turán numbers, *Combinatorica* **16** (1996), 399–406.

[22] Y. Kupitz: On pairs of disjoint segments in convex position in the plane, *Annals Discr. Math.* **20** (1984), 203–208.

[23] Y.S. Kupitz and M.A. Perles: Extremal theory for convex matchings in convex geometric graphs, *Discr. Comput. Geom.* **15** (1996), 195–220.

[24] A. Lubotzky, R. Phillips, and P. Sarnak: Ramanujan graphs, *Combinatorica* **8** (1988), 261–277.

[25] W. Moser and J. Pach: Geometric graphs, in: *New Trends in Discrete and Computational Geometry* (J. Pach, ed.), Springer, New York, 1993.

[26] J. Pach: Personal communication, 2000.

[27] J. Pach and P. Agarwal: *Combinatorial Geometry*, Wiley, New York, 1995.

[28] J. Pach, R. Pinchasi, M. Sharir, and G. Tóth: Topological graphs with no large grids, in preparation.

[29] J. Pach, R. Pinchasi, G. Tardos, and G. Tóth: Geometric graphs with no self-intersecting path of length 3, in: *Graph Drawing 2002* (M.T. Goodrich and S.G. Kobourov, eds.), Lecture Notes in Computer Science **2528**, Springer, Berlin, 2002, 2905–311.

[30] C.C. Rousseau, C.J. Jaywardene, and H. Harborth: On path convex Ramsey numbers, *Manuscript*.

[31] M. Simonovits: Extremal graph theory, in: *Selected Topics in Graph Theory 2* (L.W. Beineke, R.J. Wilson, eds.), 1983, 161–200.

About Authors

Peter Brass is at the Department of Computer Science R8/206, City College, CUNY, 138th Street at Convent Avenue, New York NY-10031, USA; *peter@cs.ccny.cuny.edu*.

Gyula Károlyi is at the Department of Algebra and Number Theory, Eötvös University, H–1518 Budapest, Pf. 120, Hungary; *karolyi@cs.elte.hu*.

Pavel Valtr is at the Department of Applied Mathematics and Institute for Theoretical Computer Science (ITI), Charles University, Malostranské nám. 25, 118 00 Praha 1, Czech Republic; *valtr@kam.ms.mff.cuni.cz*.

Acknowledgments

The authors are much indebted to János Pach for valuable and enlightening discussions on the topic of this paper. We would like to thank Rom Pinchasi for explaining us the proof of the theorem of Perles. Work by Peter Brass was done while supported by DFG grant BR 1465/5-2. Work by Gyula Károlyi was partially supported by Hungarian Scientific Research Grants AKP 2000-78 2.1, OTKA F030822, and FKFP 0151/1999. Work by Pavel Valtr was supported by project LN00A056 of The Ministry of Education of the Czech Republic, by Charles University grants No. 99/158 and 99/159 and by Czech Republic Grant GAČR 201/99/0242.

On the Inapproximability of Polynomial-programming, the Geometry of Stable Sets, and the Power of Relaxation

Andreas Brieden
Peter Gritzmann

Abstract

The present paper introduces the *geometric rank* as a measure for the quality of relaxations of certain combinatorial optimization problems in the realm of polyhedral combinatorics. In particular, this notion establishes a tight relation between the *maximum stable set* problem from combinatorial optimization, *polynomial programming* from integer non linear programming and *norm maximization*, a basic problem from convex maximization and computational convexity.

As a consequence we obtain very tight inapproximability bounds even for the largely restricted classes of polynomial programming where the polynomial is just a sum of univariate monomials of degree at most $\lceil \log n \rceil$, and it is guaranteed that the maximum is attained at a 0-1-vector. More specifically, unless $\mathbb{NP} = \mathbb{ZPP}$ this problem does not admit a polynomial-time $n^{1-\epsilon}$-approximation for any $\epsilon > 0$, and does not even admit a polynomial-time $n^{1-O(1/\sqrt{\log \log n})}$-approximation, unless $\mathbb{NP} = \mathbb{ZPTIME}(2^{O(\log n (\log \log n)^{3/2})})$. Similar results are also given for norm maximization. In addition we relate the geometric rank of a relaxation of the stable set polytope to the question whether the separation problem for the relaxation can be solved in polynomial time. Again, the results are nearly optimal.

1 Introduction and main results

Let $\mathcal{O}$ be any set packing problem, i.e. a combinatorial optimization problem whose instances consist of a collection $\mathcal{V}$ of subsets of a non empty ground set V and whose goal it is to produce a maximum number of disjoint such subsets. Let $P_{\mathcal{O}}(I)$ be the 0-1-polytope spanned by the incidence vectors of the feasible solutions for the instance $I = (V, \mathcal{V})$. Then the set packing problem asks for $\max_{x \in P_{\mathcal{O}}(I)} e^T x$, where $e = (1, \ldots, 1)^T \in \mathbb{R}^{|\mathcal{V}|}$. Since $\mathcal{O}$ is a packing problem, $P_{\mathcal{O}}(I)$ is a monotone polytope, hence contains the standard unit vectors of $\mathbb{R}^{|\mathcal{V}|}$. Now let P be a polytope with

$$P \cap \{0,1\}^{|\mathcal{V}|} \subset P_{\mathcal{O}}(I) \subset P \subset [0,1]^{|\mathcal{V}|}.$$

Such a polytope is called a *standard relaxation* of $P_{\mathcal{O}}(I)$. The *geometric rank* $g_{\mathcal{O}}(P)$ of the standard relaxation P of $P_{\mathcal{O}}(I)$ is the minimal $p \in [1, \infty]$ such that $\max_{x \in P} \|x\|_p$ is attained at an integral point of P. Note that $1 \leq g_{\mathcal{O}}(P) \leq \infty$, and that $\max_{x \in P} \|x\|_p$ is attained at an integral point of P for any $p \geq g_{\mathcal{O}}(P)$. Clearly, $g_{\mathcal{O}}(P) = 1$ means that the optimum of the linear objective function $e^T x$ over P equals that over $P_{\mathcal{O}}(I)$, i.e. P contains a 0-1-point that is optimal.

Various papers have addressed the question of how to measure the quality of relaxations in combinatorial optimization, see e.g. [LMJ94], [Goe95]. The geometric rank is a notion that allows to study the following two questions: What are the algorithmic limitations of strengthening the linear optimization subroutines in polyhedral combinatorics by considering approximative ℓ_p-norm optimization routines? What are the limitations in terms of polynomial-time separation if we consider relaxations with small geometric rank?

The first question will lead to new inapproximability results while the second will show the limits of polynomial-time separation even for quite coarse approximations.

The following three optimization tasks and their interplay are the central topics of our study. The first is the classical combinatorial optimization problem of finding the cardinality of a maximum stable set in a graph while the other two are particular restrictions of polynomial programming, the last having a specifically strong geometric flavor.

MAXSTABLESET. Given a graph $G = (V, E)$, find the maximum α so that there is a subset V^* of V with $|V^*| = \alpha$ such that no two vertices of V^* are joined by an edge in E.

Of course, MAXSTABLESET can be regarded as set packing problem. The next problem is a restricted polynomial programming problem that has been studied before by [BR95].

0-1-POLYPROG. Given $n, r, s \in \mathbb{N}$, subsets $S_1, \ldots, S_r$ of $\{1, \ldots, n\}$, a vector $b \in \mathbb{Z}^s$ and a matrix $A \in \mathbb{Z}^{s \times n}$, compute the maximum of the multivariate polynomial $f(x) = \sum_{j=1}^{r} \prod_{i \in S_j} \xi_i$, where $x = (\xi_1, \ldots, \xi_n)^T$, over the polytope $P = \{x \in [0,1]^n \colon Ax \leq b\}$ provided the existence of a 0-1-maximizer is guaranteed.

The third problem depends on a functional $\gamma : \mathbb{N} \to \mathbb{N} \cup \{\infty\}$ that is assumed to be evaluable in time that is bounded by a polynomial in n.

γ-NORMMAX. Given $n, s \in \mathbb{N}$, a rational $s \times n$-matrix A and $b \in \mathbb{Q}^s$, compute

$$N_p(P) = \max_{x \in P} \|x\|_p^p \text{ if } p \neq \infty \quad \text{and} \quad N_\infty(P) = \max_{x \in P} \|x\|_\infty \text{ else,}$$

where $p = \gamma(n)$ and $P = \{x \in \mathbb{R}^n \colon Ax \leq b\}$.

As usual, our nonapproximability results hold under the assumption that certain (unlikely) characterizations of $\mathbb{NP}$ do not hold. Therefore recall that $\mathbb{ZPP} = \mathbb{RP} \cap \text{co}\,\mathbb{RP}$ is the class of problems that can be solved in probabilistic polynomial time with zero error. As a generalization, if we replace the polynomial bound on the running time by $O(f(n))$ for some functional $f : \mathbb{N} \to \mathbb{N}$ we obtain the class $\mathbb{ZPTIME}(f(n))$. For some introduction to the theory of computational complexity see e.g. [Jan98].

As shown in [Hås96, Hås99] MaxStableSet does not admit a polynomial-time $|V|^{1-\epsilon}$-approximation for any $\epsilon > 0$, unless $\mathbb{NP} = \mathbb{ZPP}$. Also, by [EH00], MaxStableSet does not even admit a polynomial-time $n^{1-O(1/\sqrt{\log\log n})}$-approximation, unless $\mathbb{NP} = \mathbb{ZPTIME}(2^{O(\log n(\log\log n)^{3/2})})$, i.e., given that $\mathbb{NP}$ does not admit randomized algorithms with slightly super-polynomial expected running time. On the other hand, MaxStableSet can be approximated in polynomial time within $n^{1-O(\log\log n/\log n)}$, [BH92].

Using the former result and a construction of [EHdW84] that relates MaxStableSet to 0-1-PolyProg, [BR95] derived – again under the assumption $\mathbb{NP} \neq \mathbb{ZPP}$ – the inapproximability bound $n^{1/2-\epsilon}$ for 0-1-PolyProg for any $\epsilon > 0$, where the instances are such that $\Omega(\sqrt{n}) = r = O(n)$ and $|S_j| = O(n)$ for $j = 1, \dots, r$.

For constant functions $\gamma \equiv p$ the computational complexity of p-NormMax has been studied in detail in [GK93]. Of course, ∞-NormMax can be solved in polynomial time, but p-NormMax is $\mathbb{NP}$-hard for all other p. [BGK00] shows that for $p \in \mathbb{N}$ p-NormMax is even $\mathbb{APX}$-hard and a result of [Bri02] indicates that for $p \in \mathbb{N}\setminus\{1\}$ the problem is not even 'likely' to be in $\mathbb{APX}$.

Our main inapproximability result is one for γ-NormMax.

Theorem 1.1. *Let $k \in \mathbb{N}$ and $\lambda : \mathbb{N} \to [1, \infty[$ be any function, $\gamma : \mathbb{N} \to \mathbb{N}$ be a function with $1 + \log(n/k) \leq \gamma(n)$ for all $n \in \mathbb{N}$ that can be evaluated in polynomial time, and assume that there exists a polynomial-time λ-approximation algorithm for γ-NormMax. Then there exists a polynomial-time λ-approximation algorithm for MaxStableSet.*

In conjunction with the inapproximability results of [Hås96, Hås99] and [EH00] for MaxStableSet we obtain the following corollary.

Corollary 1.2. *Let $k \in \mathbb{N}$, $\epsilon > 0$, and $\gamma : \mathbb{N} \to \mathbb{N}$ be a function with $1 + \log(n/k) \leq \gamma(n)$ for all $n \in \mathbb{N}$ that can be evaluated in polynomial time. Then there does not exist a polynomial-time $n^{1-\epsilon}$-approximation algorithm for γ-NormMax, unless $\mathbb{NP} = \mathbb{ZPP}$. In addition there does not exist a polynomial-time $n^{1-O(1/\sqrt{\log\log n})}$-approximation for γ-NormMax, unless $\mathbb{NP} = \mathbb{ZPTIME}(2^{O(\log n(\log\log n)^{3/2})})$.*

This inapproximability result for γ-NormMax means that even if $p \to \infty$ (the easy case since ∞-NormMax can be solved in polynomial time) norm-maximization over polytopes 'stays pretty intractable on the way'. In the other direction, the geometric study leading to Theorem 1.1 allows to answer the question whether for 0-1-POLYPROG the degree bound for the polynomials can be further reduced without weakening the inapproximability result. In fact, as a consequence of Theorem 1.1 we obtain the following inapproximability result for 0-1-POLYPROG.

Corollary 1.3. *Let $k \in \mathbb{N}$. There is no polynomial-time approximation algorithm for* 0-1-POLYPROG *with performance ratio $n^{1-\epsilon}$ for any $\epsilon > 0$, unless $\mathbb{NP} = \mathbb{ZPP}$, even if the instances are restricted to those whose polynomial is a functional, convex on the feasible region, that consists of at most n monomials of degree at most $\lceil 1 + \log(n/k) \rceil$.*

In addition the same class of instances does not admit a polynomial-time $n^{1-O(1/\sqrt{\log\log n})}$-approximation, unless $\mathbb{NP} = \mathbb{ZPTIME}(2^{O(\log n (\log\log n)^{3/2})})$.

Note that polynomial-time approximation with error at most n is trivial for the restricted class of instances in Corollary 1.3.

The previous results can be interpreted as showing the limitation of trying to strengthen the linear programming subroutines in polyhedral combinatorics. The following will deal with polynomial-time relaxations of the stable set polytope.

In polyhedral combinatorics the polytopes are $\mathcal{H}$-presented, i.e. given in terms of systems of linear inequalities. So suppose that for each instance $G = (V, E)$ of MAXSTABLESET, $P(G)$ is a standard $\mathcal{H}$-presented relaxation of the stable set polytope $P_{\mathcal{S}}(G)$ of G, i.e.,

$$P(G) \cap \{0,1\}^{|V|} \subset P_{\mathcal{O}}(I) \subset P(G) \subset [0,1]^{|V|}.$$

Now, set

$$\mathcal{P} = \{P(G) : G \text{ is a finite graph}\}.$$

and let $\rho : \mathbb{N} \to \mathbb{N} \cup \{\infty\}$ be a functional. Then $\mathcal{P}$ is called a *ρ-relaxation* of MAXSTABLESET if $g_{\mathcal{S}}(P(G)) \leq \rho(|V|)$. Furthermore we say that the separation problem for a given *ρ-relaxation* $\mathcal{P}$ is solvable in polynomial time if, given G as input, the separation problem for $P(G) \in \mathcal{P}$ is solvable in time bounded by a polynomial in the size of G.

Obviously, having an ∞-RELAXATION is of no particular use since it simply means that the relaxations are contained in the unit-cubes. The other extreme, having a 1-RELAXATION, means that linear optimization over the relaxed polytopes solves MAXSTABLESET. As a trivial example, $g_{\mathcal{S}}(P) = 1$ for $P(G) = [0,1]^n \cap \{x : \sum_{i=1}^n \xi_i \leq \alpha\}$, where α denotes the size of a maximal stable set in the underlying graph G on n vertices. However, we have to 'pay' for the minimal geometric rank $g_{\mathcal{S}}(P) = 1$ with the fact that the separation problem for P is as hard as the original problem.

The following theorem relates the geometric rank of a polyhedral relaxation of the stable set polytope to the question whether the separation problem for the relaxation can be solved in polynomial time.

Theorem 1.4. *Let $k \in \mathbb{N}$. Then there exists a $(1 + \log(n/k))$-*RELAXATION *for which the separation problem can be solved in polynomial time. Unless $\mathbb{NP} = \mathbb{ZPP}$, this is not the case for any k-*RELAXATION, *and, unless $\mathbb{NP} = \mathbb{ZPTIME}(2^{O(\log n (\log\log n)^{3/2})})$, there is no $O(\sqrt{\log\log n})$-*RELAXATION *for which the separation problem can be solved in polynomial time.*

2 The geometric rank of relaxations of stable set polytopes

Let $k \in \mathbb{N} \setminus \{1\}$ be fixed. Given a graph $G = (V, E)$ on $n \geq k$ vertices $v_1, \ldots, v_n$, we associate a variable ξ_i with the vertex v_i, $i = 1, \ldots, n$, and consider the polytope $P_k(G) \subset \mathbb{R}^n$ that is defined by the following system of linear inequalities.

$$\begin{array}{rcll} 0 \leq & \xi_i & \leq 1 & \text{for } i = 1, \ldots, n, \\ & \xi_i + \xi_j & \leq 1 & \text{for } \{v_i, v_j\} \in E \text{ and} \\ & \xi_{i_1} + \cdots + \xi_{i_k} & \leq \alpha_{i_1,\ldots,i_k} & \text{for } \{i_1, \ldots, i_k\} \in \mathcal{I}_k \end{array}$$

where $\alpha_{i_1,\ldots,i_k}$ denotes the size of a maximal stable set in the subgraph $G_{i_1,\ldots,i_k}$ of G that is induced by the vertices $v_{i_1}, \ldots, v_{i_k}$ and $\mathcal{I}_k$ denotes the family of all k-element subsets of $\{1, \ldots, n\}$.

Note that $P_k(G)$ is a standard relaxation of the stable set polytope $P_{\mathcal{S}}(G)$.

In the following we will write P_k for $P_k(G)$ whenever there is no risk of confusion.

Lemma 2.1. *Let x^S be an integral vertex of P_k, let $S(x^S)$ denote the stable set associated with it and let $p \in \mathbb{N}$. Then we have $|S(x^S)| = \|x^S\|_p^p$.*

Proof. Since $P_k \subset [0,1]^n$, $x^S \in \{0,1\}^n$ and $\|x^S\|_p^p = \|x^S\|_1$. □

Lemma 2.2. *Let $1/2 \leq \mu \leq 1$, $q \in]0, \infty[$, and $l \in [0, 2^q - 1]$. Then*

$$\mu^q + l(1-\mu)^q \leq 1.$$

Proof. For $q = 1$ the result is trivial. For $q \neq 1$, first note that

$$\mu^q + l(1-\mu)^q \leq \mu^q + (2^q - 1)(1-\mu)^q =: f_q(\mu).$$

It is easily seen that

$$\mu^* = \left(1 + \left(\frac{1}{2^q - 1}\right)^{1/(q-1)}\right)^{-1}$$

is the unique local extremum of f_q in $[1/2, 1]$, a minimum. Hence $\mu = 1/2, 1$ are the only local maxima of f_q in $[1/2, 1]$ and the assertion follows from

$$f_q(1/2) = \left(\frac{1}{2}\right)^q + (2^q - 1)\left(\frac{1}{2}\right)^q = 2^q 2^{-q} = 1 = f_q(1).$$

□

In the following theorem we derive an upper bound for the geometric rank of the relaxation $P_k(G)$ of the associated stable set polytope $P_{\mathcal{S}}(G)$.

Theorem 2.3. *Let $G = (V, E)$ be a graph on n vertices, and let $n \geq k$. Then $g(P_k) \leq 1 + \log(n/k)$.*

Proof. Let $p = 1 + \log(n/k)$, and take any point $x = (\xi_1, \ldots, \xi_n)^T \in P_k$. Without loss of generality we may assume that $1 \geq \xi_1 \geq \cdots \geq \xi_n \geq 0$.

If $\xi_k = 0$ we conclude

$$\|x\|_p^p \leq \sum_{i=1}^{k-1} \xi_i \leq \alpha_{1,\ldots,k-1} \leq \alpha_{1,\ldots,k} \leq \alpha,$$

where α denotes the size of a maximal stable set in G.

Otherwise we set $\xi_0 = 1$ and let l^* be the maximal l with $1 \leq l \leq n-k+1$ such that $\xi_{l+k-1} > 1 - \xi_{l-1}$ and $\xi_{l-1} > 1/2$. Obviously, by the definition of ξ_0 and since $\xi_k > 0$ this maximum is well-defined.

Note that $\xi_{l^*+k-1} > 1 - \xi_{l^*-1}$ and $\xi_{l^*-1} > 1/2$ imply the following. Since for any edge $\{v_i, v_j\} \in E$ the inequality $\xi_i + \xi_j \leq 1$ is part of the given $\mathcal{H}$-presentation of P_k, the set $S^* = \{v_1, \ldots, v_{l^*-1}\}$ is stable. (Of course, if $l^* = 1$, $S^* = \emptyset$.) Further, there is no edge in G connecting a vertex of S^* with a vertex v_{l^*+m-1} for $m = 1, \ldots, k$. Hence, for any stable set I in the subgraph $G_{l^*,\ldots,l^*+k-1}$ the set $I \cup S^*$ is a stable set in G. This yields the inequality

$$\alpha_{l^*,\ldots,l^*+k-1} \leq \alpha - |S^*| = \alpha - (l^* - 1).$$

Now, suppose first that $l^* = n - k + 1$; then we have

$$\begin{aligned} \|x\|_p^p &\leq (n-k) + \sum_{i=n-k+1}^{n} \xi_i^p \leq (n-k) + \sum_{i=n-k+1}^{n} \xi_i \\ &\leq (n-k) + \alpha_{n-k+1,\ldots,n} \leq (n-k) + \alpha - (n-k) = \alpha. \end{aligned}$$

Note that $p = 1$ implies $n = k$, whence $l^* = n - k + 1$.

Now, let $l^* \leq n-k$ (and hence $p > 1$). It follows then from the maximality of l^* that $\xi_{l^*} \leq 1/2$ or $\xi_{l^*+k} \leq 1 - \xi_{l^*}$. If $\xi_{l^*} \leq 1/2$ we have

$$\begin{aligned} \|x\|_p^p &\leq (l^*-1) + \frac{n-(l^*-1)}{k} \sum_{i=l^*}^{l^*+k-1} \xi_i^p \\ &\leq (l^*-1) + \frac{n-(l^*-1)}{k} \xi_{l^*}^{p-1} \sum_{i=l^*}^{l^*+k-1} \xi_i \\ &\leq (l^*-1) + \frac{n-(l^*-1)}{k} \left(\frac{1}{2}\right)^{\log(n/k)} \alpha_{l^*,\ldots,l^*+k-1} \\ &\leq (l^*-1) + \frac{n-(l^*-1)}{k} \frac{k}{n}(\alpha - (l^*-1)) \\ &\leq (l^*-1) + \alpha - (l^*-1) = \alpha. \end{aligned}$$

So, let $\xi_{l^*} > 1/2$ but $\xi_{l^*+k} \leq 1 - \xi_{l^*}$. With the aid of Lemma 2.2 we obtain

$$\begin{aligned}
\|x\|_p^p &\leq (l^*-1) + \sum_{i=l^*}^{l^*+k-1} \xi_i^p + \sum_{i=l^*+k}^{n} \xi_i^p \\
&\leq (l^*-1) + \xi_{l^*}^{p-1} \sum_{i=l^*}^{l^*+k-1} \xi_i + \xi_{l^*+k}^{p-1} \sum_{i=l^*+k}^{n} \xi_i \\
&\leq (l^*-1) + \left(\xi_{l^*}^{p-1} + \frac{n-(l^*-1)-k}{k} (1-\xi_{l^*})^{p-1} \right) \sum_{i=l^*}^{l^*+k-1} \xi_i \\
&\leq (l^*-1) + \alpha - (l^*-1) = \alpha.
\end{aligned}$$

All together we have shown $\max_{x \in P_k} \|x\|_p^p \leq \alpha$, and by Lemma 2.1 we have actually equality. This concludes the proof of Theorem 2.3. □

Note that for a given point $x = (\xi_1, \ldots, \xi_n)^T \in P_k$ the previous proof allows to determine a stable set with cardinality at least $\|x\|_p^p$ in polynomial time.

At first sight, the $(1 + \lceil \log(n/k) \rceil)$-relaxation

$$\mathcal{P}_k = \{P_k(G) : G \text{ is a finite graph}\}$$

might seem rather weak. However since we can solve the separation problem for $\mathcal{P}_k$ in polynomial time we see from the second part of Theorem 1.4 that we cannot expect to do much better. Also note, that there are infinitely many pairs (n, k) for which $g(\mathcal{P}_k) \leq 1 + \log(n/k)$ holds with equality. In fact, for the complete graph K_n on $n \geq 2$ vertices, $x_0 = (1/2, \ldots, 1/2)^T \in P_2(K_n)$ and $\|x_0\|_{\log n}^{\log n} = 1 = \alpha(K_n)$. Hence for K_n the bound is sharp for $k = 2$. Also, if C_n is a cycle for some odd $n \in \mathbb{N}$ and $k = n-1$, $x_0 \in P_{n-1}(C_n)$ and $\|x_0\|_{1+\log(n/(n-1))}^{1+\log(n/(n-1))} = (n-1)/2 = \alpha(C_n)$. So, again, the bound is attained. It is, on the other hand, an open problem to determine for which 'interesting classes' of graphs there exist 'significantly' better estimates.

3 Reducing MaxStableSet to 0-1-PolyProg

With Theorem 2.3 at hand Theorem 1.1 and the first part of Theorem 1.4 are easy to prove.

Proof of Theorem 1.1. Let $k \in \mathbb{N}$, let $\mathcal{A}$ be a polynomial-time λ-approximation-algorithm for γ-NormMax and let $G = (V, E)$ be an instance of MaxStableSet. If $|V| < k$ a maximum stable set in G can be computed in polynomial time. Otherwise note that the $\mathcal{H}$-presentation of P_k can be determined in polynomial time and can be given as input to $\mathcal{A}$. $\mathcal{A}$ outputs a λ-approximation for the pth power of the l_p-norm-maximum where $p =$

$\gamma(n) < \infty$ that by means of Theorem 2.3 yields also a λ-approximation for the cardinality of a maximum stable set in G. □

Proof of Part 1 of Theorem 1.4. As already observed in the previous proof, P_k is given by an $\mathcal{H}$-presentation of polynomial size. Of course, the separation problem for $\mathcal{H}$-presented polytopes is solvable in polynomial time. Hence the assertion follows from Theorem 2.3. □

Now, observe that while we have monomials of high degree but consisting of just one variable in the norm-maximization problem, the polynomials we are interested in in the context of 0-1-POLYPROG are sums of monomials that are multilinear in the occurring variables. But this can be handled by reproducing the original variables suitably often.

Proof of Corollary 1.3. Assume we have a polynomial-time approximation algorithm for 0-1-POLYPROG of the kind stated in the assertion. Let I be an instance of $(1 + \lceil\log(d/k)\rceil)$-NORMMAX, let $\xi_1, \ldots, \xi_d$ denote its variables and let $d \geq k$. We take $1 + \lceil\log(d/k)\rceil$ copies of each and denote them by $\xi_{i_1}, \ldots, \xi_{i_{1+\lceil\log(d/k)\rceil}}$, for $i = 1, \ldots, d$. Adding the constraints $\xi_{i_1} = \xi_{i_2} = \cdots = \xi_{i_{1+\lceil\log(d/k)\rceil}}$ for $i = 1, \ldots, d$ and replacing $\xi_i^{1+\lceil\log(d/k)\rceil}$ by $\Pi_{j=1}^{1+\lceil\log(d/k)\rceil}\xi_{i_j}$ in the associated objective function yields an instance of 0-1-POLYPROG in dimension $n = d(1 + \lceil\log(d/k)\rceil)$. Now, since $n = d^{1+O(\log\log d/\log d)}$, it follows that for sufficiently large n,

$$n^{1-\epsilon} \leq d^{1-\epsilon/2}$$

for each fixed $\epsilon > 0$, and also that

$$\begin{aligned} n^{1-O(1/\sqrt{\log\log n})} &= d^{(1-O(1/\sqrt{\log\log n}))(1+O(\log\log d/\log d))} \\ &< d^{1-O(1/\sqrt{\log\log n})+O(\log\log d/\log d)} \\ &= d^{1-O(1/\sqrt{\log\log n})} = d^{1-O(1/\sqrt{\log\log d})}. \end{aligned}$$

Hence we obtain a polynomial-time $d^{1-\epsilon/2}$-, respectively $d^{1-O(1/\sqrt{\log\log d})}$-approximation algorithm for $(1 + \lceil\log(d/k)\rceil)$-NORMMAX. However, unless $\mathbb{NP} = \mathbb{ZPP}$ or $\mathbb{NP} = \mathbb{ZPTIME}(2^{O(\log d(\log\log d)^{3/2})})$, respectively, such an algorithm does not exist. □

In their reduction of MAXSTABLESET to 0-1-POLYPROG [BR95] assigned to each edge in the graph $G = (V, E)$ two variables and to each of the vertices v_i one subset S_i that consists of the indices of v_i's neighbors. This leads to polynomials whose degree might be of the same order as the number of variables. In contrast, the above reduction that is based on assigning variables to the vertices gives polynomials of degree at most $1 + \lceil\log(n/k)\rceil$

4 The geometric rank and polynomial time separability

Now we turn to the question whether relaxations for the stable set polytopes can be determined for which the separation problem can be solved in polynomial time and whose geometric rank does not tend to ∞.

Lemma 4.1. *Let $n, p \in \mathbb{N}$ and*

$$B_n = [0,1]^n \cap \{(\xi_1, \ldots, \xi_n)^T : (1/n^{1-1/p}) \sum_{i=1}^{n} \xi_i \leq 1\}.$$

Then

$$\mathbb{B}_p^n \cap [0,\infty[^n \subset B_n \subset (n^{1-1/p})^{1/p} \mathbb{B}_p^n \cap [0,\infty[^n,$$

where $\mathbb{B}_p^n$ denotes the l_p-unit-ball of $\mathbb{R}^n$.

Proof. For the first inclusion take any $x = (\xi_1, \ldots, \xi_n)^T \in \mathbb{B}_p^n \cap [0,\infty[^n$. By Hölder's inequality we have

$$(1/n^{1-1/p}) \sum_{i=1}^{n} \xi_i = (1/n^{1-1/p}) e^T x \leq 1/n^{1-1/p} \|e\|_{p'} \|x\|_p \leq 1,$$

where $e = (1, \ldots, 1)^T \in \mathbb{R}^n$ and $p' = p/(p-1)$.

For the second inclusion we have to show that $\max_{x \in B_n} \|x\|_p^p \leq n^{1-1/p}$. Clearly it suffices to consider vertices of B_n that are contained in the hyperplane $\{x = (\xi_1, \ldots, \xi_n)^T : (1/n^{1-1/p}) \sum_{i=1}^{n} \xi_i = 1\}$. Any such vertex v has precisely $\lfloor n^{1-1/p} \rfloor$ 1-entries, $n - \lceil n^{1-1/p} \rceil$ 0-entries and if $n^{1-1/p} \notin \mathbb{N}$ one additional entry equal to $n^{1-1/p} - \lfloor n^{1-1/p} \rfloor$. In any case,

$$\begin{aligned} \|v\|_p^p &= \lfloor n^{1-1/p} \rfloor + (n^{1-1/p} - \lfloor n^{1-1/p} \rfloor)^p \\ &\leq \lfloor n^{1-1/p} \rfloor + (n^{1-1/p} - \lfloor n^{1-1/p} \rfloor) = n^{1-1/p}. \end{aligned}$$

□

Lemma 4.2. *For any $n \in \mathbb{N}$ let Q_n be an $\mathcal{H}$-presented rational polytope with binary size bounded by a polynomial in n such that*

$$\mathbb{B}_p^n \cap [0,\infty[^n \subset Q_n \subset \lambda^{1/p} \mathbb{B}_p^n \cap [0,\infty[^n.$$

Further, for $n \in \mathbb{N}$ let P_n be a polytope in $[0,\infty[^n$. If the separation problem for $(P_n)_{n \in \mathbb{N}}$ can be solved in polynomial time then λ-approximations of $\max_{x \in P_n} \|x\|_p^p$ can be computed in polynomial time.

Proof. Let for each $n \in \mathbb{N}$, $m_n \in \mathbb{N}$, $a_{n,1}, \ldots, a_{n,m_n} \in \mathbb{Z}^n$, and $\beta_{n,1}, \ldots, \beta_{n,m} \in \mathbb{Z}$ such that $Q_n = \bigcap_{i=1}^{m_n} \{x \in \mathbb{R}^n : a_{n,i}^T x \leq \beta_{n,i}\}$. Since the separation problem for P_n can be solved in polynomial time, $\omega_n(a_{n,i}) = \max_{x \in P_n} a_{n,i}^T x$ can be computed in polynomial time; [GLS93]. With $\omega_n = \max_{i=1,\ldots,m_n} \omega_n(a_{n,i})$

this yields $P_n \subset \omega_n Q_n$ and $P_n \not\subset \omega Q_n$ for any ω with $0 \leq \omega < \omega_n$. The first relation implies $P_n \subset \omega_n \lambda^{1/p} \mathbb{B}_p^n$. Further, since $\mathbb{B}_p^n \cap [0, \infty[^n \subset Q_n$ the second yields the existence of a point $x^* \in P_n$ with $\|x^*\|_p \geq \omega_n$. Hence, ω_n^p provides a λ-approximation for $\max_{x \in P_n} \|x\|_p^p$. □

Let us mention in passing that in the realm of the algorithmic theory of convex bodies suitable, explicitly constructed polytopes approximating the l_p-unit-balls yield asymptotically optimal algorithms for the computation of norm-maxima and other radii of convex bodies [BGK+03]. In the Euclidean case this even gives optimal randomized algorithms that (in a sense surprisingly) have the same performance ratio as deterministic ones [BGK+98].

Proof of Part 2 of Theorem 1.4. Assume that for some p we have a standard ρ-RELAXATION $\mathcal{P} = \{P(G) : G \text{ is finite a graph}\}$ with $\rho(n) \leq p$ for all $n \in \mathbb{N}$ and for which the separation problem is solvable in polynomial time. Then, given a graph G on n vertices as input, we use the polytope B_n of Lemma 4.1 to play the role of Q_n in Lemma 4.2 in order to obtain an $n^{1-1/p}$-approximation for $\max_{x \in P(G)} \|x\|_p^p$. Since $\rho(n) \leq p$, we have $\max_{x \in P(G)} \|x\|_p^p = \alpha(G)$. Hence we obtain a polynomial-time $n^{1-1/p}$-approximation algorithm for MAXSTABLESET, whence with the setting $\epsilon = 1/p$, a polynomial-time $n^{1-\epsilon}$-approximation algorithm with constant ϵ or, with the setting $p = O(\sqrt{\log \log n})$, a polynomial-time $n^{1-O(1/\sqrt{\log \log n})}$-approximation. □

References

[BR95] M. Bellare and P. Rogaway, *The complexity of approximating a non-linear program*, Math. Prog. **69** (1995), 429–441.

[Bri02] A. Brieden, *On geometric optimization problems likely not contained in* $\mathbb{APX}$, Discrete Comp. Geom. **28** (2002), 201–209.

[BGK+03] A. Brieden, P. Gritzmann, R. Kannan, V. Klee, L. Lovász, and M. Simonovits, *Deterministic and randomized polynomial-time approximation of radii*, Mathematika, 2003, in print.

[BGK+98] A. Brieden, P. Gritzmann, R. Kannan, V. Klee, L. Lovász, and M. Simonovits, *Approximation of radii and norm-maxima: Randomization doesn't help*, Proc. 39th IEEE FOCS, 1998, pp. 244–251.

[BGK00] A. Brieden, P. Gritzmann, and V. Klee, *Inapproximability of some geometric and quadratic optimization problems*, In: Approximation and Complexity in Numerical Optimization: Continuous and Discrete Problems (P.M. Pardalos, ed.), Nonconvex Optimization and its Applications, vol. 42, Kluwer (Boston), 2000, pp. 96–115.

[BH92] R. Boppana and M.M. Halldórsson, *Approximating maximum independent sets by excluding subgraphs*, BIT **32** (1992), 180–196.

[EHdW84] Ch. Ebenegger, P.L. Hammer, and D. de Werra, *Pseudo-boolean functions and stability of graphs*, In: 'Algebraic and Combinatorial Methods in Operations Research' (R.E. Burkard, R.A. Cunninghame-Green, and U. Zimmermann, eds.), Ann. of Discrete Math., vol. 19, Elsevier Science Publisher B.V. (North-Holland), 1984, pp. 83–98.

[EH00] L. Engebretsen and J. Holmerin, *Clique is hard to approximate within* $n^{1-o(1)}$, Proc. 27th ICALP, 2000, pp. 2–12.

[GK93] P. Gritzmann and V. Klee, *Computational complexity of inner and outer j-radii of polytopes in finite dimensional normed spaces*, Math. Prog. **59** (1993), 163–213.

[Goe95] M.X. Goemans, *Worst-case comparison of valid inequalities for the TSP*, Math. Prog. **69** (1995), 335–349.

[GLS93] M. Grötschel, L. Lovász, and A. Schrijver, *Geometric Algorithms and Combinatorial Optimization*, Springer, Berlin, 1988, 2nd ed. 1993.

[Hås99] J. Håstad, *Clique is hard to approximate within* $n^{1-\epsilon}$, Acta Math. **182** (1999), 105–142

[Hås96] J. Håstad, *Clique is hard to approximate within* $n^{1-\epsilon}$, Proc. 37th IEEE FOCS, 1996, pp. 627–636.

[Jan98] T. Jansen, *Introduction to the theory of complexity and approximation algorithms*, In: Lectures on Proof Verification and Approximation Algorithms (E.W. Mayr, H.J. Prömel and A. Steger, eds.), Lecture Notes in Computer Science, vol. 1367, Springer (Berlin), 1998, pp. 5–28.

[LMJ94] J. Lee and W.D. Morris Jr, *Geometric comparison of combinatorial polytopes*, Discrete Appl. Math. **55** (1994), 163–182.

About Authors

Andreas Brieden and Peter Gritzmann are at the Center for Mathematical Sciences, Technische Universität München, 80290 München, Germany; *brieden@ma.tum.de, gritzman@ma.tum.de.*

Acknowledgments

The authors gratefully acknowledge support by the Deutsche Forschungsgemeinschaft and thank an anonymous referee for valuable comments.

A Lower Bound on the Complexity of Approximate Nearest-Neighbor Searching on the Hamming Cube

Amit Chakrabarti
Bernard Chazelle
Benjamin Gum
Alexey Lvov

Abstract

We consider the nearest-neighbor problem over the d-cube: given a collection of points in $\{0,1\}^d$, find the one nearest to a query point (in the L^1 sense). We establish a lower bound of $\Omega(\log\log d/\log\log\log d)$ on the worst-case query time. This result holds in the cell probe model with (any amount of) polynomial storage and word-size $d^{O(1)}$. The same lower bound holds for the approximate version of the problem, where the answer may be any point further than the nearest neighbor by a factor as large as $2^{\lfloor(\log d)^{1-\varepsilon}\rfloor}$, for any fixed $\varepsilon > 0$.

1 Introduction

For a variety of practical reasons ranging from molecular biology to web searching, nearest-neighbor searching has been a focus of attention lately [2–4,6–10,12–15,17,19,21,22,27]. In the applications considered, the dimension of the ambient space is usually high, and predictably, classical lines of attack based on space partitioning fail. To overcome the well-known "curse of dimensionality," it is typical to relax the search by seeking only approximate answers. Curiously, no lower bound has been established — to our knowledge — on the complexity of the approximate problem in its canonical setting, i.e., points on the hypercube. Our work is an attempt to remedy this.

We note that two recent results, due to Borodin et al. [8] and Barkol and Rabani [5] do give lower bounds on the *exact* version of the problem.

Given a *database* or *key-set* $S \subseteq \{0,1\}^d$, a δ*-approximate nearest neighbor* (δ*-ANN*) of a *query point* $x \in \{0,1\}^d$ is any key $y \in S$ such that $\|x-y\|_1 \leq \delta\|x-z\|_1$, for any $z \in S$. The parameter $\delta \geq 1$ is called the *approximation factor*[1] of the problem. Given some δ, the problem is to preprocess S so as to be able to find a δ-*ANN* of any query point efficiently. The data structure consists of a table T whose entries hold $d^{O(1)}$ bits each. This means that a point can be read in constant time. This assumption might be unrealistically

[1]Most of the literature about ANN is concerned with algorithms that achieve approximation factors close to 1 and sometimes they use the term "ε-ANN" (for small positive ε) to mean what we would call a $(1+\varepsilon)$-ANN.

generous when d is large, but note that this only strengthens our lower bound result.

Theorem 1.1 *Suppose the table T, constructed from preprocessing a database S of n points in $\{0,1\}^d$, is of size polynomial in n and d and holds $d^{O(1)}$-bit entries. Then, for any algorithm using T for δ-ANN searching, there exists some S such that the query time is $\Omega(\log\log d/\log\log\log d)$. This holds for any approximation factor $\delta \leq 2^{\lfloor(\log d)^{1-\varepsilon}\rfloor}$, for any fixed $\varepsilon > 0$.*

How good is the lower bound? First, note that the problem can be trivially solved *exactly* in constant time, by using a table of size 2^d. Moreover, recent algorithmic results [15, 17, 21], when adapted to our model of computation, show that for *constant* $\delta > 1$ there is a polynomial sized table with d-bit entries and a *randomized* algorithm that enables us to answer δ-*ANN* queries using $O(\log\log d)$ probes to the table. Although there seems to be only a small gap between this upper bound and our lower bound, the two bounds are in fact incomparable because of the randomization. An important open theoretical question regarding *ANN* searching is to extend our lower bound to allow randomization.

In this context it is worth mentioning that a stronger lower bound for *exact* nearest neighbor search is now known. Recent results of Borodin et al. [8] show that even randomized algorithms for this problem require $\Omega(\log d)$ query time in our model; Barkol and Rabani [5] improve this bound to $\Omega(d/\log n)$.

2 The Cell Probe Model

Yao's *cell probe model* [26] provides a framework for measuring the number of memory accesses required by a search algorithm. Because of its generality, any lower bound in that model can be trusted to apply to virtually any conceivable sequential algorithm. In his seminal paper [1], Ajtai established a nontrivial lower bound for predecessor queries in a discrete universe (for recent improvements, see [6, 23, 25]). Our proof begins with a similar adversarial scenario. Given a key-set of n points in $\{0,1\}^d$, a table T is built in preprocessing: its size is $(dn)^c$, for fixed (but arbitrary) $c > 0$ and each entry holds $d^{O(1)}$ bits. (For simplicity, we assume that an entry consists of exactly d bits; the proof is very easily generalized if this is number d is changed to $d^{O(1)}$.) To answer queries, the algorithm has at its disposal an infinite supply of functions f_1, f_2, etc. Given a query x, the algorithm evaluates the index $f_1(x)$ and looks up the table entry $T[f_1(x)]$. If $T[f_1(x)]$ is a δ-*ANN* of x, it can stop after this single round. Otherwise, it evaluates $f_2(x, T[f_1(x)])$ and looks up the entry $T[f_2(x, T[f_1(x)])]$. Again it stops if it this entry is the desired answer (at a cost of two rounds), else it goes on in this fashion. The query time of the algorithm is defined to be the maximum number of rounds, over all queries $x \in \{0,1\}^d$, required to find a δ-*ANN* of x in the table. Note that we do not charge the algorithm for the time it takes to compute the functions f_r or for the time it takes to decide whether or not to stop. Note also that we require the last entry of T fetched by the algorithm to be the answer that it will give; this might seem like an artificial requirement but it aids our proof and adds at most one to the query time.

We couch our cell-probe arguments in a communication-complexity setting as we model our adversarial lower bound proof as a game between Bob and Alice [11, 20]. The algorithm is modeled by a sequence of functions $f_1, f_2, \ldots$. Alice starts out with a set $P_1 \subseteq \{0,1\}^d$ of candidate queries and Bob holds a collection $K_1 \subseteq 2^{\{0,1\}^d}$ of candidate key-sets, each set in K_1 being of size n. The goal of Alice and Bob is to force as many communication rounds as possible, thereby exhibiting a hard problem instance. We remark that our language differs subtly from that of Miltersen [23] and Miltersen et al. [24] who instead invent communication *problems* that can be reduced to their cell-probe data structure problems, and then prove lower bounds for the communication problems.

The possible values of $f_1(x)$ (provided by the algorithm for every x) partition P_1 into equivalence classes. Alice chooses one such class and the corresponding value of $f_1(x)$, thus restricting the set of possible queries to $P_2 \subseteq P_1$. Given this fixed value of $f_1(x)$, which Alice communicates to Bob, the entry $T[f_1(x)]$ depends only on Bob's choice of key-set. All the possible values of that entry partition K_1 into equivalence classes. Bob picks one of them and communicates the corresponding value of $T[f_1(x)]$ to Alice, thus restricting the collection of possible key-sets to $K_2 \subseteq K_1$. Alice and Bob can then iterate on this process. This produces two nested sequences of *admissible* query sets,

$$P_1 \supseteq P_2 \supseteq \cdots \supseteq P_t\,,$$

and *admissible* key-set collections,

$$K_1 \supseteq K_2 \supseteq \cdots \supseteq K_t\,.$$

An element of $P_r \times K_r$ specifies a problem instance. The set $P_r \times K_r$ is called *nontrivial* if it contains at least two problem instances with distinct answers, meaning that no point can serve as a suitable *ANN* in both instances. If $P_r \times K_r$ is nontrivial, then obviously round r is needed, and possibly others as well.

We show that for some appropriate value of $n = n(d)$, there exists an admissible starting $P_1 \times K_1$, together with a strategy for Alice and Bob, that leads to a nontrivial $P_t \times K_t$, for $t = \Theta(\log\log d / \log\log\log d)$. What makes the problem combinatorially challenging is that a greedy strategy can be shown to fail. Certainly Alice and Bob must ensure that P_r and K_r do not shrink too fast; but just as important, they must choose a low-discrepancy strategy that keeps query sets and key-sets "entangled" together. To achieve this, we adapt to our needs a technique used by Ajtai [1] for probability amplification in product spaces.[2]

3 Combinatorial Preliminaries

Before getting into the proof of our lower bound, we introduce some notation and describe a combinatorial construction that will be central to our arguments. For the rest of this paper, we assume that d is large enough and that logarithms are to the base 2. The term "distance" refers to the Hamming distance between two points in $\{0,1\}^d$. A "ball of radius r centered at x" denotes the set $\{y \in \{0,1\}^d : \mathrm{dist}(x,y) \leq r\}$. To begin with, we specify the size n of the admissible key-sets and the number t of rounds, and also define two auxiliary numbers h and β:

$$h \stackrel{\text{def}}{=} 6ct \tag{1}$$

$$\beta \stackrel{\text{def}}{=} 16 \cdot 2^{\lfloor (\log d)^{1-\varepsilon} \rfloor} \tag{2}$$

$$t = \left\lceil \frac{\varepsilon \log\log d}{2 \log\log\log d} \right\rceil \tag{3}$$

$$n = (h-1)^{t-1} d^{5t} \tag{4}$$

The significance of the above formulae will become clear in the proofs of Lemmas 3.4–4.3. Recall that the constant c parametrizes the size of the table in the cell-probe algorithm: the table has $(dn)^c$ entries, each consisting of d bits.

The main combinatorial object we construct is a hierarchy $\mathcal{H}$ of balls or, more formally, a rooted tree $\mathcal{H}$ and a family of balls, one associated with each node of $\mathcal{H}$, such that the parent–child relation in $\mathcal{H}$ translates into the inclusion relation in the family of balls. Our proof will rely crucially on three quantities being large enough: the height of $\mathcal{H}$, the degree of a node in $\mathcal{H}$, and the minimum distance between balls associated with sibling nodes in $\mathcal{H}$.

From this main tree $\mathcal{H}$, we derive additional trees $\mathcal{H}_r$, $1 \leq r \leq t$; the tree $\mathcal{H}_r$ is used in the rth round of communication between Alice and Bob. The tree $\mathcal{H}_1$ is a certain contraction (in the graph theoretic sense) of $\mathcal{H}$. The trees $\mathcal{H}_2, \ldots, \mathcal{H}_r$ are "nondeterministic" in the following sense: roughly speaking, $\mathcal{H}_r$ is obtained by picking a node v of $\mathcal{H}_{r-1}$, "uncontracting" v to obtain a subtree of $\mathcal{H}$, and then contracting this subtree in a different way. Notice that the choice of v determines the resulting $\mathcal{H}_r$. In the proof, we shall fix the node v (and thus determine $\mathcal{H}_r$) only during the rth round of the Alice–Bob game.

[2] This simple but powerful technique, which is described in §4.3, has been used elsewhere in communication complexity, for example by Karchmer and Wigderson [18].

The sets P_i, of admissible queries, and K_i, of admissible key-sets, will be constructed based on these trees $\mathcal{H}_r$.

We shall now describe the constructions of these trees precisely. We begin with a geometric fact about the hypercube.

Definition 3.1 *A family of balls is said to be γ-separated if the distance between any two points belonging to distinct balls in the family is more than γ times the distance between any two points belonging to any one ball in the family. Here γ is any positive real quantity.*

Lemma 3.2 *Let $B \subseteq \{0,1\}^d$ be a ball of radius $k \leq d$ large enough. For any $\gamma \geq 16$ there exists a $\gamma/16$-separated family of balls within B, such that the size of the family is at least $2^{k/13}$ and the radius of each ball in the family is k/γ.*

Proof: We use an argument similar to the proof of Shannon's theorem. Let V_r be the volume of (i.e. the number of points in) a ball in $\{0,1\}^d$ of radius r, centered at a point in $\{0,1\}^d$. (Notice that this number does not depend on the center). Clearly

$$V_r = \sum_{i=0}^{\lfloor r \rfloor} \binom{d}{i}.$$

Consider the ball B', concentric with B and of radius $k/3$, and call its points initially *unmarked.* We proceed to mark the points of B' as follows: while there is an unmarked point left in B', pick one and mark all the points at distance at most $k/4$ from that point. The number N of points we pick in B' satisfies

$$N \geq \frac{V_{k/3}}{V_{k/4}}.$$

We can bound N from below:

$$N \geq \frac{V_{k/3}}{V_{k/4}} = \frac{\sum_{i=0}^{\lfloor k/3 \rfloor} \binom{d}{i}}{\sum_{i=0}^{\lfloor k/4 \rfloor} \binom{d}{i}} \geq \frac{\binom{d}{\lfloor k/3 \rfloor}}{\sum_{i=0}^{\lfloor k/4 \rfloor} \binom{d}{i}}.$$

Note that in each term of the sum in the denominator i is at most $d/4$. For such i,

$$\frac{\binom{d}{i}}{\binom{d}{i-1}} = \frac{d-i+1}{i} \geq 3,$$

so

$$\sum_{i=0}^{\lfloor k/4 \rfloor} \binom{d}{i} \leq \frac{3}{2} \binom{d}{\lfloor k/4 \rfloor}.$$

This gives

$$N \geq \frac{2}{3} \cdot \frac{\binom{d}{\lfloor k/3 \rfloor}}{\binom{d}{\lfloor k/4 \rfloor}}$$

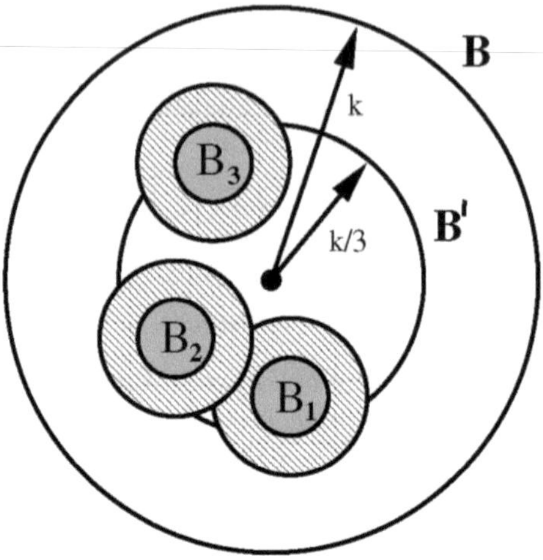

Fig. 1. Picking the separated family of balls $B_1, B_2, \ldots$. The marked points are indicated by hatching; the picked balls by solid fill.

$$
\begin{aligned}
&= \frac{2}{3} \cdot \prod_{i=\lfloor k/4 \rfloor}^{\lfloor k/3 \rfloor} \frac{d-i+1}{i} \\
&\geq \frac{2}{3} \cdot (\underbrace{2 \times 2 \times \cdots \times 2}_{k/12-2 \text{ factors}}) \\
&\geq 2^{k/12-3},
\end{aligned}
$$

and for large enough k, this implies $N \geq 2^{k/13}$.

Now pick balls of radius k/γ centered at the N picked points; their centers are in B' and their common radius is at most $k/16$, so these balls lie within B. Moreover, it is easy to see that they form a $\gamma/16$-separated family. To see why, suppose on the contrary that two points p and q in balls centered at distinct points p_0 and q_0 lie within $k/8$ of each other. Then,

$$
\begin{aligned}
\text{dist}(p_0, q_0) &\leq \text{dist}(p_0, p) + \text{dist}(p, q) + \text{dist}(q, q_0) \\
&\leq k/\gamma + k/8 + k/\gamma \\
&\leq k/4,
\end{aligned}
$$

since $\gamma \geq 16$. But this is a contradiction since, by construction, $\text{dist}(p_0, q_0) > k/4$. ∎

Corollary 3.3 *For k divisible by β, there exists a $\beta/16$-separated family of radius-(k/β) balls within B, of size $2^{k/\beta}$.* ∎

Let $\mathcal{H}$ be the tree whose root is associated with the ball of radius d centered at $(0, \ldots, 0)$. The children of the root are each associated with one of the $2^{d/\beta}$ balls specified by the above corollary.[3] Their children, grandchildren, etc., are defined similarly. In general, a node of depth k (root being of depth 0) is associated with a ball of radius d/β^k and its number of children is $2^{d/\beta^{k+1}}$. We iterate this recursive construction until the leaves of $\mathcal{H}$ are of depth h^{t-1}. Note that the balls associated with the leaves of $\mathcal{H}$ are of radius at least $d/\beta^{h^{t-1}}$, and thus, by our choice of t, large enough for the application of Lemma 3.2; specifically, its corollary.

The tree $\mathcal{H}$ is used to build other trees, each one associated with a separate round. We begin with the round-one tree $\mathcal{H}_1$. Given $v \in \mathcal{H}$, let $\mathcal{H}_1^*(v)$ denote the subtree of depth h^{t-2} rooted at v. For each node v of $\mathcal{H}$ whose depth is divisible by h^{t-2}, remove from $\mathcal{H}$ all the nodes of $\mathcal{H}_1^*(v)$, except for its leaves, which we keep in $\mathcal{H}$ and make into the children of v: these operations transform $\mathcal{H}$ into a tree $\mathcal{H}_1$ of depth h. In this way, each node v of $\mathcal{H}_1$ (together with its children) forms a contraction of the tree $\mathcal{H}_1^*(v)$. We can easily check that a node of $\mathcal{H}_1$ of depth $k < h$ has exactly

$$2^{\nu d/\beta^{kh^{t-2}+1}}$$

children, where $\nu = (1 - 1/\beta^{h^{t-2}})/(1 - 1/\beta)$.

For $1 < r < t$, we define $\mathcal{H}_r$ by induction. We pick some internal node v of $\mathcal{H}_{r-1}$ and consider the tree $\mathcal{H}_{r-1}^*(v)$ of which it is the contraction. This tree now plays the role of $\mathcal{H}$ earlier: For $z \in \mathcal{H}_{r-1}^*(v)$, we let $\mathcal{H}_r^*(z)$ denote the subtree of $\mathcal{H}_{r-1}^*(v)$ of depth h^{t-r-1} rooted at z. If the depth of z in $\mathcal{H}_{r-1}^*(v)$ is divisible by h^{t-r-1}, we turn the leaves of $\mathcal{H}_r^*(z)$ into the children of z, which transforms $\mathcal{H}_{r-1}^*(v)$ into a tree of depth h that is the desired $\mathcal{H}_r$.

For $r = t$, we define $\mathcal{H}_r$ (with respect to an internal node $v \in \mathcal{H}_{r-1}$) as simply the tree formed by v and its children. We note once again that, for any $r > 1$, the definition of $\mathcal{H}_r$ is not deterministic, since the initial choice of v is left unspecified.

Lemma 3.4 *Any internal node v of any $\mathcal{H}_r$ satisfies $2^{\sqrt{d}} < \deg(\mathcal{H}_r, v) < 2^{2d/\beta}$, where $\deg(\mathcal{T}, v)$ denotes the number of children of node v in tree $\mathcal{T}$.*

Proof: Observe that $\deg(\mathcal{H}_t, v) = \deg(\mathcal{H}_{t-1}, v)$. So, it suffices to prove the lemma for $1 \leq r \leq t-1$. Pick any such r and consider any internal node v of $\mathcal{H}_r$: $\deg(\mathcal{H}_r, v)$ is the number of leaves of $\mathcal{H}_r^*(v)$, which itself is a subtree of $\mathcal{H}$ of depth h^{t-r-1}. So, if k is the depth of v in $\mathcal{H}$, then

$$\deg(\mathcal{H}_r, v) = \prod_{i=1}^{h^{t-r-1}} 2^{d/\beta^{k+i}}.$$

[3]To simplify the notation, we shall assume that d is a large enough power of 2. Note that β is already a power of 2.

It follows that the number $\deg(\mathcal{H}_r, v)$ is largest when $r = 1$, $k = 0$, and smallest when $r = t-1$, $k = h^{t-1} - 1$. Thus

$$\deg(\mathcal{H}_r, v) \leq \prod_{i=1}^{h^{t-2}} 2^{d/\beta^i} = 2^{\nu d/\beta} < 2^{2d/\beta}.$$

On the other hand, $\deg(\mathcal{H}_r, v) \geq 2^{d/\beta^{h^{t-1}}}$, so it suffices to prove that

$$h^{t-1} \log \beta < \frac{1}{2} \log d. \tag{5}$$

But this follows after some routine algebra from (1), (3), and the fact that d is large enough. ∎

The association between balls and nodes of $\mathcal{H}_r$ is inherited from $\mathcal{H}$ in the obvious manner.

4 The Lower Bound

We now turn to the proof of the lower bound itself. Recall that the proof consists of an adversarial strategy that leads to well-structured sets P_r, of admissible queries, and K_r, of admissible key-sets. We shall require these sets to satisfy certain invariants and shall ensure that Alice's and Bob's strategies maintain these invariants. Finally, we shall show that if the invariants have been maintained for $t-1$ rounds, then the problem instance $P_t \times K_t$ is nontrivial, which would imply that round t is necessary, completing the proof.

4.1 Admissible Queries

Recall that Alice's message in round r partitions P_r into equivalence classes. This partitioning can be unwieldy, so we restrict the admissible query sets to be part of another, better-structured, nested sequence

$$P_1^* \supseteq P_2^* \supseteq \cdots \supseteq P_t^*,$$

where each $P_r^* \supseteq P_r$.

The centers of the balls at the leaves of $\mathcal{H}$ constitute the set P_1^*. For $r > 1$, we define P_r^* as the intersection of P_{r-1}^* with the balls at the leaves of $\mathcal{H}_r$. We define $P_1 = P_1^*$. For $r > 1$, Alice chooses the set P_r to be a certain subset of P_r^* according to a strategy to be specified in §4.4. For $r > 1$, we keep the set P_r of admissible queries from being too small by requiring the following:

- QUERY INVARIANT: The fraction of the leaves in $\mathcal{H}_r$ whose associated balls intersect P_r is at least $1/d$.

Note that the size of the initial collection P_1 of admissible queries is not quite as large as 2^d, although it is still a fractional power of it. Indeed,

$$|P_1| = (2^d)^{\frac{1-1/\beta^{h^{t-1}}}{\beta-1}}.$$

By our assumption on table size, the index $f_1(x)$ that Alice gives Bob during the first round can take on at most $(dn)^c$ distinct values. This subdivides P_1 into as many equivalence classes. The same is true at any around $r < t$, and so P_r is partitioned into the classes $P_{r,1}, \ldots, P_{r,(dn)^c}$. An internal node v of $\mathcal{H}_r$ is called *dense for* $P_{r,i}$ if the fraction of its children whose associated balls intersect $P_{r,i}$ is at least $1/d$. The node v is said to be dense if it is dense for at least one $P_{r,i}$.

Lemma 4.1 *The union of the balls associated with the dense non-root nodes of $\mathcal{H}_r$ contains at least a fraction $1/2d$ of the balls at the leaves.*

Proof: Consider one of the partitions $P_{r,i}$. Color the nodes of $\mathcal{H}_r$ whose associated balls intersect $P_{r,i}$. Further, mark every colored non-root node that is dense for $P_{r,i}$. Finally, mark every descendant in $\mathcal{H}_r$ of a marked node. For $1 \leq k \leq h$, let L_k be the number of leaves of $\mathcal{H}_r$ whose depth-k ancestor in $\mathcal{H}_r$ is colored and unmarked. (We include v as one of v's ancestors.) Let L be the number of leaves of $\mathcal{H}_r$. Clearly $L_1 \leq L$. For $k > 1$, an unmarked colored depth-k node is the child of a colored depth-$(k-1)$ node that is *not* dense for $P_{r,i}$. It follows that $L_k < L_{k-1}/d$ and so, for any $k \geq 1$, $L_k \leq L/d^{k-1}$.

Repeating this argument for all the $P_{r,i}$'s in the partition, we find that all the unmarked, colored nodes, at a fixed depth $k \geq 1$, are ancestors of at most $(dn)^c L/d^{k-1}$ leaves. In particular, the number of unmarked, colored leaves is at most

$$(dn)^c L/d^{h-1} < L/2d. \tag{6}$$

This last inequality follows from (1) and (4). Incidentally, the quantity h is defined the way it is precisely to make this inequality hold.

The query invariant ensures at least L/d colored leaves, so there are at least $L/2d$ colored, marked leaves. Moving up the tree $\mathcal{H}_r$, we find that the marked nodes whose parents are unmarked are ancestors of at least $L/2d$ leaves. All such nodes are dense, which completes the proof. ∎

4.2 Admissible Key-Sets

The collections K_r of admissible key-sets need not be specified explicitly. Instead, we define a probability distribution $\mathcal{D}_r$ over the set of all $\binom{2^d}{n}$ key-sets of size n and indicate a lower bound on the probability that a random key-set drawn from $\mathcal{D}_r$ is admissible, i.e., belongs to K_r. Beginning with the case $r = 1$, we define a random key-set S_1 recursively in terms of a random variable S_2, which itself depends on $S_3, \ldots, S_t$. To treat all these cases at once, we define S_r, for $1 \leq r \leq t$:

- For $r < t$, we define a *random S_r within $\mathcal{H}_r$* in two stages:

  [1] For each $k = 1, 2, \ldots, h-1$, choose d^5 nodes of $\mathcal{H}_r$ of depth k at random, uniformly without replacement among the nodes of depth k that are not descendants of chosen nodes of smaller depth. The $(h-1)d^5$ nodes chosen in this way are said to be *picked by S_r*.

  [2] For each node v picked by S_r, recursively choose a random S_{r+1} within the corresponding tree $\mathcal{H}_{r+1}$ (i.e., defined with respect to node v). Such a S_{r+1} is called the *canonical projection* of S_r on v. The union of these $(h-1)d^5$ projections S_{r+1} defines a *random S_r within $\mathcal{H}_r$*.

- For $r = t$, a random S_t within (some) $\mathcal{H}_t$ is obtained by selecting d^5 nodes at random, uniformly without replacement, among the leaves of the depth-one tree $\mathcal{H}_t$: S_t consists of the d^5 centers of the balls associated with these leaves.

Note that a random S_r consists of exactly $(h-1)^{t-r}d^{5(t-r+1)}$ points, thus satisfying the definition of n in (4) for the case of S_1. A random S_1 is admissible with probability one (since no information has been exchanged yet), and so the set of all S_1's constitutes K_1. Obviously, this cannot be true for $r > 1$, since for one thing S_r does not even have the right size, i.e., n.

Suppose we have defined the distribution $\mathcal{D}_{r-1}$, for some $r > 1$. As we shall see from Bob's strategy, this implies the choice of a specific $\mathcal{H}_{r-1}$. To define $\mathcal{D}_r$, we choose some node v in $\mathcal{H}_{r-1}$ (which immediately implies the choice of $\mathcal{H}_r$ for the next round). Any key-set S_1 whose construction involves choosing an S_r within the tree $\mathcal{H}_r$ associated with node v is called *v-based* and its subset formed by the corresponding S_r is called its *v-projection*.

By abuse of terminology, we say that S_r is admissible if it is a v-projection of at least one key-set of K_{r-1}: for each admissible S_r, choose one such key-set arbitrarily and call it the *v-extension* of S_r; for any other S_r, choose as its (unique) v-extension any v-based key-set whose v-projection is S_r (such a key-set is non-admissible). To define the distribution $\mathcal{D}_r$, we assign probability zero to any key-set S_1 that is not a v-extension; if it is, we assign it the probability of its v-projection with respect to the distribution of a random S_r. During round $r-1$, Bob gets to choose K_r among the key-sets with nonzero probability in $\mathcal{D}_r$.

We set a lower bound on the number of admissible key-sets by requiring Bob's strategy to enforce the following

- KEY-SET INVARIANT: A random S_r is admissible with probability at least 2^{-d^2}.

The underlying distribution is the one derived from the construction of S_r, which is also equivalent to $\mathcal{D}_r$.

In what follows, we shall need a tail estimate for the hypergeometric distribution. The next lemma provides it:

Lemma 4.2 *Consider a set of N of objects, a fraction $1/T$ of which are "good". Pick a random subset of size $m \leq N$ of these objects, all subsets of size m being equally likely, and let the random variable X denote the fraction of elements of this subset that are good. Then for any real $t > 0$ we have* $\text{Prob}[\frac{1}{T} - X \geq t] \leq e^{-2mt^2}$.

Proof: This follows directly from Theorems 1 and 4 in [16].

Lemma 4.3 *Fix an arbitrary $\mathcal{H}_r$ $(r < t)$. There exists some k_0 $(1 \leq k_0 < h)$ such that, with probability at least 2^{-d^2-1}, a random S_r within $\mathcal{H}_r$ is admissible and picks at least d^3 dense nodes of $\mathcal{H}_r$ of depth k_0.*

Proof: By Lemma 4.1, the dense non-root nodes of $\mathcal{H}_r$ are ancestors of at least a fraction $1/2d$ of the leaves. By the pigeonhole principle, for some k_0 with $1 \leq k_0 < h$, at least a fraction $1/2dh$ of the nodes of depth k_0 are dense. Of course, not all these nodes can be picked by S_r: only those that do not have ancestors that have been picked further up the tree are candidates. But this rules out fewer than hd^5 nodes, which by Lemma 3.4, represents a fraction at most $hd^5 2^{-\sqrt{d}}$ of all the nodes of depth k_0. This means that from among the set of depth-k_0 nodes that can be picked by S_r, the fraction $1/T_0$ that is dense satisfies

$$\frac{1}{T_0} \geq \frac{\frac{1}{2dh} - \frac{hd^5}{2^{\sqrt{d}}}}{1 - \frac{hd^5}{2^{\sqrt{d}}}} > \frac{1}{3dh}.$$

Among the d^5 nodes we pick at depth k_0, we expect at least $d^5/3dh$ of them to be dense, and thus we should exceed the lemma's target of d^3 with overwhelming probability, say, $1 - 2^{-d^2-1}$. Using Lemma 4.2 we see that this is indeed the case: choose the set of objects in the lemma to be the set of depth-k_0 nodes that are available for picking by S_r and let the "good" objects among these nodes be the dense nodes. Choose $m = d^5$, $T = T_0$ and $t = 1/T_0 - 1/d^2 > 0$. The lemma now says that the number R of dense nodes we pick satisfies

$$\text{Prob}[R \leq d^3] \; = \; \text{Prob}[R/d^5 \leq 1/d^2] \; \leq \; e^{-2d^5(\frac{1}{T_0} - \frac{1}{d^2})^2}$$

But, as observed above, $T_0 \leq 3dh$ and so, after some routine algebra, we obtain $\text{Prob}[R \leq d^3] \leq 2^{-d^2-1}$.

The key-set invariant completes the proof. ∎

4.3 Probability Amplification

In the rth round, the table entry $T[f_r(x, T[f_1(x)], \ldots)]$ that Bob returns to Alice can take on at most 2^d distinct values, and so the collection of admissible key-sets is partitioned into equivalence classes $K_{r,1}, \ldots, K_{r,2^d}$. Bob has to choose one of these classes to form the new collection K_{r+1} of admissible key-sets. Unfortunately, such a large number of classes is likely to cause a violation of the key-set invariant. To amplify the probability that a random key-set is admissible back to 2^{-d^2}, we exploit the fact that the distribution is defined over a product space, and borrowing an idea from Ajtai [1], we project the distribution on its "highest-density" coordinate.

Lemma 4.4 *For $r < t$, there exists a dense node v of $\mathcal{H}_r$ such that the conditional probability that the canonical projection on v of a random S_r is admissible, given that it picks v, is at least* $1/2$.

Proof: Let D be a subset of dense nodes of depth k_0 (referred to in Lemma 4.3). We define $\mathcal{E}_D$ to be the event that the set of dense nodes of depth k_0 picked by S_r is exactly D. Let p_D be the probability that S_r is admissible and that $\mathcal{E}_D$ occurs, and let c_D be the conditional probability that S_r is admissible, given $\mathcal{E}_D$. By Lemma 4.3, summing over all subsets D of dense depth-k_0 nodes of size at least d^3, we find that

$$\sum_D c_D \cdot \text{Prob}[\mathcal{E}_D] = \sum_D p_D \geq 2^{-d^2-1},$$

and therefore $c_{D_0} \geq 2^{-d^2-1}$, for some D_0 of size at least d^3.

Now we derive a key fact from the product construction of the probability spaces for key-sets. Consider the $|D_0|$-dimensional space, where each $v \in D_0$ defines a coordinate. Each point in this space represents an S_r and is characterized by a vector $\langle n_1, \ldots, n_{|D_0|} \rangle$, where n_i is the canonical projection of S_r onto the ith node of D_0. By the definition of admissibility for S_r's, if $\langle n_1, \ldots, n_{|D_0|} \rangle$ is an admissible S_r, then all the n_i's in its vector representation are admissible S_{r+1}'s. Let A_{n_i} be the set of *all* admissible S_{r+1}'s within the $\mathcal{H}_{r+1}$ corresponding to the ith node of D_0. Clearly the admissible S_r's that belong to the $|D_0|$-dimensional space are all contained in

$$A_{n_1} \times \cdots \times A_{n_{|D_0|}}$$

the size of which is a fraction $\prod_{v \in D_0} c_v$ of the S_r's for which D_0 is exactly the set of dense nodes of depth k_0 picked by S_r, where c_v is the probability that a random S_{r+1} within the $\mathcal{H}_{r+1}$ corresponding to v is admissible. Because within S_r the random construction of any S_{r+1} is independent of $S_r \setminus S_{r+1}$, c_v is also the conditional probability that the canonical projection on v of a random S_r is admissible, given that it picks v. Thus we see that

$$c_{D_0} \leq \prod_{v \in D_0} c_v.$$

Since $|D_0| \geq d^3$, it follows that

$$c_v \geq \left(2^{-d^2-1}\right)^{1/|D_0|} \geq \frac{1}{2},$$

for some $v \in D_0$. ∎

4.4 Maintaining the Invariants

We summarize the strategies of Alice and Bob and discuss the enforcement of the two invariants. Skipping the trivial case $r = 1$, we show that if the invariants hold at the beginning of round $r < t$, they also hold at the beginning of round $r+1$. Prior to round r, consider the node v from $\mathcal{H}_r$ described in Lemma 4.4. Since v is dense there is some $P_{r,i}$ such that the fraction of v's children whose associated balls intersect $P_{r,i}$ is at least $1/d$. Alice chooses such a $P_{r,i}$ and defines $P_{r,i} \cap P^*_{r+1}$ to be P_{r+1}, the set of admissible queries prior to round $r+1$. The tree $\mathcal{H}_{r+1}$ is then rooted at v, and its leaves coincide with the children of v in $\mathcal{H}_r$. Thus, the fraction of the leaves of $\mathcal{H}_{r+1}$ whose associated balls intersect P_{r+1} is at least $1/d$, and the query invariant holds.

Turning now to the key-set invariant, recall that during round r, Bob is presented with a table entry, which holds one of 2^d distinct values. By the choice of v in Lemma 4.4, the probability that a random S_{r+1} at v is admissible is at least a half. A key observation is that this is the same probability that a random key-set from $\mathcal{D}_{r+1}$ is in K_r. By the pigeonhole principle, there is a value of the table entry for which, with probability at least $(1/2)2^{-d}$, a random key-set from $\mathcal{D}_{r+1}$ is in K_r and produces a table with that specific entry value. Since $2^{-d-1} > 2^{-d^2}$, the key-set invariant holds after round r.

4.5 Forcing t Rounds

To complete the proof of Theorem 1.1, we must show that the invariants on query-sets and key-sets are strong enough to guarantee that $P_t \times K_t$ is nontrivial, i.e. that after $t-1$ rounds, we still have at least two admissible problem instances which produce different answers. We shall soon prove that there exists at least one key-set $S \in K_t$ which picks two distinct leaves v_1 and v_2 of the tree $\mathcal{H}_t$ whose associated balls contain queries q_1 and q_2, respectively, in P_t. Notice that by construction, the family of balls associated with the leaves of $\mathcal{H}_t$ is a $\beta/16$-separated family. Since any key must lie within some ball in this family, no key can be a $\beta/16$-*ANN* for both q_1 and q_2. But (2) says that $\beta/16 = 2^{\lfloor (\log d)^{1-\varepsilon} \rfloor}$ which concludes the argument.

We prove the existence of such an S by contradiction. For any S_t let $\nu(S_t)$ denote the number of queries in P_t that it picks (which is shorthand for "the number of nodes it picks each of whose balls contains at least one query in

P_t"). Suppose that no admissible S_t picks more than one query. Then the probability p that a random S_t is admissible satisfies

$$p \leq \mathrm{Prob}[\nu(S_t) = 0] + \mathrm{Prob}[\nu(S_t) = 1].$$

To form a random S_t we pick d^5 leaves of $\mathcal{H}_t$ at random, uniformly. By the query invariant, at least $1/d$ of them belong to P_t. So,

$$p < \left(1 - \frac{1}{d}\right)^{d^5} + 2^d \left(1 - \frac{1}{d}\right)^{d^5 - 1} < e^{-d^4} + 2^{d+1} e^{-d^4} < e^{-d^3}.$$

By the key-set invariant, we must have $p > 2^{-d^2}$, hence a contradiction. This concludes the proof of Theorem 1.1. ∎

References

[1] M. Ajtai. A lower bound for finding predecessors in Yao's cell probe model. *Combinatorica*, 8:235–247, 1988.

[2] S. Arya and D. M. Mount. Approximate nearest neighbor searching. In *Proc. 4th Annu. ACM-SIAM Symp. Disc. Alg.*, pages 271–280, 1993.

[3] S. Arya, D. M. Mount, and O. Narayan. Accounting for boundary effects in nearest-neighbor searching. *Disc. Comput. Geom.*, 16(2):155–176, 1996.

[4] S. Arya, D. M. Mount, N. S. Netanyahu, R. Silverman, and A. Wu. An optimal algorithm for approximate nearest neighbor searching. *J. ACM*, 45(6):891–923, 1998. Preliminary version in *Proc. 5th Annu. ACM-SIAM Symp. Disc. Alg.*, pages 573–582, 1994.

[5] O. Barkol and Y. Rabani. Tighter lower bounds for nearest neighbor search and related problems in the cell probe model. In *Proc. 32nd Annu. ACM Symp. Theory Comput.*, pages 388–396, 2000.

[6] P. Beame and F. Fich. Optimal bounds for the predecessor problem. In *Proc. 31st Annu. ACM Symp. Theory Comput.*, pages 295–304, 1999.

[7] M. Bern. Approximate closest-point queries in high dimensions. *Inform. Process. Lett.*, 45(2):95–99, 1993.

[8] A. Borodin, R. Ostrovsky, and Y. Rabani. Lower bounds for high dimensional nearest neighbor search and related problems. In *Proc. 31st Annu. ACM Symp. Theory Comput.*, pages 312–321, 1999.

[9] F. Cazals. Effective nearest neighbors searching on the hyper-cube, with applications to molecular clustering. In *Proc. 14th Annu. ACM Symp. Comput. Geom.*, pages 222–230, 1998.

[10] T. Chan. Approximate nearest neighbor queries revisited. *Disc. Comput. Geom.*, 20(3):359–373, 1998. Preliminary version in *Proc. 13th Annu. ACM Symp. Comput. Geom.*, pages 352–358, 1997.

[11] B. Chazelle. *The Discrepancy Method: Randomness and Complexity*. Cambridge University Press, Cambridge, 2000.

[12] K. L. Clarkson. A probabilistic algorithm for the post office problem. In *Proc. 17th Annu. ACM Symp. Theory Comput.*, pages 175–184, 1985.

[13] K. L. Clarkson. A randomized algorithm for closest-point queries. *SIAM J. Comput.*, 17(4):830–847, 1988.

[14] K. L. Clarkson. An algorithm for approximate closest-point queries. In *Proc. 10th Annu. ACM Symp. Comput. Geom.*, pages 160–164, 1994.

[15] S. Har-Peled. A replacement for voronoi diagrams of near linear size. In *Proc. 42nd Annu. IEEE Symp. Found. Comput. Sci.*, pages 94–103, 2001.

[16] W. Hoeffding. Probability inequalities for sums of bounded random variables. *J. Amer. Stat. Assoc.*, 58(301):13–30, 1963.

[17] P. Indyk and R. Motwani. Approximate nearest neighbors: towards removing the curse of dimensionality. In *Proc. 30th ACM Symp. Theory Comput.*, pages 604–613, 1998.

[18] M. Karchmer and A. Wigderson. Monotone circuits for connectivity require super-logarithmic depth. *SIAM J. Disc. Math.*, 3(2):255–265, 1990.

[19] J. M. Kleinberg. Two algorithms for nearest neighbor search in high dimensions. In *Proc. 29th Annu. ACM Symp. Theory Comput.*, pages 599–608, 1997.

[20] E. Kushilevitz and N. Nisan. *Communication Complexity.* Cambridge University Press, Cambridge, 1997.

[21] E. Kushilevitz, R. Ostrovsky, and Y. Rabani. Efficient search for approximate nearest neighbor in high-dimensional spaces. *SIAM J. Comput.*, 30(2):457–474, 2000. Preliminary version in *Proc. 30th Annu. ACM Symp. Theory Comput.*, pages 614–623, 1998.

[22] N. Linial, E. London, and Y. Rabinovich. The geometry of graphs and some of its algorithmic applications. *Combinatorica*, 15(2):215–245, 1995. Preliminary version in *Proc. 35th Annu. IEEE Symp. Found. Comput. Sci.*, pages 577–591, 1994.

[23] P. B. Miltersen. Lower bounds for union-split-find related problems on random access machines. In *Proc. 26th Annu. ACM Symp. Theory Comput.*, pages 625–634, 1994.

[24] P. B. Miltersen, N. Nisan, S. Safra, and A. Wigderson. On data structures and asymmetric communication complexity. *J. Comput. Syst. Sci.*, 57(1):37–49, 1998. Preliminary version in *Proc. 27th Annu. ACM Symp. Theory Comput.*, pages 103–111, 1995.

[25] B. Xiao. *New Bounds in Cell Probe Model.* PhD thesis, UC San Diego, 1992.

[26] A. C. Yao. Should tables be sorted? *J. ACM*, 28(3):615–628, 1981.

[27] P. N. Yianilos. Data structures and algorithms for nearest neighbor search in general metric spaces. In *Proc. 4th Annu. ACM-SIAM Symp. Disc. Alg.*, pages 311–321, 1993.

About Authors

Amit Chakrabarti is at the Department of Computer Science, Princeton University, Princeton, NJ 08544, USA; *amitc@cs.princeton.edu*.

Bernard Chazelle is at the Department of Computer Science, Princeton University, Princeton, NJ 08544, USA; *chazelle@cs.princeton.edu.*

Benjamin Gum is at the Department of Mathematics and Computer Science, Grinnell College, Grinnell, IA 50112, USA; *gum@cs.grinnell.edu.*

Alexey Lvov is at the IBM T. J. Watson Research Center, Yorktown Heights, NY 10598, USA; *lvov@us.ibm.com.*

Acknowledgments

This work was supported in part by NSF Grant CCR-93-01254, NSF Grant CCR-96-23768, ARO Grant DAAH04-96-1-0181, and NEC Research Institute. Amit Chakrabarti's work was supported in part by a DIMACS Summer Fellowship. Benjamin Gum's work was supported in part by a National Science Foundation Graduate Fellowship.

The authors wish to thank Satish B. Rao and Warren D. Smith for interesting discussions on *ANN* searching, which inspired them to look at lower bounds for this problem. They also thank Piotr Indyk, Allan Borodin, Eyal Kushilevitz, Rafail Ostrovsky and Yuval Rabani for interesting comments, suggestions and clarifications.

Detecting Undersampling in Surface Reconstruction

Tamal K. Dey
Joachim Giesen

Abstract

Current surface reconstruction algorithms perform satisfactorily on well sampled, smooth surfaces without boundaries. However, these algorithms face difficulties with undersampling. Cases of undersampling are prevalent in real data since often these data sample a part of the boundary of an object, or are derived from a surface with high curvature or non-smoothness. In this paper we present an algorithm to detect the boundaries where dense sampling stops and undersampling begins. This information can be used to reconstruct surfaces with boundaries, and also to localize small and sharp features where usually undersampling happens. We report the effectiveness of the algorithm with a number of experimental results. Theoretically, we justify the algorithm with some mild assumptions that are valid for most practical data.

1 Introduction

Many applications in CAD, computer graphics and scientific computations involve approximating a surface from its samples. Piecewise linear approximations to the surface which are sought in *surface reconstruction* are often appropriate for visual aids. They may be used also as the control net for generating limit surfaces with subdivision methods [19]. These surfaces have higher-order continuity which is useful in many CAD applications.

Among the algorithms proposed in the literature for surface reconstruction, some [1, 7, 8, 10, 14, 18] concentrated on the empirical results and did not focus so much on the theoretical guarantees. Edelsbrunner [13] reports on the development of a commercial software under proprietary rights which is based on the ideas of α-shapes [14]. Very recently, starting with the algorithm of [2] four algorithms have been proposed with the guarantee that the output surface is homeomorphic and geometrically close to the sampled surface. They are the Crust algorithm by Amenta and Bern [2], the Cocone algorithm by Amenta, Choi, Dey and Leekha [4], the PowerCrust by Amenta, Choi, Kolluri [5], and the natural neighbor algorithm by Boissonat and Cazals [9]. The theoretical guarantee provided by these algorithms requires that the given data sample a surface densely. All these algorithms run into trouble if this condition is not met. In practice, the data may sample only part of a surface densely. This may be intentional for introducing boundaries, may be accidental in high curvature regions, or may be unavoidable due to non-smoothness. Currently, a systematic treatment of these cases of undersampling is missing from the literature.

In this paper we present an algorithm that detects the regions of undersampling with some assumptions that hold for most practical data. When a sample P undersamples a surface F, we have surface patches of F that are well-sampled by P leaving out others that are not. The boundaries of the well-sampled surface patches demark the regions where dense sampling stops and undersampling begins. The main task in detecting undersampling is to approximate the boundaries of these surface patches, see Figure 1. This detection is accomplished by first identifying the sample points that represent the boundaries and then using this information to reconstruct well-sampled surface patches with the COCONE algorithm of [4].

The main idea of detecting the *boundary sample points*, a representative of the boundaries, is to consider the structure of the Voronoi cells as indicated in [3] and used in two dimensions for curve reconstruction [11]. In [12] a first attempt was made to identify the boundary sample points via a *flatness condition* for each Voronoi cell that captures its elongation. Unfortunately, this simple minded algorithm could not be proved to detect all and only boundary sample points. In this paper we mature this idea with a new definition of flatness and present a new algorithm that finds all and only boundary sample points for a surface patch under some mild assumptions. In contrast to the single pass boundary detection of [12], this algorithm works in two phases. In the first phase it uses a conservative condition to identify a set of interior sample points with well-sampled neighborhoods. In the second phase it expands this initial set of interior sample points by relaxing the condition and looking at their neighbors. In the end we argue that the algorithm captures all interior sample points leaving only the boundary ones.

Detection of boundary sample points finds immediate applications in reconstructing surfaces with boundaries and also in scientific analysis where localizing undersampled regions provides information about discontinuities in sampled functions [16]. We present experimental results with several data sets. The algorithm detects the boundaries in these test cases and also points out the high curvature regions including non-smooth ones where the surface is undersampled. In some cases, this detection aids reconstruction of non-smooth edges, in others it leaves 'holes'. A repair phase can follow to fill up these 'holes' either algorithmically or by scanning more sample points manually in the detected undersampled regions.

The rest of the paper is organized as follows. Section 2 introduces some basic definitions that are used for the boundary detection. Section 3 establishes some facts about the geometry of the Voronoi cells based on which the boundary detection algorithm is designed in Section 4. Section 5 details the COCONE reconstruction of a number of data sets with the boundary detection. We conclude in Section 6.

2 Definitions

Let F be a surface sampled by a point set P. The set P does not necessarily sample F well, but we assume that it does so for patches of F which we call S. (The definition of a well-sampled patch is given in section 2.1) The boundaries of S coincide with the boundaries of the undersampled regions in F, and our goal is to reconstruct them from the sample P. Since only P is known, we have to define the notion of boundary with respect only to P and not to F or S. This is the topic of sections 2.2 and 2.3.

2.1 Sampling

We base our algorithm on the notion of ε-sampling of smooth surfaces as introduced in [2]. This definition builds on the notion of *medial axis* and *local feature size*. The *medial axis* of a smooth surface $F \subset \mathbb{R}^3$ is the closure of the set of points that have more than one closest point on F. The *local feature size* $f_F(x)$ at a point $x \in F$ is the least distance of x to the medial axis. The *medial balls* at x are defined as the balls that touch F tangentially at x and are centered on the medial axis. A point set $P \subset F$ is called an ε-sample of a surface F if each point $x \in F$ has a sample point within the distance of $\varepsilon f_F(x)$.

Well-sampled patch: Let $S \subseteq F$ be a surface patch such that each point $x \in S$ has a sample point within $\varepsilon f_F(x)$ distance and S is maximal in the sense that no other point $y \in F \setminus S$ has this property. Notice that S may have several components. It is essential for the rest of the paper to make the distinction between S and F. The input P does not necessarily well-samples F, but does so for $S \subseteq F$ by definition. In Figure 1, the shaded patches are sampled densely by the sample points. The rest of the surface is undersampled. Our goal is to approximate the *boundaries* (shown with dotted curves) between these well-sampled and undersampled patches.

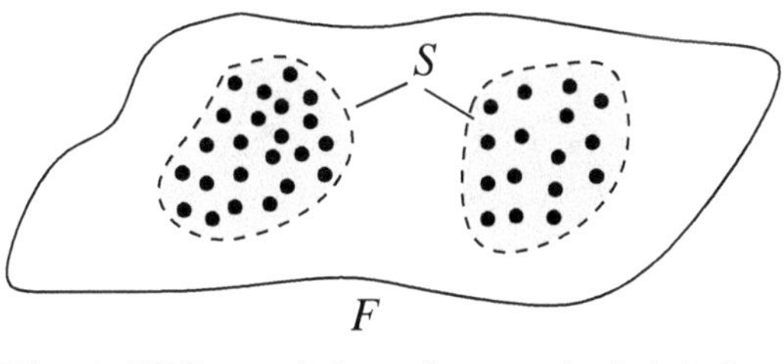

Fig. 1. Well-sampled patches are shaded darker.

2.2 Boundaries

In any compact surface S, interior points are distinguished from boundary points by their neighborhoods. An interior point has a neighborhood homeomorphic to the plane $\mathbb{R}^2$. A boundary point, on the other hand, has a neighborhood homeomorphic to the halfplane $\mathbb{H}^2_+ = \{(x,y) \in \mathbb{R}^2 : x \geq 0\}$. Even though all sample points in P may be interior points in the well-sampled patch S, the existence of a non empty boundary should be recognizable from the sample points. We are aiming for a classification of interior and boundary sample points that captures the intuitive difference between interior and boundary points. In [12] *restricted Delaunay triangulation* was used to make this distinction. In this paper we find it more appropriate for our proofs to define these sample points more directly from the *restricted Voronoi diagram* [15].

Restricted Voronoi diagram: Let P be a sample of a compact surface S with or without boundary embedded in $\mathbb{R}^3$. Denote the Voronoi diagram of P by V_P and the Voronoi cell for a point p by V_p. The restriction of V_P on the surface S defines the *restricted Voronoi diagram* containing the *restricted Voronoi cells* $V_{p,S} = V_p \cap S$.

The restriction of V_p on S, namely $V_{p,S}$ consists of the closure of all points in S that have p as their closest sample point. In other words, p is a discrete representative of the surface patch $V_{p,S}$. This leads us to define the *neighborhood* of a sample point p as $V_{p,S}$. Using this definition of neighborhood we define *interior* and *boundary* sample points.

Definition 2.1. (*Interior and boundary sample points*) A sample point p from a sample P of S is called interior if $V_{p,S}$ does not contain a boundary point of S. Sample points that are not interior are called boundary.

Arguments of [2] can be used to show that the neighborhoods of interior sample points are homeomorphic to disks if $\varepsilon < 0.1$.

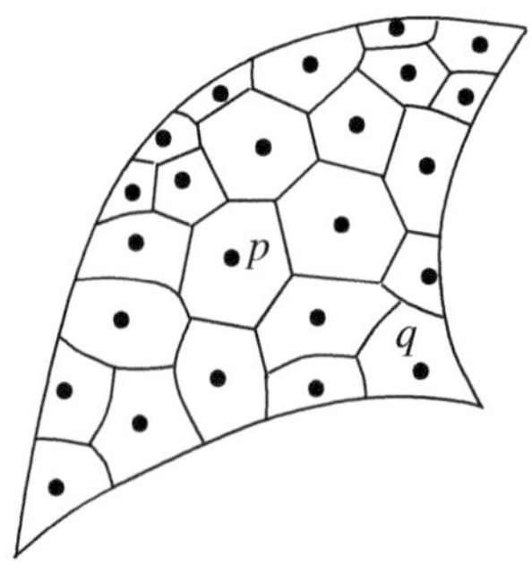

Fig. 2. Intersection of V_P with the surface S; p is an interior sample point, q is a boundary sample point.

2.3 Flat sample points

We cannot compute the restricted Voronoi diagram $V_{P,S}$ only from P when S is unknown. Therefore we need a characterization of boundary sample points that can be exploited algorithmically.

To this end we define a *flatness* condition that can be checked algorithmically and show that boundary sample points cannot be flat whereas interior sample points with well-sampled neighborhoods are necessarily flat. The flatness condition defined in [12] did not have this property and as a result could not be used to detect all and only boundary sample points. We need few other notations and definitions from earlier works. We borrow the definition of poles from [2], cocones from [4], cocone neighbors from [12]. We use the notation $\angle(\mathbf{u}, \mathbf{v})$ to denote the acute angle between the lines supporting two vectors $\mathbf{u}$ and $\mathbf{v}$.

Poles: The farthest Voronoi point p^+ from p in V_p is called the positive *pole* of p. The negative pole of p is the farthest point $p^- \in V_p$ from p such that the two vectors from p to p^+ and p^- make an angle more than $\frac{\pi}{2}$. We call $\mathbf{v}_p = p^+ - p$, the *pole vector* for p. If V_p is unbounded, p^+ is taken at infinity, and the direction of $\mathbf{v}_p$ is taken as the average of all directions given by unbounded Voronoi edges.

Cocone: The set $C_p = \{y \in V_p : \angle((y-p), \mathbf{v}_p) \geq \frac{3\pi}{8}\}$ is called the cocone of p. In words, C_p is the complement of a double cone (clipped within V_p) centered at p with an opening angle $\frac{3\pi}{8}$ around the axis aligned with $\mathbf{v}_p$. See Figure 3 for an example of a cocone.

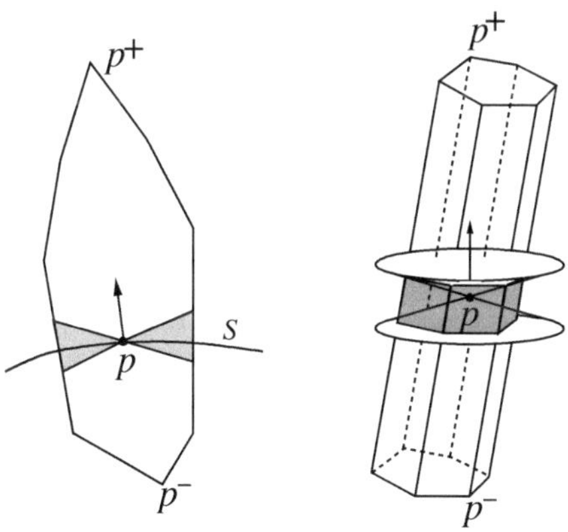

Fig. 3. A Voronoi cell together with the normalized pole vector and the cocone (shaded) in two dimensions (left) and three dimensions (right).

Cocone neighbors: The set $N_p = \{q \in P : C_p \cap V_q \neq \emptyset\}$ is called the set of cocone neighbors of p.

Radius and height: The *radius* r_p of the Voronoi cell V_p is defined as the radius of C_p, i.e., $r_p = \max\{||y - p|| : y \in C_p\}$. Define the *height* h_p of V_p as the distance $||p^- - p||$.

The radius captures how 'fat' the Voronoi cell is, whereas the height captures how 'long' it is. It is important for our proofs that the height be defined as the distance of a sample point to its negative pole rather than the positive one. The width and height defined in [12] also capture similar measures, but the definition of width is slightly different from that of radius and we need to incorporate this adjustment for our proofs, specifically for Lemma 3.10.

Now, we are prepared to define a new flatness condition, which depends on two parameters ρ and α as opposed to a single parameter ρ in [12]. (The reader should note that the common parameter ρ does *not* have the same meaning as in [12]).

Definition 2.2. (*Flatness*) A sample point $p \in P$ is called flat with respect to ρ and α if the following two conditions hold:

1. Ratio condition: $r_p \leq \rho h_p$
2. Normal condition: $\forall q$ *with* $p \in N_q$, $\angle(\mathbf{v}_p, \mathbf{v}_q) \leq \alpha$.

Ratio condition imposes that the Voronoi cell V_p is long and thin in the direction of the line supporting $\mathbf{v}_p$. The normal condition stipulates that the direction of elongation of V_p matches with that of any sample point whose cocone neighbor is p. For theoretical guarantees, we use $\rho = 1.3\varepsilon$ and $\alpha = 0.14$ radians.

3 Voronoi cell geometry

Our goal is to exploit the definition of flat sample points in a boundary detection algorithm. In Theorem 3.8 we prove that interior sample points with well-sampled neighborhoods are flat. In Theorem 3.11 we prove that the boundary sample points cannot be flat. These two theorems form the basis of our boundary detection algorithm.

Let us state two useful lemmas from [2] that are proved for smooth compact surfaces without boundaries. Here we need to extend them for surfaces that may have a non-empty boundary such as the well-sampled patch $S \subset F$. We denote the normal to F at a point p as $\mathbf{n}_p$. The first lemma taken from [2] states that the local feature size defined on F is 1-Lipschitz continuous.

Lemma 3.1. *For any two points p and q on F we have $f_F(p) \leq f_F(q) + ||q - p||$.*

The next lemma is proved in [12]. We include it here for completeness. Recall that by our notation, the angle $\angle((y-p), \mathbf{n}_p)$ denotes the smaller angle between $(y-p)$ and $\mathbf{n}_p$ or $-\mathbf{n}_p$.

Lemma 3.2. *Let p be an interior sample point in an ε-sample P of a surface S with the surface normal $\mathbf{n}_p$ at p. Let y be any point in the Voronoi cell V_p such that $||y-p|| > \nu f_F(p)$ for some $\nu > 0$. Then*

$$\angle((y-p), \mathbf{n}_p) \leq \sin^{-1}\left(\frac{\varepsilon}{\nu(1-\varepsilon)}\right) + \sin^{-1}\left(\frac{\varepsilon}{1-\varepsilon}\right).$$

Proof. We can copy almost the entire proof given in [2] but some minor adjustments are necessary to incorporate surfaces with non-empty boundary.

The proof in [2] makes use of the fact that a smooth compact surface S with empty boundary divides the space $\mathbb{R}^3$ in two components that are not connected. One of these components is bounded, the other one is unbounded. In fact, the proof needs that, given two points in a Voronoi cell such that one lies in the bounded and the other one in the unbounded component of $\mathbb{R}^3$, the line segment connecting these points has to intersect the surface S at least once.

Here we are only concerned with interior sample points. The arguments in [2] apply to show that $V_{p,S}$ is a disk if p is an interior sample point and $\varepsilon \leq 0.1$. This means that the two sides of the surface S in the Voronoi cell V_p are well defined. The convexity of V_p implies that a line segment connecting two points on opposite sides of S in the Voronoi cell V_p has to lie entirely in V_p. Thus this line segment intersects $V_{p,S}$ at least once and we can apply the same reasoning as in the proof in [2]. □

We have $||p^+ - p|| > f_F(p)$, because the two medial balls have radii at least $f_F(p)$ and the centers of these medial balls must lie in V_p. Thus we have the following corollary from Lemma 3.2.

Corollary 3.3. *For an interior sample point p we have*

$$\angle(\mathbf{v}_p, \mathbf{n}_p) \leq 2\sin^{-1}\frac{\varepsilon}{1-\varepsilon}.$$

The next lemma is proved in [2] and Lemma 3.5 follows from a result in [4].

Lemma 3.4. *Let p, q be two points on F so that $||p-q|| < \nu \min\{f_F(p), f_F(q)\}$ with $\nu < 1/3$. Then $\angle(\mathbf{n}_p, \mathbf{n}_q) \leq \frac{\nu}{1-3\nu}$.*

Lemma 3.5. *Let y be any point at a distance $\nu f_F(p)$ from p with $\nu < 0.1$. Then, $\angle((y-p), \mathbf{n}_p) \geq \frac{\pi}{2} - \nu$.*

Now we prepare for the two main theorems, Theorem 3.8 and Theorem 3.11. For the rest of the proofs we assume $\varepsilon \leq 0.01$, $\rho = 1.3\varepsilon$ and $\alpha = 0.14$ radians.

Lemma 3.6. *Interior sample points satisfy the ratio condition.*

Proof. Let p be any interior sample point. From Corollary 3.3 we get

$$\begin{aligned} \angle(\mathbf{v}_p, \mathbf{n}_p) &= \phi \\ &\leq 2\sin^{-1}\left(\frac{\varepsilon}{1-\varepsilon}\right). \end{aligned}$$

Let y be any point in C_p. By definition $\angle(\mathbf{v}_p, (y-p)) \geq \frac{3\pi}{8}$. From Lemma 3.2 (applying the contrapositive of the implication stated there) we get $||y-p|| \leq \nu f_F(p)$ where ν fulfills the inequality

$$\sin^{-1}\left(\frac{\varepsilon}{\nu(1-\varepsilon)}\right) + \sin^{-1}\left(\frac{\varepsilon}{1-\varepsilon}\right) + \phi \geq \frac{3\pi}{8}.$$

Solving for ν we get $\nu \leq 1.3\varepsilon$. That is, the radius of the Voronoi cell V_p is at most $1.3\varepsilon f_F(p)$, i.e., $r_p \leq 1.3\varepsilon f_F(p)$.

Next we show that the height $h_p = ||p^- - p||$ is at least $f_F(p)$. Recall that p^- is the farthest point in V_p from p so that $(p^- - p)\cdot \mathbf{v}_p < 0$. Since $\mathbf{n}_p$ make a small angle upto orientation with $\mathbf{v}_p$ (Corollary 3.3), one of the two medial balls going through p has its center m such that the vector $m-p$ does not point in the same direction as $\mathbf{v}_p$, i.e., $(m-p)\cdot \mathbf{v}_p < 0$. We know that $||m-p|| \geq f_F(p)$ and $m \in V_p$. This immediately implies that p^- is at least $f_F(p)$ away from p. Therefore, $h_p \geq f_F(p) \geq \frac{r_p}{1.3\varepsilon}$. Thus the ratio condition in Definition 2.2 is fulfilled for $\rho = 1.3\varepsilon$. □

Although the ratio condition holds for all interior sample points, the normal condition may not hold for all of them. Nevertheless, we can show that interior sample points with well-sampled neighborhoods satisfy the normal condition. To be precise we introduce the following definition.

Definition 3.7. (*Deep interior sample point*) An interior sample point p is called deep if there is no boundary sample point with p as its cocone neighbor.

Theorem 3.8. *All deep interior sample points are flat.*

Proof. It follows from Lemma 3.6 that deep interior sample points satisfy the ratio condition. We show that they satisfy the normal condition as well. Let q be any Voronoi neighbor of p so that $p \in N_q$. The sample point q is interior by definition. Therefore, we can apply Corollary 3.3 to assert that $\angle(\mathbf{v}_q, \mathbf{n}_q) \leq 2\sin^{-1}\frac{\varepsilon}{1-\varepsilon}$. Also, since $p \in N_q$ there is a point $x \in C_q$ with $||x-q|| \leq 1.3\varepsilon f(q)$ and $x \in V_p \cap V_q$ implying $||p-q|| \leq 2.6\varepsilon f(q)$. For small $\varepsilon \leq 0.01$ we can apply Lemma 3.4 to deduce $\angle(\mathbf{n}_p, \mathbf{n}_q) \leq 0.03$ radians. Thus, we have

$$\begin{aligned} \angle(\mathbf{v}_q, \mathbf{v}_p) &\leq \angle(\mathbf{v}_q, \mathbf{n}_q) + \angle(\mathbf{n}_q, \mathbf{n}_p) + \angle(\mathbf{n}_p, \mathbf{v}_p) \\ &\leq 0.14 \; \textit{radians} \end{aligned}$$

which satisfies the normal condition for $\alpha = 0.14$ radians. □

For Theorem 3.11 we need some boundary assumptions. The first assumption (i) says that boundary sample points remain as boundary even if S is expanded with a small collar around its boundary, and the assumption (ii) stipulates that the boundaries are 'well-separated'.

Assumption 3.9. *(*Boundary assumption*)*

i. Let $S' \subseteq S$ be the maximal surface patch such that any $x \in S'$ has a sample point within distance $\delta f(x)$, for some $\delta > \varepsilon$. The S and S' define the same set of boundary sample points (In our proofs, we will take $\delta = 1.3\Delta\varepsilon$, where Δ is some global quantity of S defined in the following lemma).

ii. Neighborhood of each boundary sample point intersects the neighborhood of at least one interior sample point.

Lemma 3.10. *Let p be a sample point for which the ratio condition holds. If $\angle(\mathbf{v}_p, \mathbf{n}_p) \leq 0.2$, then p must be an interior sample point.*

Proof. Suppose, on the contrary, p is a boundary sample point. Let $\Delta_p = \frac{h_p}{f_F(p)}$. Since the ratio condition holds, we have $||x - p|| \leq \rho h_p = \rho \Delta_p f_F(p)$ for any point $x \in C_p$. Since $h_p = ||p^- - p||$ we have an upper bound on h_p assuming not all data points are lying on a plane. Also $f_F(p) > 0$ if F is assumed to be smooth. Thus, we have an upper bound, say Δ on Δ_p. With $\rho = 1.3\varepsilon$, we have $||x - p|| < \delta f_F(p)$ where $\delta = \Delta\rho = 1.3\Delta\varepsilon$. Let y be any point on F with $||y - p|| < \delta f_F(p)$. Assuming ε to be small enough so that $\delta < 0.1$ we can apply Lemma 3.5 to have $\angle((y - p), \mathbf{n}_p) \geq \frac{\pi}{2} - \delta$. Since $\angle(\mathbf{v}_p, \mathbf{n}_p) \leq 0.2$ by condition of the lemma, we have

$$\begin{aligned} \angle(\mathbf{v}_p, (y - p)) &\geq \frac{\pi}{2} - \delta - 0.2 \\ &> \frac{3\pi}{8}. \end{aligned}$$

It implies that any point $y \in F$ with $||y - p|| < \delta f_F(p)$ cannot lie on the boundary of the double cone defining C_p. In other words, $F \cap V_p \in C_p$. Therefore, any point $y \in F \cap V_p$ satisfies $||y - p|| < \delta f_F(p)$. According to the boundary assumption the surface $S' \supset S$ must define p as a boundary sample point. But, that would require the existence of a point y with $||y - p|| = \delta f_F(p)$ to be inside V_p. This contradicts our boundary assumption 3.9(i) with $\delta = 1.3\Delta\varepsilon$. □

Remark. Even though Δ_p has an upper bound with our assumptions, it may seem that this bound could be catastrophically large. However, we are saved from this blow-up due to the fact that when $h(p)$ becomes really large, so does the local feature size $f_F(p)$. Indeed, when the points become more and more dense, we can expect that the Voronoi cells converge to a small cylinder of height $f_F(p)$ (notice that it is crucial to define height as $\|p - p^-\|$). Hence Δ will be almost 1 everywhere if the sample is sufficiently dense.

Theorem 3.11. *Boundary sample points cannot be flat.*

Proof. Let p be a boundary sample point. Suppose that, on the contrary, p is flat. Consider an interior sample point q so that $V_{q,S} \cap V_{p,S} \neq \emptyset$ (Assumption 3.9 (ii)). The sample point p is a cocone neighbor of q since $C_q \cap S = V_q \cap S$. Normal condition requires that $\angle(\mathbf{v}_p, \mathbf{v}_q) \leq 0.14$. Also, $||q - p|| \leq 2.6\varepsilon f_F(q)$. It implies that $\angle(\mathbf{n}_p, \mathbf{n}_q) \leq 0.03$ (Lemma 3.4). Thus,

$$\begin{aligned}\angle(\mathbf{v}_p, \mathbf{n}_p) &\leq \angle(\mathbf{v}_p, \mathbf{v}_q) + \angle(\mathbf{v}_q, \mathbf{n}_q) + \angle(\mathbf{n}_q, \mathbf{n}_p) \\ &\leq 0.14 + 0.021 + 0.03 = 0.191.\end{aligned}$$

Thus, p satisfies the conditions of Lemma 3.10 and hence is an interior sample point reaching a contradiction. □

Remark. It is interesting to note that one can assume the surface F to have small features around any sample point to render it as a boundary sample point in our definition. After all, the surface F is unknown to us. Notice that the algorithm can classify any sample point as boundary with an appropriate small value of ρ. On the other hand, we argue that an appropriate small value of ρ is *necessary* for classifying any sample point as a boundary one. Observe that, the smaller the feature size gets to render a sample point as boundary, the larger Δ gets with a fixed sample since $\Delta = \max\{\frac{h_p}{f_F(p)}\}$. In order for the boundary assumption to be valid, we need δ to be fixed even though Δ increases. This requires ε to decrease as $\delta = 1.3\Delta\varepsilon$. But, decreasing ε would indeed require decreasing ρ.

4 Boundary detection

The algorithm for boundary detection first computes the set of interior sample points, R, that are flat. It uses two parameters ρ and α to check the ratio and normal conditions. Theorem 3.8 and Assumption 4.1 guarantee that R is not empty. In a subsequent phase R is expanded to include all interior sample points in an iterative procedure. A generic iteration proceeds as follows. Let p be any cocone neighbor of a sample point $q \in R$ so that $p \notin R$ and V_p satisfies the ratio condition. If $\mathbf{v}_p$ and $\mathbf{v}_q$ make small angle up to orientation, i.e., if $\angle(\mathbf{v}_p, \mathbf{v}_q) \leq \alpha$, we include p in R. If no such sample point can be found, the iteration stops. We argue that R includes all and only interior sample points at the end. The rest of the sample points are detected as boundary ones.

The following routine ISFLAT checks the conditions stated in Definition 2.2 to detect flat sample points. The input is a sample point $p \in P$ with two parameters ρ and α. The return value is `true` if p is a flat sample point, and `false` otherwise. The routine BOUNDARY uses ISFLAT to detect the boundary sample points.

IsFlat ($p \in P$, α, ρ)
1 compute the radius r_p and the height h_p
2 **if** $r_p \leq \rho h_p$
3   **if** $\forall q$ with $p \in N_q : \angle(\mathbf{v}_p, \mathbf{v}_q) \leq \alpha$
4   **return** `true`
5 **return** `false`

Boundary (P, α, ρ)
1 $R := \emptyset$
2 **for** all $p \in P$
3   **if** IsFlat(p) $R := R \cup p$
4 **endfor**
5 **while** $\exists p \notin R$ and $\exists q \in R$ with $p \in N_q$,
    and $r_p \leq \rho h_p$ and $\angle(\mathbf{v}_p, \mathbf{v}_q) \leq \alpha$
6     $R := R \cup p$
7 **endwhile**
8 **return** $P \setminus R$

4.1 Justification

Now we argue that Boundary outputs all and only boundary sample points. We need an interior assumption that says that all interior sample points have well-sampled neighborhoods.

Assumption 4.1. (Interior assumption) *Each interior sample point is path connected to a deep interior sample point where the path lies only inside the restricted Voronoi cells of the interior sample points.*

Claim 4.2. Boundary *outputs all and only boundary sample points.*

Proof. Inductively assume that the set R computed by Boundary contains only interior sample points. Initially, the assumption is valid since steps 2 and 3 compute the set of flat sample points, R, which must be interior due to Theorem 3.11. Assumption 3.9 (ii) and Assumption 4.1 imply that each component of S must have a deep interior sample point. Thus, R cannot be empty initially. In the *while* loop if a sample point p is included in the set R, it must satisfy the ratio condition. Also, there exists $q \in R$ so that $\angle(\mathbf{v}_p, \mathbf{v}_q) \leq 0.14$ (assuming $\alpha = 0.14$ radians). Since q is an interior sample point by inductive assumption $\angle(\mathbf{v}_q, \mathbf{n}_q) \leq 2\sin^{-1}\frac{\varepsilon}{1-\varepsilon}$. It follows that $\angle(\mathbf{v}_p, \mathbf{n}_q) \leq 0.161$. Since p is a cocone neighbor of q, we have $||q - p|| \leq 2.6\varepsilon f_F(q)$. Applying Lemma 3.4 we get $\angle(\mathbf{n}_q, \mathbf{n}_p) \leq 0.03$ for $\varepsilon \leq 0.01$. Therefore,

$$\angle(\mathbf{v}_p, \mathbf{n}_p) \leq \angle(\mathbf{v}_p, \mathbf{n}_q) + \angle(\mathbf{n}_q, \mathbf{n}_p) \leq 0.2.$$

It follows from Lemma 3.10 that q is an interior sample point proving the inductive hypothesis.

Now we argue that each interior sample point p is included in R at the end of the *while* loop. Assumption 4.1 implies that one can reach p walking through adjacent cocones from a deep interior sample point. Any interior sample point that is a cocone neighbor of a sample point in R satisfies the condition of the *while* loop (Lemma 3.4 and Corollary 3.3). It follows that p is encountered in the *while* loop during some iteration and is included in R. □

4.2 Non-smoothness

Recall that our theory is based on the assumption that the sampled surface F is smooth. However, we observe that the boundary detection algorithm also detects undersampling in non-smooth surfaces, see Figure 5(a) and (b). The ability to handle non-smooth surfaces owes to the fact that non-smooth surfaces may be approximated with a smooth one that interpolates the sample points. For example, one can resort to the implicit surface that is C^1-smooth and interpolates the sample points using natural co-ordinates as explained in [9]. For higher order continuity results of [17] can be called upon. These smooth surfaces have high curvatures near the sharp features of the original non-smooth surface. Our theory can be applied to the approximating smooth surface to ascertain that the sample points in the vicinity of sharp features act as boundary sample points in the vicinity of high curvatures for the smooth surface.

5 Reconstruction

We use the cocone algorithm of [4] to complete the surface reconstruction after we detect the boundary sample points. The original cocone algorithm chooses all triangles incident to a sample point p whose dual Voronoi edges are intersected by the cocone C_p. But, this causes the boundary sample points to choose undesirable triangles since the estimated normals at these sample points are not correct. See Figure 4(a) for an example. In [12], the boundary detection phase checked if a sample point satisfies the ratio condition. If it does not, then that sample point is not allowed to choose any triangle. Since it is unknown if only the ratio condition distinguishes boundary sample points from interior ones, it could not be established that only boundary sample points are disallowed to choose triangles. The boundary detection as presented in this paper detects all and only boundary sample points, at least theoretically under the assumptions listed earlier. The desired set of triangles incident to boundary sample points is chosen by some interior sample points. As a result 'garbage' triangles are eliminated and clean holes appear at the undersampled regions. Figure 4(b) illustrates this effect. The complete cocone algorithm with the new boundary detection algorithm has the following steps.

```
Cocone (P, α, ρ)
    compute V_P
    B :=Boundary(P, α, ρ)
    for each p ∈ P \ B, mark the set of triangles
        that are dual to Voronoi edges intersecting C_p
    Choose the set of marked triangles, T, that are
        marked by all its vertices which are not in B
    Extract a manifold from T using pruning and walking
        as mentioned in [2] and detailed in [12].
```

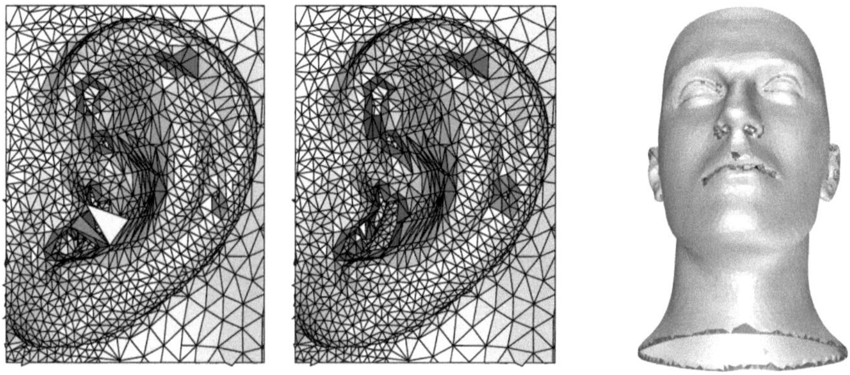

a b c

Fig. 4. Reconstruction of the dataset Mannequin (triangles incident to boundary sample points are shaded darker) (a) without boundary detection 'garbage' triangles appear at the high curvature region in the ear, (b) clean holes appear after boundary detection is applied with Boundary, (c) Undersampled regions are detected near small features at the ears, lips, eyes and the true boundary at the bottom of the neck.

5.1 Implementation

We implemented the above cocone algorithm with CGAL 2.3 [21] library together with filtered predicates. The software is available from [20]. Computation of filtered predicates simulates exact evaluation on a demand basis and thus runs faster than predicate evaluations with exact arithmetic. Instead of filtered predicates if we use floating point arithmetic the running time decreases by a factor of two. However, the results may not be correct. Table 1 shows the running times with filtered predicates on a 933 MHz PC with 512 MB memory. The code was compiled with `g++` compiler with `O1` optimization.

In the proofs we choose the ratio ρ to be at most 1.3ε. In practice a value between 0.66 and 0.99 gives good results. Although α is 0.14 radians

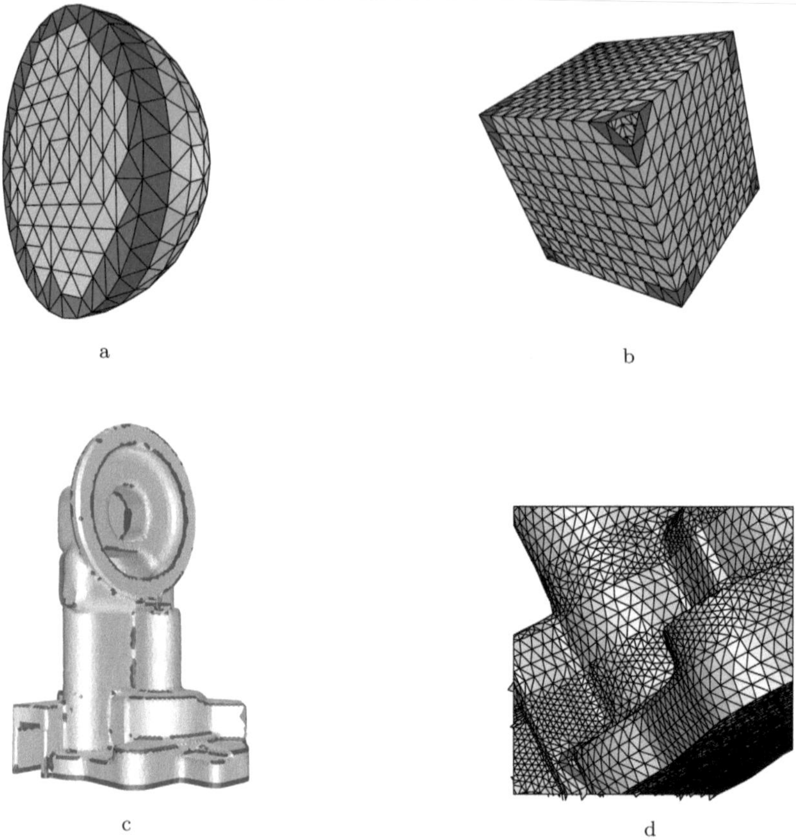

Fig. 5. Reconstruction of the data sets HALFSPHERE, CUBE and OILPUMP (a) sample points on the sharp edge are detected and they are reconstructed as neighbor sample points choose correct triangles for them, (b) The sample points on the sharp edges of the cube are also detected, and they are also reconstructed for the same reason. However, the corners (darkly shaded) are not reconstructed properly since the triangles incident to a corner vertex may have all three vertices on sharp edges which do not choose any triangle as they are detected as boundary sample points, (c) undersampled regions around sharp edges and high curvature regions of OILPUMP as detected by the algorithm are darkly shaded, (d) a zoom on the reconstruction of OILPUMP to show the reconstructed sharp edges between the flat base and the upright the structure, the darkly shaded region (right of lower left corner) around a sharp concave point cannot be reconstructed properly as the missing triangle is incident to three vertices on sharp edges.

Table 1. Experimental data.

object	number of points	number of triangles	running time(sec.)
HALFSPHERE	245	486	0.11
CUBE	602	1188	1.44
MANNEQUIN	12772	25339	15.79
FOOT	20021	39995	26.01
OILPUMP	30931	61548	42.49
MONKEYSADDLE	10000	19596	122.36

in theory, a value as large as $\frac{\pi}{6}$ gives good results as shown in Figures 4, 5 and 6. The triangles incident to boundary samples are shaded darkly.

Figure 5 illustrates the effective use of the boundary detection for reconstructing non-smooth surfaces. In these cases non-smooth edges are sampled densely. The sample points lying on these edges are detected as boundary sample points are disallowed to choose any triangle. The neighbor sample points choose correct triangles for them. Figure 5 show such reconstruction of sharp edges for the HALFSPHERE, CUBE and OILPUMP. Figure 6 shows two examples where BOUNDARY is used to detect real boundaries in the surface.

6 Conclusions

In this paper we present an algorithm to detect the regions of undersampling in data that are sampled from some surface. This provides a unified approach to reconstruct surfaces with boundaries and to identify the regions of non-smoothness or high curvature where undersampling does occur. As exhibited by our empirical results, our boundary detection algorithm correctly identifies the sample points that are visibly lying on the boundary. Also, it identifies the regions of non-smoothness and high curvature effectively in practice.

A probable follow up of this work is to investigate how this algorithm can be used to reconstruct non-smooth surfaces. After detecting the regions of non-smoothness, can we repair the surface to fill up the 'holes'? Currently research on this question is under progress.

References

[1] U. Adamy, J. Giesen, M. John. Surface reconstruction using umbrella filters. *Proc. IEEE Visualization*, (2000), 373–380.

[2] N. Amenta and M. Bern. Surface reconstruction by Voronoi filtering. *Discr. Comput. Geom.*, **22** (1999), 481–504.

[3] N. Amenta, M. Bern and M. Kamvysselis. A new Voronoi-based surface reconstruction algorithm. *SIGGRAPH 98*, (1998), 415-421.

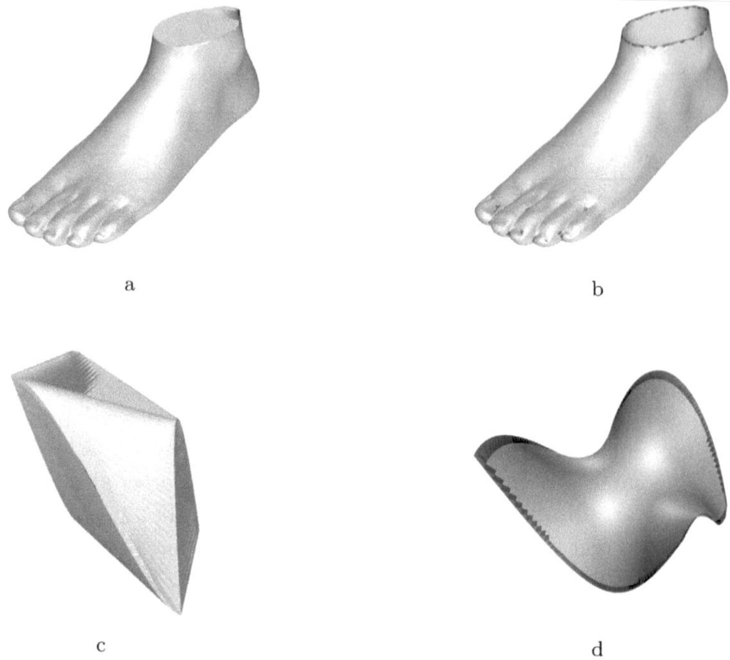

Fig. 6. Reconstruction of the dataset FOOT (a) without boundary detection; the big hole above the ankle is covered with triangles, (b) with boundary detection using the algorithm BOUNDARY; the hole above the ankle is well detected. MONKEY saddle (c) without boundary detection; the triangles between boundary vertices appear which covers the actual monkey saddle surface, (d) with boundary detection these undesirable triangles are removed.

[4] N. Amenta, S. Choi, T. K. Dey and N. Leekha. A simple algorithm for homeomorphic surface reconstruction. *Internat. J. Comput. Geom. & Appl.*, **12** (2002), 125–141.

[5] N. Amenta, S. Choi and R. K. Kolluri. The power crust, unions of balls, and the medial axis transform. *Internat. J. Comput. Geom. & Appl.*, **19** (2000), 127–153.

[6] D. Attali. r-regular shape reconstruction from unorganized points. *Proc. 13th Ann. Sympos. Comput. Geom.*, (1997), 248–253.

[7] C. Bajaj, F. Bernardini and G. Xu. Automatic reconstruction of surfaces and scalar fields from 3D scans. *SIGGRAPH 95*, (1995), 109–118.

[8] J. D. Boissonnat. Geometric structures for three dimensional shape representation. *ACM Transact. on Graphics*, **3** (1984) 266–286.

[9] J. D. Boissonnat and F. Cazals. Smooth surface reconstruction via natural neighbor interpolation of distance functions. *Proc. 16th. Ann. Sympos. Comput. Geom.*, (2000), 223–232.

[10] B. Curless and M. Levoy. A volumetric method for building complex models from range images. *SIGGRAPH 96*, (1996), 303-312.

[11] T. K. Dey, K. Mehlhorn and E. A. Ramos. Curve reconstruction: connecting dots with good reason. *Comput. Geom. Theory Appl.*, **15** (2000), 229–244.

[12] T. K. Dey, J. Giesen, N. Leekha and R. Wenger. Detecting boundaries for surface reconstruction using co-cones. *Internat. J. Comput. Graphics & CAD/CAM*, **16** (2001), 141–159.

[13] H. Edelsbrunner. Shape reconstruction with Delaunay complex. *LNCS 1380, LATIN'98: Theoretical Informatics*, (1998), 119–132.

[14] H. Edelsbrunner and E. P. Mücke. Three-dimensional alpha shapes. *ACM Trans. Graphics*, **13** (1994), 43–72.

[15] H. Edelsbrunner and N. Shah. Triangulating topological spaces. *Internat. J. Comput. Geom. & Appl.*, **7** (1997), 365–378.

[16] T. Gutzmer and A. Iske. Detection of discontinuities in scattered data approximation. *Numerical Algorithms*, **16** (1997), 155–170.

[17] H. Hiyoshi and K. Sugihara. Voronoi-based interpolation with higher continuity. *Proc. 16th. Ann. Sympos. Comput. Geom.*, (2000), 242–250.

[18] H. Hoppe, T. DeRose, T. Duchamp, J. McDonald and W. Stuetzle. Surface reconstruction from unorganized points. *SIGGRAPH 92*, (1992), 71-78.

[19] D. Zorin and P. Schröder. Subdivision for modeling and animation. *SIGGRAPH 99 Course Notes.*

[20] http://www.cis.ohio-state.edu/~tamaldey/cocone.html

[21] http://www.cgal.org

About Authors

Tamal K. Dey is at the Department of Computer and Information Science, Ohio State University, Columbus, OH 43210, USA; *tamaldey@cis.ohio-state.edu.*

Joachim Giesen was at the Department of Computer and Information Science, Ohio State University, Columbus, OH 43210, USA, when this work was done. Currently, he is at Institut für Theoretische Informatik, ETH Zentrum, CH-8092 Zürich, Switzerland; *giesen@inf.ethz.ch.*

Acknowledgments

Work on this paper has been supported by a grant CCR-9988216 under NSF.

The authors would also like to thank the referee who suggested improvements in presentation of the paper.

A Survey of the Hadwiger-Debrunner (p, q)-problem

Jürgen Eckhoff

1 Introduction

At the annual meeting of the Swiss Mathematical Society held in September 1956 in Basel, H. Hadwiger presented the following theorem. It was published one year later in a joint paper with H. Debrunner, his colleague at the University of Bern.

Theorem. (Hadwiger and Debrunner [47]) *Let p and q be integers with $p \geq q \geq d+1$ and $(d-1)p < d(q-1)$, and let $\mathcal{F}$ be a finite family of convex sets in R^d. Suppose $\mathcal{F}$ has at least p members, and among any p members of $\mathcal{F}$ some q have a common point. Then there exist $p-q+1$ points of R^d such that each member of $\mathcal{F}$ contains at least one of them.*

The primary interest of the authors was to extend the famous theorem of Helly (which is the case $p = q = d+1$) by relaxing its characteristic hypothesis in a combinatorial manner and varying its conclusion accordingly. The "(p, q)-theorem" they established stands at the origin of a branch of combinatorial geometry in which "(p, q)-problems" analogous to the one solved above are studied in various settings. The object is always to "pierce" a family of sets having the "(p, q)-property" as economically as possible, that is, to find as few points as possible so that each set in the family includes at least one of them. In the Hadwiger-Debrunner theorem, the "piercing number" $p - q + 1$ is the smallest which has the stated property. It will be seen, however, that exact results such as this are rare, and the focus will then be on finding good approximations. In some instances, one can do little more than proving that a finite piercing number exists at all. (This is the case, for example, when the integers p and q do *not* satisfy the restrictions imposed by the Hadwiger-Debrunner theorem.) As Section 1 below will demonstrate, even that can be a demanding task.

In order to formulate the (p, q)-problem in a general setting, we begin by introducing some terminology. Let $\mathcal{G}$ be a family of sets which are subsets of a ground set X, and let $\mathcal{F}$ be a subfamily of $\mathcal{G}$. A *transversal* of $\mathcal{F}$ is a subset of X that intersects each member of $\mathcal{F}$. The smallest cardinality of such a transversal is called the *transversal number* of $\mathcal{F}$ and denoted by $\tau(\mathcal{F})$. In geometrical contexts, the term *piercing number* is also used. If $\tau(\mathcal{F}) \leq k$, then $\mathcal{F}$ can be split into k subfamilies, each having a nonempty intersection. In this case, $\mathcal{F}$ is said to be *k-pierceable* or *k-fixable.*

A *matching* in $\mathcal{F}$ is a set of pairwise disjoint members of $\mathcal{F}$. The *matching number* of $\mathcal{F}$, also called the *packing number*, is the largest cardinality of such a matching. It is denoted by $\nu(\mathcal{F})$. Clearly, $\tau(\mathcal{F}) \geq \nu(\mathcal{F})$ although both numbers need not be finite. In general, no upper bound on $\tau(\mathcal{F})$ in terms of $\nu(\mathcal{F})$ exists. To derive such a bound is an important special case of the (p,q)-problem.

Finally, let p and q be integers with $p \geq q \geq 2$. Then $\mathcal{F}$ is said to have the *(p,q)-property* if $\mathcal{F}$ contains at least p members and among every p members of $\mathcal{F}$, some q have a common point. If this condition holds, we write $\mathcal{F} \in \Pi(p,q)$. As customary, $\Pi(p,p)$ will be shortened to $\Pi(p)$.

The (p,q)-problem for $\mathcal{G}$ consists in determining (or at least estimating) the *piercing numbers*

$$J(p,q;\mathcal{G}) := \sup\{\tau(\mathcal{F}) \,:\, \mathcal{F} \subset \mathcal{G},\, |\mathcal{F}| < \infty,\, \mathcal{F} \in \Pi(p,q)\}.$$

In words, $J(p,q;\mathcal{G})$ is the smallest number k such that every finite subfamily of $\mathcal{G}$ having the (p,q)-property is k-pierceable. We set $J(p,q;\mathcal{G}) := \infty$ if no finite such number exists.

In spite of the generality of these definitions, the present survey will mostly deal with the case in which X is the d-dimensional real space R^d and $\mathcal{G}$ is a suitable family of convex sets in R^d. When $\mathcal{G}$ consists of *all* convex sets in R^d, the (p,q)-problem is the original Hadwiger-Debrunner problem stated in the beginning. It is the topic of Section 2. Sections 3 and 4 will consider families of axis-parallel boxes in R^d, resp., families of homothets or translates of a given convex set in R^d. Compared to these 'classical' problems, the (p,q)-problem for d-intervals and homogeneous d-intervals considered in Section 5 is of a different nature. It has recently seen remarkable progress. Finally, in Section 6, the (p,q)-problem will be studied in an abstract, or combinatorial, setting. This part of the theory is strongly tied to the original Hadwiger-Debrunner problem in Section 2.

We point out that the two 'extremal' instances of the (p,q)-problem, namely, the cases $p = q$ and $q = 2$, have received special attention in the literature. Recall that $\mathcal{F} \in \Pi(p)$ simply means that any p or fewer members of $\mathcal{F}$ intersect. The (p,p)-problem is said to be a problem of *Gallai-type* (or of *Helly-type* if the piercing number turns out to be 1); it has been considered even before the general (p,q)-problem was introduced. The question of T. Gallai which motivated this terminology asks for the value of $J(2,2;\mathcal{H}(B^2))$, where $\mathcal{H}(B^2)$ is the family of circular disks in the plane. (For the answer, see (4.6) below.)

If $p > q$ holds, then the (p, q)-problem has a rather different flavor. One of the intricacies—and challenges—in this case is the fact that although it is known that *some* q out of every p sets in $\mathcal{F}$ intersect, it is a priori not known *which* q sets do. In this respect, the weakest assumption one can make is $\mathcal{F} \in \Pi(p, 2)$, i.e., $\nu(\mathcal{F}) < p$. The $(p, 2)$-problem thus amounts to estimating $\tau(\mathcal{F})$ in terms of $\nu(\mathcal{F})$, a paradigmatic problem in discrete geometry, combinatorics and graph theory.

In what follows, we will be using to some extent material from hypergraph (and graph) theory. Rather than developing this subject in any detail, we shall refer the reader to the excellent survey on hypergraphs by Duchet [26] where all the relevant results are described and appropriate references are given, and to the book by Lovász and Plummer [67].

2 The (p, q)-problem for convex sets in R^d

In this section, the ground family is the family $\mathcal{C}^d$ of all convex sets in R^d. Define

$$M(p, q; d) := J(p, q; \mathcal{C}^d), \quad p \geq q \geq 2.$$

This is the function originally studied by Hadwiger and Debrunner [47]. We begin with some easy observations:

$$M(p, q; d) \geq p - q + 1, \tag{2.1}$$
$$M(p, q; 1) = p - q + 1, \tag{2.2}$$
$$M(p, p; d) = 1, \ p \geq d + 1, \tag{2.3}$$
$$M(p, q; d) = \infty, \ q \leq d. \tag{2.4}$$

Here (2.1) is a universal lower bound which also applies to the other (p, q)-problems considered in this survey. It is demonstrated by any family of $p - q + 1$ pairwise disjoint sets of which one is taken with multiplicity q. Note that (2.1) is not meant to imply that $M(p, q; d)$ is finite. Proofs of (2.2) can be found in [47], [48] and [49]; a particularly nice argument appears in Károlyi [57]. The case $q = 2$ is equivalent to saying that $\tau(\mathcal{F}) = \nu(\mathcal{F})$ for any finite family $\mathcal{F}$ of intervals on the line. This is usually attributed to Gallai (see [51], where a more general graph-theoretic result is proved, and [42]). Assertion (2.3) is Helly's theorem. To see that (2.4) is true, take any family $\mathcal{F}$ of n hyperplanes in general position in R^d. Then $\mathcal{F} \in \Pi(d)$ and $\tau(\mathcal{F}) = \lceil n/d \rceil$. As n can be made arbitrarily large this shows that $M(d, d; d) = \infty$, which is enough.

Throughout the section, we therefore assume that $p \geq q \geq d+1$. In terms of $M(p, q; d)$, the Hadwiger-Debrunner (p, q)-theorem reads

$$M(p, q; d) = p - q + 1, \quad p < \frac{d}{d-1}(q - 1), \tag{2.5}$$

where it is understood that the right-hand condition is empty when $d = 1$. Thus $M(p,q;d)$ attains the bound in (2.1) if $d = 1$, or if (p,q) lies in a certain sector of the p,q-plane (which gets smaller as d increases).

For an application of (2.5) to separation and illumination properties of sets in R^d, see Netrebin [72]. Results inspired by (2.5) can be found in Trotter [82] and Dol'nikov [25].

For $d = 2$, the proof of (2.5) is reproduced in [48] and [49]. A sketch of the reasoning in R^d is as follows.

Call a pair (p,q) *admissible* if p and q satisfy the restrictions in (2.5). The proof is by induction on p. If $p = d+1$ or more generally, $p = q$, then (2.3) applies. Hence suppose $p > q$. If $(d-1)p < d(q-1)-1$, then $(p-1,q-1)$ is also admissible; since $\Pi(p,q)$ implies $\Pi(p-1,q-1)$, the assertion follows. Assume, then, that $(d-1)p = d(q-1)-1$. In this case, both $(p-1,q)$ and $(p-d,q-d+1)$ are admissible. Let $\mathcal{F}$ be a finite family of convex sets in R^d with $\mathcal{F} \in \Pi(p,q)$. Suppose, without loss of generality, that the members of $\mathcal{F}$ are compact. There is a convex polytope $P \subset R^d$ of minimum volume which meets every member and every nonempty intersection of members of $\mathcal{F}$. Set $\mathcal{F}_0 = \{K \cap P : K \in \mathcal{F}\}$. Then $\mathcal{F}_0 \in \Pi(p,q)$, and so we may as well assume that $\mathcal{F} = \mathcal{F}_0$ and P is d-dimensional. Choose a vertex v of P and define $\mathcal{F}' = \{K \in \mathcal{F} : v \in K\}$ and $\mathcal{F}'' = \mathcal{F} \setminus \mathcal{F}'$. If $\{v\} \in \mathcal{F}'$, then $\mathcal{F}'' \in \Pi(p-1,q)$ and by the induction hypothesis, $\tau(\mathcal{F}') \leq p-q$. If, on the other hand, $\{v\} \notin \mathcal{F}'$, then there is a hyperplane $H \subset R^d$ strictly separating v from the sets in $\mathcal{F}''$ and intersecting each set in $\mathcal{F}'$. Clearly, the fact that P has minimum volume implies that the sets $K \cap H$ with $K \in \mathcal{F}'$ have empty intersection. By Helly's theorem applied to H, some d or fewer such sets have empty intersection. From this it can be deduced that $\mathcal{F}'' \in \Pi(p-d,q-d+1)$ and therefore, again by the induction hypothesis, $\tau(\mathcal{F}'') \leq p-q$. Since $\mathcal{F}'$ can be pierced by v, it follows in both cases that $\tau(\mathcal{F}) \leq p-q+1$ which in view of (2.1) is enough. □

Before proceeding we like to mention that the conclusion of theorem (2.5) can sometimes be strenghtened in a certain direction. For example, if $\mathcal{F}$ has the $(p,p-1)$-property with $p = \lfloor (d+3)^2/4 \rfloor$, then $\mathcal{F}$ is not only 2-pierceable but with at most one exception all the members of $\mathcal{F}$ intersect. Moreover, the above p is the smallest for which the statement is true. This somewhat mysterious result is due to Perles [74] and (in a special case) Nadler [70] but follows in fact from a combinatorial theorem of Erdős and Gallai [31]. The broader context to which the result belongs will be discussed in Section 6.

Returning to the Hadwiger-Debrunner function we point out that, as of this writing, no other values of $M(p,q;d)$ except those appearing in (2.5) have been determined. Hadwiger [44] had asked (led by the corresponding result $N(4,3;2) = 2$ for boxes in Section 3) whether $M(4,3;2) = 2$ was true. This was immediately refuted by Danzer [loc.cit.] who presented a family of six congruent triangles in the plane to the effect that $M(4,3;2) \geq 3$, if indeed this number exists. Another such family consisting of nine congruent

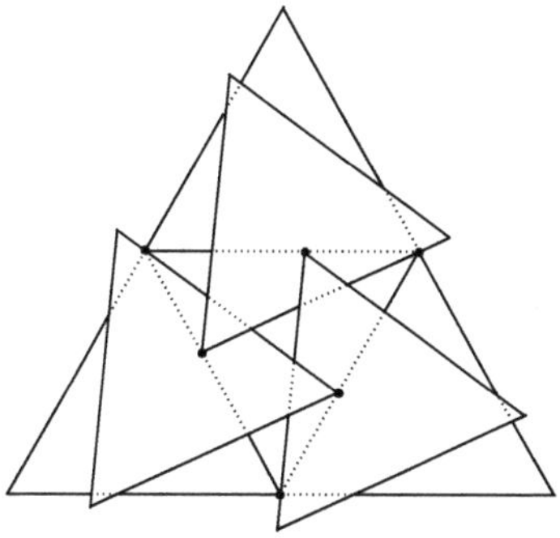

Fig. 1. Danzer's example

disks appears in Grünbaum [38] (see Fig. 11 in [49]). Danzer's example is shown in [44], [48] and (incorrectly described) [49]. A version with equilateral triangles is depicted in Fig. 1. Since no four of the triangles intersect, one simply checks that no two of the marked points where three triangles meet pierce them all. It is now known that $M(4,3;2)$ is indeed finite; we shall see upper bounds on it at the end of this section.

In any case, it is not true that $M(p,q;d) = p-q+1$ whenever the number exists. But more important, the very existence of $M(p,q;d)$ in all cases not covered by (2.4) or (2.5) remained undecided for 35 years.

A first attack on the existence question was made by Fullbright [36] in 1974. He considered finite families $\mathcal{F} \subset \mathcal{C}^d$ having the (p,q)-property and an additional property measured by α, with $0 \leq \alpha \leq 1$. The latter requires that for each $K \in \mathcal{F}$ there exist a point $a \in K$ and numbers $\alpha_1, \alpha_2 \in \mathbb{R}_+$ such that $\alpha_1 C^d + a \subset K \subset \alpha_2 C^d + a$ and $\alpha_1/\alpha_2 \geq \alpha$. Here C^d is the unit cube in R^d; for simplicity, it is assumed that K is compact and has interior points. In a sense, α measures the "cubeness" of the members of $\mathcal{F}$. Set $M_\alpha(p,q;d) := sup_{\mathcal{F}}\, \tau(\mathcal{F})$, where the supremum is over the above families. It is easily seen that $M_0(p,q;d) = M(p,q;d)$. Also, as a function of α to the positive integers union ∞, $M_\alpha(p,q;d)$ is decreasing and continuous from the right at each $\alpha < 1$. The main result in [36] asserts that $M_\alpha(p,q;d)$ is finite if $p \geq q \geq 2$ and $\alpha > 0$. The case $\alpha = 0$ had to be left open.

In 1992, Alon and Kleitman [6] announced that they had established the existence of $M(p,q;d)$ for all $p \geq q \geq d+1$, that is,

$$M(p,q;d) < \infty, \quad p \geq q \geq d+1. \tag{2.6}$$

Their proof, presented in detail in Alon and Kleitman [7], is a striking achievement and, although not long, a veritable tour de force. It has started a new chapter in the history of the (p,q)-problem.

We shall briefly describe the main ingredients in the proof of (2.6) and the way they are assembled. (This is probably not the order in which they were discovered.) It is sufficient to consider the case $q = d+1$. Let $\mathcal{F}$ be a finite family of convex sets in R^d. One of the central ideas in the proof is to work with families $\mathcal{G}$ obtained from $\mathcal{F}$ by "cloning", that is, by duplicating some of the members. Let $|\mathcal{G}|$ denote the total number of copies of members of $\mathcal{F}$ in $\mathcal{G}$. Similarly, if Y is a finite multiset with underlying set $X \subset R^d$, then $|Y|$ will be the number of points of X counted with multiplicities. The first step is to establish

(A) There is a positive real number $\beta = \beta(p,d)$ such that the following holds: If $\mathcal{F} \in \Pi(p, d+1)$ and $\mathcal{G}$ arises from $\mathcal{F}$ by cloning, then there is a point $x \in R^d$ which lies in at least $\beta\,|\mathcal{G}|$ members of $\mathcal{G}$.

In proving (A), the authors make use of the "fractional" Helly theorem due to Katchalski and Liu [59]. They supply an explicit value for β but make no attempt, here and in the sequel, to optimize the constants. Without cloning, a simple combinatorial argument shows that at least $\binom{m}{d+1}/\binom{p}{d+1}$ subfamilies of $\mathcal{F}$ of size $d+1$ intersect, where $m = |\mathcal{F}|$. Hence, by [59], β can be taken to be $((d+1)\binom{p}{d+1})^{-1}$. After cloning, $\mathcal{G} \in \Pi(p, d+1)$ need no longer hold but it is readily seen that $(2(d+1)\binom{p}{d+1})^{-1}$ will still do. (This improves a bit on [7].) The next step is

(B) Let $\beta = \beta(p,d)$ be the constant in (A). If $\mathcal{F} \in \Pi(p, d+1)$, then there is a finite multiset $Y \subset R^d$ such that for all $K \in \mathcal{F}$, $|K \cap Y| \geq \beta\,|Y|$.

It turns out that (A) and (B) are equivalent in the following sense. Associate with $\mathcal{F}$ a hypergraph $H = (X, Y)$ by choosing a finite set $X \subset R^d$ which contains one point from each nonempty intersection of members of $\mathcal{F}$ and letting $Y = \{K \cap X : K \in \mathcal{F}\}$. Then (A), (B) can be expressed as saying

$$\forall g : Y \to \mathbb{Z}_+ \ \exists x \in X : \sum_{x \in y} g(y) \geq \beta \sum_{y} g(y)\,,$$

$$\exists f : X \to \mathbb{Z}_+ \ \forall y \in Y : \sum_{x \in y} f(x) \geq \beta \sum_{x} f(x)\,.$$

Here $g(y)$ is the number of clones of the set $K \in \mathcal{F}$ corresponding to $y \in Y$ and $f(x)$ is the multiplicity of $x \in X$ in Y (so X underlies Y). Clearly, $\mathbb{Z}_+$ can be replaced by $\mathbb{Q}_+$ and even by $\mathbb{R}_+$ in these relations, the latter by continuity. In hypergraph language, the first relation states that $\nu^*(H) \leq 1/\beta$ and the second that $\tau^*(H) \leq 1/\beta$, where $\nu^*(H)$ and $\tau^*(H)$ are the *fractional* matching and transversal number of H, respectively. (See Duchet [26], Section 4.) That the assertions are equivalent follows from the fact that $\nu^*(H) = \tau^*(H)$. This in turn is a consequence of the duality theorem of linear programming.

For the last step of the proof, assume that $0 < \epsilon < 1$.

(C) There is a positive integer $k = k(\epsilon, d)$ such that the following is true: If Y is a finite multiset in R^d, then the family $\{\mathrm{conv} A : A \subset Y, |A| \geq \epsilon\,|Y|\}$ is k-pierceable.

Any transversal of the family described in (C) is called a *weak ϵ-net* for Y. What is remarkable here is that its size is independent of the size of Y. Several proofs of (C) leading to various upper bounds on $k(\epsilon,d)$ are given in Alon et al. [3]. One is based on a "selection theorem" of Bárány [9] which itself employs the celebrated Tverberg theorem. It yields a weak ϵ-net of size $O(\epsilon^{-(d+1)})$ for every fixed d. A slightly better bound is obtained in [3] by using a more sophisticated selection theorem. In turn, this has been improved for $d \geq 3$ by Chazelle et al. [16] who establish a bound close to $O(\epsilon^{-d})$. In the plane, a separate argument in [3] yields $k(\epsilon,2) \leq 7/\epsilon^2$.

The finiteness of $M(p,q;d)$ is now an immediate consequence of (A), (B) and (C) since we clearly have $M(p,q;d) \leq M(p,d+1;d) \leq k(\beta(p,d),d)$, $p \geq q \geq d+1$. □

A lucid explanation of the proof is given in Section 10.6 of the recent book by Matoušek [69].

It is almost safe to say that the bound on $M(p,q;d)$ obtained above is much too large. Even slightly smaller ϵ-nets could reduce it substantially. However, both $k(\epsilon,d)$ and $\beta(p,d)$ are far from being well known.

In 1996, Alon and Kleitman [8] published a more elementary and self-contained version of their proof. They replaced the rather sophisticated tools from convex geometry by purely combinatorial methods. For example, Bárány's theorem quoted above is replaced by a simple argument which leads to an ϵ-net of size $O(\epsilon^{-2^d})$ for $d \geq 3$. To derive (B) from (A) without using duality theory, they employ a combinatorial cloning argument due to Welzl (also described in [7]). While the β in (B) has to be lowered to $1-\log_2(2-\beta)$, the advantage is that there is now a constructive procedure for obtaining Y.

Already in [7], Alon and Kleitman had indicated how their estimate for $M(p,q;d)$ could be improved in a few isolated cases, thanks to known bounds for Turán's hypergraph problem. Such a case is the notorious $M(4,3;2)$ for which the general proof yields $M(4,3;2) \leq 4032$. Recall that the Turán number $T(m,4,3)$ is the smallest possible number of 3-element subsets of a set of size m such that every 4-element subset contains at least one of them. The sequence $T(m,4,3)/\binom{m}{3}$ increases as $m \to \infty$, and Turán conjectured that its limit (which exists) is $4/9$. The strongest partial result known at present is $T(m,4,3)/\binom{m}{3} \geq 0.4064$ for large m due to Chung and Lu [17].

Now let $\mathcal{F}$ be a family of convex sets in the plane having the $(4,3)$-property. If, in addition, $\mathcal{F} \in \Pi(2)$, then any family $\mathcal{G}$ obtained from $\mathcal{F}$ by cloning also has the $(4,3)$-property. This shows that if $m := |\mathcal{G}|$ is large (which may be assumed), then $0.4064\binom{m}{3}$ triples of sets in $\mathcal{G}$ are intersecting. This allows us to bypass the proof of (A) and directly apply the sharp quantitative version of the fractional Helly theorem due to Kalai [55]. We deduce that some point of R^2 lies in at least βm members of $\mathcal{G}$, where $\beta = 1-(1-0.4064)^{1/3} > 0.1595$. Of course, $\mathcal{F} \in \Pi(2)$ need not hold but after deleting two disjoint members, the remaining sets are pairwise intersecting, thanks to $\Pi(4,3)$. In any case, the conclusion is that $M(4,3;2) \leq \lfloor 7/0.1595^2 \rfloor + 2 = 276$.

If Turán's conjecture turned out to be correct, then 276 could be replaced by 223. But much more can now be said. Using a clever and completely elementary approach, Kleitman, Gyárfás and Tóth [61] have recently proved that $M(4,3;2) \leq 13$. This is a remarkable break-through and may well be close to the truth. So we have

$$3 \leq M(4,3;2) \leq 13\,. \tag{2.7}$$

The reasoning in [61] seems taylor-made for dealing just with the $(4,3)$-problem. It remains to be seen whether the approach can be exploited for other instances of $M(p,q;d)$ as well.

To conclude, we remark that the Alon-Kleitman technique has been applied to various analogs and generalizations of theorem (2.6). Alon [1] has used it in the case of homogeneous d-intervals which will be treated in Section 5. Alon and Kalai [4] have extended (2.6) to families of "polyconvex" sets. They show that if $\mathcal{C}^{d,n}$ is the family of all unions of at most n convex sets in R^d, where n is a positive integer, then $J(p,q;\mathcal{C}^{d,n}) < \infty$, $p \geq q \geq d+1$. They also prove a variant of (2.6) for common hyperplane transversals in R^d. (See also Matoušek [69].) More recently, Alon et al. [5] have examined the principles which underly the proof of theorem (2.6) in an abstract setting. They show that an appropriate "fractional Helly theorem" is sufficient for establishing the existence of weak ϵ-nets and hence the validity of a (p,q)-theorem. One consequence they derive is a topological (p,q)-theorem which states that (2.6) continues to hold for finite good covers in R^d. A family of subsets of R^d is called a *good cover* if its members are all open or all closed and each nonempty intersection of members is contractible. The required fractional Helly theorem directly generalizes that of Kalai [55] but its proof is much deeper. Another result in [5] is that (2.6) remains valid for finite families of convex lattice sets in R^d. These are the sets of the form $K \cap \mathbb{Z}^d$, where $K \in \mathcal{C}^d$ and $\mathbb{Z}^d$ is the d-dimensional integer lattice. To be precise, the authors had to assume that $p \geq q \geq 2^d$ holds but conjectured that $p \geq q \geq d+1$ would do (and then be best possible). Meanwhile, Bárány and Matoušek [10] have confirmed this conjecture by showing that $d+1$ is the correct "fractional Helly number" for convex lattice sets in R^d. Note that two "colored" (p,q)-theorems (in the sense of [9]) for convex sets and for convex lattice sets in R^d are also proved in [10].

3 The (p,q)-problem for boxes in R^d

Let $\mathcal{Q}^d$ denote the family of all boxes in R^d with edges parallel to the coordinate axes. For brevity, the members of $\mathcal{Q}^d$ will be called axis-parallel boxes, or simply *boxes*.

The (p,q)-problem for boxes in R^d was first studied in Hadwiger and Debrunner [48] and Hadwiger, Debrunner and Klee [49] (strictly speaking, for $d = 2$ only). Letting

$$N(p,q;d) := J(p,q,\mathcal{Q}^d), \quad p \geq q \geq 2,$$

we have the basic results

$$N(p,q;d) \geq p - q + 1, \tag{3.1}$$
$$N(p,q;1) = p - q + 1, \tag{3.2}$$
$$N(p,p;d) = 1. \tag{3.3}$$

Here (3.2) is the same as (2.2), while (3.3) (more precisely, $N(2,2;d) = 1$) is Helly's theorem for boxes in R^d. It shows that the intersection properties of a family of boxes are completely determined by its intersection graph.

Due to the simple shape of boxes, and very much opposed to the effort required in the corresponding situation in Section 2, it is not hard to show that $N(p,q;d)$ is finite for all $p \geq q \geq 2$. In fact, we have

$$N(p,q;d) \leq \binom{p-q+d}{d}, \quad p \geq q \geq 2. \tag{3.4}$$

The upper bound here is achieved for $d = 1$, for $p = q$, and possibly for $p = 3$, $q = 2$. (See (3.14) for the latter case.) In most instances, however, it is very crude.

The proof of theorem (3.4) in [48] and [49] is given for the planar case but can easily be extended to higher dimensions. It proceeds by induction on d and p. Let $\mathcal{F} \subset \mathcal{Q}^d$ be a finite family of boxes having the (p,q)-property. Assume, by (3.2) and (3.3), that $d > 1$ and $p > q$. Choose a closed half-space $H \subset R^d$ whose bounding hyperplane E is parallel to some coordinate hyperplane and such that H supports some member of $\mathcal{F}$, while no member of $\mathcal{F}$ lies in the interior of H. Set $\mathcal{F}' = \{Q \in \mathcal{F} : Q \cap E \neq \emptyset\}$ and $\mathcal{F}'' = \mathcal{F} \setminus \mathcal{F}'$. Then $\mathcal{F}''$ has the $(p-1,q)$-property and $\mathcal{F}'$ can be regarded as a family of boxes in R^{d-1}, since its intersection graph is the same as that of $\{Q \cap E : Q \in \mathcal{F}'\}$. This implies $\tau(\mathcal{F}') \leq N(p,q;d-1)$ and $\tau(\mathcal{F}'') \leq N(p-1,q;d)$. Hence, by the induction hypothesis and the recurrence relation for binomial coefficients, the assertion follows. (Strictly speaking, the argument requires that $|\mathcal{F}'| \geq p$ and $|\mathcal{F}''| \geq p-1$ but this can be assumed thanks to (3.1).) □

Now that the existence of $N(p,q;d)$ in all cases is guaranteed, the focus is on evaluating these numbers or obtaining good lower and upper bounds. In [48] and [49] it is proved that

$$N(p,q;d) = p - q + 1, \quad 2 \leq q \leq p \leq 2q - 2, \tag{3.5}$$

independently of d. For a shorter proof, see Grünbaum [39]. Actually, this is essentially a combinatorial result (see (6.1) below) and has been established by many authors.

After the pioneering work of Hadwiger and Debrunner, the (p, q)-problem for boxes was taken up and intensively studied by Wegner [85, 87] and independently by Dol'nikov [23]. More recently, major contributions have come from Károlyi [57], Fon-Der-Flaas and Kostochka [35] and Scheller [76]. These authors rediscovered or slightly improved on several of Wegner's results (some of which have not been published).

Let us first deal with the case $q = 2$, i.e., the problem of bounding $\tau(\mathcal{F})$ in terms of $\nu(\mathcal{F})$. Theorem (3.4) shows that $N(p, 2; d) = O(p^d)$ for any fixed d but this can be considerably improved. At present, the best known upper bound on $N(p, 2; d)$ results from the following recurrence relation which was found independently by Wegner [87] (for $d = 2$) and Fon-Der-Flaas and Kostochka [35]:

$$N(p, 2; d) \leq N(\lfloor p/2 \rfloor, 2; d) + N(\lceil p/2 \rceil, 2; d) + N(p, 2; d-1). \tag{3.6}$$

Here we assume $d \geq 2$ and set $N(1, 2; d) = 0$. An outline of the proof is as follows.

Suppose $\mathcal{F} \subset \mathcal{Q}^d$ is a finite family with $\mathcal{F} \in \Pi(p, 2)$, i.e., $\nu(\mathcal{F}) < p$. For a given real number a, let $\mathcal{F}'_a$ and $\mathcal{F}''_a$ be the subfamilies of all members of $\mathcal{F}$ which lie in the open half-spaces of R^d defined by $x_1 < a$, resp., $x_1 > a$. (Here x_1 is one of the coordinates of R^d.) Now either $\nu(\mathcal{F}) < \lfloor p/2 \rfloor$, in which case $\tau(\mathcal{F}) \leq N(\lfloor p/2 \rfloor, 2; d)$ holds, or $\nu(\mathcal{F}) \geq \lfloor p/2 \rfloor$. In the latter case, there is a smallest value a_0 such that the closed half-space given by $x_1 \leq a_0$ contains $\lfloor p/2 \rfloor$ pairwise disjoint members of $\mathcal{F}$. This in turn implies $\nu(\mathcal{F}'_{a_0}) < \lfloor p/2 \rfloor$ and $\nu(\mathcal{F}''_{a_0}) < \lceil p/2 \rceil$, i.e., $\tau(\mathcal{F}'_{a_0}) \leq N(\lfloor p/2 \rfloor, 2; d)$ and $\tau(\mathcal{F}''_{a_0}) \leq N(\lceil p/2 \rceil, 2; d)$. Finally, let $\mathcal{F}_{a_0}$ be the subfamily whose members intersect the hyperplane $x_1 = a_0$. As far as their intersection properties are concerned, the sets in $\mathcal{F}_{a_0}$ can be regarded as if lying in the hyperplane. Hence $\tau(\mathcal{F}_{a_0}) \leq N(p, 2; d-1)$, and the proof is complete. Again by (3.1), we need not worry whether the subfamilies in question have sufficiently many members. □

From (3.6) it follows that $N(p, 2; d) = O(p \log_2^{d-1} p)$ as a function of p, for any given d. Using a completely different approach, Károlyi [57] has obtained the same order of magnitude (with slightly weaker constants).

In the well-studied case of parallel rectangles in the plane, the recurrence in (3.6) has the following solution, by virtue of (3.2):

$$N(p, 2; 2) \leq p\lceil \log_2 p \rceil - 2^{\lceil \log_2 p \rceil} + 1, \quad p \geq 2\,. \tag{3.7}$$

(Note that Fon-Der-Flaas and Kostochka [35] fail to draw the best possible consequences from (3.6).) We will come back to (3.7) in a moment.

Turning to the case $q > 2$, it appears that the strongest bounds on $N(p, q; d)$ are currently obtained by exploiting a recursive algorithm due to

Wegner [85] and Dol'nikov [23]. The method is combinatorial in nature . We content ourselves with describing the planar case. The bound can be written as

$$N(p, q; 2) \leq \max_{1 \leq \beta \leq p-q+1} \min(X_1, X_2, X_3), \tag{3.8}$$

where $X_1 = N(\beta+1, 2; 2)$, $X_2 = N(p-\beta, q-1; 2) + \beta - 1$, and X_3 is itself a max-min term, namely,

$$X_3 = \max_{\substack{1 \leq \alpha \leq p-q-\beta+2 \\ \alpha \leq \beta}} \min(N(\alpha+1, 2; 2) + \beta - 1, N(p-\beta-\alpha, q-2; 2) + \beta).$$

A short sketch of (3.8) is as follows. Suppose $\mathcal{F} \subset \mathcal{Q}^2$ has the (p, q)-property, and let $\beta = \nu(\mathcal{F})$. Clearly, $\tau(\mathcal{F}) \leq X_1$. If $\mathcal{F}'$ arises from $\mathcal{F}$ by removing β pairwise disjoint members, then $\mathcal{F}'$ has the $(p-\beta, q-1)$-property and so $\tau(\mathcal{F}') \leq N(p-\beta, q-1; 2)$. As $N(2, 2; 2) = 1$ and each set in $\mathcal{F}'$ intersects at least one of the discarded sets, one of the points piercing $\mathcal{F}'$ is easily seen to be redundant. This gives $\tau(\mathcal{F}) \leq X_2$. Now repeat the procedure by removing α pairwise disjoint members of $\mathcal{F}'$, where $\alpha = \nu(\mathcal{F}')$. Clearly, $\alpha \leq \min(\beta, p-q-\beta+2)$, and the resulting family $\mathcal{F}''$ has the $(p-\beta-\alpha, q-2)$-property. The $\alpha+\beta$ sets removed altogether can be pierced by β points. This is a consequence of Hall's marriage theorem. Indeed, the intersection graph of $\mathcal{F} \setminus \mathcal{F}''$ is bipartite and so its stability number β equals its edge covering number, i.e., the minimum number of edges required to cover all vertices (see Lovász and Plummer [67], Cor. 1.1.7). If follows that $\tau(\mathcal{F}) \leq N(p-\beta-\alpha, q-2; 2) + \beta$. This in turn proves $\tau(\mathcal{F}) \leq X_3$ and establishes the assertion. □

We shall call (3.8) the *reduction lemma* in what follows. It can obviously be applied to other (p, q)-functions treated in this survey, with one modification. Since the analog of $N(2, 2; 2) = 1$ is then no longer true, the terms $\beta - 1$ in X_2 and X_3 must be replaced by β.

The reduction lemma is a powerful tool although difficult to work with computationally. We demonstrate its usefulness with a few examples. Wegner [85] (in a slightly weaker form) and Dol'nikov [23] have shown that if $p < \binom{q+1}{2}$, then $N(p, q; 2) = p-q+1$. This follows from $N(p, 2; 2) \leq \binom{p}{2}$ (see (3.4)) and improves on the Hadwiger-Debrunner result (3.5). But (3.7) and (3.8) now yield a considerably larger (p, q)-sector in which the same conclusion holds. Define $s(q)$ to be the largest integer for which the reduction lemma implies $N(p, q; 2) = p - q + 1$, $2 \leq q \leq p \leq s(q)$. Then the following table is obtained (see Scheller [76]):

q	2	3	4	5	6	7	8	9	10	11	12
$s(q)$	2	5	14	33	81	169	360	684	1347	2466	4588

It would be interesting to estimate $s(q)$ analytically or determine its asymptotic behavior as $q \to \infty$. Supplementing the table we have

$$N(6, 3; 2) = 4, \tag{3.9}$$

as was shown by Dol'nikov [23] in a more general graph-theoretic setting. The nice geometric proof of (3.9) in Grünbaum [39] seems flawed but can easily be repaired.

We refer to Scheller [76] for an extensive numerical treatment of the consequences of the reduction lemma. Two further examples may suffice. If we abbreviate the right-hand side of (3.7) by $f(p)$, we get

$$N(p,3;2) \le \max_{1\le\beta\le p-2} \min(f(\beta+1), f(p-\beta)+\beta-1) \tag{3.10}$$

and, after some simplification,

$$N(p,4;2) \le \max_{1<\gamma\le\lfloor p/2\rfloor} \min(f(p-2\gamma+2), f(\gamma)+p-2\gamma). \tag{3.11}$$

For instance, this yields $N(100,3;2) \le 266$ and $N(100,4;2) \le 165$ which should be compared with the bounds obtainable from (3.4).

It is clear that better estimates for $N(p,2;2)$ will produce a larger sector function $s(q)$ and smaller upper bounds for $N(p,q;2)$ in general. There is a longstanding belief, and some evidence, that $N(p,2;2) = O(p)$ (see Wegner [85], Gyárfás and Lehel [42]). More precisely, Wegner raised the possibility that the following might hold:

$$\textit{Conjecture:}\quad N(p,2;2) \le 2p-3, \quad p \ge 2\,. \tag{3.12}$$

If true, (3.12) would have very strong implications, namely, $N(p,3;2) \le \lfloor 4p/3 \rfloor - 3$, $p \ge 3$, and $N(p,q;2) = p-q+1$, $p \ge q \ge 4$. For the latter, see [85] (where the conclusion is slightly weaker) and [23]. These assertions follow at once by applying (3.10) and (3.11) to $f(p) = 2p-3$. For example, $N(p,4;2) \le f(\gamma)+p-2\gamma = p-3$, and since $\Pi(p,q)$ implies $\Pi(p-1,q-1)$, the rest is clear.

As for evidence supporting (3.12), we remark that the analogous result for axis-parallel squares is true (see [23]). Also, as announced in footnote (6) of Wegner [85], we have

$$N(4,2;2) = 5. \tag{3.13}$$

Here $N(4,2;2) \le 5$ follows from (3.7). To show that the bound is sharp, Wegner [87] uses the example illustrated in Fig. 2. Consider the square $Q = \{(x_1,x_2) : |x_1|,|x_2| \le 1\}$ in the x_1,x_2-plane. Let the points $a_1,\ldots,a_{20}$ divide the boundary of Q into equal parts, starting with a vertex and going in clockwise direction, say. Define A_i as the smallest axis-parallel rectangle containing a_i and a_{i+5} (subscripts mod 20) and define $B = \{(x_1,0) : |x_1| \le 1\}$, $C = \{(0,x_2) : |x_2| \le 1\}$, and $D = \lambda Q$ with some fixed $\lambda \in (\frac{3}{5},1)$. A few of these sets are indicated in the figure. Note that $A_1, A_6, A_{11}, A_{16}, B$ and C are degenerate rectangles. Setting $\mathcal{F} = \{A_1,\ldots,A_{20},B,C,D\}$, it can be shown that $\mathcal{F} \in \Pi(4,2)$ and $\tau(\mathcal{F}) = 4$. Under a slightly different guise, this example was found independently by Fon-Der-Flaas and Kostochka [35].

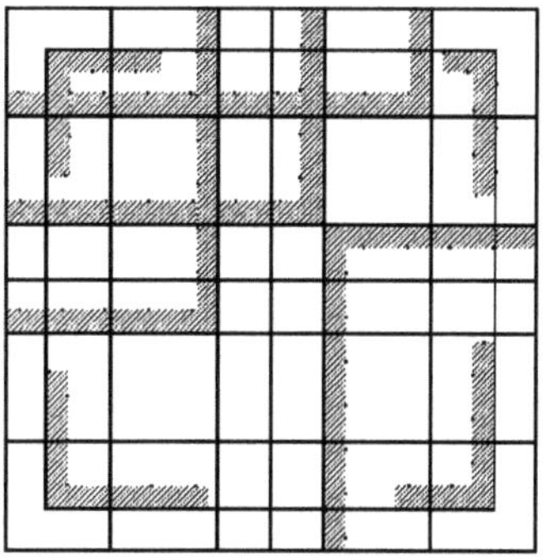

Fig. 2. Wegner's example

In passing, we like to mention yet another consequence of Wegner's conjecture. Larman et al. [62] have shown (among others) that if $\mathcal{F}$ is a family of n parallel rectangles in the plane, then one can always choose $\sqrt{n/(2\log_2 n)}$ of them which are either pairwise disjoint or pairwise intersecting. Using the upper bound in (3.7), this number can be doubled. If (3.12) were true, it could be increased to $\sqrt{n/2}$.

To conclude, let us briefly comment on two aspects of the (p, q)-problem for boxes in higher dimensions. It follows from (3.6) or already from (3.4) that

$$N(3, 2; d) \leq d + 1. \tag{3.14}$$

Here equality holds at least for $d \leq 4$. For $d = 2$ see [48], [49] and [42], for $d = 3$ see [85], [39] and for $d = 3, 4$ see [35]. The necessary examples are constructed in a uniform fashion by representing the complement of the well-known $(d + 1)$-chromatic Mycielski graph M_{d+1} as the intersection graph of a family of boxes in R^d $(d \leq 4)$. Since M_{d+1} is triangle-free, its complement has the (3,2)-property. However, it is unknown whether such a representation exists for $d > 4$. It is quite possible that equality holds in (3.14), or that $N(3, 2; d) = o(d)$ as $d \to \infty$. The best known lower bound is

$$N(3, 2; d) \geq c\sqrt{d}/\log d \tag{3.15}$$

for some positive constant c, derived from Erdős' classical result that there exist k-chromatic triangle-free graphs on $O(k^2(\log k)^2)$ vertices (see [35]).

Another interesting question is whether, for any given d, $N(p, 2; d) = O(p)$ is true. This seems a bold conjecture in view of what is known today but there is a comparable geometric situation in which it has been possible to bridge the gap (see (5.3) below). The answer is "yes" when families of homothets of a given box are considered (see [23]). According to Dol'nikov, an affirmative answer implies that $N(p, q; d) = p - q + 1$, $p \geq q \geq q(d)$, where $q(d)$ is some 'threshold' function. The proof given in [23] appears to be incorrect, but the result itself is most likely true.

4 The (p,q)-problem for homothets and translates in R^d

In this section, the (p,q)-problem will be studied for set families in R^d whose members are homothets, or translates, of a given convex set K. These families are sufficiently well-behaved so that the corresponding (p,q)-numbers turn out to be finite for all $p \geq q \geq 2$. But unlike the situation encountered in the preceding section, reasonable bounds are not easily obtained and depend, of course, on the set K. The literature deals mainly with Gallai-type problems, that is, the case $p = q$. This part is older than the general (p,q)-theory. Even so, significant results exist only for special convex sets K such as the d-ball or the d-cube.

Let K be a compact convex set in R^d. A (positive) *homothet* of K is a set of the form $\alpha K + a$ with $a \in R^d$ and $\alpha \in \mathbb{R}, \alpha > 0$. A *translate* of K is a homothet with $\alpha = 1$. The family of all homothets and all translates of K will be denoted by $\mathcal{H}(K)$ and $\mathcal{T}(K)$, respectively. The dependence on d is tacitly understood. Define, for $p \geq q \geq 2$,

$$H(p,q;K) := J(p,q;\mathcal{H}(K)),$$

$$T(p,q;K) := J(p,q;\mathcal{T}(K)).$$

Clearly, $T(p,q;K) \leq H(p,q;K)$. It will be seen shortly that all these numbers are finite. In the literature, $H(p,p;K)$ and $T(p,p;K)$ are called the pth *Gallai number* of $\mathcal{H}(K)$ and $\mathcal{T}(K)$, respectively. In view of Helly's theorem in R^d, they are of interest only if $p \leq d$. Grünbaum [38] writes $h(K)$ and $H(K)$ for $T(2,2;K)$ and $H(2,2;K)$ and derives upper bounds for these numbers depending only on d. The same abbreviations are used in [12], [13], [14], [15] and [65].

Some special notation is convenient. For a bounded set $M \subset R^d$, define $[M/K]$ to be the smallest number of translates of K needed to cover M. The frequently used $[(K-K)/K]$ is written as $\bar{\gamma}(K)$. It is not hard to show that $\bar{\gamma}(K)$ equals the transversal number, in the worst case, of a family of translates of K in which some member meets every other member (see [22]). Upper bounds on $\bar{\gamma}(K)$ in terms of d alone were established by Grünbaum [38] and Danzer [18,19] and are collected in [22]. However, in most cases they are outdated by the probabilistic method of Erdős and Rogers [33] which yields

$$\bar{\gamma}(K) \leq 3^{d+1}2^d(d+1)^{-1}d(\log d + \log\log d + 4), \tag{4.1}$$

provided d is sufficiently large. Here $d(\log d + \log\log d + 4)$ is the density of an economical covering of R^d by translates of K, for each given K. In unpublished lecture notes, Danzer has replaced the constant 4 by $2 + r(d)$, where $r(d) \leq \log 2$ and $r(d) \to 0$ for $d \to \infty$. If K is centrally symmetric, better bounds are available.

The (p, q)-problem for families of *homothets* will be treated first. The basic observations here are

$$H(p,q;K) \leq H(q,q;K) + (p-q)\bar{\gamma}(K), \quad p > q \geq 2, \tag{4.2}$$
$$H(p,p;K) \leq H(2,2;K) \leq \bar{\gamma}(K), \quad p \geq 2. \tag{4.3}$$

The first is due to Wegner [86] and the second appears in [22]. Together they establish the existence of all piercing numbers considered in this section. Of course, the universal bound based on (4.1) is very rough. A lot more can be said in special cases such as $K = B^d$, where B^d is the unit ball in R^d. This case was thoroughly investigated by Danzer [18,19]. What is required are good estimates for $\bar{\gamma}(B^d) = [2B^d/B^d]$. Danzer observed that $\bar{\gamma}(B^d) \leq 1 + \kappa_d$, where κ_d is the minimum number of spherical caps of radius $\pi/6$ that will cover the unit sphere S^{d-1} in R^d. For example, he shows that $\kappa_2 = 6$, $\kappa_3 \leq 20$ and $\kappa_4 \leq 70$. His upper bound on $H(2,2;B^d)$ is of order $(1+\sqrt{3})^d$ as $d \to \infty$. For large d, this can be replaced by the Erdős-Rogers estimate

$$H(2,2;B^d) \leq 2^d \, d^{3/2} \sqrt{3\pi/2}\,(\log d + \log\log d + 1). \tag{4.4}$$

The idea of studying $H(2,2;B^d)$ by means of $\bar{\gamma}(B^d)$ was extended by Danzer [19] to $H(p,p;B^d)$. His main tool is a lemma which states (in the Euclidean case) that a family of balls having nonempty intersection covers the convex hull of the set of its centers. This enables him to intersect the given family with a suitable flat of dimension $d-p+2$ and thus reduce the dimension by $p-2$. Consequently,

$$H(p,p;B^d) \leq H(2,2;B^{d-p+2}) \leq [2\,B^{d-p+2}/B^{d-p+2}], \quad 2 \leq p \leq d. \tag{4.5}$$

This also holds under a certain condition weaker than $\Pi(p)$, see Proposition (7.12) in [22].

We now turn to the problem of evaluating $H(2,2;B^2)$. This is the original "Gallai's problem" after which this part of the (p, q)-theory is named. It was first published in the book by Fejes Tóth [34] and later in Hadwiger [45], Hadwiger and Debrunner [48] and Hadwiger, Debrunner and Klee [49]. Plainly, $H(2,2;B^2) \leq 7$. In fact, any family of pairwise intersecting disks in the plane can be pierced by the center of a smallest disk (of radius 1, say) and six equally spaced points at distance $\sqrt{3}$ from the center (see Fig. 7 in [22]). This was observed by Ungar and Szekeres. Later Heppes (unpublished) reduced the bound to 6, and then Stachó [78] to 5, the value that Gallai had expected. But already in a colloquium talk at the University of Munich in 1954, and again in July 1956 in Oberwolfach, Danzer had announced that he could lower the bound to 4. That a further reduction to 3 is not possible is shown by a family of 21 disks constructed in Grünbaum [38]. Hence

$$H(2,2;B^2) = 4. \tag{4.6}$$

This is still one of the most spectacular results in the area. The first published proof was given by Stachó [79], who used a refinement of his method in [78]. Then Danzer [21] also published a proof—not the original one—which showed a bit more, namely, that every family of pairwise intersecting spherical caps on the 2-shere is 4-pierceable. Both proofs are far too long and involved to be sketched in this survey. We refer to Danzer [21] for further details and a short history of the problem (see also Eckhoff [28]). Danzer's example demonstrating $H(2,2;B^2) \geq 4$ consists of only 10 disks. Since among every five pairwise intersecting homothets some three have a common point (see [86], [78]), any such example must have at leas nine members.

What can be said about the numbers $H(p,q;B^2)$? For $q=2$, this question was also asked by Gallai (see Erdős [30]). Using (4.6), Wegner [89] obtained

$$4p-4 \leq H(p,2;B^2) \leq 7p-10, \quad p \geq 2, \tag{4.7}$$

by elementary arguments. The planar case of theorem (2.5) yields $H(p,q;B^2) = p-q+1$, if $p \leq 2q-3$, and the reduction lemma of Section 3 leads to reasonably small upper bounds if $p \geq 2q-2$. One result may suffice, namely, $H(p,q;B^2) \leq p-q+2$ for $q \geq 14$.

We conclude this subsection by mentioning another isolated result, this time in R^d, due to Dol'nikov [23]. Let C^d denote the unit cube in R^d. Then

$$H(p,2;C^d) \leq 2^d(p-1)+1, \quad d \geq 2. \tag{4.8}$$

This is very easy to prove and probably far from being a good estimate, but it does show that $H(p,2;C^d) = O(p)$ for any fixed d. Hence by Theorem 3 in [23], there is a 'threshold' $q(d)$ such that $H(p,q;C^d) = p-q+1$, $2 \leq q \leq p \leq q(d)$.

Let us now consider the (p,q)-problem for families of *translates* of K which is equivalent (not just related, as above) to a covering problem. In fact, the pth Gallai number $T(p,p;K)$ equals the smallest number of translates of K needed to cover a set of points in R^d any p of which lie in some translate of K. See Proposition 7.6 in [22]; for centrally symmetric sets, see also Hadwiger and Schaer [50]. Again, most of the known results deal with the Euclidean case $K = B^d$. We refrain from reproducing the various bounds listed in [22] since they are outdated by the covering method of Erdős and Rogers. The most important results concern $T(2,2;B^d)$. This is the minimum number of unit balls in R^d that can cover, suitably translated, every set of diameter 2. The oldest result here is

$$T(2,2;B^2) = 3 \tag{4.9}$$

which appeared in [48] and [49]. A non-2-pierceable family of nine pairwise intersecting unit disks in R^2 constructed in [38] is also shown there. Incidentally, (4.9) is the only Gallai-type result in Hadwiger and Debrunner [46], the predecessor of [48]. For a strengthening, see Schopp [77]. As for higher dimensions, Grünbaum [38] had speculated whether $T(2,2;B^d) = d+1$ was always true. This was confirmed by Katzarowa-Karanowa [60] for $d=3$ but refuted by Danzer [20] for large d. Danzer's counterexample shows that $T(2,2;B^d) > 1.003^d$; hence $T(2,2;B^d)$ grows exponentially with d. More recently, Bourgain and Lindenstrauss [11] replaced the lower bound by $T(2,2;B^d) \geq 1.0645^d$ and established an upper bound of the form

$$T(2,2;B^d) \leq (\sqrt{3/2}+\epsilon)^d \tag{4.10}$$

for every $\epsilon > 0$ and every $d \geq d(\epsilon)$. The upper bound is much more demanding than the lower one. Returning to the planar case, we remark that Wegner [89] made some progress in estimating $T(p,q;B^2)$. He notes that (4.9) implies $T(p,2;B^2) \geq 3p-3$ and then proves the following:

$$T(p,2;B^2) \leq 4p-6, \quad p \geq 3, \tag{4.11}$$

$$T(p,q;B^2) = p-q+1, \quad 3 \leq q \leq p \leq 3q-6. \tag{4.12}$$

Here (4.11) shows that $T(3,2;B^2) = 6$, and (4.12) improves on theorem (2.5). Another result is $T(4,3;B^2) = 3$. Exploiting the fact that $T(p,2;B^2)$ is bounded by a linear function of p, the reduction lemma of Section 3 yields very good estimates for $q > 2$. In particular, $T(p,q;B^2) \leq p-q+2$ if $q \geq 8$.

So far we have dealt exclusively with families of translates of a ball. For more general sets, much less is known. Grünbaum [37] extended (4.9) above to $T(2,2;K) \leq 3$ for any centrally symmetric K. He made the natural conjecture that this bound is valid without the central symmetry, that is, every family of pairwise intersecting translates of a planar convex set should be 3-pierceable. This was later confirmed by Chakerian and Stein [15], Chakerian and Sallee [13] and Loomis [65] in a number of special cases. See the reports by Chakerian [12] and Eckhoff [28]. In [15] it was also shown that $T(2,2,K) \leq 4$. A more general conjecture of Grünbaum [38] claiming that $T(2,2;K) \leq d+1$ for all $K \subset R^d$ is false already for $d=3$; in fact, $T(2,2;K) \geq 7$ for all $K \subset R^3$ according to [14]. In the plane, however, a full proof of Grünbaum's conjecture has now been given by Karasev [56]:

$$T(2,2;K) \leq 3, \quad K \subset R^2. \tag{4.13}$$

Karasev notes that it is enough to show $[M/K] \leq 3$ for any bounded convex set $M \subset R^2$ satisfying $M - M \subset K - K$. The latter condition says that the width of M, measured in the normed space with unit ball $K-K$, is at most 1 in every direction. The three translates of K needed to cover M are explicitly constructed, using a continuity argument.

To end this section, let C^d be the unit cube in R^d. Using the same technique as in the proof of theorem (3.4), Dol'nikov [23] showed that

$$T(p,2;C^d) \leq 2^{d-1}(p-1)+1, \quad d \geq 2\,. \tag{4.14}$$

Since the right-hand side is linear in p, for fixed d, it follows that $T(p,q;C^d) = p-q+1\,, p \geq q \geq q(d)$ for some function $q(d)$. In particular, applying the reduction lemma of Section 3, one finds that $T(p,q;C^2) = p-q+1$, $p \geq q \geq 4$.

5 The (p, q)-problem for d-intervals

Compared with the well-studied (p, q)-problems considered in the three preceding sections, the (p, q)-problem for d-intervals and homogeneous d-intervals to be treated in the present section is more recent. It also differs from the former ones in that the ground family $\mathcal{G}$ is not a family of convex sets in R^d. Nevertheless, the problem bears some resemblance to the (p, q)-problem for boxes in R^d.

There are two closely related concepts of "multiple" intervals that will be reviewed here. We first deal with the case of d-intervals.

Let $L_1, \ldots, L_d$ be distinct parallel lines in the plane. A *d-interval* I is a union of d intervals, one in each of the given lines. If $I = I_1 \cup \cdots \cup I_d$, where $I_j \subset L_j$ for $j = 1, \ldots, d$, then I_j is called the jth component of I. It is usually assumed that the components of I are nonempty.

Incidentally, a d-interval can be regarded, in a natural way, as an axis-parallel box in R^d. A point piercing a family of d-intervals then becomes a hyperplane transversal of the corresponding family of boxes perpendicular to some coordinate axis of R^d. This interpretation is used in Grünbaum [39] (see also [88], [40] and [61]).

According to Gyárfás and Lehel [41], Gallai raised the problem of bounding the transversal number of a family of pairwise intersecting d-intervals by a function of d. More generally, letting $\mathcal{I}^d$ denote the family of all d-intervals, the problem is to determine

$$I(p, q; d) := J(p, q; \mathcal{I}^d), \quad p \geq q \geq 2\,.$$

We point out, however, that the literature deals almost exclusively with the case $q = 2$, i.e., Gallai's problem just mentioned. Consequently, the emphasis here is also on this case. The general case will be briefly touched by invoking the reduction lemma of Section 3. Clearly, $I(p, q; 1) = p - q + 1$ (see (2.2) or (3.2)). That $I(p, q; d)$ exists for all p, q and d was first proved by Gyárfás and Lehel [41] and independently by Wegner [88]. Note that it suffices to consider the case $q = 2$.

The bounds on $I(p, q; d)$ obtained in these papers are enormous. For example, the proof in [41] shows that the order of $I(p, 2; d)$ for fixed d is roughly $O(p^{d!})$. A much better estimate results from the recursion

$$I(p, 2; d) \leq I(p-1, 2; d) + I((p-1)(d-1)+1, 2; d-1) + (p-1)(d-1)^2 + 1$$

proved by Károlyi and Tardos [58]. It yields $I(p, 2; d) = O(p^d)$ for fixed d and even $O(p^{d-1})$, in view of (5.2) below. Using an ingenious topological approach, Kaiser [52] has now shown that the correct order of magnitude of $I(p, 2; d)$ as a function of p is $O(p)$. We shall report on this in a moment.

Let us start with a nice result of Gyárfás and Lehel [41] which asserts

$$I(2, 2; 3) = 4\,. \tag{5.1}$$

Here $I(2,2;3) \geq 4$ is demonstrated by a family of ten 3-intervals. A different proof of (5.2) is due to Tardos [80]. For an application of (5.2) to a problem on common transversals in R^3, see Wegner [88].

We next deal with families of 2-intervals. The recurrence in [41] yields $I(p,2;2) \leq p(p-1)$. The problem was taken up by Tardos [80] who was the first to introduce topological methods into the field. He succeeded in finding the exact solution, namely,

$$I(p,2;2) = 2p - 2. \tag{5.2}$$

That the right-hand side here is best possible can be seen by taking $p-1$ "disjoint" copies of the family showing equality in $I(2,2;2) = 2$. The latter result was first obtained in Gyárfás and Lehel [41] and in Wegner [88]. For proofs using common transversals in the plane, see Grünbaum [39] (where the argument seems incomplete) and Eckhoff [27]. For a recent application to obtaining an upper bound on $M(4,3;2)$, see Kleitman, Gyárfás and Tóth [61].

Replacing a bound of magnitude $O(p^2)$ by an $O(p)$ bound may appear like a modest improvement but is, in fact, a huge step. Tardos [80] established (5.1) and (5.2) in a uniform manner by applying homology and homotopy theory to certain simplicial complexes associated with the families in question. Rather than explaining his approach here, we shall describe the core of the topological method in connection with Kaiser's proof of (5.8) below.

At the end of his paper, Tardos conjectured that $I(p,2;d)$ is bounded by a linear function of p, for any given d. This was confirmed by Kaiser [52] who showed that

$$I(p,2;d) \leq d(d-1)(p-1), \quad d \geq 2. \tag{5.3}$$

Despite the fact that this is a striking result, with an outstanding proof, there is still room for improvement. The problem of how $I(p,2;d)$ might grow as a function of d remains open. Károlyi and Tardos [58] noted that a lower bound for $I(p,2;d)$ is $d(p-1)$, and Kaiser [53] increased this to $2(d-1)(p-1)$. In view of (5.1) and (5.2), one is tempted to conjecture that the latter is always tight; this would also settle Gallai's original question. However, a recent construction of Matoušek [68] establishes a near-quadratic lower bound of the form $I(p,2;d) \geq c(d/\log d)^2(p-1)$, where c is a positive constant. We will return to (5.3) at the end of this section.

Before proceeding we briefly consider the case $q > 2$. Except for $I(p,q;1) = p-q+1$ and a few sporadic Gallai-type results such as $I(3,3;3) = 2$ and the trivial $I(p,p;d) = 1$ for $p \geq 2d$ (see [41], [88]), the literature is silent. Fortunately, $I(p,2;d)$ is bounded by a linear function of p, and so the reduction lemma of Section 3 yields nearly optimal estimates for $q > 2$. Assume, for simplicity, that $d = 2$. It follows from $I(2,2;2) = 2$ and assertion (6.1) below that $I(p,q;2) \leq p-q+2$ for $p \leq 2q-2$. The reduction lemma shows, in addition, that $I(p,3;2) \leq \lfloor \frac{4}{3}(p-1) \rfloor$, $I(p,4;2) \leq p-1$ and

$$I(p,q;2) \leq p-q+2, \quad p > q \geq 5. \tag{5.4}$$

We point out that Kaiser and Rabinovich [54] have studied intersection properties of d-intervals (and more general "(d,n)-convex" sets) which are loosely related to the (p,q)-property. Among others, they showed that if $\mathcal{F} \subset \mathcal{I}^d$ satisfies $\Pi(\lfloor \log_2(d+1) \rfloor + 1)$, then $\mathcal{F}$ has a transversal $\{x_1, \ldots, x_d\}$ with $x_j \in L_j$, $j = 1, \ldots, d$. For $d = 2$, this is also proved in Eckhoff [27] and slightly stronger than $I(2,2;2) = 2$.

We now turn to the (p,q)-problem for families of *homogeneous d-intervals.* By definition, a homogeneous d-interval is a subset of the line expressible as the union of d or fewer intervals. Denote by $\mathcal{I}_h^d$ the family of all such sets and define

$$I_h(p,q;d) := J(p,q;\mathcal{I}_h^d), \quad p \geq q \geq 2.$$

Since a homogeneous d-interval arises from a d-interval by letting the lines $L_1, \ldots, L_d$ coincide, it is easily seen that

$$I(p,q;d) \leq I_h(p,q;d) \tag{5.5}$$

for all p, q and d (see [58], [81]).

The existence of $I_h(p,2;d)$, and hence of $I_h(p,q;d)$ for all $p \geq q \geq 2$ was first established by Gyárfás and Lehel [41] (see also [42]), yielding an upper bound of order $O(p^{(d+1)!/2})$ for any fixed d. Later Károlyi and Tardos [58] used a stronger recurrence relation to obtain a bound of order $O(p^d)$. For a different existence proof, see Pach [73].

An explicit upper bound of the form

$$I_h(p,2;d) \leq I(2d(d-1)(p-1)+1,2;d) \tag{5.6}$$

appears in Tardos [80]. By (5.2), it implies that $I_h(p,2;2) \leq 8p-8$. The latter was then reduced to $I_h(p,2;2) \leq 3p-3$ by Kaiser [52]. Now $I_h(2,2;2) = 3$ had already been proved by Gyárfás and Lehel [41,42]. Using $p-1$ "disjoint" copies of the family in [41] showing equality for $p = 2$, one gets

$$I_h(p,2;2) = 3p-3, \quad p \geq 2\,. \tag{5.7}$$

Actually, Kaiser established a much stronger result which shows that $I_h(p,2;d)$ is bounded by a linear function of p, for every given value of d. More precisely, he proved the following:

$$I_h(p,2;d) \leq (d^2-d+1)(p-1), \quad d \geq 2. \tag{5.8}$$

This result should be compared with (5.3). It is best possible as far as the dependence on p is concerned. If $d \geq 3$ and no projective plane of order $d-1$ exists, then the right-hand side can be lowered to $d(d-1)(p-1)$. On the other hand, $I_h(p,2;d)$ is certainly not a linear function of d. This is again demonstrated in the recent paper by Matoušek [68] where it is shown that $I_h(p,2;d) \geq c(d^2/\log d)(p-1)$ for some constant $c > 0$. This comes close to matching the bound in (5.8).

We mention that Alon [1] has given a very short und elementary argument for proving $I_h(p,2;d) \le 2d^2(p-1)$, a bound differing from (5.8) only by a factor. The proof uses the technique of Alon and Kleitman [7].

Here is an outline of Kaiser's remarkable proof of (5.8).

Let $\mathcal{H}$ be a finite family of homogeneous d-intervals having the $(p,2)$-property. Assume, without loss of generality, that the sets in $\mathcal{H}$ are closed and contained in the unit interval (0,1). Fix an integer n (to be assigned a specific value later) and consider the n-simplex

$$\Delta^n = \{(x_1,\dots,x_n) \in R^n : 0 \le x_1 \le \dots \le x_n \le 1\}$$

which may be regarded as the space of candidates of transversals of size n piercing $\mathcal{H}$. This space is parametrized by the n-sphere S^n in R^{n+1} by mapping the point $z = (z_0,\dots,z_n) \in S^n$ to $x = (x_1,\dots,x_n) \in \Delta^n$, where

$$x_i = z_0^2 + z_1^2 + \dots + z_{i-1}^2, \quad i = 1,\dots,n.$$

Now fix a point $z \in S^n$ and hence $x \in \Delta^n$. The unit interval is broken into $n+1$ open subintervals $U_i = (x_i, x_{i+1})$, $i = 0,\dots,n$, where $x_0 = 0$ and $x_{n+1} = 1$. Note that some of the U_i may be empty. Set $V = \{0,1,\dots,n\}$ and define, for any $E \subset V$,

$$w_E(z) := max_K\, dist(K, \{x_1,\dots,x_n\}).$$

Here dist is the usual distance on the line and the maximum is taken over all $K \in \mathcal{H}$ satisfying

$$K \subset \bigcup_{i\in E} U_i \quad \text{and} \quad K \cap U_i \ne \emptyset,\ i \in E.$$

If no such K exists, set $w_E(z) = 0$. Also define, for any $i \in V$, $w_i(z) := \sum_{i\in E} w_E(z)$. The functions $w_E,\ w_i : S^n \to \mathbb{R}_+$ are clearly continuous. We seek a point $\bar{z} \in S^n$ for which $w_0(\bar{z}) = \dots = w_n(\bar{z})$. To this end, define $h_i : S^n \to \mathbb{R}\ (i = 1,\dots,n)$ by letting $h_i(z) = sgn(z_i)w_i(z) - sgn(z_0)w_0(z)$, where $sgn(z_i)$ is the signature of z_i. In spite of the fact that $sgn(z_i)$ has a jump at $z_i = 0$, the $h(i)$ are continuous. Indeed, $z_i = 0$ means $U_i = \emptyset$, hence $w_E(z) = 0$ if $i \in E$ and so $w_i(z) = 0$. Since $w_E(-z) = w_E(z)$ and $h_i(-z) = -h_i(z)$ for all E and i, the Borsuk-Ulam theorem yields a point $\bar{z} \in S^n$ such that $h_i(\bar{z}) = 0$, $i = 1,\dots,n$. This is the point we are looking for.

Set $w = w_0(\bar{z})$ and assume, for a contradiction, that $w > 0$. Let H be the hypergraph whose vertex set is V and whose edges are the sets E with $w_E(\bar{z}) > 0$. Clearly, pairwise disjoint edges of H correspond to pairwise disjoint members of $\mathcal{H}$, whence $\nu(\mathcal{H}) \le p-1$. By the definition of w, the function mapping $E \subset H$ to $w_E(\bar{z})/w$ is a fractional matching of H. This implies

$$\nu^*(H) \ge \frac{1}{w} \sum_E w_E(\bar{z}).$$

On the other hand, double counting yields

$$d \sum_{E} w_e(\bar{z}) \geq (n+1)w,$$

simply because $|V| = n+1$ and $|E| \leq d$ for all $E \subset H$. Hence, $\nu^*(H) \geq (n+1)d$. The last step is to apply a powerful result of Füredi, proved for $d = 2$ by Lovász, which asserts that $\nu^*(H) \leq (d - 1 + 1/d)\,\nu(H)$ for any hypergraph H of rank at most d (see Duchet [26], Thm. 4.29). It follows that $n+1 \leq (d^2-d+1)(p-1)$. If we now specialize n to be $(d^2-d+1)(p-1)$, this inequality is violated. The conclusion is that $w = 0$, i.e., $w_E(\bar{z}) = 0$ for all $E \subset V$. This in turn implies that $\{\bar{x}_1, \ldots, \bar{x}_n\}$ pierces $\mathcal{H}$. Hence $\tau(\mathcal{H}) \leq n$, and the proof is complete. □

We point out that Kaiser's proof of theorem (5.3) for d-intervals uses a scheme very similar to that outlined above. The main difference is that the space of candidates of transversals of $\mathcal{H}$ is then parametrized by the product of d copies of S^n, one for each component line. This requires an extension of the Borsuk-Ulam theorem adapted to the situation which had been proved shortly before by Ramos [75]. The details are much more complicated.

Let us briefly comment on the case $q > 2$. Since $I_h(p, 2; 2)$ is bounded by a linear function of p, the reduction lemma of Section 3 produces quite good upper bounds for $I_h(p, q; 2)$. We content ourselves with just one implication of (5.7), namely,

$$I_h(p, q; 2) \leq p - q + 4, \quad p \geq q \geq 6. \tag{5.9}$$

If $p \leq 2q-2$, then even $I_h(p, q; 2) \leq p-q+3$ holds, thanks to $I_h(2, 2; 2) = 3$ and (6.1) below.

We remark that the literature contains several generalizations of the notion of "multiple interval family". The extensions to polyconvex sets (Alon and Kalai [4]) and to (d, n)-convex sets (Kaiser and Rabinovich [54]) have already been mentioned. Lehel [63] proved that if $\mathcal{Q}^{d,n}$ is the family of all unions of at most n parallel boxes in R^d, then $J(p, q; \mathcal{Q}^{d,n}) < \infty$ if and only if $p \geq q \geq \min(d, n) + 1$. For general existence proofs in a combinatorial framework, see Gyárfás [40] and Pach [73]. Another possibility is to consider graph-theoretical extensions in which lines are replaced by trees, intervals by subtrees and homogeneous d-intervals by subforests with at most d components. For instance, a family of d-subforests of a tree in which every $d+1$ members intersect can be pierced by d vertices (see [63]). The analog of $I_h(2, 2; 2) = 3$ is also valid (see Gyárfás and Lehel [41]) but the proof is now much harder. More recently, Alon [2] has shown that the elementary bound $I_h(p, 2; d) \leq 2d^2(p-1)$ which he obtained in [1] continues to hold in the graph setting.

6 The combinatorial (p,q)-problem

In this final section, the Hadwiger-Debrunner (p,q)-problem will be studied in a combinatorial, or abstract, setting. This seems justified in view of the apparent combinatorial overtones present in some of the material in the preceding sections. The approach helps to clarify the role played by combinatorial arguments and to separate them from the genuine geometrical aspects of the subject. Of course, the abstract (p,q)-property can also be considered in its own right.

This line of research was started by Wegner [85, 87] although traces of it in graph and hypergraph theory can be found in the earlier literature. Wegner's motivation was to uncover the extent to which the Hadwiger-Debrunner result (2.5) is a purely combinatorial consequence of Helly's theorem. Apart from a small "missing link" this goal has been achieved, but other intriguing questions remain.

It is convenient to use the language of (finite, abstract) simplicial complexes, or *complexes*, for short. The complex K is said to satisfy the *d-Helly condition* if the minimal non-faces of K have at most $d+1$ vertices. Equivalently, a set of $d+2$ or more vertices of K spans a face if, and only if, each of its $(d+1)$-element subsets does. We write $K \in H(d)$ when this condition holds. Note that K is then completely determined by its d-skeleton, i.e., the subcomplex of all faces with $d+1$ or fewer vertices. We say that K has the *(p,q)-property*, written $K \in \Pi(p,q)$, if K has at least p vertices and from any p vertices, some q span a face of K. Here, as always, $p \geq q \geq 2$. Finally, let $\rho(K)$ denote the minimum number of faces of K required to cover all vertices of K. Define

$$J(p,q;d) := \sup\{\rho(K) : K \in \Pi(p,q) \cap H(d)\}$$

and set $J(p,q;d) = \infty$ if no finite supremum exists. This definition is due to Wegner [85] and independently to Dol'nikov [24]. Notice, however, that Dol'nikov uses the language of uniform hypergraphs and considers a dualized version of the problem. In a sense, $J(p,q;d)$ can be regarded as the abstract counterpart of the function $M(p,q;d)$ studied in Section 2. The latter is recovered when only *d-representable* complexes are admitted, that is, nerve complexes of finite families of convex sets in R^d.

Let us begin with the case $d = 1$ which has received much attention in the literature. By virtue of the 1-Helly condition, the (p,q)-problem is then a problem on graphs. If $K \in H(1)$ and G is the 1-skeleton, or edge-graph, of K, then the faces of K correspond to the *cliques* (complete subgraphs) of G and $\rho(K)$ becomes the *clique covering number* of G. We have

$$J(p,q;1) = \begin{cases} p-q+1, & p \leq 2q-2, \\ \infty, & p \geq 2q-1. \end{cases} \tag{6.1}$$

Expressed in graph-theoretic terms, (6.1) states that if $p \leq 2q-2$ and among any p vertices of a graph G some q vertices span a clique, then all the vertices of G can be covered by $p-q+1$ cliques. If $p \geq 2q-1$, no such number exists.

This result has been rediscovered many times, often under geometrical guises (see [49], [85], [39] and [23]). Since the nerve complex of a family of boxes satisfies the $H(1)$-condition, assertion (3.5) in Section 3 is a special case of (6.1). Linial and Rabinovich [64] established a stronger result, namely, that $p-q$ of the $p-q+1$ cliques covering G can be assumed to consist of a single vertex. In an equivalent form, however, this was already shown by Erdős and Gallai [31], Thm. 3.5. The latter seems to be the first occurence of the result in the literature.

That $J(p,q;1)=\infty$ for $p \geq 2q-1$ can be seen as follows. Erdős [29] had proved that there exist graphs having girth at least $2q$ and arbitrary high chromatic number. In such a graph G, every subgraph on $2q-1$ vertices is cycle-free and thus 2-colorable, and so the complement $\overline{G}$ has a clique of size q. This means that $\overline{G}$ has the (p,q)-property while its clique covering number can be made arbitrarily large. This proves the assertion.

For connections with Ramsey's theorem, see Gyárfás [40], Dol'nikov [24] and Linial and Rabinovich [64].

We now turn to the general case which was investigated by Wegner [85]. It follows from (2.4) that $J(p,q;d)=\infty$ if $q \leq d$. Therefore, we assume that $p \geq q \geq d+1$. There is ample evidence that the following conjecture is true.

$$\textit{Conjecture:} \quad J(p,q;d) = p-q+1, \quad p \leq \frac{d+1}{d}(q-1). \tag{6.2}$$

In fact, Wegner's "Satz 1" stated this result in categorical form. Unfortunately, the proof in [85] is not complete, as the case analysis for $d>1$ does not cover all possibilities (see Wegner [87]). To some extent, however, the gap can be filled. It is shown in [87] that (6.2) holds with $p-q+1$ replaced by $p-q+2$ and that

$$J(p,q;d) \leq p-q+1, \quad p \leq \frac{d+1}{d}q-2; \tag{6.3}$$

the latter is slightly weaker than (6.2) when q is not divisible by d.

The effort should of course be aimed at verifying (6.2). Wegner [87] proved that in order to do so, it is enough to settle the special case $J(2d+2, 2d+1; d)=2$. This harmless looking assertion is actually quite intriguing. Wegner has verified it for $d \leq 4$ and Klaus-Peter Nischke independently for $d \leq 5$ (private communications).

The importance of establishing (6.2) lies (among others) in the fact that the resulting theorem would be best possible. Wegner [85] describes an example of Danzer to the effect that $J(d+2,d+1;d)=\infty$ but in fact the following is true:

$$J(p, q; d) = \infty, \quad p > \frac{d+1}{d}(q-1). \tag{6.4}$$

Thus if the argument in [85] could be repaired, the function $J(p, q; d)$ would be completely known and hence the degree to which theorem (2.5) is a combinatorial consequence of Helly's theorem alone.

Let us sketch a proof of (6.4). (It seems that the details appear here for the first time.) Erdős and Hajnal [32] proved that for all positive integers d, k and p, there exists a $(d+1)$-uniform hypergraph H with $\chi(H) \geq k$ in which no circuit has length p or less. (See Duchet [26], Thm. 5.25.) Here $\chi(H)$ is the chromatic number of H. The proof in [32] is non-constructive, but constructive proofs were later given by Lovász [66] and Nešetřil and Rödl [71]. Now a theorem of Berge and Las Vergnas asserts that for hypergraphs without odd circuits, the transversal number is equal to the matching number (see Duchet [26], Thm. 3.5, Lovász and Plummer [67], Thm. 12.3.3). In particular, if H' is a partial subhypergraph of H on p vertices, then $\tau(H') = \nu(H')$ holds. Since clearly $\nu(H') \leq \lfloor \frac{p}{d+1} \rfloor$, we have $\tau(H') \leq \lfloor \frac{p}{d+1} \rfloor$. Next define a simplicial complex K on the vertex set of H by declaring a subset to be a face of K if it doesn't contain any edge of H. Then $K \in H(d)$, and $\chi(H) \geq k$ translates into $\rho(K) \geq k$. Because $p > \frac{d+1}{d}(q-1)$ is equivalent to $p - q \geq \lfloor \frac{p}{d+1} \rfloor$, we get $\tau(H') \leq p - q$. This means that some set of q vertices of H' fails to contain an edge of H and thus is a face of K. In other words, $K \in \Pi(p, q)$, and as k can be arbitrarily large the assertion follows. □

The remainder of this section is devoted to a Helly-type problem for complexes closely related to the (p, q)-problem treated above. Recall from (6.1) that a graph having the (p, q)-property with $p \leq 2q - 2$ has clique covering number at most $p - q + 1$. Moreover, it can be covered by $p - q$ one-vertex cliques and one "big" clique containing the remaining vertices. The question is whether the latter result has an analog for $d > 1$. Theorem 1 in Dol'nikov [24] claimed that this was indeed the case if $p \leq \frac{d+1}{d}(q-1)$. This would generalize the result conjectured in (6.2). However, a proof is not given in [24] and the assertion is in fact far from true.

We find it convenient to write $k = p - q$ in what follows. Define $h(d, k)$ to be the smallest integer p for which the following statement is true:

If $K \in \Pi(p, p-k) \cap H(d)$, then all but at most k vertices of K span a face of K.

Since $K \in \Pi(p, p-k)$ means that all but at most k of any p given vertices of K span a face, the evaluation of $h(d, k)$ is indeed a Helly-type problem. Dol'nikov's claim in [24] would amount to $h(d, k) = (d+1)(k+1)$ but apart from $k = 0$; $d = 1$; and $d = 2$, $k = 1$, this is quite false. By Theorem 3.5 of Erdős and Gallai [31] quoted earlier and Theorem 5.6 in the same paper, we have

$$h(1,k) = 2(k+1), \quad (6.5)$$
$$h(d,1) = \lfloor (d+3)^2/4 \rfloor. \quad (6.6)$$

The general case was studied by Gyárfás, Lehel and Tuza [43] and Tuza [83, 84] in a hypergraph setting. The following results were obtained:

$$h(d,k) < \binom{d+k+1}{d} + \binom{d+k}{d}, \quad (6.7)$$
$$h(d,k) \geq t\binom{k+d+1-t}{k} + k + d + 1 - t, \quad t := \left\lceil \frac{d+1}{k+1} \right\rceil. \quad (6.8)$$

It is conjectured that the lower bound is always tight. For instance, this would mean that $h(2,k) = \binom{k+3}{2}$. Since the lower and the upper bound differ by a factor of less than 4, it follows that for fixed d and $k \to \infty$, $h(d,k) = \Theta(k^d)$.

The foregoing results have a geometric application which leads us back to the original (p,q)-problem of Hadwiger and Debrunner. For a certain sector of the p,q-plane which is smaller than that described in Sections 1 and 2, the conclusion of theorem (2.5) can be strengthened as follows:

Let $\mathcal{F}$ be a finite family of convex sets in R^d such that among every $h(d,k)$ members of $\mathcal{F}$, all but at most k members have a common point. Then there is a point common to all but at most k members of $\mathcal{F}$.

This follows at once from the definition of $h(d,k)$ by specializing K to be the nerve complex of $\mathcal{F}$. For $k = 0$ this is merely Helly's theorem, in view of $h(d,0) = d+1$. The result is also best possible for $k = 1$, as shown by Perles [74] and Nadler [70] (see Section 2), and trivially for $d = 1$. It is somewhat unlikely that $h(d,k)$ should always be optimal, even if Tuza's conjecture were correct. The question is whether the combinatorial examples such as those proving (6.8) can be realized geometrically.

References

[1] N. Alon, Piercing d-intervals, Discrete Comput. Geom. 19 (1998) 333-334.

[2] N. Alon, Covering a hypergraph of subgraphs, Discrete Math., to appear.

[3] N. Alon, I. Bárány, Z. Füredi and D. J. Kleitman, Point selection and weak ϵ-nets for convex hulls, Combin. Probab. Comput. 1 (1992) 189-200.

[4] N. Alon and G. Kalai, Bounding the piercing number, Discrete Comput. Geom. 13 (1995) 245-256.

[5] N. Alon, G. Kalai, J. Matoušek and R. Meshulam, Transversal numbers for hypergraphs arising in geometry, Adv. Appl. Math. 29 (2002) 79-101.

[6] N. Alon and D. J. Kleitman, Piercing convex sets, Bull. Amer. Math. Soc. (N.S.) 27 (1992) 252-256.

[7] N. Alon and D. J. Kleitman, Piercing convex sets and the Hadwiger-Debrunner (p, q)-problem, Adv. Math. 96 (1992) 103-112.

[8] N. Alon and D. J. Kleitman, A purely combinatorial proof of the Hadwiger Debrunner (p, q) conjecture, Electron. J. Combin. 4 (1997) #R1.

[9] I. Bárány, A generalization of Carathéodory's theorem, Discrete Math. 40 (1982) 141-152.

[10] I. Bárány and J. Matoušek, A fractional Helly theorem for convex lattice sets, Adv. Math., to appear.

[11] J. Bourgain and J. Lindenstrauss, On covering a set in R^N by balls of the same diameter, in: Geometric Aspects of Functional Analysis (1989-90), pp. 138-144, Lecture Notes in Mathematics 1469, Springer-Verlag, Berlin, 1991.

[12] G. D. Chakerian, Intersection and covering properties of convex sets, Amer. Math. Monthly 76 (1969) 753-766.

[13] G. D. Chakerian and G. T. Sallee, An intersection theorem for sets of constant width, Duke Math. J. 36 (1969) 165-170.

[14] G. D. Chakerian and S. K. Stein, On measures of symmetry of convex bodies, Canad. J. Math. 17(1965) 497-504.

[15] G. D. Chakerian and S. K. Stein, Some intersection properties of convex bodies, Proc. Amer. Math. Soc. 18 (1967) 109-112.

[16] B. Chazelle, H. Edelsbrunner, M. Grigni, L. Guibas, M. Sharir and E. Welzl, Improved bounds on weak ϵ-nets for convex sets, Discrete Comput. Geom. 13 (1995) 1-15.

[17] F. Chung and L. Lu, An upper bound for the Turán number $t_3(n, 4)$, J. Combin. Theory Ser. A 87 (1999) 381-385.

[18] L. Danzer, Über zwei Lagerungsprobleme; Abwandlungen einer Vermutung von T. Gallai, Dissertation, Techn. Hochschule München, 1960.

[19] L. Danzer, Über Durchschnittseigenschaften n-dimensionaler Kugelfamilien, J. Reine Angew. Math. 208 (1961) 181-203.

[20] L. Danzer, On the k-th diameter in E^d and a problem of Grünbaum, in: Proc. Colloq. Convexity Copenhagen 1965, p. 41, Københavns Universitets Matematiske Institut, Copenhagen, 1967.

[21] L. Danzer, Zur Lösung des Gallaischen Problems über Kreisscheiben in der euklidischen Ebene, Studia Sci. Math. Hungar. 21 (1986) 111-134.

[22] L. Danzer, B. Grünbaum and V. Klee, Helly's theorem and its relatives, in: Convexity, Proc. Symp. Pure Math. 7 (V. L. Klee, ed.), pp. 101-180, Amer. Math. Soc., Providence, R. I., 1963.

[23] V. L. Dol'nikov, A coloring problem, Siberian Math. J. 13 (1972) 886-894. Translation of Sibirsk. Mat. Zh. 13 (1972) 1272-1283.

[24] V. L. Dol'nikov, On a generalization of Ramsey's theorem, Soviet Math. Dokl. 18 (1977) 223-226. Translation of Dokl. Akad. Nauk SSSR 232 (1977) 1241-1244.

[25] V. L. Dol'nikov, A theorem of Helly type for sets defined by systems of equations, Math. Notes 46 (1989) 837-840. Translation of Mat. Zametki 46 (1989), no. 5, 13-16.

[26] P. Duchet, Hypergraphs, in: Handbook of Combinatorics (R. L. Graham et al., eds.), Vol. I, Chapter 7, pp. 381-432, North-Holland, Amsterdam, 1995.

[27] J. Eckhoff, Transversalenprobleme in der Ebene, Arch. Math. (Basel) 24 (1973) 195-202.

[28] J. Eckhoff, Helly, Radon and Carathéodory type theorems, in: Handbook of Convex Geometry (P. M. Gruber and J. M. Wills, eds.), Vol. A, Chapter 2.1, pp. 389-448, North-Holland, Amsterdam, 1993.

[29] P. Erdős, Graph theory and probability, Canad. J. Math. 11 (1959) 34-38.

[30] P. Erdős, Personal reminiscences and remarks on the mathematical work of Tibor Gallai, Combinatorica 2 (1982) 207-212.

[31] P. Erdős and T. Gallai, On the minimal number of vertices representing the edges of a graph, Publ. Math. Inst. Hungar. 6 (1961) 181-203.

[32] P. Erdős and A. Hajnal, On chromatic number of graphs and set-systems, Acta Math. Hungar. 17 (1966) 61-99.

[33] P. Erdős and C. A. Rogers, Covering space with convex bodies, Acta Arith. 7 (1962) 281-285.

[34] L. Fejes Tóth, Lagerungen in der Ebene, auf der Kugel und im Raum, Springer-Verlag, Berlin, 1953. [Second edition 1972]

[35] D. G. Fon-Der-Flaas and A. V. Kostochka, Covering boxes by points, Discrete Math. 120 (1993) 269-275.

[36] B. E. Fullbright, Intersectional properties of certain families of compact convex sets, Pacific J. Math. 50 (1974) 57-62.

[37] B. Grünbaum, Borsuk's partition conjecture in Minkowski planes, Bull. Research Council Israel, Sect. F, 7 (1957) 25-30.

[38] B. Grünbaum, On intersections of similar sets, Portugal. Math. 18 (1959), 155-164.

[39] B. Grünbaum, Lectures on Combinatorial Geometry, Mimeographed notes, University of Washington, Seattle, 1974.

[40] A. Gyárfás, A Ramsey-type theorem and its applications to relatives of Helly's theorem, Period. Math. Hungar. 3 (1976) 261-270.

[41] A. Gyárfás and J. Lehel, A Helly-type problem in trees, in: Combinatorial Theory and its Applications II (P. Erdős et al., eds.), Colloq. Math. Soc. János Bolyai 4, pp. 571-584, North-Holland, Amsterdam, 1970.

[42] A. Gyárfás and J. Lehel, Covering and coloring problems for relatives of intervals, Discrete Math. 55 (1985) 167-180.

[43] A. Gyárfás, J. Lehel and Z. Tuza, Upper bounds on the order of τ-critical hypergraphs, J. Combin. Theory Ser. B 33 (1982) 161-165.

[44] H. Hadwiger, Ungelöste Probleme, Nr. 15, Elem. Math. 12 (1957), 10-11; Nachtrag ibid. p. 62.

[45] H. Hadwiger, Ungelöste Probleme, Nr. 19, Elem. Math. 12 (1957) 109-110.

[46] H. Hadwiger and H. Debrunner, Ausgewählte Einzelprobleme der kombinatorischen Geometrie in der Ebene, Enseign. Math. (2) 1 (1955) 56-89.

[47] H. Hadwiger and H. Debrunner, Über eine Variante zum Hellyschen Satz, Arch. Math. (Basel) 8 (1957) 309-313.

[48] H. Hadwiger and H. Debrunner, Kombinatorische Geometrie in der Ebene, Monogr. Enseign. Math., No. 2, Université, Genève, 1960.

[49] H. Hadwiger, H. Debrunner and V. Klee, Combinatorial Geometry in the Plane, Holt, Rinehart and Winston, New York, 1964.

[50] H. Hadwiger and J. Schaer, Studie zur kombinatorischen Geometrie zentralsymmetrischer Eikörper, Portugal. Math. 30 (1971) 145-152.

[51] A. Hajnal and J. Surányi, Über die Auflösung von Graphen in vollständige Teilgraphen, Ann. Univ. Sci. Budapest. Rolando Eötvös 1 (1958) 113-121.

[52] T. Kaiser, Transversals of d-intervals, Discrete Comput. Geom. 18 (1997) 195-203.

[53] T. Kaiser, Piercing problems and topological methods, Doctoral dissertation, Charles University, Prague, 1998.

[54] T. Kaiser and Y. Rabinovich, Intersection properties of families of convex (n, d)-bodies, Discrete Comput. Geom. 21 (1999) 275-287.

[55] G. Kalai, Intersection patterns of convex sets, Israel J. Math. 48 (1984) 161-174.

[56] R. N. Karasev, Transversals for families of translates of a two-dimensional convex compact set, Discrete Comput. Geom. 24 (2000) 345-353.

[57] G. Károlyi, On point covers on parallel rectangles, Period. Math. Hungar. 23 (1991) 105-107.

[58] G. Károlyi and G. Tardos, On point covers of multiple intervals and axis-parallel rectangles, Combinatorica 16 (1996) 213-222.

[59] M. Katchalski and A. Liu, A problem of geometry in R^n, Proc. Amer. Math. Soc. 75 (1979) 284-288.

[60] P. Katzarowa-Karanowa, Über ein euklidisch-geometrisches Problem von B. Grünbaum, Arch. Math. (Basel) 18 (1967) 663-672.

[61] D. J. Kleitman, A. Gyárfás and G. Tóth, Convex sets in the plane with three of every four meeting, Combinatorica 21 (2001) 221-232.

[62] D. Larman, J. Matoušek, J. Pach and J. Törőcsik, A Ramsey-type result for convex sets, Bull. London Math. Soc. 26 (1994) 132-136.

[63] J. Lehel, Gallai-type results for multiple boxes and forests, European J. Combin. 9 (1988) 113-120.

[64] N. Linial and Y. Rabinovich, Local and global clique numbers, J. Combin. Theory Ser. B 61 (1994) 5-15.

[65] P. Loomis, Covering among constant relative width bodies in the plane, Ph. D. thesis, University of California, Davis, 1980.

[66] L. Lovász, On chromatic number of finite set-systems, Acta Math. Hungar. 19 (1968) 59-67.

[67] L. Lovász and M. D. Plummer, Matching Theory, Ann. of Discrete Math., No. 29, North-Holland, Amsterdam, 1986.

[68] J. Matoušek, Lower bounds on the transversal numbers of d-intervals, Discrete Comput. Geom. 26 (2001) 283-287.

[69] J. Matoušek, Lectures on Discrete Geometry, Springer-Verlag, New York, 2002.

[70] D. Nadler, Minimal 2-fold coverings of E^d, Geom. Dedicata 65 (1997) 305-312.

[71] J. Nešetřil and V. Rödl, A short proof of the existence of highly chromatic hypergraphs without short cycles, J. Combin. Theory Ser. B 27 (1979) 225-227.

[72] A. G. Netrebin, Separability of convex sets and the concept of convexity on the surface of a convex body, Math. Notes 27 (1980) 311-316. Translation of Mat. Zametki 25 (1980) 603-618.

[73] J. Pach, A remark on transversal numbers, in: The Mathematics of Paul Erdős II (R. L. Graham and J. Nešetřil, eds.), pp. 310-317, Springer-Verlag, Berlin, 1997.

[74] M. A. Perles, A Helly type theorem for almost intersecting families, Talk at the Convex Geometry meeting, Oberwolfach, June 1993.

[75] E. A. Ramos, Equipartition of mass distributions by hyperplanes, Discrete Comput. Geom. 15 (1996) 147-167.

[76] N. Scheller, (p, q)-Probleme für Quaderfamilien, Diplomarbeit, Universität Dortmund, 1996.

[77] J. Schopp, Verschärfung eines Kreisabdeckungssatzes, Elem. Math. 17 (1962) 12-14.

[78] L. Stachó, Über ein Problem für Kreisscheibenfamilien, Acta Sci. Math. (Szeged) 26 (1965) 273-282.

[79] L. Stachó, A solution of Gallai's problem on pinning down circles. (Hungarian), Mat. Lapok 32 (1981/84) 19-47.

[80] G. Tardos, Transversals of 2-intervals, a topological approach, Combinatorica 15 (1995) 123-134.

[81] G. Tardos, Transversals of d-intervals—comparing three approaches, in: European Congress of Mathematics, Budapest 1996 (A. Balog et al., eds.), Vol. II, pp. 234-243, Progress in Mathematics, Vol. 169, Birkhäuser-Verlag, Basel, 1998.

[82] W. T. Trotter, Jr., A decomposition theorem for collections of universal subcontinua, Colloq. Math. 23 (1971) 233-239.

[83] Z. Tuza, Critical hypergraphs and intersecting set-pair systems, J. Combin. Theory Ser. B 39 (1985) 134-145.

[84] Z. Tuza, Minimum number of elements representing a set system of given rank, J. Combin. Theory Ser. A 52 (1989) 84-89.

[85] G. Wegner, Über eine kombinatorisch-geometrische Frage von Hadwiger und Debrunner, Israel J. Math. 3 (1965) 187-198.

[86] G. Wegner, Eigenschaften der Nerven homologisch-einfacher Familien im R^n, Dissertation, Universität Göttingen, 1967.

[87] G. Wegner, Anmerkungen zu 'Über eine kombinatorisch-geometrische Frage von Hadwiger und Debrunner', Unpublished notes (4 pages), Göttingen, 1968.

[88] G. Wegner, Ein ebenes Transversalenproblem, Monatsh. Math. 77 (1973) 72-81.

[89] G. Wegner, Über Helly-Gallaische Stichzahlprobleme, in: 3. Kolloquium über Diskrete Geometrie, pp. 277-282, Salzburg, 1985.

Surface Reconstruction by Wrapping Finite Sets in Space

Herbert Edelsbrunner

Abstract

Given a finite point set in $\mathbb{R}^3$, the surface reconstruction problem asks for a surface that passes through many but not necessarily all points. We describe an unambiguous definition of such a surface in geometric and topological terms, and sketch a fast algorithm for constructing it. Our solution overcomes past limitations to special point distributions and heuristic design decisions.

1 Introduction

The original version of this paper was written in 1995. To preserve that version, we have limited modifications to minor stylistic changes and to the addition of a paragraph that accounts for the new and related work during the years from 1996 to 2001. All citations of work during these five years use letters rather than numbers in the citation.

Problem and solution. The input to the *surface reconstruction problem* is a finite set of points scattered in three-dimensional Euclidean space. The general task is to find a surface passing through the points. There are of course many possible such surfaces, and we would want one that in some sense is most reasonable and best represents the way the input points are distributed. We allow for the case that some points lie off the surface inside the bounded volume. We propose a solution that provides structural information in terms of a mesh or complex connecting the points. Section 2 will be more specific about what exactly we mean by this and what properties we expect from the mesh.

The first part of our solution consists of a description of the surface in geometric and topological terms. There are minimum distance functions and ideas from Morse theory turning these functions into vector fields and cell decompositions. For generic data sets, this description is unambiguous and completely determines the surface. The second part of our solution is an efficient algorithm that constructs the defined surface. The algorithm is based on Delaunay complexes and extracts a subcomplex through repeated collapsing.

All ideas and results generalize to any arbitrary fixed number of dimensions. For reasons of specificity, the discussion in this paper is exclusively three-dimensional.

Work prior to 1995. The surface reconstruction problem has a long history. Most of the previous work assumes some kind of additional structure given along with the data points. A common assumption is that the points lie on curves defined by slicing a surface with a collection of parallel planes [9, 13]. The surface reconstruction is reduced to a sequence of steps, each connecting two curves in contiguous planes. Another common assumption is differentiability [11]. The surface is constructed from patches defining diffeomorphisms between $\mathbb{R}^2$ and local neighborhoods on the surface. Fairly dense point distributions are required to allow the reconstruction of tangents and normals.

We are interested in the *general* surface reconstruction problem that admits no assumption other than that the input consists of finitely many points in $\mathbb{R}^3$. At the time of writing this paper in 1995, we found only three pieces of work studying the general problem. In two cases, the surface or shape is obtained from the three-dimensional Delaunay complex of the input points. This is also the approach followed in this paper. Boissonnat [1] compromises the global nature of the approach by using local rules for removing simplices from the Delaunay complex. The resulting surfaces are somewhat unpredictable and not amenable to rational analysis. Edelsbrunner and Mücke [6] use distance relationships to identify certain subcomplexes of the Delaunay complex as alpha shapes of the given data set. For uneven densities, these shapes tend to either exhibit a lack of detail in dense regions or gaps and holes in sparse regions. Rather than subcomplexes of the Delaunay complex, Veltkamp [15] uses two-parameter neighborhood graphs to form surfaces from points in space. Depending on the choice of the parameters the graphs may exhibit self-intersections or poor shape representation.

Development after 1995. The algorithm described in this paper has been implemented in 1996 at Raindrop Geomagic, which successfully commercialized it as `geomagic Wrap`. It is also described in U. S. Patent No. 6,3777,865, which has issued on April 23, 2002.

The surface reconstruction problem has enjoyed increasing popularity over the last few years, both in computer graphics and in computational geometry. A number of essentially two-dimensional algorithms that rely primarily on density and smoothness assumptions of the data have been developed [C, E, F, J]. In parallel, the use of three-dimensional Delaunay complexes has been refined [A, B, D, I]. The focus of that work is the detailed study of Delaunay complexes for data sets that satisfy density and smoothness assumptions, and to exploit their special structure for surface reconstruction. Possibly surprisingly, neither development has come close to reproducing the ideas presented in this paper.

From a completely different angle, Robin Forman's development of a discrete Morse theory for simplicial complexes [H] is related to work in this paper. According to Forman, a *discrete Morse function* is a map f from the collection of simplices to the real numbers such that for every simplex τ the following two conditions hold:

(1) there is at most one face $\upsilon \leq \tau$ with $\dim \upsilon = \dim \tau - 1$ and $f(\upsilon) \geq f(\tau)$, and

(2) there is at most one coface $\sigma \geq \tau$ with $\dim \sigma = \dim \tau + 1$ and $f(\sigma) \leq f(\tau)$.

The theory developed in this paper uses relations that correspond to functions violating these conditions and thus does not seem to fit into Forman's framework. It would be interesting to elucidate the connection between the two approaches to a discrete Morse theory. Another recent development that resonates with the work in this paper is the introduction of persistent Betti numbers [G]. It relates to the discussion of granularity in Section 9, in which it is suggested to construct coarse-grained decompositions of the Delaunay complex by merging discrete stable manifolds, possibly by suppressing some of the less persistent critical points. Maybe the time has come to integrate all these ideas and to develop a hierarchical approach to surface reconstruction based on a more extensive use of algebraic structures developed in topology.

Outline. Section 2 displays sample results obtained with software implemented at Raindrop Geomagic. Section 3 reviews Delaunay complexes for finite point sets in $\mathbb{R}^3$. Section 4 reviews notions from Morse theory and constructs a family of Morse functions from local distance information. Section 5 derives an ordering principle for the Delaunay simplices from the Morse functions. Section 6 studies mechanisms to cluster simplices based on the ordering. Section 7 defines the basic surface construction as a sequence of collapses. Section 8 discusses generalizations of the basic construction with and without interactive surface modification. Section 9 mentions possible extensions of the presented results and related open questions.

2 Examples and Properties

We use notation and terminology from combinatorial topology [10] to describe the surfaces and the algorithm that constructs them. In a nut-shell, that algorithm starts with the Delaunay complex of the input set and constructs an acyclic partial order over its simplices. This order is motivated by a continuous flow field in which every point is attracted by the Voronoi vertex that is nearest in a weighted sense. The Voronoi vertex at infinity corresponds to a dummy simplex in the relation, and the reconstructed surface is obtained by sculpting away all predecessors of that dummy simplex. We begin by giving a few examples constructed by software implementing the algorithm.

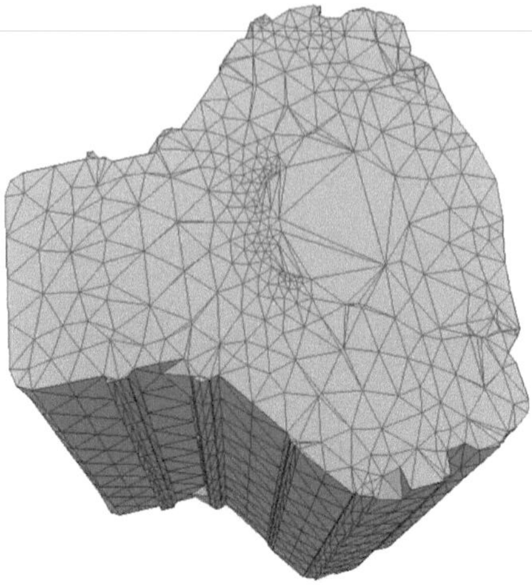

Fig. 1. An engine block surface homeomorphic to a sphere.

Sample surfaces. In the simplest and possibly most common case, the surface constructed from a finite point set in $\mathbb{R}^3$ is connected like a sphere, as in Figure 1. With our approach, it is possible to modify the construction and to introduce tunnels, as illustrated in Figure 2. There are also cases, in which the surface cannot be naturally closed and remains connected like a disk, as in Figure 3. Finally, if the points are lined up, it is possible that the surface reconstructed by our algorithm degenerates to a curve, as in Figure 4.

In technical terms, the constructed surface is a simplicial complex of dimension 2 with the topology of a possibly pinched sphere. If forced by the distribution of the data points, the complex can be one- or zero-dimensional. We first introduce the relevant terminology and then describe the reconstructed surface in more detail.

Spaces and maps. All topological spaces in this paper are subsets of Euclidean space of some dimension k, denoted by $\mathbb{R}^k$. Without exception, we use the topology induced by the Euclidean metric in $\mathbb{R}^k$. The Euclidean distance between points x and y is denoted by $\|x - y\|$, and the *norm* of x is the distance from the origin, which is $\|x\| = \|x - 0\|$. Other than for Euclidean k-dimensional space, we need short notation for the k-dimensional sphere and the k-dimensional ball,

$$\begin{aligned} \mathbb{S}^k &= \{x \in \mathbb{R}^{k+1} \mid \|x\| = 1\}, \\ \mathbb{B}^k &= \{x \in \mathbb{R}^k \mid \|x\| \leq 1\}. \end{aligned}$$

We refer to $\mathbb{S}^k$ as the *k-sphere* and to $\mathbb{B}^k$ as the *k-ball.* For example, the 1-sphere is a circle, the 0-sphere is a pair of points, and the (-1)-sphere is

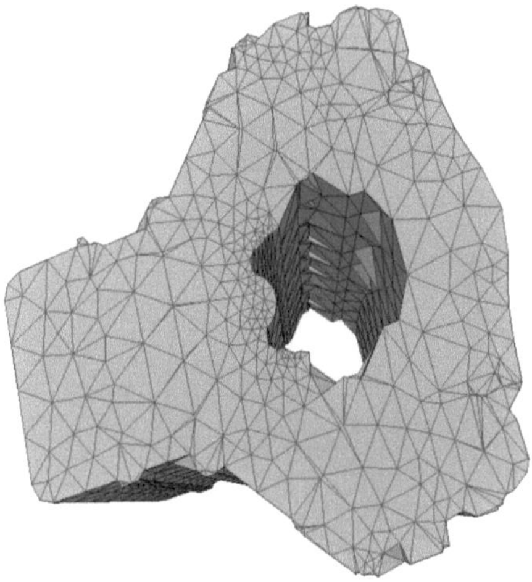

Fig. 2. The engine block surface in Figure 1 after pushing open a tunnel.

the empty set. The 2-sphere is what we ordinarily call a sphere. The 2-ball is a closed disk, the 1-ball is a closed interval, and the 0-ball is a point.

Topological spaces are compared via continuous functions referred to as *maps*. A *homeomorphism*, $\beta : \mathbb{X} \to \mathbb{Y}$, is a continuous bijection with continuous inverse. The inverse of a homeomorphism is again a homeomorphism. $\mathbb{X}$ and $\mathbb{Y}$ are *homeomorphic* or *topologically equivalent*, denoted $\mathbb{X} \approx \mathbb{Y}$, if there is a homeomorphism between them. An *embedding* is an injective map $\iota : \mathbb{X} \to \mathbb{Y}$ whose restriction to the image, $\iota(\mathbb{X})$, is a homeomorphism.

A *homotopy* between two maps $a, b : \mathbb{X} \to \mathbb{Y}$ is a continuous function $F : \mathbb{X} \times [0,1] \to \mathbb{Y}$ with $F(x, 0) = a(x)$ and $F(x, 1) = b(x)$ for all $x \in \mathbb{X}$. $\mathbb{X}$ and $\mathbb{Y}$ are *homotopy equivalent*, denoted by $\mathbb{X} \simeq \mathbb{Y}$, if there are maps $f : \mathbb{X} \to \mathbb{Y}$ and $g : \mathbb{Y} \to \mathbb{X}$ and homotopies between $g \circ f$ and the identity for $\mathbb{X}$ and between $f \circ g$ and the identity for $\mathbb{Y}$. Two spaces have the same *homotopy type* if they are homotopy equivalent. A space is *contractible* if it is homotopy equivalent to a point. Homotopy equivalence is weaker than topological equivalence, in the sense that $\mathbb{X} \approx \mathbb{Y}$ implies $\mathbb{X} \simeq \mathbb{Y}$.

Simplicial complexes. We use combinatorial structures to represent topological spaces in the computer. We begin by introducing the geometric elements that make up these structures. The convex hull of a finite collection of points U is denoted as $\operatorname{conv} U$. A *k-simplex*, σ, is the convex hull of $k+1$ affinely independent points. The *dimension* of σ is $\dim \sigma = k$. At most four points can be affinely independent in $\mathbb{R}^3$ and we have four types of simplices: *vertices* or 0-simplices, *edges* or 1-simplices, *triangles* or 2-simplices, and *tetrahedra* or 3-simplices. A simplex $\tau = \operatorname{conv} T$ is a *face* of another simplex $\sigma = \operatorname{conv} U$, and σ is a *coface* of τ, if $T \subseteq U$. We denote this relationship by $\tau \leq \sigma$. The *boundary* of σ, $\operatorname{bd} \sigma$, is the union of all proper faces, and the

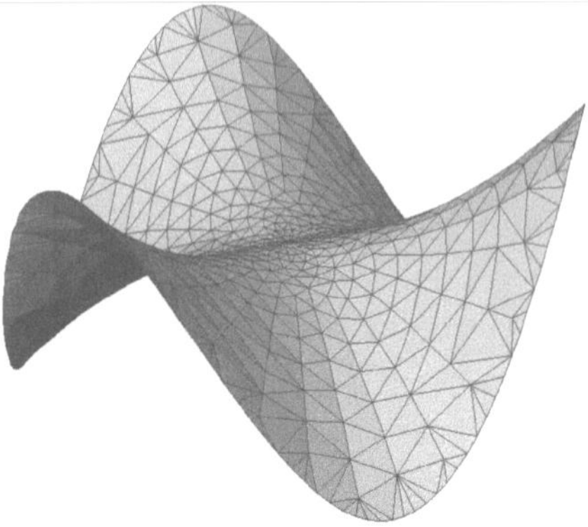

Fig. 3. Points on a saddle surface triangulated to form a patch homeomorphic to a disk.

Fig. 4. Points on the moment curve connected to form a curve homeomorphic to an interval.

interior is $\operatorname{int}\sigma = \sigma - \operatorname{bd}\sigma$. For example, the boundary of an edge consists of its two endpoints and the interior is the open edge, without endpoints. The boundary of a vertex is empty and the interior is the vertex itself.

A *simplicial complex*, $\mathcal{K}$, is a finite collection of simplices such that $\sigma \in \mathcal{K}$ and $\tau \leq \sigma$ implies $\tau \in \mathcal{K}$, and $\sigma, \sigma' \in \mathcal{K}$ implies that $\sigma \cap \sigma'$ is either empty or a face of both. The *dimension* of $\mathcal{K}$ is the maximum dimension of any of its simplices. A *principal simplex* has no proper coface in $\mathcal{K}$. For example, if $\dim \mathcal{K} = 3$ then every tetrahedron in $\mathcal{K}$ is a principal simplex. A *subcomplex* is a simplicial complex $\mathcal{L} \subseteq \mathcal{K}$. The *vertex set* of $\mathcal{K}$ is $\operatorname{Vert}\mathcal{K} = \{\sigma \in \mathcal{K} \mid \dim \sigma = 0\}$. $\mathcal{K}$ is *connected* if for every non-trivial partition $\operatorname{Vert}\mathcal{K} = V_1 \dot{\cup} V_2$ there is a simplex $\sigma \in \mathcal{K}$ that has vertices in V_1 and in V_2. The *underlying space* is $|\mathcal{K}| = \bigcup_{\sigma \in \mathcal{K}} \sigma$. The interiors of simplices partition the underlying space. In other words, for every $x \in |\mathcal{K}|$ there is a unique $\sigma \in \mathcal{K}$ with $x \in \operatorname{int}\sigma$.

Special subsets and subcomplexes. The *star* of a simplex τ in a simplicial complex $\mathcal{K}$ is the set of cofaces, the *closure* of a subset $L \subseteq \mathcal{K}$ is the smallest subcomplex that contains L, and the *link* of a simplex $\tau \in \mathcal{K}$ is the set of simplices in the closed star that are disjoint from τ:

$$\begin{aligned} \operatorname{St}\tau &= \{\sigma \in \mathcal{K} \mid \tau \leq \sigma\}, \\ \operatorname{Cl}L &= \{\tau \in \mathcal{K} \mid \tau \leq \sigma \in L\}, \\ \operatorname{Lk}\tau &= \{\sigma \in \operatorname{Cl}\operatorname{St}\tau \mid \sigma \cap \tau = \emptyset\}. \end{aligned}$$

In other words, the link consists of all simplices in the closed star of τ that do not belong to the (open) star of any face of τ. Links can be used to introduce combinatorial notions of interior and boundary. They are defined relative to the space that contains the complex, which in this paper is $\mathbb{R}^3$. The *interior* of $\mathcal{K}$ is the set of simplices, σ, whose links are homeomorphic to spheres of appropriate dimension, and the *boundary* consists of all other simplices:

$$\begin{aligned} \operatorname{Int}\mathcal{K} &= \{\sigma \in \mathcal{K} \mid |\operatorname{Lk}\sigma| \approx \mathbb{S}^{2-\dim\sigma}\}, \\ \operatorname{Bd}\mathcal{K} &= \mathcal{K} - \operatorname{Int}\mathcal{K}. \end{aligned}$$

In $\mathbb{R}^3$, the boundary of a simplicial complex consists of all simplices of dimension 2 or less that are not completely surrounded by tetrahedra.

Surface properties. Let $S \subseteq \mathbb{R}^3$ be finite. The solution to the surface reconstruction problem proposed in this paper is a simplicial complex, $\mathcal{W}$. Its underlying space is what we call a *pinched 2-sphere*, that is, $|\mathcal{W}|$ is the image of a map $\varphi : \mathbb{S}^2 \to \mathbb{R}^3$ and every neighborhood of φ contains an embedding of $\mathbb{S}^2$ in $\mathbb{R}^3$. Another way to express the latter condition is that for every real $\varepsilon > 0$ there is an embedding $\iota : \mathbb{S}^2 \to \mathbb{R}^3$ with $\|\varphi(x) - \iota(x)\| < \varepsilon$ for every $x \in \mathbb{S}^2$. As we will see in Section 6 — which contains the formal definition of $\mathcal{W}$ — not every topological type of a pinched 2-sphere can be realized by $\mathcal{W}$. Here we just list a few not necessarily independent properties of $\mathcal{W}$:

(P1) $\operatorname{Vert}\mathcal{W} \subseteq S$.

(P2) $\mathcal{W}$ is connected.

(P3) $\mathbb{R}^3 - |\mathcal{W}|$ consists of $c + 1 \geq 1$ open components exactly one of which is unbounded.

Let Ω be the unbounded component, and let $\mathcal{X} \supseteq \mathcal{W}$ be a complex triangulating the complement of Ω, that is, $|\mathcal{X}| = \mathbb{R}^3 - \Omega$. Our algorithm implicitly constructs such a complex $\mathcal{X}$ that satisfies the following again not necessarily independent properties:

(P4) $\operatorname{Vert}\mathcal{X} = S$.

(P5) If $c = 0$ then $\mathcal{X} = \mathcal{W}$.

(P6) $|\mathcal{X}|$ is contractible.

(P7) $\mathcal{W} = \operatorname{Bd}\mathcal{X}$.

(P8) $\mathcal{X}$ may have principal simplices of dimension less than 3.

We also consider variants of the basic construction yielding complexes $\mathcal{W}$ that violate property (P2) and complexes $\mathcal{X}$ that violate (P4) and (P6).

3 Delaunay Complexes

The complexes $\mathcal{W}$ and $\mathcal{X}$ of Section 2 are both constructed as subcomplexes of the Delaunay complex of the data set. This section introduces Delaunay complexes and mentions properties relevant to the discussions in this paper.

Voronoi cells and Delaunay simplices. Let S be a finite set in $\mathbb{R}^3$. The *Voronoi cell* of $p \in S$ is

$$V_p = \{x \in \mathbb{R}^3 \mid \|x - p\| \leq \|x - q\| \text{ for all } q \in S\}.$$

Let $V_T = \{V_p \mid p \in T\}$ for every $T \subseteq S$. Any two Voronoi cells have disjoint interiors and the collection of all Voronoi cells, V_S, covers the entire $\mathbb{R}^3$. Throughout this paper, we assume the generic case, in which no four points lie on a common plane and no five points lie on a common sphere. The algorithmic justification of this assumption is a simulated perturbation, as described in [5]. In the generic case, two Voronoi cells are either disjoint or they meet along a common two-dimensional face. Three cells either have no points in common or they meet along a common line segment or half-line. Four cells either have no points in common or they meet in a common point. Five or more cells have no points in common.

The intersection pattern among the Voronoi cells can be recorded using a simplicial complex. More specifically, the *Delaunay complex* of S, defined as

$$\operatorname{Del} S = \{\operatorname{conv} T \mid \bigcap V_T \neq \emptyset\},$$

is such a record. The non-degeneracy assumption implies that $\operatorname{Del} S$ is a simplicial complex in $\mathbb{R}^3$. Its underlying space is $|\operatorname{Del} S| = \operatorname{conv} S$, and its vertex set is $\operatorname{Vert}\operatorname{Del} S = S$. Whether or not a simplex belongs to $\operatorname{Del} S$ can be decided by a local geometric test. Call a sphere in $\mathbb{R}^3$ *empty* if all points of S lie on or outside the sphere.

FACT 1. $\sigma = \operatorname{conv} T \in \operatorname{Del} S$ iff there is an empty sphere that passes through all points of T.

If σ is a tetrahedron then $\operatorname{card} T = 4$ and there is a unique sphere $\Sigma = \Sigma(\sigma)$ passing through the four points. We call Σ the *orthosphere* of σ or T. $\operatorname{Del} S$ contains exactly all tetrahedra whose orthospheres are empty. The triangles, edges, and vertices in $\operatorname{Del} S$ are the faces of these tetrahedra.

Weighted points. In some circumstances, it is useful to assign real weights to the points in S. With the proper generalization of definitions, all results extend from the unweighted to the more general weighted case, in which S is a finite subset of $\mathbb{R}^3 \times \mathbb{R}$. It is convenient to be ambiguous about the meaning of a point p in S: it can either be the weighted point $p \in \mathbb{R}^3 \times \mathbb{R}$ or its projection to $\mathbb{R}^3$. In either case, the *weight* is denoted by $w_p \in \mathbb{R}$. The *weighted square distance* of a point $x \in \mathbb{R}^3$ from the weighted point p is $\pi_p(x) = \|x - p\|^2 - w_p$. Voronoi cells and Delaunay simplices can be defined as before, substituting weighted square distance for Euclidean distance. We still have $|\operatorname{Del} S| = \operatorname{conv} S$. It is possible that $\operatorname{Vert} \operatorname{Del} S$ is not equal to but rather a proper subset of S. Specifically, $p \in S$ is not a vertex of $\operatorname{Del} S$ iff its Voronoi cell is empty.

To extend the local criterion expressed by Fact 1, we need to generalize the notion of orthosphere. A sphere with center $z \in \mathbb{R}^3$ and radius r is *orthogonal* to a weighted point p if $\|p - z\|^2 = w_p + r^2$, and it is *further than orthogonal* if $\|p - z\|^2$ exceeds $w_p + r^2$. Here, $r^2 \in \mathbb{R}$ and it is convenient to choose r from the set of non-negative multiples of the real and the imaginary units, 1 and $\sqrt{-1}$. We call a sphere *empty* if it is orthogonal to or further than orthogonal from all weighted points in S.

FACT 2. $\sigma = \operatorname{conv} T \in \operatorname{Del} S$ iff there is an empty sphere orthogonal to all weighted points in T.

Again we assume non-degeneracy. In the weighted case, this means that every four weighted points have a unique sphere orthogonal to all of them, and no five points have such a sphere. The *orthosphere* of a tetrahedron $\sigma = \operatorname{conv} T$ is the sphere orthogonal to the four weighted points in T. $\operatorname{Del} S$ contains exactly all tetrahedra with empty orthosphere. The triangles, edges, and vertices in $\operatorname{Del} S$ are the faces of these tetrahedra.

Relative position. Call the non-empty intersection of $k+1$ Voronoi cells an *ℓ-cell*, for $0 \le k \le 3$ and $\ell = 3 - k$. An intersection of Voronoi cells $\nu = \bigcap V_T$ is an ℓ-cell iff $\sigma = \operatorname{conv} T$ is a k-simplex in $\operatorname{Del} S$. We are interested in the position of ν and σ relative to each other in space. Their affine hulls are orthogonal flats of complementary dimension, $\ell + k = 3$, that intersect at a point y. In the assumed generic case, y is either contained in the interior of ν or it lies outside ν. Similarly, y is either contained in $\operatorname{int} \sigma$ or $y \notin \sigma$. We distinguish four mutually exclusive cases:

(R1) $\operatorname{int} \nu \cap \operatorname{int} \sigma \neq \emptyset$,

(R2) $\operatorname{int} \nu \cap \operatorname{int} \sigma = \emptyset$ and $\operatorname{int} \nu \cap \operatorname{aff} \sigma \neq \emptyset$,

(R3) $\operatorname{int} \nu \cap \operatorname{int} \sigma = \emptyset$ and $\operatorname{aff} \nu \cap \operatorname{int} \sigma \neq \emptyset$,

(R4) $\operatorname{int} \nu \cap \operatorname{aff} \sigma = \operatorname{aff} \nu \cap \operatorname{int} \sigma = \emptyset$.

Figure 5 illustrates all cases for all values of k. For $k = 0$, only Cases (R1) and (R3) and for $k = 3$ only Cases (R1) and (R2) are possible. The ℓ-cell ν is

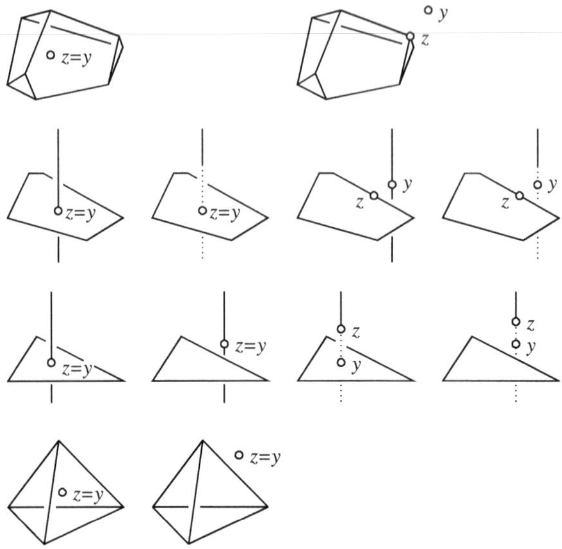

Fig. 5. From left to right: Cases (R1), (R2), (R3), and (R4), and from top to bottom: $k = 0$, 1, 2, and 3. In each case, the center of the smallest empty sphere orthogonal to all $p \in T$ is z and the intersection of the two affine hulls is y.

the set of points $x \in \mathbb{R}^3$ for which the sphere with center x and radius $\pi_p(x)$, with $p \in T$, is empty and orthogonal to all points in T. The smallest such sphere is centered at the point $z \in \nu$ closest to the point $y = \operatorname{aff} \nu \cap \operatorname{aff} \sigma$. This implies that for $k = 1$, the Cases (R2) and (R4) cannot occur unless the points are weighted. For $k = 2$, all four cases are possible even in the unweighted case.

4 Morse Functions

The complex $\mathcal{X}$ of Section 2 is a subcomplex of Del S constructed by collapsing Delaunay simplices from the outside in. To decide which simplices to collapse and which not, we construct an acyclic relation among the Delaunay simplices, which is motivated by a particular family of Morse functions. This section constructs these functions after reviewing relevant concepts from Morse theory. The reader interested in a more complete account of that theory is referred to Milnor [14] or Wallace [16].

Vector fields and flow curves. We are interested in smooth functions $f : \mathbb{R}^3 \to \mathbb{R}$ that satisfy a few genericity assumptions. Smoothness means that f is continuous and infinitely often differentiable, but we will see later that this can be weakened to twice differentiable or even only once differentiable and almost everywhere twice differentiable. The *gradient* of f is a vector field

$\nabla f : \mathbb{R}^3 \to \mathbb{R}^3$ defined by

$$\nabla f(x) = \left(\frac{\partial f}{\partial x_1}(x), \frac{\partial f}{\partial x_2}(x), \frac{\partial f}{\partial x_3}(x) \right).$$

It is smooth because f is smooth. A point $y \in \mathbb{R}^3$ is *critical* if $\nabla f(y) = (0,0,0)$. The *Hessian* at y is the three-by-three matrix of partial second derivatives. A critical point is *non-degenerate* if the Hessian at that point has full rank. Non-degenerate critical points are necessarily isolated. A *Morse function* is a smooth function, f, whose critical points are non-degenerate. The fundamental Morse lemma asserts that for every critical point y there are local coordinates originating from y so that

$$f(x) = f(y) \pm x_1^2 \pm x_2^2 \pm x_3^2$$

for all x in a neighborhood of y. The number of minus signs is the *index* of y. For example, critical points of index 0 are local minima, and critical points of index 3 are local maxima. Critical points of index 1 and 2 are different types of saddle points.

The gradient of f defines a first-order differential equation. A solution is a maximal embedding $\iota : \mathbb{R} \to \mathbb{R}^3$ whose tangent vectors agree with the gradient of f. For each non-critical point $x \in \mathbb{R}^3$, there is a unique solution or *flow curve* ι_x that contains x, that is, $x = \iota_x(t)$ for some $t \in \mathbb{R}$. It follows that two flow curves are either the same or disjoint. The orientation of $\mathbb{R}$ from $-\infty$ to $+\infty$ imposes an orientation on the flow curve. It is convenient to compactify $\mathbb{R}^3$ to $\mathbb{S}^3$ by stipulating a critical point ω at infinity. Then every flow curve starts at a critical point and ends at a critical point. In this paper, ω will only be an endpoint of flow curves and we define its index to be 3. Let $C(f)$ be the collection of critical points, including ω. The *stable manifold* of $y \in C(f)$ is

$$M_y = \{y\} \cup \{x \in \mathbb{R}^3 \mid \iota_x(t) \to y \text{ as } t \to \infty\}.$$

If the index of y is j, then the stable manifold consists of y and a $(j-1)$-dimensional sphere of flow curves. For all $y \neq \omega$, M_y is the image of an injective map from $\mathbb{R}^j$ to $\mathbb{R}^3$. If M_y fails to be homeomorphic to $\mathbb{R}^j$ that is only because it is possible that flow curves ending at y share the same starting point. M_ω is the image of an injective map from $\mathbb{R}^3 - \{0\}$, the punctured three-dimensional space, to $\mathbb{R}^3$. The stable manifolds are mutually disjoint open sets that cover the entire $\mathbb{R}^3$. We call this the *complex of stable manifolds* and write

$$\operatorname{Sm} f = \{M_y \mid y \in C(f)\}.$$

The *j-cells* of $\operatorname{Sm} f$ are the stable manifolds of critical points of index j.

Distance from empty spheres. Given a finite set S of weighted or unweighted points in $\mathbb{R}^3$, we construct a real-valued function by considering empty spheres. Think of a sphere $\Sigma = (z, r)$ as a weighted point $(z, r^2) \in \mathbb{R}^3 \times \mathbb{R}$. With this interpretation, the weighted square distance of a point $x \in \mathbb{R}^3$ from Σ is well-defined as $\pi_\Sigma(x) = \|x - z\|^2 - r^2$. Consider the function that maps every point $x \in \mathbb{R}^3$ to the minimum weighted square distance from any empty sphere. For a point inside the convex hull of S, the minimum weighted square distance is defined by the orthosphere of a Delaunay tetrahedron. For a point outside the convex hull, the minimum weighted square distance is defined by an infinitely large sphere or hyperplane that supports the convex hull in a triangle. This sphere can be interpreted as the orthosphere of an infinitely large tetrahedron spanned by the triangle and a point at infinity.

The exact shape of this infinitely large tetrahedron can be obtained by constructing the Voronoi cells for the set of orthospheres, including the ones of infinite size. To eliminate the remaining ambiguity, we approximate each infinitely large sphere by a sphere of radius $\frac{1}{\delta}$ that is orthogonal to the same three weighted points. As δ goes to zero, some of the Voronoi cells do not change, some grow and eventually become unbounded, and some cells disappear to infinity. The first kind are the original Delaunay tetrahedra, and the second kind are the infinitely large tetrahedra defined by convex hull triangles. Together, the two types of tetrahedra cover the entire $\mathbb{R}^3$. Let D^3 be the set of tetrahedra of both types. We construct $g : \mathbb{R}^3 \to \mathbb{R}$ by defining $g(x) = \max\{-\pi_{\Sigma(\sigma)}(x) \mid \sigma \in D^3\}$. For a point x, the relevant orthosphere is defined by the tetrahedron that contains the point x.

FACT 3. If $x \in \sigma \in D^3$ then $g(x) = -\pi_{\Sigma(\sigma)}(x)$.

Note that $g(x) = +\infty$ if $x \notin \operatorname{conv} S$. This is a slight inconvenience in the subsequent discussion. All difficulties can be finessed by again approximating each infinitely large sphere by a sphere of radius $\frac{1}{\delta}$. To simplify the discussion, we do not explicate this approximation, but we do pretend that g is a continuous map from $\mathbb{R}^3$ to $\mathbb{R}$.

Smoothing. The function g is continuous but does not quite qualify as a Morse function because it is not everywhere smooth. Smooth functions can be derived from g using appropriate cut-off functions blending between adjacent Delaunay tetrahedra. Figure 6 illustrates the effect of smoothing on Delaunay edges in $\mathbb{R}^2$.

We construct an infinite family of smooth functions f_ε approximating g. Consider the graph of $g : \mathbb{R}^3 \to \mathbb{R}$, which is $G = \{(x, g(x)) \mid x \in \mathbb{R}^3\} \subseteq \mathbb{R}^4$. It is a three-dimensional manifold that consists of finitely many smooth patches, which fit together in a continuous but non-differentiable manner. Let $\varepsilon > 0$ be real, let $b_\varepsilon = \{u \in \mathbb{R}^4 \mid \|u\| \leq \varepsilon\}$, and consider $G + b_\varepsilon = \{z + u \mid z \in G, u \in b_\varepsilon\}$. For sufficiently small positive ε, the lower boundary of $G + b_\varepsilon$ is

Fig. 6. From top to bottom: the gradient of f_ε on the left and its limit, ∇g, on the right for a centered, a confident, an equivocal edge in $\mathbb{R}^2$. The centered and confident edges repel a flow curve unless it lies exactly on the edge. The equivocal edge is crossed by an interval of bending flow curves.

the graph of a differentiable function

$$f_\varepsilon(x) = \min\{r \mid (x, r) \in G + b_\varepsilon\}.$$

The f_ε are not smooth in the sense of being infinitely often differentiable. Still, they are everywhere differentiable and almost everywhere twice differentiable, which suffices for the purposes of this paper. Most importantly, the notions of gradient, critical point, and flow curve are defined. For example, outside the convex hull of S, g and therefore the f_ε are approximately infinitely steep and the flow curves go quickly to infinity. By assumption of non-degeneracy, the f_ε are twice differentiable at all critical points, and we can define indices and stable manifolds as before.

Limit construction. We use limit considerations to construct a vector field for g. For every point $x \in \mathbb{R}^3$, define $\nabla g(x) = \lim_{\varepsilon \to 0} \nabla f_\varepsilon(x)$. ∇g is a vector field albeit not continuous because g is not differentiable, as seen in Figure 6. Observe that the f_ε have identical critical points. In other words, $C(f_\varepsilon) = C(f_\delta)$ for any two sufficiently small $0 < \varepsilon < \delta$. The following relation between g and the f_ε is fairly straightforward to prove.

FACT 4. $y \in \mathbb{R}^3$ is a critical point of f_ε iff $\nabla g(y) = (0, 0, 0)$, and the index of y is j iff the Delaunay simplex σ that contains y in its interior has dimension j.

The vector field ∇g does not enjoy some of the nice properties of the ∇f_ε. In particular, ∇g is not continuous. We finesse some of the resulting difficulties with limit considerations. As an example, consider a non-critical point $x \in \mathbb{R}^3$. For each sufficiently small $\varepsilon > 0$, there is a unique flow curve $\iota_{x,\varepsilon}$ of ∇f_ε that contains x. Define the *limit curve* for x as $\lambda_x = \lim_{\varepsilon \to 0} \iota_{x,\varepsilon}$. The curve λ_x is a continuous though generally not a smooth embedding of $\mathbb{R}$ in $\mathbb{R}^3$. Indeed, λ_x is piecewise linear, and for each simplex σ, $\lambda_x \cap \operatorname{int} \sigma$ is either empty, a point, or a line segment. While two flow curves are either disjoint or the same, two limit curves can partially overlap. However, once they separate they stay apart. In other words, if $x \in \lambda_u \cap \lambda_v$ then the portions of λ_u and λ_v preceding x are the same. The reason for this is the repulsion of nearby flow curves by centered and confident simplices. Only equivocal simplices attract nearby flow curves, but these curves pass right through the simplex, without ambiguity or merging of curves.

Based on the definition of the λ_x, we can consider limits of stable manifolds and the complex they form. These limits provide the guiding principle motivating our surface construction method. At his moment, we recall that $\mathcal{W}$ refers to the surface and $\mathcal{X}$ refers to a triangulation of the portion of $\mathbb{R}^3$ bounded by $\mathcal{W}$.

INTUITION. In the limit, $\mathcal{X}$ triangulates the complement of the stable manifold of ω and $\mathcal{W}$ is the boundary of $\mathcal{X}$.

We will take small liberties in translating this intuition into an unambiguous construction, which will be given in Section 6.

5 Ordering Simplices

The flow and limit curves motivate the construction of an acyclic relation over the set of Delaunay simplices. The complexes $\mathcal{W}$ and $\mathcal{X}$ of Section 2 are then constructed by collapsing simplices from back to front in the relation.

Flow relation. Introduce a dummy simplex, ω, that represents the outside, or complement of $|\operatorname{Del} S|$. It replaces the collection of infinitely large tetrahedra introduced in Section 4. All tetrahedra in this collection have the same flow behavior and can be treated uniformly. We deliberately choose the same name, ω, for the dummy simplex and the dummy critical point, and this will not cause any confusion. By definition, the faces of ω are the convex hull faces, which are the simplices in $\operatorname{Bd} \operatorname{Del} S$. Let $D = \operatorname{Del} S \cup \{\omega\}$. The *flow relation*, $\prec \subseteq D \times D$, is constructed to mimic the behavior of the limit curves. Specifically, $\tau \prec \upsilon \prec \sigma$ if υ is a proper face of τ and of σ and there is a point $x \in \operatorname{int} \upsilon$ with λ_x passing from $\operatorname{int} \tau$ through x to $\operatorname{int} \sigma$. We pronounce this as τ precedes υ and υ precedes σ. The condition implies that every neighborhood of x contains a non-empty subset of $\lambda_x \cap \operatorname{int} \tau$ and a non-empty subset of $\lambda_x \cap \operatorname{int} \sigma$. We call τ a *predecessor* and σ a *successor* of

υ. The sets of *descendents* and *ancestors* are

$$\begin{aligned} \operatorname{Des} \upsilon &= \{\upsilon\} \cup \bigcup_{\upsilon \prec \sigma} \operatorname{Des} \sigma, \\ \operatorname{Anc} \upsilon &= \{\upsilon\} \cup \bigcup_{\tau \prec \upsilon} \operatorname{Anc} \tau. \end{aligned}$$

It is convenient to study the flow relation by distinguishing three types of Delaunay simplices: centered, confident, and equivocal. These are related to the classification of Delaunay simplices introduced in Section 3 and illustrated in Figure 5. The three types are mutually exclusive and exhaust all possible Delaunay simplices. For each type, we specify predecessors and successors in terms of their local geometric properties.

Centered simplices. A simplex $\sigma \in D$ is *centered* if, for every $x \in \operatorname{int} \sigma$, the portion of λ_x succeeding x is contained in $\operatorname{int} \sigma$. Illustrations of a centered edge and a centered triangle can be seen in Figures 6 and 7. In words, limit curves enter but do not exit $\operatorname{int} \sigma$. In the case of $\sigma = \omega$, the limit curves

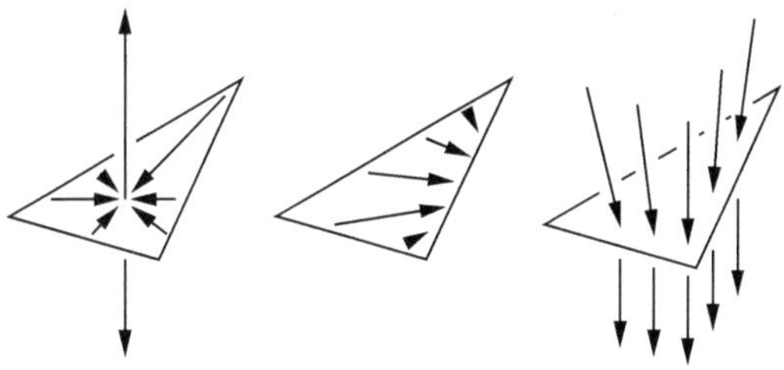

Fig. 7. From left to right: a centered, a confident, an equivocal triangle in $\mathbb{R}^3$. We remember this terminology by thinking of a triangle that contains its own flow as confident. If on top of its own flow it also contains the limit point, we think of it as overly confident or (self-)centered. We think of a triangle that is too weak to contain its flow as equivocal.

diverge, and in the case of $\sigma \in \operatorname{Del} S$, they converge towards a critical point y in the interior of $\sigma = \operatorname{conv} T$. This point is also contained in the interior of the corresponding ℓ-cell $\nu = \bigcap V_T$. It follows that ν and σ fall into Case (R1), which is illustrated by the leftmost column in Figure 5. The index of y is the dimension of σ.

FACT 5. A Delaunay simplex $\sigma \in \operatorname{Del} S$ with dual Voronoi cell ν is centered iff $\operatorname{int} \nu \cap \operatorname{int} \sigma \neq \emptyset$. The intersection is a point $y \in \mathbb{R}^3$, y is critical for all f_ε, and the index of y is $\dim \sigma$.

Since the limit curves do not exit its interior, σ has no successors in the flow relation. All predecessors are faces, but in general not all faces are predecessors of σ.

Confident simplices. A simplex $\tau \in D$ is *confident* if it is not centered and for every $x \in \text{int}\,\tau$ there is a sufficiently small neighborhood N of x with $\lambda_x \cap N \subseteq \text{int}\,\tau$. Illustrations of a confident edge and a confident triangle can be seen in Figures 6 and 7. Confident simplices are quite similar to centered ones, in the sense that they would be centered if they covered a large enough part of their affine hull. In other words, the affine hull of $\tau = \text{conv}\,T$ intersects the interior of the corresponding ℓ-cell $\nu = \bigcap V_T$. Thus, ν and τ fall into Case (R2), which is illustrated by the second column from the left in Figure 5. All tetrahedra that are not centered are confident.

FACT 6. A Delaunay simplex $\tau \in \text{Del}\,S$ with dual Voronoi cell ν is confident iff $\text{int}\,\nu \cap \text{int}\,\tau = \emptyset$ and $\text{int}\,\nu \cap \text{aff}\,\tau \neq \emptyset$.

All predecessors and successors of a confident τ are faces of τ. To determine which faces are successors, consider the center of the smallest orthosphere of τ. This is the point $z = \text{int}\,\nu \cap \text{aff}\,\tau$, which lies outside $\text{int}\,\tau$. Let $k = \dim \tau$ and consider each $(k-1)$-dimensional face ξ of τ. By assumption of general position, the affine hull of ξ does not contain z. Either $\text{aff}\,\xi$ separates z and $\text{int}\,\tau$ within $\text{aff}\,\tau$ or z and $\text{int}\,\tau$ lie on the same side of $\text{aff}\,\xi$. A proper face υ of τ is a successor iff the affine hulls of all $(k-1)$-dimensional faces ξ of τ that contain υ separate z and $\text{int}\,\tau$. Observe that there is a unique lowest-dimensional successor υ. It has the property that the successors of τ are precisely the proper faces of τ that are cofaces of υ. Every other proper face of τ is either a predecessor of τ or neither a successor nor a predecessor.

Equivocal simplices. A simplex $\upsilon \in D$ is *equivocal* if $\lambda_x \cap \text{int}\,\upsilon = x$ for every point $x \in \text{int}\,\upsilon$. Illustrations of an equivocal edge and an equivocal triangle can be seen in Figures 6 and 7. The center z of the smallest empty sphere orthogonal to all $p \in T$, with $\upsilon = \text{conv}\,T$, lies outside the affine hull of υ. In other words, the affine hull of υ misses the interior of the corresponding ℓ-cell, $\nu = \bigcap V_T$. Thus ν and υ fall into Cases (R3) or (R4), which are illustrated by the right two columns of Figure 5.

FACT 7. A Delaunay simplex $\upsilon \in \text{Del}\,S$ with dual Voronoi cell ν is equivocal iff $\text{int}\,\nu \cap \text{aff}\,\upsilon = \emptyset$.

All predecessors and successors are cofaces of υ. For example, an equivocal triangle has exactly one predecessor, a tetrahedral coface, and exactly one successor, the other tetrahedral coface. The second coface can also be ω. All predecessors and successors of an equivocal simplex are confident or centered. Symmetrically, all predecessors and successors of confident and centered simplices are equivocal. An equivocal edge or vertex can have an arbitrary number of successors, but there is always only one predecessor. This fact is important and deserves a proof.

CLAIM 8. Every equivocal simplex has exactly one predecessor.

PROOF. Let υ be equivocal and $x \in \operatorname{int} \upsilon$. Consider the collection of limit curves λ_u that pass through x. As mentioned earlier, all λ_u approach x from the same direction although they possibly leave x in different directions. Since before x all λ_u are the same, we consider only λ_x and in particular a sufficiently small portion of λ_x immediately preceding x. This portion is a line segment contained in the interior of a simplex $\tau = \operatorname{conv} U$. We have $\tau \prec \upsilon$ and τ is confident. Every limit curve that intersects $\operatorname{int} \tau$ does so in a portion that lies on a line passing through the center z of the smallest empty sphere orthogonal to all $p \in U$. It follows that for every point $x \in \operatorname{int} \upsilon$, a sufficiently small portion of λ_x immediately preceding x lies in the affine hull of υ and z and therefore in $\operatorname{int} \tau$. In other words, each x identifies the same simplex τ, which implies that τ is the only predecessor of υ. □

The unique predecessor τ of the equivocal υ can be determined through local geometric considerations. Recall the definitions of z and $y = \operatorname{aff} \nu \cap \operatorname{aff} \upsilon$. By Fact 7, we have $y \notin \nu$, and since $z \in \nu$ is closest to y, it lies on the boundary. Let $U \subseteq S$ be maximal with $z \in \bigcap V_U$. We have $T \subseteq U$ and $T \neq U$ because $z \in \operatorname{bd} \nu$. The predecessor of υ is $\tau = \operatorname{conv} U$.

6 Clustering Simplices

A *sink* is a simplex $\sigma \in D$ without successor in the flow relation. By construction, the sinks are exactly the centered simplices together with ω. We use $\prec$ to define for each sink σ a set of simplices gravitating towards σ. Think of σ analogous to a critical point and of this set analogous to a stable manifold.

Acyclicity. We show that the flow relation is acyclic. This is plausible since the value of the function g strictly increases along limit curves. A *cycle* is a sequence of simplices $\sigma_1 \prec \sigma_2 \prec \ldots \prec \sigma_\ell$, with $\ell \geq 3$ and $\sigma_1 = \sigma_\ell$. The algorithm in Section 8 relies on the absence of cycles in the flow relation.

CLAIM 9. The relation $\prec$ is acyclic.

PROOF. Let $\sigma_i = \operatorname{conv} T \in \operatorname{Del} S$, and let $\Sigma_i = (z_i, r_i)$ be the smallest empty sphere orthogonal to all $p \in T$. Consider $\sigma_i \prec \sigma_j$ and note that σ_i cannot be centered since otherwise it has no successor. If σ_i is confident then σ_j is equivocal and we have $\Sigma_i = \Sigma_j$ and $\dim \sigma_i > \dim \sigma_j$. If σ_i is equivocal then σ_j is centered or confident. Hence, $z_i \neq z_j$ and by assumption of non-degeneracy we have $r_i^2 < r_j^2$. To prove a cycle cannot exist, we assign to each σ_i the pair $(r_i^2, -\dim \sigma_i)$. The pairs increase lexicographically along a chain, which implies the chain cannot come back to where it started. □

Ancestor sets. The analogy between stable manifolds and ancestor sets of centered simplices is generally correct but troubled by inconsistencies in the details. The source of the trouble are simplices with more than one successor. Their existence implies the possibility of non-disjoint ancestor sets. This is in contrast to stable manifolds of a smooth Morse function, which are pairwise disjoint, but not unlike the closures of stable manifolds, which can overlap. There are two types of simplices which may have more than one successor:

(S1) equivocal edges and vertices,

(S2) confident tetrahedra and triangles.

Type (S1) simplices relate to the pinching or flattening of stable manifolds that occurs in the limit. Type (S2) simplices are a result of the non-continuous dependence of the stable manifolds from the input points. There are no type (S2) simplices in the two-dimensional unweighted case, where the limit of the complex of stable manifolds is the Gabriel graph [12]. Nonetheless, type (S2) simplices appear already in the two-dimensional weighted and the three-dimensional unweighted cases.

Despite the possibility of type (S1) and (S2) simplices, ancestor sets retain the containment relation of the closures of stable manifolds. Let y and z be critical points of a differentiable map f_ε, for a sufficiently small $\varepsilon > 0$, and recall that M_y and M_z denote their stable manifolds, as defined in Section 4. Let σ and τ be the centered simplices with $y \in \operatorname{int}\sigma$ and $z \in \operatorname{int}\tau$, and recall that $\operatorname{Anc}\sigma$ and $\operatorname{Anc}\tau$ are their ancestor sets.

CLAIM 10. $M_z \subseteq \operatorname{cl} M_y$ implies $\operatorname{Anc}\tau \subseteq \operatorname{Cl}\operatorname{Anc}\sigma$.

PROOF. Assume first that there are no type (S2) simplices. Then the dimensions of the confident simplices along a chain of the flow relation cannot decrease. If follows that $\dim\tau$ is the maximum dimension of any simplex in the ancestor set of τ, $\dim\operatorname{Cl}\operatorname{Anc}\tau = \dim\tau$. The dimension of τ is also the index of z. The claimed subset relation follows because limit curves are approximated by the flow curves as ε goes to 0.

If D contains type (S2) simplices, the dimension of $\operatorname{Cl}\operatorname{Anc}\tau$ may exceed the dimension of τ. The claimed subset relation still holds because the simplices $\upsilon \in \operatorname{Anc}\tau$ with $\dim\upsilon > \dim\tau$ have descendents outside $\operatorname{Anc}\tau$ and in particular σ is a descendent of υ. ◻

The face-coface relation over the set of stable manifolds is transitive and induces a partial order over the collection of centered simplices. This relation will be used in Section 8.

Definition of $\mathcal{X}$ and $\mathcal{W}$. Ancestor sets seem slightly too large to faithfully represent stable manifolds. We introduce a more conservative notion that admits only simplices whose cofaces have descendent sets contained in ancestor sets. Let $C \subseteq D$ be the set of sinks, including ω. For a subset

$B \subseteq C$, define its ancestor set as the union of ancestor sets of its members, $\operatorname{Anc} B = \bigcup_{\sigma \in B} \operatorname{Anc} \sigma$. The *conservative ancestor set* of B is

$$\begin{aligned} \operatorname{Cnc} B &= \operatorname{Int} \{\tau \in D \mid \operatorname{Des} \tau \subseteq \operatorname{Anc} B\} \\ &= \{\tau \in D \mid \operatorname{Des} \sigma \subseteq \operatorname{Anc} B \text{ for all } \sigma \in \operatorname{St} \tau\}. \end{aligned}$$

Observe that the sets $\operatorname{Cnc} \sigma = \operatorname{Cnc} \{\sigma\}$, $\sigma \in C$, do not necessarily cover D. Indeed, a simplex is not covered by any set $\operatorname{Cnc} \sigma$ if it belongs to more than one set $\operatorname{Anc} \sigma$. On the other hand, $\operatorname{Cnc} C = D$ which shows that $\operatorname{Cnc} B$ is generally not equal to the union of conservative ancestor sets of its members. In this paper, we are only interested in

$$\begin{aligned} \Omega &= \operatorname{Cnc} \omega \\ &= \operatorname{Int} \{\tau \in D \mid \operatorname{Des} \tau \subseteq \operatorname{Anc} \omega\} \\ &= D - \operatorname{Cl} \bigcup \operatorname{Anc} \sigma \\ &= D - \bigcup \operatorname{Cl} \operatorname{Anc} \sigma, \end{aligned}$$

where the union is taken over all sinks $\sigma \neq \omega$. The wrapping surface, $\mathcal{W}$, is the boundary of

$$\begin{aligned} \mathcal{X} &= D - \Omega \\ &= \bigcup \operatorname{Cl} \operatorname{Anc} \sigma, \end{aligned}$$

where the union is again taken over all sinks $\sigma \neq \omega$. We have $\Omega = \operatorname{Int} \Omega$, by definition, and therefore $\mathcal{X} = \operatorname{Cl} \mathcal{X}$. In words, $\mathcal{X}$ is a simplicial complex. We will see shortly that $|\mathcal{X}|$ is contractible. In summary, Ω and $\mathcal{X}$ have the topological properties suggested by the analogy between Ω and the stable manifold of ω.

7 Collapsing Simplices

In this section, we show how to construct the complex $\mathcal{X}$ by collapsing simplices of the Delaunay complex. We refer to this algorithm as the *basic construction.*

Collapses. Consider a simplicial complex $\mathcal{K}$, and let υ be a simplex with exactly one proper coface $\tau \in \mathcal{K}$. In this case, $\dim \upsilon = \dim \tau - 1$ and τ is a principal simplex in $\mathcal{K}$. The operation that removes τ and υ from $\mathcal{K}$ is called an *elementary collapse*, and we write $\mathcal{K} \searrow \mathcal{K}_1$, where $\mathcal{K}_1 = \mathcal{K} - \{\tau, \upsilon\}$. An elementary collapse maintains the homotopy type of the complex.

FACT 11. $|\mathcal{K}_1|$ is homotopy equivalent to $|\mathcal{K}|$.

The homotopy equivalence can be established by constructing a deformation retraction of $|\mathcal{K}|$ to $|\mathcal{K}_1|$. This is a homotopy $F : |\mathcal{K}| \times [0,1] \to |\mathcal{K}|$ between

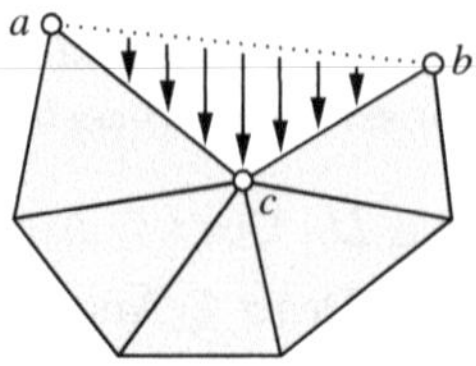

Fig. 8. The elementary collapse removes the edge ab together with the triangle abc. The corresponding deformation retraction moves all points of abc parallel to the direction from the barycenter of ab to c.

the identity map on $|\mathcal{K}|$ and a map from $|\mathcal{K}|$ to $|\mathcal{K}_1|$ that keeps all points $x \in |\mathcal{K}_1|$ fixed for all $t \in [0,1]$. Such a homotopy is indicated in Figure 8.

An ℓ-simplex υ is *free* if there is a $k > \ell$ and a k-simplex $\tau \in \mathcal{K}$ such that all cofaces of υ are faces of τ. It follows that all cofaces of a free υ are free, except for τ, which is a principal simplex in $\mathcal{K}$. The operation that removes all cofaces of the free υ is called a *(k,ℓ)-collapse.* The number of simplices removed is $2m = 2^{k-\ell}$, and the (k,ℓ)-collapse can be written as a composition of m elementary collapses:

$$\mathcal{K} \searrow \mathcal{K}_1 \searrow \mathcal{K}_2 \searrow \ldots \searrow \mathcal{K}_m.$$

In $\mathbb{R}^3$, the (k,ℓ)-collapses satisfy $0 \leq \ell < k \leq 3$, and there are six different cases all illustrated in Figure 9. We use collapses to shrink a subcomplex $\mathcal{Y}$

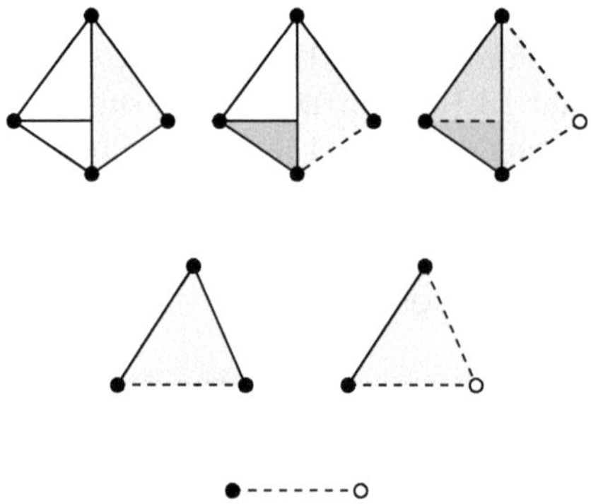

Fig. 9. From left to right, top to bottom: collapsing a tetrahedron from a triangle, an edge, a vertex, collapsing a triangle from an edge, a vertex, and collapsing an edge from a vertex. In each case, the collapse removes the tetrahedron, the shaded triangles, the dashed edges, and the hollow vertices, if any.

of the Delaunay complex. Call a pair $\upsilon \leq \tau$ *collapsible* if

(i) $\upsilon, \tau \in \mathcal{Y}$, υ is free, υ is equivocal, and

(ii) $\tau \prec \upsilon$, τ is the highest-dimensional coface of υ in $\mathcal{Y}$, υ is the lowest-dimensional successor of τ.

Observe that $\upsilon \in \mathrm{Bd}\,\mathcal{Y}$ because υ is free. Its coface τ may have several successors, all of which are free because they contain υ, which is free.

Correctness. A *constructive retraction* is an algorithm A that starts with the Delaunay complex and removes simplices by collapsing as long as there are collapsible pairs. Let $\mathcal{Y}_{\mathsf{A}}$ be the remaining subcomplex. We claim that every constructive retraction correctly constructs $\mathcal{X}$, no matter what sequence of collapses it chooses.

THEOREM. $\mathcal{Y}_{\mathsf{A}} = \mathcal{X}$ for every constructive retraction A.

PROOF. We prove $\mathcal{X} \subseteq \mathcal{Y} = \mathcal{Y}_{\mathsf{A}}$ by induction in the order of increasing descendent sets. Let $\xi \in D - \mathcal{Y}$. To show $\xi \in D - \mathcal{X}$, recall that $\xi \in \Omega = D - \mathcal{X}$ iff $\mathrm{Des}\,\eta \subseteq \mathrm{Anc}\,\omega$ for all cofaces η of ξ. Let $\upsilon \leq \tau$ be the pair whose collapse removes ξ from $\mathrm{Del}\,S$, and note that $\upsilon \leq \xi \leq \tau$. We begin by proving $\mathrm{Des}\,\xi \subseteq \mathrm{Anc}\,\omega$. If $\xi \neq \tau$ then ξ is equivocal and all successors σ are cofaces that have already been removed. Then $\mathrm{Des}\,\sigma \subseteq \mathrm{Anc}\,\omega$ by induction and $\mathrm{Des}\,\xi \subseteq \mathrm{Anc}\,\omega$ follows. If $\xi = \tau$ then all successors σ are proper faces of τ and cofaces of υ. We just proved $\mathrm{Des}\,\sigma \subseteq \mathrm{Anc}\,\omega$ for all such σ and $\mathrm{Des}\,\xi \subseteq \mathrm{Anc}\,\omega$ again follows. Finally observe that every coface η of ξ has either already been removed or $\xi \leq \eta \leq \tau$. In both cases we have $\mathrm{Des}\,\eta \subseteq \mathrm{Anc}\,\omega$ and therefore $\xi \in \mathrm{Cnc}\,\omega = D - \mathcal{X}$.

We prove $\mathcal{Y} \subseteq \mathcal{X}$ by contradiction. Assume $\mathcal{Y} - \mathcal{X} \neq \emptyset$. For each simplex $\sigma = \mathrm{conv}\,T$, consider the pair $(r^2, -k)$, where $k = \dim\sigma$ and r is the radius of the smallest empty sphere orthogonal to all $p \in T$. By assumption of genericity, all pairs are different and, as shown in the proof of Claim 9, they lexicographically increase along chains in the flow relation. Let υ be the simplex in $\mathcal{Y} - \mathcal{X}$ with lexicographically largest pair. Since $\upsilon \in \Omega$, each successor σ of υ belongs to Ω. Furthermore, the pair of σ is lexicographically larger than that of υ. It follows that υ has no successor in $\mathcal{Y}$ and is therefore either a tetrahedron or equivocal. To contradict the first possibility, observe that for every tetrahedron υ with $\mathrm{Des}\,\upsilon \subseteq \mathrm{Anc}\,\omega$, there is an alternating sequence of tetrahedra and triangles,

$$\upsilon \prec \tau_0 \prec \upsilon_1 \prec \tau_1 \prec \ldots \prec \tau_j \prec \omega,$$

connecting υ to ω. So $\upsilon \notin \mathcal{Y}$ implies $\tau_0 \notin \mathcal{Y}$, contradicting the choice of υ. Second, consider the case in which υ is an equivocal simplex. Let τ be the predecessor of υ, which is unique and confident. The predecessor τ and its faces are the only cofaces of υ in $\mathcal{Y}$, since all others have larger r^2 value than υ. It follows that υ is free and $\upsilon \leq \tau$ is collapsible, which contradicts $\upsilon \in \mathcal{Y}$. ◻

The construction of $\mathcal{X}$ starts with $\mathrm{Del}\,S$, which is a contractible simplicial complex. The collapses maintain the homotopy type, which implies that $|\mathcal{X}|$ is indeed contractible, as claimed at the end of Section 6.

8 Deleting Simplices

A strength of the basic construction in Section 7 is that the wrapping surface is unique and its computation is fully automatic. A complementary weakness is the lack of variability in the result. This section generalizes the basic construction so the shape of the surface is influenced by the choice of additional parameters. The surface may wrap tighter around input points or develop holes and change its topology. We first describe a simplex removing operation that changes the homotopy type and then use this operation to modify the surface.

Discriminating by size. The idea is to collapse simplices not only from ω but more generally from all significant sinks. Recall that each sink, or centered simplex $\sigma = \operatorname{conv} T$, with $T \subseteq S$, corresponds to a critical point $y \in \operatorname{int} \sigma$. Call $|\sigma| = g(y)$ the *size* of σ, where g is the same as in Section 4. By definition of g, the sphere $\Sigma = (y, \sqrt{|\sigma|})$ is empty and the smallest sphere orthogonal to all $p \in T$. It is intuitively plausible that large size is indicative of space through which the wrapping surface may want to be pushed.

The non-degeneracy assumption on the input points implies that all sizes are different. Sort the sinks in order of decreasing size $|\sigma_0| > |\sigma_1| > \ldots > |\sigma_m|$, where $\sigma_0 = \omega$ and $|\omega| = +\infty$. For each index $0 \leq j \leq m$, define

$$\begin{aligned} \mathcal{X}_j &= D - \operatorname{Cnc} B_j \\ &= \bigcup_{i=j+1}^{m} \operatorname{Cl} \operatorname{Anc} \sigma_i, \end{aligned}$$

where $B_j = \{\sigma_0, \sigma_1, \ldots, \sigma_j\}$ and $\operatorname{Cnc} B_j$ is the conservative ancestor set, as defined in Section 6. Define $\mathcal{W}_j = \operatorname{Bd} \mathcal{X}_j$. The $\mathcal{X}_j$ form a nested sequence of subcomplexes:

$$\mathcal{X} = \mathcal{X}_0 \supseteq \mathcal{X}_1 \supseteq \ldots \supseteq \mathcal{X}_m = \emptyset.$$

Correspondingly, the $\mathcal{W}_j$ form a nested sequence of wrapping surfaces.

An operation that removes a principal simplex σ from a complex $\mathcal{K}$ is called a *deletion*. In contrast to a collapse, a deletion alters the homotopy type of $|\mathcal{K}|$. A particular $\mathcal{X}_j$ is constructed from D by a succession of deletions and collapses. Each deletion is followed by collapses until no collapsible pairs remain. We refer to such an exhaustive sequence of collapses as a *retraction*. For example the basic construction computes $\mathcal{X} = \mathcal{X}_0$ from D by first deleting ω and then performing a retraction. The complex $\mathcal{X}_j$ is computed by repeating these two operations $j+1$ times, once each for $\sigma_0, \sigma_1, \ldots, \sigma_j$.

Local modifications. There is no reason other than convenience that requires a total order of the retractions. Indeed, it is possibly to perform retractions in any order consistent with the face-coface relation of stable manifolds. Recall that $C \subseteq D$ is the set of sinks, including ω. Let τ and σ

be centered simplices and $z \in \text{int}\,\tau$ and $y \in \text{int}\,\sigma$ the corresponding critical points. The pair $\tau \vdash \sigma$ is in the *sink relation* $\vdash \subseteq C \times C$ if $M_z \subseteq \text{cl}\,M_y$, as discussed in Claim 10. We call σ a *cosink* of τ and write $\text{Cos}\,\tau$ for the set of cosinks, including τ. The relation is acyclic and transitive and therefore a partial order. It can be used to locally change or refine the wrapping surface. To describe how this may work, let B and B' be disjoint sets of centered simplices with $\mathcal{X} = D - \text{Cnc}\,B$ and $B' \subseteq \mathcal{W} = \text{Bd}\,\mathcal{X}$. We call the set of simplices in the conservative ancestor set of $B \cup B'$ that belong to $\mathcal{W}$ a *front*: $F = \mathcal{W} \cap \text{Cnc}\,(B \cup B')$. Locality is understood in terms of F, that is, changes to the surface are triggered only from simplices in F. We exemplify this idea by setting a size threshold δ and removing simplices that have descendents σ with size exceeding δ. It suffices to consider sinks and we restrict attention to simplices τ in $F_\delta = \{\sigma \in F \cap C \mid |\sigma| > \delta\}$. To remove a simplex τ, we remove the entire conservative ancestor set of its cosinks. This is repeated for every $\tau \in F_\delta$. The local modification of $\mathcal{X}$ is completely specified by F and δ and creates $\mathcal{X}' = D - \text{Cnc}\,(B \cup \text{Cos}\,F_\delta)$. It is possible that $\mathcal{X}'$ contains a simplex $\upsilon \in F$ even though all centered descendents of υ and their cosinks have been removed. This is the case if υ has a coface ξ with at least one centered descendent remaining in $\mathcal{X}'$. The construction of $\mathcal{X}'$ is again reduced to a sequence of deletions and retractions: for each $\tau \in F_\delta$, we find all $\sigma \in \text{Cos}\,\tau$ and repeatedly delete the σ without remaining cosinks. Each deletion is followed by a retraction. Similar to the basic construction, the sequence in which the $\tau \in F_\delta$ are removed is irrelevant, since any one results in the same $\mathcal{X}'$.

An interesting variant of the above described local deformation expands F_δ to include sinks υ that become part of $\mathcal{W}$ during the process. Only sinks υ with size $|\upsilon| > \delta$ are considered. This recursive construction amounts to replacing F_δ by G_δ, which is the maximal set of sinks $\upsilon \in \mathcal{X}$ with $|\upsilon| > \delta$ so that every component of $\text{Cnc}\,G_\delta$ contains a simplex in F_δ.

9 Discussion

The solution to the surface reconstruction problem presented in this paper is based on discrete methods inspired by concepts in continuous mathematics. In particular, we construct subcomplexes of the Delaunay complex of a finite point set by collapsing and occasionally deleting simplices. Continuous mathematics enters in the form of Morse functions and their gradient fields, which constitute the rationale for the rules that decide when to collapse and when not. The remainder of this section briefly discusses possible extensions of the ideas in this paper and formulates open questions.

Adjusting granularity. The discrete version of the complex of stable manifolds can be interpreted as a coarse-grained view of the finer Delaunay complex. Each stable manifold is represented by a cluster of Delaunay simplices glued together by the flow relation. We can imagine an extension to a 1-parameter family of flow relations. The *granularity* parameter $\gamma \in \mathbb{R}$ controls the coarseness of the clustering. We aim at a parametrization with $\gamma = -\infty$, 0, and $+\infty$ corresponding to the Delaunay complex, the complex of stable manifolds, and $\{\mathbb{R}^3\}$. If $\gamma_1 < \gamma_2$, then the clustering for γ_1 should be a refinement of the clustering for γ_2. The present discussion conveniently ignores that ancestor sets representing stable manifolds can overlap and do not exactly partition the set of Delaunay simplices. Eventually, this set-theoretic inconvenience will have to be dealt with, possibly with concepts similar to conservative ancestor sets.

A mathematical formulation of granularity will have to be based on size and the comparison of sizes. Consider for example a centered triangle τ shared by tetrahedra σ_1 and σ_2. Then $|\tau| < |\sigma_1|$ and $|\tau| < |\sigma_2|$, and we suppose $|\sigma_1| < |\sigma_2|$. It is plausible to stipulate that for $\gamma > |\sigma_1| - |\tau|$, the triangle ought to change its behavior and act like an equivocal triangle with flow from smaller to larger size: $\sigma_1 \prec_\gamma \tau \prec_\gamma \sigma_2$. The intuition for this stipulation is to permit a limited amount of downward flow, namely from σ_1 to τ. The permitted amount is bounded from above by γ. The idea of limited downward flow can be generalized to simplices of all dimensions. For $\gamma > 0$, we cannot expect that the resulting cells are necessarily contractible. For negative γ, we get fewer pairs than in $\prec \; = \; \prec_0$ and therefore a finer partition of D than for $\gamma = 0$. Variants substituting $|\sigma|/|\tau|$ for $|\sigma| - |\tau|$ or cosinks for cofaces σ are conceivable.

Final remarks and open questions. An interesting variant of the basic construction maintains the wrapping surface $\mathcal{W} \subseteq \operatorname{Del} S$ while adding one point at a time to S. The Delaunay complex can be constructed in randomized time between constant and logarithmic per simplex [3, 7]. Is it possible to maintain $\mathcal{W}$ in about the same or possibly less time? More generally, we ask for an algorithm that maintains $\mathcal{W}$ through a sequence of point insertions, point deletions, point motions, and weight changes. An efficient such algorithm would be useful in conjunction with a fast algorithm for maintaining the Delaunay complex under such operations. The implementation of such an algorithm for three-dimensional Delaunay complexes is described in [8].

Given a viewing point $x \in \mathbb{R}^3$, a *depth-ordering* of the simplices in $\mathcal{W}$ is a linear extension of the in front-behind relation defined for x. Every Delaunay complex has a depth-ordering for every viewpoint in space [4], and since $\mathcal{W} \subseteq \operatorname{Del} S$, this is also true for $\mathcal{W}$. The depth-ordering opens up the possibility to use hidden-surface algorithms other than z-buffering to generate pleasing graphical representations.

The ideas presented in this paper can be extended to $\mathbb{R}^4$ and higher dimensions. It might be worthwhile to develop and implement such an extension, which could then be used in the analysis of point data beyond three dimensions. Such data is common in studies of time series, dynamical systems, and other areas of science and applied mathematics. The independence of the algorithm from assumptions about the data makes it an attractive approach to discovering structure in point data. The examples in Section 2 show that the algorithm has the ability to adapt to the dimension of the data, which is a useful feature in data exploration [2].

We conclude this paper with a question about the stability of the complex of stable manifolds. Small motion in the input data can cause critical points to appear or disappear. This causes non-continuous changes in the complex and possibly in the wrapping surface. It would be interesting to understand how exactly this lack of stability is related to the phenomenon of overlapping ancestor sets, or more precisely to the existence of type (S2) simplices, which are discussed in Section 6.

References

[1] J.-D. Boissonnat. Geometric structures for three-dimensional shape representation. *ACM Trans. Graphics* **3** (1984), 266–286.

[2] C. Bregler and S. M. Omohundro. Nonlinear manifold learning for visual speech recognition. *In* "Proc. 5th Internat. Conf. Comput. Vision, 1995", 494-499.

[3] K. L. Clarkson, K. Mehlhorn and R. Seidel. Four results on randomized incremental constructions. *In* "Proc. 9th Ann. Sympos. Theoret. Aspects Comput. Sci., 1992", 463–474, Lecture Notes in Comput. Sci. **577**, Springer-Verlag.

[4] H. Edelsbrunner. An acyclicity theorem for cell complexes in d dimensions. *Combinatorica* **10** (1990), 251–260.

[5] H. Edelsbrunner and E. P. Mücke. Simulation of Simplicity: a technique to cope with degenerate cases in geometric algorithms. *ACM Trans. Graphics* **9** (1990), 66–104.

[6] H. Edelsbrunner and E. P. Mücke. Three-dimensional alpha shapes. *ACM Trans. Graphics* **13** (1994), 43–72.

[7] H. Edelsbrunner and N. R. Shah. Incremental topological flipping works for regular triangulations. *In* "Proc. 8th Ann. Sympos. Comput. Geom., 1992", 43–52.

[8] M. Facello. *Geometric Techniques for Molecular Shape Analysis.* Ph. D. Thesis, Dept. Comput. Sci., Univ. Illinois, Urbana, 1996.

[9] H. Fuchs, Z. Kedem, and S. P. Uselton. Optimal surface reconstruction from planar contours. *Comm. ACM* **20** (1977), 693–702.

[10] P. J. Giblin. *Graphs, Surfaces, and Homology.* Second edition, Chapman and Hall, London, 1981.

[11] H. Hoppe, T. de Rose, T. Duchamp, J. McDonald, and W. Stützle. Surface reconstruction from unorganized points. *Computer Graphics*, Proc. SIGGRAPH 1992, 71–78.

[12] D. W. MATULA AND R. R. SOKAL. Properties of Gabriel graphs relevant to geographic variation research and the clustering of points in the plane. *Geographic Analysis* **12** (1980), 205–222.

[13] D. MEYERS, S. SKINNER AND K. SLOAN. Surfaces from contours. *ACM Trans. Graphics* **11** (1992), 228–258.

[14] J. MILNOR. *Morse Theory.* Annals Math. Studies, Princeton Univ. Press, 1963.

[15] R. C. VELTKAMP. *Closed Object Boundaries from Scattered Points.* Springer-Verlag, Berlin, 1994.

[16] A. WALLACE. *Differential Topology. First Steps.* Benjamin, New York, 1968.

References

[A] N. AMENTA AND M. BERN. Surface reconstruction by Voronoi filtering. *Discrete Comput. Geom.* **22** (1999), 481–504.

[B] N. AMENTA, S. CHOI AND R. KOLLURI. The power crust, unions of balls, and the medial axis transform. *Comput. Geom. Theory Appl.* **19** (2001), 127–153.

[C] F. BERNARDINI, J. MITTLEMAN, H. RUSHMEIER, C. SILVA AND G. TAUBIN. The ball-pivoting algorithm for surface reconstruction. *IEEE Trans. Visual. Comput. Graphics* **5**(1999), 349–359.

[D] F. BERNARDINI AND C. L. BAJAJ. Sampling and reconstructing manifolds using α-shapes. *In* "Proc. 9th Canadian Conf. Comput. Geom., 1997", 193–198.

[E] J.-D. BOISSONNAT AND F. CAZALS. Smooth surface reconstruction via natural neighbour interpolation of distance functions. *In* "Proc. 16th Ann. Sympos. Comput. Geom. 2000", 223–232.

[F] B. CURLESS AND M. LEVOY. A volumetric method for building complex models from range images. *Comput. Graphics*, Proc. SIGGRAPH 1996, 303–312.

[G] H. EDELSBRUNNER, D. LETSCHER AND A. ZOMORODIAN. Topological persistence and simplification. *Discrete Comput. Geom.*, to appear.

[H] R. FORMAN. Combinatorial differential topology and geometry. In *New Perspective in Geometric Combinatorics*, MSRI Publication **8**, 1999, 177–206.

[I] S. FUNKE AND E. A. RAMOS. Smooth surface reconstruction in near-linear time. *In* "Proc. 13th Ann. SIAM-ACM Sympos. Discrete Algorithms, 2002", 781–790.

[J] G. MEENAKSHISUNDARAM. *Theory and Practice of Sampling and Reconstruction for Manifolds with Boundaries.* Ph. D. Thesis, Dept. Comput. Sci., Univ. North Carolina, Chapel Hill, 2001.

About Authors

Herbert Edelsbrunner is with the Departments of Computer Science and Mathematics at Duke University, Durham, North Carolina, and with Raindrop Geomagic, Research Triangle Park, North Carolina, USA.

Infeasibility of Systems of Halfspaces

Stefan Felsner
Nicole Morawe

Abstract

An oriented hyperplane is a hyperplane with designated good and bad sides. The infeasibility of a cell in an arrangement $\vec{\mathcal{A}}$ of oriented hyperplanes is the number of hyperplanes with this cell on the bad side. With $\text{MinInf}(\vec{\mathcal{A}})$ we denote the minimum infeasibility of a cell in the arrangement. A subset of hyperplanes of $\vec{\mathcal{A}}$ is called an infeasible subsystem if every cell in the induced subarrangement has positive infeasibility. With $\text{MaxDis}(\vec{\mathcal{A}})$ we denote the maximal number of disjoint infeasible subsystems of $\vec{\mathcal{A}}$. For every arrangement $\vec{\mathcal{A}}$ of oriented hyperplanes

$$\text{MinInf}(\vec{\mathcal{A}}) \geq \text{MaxDis}(\vec{\mathcal{A}}).$$

In this paper we investigate bounds for the ratio of the LHS over the RHS in the above inequality. The main contribution is a detailed discussion of the problem in the case $d = 2$, i.e., for 2-dimensional arrangements. We prove that $\text{MinInf}(\vec{\mathcal{A}}) \leq 2 \cdot \text{MaxDis}(\vec{\mathcal{A}})$, in this case. An example shows that the factor 2 is best possible. If an arrangement $\vec{\mathcal{A}}$ of n lines contains a cell of infeasibility n, then the factor can be improved to $3/2$, which is again best possible. We also consider the problem for arrangements of pseudolines in the Euclidean plane and show that the factor of 2 suffices in this more general situation.

1 Introduction

Let $Mx < b$ be a system of n linear inequalities in d variables. Think of (M, b) as an arrangement $\mathcal{A}_M$ of affine hyperplanes in $\mathbb{R}^d$. From the two halfspaces defined by a hyperplane $(m_i, x) = b_i$ we call the halfspace $(m_i, x) < b_i$ the *good* side and the halfspace $(m_i, x) > b_i$ the *bad* side of the hyperplane. The *infeasibility* of a point $x \in \mathbb{R}^d$ is the number of inequalities violated by x, i.e., the number of hyperplanes with x on the bad side. We denote the infeasibility of x by $\text{Inf}(x)$. Given a cell c in $\mathcal{A}_M$ then $\text{Inf}(x) = \text{Inf}(y)$ for all $x, y \in c$, therefore, we can define the infeasibility of cell c as the infeasibility of any point in c. A cell c which is on the good side of all hyperplanes, i.e., $\text{Inf}(c) = 0$, it is called *feasible*. An arrangement is called *feasible* if it contains a feasible cell and *infeasible* otherwise.

Let $\vec{\mathcal{A}}$ be an arrangement of oriented hyperplanes, i.e., of hyperplanes with designated good and bad sides. The cells of minimum infeasibility of $\vec{\mathcal{A}}$ are of some interest. These cells can be made feasible by removing or reverting the least number of inequalities. We define the *minimum infeasibility* $\text{MinInf}(\vec{\mathcal{A}})$ as the minimum of $\text{Inf}(c)$ over all cells c of $\vec{\mathcal{A}}$.

A *subarrangement* of $\vec{\mathcal{A}}$ is a subset $\vec{\mathcal{B}}$ of the hyperplanes of $\vec{\mathcal{A}}$, with the same orientations. We denote by $\text{MaxDis}(\vec{\mathcal{A}})$ the maximum number of disjoint infeasible subarrangements of $\vec{\mathcal{A}}$. Given a cell c of $\vec{\mathcal{A}}$, a point $x \in c$ and disjoint infeasible subarrangements $\vec{\mathcal{B}}_1, \ldots, \vec{\mathcal{B}}_k$ we note that each $\vec{\mathcal{B}}_i$ contributes at least one to the infeasibility of x in $\vec{\mathcal{A}}$. From this observation we obtain the following inequality.

Proposition 1.1. *For every arrangement $\vec{\mathcal{A}}$ of oriented hyperplanes*

$$\text{MinInf}(\vec{\mathcal{A}}) \geq \text{MaxDis}(\vec{\mathcal{A}}).$$

In this paper we investigate bounds for the ratio of the LHS over the RHS in the above inequality. The existence of such a ratio follows from Helly's Theorem, see Proposition 2.4.

The main contribution of the paper is a detailed discussion of the problem in the case $d = 2$, i.e., for 2-dimensional arrangements. We prove that $\text{MinInf}(\vec{\mathcal{A}}) \leq 2 \cdot \text{MaxDis}(\vec{\mathcal{A}})$, in this case. An example shows that the factor 2 is best possible. If an arrangement $\vec{\mathcal{A}}$ of n lines contains a cell of infeasibility n, then the factor can be improved to $3/2$, which is again best possible. Finally, we consider the problem for arrangements of pseudolines in the Euclidean plane. A completely independent proof is required to show that the factor of 2 suffices in this more general situation.

The problem of investigating the gap in Proposition 1.1 came to our attention through a question posed by Komei Fukuda at the Monte Verità Conference on Discrete and Computational Geometry 1998. Fukuda remarked that he was led to the problem by the observation that some well studied graph problems can be viewed as problems about minimum infeasibility of certain arrangements. We briefly discuss two such instances.

The *Maximum Acyclic Subgraph* problem asks for the minimum number of arcs of a directed graph $D = (V, A)$ whose removal makes the graph acyclic (see e.g. [4]). Associate a variable x_v with vertex v and an inequality $x_u - x_v \geq 1$ with each arc (u, v). The Maximum Acyclic Subgraph problem is the problem of determining $\text{Inf}(\vec{\mathcal{A}})$ for the corresponding arrangement. A minimal infeasible subarrangement corresponds to a simple cycle, hence, Proposition 1.1 translates to the lower bound given by a maximum collection of disjoint cycles.

The *Lucchesi-Younger Theorem* for directed graphs is a MinMax result which can be interpreted as a situation in which equality in Proposition 1.1 holds (for proofs see [2, 6, 7]). Let $D = (V, A)$ be a directed graph with a 2-connected underlying graph G_D. A *dicut* of D is a set B of arcs which is a cut (X, Y) for G_D such that all arcs of B are oriented from X to Y.

The Lucchesi-Younger Theorem states that the minimum number of arcs of D that have to be reversed to make D strongly connected equals the maximum number of disjoint dicuts of D. For the translation provide a variable x_a for each arc of D, now consider the flow constraint for each vertex $v \in V$, i.e., $\sum_{a \in in(v)} x_a - \sum_{a \in out(v)} x_a = 0$, and positivity inequalities $x_a \geq 1$. The theorem is equivalent to $\text{MININF}(\vec{\mathcal{A}}) = \text{MAXDIS}(\vec{\mathcal{A}})$ when $\vec{\mathcal{A}}$ is the arrangement induced by the inequalities on the $(|A| - |V|)$-dimensional affine space defined by the flow constraints.

2 Introductory Observations and Examples

In this section we collect some general observations about infeasibility in arrangements of oriented hyperplanes. For the sake of simplicity we assume throughout the paper that the arrangements we consider are '*in general position*'. By this we mean that the intersection of every set of d hyperplanes of the arrangement is a point and no point is contained in more than d of the hyperplanes. Our results (except Theorem 4.1) hold without this assumption, but the proofs would be broadened by additional details.

Without proof we state the following two observations.

Observation 2.1. *An infeasible arrangement in $\mathbb{R}^d$ contains at least $d+1$ hyperplanes.*

Observation 2.2. *An arrangement in $\mathbb{R}^d$ consisting of $d+1$ hyperplanes is infeasible iff its (unique) bounded cell has infeasibility $d+1$.*

An infeasible arrangement in $\mathbb{R}^d$ consisting of $d+1$ hyperplanes is called a *infeasible base arrangement.*

Lemma 2.3. *Every infeasible arrangement contains an infeasible base arrangement.*

Proof. Let $\vec{\mathcal{A}}$ be an infeasible arrangement. Suppose that $\vec{\mathcal{A}}$ contains no infeasible base arrangement. This means that for each set of $d+1$ good halfspaces there is a point being contained in all of them. Helly's Theorem implies the existence of a point x which is contained in all the good halfspaces. By definition this point and, hence, his cell has infeasibility 0. This means that $\vec{\mathcal{A}}$ is feasible, contradiction. □

Proposition 2.4. *For every arrangement $\vec{\mathcal{A}}$ in $\mathbb{R}^d$*

$$\text{MININF}(\vec{\mathcal{A}}) \leq (d+1) \cdot \text{MAXDIS}(\vec{\mathcal{A}}).$$

Proof. Recursively remove infeasible base arrangements from $\vec{\mathcal{A}}$ until the arrangement becomes feasible. This is possible by Lemma 2.3. Let x be a point in the feasible cell. The removal of each infeasible base arrangements has decreased the infeasibility of x by at most $d+1$. Therefore, $\text{MININF}(\vec{\mathcal{A}}) \leq \text{MININF}(x) \leq (d+1) \cdot \#$ removed subarrangements $\leq (d+1) \cdot \text{MAXDIS}(\vec{\mathcal{A}})$. □

We turn to a lower bound construction. Choose n points uniformly at random from the unit sphere in $\mathbb{R}^d$. With each point consider the tangent hyperplane and let the bad halfspace of the hyperplane be the side containing the sphere. The infeasibility of a point x in the interior of the sphere is n. Along every ray starting at x the infeasibility is monotonically decreasing. Therefore, every cell of minimum infeasibility is an unbounded cell. The infeasibility of an unbounded cell of the arrangement equals the number of random points on the averted hemisphere. By construction this number will be approximately $n/2$. Since a infeasible subsystem consists of $d+1$ halfspaces, there are at most $n/(d+1)$ of them. On the basis of this idea there is a rigorous proof for the next proposition.

Proposition 2.5. *For every $\epsilon > 0$ there are arrangements $\vec{\mathcal{A}}$ in $\mathbb{R}^d$ with*

$$\text{MININF}(\vec{\mathcal{A}}) \geq \left(\frac{d+1}{2} - \epsilon\right) \cdot \text{MAXDIS}(\vec{\mathcal{A}}).$$

As a warmup we consider the very simple case of 1-dimensional arrangements. A hyperplane in dimension $d = 1$ is just a point p in $\mathbb{R}$ and its halfspaces are the two open intervals $]-\infty, p\,[$ and $]\,p, \infty[$. A cell of an arrangement corresponds to an (open) interval. An infeasible base arrangement consists of two points with disjoint good halfspaces, see Figure[1] 1.

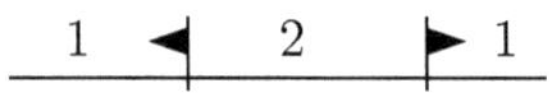

Fig. 1. An infeasible base arrangement in one dimension, cells are labeled with their infeasibility.

Let $\vec{\mathcal{A}}$ be a 1-dimensional arrangement. Since we are only interested in the infeasibility of the cells of $\vec{\mathcal{A}}$, we disregard the exact position of the points (hyperplanes) and restrict the attention to their ordering and the orientation information, i.e., whether the left or the right side is the good one. From $\vec{\mathcal{A}}$ construct a $\{+,-\}$-sequence containing n signs: Scan the points from left to right. If the bad side is to the right of the point write a plus-sign, otherwise, write a minus-sign. Infeasible base arrangements of $\vec{\mathcal{A}}$ are in one-to-one correspondence to the $+-$ subsequences of this sequence. A $+-$ subsequence of consecutive signs will be called *tight subarrangement.*

Observation 2.6. *A 1-dimensional infeasible arrangement contains a tight subarrangement.*

Observation 2.7. *The bounded cell of a tight subarrangement of $\vec{\mathcal{A}}$ is not a cell of minimal infeasibility in $\vec{\mathcal{A}}$.*

[1]In figures we mark the good side of a hyperplane with a flag.

Observation 2.7 is proven by considering the infeasibility of one of the two cells adjacent to the cell between the points of the tight subarrangement. From this observation it follows that removing a tight subarrangement from $\vec{\mathcal{A}}$ decreases the infeasibility of the arrangement by one.

Given an infeasible arrangement $\vec{\mathcal{A}}$, recursively remove tight subarrangements, as long as possible. When the procedure stops the arrangement is feasible (Observation 2.6). This shows

Proposition 2.8. *For every 1-dimensional arrangement* $\vec{\mathcal{A}}$

$$\textsc{MinInf}(\vec{\mathcal{A}}) = \textsc{MaxDis}(\vec{\mathcal{A}}).$$

3 Arrangements of Lines

In this section we discuss arrangements of lines in 2-dimensional space. With every line we assume that a good and a bad halfplane are specified. An infeasible base arrangement is shown in Figure 2.

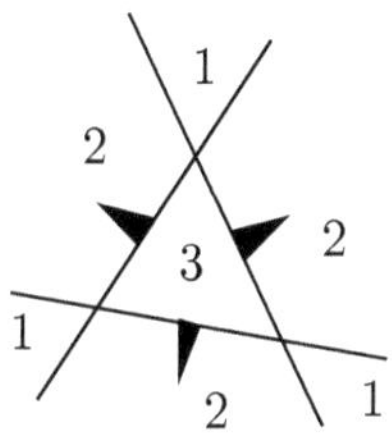

Fig. 2. An infeasible base arrangement in $\mathbb{R}^2$, cells are labeled with their infeasibility.

It is easy to give a more explicit example for the lower bound of Proposition 2.5 in dimension two. Consider the arrangement of lines supported by the edges of a regular n-gon with orientations such that all halfspaces containing the n-gon are bad. In the even case small perturbations are required to avoid parallel lines. It is easy to see that $\textsc{MaxDis}(\vec{\mathcal{A}}) = \lfloor n/3 \rfloor$ and $\textsc{MinInf}(\vec{\mathcal{A}}) = \lfloor (n-1)/2 \rfloor$. An example with $n = 5$ is shown in Figure 3.

Theorem 3.1. *For every arrangement* $\vec{\mathcal{A}}$ *of oriented lines*

$$\textsc{MinInf}(\vec{\mathcal{A}}) \leq 2 \cdot \textsc{MaxDis}(\vec{\mathcal{A}}).$$

This bound is best possible.

There is a class of arrangements which allows to prove a smaller factor.

Theorem 3.2. *For every arrangement* $\vec{\mathcal{A}}$ *of oriented lines with a cell which is on the bad side of every line*

$$\textsc{MinInf}(\vec{\mathcal{A}}) \leq \lceil (3/2) \cdot \textsc{MaxDis}(\vec{\mathcal{A}}) \rceil.$$

This bound is best possible.

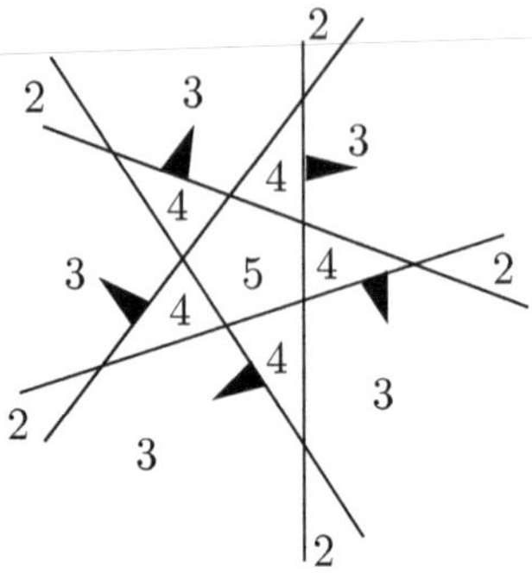

Fig. 3. Arrangement with MAXDIS($\vec{\mathcal{A}}$) = 1 and MININF($\vec{\mathcal{A}}$) = 2

3.1 Proof of Theorem 3.1

The proof is based on a recursive removal of infeasible base arrangements. The goal is to find a subarrangement whose removal decreases the infeasibility of the remaining arrangement by at most two, we call such a infeasible base subarrangement a *tight subarrangement.*

For a cell c of minimum infeasibility let $\mathcal{A}(c)$ be the arrangement formed by all lines having c on their bad side. Let $z(c)$ be the cell in the arrangement $\mathcal{A}(c)$ containing c. With d_c we denote be the minimal distance between a point from cell c (including its boundary) and a line of $\vec{\mathcal{A}}(c)$. The minimal distance d_c is greater than 0, since the boundary of c is not contained in $\vec{\mathcal{A}}(c)$. Note that d_c is achieved at a corner of c. Among all cells of minimum infeasibility choose the cell c which minimizes d_c. Let p be the corner of c and l be the line of $\mathcal{A}(c)$ such that $d_c = dist(p, l)$. With the following two lemmas we show that line l and the two lines l_1, l_2 crossing at p form a tight subarrangement. Figure 4 illustrates the situation.

Lemma 3.3. *The subarrangement l, l_1, l_2 is infeasible.*

Proof. Lines l_1 and l_2 divide the plane into four open unbounded sectors. Let R be the sector which is on the good side of both lines, sector R is opposite to the sector containing c, see Figure 5. Since $d_c = dist(p, l)$ the point p is the closest point to l in cell c. Therefore, line l has to cross region R and the subsystem formed by the three lines is infeasible. □

Lemma 3.4. *Removing the subarrangement l, l_1, l_2 decreases the infeasibility of the remaining arrangement by at most two.*

Proof. We have to show that there is no cell in the triangular cell formed by l, l_1, l_2 which is of minimal infeasibility.

Suppose $\bar{c}$ is a cell with minimal infeasibility contained in the triangular region of l, l_1, l_2. Define $\vec{\mathcal{A}}(\bar{c})$ as above. Then l belongs to $\vec{\mathcal{A}}(\bar{c})$ and we get $d_{\bar{c}} \leq dist(\bar{c}, l) < d_c$. Since we have chosen c as the cell of minimal infeasibility minimizing d_c, the existence of such a cell $\bar{c}$ is impossible. □

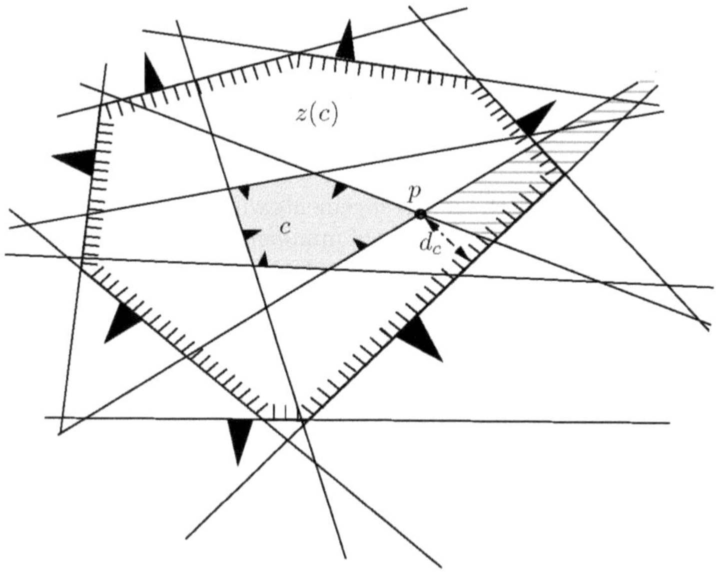

Fig. 4. The construction: The large 7-gon is $z(c)$ and the fasciated region is the triangle of the subarrangement l, l_1, l_2.

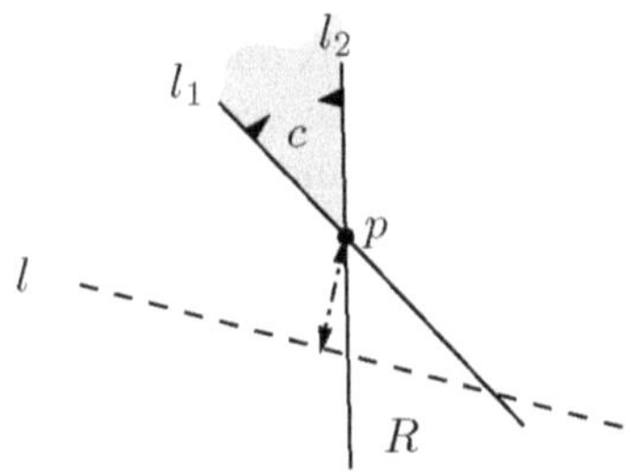

Fig. 5. The relative positions of l_1, l_2 and l.

Since removing a tight subarrangement decreases the infeasibility of every cell of minimal infeasibility by at most two we get $\text{MinInf}(\vec{\mathcal{A}}) \leq 2 \cdot \#\text{removed subarrangements} \leq 2 \cdot \text{MaxDis}(\vec{\mathcal{A}})$. This completes the proof of the inequality of Theorem 3.1.

To complete the proof of Theorem 3.1 we need a construction showing that the factor 2 is best possible. Such a construction will be given in Subsection 4.2.

Remark. The idea of this proof generalizes to higher dimensions. An analogous distance argument allows to detect an infeasible base subarrangement whose removal decreases the infeasibility of every minimal infeasible cell by at most d. This improves the factor of Proposition 2.4 and gives: For every arrangement $\vec{\mathcal{A}}$ in $\mathbb{R}^d$

$$\textsc{MinInf}(\vec{\mathcal{A}}) \leq d \cdot \textsc{MaxDis}(\vec{\mathcal{A}}).$$

3.2 Proof of Theorem 3.2

Recall that now we deal with line-arrangements with one cell being on the bad side of every line. In this case all cells of minimal infeasibility are unbounded and we restrict the attention to those cells which intersect the circle at infinity.

We encode the infeasibility of all unbounded cells in a $\{+1, -1\}$-sequence of length n: Let k be the minimum infeasibility of the arrangement and let x be a point with infeasibility k on the circle at infinity. Let y be the antipodal point of x. Note that the infeasibility of y is $n - k$. Traversing the lines at infinity clockwise from x to y we write $+1$, if x is on the good side of a line, and we write -1 if x is on the bad side.

The infeasibility of every unbounded cell c is encoded by this sequence. Let the unbounded cells be numbered $0, 1, \ldots, 2n-1$ in clockwise order, such that cell 0 contains x. If c is the i-th cell with $i \leq n$, then the infeasibility of c is k plus the sum of the first i numbers of the sequence. Otherwise, if $i > n$, use its antipodal cell c^* to calculate the infeasibility as $\textsc{Inf}(c) = n - \textsc{Inf}(c^*)$.

By the choice of x as a point of minimal infeasibility (and thereby y as a point of maximal infeasibility) the resulting sequence has a total sum of $n-2k$ and the property that every prefix and postfix has a non-negative sum.

The next lemma describes how the sequence represents infeasible base subarrangements. We abbreviate $+1$ and -1 with $+$ and $-$.

Lemma 3.5. *Infeasible base arrangements are represented in the $+,-$ sequence by $+-+$ or $-+-$ subsequences.*

Proof. Fix an infeasible base arrangement. Then x is either contained in one of its regions with infeasibility 1 or infeasibility 2. Constructing the sequences transforms the base arrangement to $+-+$ in case of x lying in a region of infeasibility 1 and $-+-$ otherwise.

Conversely, let $+-+$ be a subsequence of the sequence. With respect to x there are only two possibilities for the relative position of its three lines. The induced orientation of $+-+$ leads to the two configurations show in Figure 6. The left configurations in fact is an infeasible base arrangement. The right configuration is not possible, since no point lies in all three bad halfplanes. The same reasoning applies to $-+-$ subsequences. □

To complete the proof of Theorem 3.2 we need a combinatorial lemma about decompositions of sequences of plus- and minus-signs.

Lemma 3.6. *Let S be a $\{+,-\}$-sequence containing $n-k$ plus-signs and k minus-signs with the property that every prefix and postfix of S has a non-negative sum. Then S contains at least $\lfloor \frac{2k+1}{3} \rfloor$ disjoint subsequences of the form $+-+$ or $-+-$.*

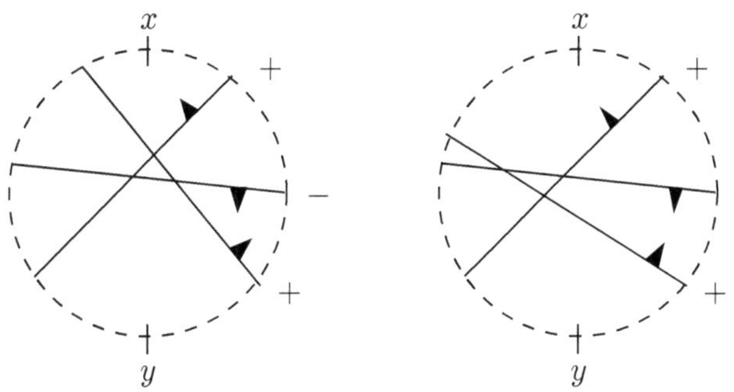

Fig. 6. The two arrangements defined by $+-+$.

Proof. We abstract the string S to a a zig-zag path with steps $(1,1)$ and $(1,-1)$ starting at $(0,k)$ and ending at $(n,n-k)$, always staying in the y-interval $[k,n-k]$, see Figure 7.

Fig. 7. The zig-zag path for the string $+-++-+--+++-++$.

A level of S is the set of symbols corresponding to a step in the zig-zag path with y-coordinates between i and $i+1$, for some i. Note that each level contains either zero or an odd number of symbols.

We prove the lemma by induction on the length of the string. If the string contains only one symbol it is $+$ and thus

$$\#\text{disjoint alternating triples} = 0 = \left\lfloor \frac{2\cdot 0+1}{3} \right\rfloor.$$

We distinguish several reducing operations on the sequence.

1. There is a level which contains exactly one symbol:

   This symbol must be $+$. Removing it leads to a shorter valid string with the same number of negative entries. By induction this shorter sequence contains $\lfloor \frac{2k+1}{3} \rfloor$ disjoint alternating triples.

2. There is a level which contains exactly three symbols:

   These three symbols form the substring $+-+$. Removing them leaves a shorter valid string with $k-1$ negative symbols. By induction

$$\#\text{disjoint alternating triples} \geq \left\lfloor \frac{2(k-1)+1}{3} \right\rfloor + 1 \geq \left\lfloor \frac{2k+1}{3} \right\rfloor.$$

3. There is a level containing at least six symbols:

   The first six symbols of this level form the substring $+-+-+-$, which consists of two alternating triples. Removing all six symbols leaves a shorter valid string with $k-3$ negative symbols. Thus

$$\#\text{disjoint alternating triples} \geq \left\lfloor \frac{2(k-3)+1}{3} \right\rfloor + 2 = \left\lfloor \frac{2k+1}{3} \right\rfloor .$$

4. If non of the above cases holds, then every level contains exactly five symbols.

   • If there is only one level, then $n = 5$, $k = 2$ and the first three symbols form a substring $+-+$. Hence,

$$\#\text{disjoint alternating triples} = 1 = \left\lfloor \frac{2 \cdot 2+1}{3} \right\rfloor .$$

   • If there are at least two levels, then consider the 10 symbols of height k and $k+1$. This substring has the shape of one of the 6 cases shown in Figure 8. The figure indicates how the first nine symbols of each string can be decomposed into three disjoint alternating triples. Removing the ten symbols leaves a valid string with $k-4$ negative symbols. Thus

$$\#\text{disjoint alternating triples} \geq \left\lfloor \frac{2(k-4)+1}{3} \right\rfloor + 3 \geq \left\lfloor \frac{2k+1}{3} \right\rfloor .$$

□

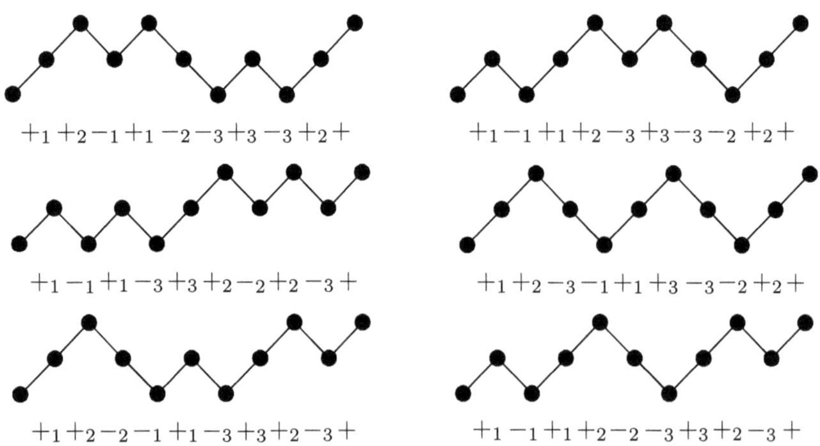

Fig. 8. The six possible patterns of two levels, each of 5 symbols

Since, by construction the parameter k was the infeasibility of the arrangement, the lemma gives $\textsc{MaxDis}(\vec{\mathcal{A}}) \geq \lfloor \frac{2}{3}\textsc{MinInf}(\vec{\mathcal{A}}) + \frac{1}{3} \rfloor$. This implies $\lceil \frac{3}{2}\textsc{MaxDis}(\vec{\mathcal{A}}) \rceil \geq \textsc{MinInf}(\vec{\mathcal{A}})$, i.e., the formula stated in the theorem.

The tightness of the factor 3/2 is shown via the arrangement of the n-gon. Consider the arrangement of lines supported by the edges of a regular $(6r+5)$-gon oriented so that all halfspaces containing the polygon are bad. Then $\textsc{MaxDis}(\vec{\mathcal{A}}) = 2r+1$ and $\textsc{MinInf} = 3r+2 = \lceil \frac{3}{2}\textsc{MaxDis}(\vec{\mathcal{A}}) \rceil$.

4 Arrangements of Pseudolines

A *pseudoline* is a curve in $\mathbb{R}^2$ whose removal from the plane leaves two unbounded connected components. In other words a pseudoline is a simple curve which goes to infinity on both sides. An *arrangement of pseudolines* is a family of pseudolines with the property that each pair of pseudolines has a unique point of intersection, where the two pseudolines cross. Arrangements of pseudolines are a natural generalization of arrangements of straight lines, they have been studied in a wide variety of contexts. Grünbaum [5] is a nice little monograph collecting results and many problems about arrangements of both straight-lines and pseudolines. A more recent overview is given by Goodman [3].

Let $\mathcal{A}$ be an arrangement of pseudolines. Choose an orientation for each pseudoline, or equivalently, assign a good and a bad side to each pseudoline. For the oriented arrangement $\vec{\mathcal{A}}$ there are natural notions of infeasibility for points, cells and subarrangements. Hence, questions regarding the relation between $\textsc{MinInf}(\vec{\mathcal{A}})$ and $\textsc{MaxDis}(\vec{\mathcal{A}})$ can be asked in this more general setting. Let us review the results for 2-dimensional arrangements of lines and see what can be adapted to pseudolines. The validity of Propositions 1.1 in the new context is obvious. The upper bound of Proposition 2.4, i.e., $\textsc{MinInf}(\vec{\mathcal{A}}) \leq 3 \cdot \textsc{MaxDis}(\vec{\mathcal{A}})$, holds, because of an analog of Helly's Theorem for pseudo-halfspaces. What about pseudoline analogs of Theorems 3.1 and 3.2? In the case of Theorem 3.2 the answer is easy. The proof given in Subsection 3.2 is completely combinatorial, it makes no use of straightness. Therefore, the result remains valid in the setting of arrangements of pseudolines:

Theorem 4.1. *For every arrangement $\vec{\mathcal{A}}$ of pairwise crossing oriented pseudolines with a cell which is on the bad side of every pseudoline*

$$\textsc{MinInf}(\vec{\mathcal{A}}) \leq \lceil (3/2) \cdot \textsc{MaxDis}(\vec{\mathcal{A}}) \rceil.$$

This bound is best possible.

The statement of this theorem gets false without the assumption that the pseudolines are pairwise crossing. There are sets $\vec{\mathcal{A}}$ of n pairwise parallel oriented pseudolines such that $\textsc{MinInf}(\vec{\mathcal{A}}) = n-1$ and $\textsc{MaxDis}(\vec{\mathcal{A}}) = \lfloor n/2 \rfloor$.

The proof of Theorem 3.1 in Subsection 3.1 makes use of a metric argument. Therefore, the question whether the factor of 2 remains valid for arrangements of pseudolines cannot be answered on the basis of the old proof. The main result of this section is a proof of the following generalization of Theorem 3.1.

Theorem 4.2. *For every arrangement $\vec{\mathcal{A}}$ of oriented pseudolines*

$$\text{MinInf}(\vec{\mathcal{A}}) \leq 2 \cdot \text{MaxDis}(\vec{\mathcal{A}}).$$

This bound is best possible.

In the next subsection we collect preparing facts about arrangements of pseudolines. Along the way we learn some combinatorics of cyclic arrangements and use this to produce the lower bound example for Theorem 3.1. Subsection 4.3 contains the main body of the proof for Theorem 4.2.

4.1 Basic Facts

Given an arrangement $\mathcal{A}$ of pseudolines, choose a unbounded cell $\hat{c}$ and imagine that $\hat{c}$ contains the northpole. The complementary cell $\check{c}$ is the unbounded cell separated from $\hat{c}$ by all the pseudolines of the arrangement. Label the pseudolines so that traversing the circle at infinity counterclockwise from $\hat{c}$ to $\check{c}$ they are met in the order $1, 2, \ldots, n$. This results in a *marked arrangement of pseudolines.*

On the set of all (combinatorially different) marked arrangements of n pseudolines consider a graph $\mathcal{G}_n$ whose edges correspond to *triangular flips*, see Figure 9. To be precise, the pair $(\vec{\mathcal{A}}, \vec{\mathcal{B}})$ is an edge of $\mathcal{G}_n$ iff there are three

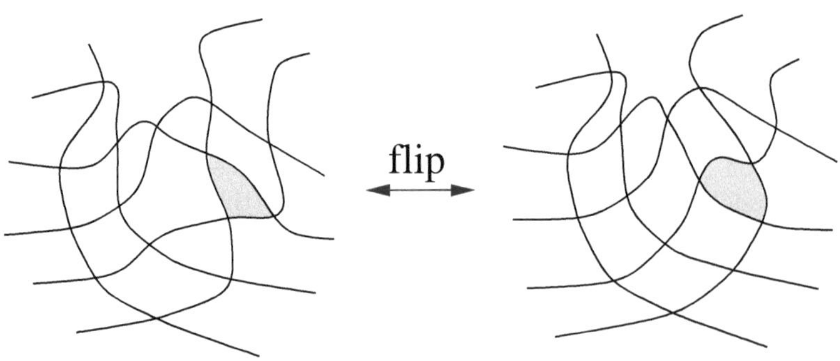

Fig. 9. A flip at the shaded triangular cell.

pseudolines $i < j < k$ in $\vec{\mathcal{A}}$ such that:

- There is a triangular cell bounded by the tree pseudolines i, j, k.
- The northpole is separated from the crossing of pseudolines i and k by pseudoline j.

- Arrangement $\vec{\mathcal{B}}$ is equivalent to the arrangement obtained from $\vec{\mathcal{A}}$ by pulling pseudoline j below the crossing of i and k.

Various aspects of the graph $\mathcal{G}_n$ have been studied [1, 10]. We will need the following properties of $\mathcal{G}_n$.

(1) There is a unique arrangement $\mathcal{C}$ of n pseudolines, so that the indegree of $\mathcal{C}$ in $\mathcal{G}_n$ is zero. The arrangement $\mathcal{C}$ is the *cyclic arrangement*, it will be studied in more detail in the next subsection.

(2) $\mathcal{G}_n$ is the diagram of a ranked poset. (This poset is the "higher Bruhat order" $B(n, 2)$ introduced in [8] and further studied in [10].)

The general idea for the proof of Theorem 4.2 is as follows. In a first step we show the factor of 2 for all orientations of the cyclic arrangement. This is done in Subsection 4.2. To prove the same factor for a marked oriented arrangement $\vec{\mathcal{A}}$ consider a path $\mathcal{C} = \mathcal{A}_0, \mathcal{A}_1, \ldots, \mathcal{A}_r = \mathcal{A}$ in $\mathcal{G}_n$, which starts in the cyclic arrangement $\mathcal{C}$ and ends in the unoriented version $\mathcal{A}$ of $\vec{\mathcal{A}}$. The existence of such a path follows from the two properties of $\mathcal{G}_n$ mentioned above, the original source for the connectivity of $\mathcal{G}_n$ is Ringel [9].

Let $\vec{\mathcal{A}}_i$ be the orientation of $\mathcal{A}_i$ where line j is oriented as in $\vec{\mathcal{A}}$. The idea for the inductive proof is to show that $\text{MinInf}(\vec{\mathcal{A}}_i) \leq 2 \cdot \text{MaxDis}(\vec{\mathcal{A}}_i)$ implies $\text{MinInf}(\vec{\mathcal{A}}_{i+1}) \leq 2 \cdot \text{MaxDis}(\vec{\mathcal{A}}_{i+1})$. Actually the proof given in Subsection 4.3 is a bit more complicated.

4.2 Cyclic Arrangements

The marked cyclic arrangement $\mathcal{C}$ of n lines is characterized by the property that for any three lines $i < j < k$ the crossing of i and k is separated from the northpole by line j. A straight-line realization of this arrangement can be obtained by choosing n different tangents to the parabola $y = x^2$, the special north-cell $\hat{c}$ is the cell containing the parabola. Figures 11 and 12 show cyclic arrangements.

An orientation of the cyclic arrangement, i.e., an assignment of good and bad sides to all lines, can be encoded by a sequence S of $+$ and $-$ signs. Take the lines in the order of their labels and write a $+$ if the northern side is bad and $-$ if the northern side is good. Figure 10 illustrates the following observation.

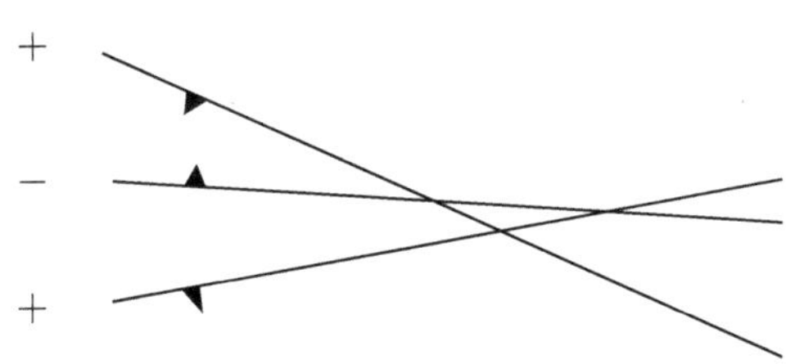

Fig. 10. Infeasible base arrangement in a cyclic arrangement.

Observation 4.3. *Let $\vec{\mathcal{C}}$ be an oriented cyclic arrangement. Infeasible base arrangements of $\vec{\mathcal{C}}$ correspond one-to-one to the $+-+$ subsequences in the sequence S of the orientation.*

Given the sequence S of an orientation of $\vec{\mathcal{C}}$ we want to calculate the infeasibility of a cell c. The infeasibility of the north-cell $\hat{c}$ equals the number of plus-signs in S. The infeasibility of a cell c different from the north-cell is obtained by reverting in S the signs of all the lines above c and then counting the number of plus-signs.

Associate with every cell c of the cyclic arrangement $\mathcal{C}$ of n lines associate the set of lines above c, i.e., the set of those lines separating c from the northpole. This gives a bijective mapping from the cells of $\vec{\mathcal{C}}$ to intervals $[i, \ldots, j]$ with $1 \leq i \leq j \leq n$, together with $\emptyset$. Figure 11 exemplifies the correspondence.

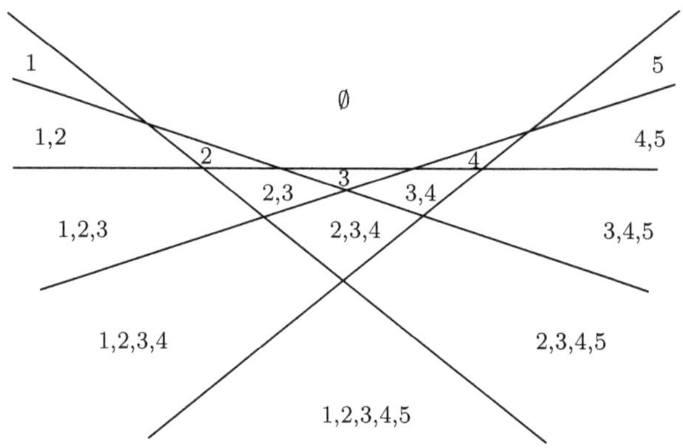

Fig. 11. Cyclic arrangement, $n = 5$, cells are labeled by lines above them.

We are ready, now, to provide the lower bound example for Theorems 3.1 and 4.2.

Proposition 4.4. *There is an orientation $\vec{\mathcal{C}}$ of the cyclic arrangement $\mathcal{C}$ with $4k+1$ lines, such that* $\textsc{MaxDis}(\vec{\mathcal{C}}) = k$ *and* $\textsc{MinInf}(\vec{\mathcal{C}}) = 2k$.

Proof. Consider the orientation $\vec{\mathcal{C}}$ of the cyclic arrangement corresponding to the sequence $+[-+-+]^k$ of length $4k+1$. Figure 12 shows an example with $k = 2$.

There are only $2k+1$ plus-signs in the sequence and every $+-+$ subsequence requires two, therefore, there can be at most k disjoint $+-+$ subsequences. Obviously, there are as many and $\textsc{MaxDis}(\vec{\mathcal{C}}) = k$.

Reverting the signs in an interval of S can decrease the number of plus-signs by at most one. Therefore, the infeasibility of every cell in that arrangement is at least $2k$. □

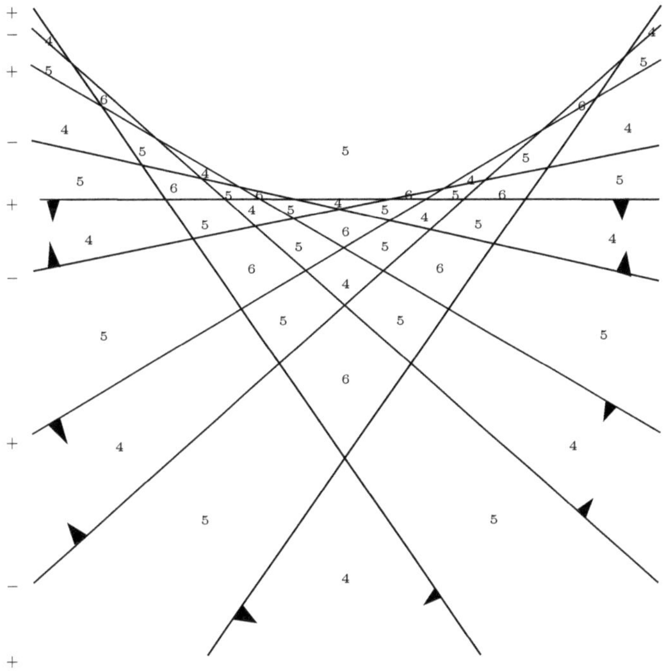

Fig. 12. Cyclic arrangement with $\textsc{MaxDis}(\vec{\mathcal{C}}) = 2$ and $\textsc{MinInf}(\vec{\mathcal{C}}) = 4$.

Cyclic arrangements can be realized as straight-line arrangements, therefore, the statement of the next proposition is a consequence of Theorem 3.1. Here we give a new proof which has the virtue of being purely combinatorial.

Proposition 4.5. *Given an orientation $\vec{\mathcal{C}}$ of the cyclic arrangement with $\textsc{MaxDis}(\vec{\mathcal{C}}) = k$, then there is a cell c with $\textsc{Inf}(c) \leq 2k$.*

Proof. The orientation $\vec{\mathcal{C}}$ is encoded by a sequence S with no more then k disjoint $+-+$ subsequences. We have to show that there is an interval such that reverting the signs of this interval results in a sequence with at most $2k$ plus-signs.

Define the *span* of a subset T of signs of S as the length of the shortest interval containing all the signs of T. The span a family of k disjoint $+-+$ subsequences of S is the sum of the spans of its k triples.

Let F be a family of k disjoint $+-+$ triples of S such that the span of F if minimal among all such families. In S color the $3k$ signs from the triples in F red and all the remaining signs blue. Since, the blue signs are not allowed to contain an additional $+-+$ subsequence, the induced blue sequence is described by the regular expression $[-]^*[+]^*[-]^*$.

For the reversal we choose the smallest interval containing all the blue plus-signs. Let S' be the sequence after reversal of the signs in this interval. In S' there is no blue plus-sign. Let T be one of the k red $+-+$ triples of

F and let T' be the corresponding triple in S'. We claim that T' contains at most two plus-signs. Since all plus-signs of S' are red the claim immediately yields an upper bound of $2k$ for the number of plus-signs in S', which proves the proposition.

Suppose T is a triple in F violating the claim. This means, that the middle symbol, the $-$, of T has been reversed, while the other two symbols remained unchanged. By the choice of the interval for reversal we find that the first and the last blue plus-sign together with the $-$ from T form a $+-+$ subsequence T^* in S. The span of T^* is less then the span of T. Therefore, $F - T + T^*$ is a family of k disjoint $+-+$ subsequences and span less then the span of F, contradiction. □

Remark. The notion of a cyclic arrangement is not restricted to the plane. Let $\mathcal{C}_o$ be the orientation of such an arrangement given by alternating sequence of plus- and minus-signs. In even dimension d it can be shown that $\text{MinInf}(\vec{\mathcal{C}}) \geq \frac{d+2}{2}\text{MaxDis}(\vec{\mathcal{C}}) - \frac{d}{2}$. This is a slight improvement over the lower bound of Proposition 2.5. In odd dimensions the gain is smaller. Compared to Proposition 2.5 $\vec{\mathcal{C}}$ has the advantage of providing a deterministic construction for the factor $(d+1)/2$ and saving the ϵ.

4.3 Proof of Theorem 4.2

By the sketchy outline of the proof given on page 417 it is sufficient to prove the claim: *If $\vec{\mathcal{A}}$ and $\vec{\mathcal{B}}$ are related by a triangular flip as in Figure 9 the inequality for $\vec{\mathcal{A}}$ implies the inequality for $\vec{\mathcal{B}}$.* In the course of the argument we need induction, so we assume the the inequality $\text{MinInf}(\vec{\mathcal{A}}) \leq 2\text{MaxDis}(\vec{\mathcal{A}})$ holds for $\vec{\mathcal{A}}$ and also for all arrangements with fewer lines than $\vec{\mathcal{A}}$.

Given that the implication holds for every pair of flip-adjacent arrangements we can apply this to all pairs of a path $C = A_0A_1...A_r = A$ connecting the cyclic arrangement with A. The proof of the above claim follows.

Let c be the triangular cell in $\vec{\mathcal{A}}$ which is flipped to get $\vec{\mathcal{B}}$ and let T be the triple of pseudolines bounding c. We distinguish three cases:

- Cell c is on the good side of the three pseudolines of T in $\vec{\mathcal{A}}$, abusing terminology we characterize this case by saying T *is feasible.*
- The subsystem induced by T is infeasible, i.e., c is on the bad side of all three pseudolines in $\vec{\mathcal{A}}$.
- The subsystem induced by T in $\vec{\mathcal{A}}$ is neither feasible nor infeasible, then we call it *neutral.*

First consider the case that subsystem induced by T is neutral. Since rotations and careful local deformations do not affect the combinatorial nature the flip is described by th left to right arrow or by the right to left arrow in Figure 13. in either case after performing the flip the subsystem remains neutral. Since $\text{MinInf}(\vec{\mathcal{A}})$ and $\text{MinInf}(\vec{\mathcal{B}})$ are both at most $i-2$, this value is not affected by the flip. Moreover, a subsystem of $\vec{\mathcal{A}}$ is infeasible iff it is infeasible in $\vec{\mathcal{B}}$. Therefore, $\textsc{MaxDis}(\vec{\mathcal{A}}) = \textsc{MaxDis}(\vec{\mathcal{B}})$ and the inequality $\textsc{MinInf}(\vec{\mathcal{B}}) \leq 2 \cdot \textsc{MaxDis}(\vec{\mathcal{B}})$ follows from the corresponding inequality for $\vec{\mathcal{A}}$.

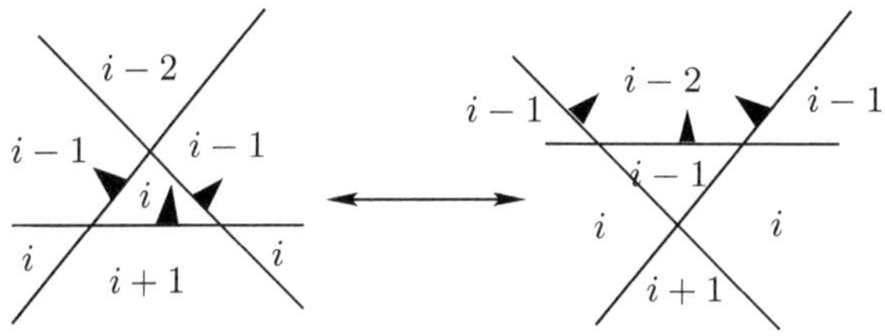

Fig. 13. A neutral flip.

Next suppose that the subsystem induced by T is feasible, i.e., in $\vec{\mathcal{A}}$ cell c is on the good side of the lines in T. Let $\vec{\mathcal{B}}^*$ be the arrangement obtained from $\vec{\mathcal{B}}$ by deleting the three pseudolines of T. Recall that by induction

$$\textsc{MinInf}(\vec{\mathcal{B}}^*) \leq 2 \cdot \textsc{MaxDis}(\vec{\mathcal{B}}^*).$$

Since T is an infeasible base system of $\vec{\mathcal{B}}$ we have

$$\textsc{MaxDis}(\vec{\mathcal{B}}^*) + 1 \leq \textsc{MaxDis}(\vec{\mathcal{B}}).$$

To relate the minimum infeasibility of $\vec{\mathcal{B}}$ and $\vec{\mathcal{B}}^*$ consider a cell $c_{\min}$ of $\vec{\mathcal{B}}$ with $\textsc{Inf}(c_{\min}) = \textsc{MinInf}(\vec{\mathcal{B}})$. Let c' be the triangular cell of $\mathcal{B}$ bounded by the lines in T. Since $c_{\min} \neq c'$ the infeasibility of all points in $c_{\min}$ can decrease by at most two upon removal of the three lines of T. Therefore,

$$\textsc{MinInf}(\vec{\mathcal{B}}) \leq \textsc{MinInf}(\vec{\mathcal{B}}^*) + 2.$$

Combining the inequalities leads to the desired result for $\vec{\mathcal{B}}$:

$$\textsc{MinInf}(\vec{\mathcal{B}}) \leq \textsc{MinInf}(\vec{\mathcal{B}}^*) + 2 \leq 2 \cdot \textsc{MaxDis}(\vec{\mathcal{B}}^*) + 2 \leq 2 \cdot \textsc{MaxDis}(\vec{\mathcal{B}}).$$

The last case is that the subsystem induced by T is infeasible. After the flip the subsystem becomes a feasible subsystem of $\vec{\mathcal{B}}$. Clearly $\textsc{MinInf}(\vec{\mathcal{A}}) \geq \textsc{MinInf}(\vec{\mathcal{B}})$. If $\textsc{MaxDis}(\vec{\mathcal{A}}) = \textsc{MaxDis}(\vec{\mathcal{B}})$ we immediately obtain the result:

$$\textsc{MinInf}(\vec{\mathcal{B}}) \leq \textsc{MinInf}(\vec{\mathcal{A}}) \leq 2 \cdot \textsc{MaxDis}(\vec{\mathcal{A}}) = 2 \cdot \textsc{MaxDis}(\vec{\mathcal{B}}).$$

The more complex situation is when $\textsc{MaxDis}(\vec{\mathcal{A}}) - 1 = \textsc{MaxDis}(\vec{\mathcal{B}})$. In this case the three lines of T form an infeasible subsystem of $\vec{\mathcal{A}}$ which is contained in every maximal set of disjoint infeasible subsystems of $\vec{\mathcal{A}}$. To deal with this situation we need the following lemma:

Lemma 4.6. *Let $\vec{\mathcal{A}}$ be an oriented arrangement of pseudolines with an infeasible triangular cell c whose boundary lines form an infeasible system T which is contained in every maximal set of disjoint infeasible subsystems of $\vec{\mathcal{A}}$. Then*

(1) every pseudoline g with c on the bad side is contained in every maximal set of disjoint infeasible subsystems of $\vec{\mathcal{A}}$, and

(2) there is no infeasible base system R disjoint from T in $\vec{\mathcal{A}}$ such that c is contained in the bounded region of R.

Proof. Let l_1, l_2, l_3 be the pseudolines of the system T. Suppose there is a pseudoline g which has c on the bad side and is not used by some maximal system of disjoint infeasible subsystems. Up to relabeling the situation is as shown in Figure 14. Replacing the triple $T = l_1, l_2, l_3$ by g, l_2, l_3 gives a maximal system of disjoint infeasible subsystems not containing T, a contradiction.

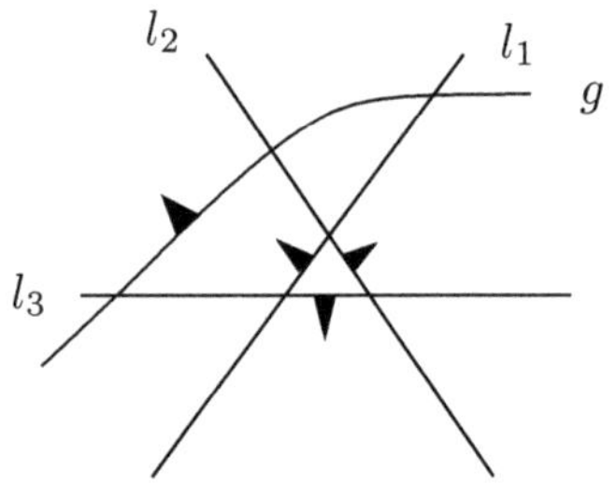

Fig. 14. The infeasible triangular cell c and an extra line g with c on its bad side.

For the proof of part 2, suppose there is an infeasible base subarrangement R containing c in its bounded region. Let g_1, g_2, g_3 be the pseudolines of R. We may assume that the arrangement induced by l_1, l_2, l_3 and g_1 is again as shown in Figure 14.

Suppose that the crossing of g_2 and g_3 is on the bad side of l_1. In this case the triples l_2, l_3, g_1 and g_2, g_3, l_1 form two disjoint infeasible subarrangements from the six lines of T and R.
If g_2 and g_3 cross on the good side of l_1 then the situation is as described by Figure 15.
Suppose that the crossing of l_2 and g_3 is on the good side of l_1 and hence on the bad side of l_3. In this case the triples l_2, l_3, g_3 and l_1, g_1, g_2 form two disjoint infeasible subarrangements. Otherwise, the crossing of l_2 and g_3 is on the bad side of l_1. In this case the triples l_1, l_2, g_3 and l_3, g_1, g_2 form two disjoint infeasible subarrangements.

In all cases we can form two new infeasible base arrangements from the six lines of T and R. This contradicts the assumption that T is contained in every maximal set of disjoint infeasible subsystems of $\vec{\mathcal{A}}$. □

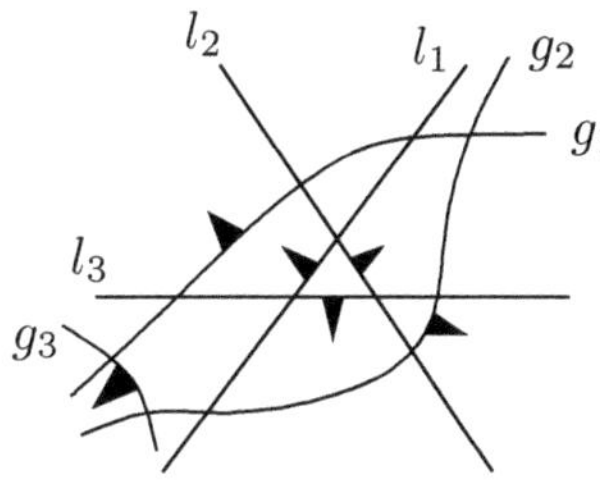

Fig. 15. Lines g_1 and g_2 cross on the good side of l_1.

We now estimate the infeasibility of the the triangular cell c defined by T in $\vec{\mathcal{A}}$. Let F be a maximal system of disjoint infeasible subsystems in $\vec{\mathcal{A}}$. Because of Lemma 4.6.(2) every subarrangement $T' \neq T$ in F contains c in a region of infeasibility at most 2. By Lemma 4.6.(1) this accounts for all pseudolines having c on the bad side. Therefore,

$$\text{INF}(c) \leq 2 \cdot (\text{MAXDIS}(\vec{\mathcal{A}}) - 1) + 3$$

Since the infeasibility of cell c is decreased by 3 through the flip, we have

$$\text{MININF}(\vec{\mathcal{B}}) \leq \text{INF}(c) - 3 \leq 2 \cdot (\text{MAXDIS}(\vec{\mathcal{A}}) - 1).$$

This completes the proof for the case where $\text{MAXDIS}(\vec{\mathcal{B}}) = \text{MAXDIS}(\vec{\mathcal{A}}) - 1$ and, hence, the proof of Theorem 4.2. □

Conclusion

We have investigated bounds of the LHS over the RHS in the inequality

$$\text{MININF}(\vec{\mathcal{A}}) \geq \text{MAXDIS}(\vec{\mathcal{A}}).$$

For the 2-dimensional case we gave tight results both for arrangements of lines and pseudolines. In d dimensions we have seen that the best possible factor is between $(d+1)/2$ and $d+1$. Based on the ideas of this paper these bounds can be improved to $(d+2)/2$ and d. For $d = 2$ these lower and upper bounds match, but for higher d there remains a wide gap. It would be interesting to know on which side of the interval the truth hides.

References

[1] S. FELSNER AND H. WEIL, *Sweeps, arrangements and signotopes*, Discrete Applied Mathematics, 109 (2001), pp. 67–94.

[2] A. FRANK, *How to make a digraph strongly connected.*, Combinatorica, 1 (1981), pp. 145–153.

[3] J. E. GOODMAN, *Pseudoline arrangements*, in Handbook of Discrete and Computational Geometry, Goodman and O'Rourke, eds., CRC Press, 1997, pp. 83–109.

[4] M. Grötschel, M. Jünger, and G. Reinelt, *Acyclic subdigraphs and linear orderings*, in Graphs and Order, Rival, ed., NATO ASI Ser., Ser. C 147, Reidel, 1985, pp. 217–264.

[5] B. Grünbaum, *Arrangements and Spreads*, Regional Conf. Ser. Math., Amer. Math. Soc., 1972.

[6] L. Lovász, *On two minimax theorems in graph.*, J. Comb. Theory, Ser. B, 21 (1976), pp. 96–103.

[7] C. Lucchesi and D. Younger, *A minimax theorem for directed graphs.*, J. London Math. Soc., II. Ser., 17 (1978), pp. 369–374.

[8] Y. Manin and V. Schechtman, *Arrangements of hyperplanes, higher braid groups and higher Bruhat orders*, in Algebraic Number Theory – in honour of K. Iwasawa, J. C. et al., ed., vol. 17 of Advanced Studies in Pure Mathematics, Kinokuniya Company/Academic Press, 1989, pp. 289–308.

[9] G. Ringel, *Teilungen der Ebenen durch Geraden oder topologische Geraden.*, Math. Z., 64 (1956), pp. 79–102.

[10] G. Ziegler, *Higher Bruhat orders and cyclic hyperplane arrangements*, Topology, 32 (1993), pp. 259–279.

About Authors

Stefan Felsner and Nicole Morawe are at the Fachbereich Mathematik und Informatik, Freie Universität Berlin, Takustr. 9, 14195 Berlin, Germany; *felsner@inf.fu-berlin.de*.

Combinatorial Generation of Small Point Configurations and Hyperplane Arrangements

Lukas Finschi
Komei Fukuda

Abstract

A recent progress on the complete enumeration of oriented matroids enables us to generate all combinatorial types of small point configurations and hyperplane arrangements in general dimension, including degenerate ones. This extends a number of former works which concentrated on the non-degenerate case and are usually limited to dimension 2 or 3. Our initial study on the complete list for small cases has shown its potential in resolving geometric conjectures.

1 Introduction

The generation of combinatorial types of point configurations and hyperplane arrangements has long been an outstanding problem of combinatorial geometry. A point configuration is a set of n points in the real Euclidean space $\mathbb{R}^d$. Its combinatorial type, called *order type*, is determined by the relative positions of the points, more formally by the set of all partitions of the n points by hyperplanes, where the points may be arbitrarily relabeled. Similarly, a hyperplane arrangement is a set of n affine hyperplanes in $\mathbb{R}^d$, and its combinatorial type, which we call its *dissection type*, is determined by the relative positions of all cells. For the generation of these combinatorial types no direct method is known, and it appears to be necessary to use combinatorial abstractions — allowable sequences of permutations, λ-functions, chirotopes, combinatorial geometries, or oriented matroids; in our work we will use *oriented matroids* [BLVS+99]. These abstractions are more general than their geometric counterparts, e.g., there exist oriented matroids which cannot be *realized* by any point configuration. Although it is NP-hard to decide whether a given oriented matroid is realizable or not [Mnë88, Sho91] the realizability problem is decidable and there are practical methods which work satisfactory for small instances, at least in the non-degenerate case (e.g., see [RG92]).

The former work on generation of point configurations and related structures (e.g., see [GP83, GP84, BGdO00, AAK01]) concentrated on the special cases of non-degenerate configurations (i.e., the cases where e.g. no three points lie on a line) and low dimensions (i.e., $d = 2$ or $d = 3$). We generate the entire list of all *abstract* order types and dissection types (i.e., we also list non-realizable cases) for small n, including also degenerate cases in arbitrary dimension d. The realizability problem in general is not addressed in this article; however, if n is small enough then all instances are realizable (cf. Section 3.4). Although the realizability problem is very hard and needs further investigation, the complete generation of the abstract combinatorial types of point configurations and hyperplane arrangements offers a powerful database for various investigations (e.g., see Section 4).

This article is the full paper version of the extended abstract presented in [FF01]. A database of computational results can be accessed via the Internet on http://www.om.math.ethz.ch.

2 Combinatorial Types and Oriented Matroids

This section discusses order types and dissection types and their relation to oriented matroids.

2.1 Order Types of Point Configurations

Consider a point configuration $\mathcal{P} = \{v^1, \ldots, v^n\}$ of n points in $\mathbb{R}^d$. An oriented hyperplane H partitions $\mathcal{P}$ into three (possibly empty) sets $\mathcal{P}^+$, $\mathcal{P}^-$, $\mathcal{P}^0$ and defines a corresponding sign vector in $\{-,+,0\}^n$, see Figure 1. The collection of all possible sign vectors obtained by hyperplanes from $\mathcal{P}$

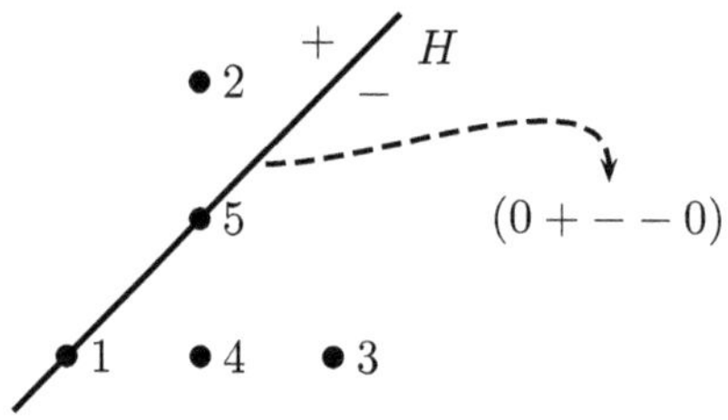

Fig. 1. Sign vector defined by a hyperplane in a point configuration

defines the order type of $\mathcal{P}$. More formally, for $x \in \mathbb{R}^{d+1}$ define a sign vector X by $X_e = \text{sign}(\sum_{i=1}^{d+1} v_i^e x_i)$ for $e \in E := \{1, \ldots, n\}$, where $v_{d+1}^e := 1$. Let A be the matrix of the $n = |E|$ row vectors $v^e \in \mathbb{R}^{d+1}$, $e \in E$, and set $\mathcal{F}(\mathcal{P}) := \{\text{sign}(Ax) \mid x \in \mathbb{R}^{d+1}\}$. Two point configurations $\mathcal{P}$ and $\mathcal{P}'$ are *combinatorially equivalent* if the sets of sign vectors $\mathcal{F}(\mathcal{P})$ and $\mathcal{F}(\mathcal{P}')$ are the same after possibly relabeling the points.

Definition 1 (Order Type of a Point Configuration). Consider a point configuration $\mathcal{P} = \{v^e \mid e \in E\}$ and $\mathcal{F}(\mathcal{P})$ as described above. The *order type of* $\mathcal{P}$ is defined as the *relabeling class* $\text{LC}(E, \mathcal{F}(\mathcal{P}))$ of $(E, \mathcal{F}(\mathcal{P}))$, which is the set of all $(E', \mathcal{F}')$ that are obtained from $(E, \mathcal{F}(\mathcal{P}))$ by relabeling the elements in E.

The set $\mathcal{F}(\mathcal{P})$ is known as (the set of covectors of) a realizable oriented matroid: A pair $(E, \mathcal{F})$ is called a *realizable oriented matroid* if E is a finite set and $\mathcal{F} = \{\text{sign}(y) \mid y \in V\}$ for some vector subspace V of $\mathbb{R}^E$. In general, *oriented matroids* are defined by a set of properties which hold for any $\mathcal{F}(\mathcal{P})$ (see Section 3). Note that for any point configuration the sign vector $(+\cdots+)$ is in $\mathcal{F}(\mathcal{P})$. An oriented matroid $(E, \mathcal{F})$ such that $(+\cdots+) \in \mathcal{F}$ is called an *acyclic oriented matroid.*

The above definition of an order type is what we initially have described. The notion of *order types* has been introduced by Goodman and Pollack [GP83] using λ-functions, first for labeled point configurations and in the further discussion also for unlabeled point configurations. It was pointed out in [GP83] that the definitions of order types by λ-functions and oriented matroids are equivalent.

The one-to-one correspondence between all order types and all relabeling classes of realizable acyclic oriented matroids can be illustrated by use of oriented *sphere arrangements* which represent realizable oriented matroids. As before, we embed $\mathcal{P}$ in $\mathbb{R}^{d+1}$ by adding $v^e_{d+1} = 1$ to every v^e. The extended vectors from $\mathcal{P}$ are the normal vectors of a central arrangement of hyperplanes. Its intersection with the unit sphere S^d defines a sphere arrangement $\mathcal{S}$ as depicted in Figure 2, where every sphere in $\mathcal{S}$ is oriented

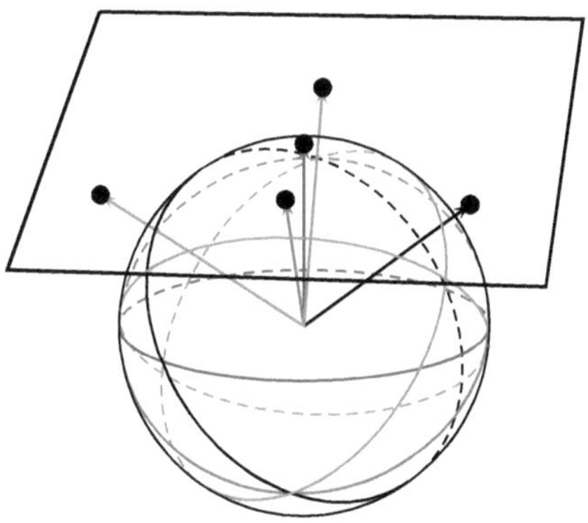

Fig. 2. Point configuration and sphere arrangement

according to its normal vector. Every cell in the sphere arrangement $\mathcal{S}$ has a one-to-one relation to a sign vector in $\mathcal{F}(\mathcal{P})$ as introduced above. The cell containing $v = (0, \ldots, 0, 1)$ corresponds to the sign vector $(+\cdots+) \in \mathcal{F}(\mathcal{P})$.

A point configuration is non-degenerate, i.e., the points are in general position, if and only if the corresponding acyclic oriented matroid is *uniform* (in a corresponding sphere arrangement in $\mathbb{R}^d$ this means that no d spheres intersect in a common point).

2.2 Dissection Types of Hyperplane Arrangements

Let $\mathcal{Q} = \{h^1, \dots, h^n\}$ be a hyperplane arrangement in $\mathbb{R}^d$. For $e \in \{1, \dots, n\}$ let h^e be described by $v^e \in \mathbb{R}^{d+1}$ such that h^e is the set of points $x \in \mathbb{R}^d$ for which $\sum_{i=1}^{d+1} v_i^e x_i = 0$, where $x_{d+1} := 1$. Let A be the matrix of the $n+1$ row vectors $v^e \in \mathbb{R}^{d+1}$ for $e \in E := \{1, \dots, n\} \cup \{g\}$, where $v^g \in \mathbb{R}^{d+1}$ is defined by $v_{d+1}^g := 1$ and $v_i^g := 0$ otherwise; $g \in E$ is a new index element which is called the *infinity element.* We define $\mathcal{F}(\mathcal{Q}) := \{\text{sign}(Ax) \mid x \in \mathbb{R}^{d+1}\}$ which is the set of covectors of a realizable oriented matroid (cf. Section 2.1).

The orientation of the hyperplanes H_e has been arbitrarily determined by the choice of v^e. Therefore, two hyperplane arrangements $\mathcal{Q}$ and $\mathcal{Q}'$ are *combinatorially equivalent* if $\mathcal{F}(\mathcal{Q})$ and $\mathcal{F}(\mathcal{Q}')$ are the same after possibly reorienting and relabeling the elements such that the infinity elements are identified.

Definition 2 (Dissection Type of a Hyperplane Arrangement). Let $\mathcal{Q}$ be a hyperplane arrangement, $g \in E$ the infinity element, and $\mathcal{F}(\mathcal{Q})$ as above. The *dissection type of* $\mathcal{Q}$ is the *affine isomorphism class* $\text{AC}(E, \mathcal{F}(\mathcal{Q}), g)$, which is the set of all $(E', \mathcal{F}', g')$ that are obtained from $(E, \mathcal{F}(\mathcal{Q}), g)$ by isomorphisms which map g to g', where an isomorphism is the composition of a reorientation (i.e., for some $S \subseteq E$ each $X \in \mathcal{F}$ is replaced by X' for which $X'_e = -X_e$ if $e \in S$ and $X'_e = X_e$ otherwise) and relabeling of elements.

We introduce the notion of *dissection types* of hyperplane arrangements in analogy to the notion of *order types* of point configurations.

For an oriented matroid $(E, \mathcal{F})$ and $g \in E$ we call the triple $(E, \mathcal{F}, g)$ an *affine oriented matroid.* The correspondence between dissection types and realizable affine oriented matroids can be illustrated similarly as for point configurations (see Section 2.1). The hyperplane arrangement $\mathcal{Q}$, embedded in $\mathbb{R}^{d+1}$ by adding a coordinate $x_{d+1} = 1$, defines a sphere arrangement $\mathcal{S}$ on the unit sphere, where an extra sphere with normal vector $(0, \dots, 0, 1)$ is specially marked and corresponds to the infinity element (see Figure 3).

A hyperplane arrangement is non-degenerate, i.e., the hyperplanes are in general position (which also means that there are no parallel hyperplanes), if and only if the corresponding affine oriented matroid is uniform.

2.3 Oriented Matroids and Abstract Combinatorial Types

Realizable oriented matroids correspond to sets of sign vectors defined by real vector subspaces or sphere arrangements. The following set (F0) to (F3) of properties are satisfied by realizable oriented matroids as introduced above; they are used as the axioms which define oriented matroids. Then oriented matroids are representable by *topological* sphere arrangements (also called *pseudosphere* arrangements) [FL78, Man82].

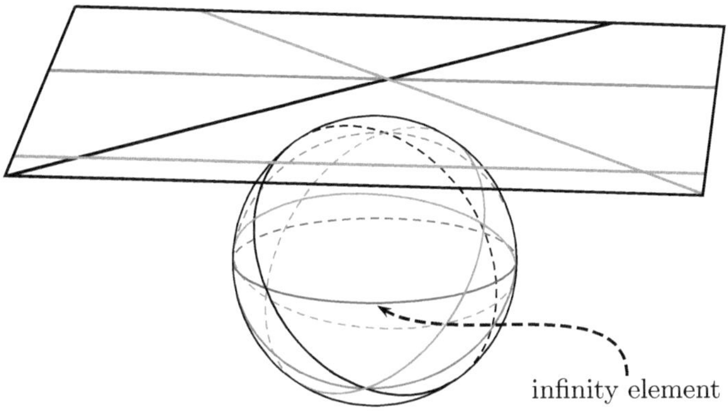

Fig. 3. Hyperplane arrangement and sphere arrangement

Definition 3 (Oriented Matroid). An *oriented matroid* $\mathcal{M}$ is a pair $(E, \mathcal{F})$ of a finite set E and a set $\mathcal{F} \subseteq \{-,+,0\}^E$ of sign vectors (called *covectors*) for which the following *covector axioms* (F0) to (F3) are valid:

(F0) $(0 \cdots 0) \in \mathcal{F}$.

(F1) If $X \in \mathcal{F}$ then $-X \in \mathcal{F}$.

(F2) If $X, Y \in \mathcal{F}$ then $X \circ Y \in \mathcal{F}$, where $(X \circ Y)_e := X_e$ if $X_e \neq 0$ and $(X \circ Y)_e := Y_e$ otherwise.

(F3) For all $X, Y \in \mathcal{F}$ and $e \in D(X,Y) := \{e \in E \mid X_e = -Y_e \neq 0\}$ there exists $Z \in \mathcal{F}$ such that $Z_e = 0$ and $Z_f = (X \circ Y)_f$ for all $f \in E \setminus D(X,Y)$.

We have seen above that the order types of point configurations correspond to the relabeling classes of realizable acyclic oriented matroids, and the dissection types of hyperplane arrangements to the affine isomorphism classes of realizable affine oriented matroids. It is straightforward to generalize these notions (cf. Definitions 1 and 2) to oriented matroids as follows:

Definition 4 (Abstract Order Type, Abstract Dissection Type). We call the relabeling class of an acyclic oriented matroid an *abstract order type.* For an oriented matroid $(E, \mathcal{F})$ and $g \in E$ we call the affine isomorphism class $\mathrm{AC}(E, \mathcal{F}, g)$ an *abstract dissection type.* If the oriented matroid is uniform then the corresponding abstract order type (or abstract dissection type) is called *non-degenerate.*

3 Generation of Combinatorial Types

This section discusses the generation of complete listings of (abstract) order types and dissection types. For this generation we use the relations to oriented matroids as discussed in Section 2. It is not evident how one can design any finite method to generate all order types and dissection types directly, i.e., without an axiomatic abstraction such as oriented matroids.

The listings of combinatorial types are organized by number of elements (e.g., points) n and dimension d. In order to avoid unnecessary redundancies, a point configuration $\mathcal{P}$ in $\mathbb{R}^d$ is listed only if all points are distinct and $\mathcal{P}$ is not contained in a $(d-1)$-dimensional affine subspace. Similarly, a hyperplane arrangement is listed only if all hyperplanes are distinct and their normal vectors are not contained in a $(d-1)$-dimensional affine subspace. Equivalently, we will only consider oriented matroids $(E, \mathcal{F})$ which are so-called *simple*, i.e., there are no *loops* (elements $e \in E$ such that $X_e = 0$ for all $X \in \mathcal{F}$) and no *parallel elements* (elements $e, f \in E$ such that $X_e = X_f$ for all $X \in \mathcal{F}$ or $X_e = -X_f$ for all $X \in \mathcal{F}$). The dimension of a point configuration or hyperplane arrangement is reflected by the *dimension* of the corresponding oriented matroid which is equal to the dimension spanned by a corresponding (topological) sphere arrangement.

3.1 Generation of Oriented Matroids

For given integers n and d consider the complete list $\mathrm{IC}(n, d)$ of all simple oriented matroids on n elements of dimension d up to isomorphism, i.e., up to reorientation and relabeling. We may think of $\mathrm{IC}(n, d)$ as a list containing all types of unlabeled and unoriented topological sphere arrangements with n spheres on S^d. We described in [FF02] several methods how $\mathrm{IC}(n, d)$ can be generated. The following is a short sketch of the main steps using (topological) sphere arrangements for illustration.

Every arrangement of n (topological) spheres can be generated from some arrangement $\mathcal{S}$ of $n-1$ spheres by inserting a new sphere S_n; this is called a *single element extension*. The oriented sphere S_n defines for every vertex (i.e., 0-dimensional intersection of spheres) of $\mathcal{S}$ a corresponding sign ($-$, $+$, or 0); the collection of all signs defined by a single element extension is called a *localization*. Localizations (and by this single element extensions) can be characterized by a result of Las Vergnas [LV78] using *coline cycles*, which are the 1-dimensional intersections of spheres in $\mathcal{S}$: a collection of signs on the vertices of $\mathcal{S}$ is a localization if and only if the sign pattern of every coline cycle has one of the three types that are indicated in Figure 4. While the only-if

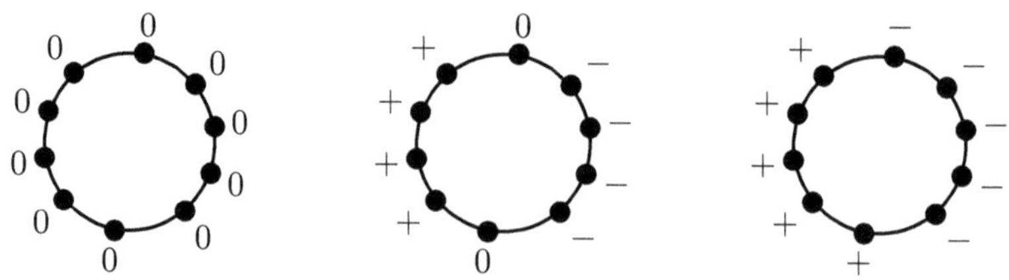

Fig. 4. Sign patterns of coline cycles characterizing single element extensions

part is obvious, the if part is nontrivial. This characterization can be used in a backtracking method which enumerates all single element extensions; there are efficient implementations of the algorithm using a dynamic ordering of the coline cycles in the backtracking process and a special data structure which encodes the mutual intersections of coline cycles.

By use of special canonical representations of isomorphism classes of oriented matroids (based on *chirotopes* representing basis orientations) and applying a *reverse search* technique [AF96], the generation methods can be designed so that they list the isomorphism classes without repetition in a canonical ordering (see [Fin01]). The expensive operation is the computation of canonical representations; our implementations have a worst case complexity of $O(n!)$. We hope that this can be improved significantly in the future. However, our methods reduce redundancies already while backtracking (using a coarse identity induced by symmetries of the given arrangement $\mathcal{S}$), so only relatively few canonical representations have to be computed.

Table 1 on the left shows the numbers $|\operatorname{IC}(n,d)|$ of oriented matroids up to isomorphism, and for comparison we give on the right hand side of Table 1 the corresponding numbers for the non-degenerate (i.e., uniform) cases only (these have been generated and discussed in former work; see Table 6 in [Bok93]). For $d \geq n$ there are blanks in the tables as there are too few elements to span the dimension. We will use the lists $\operatorname{IC}(n,d)$ as the input for the following.

Table 1. Number of oriented matroids up to isomorphism

	all oriented matroids							uniform only						
$n=$	3	4	5	6	7	8	9	3	4	5	6	7	8	9
$d=2$	1	2	4	17	143	4 890	461 053	1	1	1	4	11	135	4 382
$d=3$		1	3	12	206	181 472	?		1	1	1	11	2 628	?
$d=4$			1	4	25	6 029	?			1	1	1	135	?
$d=5$				1	5	50	508 321				1	1	1	4 382
$d=6$					1	6	91					1	1	1
$d=7$						1	7						1	1
$d=8$							1							1

3.2 Generation of Abstract Order Types

Consider an oriented sphere arrangement $\mathcal{S}$ in $\operatorname{IC}(n,d)$. The corresponding oriented matroid is acyclic if some cell c of maximal dimension corresponds to the sign vector $X(c) = (+\cdots+)$; geometrically, we can assume (after an appropriate rotation of $\mathcal{S}$) that c contains the vector $(0,\ldots,0,1)$ (cf. Section 2.1). For an *arbitrary* cell c of maximal dimension $\mathcal{S}$ in $\operatorname{IC}(n,d)$ a reorientation of $\mathcal{S}$ according to $X(c)$ will let c correspond to $(+\cdots+)$. Hence the list of all sign vectors corresponding to cells of maximal dimension in $\mathcal{S}$ (known as *topes*), which we can compute efficiently, is sufficient to find all abstract order types isomorphic to $\mathcal{S}$. See Pseudo-Code 1 for an outline of the algorithm. Note that every abstract order type belongs to a unique isomorphism class of oriented matroids, hence every abstract order type is generated exactly once.

Input: n, d.
Output: all abstract order types with n points and dimension d.

for every representative $\mathcal{M} \in \mathrm{IC}(n,d)$ **do**
  compute the set of topes $\mathcal{T}$ of $\mathcal{M}$;
  for every tope $X \in \mathcal{T}$ **do**
    $\mathcal{M}^X :=$ reorientation of $\mathcal{M}$ according to X;
    $\mathcal{M}^X_{\mathrm{rep}} :=$ canonical representative in the relabeling class $\mathrm{LC}(\mathcal{M}^X)$;
  endfor;
  after removing multiple entries, output the set $\{\mathcal{M}^X_{\mathrm{rep}} \mid X \in \mathcal{T}\}$;
endfor.

Pseudo-Code 1: Generation of abstract order types

The left hand side of Table 2 shows the numbers of abstract order types obtained by computations. Note that there are considerably fewer *non-degenerate* abstract order types, i.e., abstract order types corresponding to uniform oriented matroids (see right hand side of Table 2); the numbers for $d = 2$ coincide with the known results [Knu92, AAK01].

Table 2. Number of abstract order types

	all abstract order types							non-degenerate only						
$n=$	3	4	5	6	7	8	9	3	4	5	6	7	8	9
$d=2$	1	3	11	93	2121	122508	15296266	1	2	3	16	135	3315	158830
$d=3$		1	5	55	5083	10775236	?		1	2	4	246	160020	?
$d=4$			1	8	204	505336	?			1	3	8	11174	?
$d=5$				1	11	705	?				1	3	11	938513
$d=6$					1	15	2293					1	4	22
$d=7$						1	19						1	4
$d=8$							1							1

3.3 Generation of Abstract Dissection Types

If in a realizable oriented matroid $\mathcal{M}$ an element g is marked as the infinity element, then there exists a representation by a sphere arrangement $\mathcal{S}$ where the sphere corresponding to g has the normal vector $(0,\ldots,0,1)$. Analogously, the complete list of abstract dissection types for n hyperplanes in $\mathbb{R}^d$ is obtained from $\mathrm{IC}(n+1,d)$ by marking infinity elements in all possible ways and by identifying affine isomorphic instances. Pseudo-Code 2 summarizes the algorithm. Every abstract dissection type belongs to a unique isomorphism class of oriented matroids, hence every abstract dissection type is generated exactly once.

> *Input:* n, d.
> *Output:* all abstract dissection types with n hyperplanes and dimension d.
>
> **for** every representative $\mathcal{M} = (E, \mathcal{F}) \in \mathrm{IC}(n+1, d)$ **do**
>   **for** every element $g \in E = \{1, \ldots, n+1\}$ **do**
> $\mathcal{M}^g_{\mathrm{rep}}$:= canonical representative in the affine isomorphism class $\mathrm{AC}(E, \mathcal{F}, g)$;
>   **endfor**;
>   after removing multiple entries, output the set $\{\mathcal{M}^g_{\mathrm{rep}} \mid g \in E\}$;
> **endfor**.

Pseudo-Code 2: Generation of abstract dissection types

The numbers of abstract dissection types obtained by computations can be found in Table 3 on the left. For comparison, the right hand side of Table 3 shows corresponding numbers for *non-degenerate* dissection types; the known numbers (see [Rin56]) for $d = 2$ and $n \leq 7$ coincide with the numbers obtained by our programs.

Table 3. Number of abstract dissection types

	all abstract dissection types							non-degenerate only						
$n+1=$	3	4	5	6	7	8	9	3	4	5	6	7	8	9
$d=2$	1	3	8	46	790	37 829	4 134 939	1	1	1	6	43	922	38 612
$d=3$		1	5	27	1 063	1 434 219	?		1	1	1	43	20 008	?
$d=4$			1	7	71	44 956	?			1	1	1	922	?
$d=5$				1	9	156	?				1	1	1	38 612
$d=6$					1	11	325					1	1	1
$d=7$						1	13						1	1
$d=8$							1							1

3.4 Notes on the Realizability

The generation of combinatorial types often was considered together with the realizability problem, the problem of classifying which abstract types can be realized by coordinates in Euclidean space and which are not realizable.

For $d = 2$ the classification of the *uniform* cases is due to Grünbaum [Grü67, Grü72] for $n = 7$, Goodman and Pollack [GP80a] for $n = 8$, Richter-

Gebert [Ric88] and Gonzalez-Sprinberg and Laffaille [GSL89] for $n = 9$, finally Bokowski, Laffaille, and Richter-Gebert (unpublished) for $n = 10$; for $d = 3$ and $n = 8$ the classification is due to Bokowski and Richter-Gebert [BRG90]. The realizability problem is attacked from two sides: (i) finding realizations (using randomly generated points, various insertion or perturbation techniques) and (ii) proving that no realization can exist (e.g., with final polynomials [RG92]).

In the *general* case (which includes degenerate configurations) important questions are still open. It is not clear how the methods to detect non-realizability of uniform instances can be generalized to the degenerate case. Furthermore, finding coordinates for realizable instances has the additional difficulty that no rational solution may exist. The classification problem for the general case is solved for $d = 2$ and $n \leq 8$ due to Goodman and Pollack [GP80b] (all cases are realizable, which was a conjecture of Grünbaum [Grü72]). By duality, all cases are realizable for $n \leq 8$ in any dimension except $d = 3$, where the first non-realizable cases occur already for $n = 8$. Also by duality, all instances with $n - d \leq 3$ are realizable. Hence, our computational results contain non-realizable cases only in the lists for (n, d) equal to $(8, 3)$, $(9, 2)$, and $(9, 5)$; here, the number of non-realizable *uniform* isomorphism classes of oriented matroids are 24, 1, and 1, respectively. Furthermore, it is known that for $(n, d) = (10, 2)$ there are 312 356 isomorphism classes whereof 242 are non-realizable. W.r.t. non-degenerate abstract order types it is known that for $(n, d) = (9, 2)$ there are 13 non-realizable types, and for $(n, d) = (10, 2)$ there are 10 635 non-realizable types out of 14 320 182 abstract types [AAK01]. Corresponding numbers for degenerate non-realizable types are not known.

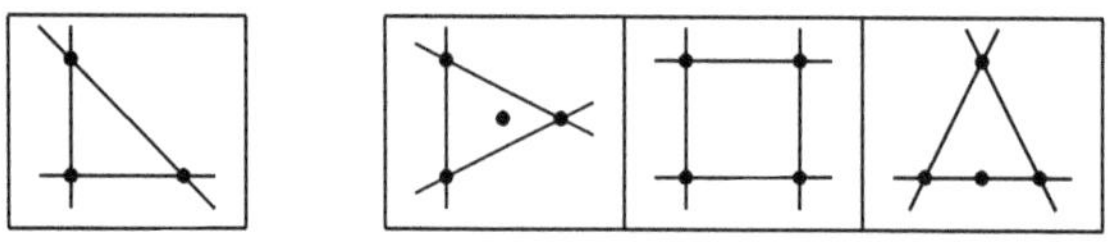

Fig. 5. The order types with 3 and 4 non-collinear points in $\mathbb{R}^2$

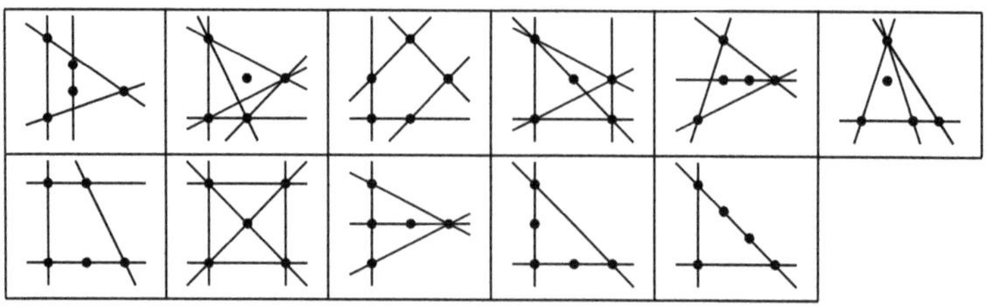

Fig. 6. The 11 order types with 5 non-collinear points in $\mathbb{R}^2$; only the first 3 are non-degenerate

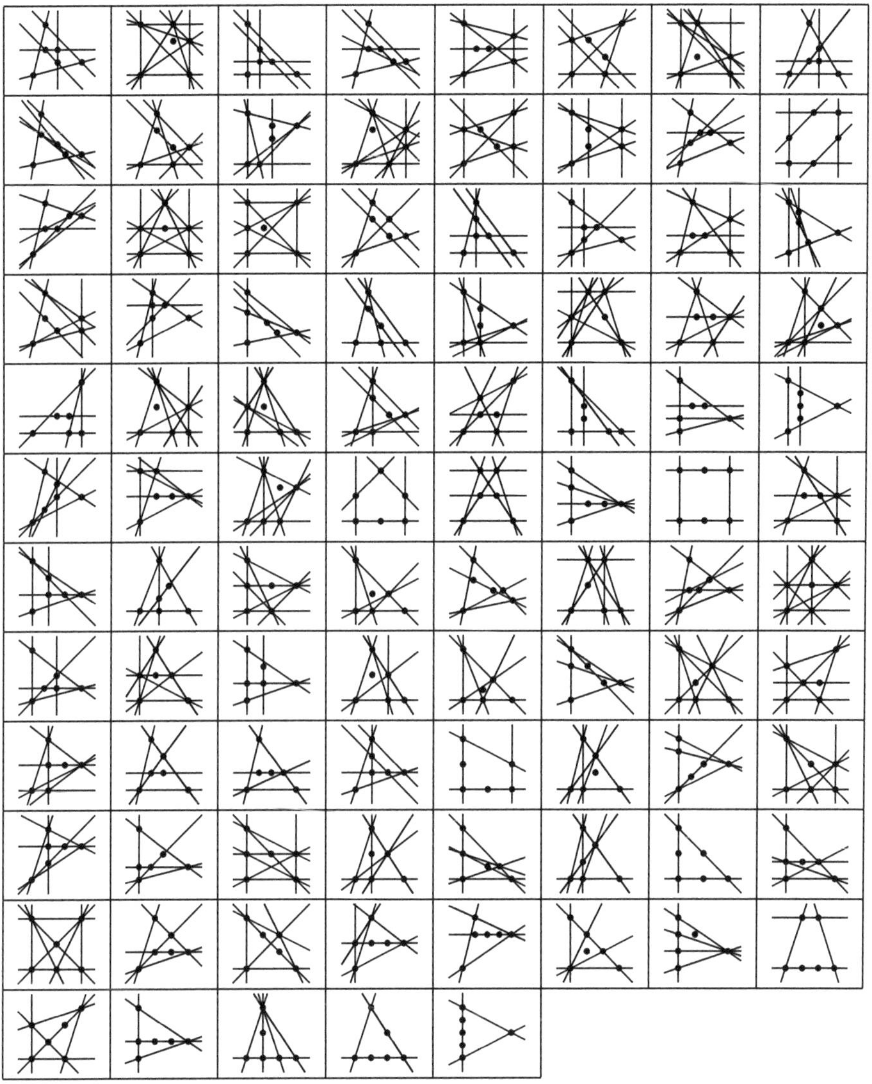

Fig. 7. The 93 order types with 6 non-collinear points in $\mathbb{R}^2$; only the first 16 are non-degenerate

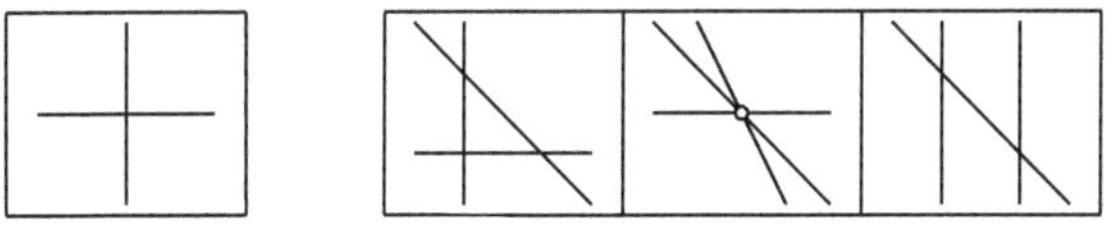

Fig. 8. The dissection types with 2 and 3 non-parallel hyperplanes in $\mathbb{R}^2$

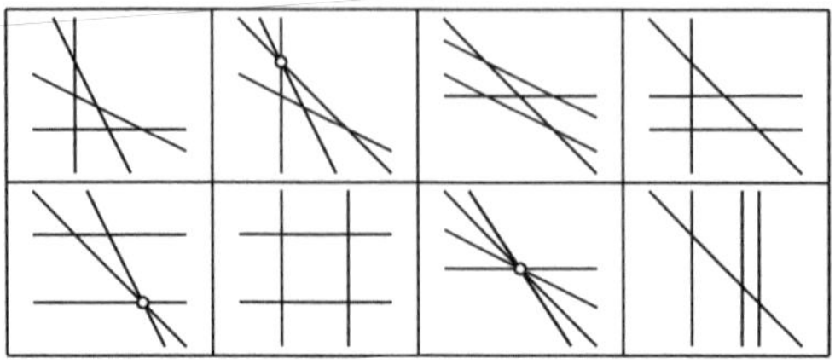

Fig. 9. The 8 dissection types with 4 non-parallel hyperplanes in $\mathbb{R}^2$; only the first is non-degenerate

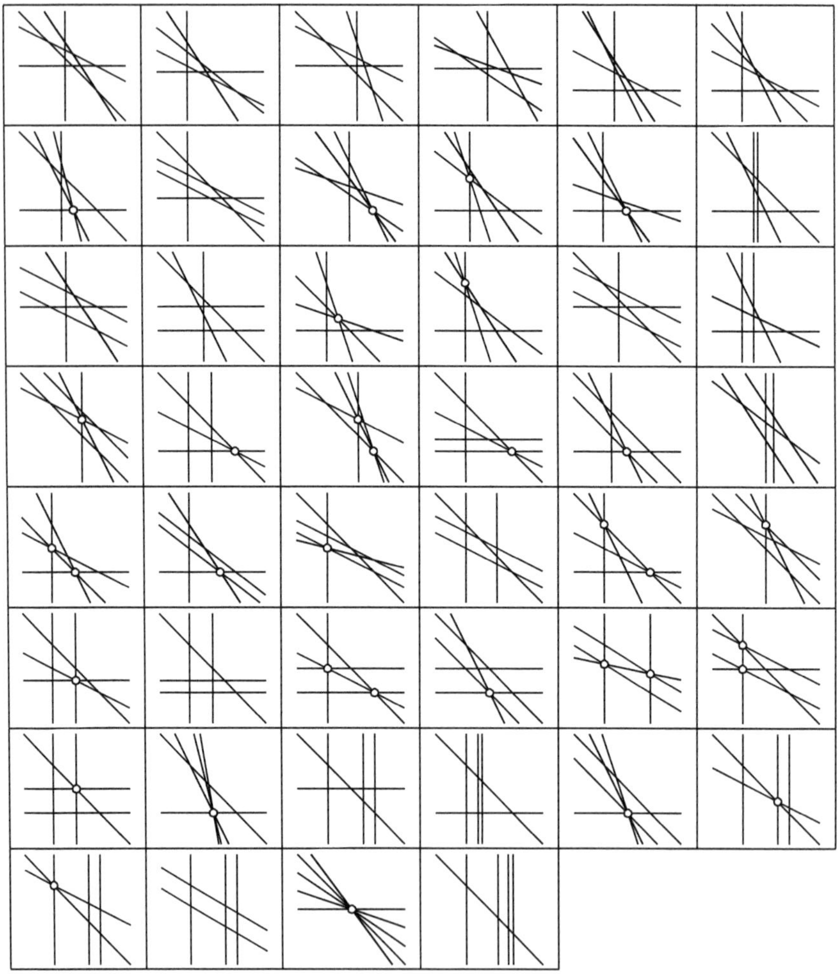

Fig. 10. The 46 dissection types with 5 non-parallel hyperplanes in $\mathbb{R}^2$; only the first 6 are non-degenerate

We present in Figures 5 to 7 realizations of abstract order types for small instances in $\mathbb{R}^2$, i.e., for configurations of 3 to 6 points. The trivial types of collinear points (i.e., all points on a line) correspond to combinatorial types in $\mathbb{R}^1$ and are not counted in $\mathbb{R}^2$. We draw the point configurations in the figures with some lines which may be helpful when reading the picture. The lines which we added are marking a non-degeneracy (i.e., three or more collinear points) and the convex hull, and if there are points in the interior of the convex hull, all lines through the points on the boundary of the convex hull (*extreme points*) are shown; the rule is applied recursively after removing the extreme points.

We present in Figures 8 to 10 realizations of abstract dissection types for small instances in $\mathbb{R}^2$, i.e., for arrangements of 2 to 5 hyperplanes. Degenerate intersections (i.e., points where three or more lines intersect) are marked; lines without intersection in the drawing are parallel. The trivial types of all lines parallel correspond to combinatorial types in $\mathbb{R}^1$ and are not counted in $\mathbb{R}^2$.

4 Applications

Before we discuss one example in more detail below, the following few remarks may hint on some possible impacts of our database of combinatorial types. There has been a strong interest in the number of faces (f-vectors) and specially the number of simplicial topes (mutations) of arrangements and oriented matroids, or in k-sets and extremal properties of point configurations; our database provides all these data for further investigations. For previous results based on listings of (non-degenerate) order types see also [AK01].

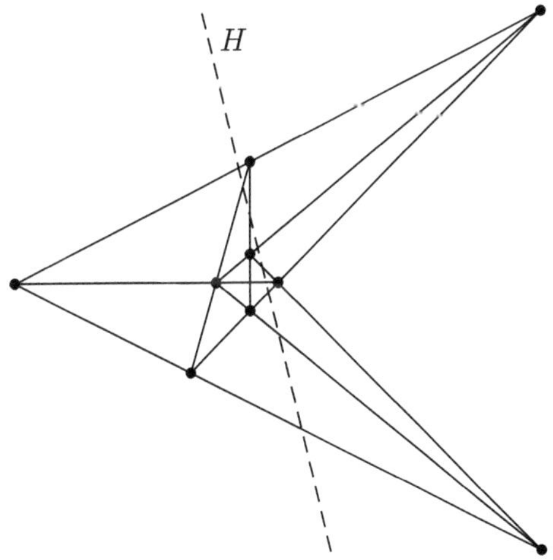

Coordinates	
x	y
1	1
$\frac{1}{2}$	0
$\frac{1}{\sqrt{5}}$	$-1+\frac{2}{\sqrt{5}}$
$\frac{1}{3}$	$-\frac{1}{3}$
$\frac{3}{2}-\frac{\sqrt{5}}{2}$	0
$\frac{1}{\sqrt{5}}$	$1-\frac{2}{\sqrt{5}}$
0	0
$\frac{1}{\sqrt{5}}$	$\frac{1}{\sqrt{5}}$
1	-1

Fig. 11. The counter-example to the conjecture of da Silva and Fukuda with 9 points

The list of abstract order types has been used to compute all (abstract) types of convex polytopes which coincide with the known numbers of combinatorial types of polytopes (e.g., see [KK95]), by this providing an independent proof of these results.

Consider the following conjecture of da Silva and Fukuda (Conjecture 4.2 in [dSF98]), which is a strong version of the Sylvester-Gallai Theorem: *Let $\mathcal{P}$ be a point configuration in $\mathbb{R}^2$, not all points on a line. Let H be a line which does not contain a point from $\mathcal{P}$ but separates $\mathcal{P}$ into two parts $\mathcal{P}^-$ and $\mathcal{P}^+$ such that $|\mathcal{P}^-|$ and $|\mathcal{P}^+|$ differ by at most 1. Then there exists a line $\tilde{H}$ which contains exactly two points of $\mathcal{P}$, one from $\mathcal{P}^-$ and one from $\mathcal{P}^+$.*

Some weaker versions of this conjecture have been proved by Pach and Pinchasi [PP00]. We have tested the conjecture itself against our database of abstract order types: It is valid for $n \leq 8$ and for $n = 10$ points, but for $n = 9$ points the list of **15 296 266** abstract order types contains one counter-example, and this is the only one for $n = 9$. Moreover this abstract order type has been found to be realizable; a picture of the counter-example is given in Figure 11.

References

[AAK01] O. Aichholzer, F. Aurenhammer, and H. Krasser. Enumerating order types for small point sets with applications. In *Proc. 17th Ann. ACM Symp. Comput. Geometry*, pages 11–18. Medford, MA, USA, 2001.

[AF96] D. Avis and K. Fukuda. Reverse search for enumeration. *Discrete Appl. Math.*, 65(1-3):21–46, 1996.

[AK01] O. Aichholzer and H. Krasser. The point set order type data base: a collection of applications and results. In *Thirteenth Canadian Conference on Computational Geometry*, pages 17–20. University of Waterloo, 2001.

[BGdO00] J. Bokowski and A. Guedes de Oliveira. On the generation of oriented matroids. *Discrete Comput. Geom.*, 24(2-3):197–208, 2000.

[BLVS+99] A. Björner, M. Las Vergnas, B. Sturmfels, N. White, and G. M. Ziegler. *Oriented matroids.* Cambridge University Press, Cambridge, second edition, 1999.

[Bok93] J. Bokowski. Oriented matroids. In *Handbook of convex geometry, Vol. A, B*, pages 555–602. North-Holland, Amsterdam, 1993.

[BRG90] J. Bokowski and J. Richter-Gebert. On the classification of non-realizable oriented matroids. Preprint Nr. 1345, Technische Hochschule, Darmstadt.

[dSF98] I. P. F. da Silva and K. Fukuda. Isolating points by lines in the plane. *J. Geom.*, 62(1-2):48–65, 1998.

[FF01] L. Finschi and K. Fukuda. Complete combinatorial generation of small point configurations and hyperplane arrangements. In *Thirteenth Canadian Conference on Computational Geometry*, pages 97–100. University of Waterloo, 2001.

[FF02] L. Finschi and K. Fukuda. Generation of oriented matroids — a graph theoretical approach. *Discrete Comput. Geom.*, 27:117–136, 2002.

[Fin01] L. Finschi. *A Graph Theoretical Approach for Reconstruction and Generation of Oriented Matroids*. Ph.D. thesis, Swiss Federal Institute of Technology ETH Zürich, 2001.

[FL78] J. Folkman and J. Lawrence. Oriented matroids. *J. Combin. Theory Ser. B*, 25(2):199–236, 1978.

[GP80a] J. E. Goodman and R. Pollack. On the combinatorial classification of nondegenerate configurations in the plane. *J. Combin. Theory Ser. A*, 29(2):220–235, 1980.

[GP80b] J. E. Goodman and R. Pollack. Proof of Grünbaum's conjecture on the stretchability of certain arrangements of pseudolines. *J. Combin. Theory Ser. A*, 29(3):385–390, 1980.

[GP83] J. E. Goodman and R. Pollack. Multidimensional sorting. *SIAM J. Comput.*, 12(3):484–507, 1983.

[GP84] J. E. Goodman and R. Pollack. Semispaces of configurations, cell complexes of arrangements. *J. Combin. Theory Ser. A*, 37(3):257–293, 1984.

[Grü67] B. Grünbaum. *Convex polytopes*. Interscience Publishers John Wiley & Sons, Inc., New York, 1967.

[Grü72] B. Grünbaum. *Arrangements and spreads*. American Mathematical Society Providence, R.I., 1972.

[GSL89] G. Gonzalez-Sprinberg and G. Laffaille. Sur les arrangements simples de huit droites dans $\mathbf{RP}^2$. *C. R. Acad. Sci. Paris Sér. I Math.*, 309(6):341–344, 1989.

[KK95] V. Klee and P. Kleinschmidt. Convex polytopes and related complexes. In *Handbook of combinatorics, Vol. 1, 2*, pages 875–917. Elsevier, Amsterdam, 1995.

[Knu92] D. E. Knuth. *Axioms and hulls*, volume 606 of *Lecture Notes in Computer Science*. Springer-Verlag, Berlin, 1992.

[LV78] M. Las Vergnas. Extensions ponctuelles d'une géométrie combinatoire orientée. In *Problèmes combinatoires et théorie des graphes (Colloq. Internat. CNRS, Univ. Orsay, Orsay, 1976)*, pages 265–270. CNRS, Paris, 1978.

[Man82] A. Mandel. *Topology of oriented matroid*. Ph.D. thesis, University of Waterloo, 1982.

[Mnë88] N. E. Mnëv. The universality theorems on the classification problem of configuration varieties and convex polytopes varieties. In *Topology and geometry—Rohlin Seminar*, volume 1346 of *Lecture Notes in Math.*, pages 527–543. Springer, Berlin, 1988.

[PP00] J. Pach and R. Pinchasi. Bichromatic lines with few points. *J. Combin. Theory Ser. A*, 90(2):326–335, 2000.

[RG92] J. Richter-Gebert. *On the realizability problem of combinatorial geometries decision methods*. Ph.D. thesis, TU Darmstadt, 1992.

[Ric88] J. Richter. *Kombinatorische Realisierbarkeitskriterien für orientierte Matroide*. Master's thesis, Technische Hochschule, Darmstadt, 1988. published in Mitteilungen Mathem. Seminar Giessen, Heft 194, Giessen 1989.

[Rin56] G. Ringel. Teilungen der Ebene durch Geraden oder topologische Geraden. *Math. Z.*, 64:79–102, 1956.

[Sho91] P. W. Shor. Stretchability of pseudolines is NP-hard. In P. Gritzmann and B. Sturmfels, editors, *Applied geometry and discrete mathematics: the Victor Klee Festschrift*, volume 4 of *DIMACS series in discrete mathematics and theoretical computer science*, pages 531–554. Amer. Math. Soc., Providence, RI, 1991.

About Authors

Lukas Finschi is at the Institute for Operations Research, Swiss Federal Institute of Technology, Zurich, Switzerland; *finschi@ifor.math.ethz.ch*.

Komei Fukuda is at the Institute for Operations Research, Swiss Federal Institute of Technology, Zurich, Switzerland and Department of Mathematics, Swiss Federal Institute of Technology, Lausanne; *fukuda@ifor.math.ethz.ch*.

Acknowledgments

Work on this paper has been partially supported by the Swiss National Science Foundation Grant 21-58977.99.

Relative Closure and the Complexity of Pfaffian Elimination

Andrei Gabrielov

Abstract

We introduce the "relative closure" operation on one-parametric families of semi-Pfaffian sets. We show that finite unions of sets obtained with this operation ("limit sets") constitute a structure, i.e., a Boolean algebra closed under projections. Any Pfaffian expression, i.e., an expression with Boolean operations, quantifiers, equations and inequalities between Pfaffian functions, defines a limit set. The structure of limit sets is effectively o-minimal: there is an upper bound on the complexity of a limit set defined by a Pfaffian expression, in terms of the complexities of the expression and the Pfaffian functions in it.

1 Introduction

Pfaffian functions [14, 15] are solutions of a triangular system of first-order partial differential equations with polynomial coefficients (see Definition 2.1 below). A semi-Pfaffian set, defined by a Boolean formula with equations and inequalities between Pfaffian functions, is characterized by global finiteness properties. This means that the geometric and topological complexity of a semi-Pfaffian set admits an upper bound in terms of the complexity of its defining formula.

A sub-Pfaffian set Y is the image of a projection of a semi-Pfaffian set X into a subspace. Many finiteness properties of Y can be derived from the corresponding properties of X. These finiteness properties make semi- and sub-Pfaffian sets one of the favorite objects in the theory of o-minimal structures (see [2, 3]).

Upper bounds on the topological complexity of semi-Pfaffian sets were established in [15]. Different aspects of the geometric complexity of semi-Pfaffian and sub-Pfaffian sets, such as the order of tangency (Lojasiewicz inequality), stratification, frontier and closure, were addressed in [4–8].

For a restricted sub-Pfaffian set Y (projection of a restricted semi-Pfaffian set, see Definition 2.4) the complement of Y is sub-Pfaffian [5, 8, 22]). The algorithm in [8] provides an upper bound on the complexity of an existential expression for the complement of Y in terms of the complexity of an existential expression for Y.

For non-restricted semi-Pfaffian sets, Charbonnel [1] and Wilkie [23] introduced the "closure at infinity" operation. Charbonnel-Wilkie theorem ([23], see also [13, 18, 21]) implies that the sets constructed from non-restricted semi-Pfaffian sets by a finite sequence of projections and closures at infinity constitute an o-minimal structure.

In this paper, we introduce the "relative closure" operation (see Definition 3.5 below) on one-parametric families of semi-Pfaffian sets. A "limit set" is a finite union of the relative closures of semi-Pfaffian families. Every semi-Pfaffian set is a limit set. The main results of this paper (Theorems 3.10 and 6.1) state that limit sets constitute an effectively o-minimal structure, i.e., any expression with limit sets defines a limit set, with an upper bound on the complexity of the resulting limit set in terms of the complexity of the expression and of the limit sets in it. Since the number of connected components of a limit set admits an upper bound in terms of its complexity (Theorem 3.13) this provides an efficient version of the Charbonnel-Wilkie theorem for Pfaffian expressions.

2 Pfaffian functions and semi-Pfaffian sets

For a set $X \subset \mathbb{R}^n$, let $\overline{X}$ and $\partial X = \overline{X} \setminus X$ denote its closure and frontier. We assume that the closure points of X at infinity are included in $\overline{X}$ and ∂X. To avoid the separate treatment of infinity, we assume that $\mathbb{R}^n$ is embedded in the projective space, and all constructions are performed in an affine chart U such that X is relatively compact in U. To achieve this, it may be necessary to subdivide X into smaller pieces, each of them relatively compact in its own chart.

Definition 2.1 (See [15]). A *Pfaffian chain* of *order* $r \geq 0$ and *degree* $p \geq 1$ in $\mathbb{R}^n$ is a sequence of functions $y(x) = (y_1(x), \ldots, y_r(x))$, each y_i defined and analytic in an open domain $G_i \subset \mathbb{R}^n$, satisfying a system of Pfaffian equations

$$dy_i(x) = \sum_{j=1}^{n} P_{ij}\big(x, y_1(x), \ldots, y_i(x)\big)dx_j, \text{ for } x \in G_i,\ i = 1, \ldots, r. \quad (1)$$

Here $P_{ij}(x, y_1, \ldots, y_i)$ are polynomials of degree at most p. The system (1) is triangular: P_{ij} does not depend on y_k with $k > i$.

Each domain G_i should satisfy the following conditions:

(i) The graph $\Gamma_i = \{x \in G_i,\ t = y_i(x)\}$ of $y_i(x)$ belongs to an open domain

$$\Omega_i = \{x, t : x \in G_{i-1},\ S_{i\nu}(x, y_1(x), \ldots, y_{i-1}(x), t) > 0, \text{ for } \nu = 1, \ldots, N_i\}$$

with $S_{i\nu}$ polynomial in $x, y_1, \ldots, y_{i-1}, t$, and $\partial\Gamma_i \subset \partial\Omega_i$.

(ii) Γ_i is a *separating submanifold* ("Rolle leaf") in Ω_i, i.e., Ω_i is a disjoint union of Γ_i and two open domains Ω_i^- and Ω_i^+. This is true, for example, when G_i is connected and Ω_i simply connected ([15], p.38).

A *Pfaffian function* of degree $d > 0$ with the Pfaffian chain $y(x)$ is a function $q(x) = Q(x, y(x))$, where $Q(x, y)$ is a polynomial of degree at most d. The function $q(x)$ is defined in a *semi-Pfaffian domain*

$$G = \bigcap G_i = \{S_{i\nu}(x, y_1(x), \ldots, y_i(x)) > 0, \text{ for } i = 1, \ldots, r, \ \nu = 1, \ldots, N_i\}. \tag{2}$$

Remark 2.2. The above definition of a Pfaffian chain corresponds to the definition of a special Pfaffian chain in [4] (see also [7]). It is more restrictive than definitions in [15] and [4] where Pfaffian chains are defined as sequences of nested integral manifolds of polynomial 1-forms. Both definitions lead to (locally) the same class of Pfaffian functions.

More general definitions of Pfaffian functions, where the coefficients of equations (1) can be non-polynomial, are considered in [LR] and [MS]. Most of our constructions can be adjusted to this more general definition. However, efficient upper bounds on the complexity do not hold in this case.

Example 2.3 (Iterated exponential and logarithmic functions). For $r = 1, 2, \ldots$, let $e_r(t) = \exp(e_{r-1}(t))$, with $e_0(t) = t$. The functions $e_1, \ldots, e_r$ constitute a Pfaffian chain of order r and degree r, since $de_r = e_r \cdots e_1\, dt$.

For $r = 1, 2, \ldots$, let $l_r(t) = \ln(l_{r-1}(t))$ for $t > e_{r-1}(0)$, with $l_0(t) = t$. Define

$$\eta_r(\lambda) = 1/l_r(1/\lambda). \tag{3}$$

The function $\eta_r(\lambda)$ is defined in $G_r = \{0 < \lambda < 1/e_r(0)\}$. The functions $\eta_0, \ldots, \eta_r$ constitute a Pfaffian chain of order $r+1$ and degree $r+2$, since

$$d\eta_0 = -\eta_0^2\, d\lambda, \ d\eta_1 = \eta_0\eta_1^2\, d\lambda, \ldots, d\eta_r = (-1)^{r-1}\eta_0 \cdots \eta_{r-1}\eta_r^2\, d\lambda.$$

In the following, we fix a Pfaffian chain $y(x) = (y_1(x), \ldots, y_r(x))$ and, if not explicitly stated otherwise, consider only Pfaffian functions with this particular Pfaffian chain, without explicit reference to the functions $y_i(x)$ and their domains of definition G_i.

Definition 2.4. A *basic semi-Pfaffian set* X of the *format* (I, J, n, r, p, d) in a semi-Pfaffian domain $G \subset \mathbb{R}^n$ is defined by a system of equations and inequalities

$$X = \{x \in G, \ \phi_i(x) = 0, \ \psi_j(x) > 0, \text{ for } i = 1, \ldots, I, \ j = 1, \ldots, J\} \tag{4}$$

where ϕ_i and ψ_j are Pfaffian functions in G of degree not exceeding d, with a common Pfaffian chain of order r and degree p. We assume that G satisfies conditions (i) and (ii) of Definition 2.1, and the inequalities (2) for G are included in the definition of X.

The set X is *restricted* in G if $\overline{X} \subset G$.

A *semi-Pfaffian set* of the *format* (N, I, J, n, r, p, d) is a finite union of at most N basic semi-Pfaffian sets of the formats not exceeding (I, J, n, r, p, d) component-wise, all with the same Pfaffian chain. A semi-Pfaffian set X is *restricted* if it is a finite union of restricted basic semi-Pfaffian sets.

We need the following properties of semi-Pfaffian sets.

Proposition 2.5. *Semi-Pfaffian sets in G constitute a Boolean algebra. The format of a set defined by a Boolean formula with semi-Pfaffian sets admits an upper bound in terms of the formats of these sets and the complexity of the Boolean formula.*

Theorem 2.6 (Khovanskii [15], see also [24]). *The number of connected components of a semi-Pfaffian set X is finite, and admits an upper bound in terms of the format of X.*

Definition 2.7. A semi-Pfaffian set X is *nonsingular* of codimension k if, in a neighborhood of any point $x^0 \in X$, it coincides with a basic semi-Pfaffian set $\{\phi_1(x) = \cdots = \phi_k(x) = 0\}$ with the differentials of the functions $\phi_1, \ldots, \phi_k$ independent at x^0.

Proposition 2.8 (See [7]). *Every semi-Pfaffian set X can be represented as a disjoint union of semi-Pfaffian subsets X^k, nonsingular of codimension k. For each k, $\bigcup_{l \geq k} X^l$ is relatively closed in X. The formats of X^k admit upper bounds in terms of the format of X.*

Definition 2.9. *Dimension* of a semi-Pfaffian set X is the maximum d such that X^{n-d} in Proposition 2.8 is nonempty.

Proposition 2.10. *Let X be a semi-Pfaffian set in a semi-Pfaffian domain G. Then $\overline{X} \cap G$ and $\partial X \cap G$ are semi-Pfaffian sets. The formats of these sets admit upper bounds in terms of the format of X.*

Proof. This follows from the algorithm [6] for the frontier and closure of a semi-Pfaffian set, and from the complexity estimates in [8]. □

Lemma 2.11 (Curve selection). *Let X be a semi-Pfaffian set in a semi-Pfaffian domain G such that $0 \in \overline{X \setminus \{0\}}$. There exists a one-dimensional nonsingular semi-Pfaffian subset γ of $X \setminus \{0\}$ such that $0 \in \overline{\gamma}$. The format of γ admits an upper bound in terms of the format of X.*

Proof. Due to Proposition 2.8, we can suppose X to be a nonsingular basic semi-Pfaffian set of codimension k such that the differentials of $\phi_1, \ldots, \phi_k$ in (4) are independent at each point of X. Let ψ be the the product of all functions ψ_j in (4) multiplied by $1 + (c, x)$, with a generic vector c. If there are no inequalities in (4), we set $\psi = 1 + (c, x)$.

We assume (see Definition 2.4) that the functions ψ_j include the inequalities for G. In particular, ψ vanishes on ∂X.

Consider the set where $|\psi|$ is maximal over $X_\epsilon = \{x \in X : |x| = \epsilon\}$. This set is contained in the set γ_ϵ of critical points of $\psi|_{X_\epsilon}$. It follows from Lemma 2.15 below that, for a generic c, these critical points are non-degenerate, for small $\epsilon > 0$. Hence, for a small $\delta > 0$, the set $\gamma = \{(\epsilon, \gamma_\epsilon) : 0 < \epsilon < \delta\}$ is nonsingular one-dimensional. It is clear that γ is semi-Pfaffian and $0 \in \overline{\gamma}$. □

Proposition 2.12 (Exponential Lojasiewicz inequality, [12, 16, 17]). *Let X be a semi-Pfaffian set in $G \subset \mathbb{R}^n$ with a Pfaffian chain of order r, and let $q(x)$ be a Pfaffian function in $\mathbb{R}^n$. Suppose that 0 belongs to the closure of $X \cap \{q(x) > 0\}$. Then 0 belongs to the closure of*

$$x \in X : q(x) \geq 1/e_r(|x|^{-N}), \tag{5}$$

for some $N > 0$. Here $e_r(\lambda)$ is the iterated exponential function from Example 2.3.

Proof. Let $X_\epsilon = X \cap \{|x| = \epsilon\}$. Due to Lemma 2.11, we can suppose that $X \cap \{q > 0\}$ is a nonsingular curve. Let us choose a branch γ of this curve such that $0 \in \overline{\gamma}$. Let $y(x) = (y_1(x), \ldots, y_r(x))$ be the Pfaffian chain for X. We have $\gamma \subset \{\phi_1(x) = \cdots = \phi_{n-1}(x) = 0\}$ where $\phi_j(x) = Q_j(x, y(x))$ are Pfaffian functions, with Q_j polynomial in (x, y), and the differentials of $\phi_j(x)$ are independent on γ. This implies that the differentials of $Q_1(x, y), \ldots, Q_{n-1}(x, y)$ are independent on $\Gamma = \{x \in \gamma,\ y = y(x)\}$. In particular, there is a $(r+1)$-dimensional irreducible component Z of the algebraic set $\{Q_1(x, y) = \cdots = Q_{n-1}(x, y) = 0\}$ in $\mathbb{R}^{n+r}$ such that $\Gamma \subset Z$. After a linear change of variables in $\mathbb{R}^n$, we can suppose that $|x_n| = \max_i |x_i|$ on γ in the neighborhood of 0. Since $Z \not\subset \{x_n = 0\}$, there exist linear functions $l_1(x, y), \ldots, l_r(x, y)$ in $\mathbb{R}^{n+r}$ such that $\mathbb{R}[x, y]/I(Z)$ is algebraic over $\mathbb{R}[x_n, l_1, \ldots, l_r]$. In particular, functions $x_1, \ldots, x_{n-1}$ and $y_1(x), \ldots, y_r(x)$ restricted to γ are algebraic over the field generated by $r+1$ functions $x_n, l_1(x, y(x)), \ldots, l_r(x, y(x))$ restricted to γ.

Consider $t = 1/|x_n|$ as a parameter on γ in the neighborhood of $x = 0$. Restrictions of Pfaffian functions to γ can be considered as functions in t defined for large t. Due to the finiteness properties of Pfaffian functions [15], germs at $t = \infty$ of these functions generate a Hardy field H. The above arguments imply that H has transcendence degree at most r over $\mathbb{R}(t)$. Due to Proposition 5 of [20], rank of H does not exceed $r+1$. From Theorem 2 of [20], any function $h(t)$ in H is dominated by an iterated exponential function e_r (see Example 2.3 above): $|h(t)| < e_r(t^N)$ for some $N > 0$ as $t \to \infty$. Our statement follows from this inequality applied to $h = (1/q)|_\gamma$, since $|x_n| = \max_i |x_i| \geq |x|/\sqrt{n}$ on γ in the neighborhood of 0. □

Lemma 2.13. *Let X be a smooth manifold in $\mathbb{R}^n$. Let $f_c(x) = f(x) - \sum_\alpha c_\alpha g_\alpha(x)$ be a family of smooth functions on X depending on parameters $c \in \mathbb{R}^m$. Suppose that, for any $x \in X$, the differentials of g_α generate* *the cotangent space to X at x. Then, for a generic c, $f_c(x)$ has only non-degenerate critical points. More precisely, the values of c such that $f_c(x)$ has a degenerate critical point constitute a zero measure set $S \subset \mathbb{R}^m$.*

Proof. This is a variant of Thom's transversality theorem (See, e.g., [11], Ch. II). For convenience, we give a proof here. Let $d = \dim X$. Fix $x^0 \in X$. One can renumber g_α so that the differentials of $g_1, \dots, g_d$ generate the cotangent space to X at x^0. Let us change coordinates in a neighborhood U of x^0 so that $g_i(x) = x_i - a_i$, for $x \in U$, $i = 1, \dots, d$. Consider the mapping $df : U \to \mathbb{R}^d$ in these coordinates. The set of critical points of f_c in U coincides with $df^{-1}(c)$, and all these points are non-degenerate when c is not a critical value of df. From Sard's theorem, the set S_U of critical values of df has zero measure. Since the sets U selected for different points x^0 cover X, a countable covering of X by these sets can be found. Accordingly, the set S, a countable union of the sets S_U, has zero measure. □

Lemma 2.14. *Let X be a smooth manifold in $\mathbb{R}^n$, and $f(x)$ a smooth non-vanishing function on X. For a generic $c = (c_1, \dots, c_n)$, all critical points of a function $f(x)(1 + (c, x))$ are non-degenerate. More precisely, the values of c such that $f(x)(1 + (c, x))$ has a degenerate critical point constitute a zero measure set $V \subset \mathbb{R}^n$.*

Proof. Consider the following family: $f_{a,c} = f(x) - af(x) + (c, x)f(x)$. It is easy to see that the differentials of $f(x)$ and $x_i f(x)$ generate the cotangent space to X at each point $x^0 \in X$. Lemma 2.13 implies that the set

$$S = \{(a, c) : f_{a,c} \text{ has a degenerate critical point}\}$$

has zero measure in $\mathbb{R}^{n+1}$. Since multiplication by a constant does not change critical points and their degeneracy, $S \cap \{a \neq 1\}$ is a cylinder over the set V. Hence V has zero measure in $\mathbb{R}^n$. □

Lemma 2.15. *Let X be a smooth manifold in $\mathbb{R}^n$, and $F(x, \lambda)$ a smooth non-vanishing function on $X \times \mathbb{R}^d$. For a fixed λ, consider $f_\lambda(x) = F(x, \lambda)$ as a function on X. For a generic c, the set*

$$W_c = \{\lambda : f_\lambda(x)(1 + (c, x)) \textit{ has a degenerate critical point}\}$$

has zero measure in $\mathbb{R}^d$.

Proof. Lemma 2.14 implies that, for each λ, the set

$$S_\lambda = \{c : f_\lambda(x)(1 + (c, x)) \text{ has a degenerate critical point}\}$$

has zero measure in $\mathbb{R}^n$. Let $S = \cup_\lambda (S_\lambda, \lambda) \subset \mathbb{R}^n \times \mathbb{R}^d$. Due to Fubini theorem, S has measure zero in $\mathbb{R}^n \times \mathbb{R}^d$. This implies that, for a generic c, the set $W_c = S \cap \{c = \text{const}\}$ has zero measure in $\mathbb{R}^d$. □

3 Relative closure and limit sets

Let $\mathbb{R}^n \times \mathbb{R}$ be $(n+1)$-dimensional space, with coordinates $x = (x_1, \ldots, x_n)$ and λ. For a set $X \subset \mathbb{R}^n \times \mathbb{R}$, we define $X_+ = X \cap \{\lambda > 0\}$, $X_\lambda = X \cap \{\lambda = \text{const}\}$, and $\check{X} = \overline{X_+} \cap \{\lambda = 0\}$. Coordinate λ is considered as a *parameter*, and the set X is considered as a *family* of sets X_λ in $\mathbb{R}^n$.

Definition 3.1. Let G be a semi-Pfaffian domain (see Definition 2.1) in $\mathbb{R}^n \times \mathbb{R}$. A subset $X \subset G$ is a *semi-Pfaffian family* if X is a semi-Pfaffian set with a Pfaffian chain defined in G and, for any $\epsilon > 0$, the set $X \cap \{\lambda > \epsilon\}$ is restricted in G. The *format* of X is defined as the format of a semi-Pfaffian set X_λ for a small $\lambda > 0$.

Remark 3.2. In all constructions below, upper bounds on the complexity can be established for semi-Pfaffian families considered as semi-Pfaffian sets in $\mathbb{R}^n \times \mathbb{R}$. However, the upper bounds in terms of the format of a family (i.e., the complexity of the fibers X_λ) are more important in applications, since they provide better estimates for the geometric and topological complexity of limit sets.

Proposition 3.3. *Let X be a semi-Pfaffian family. Then $\overline{X}_+$ and $(\partial X)_+$ are semi-Pfaffian families. The formats of these families admit upper bounds in terms of the format of X.*

Proof. Since $X \cap \{\lambda > \epsilon\}$ is restricted in G, for any $\epsilon > 0$, the set $\overline{X}_+$ is contained in G. Proposition 3.3 implies that $\overline{X}_+$ and $(\partial X)_+$ are semi-Pfaffian sets in G. The sets $\overline{X}_+ \cap \{\lambda > \epsilon\}$ and $(\partial X)_+ \cap \{\lambda > \epsilon\}$ are restricted in G, for any $\epsilon > 0$, since this is true for X.

The statement on the formats follows from Proposition 2.10, since $(\overline{X})_\lambda = \overline{X_\lambda}$ and $(\partial X)_\lambda = \partial(X_\lambda)$ for a generic $\lambda > 0$. These equalities can be derived from Proposition 2.8, Sard's theorem, and the finiteness properties of semi-Pfaffian sets. □

Definition 3.4. Two semi-Pfaffian families X and Y form a *semi-Pfaffian couple* (X, Y) if Y is relatively closed in $\{\lambda > 0\}$ (i.e., $\overline{Y}_+ = Y_+$) and contains $(\partial X)_+$. The format of the couple (X, Y) is defined as the component-wise maximum of the formats of X and Y.

Definition 3.5. Let (X, Y) be a semi-Pfaffian couple in $G \subset \mathbb{R}^n \times \mathbb{R}$. The *relative closure* of (X, Y) is defined as

$$(X, Y)_0 = \check{X} \setminus \check{Y} \subset \check{G} \subset \mathbb{R}^n. \tag{6}$$

If $Y = (\overline{\partial X})_+$, we write X_0, the *relative closure* of X, instead of $(X, Y)_0$. The format of $(X, Y)_0$ is defined as the format of the couple (X, Y).

Definition 3.6. A *limit set* in $\Omega \subset \mathbb{R}^n$ is a finite union of the relative closures $(X_i, Y_i)_0$ of semi-Pfaffian couples (X_i, Y_i) in $G_i \subset \mathbb{R}^n \times \mathbb{R}$, such that

$\check{G}_i = \Omega$ for all i. The format of a limit set is defined as (K, N, I, J, n, r, p, d) where (N, I, J, n, r, p, d) is the component-wise maximum of the formats of the couples (X_i, Y_i), and K is the number of these couples.

Proposition 3.7 (Complement of a limit set). *Let (X, Y) be a semi-Pfaffian couple in $G \subset \mathbb{R}^n \times \mathbb{R}$. Then the complement $\check{G} \setminus (X, Y)_0$ of $(X, Y)_0$ in $\check{G}$ is a limit set. The format of this limit set admits an upper bound in terms of the format of (X, Y).*

Proof. We assume that inequalities $s_{i\nu}(x, \lambda) = S_{i\nu}(x, y_1(x, \lambda), \ldots, y_i(x, \lambda)) > 0$ (see (2)) defining G are included in the definition of X. Let $s(x, \lambda)$ be the product of all functions $s_\nu(x, \lambda)$ in these inequalities, so that $s > 0$ in G and $s = 0$ on ∂G. Let $G' = G \cap \{\lambda > 0,\ s(x, \lambda) \geq 1/e_r(\lambda^{-N})\}$. Here r is the order of the Pfaffian chain for X and N is a positive integer. Let $Z = G \setminus X$ and $Z' = Z \cap G'$. It is clear that Z' is a semi-Pfaffian family in G, and its format (as a family) does not depend on N (since $1/e_r(\lambda^{-N})$, for a fixed λ, is a constant). It follows from Proposition 2.12 that $\check{Z} = \check{Z}'$ for large N. We are going to prove that

$$\check{G} \setminus (X, Y)_0 = (Z', \overline{X}_+)_0 \cup (Y, \emptyset)_0. \tag{7}$$

By definition of the relative closure, the right side of (7) equals $(\check{Z}' \setminus \check{X}) \cup \check{Y} = (\check{Z} \setminus \check{X}) \cup \check{Y}$. Since $(X, Y)_0 \cap (\check{Z} \setminus \check{X}) = \emptyset$ and $(X, Y)_0 \cap \check{Y} = \emptyset$, the left side of (7) contains its right side. Let now $x \in \check{G} \setminus (X, Y)_0$. Note that x belongs either to $\check{X}$ or to $\check{Z}$ (or to both). If $x \in \check{X}$ then $x \in \check{Y}$. Otherwise, $x \in \check{Z} \setminus \check{X}$. This implies that the right side of (7) contains its left side. □

Proposition 3.8 (Product of limit sets). *Let (X, Y) and (X', Y') be two semi-Pfaffian couples in $G \subset \mathbb{R}^n \times \mathbb{R}$ and $G' \subset \mathbb{R}^m \times \mathbb{R}$, respectively. Then the product of $(X, Y)_0$ and $(X', Y')_0$ is a limit set in $\check{G} \times \check{G}' \subset \mathbb{R}^n \times \mathbb{R}^m$:*

$$(X, Y)_0 \times (X', Y')_0 = (X \times_{\mathbb{R}} X', Z)_0, \text{ where } Z = (\overline{X}_+ \times_{\mathbb{R}} Y') \cup (Y \times_{\mathbb{R}} \overline{X'}_+). \tag{8}$$

Here $X \times_{\mathbb{R}} X' = \{(x, x', \lambda) : (x, \lambda) \in X,\ (x', \lambda) \in X'\}$ is the fibered product over $\mathbb{R}$.

Proof. Let $z \in \check{X}$ and $z' \in \check{X}'$. From Lemma 2.11, one can find continuous functions $x = x(\lambda)$ and $x' = x'(\lambda)$ defined for small $\lambda > 0$ such that $(x(\lambda), \lambda) \in X$, $(x'(\lambda), \lambda) \in X'$, and $\lim_{\lambda \searrow 0}(x(\lambda), x'(\lambda)) = (z, z')$. Hence $(z, z') \in (X \times_{\mathbb{R}} X')\check{}$. This implies $\check{X} \times \check{X}' = (X \times_{\mathbb{R}} X')\check{}$. Similarly, $(\check{X} \times \check{Y}') \cup (\check{Y} \times (\check{X}') = \check{Z}$. The statement then follows from standard set-theoretic arguments. □

Proposition 3.9 (Intersection of limit sets). *Let (X, Y) and (X', Y') be two semi-Pfaffian couples. Then $(X, Y)_0 \cap (X', Y')_0$ is a limit set. The format of this limit set admits an upper bound in terms of the formats of the couples (X, Y) and (X', Y').*

Proof. We are going to prove that, for large integer N,

$$(X,Y)_0 \cap (X',Y')_0 = ((X \times_{\mathbb{R}} X') \cap W_N, Z)_0 \tag{9}$$

where Z is defined in (8) and

$$W_N = \{(x,x',\lambda) : |x-x'| \leq \eta_r(\lambda)^{1/N}\}. \tag{10}$$

Here r is the order of the Pfaffian chain for X, Y, X', Y', η_r is the iterated logarithmic function defined in (3), and $\check{G}$ is identified with its diagonal embedding in $\mathbb{R}^n \times \mathbb{R}^n$. The statement follows from Propositions 2.12 and 3.8, and the identity

$$(X,Y)_0 \cap (X',Y')_0 = [(X,Y)_0 \times (X',Y')_0] \cap \{x = x'\}.$$

We only have to show that

$$\check{}((X \times_{\mathbb{R}} X') \cap W_N) = \check{}(X \times_{\mathbb{R}} X') \cap \{x = x'\}.$$

Due to Lemma 2.11, a point (z,z) belongs to $\check{}(X \times_{\mathbb{R}} X')$ if and only if z belongs to $\check{X} \cap \check{X}'$. From (5) applied to $q \equiv \lambda$, the point $(z,0)$ belongs to the closures of $X \cap \{(x,\lambda) : \eta_r(\lambda) \geq |x-z|^N\}$ and $X' \cap \{(x',\lambda) : \eta_r(\lambda) \geq |x'-z|^N\}$, for large enough N. Let (x,λ) and (x',λ) be two points in X and X', respectively, satisfying these two inequalities. Then $|x-x'| \leq |x-z| + |x'-z| \leq 2(\eta_r(\lambda))^{1/N}$. For small λ, this implies $|x-x'|^{N+1} \leq \eta_r(\lambda)$, hence $(z,z,0)$ belongs to the closure of $(X_+ \times_{\mathbb{R}} X'_+) \cap W_{N+1}$, q.e.d.

To derive an upper bound for the format of $(X,Y)_0 \cap (X',Y')_0$, note that, for a fixed λ, $\eta_r(\lambda)^{1/N}$ is a constant, and $(W_N)_\lambda$ is a semialgebraic set of degree 2. □

Theorem 3.10. *Limit sets constitute a Boolean algebra. The format of a limit set defined by a Boolean formula with limit sets $X_1, \ldots, X_N$ admits an upper bound in terms of the complexity of the formula and the formats of $X_1, \ldots, X_N$.*

Proof. This follows from Propositions 3.7 and 3.9. □

Proposition 3.11. *Let (X,Y) be a semi-Pfaffian couple, and X' a semi-Pfaffian family such that X' is a relatively closed subset of X. Then $(X \setminus X', Y \cup X')$ and (X',Y) are semi-Pfaffian couples, and $(X,Y)_0$ is a disjoint union of $(X \setminus X', Y \cup X')_0$ and $(X',Y)_0$.*

Proof. Since X' is relatively closed in X, we have $(\partial X')_+ \subset (\partial X)_+ \subset Y$. In particular, (X',Y) is a semi-Pfaffian couple, and $Y \cup X'$ is relatively closed in $\{\lambda > 0\}$. Since a point in $\partial(X \setminus X')$ belongs either to ∂X or to X', we have $\left(\partial(X \setminus X')\right)_+ \subset Y \cup X'$, hence $(X \setminus X', Y \cup X')$ is a semi-Pfaffian couple.

It is clear that $(X \setminus X', Y \cup X')_0$ and $(X',Y)_0$ are disjoint subsets of $(X,Y)_0$. If a point $x^0 \in (X,Y)_0$ belongs to $\check{X}'$, then $x^0 \in (X',Y)_0$. Otherwise, x^0 belongs to $(X \setminus X', Y \cup X')_0$. □

Proposition 3.12. *Let (X, Y) be a semi-Pfaffian couple. Then $(X, Y)_0$ is a disjoint union of sets $(X^k, Y^k)_0$ with nonsingular k-dimensional sets X^k. Here $k = 0, \ldots, \dim X$. The formats of the semi-Pfaffian couples (X^k, Y^k) admit upper bounds in terms of the format of (X, Y).*

Proof. This follows from Propositions 2.8 and 3.11. □

Theorem 3.13 (See also [10]). *Let (X, Y) be a semi-Pfaffian couple. Then the number of connected components of $(X, Y)_0$ is finite, and admits an upper bound in terms of the format of (X, Y).*

Proof. Let $\Psi(x) = \min_{x' \in \check{Y}} (x - x')^2$ be the (squared) distance from x to $\check{Y}$ and, for $\lambda > 0$, let $\Psi_\lambda(x) = \min_{y \in Y_\lambda} (x - y)^2$ be the distance from x to Y_λ. Let Z_λ be the set of local maxima of $\Psi_\lambda|_{X_\lambda}$.

For every connected component C of $(X, Y)_0$, the function $\Psi(x)$ is positive on C and vanishes on ∂C, hence Ψ has a local maximum $x_0 \in C$. For small $\lambda > 0$, there exist $x_\lambda \in X_\lambda$ such that $|x_\lambda - x_0| \to 0$ as $\lambda \searrow 0$. This implies $\lim_{\lambda \searrow 0} \Psi_\lambda(x_\lambda) = \Psi(x_0) > 0$. In particular, there exists a positive constant ϵ such that $\Psi_\lambda(x_\lambda) > \epsilon$ for small $\lambda > 0$. Let $W_{\lambda,\epsilon} = \{x \in X_\lambda,\ \Psi_\lambda(x) > \epsilon$, and let C_λ be the connected component of x_λ in $W_{\lambda,\epsilon}$. Since $\Psi_\lambda(x) > \epsilon$ for any $x \in C_\lambda$, the sets C_λ are close to C for small positive λ, i.e., the closure of $\bigcup_{\lambda > 0} C_\lambda$ intersected with $\{\lambda = 0\}$ is a connected subset of $(X, Y)_0$ containing x_0, hence a subset of C. From the definition of C_λ, there exists a local maximum z_λ of $\Psi_\lambda|_{X_\lambda}$ in C_λ, and a connected component V_λ of Z_λ containing z_λ belongs to C_λ. Hence V_λ is close to C for small positive λ. This implies that the number of connected components of $(X, Y)_0$ does not exceed the number of connected components of Z_λ, for small positive λ.

Since Z_λ is a restricted sub-Pfaffian set, an upper bound on the number of its connected components in terms of the format of (X, Y) can be obtained either from [8] or from the bounds on the Betti numbers of restricted sub-Pfaffian sets in [9]. □

4 Regular families and dimension of limit sets

We consider $\mathbb{R}^n$ equipped with the standard Euclidean metric $|x|^2 = \sum x_i^2$. For a linear subspace $L \subset \mathbb{R}^n$, we define $L^\perp$ to be its orthogonal complement in $\mathbb{R}^n$. Let $\pi_L : \mathbb{R}^n \to L^\perp$ be a projection along L. For $x \in \mathbb{R}^n$ or $z = \pi_L x \in L^\perp$, let $L + x = L + z$ denote an affine subspace of $\mathbb{R}^n$ through x parallel to L.

For $I = \{i_1, \ldots, i_d\} \subset \{1, \ldots, n\}$, let $\mathbb{R}_I$ be the $(n - d)$-dimensional coordinate subspace of $\mathbb{R}^n$ defined by $x_I = (x_{i_1}, \ldots, x_{i_d}) = 0$. Let π_I denote a projection along $\mathbb{R}_I$.

Definition 4.1. Let L and T be two linear subspaces in $\mathbb{R}^n$. We define *internal* distance from L to T as

$$\operatorname{dist}_i(L, T) = \sup_{x \in L,\, |x|=1} \ \inf_{y \in T,\, |y|=1} |x - y|. \tag{11}$$

Note that $\operatorname{dist}_i(L,T) \neq \operatorname{dist}_i(T,L)$, and $\operatorname{dist}_i(L,T) > 0$ if and only if $L \not\subset T$. *External* distance between L and T in $\mathbb{R}^n$ is defined as

$$\operatorname{dist}_e(L,T) = \inf_{x\in T^\perp,\,|x|=1}\ \inf_{y\in L^\perp,\,|y|=1} |x-y|. \tag{12}$$

Note that dist_e depends on the ambient space $\mathbb{R}^n$. When it is necessary to specify the ambient space, we write $\operatorname{dist}_e(L,T;\mathbb{R}^n)$ instead of $\operatorname{dist}_e(L,T)$. We have $\operatorname{dist}_e(L,T;\mathbb{R}^n) > 0$ if and only if L and T are transversal: $L+T=\mathbb{R}^n$.

Lemma 4.2. *For fixed dimensions, d and k, of L and T, both* $\operatorname{dist}_i(L,T)$ *and* $\operatorname{dist}_e(L,T)$ *are continuous nonnegative semialgebraic functions on $G_{d,n}\times G_{k,n}$, where $G_{d,n}$ denotes the Grassmannian of d-dimensional subspaces in $\mathbb{R}^n$.*

Proof. This follows from Definition 4.1 and the Tarski-Seidenberg principle. □

Lemma 4.3. *Let d and n be two positive integers, $d<n$. There exists a constant $C_{d,n}>0$ such that, for any d-dimensional subspace L of $\mathbb{R}^n$, there is a subset $I=\{i_1,\dots,i_d\}\subset\{1,\dots,n\}$ with* $\operatorname{dist}_e(L,\mathbb{R}_I) > C_{d,n}$.

Proof. For any d-dimensional subspace L of $\mathbb{R}^n$, there exists $I=\{i_1,\dots,i_d\}\subset\{1,\dots,n\}$ such that L is transversal to $\mathbb{R}_I$. This implies that

$$\rho(L) = \max_{I:|I|=d} \operatorname{dist}_e(L,\mathbb{R}_I)$$

is positive. Since ρ is a continuous function on $G_{d,n}$, its minimum value $C_{d,n}$ is positive. □

Definition 4.4. Let X be a semi-Pfaffian family in $\mathbb{R}^n\times\mathbb{R}$, and L a linear subspace of $\mathbb{R}^n$. We say that X is *L-regular* at $x^0\in\mathbb{R}^n$ if there exists a neighborhood Ω of x^0 and a constant $C>0$ such that, for small $\lambda>0$, the set $X_\lambda\cap\Omega$ is nonsingular and

$$\operatorname{dist}_e(L,T_xX_\lambda;\mathbb{R}^n) > C. \tag{13}$$

for all $x\in X_\lambda\cap\Omega$. In other words, for any sequence $(x^\nu,\lambda^\nu)\in X_+$ converging to $(x^0,0)$, the limit of $T_{x^\nu}X_{\lambda^\nu}$, if exists, is transversal to L.

A couple (X,Y) is *L-regular* if X is L-regular at each point $x^0\in(X,Y)_0$. For $L=\mathbb{R}_I$, an L-regular couple is called I-regular.

Proposition 4.5. *Let (X,Y) be a semi-Pfaffian couple in $G\subset\mathbb{R}^n\times\mathbb{R}$. Let L be a linear subspace in $\mathbb{R}^n$. Suppose that (X,Y) is L-regular at $x^0\in(X,Y)_0$. Let $T=(L+x^0)\times\mathbb{R}$. Then $x^0\in(X\cap T,Y)_0$.*

Proof. From the definition of L-regularity, there exists a neighborhood Ω of x^0 and a constant $C>0$ such that (13) holds for small $\lambda>0$ and $x\in X_\lambda\cap\Omega$. One can choose Ω a cylinder over a neighborhood U of $z^0=\pi_Lx^0$ in $L^\perp$.

Let (x^ν, λ^ν) be a sequence of points in X_+ converging to $(x^0, 0)$. We have $x^\nu \in X_{\lambda^\nu} \cap \Omega$ for large ν. Since $x^0 \in (X,Y)_0$, we have also $\partial X_{\lambda^\nu} \cap \Omega = \emptyset$ for large ν. Let $z^\nu = \pi_L x^\nu$. Let us connect (z^0, λ^ν) with (z^ν, λ^ν) by a line segment S_ν of the length $s_\nu = |z^\nu - z^0|$. We have $S_\nu \subset U$ for large ν. Let us parametrize S_ν by $t \in [0, s_\nu]$, with $t = 0$ corresponding to z^ν and $t = s_\nu$ to z^0. Let $\xi_\nu = \partial/\partial t$ be a unit tangent vector field to S_ν. For large ν the set $Z_\nu = X_{\lambda^\nu} \cap \pi_L^{-1} S_\nu \cap \Omega$ is nonsingular, and there is a unique smooth vector field ζ_ν on Z_ν orthogonal to $Z_\nu \cap (L + x^\nu)$ such that $\pi_L \zeta_\nu = \xi_\nu$. Due to (13), $sup_{Z_\nu}|\zeta_\nu|$ is bounded uniformly in ν.

Let γ_ν be a trajectory of ζ_ν starting at (x^ν, λ^ν). Since ζ_ν is uniformly bounded, we can assume, taking U small enough, that γ_ν cannot escape Ω at a point $x \in \partial\Omega$ such that $\pi_L x \in U$. Since $X \cap \{\lambda > \epsilon\}$ is restricted in G, for every $\epsilon > 0$, γ_ν cannot escape G other than through ∂X. Since $\partial X \subset Y$ and $Y_{\lambda^\nu} \cap \Omega = \emptyset$ for large ν, the only possibility for γ_ν is to end at a point $(u^\nu, \lambda^\nu) \in X_{\lambda^\nu} \cap \Omega$ such that $\pi_L u^\nu = z^0$, hence $(u^\nu, \lambda^\nu) \in X_+ \cap T$. Since $x^\nu \to x^0$ and ζ^ν is uniformly bounded, we have $u^\nu \to x^0$ as $\nu \to \infty$. This implies $x^0 \in \check{(}X \cap T)$. Since $x^0 \notin \check{Y}$, we have $x^0 \in (X \cap T, Y)_0$. □

Definition 4.6. Let L be a linear subspace in $\mathbb{R}^n$. A subset Z of $\mathbb{R}^n$ is *L-Lipschitz* if, in a neighborhood of each point $x^0 \in Z$, the set Z coincides with a finite union of graphs of Lipschitz functions $f_\nu : L^\perp \to L$. For $L = \mathbb{R}_I$, L-Lipschitz sets are called I-Lipschitz.

Proposition 4.7. *Let L be a linear subspace of $\mathbb{R}^n$ of codimension d. Let (X,Y) be a L-regular semi-Pfaffian couple in $\mathbb{R}^n \times \mathbb{R}$ with* $\dim X = d + 1$. *Then $(X,Y)_0$ is an L-Lipschitz set.*

Proof. Let $x^0 \in (X,Y)_0$. Due to Proposition 4.5, $(x^0, 0)$ belongs to the closure of $\Gamma = X_+ \cap T$ where $T = (L + x^0) \times \mathbb{R}$. The set Γ is nonsingular one-dimensional in the neighborhood of $(x^0, 0)$. Let Γ_k be distinct branches of Γ such that $(x^0, 0) \in \overline{\Gamma_k}$.

Let Ω be a neighborhood of x^0 in $\mathbb{R}^n$ such that, for small $\lambda > 0$, we have $Y_\lambda \cap \Omega = \emptyset$ and (13) holds at each point of $X_\lambda \cap \Omega$. We can choose Ω a cylinder over $U \subset L^\perp$ where U is a small neighborhood of $z^0 = \pi_L x^0$ in $L^\perp$. With the same arguments as in the proof of Proposition 4.5, one can show that, for small $\lambda > 0$, the set $X_\lambda \cap \Omega$ is a finite union of graphs of smooth functions $f_{k,\lambda}$ on U with values in L, with the graph of $f_{k,\lambda}$ passing Γ_k.

Since X is L-regular at x^0, the gradients of $f_{k,\lambda}$ are uniformly bounded, independent of λ. For a fixed $z \in U$ and a fixed k, the values $f_{k,\lambda}(z)$ are bounded and depend monotonously on λ as $\lambda \to 0$. Let X_k be the union over $\lambda > 0$ of the graphs of $f_{k,\lambda}$. Then $Z_k = \check{X}_k \subset \check{X} \cap \Omega$ is a graph of a Lipschitz function in U with values in L, and $(X,Y)_0 \cap \Omega = \cup_k Z_k$. □

Proposition 4.8. *Let (X,Y) be a semi-Pfaffian couple in $G \subset \mathbb{R}^n \times \mathbb{R}$ with* $\dim X = d + 1$. *Then*

$$(X,Y)_0 = \bigcup (X_I, Y_I)_0, \tag{14}$$

union over $I \subset \{1, \dots, n\}$ with $|I| \le d$, so that

(a) *(X_I, Y_I) is an I-regular semi-Pfaffian couple in G,*

(b) *$X_I \subset X$ is either empty or $(|I|+1)$-dimensional, and* $\dim Y_I \le \max(\dim Y, d)$. *The formats of (X_I, Y_I) admit upper bounds in terms of the format of (X, Y).*

Proof. For $d = 0$, we can suppose X to be nonsingular 1-dimensional. Then (X, Y) is I-regular for $I = \emptyset$.

Due to Proposition 2.8, there exists a relatively closed subset $V \subset X$ such that $X \setminus V$ is nonsingular $(d+1)$-dimensional, and $\dim V \le d$. For $I \subset \{1, \dots, n\}$ with $|I| = d$, let $X_I = \{(x, \lambda) \in X \setminus V : \text{dist}_e(\mathbb{R}_I, T_x X_\lambda) > C_{d,n}\}$, where $C_{d,n}$ is defined in Lemma 3.3. Then $X \setminus V = \bigcup_{|I|=d} X_I$ and ∂X_I is relatively closed in $X \setminus V$. Due to Proposition 3.11,

$$(X, Y)_0 = \bigcup_{|I|=d} (X_I, Y_I)_0 \cup (W, Y)_0, \text{ where } Y_I = Y \cup V \cup \partial X_I \text{ and}$$

$$W = V \bigcup_{|I|=d} (X \cap \partial X_I).$$

Note that each couple (X_I, Y_I) is I-regular, and $\dim W \le d$. The statement follows now from the induction hypothesis. □

Definition 4.9. For a semi-Pfaffian couple (X, Y) in $\mathbb{R}^n \times \mathbb{R}$, *dimension* $\dim(X, Y)_0$ is defined as maximum of $|I|$ over $I \subset \{1, \dots, n\}$ such that $(X_I, Y_I)_0 \ne \emptyset$ in (14).

Proposition 4.10. *Let $K \subset \{1, \dots, n\}$. Suppose that (X, Y) in Proposition 4.8 satisfies the following property: $X \subset Z$ where Z is a $(|K|+1)$-dimensional semi-Pfaffian family, K-regular at all $x \in (X, Y)_0$. Then the union in (14) can be taken over $I \subset K$.*

Proof. We repeat the arguments in the proof of Proposition 4.8, replacing the condition on $T_x X_\lambda$ in the definition of X_I by the corresponding condition on $\pi_K T_x X_\lambda$. Let $d = \dim X - 1$ and $k = |K|$. For $I \subset K$ with $|I| = d$, let

$$X_I = \{(x, \lambda) \in X \setminus V : \text{dist}_e(\pi_K \mathbb{R}_I, \pi_K(T_x X_\lambda); \mathbb{R}_K^\perp) > C_{d,k}\},$$

where V is the singular set of X and $C_{d,k}$ is defined in Lemma 3.3. Then

$$X \setminus V = \bigcup_{I \subset K, |I|=d} X_I \text{ and } (X, Y)_0 = \bigcup_{I \subset K, |I|=d} (X_I, Y_I)_0 \bigcup (W, Y)_0,$$

where $Y_I = Y \cup V \cup \partial X_I$ and $W = V \bigcup_{I \subset K, |I|=d} (X \cap \partial X_I)$. □

5 L-tangent families and projections of limit sets

Definition 5.1. Let L be a linear subspace in $\mathbb{R}^n$. A nonsingular family X in $\mathbb{R}^n \times \mathbb{R}$ is *L-tangent* at $x^0 \in \mathbb{R}^n$ if, for any sequence (x^ν, λ^ν) of points in X_+ converging to $(x^0, 0)$, we have $\lim_{\nu \to \infty} \operatorname{dist}_i(L, T_{x^\nu} X_{\lambda^\nu}) = 0$. In other words, the limit of $T_{x^\nu} X_{\lambda^\nu}$, if exists, is contained in L. A couple (X, Y) is L-tangent if X is L-tangent at each point of $(X, Y)_0$. For $L = \mathbb{R}_I$, an L-tangent couple is called I-tangent.

Proposition 5.2. *Let (X, Y) be an L-tangent semi-Pfaffian couple in $\mathbb{R}^n \times \mathbb{R}$. Then $(X, Y)_0$ is contained in a finite number of affine subspaces parallel to L. The number of these planes admits an upper bound in terms of the format of (X, Y).*

Proof. One can assume $L = \mathbb{R}_K$ where $K = \{1, \ldots, k\}$. Due to Proposition 4.8, $(X, Y)_0 = \cup_I (X_I, Y_I)_0$ with $X_I \subset X$ either empty or $(|I|+1)$-dimensional, and (X_I, Y_I) I-regular, for each $I \subset \{1, \ldots, n\}$. Let $x^0 \in (X_I, Y_I)_0$, for some I. In particular, $(X_I, Y_I)_0 \neq \emptyset$. Since X is K-tangent at x^0 and $X_I \subset X$, X_I is K-tangent at x^0. This is only possible when $I \cap K = \emptyset$, i.e., $\mathbb{R}_K^\perp \subset \mathbb{R}_I$. According to Proposition 4.7, $(X_I, Y_I)_0$ is an I-Lipschitz set. In the neighborhood of x^0, it is a finite union of graphs of Lipschitz functions $f_\nu : \mathbb{R}_I^\perp \to \mathbb{R}_I$. Since X_I is K-tangent, the first k components of each f_ν are constants. This implies that $(X_I, Y_I)_0$ is contained in at most countable set of affine planes parallel to $\mathbb{R}_K$. The number of these planes does not exceed the number of connected components of $(X_I, Y_I)_0$, which admits an upper bound in terms of the format of (X, Y) (Theorem 3.13 and Proposition 4.8). □

Proposition 5.3. *Let (X, Y) be a semi-Pfaffian couple in $\mathbb{R}^n \times \mathbb{R}$ with $\dim X = d + 1$, and $J \subset \{1, \ldots, n\}$. Then*

$$(X, Y)_0 = \bigcup (X_I, Y_I)_0, \tag{15}$$

union over $I \subset \{1, \ldots, n\}$ with $|I| \leq d$, so that

(a) *(X_I, Y_I) is an I-regular semi-Pfaffian couple in $\mathbb{R}^n \times \mathbb{R}$,*

(b) *$X_I \subset X$ is either empty or $(|I|+1)$-dimensional, and* $\dim Y_I \leq \max(\dim Y, d)$.

(c) *for any affine space $T \subset \mathbb{R}^n \times \mathbb{R}$ parallel to $\mathbb{R}_{I \cap J} \times \mathbb{R}$, $(X_I \cap T, Y_I)$ is J-tangent.*

Proof. We use induction on d, as in the proof of Proposition 4.8. For $d = 0$, the set X is 1-dimensional. Then (X, Y) is I-regular for $I = \emptyset$ and J-tangent for any J.

Let V be a relatively closed subset in X such that $X \setminus V$ is nonsingular $(d+1)$-dimensional, and $\dim V \leq d$. Let $I \subset \{1, \ldots, n\}$ with $|I| = d$. Let

$K = I \cap J$, $k = |K|$, $m = |J|$. Define

$$X_I = \{(x,\lambda) \in X \setminus V : \quad \begin{array}{l} \text{dist}_e(\mathbb{R}_K, T_x X_\lambda; \mathbb{R}^n) > C_{k,m}, \\ \text{dist}_e(\mathbb{R}_I, \mathbb{R}_K \cap T_x X_\lambda; \mathbb{R}_K) > C_{d-k,n-m}, \\ \text{dist}_i(\mathbb{R}_J, \mathbb{R}_K \cap T_x X_\lambda) < \eta_r(\lambda^N)\} \end{array}$$

where N is a large number, r is the order of the Pfaffian chain for X, and η_r is defined in (3). The constants $C_{k,m}$ and $C_{d-k,n-m}$ are defined in Lemma 3.3.

It can be shown, using Proposition 2.12, that

$$(X,Y)_0 = \bigcup_{|I|=d} (X_I, Y_I)_0 \bigcup (W,Y)_0 \text{ where } Y_I = Y \cup V \cup \partial X_I \text{ and}$$

$$W = V \bigcup_{|I|=d} (X \cap \partial X_I).$$

The statement follows from the induction hypothesis, since (X_I, Y_I) satisfy conditions (a)–(c) and $\dim W \leq d$. □

Definition 5.4. For $J \subset \{1,\ldots,n\}$, let $\pi_J : \mathbb{R}^n \to \mathbb{R}_J^\perp$ be a natural projection along $\mathbb{R}_J$. For a semi-Pfaffian couple (X,Y) in $\mathbb{R}^n \times \mathbb{R}$, *dimension* $\dim \pi_J(X,Y)_0$ is defined as maximum of $|I \cap J|$ over $I \subset \{1,\ldots,n\}$ such that $(X_I,Y_I)_0 \neq \emptyset$ in a decomposition (15) satisfying conditions (a)–(c) of Proposition 5.3.

Proposition 5.5. *Let $K, J \subset \{1,\ldots,n\}$. Suppose that (X,Y) in Proposition 5.3 satisfies the following property: $X \subset Z$ where Z is a semi-Pfaffian family in $\mathbb{R}^n \times \mathbb{R}$ such that*

(i) $\dim Z = |K| + 1$,

(ii) *Z is K-regular at all $x \in (X,Y)_0$,*

(iii) *for any affine space $T \subset \mathbb{R}^n \times \mathbb{R}$ parallel to $\mathbb{R}_{K\cap J} \times \mathbb{R}$, $Z \cap T$ is J-tangent at all $x \in (X,Y)_0$. Then the union in (15) can be taken over $I \subset K$.*

Proof. The proof is similar to the proof of Proposition 4.10. □

Lemma 5.6 (Fiber cutting). *Let (X,Y) be a semi-Pfaffian couple in $\mathbb{R}^n \times \mathbb{R}$. Let $K, J \subset \{1,\ldots,n\}$ and $\pi = \pi_J$. Suppose that (X,Y) is K-regular and, for any affine subspace $T \subset \mathbb{R}^n \times \mathbb{R}$ parallel to $\mathbb{R}_{J\cap K} \times \mathbb{R}$, the couple $(X \cap T, Y)$ is J-tangent. In particular, $d = \dim \pi(X,Y)_0 = |J \cap K|$. Let $\mathbb{R}^{2n} = \mathbb{R}^n \times \mathbb{R}^n$, and $\rho : \mathbb{R}^{2n} \to \mathbb{R}^n$ a projection to the first factor.*

There exist semi-Pfaffian couples (V,W) and (V',W') in $\mathbb{R}^{2n} \times \mathbb{R}$ such that

(i)$\rho V, \rho V' \subset X$,

(ii)$\pi(X,Y)_0 = \pi\rho(V,W)_0 \cup \pi\rho(V',W')_0$,

(iii)$\dim \pi\rho(V', W')_0 < d$,

(iv)(V, W) *is* $(J \cap K)$*-regular,*

(v) V *is* $(d+1)$*-dimensional,*

(vi)*for any* $\lambda > 0$ *and any affine subspace* L *of* $\mathbb{R}^{2n}$ *parallel to* $\mathbb{R}_{J\cap K} \times \mathbb{R}^n$*, the set* $V_\lambda \cap L$ *is finite. The formats of* (V, W) *and* (V', W') *admit upper bounds in terms of the format of* (X, Y)*.*

Proof. Due to Proposition 4.5, $(X, Y)_0$ is the union of $(X \cap T, Y)_0$ over all affine T parallel to $\mathbb{R}_{K\cap J} \times \mathbb{R}$. Due to Proposition 5.2, $\pi(X \cap T, Y)_0$ is finite, for any such T.

We want to apply the arguments in the proof of Theorem 3.13 to each couple $(X \cap T, Y)$. Let $Y = \cup_k Y^k$ be a weak stratification of Y (see Proposition 2.8). For a generic $2n$-vector (c, c'), consider a "distance" function $\Phi(x, x') = [1 + (c, x) + (c', x')](x - x')^2$ on $\mathbb{R}^{2n}$. Suppose that (c, c') is chosen so that $1 + (c, x) + (c', x')$ is positive on $X \times Y$. For $z \in \mathbb{R}^{\perp}_{K\cap J}$, let $T_z = \{x \in \mathbb{R}^n : \pi_{J\cap K} x = z\}$ be an affine subspace parallel to $\mathbb{R}_{J\cap K}$. Define semi-Pfaffian families $V^* = \bigcup_{z,\lambda} V_{z,\lambda}$ and $W^* = \bigcup_{z,\lambda} W_{z,\lambda}$ in $\mathbb{R}^{2n} \times \mathbb{R}$ as follows:

$$V^k_{z,\lambda} = \{x \in X_\lambda \cap T_z,\ x' \in Y^k_\lambda,\ (x, x') \text{ is a critical point of } \Phi|_{(X_\lambda \cap T_z)\times Y^k_\lambda}\};$$

$$V_{z,\lambda} = \bigcup_k V^k_{z,\lambda};$$

$$W_{z,\lambda} = \bigcup_k \{(x, x') \in V^k_{z,\lambda} \text{ a degenerate critical point of } \Phi|_{(X_\lambda \cap T_z)\times Y^k_\lambda}\}.$$

Note that V^* and W^* are relatively closed in $X \times_{\mathbb{R}} Y$. The set $V_{z,\lambda}$ contains all points $x^0 \in (X_\lambda \cap T_z) \setminus Y$ where $\Psi_\lambda(x) = \min_{x' \in Y_\lambda} \Phi(x, x')$ has a local maximum on $X_\lambda \cap T_z$ at x^0. This implies $\pi\rho(V^*, Y \times_{\mathbb{R}} Y)_0 = \pi(X, Y)_0$.

Let S be the set of those (z, λ) for which $W_{z,\lambda}$ is non-empty. Due to Lemma 2.15, for a generic (c, c') the set S has zero measure in $\mathbb{R}_{J\cap K} \times \mathbb{R}$. Since $W^* \subset X \times_{\mathbb{R}} Y$ and (X, Y) is K-regular, Proposition 5.5 implies that $\dim \pi(W^*, Y \times_{\mathbb{R}} Y)_0 < d$.

For $(z, \lambda) \notin S$, the set $V_{z,\lambda}$ is discrete. Since $V^* \subset X \times_{\mathbb{R}} Y$, proposition 5.5 applied to $Z = X \times \mathbb{R}^n$ and $K' = K \cup \{n+1, \dots, 2n\}$ implies that

$$(V^*, (Y \times_{\mathbb{R}} Y) \cup W^*)_0 = \bigcup (V_I, W_I)_0,$$

union over $I \subset K'$, where (V_I, W_I) satisfy conditions (a)–(c) of Proposition 5.3 and the sets $(V_I, W_I)_0$ are empty for all $I \supset J \cap K$ unless $I = J \cap K$.

Let $V = V_{J\cap K}$ and $W = W_{J\cap K}$. The set $W_{J\cap K}$ in the proof of Proposition 5.5 can be chosen so that $(V^*, (Y \times_{\mathbb{R}} Y) \cup W^*)_0 = (V, W)_0 \cup (W, (Y \times_{\mathbb{R}} Y) \cup W^*)_0$. and $\dim \pi\rho(W, (Y \times_{\mathbb{R}} Y) \cup W^*)_0 < d$ Let $V' = W^* \cup W$ and $W' = Y \times_{\mathbb{R}} Y$. Then the couples (V, W) and (V', W') satisfy conditions of Lemma 5.6. □

6 Projection theorem

In this section, we fix $J = \{1, \ldots, m\} \subset \{1, \ldots, n\}$ and denote $\pi = \pi_J : \mathbb{R}^n \to \mathbb{R}^m$. For $x \in \mathbb{R}^n$, let $x = (y, z)$ where $y = (x_1, \ldots, x_m)$ and $z = (x_{m+1}, \ldots, x_n)$.

Theorem 6.1 (Projection of a limit set). *Let (X, Y) be a semi-Pfaffian couple in $G \subset \mathbb{R}^n \times \mathbb{R}$. Then $\pi(X, Y)_0$ is a limit set in $\pi\check{G} \subset \mathbb{R}^m$, and its format admits an upper bound in terms of the format of (X, Y).*

Proof of this theorem will be given at the end of this section. First, we prove it for X relatively closed in $\{\lambda > 0\}$, in two special cases: when Y is empty, and when each fiber of π restricted to $\check{X}$ contains at most one point. Next, we reduce the case of finite fibers to the case of one-point fibers. Finally, general case is reduced to the case of finite fibers by fiber-cutting.

Proposition 6.2. *Let X be a semi-Pfaffian family in $G \subset \mathbb{R}^n \times \mathbb{R}$. Suppose that X is relatively closed in $\{\lambda > 0\}$, i.e., $X_+ = \overline{X} \cap \{\lambda > 0\}$. Then $\pi\check{X}$ is a limit set in $\pi\check{G} \subset \mathbb{R}^m$.*

Proof. Let $f_1(x, \lambda), \ldots, f_r(x, \lambda)$ be a Pfaffian chain for X. Define a "z-cone over X" as

$$CX = \{(y, z, \lambda) \in \mathbb{R}^n \times \mathbb{R} : \lambda > 0,\ (y, \frac{z}{\lambda}, \lambda) \in X\}.$$

This is a semi-Pfaffian family in the z-cone CG over G, with the Pfaffian chain

$$f_1(y, \frac{z}{\lambda}, \lambda), \ldots, f_r(y, \frac{z}{\lambda}, \lambda).$$

Note that $\check{}(CG) = \pi\check{G}$. We have $\pi\check{X} = \check{}(CX) = (CX, \emptyset)_0$. □

Proposition 6.3. *Let (X, Y) be a semi-Pfaffian couple in $G \subset \mathbb{R}^n \times \mathbb{R}$ and Z a limit set in $\pi\check{G} \subset \mathbb{R}^m$. Suppose that X is relatively closed in $\{\lambda > 0\}$ and, for each $y \in \pi(X, Y)_0 \setminus Z$, the set $\check{X} \cap \pi^{-1}y$ contains at most one point. Then $\pi(X, Y)_0 \setminus Z$ is a limit set in $\pi\check{G}$.*

Proof. Due to Proposition 6.2, $\pi\check{X}$ is a limit set in $\mathbb{R}^m$. Let $y = \pi x \in \pi\check{X} \setminus Z$, where $x = (y, z) \in \check{X}$. If $y \notin \pi(X, Y)_0$ then $x \in \check{Y}$, hence $y \in \pi(\check{X} \cap \check{Y})$. Conversely, if $y \in \pi(\check{X} \cap \check{Y})$ then $y \notin \pi(X, Y)_0$. Otherwise, x would be a unique point in $\check{X} \cap \pi^{-1}y$, hence $x \in \check{X} \cap \check{Y}$, and $y = \pi x \notin \pi(X, Y)_0$. This implies $\pi(X, Y)_0 \setminus Z = \pi\check{X} \setminus (\pi(\check{X} \cap \check{Y}) \cup Z)$. From (8) and (9) follows that $\check{X} \cap \check{Y} = \check{}((X \times_{\mathbb{R}} Y) \cap W)$, for a closed semi-Pfaffian family $W \subset \mathbb{R}^n \times \mathbb{R}^n \times \mathbb{R}$. Due to Proposition 6.2, $\pi(\check{X} \cap \check{Y})$ is a limit set. Hence $\pi(X, Y)_0$ is a limit set. □

Proof of Theorem 6.1. We proceed by induction on $d = \dim \pi(X, Y)_0$. Due to Proposition 5.3, we can suppose that, for some $I \subset \{1, \ldots, n\}$, the couple (X, Y) is I-regular, X is $(|I| + 1)$-dimensional, and $(X \cap (T \times \mathbb{R}), Y)$

is J-tangent for any affine space T parallel to $\mathbb{R}_{I\cap J}$. Due to the induction hypothesis, we can consider only those I for which $|I \cap J| = d$.

Due to Lemma 5.6 (with $K = I$) we can replace $(X,Y)_0$ by $(V,W)_0 \cup (V',W')_0$, where V is $(d+1)$-dimensional and projection of $(V',W')_0$ to $\mathbb{R}^m$ is less than d-dimensional. Due to the induction hypothesis, projection of $(V',W')_0$ to $\mathbb{R}^m$ is a limit set, hence it is enough to prove that projection of $(V,W)_0$ to $\mathbb{R}^m$ is a limit set. Accordingly, we can suppose from the very beginning that X is $(d+1)$-dimensional. Applying Proposition 5.3 to $(X, \partial X)$, we can suppose that, for a semi-Pfaffian family $S \supset \partial X$ with $\dim S \le d$, the semi-Pfaffian couple (X,S) is K-regular, for $K \subset J$ with $|K| = d$. Let Δ be projection of S_0 to $\mathbb{R}^d$. Due to the induction hypothesis, Δ is a limit set.

Let ρ denote projection from $\mathbb{R}^m$ to $\mathbb{R}^d$, and $y = (u,v)$ where $u = (x_1,\dots,x_d)$ and $v = (x_{d+1},\dots,x_m)$. For $u \in \mathbb{R}^d \setminus \Delta$ and $\lambda > 0$, the sets $X_{u,\lambda} = X_\lambda \cap \{(x_1,\dots,x_d) = u\}$ are finite. Let $N_{\max}$ be the maximum, over $u \in \mathbb{R}^d \setminus \Delta$ and $\lambda > 0$, number of points in $X_{u,\lambda}$. For $N = 1,\dots,N_{\max}$, let

$$\begin{aligned} X_N = \{u, v_1, z_1, \dots, v_N, z_N, \lambda : \lambda > 0,\\ (u, v_1, z_1) \in X_\lambda, \dots, (u, v_N, z_N) \in X_\lambda,\\ (v_1, z_1) < \dots < (v_N, z_N)\}. \end{aligned}$$

Here "$<$" is the lexicographic order. Each X_N is a semi-Pfaffian family in $\mathbb{R}^{d+N(n-d)} \times \mathbb{R}$, such that $X_{N,u,\lambda} = X_N \cap \{(u,\lambda) = \text{const}\}$ contains exactly one point when $X_{u,\lambda}$ contains exactly N points, and $X_{N,u,\lambda}$ is empty when $X_{u,\lambda}$ contains less than N points. For $j = 1,\dots,N$, let

$$\pi_{N,j}(u, v_1, z_1, \dots, u_N, z_N) = (u, v_j).$$

Let $Z = \rho^{-1}\Delta \cap \pi(X,Y)_0$. It is easy to show that Z is a projection of a limit set, and $\dim Z < d$. Due to the induction hypothesis, Z is a limit set.

For $y^0 = (u^0, v^0) \in \pi(X,Y)_0 \setminus Z$, let N be the maximum number such that $(y^0, 0) \in \pi_{N,j}\overline{X_N}$, for some j. Let $x^0 = (y^0, z^0)$ be a point in $(X,Y)_0 \cap \pi^{-1} y^0$. Since X is K-regular at each point of $\pi^{-1}\rho^{-1}u^0$, the point $(x^0, 0)$ belongs to the closure of $X \cap \{u = u^0\}$, due to Proposition 4.5. Since $(y^0, 0)$ does not belong to $\pi_{N+1,j}\overline{X_{N+1}}$, for all j, the set $X_{u^0,\lambda}$ contains exactly N points, for small $\lambda > 0$. Hence $X_{N,u^0,\lambda}$ contains exactly one point, for small $\lambda > 0$. This implies that $\overline{X_{N,u^0}} \cap \{\lambda = 0\}$ contains exactly one point. It is easy to see that X_N is K-regular at each point of $\pi^{-1}\rho^{-1}u^0$. Hence, $\overline{X_N} \cap \pi_{N,j}^{-1}(y^0, 0) = \overline{X_{N,u^0}} \cap \pi_{N,j}^{-1}(y^0, 0)$ contains exactly one point.

Let

$$Y_{N,j} = \partial X_N \bigcup \{u, v_1, z_1, \dots, u_N, z_N, \lambda : (u, v_j, z_j, \lambda) \in Y\},$$

and $Z_N = \cup_j \pi_{N,j}(X_N, Y_{N,j})_0$. Here we consider all sets $\pi_{N,j}(X_N, Y_{N,j})_0$ as subsets of the same space $\mathbb{R}^m$.

For $N = N_{\max}$, each set $\pi_{N,j}(X_N, Y_{N,j})_0 \setminus Z$ is a limit set due to Proposition 6.3. In particular, $Z_{N_{\max}} \setminus Z$ is a limit set. Applying the same arguments to $N = N_{\max} - 1$ and $Z \cup Z_{N_{\max}}$ instead of Z, we prove that each set $\pi N, j(X_N, Y_{N,j})_0 \setminus (Z \cup Z_N)$ is a limit set, for $N = N_{\max} - 1$, hence $Z_{N_{\max}-1} \setminus (Z \cup Z_{N_{\max}})$ is a limit set. Repeating these arguments for decreasing N, we prove that each set $Z_N \setminus (Z \cup Z_{N+1})$ is a limit set. Finally, $\pi(X,Y)_0 = (\pi(X,Y)_0 \cap Z) \cup_N (Z_N \setminus (Z \cup Z_{N+1})$ is a limit set, since $\pi(X,Y)_0 \cap Z = \pi(X,Y)_0 \cap \rho^{-1}\Delta$ is a projection of a limit set and its dimension is less than d. □

References

[1] J.-Y. Charbonnel, Sur certains sous-ensembles de l'espace euclidien, *Ann. Inst. Fourier (Grenoble)*, 41 (1991), 679–717.

[2] L. van den Dries, Tame Topology and O-minimal Structures, *Cambridge University Press*, Cambridge, UK (1998).

[3] L. van den Dries and C. Miller, Geometric categories and o-minimal structures, *Duke Math. J.*, 84 (1996), 497–540.

[4] A. Gabrielov, Multiplicities of Pfaffian Intersections and the Lojasiewicz Inequality, *Selecta Mathematica, New Series*, 1 (1995), 113–127.

[5] A. Gabrielov, Complements of subanalytic sets and existential formulas for analytic functions, *Inventiones Math.*, 125 (1996), 1–12.

[6] A. Gabrielov, Frontier and closure of a semi-Pfaffian set, *J. Discr. Comput. Geometry*, 19 (1998), 605–617.

[7] A. Gabrielov and N. Vorobjov, Complexity of stratifications of semi-Pfaffian sets, *J. Discr. Comput. Geometry*, 14 (1995), 71–91.

[8] A. Gabrielov and N. Vorobjov, Complexity of cylindrical decompositions of sub-Pfaffian sets, *J. Pure Appl. Algebra*, 164 (2001), 179-197.

[9] A. Gabrielov, N. Vorobjov, and T. Zell, Betti numbers of semialgebraic and sub-Pfaffian sets, *J. London Math. Soc.*, (2003), to appear.

[10] A. Gabrielov and T. Zell, On the number of connected components of the relative closure of a semi-Pfaffian family, *Algorithmic and Quantitative Real Algebraic Geometry, AMS-DIMACS*, (2003), to appear.

[11] M. Golubitsky and V. Guillemin, Stable Mappings and Their Singularities, *Springer-Verlag*, New York, (1973).

[12] D. Grigoriev, Deviation theorems for Pfaffian sigmoids, *Algebra i Analiz*, 6 (1994), 127–131. English transl. in *St. Petersburg Math. J.*, 6 (1995), 107–111.

[13] M. Karpinski and A. Macintyre, A generalization of Wilkie's theorem of the complement, and an application to Pfaffian closure, *Selecta Mathematica, New Series*, 5 (1999), 507–515.

[14] A.G. Khovanskii, On a class of systems of transcendental equations, *Soviet Math. Dokl.*, 22 (1980), 762–765.

[15] A.G. Khovanskii, Fewnomials, AMS Translation of mathematical monographs 88, *AMS*, Providence, RI, (1991).

[16] J.-M. Lion, Inégalité de Lojasiewicz en géométrie pfaffienne, *Illinois J. Math.*, 44 (2000), 889–900.

[17] J.-M. Lion, C. Miller, and P. Speissegger, Differential equations over polynomially bounded o-minimal structures, *Proc. AMS*, 131 (2003), 175–183.

[18] J.-M. Lion and J.-P. Rolin, Volumes, feuilles de Rolle de feuilletages analytiques et théorème de Wilkie, *Ann. Fac. Sci. Toulouse Math.*, 7 (1998), 93–112.

[19] C. Miller and P. Speissegger, Pfaffian differential equations over exponential o-minimal structures, *J. Symbolic Logic*, 67 (2002), 438–448.

[20] M. Rosenlicht, The rank of a Hardy field, *Trans. AMS*, 280 (1983), 659–671.

[21] P. Speissegger, The Pfaffian closure of an o-minimal structure, *J. Reine Angew. Math.*, 508 (1999), 189–211.

[22] A.J. Wilkie, Model completeness results for expansions of the ordered field of real numbers by restricted Pfaffian functions and the exponential function, *J. Amer. Math. Soc.*, 9 (1996), 1051–1094.

[23] A.J. Wilkie, A theorem of the complement and some new o-minimal structures *Selecta Mathematica, New Series*, 5 (1999), 397–422.

[24] T. Zell, Betti numbers of semi-Pfaffian sets, *J. Pure Appl. Algebra*, 139 (1999), 323-338.

About Author

Andrei Gabrielov is at the Department of Mathematics, Purdue University, W. Lafayette, IN 47907-1395, USA; *agabriel@math.purdue.edu*, *www.math.purdue.edu/~agabriel*

Acknowledgments

Supported by NSF Grant # DMS-0070666 and James S. McDonnell Foundation. Part of this work was done when the author was visiting MSRI at Berkeley, CA.